W9-ANQ-909

Scanning Electron Microscopy and X-Ray Microanalysis

A Text for Biologists, Materials Scientists, and Geologists

SECOND EDITION

Scanning Electron Microscopy and X-Ray Microanalysis

A Text for Biologists, Materials Scientists, and Geologists

SECOND EDITION

Joseph I. Goldstein
Lehigh University
Bethlehem, Pennsylvania

Dale E. Newbury
National Institute of
Standards and Technology
Gaithersburg, Maryland

Patrick Echlin
University of Cambridge
Cambridge, England

David C. Joy
University of Tennessee
Knoxville, Tennessee

A. D. Romig, Jr.
Sandia National Laboratories
Albuquerque, New Mexico

Charles E. Lyman
Lehigh University
Bethlehem, Pennsylvania

Charles Fiori
National Institute of
Standards and Technology
Gaithersburg, Maryland

Eric Lifshin
General Electric Corporate Research
and Development
Schenectady, New York

PLENUM PRESS • NEW YORK AND LONDON

Library of Congress Cataloging-in-Publication Data

Scanning electron microscopy and x-ray microanalysis : a text for biologists, materials scientists, and geologists / Joseph I. Goldstein ... [et al.]. -- 2nd ed.
p. cm.
Includes bibliographical references and index.
ISBN 0-306-44175-6
1. Scanning electron microscopy. 2. X-ray microanalysis.
I. Goldstein, Joseph, 1939-
QH212.S3S29 1992
502'.8'25--dc20 92-9840
CIP

10 9 8 7 6 5

ISBN 0-306-44175-6

© 1992, 1981 Plenum Press, New York
A Division of Plenum Publishing Corporation
233 Spring Street, New York, N.Y. 10013

All rights reserved

No part of this book may be reproduced, stored in a retrieval system, or transmitted in any form or by any means, electronic, mechanical, photocopying, microfilming, recording, or otherwise, without written permission from the Publisher

Printed in the United States of America

Preface

In the last decade, since the publication of the first edition of *Scanning Electron Microscopy and X-ray Microanalysis,* there has been a great expansion in the capabilities of the basic SEM and EPMA. High-resolution imaging has been developed with the aid of an extensive range of field emission gun (FEG) microscopes. The magnification ranges of these instruments now overlap those of the transmission electron microscope. Low-voltage microscopy using the FEG now allows for the observation of noncoated samples. In addition, advances in the development of x-ray wavelength and energy dispersive spectrometers allow for the measurement of low-energy x-rays, particularly from the light elements (B, C, N, O). In the area of x-ray microanalysis, great advances have been made, particularly with the "phi rho z" [$\phi(\rho z)$] technique for solid samples, and with other quantitation methods for thin films, particles, rough surfaces, and the light elements. In addition, x-ray imaging has advanced from the conventional technique of "dot mapping" to the method of quantitative compositional imaging. Beyond this, new software has allowed the development of much more meaningful displays for both imaging and quantitative analysis results and the capability for integrating the data to obtain specific information such as precipitate size, chemical analysis in designated areas or along specific directions, and local chemical inhomogeneities.

During these 10 years we have taught over 1500 students in our Lehigh SEM short course in basic SEM and x-ray microanalysis and have updated our notes to the point that the instructors felt that a completely rewritten book was necessary. In this book we have incorporated information about the new capabilities listed above and added new material on specimen preparation for polymers, a growing area for the use of the SEM. On the other hand, we have retained the features of the First Edition, including the same general chapter headings that have been so well accepted. The authors have noticed that there are generally two groups of students who use this textbook and who attend our course, the real introductory or novice student and the experienced student who is looking to sharpen his or her basic skills and to delve into the newer techniques. Therefore, we have decided to highlight in the left margin the material which is essentially basic and should be read by every student who is a novice in the field. We have also added a new introductory chapter on quantitative x-ray microanalysis of bulk samples which will serve as a beginning for those readers interested in quantitation but overwhelmed at first by the physics and the mathematical expressions. This introductory chapter is descriptive in nature with a minimum of equations and should help those readers who want to understand the basic features of the quantitative analysis approach.

The authors wish to thank their many colleagues who have contributed to this volume by allowing us to use material from their research, by their criticism of drafts of the chapters, and by their general support. One of the authors (J. I. G.) wishes to acknowledge the research support and encouragement from the Extraterrestrial Materials Program of the National Aeronautics and Space Administration. Special thanks go to Ms. Sharon Coe for her efforts with the manuscript, to Dr. John Friel of Princeton Gamma Tech and Dr. Bill Bastin of the Technical University of Eindhoven for their contributions to the chapters on quantitative x-ray microanalysis, and to Dr. David Williams of Lehigh University for continuous and helpful advice as the textbook was developed.

J. I. Goldstein
D. E. Newbury

Contents

1. Introduction 1

1.1. Evolution of the Scanning Electron Microscope 2
1.2. Evolution of the Electron Probe Microanalyzer 10
1.3. Outline of This Book 17

2. Electron Optics 21

2.1. How the SEM Works 21
2.1.1. Functions of the SEM Subsystems 21
2.1.2. Why Learn about Electron Optics? 24
2.2. Electron Guns 25
2.2.1. Thermionic Electron Emission 25
2.2.2. Conventional Triode Electron Guns 26
2.2.3. Brightness 29
2.2.4. Tungsten Hairpin Electron Gun 31
2.2.5. Lanthanum Hexaboride (LaB_6) Electron Guns . . . 35
2.2.6. Field Emission Electron Guns 38
2.3. Electron Lenses 43
2.3.1. Properties of Magnetic Lenses 43
2.3.2. Lenses in SEMs 46
2.3.3. Producing Minimum Spot Size 48
2.3.4. Lens Aberrations 53
2.4. Electron Probe Diameter versus Electron Probe Current . . 57
2.4.1. Calculation of d_{min} and i_{max} 57
2.4.2. Comparison of Electron Sources 60
2.4.3. Measurement of Microscope Parameters 65
2.5. Summary of SEM Microscopy Modes 67

3. Electron–Specimen Interactions 69

3.1. Introduction 69
3.2. Electron Scattering 70
3.2.1. Elastic Scattering 71
3.2.2. Inelastic Scattering 73

3.3. Interaction Volume 79
3.3.1. Experimental Evidence 79
3.3.2. Monte Carlo Electron-Trajectory Simulation 81
3.3.2.1. Influence of Beam Energy on Interaction Volume 83
3.3.2.2. Influence of Atomic Number on Interaction Volume 84
3.3.2.3. Influence of Specimen Surface Tilt on Interaction Volume 86
3.3.3. Measures of Interaction Volume—Electron Range . . 87
3.3.3.1. Bethe Range 88
3.3.3.2. Kanaya–Okayama Range 89
3.3.3.3. Range for a Tilted Specimen 89
3.3.3.4. Comparison of Ranges 89
3.3.3.5. Range at Low Beam Energy 90
3.4. Signals from Elastic Scattering 90
3.4.1. Backscattered Electrons 91
3.4.1.1. Atomic Number Dependence 92
3.4.1.2. Beam-Energy Dependence 94
3.4.1.3. Tilt Dependence 95
3.4.1.4. Angular Distribution 97
3.4.1.5. Energy Distribution 100
3.4.1.6. Lateral Spatial Distribution 101
3.4.1.7. Sampling Depth of Backscattered Electrons 104
3.5. Signals from Inelastic Scattering 106
3.5.1. Secondary Electrons 107
3.5.1.1. Definition and Origin 107
3.5.1.2. Energy Distribution 108
3.5.1.3. Specimen Composition Dependence . . . 108
3.5.1.4. Beam-Energy Dependence 110
3.5.1.5. Specimen Tilt Dependence 111
3.5.1.6. Angular Distribution of Secondary Electrons 112
3.5.1.7. Range and Escape Depth of Secondary Electrons 113
3.5.1.8. Relative Contributions of SE_I and SE_{II} . . 115
3.5.2. X-Rays 116
3.5.2.1. Continuum X-Ray Production 117
3.5.2.2. Inner-Shell Ionization 119
3.5.2.3. X-Ray Absorption 135
3.5.2.4. X-Ray Fluorescence 139
3.5.3. Auger Electrons 142
3.5.4. Cathodoluminescence 144
3.5.5. Specimen Heating 146
3.6. Summary 146

4. Image Formation and Interpretation 149

4.1. Introduction 149
4.2. The Basic SEM Imaging Process 150

4.2.1. Scanning Action 150
4.2.2. Image Construction (Mapping) 152
4.2.2.1. Line Scans 153
4.2.2.2. Image (Area) Scanning 154
4.2.2.3. Digital Imaging: Collection and Display . . 156
4.2.3. Magnification 157
4.2.4. Picture Element (Pixel) Size 159
4.2.5. Low-Magnification Operation 163
4.2.6. Depth of Field (Focus) 163
4.2.7. Image Distortions 166
4.2.7.1. Projection Distortion: Gnomonic Projection 166
4.2.7.2. Projection Distortion: Image Foreshortening of Tilted Objects 167
4.2.7.3. Corrections for Tilted Flat Surfaces 170
4.2.7.4. Scan Distortion: Pathological 170
4.2.7.5. Moiré Effects 174
4.3. Detectors 174
4.3.1. Electron Detectors 176
4.3.1.1. Everhart–Thornley Detector 177
4.3.1.2. Dedicated Backscattered-Electron Detectors 181
4.3.1.3. Specimen Current (The Specimen As Detector) 186
4.3.2. Cathodoluminescence Detector 188
4.4. Image Contrast at Low Magnification (<10,000×) 189
4.4.1. Contrast 190
4.4.2. Compositional (Atomic Number) Contrast 191
4.4.2.1. Compositional Contrast with Backscattered Electrons 191
4.4.2.2. Compositional Contrast with Secondary Electrons 195
4.4.2.3. Compositional Contrast with Specimen Current 197
4.4.3. Topographic Contrast 198
4.4.3.1. Origin 198
4.4.3.2. Topographic Contrast with the Everhart–Thornley Detector 200
4.4.3.3. Light–Optical Analogy 203
4.4.3.4. Topographic Contrast with Other Detectors 205
4.4.3.5. Separation of Contrast Components 210
4.4.3.6. Other Contrast Mechanisms 214
4.5. Image Quality 215
4.6. High-Resolution Microscopy: Intermediate (10,000–100,000×) and High Magnification (>100,000×) 219
4.6.1. Electron–Specimen Interactions in High-Resolution Microscopy 220
4.6.1.1. Backscattered Electrons 220
4.6.1.2. Secondary Electrons 223
4.6.2. High-Resolution Imaging at High Voltage 224
4.6.3. High-Resolution Imaging at Low Voltage 226

4.6.4. Resolution Improvements: The Secondary-Electron Signal . . . 227
4.6.5. Image Interpretation at High Resolution . . . 229
4.7. Image Processing for the Display of Contrast Information . . 231
4.7.1. The Visibility Problem . . . 232
4.7.2. Analog Signal Processing . . . 233
4.7.2.1. Display of Weak Contrast (Differential Amplification) . . . 234
4.7.2.2. Enhancement of a Selected Contrast Range (Gamma Processing) . . . 237
4.7.2.3. Enhancement of Selected Spatial Frequencies (Derivative Processing) . . . 238
4.7.2.4. Signal Mixing . . . 242
4.7.2.5. Contrast Reversal . . . 243
4.7.2.6. *Y*-Modulation . . . 243
4.7.3. Digital Image Processing . . . 244
4.7.3.1. Real Time Digital Imaging . . . 244
4.7.3.2. Off-Line Digital Image Processing . . . 245
4.7.3.3. Digital Imaging for Minimum-Dose Microscopy . . . 246
4.8. Defects of the SEM Imaging Process . . . 247
4.8.1. Contamination . . . 247
4.8.2. Charging . . . 249
4.8.2.1. Incipient Charging . . . 253
4.8.2.2. Severe Charging . . . 254
4.9. Special Topics in SEM Imaging . . . 255
4.9.1. SEM at Elevated Pressures (Environmental SEM) . . 255
4.9.1.1. The Vacuum Environment . . . 255
4.9.1.2. Detectors for Elevated-Pressure Microscopy . . . 256
4.9.1.3. Contrast in Elevated-Pressure Microscopy 257
4.9.1.4. Resolution . . . 258
4.9.1.5. Benefits of SEM at Elevated Pressures . . 258
4.9.2. Stereo Microscopy . . . 260
4.9.2.1. Qualitative Stereo Microscopy . . . 260
4.9.2.2. Quantitative Stereo Microscopy . . . 263
4.9.3. STEM in SEM . . . 267
4.10. Developing a Comprehensive Imaging Strategy . . . 270

5. X-Ray Spectral Measurement: WDS and EDS . . . 273

5.1. Introduction . . . 273
5.2. Wavelength-Dispersive Spectrometer . . . 273
5.2.1. Basic Design . . . 273
5.2.2. The X-Ray Detector . . . 280
5.2.3. Detector Electronics . . . 283
5.3. Energy-Dispersive X-Ray Spectrometer . . . 292
5.3.1. Operating Principles . . . 292
5.3.2. The Detection Process . . . 296
5.3.3. Charge-to-Voltage Conversion . . . 297

5.3.4. Pulse-Shaping Linear Amplifier and Pileup Rejection Circuitry 298
5.3.5. The Computer X-Ray Analyzer 304
5.3.6. Artifacts of the Detection Process 310
5.3.6.1. Peak Broadening 310
5.3.6.2. Peak Distortion 313
5.3.6.3. Silicon X-Ray Escape Peaks 315
5.3.6.4. Absorption Edges 316
5.3.6.5. Internal Fluorescence Peak of Silicon . . . 319
5.3.7. Artifacts from the Detector Environment 319
5.3.7.1. Microphony 320
5.3.7.2. Ground Loops 321
5.3.7.3. Ice–Oil Accumulation 323
5.3.7.4. Sensitivity to Stray Radiation 325
5.3.8. Summary of EDS Operation and Artifacts 330
5.4. Comparison of WDS and EDS 331
5.4.1. Geometrical Collection Efficiency 331
5.4.2. Quantum Efficiency 332
5.4.3. Resolution 332
5.4.4. Spectral Acceptance Range 334
5.4.5. Maximum Count Rate 334
5.4.6. Minimum Probe Size 334
5.4.7. Speed of Analysis 336
5.4.8. Spectral Artifacts 336
Appendix: Initial Detector Setup and Testing 337

6. Qualitative X-Ray Analysis 341

6.1. Introduction 341
6.2. EDS Qualitative Analysis 343
6.2.1. X-Ray Lines 343
6.2.2. Guidelines for EDS Qualitative Analysis 348
6.2.2.1. General Guidelines for EDS Qualitative Analysis 348
6.2.2.2. Specific Guidelines for EDS Qualitative Analysis 349
6.2.3. Pathological Overlaps in EDS Qualitative Analysis 353
6.2.4. Examples of EDS Qualitative Analysis 355
6.3. WDS Qualitative Analysis 357
6.3.1. Measurement of X-Ray Lines 357
6.3.2. Guidelines for WDS Qualitative Analysis 361
6.4. Automatic Qualitative EDS Analysis 363

7. X-Ray Peak and Background Measurements 365

7.1. General Considerations for X-Ray Data Handling 365
7.2. Background Correction 366
7.2.1. Background Correction for EDS 366
7.2.1.1. Background Modeling 368
7.2.1.2. Background Filtering 373

7.2.2. Background Correction for WDS 376
7.2.2.1. Interpolation 376
7.2.2.2. Substitute Material Method 377
7.3. Peak Overlap Correction 377
7.3.1. EDS Peak Overlap Correction 377
7.3.1.1. Linearity 379
7.3.1.2. Goodness of Fit 380
7.3.1.3. The Linear Methods 381
7.3.1.4. The Nonlinear Methods 383
7.3.1.5. Error Estimation 386
7.3.2. WDS Peak-Overlap Correction 391

8. Quantitative X-Ray Analysis: The Basics 395

8.1. Introduction 395
8.2. Advantages of Quantitative X-Ray Microanalysis in the SEM/EPMA 396
8.3. Quantitative Analysis Procedures 397
8.4. The Approach to X-Ray Quantitation: The Need for Matrix Corrections 399
8.5. The Physical Origin of Matrix Effects 400
8.6. X-Ray Production 401
8.6.1. Effect of Atomic Number 401
8.6.2. X-Ray Generation with Depth, $\phi(\rho z)$ 403
8.7. ZAF Factors in Microanalysis 405
8.7.1. Atomic Number Effect 405
8.7.2. X-Ray Absorption Effect 407
8.7.3. X-Ray Fluorescence 412
8.8. Types of Matrix Correction Schemes 413
8.9. Caveats 415

9. Quantitative X-Ray Analysis: Theory and Practice 417

9.1. Introduction 417
9.2. ZAF Technique 417
9.2.1. Introduction 417
9.2.2. The Atomic Number Correction Z 419
9.2.3. The Absorption Correction A 424
9.2.3.1. Formulation 424
9.2.3.2. Expressions for $f(\chi)$ 425
9.2.3.3. Practical Considerations 426
9.2.3.4. Calculations of the Absorption Factor A . . 427
9.2.4. The Characteristic Fluorescence Correction F 429
9.2.5. The Continuum Fluorescence Correction 432
9.2.6. Summary Discussion of the ZAF Method 434
9.3. $\phi(\rho z)$ Technique 436
9.3.1. Introduction 436
9.3.2. The $\phi(\rho z)$ Curves 437
9.3.2.1. Definition 437
9.3.2.2. Measurement of $\phi(\rho z)$ Curves 439

9.3.3. Calculation of $\phi(\rho z)$ Curves 440
9.3.4. Atomic Number Correction Z_i 443
9.3.5. Absorption Correction A_i. 445
9.3.6. Summary Discussion of the $\phi(\rho z)$ Method 448
9.4. Quantitative Analysis with Nonnormal Electron-Beam Incidence 453
9.5. Standardless Analysis 456
9.6. The Biological or Polymer Specimen: Special Procedures . . . 460
9.6.1. The Characteristic Signal 461
9.6.2. The Continuum Signal 462
9.6.3. Derivation of the Hall Procedure in Terms of X-Ray Cross-Sections 463
9.6.4. Error Analysis 465
9.6.5. Bulk Targets and Analysis of a Minor Element . . . 465
9.7. Special Procedures for Geological Analysis 466
9.7.1. Introduction 466
9.7.2. Formulation of the Bence–Albee Procedure 467
9.7.3. Application of Bence–Albee Procedure 468
9.7.4. Specimen Conductivity 470
9.8. Special Sample Analysis 471
9.8.1. Introduction: The Analytical Total 471
9.8.2. Films on Substrates 472
9.8.3. Foils 476
9.8.4. Particles and Rough Surfaces 479
9.8.4.1. Mass Effect 479
9.8.4.2. Absorption Effect 480
9.8.4.3. Fluorescence Effect 486
9.8.4.4. Compensating for Geometric Effects in Quantitative Analysis 486
9.9. Precision and Sensitivity in X-Ray Analysis 493
9.9.1. Statistical Basis for Calculating Precision and Sensitivity 493
9.9.2. Sample Homogeneity 496
9.9.3. Analytical Sensitivity 497
9.9.4. Trace-Element Analysis 499
9.9.5. Variance under Peak Overlap Conditions in the EDS 502
9.10. Light-Element Analysis 503
9.10.1. Introduction 503
9.10.2. Operating Conditions for Light-Element Analysis . . 504
9.10.3. X-Ray Spectrometers 505
9.10.3.1. WDS 505
9.10.3.2. EDS 509
9.10.4. Chemical Bonding Shifts 510
9.10.4.1. WDS 510
9.10.4.2. EDS 512
9.10.5. Standards for Light-Element Analysis 513
9.10.6. Surface Contamination 514
9.10.7. Measurement of Background Intensities 516
9.10.7.1. WDS 516
9.10.7.2. EDS 517
9.10.8. Quantitation Procedures for the Light Elements . . . 517

Appendix 9.1. Equations for the α, β, γ, and $\phi(0)$ Terms of the Packwood–Brown $\phi(\rho z)$ Equation 520
Appendix 9.2. Solutions for the Atomic Number and Absorption Corrections 522

10. Compositional Imaging 525

10.1. Introduction 525
10.2. Analog X-Ray Area Scanning (Dot Mapping) 526
10.2.1. Procedure 526
10.2.2. Limitations and Artifacts 529
10.3. Digital Compositional Mapping 535
10.3.1. Principles 535
10.3.1.1. Data Collection 536
10.3.1.2. Dead-time Correction 536
10.3.1.3. Defocusing or Decollimation Correction . . 536
10.3.1.4. Background Correction 537
10.3.1.5. Standardization (k-Value) 538
10.3.1.6. Matrix Correction 538
10.3.1.7. Combined EDS–WDS Strategy 538
10.3.1.8. Statistics in Compositional Mapping 539
10.3.2. Advantages 539

11. Specimen Preparation for Inorganic Materials: Microstructural and Microchemical Analysis 547

11.1. Metals 547
11.1.1. Specimen Preparation for Surface Topography . . . 549
11.1.2. Specimen Preparation for Microstructural and Microchemical Analysis 550
11.1.2.1. Final Specimen Preparation for Microstructural Analysis 550
11.1.2.2. Final Specimen Preparation for Microchemical Analysis 552
11.1.3. Preparation of Standards for X-Ray Microanalysis . . 554
11.2. Ceramics and Geological Specimens 556
11.2.1. Initial Specimen Preparation 556
11.2.2. Mounting 556
11.2.3. Polishing 557
11.2.4. Final Specimen Preparation 557
11.3. Electronic Devices and Packages 557
11.3.1. Initial Specimen Preparation 558
11.3.2. Mounting 559
11.3.3. Polishing 560
11.3.4. Final Preparation 560
11.4. Semiconductors 560
11.4.1. Voltage Contrast 560
11.4.2. Charge Collection 561
11.4.3. Electron Channeling 562
11.5. Sands, Soils, and Clays 562
11.6. Particles and Fibers 564

11.6.1. Particle Substrates 565
11.6.1.1. Bulk Substrates 565
11.6.1.2. Thin-Foil Substrate 566
11.6.2. Transfer and Attachment of Particles to Substrates 566
11.6.2.1. Abundant, Loose Particles 567
11.6.2.2. Particle Transfer from a Filter 568
11.6.2.3. Particles in a Solid Matrix 568
11.6.3. Transfer of Individual Particles 569

12. Sample Preparation for Biological, Organic, Polymeric, and Hydrated Materials 571

12.1. Introduction 571
12.2. Compromising the Electron-Beam Instrument 572
12.2.1. Environmental Stages 572
12.2.2. Environmental Microscopes 572
12.2.3. Nonoptimal Microscope Performance 573
12.3. Compromising the Sample 574
12.4. Correlative Microscopy 576
12.5. Techniques for Structural Studies 576
12.5.1. Specimen Selection 578
12.5.2. Specimen Cleaning 579
12.5.3. Specimen Stabilization 582
12.5.4. Specimen Dehydration 585
12.5.4.1. Chemical Dehydration 585
12.5.4.2. Critical-Point Drying 590
12.5.4.3. Low-Temperature Drying 592
12.5.4.4. Ambient-Temperature Sublimation 594
12.5.5. Exposure of Internal Surfaces 594
12.5.5.1. Sectioning 596
12.5.5.2. Fracturing 599
12.5.5.3. Replication 604
12.5.5.4. Surface Etching 606
12.5.6. Specimen Supports 606
12.5.7. High-Resolution Scanning Microscopy 609
12.5.7.1. Isolation of Object of Interest 609
12.5.7.2. Stabilization and Conductive Staining 611
12.5.7.3. Specimen Coating 611
12.6. Specimen Preparation for Localization of Metabolic Activity and Chemical Specificity 611
12.6.1. Introduction 611
12.6.2. The Nature of the Problem 612
12.6.2.1. The Form of the Substance Being Analyzed 612
12.6.2.2. Precision of Analytical Investigation 612
12.6.2.3. Types of Specimens 614
12.6.2.4. Types of Instrumentation 616
12.6.2.5. Types of Analytical Applications 616
12.6.3. General Preparative Procedures 617
12.6.3.1. Before Fixation 617
12.6.3.2. Fixation 617
12.6.3.3. Histochemical Techniques 619

12.6.3.4. Precipitation Techniques 620
12.6.3.5. Dehydration 621
12.6.3.6. Embedding 622
12.6.3.7. Sectioning and Fracturing 623
12.6.3.8. Specimen Supports 623
12.6.3.9. Specimen Staining 624
12.6.4. Localizing Regions of Biological Activity and Chemical Specificity 624
12.6.4.1. Backscattered-Electron Cytochemical Methods 624
12.6.4.2. Radioactive Labelling Methods 628
12.6.4.3. Immunocytochemical Methods 629
12.6.5. Criteria for Satisfactory Specimen Preparation . . . 633
12.7. Preparative Procedures for Organic Samples Such as Polymers, Plastics, and Paints 635
12.7.1. Introduction 635
12.7.2. Examination of the Surface of Polymers and Fibers 635
12.7.3. Examination of the Interior of Polymers and Fibers 635
12.7.3.1. Sectioning 636
12.7.3.2. Polished Cut Surfaces 636
12.7.3.3. Peelback Procedure 637
12.7.3.4. Fracturing 638
12.7.3.5. Replicas 639
12.7.4. Surface Etching of Polymers 639
12.7.4.1. High-Energy Beam Bombardment 640
12.7.4.2. Argon Ion–Beam Etching 640
12.7.4.3. Oxygen Plasma Etching 640
12.7.4.4. Chemical Dissolution 640
12.7.4.5. Chemical Attack 641
12.7.4.6. Enzymatic Digestion 641
12.7.5. Staining of Polymers and Plastics 641
12.7.5.1. Osmium Tetroxide 641
12.7.5.2. Chlorosulphonic Acid 641
12.7.5.3. Phosphotungstic Acid 642
12.7.5.4. Ruthenium Tetroxide 642
12.7.5.5. Silver Salts 642
12.7.6. Specialized Preparative Methods 643
12.8. Low-Temperature Specimen Preparation for Structural and Analytical Studies 644
12.8.1. Introduction 644
12.8.2. Water: Properties 645
12.8.3. Ice 646
12.8.4. Rapid Cooling Procedures 647
12.8.5. Cryosectioning 649
12.8.6. Cryofracturing 649
12.8.7. Freeze Drying 651
12.8.8. Freeze Substitution and Low-Temperature Embedding 652
12.8.9. Low-Temperature SEM 654
12.8.10. Low-Temperature X-Ray Microanalysis 654
12.9. Damage, Artifact, and Interpretation 658
12.9.1. Introduction 658

12.9.2. Sample Damage during Preparation 658
12.9.3. Specimen Damage during Examination and Analysis 659
12.9.3.1. Observable Damage 659
12.9.3.2. Nonobservable Damage 662
12.9.4. Artifacts 668
12.9.5. Interpretation 669
12.10. Specific Preparative Procedures: A Bibliography 669

13. Coating and Conductivity Techniques for SEM and Microanalysis 671

13.1. Introduction 671
13.2. Specimen Characteristics 672
13.2.1. Conductivity 672
13.2.2. Thermal Damage 672
13.2.3. Secondary- and Backscattered-Electron Emission . . 674
13.2.4. X-Ray and Cathodoluminescence Emission 674
13.2.5. Mechanical Stability 675
13.3. Untreated Specimens 675
13.4. Bulk Conductivity Staining Methods 678
13.5. Specimen Mounting Procedures 684
13.6. Thin-Film Methods 685
13.7. Thermal Evaporation 686
13.7.1. High-Vacuum Evaporation 688
13.7.2. The Apparatus 692
13.7.3. Choice of Evaporant 694
13.7.4. Evaporation Techniques 696
13.7.5. Artifacts Associated with Evaporative Coating . . . 700
13.7.6. Low-Vacuum Evaporation 701
13.8. Sputter Coating 701
13.8.1. Diode or Direct-Current Sputtering 703
13.8.2. Plasma–Magnetron Sputtering 705
13.8.3. Ion-Beam Sputtering 707
13.8.4. Penning Sputtering 711
13.8.5. Sputtering Techniques 715
13.8.6. Choice of Target Material 716
13.8.7. Coating Thickness 716
13.8.8. Advantages of Sputter Coating 717
13.8.9. Artifacts Associated with Sputter Coating 719
13.9. Specialized Coating Methods 720
13.9.1. High-Resolution Coating 720
13.9.2. Coating Samples Maintained at Low Specimen Temperatures 723
13.9.3. Coating Frozen-Hydrated Material Maintained at Low Temperatures 726
13.9.4. Coating Techniques for X-Ray Microanalysis 729
13.10. Determination of Coating Thickness 732
13.10.1. Estimation of Coating Thickness 732
13.10.2. Measurement during Coating 733
13.10.3. Measurement after Coating 735
13.10.4. Removing Coating Layers 736

13.11. Artifacts Related to Coating and Bulk-Conductivity Procedures 737
13.12. Conclusions 739

14 Data Base 741

Table 14.1. Atomic Number, Atomic Weight, and Density of Elements 741
Table 14.2. Common Oxides of the Elements 743
Table 14.3. Mass Absorption Coefficients for $K\alpha$ Lines 744
Table 14.4. Mass Absorption Coefficients for $L\alpha$ Lines 752
Table 14.5. Mass Absorption Coefficients for $M\alpha$ Lines 768
Table 14.6. K Series X-Ray Wavelengths and Energies 778
Table 14.7. L Series X-Ray Wavelengths and Energies 780
Table 14.8. M Series X-Ray Wavelengths and Energies 781
Table 14.9. J and Fluorescent Yield (ω) by Atomic Number 782
Table 14.10. Important Properties of Selected Coating Elements 784

References 787

Index 807

1 Introduction

In our era of rapidly expanding technology, the scientist must observe, analyze, and correctly explain phenomena occurring on a micrometer (μm) or submicrometer scale. The scanning electron microscope and electron microprobe are powerful instruments which permit the observation and characterization of heterogeneous organic and inorganic materials and surfaces on such a local scale. In both instruments, the area to be examined or the microvolume to be analyzed is irradiated with a finely focused electron beam, which may be static or swept in a raster across the surface of the specimen. The types of signals produced when the electron beam impinges on a specimen surface include secondary electrons, backscattered electrons, Auger electrons, characteristic x rays, and photons of various energies. These signals are obtained from specific emission volumes within the sample and can be used to examine many characteristics of the sample (composition, surface topography, crystallography, etc.).

In the scanning electron microscope (SEM), the signals of greatest interest are the secondary and backscattered electrons, since these vary according to differences in surface topography as the electron beam sweeps across the specimen. The secondary-electron emission is confined to a volume near the beam's impact area, permitting images to be obtained at relatively high resolution. Other signals are available, which prove similarly useful in many cases.

In the electron probe microanalyzer (EPMA), frequently referred to as the electron microprobe, the primary radiation of interest is the characteristic x rays emitted as a result of the electron bombardment. The analysis of the characteristic x radiation can yield both qualitative identification and quantitative compositional information from regions of a specimen as small as a micrometer in diameter.

Historically, the scanning electron microscope and electron microprobe evolved as separate instruments. It is obvious on inspection, however, that these two instruments are quite similar, differing mainly in the way in which they are used. The development of each of these instruments (SEM and EPMA) and the differences and similarities among modern commercial instruments are discussed in this chapter.

1.1. Evolution of the Scanning Electron Microscope

The SEM is one of the most versatile instruments available for the examination and analysis of the microstructural characteristics of solid objects. The primary reason for the SEM's usefulness is the high resolution that can be obtained when bulk objects are examined; values on the order of 2 to 5 nm (20–50 Å) are now usually quoted for commercial instruments, while advanced research instruments are available that have achieved resolutions of better than 1 nm (10 Å) (Nagatani *et al.*, 1987). The high-resolution micrograph shown in Fig. 1.1 was taken with an advanced, field-emission-gun, commercial SEM under routine operating conditions.

Another important feature of the SEM is the three-dimensional appearance of the specimen image, a direct result of the large depth of field, as well as to the shadow-relief effect of the secondary and backscattered electron contrast. Figure 1.2a shows the skeleton of a

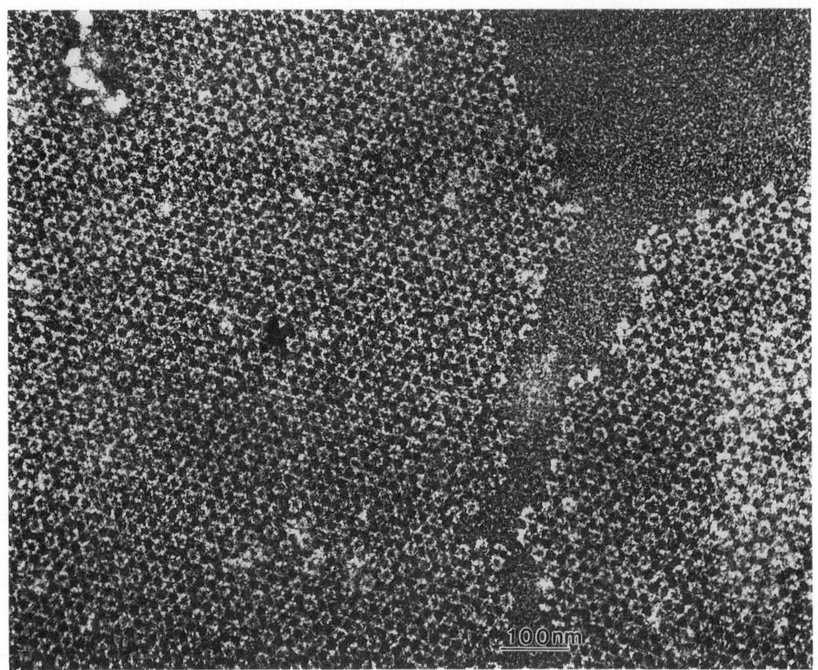

Figure 1.1. High-resolution secondary-electron image of crystalline HPI-layer (polyhead) shadowed with Pt–Ir–C. Image recorded at 30 keV on a Hitachi S-900 FEG SEM. (Courtesy of P. Walther, University of Wisconsin, Madison.)

Optical Micrograph

25µm

Figure 1.2. (a) Optical micrograph of the radiolarian *Trochodiscus longispinus*. (b) SEM micrograph of same radiolarian. The greater depth of focus and superior resolving capability are apparent.

small marine organism (the radiolarian *Trochodiscus longispinus*) viewed with a light microscope, and Fig. 1.2b shows the same object viewed with the SEM. The greater depth of field of the SEM provides much more information about the specimen. In fact, the SEM literature indicates that it is this feature which is of the most value to the SEM user. Most SEM micrographs have been produced with magnifications below 8000 × (8,000 diameters). At these magnifications the SEM is operating well within its resolution capabilities. Figure 1.3, a micrograph of a pollen grain of *Ipomoea purpurea* L. (morning glory), shows in one picture the complex surface topography of the wall of this single plant cell. The only other ways to obtain this type of detailed information would be to painstakingly reconstruct the three-dimensional structure from planar serial sections observed in the transmission electron microscope or to reconstruct the image from a through-focal series of images formed in a scanning confocal optical microscope, but at the penalty of greatly reduced spatial resolution. It would be difficult to make a faithful replica of such a detailed and irregular surface.

The SEM is also capable of examining objects at very low magnification. This feature is useful in forensic studies as well as other fields because the electron image complements the information available from the optical image. A low-magnification micrograph of an archaeological subject is shown in Fig. 1.4.

The basic components of the SEM are the lens system, electron gun, electron collector, visual and recording cathode ray tubes

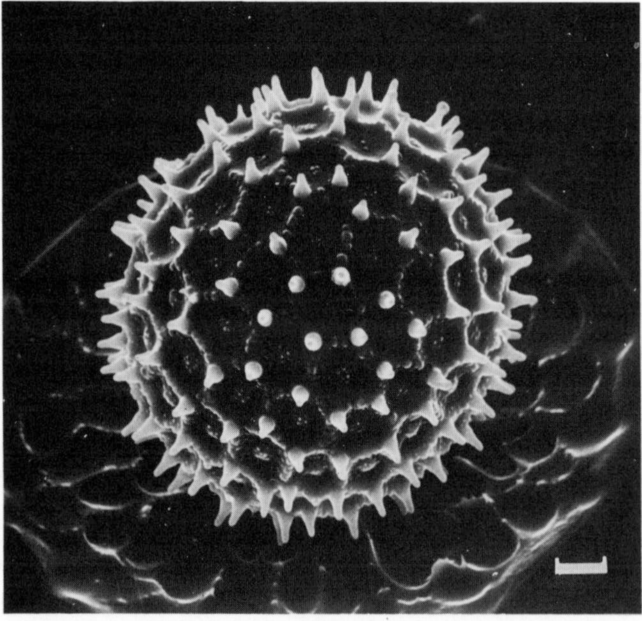

Figure 1.3. A whole pollen grain from *Ipomoea purpurea* L. (morning glory). Note the large spines, some of which are as long as 5 μm. Marker = 6 μm.

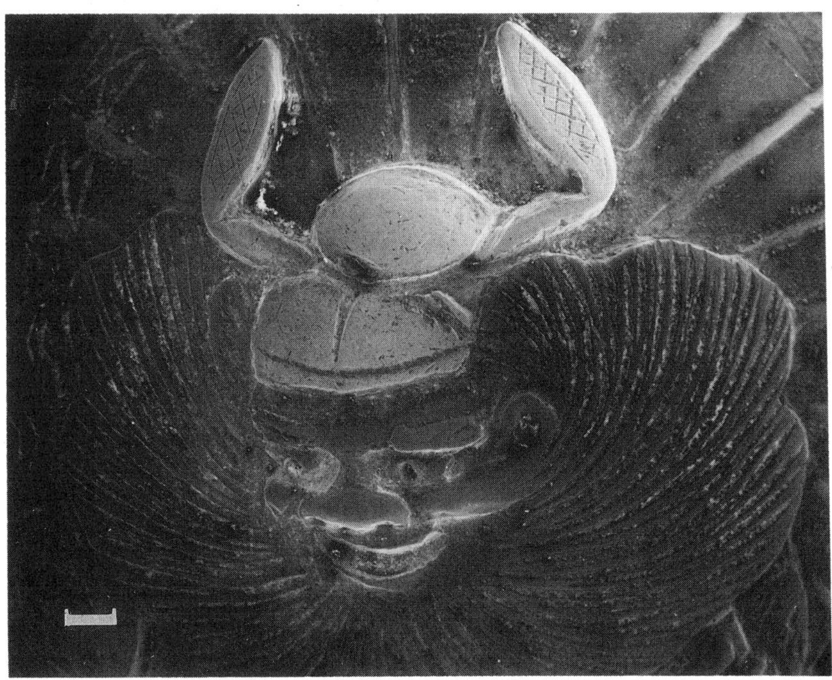

Figure 1.4. SEM image of face of a helmeted, demonlike warrior from the rear of the handle of an eighteenth-century Japanese sword. The region of the helmet is gilt. (Courtesy of M. Notis, Lehigh University.) Marker = 1 mm.

(CRTs), and the electronics associated with them. The first successful commercial packaging of these components (the Cambridge Scientific Instruments Mark I) was offered in 1965. This instrument was the culmination of a period of development extending over more than thirty years and extending over two continents. The purpose of this brief historical introduction is to point out the pioneers of scanning electron microscopy and in the process to trace the evolution of the instrument. For a detailed history, see Oatley (1982).

The earliest recognized work describing the concept of an SEM is that of Knoll (1935) who, along with other pioneers in the field of electron optics, was working in Germany. Subsequently von Ardenne (1938) constructed a scanning transmission electron microscope (STEM) by adding scan coils to a transmission electron microscope (TEM). Both the theoretical basis and the practical aspects of STEM were discussed in fairly complete detail by von Ardenne (1938), and his instrument incorporated many features which have since become standard. The first STEM micrograph was of a ZnO crystal imaged at an operating voltage of 23 kV at a magnification of 8000 diameters and with a spatial resolution between 50 and 100 nm. The photograph contained 400 × 400 scan lines and took 20 min to record because the film was mechanically scanned in synchronism with the beam. The instrument had two electrostatic condenser lenses, with the scan coils

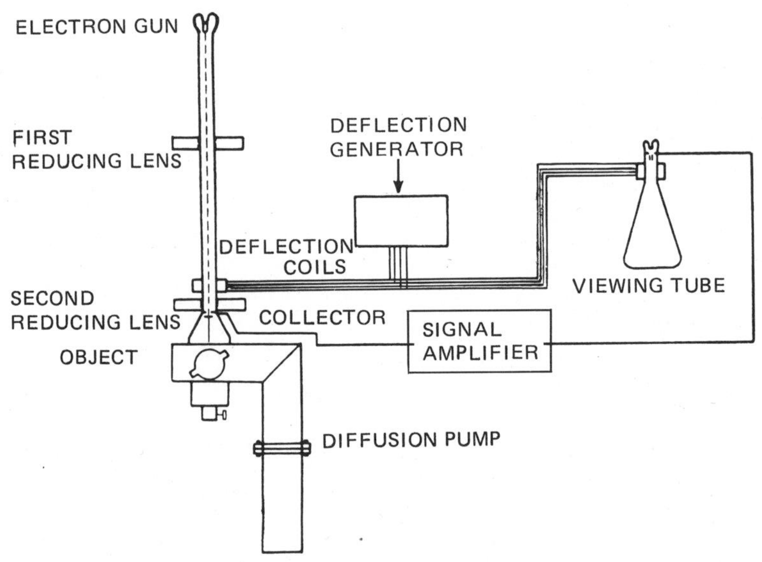

Figure 1.5. Schematic diagram of an early SEM (Zworykin *et al.*, 1942).

being placed between them. The instrument possessed a CRT for viewing, but it was not used to photograph the image.

The first SEM used to examine thick specimens was described by Zworykin *et al.* (1942), working at the RCA Laboratories in the United States. The authors recognized that secondary-electron emission would be responsible for topographic contrast and accordingly constructed the design shown in Fig. 1.5. The collector was biased positively relative to the specimen by 50 V, and the secondary-electron current collected on it produced a voltage drop across a resistor. This voltage drop was sent to a television set to produce the image; however, a resolution of only 1 μm was attained. This was considered unsatisfactory by the authors since they sought to achieve better resolution than that obtainable with the light microscope (i.e., about 0.5 μm). They therefore decided to reduce the electron beam spot size, as well as to improve the signal-to-noise ratio, and so produce a better instrument. A detailed analysis of the interrelationship of lens aberrations, gun brightness, and spot size resulted in a method for determining the minimum spot size as a function of beam current (Zworykin *et al.*, 1942). They also sought to improve gun brightness by using a field-emitter source, but instability in these cold-cathode sources forced a return to thermionic emission sources although even in 1942 they were able to produce high-magnification, high-resolution images. Their next contribution was the use of an electron multiplier tube as a preamplifier for the secondary-emission current from the specimen. Useful but noisy (by today's standards) photomicrographs were obtained. The electron optics of the instrument consisted of three electrostatic lenses with scan coils placed between the second and third lenses. The electron gun was at the bottom so that the specimen chamber was at a comfortable height for the operator. Resolutions of about 50 nm (500 Å) were achieved with this first modern SEM but, by comparison with the performance then obtainable from the rapidly developing TEM, this figure was considered unexciting and further development languished.

In 1948 C. W. Oatley, then a lecturer at the University of Cambridge, became interested in building electron microscopes and decided to reinvestigate the SEM as a complement to the work being done on the TEM by V. E. Cosslett, also in Cambridge. During the next few years, Oatley and his student, D. McMullan, built their first SEM, and by 1952 this unit had achieved a resolution of 50 nm (500 Å) (McMullan, 1952). McMullan was followed by K. C. A. Smith (1956) who, recognizing that signal processing could improve micrographs, introduced nonlinear signal amplification (γ processing). He also replaced electrostatic with electromagnetic lenses and improved the scanning system by introducing double-deflection scanning. He was also the first to insert a stigmator into the SEM (Smith, 1956).

The next step forward was the improvement of the secondary-electron detector described by Zworykin *et al.* (1942). This was accomplished by Everhart and Thornley (1960), who employed a scintillator to convert the electrons to light, which was then transmitted by a light pipe directly to the face of a photomultiplier. Replacement of the electron multiplier by the more efficient photomultiplier increased the amount of signal collected and resulted in an improvement in signal-to-noise ratio. Hence, weak contrast mechanisms, such as the phenomenon of voltage contrast discovered by Oatley and Everhart (1957), could be better investigated. Image interpretation was also improved when Wells (1959) made the first studies of the effects of beam penetration on image formation in the SEM and was the first to use stereographic pairs to produce SEM micrographs with quantifiable depth information.

Pease (1963) built a system, known as SEM V, with three magnetic lenses and the gun at the bottom, using the Everhart–Thornley detector system. This instrument became the prototype of the Cambridge Scientific Instruments Mark I "Stereoscan" (Pease, 1963; Pease and Nixon, 1965). A. D. G. Stewart and co-workers at the Cambridge Scientific Instrument Co. carried out the commercial design and packing of the instrument. In the ensuing 30 yr, more than 15,000 SEM units have been sold by a dozen or more manufacturers (from the United States, the United Kingdom, the Netherlands, Japan, Germany, and France), which are actively developing new, improved instruments.

Since the first commercial instrument of 1965, many advances have been made. One of these was the development of the lanthanum hexaboride (LaB_6) cathode by Broers (1969a). This material provides a high-brightness electron gun; hence, more electron current can be concentrated into a smaller beam spot and an effective improvement in resolution can be obtained. The field-emission electron source, first used in the SEM in 1942, was revived by Crewe (1969) and has now been developed to the point where it can be routinely used for high-resolution imaging. The advantage of the field-emission gun is that the source is very small, so only simple optics are required to produce a probe of nanometer size, whereas the brightness is very high, so that current densities of thousands of amperes per square centimeter are available even when the beam current is on the order of 0.1 nA. Field-emission sources, however, require clean, high vacuums on the order of 10^{-8} Pa (10^{-10} torr) or better to operate reliably, and such pressures require special attention to vacuum technology. Field-emission-gun SEMs are now commercially available from several companies and are rapidly gaining acceptance because of their high resolution and ability to operate efficiently at low beam energies.

Often the contrast of the features one is examining is so low as to be invisible to the eye. Therefore, contrast enhancement by processing of the signal is needed. Early signal processing included nonlinear amplification, as noted, and differential amplification (black level suppression), both incorporated into SEMs at Cambridge University. Derivative signal processing (differentiation) to enhance small details was introduced later (Lander *et al.*, 1963; Heinrich *et al.*, 1970; Fiori *et al.*, 1974). While most commercial SEMs today are still provided with these simple analog signal-processing capabilities, increasingly the SEM is being coupled to a computer to provide a digital signal, as well as processing and storage capability.

Many SEMs are now equipped with digital framestores which permit images to be stored temporarily for observation or, after transfer to a magnetic disc, permanently for the record. Once in digital form, the image may then be processed to achieve all of the effects attainable by the older analog methods, as well as in many new and productive ways. With the advent of powerful and inexpensive personal computers equipped with high-resolution graphical displays, software packages capable of a full range of processing and quantitative functions on digital images have become available, giving the microscopist an unprecedented degree of flexibility and convenience in using the output of the SEM.

Other advances involve contrast mechanisms not readily available in other types of instrumentation. Electron-channeling contrast, produced by crystal orientation and lattice interactions with the primary beam, was first observed by Coates (1967) and has been developed to permit a full range of crystallographic measurements in the SEM (Joy *et al.*, 1982). Contrast from magnetic domains in uniaxial materials was observed by Joy and Jakubovics (1968). Magnetic contrast in cubic materials was first observed by Philibert and Tixier (1969); this contrast mechanism was explained later by Fathers *et al.* (1973, 1974).

Some of the first scanning micrographs were of selected biological materials, such as hard-shell insects and wood fibers, which were strong enough to withstand without distortion the process of being dried in air (Smith and Oatley, 1955). Later Thornley (1960) showed SEM images of freeze-dried biological material, examined at 1 keV to avoid charging. Boyde and Stewart (1962) published one of the first papers in which SEM was used specifically to study a biological problem.

Advances in the biological field have been regulated to some extent by necessary advances in specimen preparation. Most biological specimens are wet, radiation sensitive, thermolabile samples of low contrast and weak emissivity, and they are invariably poor conductors. Much attention has been paid to stabilizing the delicate organic materials, removing or immobilizing the cell fluids, and coating the samples with a thin layer of a conducting film. Ways have been devised to selectively increase the contrast of specimens and to reveal their contents at predetermined planes within the sample. The development of low-temperature stages has helped to reduce the mass loss and thermal damage in sensitive specimens (Echlin, 1978). The preparative techniques and associated instrumentation have now advanced to the point where any biological material may be examined in the SEM with the certainty that the images obtained are a reasonable representation of the once-living state.

The large depth of field available in the SEM makes it possible to observe three-dimensional objects using stereoscopy (Wells, 1960). This is particularly important in the examination of biological samples, which—unlike many metals, rocks, polymers, and ceramics—are morphologically very heterogeneous and must be examined in depth as well as at a given surface. The three-dimensional images allow various morphological features to be correctly interrelated and definitively measured. Equipment has been developed that allows quantitative evaluation of surface topography making use of this feature (Boyde, 1974a: Boyde and Howell, 1977; Boyde, 1979). Provisions for direct, real-time, high-resolution stereo viewing on the SEM have been described as well (Pawley, 1988).

The addition of an energy-dispersive x-ray detector to an electron probe microanalyzer (Fitzgerald *et al.*, 1968) signaled the eventual coupling of such instrumentation to the SEM. Today, it is estimated that well over half of all SEMs are now equipped with x-ray analytical capabilities. Thus, topographic, crystallographic, and compositional information can be obtained rapidly, efficiently, and simultaneously from the same area.

In its current form, the SEM is both competitive with and complementary to the capabilities offered by other microscopes. It offers much of the use and image interpretation found in the conventional light microscope while providing an improved depth of field and the benefits of image processing. While the SEM lacks the three-dimensional sectioning abilities of scanning beam confocal microscopes, the analytical capabilities of the SEM provide information that cannot be obtained on the light-optical instruments, and the ultimate resolution is far superior. In fact, the best modern SEMs now have spatial resolutions in the nanometer range and are thus directly comparable in their performance with TEMs in many situations, but the SEM has the added advantage that the specimen need not be made thin enough to transmit electrons. The lateral resolution of a high-performance SEM on solid specimens is even comparable with that obtained by the scanning tunneling microscope (STM) or atomic force microscope (AFM), and although the resolution in the vertical direction is much poorer than for the STM or AFM, the image formation mechanisms are much better understood. The SEM is better able to deal with specimens having pronounced topography than the STM or AFM, where probe interactions become difficult to control and understand when multiple surfaces are involved. The SEM is thus a versatile and powerful machine and consequently a major tool in research and technology.

1.2. Evolution of the Electron Probe Microanalyzer

The EPMA is one of the most powerful instruments for the microanalysis of inorganic and organic materials. The primary reason for the EPMA's usefulness is that compositional information, using characteristic x-ray lines, with a spatial resolution on the order of 1 μm can be obtained from a sample. The sample is analyzed nondestructively, and quantitative analysis can be obtained in many cases with an accuracy of the order of 1–2% of the amount present for a given element (5–10% in biological materials). In the case of some labile biological and organic substances, there can be a substantial (i.e., up to 90%) mass loss during the period of analysis. Elemental losses can also occur, particularly in the case of light elements in an organic matrix. Both mass loss and elemental loss can be reduced to negligible properties by maintaining the specimen at low temperature.

Figure 1.6 shows a microprobe analysis of C, Fe, Ni, and Co across a carbide (cohenite) phase which has nucleated and grown by a diffusion-controlled process in a ferrite α-bcc phase matrix during cooling of a lunar metal particle on the moon's surface (Goldstein *et al.*, 1976). The microchemical analyses were taken with a commercial EPMA and illustrate the resolution of the quantitative x-ray analysis which can be obtained.

Another important feature of the EPMA is its capability of obtaining compositional mapping with characteristic x rays. The x ray images show the elemental distribution in the area of interest. Figure 1.7 shows the distribution of Fe and P in the various phases of a metal grain from lunar soil 68501 (Hewins *et al.*, 1976). Magnifications up to 2500X are possible without exceeding the x-ray spatial resolution of the instrument. The attractiveness of this form of data gathering is that detailed microcompositional information can be directly correlated with light-optical and electron metallography. In addition, reduced-

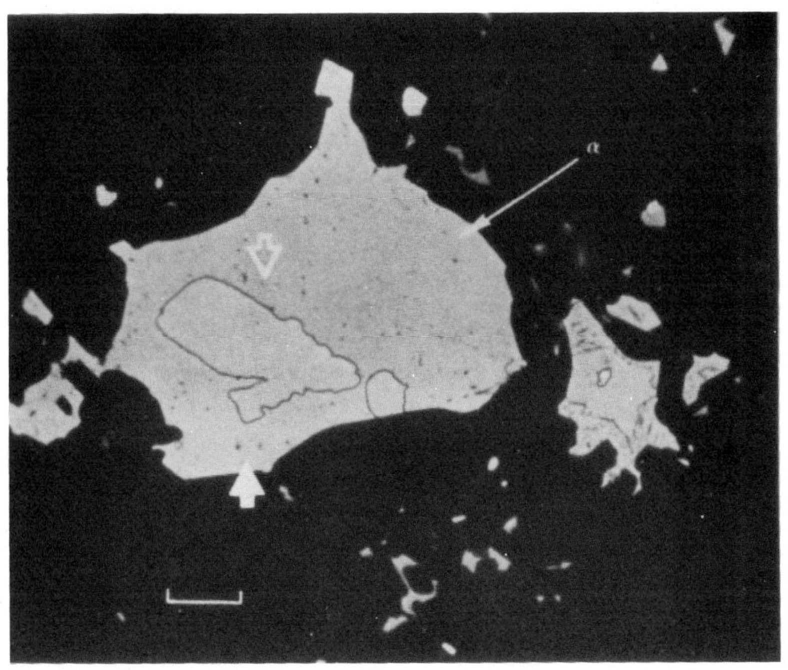

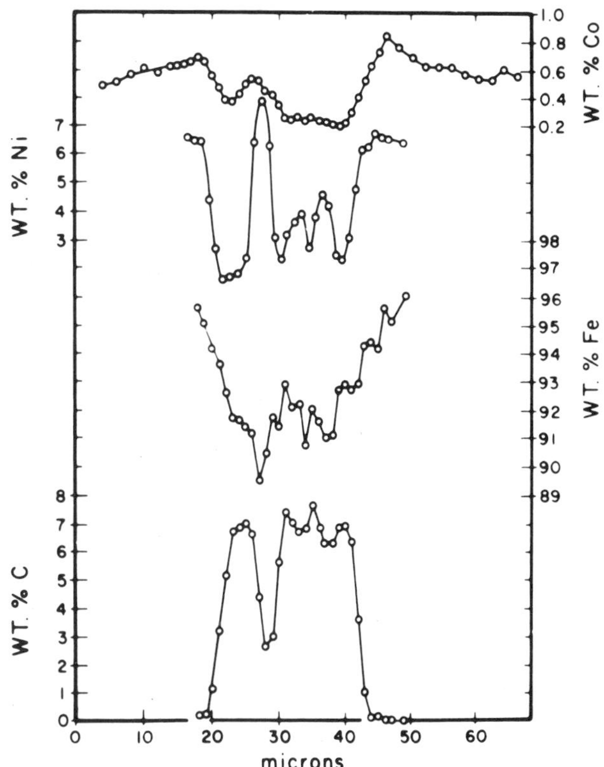

Figure 1.6. (a) Micrograph in reflected light of lunar rock 73275,68. The α phase surrounds the cohenite, and the arrows indicate the extent of the microprobe scan shown in (b) (Goldstein *et al.*, 1976). Marker = 16 μm. (b) Variation of C, Fe, Ni, and Co across α-cohenite (Fe_3C) particle in lunar rock 73275 shown in (a).

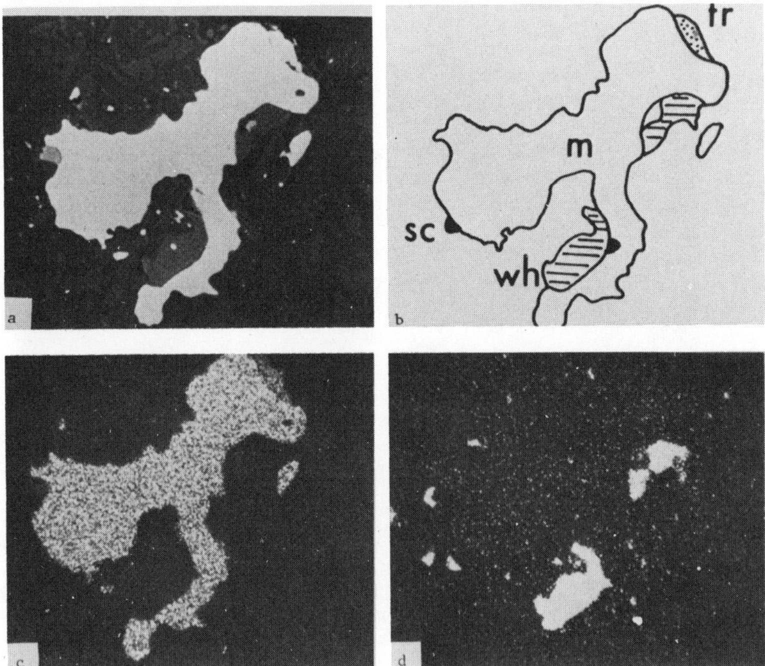

Figure 1.7. A metal grain about 105 μm long occurring with a phosphate mineral in a lunar melt rock from soil 68501 (Hewins *et al.*, 1976). (a) Backscattered-electron image showing metal (white), phosphate and pyroxene (medium gray), and plagioclase (dark gray). (b) Sketch of (a) showing grains of metal (m), trolite, FeS (tr), phosphide $(FeNi)_3P$ (sc), and phosphate (wh). (c) Fe Kα image of (a). (d) P $K\alpha$ image of (a).

area scans can be used to compare the relative compositions of different areas of interest. Figure 1.8 shows a simple biological application of EPMA in which differences in elemental ratios have been measured in developing plant roots. In this example, chlorine is present in the nucleus but not detected in the cytoplasm. This analysis was carried out using a reduced raster measuring approximately 1 μm^2. The EPMA literature indicates that the x-ray scanning feature of the EPMA is of great value to the user. In addition, the variety of signals available in the EPMA (emitted electrons, photons, etc.) can provide useful information about surface topography and composition in small regions of the specimen. The early development of EPMA is reviewed in the following paragraphs.

In 1913, Moseley found that the frequency of emitted characteristic x-ray radiation is a function of the atomic number of the emitting element (Moseley, 1913). This discovery led to the technique of x-ray spectrochemical analysis, by which the elements present in a specimen could be identified by the examination of the directly or indirectly excited x-ray spectra. The area analyzed, however, was quite large (>1 mm^2). The idea of the electron microanalyzer, in which a focused electron beam was used to excite a small area on a specimen ($\sim 1\ \mu m^2$) and which included a light-optical microscope for locating the area of interest, was first patented in the 1940s (Marton, 1941; Hillier, 1947).

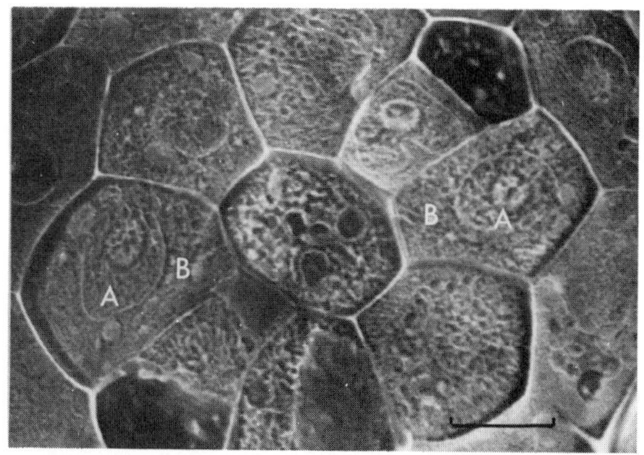

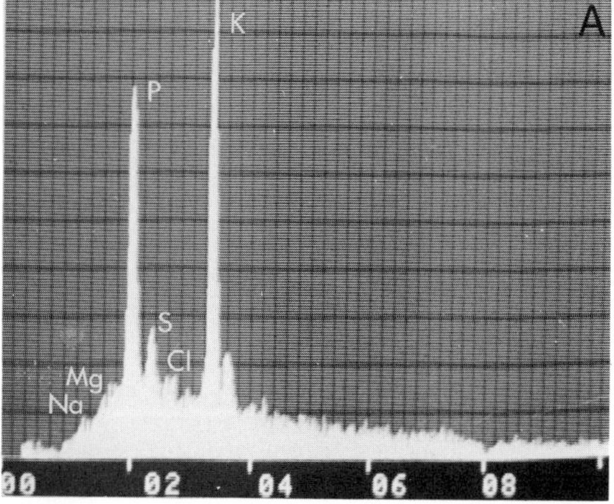

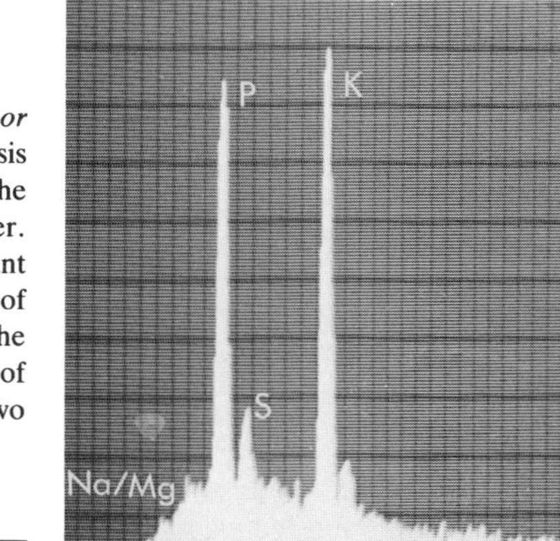

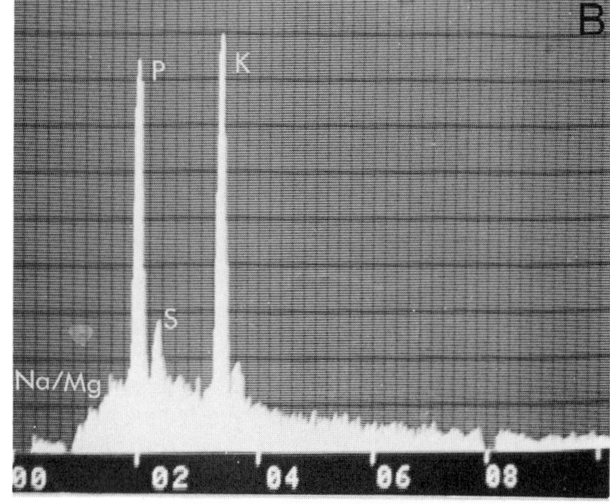

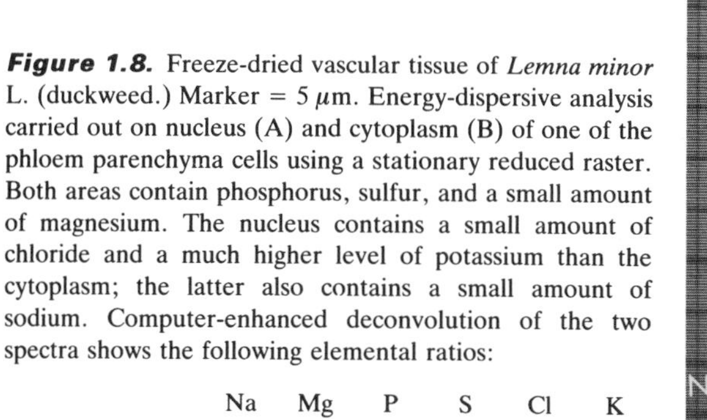

Figure 1.8. Freeze-dried vascular tissue of *Lemna minor* L. (duckweed.) Marker = 5 μm. Energy-dispersive analysis carried out on nucleus (A) and cytoplasm (B) of one of the phloem parenchyma cells using a stationary reduced raster. Both areas contain phosphorus, sulfur, and a small amount of magnesium. The nucleus contains a small amount of chloride and a much higher level of potassium than the cytoplasm; the latter also contains a small amount of sodium. Computer-enhanced deconvolution of the two spectra shows the following elemental ratios:

	Na	Mg	P	S	Cl	K
Nucleus	0	6	35	6	2	42
Cytoplasm	2	4	31	5	0	29

It was not until 1949, however, that R. Castaing, under the direction of Guinier, described and built an instrument called the "Microsonde electronique," or electron microprobe (Castaing and Guinier, 1950; Castaing, 1951; Castaing, 1960). In his doctoral thesis, Castaing demonstrated that a localized chemical analysis could be made on the surface of a specimen and outlined the approach by which this information could be quantified. Concurrently with the work of Castaing in France, Borovskii in the U.S.S.R. developed an EPMA quite dissimilar in design (Borovskii and Ilin, 1953).

During the early 1950s, several EPMA instruments were developed in laboratories in both Europe and the United States (Haine and Mulvey, 1959; Birks and Brooks, 1957; Fisher and Schwarts, 1957; Wittry, 1957; Cuthill *et al.*, 1963). The first commercial EPMA instrument, introduced by CAMECA in France in 1956, was based on the design of an EPMA built by Castaing in the laboratories of the Recherches Aeronautiques (Castaing, 1960). Figure 1.9 shows diagrammatically the design of the apparatus. The electron optics consisted of an electron gun followed by reducing lenses which formed an electron probe with a diameter of approximately 0.1 to 1 μm on the specimen. Since electrons produce x-rays from a region often exceeding 1 μm wide and 1 μm deep, it is usually unnecessary to use probes of very small diameter. A light microscope for accurately choosing the point to be analyzed and a set of wavelength dispersive spectrometers

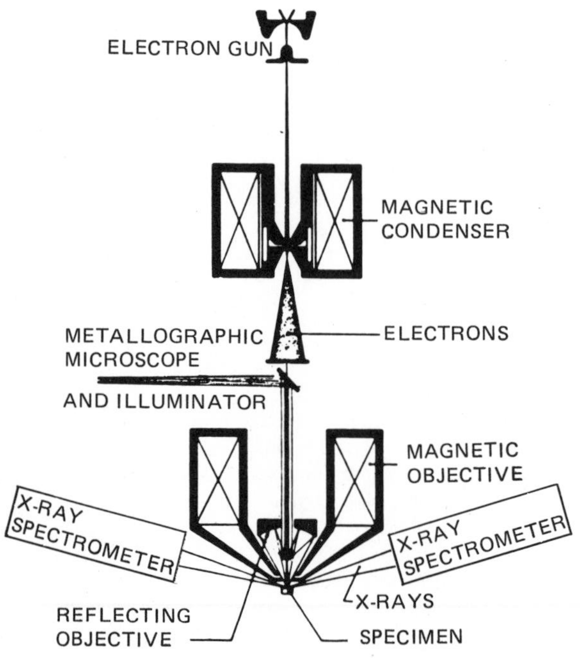

Figure 1.9. Schematic diagram of the French microanalyzer (adapted from Castaing, 1960).

for analyzing the intensity of x-ray radiation emitted as a function of energy are also part of the instrument.

Cosslett and Duncumb (1956) designed and built the first scanning electron microprobe at the Cavendish Laboratories in Cambridge, England. Whereas all previous electron microprobes had operated with a static electron probe, Cosslett and Duncumb swept the beam across the surface of a specimen in a raster, as is done in current SEMs. They used the backscattered electron signal to modulate the brightness of a cathode ray tube sweeping in synchronism with the electron probe. They also used the x-ray signal to modulate the brightness, permitting a scanned image to be obtained showing the lateral distribution of a particular element. Although the concept of a local x-ray analysis is in itself a strong incentive for the use of a microprobe, the addition of the scanning concept was an extremely significant contribution and probably accounted for the subsequent increased popularity of the electron microprobe.

In his doctoral thesis, Castaing developed the physical theory enabling the analyst to convert measured x-ray intensities to chemical composition (Castaing, 1951). Castaing proposed a method of analysis based on comparisons of intensity, i.e., the comparison of the intensity I_i of a characteristic line of a given element emitted by a specimen under one set of conditions of electron bombardment and the intensity $I_{(i)}$ of the same characteristic radiation when emitted by a standard containing the same element under the same electron bombardment conditions. The ratio of the two readings (the so-called k ratio) is proportional to the mass concentration of a given element in the region analyzed. Recognition of the complexity of converting x-ray intensities to chemical composition has led numerous investigators (Wittry, 1963; Philibert, 1963; Duncumb and Shields, 1966) to expand the theoretical treatment of quantitative analysis proposed by Castaing.

In the years since the development of the first scanning EPMA instrument, many advances have been made. Of particular importance was the development of diffracting crystals having large interplanar spacings (Henke, 1964; 1965). These crystals enable long-wavelength x-rays from the light elements to be measured with wavelength-dispersive spectrometers. The ability to detect fluorine, oxygen, nitrogen, carbon, and boron enabled users of the EPMA to investigate many new types of problems with the instrument (note Fig. 1.6). Techniques of soft x-ray spectroscopy have now been applied in the EPMA to establish how each element is chemically combined in the sample. The chemical effect may be observed as changes in wavelength, shape, or relative intensity of emission and absorption spectra.

The x-ray microanalysis of biological material is beset by the same problems associated with the examination of the sample in the SEM. The experimenter has to be very careful that the elements being measured remain in the specimen and are not removed or relocated by the preparative procedure. While specimen preparation for biological material is more exacting than the procedures used in the materials sciences, the quantitation methods are in principle somewhat simpler because they are based on the analysis of thin sections, so that absorption effects may often be neglected. However, such thin sections are all too easily damaged by the electron beam and much effort in the past decade has gone into devising instrumentation as well as preparative and analytical procedures to limit specimen damage in organic samples.

When the application of the EPMA was extended to nonmetallic specimens, it became apparent that other types of excitation phenomena might also be useful. For example, the color of visible light (cathodoluminescence) produced by the interaction of the electron probe with the specimen has been associated with the presence of certain impurities in minerals (Long and Agrell, 1965). In addition, photon radiation produced by the recombination of excess hole–electron pairs in a semiconductor can be studied (Kyser and Wittry, 1966). Measurement of cathodoluminescence in the EPMA has now been developed as another important use of this instrument (Yacobi and Holt, 1990).

Increased use of the computer in conjunction with the EPMA has greatly improved the quality and quantity of the data obtained. Many computer programs have been developed to convert the x-ray intensity ratios to chemical compositions, primarily because some of the correction parameters are functions of concentration and hence make successive approximations necessary. These programs can be run on a minicomputer or a fast personal computer and the compositions calculated directly from recorded digital data. The advantage of rapid calculation of chemical compositions is that the operator has greater flexibility in carrying out analyses. In addition, computer automation of the EPMA has been developed to varying degrees of complexity. Dedicated computers and the accompanying software to control the electron beam, specimen stage, and spectrometers are now commercially available. The advantages of automation are many, but in particular it greatly facilitates repetitive analysis, increases the amount of quantitative analysis performed, and leaves the operator free to concentrate on evaluating the analysis and designing further experiments to be performed.

The development of the energy-resolving x-ray spectrometer based upon the silicon (lithium-drifted) Si (Li) solid-state detector (Fitzgerald *et al.*, 1968), has revolutionized x-ray microanalysis. The energy-dispersive spectrometer (EDS) system is now the most common x-ray measurement system to be found in the SEM lab. Even in classical EPMA configurations, the Si (Li) detector plays an important part alongside the wavelength-dispersive spectrometer. The EDS system offers a means of rapidly evaluating the elemental constituents of a sample; major constituents (10 wt % or more) can be identified in only 10 s, and a 100-s accumulation is often sufficient for identification of minor elements (on the order of 1 wt %). In addition to rapid qualitative analysis, accurate quantitative analysis can also be achieved with EDS x-ray spectrometry. These great advantages are tempered by the relatively poor energy resolution of the EDS (140 eV at Mn $K\alpha$, compared with 5–10 eV for a wavelength spectrometer) which leads to frequent unresolved spectral interferences (e.g., S $K\alpha, \beta$, Mo $L\alpha, \beta$, Pb $M\alpha, \beta$), poor peak-to-background values, and the resultant poor limit of detection (typically ~0.1 wt % compared to ~0.01 wt % for a wavelength spectrometer).

The development of quantitative compositional mapping provides another strong link between the imaging of the SEM and the quantitative elemental analysis of the EPMA. In quantitative compositional mapping, a complete quantitative analysis is carried out under computer control at every discrete beam location in a scanned field (Newbury *et al.*, 1990a,b; 1991). The x–y arrays of numerical concentration values corresponding to the beam positions on the specimen are assembled into images with a computer digital image processor by encoding the concentration axes with an appropriate gray or color scale. The resulting images, or compositional maps, are supported at every picture element (pixel) by the complete numerical concentration values, so the analyst can readily recover the analysis corresponding to any single pixel or array of pixels, and the compositional maps can be correlated with SEM images prepared from any of the available signals. An example of a compositional map of zinc at the grain boundaries of copper following diffusion-induced grain boundary migration is shown in Fig. 1.10.

1.3. Outline of This Book

The SEM and EPMA are in reality very similar instruments. Therefore several manufacturers have constructed instruments each

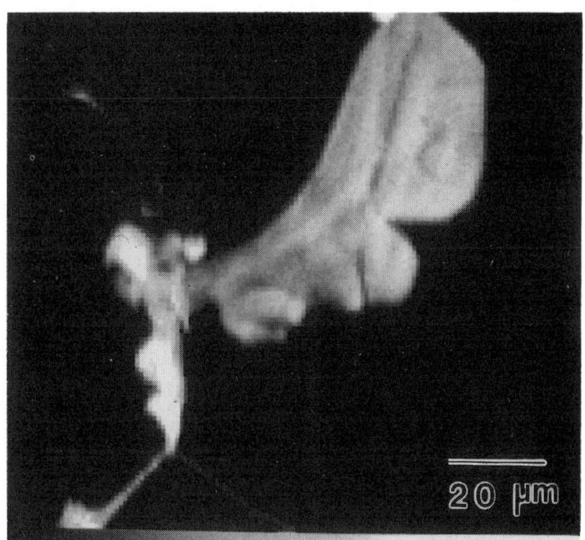

Figure 1.10. Compositional map of zinc at the grain boundaries of copper following diffusion-induced grain boundary migration. The gray scale depicts the concentration range from 0–10 wt % zinc. Field width = 100 μm (Newbury *et al.*, 1988).

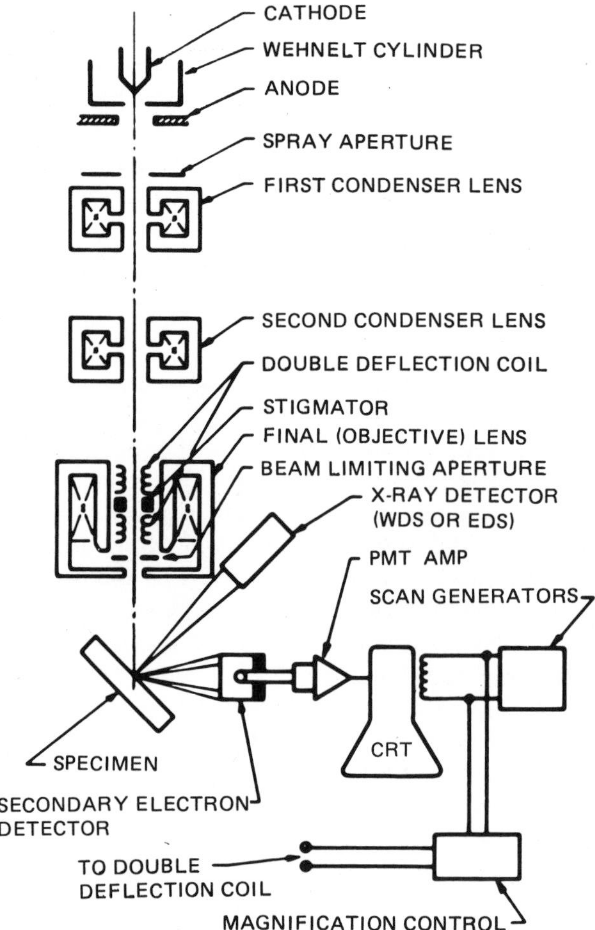

Figure 1.11. Schematic drawing of the electron and x-ray optics of a combined SEM–EPMA.

capable of being operated as an electron microprobe and as a high-resolution SEM. Figure 1.11 shows a schematic of the electron and x-ray optics of such a combination instrument. Both a secondary electron detector and an x-ray detector or detectors are placed below the final lens. For quantitative x-ray analysis and for the measurement of x rays from the light elements, at least two wavelength-dispersive spectrometers and an energy-dispersive spectrometer are desirable. In addition, provision for a broad useful current range, 1 pA to 1 μA, must be made. The scanning unit allows both electron and x-ray signals to be measured and displayed on the CRT. It is logical, therefore, that SEM–EPMA be considered one instrument in this textbook as well.

We recognize that the SEM community consists of users with an enormous diversity in technical backgrounds. The material of this book is comprehensive, ranging from introductory through intermediate levels. Persons new to the SEM–EPMA field may wish to skip the intermediate material in a first reading of this text. We have therefore distinguished the introductory and intermediate material by slightly indenting the former and placing a thick line along its left margin, as throughout this first chapter.

The electron optics system and the signals produced during electron bombardment in the SEM–EPMA are discussed in Chapters 2 and 3. The remainder of the book is devoted to the details of measuring the available signals, to the techniques of using these signals to determine particular types of information about organic and inorganic samples, and to the preparation of the specimens. The emphasis is on the selection and use of techniques appropriate to the solution of problems often presented to the SEM and EPMA analysis staff by their clients.

Chapters 4 and 5 consider the detection and processing of secondary-electron, backscattered-electron, cathodoluminescence, and x-ray signals as obtained from the SEM-EPMA. Following this material, Chapters 6, 7, 8, and 9 discuss various methods for qualitative and quantitative x-ray analysis. Chapter 10 covers qualitative and quantitative compositional mapping. The methods for preparation of solid materials such as rocks, metals, and ceramics for SEM and x-ray microanalysis are given in Chapter 11, and the methods for preparation of biological samples for SEM are discussed in Chapter 12. Coating techniques for nonconducting samples are considered in Chapter 13. Chapter 14 is a compendium of critical data useful to the analyst.

2

Electron Optics

The final electron beam diameter, often called the spot size or probe size, limits the best possible image resolution in the SEM and the EPMA. The amount of current in the final probe determines the intensity of emitted signals such as secondary electrons, backscattered electrons, or x rays. Unfortunately, the smaller the electron probe, the smaller the available probe current. Therefore, the operator must adjust the microscope controls to produce the desired result in each microscopy mode: high resolution, high depth-of-field, and microanalysis. Intelligent use of these instruments requires an understanding of how the components of the electron column control probe size and probe current. In this chapter, we will describe the electron optical column and develop the relationship between electron-probe current and electron-probe diameter (spot size).

2.1. How the SEM Works

This section provides a brief outline of the operating principles of the SEM. The basic principles given here apply to all SEMs, but the details of electron-optical design vary from manufacturer to manufacturer.

2.1.1. Functions of the SEM Subsystems

Electron Lenses Produce a Small Spot. The electron column consists of an electron gun and two or more electron lenses, operating in a vacuum. The electron gun produces a source of electrons and accelerates these electrons to an energy in the range 1–40 keV. The beam diameter produced directly by the conventional electron gun is too large to generate a sharp image at high magnification. Electron lenses are used to reduce the diameter of this source of electrons and place a small, focused electron beam on the specimen, as shown

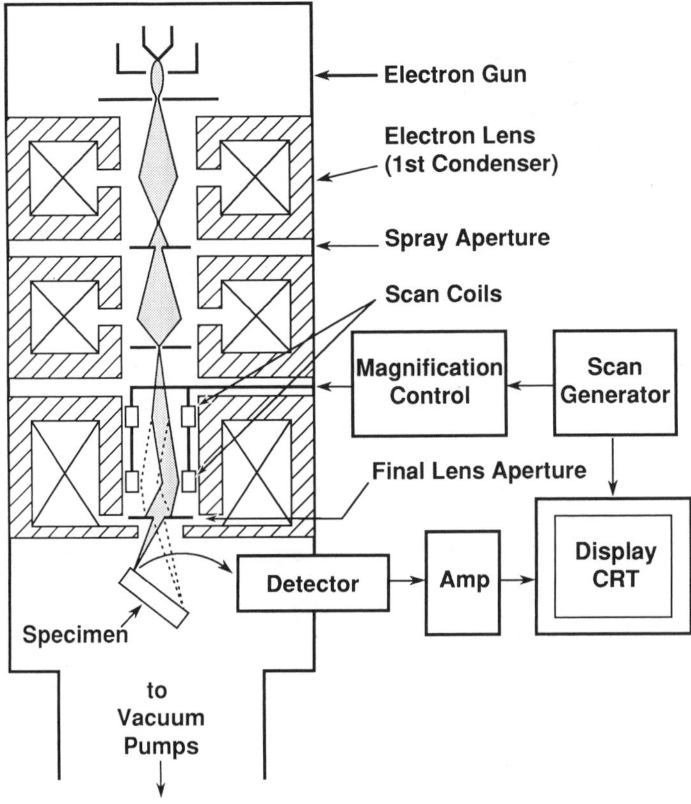

Figure 2.1. Schematic drawing showing the electron column, the deflection system, and the electron detectors.

schematically in Fig. 2.1. Most SEMs can generate an electron beam at the specimen surface with a spot size less than 10 nm (100 Å) while still carrying sufficient current to form an acceptable image. In most SEMs, the electron beam emerges from the final lens into the specimen chamber, where it interacts with the near-surface region of the specimen to a depth of approximately 1 μm and generates signals used to form an image. The actual formation of an image requires a scanning system to construct the image point by point.

Deflection System Controls Magnification. In order to produce contrast in an image, the signal intensity from the beam–specimen interaction must be measured from point to point across the specimen surface. The function of the deflection system is to scan the beam along a line and then displace the line position for the next scan so that a rectangular raster is generated on both the specimen and the viewing screen. Two pairs of electromagnetic deflection coils (scan coils) are used to control the raster of the beam. The first pair of scan coils bends the beam off the optical axis of the microscope and the second pair bends the beam back onto the axis at the pivot point of the scan (Fig. 2.1). The magnification M of the specimen image is the ratio of the linear size of the viewing screen, known as the cathode ray tube (CRT), to the linear size of the raster on the specimen. Thus, increased magnification may be obtained by exciting the scan coils less strongly so that the beam deflects a smaller distance on the specimen. Higher magnification is obtained because a smaller raster width on the specimen appears larger when displayed on the viewing screen. Thus, a 100-μm-wide raster on the specimen becomes an image of 1000X magnification when the image is displayed on a 10-cm-wide viewing screen.

Electron Detectors Collect the Signal. The interaction of the electron beam with the specimen causes the generation of many signals (see Chapter 3 for a detailed discussion of beam–specimen interactions), which may be used to modulate the intensity of the viewing CRT and produce an image (see Chapter 4 for a detailed discussion of image formation). The two signals most often used to produce images are secondary electrons (SEs) and backscattered electrons (BSEs), as shown in Fig. 2.2a. Secondary and backscattered electrons are collected by the Everhart–Thornley (E–T) electron detector. The E–T detector consists of a scintillator, a light pipe, and a photomultiplier tube. This detector is located to the side of the specimen, and the specimen is usually tilted toward it. The E–T detector is electrically isolated from the rest of the microscope and has a wire mesh screen in front of it at a potential of about +300 V. This positively charged screen draws the low-energy secondary electrons into the detector from anywhere in the specimen chamber, but to form a high quality SEM image the strongest source of SEs should be located at the point where the beam enters the specimen surface. A small fraction of the BSEs from the specimen also directly enter the E–T detector. All electrons entering the E–T detector are accelerated by a voltage of

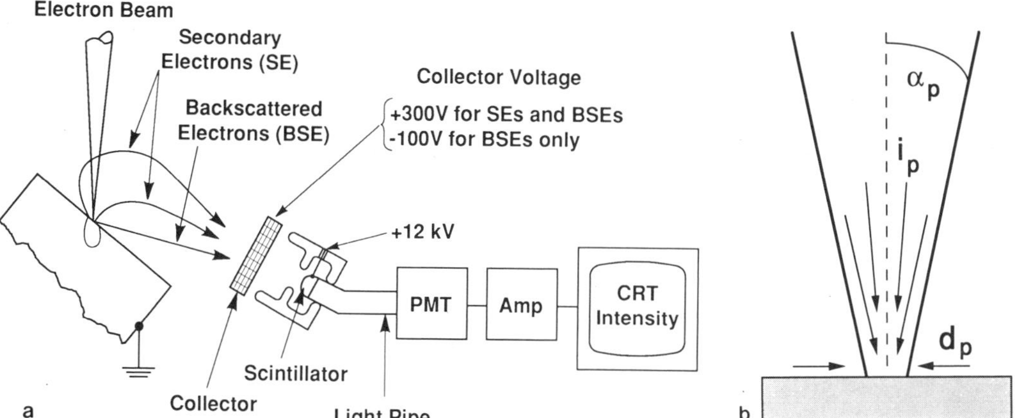

Figure 2.2. (a) Diagram showing backscattered and secondary-electron emission and Everhart–Thornley (E–T) detector. (b) Diagram showing the point where the electron beam meets the specimen. The three major electron-beam parameters are defined: electron probe diameter d_p, electron probe current i_p, and electron probe convergence α_p.

+12 kV placed on the aluminum coating of the scintillator. The now-energetic SEs strike the scintillator material and produce light that travels down a light pipe to the photomultiplier tube (PMT), which converts the light to an amplified electrical signal. Variations in signal occurring as the beam moves over the specimen surface provide the intensity changes we see on the viewing screen as an image.

Backscattered electrons are most efficiently and selectively collected with an "overhead" backscatter detector (not shown in Fig. 2.2a). A large fraction of the BSEs can be collected with this arrangement, using either a semiconductor or a scintillator–photomultiplier detector. As mentioned above, the E–T detector also detects BSEs. A small fraction of the backscattered signal enters the E–T detector by line of sight from the specimen. However, BSEs emitted in other directions may be sensed with a positively biased (+300 volts) E–T detector by capturing SEs emitted when those BSEs strike the polepiece and chamber walls. A pure BSE signal can be collected with the E–T detector by repelling the SEs with a −100-V potential on the screen in front of the detector. While only SEs and BSEs have been discussed here, it should be clear that any signal collected and suitably amplified may be used to form an image in the SEM.

Camera Records the Image. Most SEMs have a separate CRT viewed by a camera to record slow-scan images on Polaroid or conventional wet-processed film. Alternatively, a video printer may be used to convert the record CRT signal to a positive image on special printing paper. For more detail about signal generation, detectors, and image formation, see Chapter 4.

2.1.2. Why Learn about Electron Optics?

The sharpness and contrast of SEM micrographs and the depth of field in images are dependent upon the three major electron beam parameters: electron-probe size d_p, electron-probe current i_p, and the electron probe convergence angle α_p, as shown in Fig. 2.2b. The electron probe size d_p, often called the spot size, is defined as the diameter of the final beam. In this book, the term *electron beam* refers to electrons at any point in the column outside the electron gun, whereas the term *electron probe* refers to the focused electron beam at the specimen. The electron probe current i_p is defined as the current in the final probe which impinges upon the specimen and generates the various imaging signals. The electron probe convergence angle α_p is the half-angle that the cone of electrons at the specimen makes with the centerline of the beam. This angle is also called the *divergence angle.* We will use the term *convergence angle* when we are specifically dealing with the beam at the specimen, where it converges. For the highest resolution image, d_p must be as small as possible. For the best image quality and for x-ray microanalysis, i_p must be as large as possible. For the best depth of field, where a large range of heights on the specimen appear in focus at the same time, α_p must be as small as possible. We would like to have all these conditions met simultaneously; however, i_p is inescapably reduced when d_p and α_p are made small. The operator must select these three parameters intelligently to obtain appropriate imaging conditions in each situation. Therefore, an elementary understanding of electron optics is necessary for productive SEM operation.

2.2. Electron Guns

The purpose of the electron gun is to provide a large, stable current in a small electron beam. Several types of electron guns are used on current SEMs. These guns vary in the amount of current they can produce, the size of the source, the stability of the emitted current, and the lifetime of the source. Since most SEM electron sources are of the tungsten thermionic type, most of this section will deal with this type of electron gun. The newer LaB_6 and field-emission guns offer considerably better SEM performance, and these sources will also be described with appropriate comparisons to the conventional tungsten source.

2.2.1. Thermionic Electron Emission

Thermionic emission occurs when enough heat is supplied to the emitter so that electrons can overcome the work-function energy barrier E_w of the material and escape from the material. Figure 2.3 illustrates the concept of overcoming the work function for the

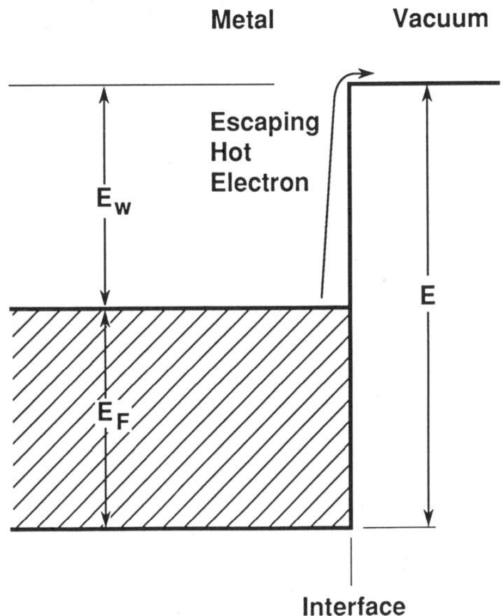

Figure 2.3. Energy model for electron emission from tungsten metal with no electric field applied (adapted from Kittel, 1966).

situation when no strong extraction electric field is present in the gun. The symbol E represents the energy (work) necessary to place an electron in the vacuum from the lowest energy state in the metal. Electrons have a range of energies, and the highest energy state in the metal is called the Fermi level, E_F. When the emitter material is heated to a high temperature, a small fraction of the electrons at the Fermi level acquire enough energy to overcome E_w and escape into the vacuum. The cathode current density J_c obtained from an emitter by thermionic emission is expressed by the Richardson equation

$$J_c = A_c T^2 \exp(-E_w/kT) \tag{2.1}$$

A/cm^2, where $A_c = 120\,\text{A/cm}^2\text{K}^2$ is a constant for all thermionic emitters (Heinrich, 1981), T(K) is the absolute emission temperature, E_w(eV) is the work function of the filament material, and k is Boltzmann's constant (8.6×10^{-5} eV/K). Both the temperature T and the work function E_w have a strong effect on the emission current density J_c that can be obtained from the filament, whereas the constant A_c has only a minor effect. Because it is desirable to operate the electron gun at the lowest possible temperature to reduce evaporation of the filament, materials of low work function are desired. Tungsten wire is often the material of choice because it has a low E_w and produces a high value of J_c at temperatures well below its melting point. A sample calculation using Eq. 2.1 may be obtained by substituting typical values for tungsten: $T = 2700$ K and $E_w = 4.5$ eV. The resulting cathode current density is $J_c = 3.4\,\text{A/cm}^2$.

2.2.2. Conventional Triode Electron Guns

The most common electron gun consists of three components (a triode): a tungsten wire filament serving as the cathode, the Wehnelt cylinder or grid cap, and the anode, as shown schematically in Fig. 2.4. These components are maintained at different electrical potentials (voltages) by appropriate connections to the high voltage supply, which is usually variable in the range 1–40 kV (1000 to 40,000 volts).

Filament. The typical tungsten filament is a bent wire that is heated resistively using a filament-heating power supply. During operation the filament and its heating supply are maintained at a high negative potential by the high voltage supply. At the operating temperature, electrons are emitted from the V-shaped filament.

Grid Cap. Electrons are emitted from the heated filament in all directions. The grid cap (Wehnelt cylinder) acts to focus electrons inside the gun and to control the amount of electron emission. This electrode is maintained at a slightly more negative potential than the filament itself, using the variable bias resistor as a voltage divider. The bias resistor provides a bias voltage on the grid cap. This bias voltage varies with the emission current and gives rise to the term "self-biased electron gun." In Fig. 2.4, lines of constant electrostatic field potential

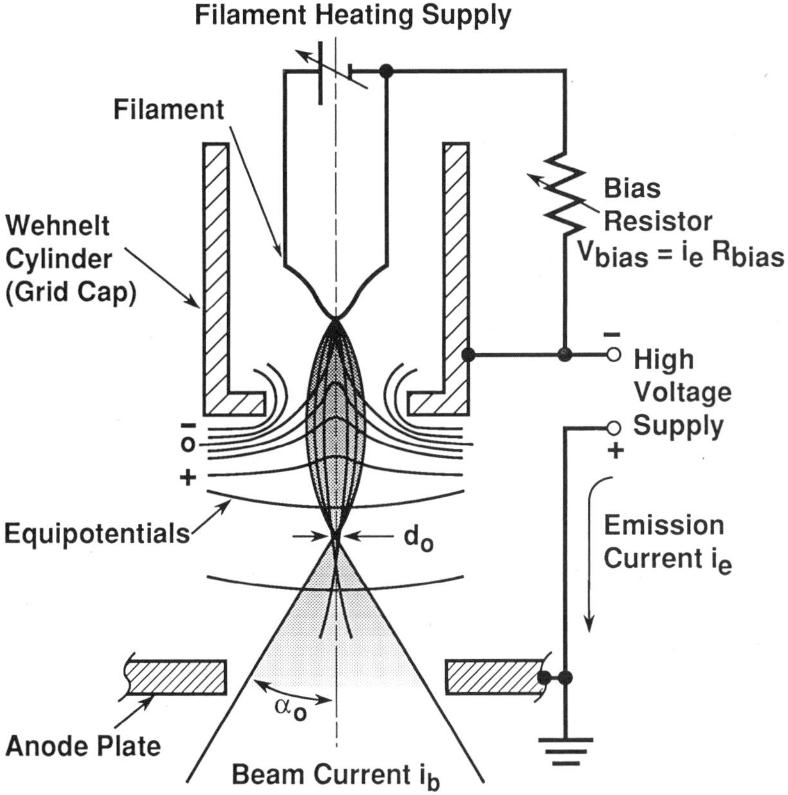

Figure 2.4. Schematic diagram of the conventional self-biased thermionic (triode) electron gun (adapted from Hall, 1966).

(equipotentials) are shown that vary from negative to zero to positive. Negatively charged electrons move only toward positive equipotentials and electrodes; they are repelled by negative equipotentials. Thus, electrons leave the filament only where the positive electrostatic field meets the surface of the filament. Electrons emitted into this region of positive potential tend to follow the maximum gradient in electrostatic potential from the filament to the anode. Note the strong focusing action as electrons are repelled by the negative field lines around the grid cap. This focusing action forces the electrons to a crossover of diameter d_0 and divergence angle α_0 between the grid cap and the anode, as shown in Fig. 2.4. The intensity distribution of electrons at the gun crossover is usually assumed to be Gaussian. As we shall see, the condenser and objective lenses produce a demagnified image of this crossover, which is the final electron probe d_p on the specimen.

Anode. Electrons are accelerated from the high negative potential of the filament, e.g., −20,000 V, to ground potential (0 volts) at the anode. A hole in the anode allows a fraction of these electrons to continue down the column towards the lenses. Electrons collected on the anode return via ground to the high voltage power supply.

Emission Current and Beam Current. Two important parameters for any electron gun are the amount of current it produces and the current stability. In electron guns using thermionic emitters (W, LaB_6), the filament heating current i_f is used to raise the temperature of the emitter by resistive heating to the point at which electron emission occurs. Most of the electrons emitted in this way from the filament are captured by the anode and return to the high voltage supply where they are measured as the "emission current," i_e. The portion of the electron current that leaves the gun through the hole in the anode is called the "beam current," i_b. At each lens and aperture along the column the beam current becomes smaller, and it is several orders of magnitude smaller when it is measured at the specimen as the "probe current," i_p.

Saturation. A constant beam current is required for microscopy and microanalysis since all information (the image, the x-ray spectrum, etc.) is recorded as a function of time. For stable operation a condition of saturation must be established in which small increases or decreases in the filament heating current i_f do not change the electron beam current. Saturation of the filament is an important task for the operator. Figure 2.5 shows the emission characteristics of a self-biased gun in which the beam current i_b is plotted versus the filament heating current i_f. As the filament current is increased, the beam current eventually reaches a point where there is no further increase, a condition called *saturation*. Saturation may be understood as evidence that a self-regulating negative feedback circuit is operating to produce a stable beam current. Suppose that a chance fluctuation in filament heating current i_f occurs that produces a momentary increase in emission. This increased emission current flows back through the high voltage supply and through the bias resistor, generating an increased negative voltage on the grid cap. This slightly higher bias voltage automatically counters any further increases in emission.

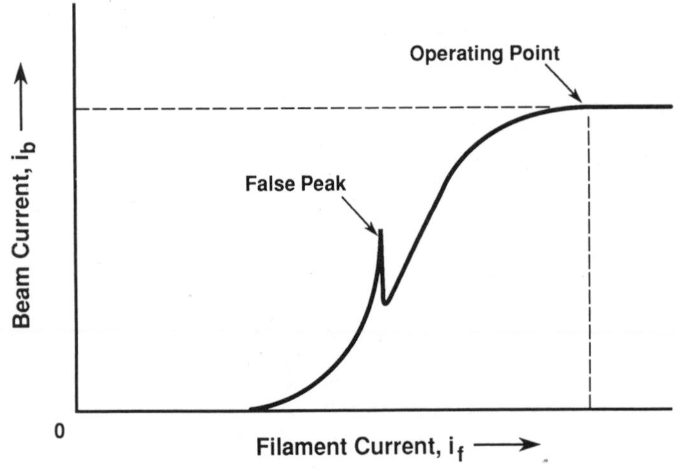

Figure 2.5. Saturation of a tungsten hairpin electron gun. Operating point is the level of heating current for which no further increase in beam current can be obtained. Usually a false peak is observed even with a well-aligned gun. A misaligned gun exhibits a maximum emission with increased filament heating current.

As the operator turns up the filament heating circuit, a false peak (Fig. 2.5) may be observed where the beam current rises, and then falls, before the true saturation condition is established. This false peak appears when some other part of the filament surface reaches emission temperature before the filament tip. Since this lobe of emission usually passes through the grid cap aperture before emission from the tip itself, it appears to the operator as a false peak. When saturation is complete, however, all the lobes of intensity collapse into a small, tight beam that is very stable because of the self-regulating nature of the bias voltage.

2.2.3. Brightness

Two important measures of electron gun performance are the electron current density at various locations along the column and the electron optical brightness of the gun. For example, the current density inside the gun is simply the emission current i_e divided by the cross-sectional area of the gun crossover d_0. This emission current i_e inside the gun is typically about 100 μA and is the total current emitted from the filament that returns to the high voltage supply through the electrical ground. In fact, all the current in the gun crossover would be concentrated into a focused spot on the specimen if most of it were not intercepted by various apertures in the column, especially the anode.

Current Density. Only a small portion of the gun emission current escapes through the anode aperture and proceeds further down the column as the beam current i_b. The beam is often characterized by the current density, so that the relative intensity of the beam may be measured at any point in the column, regardless of the fraction of the beam sampled. Current density in the beam is expressed as

$$J_b = \frac{\text{current}}{\text{area}} = \frac{i_b}{\pi\left(\dfrac{d}{2}\right)^2}, \tag{2.2}$$

where i_b is the beam current and d the diameter of the beam at some point in the column. The value of J_b is different after each aperture and lens in the column because the angular spread (divergence) of the beam is such that a fraction of the beam is intercepted by microscope components and lost. The concept of current density ignores the beam divergence. A parameter that indicates gun performance specifically is the electron optical brightness β.

Brightness Equation. The parameter of brightness incorporates the current density but also recognizes the changes in the angular spread of the electrons as they are focused to form a beam crossover at various points in the column. Brightness is defined as the current density per solid angle and is given in an important relationship known as the brightness equation:

$$\beta = \frac{\text{current}}{(\text{area})(\text{solid angle})} = \frac{4i_b}{\pi^2 d^2 \alpha^2} \quad (2.3)$$

A/cm² sr, where i_b is the beam current at some point in the electron column outside the gun, d is the diameter of the beam at this point, and α is the beam convergence (or divergence) angle at this point. The dimensionless unit of solid angle is the steradian (sr); a solid angle may be thought of as an ice-cream-cone-like section cut from a sphere. The solid angle in steradians may be calculated as $\pi\alpha^2$, where α is the angle from the cone's surface to its centerline. Equation (2.3) is the first of two critical equations which the microscopist must use to develop a practical understanding of electron-optical limitations on images. The second is the threshold equation given in Chapter 4.

Neglecting lens aberrations, electron beam brightness is a constant throughout the electron column as the individual values of i_b, d, and α change. Measurements of brightness made at the specimen level are only estimates of the actual gun brightness, because of the action of lens aberrations. Since brightness is the most important performance indicator for electron guns, even relative estimates are valuable.

Maximum Theoretical Brightness. For thermionic emitters at high voltages, the maximum theoretical brightness according to Langmuir (1937) is

$$\beta_{max} = \frac{J_c e V_0}{\pi k T} \quad (2.4)$$

A/cm²sr, where J_c is the current density at the cathode surface [given by Eq. 2.1], V_0 is the accelerating voltage, e is the electronic charge [1.59×10^{-19} C, (coulomb)], k is Boltzmann's constant (8.6×10^{-5} eV/K), and T is the absolute temperature (K). An important result of Eq. 2.4 is that brightness increases linearly with accelerating voltage and inversely with filament temperature. Substituting typical values for a tungsten gun operated at 2700 K and 20 kV accelerating voltage, the maximum theoretical brightness is 9.2×10^4 A/cm²sr. Uncertainties in J_c from Eq. 2.1 can lead to errors in β_{max} of a factor of two or more.

Effect of Bias Voltage. To achieve the highest brightness for a given triode electron gun, the bias voltage between the filament and the grid cap must be optimized by adjusting the bias resistor. At low bias, the negative field in front of the filament becomes weak, providing a relatively poor focusing action (Fig. 2.6). The electrons respond to a strong positive field toward the anode, and therefore the emission current i_e is high, as shown in Fig. 2.7 for a bias voltage of −300 V. Because of the weak focusing at low bias, the crossover d_0 is large and the brightness β is not optimum. At very high bias, the negative field potential in front of the filament is so strong that electrons emitted are forced to return to the filament (Fig. 2.6). Thus, under high bias conditions (cut-off) both the emission current and the brightness decrease to zero. Somewhere in between these two extremes lies an optimum bias condition (Fig. 2.6), which provides good emission, good focusing, and the maximum in brightness (about −400 V in Fig. 2.7). Haine and Cosslett (1961) showed that this optimum brightness is very close to the maximum theoretical brightness for T = 2700 K. The optimum brightness condition can also be altered by changing the filament-to-grid cap distance. Often the filament-to-grid spacing is fixed, leaving only the bias voltage adjustment to the operator.

2.2.4. Tungsten Hairpin Electron Gun

While high brightness electron guns are of increasing interest for high-resolution SEM applications, the tungsten filament has served the TEM–SEM community well for over 50 years. It is reliable, its

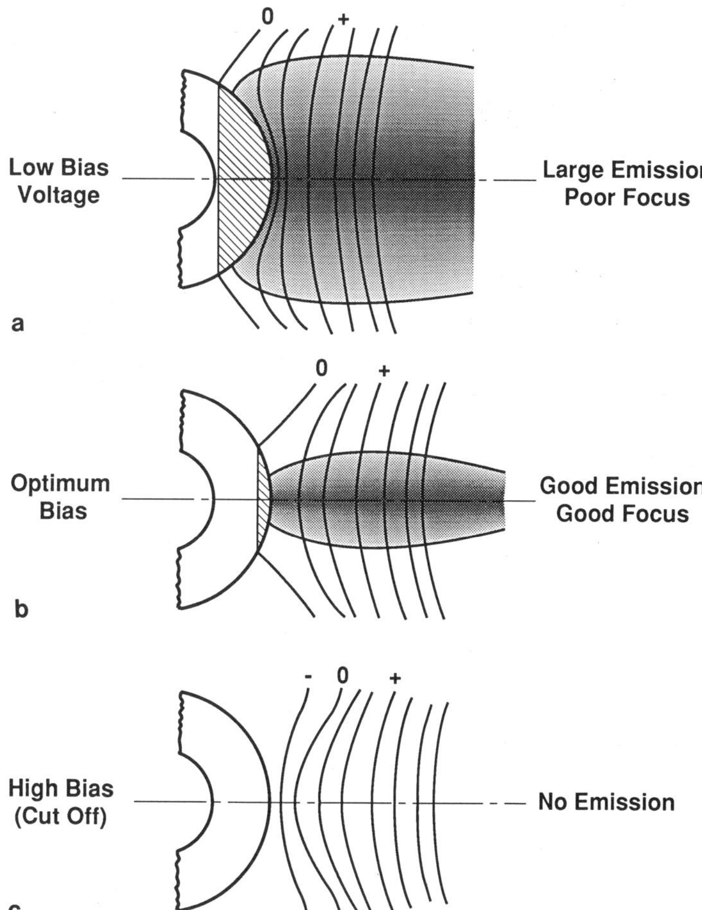

Figure 2.6. Typical emission distributions from the tip of a tungsten hairpin filament for (a) low bias voltage producing high emission but poor focusing, (b) optimum bias voltage producing good focus and good emission, and (c) high bias voltage (cut off) emitting no current (adapted from Haine and Cosslett, 1961).

properties are well understood, and it is relatively inexpensive. Therefore, for many SEM applications where high brightness is not necessary, such as low magnification, or where stable high currents are required, as in x-ray microanalysis, the thermionic tungsten filament may be used without loss of performance, and may, in fact, be the best choice.

The typical tungsten cathode is a wire filament about 100 μm in diameter that is bent to a V-shaped hairpin tip with a radius also about 100 μm, as shown in Fig. 2.8. To achieve thermionic emission, the cathode is heated resistively by the filament heating current i_f. Under normal operating conditions, the electron emission area is about 100 μm $\times$ 150 μm. For tungsten, as for all cathodes, there are several important performance parameters: brightness, lifetime, source size, energy spread, and stability. The performance characteristics for thermionic tungsten hairpin guns have been carefully investigated (Haine and Einstein, 1952; Haine and Cosslett, 1961).

Brightness. To calculate the brightness of a conventional SEM electron gun, one might consider using the brightness equation [Eq. (2.3)] and substituting typical values for the gun crossover of $d_0 = 30$ to 100 μm and $\alpha_0 = 3 \times 10^{-3}$ to 8×10^{-3} rad. Unfortunately, we usually do not know the beam current at the point where it leaves the gun. Thus, the maximum calculated brightness [Eq. (2.4)] or a brightness estimated from measurements at the specimen is quoted. An important result from Eq. (2.4) is that brightness increases linearly with accelerating voltage. Thus, if β_{max} is 9.2×10^4 A/cm^{2}sr for 20 kV, then for 30 kV, β_{max} is 1.4×10^5 A/cm^{2}sr.

Lifetime. The lifetime of the thermionic tungsten filament decreases as the temperature increases. At a temperature of 2700 K, and a brightness of $\beta = 10^5$ A/cm^{2}sr, a typical lifetime is about 30 to 100 h in a reasonably good vacuum of approximately 10^{-3} Pa (10^{-5} torr). In

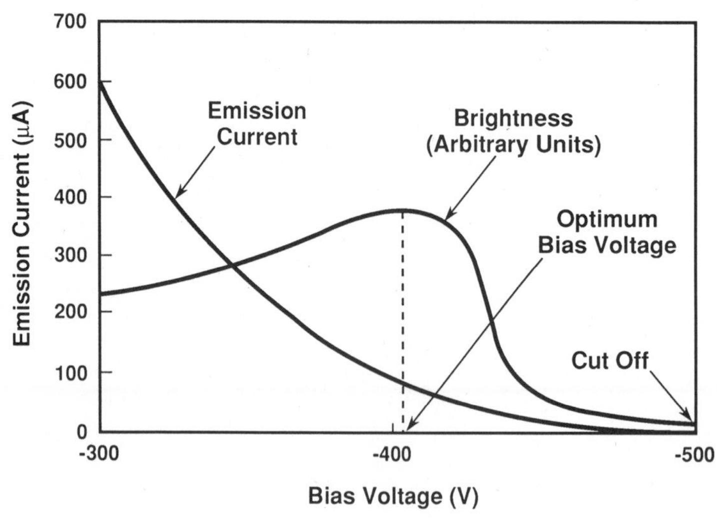

Figure 2.7. Relationship of emission current and brightness to bias voltage. While a brightness maximum should be obtainable in any gun, this schematic diagram shows values and curves for a hypothetical system.

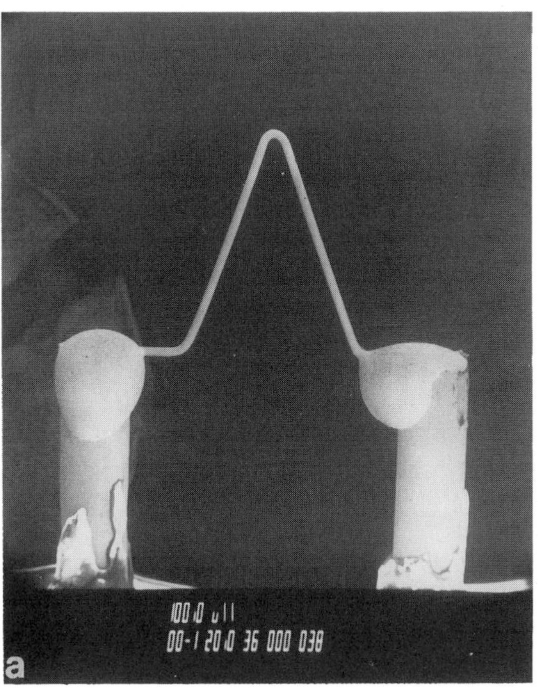

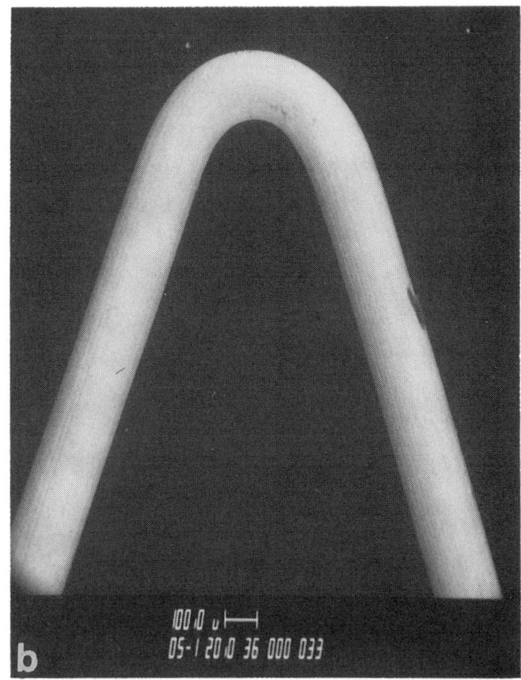

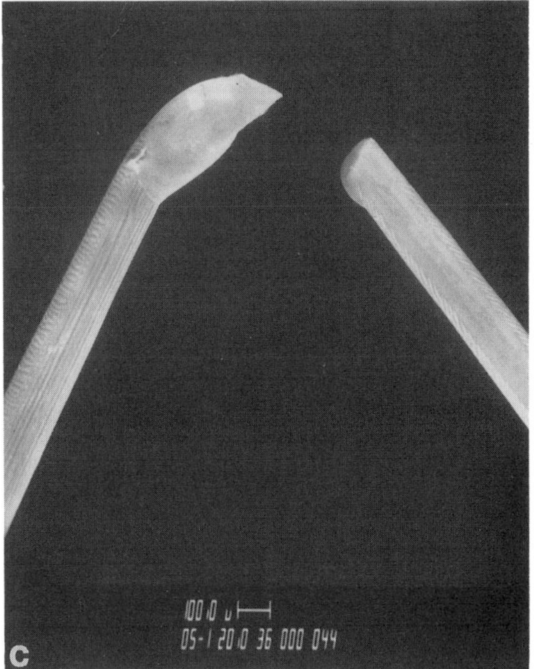

Figure 2.8. SEM micrographs of the tungsten hairpin filament. (a) Filament wire spot-welded to support posts. (b) Tip of wire showing fine wire drawing marks. (c) Filament failure by overheating (oversaturation).

such a vacuum the common mode of filament failure is tungsten evaporation. Evaporation is slow because of the high melting point of tungsten, 3683 K (3410°C). While it is true that brightness may be increased by raising the temperature of the filament (oversaturating), this causes excessive evaporation in a local hot spot, at which the filament fails in perhaps 1 h or less (Fig. 2.8c). In a poor vacuum the filament usually fails after erosion by ion bombardment from the residual gases in the gun.

Source Size, Energy Spread, Stability. The electron source size at the gun crossover ranges between $d_0 \sim 30$ to 100 μm depending on the gun configuration and operating conditions. This relatively large size means that considerable electron-optical demagnification is necessary to achieve a small electron probe needed for reasonable resolution in the SEM. The electron beam energy spread ΔE, which is the variation of electron energies leaving the filament to form the beam, is relatively large for a thermionic tungsten filament at about 1 to 3 eV (Troyon, 1987). This parameter is important for low voltage SEM operation, to be discussed later. The 1% current stability of the thermionic source is the parameter that recommends its use in the electron probe microanalyzer, where precision of measurement is paramount. The parameters discussed above are listed in Table 2.1 for comparison with higher-brightness sources to be discussed below.

To improve image resolution in the SEM, it is essential to reduce the electron-probe size without causing a loss of current in the probe. The only practical way to accomplish this is to increase the electron gun brightness. Pointed tungsten filaments have been used to produce somewhat higher brightness (Hibi *et al.* 1962), but these filaments are more difficult to use than the bent hairpin configuration. Changing the filament material or the mechanism of emission can improve source brightness by an order of magnitude (LaB_6) or more (field emission).

Table 2.1. Comparison of Electron Sources at 20 kV

Source	Brightness	Lifetime	Source size	Energy spread ΔE	Beam current Stability	References
Tungsten hairpin	10^5 A/cm^{2}sr	40–100 h	30–100 μm	1–3 eV	1%	[a,b]
LaB_6	10^6	200–1000	5–50 μm	1–2	1%	[b,c]
Field Emission						
Cold	10^8	>1000	<5 nm	0.3	5%	[d,e]
Thermal	10^8	>1000	<5 nm	1	5%	[e]
Schottky	10^8	>1000	15–30 nm	0.3–1.0	2%	[e]

[a] Haine and Cosslett (1961).
[b] Troyon (1987).
[c] Broers (1974).
[d] Crewe *et al.* (1971).
[e] Tuggle *et al.* (1985).

2.2.5. Lanthanum Hexaboride (LaB_6) Electron Guns

A block of lanthanum hexaboride (LaB_6) heated to thermionic emission (Fig. 2.9) is the most common high-brightness source. This source offers about 5–10 times more brightness and a longer lifetime than tungsten, but the required vacuum conditions are more stringent.

Higher Current, Longer Life. From Eq. (2.1), it can be seen that the cathode current density, and hence the brightness [Eq. (2.4)] may be increased by lowering the work function E_w. Because E_w is in the exponent, the effect can be dramatic. For example, at the tungsten operating temperature, each 0.1 eV reduction in E_w increases J_c by about 1.5 times. Of the many oxides and borides that exhibit a low value of E_w, the most successful has been lanthanum hexaboride, first investigated by Lafferty (1951). The measured work function of $\langle 100 \rangle$ single crystal LaB_6 is about 2.5 eV (Swanson *et al.*, 1981), compared with 4.5 eV for tungsten. Since the constant in Eq. (2.1), $A_c = 120\ \text{A/cm}^2\text{K}^2$, remains the same as for tungsten, a current density equal to that produced by a conventional tungsten filament (about $3\ \text{A/cm}^2$) is available at an operating temperature under 1600 K. At the typical LaB_6 operating temperature of 1800 K, the current density is nearly $40\ \text{A/cm}^2$. These low operating temperatures increase emitter lifetime by reducing the material evaporation rate, even at significantly improved brightness levels.

These substantial advantages, however, are accompanied by a practical difficulty. LaB_6 is extremely chemically reactive when hot, and it readily forms compounds with elements in the gas phase that "poisson" the cathode and reduce its emission. In addition, the volatile oxides of La and B which form in the presence of water vapor or oxygen may cause a significant loss of emitter material. Thus, the vacuum in the gun chamber must be better than 10^{-5} Pa (10^{-7} torr) or about two orders of magnitude better than the maximum operating pressure required for a tungsten cathode.

Commercial LaB_6. The first practical LaB_6 gun employed indirect heating of a sharpened rod of pressed and sintered LaB_6 powder (Broers, 1969). Development of LaB_6 guns for commercial SEMs departed from the original Broers design and emphasized directly heated LaB_6 emitter systems as direct plug-in replacements for the conventional tungsten hairpin, as shown in Fig. 2.10. The emitter is usually fashioned from a tiny block of single-crystal LaB_6 about 100 μm in diameter (or square cross-section) and about 0.5 mm long. The LaB_6 crystal is both supported and resistively heated by either carbon or rhenium, two materials that do not react to form a compound with LaB_6. For example, the resistance of graphite supports may be chosen so that a current of one or two amperes provides enough heat to raise the temperature of the LaB_6 to emission. Assuming the vacuum in the gun chamber is adequate, only the bias voltage and the grid cap shape need be modified to replace the usual tungsten filament.

Schottky Effect. The sharpness of the tip can also affect the emission. A small radius (1–10 μm) at the cathode tip in combination

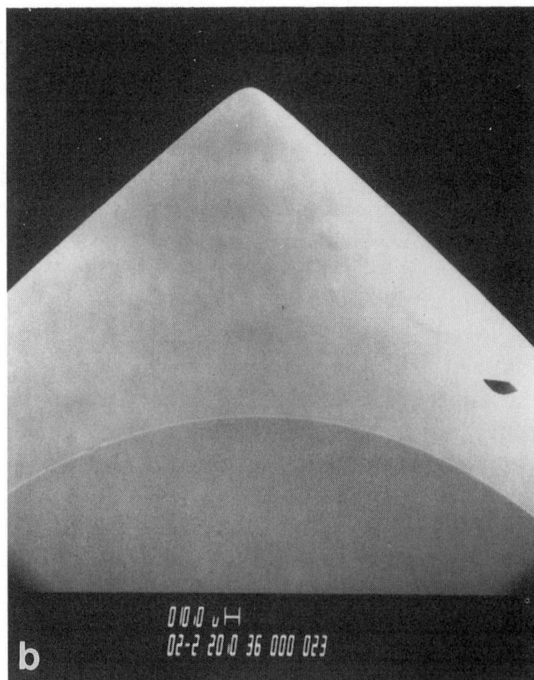

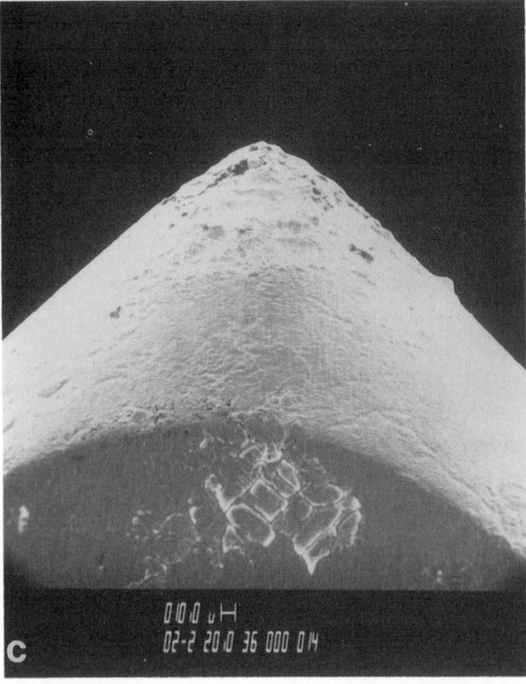

Figure 2.9. SEM micrographs of the lanthanum hexaboride source. (a) LaB_6 block mounted on support. (b) Higher magnification, showing finely ground tip with 10-μm radius. (c) Failed LaB_6 tip with evidence of evaporation and oxide formation.

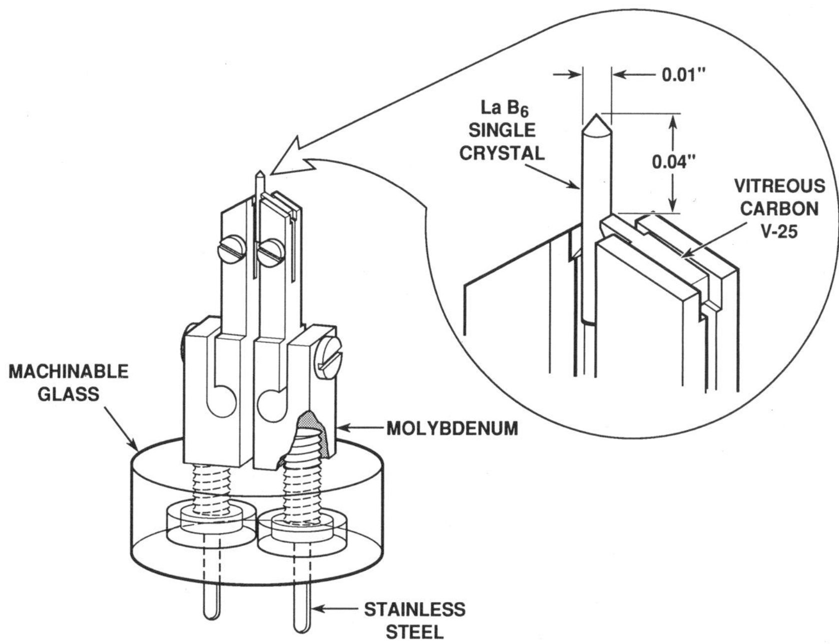

Figure 2.10. Schematic diagram of a directly heated LaB_6 emitter that directly replaces a tungsten filament assembly.

with a high bias voltage causes a high local electric field at the tip. This field, which may be 10^6 V/cm or more, lowers the potential barrier (the work function E_w) confronting an electron escaping from the cathode. The amount by which this "Schottky effect" can reduce E_w is typically 0.1 V or more (Broers, 1975), leading to a doubling of the cathode current density.

Performance Parameters. Typical brightness values for LaB_6 emitters range from 5×10^5 to 5×10^6 A/cm^{2}sr, depending upon the sharpness of the tip. Sharp emitters have the highest brightness but the shortest lifetime. Emitters with blunt or truncated tips exhibit lower brightness but can last in excess of 1000 h in an ideal vacuum environment. Pressures better than 10^{-5} Pa (10^{-7} torr) in the gun chamber allow emitter lifetimes several times longer than pressures of about 10^{-4} Pa (10^{-6} torr). Vacuum is critical since poor vacuum also causes buildup of lanthanum oxide on the grid cap and anode. The energy spread is 1–2 eV, about the same as tungsten, but varies with operating conditions (Troyon, 1987). The characteristics of LaB_6 sources are listed in Table 2.1.

Operation. The typical operating temperature is just above 1800 K, but it is not easy to tell when this has been reached. The high electric field at the tip of an LaB_6 cathode means that the type of saturation shown in Fig. 2.5 usually does not occur. When correctly biased, the emission will rise to a shallow "knee" which appears to the operator as the first of two separate saturation levels. The best operating condition is just below the second saturation level. Care must be taken to avoid overheating the cathode in an attempt to reach saturation. It should also be noted that when first operated, or after exposure to the atmosphere, an LaB_6 emitter may require some time to activate, and the heater current should be increased very slowly, over 30–60 min. During the activation period, contaminants on the surface are removed by evaporation. Any attempt to reach full output before activation has occurred can lead to the destruction of the cathode by overheating. When activation is complete, source emission will rise rapidly to its usual value. Since LaB_6 evaporation ceases below 1670 K, some manufacturers recommend leaving the desaturated filament on 24 h a day even when not in use. This procedure avoids thermal shock from heating/cooling cycles, reduces thermal drift, improves the vacuum, and extends emitter life.

An LaB_6 gun is generally more expensive to operate than the conventional tungsten hairpin gun. Vacuum requirements may require differential pumping of the gun region with an additional vacuum pump such as an ion pump. LaB_6 filaments are more costly, typically ten times as expensive as tungsten filaments. Nevertheless the increased current in a given probe size and the significantly longer filament life ($\geq$1000 h) clearly justify the increased cost for many applications.

2.2.6. Field Emission Electron Guns

The electron sources described so far rely on high temperatures to enable some of the free electrons in the cathode material to overcome the work-function energy barrier and escape into the vacuum. These thermionic sources have the disadvantages of relatively low brightness, evaporation of the cathode material, and thermal drift during operation. Field emission is another way of generating electrons that is free from these disadvantages.

Electrons from a Sharp Tip. The field emission cathode is usually a wire of single-crystal tungsten fashioned into a sharp point and spot welded to a tungsten hairpin, as shown in Fig. 2.11. The significance of the small tip radius, about 100 nm or less, is that an electric field can be concentrated to an extreme level. If the tip is held at a negative 3–5 kV relative to the first anode (Fig. 2.12), the applied electric field F at the tip is so strong ($>10^7$ V/cm) that the potential barrier for electrons becomes narrow in width as well as reduced in height by the Schottky effect (see Fig. 2.13). This narrow barrier allows electrons to "tunnel" directly through the barrier and leave the cathode without requiring any thermal energy to lift them over the work function barrier (Gomer, 1961). Tungsten is the cathode material of choice, since only very strong materials can withstand the high mechanical stress placed on the tip in such a high electrical field. A cathode current density as high as 10^5 A/cm^2 may be obtained from a field emitter, compared with about 3 A/cm^2 from a tungsten hairpin filament. Electrons emanate from a very small virtual source ($\sim$10 nm) behind the sharp tip into a large semi-angle (nearly 20°, or about 0.3 rad), which gives a high current per solid angle and thus a high brightness. A second anode is used to accelerate the electrons to the operating voltage (see Fig. 2.12).

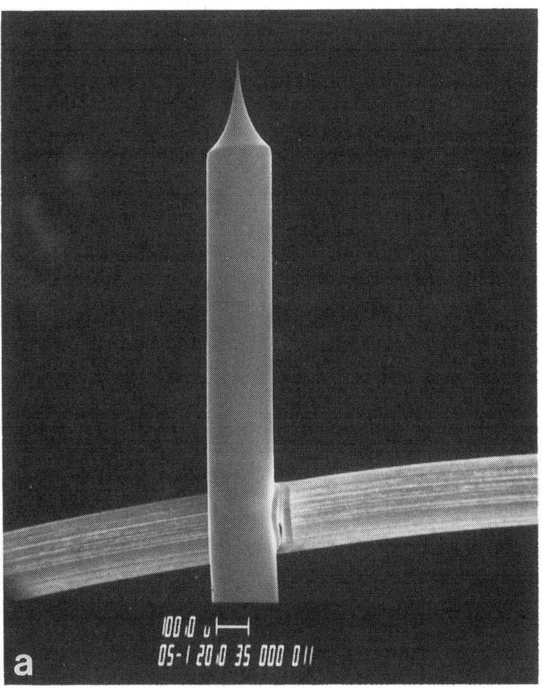

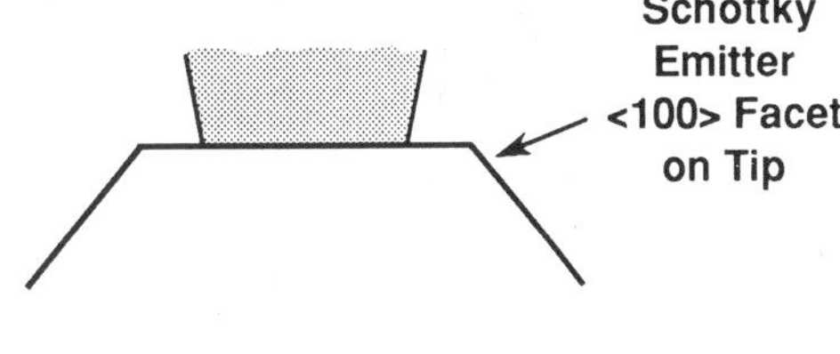

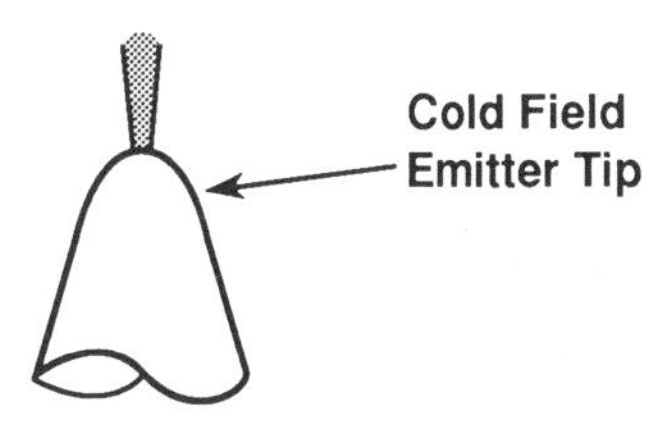

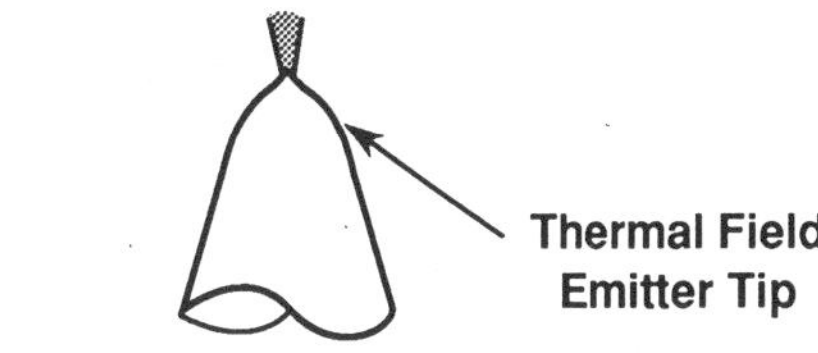

Figure 2.11. Examples of field emission sources. (a) SEM micrograph of ⟨310⟩ single crystal wire spot welded to a tungsten wire. (b) Higher-magnification image of the tip. (c) Schematic diagram comparing tip shapes for the Schottky emitter, cold field emitter, and thermal field emitter (adapted from Tuggle *et al.*, 1985).

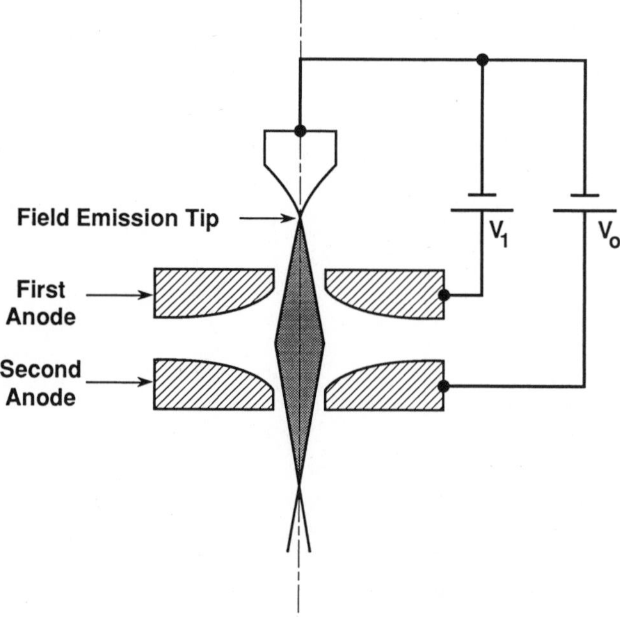

Figure 2.12. Schematic diagram of the Butler triode field emission source. V_1 is the extraction voltage, typically a few kV, and V_0 is the accelerating voltage (adapted from Crewe, 1969).

Brightness and Probe Current. The cathode current density J_c for a cold field emission source strongly depends on the applied field strength, similar to the way thermionic emission is dependent upon temperature. Expressions derived to calculate J_c (Fowler and Nordheim, 1928; Good and Muller, 1956) yield values in the range $J_c = 10^4$–

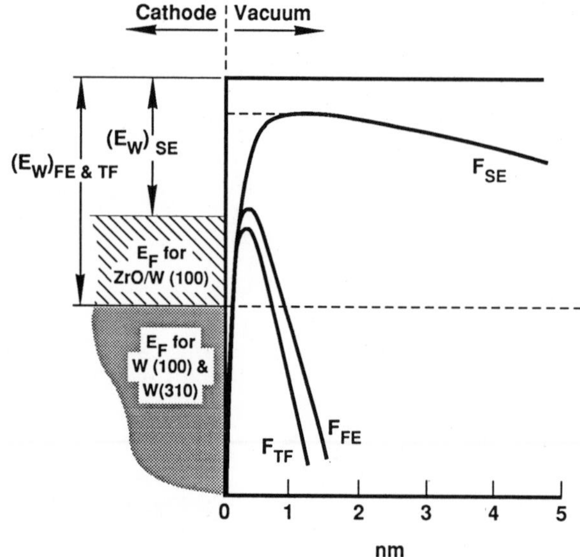

Figure 2.13. Energy-level diagram for cold field emission (FE), thermal field emission (TF), and Schottky emission (SE). Electrons tunnel through the narrow barriers in the FE and TF cases. For the SE case, the sharp tip covers the barrier and the zirconium oxide coating reduces the work function so that electrons can escape over the barrier (adapted from Tuggle *et al.*, 1985).

10^5 A/cm^2. The maximum theoretical brightness of a field emission gun is given (Troyon, 1984) by

$$\beta_{max} = \frac{J_c e V_0}{\pi \Delta E} \tag{2.5}$$

A/cm^{2}sr, where ΔE, the energy spread of the beam, is only about 0.3 eV for cold field emission (Crewe *et al.*, 1971). Substituting $\Delta E = 0.3$ eV, $J_c = 10^5$ A/cm^2, and $V_0 = 20$ kV into Equation 2.5 yields $\beta_{max} \sim 2 \times 10^9$ A/cm^{2}sr, which is 10^2–10^3 times greater than the brightnesses of thermionic sources. Working back through the brightness equation [Eq. (2.3)], ignoring lens aberrations, and assuming typical values for the beam parameters at the specimen ($d_p = 2$ nm and $\alpha_p = 7 \times 10^{-3}$ rad) leads to an estimated probe current of $(i_p)_{max} \sim 10$ nA. The practical brightness for field emission sources is difficult to measure, but it is expected to be about 10^8 A/cm^{2}sr at 20 kV, so practical values of i_p are typically less than a few nA.

In order to increase the current that actually gets through the first anode aperture, several methods have been developed to direct more current into a smaller angle. The orientation of the single-crystal wire is usually ⟨310⟩, so that the direction of minimum work function (E_w = 4.35 eV) is directed toward the anode (Orloff, 1984). Several other methods have been used to confine the angle of emission from the tip: reshaping the tip by heating while applying a field (Crewe *et al.*, 1968; Swanson and Crouser, 1969), adsorption of oxygen to induce reshaping by evaporating WO_3 (Veneklasen and Siegel, 1972), and adsorption of ZrO on the tip to reduce the work function (Swanson, 1975) all have been used to confine the angle of emission from the tip. In addition, electrostatic (Butler, 1966) and magnetic (Kuo, 1976) focusing elements have been placed between the first and second anode to increase the current through the second anode aperture.

Triode Gun. The first practical triode cold field emission gun employed shaped anodes for electrostatic focusing (Butler, 1966), as shown in Fig. 2.12, but some subsequent designs use planar electrodes. A voltage V_1, usually a few kV, between the field emission tip and the first anode, called the *extraction voltage,* controls the emission current. Emission currents are in the range 1–20 μA. Field emission tips with larger radii require a higher extraction voltage to produce the same emission current. A second voltage V_0 between the tip and the second anode determines the accelerating voltage, which usually ranges from about 1 to 30 kV. The anodes act as a pair of electrostatic lenses and form a real crossover at some distance beyond the second anode. However, as the tip blunts during long-term operation, more extraction voltage is required, and the position of the crossover changes. For gun designs with planar electrodes, focusing characteristics do not change much as the extraction voltage is increased. However, the electron beam diverges as it leaves the gun, and an additional electron lens further down the column or incorporated into the gun must be used to converge the beam to a crossover.

Cold Field Emission Operation. Cold field emission cathodes must be operated in clean vacuum better than 10^{-8} Pa ($<10^{-10}$ torr) in order to obtain reasonably stable emission. A monolayer or thicker coating of foreign gas atoms on the tip's surface tends to reduce emission. A single atom adsorbing to the monolayer can individually field emit, causing brief periods of unstable high emission before the gas atom desorbs. This gas adsorption causes the emission to become low and unstable after some hours or days. Higher and more stable emission may be restored by briefly heating the tip to about 2500 K. This process known as "flashing," cleans off adsorbed gas atoms. Following a flash, the high emission from the clean tip decays exponentially for about 10–15 min, after which there is only a modest decrease with time. Instabilities serious enough to cause streaks in the scanned image may be reduced by using a feedback circuit for compensation.

There is no fixed maximum emission from a field emission gun. Higher emission can be drawn from the gun until the additional electron flux desorbs so many gas atoms from the metal housing of the gun that the emission becomes unstable. Field emitter tips do not "burn out," so the lifetime for field emitters is determined primarily by the vacuum level in the gun, and thus the total number of flashes. The more often the tip is flashed, the blunter it gets. After many months (or even years) of operation, the extraction voltage power supply reaches its maximum rated voltage, at which time the tip must be exchanged for a new, sharper tip. The tips are also vulnerable to catastrophic failure due to high-voltage arc discharges. The greatest advantages of cold field emission are that

1. The virtual source is so small that little demagnification of the beam is necessary to form a small spot on the order of 1–2 nm, and
2. The energy spread is very low, which improves performance for low-voltage operation (see Table 2.1).

Thermal Field Emission Operation. By heating a ⟨100⟩ single-crystal tungsten field emitter to 1800 K in a strong electric field, buildup of the tip occurs, which effectively sharpens the tip (see Fig. 2.11c). Continuous heating of the tip to temperatures in the range 1300–2000 K prevents most of the gas molecules from alighting on the tip's surface. Thermal field emission thus alleviates the need for flashing the tip. In addition to maintaining better emission stability over time, a thermal field emitter may be operated in the somewhat poorer vacuum of 10^{-7} Pa (10^{-9} torr). While the brightness of thermal field emitters is similar to that of cold field emitters, the energy spread of the electrons is larger by a factor of 3–5.

Schottky Emitter Operation. The Schottky emission cathode operates at 1800 K and also uses a ZrO coating on a ⟨100⟩ tungsten facet at the tip to reduce the work function (Tuggle *et al.*, 1985). The ZrO lowers the work function from 4.5 eV to about 2.8 eV to provide emission, and the flat emitting area on the ⟨100⟩ facet provides good emission stability.

Note in Fig. 2.13 that Schottky emission electrons must still be thermally assisted to overcome the energy barrier. The Schottky emitter exhibits an energy spread ΔE that in some cases is almost as small as for cold field emission, which is a major consideration in selecting sources for low-voltage SEM. Since the Schottky virtual source size is larger than that for the cold field emission source, greater demagnification must be used to produce a 1–2 nm spot. Hence, the major advantage of this source is greater emission stability, but this advantage is somewhat offset by the larger demagnification required, which reduces the current available in the smallest electron probes.

Stability and Total Current. Field-emission sources provide high current (1 nA) into small electron probes (1–2 nm), and this produces excellent SEM images, even though beam instabilities must be compensated for by a feedback loop. However, both the long- and short-term current instabilities of cold field emission hamper quantitative x-ray microanalysis. Moreover, field emission sources are not suited to all applications. For probe sizes of 200 nm or larger, the tungsten hairpin and the LaB_6 emitter deliver more total current into the probe than a field emitter (Cleaver and Smith, 1973).

2.3. Electron Lenses

Electron lenses beyond the electron gun are used to demagnify the image of the crossover in the electron gun ($d_o \sim$ 10–50 μm for a thermionic gun) to the final spot size on the specimen (1 nm–1 μm). This represents a demagnification of as much as 10,000 for a thermionic source. Since the source size in a field emission system is already small, a 1–2 nm probe size in this case requires only a demagnification of 10–100. Electrons can be focused by either electrostatic or magnetic fields. We have already encountered electrostatic focusing inside the electron gun; however, in the rest of the column, most electron microscopes employ electromagnetic lenses because of their lower inherent aberrations. Even with magnetic focusing, the electron lens is a notoriously poor lens compared to the typical glass lens for focusing light since some aberrations that can be corrected in glass lenses cannot be corrected in electron lenses.

2.3.1. Properties of Magnetic Lenses

A simple iron electromagnet producing a magnetic field is shown in Fig. 2.14a, where the field across the gap is created by a current I (usually a few amps) energizing a coil of N turns. This device produces a homogeneous magnetic field in the center of the gap with the direction of the field from north to south. An electron passing through this field perpendicular to the field lines moves in a curved trajectory out of the plane of the paper because there is a force on the electron created by the electron's movement in the magnetic field. At the edge of the gap, the field lines are bent. These field lines are called the *fringing field,* and it is this field that is important in electron focusing.

Fringing Field in the Gap. An electron lens is a rotationally symmetric electromagnet with the coil windings on the inside, as shown in Fig. 2.14b. Figure 2.15a shows the fringing magnetic field crossing the gap inside a schematic cross-section of a magnetic lens. The magnetic flux density **B** denotes the intensity of the field. (The symbol **H** is often used for magnetic field strength, but in vacuum **H** = **B**.) Along each flux line, **B** is represented by a vector quantity having both magnitude and direction. To understand how moving electrons interact with **B**, it is useful to separate **B** into its component vectors $\mathbf{B_r}$ in the radial direction and $\mathbf{B_z}$ along the optic axis. Figure 2.15b shows that $\mathbf{B_r}$ reverses in direction as the flux line curves back to the lens, whereas $\mathbf{B_z}$ goes through a maximum.

Electron Focusing. The equation that relates the force **F** on the electron to its velocity **v** and the magnetic flux density **B** is

$$\mathbf{F} = -e(\mathbf{v} \times \mathbf{B}), \tag{2.6}$$

where *e* is the charge on the electron and the operation is the vector cross product of **v** and **B**, which follows the "right hand rule" shown in Fig. 2.15c. The minus sign indicates that the force on an electron is opposite that on a positive charge. Figure 2.15c shows the force on a positive charge along the thumb when the index finger represents the particle velocity **v** and the second finger represents the magnetic flux density **B**. The actual force on the electron is in the direction opposite that of the thumb because the electron is negatively charged.

Electron focusing occurs because the electron interacts with $\mathbf{B_r}$ and $\mathbf{B_z}$ separately, as shown in Fig. 2.15d. An electron of velocity $\mathbf{v_z}$ enters

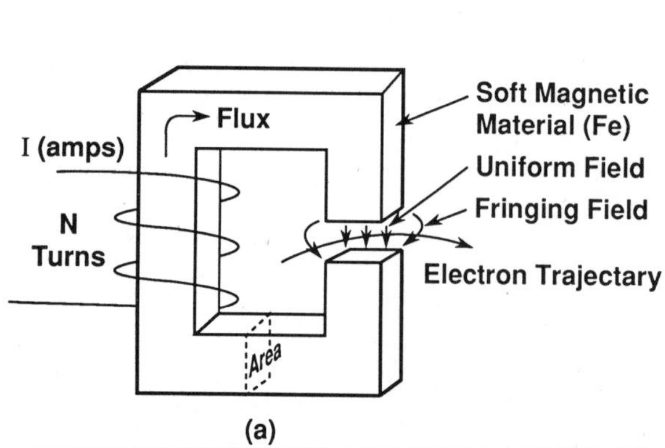

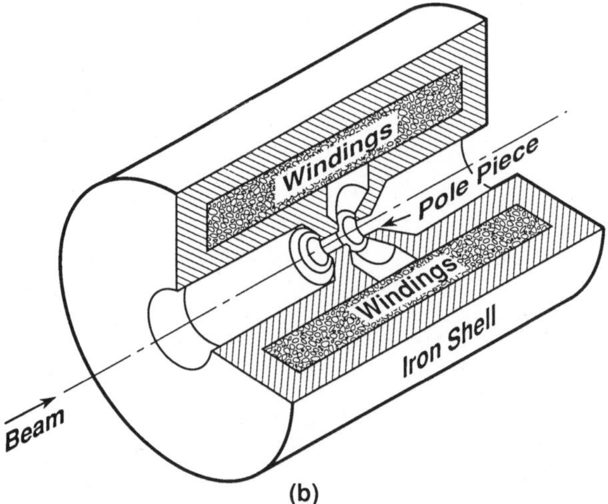

Figure 2.14. Production of a magnetic field inside an electromagnetic lens. (a) A coil of wire energizing a simple magnetic circuit to produce a magnetic field across a gap in the iron circuit. (b) A rotationally symmetric electron lens where the coil windings are inside the iron shroud and the field is produced across the lens gap between polepieces (adapted from Hall, 1966).

the lens parallel to the optic axis and interacts with the radial component $\mathbf{B_r}$. According to the right hand rule of Fig. 2.15c, the vector product $-e(\mathbf{v_z} \times \mathbf{B_r})$ produces a rotational force $\mathbf{F_{\theta in}}$, which in turn gives the electron a rotational velocity $\mathbf{v_{\theta in}}$. This rotational velocity then interacts with the axial component of the $\mathbf{B_z}$ of the field to produce a radial force $\mathbf{F_r} = -\mathbf{e}(\mathbf{v_{\theta in}} \times \mathbf{B_z})$ on the electron. This radial force causes the electron's trajectory to curve toward the optic axis and cross it. The focal length f of the lens is the distance along the optic axis from the point where an electron first changes direction to the point where the electron

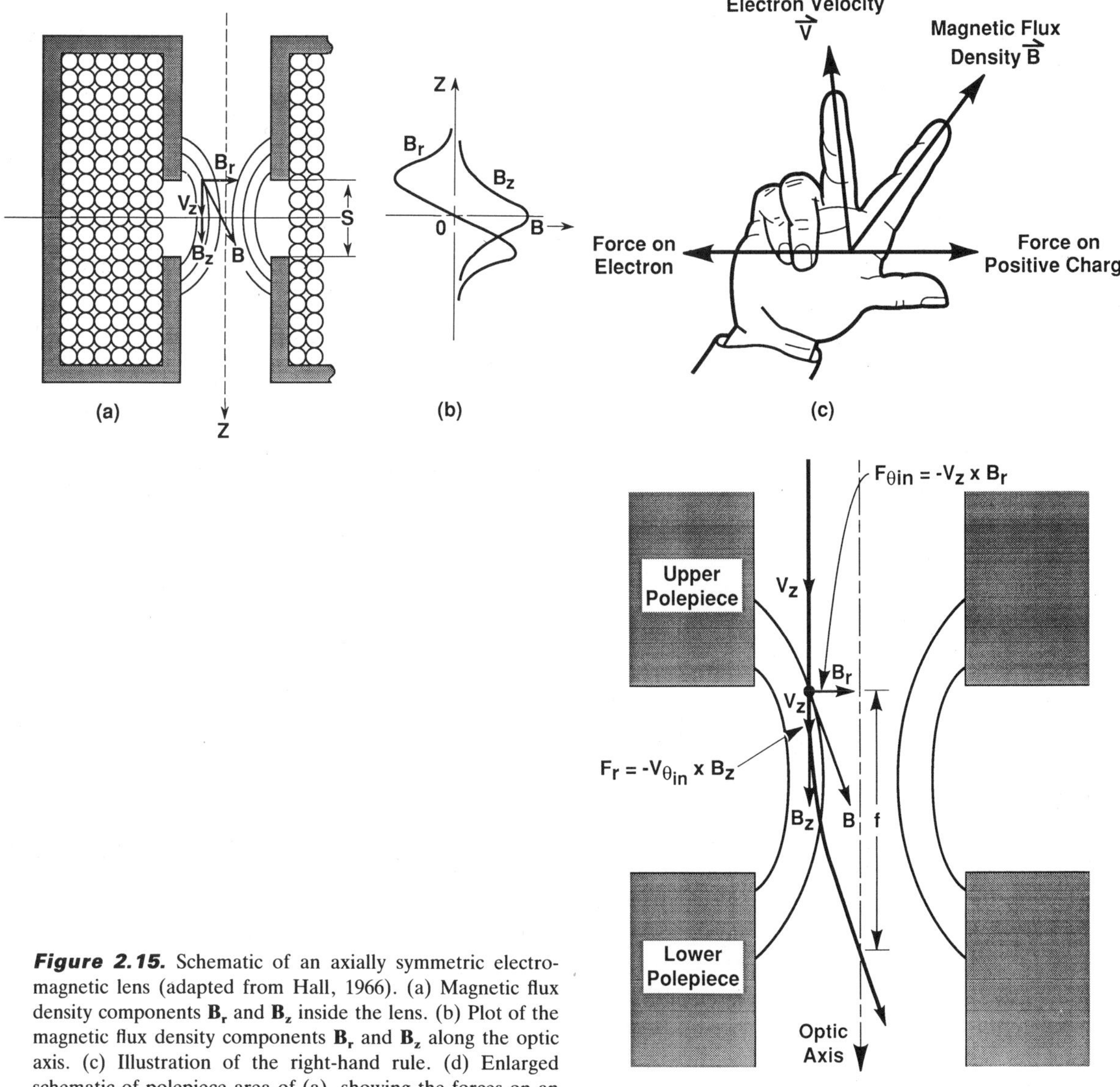

Figure 2.15. Schematic of an axially symmetric electromagnetic lens (adapted from Hall, 1966). (a) Magnetic flux density components $\mathbf{B_r}$ and $\mathbf{B_z}$ inside the lens. (b) Plot of the magnetic flux density components $\mathbf{B_r}$ and $\mathbf{B_z}$ along the optic axis. (c) Illustration of the right-hand rule. (d) Enlarged schematic of polepiece area of (a), showing the forces on an electron that cause it to be focused.

crosses the axis. Note that the actual trajectory of the electron as it traverses the lens is a spiral and that the final image will show this as a rotation of the image when the strength of the objective lens is changed.

Lens Current Changes Focal Length. We can control the lens by changing the current I in the lens coil, which changes the focal length f. For most lenses, the focal length f is nearly proportional to $V_0/(NI)^2$, where N is the number of turns in the lens coil and V_0 is the accelerating voltage (Liebmann, 1955). Thus, the focal length f decreases as the current I increases, making the lens stronger. Note also that the focal length increases at higher accelerating voltages (higher kV) for the same lens excitation, since the velocity of the electrons increases with increasing beam voltage. All modern SEMs automatically change I as a function of accelerating voltage to compensate for the change in focal length.

2.3.2. Lenses in SEMs

Condenser Lenses. SEMs employ one to three condenser lenses to demagnify the electron-beam crossover diameter in the gun to a smaller size. The first condenser lens controls the amount of demagnification. In microscopes equipped with a second condenser lens, the control of both condensers is usually ganged so that a single knob adjusts both. This knob is usually labeled *spot size, condenser, C*1, or *resolution.* As described below, the more current I flowing through the condenser, the smaller the final probe size and the smaller the beam current that reaches the specimen. The condenser is usually air cooled since it is a relatively weak lens and the heat generated by the modest current flowing through it is easily dissipated.

Objective Lenses. The final lens in the column, called the *probe-forming* or *objective* lens, focuses the image by controlling the movement of the probe crossover along the optic axis (Z-axis) of the column. The knob controlling the current flowing through the winding of this lens is usually labeled *objective* or *focus.* Since the objective lens is the strongest lens in the SEM, with the largest current I flowing through its windings, it must usually be liquid cooled. The design of this lens often incorporates space for the scanning coils, the stigmator, and the beam limiting aperture. There are two basic designs of the objective lens: one for large specimens, where the specimen must be outside the lens, and one for specimens small enough to be placed inside the lens.

The most common objective lens, called the *asymmetrical pinhole lens* or *conical lens,* is shown in Fig. 2.16a. The final polepiece has a very small bore that keeps most of the magnetic field within the lens, allowing for collection of secondary electrons from the specimen, undisturbed by magnetic fields. At the same time, a relatively large specimen may be positioned close to the lens gap, where $\mathbf{B}_z$ is a maximum and the focal length is short. For high image resolution, it is important to keep the specimen close to the gap because lens aberrations that enlarge the final probe size increase rapidly with focal length. The focal length of the objective lens increases as the distance from the lower surface of the polepiece to the specimen, known as the *working distance W,* increases. Typical focal lengths are between 10 mm and 40 mm, while typical working distances are from 5 mm to 40 mm. In some microscopes the final lens aperture, which controls the beam convergence angle, is located in the objective lens gap. There are two main operational advantages of the pinhole lens. First, specimen size is limited not by the lens but by the size of the specimen chamber below the lens. Second, the large variability in working distance W allows excellent depth of field at long working distances.

For the second type of objective lens, the immersion lens, a small specimen is placed directly inside the lens gap (Fig. 2.16b). Since the specimen is immersed in the magnetic field, this lens provides focal lengths in the range 2 to 5 mm, similar to those used in TEM. Because lens aberrations scale with focal length, this type of lens yields the lowest aberrations, the smallest probe size, and the highest image resolution. Collection of secondary electrons in this lens takes advantage of the fact that secondaries spiral upward in the strong magnetic field to a detector

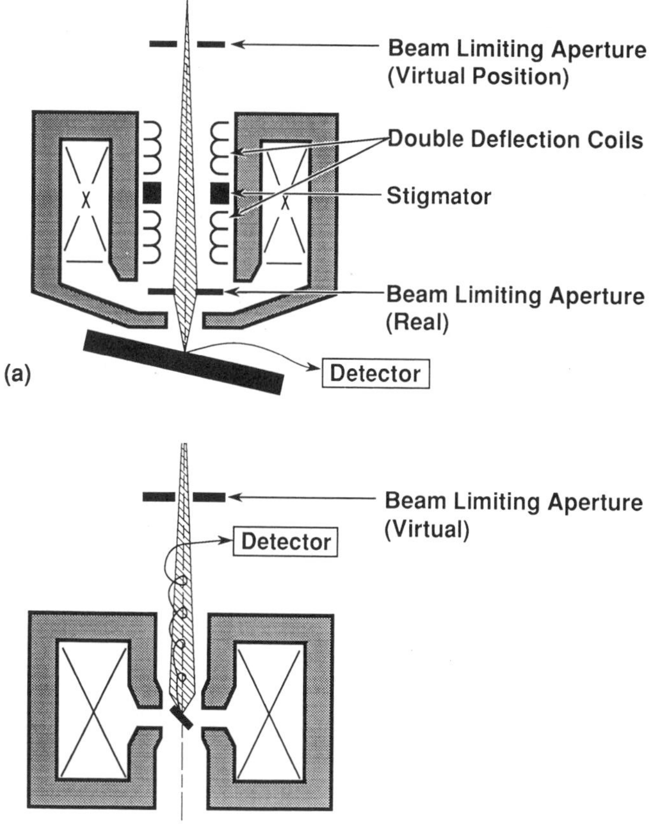

Figure 2.16. Two objective lens configurations. (a) Asymmetrical pinhole lens or conical lens, allowing a large specimen to be placed outside the lens. (b) Symmetrical immersion lens, where a small specimen is placed inside the lens.

above the lens (Koike *et al.*, 1971). This arrangement has the advantage that only secondaries are collected, because they are effectively confined by the strong magnetic field inside the lens while the highly energetic backscattered electrons strike the walls of the lens and do not reach the collector. For immersion lenses the objective aperture must be located outside the lens in the so-called virtual position. Since immersion objective lenses are standard on TEMs, a TEM with a scanning attachment and an SE detector gives the highest resolution possible for a thermionic source. The main advantages of the immersion lens are higher resolution and selective separation of secondary electrons from backscattered electrons.

Real and Virtual Objective Apertures. An objective aperture is called a *real* objective aperture if it is located just before the specimen in the gap of the probe-forming lens (Fig. 2.16a). However, beam convergence may be limited at any point between the crossover after the last condenser lens and the objective lens gap. Thus, in some SEMs the final aperture is located between the last condenser lens and the objective lens in a position known as the *virtual objective aperture* position (Figs. 2.16a–b). The sizes of virtual apertures are smaller in order to limit the beam to the same degree as a real aperture. The divergence angle of the virtual aperture is multiplied by the demagnification factor ($m \geq 1$) of the objective lens to yield the equivalent real objective aperture size (see next section). One advantage of this configuration is that the apertures tend to stay clean since they are removed from the specimen region where contaminants from the specimen may be desorbed by the electron beam. For either type of objective aperture, centering is important, either to prevent the image from sweeping across the screen during scanning (real objective aperture alignment) or to produce a uniform beam intensity across the scanned field (virtual objective aperture alignment).

2.3.3. Producing Minimum Spot Size

Neglecting lens aberrations (see Section 2.3.5), demagnification in an electron lens may be described by geometrical optics because the situation is similar to that of a light-optical lens. For weak lenses, when the lens thickness is negligible compared to both the object distance and the image distance, the focal length f may be calculated from the same thin-lens equation used in light optics:

$$\frac{1}{f} = \frac{1}{p} + \frac{1}{q}, \tag{2.7}$$

where p is the distance from the object to the center of the lens and q is the distance from the center of the lens to the image, as shown in Fig. 2.17. Note that the image is inverted compared to the object, and, for magnetic lenses, there is an additional rotation of the image (not shown). The focal length f is shown as the distance from the center of the lens to the point where a ray parallel to the axis crosses the axis (compare with Fig. 2.15d). While Eq. (2.7) relates f, p, and q for lenses used to magnify or demagnify, our interest is in the demagnification of the gun crossover d_0.

Demagnification (Reducing the Spot Size). The demagnification for the lens in Fig. 2.17 is given by $m = p/q = 3.35$, a number greater than 1. Figure 2.18 shows a schematic two-lens SEM that illustrates the operator's control over microscope parameters. The ray traces of electrons in this and the following figures are shown with highly exaggerated α angles compared to actual α angles in the SEM, which are typically 0.001–0.02 radians (0.05°–1°). The crossover at the electron gun, of diameter d_0 and divergence angle α_0, passes through the condenser lens and is reduced to a diameter d_1 with an increased convergence angle α_1. Usually most of the beam leaving the gun enters the first condenser lens, but in a real SEM there would be an aperture in this lens that is not shown in Fig. 2.18. The distance p_1 from the gun crossover point to the condenser lens gap is constant, whereas the distance q_1 between the condenser lens and the next crossover can be varied by changing the strength of the condenser lens. As the current in the condenser winding increases, the focal length f_1 decreases. Therefore, according to Eq. (2.7), since p_1 is constant, q_1 decreases. Also, as the strength of the lens increases, the demagnification m_1 will increase, reducing the spot size d_1 and increasing the divergence angle α_1 of electrons at the crossover below the condenser. The diameter of the intermediate crossover $d_1 = d_0/m_1 = d_0 q_1/p_1$. The divergence angle α_1 is equal to α_0 times the demagnification m_1. For the two-lens system shown in Fig. 2.18, Eq.

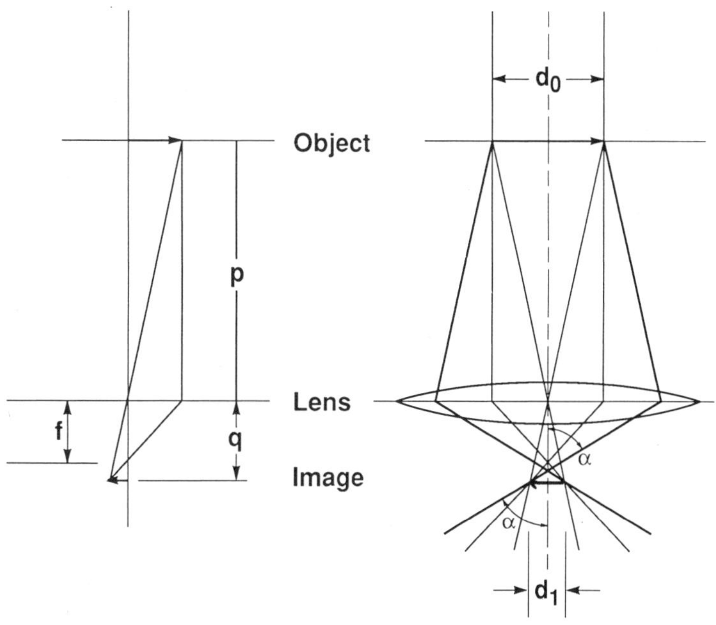

Figure 2.17. Geometric optics for a thin lens operated in a demagnifying mode. (a) Geometrical ray diagram showing the construction to determine f and (b) the demagnification of the gun crossover d_0 to form the intermediate crossover d_1 after the first condenser lens.

(2.7) may be applied for each lens. Since the variable distance from the intermediate crossover to the objective lens gap is p_2 and the distance from the objective lens to the focused spot on the specimen is q_2, the demagnification for the objective lens is $m_2 = p_2/q_2$. Thus, the final geometrical probe size in Fig. 2.18 is given by $d_2 = d_0/m_1m_2$. When aberrations are neglected, as in this case, the final probe size is called the Gaussian probe size.

Effect of Final Aperture Size. A real objective aperture placed in the gap of the probe-forming lens (usually 50 to 300 μm in diameter) is shown in Fig. 2.18. This aperture decreases the divergence angle α_1 from the condenser lens to a smaller angle α_a for the electrons entering the final lens. The final aperture has three important effects on the final probe. First, there is an optimum aperture angle that minimizes the detrimental effects of aberrations on final probe size, as will be shown below. Second, the current in the final probe is controlled by the size of the aperture since only a fraction of the

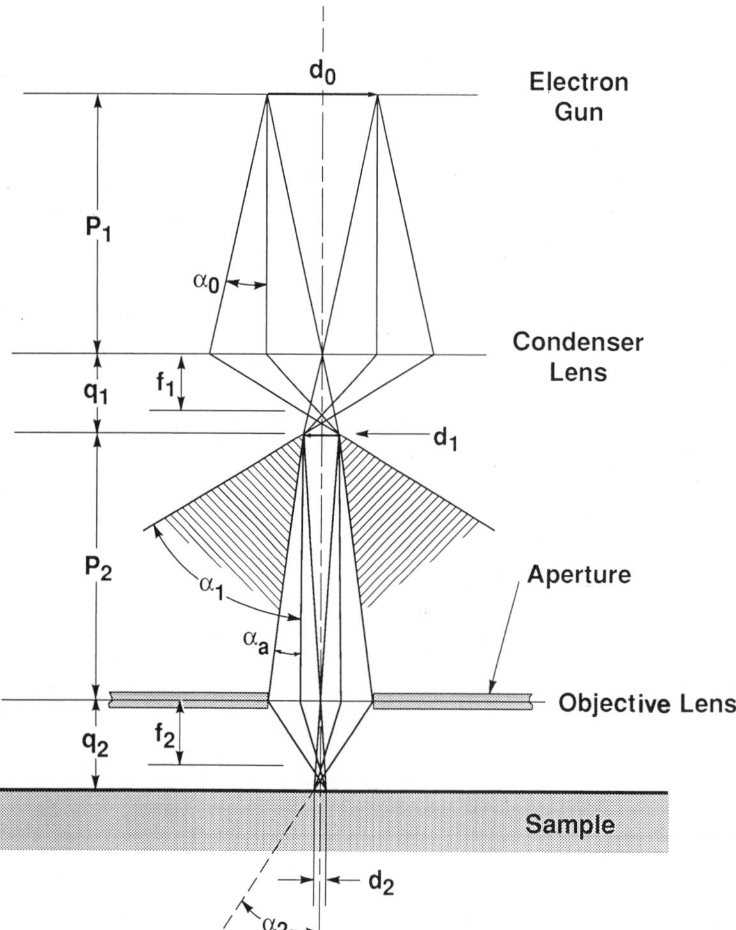

Figure 2.18. Ray traces in a schematic two-lens SEM column with a single condenser lens and a probe-forming or objective lens.

current sprayed out to angles α_1 passes within the aperture angle α_a. Third, the final convergence angle α_2 $(=\alpha_p)$, the probe convergence angle, controls the depth of field, that is, the range of heights on the specimen surface for which the image is in focus (see Chapter 4). Smaller α_2 angles produce greater depth of field.

Effect of Working Distance. The objective lens can focus the final probe at various specimen working distances, as illustrated in Fig. 2.19. In both diagrams of Fig. 2.19 the condenser lens strength is the same, producing the same demagnification $m_1 = p_1/q_1$. Also, the same aperture size is used, so that α_a is the same in both cases. The working distance W increases and q_2 increases as the specimen is moved further from the lens, as in Fig. 2.19b. Since the demagnification $m_2 = p_2/q_2$ decreases, the spot size on the specimen increases and the image resolution degrades. However, the convergence angle α_2

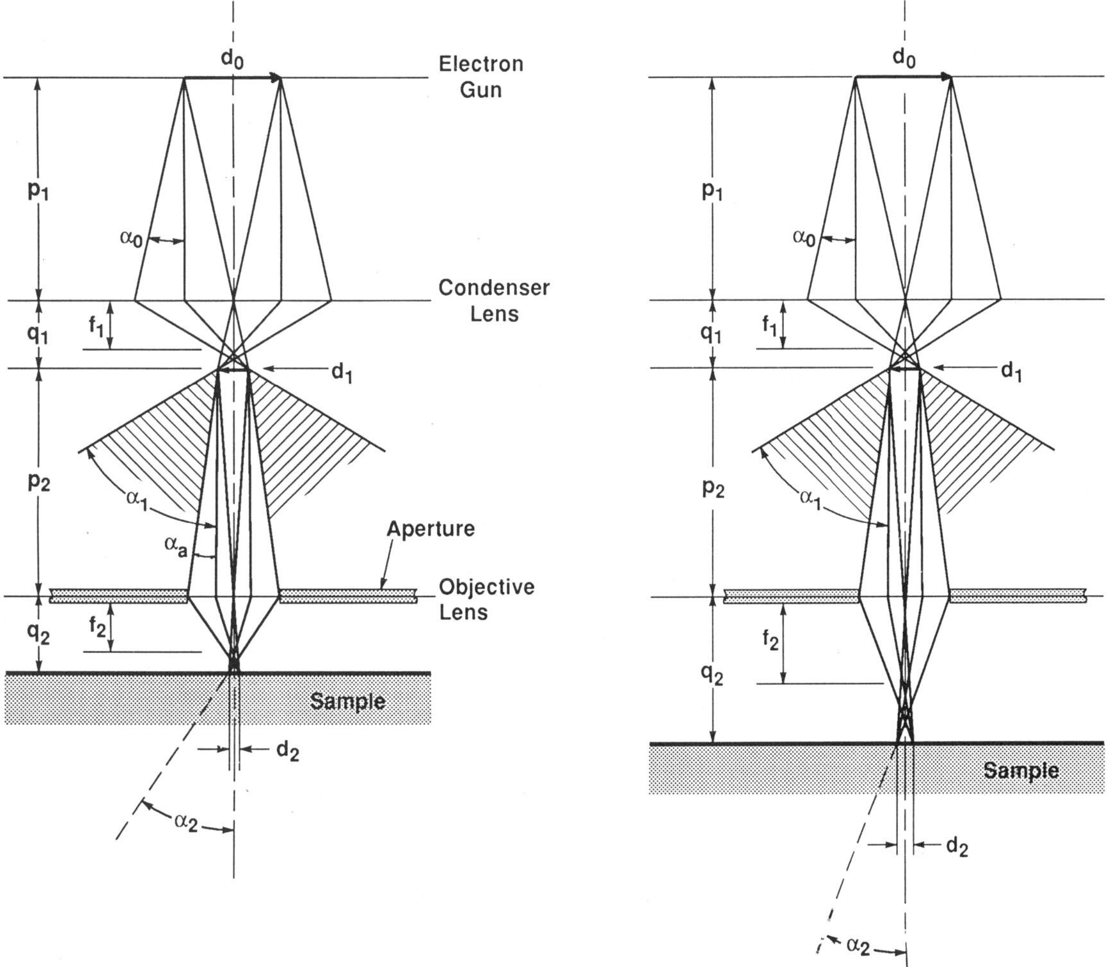

Figure 2.19. Schematic of ray traces for two-lens probe forming system. (a) Small working distance. (b) Large working distance.

decreases, giving rise to an improved depth of field. To obtain this focusing condition at large W, the objective lens current must be decreased, which in turn increases the focal length f_2 of the lens. This increased working distance also increases the scan length that the beam traverses on the specimen. Thus, long working distances can be used to obtain very low magnifications. Finally, by considering Fig. 2.19, one can understand the alternate method of focusing whereby the working distance is selected by setting the current in the objective lens and the operator physically moves the specimen vertically along the z-axis until it comes into focus on the screen.

Effect of Condenser Lens Strength. Increasing the strength of the condenser lens decreases both the final probe size $d_p = d_2$ and the amount of current i_p in the final probe, as shown in Fig. 2.20. With a constant working distance and objective lens aperture size, an increase

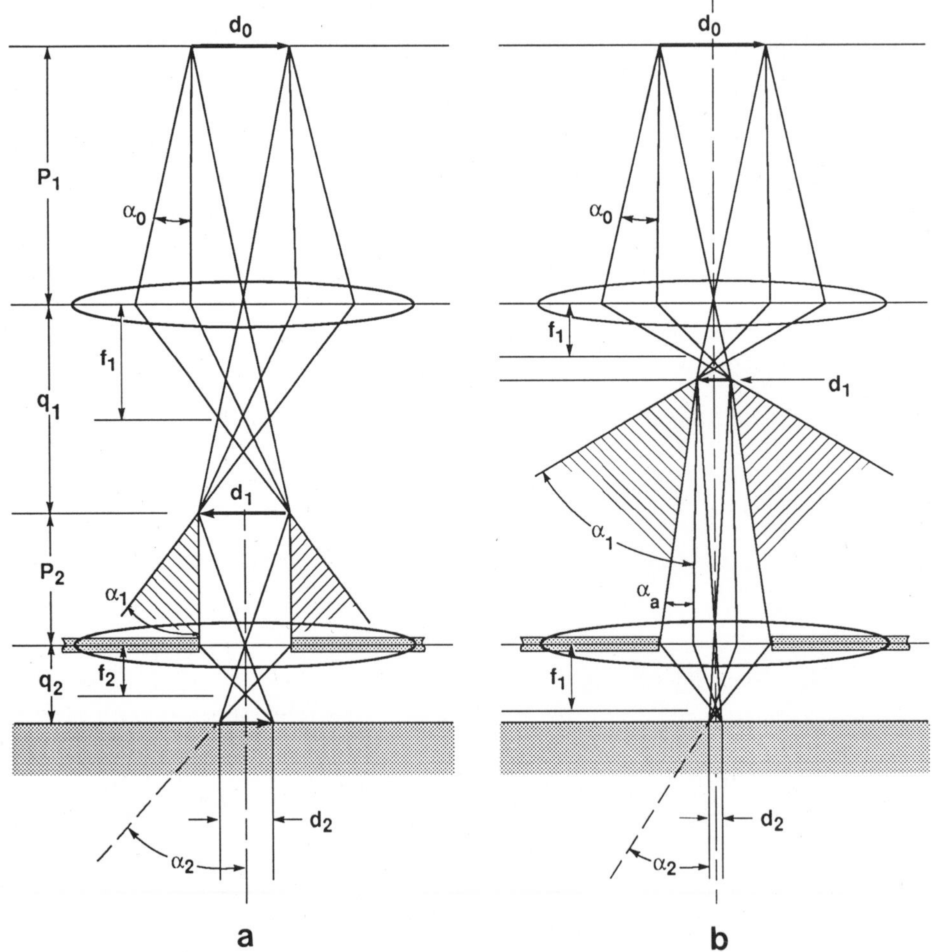

Figure 2.20. Schematic of ray traces for two-lens probe forming system: (a) weak condenser lens, (b) strong condenser lens.

in condenser lens strength causes an increase in demagnification of the probe, because m_1 and m_2 both increase as f_1 becomes shorter. However, as the intermediate crossover image d_1 decreases, α_1 increases. As the cross-hatched region in Fig. 2.20b shows, when the condenser lens is set to a shorter focal length, d_1 and d_2 are smaller, but a smaller portion of the beam passes through the final lens aperture. The current reaching the specimen may be calculated using the ratio of the solid angles of the cones subtended by α_a and by α_1. The beam current which enters the final lens is given by the ratio $(\alpha_a/\alpha_1)^2$ multiplied by the current available in the intermediate image d_1. To increase the current i_p in the probe, it is necessary to weaken the condenser lens, as in Fig. 2.20a so more current passes through the aperture. *Therefore, one can minimize spot size only at the expense of current in the final probe.* The operator must make a conscious decision as to whether minimum probe size or maximum probe current and signal is desired.

Gaussian Probe Diameter. Most calculations of electron probe diameter start with a calculation of the Gaussian probe diameter d_g. Strictly, the value calculated is the full width at half maximum (FWHM) of the Gaussian intensity distribution of d_g. If a real crossover of known diameter exists in the electron gun (thermionic sources), the Gaussian probe size may be calculated from the total demagnification, as described above. However, to compare field-emission and thermionic electron sources, a method to estimate d_g appropriate to all cases would be useful. In Section 2.2.3 there is a statement that, in the absence of aberrations, the electron optical brightness is constant throughout the electron column. Thus, we can use the gun brightness to estimate d_g (FWHM) of the electron probe intensity profile from a rearrangement of Eq. (2.3) as applied at the final probe spot at the specimen surface:

$$d_g = \sqrt{\frac{4i_p}{\beta\pi^2\alpha_2^2}}. \tag{2.8}$$

The current in the final probe can also be estimated by rearranging this equation:

$$i_p = \frac{\beta\pi^2\alpha_2^2 d_g^2}{4}. \tag{2.9}$$

If there were no aberrations in the system it would be necessary only to increase α_2, the probe convergence angle, in order to increase the current at a constant probe diameter. However, because of the several aberrations present in the electron-optical system, α_2 must be kept small and the current available for a given probe diameter is limited.

2.3.4. Lens Aberrations

Electron rays that are very close to the optic axis ($\alpha \cong 0$) represent the object as an image with the correct position, shape, and size, as outlined in the previous section. Electron rays that are inclined to the axis ($\alpha > 0$) do not come to focus at the correct location, causing a blur in the image called an *aberration*. Since the amount of current in the probe increases with α, it is clear that a compromise must be made. While most aberrations can be corrected in glass lenses that focus light, this is not true in electron lenses.

Spherical Aberration. Spherical aberration arises because electrons in trajectories which are further away from the optic axis are bent more strongly by the magnetic field than those near the axis. In other words, the strength of the lens is greater for rays passing through the outside of the lens than for those passing through the center of the lens. This aberration is illustrated in Fig. 2.21a. In this figure, the z-axis of the

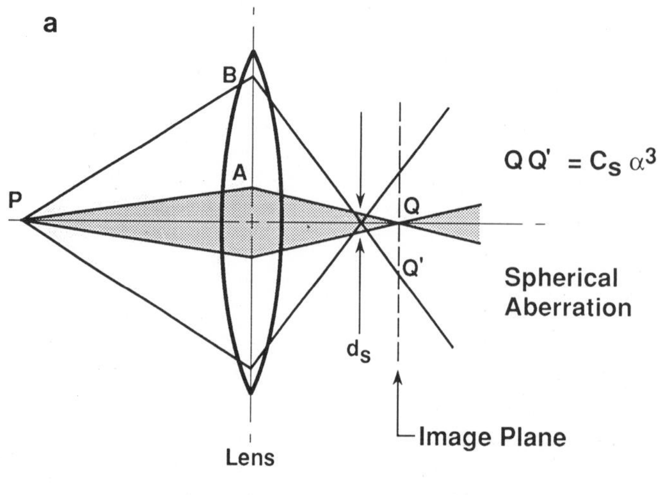

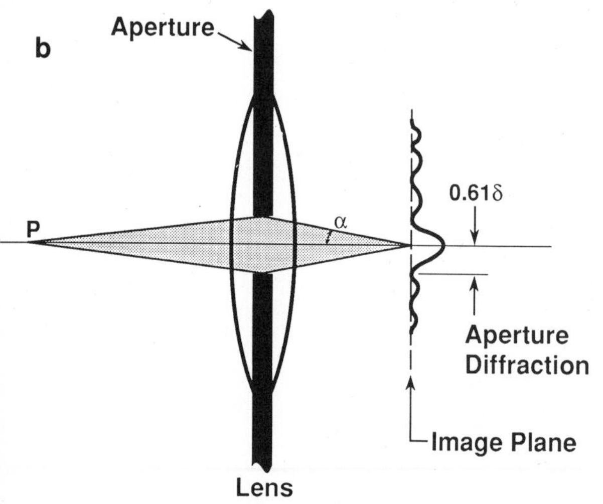

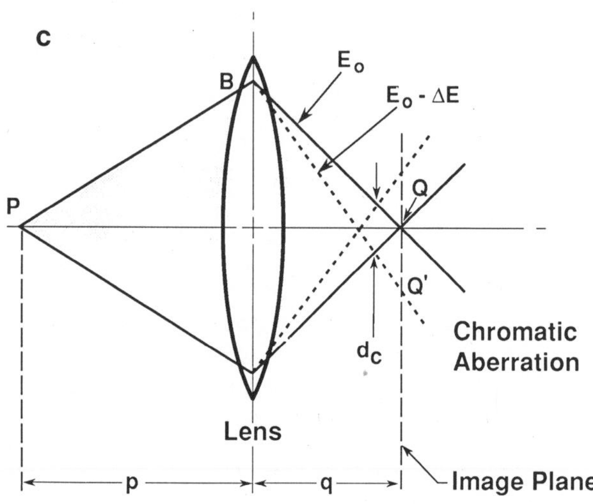

Figure 2.21. Schematic drawings showing how (a) spherical aberration, (b) aperture diffraction, and (c) chromatic aberration in a lens cause a point object at P to blur into a broadened image at Q. The drawings are adapted from Hall (1966) and Oatley (1972).

electron optical column is drawn horizontally instead of vertically. Electrons emerging from point P, and following a path close to the axis, such as PA, are focused to a point Q at the dotted line known as the Gaussian image plane. Electrons following the path PB, which is the maximum divergence allowed by the aperture of the lens, are focused more strongly and cross the optic axis closer to the lens. Thus, electrons of path PB are focused to a point Q′ rather than to point Q at the Gaussian image plane. This is because the magnetic field between the lens polepieces is not uniform. This process causes an enlarged image of point P in the Gaussian image plane of diameter 2QQ′. The smallest image of point P occurs just in front of QQ′ and is often called the *spherical aberration disk of minimum confusion.* The diameter d_s of this disk can be written as

$$d_s = \tfrac{1}{2}C_s\alpha^3, \tag{2.10}$$

where C_s is the spherical aberration coefficient, a lens parameter related to the focal length f and the lens excitation $V_0/(NI)^2$, and α is the convergence angle formed between BQ and the optic axis. For pinhole objective lenses, the value of C_s is about 2 cm, while for immersion lenses C_s is a few millimeters.

Since electron lenses are always convergent (positive) lenses, spherical aberration cannot be removed by a combination of positive and negative lenses as in light optics. The contribution of d_s to the final electron-probe diameter can be reduced only by decreasing α with an objective lens aperture. Unfortunately, a small aperture reduces the current in the probe and introduces aperture diffraction.

Aperture Diffraction. For very small apertures, the wave nature of electrons gives rise to a diffraction pattern instead of a point at the Gaussian image plane. This is not actually a lens aberration but a consequence of the small aperture used to reduce spherical aberration. Electrons emerging from point P diffract at the small aperture and appear in the Gaussian image plane as a broad "Airy disk" intensity distribution surrounded by small subsidiary maxima, as shown in Fig. 2.21b. Following Wells (1974a) we take half the diameter of the Airy disk as the diffraction contribution to the spot size d_d given by

$$d_d = \frac{0.61\,\lambda}{\alpha}, \tag{2.11}$$

where λ is the wavelength of the electrons, given nonrelativistically by $\lambda = 1.24/(E_0)^{1/2}$, with λ in nanometers, and E_0 in electron volts and α is the angle between the converging ray and the electron optic axis, in radians. The term δ shown in Fig. 2.21b is λ/α. For this effect, the larger the value of α, the smaller the contribution of d_d. Thus, spherical aberration and aperture diffraction vary in opposite directions with α. This leads to the need to find an optimum aperture angle α_{opt} which is a balance between these two effects.

Chromatic Aberration. A variation in the electron beam energy E_0 is the lens current I also changes the point at which electrons from point

P are focused in the Gaussian image plane. For example, Fig. 2.21c shows that if two electrons in the ray PB have different energies, say E_0 and $E_0 - \Delta E$, they focus at different points in the image plane, namely Q and Q′, respectively. This process causes the image of point P to be enlarged to 2QQ′ at the Gaussian image plane. The diameter d_c of the disk of least confusion which forms in front of QQ′ is usually written as

$$d_c = C_c \alpha \frac{\Delta E}{E_0}, \tag{2.12}$$

where C_c is the chromatic aberration coefficient, α is the convergence angle, and $\Delta E / E_0$ is the fractional variation in the electron-beam energy. The chromatic aberration coefficient is directly related to the focal length f of the lens and of about the same magnitude. For most SEM objective lenses, the value of C_c is similar to that of C_s but may be higher or lower than C_s depending on operating conditions.

Contributions to d_c may occur from variations in both accelerating voltage and lens currents resulting from imperfect stabilization of the respective power supplies. However, if the high voltage and lens currents are stabilized to one part in 10^6 per minute, the effects of these electrical instabilities will be unimportant (Oatley, 1972). Nevertheless, there is a variation ΔE in the energy of the electrons due to the spread of the initial energies of the electrons leaving the cathode. As shown in Table 2.1, typical values of this energy spread ΔE are 1–3 eV for the tungsten hairpin filament, 1–2 eV for LaB_6, depending on the bluntness of the tip and the current drawn; and 0.2–0.3 eV for cold field emission. The value of d_c can also be minimized by using a lens with low C_c, such as an immersion lens, and by reducing the convergence angle α.

Astigmatism. Although it has been tacitly assumed that magnetic lenses have perfect rotational symmetry, as suggested by Fig. 2.14b, the perfection required for small probe formation is not generally possible. Machining errors, inhomogeneities in the iron of the lens, asymmetry in its windings, and dirty apertures lead to a lens that is not perfectly round. For example, if a lens has elliptical rather than circular symmetry, electrons diverging from a point object will come to focus at the image as two separate line foci at right angles to each other, as shown in Fig. 2.22. Thus, astigmatism would enlarge the size of the final electron probe diameter to a diameter d_a even if all other aberrations were negligible. This effect can be recognized by the stretching of the image in two perpendicular directions when the objective lens is first underfocused and then overfocused, as shown in Fig. 2.23. The stretching in the image vanishes at exact focus (Figure 2.23a) but the effective probe size still may be many times the smallest possible probe, and the image appears blurred.

Astigmatism can be corrected with the stigmator, a device that applies a weak supplemental magnetic field to make the lens appear symmetrical to the electron beam. The stigmator is usually an octupole of small electromagnetic coils that apply the required level of additional field in the appropriate direction. One can usually correct for astigmatism in the objective lens by adjusting the two stigmator controls alternately while focusing an image with fine random detail at moderately high magnifications (10,000X or higher). This cycle (adjusting x-stigmator, focus, y-stigmator, focus) should be repeated until the sharpest image is obtained (Fig. 2.23d). If the stigmator cannot correct the astigmatism of the image, the apertures or liner tube should be cleaned and the SEM column should be aligned.

Effects of Aberrations on Probe Size. The effects of the above aberrations are most significant in the objective lens since the amount of blurring they cause in preceding lenses is small when compared to the size of the intermediate crossovers formed by those lenses. While astigmatism can be completely corrected in the final lens of a properly maintained SEM, the effects of spherical aberration, aperture diffraction, and chromatic aberration remain.

2.4. Electron Probe Diameter versus Electron Probe Current

The general approach to handling aberrations in SEM lenses is to determine the objective lens aperture angle α_{opt} that minimizes the effect of the lens aberrations on the final probe size. Use of α_{opt} will allow one to obtain the maximum amount of current at the minimum probe size.

2.4.1. Calculation of d_{min} and i_{max}

The most important figure of merit for an SEM is the amount of current that can be placed in small electron probes used for imaging and analysis. Following Smith (1956), it is possible to calculate the diameter d_p of the electron probe impinging on the sample carrying a given current i_p. Calculations of d_p are usually made for the FWHM of a Gaussian intensity distribution that contains about half of the total probe current.

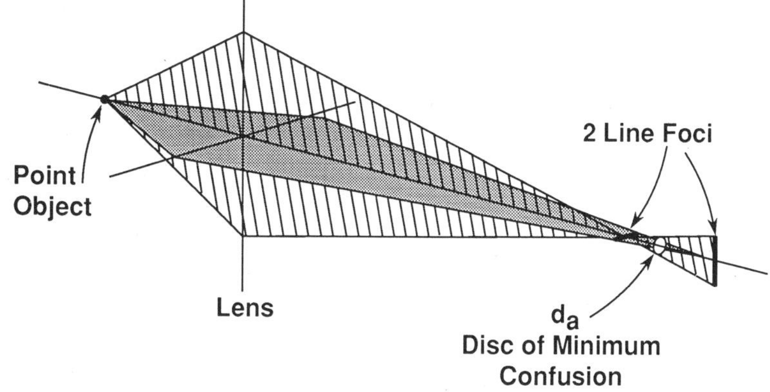

Figure 2.22. Schematic diagram showing the origin of astigmatism. A point object is focused to two line foci at the image, and the desired small focused beam can be obtained only by forcing the two line foci to coincide using the stigmator (adapted from Hall, 1966).

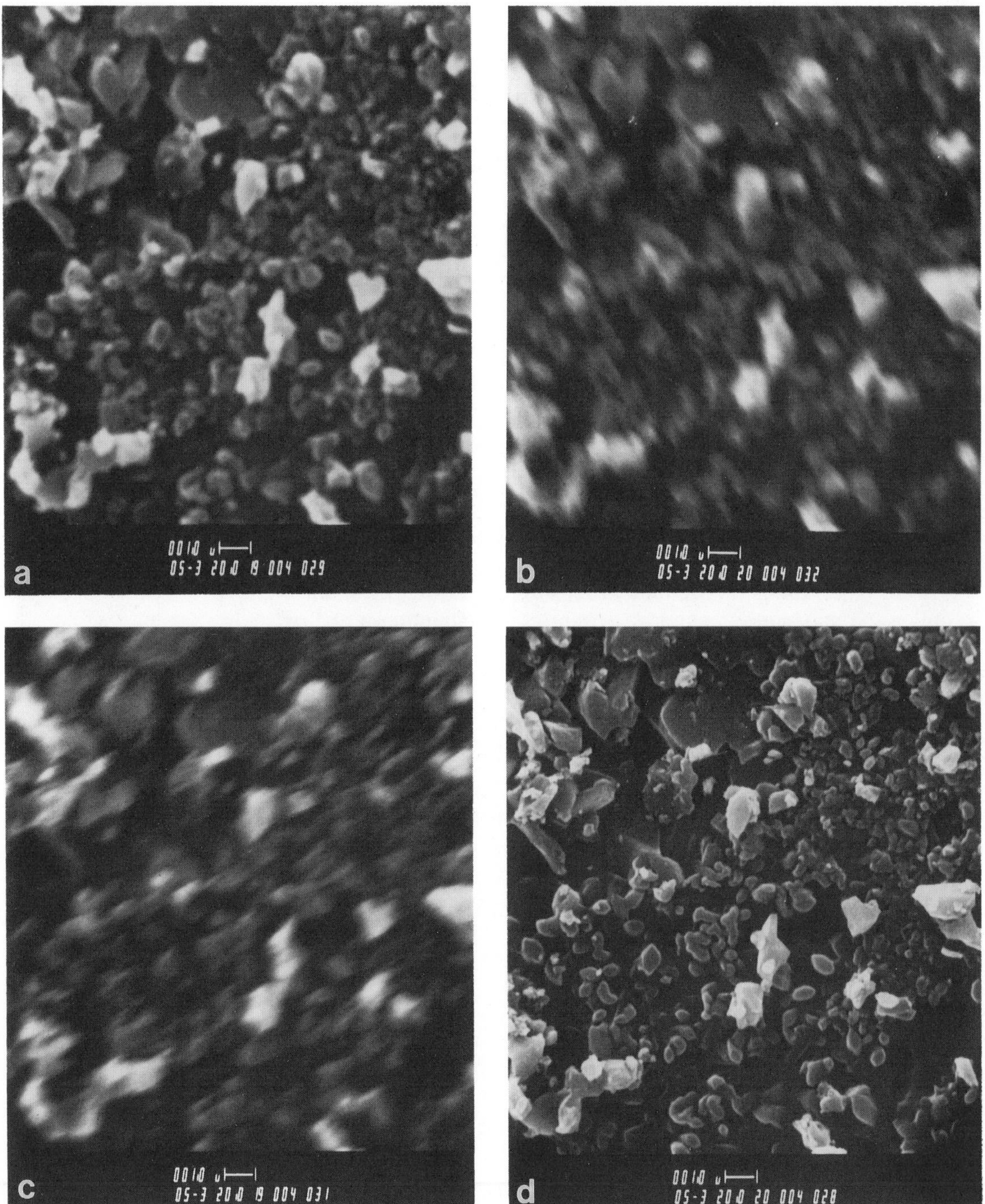

Figure 2.23. Effect of astigmatism in the probe-forming lens: (a) initial image before astigmatism correction, (b) underfocus, (c) overfocus, and (d) image corrected for astigmatism.

To obtain d_p (FWHM), it is usually assumed that all the significant aberrations are caused by the final (objective) lens. To calculate the final probe size, the diameter of the Gaussian probe d_g and the various aberration disks are assumed to be error functions. The effective final spot size d_p(FWHM) is considered equal to the square root of the sum of the squares of the separate diameters,

$$d_p = (d_g^2 + d_s^2 + d_d^2 + d_c^2)^{1/2}, \tag{2.13}$$

where d_g is the Gaussian probe size at the specimen, d_s is the spherical aberration disk, d_d is the aperture diffraction disk, and d_c is the chromatic aberration disk. Substituting from Eqs. (2.8) and (2.10)–(2.12), we obtain

$$d_p = \left[\frac{4i_p}{\beta\pi^2\alpha^2} + (\tfrac{1}{2}C_s)^2\alpha^6 + \frac{(0.61\lambda)^2}{\alpha^2} + \left(\frac{\Delta E}{E_0}C_c\right)^2\alpha^2\right]^{1/2}. \tag{2.14}$$

In Eq. (2.14), lens instabilities in the chromatic aberration term are assumed to be negligible.

Approximation for Normal Accelerating Voltages. For typical 10 to 30 kV operation, it is possible to understand the relationship between probe size and probe current by considering the simplified case of a probe limited by spherical aberration and diffraction effects, but neglecting chromatic effects. The appropriate expression for the Gaussian probe diameter to substitute into the brightness equation, Eq. (2.9), can be obtained from

$$d_g^2 = d_p^2 - d_s^2 - d_d^2 \tag{2.15}$$

By substituting Eqs. (2.10) and (2.15) into Eq. (2.9) and differentiating with respect to α, we obtain an expression for $dI/d\alpha$. The aperture angle α goes through a minimum, which we designate the *optimum* aperture angle, α_{opt}, when $dI/d\alpha$ is zero. Solving the differentiated equation for the condition $dI/d\alpha = 0$ gives the result

$$\alpha_{\text{opt}} = \left(\frac{d_p}{C_s}\right)^{1/3}. \tag{2.16}$$

Substituting Eq. (2.16) into both Eq. (2.14) (neglecting chromatic aberration for the range 10 to 30 kV) and Eq. (2.9) yields the following results:

$$d_{\text{min}} = KC_s^{1/4}\lambda^{3/4}\left(\frac{i_p}{\beta\lambda^2} + 1\right)^{3/8} \tag{2.17}$$

$$i_{\text{max}} = \frac{3\pi^2}{16}\beta\frac{d_p^{8/3}}{C_s^{2/3}}, \tag{2.18}$$

where K is a constant close to unity. The value of d_{min} is given in nanometers when C_s and λ are in nanometers, and i is given in amperes (A). Equation 2.17 shows that the minimum probe size d_{min} decreases as the brightness β increases and as the electron wavelength λ and spherical aberration coefficient C_s decrease. In fact, in the limit of zero probe current, d_{min} reaches a value of approximately $C_s^{1/4}\lambda^{3/4}$, which can be regarded as a measure of the theoretical resolution of the microscope. From Eq. (2.18) we can see that the maximum probe current varies as the 8/3 power of the probe diameter. Since backscattered-electron, secondary-electron, and x-ray emission vary directly with probe current, these signals fall off very rapidly as the probe diameter is reduced. More detailed expressions equivalent to Eqs. (2.17) and (2.18) have been published (Pease and Nixon, 1965; Wells, 1974a).

Increasing i_{max}. We can see from Eq. (2.17) that there are several ways in which i_{max} can be increased. First, the brightness may be increased either by changing to a high-brightness source or by increasing the accelerating voltage (higher kV). However, to keep the x-ray emission volume relatively small, as discussed in Chapter 3, the maximum voltage that can successfully be applied for x-ray analysis is about 30 kV. Second, the value of i_{max} can be increased if C_s is lowered by using a short working distance to reduce the focal length of the objective lens, and thus C_s. However, for a pinhole objective lens the working distance must be large enough to provide adequate tilting of the specimen, so the focal length cannot be greatly reduced. Nevertheless, changes in lens design can decrease C_s and hence increase i_{max}. For example, a reduction in C_s by a factor of 10, as in an immersion objective lens (Section 2.3.2), increases i_{max} by a factor of 5 and reduces d_{min} by a factor of 2.

Low Voltage Operation. Low accelerating voltages can eliminate the need for metal coating of insulators (see Chapters 3 and 4). However, at accelerating voltages near 1 kV, $\Delta E/E_0$ in the chromatic aberration term of Eq. (2.14) can no longer be neglected, and in fact the chromatic aberration term becomes larger than the spherical aberration term and dominates the probe size for high-resolution operation. The situation at low voltage is also more complex because the chromatic term is not Gaussian like the other aberrations. One way to calculate d_{min} and i_{max} including the chromatic effect is to use the method of Shao and Crewe (1989) to calculate α_{opt} for a given set of microscopic parameters C_s, C_c, ΔE, and E_0. The values of d_{min} and i_{max} may then be calculated from Eqs. (2.14) and (2.9), respectively.

2.4.2. Comparison of Electron Sources

The calculated minimum probe size d_{min} as a function of probe current for the three major electron sources is shown in Fig. 2.24. The 10-kV and 30-kV curves for tungsten and LaB_6, shown in Figs 2.24a and b, were calculated using Eq. (2.18) with brightness values of 5×10^4 A/cm²sr (10 kV) and 1.5×10^5 A/cm²sr (30 kV) for tungsten and 5×10^5 A/cm²sr (10 kV) and 1.5×10^6 A/cm²sr (30 kV) for LaB_6. The value of C_s for the pinhole lens was 2 cm. For probe sizes over 50 nm, the effect of chromatic aberration was neglected. For probe sizes under 50 nm, chromatic aberration becomes significant. Thus, for the smallest probe diameters, d_c was also calculated, using $\alpha = \alpha_{opt}$ from Eq. (2.16). Clearly, this is an approximation, since α_{opt} is calculated without considering chromatic aberration, but the procedure yields values close to those from a more rigorous approach. The effect of chromatic aberration, d_c, was added to d_p in quadrature. The field emission curves in Fig. 2.24a and b were plotted by substituting α_{opt} from the Shao and Crewe equation (Shao and Crewe, 1989) into Eq. (2.14) to calculate d_p. The appropriate α_{opt} and d_p values were then substituted into Eq. (2.9) to find the probe current i_p incident on the specimen for various geometrical probe sizes d_g. The lens aberrations coefficients were $C_s = 2$ cm and $C_c = 1$ cm for the normal pinhole objective lens and $C_s = 0.2$ cm and $C_c = 0.25$ cm for the immersion objective lens. Values for energy spread ΔE typical of the three sources were taken as 3 eV for tungsten, 1.5 eV for LaB_6, and 0.3 eV for cold field emission (Table 2.1). It was assumed that the gun brightness β decreased linearly with beam energy, so that at 1 kV the values were 5×10^3 A/cm^2 sr for tungsten, 5×10^4 A/cm^2 sr for LaB_6, and 3×10^6 A/cm^2 sr for cold field emission. Because of the assumptions made about lens aberrations, brightness, system magnification, and probe shape, the curves in Fig. 2.24 may differ by a factor of 2 when compared to actual observations. The probe-diameter-vs-probe-current plots given in Fig. 2.24 can be used as a guide to the operation of the electron optics.

High Resolution SEM. Figure 2.24a shows the probe size and probe current for high resolution imaging using a pinhole objective lens with tungsten, LaB_6, and cold field emission sources at 30 kV. The bottom curve is for a field emission source combined with an immersion objective lens. At 30 kV each source can provide a probe size under 10 nm with enough probe current to take a picture (>0.01 nA); however, there is a systematic improvement in probe size (and thus in instrumental resolution) from the tungsten hairpin, to the LaB_6, to the field-emission gun with the pinhole lens, to the field-emission gun with the immersion lens. For example, Fig. 2.24a shows that the smallest probe size for a tungsten gun is about 7 nm (70 Å), whereas a field-emission gun system employing an immersion lens produces a probe size close to 1 nm (10 Å). It may appear that the LaB_6 source produces the same resolution as the field emission (FE) source with a pinhole lens. However, field emission provides about 100 times more current, nearly 1 nA, in the smallest probe. A probe current of 1 nA cannot be obtained with a tungsten hairpin source until the first condenser lens is weakened enough to make the probe size 40 nm (400 Å). When the accelerating voltage is lowered to 10 kV, the probe size must be approximately double that obtained at 30 kV to achieve the same levels of probe current, as shown in Fig. 2.24b.

Low-Voltage SEM. For operation at 1 kV, there is a dramatic enlargement of the probe diameter for all electron sources, as shown in Fig. 2.24c. This enlargement is due to the chromatic aberration term, which dominates the probe size [Eq. (2.14)]. The values of energy spread ΔE for thermionic sources make the factor $\Delta E/E_0$ relatively large at 1 kV. Thus, with these assumptions, the smallest practical probe size obtainable with a tungsten hairpin source is about 100 nm (1000 Å). However, for the field-emission source, ΔE is smaller by an order of magnitude, and the factor $\Delta E/E_0$ is reduced proportionally, allowing the field-emission SEM with the pinhole lens to produce a probe size of about 20 nm (200 Å) at 1 kV. Since C_c is about an order of magnitude lower for an immersion objective lens, an additional factor-of-two improvement can be expected, to about 10 nm (100 Å) with this type of system. It is clear from Fig. 2.24c that the best way to obtain SEM images with even moderate resolution at 1 kV is to use a field emission gun. The assumptions made for the various

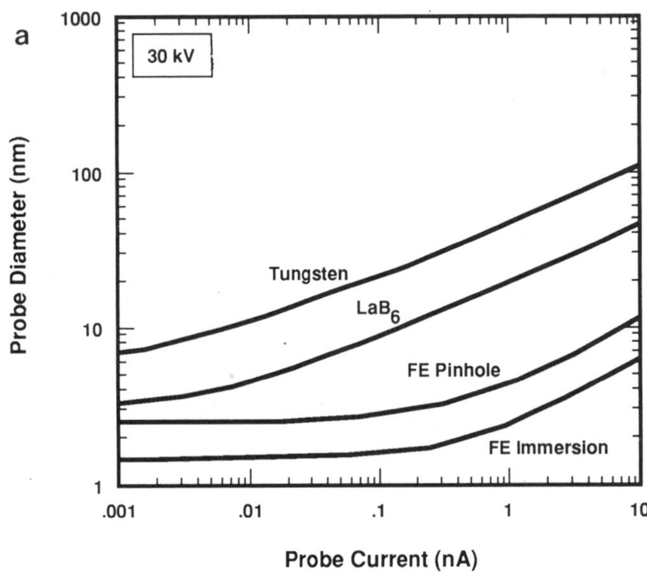

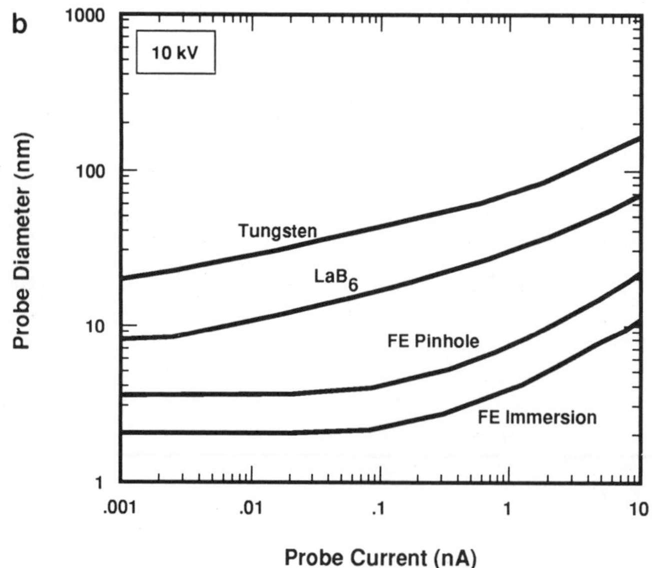

Figure 2.24. Relationship between probe diameter d_p and probe current i_p calculated for tungsten hairpin, LaB_6, and field emission (FE) sources used with a pinhole lens (C_s = 20 mm, C_c = 10 mm). Curves are also shown for the FE source with an immersion lens (C_s = 2 mm, C_c = 2.5 mm). (a) Normal SEM imaging range at 30 kV. (b) SEM range at 10 kV. (c) Low-voltage SEM at 1 kV. (d) Microanalysis range at 10 kV and 30 kV. (FE does not produce large currents in this range and is not shown.)

parameters used to calculate the 1 kV curves are believed to be conservative. Actual FE–SEM results at low accelerating voltage appear to be better than would be predicted from Fig. 2.24c. For example, Fig. 4.45b shows a structure with a 12-nm repeat spacing clearly resolved with a 2-kV beam.

Microanalysis. X-ray microanalysis usually requires 15 kV or more to efficiently excite x rays from heavier elements and 10 kV or less for efficient excitation of the light elements (see Chapter 3).

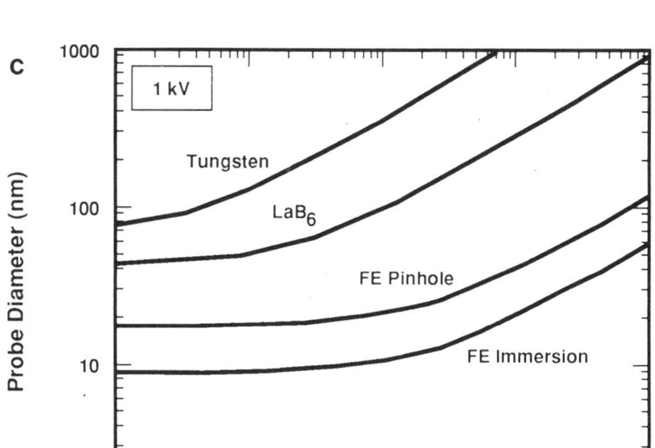

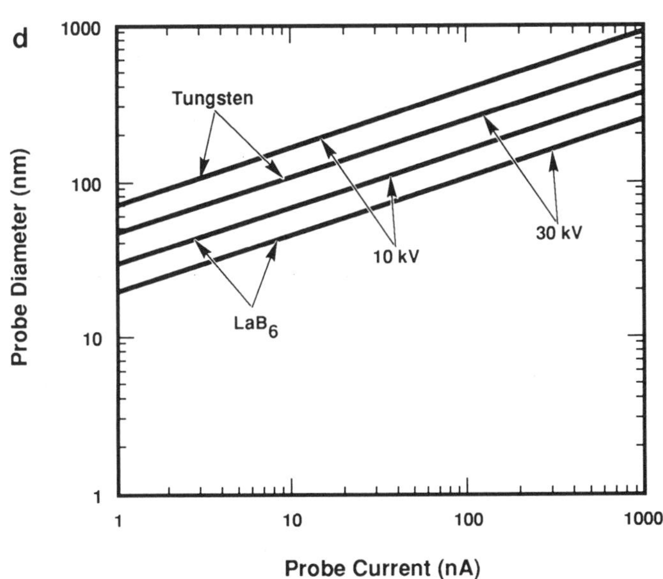

Figure 2.24. (*Continued*)

Figure 2.24d shows the 30-kV and 10-kV probe sizes and probe currents typical of the EPMA microanalysis range calculated using Eq. (2.18). Even at an electron-probe current of 1 μA (1000 nA) at 30 kV, the probe diameter is smaller than 1 μm for a conventional tungsten source. A minimum current of about 10 nA (10^{-8} A) is usually needed to perform quantitative x-ray analyses with wavelength-dispersive spectrometers (WDS). From Fig. 2.24d, WDS x-ray analysis can be performed at 10 nA using a 30-kV tungsten source with electron beam sizes as small as 0.1 μm (100 nm). However, as we will see in Chapter 3, at 30 kV the diameter of the region of x-ray emission will still be on the order of 1 μm, so there is no advantage in sacrificing probe current by using a smaller probe size.

For SEM beam sizes of approximately 10 nm (100 Å), which would be desirable for high resolution imaging, the beam current shown in Fig. 2.24a for the tungsten filament at 30 kV is 0.01 nA. This current is an order of magnitude lower than the 0.1 nA (10^{-10} A) necessary for reasonable energy-dispersive x-ray spectrometry (EDS), as discussed in Chapter 5. The LaB_6 gun, however, provides 0.1 nA at 30 kV, enough current for EDS analysis, with a nominal 10-nm probe diameter. Field emission sources are even better in this range of small probe sizes.

No field emission curve was drawn in Fig. 2.24d for the microanalysis range of operation. Above about 200 nm (2000 Å), the tungsten thermionic gun can deliver more current than a field emission gun (Cleaver and Smith, 1973) because the aberrations of the field emission gun itself become significant in this range and limit the available current. One might expect that an increased extraction voltage would lead to more total current from a field emission gun, but the beam becomes unstable in such a case because the high emission desorbs considerable gas from the gun metalwork.

Environmental Barriers to High-Resolution Imaging. The curves of Fig. 2.24 give optimum probe sizes assuming that the microscope is clean and well-aligned and that external effects such as mechanical vibration and AC magnetic field interference are minimal. In many practical cases the resolution obtained is more a function of the room environment than of the electron optics of the microscope. Sources of vibration may be motors (over a large range of frequencies) or low-frequency vibrations (2–10 Hz) related to the natural vibrations of the building. The effects of most vibrations can be minimized with an antivibration table under the microscope column. Oscillating magnetic fields may be caused by power supplies (60 Hz or higher) in the microscope room or an adjacent room. Sensitivity of the electron beam to stray magnetic fields is greater at low accelerating voltages, and extra magnetic shielding of the column becomes critical. In most cases the microscope manufacturer advises customers concerning the room environment. Even when the microscope environment is ideal, high-resolution imaging may be limited by contamination of specimens either before or after insertion into the microscope.

2.4.3. Measurement of Microscope Parameters

All of the parameters characterizing the electron beam incident on the sample (i.e., the incident probe current i_p, the probe diameter d_p, and the convergence angle α_p) can be experimentally determined. While there is no need to monitor such quantities continuously, it is valuable to associate particular values of i_p, d_p, and α_p with specific operating conditions, both as a means of setting desired operating parameters and as a diagnostic device in the event of problems with the microscope.

Probe Current. The most straightforward quantity to measure is the incident beam current i_p, since this can be done with a 'Faraday cup.' which is simply a container completely closed except for a small entrance aperture (see Fig. 2.25). An electron microscope aperture (3-mm diameter) with a hole 25–100 μm in diameter is convenient for this purpose. The container is made from a material (Ti or C) different from that used to fabricate the microscope stage. In this way any x rays produced by stray electrons outside the focused beam which strike the Faraday cup can easily be detected. The Faraday cup does not allow the backscattered and secondary electrons generated by the incident beam to escape. The current flowing to ground is therefore exactly equal to the incident beam current i_p, and it can conveniently be measured with a DC picoammeter or a calibrated specimen-current amplifier. For cases where the highest accuracy is not required, a flat carbon block may be substituted for the Faraday cup. In this case the measured specimen current i_{sc} and incident beam current i_p are related as $i_p = i_{sc}/[1 - (\eta + \delta)]$ where η and δ are the backscatter and secondary electron yields (Chapter 3), respectively. For a carbon sample normal to the beam, both η and δ are small, so the measured specimen current is about 90% of i_p.

Probe Size. The probe diameter, as previously defined, is measured by sweeping the beam across a sharp, electron-opaque edge and

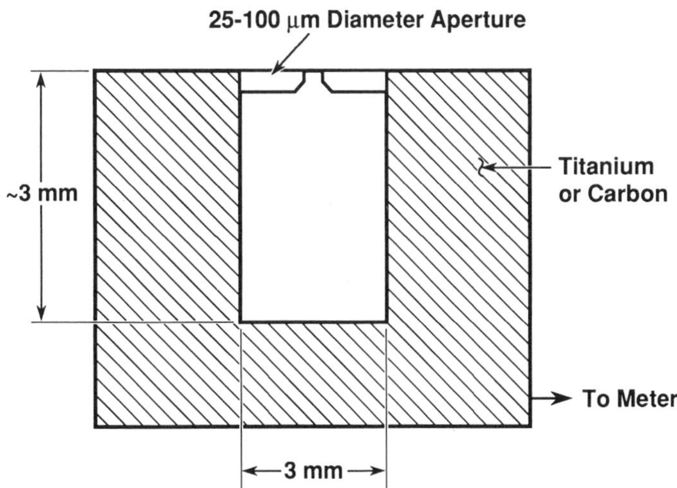

Figure 2.25. Schematic drawing of a cross-section through a Faraday cup suitable for measuring incident probe current i_p.

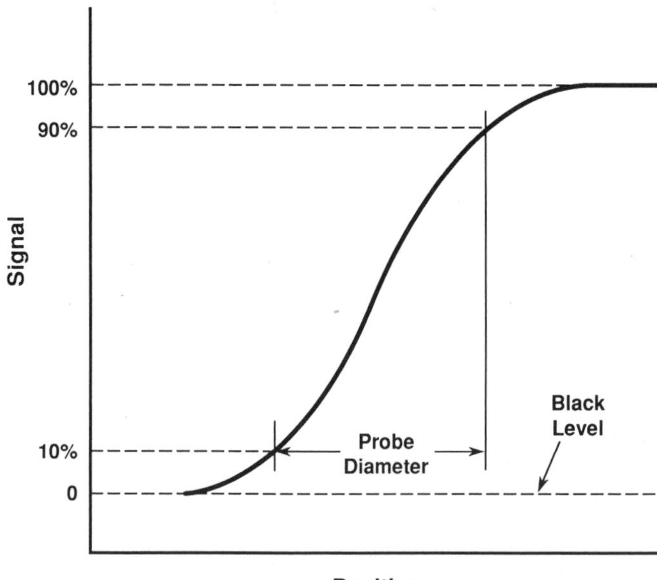

Figure 2.26. Schematic drawing of the electron signal emitted when a Gaussian electron beam is swept across a sharp edge of a specimen to measure the probe diameter d_p.

observing the change in signal as a function of the beam position. The profile has the form shown in Fig. 2.26. Typically the diameter is taken as the distance between the 10% and 90% signal levels. Suitable sharp edges include clean razor blades, cleavage edges in materials such as silicon, and fine-drawn wires such as fuse wire. The edge must be clean, smooth, and nontransmitting to electrons. While these conditions are easy to satisfy for large probes (~1 μm diameter), it is difficult to fabricate suitable edges for smaller probe diameters. Ideally the portion of the edge scanned should be over the entrance to a Faraday cup so that none of the scatter of the incident beam modifies the profile.

Probe Convergence Angle. The probe convergence angle (total angle 2α) can be measured by using the same technique as described above for the probe diameter. The beam is focused on a sharp edge, and the "initial" probe diameter d_i is measured. Without changing the focus, the test edge is moved vertically upward a distance z using the z control of the SEM stage, and the larger "final" diameter d_f is measured. Thus, as shown in Fig. 2.27, the convergence angle α may be found from

$$\alpha_p = \frac{d_f - d_i}{2z}. \tag{2.22}$$

Equation (2.22) works whether the aperture is a real objective-lens aperture or a virtual objective aperture. For the case of the real objective aperture in the gap of the pinhole lens of Fig. 2.16a, the convergence may also be obtained using the diameter D_A of the aperture in the final lens and the distance q from this aperture to the specimen (see Fig. 2.18),

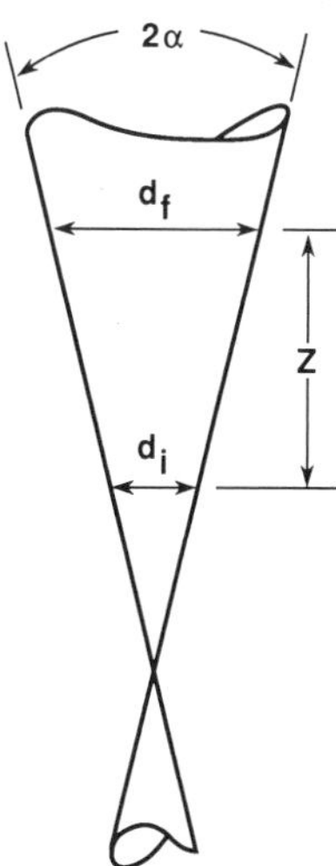

Figure 2.27. Schematic drawing of the cross-section of the electron beam. The convergence α_p of the probe can be calculated by measuring the beam diameter at two well-separated points along the optic axis.

that is,

$$\alpha_p = \frac{D_A}{2q}. \tag{2.23}$$

In most cases the working distance W may be substituted for q.

Estimation of Beam Brightness. By measuring i_p, d_p, and α_p under the same set of conditions (kV, C1 lens setting, final aperture size), the brightness may be estimated using Eq. (2.3). Since the brightness is constant throughout the electron column, neglecting aberrations, this estimate of the brightness is also an indication of the gun brightness. However, since measurement of the probe size is difficult below 1 μm and since convergence angles depend upon probe-size measurements for systems with a virtual objective aperture, brightness measurements under small probe conditions are only approximate.

2.5. Summary of SEM Microscopy Modes

This chapter provides some practical knowledge of electron sources and lenses. From this understanding the SEM operator should be able to select the proper operating conditions for the major modes of microscopy.

High Depth-of-Field Images. This is the SEM capability most often used in routine microscopy. High depth of field is attained when different heights in the image of a rough surface are all in focus at the same time. This mode requires a small convergence angle α_p so that the beam appears small over large height differences on the specimen.

This small beam angle can be obtained by using a small objective lens aperture, a long working distance, or both.

High-Resolution Images. High-resolution images require a small probe size, adequate probe current, and minimal interference from external vibration and stray AC magnetic fields. Electron-optically, the smallest probe is obtained by selecting a high operating voltage, strengthening the first condenser lens, using the optimum objective-lens aperture size, and using a short working distance. The penalty for using a very small probe is typically a very low probe current. High brightness sources and short focal length immersion objective lenses can improve resolution by a total factor of 5–10 times.

High Beam Current for Image Quality and X-Ray Microanalysis. While a probe current of at least 0.001 nA (10^{-12} A) is required to produce a photographic image, the image may be so noisy that image detail is lost. Large probe currents are obtained by weakening the first condenser lens. As the first condenser lens becomes weak, the probe contains more current and image quality improves, but the probe size increases dramatically. At low magnification there is no reason to image with a probe much smaller than the image detail we can see, so high-current, large probes provide high quality images. Currents of at least 0.1 nA (10^{-10} A) are needed for x-ray detection by the energy-dispersive spectrometer (EDS) while the wavelength-dispersive spectrometer (WDS) requires at least 10 nA (10^{-8} A). Often the probe diameter must be intentionally enlarged to obtain an adequate signal for microanalysis.

3

Electron–Specimen Interactions

3.1. Introduction

The versatility of scanning electron microscopy and of x-ray microanalysis is derived in large measure from the rich variety of interactions that the beam electrons undergo in a specimen. These interactions can reveal information on the specimen's composition, topography, crystallography, electrical potential, local magnetic field, and other properties. The electron–specimen interactions can be divided into two classes.

1. Elastic scattering events affect the trajectories of the beam electrons inside the specimen without altering the kinetic energy of the electron. Elastic scattering is responsible for the phenomenon of electron backscattering that forms an important imaging signal in scanning electron microscopy.
2. Inelastic scattering events of several types result in a transfer of energy from the beam electrons to the atoms of the specimen, leading to the generation of secondary electrons; Auger electrons; characteristic and bremsstrahlung (continuum) x rays; electron–hole pairs in semiconductors and insulators; long-wavelength electromagnetic radiation in the visible, ultraviolet, and infrared regions of the spectrum (cathodoluminescence); lattice vibrations (phonons); and electron oscillations in metals (plasmons).

In principle, all of these products of the primary beam interaction can be used to derive information on the nature of the specimen, including what atomic species are present within the region excited by the beam.

To derive the full benefit of this rich variety of interactions as observed in SEM images and x-ray spectra, the microscopist or analyst needs a working knowledge of electron–specimen interactions. This knowledge base should be at least broadly qualitative where an understanding of the character of the variables involved will suffice, but in certain critical cases, such as the generation and propagation of characteristic x rays, the level of understanding must be quantitative.

This chapter provides an overview of the complex field of electron–specimen interactions, with the goal of providing the information necessary to support the interpretation of images and an understanding of the physics of characteristic x rays as the basis for qualitative and quantitative x-ray microanalysis. Additional information on the physics of x rays will be included in subsequent chapters on x-ray microanalysis.

3.2. Electron Scattering

The electron-optical column which precedes the specimen has as its function the definition and control of the beam through manipulation of three beam parameters: diameter, current, and convergence. Typical electron beams used in microscopy and microanalysis consist of electrons with a precisely defined incident energy in the range 1–40 keV. The electrons within the beam follow paths that are nearly parallel, with a convergence in the range 2×10^{-2} to 10^{-3} radians (1°–0.05°). The beam is focused to form a final probe diameter at the specimen plane in the range 1 nm to 1 μm. This probe carries a current in the range 1 pA–1 μA. To successfully address a wide range of problems in imaging and microanalysis, we are compelled to utilize beam parameters selected appropriately throughout these ranges. However, it is critical to realize that while we have great control of the beam electrons before they reach the specimen, once the electrons enter the specimen, the scattering processes control subsequent behavior, with many important consequences. For example, we are often interested in obtaining information on fine spatial details of a specimen. Since the SEM image is constructed from information derived from a matrix of closely spaced probe locations produced by scanning, a probe diameter at least as small as the features of interest is obviously a first requirement for high resolution imaging. The electron-optical column is clearly capable of achieving a finely focused probe, with aberration-limited probes as small as 1 nm in diameter possible in the field emission SEM, as discussed in Chapter 2. An important question to ask is: Are the signals generated in the specimen confined to the area of the beam impact on the specimen's surface, since this would obviously produce images of the highest resolution? Unfortunately, the answer to this question is negative in many cases. This is a direct consequence of the effects of electron scattering, which leads to degraded spatial resolution.

Electron scattering is an interaction between the probe electron and the specimen atoms that results in a change in the electron trajectory (direction of travel) and/or energy. In discussing scattering, a key concept is that of the cross section, which is the measure of the probability that an event will take place. The cross section, usually denoted Q or σ, is defined in its most general terms as (Considine, 1976)

$$Q = N/n_i n_t \tag{3.1}$$

cm^2, where N is the number of events of a certain type per unit volume (events/cm^3), n_t is the number of target sites per unit volume (sites/cm^3), and n_i is the number of incident particles per unit area (particles/cm^2). The full dimensionality of the cross section for electrons (denoted e^-) as the incident particles scattering with the atoms of the sample is thus

$$Q(\text{events/cm}^3)/[(e^-/\text{cm}^2)(\text{atom/cm}^3). \tag{3.2}$$

In Eq. (3.1), the terms "events," "e^-," and "atoms" are usually thought of as dimensionless. Cancelling the cm^3 terms reduces the apparent dimensions of the cross section to those of area, cm^2. Cross sections are sometimes referred to as the effective "size" which the atom presents as a target to the incident particle. It should be noted, however, that the cross section is not the physical size of the atom. Atomic radii are of the order of 10^{-8} cm, so that the cross-sectional area of an atom is approximately 10^{-16} cm^2. For many processes of interest in the SEM, the cross section for the process of interest is much smaller than the physical size of an atom. For example, inner shell ionization, which gives rise to the characteristic x rays used for x-ray microanalysis, typically has a cross section in the range 10^{-22} to 10^{-24} cm^2.

While the concept of a probability or cross section for a process may seem somewhat abstract, a closely related and more readily understandable concept is that of the mean free path λ, which is the average distance the electron travels in the specimen between events of a certain type. The mean free path can be calculated from the cross section by the equation

$$\lambda = A/(N_A \rho Q) \tag{3.3}$$

cm, where A is the atomic weight (g/mol), N_A is Avogadro's number (6.02×10^{23} atoms/mol), and ρ is the density (g/cm^3). To determine the mean free path for a particular type of event, λ_j, the specific cross section for that type of event, Q_j, is substituted in Eq. (3.3). The mean free path for all possible types of events is found by considering the individual scattering processes. The total mean free path, λ_{tot}, is calculated according to the equation

$$\frac{1}{\lambda_{\text{tot}}} = \frac{1}{\lambda_a} + \frac{1}{\lambda_b} + \frac{1}{\lambda_c} + \cdots. \tag{3.4}$$

Note that the total mean free path must always be less than the smallest value among the mean free paths of the various possible processes.

3.2.1. Elastic Scattering

Elastic scattering, which is illustrated in Fig. 3.1(a), changes the direction component of the electron vector velocity $\mathbf{v}$, but the

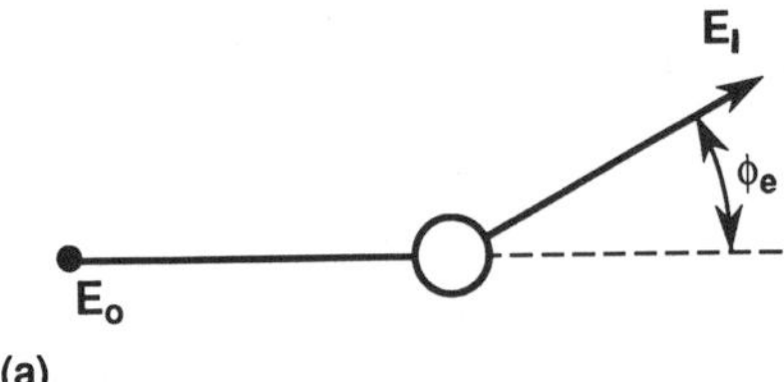

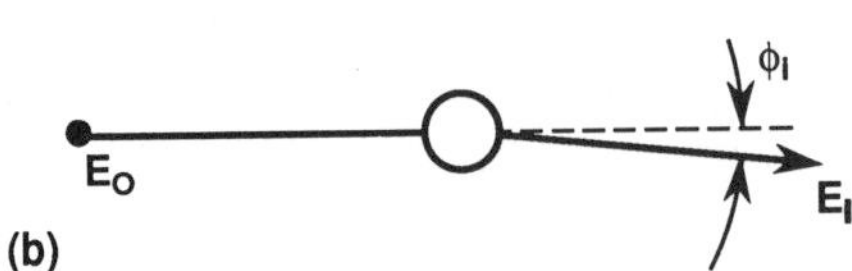

Figure 3.1. Schematic illustration of scattering processes that occur when an energetic electron of energy E_0 interacts with an atom: (a) Elastic scattering, in which the energy E_I after scattering equals E_0. (b) Inelastic scattering, in which E_I is less than E_0. Note that the elastic scattering angle $\phi_e \gg \phi_i$.

magnitude of the velocity, v, remains virtually constant, so that the electron's kinetic energy, $E = m_e v^2/2$, where m_e is the electron mass, is unchanged. The electron is scattered by Coulombic interaction with the charge of the atomic nucleus (Rutherford scattering), as partially screened by the atomic orbital electrons. Less than 1 eV of energy is transferred from the beam electron to the specimen; this loss is negligible compared to the incident energy, which is typically greater than 1 keV. As a result of this elastic scattering, the electron deviates from its initial path by an angle ϕ_e, where the subscript denotes "elastic". The elastic scattering angle can take any value in the range 0–2π radians (0°–180°), with an average value per interaction of about 2°–5°. Thus, for most elastic scattering events, the electron continues to propagate in approximately the same direction, but occasionally an elastic scattering event occurs that causes the trajectory to deviate sharply, and very rarely the electron can scatter through nearly 180° in a single event, reversing its direction of travel.

The cross section for elastic scattering can be conveniently described by the screened Rutherford expression (Evans, 1955):

$$Q(>\phi_0) = 1.62 \times 10^{-20} \frac{Z^2}{E^2} \cot^2 \frac{\phi_0}{2} \tag{3.5}$$

events/e^- (atom/cm^2), where $Q(>\phi_0)$ is the cross section for an elastic scattering event exceeding a specified angle ϕ_0, Z is the atomic number of the scattering atom, and E is the electron energy (keV). Equation (3.5) is plotted for various scattering atoms and incident electron energies in Fig. 3.2; note the logarithmic scale for the cross section. Inspection of Eq. (3.5) and these plots reveals a strong dependence of elastic scattering upon atomic number and incident electron energy, with the cross section increasing with the square of the atomic number and decreasing with the inverse square of the electron energy. Also, as smaller and smaller scattering angles ϕ_0 are considered, the cross section increases rapidly toward infinity.

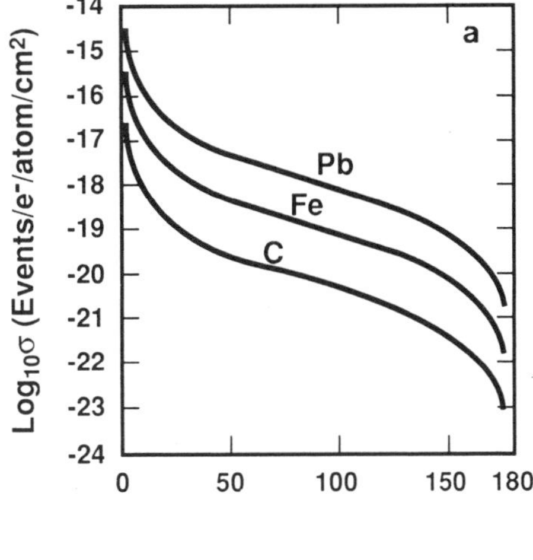

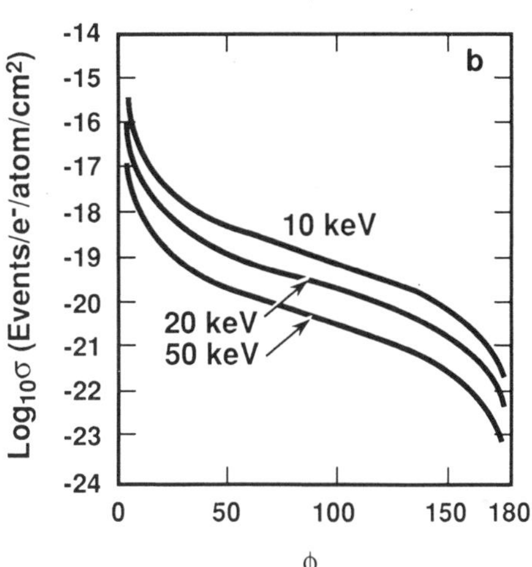

Figure 3.2. Plot of elastic scattering cross-section from (3.5), (a) as a function of atomic number for $E_0 = 10\,\text{keV}$, (b) as a function of beam energy (10, 20, and 30 keV) for an iron target.

If we consider elastic events greater than 2°, the elastic mean free path can be calculated from Eqs. (3.3) and (3.5), as listed in Table 3.1. Inspection of Table 3.1 reveals that the mean free path increases with increasing electron energy and decreasing atomic number. In traversing a specific thickness of various materials, elastic scattering is more probable for high atomic number materials and for low beam energy.

3.2.2. Inelastic Scattering

The second general category of scattering is inelastic scattering. During an inelastic scattering event, illustrated schematically in Fig. 3.1b, energy is transferred to the tightly bound inner-shell electrons and loosely bound outer-shell electrons of the atoms, and the kinetic energy of the energetic electron involved decreases. Depending on the type of process, a single inelastic event can transfer any amount of

Table 3.1 Elastic Mean Free Path in Nanometers (Scattering Angle >2°)

	Energy (keV)				
Element	10	20	30	40	50
C	5.5	22	49	89	140
Al	1.8	7.4	17	29	46
Fe	0.3	1.3	2.9	5.2	8.2
Ag	0.15	0.6	1.3	2.3	3.6
Pb	0.08	0.34	0.76	1.4	2.1
U	0.05	0.19	0.42	0.75	1.2

energy from the beam electron, ranging from a fraction of an electron volt (e.g., phonon excitation) to the entire energy carried by an incident probe electron carrying many kiloelectron volts (e.g., bremsstrahlung). Despite the loss of energy, the electron trajectory deviates only by a small angle, on the order of 0.1° or less. There are many types of inelastic scattering; we shall consider only the principal processes of interest in scanning electron microscopy and microanalysis (for a basic treatment of the following terms, see Kittel, 1956). In order of increasing energy loss, the main processes are summarized as follows:

Phonon Excitation. A substantial portion of the energy deposited in the solid by the beam electrons is transferred by small (<1 eV) energy loss events that cause the excitation of lattice oscillations (phonons); this process is manifested as heating of the target. A question inevitably arises as to the temperature rise which takes place in the specimen as a result of this energy transfer. For the case of an electron beam incident on a bulk target, the region in which the beam electrons deposit energy is in intimate contact with the surrounding matter so that good thermal conductivity exists, even for nonmetals. A maximum temperature rise of only a few degrees Celsius is observed for bulk targets with a typical SEM beam current of 1 nA. For thin specimens (<100 nm) or for high beam currents (> 1 μA) in bulk nonconducting specimens, significant heating with temperature rises of several hundred degrees Celsius are possible (Reimer, 1976). The acoustic waves that result from conversion of the thermal energy deposited by a scanned beam can be detected with a transducer as the basis of "thermal wave microscopy" (Davies, 1983).

Plasmon Excitation. For metallic species such as aluminum or copper, the outermost atomic electrons are so loosely bound that in a solid these electrons are not localized to a specific atom but rather form a "free-electron gas" or "sea," which permeates the ion cores. The moving energetic electron can excite waves in this free-electron sea. Because of the regularity of the atomic arrangement, the plasmon energy tends to have a specific value; in aluminum, the excitation of a plasmon involves a transfer of about 15 eV from the energetic electron to the free electron gas of the solid.

Secondary Electron Excitation. Inelastic scattering of the energetic electron can lead to the promotion of loosely bound electrons from the valence band to the conduction band in a semiconductor or insulator ("electron–hole pair production") with enough kinetic energy for subsequent motion through the solid. Such promotion can occur directly from the conduction-band electrons in a metal. This specimen electron, designated a *secondary* electron, propagates through the specimen and is itself subject to inelastic scattering and energy loss. If the secondary electron retains sufficient energy when it reaches the surface to overcome the surface barrier energy, it will escape from the solid as a secondary electron. The majority of ejected secondary electrons have energies <10 eV, and by arbitrary definition, all "slow" secondary electrons are ejected with <50 eV. More tightly bound atomic electrons can be ejected with higher kinetic energies, creating the so-called "fast secondary electrons," which are far less numerous than slow secondaries.

Bremsstrahlung or Continuum X-ray Generation. An energetic beam electron can undergo deceleration in the Coulombic field of the specimen atoms. The energy lost by the beam electron in this deceleration process is converted into a photon of electromagnetic energy; this radiation is known as *bremsstrahlung* ("braking radiation"). The energy loss can take on any value from a fraction of an electron volt to the total energy carried by the incident electron, which may be 10 keV or more. The bremsstrahlung x rays thus form a continuous spectrum from zero energy up to the beam energy. Because the formation of bremsstrahlung is dependent on the direction of flight of the beam electron relative to the scattering center (atom), the angular distribution of the intensity of the bremsstrahlung depends on the trajectory of the beam electron relative to the sample atoms. For a single trajectory, the distribution is anisotropic and favors a forward direction along the trajectory. In a solid sample, the effect of elastic scattering is to randomize the trajectories, so that the effective bremsstrahlung generation, as viewed by an external observer, is nearly isotropic.

Ionization of Inner Shells. A sufficiently energetic electron can interact with and cause the ejection of a tightly bound inner-shell atomic electron, leaving the atom in an excited state. Subsequent decay of this excited state results in the emission of a characteristic x ray or an Auger electron.

The cross sections for several of these processes have been calculated after the work of Shimizu *et al.* (1976) for an aluminum target over a range of electron energies. As shown in Fig. 3.3, all of the cross sections decrease with increasing energy.

Continuous Energy Loss Approximation. Inelastic scattering occurs by a variety of discrete processes, with a variable amount of energy transferred to the target atoms, depending on the strength of each interaction. Cross sections for the individual processes are

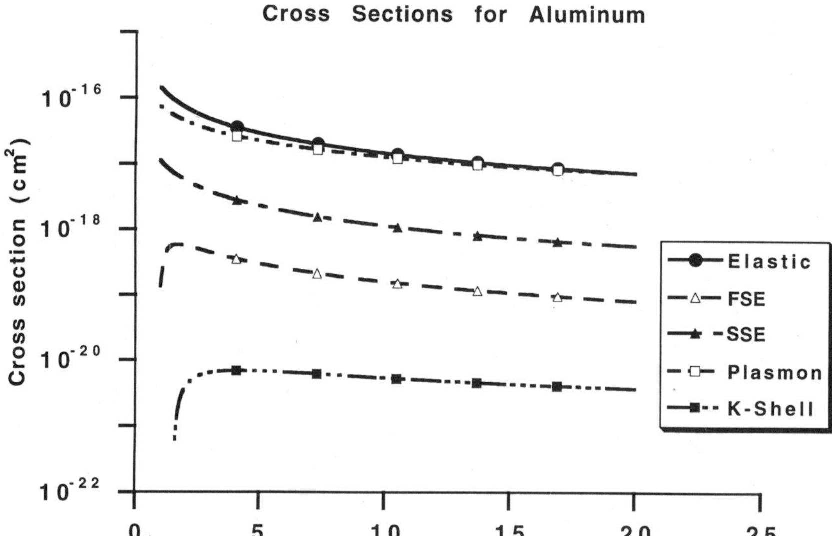

Figure 3.3. Plot of total cross-sections for elastic scattering and several inelastic scattering processes: plasmon, conduction-electron excitation, and inner-shell ionization (adapted from Shimizu *et al.*, 1976).

difficult to obtain for all targets of interest. It is useful in many calculations to consider all inelastic processes grouped together to give an average rate of energy loss, dE/ds, where s represents the distance travelled in the specimen. The continuous energy loss expression of Bethe (1933) provides a useful approximation for dE/ds for energies greater than approximately 3–5 keV, where the cutoff depends on the material:

$$\frac{dE}{ds} = -2\pi e^4 N_0 \frac{Z\rho}{AE_m} \ln \frac{1.166E_m}{J}, \tag{3.6}$$

where e is the electronic charge, N_0 is Avogadro's number, Z is the atomic weight (g/mol), ρ is the density, E_m (keV) is the average energy along the path segment s, and J is the mean ionization potential. Combining the constants and expressing in units which give dE/ds in terms of keV/cm,

$$\frac{dE}{ds} = -7.85 \times 10^4 \frac{Z\rho}{AE_m} \ln \frac{1.166E_m}{J}. \tag{3.7}$$

The mean ionization potential is the average energy loss per inelastic interaction considering all the energy loss processes. J can be approximated as $0.115Z$ (keV), but a more exact and widely used expression is that described by Berger and Seltzer (1964):

$$J\,(\text{keV}) = (9.76Z + 58.5Z^{-0.19}) \times 10^{-3}. \tag{3.8}$$

The behavior predicted by Eq. (3.7) for Si, Fe and Au is plotted in Fig. 3.4 for the energy range 2–50 keV.

Recent studies by Joy and Luo (1989) suggest that the Bethe expression is satisfactory over a range extending from approximately $7J$ to a maximum of approximately 50 keV. The value of 50 keV is a convenient upper bound, since this value is higher than the maximum initial beam energy encountered in modern SEM–EPMA instrumentation, which is typically 40 keV or lower. Higher beam energies, such as those used in analytical electron microscopy, require the use of relativistic corrections (Bethe, 1933). The low end of the range of valid application of the Bethe expression is another matter. The advent of high-performance low-voltage SEM (LVSEM) has made operation with beam energies below 5 keV routine. If the criterion $7J$ is used for the lower limit of applicability of the Bethe formulation for dE/ds, this limit is frequently encountered for operation under LVSEM conditions, especially for targets of intermediate and high atomic number: C (0.7 keV), Al (1.1 keV), Cu (2.2 keV), Au (5.6 keV).

Joy and Luo (1989) have noted that at low beam energies the number of processes contributing to the energy loss decreases; they

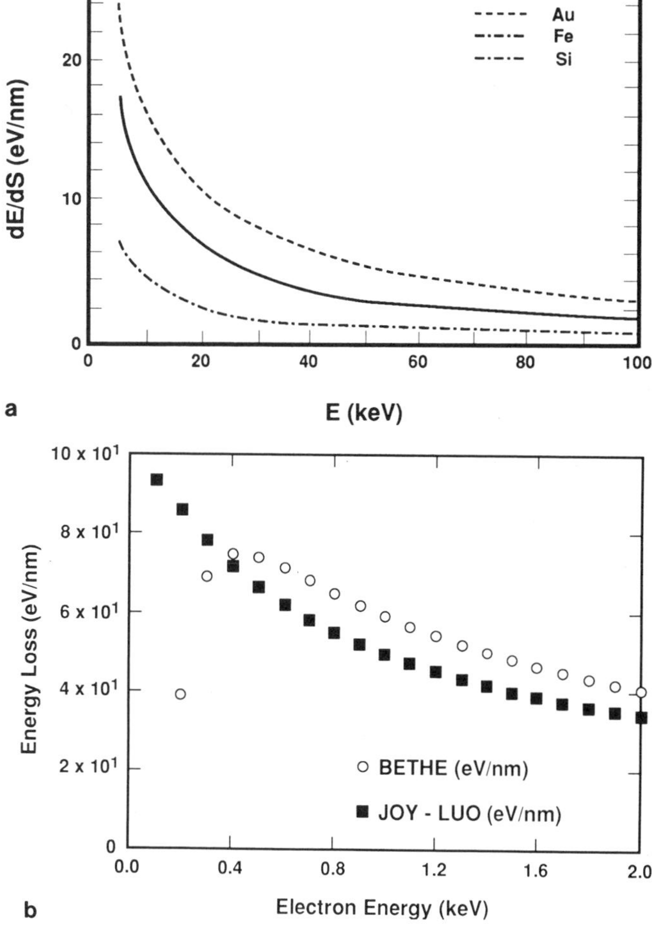

Figure 3.4. (a) Energy loss (eV/nm) due to all inelastic scattering processes for gold, iron, and silicon as a function of electron energy at high energies using the Bethe (1933) formulation of dE/ds. (b) Energy loss at low electron energy ($E < 2$ keV) using the Bethe (1933) and Joy and Luo (1989) formulations of dE/ds for silicon.

propose a modified form of the Bethe energy-loss relation which is appropriate at low beam energies, below $7J$:

$$\frac{dE}{ds} = -7.85 \times 10^4 \frac{Z\rho}{AE_m} \ln \frac{1.166E_m}{J^*} \tag{3.9}$$

(keV/cm), where the modified mean ionization potential is given by:

$$J^* = \frac{J}{1 + \dfrac{kJ}{E}}. \tag{3.10}$$

In this expression, k is a variable, dependent on atomic number, which is always close to but less than unity. The data for k presented by Joy and Luo (1989) fit an equation of the form

$$k = 0.731 + 0.0688 \log_{10} Z. \tag{3.11}$$

Figure 3.4b shows the behavior of the Bethe and Joy–Luo expressions as a function of electron energy for a copper target. The Joy–Luo expression shows a progressive increase in the rate of energy loss with decreasing electron energy, while the Bethe expression goes through a peak and then decreases, eventually becoming negative, which is physically unrealistic.

The Bethe and Joy–Luo expressions for dE/ds provide convenient and readily calculable relations for determining the amount of energy lost as the beam electron traverses a specimen. Note that the distance s in Eqs. (3.6) and (3.9) is the distance *along* the trajectory. Because of elastic scattering, the trajectory deviates significantly from a straight line, especially in targets of high atomic number. Hence the energy loss can be directly calculated only for the case of a thin foil or supported film with a thickness less than an elastic mean free path. For thicker specimens, a correction for the added path length due to elastic scattering must be made to calculate the energy loss.

An important parameter in quantitative x-ray analysis calculations directly related to the continuous energy loss approximation is that of the *stopping power S*, defined as

$$S = -\frac{1}{\rho}\frac{dE}{ds}. \tag{3.12}$$

The stopping power is simply the continuous energy loss approximation expressed in terms of mass distance units (g/cm^2) rather than the more familiar linear distance units (cm). Density is generally a strong function of atomic number. In terms of mass distance units, the stopping power per unit mass distance is actually greater for low atomic number materials than for high. To see that this is true, consider the terms in Eqs. (3.7) or (3.9) and (3.12). The mean ionization potential J is an increasing function of Z. At a specific energy E, the stopping power is therefore proportional to $(Z/A) \ln (E/f(Z))$. Both (Z/A) and $\ln (E/f(Z))$ decrease with increasing atomic number and, since ρ increases with Z, both terms in Eq.

(3.12) contribute to a decrease in S with increasing atomic number. In terms of mass distance, the stopping power is approximately 50% greater in aluminum than it is in gold at 20 keV (Cosslett and Thomas, 1966).

The processes of elastic and inelastic scattering operate concurrently. Elastic scattering causes the beam electrons to deviate from their original direction of travel, causing the initially well-defined trajectories of the electron probe to become diffuse in the solid. Inelastic scattering progressively reduces the energy of the energetic electron until it is captured by the solid, thus limiting the range of travel of the electrons within the solid. The combined effect of elastic and inelastic scattering is to limit the penetration of the beam into the solid. The resulting region over which the energetic electrons interact with the solid, depositing energy and producing those forms of secondary radiation which we measure, is known as the *interaction volume*. An understanding of the size and shape of the interaction volume as a function of specimen and beam parameters is vital for proper interpretation of features of SEM images and the spatial resolution of x-ray microanalysis.

3.3. Interaction Volume

3.3.1. Experimental Evidence

The interaction volume of the beam electrons in the target can be directly observed in a few special cases. Certain plastics, such as polymethylmethacrylate (PMMA), undergo a chemical change during electron bombardment that renders the material sensitive to etching in a suitable solvent. This phenomenon is the basis for important steps in the lithographic fabrication of semiconductor microcircuits. The same effect has been used to directly reveal the size and shape of the interaction volume (Everhart *et al.*, 1972). Figure 3.5 shows the result of an experiment in which the interaction volume is revealed in a series of successively longer chemical etchings. The etching rate depends upon the damage, which is proportional to the deposited energy and hence to the electron dose per unit volume, e^-/cm^3. Initially, the most damaged material etches out in the solvent. Etching for increasing time periods reveals contours of progressively decreasing electron energy deposition. To produce the series of images shown in Fig. 3.5, a static, 20-keV electron beam focused to less than 0.5 μm in diameter was placed for a fixed time at a series of locations on PMMA. These constant-dose locations were then subjected to chemical etching for increasing time periods. Following etching, the PMMA was cleaved to reveal the etched volume. The images in Fig. 3.5 show contours of successively decreasing energy deposition; these contours are plotted numerically in Fig. 3.6. The etched structures allow us to directly visualize electron penetration and the interaction volume in a

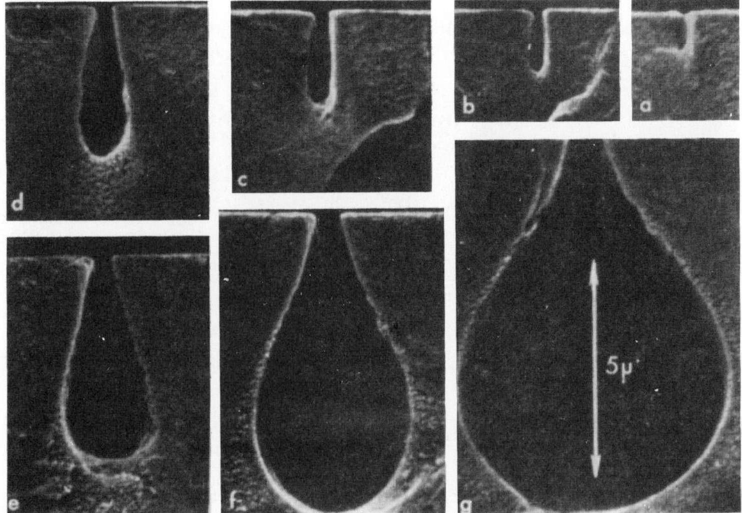

Figure 3.5. Direct visualization of the electron volume in polymethylmethacrylate. In (a) through (g), the electron dose is the same, but the etching time is increased progressively to reveal successively lower energy deposition (radiation damage) levels (from Everhart *et al.*, 1972).

low atomic number matrix similar in average atomic number to biological materials and polymers. Several points stand out:

1. Despite the fact that the incident beam diameter was well under a micrometer in diameter, the *interaction volume in a low-density, low-atomic-number target has overall dimensions of micrometers.*
2. The energy deposition rate varies rapidly throughout the interaction volume, being greatest near the beam impact point.
3. The interaction volume for a low-density, low-atomic-number target has a distinct pear shape.

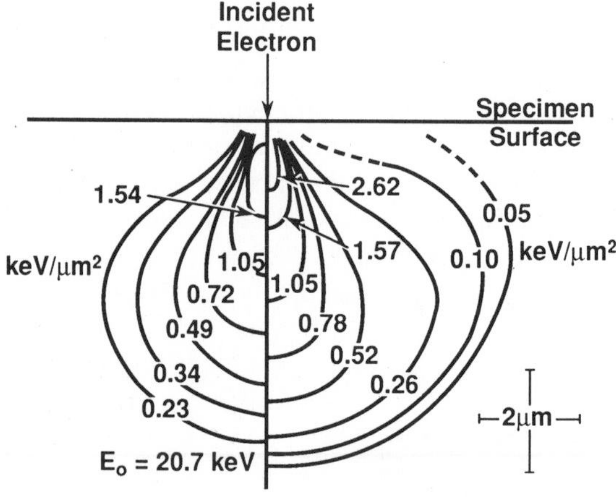

Figure 3.6. Contours of energy deposition in a low-atomic-number solid (polymethylmethacrylate) as a function of position, as measured experimentally by etching (Fig. 3.5) and as calculated by Monte Carlo electron-trajectory simulation (Everhart *et al.*, 1972).

The origin of this pear shape can be understood in terms of the characteristics of elastic and inelastic scattering. For this low-atomic-number matrix, elastic scattering is relatively weak, so that the beam initially tends to penetrate into the target, forming the narrow neck of the pear shape. Inelastic scattering is relatively more probable in this low-atomic-number material, so that the electron energy decreases. The probability for elastic scattering increases rapidly with decreasing electron energy. With further penetration and reduction in electron energy, the cumulative effect of elastic scattering tends to cause the electron trajectories to deviate, creating the bulbous portion of the pear shape. Inelastic scattering limits the eventual range of the electrons in the target.

3.3.2. Monte Carlo Electron-Trajectory Simulation

While the PMMA etching experiment provides a good idea of the size and shape of the interaction volume in the important class of materials that have a low atomic number, there are no similar damage–etching experiments which can directly reveal the interaction volume in intermediate- and high-atomic-number materials such as metals. The mathematical technique of Monte Carlo electron-trajectory simulation provides an indirect method to visualize the interaction volume in such cases (Berger, 1963; Shimizu and Murata, 1971; Heinrich *et al.*, 1976; Newbury *et al.*, 1986; Heinrich and Newbury, 1991). In the Monte Carlo electron-trajectory simulation, the effects of elastic and inelastic scattering are calculated from appropriate models to determine reasonable scattering angles, distances between scattering sites, and energy loss rates. From these parameters and equations of analytical geometry, the electron trajectory can be simulated in a stepwise fashion from the location at which it enters the specimen to its final fate when it either escapes the specimen or loses all of its energy and is captured by the specimen.

The basic principles of constructing a Monte Carlo simulation are as follows: The length of the basic repetitive step in such a calculation is usually set equal to the mean free path for elastic scattering ("single scattering Monte Carlo") or a multiple thereof ("multiple scattering Monte Carlo"). From the mean free path and the rate of energy loss due to inelastic scattering as calculated with the Bethe or Joy–Luo expressions, the decreasing energy can be calculated along the path of the electron. After the electron travels a distance equal to a mean free path, the next scattering site is reached, and a new scattering angle is chosen for the next step based upon the new value of the energy. The scattering angle is calculated only from expressions based upon elastic scattering, since inelastic scattering causes negligible angular deviation. Since the elastic scattering angle can take on any value over a wide range from 0–180°, random numbers are used with an appropriate weighting factor to produce the appropriate statistical distribution of scattering. Because of the extensive use of random numbers in the simulation, the name

"Monte Carlo" is applied to this calculation technique. The electron trajectory is followed until either the energy has decreased due to inelastic scattering to an arbitrary cutoff energy, usually chosen as the energy at which a process of interest, such as inner shell ionization, can no longer be activated or the electron passes through a surface of the target and escapes. The spatial distribution of production of secondary radiation, such as the generation of characteristic x rays or secondary electrons, can be calculated with the appropriate cross sections and the step length. Extensive testing of Monte Carlo procedures by comparison to experimental values has established the accuracy of the simulations and the limits of applicability (Newbury and Myklebust, 1984). A more complete description of the Monte Carlo simulation technique can be found in several references (Newbury and Myklebust, 1979; Newbury *et al.*, 1986; Henoc and Maurice, 1991).

An example of individual trajectories calculated by a Monte Carlo simulation is shown in Fig. 3.7a, where it can be seen that the individual trajectories differ sharply from each other because of the random nature of individual scattering events. A single trajectory, while it can be realistically simulated, is not representative of the interaction volume. A large number of trajectories, typically 10,000 to 100,000, must be calculated to achieve statistical significance. When several hundred trajectories are plotted, the shape of the interaction volume can be

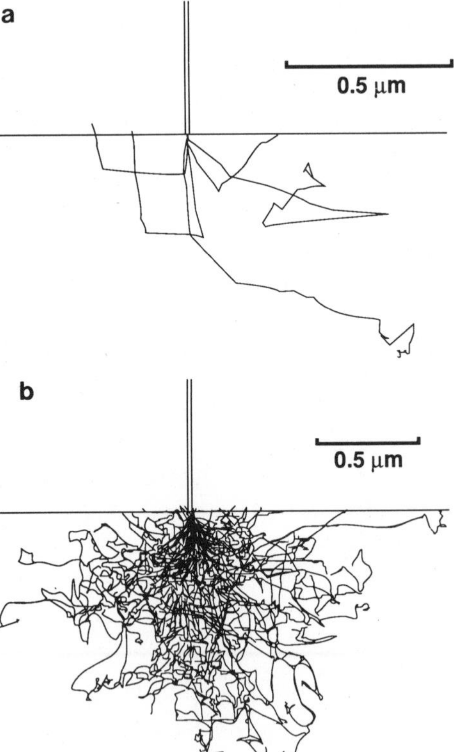

Figure 3.7. Monte Carlo electron-trajectory simulation of the beam interaction in iron, $E_0 = 20$ keV, tilt = 0°. (a) Plot of five trajectories, showing random variations. (b) Plot of 100 trajectories projected on a plane perpendicular to the surface, giving a visual impression of the interaction volume.

visualized, as shown in Fig. 3.7b. [Note that the interaction volume is three-dimensional, and the drawings are constructed by projection onto a two-dimensional plane; three-dimensional visualization is possible by stereo plotting techniques as described by Bright *et al.* (1984)].

In the following sections, Monte Carlo simulations will be used extensively to illustrate the characteristics of the interaction volume as a function of beam energy, specimen composition, and surface tilt. It must be remembered throughout this discussion that the numerical values used to describe the interaction volume are only approximate; it is useful to regard these values as "gray" numbers. Examination of a plot of many electron trajectories reveals that the boundaries of the interaction volume are not sharply defined. A new calculation with a different series of random numbers would produce a slightly different boundary contour. More important, while we like to use one or two numbers to describe the size of the interaction volume, it is clear from Figs. 3.5–3.7 that the densities of electron trajectories and consequent energy deposition vary greatly throughout the interaction volume. The interaction volume has a dense core of energy deposition near the beam impact point, and the density decreases markedly away from this core. The density of trajectories gradually approaches zero as the "limit" of the interaction volume envelope is reached. Therefore, using a single number to describe the size of the interaction volume is, at best, only approximate. A more complete numerical representation of the distributions of energy deposition and the generation of radiation products is necessary for a fully quantitative description of electron–solid interactions. Nevertheless, Monte Carlo plots are extremely useful to visualize the interaction volume and gain at least a qualitative perspective.

3.3.2.1. Influence of Beam Energy on Interaction Volume

The size of the interaction volume is a strong function of the energy with which the beam electrons interact with the target. The interaction volume in iron is shown as a function of beam energy for the range 10–30 keV in Fig. 3.8. The increase in size with beam energy can be understood from an examination of Eqs. (3.5) and (3.6). First, the cross section for elastic scattering has an inverse dependence on the square of the energy, $Q \sim 1/E^2$. Thus, as the beam energy increases, the electron trajectories near the surface become straighter and the electrons penetrate more deeply into the solid before the cumulative effects of multiple elastic scattering events cause some of the electrons to propagate back toward the surface. Second, the rate of energy loss with distance travelled, as given by the Bethe expression, is inversely related to the energy, $dE/ds \sim 1/E$. As the beam energy is increased, the electrons can penetrate to greater depths, since they enter the specimen with more energy and lose it at a lower rate. The lateral dimensions of the interaction volume are seen to scale with energy in a similar fashion to the depth. The shape of the interaction volume does not change significantly with beam energy.

(a)

(b)

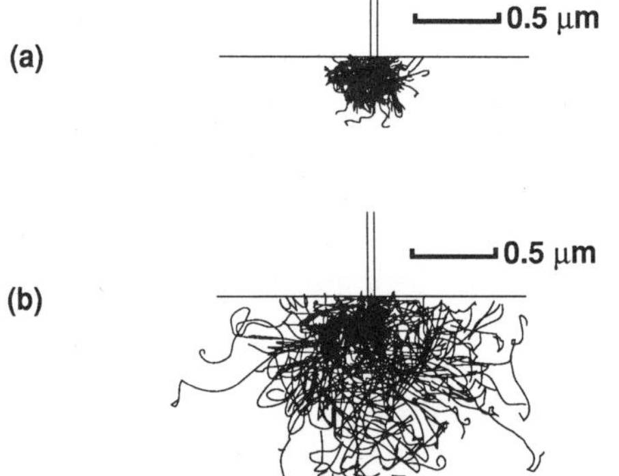

(c)

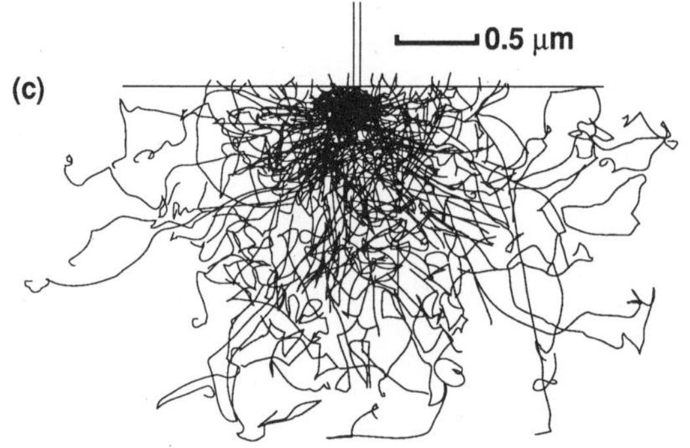

Figure 3.8. Monte Carlo electron-trajectory simulations of the interaction volume in iron as a function of beam energy: (a) 10 keV, (b) 20 keV, (c) 30 keV.

3.3.2.2. Influence of Atomic Number on Interaction Volume

Monte Carlo calculations, shown in Fig. 3.9 for targets of carbon ($Z = 6$), iron ($Z = 26$), silver ($Z = 47$), and uranium ($Z = 92$) at a beam energy of 20 keV, reveal that the linear dimensions of the interaction volume decrease with increasing atomic number at a fixed beam energy. This is a direct consequence of the increase in the cross section for elastic scattering with atomic number, $Q \sim Z^2$. In targets of high atomic number, the electrons undergo more elastic scattering per unit distance and the average scattering angle is greater, as compared to low-atomic-number targets. The electron trajectories in high-atomic-number materials thus tend to deviate out of the initial direction of travel more quickly and reduce the penetration into the solid. In low-atomic-number materials, elastic scattering is less likely and the trajectories deviate less from the initial beam path, allowing for deeper penetration into the solid. The shape of the interaction volume also changes significantly as a function of atomic number. The dense region of trajectories changes from the pear shape seen in low-atomic-number materials to a more nearly spherical shape truncated by the plane of the surface for high atomic number materials.

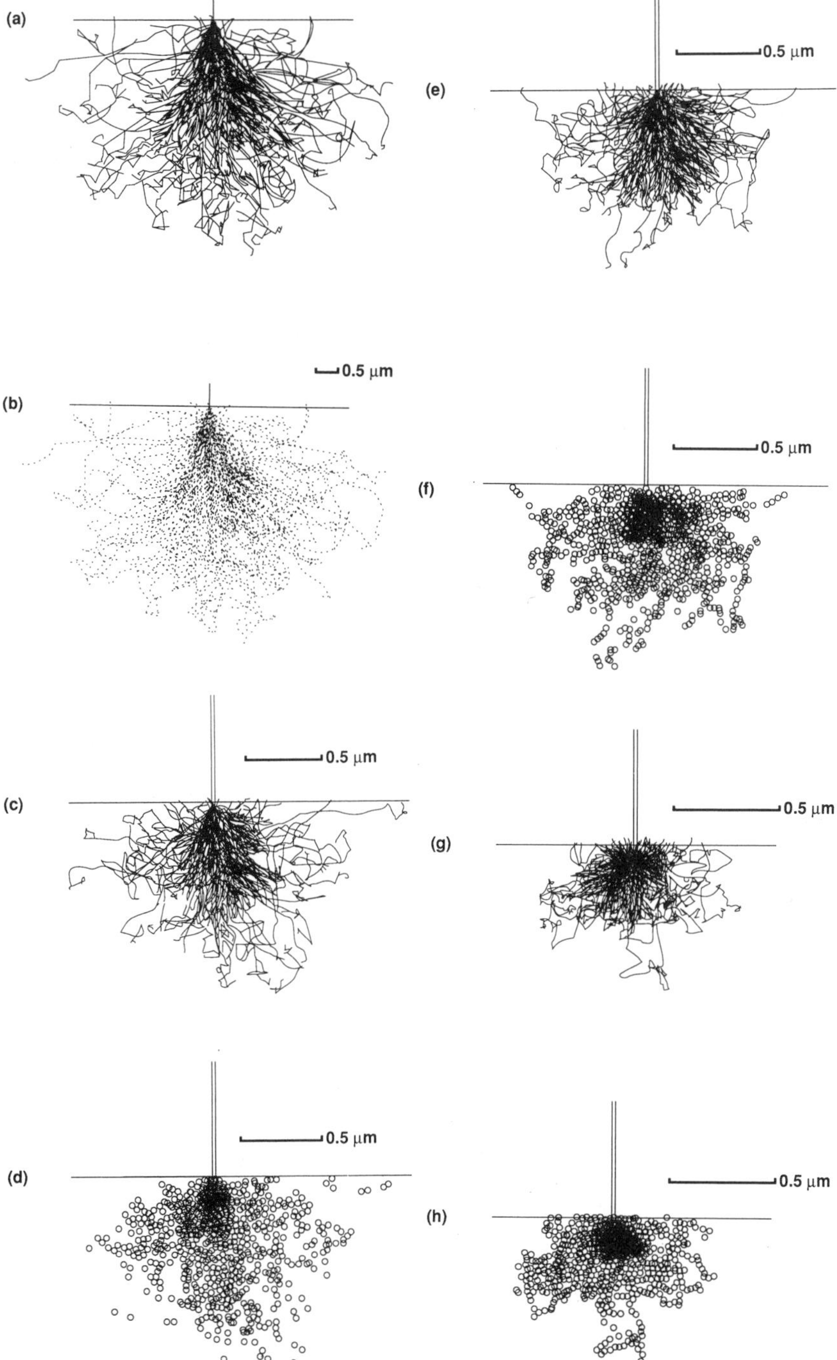

Figure 3.9. Monte Carlo electron-trajectory simulations of the interaction volume at 20 keV and 0° tilt in various targets; both trajectory plots and sites of inner-shell ionization are shown: (a), (b) carbon *K*-shell; (c), (d) iron *K*-shell; (e), (f) silver *L*-shell; (g), (h) uranium *M*-shell.

3.3.2.3. Influence of Specimen Surface Tilt on Interaction Volume

As the angle of tilt of a specimen surface increases (i.e., the angle of the beam relative to the surface decreases), the interaction volume becomes smaller and asymmetric, as shown in the Monte Carlo plots of Fig. 3.10. This behavior can be understood with the aid of Fig. 3.11, which depicts the scattering cone for an elastic event. Assume that the semicone angle is equal to the most probable value of the elastic scattering angle, which is typically in the range 3°–5°. Most elastic scattering angles are so small that the electron tends to continue in the same general direction after scattering ("forward scattering"). The electron travels with equal probability to some point on the circumference of the base of the cone. At 0° tilt, where the beam is

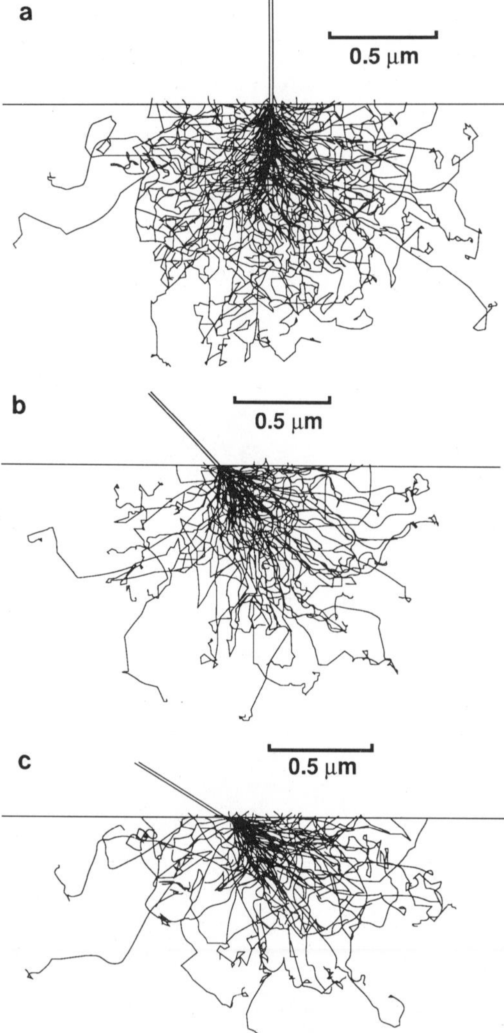

Figure 3.10. Monte Carlo electron-trajectory simulations of the interaction volume in iron at $E_0 = 20$ keV for various tilts: (a) 0° tilt, (b) 45° tilt, (c) 60° tilt.

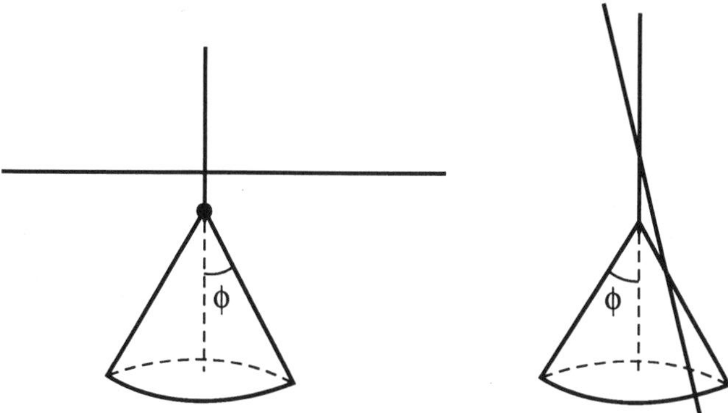

Figure 3.11. Schematic illustration of the origin of increased backscattering from tilted specimens. Consider a given average elastic scattering angle ϕ which produces the scattering cone indicated. The electron may land at any point on the base of the cone with equal probability. At normal incidence, no matter where on the base of the cone the electron lands, it tends to continue propagating into the solid. When the specimen is tilted, some of the possible locations on the base of the scattering cone actually carry the electron out of the solid immediately.

perpendicular to the surface, the tendency for forward scattering causes most of the electrons to propagate down into the specimen, as shown in Fig. 3.11a. For a tilted specimen, Fig. 3.11b, despite the tendency for forward scattering, at least some of the beam electrons propagate nearer to the surface; indeed, even with a small scattering angle, some electrons can escape through the specimen surface after a single small-angle scattering event. The electrons do not penetrate as deeply into the specimen at high tilts, and the interaction volume thus has a reduced depth dimension. The lateral dimensions of the interaction volume from a tilted specimen present a different situation. The dimension parallel to the surface and perpendicular to the axis of tilt ("downwind" from the beam impact) increases as compared to normal incidence, while the dimension parallel to the axis of tilt remains nearly the same as that for normal beam incidence.

3.3.3. Measures of Interaction Volume—Electron Range

For many applications in microscopy and microanalysis, it is useful to have a numerical estimate of the size of the interaction volume, even with the necessary caveats on the imprecision of such a number and the reality of the nonuniformity of energy deposition within the envelope of the interaction volume. To obtain such an estimate, we encounter the concept of *electron range*, which is the distance travelled by the beam electrons within the solid. Because of the complex nature of the interaction volume, a number of different definitions of the electron range exists in the literature: Bethe range, maximum range, experimental range, Kanaya–Okayama range, etc.

The discussion here will be restricted to only two range definitions, the Bethe range used in x-ray calculations and the Kanaya–Okayama range, which directly describes the overall interaction volume.

3.3.3.1. Bethe Range

If a suitable expression is available for the rate of energy loss with distance travelled, dE/ds, then a rigorous definition of the total distance traveled by an "average" electron is given by

$$R = \int_{E_0}^{E=0} \frac{1}{dE/ds}\, dE. \tag{3.13}$$

The Bethe expression for dE/ds, Eq. (3.7), can be substituted in Eq. (3.13) so that the integral gives the total distance along an electron trajectory, the so-called Bethe range. When Eq. (3.7) is substituted in Eq. (3.13), the integral contains the expression $E/\log(kE)$, which must be integrated numerically. An approximation to the integrated form of the equation is given by Henoc and Maurice (1976):

$$R\,(\mathrm{cm}) = \frac{J^2 A}{7.85 \times 10^4 \rho Z}\left[\mathrm{EI}\left(2\log_e \frac{1.166E_0}{J}\right) - \mathrm{EI}\left(2\log_e \frac{1.166E_i}{J}\right)\right]. \tag{3.14}$$

$E_i = 1.03j$, where J is the mean ionization potential and the value 1.03 is chosen to gain convergence of the integral; alternatively, the range can be calculated to a limit corresponding to a particular inner-shell ionization energy with substitution of E_c for E_i. In Eq. (3.14), EI represents the exponential integral, which may be approximated for any numerical argument x as:

$$\mathrm{EI}(x) = 0.5772 + \log_e(x) + \sum_{n=1}^{\infty} \frac{x^n}{nn!}. \tag{3.15}$$

The product $(nn!)$ rises rapidly and for adequate accuracy in calculations for arguments of $2\log_e(1.166\,E_0/J)$, the series converges in approximately 20 terms.

The Bethe range is the total distance that the beam electron travels in the target while losing all of its energy. However, the Bethe range is not a good description of the maximum dimension of the interaction volume because it does not take into account the effects of elastic scattering, which cause significant deviations of the electron trajectories, as shown in the Monte Carlo plots of Figs. 3.8 and 3.9. Such elastic scattering causes the electron trajectories to "curl up," and the depth from the surface that an electron reaches becomes significantly less than the value suggested by the Bethe range, which is the total distance along the trajectory. The Bethe range overestimates the maximum depth of the interaction volume. The difference between the Bethe range and the maximum dimension of the interaction volume becomes greater as the atomic number of the target, and therefore the amount of elastic scattering, increases. Nevertheless, the Bethe range has important applications in x-ray generation calculations.

3.3.3.2. Kanaya–Okayama Range

Kanaya and Okayama (1972) have considered the combined effects of elastic scattering and energy loss due to inelastic scattering to derive an electron range that more closely approximates the depth dimension of the interaction volume:

$$R_{\mathrm{KO}} = \frac{0.0276 A E_0^{1.67}}{Z^{0.89}\rho} \tag{3.16}$$

(μm), where E_0 is the incident beam energy in keV, A is the atomic weight in g/mole, ρ is the density in g/cm^3, and Z is the atomic number; the beam is incident at a right angle to the specimen surface. The Kanaya–Okayama range is equivalent to the radius of a circle centered on the surface at the beam impact point whose circumference encompasses the limiting envelope of the interaction volume.

3.3.3.3. Range for a Tilted Specimen

The maximum depth dimension of the interaction volume decreases as the tilt angle of the specimen increases (or, equivalently, the angle of beam incidence relative to the surface decreases). A correction for the effect of tilt over the range from 0 to 70° can be derived from Monte Carlo calculations, and a good approximation is given by

$$R_{\mathrm{KO}}(\theta) = R_{\mathrm{KO}}(0)\cos\theta, \tag{3.17}$$

where $R_{\mathrm{KO}}(0)$ is the range for 0 tilt and θ is the tilt angle.

3.3.3.4. Comparison of Ranges

Table 3.2 compares the Bethe and the Kanaya–Okayama ranges for a variety of beam energies and atomic numbers. The greatest differences

Table 3.2. Comparison of Ranges[a]

	Bethe range (μm)				Kanaya–Okayama range (μm)			
	Beam energy (keV)				Beam energy (keV)			
Target	5	10	20	30	5	10	20	30
C	0.61	2.1	7.5	13.0	0.52	1.7	5.3	10.4
Al	0.52	1.8	6.0	12.4	0.41	1.3	4.2	8.2
Cu	0.21	0.69	2.3	4.6	0.14	0.46	1.5	2.9
Au	0.19	0.54	1.6	3.2	0.085	0.27	0.86	1.7

[a] Tilt angle = 0° throughout.

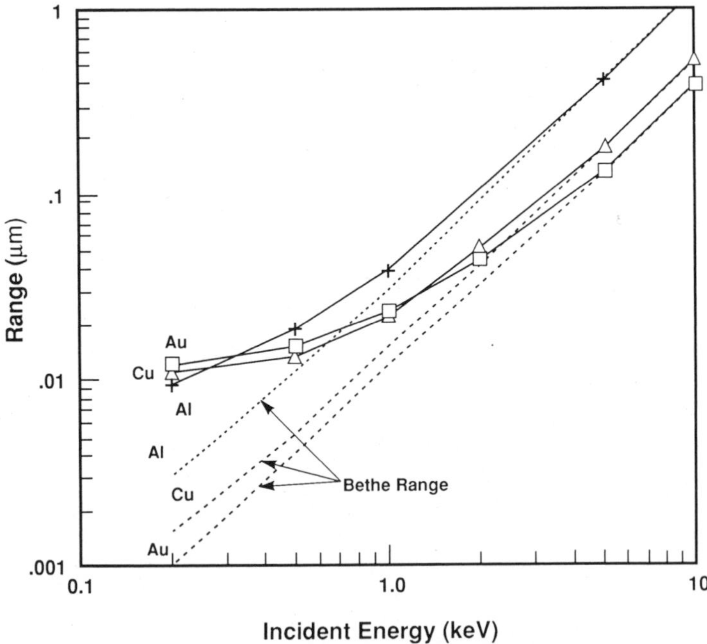

Figure 3.12. Electron range at low incident-beam energy calculated by the Bethe expression as extended by Rao–Sahib and Wittry (1972) (dotted lines) and as modified by Joy and Luo (1989) for Al, Cu, and Au (solid lines).

between these two descriptions of the electron range occur for high-atomic-number targets.

3.3.3.5. Range at Low Beam Energy

As discussed by Joy and Luo (1989), the model chosen for the stopping power has a strong influence on the calculation of the range at low beam energy. As noted above, the approximations used in the Bethe formula eventually fail at low beam energy, and an alternative value of the stopping power must be used. Recent calculations (Joy and Luo, 1989) suggest that earlier expressions for the low-energy regime, such as that of Rao–Sahib and Wittry (1972), tend to overestimate the stopping power, leading to underestimates of the electron range. Figure 3.12 shows the behavior of the range at low beam energy as calculated with the Joy–Luo formula. As the beam energy decreases below 5 keV, the range initially falls with a slope of approximatley 1.7, but at lower energies, as the electron energy falls below the critical value for various interaction processes, e.g., atomic inner shells, the curve begins to flatten and the range tends toward a nearly constant value.

3.4. Signals from Elastic Scattering

The various signals used to form images and perform analysis are generated by events within the interaction volume. Knowledge of the characteristics of these signals is vital for proper image interpretation and analysis procedures. Signals can be categorized according to whether they are primarily influenced by elastic or by inelastic scattering. The backscattered electron signal arises because of elastic scattering, but the characteristics of backscattered electrons are also heavily influenced by inelastic scattering.

3.4.1. Backscattered Electrons

It is found experimentally that a significant fraction of the incident electrons that strike a flat, bulk target placed normal to the probe subsequently escape through the same surface that the electrons entered. For example, if the beam current is first measured in a Faraday cup (see Chapter 2) and the beam is then made to impinge on a flat copper target set normal to the beam and biased positively by means of a battery to exclude the effects of secondary electrons, only about 70% of the beam current is recorded flowing from the copper as a specimen current to ground. The balance of the current, 30%, represents beam electrons which emerge from the specimen. These re-emergent beam electrons are called backscattered electrons. Strictly speaking, backscattering refers to single elastic scattering events in which the electron trajectory is changed by more than 90° from the forward direction of motion, so that the scattered electron propagates back into the same hemisphere that contains the original beam. Close examination of the Monte Carlo electron trajectory plots in Figs. 3.7–3.10 reveals that the beam electrons that follow trajectories which intersect the surface to escape as backscattered electrons actually suffer numerous elastic scattering events, most of which involve angles of less than 90° but whose cumulative effect is to result in escape from the solid. Backscattering from solid specimens is thus mostly an effect of multiple elastic scattering, although a small fraction of the backscattered electrons escape as a result of a single high-angle event.

From the result of the experiment on copper, backscattering is clearly a strong effect, resulting in a significant fraction of the incident beam current being scattered out of the specimen. The backscattered fraction is quantified by the backscatter coefficient η, which is defined as

$$\eta = \frac{n_{\mathrm{BSE}}}{n_{\mathrm{B}}} = \frac{i_{\mathrm{BSE}}}{i_{\mathrm{B}}}, \tag{3.18}$$

where n_{B} is the number of beam electrons incident on the specimen and n_{BSE} is the number of backscattered electrons (BSE). The backscatter coefficient can also be expressed in terms of currents, where i_{B} refers to the beam current injected into the specimen and i_{BSE} to the backscattered electron current passing out of the specimen.

Backscattered electrons provide an extremely useful signal for imaging in scanning electron microscopy. Backscattered electrons respond to composition (atomic number or compositional contrast), local specimen surface inclination (topographic or shape contrast), crystallography (electron channeling), and internal magnetic fields (magnetic contrast). Backscattered electrons remove a significant amount of the total energy of the primary beam, which in the absence of backscattering would contribute to the production of secondary radiation products such as the characteristic x rays measured in quantitative x-ray microanalysis. Knowledge of the properties of backscattered electrons is needed for proper image interpretation and understanding of x-ray microanalysis correction procedures. The significant properties are surveyed in the following sections; for a more detailed survey, particularly for thin films, see Nedrig (1978).

3.4.1.1. Atomic Number Dependence

The Monte Carlo trajectory plots for different elements shown in Fig. 3.9 suggest that backscattering increases with increasing atomic number. If the experiment described above for measuring the backscattering coefficient from copper is repeated for many pure elements ranging across the periodic table, the results shown in Fig. 3.13 are found. Several important points about this plot should be noted:

1. The plot shows a general, monotonic increase in the backscattering coefficient with atomic number. Whenever a sensible relationship is found between a specimen property, such as composition, and a measurable signal in the SEM, such as BSE, the basis for a *contrast mechanism* exists. In this case, the monotonic increase of η vs Z forms the basis for *atomic number contrast* (also called *compositional contrast* or *Z contrast*).

2. The slope of η vs Z is initially steep, but decreases with

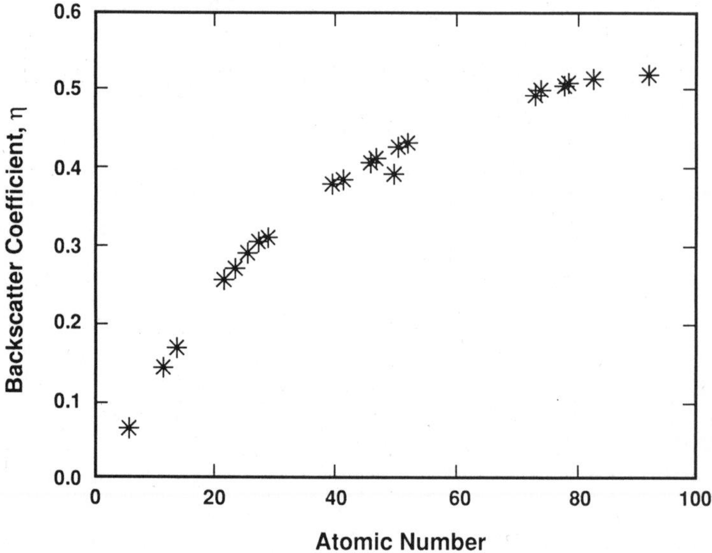

Figure 3.13. Backscattered-electron coefficient η as a function of atomic number at $E_0 = 20\,\text{keV}$ (data of Heinrich, 1966a).

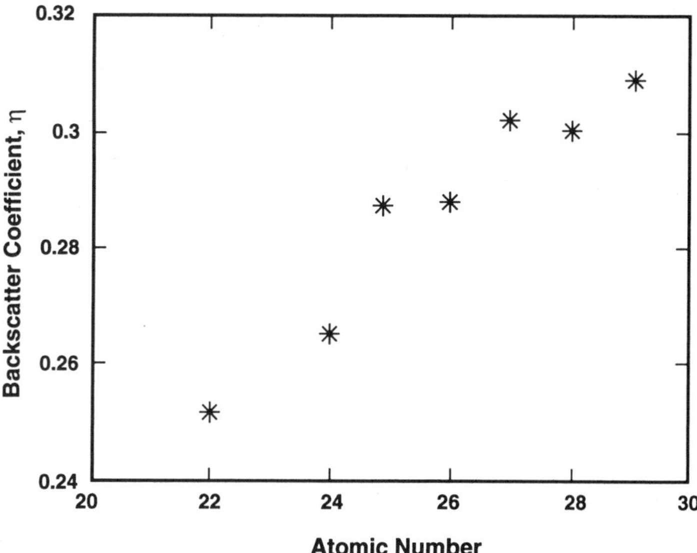

Figure 3.14. Expansion of the plot in Fig. 3.13 showing local deviations from monotonic behavior (data of Heinrich, 1966a).

increasing Z, becoming very shallow above $Z = 50$. The practical effect of this behavior is that atomic number contrast between adjacent pairs of elements is strong at low atomic number and weak at high atomic number.

3. Although the η-vs-Z curve is generally considered monotonic, close examination of values of η determined with high precision, as shown in Fig. 3.14, reveals fine-scale structure in the curve so that adjacent element pairs, as in the first transition series, may occasionally not follow the general upward trend of the overall curve (Heinrich, 1966a). This means that atomic number contrast may not always have the expected sense when small changes in Z are involved.

The curve of η vs Z can be conveniently fit with an expression obtained by Reuter (1972):

$$\eta = -0.0254 + 0.016Z - 1.86 \times 10^{-4}Z^2 + 8.3 \times 10^{-7}Z^3 \quad (3.19a)$$

This expression is useful for estimating values of η when contrast calculations must be performed (see Chapter 4). Note that while this expression fits the general trend of the data very well, the fine-scale detail shown in Fig. 3.14 is lost in the fit.

When a target is a mixture of elements that is homogeneous on an atomic scale, e.g., a solid solution, then the backscattering coefficient follows a simple rule of mixtures based on the weight (mass) concentrations C_i of the individual constituents (Heinrich, 1966a):

$$\eta = \sum_i C_i \eta_i, \quad (3.19b)$$

where i denotes each constituent, η_i is the pure-element backscatter coefficient, and the summation is taken over all constituents.

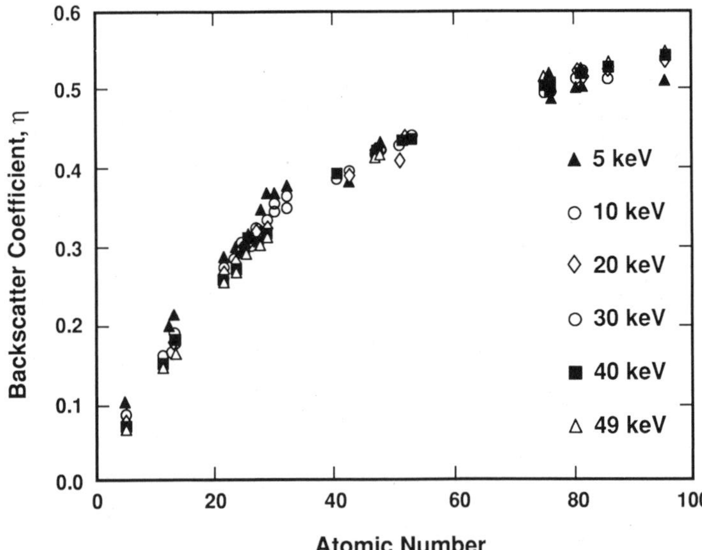

Figure 3.15. Backscattered-electron coefficient as a function of atomic number plotted for a range of beam energies from 5 keV to 49 keV [data of Bishop (1966) and Heinrich (1966a)].

3.4.1.2. Beam-Energy Dependence

The size of the interaction volume was seen in the Monte Carlo plots of Fig. 3.8 to be a strong function of the beam energy. We might reasonably expect that the backscatter coefficient would also depend strongly on beam energy. However, experimental measurements reveal that this is not the case. As shown in Fig. 3.15, there is only a small and not entirely regular change, generally less than 10%, in the backscatter coefficient as a function of beam energy for the range from 5–50 keV, which spans most of the conventional SEM/EPMA range. When data for specific elements are examined, the dependence of the backscatter coefficient is found to increase, decrease, or remain the same, depending on the particular element (Heinrich, 1966a). This unexpected result can be understood from the following qualitative argument: Range and stopping power behave inversely with electron energy. Although the range increases as approximately the 1.7 power of the beam energy, the Bethe energy-loss expression shows that the rate of energy loss decreases with increasing energy. Compare an average electron at the limit of the envelope of the interaction volume for 10-keV incident energy with an average electron at the same absolute depth in the interaction volume for 20-keV incident energy. In the 10-keV case, the electron has lost virtually all of its energy and becomes captured by the specimen. In the 20-keV case, an electron at the same depth still has at least 10 keV of its energy remaining, since it started with twice the energy and lost it at a lower rate. The electron thus has the possibility of continuing to travel and scatter, so that a significant fraction of these electrons can eventually reach the surface and escape as backscattered electrons. The backscatter coefficient is thus relatively insensitive to the beam energy.

At beam energies below 5 keV, the behavior of the backscatter coefficient is more complicated. As the beam energy is reduced toward 1 keV, the backscatter coefficients of light elements apparently increase, while those for heavy elements decrease. Hunger and Kuchler (1979) have obtained an expression for backscattering, originally intended for the beam energy range from 4–40 keV, which appears to work well when extended to energies as low as 1 keV (Joy, 1991):

$$\eta(Z, E) = E^m C, \tag{3.20a}$$

where

$$m = 0.1382 - \frac{0.9211}{Z^{1/2}} \tag{3.20b}$$

and

$$C = 0.1904 - 0.2235(\ln Z) + 0.1292(\ln Z)^2 - 0.01491(\ln Z)^3. \tag{3.20c}$$

There is a considerable practical difficulty in making low-beam energy backscatter measurements under conditions appropriate to conventional SEM. The accumulation of contamination during measurement in the vacuum environment of conventional SEMs can alter the apparent backscatter coefficient. Contamination may arise from the specimen surface or from the residual gases of the pumping system. From Fig. 3.12, the range for 1-keV electrons is only 0.01 μm (gold) to 0.04 μm (aluminum), which may be compared to a range of 0.4–1.3 μm for 10 keV electrons. If a layer of only a few nanometers of carbonaceous material builds up during a low-beam energy measurement, this layer will have a large influence on the apparent backscatter coefficient. Such contamination effects may actually dominate low-beam energy images and reduce the apparent atomic number contrast due to backscattering.

3.4.1.3. Tilt Dependence

If the backscatter coefficient is measured as a function of the tilt angle θ, which is defined as the complement of the angle between the beam and the surface plane, then a smooth, monotonic increase in backscattering with tilt is found, as demonstrated in Fig. 3.16. The slope of η vs θ is initially shallow but increases with increasing tilt. At very high tilt angles, which correspond to grazing incidence, the value of η tends toward unity. If η vs θ is plotted for a range of elements, then at high values of θ the backscatter coefficients for all elements tend to converge, as shown in Fig. 3.17. An expression suggested by Arnal *et al.* (1969) gives the backscatter coefficient as a general function of Z and θ:

$$\eta(\theta) = \frac{1}{(1 + \cos\theta)^p} \tag{3.21}$$

where $p = 9/Z^{1/2}$.

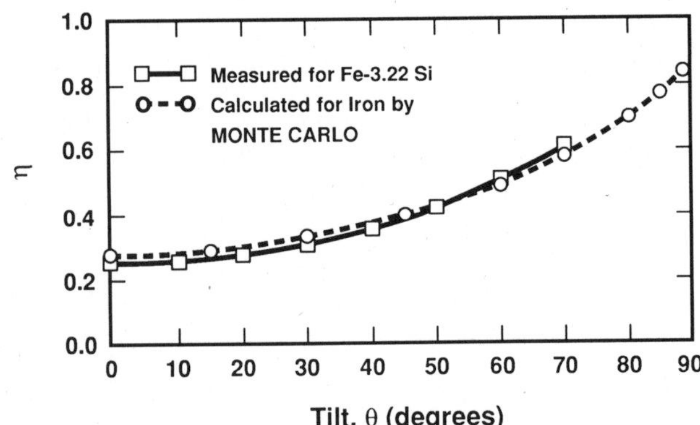

Figure 3.16. Backscattered-electron coefficient as a function of tilt as measured for iron–silicon and as calculated by Monte Carlo electron-trajectory simulation (Newbury *et al.*, 1973).

This behavior of η vs θ arises because of the dominant tendency of elastic scattering for forward scattering. That is, most elastic scattering events result in relatively small deviation angles, on the order of 5°, so the electron trajectories tend to continue in roughly the same direction after scattering as that in which they were initially traveling. When the beam is set normal to the specimen surface, i.e., $\theta = 0°$, this tendency for forward scattering means that beam electrons penetrate into the target. Only by the cumulative effects of many small scattering events and the much rarer large angle events do some of the electron trajectories reverse direction and travel back toward the surface to escape as backscattered electrons. However, if the tilt

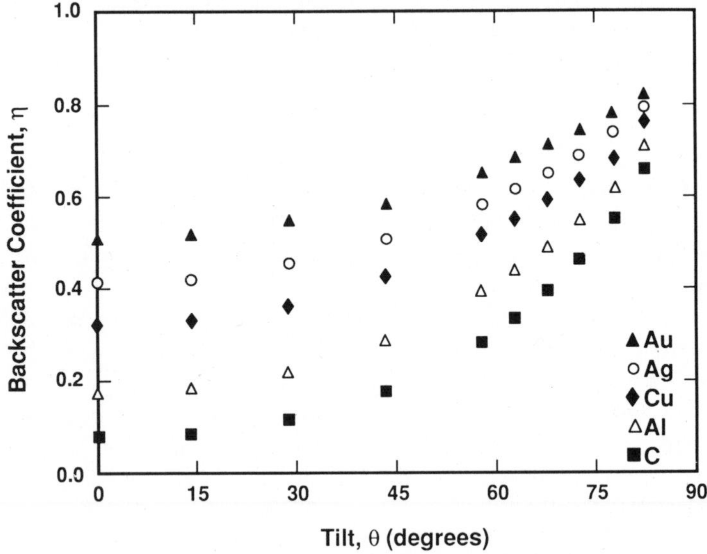

Figure 3.17. Backscattered-electron coefficient as a function of tilt as calculated for several elements by Monte Carlo electron-trajectory simulation (20 keV).

angle of the specimen surface is increased, the geometry of the situation is such that, despite the tendency for forward scattering, electrons tend to travel along trajectories near the surface, as seen in the Monte Carlo plots of Fig. 3.10. Thus, electrons can escape the surface with less total angular deviation, so the backscattering coefficient increases.

The monotonic rise of η with θ forms the basis for an important component of the mechanism of topographic contrast in the SEM, by which mechanism the shape of objects is recognized. In quantitative EPMA corrections, the enhanced loss of backscattered electrons from tilted specimens contributes to a decrease in the production of characteristic x rays as compared to a specimen at 0° tilt.

3.4.1.4. Angular Distribution

A second important consideration is the *directionality* of the backscattered electrons. The dependence of η on θ gives the total *number* of backscattered electrons that emerge at a particular tilt angle of the surface without regard to the trajectories that the backscattered electrons fcllow out of the specimen. In considering the performance of a backscattered electron detector and in order to interpret the images obtained from the BSE signal, it is necessary to understand the relationship of the detector position to the BSE trajectories as emitted from the specimen.

Normal Beam Incidence (0° tilt). The angular distribution of backscattered electrons is defined relative to the normal to the surface. Consider a specimen at 0° tilt, for which the beam is parallel to the surface normal. As shown in Fig. 3.18, an angle ϕ is defined by the vector of the surface normal **n** and a second vector **m**. If the number of backscattered electrons is measured by placing a detector with a very narrow angle of view along a specific direction **m**, then the backscatter coefficient at the angle ϕ, designated $\eta(\phi)$, follows a distribution which approximates a cosine expression:

$$\eta(\phi) = \eta_n \cos \phi, \tag{3.22}$$

where η_n is the value measured along the normal vector, **n**. In Fig. 3.18, the relative number of backscattered electrons at any angle ϕ is

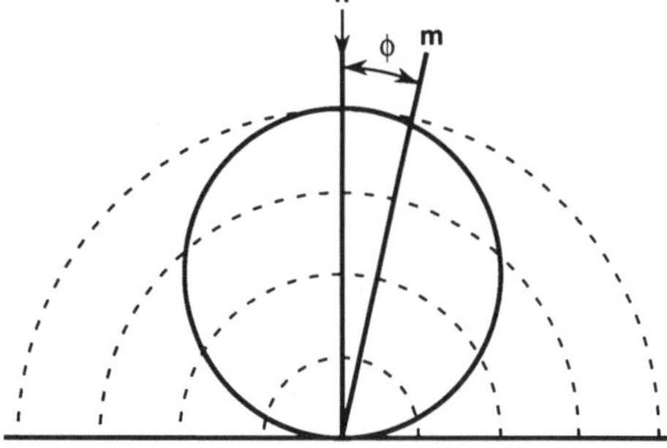

Figure 3.18. Angular distribution of backscattered electrons relative to the surface normal for a specimen surface at 0° tilt (beam perpendicular to surface).

given by the length of the line drawn from the beam impact point to intersect the solid curve. For 0° tilt, this cosine distribution is rotationally symmetric around the surface normal, so that the same angular distribution is found in any plane which contains the surface normal.

Inspection of this cosine distribution shows that the maximum number of backscattered electrons is emitted along the surface normal, $\phi = 0°$, which means that they travel back along the incident beam. As the detector is placed at larger values of ϕ away from the surface normal, the number of backscattered electrons decreases. At $\phi = 45°$, the backscattered electron intensity is approximately 0.707 of the intensity along $\phi = 0°$, while at $\phi = 60°$, the intensity has fallen to 0.50. At shallow angles just above the surface, there are virtually no backscattered electron trajectories.

The origin of the cosine distribution can be understood to a first approximation with the aid of Fig. 3.19. Consider a beam electron which has penetrated to a depth P_0 before undergoing sufficient angular deviation to begin its return to the surface. The length of the path P which the electron must travel at angle ϕ to the normal has a value of

$$P = \frac{P_0}{\cos \phi}, \tag{3.23}$$

where the angle ϕ has the same meaning that it does in Fig. 3.18. The number $N(\phi)$ of electrons which escape along a given path with angle ϕ will be inversely proportional to the path length:

$$N(\phi) \sim 1/P \tag{3.24}$$

Substituting for P from Eq. (3.23) gives the result that

$$N(\phi) \sim \cos \phi. \tag{3.25}$$

Of course the real situation is much more complicated, since the energetic electrons scattering in the solid do not follow straight-line paths out of the specimen but actually undergo many angular deviations due to multiple elastic scattering.

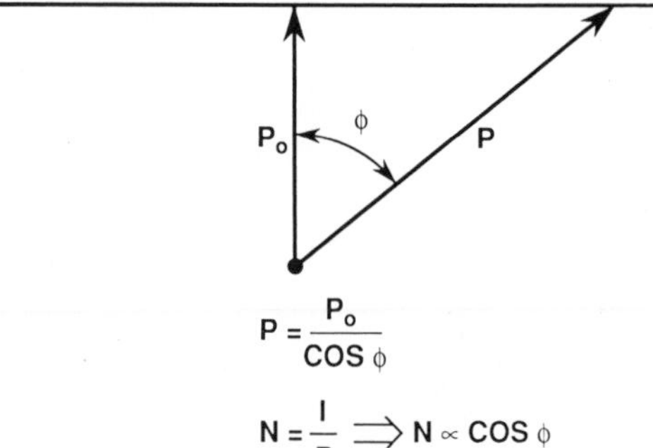

Figure 3.19. Origin of the cosine angular distribution of backscattered electrons. The path lengths P increase as the angle ϕ from the normal increases.

specimen upon reaching an energy equivalent to the equilibrium thermal energy of the specimen electrons).

2. The energy distribution shows two distinct regions: the uppermost region, denoted I, represents the high-energy hump of backscattered electrons that have lost less than 50% of E_0. For most targets of intermediate and high atomic number, most backscattered electrons will be found in region I. Region II is the broad, gradually decreasing tail of the energy distribution, representing those beam electrons which travel progressively greater distances, losing progressively more energy within the specimen prior to backscattering.

3. The energy distribution shows a peak value, which becomes more distinct for higher-atomic-number targets and whose relative energy, E/E_0, increases from approximately 0.8 for copper to 0.95 for gold (Fig. 3.21a). The distribution for carbon is extremely broad, with only a broad maximum.

4. The energy distribution of backscattered electrons also depends on the angle above the surface at which the distribution is measured. As shown in Fig. 3.21b for normal beam incidence, the peak in the energy distribution is lost at low takeoff angles, i.e., shallow backscatter emergence angles above the specimen surface. If the specimen surface is tilted to a high angle (>70°) and the energy distribution is measured for forward-scattered electrons, the distribution is shifted to higher energy. Wells (1974a) has demonstrated that for highly tilted specimens, a significant fraction of those forward-scattered electrons have lost less than 5% of the incident energy ("low loss electrons").

5. Cumulative integrals can be calculated from the backscattered electron energy distributions of Fig. 3.22. As listed in Table 3.3, most backscattered electrons retain at least 50% of the incident beam energy, with the fraction rising sharply for intermediate and high atomic number targets.

3.4.1.6. Lateral Spatial Distribution

Examination of the Monte Carlo electron trajectory simulations shown in Figs. 3.7–3.10 reveals that beam electrons can travel significant distances laterally from the beam impact point before escaping as backscattered electrons. Plots of the density of backscattered electrons per unit area have been calculated by Monte Carlo electron trajectory simulation (Murata, 1973; 1974). As shown in Fig. 3.23a and b, the spatial distribution for normal beam incidence resembles a sombrero, with a central peak and broad tails of gradually falling density. Using the Kanaya–Okayama range to provide a means of scaling the spatial distribution, the extent of the radial distribution is found to depend on the atomic number (Newbury and Myklebust, 1991). Table 3.4a gives the fraction of the Kanaya–Okayama range necessary to encompass various fractions of the total backscattering. Table 3.4b gives the constants of a power-series fit in atomic number of the cumulative radial distribution.

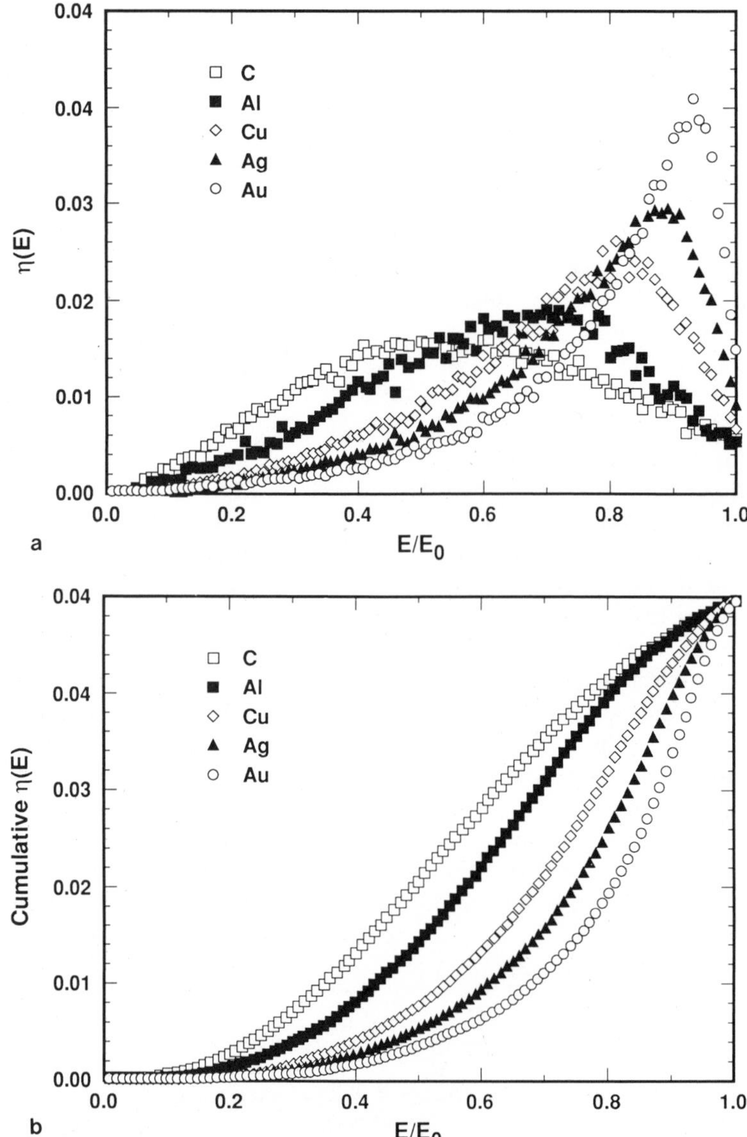

Figure 3.22. Monte Carlo electron-trajectory simulations for the energy distribution of backscattered electrons emitted into 2π steradians (Newbury and Myklebust, 1991): (a) $\eta(E)$ vs E/E_0; (b) cumulative $\eta(E)$ distribution.

Since the basic principle of SEM imaging is the measurement of information localized by focusing the beam to form a probe, a direct consequence of the lateral spreading of backscattered electrons is a decrease in the capability of the SEM to resolve fine features of the specimen on the scale of the focused probe. Examining the sombrero-shaped spatial distribution, the backscattered electrons in the peak of the distribution represent the "good" portion of the backscattered

Table 3.3. Backscattered-Electron Energy Distribution (E/E_0)

Element	Cumulative Fraction Reached		
	>0.5	>0.75	>0.9
Carbon	0.55	0.72	0.85
Aluminum	0.63	0.77	0.87
Copper	0.74	0.83	0.92
Silver	0.79	0.88	0.94
Gold	0.84	0.92	0.96

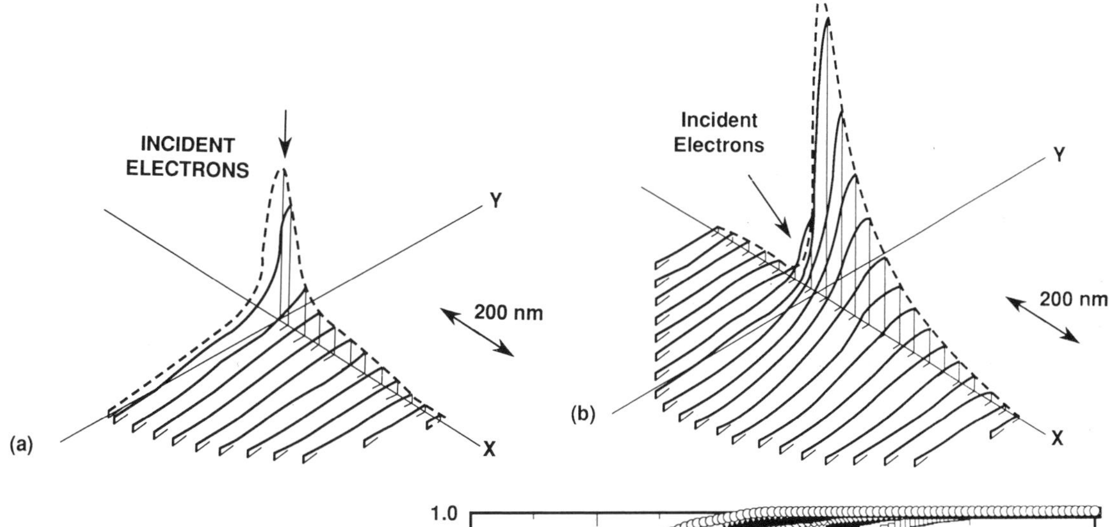

Figure 3.23. Spatial distribution of backscattered electrons from copper plotted as a histogram with the frequency axis perpendicular to the x–y surface plane: (a) copper, $E_0 = 20\,\text{keV}$, 0° tilt; (b) 45° tilt (Murata, 1973). (c) Cumulative backscattering with radial distance from beam impact, normalized to the Kanaya–Okayama range (Newbury and Myklebust, 1991).

Table 3.4. Cumulative Radial Backscattering (20 keV)

	a. Distribution Fraction		
Element	80%	90%	95%
C	0.502	0.575	0.63
Al	0.419	0.490	0.545
Cu	0.318	0.382	0.439
Ag	0.250	0.310	0.365
Au	0.195	0.248	0.295
	b. Quadratic fit ($y = M_0 + M_1Z + M_2Z^2$)		
Coefficient	80%	90%	95%
M_0	0.5453	0.6210	0.6745
M_2	−9.535E-3	−9.964E-3	−9.754E-3
M_2	6.494E-5	6.675E-5	6.304E-5

electron signal which is reasonably closely associated with the incident beam; these electrons are sensitive to fine scale features of the specimen. The backscattered electrons in the wide brim of the sombrero are emitted so far from the beam impact area that they are insensitive to the fine features in the immediate vicinity of the beam impact, and in fact contribute to the deleterious noise component of the backscattered electron signal. For low atomic number targets, the brim dominates the signal, while for high atomic number targets, the peak forms a larger fraction of the backscattered electron signal.

3.4.1.7. Sampling Depth of Backscattered Electrons

A second consequence of the finite distances traveled by the beam electrons within the specimen before backscattering concerns the sampling depth of the backscattered electron signal. The SEM is sometimes thought of as a *surface* imaging tool. If we carefully examine the Monte Carlo electron-trajectory simulations in Figs. 3.7–3.10, individual trajectories of backscattered electrons can be identified for which the beam electron actually penetrated to a significant fraction of the range before reversing its course and returning to the surface to escape as a backscattered electron. Electrons which travel along such trajectories may clearly be influenced by subsurface features of the specimen structure (e.g., inclusions of different composition, voids, etc.) and carry information on that structure upon escaping. For example, elastic scattering depends very strongly on the atomic number of the target atom. If a low-atomic-number specimen has a small inclusion of a high-atomic-number material buried below its surface, then, when the probe is positioned above the inclusion, the electron trajectories that reach the inclusion will be influenced by the larger amount of elastic scattering at that location. More trajectories will be reversed because of the increased scattering, which may produce a higher backscattering coefficient than would be observed from pure low-atomic-number material, depending on the depth of the feature. The backscattered electron signal can thus respond to subsurface details of the specimen's structure. The critical question is; How deep can such an inclusion be

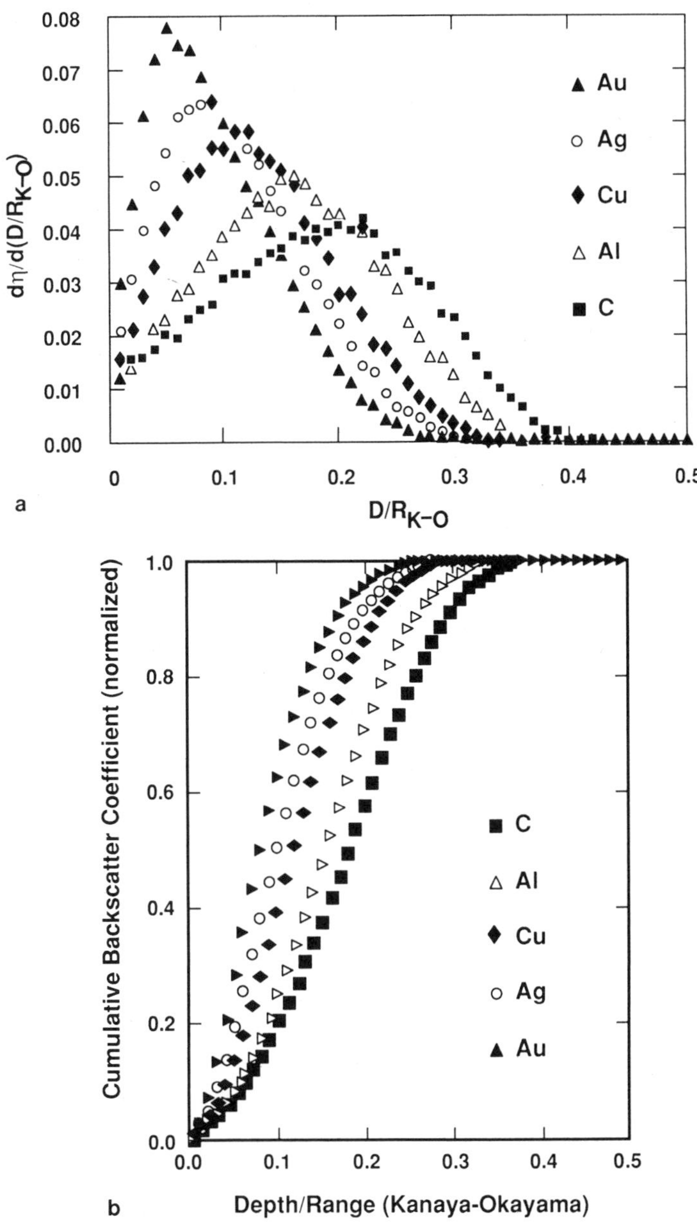

Figure 3.24. (a) Distribution of the maximum depth reached by beam electrons before returning to the surface to escape as backscattered electrons, as calculated by Monte Carlo electron-trajectory simulation. (b) Cumulative backscattering as a function of depth, derived from (a) (Newbury and Myklebust, 1991) $E_0 = 20\,\text{keV}$.

Table 3.5. Cumulative Backscattering Depth (E_0 = 20 keV D/R$_{KO}$)

a. Depth (D/R_{KO})

	0° Tilt			45° Tilt		
Element	63%	90%	95%	63%	90%	95%
C	0.21	0.29	0.32	0.15	0.23	0.27
Al	0.18	0.26	0.28	0.13	0.21	0.24
Cu	0.14	0.22	0.24	0.10	0.19	0.22
Ag	0.12	0.19	0.22	0.085	0.17	0.19
Au	0.10	0.17	0.19	0.075	0.15	0.17

b. Quadratic fit ($y = M_0 + M_1Z + M_2Z^2$)

Coefficient	63%	90%	95%	63%	90%	95%
M_0	0.226	0.309	0.333	0.163	0.238	0.280
M_1	−0.00352	−0.00377	−0.00374	−0.00263	−0.00202	−0.00268
M_2	2.459	2.559	2.469	1.935	1.148	1.630

in the specimen and still be detected in a BSE image? Monte Carlo calculations are ideally suited to answering such a question, since the history of each trajectory is followed and the maximum depth can be recorded. Figure 3.24 shows a Monte Carlo calculation of the frequency histogram of backscattering versus the maximum depth reached prior to backscattering (Newbury and Myklebust, 1991). Table 3.5a gives the BSE sampling depth obtained by integrating the distribution of Fig. 3.24, and Table 3.5b gives the parameters of a quadratic fit to this data. The strong influence of atomic number is readily apparent. For example, 90% of the total backscattering is obtained in $0.17R_{KO}$ for gold and $0.29R_{KO}$ for carbon.

Table 3.5a demonstrates that the sampling depth of backscattered electrons is a substantial fraction of the electron range, and the BSE signal is emphatically not a surface-sensitive signal at conventional SEM beam energies (>10 keV). Because of the general scaling of the characteristics of the interaction volume with beam energy, the data in Table 3.5 are widely applicable with energy. For example, as the beam energy is decreased, the range decreases rapidly, and the backscattered electron signal becomes much more surface-sensitive.

3.5. Signals from Inelastic Scattering

The energy deposited in the specimen by the beam electrons is distributed among a number of secondary processes, several of which result in signals that are important for microscopy and microanalysis.

These signals include the formation of

1. secondary electrons with a continuous energy distribution.
2. Auger electrons with characteristic energies.
3. bremsstrahlung x rays formed with a continuous energy distribution.
4. characteristic x rays.
5. cathodoluminescence radiation, which is the emission of ultraviolet, visible, or infrared radiation, and
6. lattice vibrations (heat).

The fraction of the deposited energy associated with each process depends on the exact nature of the sample, but in almost all cases, most of the energy deposition occurs as heating of the specimen.

3.5.1. Secondary Electrons

3.5.1.1. Definition and Origin

If the energy distribution of all electrons emitted from a sample is measured over the range from E_0, the incident beam energy, down to 0 keV, a curve similar to that shown schematically in Fig. 3.25a is observed. Most of the energy distribution is dominated by the backscattered electrons, which give rise to the regions denoted I, the high energy peak, and region II, the tail extending from intermediate to low energy. If this low-energy tail is extrapolated to zero energy, the yield of backscattered electrons falls smoothly to zero. However, at very low energy, below 50 eV, it is found experimentally that the number of electrons emitted from the specimen increases sharply to a

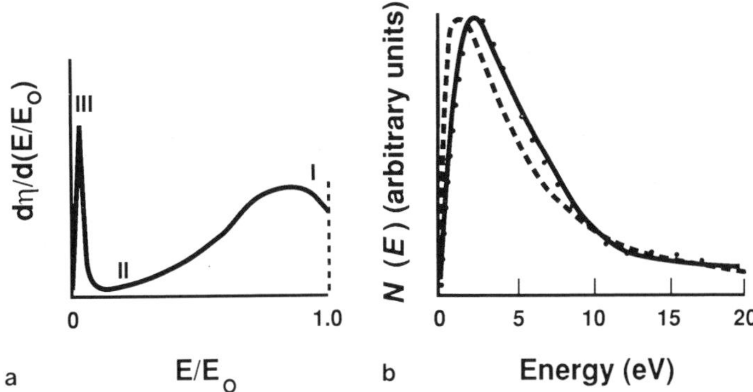

Figure 3.25. (a) Complete energy distribution of electrons emitted from a target, including backscattered electrons (regions I and II) and secondary electrons (region III). Note that the width of region III is exaggerated. (b) Secondary-electron energy distribution as measured (points) and as calculated (lines) with different assumptions on secondary propagation (Koshikawa and Shimizu, 1974).

level much greater than the expected contribution from backscattered electrons. This increase in emitted electrons forms region III in Fig. 3.25a and is due to the phenomenon of secondary electron emission (Bruining, 1954; Seiler, 1984; Cailler and Ganachaud, 1990a; b). Secondary electrons are electrons of the specimen ejected during inelastic scattering of the energetic beam electrons. Secondary electrons are defined purely on the basis of their kinetic energy; that is, all electrons emitted from the specimen with an energy less than 50 eV, an arbitrary choice, are considered secondary electrons. Although some backscattered beam electrons are obviously included in this region, their inclusion in the definition of secondary electrons introduces only a negligible effect.

The total secondary electron coefficient δ is given by

$$\delta = \frac{n_{\mathrm{SE}}}{n_{\mathrm{B}}} = \frac{i_{\mathrm{SE}}}{i_{\mathrm{B}}}, \tag{3.26}$$

where n_{SE} is the number of secondary electrons emitted from a sample bombarded by n_{B} beam electrons, and i designates the equivalent currents.

For a recent comprehensive review of the properties of secondary electrons, see Cailler and Ganachaud (1990a; b).

3.5.1.2. Energy Distribution

Secondary electrons are principally produced as a result of interactions between energetic beam electrons and weakly bound conduction-band electrons in metals or outer-shell valence electrons in semiconductors and insulators (Streitwolf, 1959). Because of the great difference in energy between the beam electrons and the specimen electrons, only a small amount of kinetic energy can be efficiently transferred to the secondary electrons. While fast secondary electrons with energies up to half the incident beam energy can be produced in some scattering events, the number of these fast secondary electrons is small compared to the low-energy slow secondaries. The energy distribution of the secondary electrons emitted from the specimen is narrow and peaks at very low energy, generally in the range 2–5 eV, as shown in Fig. 3.25b. The choice of 50 eV as the upper cutoff is conservative, since the distribution is such that more than 90% of the secondaries are emitted with less than 10 eV of energy.

3.5.1.3. Specimen Composition Dependence

Compared to the behavior of backscattered electrons, whose yield increases nearly monotonically with the atomic number of the specimen, the secondary-electron coefficient from pure-element targets is relatively insensitive to atomic number. As shown in Fig. 3.26, for a beam energy of 20 keV, δ has a value of approximately 0.1 for most elements (Wittry, 1966). Two noteworthy exceptions to this value are

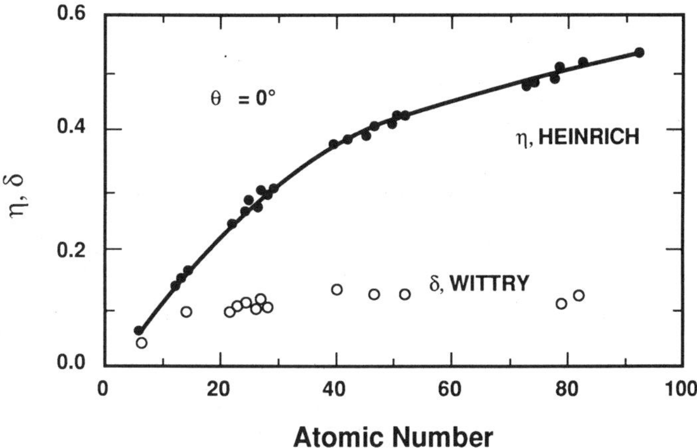

Figure 3.26. Comparison of backscattered-electron and secondary-electron coefficients as a function of atomic number, $E_0 = 30\,\text{keV}$ [data of Wittry (1966) and Heinrich (1966a)].

carbon, which is anomalously low with a value of approximately 0.05, and gold, which is anomalously high with a value of 0.2. These secondary-electron coefficients are representative of measurements made in the vacuum environment of a typical SEM/EPMA, which operates in the pressure range 10^{-3} to 10^{-6} Pa. The intense electron bombardment produced by a focused beam in such an environment results in the almost instantaneous deposition of a layer of surface contamination, generally arising from cracking of hydrocarbon deposits on the specimen surface after preparation or from the residual gases that remain in the vacuum system as a consequence of backstreaming from the oil-diffusion pump when the limiting pressure is achieved. The emission of secondary electrons is very sensitive to the condition of the surface because the low kinetic energy severely limits their range. Measurements of δ made in ultrahigh-vacuum conditions with *in situ* surface cleaning of the specimen suggest a much stronger compositional dependence (Seiler, 1983). However, while the values of δ plotted in Fig. 3.26 may not be representative of "true" material dependence, they are nevertheless appropriate to the actual operational environment of SEM/EPMA observations and measurements with conventional instruments.

There are some notable exceptions to this lack of apparent compositional dependence for δ. Under certain conditions, particularly at low beam energies, δ can approach or even exceed unity for many materials that are nominally insulators. The dependence of δ upon composition can be influenced by the nature of molecular bonds, trace elements present in semiconductors, crystal orientation, conductivity, and the perfection of the crystal surface, making it difficult to formulate a predictive description of secondary emission as a function of specimen composition. An example of a pronounced variation in δ with trace-element composition is shown in Fig. 4.32, of Chapter 4.

Table 3.6. Secondary-Electron Coefficients As a Function of Incident Energy[a]

Element	5 keV	20 keV	50 keV
Al	0.40	0.10	0.05
Au	0.70	0.20	0.10

[a] Reimer and Tollkamp (1980).

3.5.1.4. Beam-Energy Dependence

The secondary-electron coefficient generally rises as the beam energy is lowered. Table 3.6 provides experimental data for the secondary coefficients for Al and Au over a range of energy.

The behavior of the total electron emission from the sample, including both backscattering and secondary emission expressed as $\eta + \delta$, is shown schematically as a function of energy in Fig. 3.27. As discussed above, backscattering shows a stronger variability with energy at low beam energy, but the change is usually within a factor of 2 of high energy values. Most of the change in $\eta + \delta$ seen at low incident-beam energy in Fig. 3.27 really reflects the behavior of δ at low beam energy. Starting with $\delta = 0.1$, appropriate to a beam energy of 10 keV or higher, δ begins to increase significantly as the beam energy is reduced below approximately 5 keV. This increase in δ can be understood in terms of the range of the beam electrons. As discussed below (in Section 3.5.1.7.), the escape depth of secondary electrons is small, on the order of a few nanometers, so all of the secondaries created by the beam electrons at depths greater than a few nanometers are lost. When the primary beam energy is reduced below 3 keV, the primary range becomes so shallow that a much greater fraction of the production of secondaries occurs within the shallow escape depth, and δ increases.

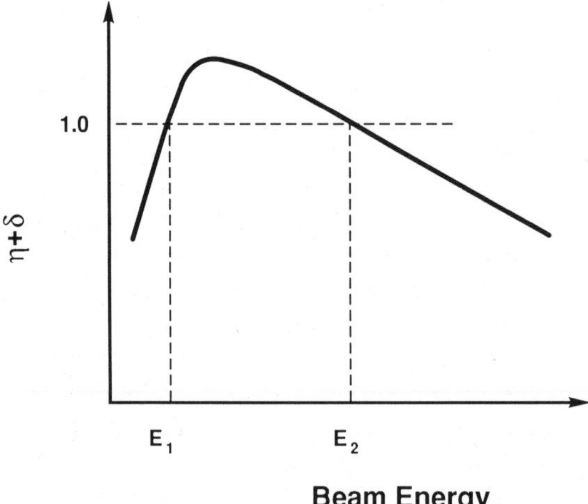

Figure 3.27. Total emitted-electron coefficient $\eta + \delta$ as a function of beam energy.

Table 3.7. Upper Crossover Energy for Various Materials (Normal Beam Incidence)

Material	E_2 (keV)	Reference
Kapton	0.4	Joy (1988)
Electron resist	0.55–0.70	Joy (1987)
Nylon	1.18	Joy (1988)
5% PB7/Nylon	1.40	Krause *et al.* (1982, 1987)
Acetal	1.65	Vaz (1986)
PVC	1.65	Vaz (1986)
Teflon	1.82	Vaz and Krause, (1986)
Glass passivation	2.0	Joy (1987)
GaAs	2.6	Joy (1987)
Quartz	3.0	Joy (1987)
Alumina	4.2	Joy (1988)

As δ increases, there comes a point, denoted E_2 and called the *upper* (or *second*) *crossover point,* where $\eta + \delta$ reaches a value of unity. As the beam energy is lowered further, $\eta + \delta$ increases above unity; that is, more electrons are emitted from the surface as a result of backscattering and secondary emission than are supplied by the beam! The peak in $\eta + \delta$ may be as high as 5–20 for certain insulators. As the beam energy is further reduced, the value of $\eta + \delta$ decreases until the *lower* (or *first*) *crossover point* E_1 is reached. Below E_1, $\eta + \delta$ decreases with decreasing beam energy. E_1 is generally below 1 keV and is difficult to measure. Values of E_2 are listed for several materials in Table 3.7. As described in detail in Chapter 4, knowledge of the E_2 point and of the behavior of $\eta + \delta$ at low beam energy forms the basis for controlling charging in SEM imaging of insulators through careful choice of the beam energy and specimen tilt.

3.5.1.5. Specimen Tilt Dependence

As the angle of specimen tilt θ is increased, δ increases following a relationship reasonably well described by a secant function (Kanter, 1961):

$$\delta(\theta) = \delta_0 \sec\theta, \tag{3.27}$$

where δ_0 is the value of δ found at 0° tilt (normal incidence). This behavior is shown in the plot in Fig. 3.28a. The origin of the secant-function behavior can be understood with the simple argument illustrated in Fig. 3.28b. Consider that all secondary electrons created along the primary beam trajectory within a distance R_0 of the surface can escape. As described in more detail in the next section, this escape depth is very shallow, so that the primary beam electrons are effectively not scattered elastically in this distance and lose so little energy that the rate of secondary production is effectively constant. When the specimen is tilted to a value of θ, the length R of the primary electron path within a distance of R_0 from the surface

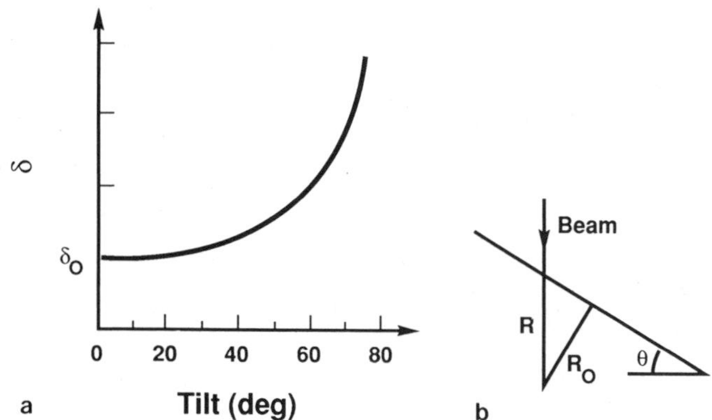

Figure 3.28. (a) Behavior of the secondary-electron coefficient δ as a function of specimen tilt θ, following a secant law. (b) Schematic diagram of the origin of the secant-law behavior.

increases as $R = R_0 \sec \theta$. Since the production and escape of secondaries is proportional to path length, the secondary-electron coefficient is found to increase following a secant relationship. This argument is not strictly complete, however, because secondary electrons can also be created by the beam electrons as they backscatter. As discussed above, backscattering also increases with tilt, which further contributes to the increase in δ at high tilt angles.

When the primary beam energy is reduced to low values on the order of E_2, the secant-law behavior described by Eq. (3.27) no longer applies. The ranges of the backscattered electrons and secondary electrons are similar, so that essentially all of the secondaries generated in the interaction volume can escape. Increasing the angle of tilt does not increase the primary electron range within the escape depth, so no added secondary production is gained within the secondary escape depth.

3.5.1.6. Angular Distribution of Secondary Electrons

In a fashion similar to the arguments used to explain the angular distribution of backscattered electrons (Figs. 3.18 and 3.19), when the specimen is placed at 0° tilt, it can be shown that the emission of secondary electrons follows a cosine distribution of an angle ϕ measured relative to the surface normal. When the specimen is tilted, however, the angular-emission behavior of secondary and backscattered electrons differs. For backscattered electrons, the angular distribution becomes asymmetrical and peaked in a forward scattering direction because elastic scattering is strongly favored in the forward scattering direction. For secondary electrons the angular distribution remains a cosine function relative to the local surface normal. This behavior is a result of the fact that secondaries are generated isotropically by the primary electrons, regardless of tilt. Although the total secondary-electron coefficient δ increases with increasng tilt, the angular distribution of emission remains the same because the distribution of path lengths out of the specimen follows a $1/\cos\phi$ distribution relative to the surface normal.

3.5.1.7. Range and Escape Depth of Secondary Electrons

An important characteristic of secondary electrons is their shallow escape depth, a direct consequence of the low kinetic energy with which they are generated. Secondary electrons are produced along the entirety of the beam electron trajectories within the specimen. However, secondary electrons are subject to inelastic scattering and energy loss during their passage through the specimen and are strongly attenuated as a result. Moreover, when the secondary electrons reach the surface, they must overcome the surface potential barrier (work function), which requires a kinetic energy of several electron volts. As a consequence of the strong attenuation of the secondaries due to inelastic scattering, the probability of escape decreases exponentially with depth:

$$p \sim \exp\frac{-z}{\lambda}, \tag{3.28}$$

where p is the probability of escape, z is the depth below the surface where the secondary-electron generation takes place, and λ is the mean free path of the secondary electrons. Seiler (1967) has determined that the maximum depth of emission is about 5λ, where λ is about 1 nm for metals and up to 10 nm for insulators. The mean free path depends on the energy of the secondary electrons, so that a range of λ actually exists over the full energy range of secondary electrons. For the purposes of rough estimation, the values of λ given above will suffice. The substantially greater range of secondary electrons in insulators is a direct consequence of the fact that inelastic scattering of secondary electrons takes place chiefly with conduction electrons, which are abundant in metals and greatly reduced in insulators. The escape probability of secondary electrons as a function of depth has been calculated by Koshikawa and Shimizu (1974).

As shown in Fig. 3.29, there is a sharp dropoff in escape probability with depth. Compared to the escape depth for backscattered electrons shown in Fig. 3.24, the escape depth for secondary electrons is about 1/100 of that for backscattered electrons for incident beam energies in the range 10–30 keV. Such a shallow escape depth seems to imply that secondary electrons collected as an imaging signal should be capable of imaging surface features and provide information about only a thin layer below the surface, but this is not true.

Secondary electrons are generated throughout the interaction volume of the beam electrons in the specimen, but only those generated within the mean escape distance from the surface carry information that can be detected by the microscopist. Such observable secondary electrons can be formed by two distinctly different processes, as shown in Fig. 3.30.

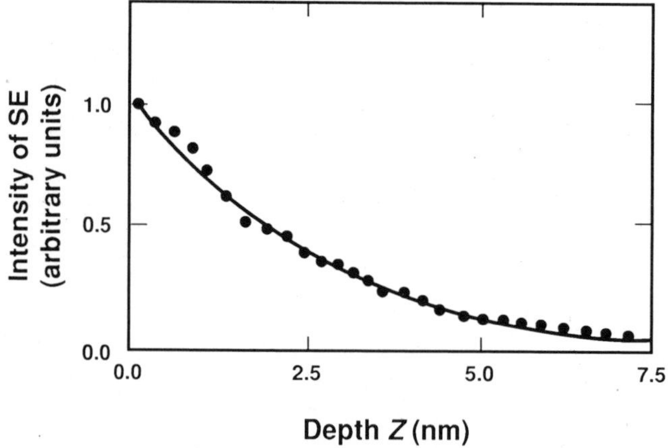

Figure 3.29. Probability of escape of secondary electrons generated at a depth Z below the surface (Koshikawa and Shimizu, 1974).

1. As the incident beam electrons pass through the specimen surface, they generate observable secondaries within the 5λ escape distance of the surface. In the nomenclature of Drescher *et al.* (1970), these secondaries associated with the incoming beam electrons are designated SE_I. The SE_I signal is inherently a high-resolution signal that preserves both the lateral spatial resolution of the focused beam and the shallow sampling depth of the secondary electrons.
2. As beam electrons scatter within the specimen and approach the surface to emerge as backscattered electrons, the secondary electrons which they generate within 5λ of the surface can also escape. These secondaries formed by the exiting backscattered electrons are designated as SE_{II} (Drescher *et al.*, 1970; Peters, 1982, 1984a).

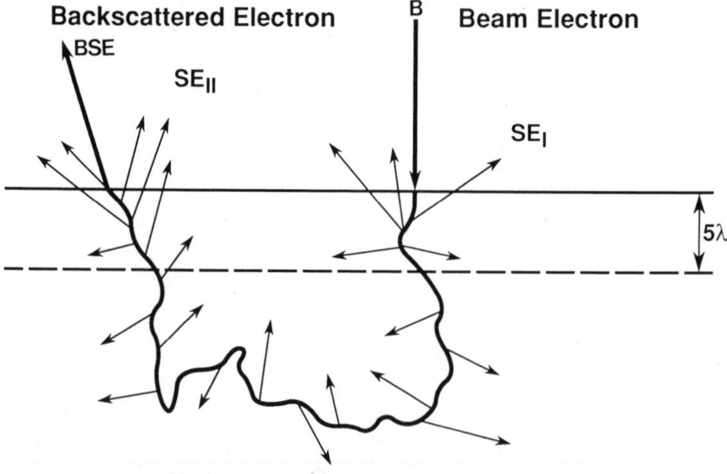

Figure 3.30. Schematic illustration of the origin of two sources of secondary electrons in the sample. Incident beam electrons (B) generate secondary electrons (SE_I) upon entering the sample. Backscattered electrons (BSE) generate secondary electrons (SE_{II}) while leaving the sample. λ is the mean free path for secondary electrons.

Because the SE_{II} signal is actually a consequence of backscattering, its characteristics as a signal carrier actually follow those of the backscattered electron signal, since changes in the SE_{II} signal correspond to changes in the backscattered signal. Both lateral and depth distribution characteristics of backscattered electrons are found in the SE_{II} signal, which is therefore inherently a low-resolution signal. Thus, the information present in the SE_I and SE_{II} signals is distinctly different.

At low beam energies, <5 keV, the range of the beam electrons decreases, and the sampling depth of the backscattered electrons is consequently reduced, while the secondary escape depth is independent of primary beam energy. At a primary beam energy equal to the upper crossover energy E_2, where $\eta + \delta = 1$, the escape depth of secondary electrons and the sampling depth of backscattered electrons are approximately equal. As noted above, the increase in δ at low beam energies reflects the increased fraction of the primary beam energy consumed in the production of secondaries within the escape depth of the primary beam.

3.5.1.8. Relative Contributions of SE_I and SE_{II}

Experimentalists have been able to distinguish the relative contributions of SE_I and SE_{II} to the total secondary emission by careful experiments on thin foils where the backscattered-electron component can be effectively eliminated. The total secondary-electron coefficient δ_T consists of two components, δ_I and δ_{II}, corresponding to SE_I and SE_{II}:

$$\delta_T = \delta_I + \delta_{II}\eta, \tag{3.29}$$

where the δ terms represent the contribution per incident beam electron. The backscatter coefficient η multiplies δ_{II} because the flux of energetic beam electrons traveling back through the entrance surface is reduced from unity to η. The values of δ_I and δ_{II} are not equal, which is to say that an incident beam electron and an average backscattered electron are not equally efficient in generating secondaries. In general, the ratio δ_{II}/δ_I is on the order of 3 to 4 (Seiler, 1967); that is, backscattered electrons are significantly more efficient at generating secondaries than those same electrons were when they first entered the specimen in the incident beam! This unexpected behavior arises because of two factors:

1. Consider a layer of 5λ thickness starting at the surface; imagine that any secondary produced within this layer will escape. For beam energies above 10 keV, the elastic-scattering mean free path is much greater than 5λ. For a condition of 0° tilt, the incident-beam electrons approach the surface at a right angle and travel through the 5λ layer without significant elastic scattering. As shown in Fig. 3.30, the length of the primary electron trajectory along which production of secondaries can lead to escape is thus only 5λ. Now consider the situation for backscattered electrons. Because of the randomizing effect of elastic scattering, a large

Table 3.8. Ratio of Secondary Electrons Produced by Beam and Backscattered Electrons

	δ_T	η	SE_{II}/SE_I
Carbon	0.05	0.06	0.18
Aluminum	0.1	0.16	0.48
Copper	0.1	0.30	0.9
Gold	0.2	0.50	1.5

fraction of the backscattered electrons is likely to approach the surface at shallower angles, and thus their path length through the 5λ escape depth is greater by the secant of the angle to the surface. This additional path length leads to the production of more secondaries that can escape from the specimen.

2. Because of inelastic scattering during their passage through the specimen, the electrons emerging from the surface as backscattered electrons have lost energy relative to the incident beam. Electrons of lower energy are more efficient at transfering kinetic energy to the weakly bound specimen electrons so as to eject them as secondaries. Due to their lower energy, backscattered electrons have greater efficiency at generating secondaries as compared to the higher-energy incident-beam electrons.

Taking $\delta_{II}/\delta_I = 3$ and using the appropriate values of δ_T and η, the ratio of the yield of secondary electrons from both processes, SE_{II}/SE_I, can be calculated for bulk targets. As listed in Table 3.8, the ratio of secondaries from the two processes is such that the SE_I signal dominates δ_T for light elements such as carbon and aluminum, the processes are approximately equal for an intermediate-atomic-number material such as copper, and the SE_{II} signal dominates δ_T for gold.

3.5.2. X Rays

During inelastic scattering of the beam electrons, x rays can be formed by two distinctly different processes.

1. the bremsstrahlung, or continuous x-ray process, and
2. the inner-shell ionization process, which can lead to the emission of characteristic x rays.

The x-ray spectrum which we observe has two distinctly different components, a characteristic component, which identifies the specific atom(s) present in the interaction volume, and a continuous component, which is nonspecific and which forms a background at all energies. In describing x rays, we shall make use of both the energy of the x ray E (keV) and its associated wavelength λ (nm), which are related by the following expression:

$$\lambda = \frac{hc}{eE} = \frac{1.2398}{E}, \tag{3.30}$$

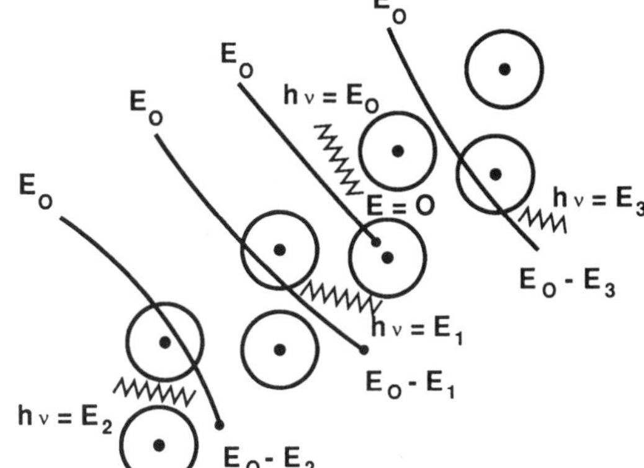

Figure 3.31. Schematic illustration of the origin of the x-ray continuum, resulting from deceleration of the beam electrons in the Coulombic field of the atoms.

(nm), where h is Planck's constant, c is the speed of light, and e is the electron charge.

3.5.2.1. Continuum X-Ray Production

Beam electrons can undergo deceleration in the Coulombic field of the atom, which is formed by the positive field of the nucleus and the negative field of the bound electrons, as shown in Fig. 3.31. The loss in energy from the electron that occurs in such a deceleration event is emitted as a photon of electromagnetic energy. This radiation is referred to as the x-ray bremsstrahlung, or "braking radiation." Because of the random nature of the interaction, the electron may lose any amount of energy in a single deceleration event. The bremsstrahlung can therefore take on any value from 0 up to the entire energy of the electron, thus forming a continuous electromagnetic spectrum. Determination of the limiting bremsstrahlung energy, known as the Duane–Hunt limit, with a calibrated x-ray spectrometer and with appropriate correction for the spectrometer response function, provides an unambiguous measure of the true electron-beam energy as the electrons actually reach the specimen (Myklebust *et al.*, 1990).

Figure 3.32 shows the x-ray spectrum, consisting of the continuous and characteristic components, generated by the electron beam within the interaction volume in a solid copper specimen. The continuum extends from virtually zero energy (infrared, visible, and ultraviolet radiation with an energy of a few electron volts), to x rays having an energy equal to the total incident energy of the beam electrons. The maximum x-ray energy corresponds to beam electrons which have lost all of their incident energy in a single event. Since x-ray wavelength is inversely proportional to energy, the most energetic x rays will have a minimum wavelength, λ_{SWL}, called the short-wavelength limit or Duane–Hunt limit, that can be related to the incident energy E_0 through Eq. (3.30).

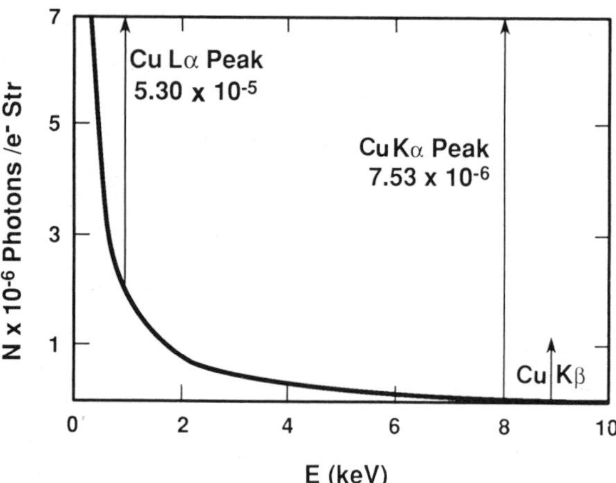

Figure 3.32. Calculated x-ray spectrum as generated in a copper target by a 20-keV electron beam. The continuum background and the Cu $K\alpha$, Cu $K\beta$, and Cu $L\alpha$ characteristic x-ray lines are shown, $k_\beta = 1.02 \times 10^{-6}$.

The intensity of the x-ray continuum I_{cm}, at any energy or wavelength has been quantified according to Kramers (1923) as

$$I_{\text{cm}} \sim i\bar{Z}\left(\frac{\lambda}{\lambda_{\text{SWL}}} - 1\right) \sim \frac{i\bar{Z}(E_0 - E_\nu)}{E_\nu}, \tag{3.31}$$

where $\bar{Z}$ is the average atomic number (based upon mass or weight fractions of the elemental constituents) of the specimen, E_0 is the incident beam energy, i is the beam current, and E_ν is the continuum photon energy. Subsequent refinements of Eq. (3.31) will be described in Chapter 9. From Eq. (3.31) and the spectrum shown in Fig. 3.32, it can be seen that the x-ray continuum intensity decreases as the photon energy increases, reaching zero intensity at the beam energy. At low photon energies, the intensity increases rapidly because of the greater probability for slight deviations in trajectory caused by the Coulombic field of the atoms. Note that Eq. (3.31) is undefined at $E_\nu = 0$. The intensity of the continuum is a function of both the atomic number and the beam energy. As the beam energy E_0 increases, the maximum continuum energy increases and the corresponding short-wavelength limit λ_{SWL} decreases. As the beam energy is increased, the intensity at a specific value of E_ν increases because, with a higher initial energy, the beam electrons have more chances to undergo a particular energy loss E_ν. The intensity of the continuum increases with the atomic number because of the increased Coulombic field of the nucleus. The intensity scales directly with the beam current (number of beam electrons per unit time) which strike the specimen.

The level of the continuum radiation plays an important role in analytical x-ray spectrometry because it forms a background under the characteristic peaks of interest. Once a photon is created with a specific energy, it is impossible to determine whether it is a continuum or characteristic x ray. Thus, the spectral background intensity due to the continuum process that occurs at the same energy as a characteristic x ray sets an eventual limit to the concentration level of the element that can be detected. The continuum is therefore usually regarded as a nuisance to the analyst. However, it should be noted that, from Eq. (3.31), the continuum carries information about the average atomic number, and hence the overall composition, of the specimen. Thus regions of different average atomic number in a specimen emit at different levels of continuum intensity at *all* x-ray energies. This fact can prove useful in developing a strategy to measure background while analyzing a specimen with wavelength-dispersive (diffraction) x-ray spectrometers, and it forms the basis for an important correction scheme (Marshall–Hall method, peak-to-background method) for quantitative analysis of particles, rough surfaces, and biological specimens that undergo mass loss due to beam damage during analysis. The atomic number dependence of the bremsstrahlung causes an important artifact in x-ray mapping of constituents at minor and trace levels, which can lead to serious misinterpretation if not recognized and corrected.

3.5.2.2. Inner-Shell Ionization

The beam electron can interact with a tightly bound inner-shell electron, ejecting the atomic electron and leaving a vacancy in that shell; the atom is left as an ion in an excited, energetic state, as shown in Fig. 3.33. The atom relaxes to its ground state (lowest energy) within approximately 1 ps through a limited set of allowed transitions of outer-shell electrons to fill the inner-shell vacancy. The energies of the electrons in the shells are sharply defined, with values characteristic of the atomic species. The energy difference of the transition is therefore also a characteristic value, and this excess energy can be released from the atom in one of two ways (Fig. 3.33, branches). In the Auger process, the difference in shell energies can be transmitted to another outer-shell electron, ejecting it from the atom as an electron with a specific kinetic energy. In the characteristic x-ray process, the difference in energy is expressed as a photon of electromagnetic radiation that is sharply defined in energy, in contrast to the bremsstrahlung process, which produces photons spanning an energy continuum. The partitioning of the de-excitation process between the x-ray and the Auger branches is described by the fluorescence yield ω and the Auger yield a:

$$\omega + a = 1. \tag{3.32}$$

For a specific shell, the Auger process is favored for low atomic numbers, e.g. $\omega \simeq 0.005$ for the carbon K-shell, while the characteristic x-ray process dominates for high atomic numbers, e.g., $\omega \simeq 0.5$ for the germanium K-shell, as shown in Fig. 3.34.

A comprehensive treatment of the properties of characteristic x

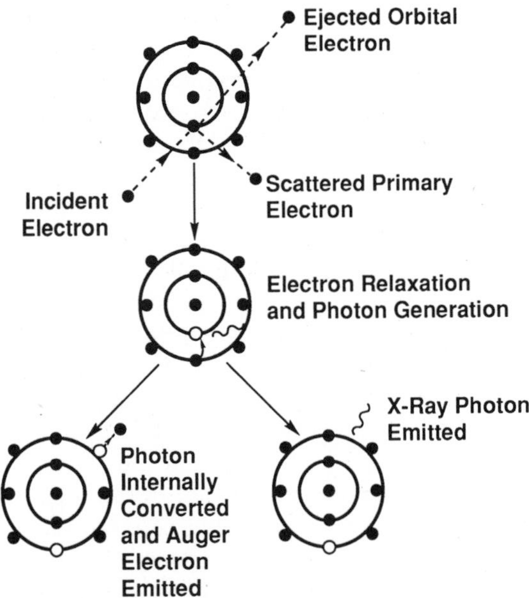

Figure 3.33. Schematic illustration of the process of inner-shell ionization and subsequent de-excitation by electron transitions. The difference in energy between the shells is expressed either by the ejection of an energetic electron with characteristic energy (Auger process) or by the emission of a characteristic x ray.

rays is beyond the scope of this book, and the interested reader is referred to the literature (e.g., Bertin, 1975). Certain basic concepts fundamental to x-ray microanalysis will be surveyed, including atomic energy levels, critical ionization energy, families of x-ray emission energies, and x-ray intensities.

Atomic Energy Levels. The electrons of an atom occupy specific energy levels, illustrated schematically in Fig. 3.35, with each electron energy level uniquely described by a set of quantum numbers (n, l, j, m_j).

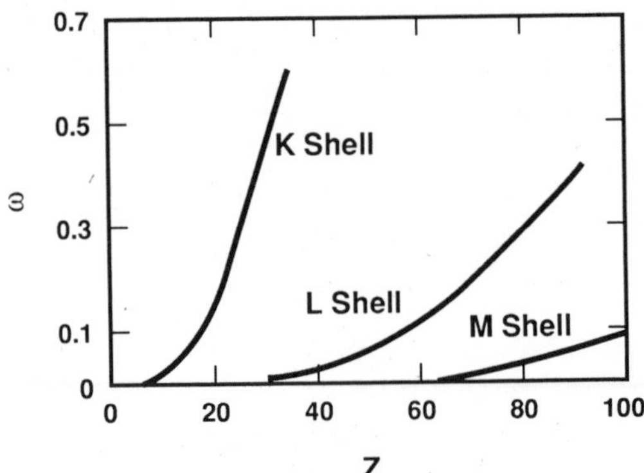

Figure 3.34. Fluorescence yield ω as a function of atomic number for several shells.

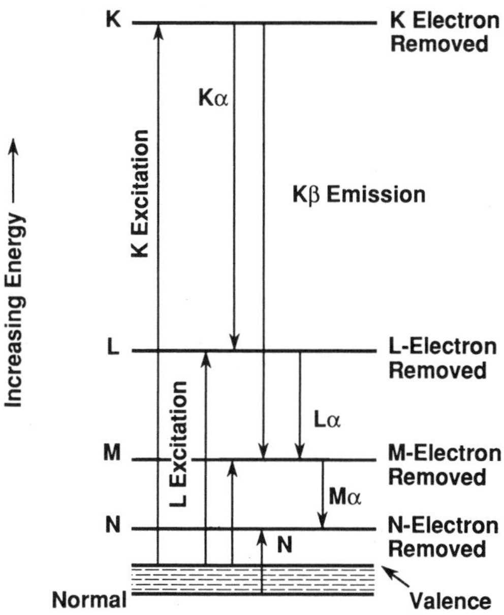

Figure 3.35. Energy-level diagram for an atom, illustrating the excitation of the *K, L, M,* and *N* shells and the formation of $K\alpha$, $K\beta$, $L\alpha$, and $M\alpha$ x rays.

1. The principal quantum number n denotes a shell in which all electrons have nearly the same energy: $n = 1$ corresponds to the shell designated *K* in x-ray terminology; $n = 2$, *L* shell; $n = 3$, *M* shell; $n = 4$, N shell; etc.
2. The orbital quantum number l characterizes the orbital angular momentum of an electron in a shell; l is restricted to values 0 to $n - 1$.
3. While orbiting, the electron is also spinning. The spin quantum number s describes that part of the total angular momentum due to the electron spinning on its own axis and is restricted to the values $\pm\frac{1}{2}$. Because magnetic coupling between the spin and orbital angular momenta occurs, the quantum number describing the total angular momentum, j, takes on values $j = l + s$.
4. Under the influence of a magnetic field, the angular momentum takes on specific directions characterized by the magnetic quantum number m_j. The values of m_j are given by $m_j \leq |j|$; e.g., for $j = \frac{5}{2}$, $m_j = \pm\frac{5}{2}$, $\pm\frac{3}{2}$, $\pm\frac{1}{2}$.

The arrangement of electrons in an atom is controlled by the Pauli exclusion principle, which imposes the restriction that no two electrons in an atom can have the exact same set of quantum numbers and therefore the same energy. Thus, each electron has a unique set of quantum numbers (n, l, j, m_j) which describes it. To the level of understanding which we need for practical x-ray spectrometry, it is sufficient to realize that the quantum number rules result in the atomic electrons being arranged in major shells (*K, L, M,* etc.) within which are minor subshells

Table 3.9. Shells and Subshells of Atoms

X-ray notation	Quantum numbers				Maximum electron population
	n	l	j	m_j	
K	1	0	$\frac{1}{2}$	$\pm\frac{1}{2}$	2
L_{I}	2	0	$\frac{1}{2}$	$\pm\frac{1}{2}$	2
L_{II}	2	1	$\frac{1}{2}$	$\pm\frac{1}{2}$	2
L_{III}	2	1	$\frac{1}{2}$	$\pm\frac{3}{2}, \pm\frac{1}{2}$	4
M_{I}	3	0	$\frac{1}{2}$	$\pm\frac{1}{2}$	2
M_{II}	3	1	$\frac{1}{2}$	$\pm\frac{1}{2}$	2
M_{III}	3	1	$\frac{3}{2}$	$\pm\frac{3}{2}, \pm\frac{1}{2}$	4
M_{IV}	3	2	$\frac{3}{2}$	$\pm\frac{3}{2}, \pm\frac{1}{2}$	4
M_{V}	3	2	$\frac{5}{2}$	$\pm\frac{5}{2}, \pm\frac{3}{2}, \pm\frac{1}{2}$	6
N_{I}	4	0	$\frac{1}{2}$	$\pm\frac{1}{2}$	2
N_{II}	4	1	$\frac{1}{2}$	$\pm\frac{1}{2}$	2
N_{III}	4	1	$\frac{3}{2}$	$\pm\frac{3}{2}, \pm\frac{1}{2}$	4
N_{IV}	4	2	$\frac{3}{2}$	$\pm\frac{3}{2}, \pm\frac{1}{2}$	4
N_{V}	4	2	$\frac{5}{2}$	$\pm\frac{5}{2}, \pm\frac{3}{2}, \pm\frac{1}{2}$	6
N_{VI}	4	3	$\frac{5}{2}$	$\pm\frac{5}{2}, \pm\frac{3}{2}, \pm\frac{1}{2}$	6
N_{VII}	4	3	$\frac{7}{2}$	$\pm\frac{7}{2}, \pm\frac{5}{2}, \pm\frac{3}{2}, \pm\frac{1}{2}$	8

(L_{I}, L_{II}, L_{III}, M_{I}, etc.). The shells and subshells, their quantum numbers, maximum electron occupation, and the corresponding x-ray notation are listed in Table 3.9.

Critical Ionization Energy. Inner-shell ionization occurs when an electron is removed from a shell and ejected from the atom. Because the energy of each shell and subshell is sharply defined, the minimum energy necessary to remove an electron from a specific shell has a sharply defined value as well, the so-called *critical ionization energy* (also known as the *critical excitation energy* or *x-ray absorption energy*). Each shell and subshell of an atom species requires a different critical ionization energy. As an example, consider the wide range in critical ionization energies for the *K*, *L*, and *M* shells and subshells of platinum ($Z = 78$), listed in Table 3.10. The critical ionization energy

Table 3.10. Critical Ionization Energies for Platinum

Shell	Critical ionization energy (keV)
K	78.39
L_{I}	13.88
L_{II}	13.27
L_{III}	11.56
M_{I}	3.296
M_{II}	3.026
M_{III}	2.645
M_{IV}	2.202
M_{V}	2.122

is an important parameter in calculating x-ray intensities. Extensive tabulations of the critical ionization energy and characteristic x-ray energies have been prepared by Bearden (1964; 1967a; 1967b).

Energy of Characteristic X-Rays. The de-excitation of an atom following ionization takes place by a mechanism involving transitions of electrons from one shell or subshell to another. The transitions may be radiative, that is, accompanied by the emission of a photon of electromagnetic radiation, or non-radiative, e.g., the Auger electron emission process.

The x rays emitted during a radiative transition are called *characteristic* x rays because their specific energies and wavelengths are characteristic of the particular element which is excited. The energy levels of the shells vary in a discrete fashion with atomic number so that the difference in energy between shells changes significantly even when the atomic number changes by only one unit. This fact was discovered by Moseley (1913; 1914) and expressed in the form of an equation:

$$\lambda = \frac{B}{(Z - C)^2}, \tag{3.33}$$

where B and C are constants which differ for each family and λ is the characteristic x-ray wavelength. Moseley's relation, which forms the basis for qualitative analysis, that is, the identification of elemental constituents, is plotted for three different x-ray lines ($K\alpha$, $L\alpha$, $M\alpha$) in Fig. 3.36.

For elements with atomic numbers ≥ 11 (sodium), the shell structure is sufficiently complex that when an ionization occurs in the innermost K shell, the transition to fill that vacancy can occur from more than one outer shell. Thus, as shown in the simplified diagram of Fig. 3.35, following ionization of a K shell, a transition to fill the

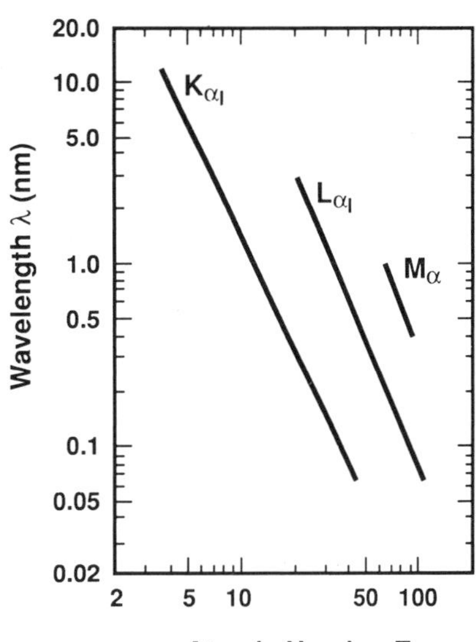

Figure 3.36. Moseley's relation between λ and Z for the $K\alpha_1$, $L\alpha_1$, and $M\alpha_1$ characteristic x-ray lines.

vacancy can occur from either the L shell or the M shell. Since the electrons in these outer shells are at different energies, the x rays created from these two possible shells must have different energies and are designated differently. In the Siegbahn notation most typically used in x-ray spectrometry, lower-case Greek letters are used to designate the x rays arising from ionizations in a particular shell in the general order of observed intensity: α, β, γ, etc., with the x ray produced from the most probable transition designated α, e.g., for the K-shell, $K\alpha$ and $K\beta$ transitions are possible. The $K\alpha$ x ray is formed from a transition from the L shell to the K shell, while a $K\beta$ x ray results from a transition from the M shell. Depending on the atomic number of the atom, these K-shell x rays can have quite different energies; in copper, for example, the $K\alpha$ x ray has an energy of 8.04 keV and the $K\beta$ has an energy of 8.90 keV. As noted in Table 3.9, the shells may be further divided into subshells of slightly different energy. These subdivisions are shown by subscripted numbers, with number 1 indicating the highest relative intensity. The $K\alpha$ x rays of copper are split into $K\alpha_1$ (8.048 keV, a transition from the L_{III} subshell) and $K\alpha_2$ (8.028 keV, a transition from the L_{II} subshell); $K\beta$ is split into $K\beta_1$ (8.904 keV, from M_{II}), $K\beta_2$ (8.977 keV, from $N_{\mathrm{II,III}}$), $K\beta_3$ (8.904 keV, from M_{II}), and $K\beta_5$ (8.976 keV, from $M_{\mathrm{IV,V}}$). Note that not all observable x-ray lines have been assigned Greek letter designations. There exists an L_l transition. A number of x-ray lines with low intensity relative to the principal lines are designated in an alternative style used in atomic physics. In this scheme, the first letter denotes the shell in which the ionization initially takes place, with the subscript indicating the specific subshell, while the second letter and number indicate the shell and subshell from which the inner shell vacancy is filled. Thus, $M_{\mathrm{II}}N_{\mathrm{IV}}$ denotes an ionization in the M_{II} subshell followed by an electron transition from the N_{IV} subshell.

The energies of the shells are so sharply defined that the uncertainty in the energy, or energy width, of a characteristic x ray is a small fraction of the energy. For example, for a Ca $K\alpha_1$ x ray, the energy is 3.692 keV $\pm$ 0.01 keV, or 0.027%, defined at the half-height of photon intensity (Salem and Lee, 1976). The narrow fundamental energy width means that x-ray peaks appear extremely narrow when measured with an appropriate high-resolution spectrometer. This narrow, line-like character of peaks superimposed on the broad, slowly changing x-ray background has led to the frequent use of the term x-ray *lines* in the literature when referring to characteristic x-ray peaks.

Since the primary ionization involves removing an electron from a bound state in a shell to an effective infinity outside the atom, whereas the characteristic x-rays are formed by transitions between bound states in shells, the energy of the characteristic x ray is *always less* than the critical ionization energy for the shell from which the original electron was removed. Thus $E_{K\alpha} = E_K - E_L$ and $E_{K\beta} = E_K - E_M$, while $E_C = E_K - 0$; therefore, $E_{K\alpha} < E_{K\beta} < E_C$.

Families of X-Ray Lines. For carbon, a low-atomic-number atom of simple structure, the shell structure consists of two electrons in the K shell and four in the L-shell. Such a structure permits only $K\alpha$ x rays with an energy of 0.277 keV to be created, because although there are L-shell electrons which can be ejected in collisions with beam electrons, there exist no M-shell electrons to fill the vacancy in the L-shell. Carbon can therefore be identified only by a single peak. For sodium ($Z = 11$), one M-shell electron exists, so that both a $K\alpha$ peak at 1.041 keV and a $K\beta$ peak at 1.071 are formed, although the $K\beta$ x ray is rare in comparison to $K\alpha$, having a ratio of $K\alpha/K\beta$ greater than 100. (A sodium L-shell x ray is also possible with an energy of 0.030 keV. X-ray energies below 0.2 keV cannot be measured under practical analytical conditions and will be ignored). When more complex atoms are considered, the detailed diagram of Fig. 3.37 demonstrates the extensive families of characteristic x rays which exist for intermediate- and high-atomic-number elements. As the atomic number of the atom increases above 10, the K-shell x rays split into $K\alpha$ and $K\beta$ pairs. Above atomic number 21, an additional family, the

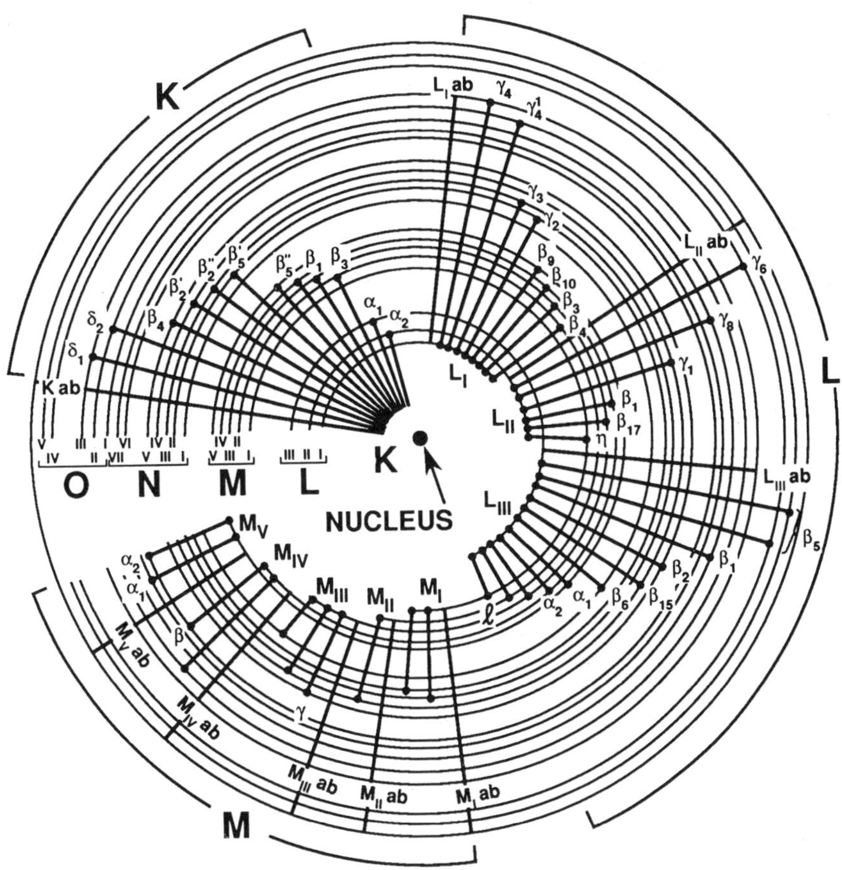

Figure 3.37. Comprehensive energy-level diagram showing all electron transitions which give rise to K, L, and M x rays (Woldseth, 1973).

L-shell x rays, begins to become measurable at 0.2 keV. When atomic number 50 is reached, the *M*-shell family begins to appear above 0.2 keV. The complexity of the shells is extended further by the splitting of lines due to the energy divisions of subshells. Examples of x-ray spectra for the energy range 0.2–20 keV taken with a low-resolution energy dispersive x-ray spectrometer are shown in Chapter 6 for carbon $K\alpha$ (Fig. 6.10a), titanium $K\alpha$, $K\beta$ (Fig. 6.2), copper $K\alpha$, $K\beta$, $L\alpha$ (Fig. 6.8), and the dysprosium *L*-family and *M*-family (Fig. 6.9). These spectra demonstrate the increase in complexity of the x-ray spectrum with atomic number. Note that in these spectra many lines are not resolved; e.g., $K\alpha_1$–$K\alpha_2$, because of the close proximity in energy of the peaks and the relatively poor resolution of the energy-dispersive x-ray spectrometer. If wavelength-dispersive spectrometry (based upon Bragg diffraction) is used, 25 or more *L*-family lines can be observed, depending on the element.

When a beam electron has sufficient energy to excite a particular excitation edge to produce the characteristic x-ray lines from that edge, all other characteristic x-rays of lower energy for the element will also be excited. This occurs because of

1. direct ionization of those lower-energy shells by beam electrons and
2. x-ray formation resulting from the propagation of a vacancy created in an inner shell to outer shells as a consequence of electron transitions.

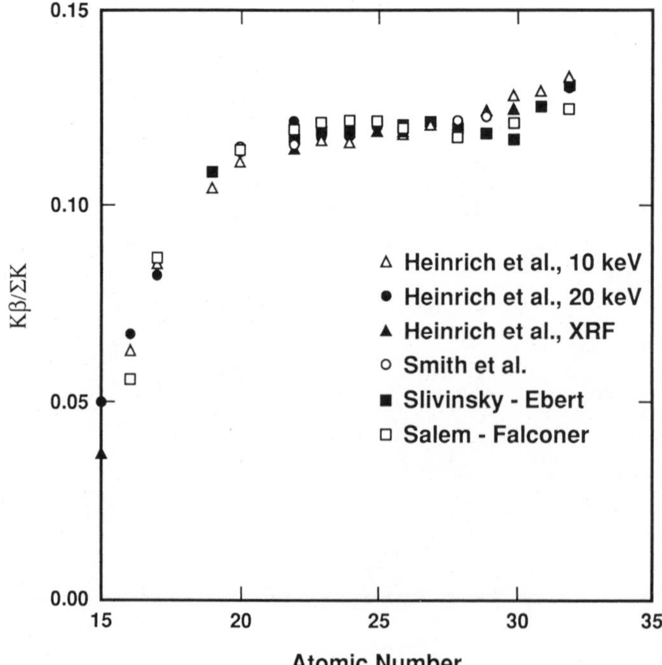

Figure 3.38. Plot of the $K\alpha$–$K\beta$ relative weights (Heinrich *et al.*, 1979).

Table 3.11. Weights of Lines

Family	Approximate intrafmily weights
K	$K\alpha = 1$ $K\beta = 0.1$
L	$L\alpha = 1$ $L\beta_1 = 0.7$ $L\beta_2 = 0.2$ $L\gamma_1 = 0.08$ $L\gamma_2 = 0.03$
	$L\gamma_3 = 0.03$ $Ll = 0.04$ $L\eta = 0.01$
M	$M\alpha = 1$ $M\beta = 0.6$ $M\zeta = 0.06$ $M\gamma = 0.05$ $M_{II}N_{IV} = 0.01$

Thus, if K-family x rays of a heavy element are excited, L- and M-family x rays will also be present at the appropriate energies in the spectrum.

Weights of Lines. Although many possible electron transitions can occur in complex atoms to fill vacancies in an inner shell, giving rise to the families of x rays, the probability for each type of transition varies considerably. The relative transition probabilities for the lines arising from an ionization of a specific shell are termed the *weights of lines*. The weights of lines are dependent upon atomic number and vary in a complex fashion. The weights of K-lines, the best known, are plotted in Fig. 3.38, while values for L- and M-family x rays are generally poorly known. General values are presented in Table 3.11 for the lines of significant intensity that can be readily observed in energy-dispersive x-ray spectra. While these values may not be exact for a specific element, these weights are a useful guide in interpreting spectra and assigning peak identifications in energy-dispersive x-ray spectrometry. Note that the values in Table 3.11 are *intra*family weights and *do not* give *inter*family weights. Interfamily weights depend strongly on the exact conditions of electron-beam excitation, i.e., the ratio of the beam energy to the critical excitation energy for each shell, as described below.

Intensity of Characteristic X-Rays:

Cross-Section for Inner Shell Ionization. Numerous cross-sections for inner shell ionization can be found in the literature; these have been reviewed by Powell (1976a,b; 1990). The basic form of the cross-section is that derived by Bethe (1930):

$$Q = 6.51 \times 10^{-20} \frac{n_s b_s}{UE_c^2} \log_e(c_s U) \tag{3.34}$$

ionizations/e^-/(atom/cm^2), where n_s is the number of electrons in a shell or subshell (e.g., $n_s = 2$ for a K shell), b_s and c_s are constants for a particular shell, E_c is the critical ionization energy (keV) of the shell, and U is the overvoltage:

$$U = \frac{E}{E_c},$$

where E is the instantaneous beam energy. The cross-section is plotted as a function of overvoltage in Fig. 3.39. Based on evaluation of published experimental data, Powell (1976a; b) recommends values of $b_K = 0.9$ and $c_K = 0.65$ for the overvoltage range $4 < U < 25$. However, in the analysis of solid specimens, the large fraction of the beam electrons which lose all of their energy because of inelastic scattering inevitably interact with inner shells at overvoltages as low as $U = 1$. It is thus important to characterize the low-overvoltage portion of the cross-section, for which relatively little data is available. For such calculations, a value of c_s must be used for which the cross-section is defined down to $U = 1$. Specifically, this condition requires that $c_s \geq 1$ (e.g., $b_k = 0.3$ and $c_k = 1$, Green and Cosslett, 1961).

X-Ray Production in Thin Foils. Using the cross-section for inner-shell ionization as a starting point, x-ray production from a thin foil can be estimated. "Thin" is defined as a thickness which is small compared to the elastic mean free path, so that the average beam electron passes through the foil along its incident trajectory without significant deflection; the energy loss due to inelastic scattering is also negligible when the foil is thin. To convert the ionization cross section given in Eq. (3.34), which has dimensions of ionizations/e^-/(atom/cm^2), to x-ray production in units of photons/e^-, n_x, the following dimensional argument is used:

$$\frac{\text{photons}}{e^-} = \frac{\text{ionizations}}{e^-(\text{atom/cm}^2)}\,\frac{\text{photons}}{\text{ionization}}\,\frac{\text{atoms}}{\text{mole}}\,\frac{\text{moles}}{\text{g}}\,\frac{\text{g}}{\text{cm}^3}\,\text{cm}$$

$$n_x = Q\omega N_0 \frac{1}{A}\rho t, \tag{3.35}$$

where ω is the fluorescence yield, N_0 is Avogadro's number, A is the atomic weight, ρ is the density, and t is the thickness.

X-Ray Production in Thick Targets (Theoretical). A thick target is one for which elastic scattering and energy loss are significant, which occurs at a multiple of a few times the elastic mean free path, or about 100 nm for average conditions. Targets greater than 10 μm thick are effectively infinite in thickness to the beam electrons for most materials

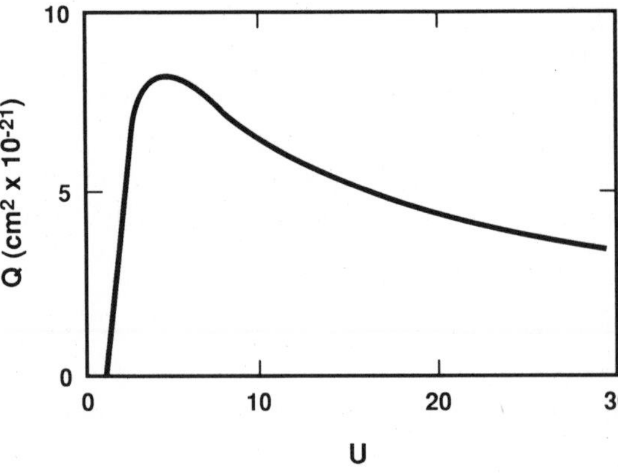

Figure 3.39. Plot of the cross-section for inner-shell ionization as a function of overvoltage $U = E/E_c$.

over the range of beam energies used in SEM/EMPA. To accurately calculate x-ray production for thick targets, the cross-section must be evaluated over the entire range of electron energy from E_0 to E_c. The thickness t in Eq. (3.35) is replaced by the infinitesimal path increment ds, and the bulk x-ray yield, I_c, is calculated by the integration along the Bethe range to the point where the electron energy falls below E_c for the line of interest:

$$I_c = \int_0^{R_B} \frac{Q_i \omega N_0 \rho}{A} ds = \frac{\omega N_0 \rho}{A} \int_{E_0}^{E_c} \frac{Q_i}{dE/ds} dE. \qquad (3.36)$$

The expression dE/ds in the denominator of the ratio in the second integral is the rate of energy loss, which can be expressed conveniently as a function by means of Eq. (3.6), the Bethe expression. A full discussion of Eq. (3.36), which forms the basis for quantitative x-ray analysis, will be given in Chapters 7 and 8. Some of the complications that arise when developing the corrections necessary for quantitative x-ray analysis can be anticipated by considering that Eq. (3.36) actually applies only to the fraction of beam electrons which lose energy down to $E = E_c$. A correction factor must be applied for those electrons which backscatter at energies $E > E_c$, carrying off energy that would have continued to contribute to the production of x rays. Since electrons can backscatter with any energy $E \leq E_0$, calculation of a backscatter correction factor itself involves a complicated integral and requires accurate values for the backscattered electron coefficient and energy distribution.

X-Ray Production in Thick Targets (Experimental). A number of workers have reported experimental measurements of the characteristic x-ray intensity I_c as generated (i.e., as produced in the specimen prior to absorption as the radiation propagates through the specimen) (Green, 1963; Lifshin *et al.*, 1980). The experimental expressions have the general form:

$$I_c = a\left(\frac{E_0 - E_c}{E_c}\right)^n = a(U - 1)^n, \qquad (3.37)$$

where a and n are constants for a particular element and shell. The results reported by Lifshin *et al.* for K-shell x-ray production ($K\alpha_{1,2} + K\beta$) are given in Table 3.12.

X-ray Peak-to-Background Ratio. An important factor in determining the limits of detection in spectrometric analysis is the presence of background, i.e., noncharacteristic radiation at the same energy as the characteristic radiation of interest. In x-ray spectrometry, the x-ray bremsstrahlung is the principal contributor to the background and is observed as a continuum from the incident beam energy down to zero energy. As described above, the characteristic intensity at a specific energy arising from an edge energy E_c is found experimentally to follow a relation of the form

$$I_c = i_B\left(\frac{E_0 - E_c}{E_c}\right)^n, \qquad (3.38)$$

Table 3.12. Experimental Measurements of X-Ray Yield[a]

Element	Z	n	Absolute efficiency (photons/e^-/sr)
Mg	12	1.42	0.114×10^{-4}
Si	14	1.35	0.316×10^{-4}
Ti	22	1.51	0.631×10^{-4}
Cr	24	1.52	0.741×10^{-4}
Ni	28	1.47	0.933×10^{-4}

[a] Lifshin *et al.* (1980).

where i_B is the beam current. The bremsstrahlung follows a relation of the form

$$I_{\mathrm{cm}} = i_B Z\left(\frac{E_0 - E_\nu}{E_\nu}\right), \tag{3.39}$$

where E_ν is the energy of the bremsstrahlung. The peak-to-background ratio is found by taking the ratio of Eqs. (3.38) and (3.39):

$$P/B = \frac{I_{\mathrm{c}}}{I_{\mathrm{cm}}} = \frac{\left(\dfrac{E_0 - E_{\mathrm{c}}}{E_{\mathrm{c}}}\right)^n}{Z\left(\dfrac{E_0 - E_\nu}{E_\nu}\right)}. \tag{3.40}$$

Assuming that $E_\nu \simeq E_{\mathrm{c}}$:

$$P/B = \frac{I_{\mathrm{c}}}{I_{\mathrm{cm}}} = \frac{1}{Z}\left(\frac{E_0 - E_{\mathrm{c}}}{E_{\mathrm{c}}}\right)^{n-1}. \tag{3.41}$$

From Eq. (3.41), the peak-to-background ratio increases, and the limit of detection, as expressed as a mass fraction, decreases, as the difference $(E_0 - E_c)$ increases. It would therefore seem to be advantageous to make E_0 as large as possible for a specific choice of E_{c}. However, as noted in the discussion of the interaction volume, the beam electrons penetrate deeper into the specimen as approximately $E_0^{1.7}$. X rays are therefore produced deeper in the specimen as the beam energy is increased. Besides degrading the depth and lateral spatial resolution of analysis, an important consequence of increasing the beam energy is an increased absorption of x rays within the specimen. This increased absorption reduces the measured x-ray intensity, degrades the limit of detection, and increases the uncertainty of the correction for absorption that must be applied in the quantitative analysis procedure. There is thus an optimum beam energy beyond which further increases in beam energy actually degrade analytical performance. This limit depends on the energy of the characteristic x rays and the composition of the specimen.

Depth of X-Ray Production (X-Ray Range). Depending on the critical excitation energy, characteristic x-rays may be generated over a substantial fraction of the interaction volume of the beam electrons in the solid, as shown in Figs. 3.7–3.9. To predict the depth of x-ray production (*x-ray range*) and the lateral x-ray source size (*x-ray spatial resolution*), the starting point is the electron penetration, as calculated with a range expression, such as the Kanaya–Okayama range. Electron range expressions have the following general form:

$$\rho R = KE_0^n, \tag{3.42}$$

where E_0 is the incident electron beam energy, ρ is the density, K depends on material parameters, and n is a constant in the range 1.2 to 1.7. This formulation of the range considers electrons which effectively lose all energy while scattering in the specimen. Characteristic x rays can be produced only within that portion of the electron trajectories for which the energy exceeds a specific value E_c while bremsstrahlung x rays continue to be produced until the photon energy of interest, E_v, is reached. The range of *direct* x-ray generation (as opposed to secondary fluorescence, which is the indirect generation of x rays by absorption of these primary x rays) is therefore always smaller than the electron range. To account for the energy limit of x-ray production, the range equation is modified to the form

$$\rho R = K(E_0^n - E_c^n). \tag{3.43}$$

If the range of production of bremsstrahlung x rays is required, the energy E_v may be substituted for E_c.

For the Kanaya–Okayama electron range, Eq. (3.16), K in the equation has the form

$$K = \frac{0.0276A}{Z^{0.889}}, \tag{3.44}$$

and thus depends on the atomic number and weight. Other authors (Reed, 1966; Anderson and Hasler, 1966; Heinrich, 1981) have set K equal to a constant without atomic number/weight dependence. The x-ray range of Andersen and Hasler (1966) has the form

$$\rho R = 0.064(E_0^{1.68} - E_c^{1.68}), \tag{3.45}$$

where R has the value of micrometers for E in keV and ρ in g/cm^3.

If we consider a Monte Carlo electron trajectory simulation, the Andersen–Hasler range is a measure of the region of x-ray generation, while the Kanaya–Okayama range gives the limiting envelope of the electron trajectories. For a beam normally incident on a surface, the maximum width of the electron interaction volume or x-ray generation volume projected on the surface is approximately equal to the appropriate range for electrons or x rays. Figure 3.40 shows the Andersen–Hasler x-ray range for the principal analytical lines of elements representing low (aluminum, 2.70 g/cm^3), intermediate (copper, 8.90 g/cm^3), and high (gold, 19.3 g/cm^3) atomic numbers. As the atomic number and density of the target increase, the depth of production for the principal line decreases. The depth of production is also a function of the critical ionization energy of the line. Figure 3.41

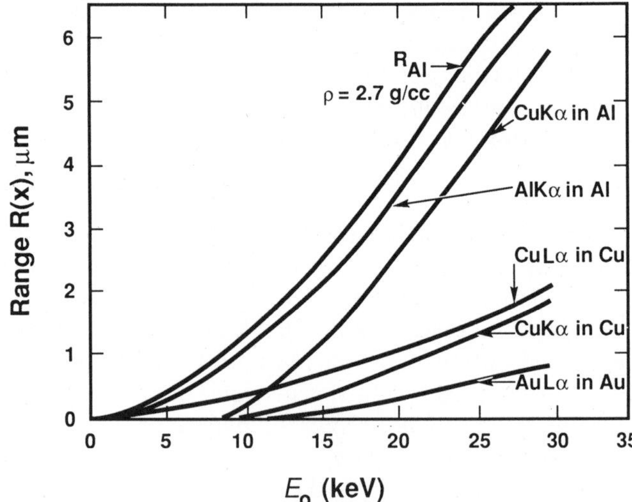

Figure 3.40. X-ray generation range for the Al $K\alpha$, Cu $K\alpha$, Cu $L\alpha$, and Au $L\alpha$ lines generated within aluminum, copper, and gold as a function of incident beam energy.

shows the electron range and the x-ray ranges for a variety of trace elements with various analytical line energies in a single matrix, iron. Note that for a specific choice for the incident electron energy, the x-ray range for high-energy x rays is only a small fraction of the electron range, while for low energy x rays, the x-ray range is almost as large as the electron range.

This discussion of concepts involved in defining the depth of x-ray production is summarized in Fig. 3.42, which shows a comparison of the depth of x-ray production at the same beam energy for aluminum $K\alpha$ and copper $K\alpha$ in a matrix of density $\sim 3\ \text{g/cm}^3$ (e.g., aluminum)

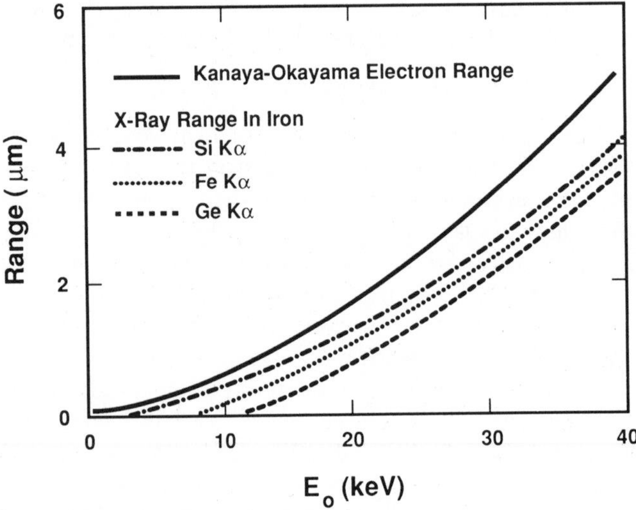

Figure 3.41. X-ray generation range (Andersen–Hasler) for Si $K\alpha$, Fe $K\alpha$, and Ge $K\alpha$ in an iron matrix as a function of beam energy.

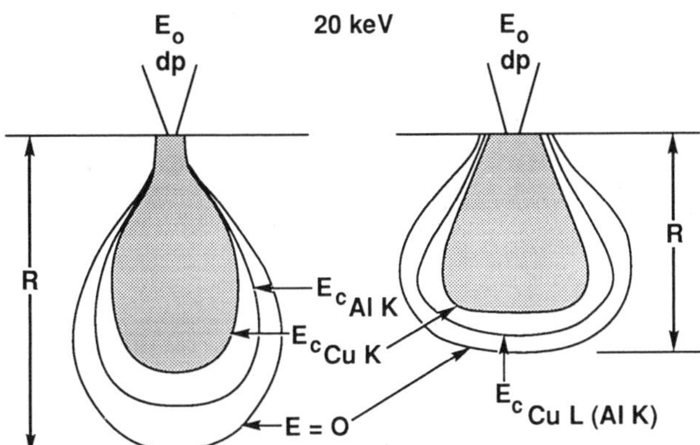

Figure 3.42. Comparison of x-ray production regions from specimens with densities of 3 (left) and 10 g/cm^3 (right).

and in a matrix of density $\sim 10\,\mathrm{g/cm^3}$ (e.g., copper, silver). Both Al Kα and Cu Kα are produced at greater depths in the low-density matrix than in the high-density matrix. The shapes of the interaction and x-ray generation volume differ considerably in the two targets, with the low-density matrix having a pear shape and the high-density matrix giving a less distinct neck.

The range of primary x-ray generation is the critical parameter in estimating the sampling volume for x-ray microanalysis. From Figs. 3.40–3.42, the sampling volume for x-ray microanalysis differs, depending on the energy of the x-ray lines measured. When two different analytical lines are available with widely differing energies, e.g., Cu $L\alpha$ (0.93 keV), and Cu $K\alpha$ (8.04 keV), the depth of the sampling volume may differ by a factor of 5 or more. If the specimen is heterogeneous in composition over a depth equivalent to the electron range, the effect of different sampling volumes depending on x-ray energy may lead to incorrect and even misleading results. Therefore, a fundamental requirement for accurate results by conventional quantitative x-ray microanalysis procedures is that the specimen must be homogeneous over the electron range.

Depth Distribution of X-Ray Production (ϕ (ρz)). While the single numerical value given by the x-ray range defines the effective spatial limit of x-ray production, even a cursory examination of the Monte Carlo electron trajectory–x-ray plots in Figs. 3.8–3.10 suggests that the depth distribution of x-ray production is not uniform within the x-ray range. The distribution is nonuniform both laterally and in depth. The lateral distribution, as projected on the surface plane, is important in defining the spatial resolution of x-ray microanalysis. The distribution in depth, which is projected on a plane perpendicular to the surface, is important because the x-rays produced below the surface must propagate through the atoms of the specimen to escape, and some fraction is lost because of photoelectric absorption, described below. The x-ray depth distribution

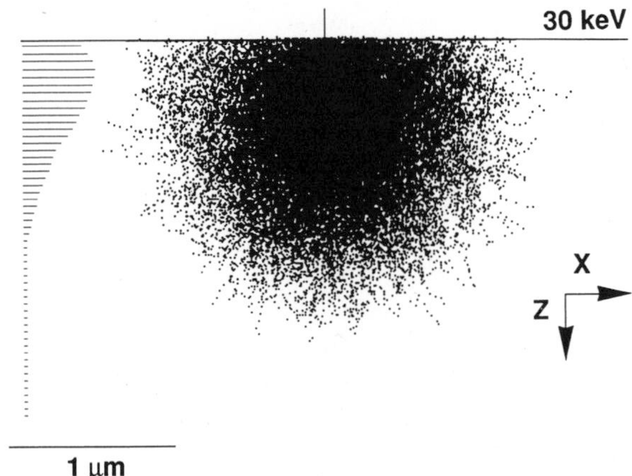

Figure 3.43. Concept of the x-ray depth distribution function $\phi(\rho z)$ as a histogram created by dividing the specimen into layers of thickness dz. The intensity in each layer is normalized by the intensity generated by the same beam in a freestanding layer dz in thickness and having the same composition.

function, an important relation in developing quantitative x-ray microanalysis correction procedures, is given the name $\phi(\rho z)$. As shown in Fig. 3.43, the $\phi(\rho z)$ function can be thought of as a histogram giving the relative number of x-rays generated in a slice of specimen of thickness dz. The histogram is normalized by the number of x rays the beam would produce in a freestanding layer of thickness equal to dz and of the same composition. $\phi(\rho z)$ curves can be calculated by Monte Carlo electron-trajectory simulation or measured experimentally by using a tracer layer buried below various thicknesses of the matrix element; results from both methods are shown in Fig. 3.44.

The $\phi(\rho z)$ function has several characteristic features. The numerical value of the first layer, which starts at the surface, is always greater than unity, since the x-ray production consists both of the contribution from the unscattered incident beam, identical to the production in the unsupported reference layer, and the x rays produced by the backscattering electrons propagating back through the surface to escape. For a specimen set normal to the beam, the shape of the $\phi(\rho z)$ function is such that, as the depth below the surface increases, the relative number of x rays produced in each layer increases, achieves a maximum, and then decreases, reaching zero at the x-ray range. The origin of this shape can be deduced from qualitative arguments. For the infinitesimal, unsupported reference layer dz, there is essentially no elastic scattering of the electrons, so the trajectory along which x rays can be generated is equal to dz. As the beam electrons penetrate into a bulk target, the effect of elastic scattering is to gradually cause the electron trajectories to deviate so that they cross through a layer of dz in thickness at an oblique angle, making the path along which x rays can be generated longer than dz. The x-ray signal, normalized to the freestanding reference layer, therefore increases. As we proceed deeper into the specimen, the cumulative

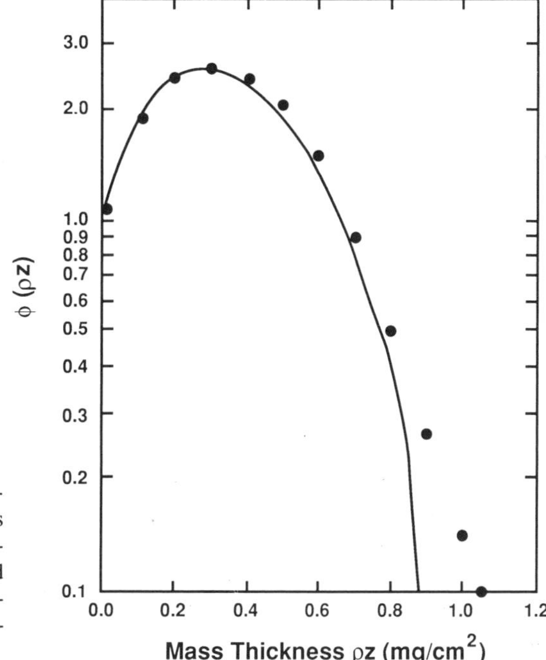

Figure 3.44. Depth distribution function $\phi(\rho z)$ of x rays produced in aluminum. Experimental data of Castaing and Henoc (1966), Monte Carlo calculation of Newbury and Yakowitz (1976).

deviation due to multiple elastic scattering events increases, and many electrons undergo sufficient change in direction to begin to propagate back toward the surface. These scattered electrons contribute extra x-ray production to the layers which they pass through for the second time, adding to the x-rays produced by the downward-propagating primary beam. However, the loss of electrons due to backscattering robs the downward-propagating beam of intensity. Moreover, the loss of energy due to inelastic scattering eventually reduces the capability for inner-shell ionization. As shown in Fig. 3.39, the cross-section for inner-shell ionization is a maximum at a value of approximately $U = 5$. From the combined effects of elastic and inelastic electron scattering and the behavior of the cross-section for inner-shell ionization, x-ray production eventually peaks and decreases at greater depths, eventually reaching zero when the electron energy falls below $U = 1$, a depth which corresponds to the x-ray range.

3.5.2.3. X-Ray Absorption

X rays as photons of electromagnetic radiation can undergo the phenomenon of photoelectric absorption upon interacting with an atom. That is, the photon is absorbed and the energy is completely transferred to an orbital electron, which is ejected with a kinetic energy equal to the photon energy minus the binding energy (critical ionization energy) by which the electron is held to the atom. X rays can also be lost due to scattering. For x rays of an incident intensity I_0

Table 3.13. X-Ray Energy and Mass Absorption Coefficients for Ni $K\alpha$ in Several Elements

Element (atomic number)	$K\alpha$	$K\beta$	E_{edge}	(μ/ρ) Ni $K\alpha$ (cm^2/g)
Mn (25)	5.895	6.492	6.537	344
Fe (26)	6.400	7.059	7.111	380
Co (27)	6.925	7.649	7.709	53
Ni (28)	7.472	8.265	8.331	59
Cu (29)	8.041	8.907	8.980	65.5

propagating through a slab of thickness t and density ρ, the intensity on the exit surface is attenuated according to the expression

$$\frac{I}{I_0} = \exp - (\mu/\rho)(\rho t), \tag{3.46}$$

where (μ/ρ) is the mass absorption coefficient (cm^2/g) of the sample atoms for the specific x-ray energy of interest. Some typical values of mass absorption coefficients in various pure elements for Ni $K\alpha$ (7.47 keV) are listed in Table 3.13. Extensive compilations are available in the literature (Heinrich 1966b; 1986). A selection of mass absorption coefficients for $K\alpha$ and $L\alpha$ characteristic lines is tabulated in Chapter 14.

Photoelectric absorption by electrons in a specific shell requires that the photon energy exceed the binding energy for that shell. When the photon energy is slightly greater than the binding energy, the probability for absorption is highest. For a specific absorber, mass absorption coefficients generally decrease in a smooth fashion with increasing x-ray energy, except in the energy region close to the critical excitation energy for each shell of the absorber. Figure 3.45 shows a plot of (μ/ρ) for nickel as an absorber of various x-ray energies. A sharp jump in (μ/ρ) occurs at the energy of the Ni K-edge, 8.331 keV; this jump is referred to as an *x-ray absorption edge.* X-rays with an energy slightly greater than the critical ionization energy can efficiently couple their energy to eject a bound electron and are therefore strongly absorbed. The mass absorption coefficient increases abruptly at the edge and behaves in a highly irregular fashion for an energy range spanning a few hundred electron volts above the edge, as shown in Fig. 3.46. This extended x-ray absorption fine structure (EXAFS) just above an absorption edge is sensitive to the electronic structure of an atom, which can reveal details of the chemical bonding and the crystallographic environment. EXAFS performed with tunable synchrotron x-ray sources is an important analytical technique (Teo and Joy, 1981). Above the EXAFS range, the mass absorption coefficient resumes a smooth decrease with increasing photon energy.

Absorption edges can be directly observed in x-ray spectra when an energy range spanning the critical excitation energy is examined. The electron-excited x-ray bremsstrahlung provides a source of radiation with

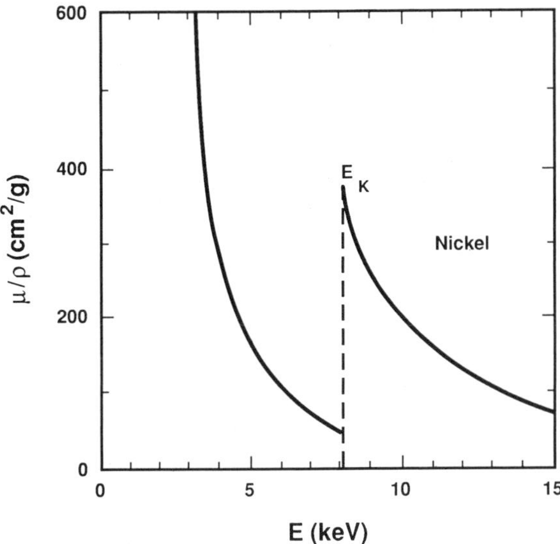

Figure 3.45. Plot of the mass absorption coefficient as a function of x-ray energy in nickel.

a continuous distribution in energy. In the region of an absorption edge, the bremsstrahlung intensity abruptly decreases for x-ray energies slightly above the edge, since the mass absorption coefficient increases. An example for nickel excited with a 40-keV electron beam is shown in Fig. 3.47. The Ni $K\alpha$ and Ni $K\beta$ lines are observed, and above the position of the critical excitation energy for the Ni K-shell, the x-ray bremsstrahlung is much lower than what would be expected if the bremsstrahlung below the peaks was extrapolated to higher energy.

The mass absorption coefficient for a sample containing a mixture of elements is found by taking the summation of the mass absorption

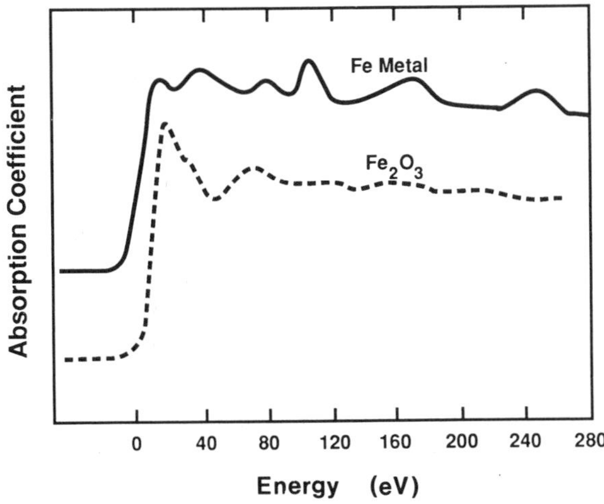

Figure 3.46. Plot of the mass absorption coefficient for iron immediately above the Fe L edge, showing the extended x-ray absorption fine structure in iron metal and in iron oxide (Fe_2O_3) (Nagel, 1968).

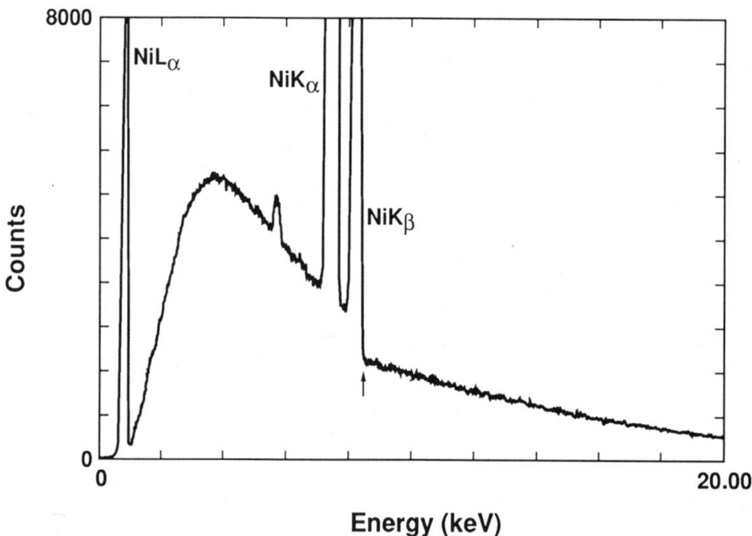

Figure 3.47. Energy-dispersive x-ray spectrum of nickel with a primary beam energy of $E_0 = 40\,\text{keV}$ showing a sharp step in the x-ray continuum background due to increased absorption just above the Ni K absorption edge.

coefficient for each element multiplied by its weight (mass) fraction:

$$\left(\frac{\mu}{\rho}\right)^i_{\text{spec}} = \sum_j \left(\frac{\mu}{\rho}\right)^i_j C_j, \tag{3.47}$$

where $(\mu/\rho)^i_j$ is the mass absorption coefficient for radiation from element i by element j and C is the weight (mass) fraction of element j.

It is important to note that x-ray photoelectric absorption is an "all or nothing" process. Either the photon is completely absorbed in a single absorption event or else it continues to propagate without modification of its energy. Characteristic x rays which escape the specimen thus retain their specific energy, which identifies the species of atom that emitted the photon. This is a critical point for analytical x-ray spectrometry.

There is a phenomenon of inelastic scattering of x-rays known as Compton scattering in which a significant amount of energy is lost by a photon in a collision with a free (i.e., not bound) electron. Because energy and momentum are conserved in the collision, both the recoil electron and the photon of lower energy propagate along new paths relative to the incident photon. For an initial wavelength λ, the Compton-shifted wavelength λ' is related by the expression:

$$\lambda' - \lambda = \frac{h}{m_e c}(1 - \cos\zeta), \tag{3.48}$$

where ζ is the angle between the original direction of propagation of the photon and the final direction after scattering. The energy loss of the scattered photon therefore depends on the angle relative to the incident photon path. Compton scattering and consequent loss of energy from the photon is a phenomenon that is readily observed in the technique of x-ray fluorescence (XRF) analysis, where a primary beam of characteristic x rays is used to excite the specimen to emit secondary x rays characteristic of the specimen atoms. The x rays emerging from the specimen are detected at a specific angle from the primary beam. Besides the x rays of interest emitted from the specimen atoms, the XRF spectrum also includes a significant peak due to elastically scattered primary x rays, and at a specific energy below this elastic peak is a "Compton peak" representing those primary photons that underwent Compton scattering in the specimen. A specific Compton energy loss is found as a result of collecting the x-ray spectrum at a specific angle from the primary x-ray beam. Since in electron-excited x-ray microanalysis the spectrum is collected at a specific angle (the take-off angle) from the specimen surface, and therefore from the primary electron beam, why then do we not find a Compton-shifted peak below every high-intensity peak in an electron-excited x-ray spectrum? Although Compton scattering of the x-rays propagating through the solid can occur, it is not an observable feature in the measured spectrum. First, the distance of propagation in the solid is only a few micrometers at most, and for the x-ray energies of interest, the cross-section for Compton scattering is much lower than for photoelectric absorption. Second, the electron-excited x-ray source in the solid is essentially isotropic with respect to the direction of emission, and the non-directionality of the source means that there is no specific Compton scattering angle selected by placing the x-ray detector at a specific take-off angle relative to the specimen surface. The Compton-scattered and energy-shifted x-ray can thus appear at any energy below the peak energy, and is thus lost in the bremsstrahlung background since Compton scattering is a low-probability event.

3.5.2.4. X-Ray Fluorescence

An inevitable consequence of photoelectric absorption of x-rays is the ejection of a bound inner-shell electron from the atom, leaving the atom in the same excited state as that produced by inelastic scattering of an energetic electron that causes inner-shell ionization. The subsequent de-excitation as the atom returns to the ground state by electron transitions is the same for both cases; the excited atom follows the same routes to de-excitation, producing either characteristic x rays or characteristic electrons (Auger electrons), with the relative abundance given by the fluoresence yield. X-ray induced emission of x rays is referred to as *x-ray fluorescence*. To distinguish the effects, *primary radiation* will refer to the direct production of x rays by the ionization of atoms by the beam electrons and *secondary radiation* will refer to the production of x rays by ionization of atoms by the higher-energy x rays. Secondary radiation can be created from two different sources, characteristic and continuum (bremsstrahlung) x rays. Since the x-ray photon causing fluorescence must have at least as much energy as the critical excitation energy of the atom in question, the energy of the secondary radiation will always be less than the energy of the photon which is responsible for ionizing the inner shell.

Characteristic Fluorescence. If a mixed sample consists of atom species A and B and the energy of the characteristic radiation from element A exceeds the critical excitation energy for element B, then characteristic fluorescence of B by the radiation of A will occur. The fluorescence effect depends on how closely the photon energy of A matches the critical excitation energy for B, with the maximum effect occurring when E_A just exceeds E_c for B. To examine this situation, consider a sample containing a sequence of transition elements, e.g., manganese, iron, cobalt, and nickel (Table 3.14). The critical excitation energy for manganese is lower than the $K\alpha$ energies for cobalt and nickel, and therefore characteristic manganese fluorescence will occur from these radiations. The $K\beta$ energies of iron, cobalt, and nickel exceed the critical excitation energy for manganese, and so these radiation sources also contribute to the secondary fluorescence of manganese. These arguments can be repeated for each element in the hypothetical specimen. As shown in Table 3.13, the situation for the generation of secondary fluorescence is different for each of the constituents.

For characteristic-induced fluorescence to be a significant effect, the primary radiation must be strongly absorbed; that is, the absorber must have a high mass absorption coefficient for the primary radiation. Examination of Tables 3.13 and 3.14 reveals how the absorption–fluorescence phenomena depend on proximity to the critical excitation energy. Ni $K\alpha$ radiation (7.47 keV) is strongly absorbed in iron ($Z = 26$, $E_c = 7.111$ keV, $\mu/\rho = 380$ cm^2/g), with consequent emission of Fe $K\alpha$ and $K\beta$, while in cobalt, which is only one atomic number higher, the absorption is reduced by more than a factor of 7 ($Z = 27$, $E_c = 7.709$ keV, $\mu/\rho = 53$ cm^2/g). The magnitude of the fluorescence effect from different radiation sources on a given element can be estimated by comparing the mass absorption coefficients. Consider a minor iron constituent in a nickel–copper alloy. The mass absorption coefficients of iron are 380 cm^2/g for Ni $K\alpha$ (7.47 keV) and 311 cm^2/g for Cu $K\alpha$ (8.04 keV). Both Ni $K\alpha$ and Cu $K\alpha$ cause fluorescence of iron, but Ni $K\alpha$ will have the greater effect.

Continuum Fluorescence. Characteristic-induced fluorescence can occur only if the radiation in question is more energetic than the

Table 3.14. Secondary Fluorescence in a Sample Containing Mn, Fe, Co, and Ni

Element	Radiation causing fluorescence
Mn	Fe $K\beta$, Co $K\alpha$, Co $K\beta$, Ni $K\alpha$, Ni $K\beta$
Fe	Co $K\beta$, Ni $K\alpha$, Ni $K\beta$
Co	Ni $K\beta$
Ni	none

critical excitation energy of the element of interest and is only significant if the characteristic energy is within approximately 3 keV of the critical excitation energy. On the other hand, the continuum (bremsstrahlung) radiation provides a source of x rays at all energies up to the incident beam energy. Continuum-induced fluorescence will therefore *always* be a component of the measured characteristic radiation from a specimen. The calculation of the intensity of continuum fluorescence involves integrating the contributions of all photon energies above the critical excitation energy for the edge of interest. Henoc (1968) has discussed the details of this calculation.

The ratio of continuum-induced fluorescence (*indirect radiation*) to the total x-ray production [electron (*direct radiation*) plus the indirect radiation] is designated S. Experimental values compiled by Green (1962; 1963) for the K and L_{III} edges are plotted in Fig. 3.48 (solid lines) and are compared with calculations by Green and Cosslett (1961) (dashed lines). The fraction of the total radiation contributed by continuum-induced fluorescence is negligible for low-atomic-number elements ($Z < 20$) but rises to approximately 10% for Cu Kα and to 30% for Au Lα.

Range of Fluorescence Radiation. The range of direct, electron-induced characteristic x-ray production is constrained to lie within the interaction volume of the electrons in the target. A good estimate of that range of production is given by Eq. (3.45). X rays have a much greater range in matter than do electrons, so the range of x-ray induced fluorescence is correspondingly much greater. Consider the case of a Ni-10% Fe alloy, where the Ni $K\alpha$ radiation can induce fluorescence of iron K-radiation. The range of Ni $K\alpha$ in the alloy can be calculated from Eq. 3.46 and 3.47 using the data in Table 3.12. Consider that the Ni $K\alpha$ radiation propagates uniformly in all

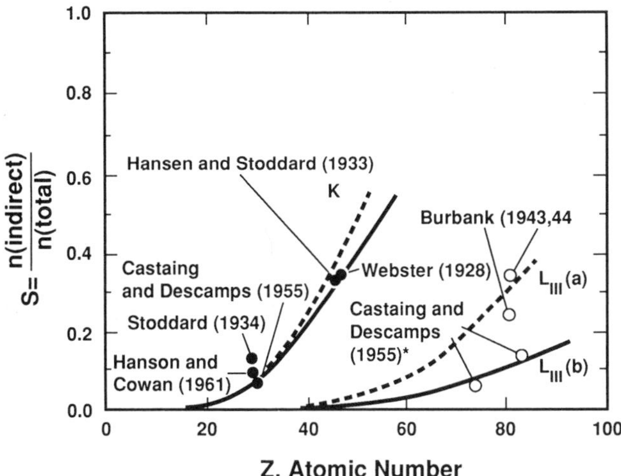

Figure 3.48. Variation of the ratio of indirect to total x-ray production with atomic number for the $K\alpha$ and $L\alpha$ lines (Green, 1963).

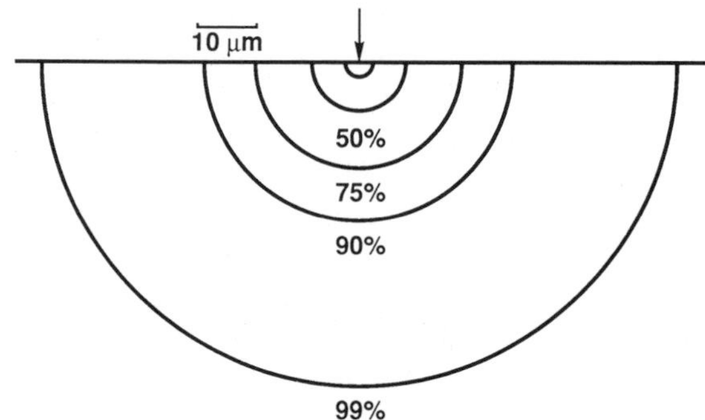

Figure 3.49. Range of secondary fluorescence of Fe $K\alpha$ by Ni $K\alpha$ in an alloy of composition Ni–10%Fe, $E_0 = 20$ keV. The innermost semicircle shows the extent of direct electron-excited production of Ni $K\alpha$ and Fe $K\alpha$ by the beam electrons.

directions from the source, which is the hemisphere containing the direct, electron-induced Ni $K\alpha$ production. Based on these calculations, the electron and x-ray ranges are compared in Fig. 3.49. The radius of the hemisphere which contains 90% of the Fe $K\alpha$ induced by Ni $K\alpha$ is approximately 10 times greater, and the volume approximately 1000 times greater, than for the hemisphere which contains all of the direct, electron-induced Fe $K\alpha$ production. The dimensions of the 50%, 75%, and 99% hemispheres are also shown in Fig. 3.49. Clearly, the x-ray induced fluorescence from both characteristic and continuum contributions originates in a much larger volume than the electron-induced characteristic radiation. This has the effect of degrading the spatial resolution of x-ray microanalysis.

3.5.3. Auger Electrons

When an atom is ionized in an inner shell, the subsequent intershell electron transitions to achieve de-excitation can result either in the emission of a characteristic x ray or in the ejection of another outer-shell electron which is designated an Auger electron. The ejected Auger electron has an energy which is characteristic of the atom, because the electron transitions occur between sharply defined levels. In Fig. 3.33, the particular Auger emission path depicted involves an original vacancy in the K-shell filled by a transition from the L-shell with subsequent ejection of another L-shell electron which is designated a KLL transition. The yield a of Auger electrons is given by $a = 1 - \omega$, where ω is the fluorescence yield for x rays. Thus, the yield of Auger electrons will be high for light elements such as carbon, where the fluorescence yield ($\omega_c < 0.01$) is low.

As an example, the Auger electron spectrum of silver is shown in Fig. 3.50 for an incident electron energy of 1 keV. In Fig. 3.50, the lowest spectrum shows the number of electrons per energy increment versus energy, $N(E)$ vs E. The features of the spectrum are a high and sharply

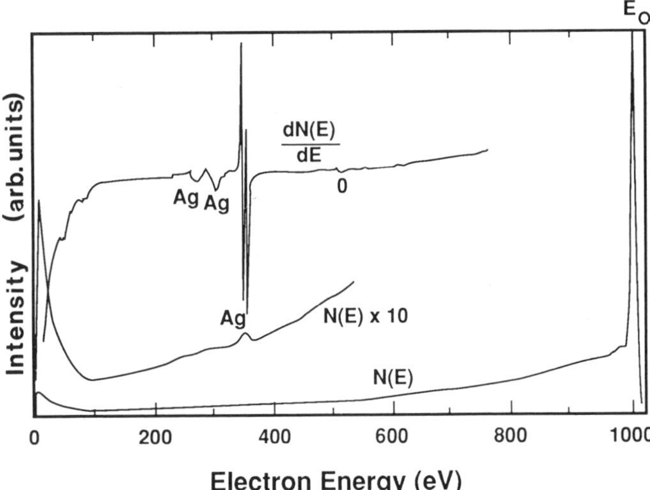

Figure 3.50. Auger electron spectrum of silver excited with an incident electron beam energy of 1 keV. Integral (shown with two different vertical expansions) and derivative representations are shown. (Courtesy of N. C. MacDonald.)

defined elastic backscattering peak at the incident energy superimposed on a broad and relatively featureless continuum of emitted electron energies. This continuum arises from backscattered electrons, fast secondary electrons, and Auger electrons that have lost energy due to inelastic scattering. A slight rise at very low energies, less than 50 eV, represents the slow secondary electrons. No Auger peak features are visible in this spectrum, displayed at a gain chosen so that the elastic peak is on scale. Expansion of the gain by a factor of 10 in the middle spectrum shows this characteristic Auger electron peak for silver associated with ionization of the silver M-shell. The peak even from a pure element only has a peak–background ratio slightly above unity. Because of this low peak-to-background ratio, it is common to differentiate the spectrum relative to energy, and this $dN(E)/dE$-versus-E spectrum is also shown in Fig. 3.50, where the Auger features in the spectrum are much more obvious. It is important to note that although the peaks are much more obvious in the $dN(E)/dE$ plot, such a differentiation operation does not actually increase the amount of information available. Unfortunately, the noise (random fluctuations in electron intensity) inherent in any counting operation is also enhanced by the differentiation operation, so that there is no additional information in the differentiated spectrum compared to the original spectrum.

Auger electrons and characteristic x rays can be measured simultaneously and in principle carry the same information about the composition of the specimen. A major difference does exist between the two characteristic signals because of profound differences in the depth of sampling. Since characteristic x-rays and Auger electrons both arise from the same inner-shell ionization events, the Monte Carlo plots (Figs. 3.7–3.10) showing the interaction volume and the sites of ionizations actually depict the distributions of the *generation sites* of both characteristic x rays and Auger electrons. The subsequent propagation of characteristic x rays and Auger electrons through the specimen to reach the surface occurs under radically different conditions. Inelastic scattering has a very low probability for characteristic x rays, and so x rays that are not totally absorbed by the photoelectric process reach the surface unchanged in energy and remain characteristic of the atoms from which they were emitted. Characteristic x rays thus sample the entire interaction volume within the specimen. Auger electrons, on the other hand, have a high probability of inelastic scattering and energy loss. For Auger electrons in the range 50 eV to 1 keV, the inelastic mean free path is on the order of 0.1–3 nm. As an Auger electron propagates through the specimen, it rapidly loses energy and therefore cannot be used to identify the emitting atom. Only Auger electrons lying within a thin surface layer (approximately 1–2 nm thick) can escape the specimen unchanged in energy and thus remain characteristic of the emitting atoms. Auger electrons formed deeper in the interaction volume may still escape the specimen, but they will have lost an indeterminate amount of energy and be unrecognizable in the background. Auger electron spectroscopy thus provides a form of surface microanalysis with a sampling depth on the order of 1 nm, regardless of the depth dimensions of the interaction volume, whereas the characteristic x-ray sampling depth for the same primary electron beam energy may be 1 μm or greater. The minimum lateral spatial resolution of Auger microanalysis is determined in part by the size of the surface region through which the backscattered electrons are emitted, since the backscattered electrons can create measurable Auger electrons as they exit.

3.5.4. Cathodoluminescence

When certain materials, such as insulators and semiconductors, are bombarded by energetic electrons, long-wavelength photons are emitted in the ultraviolet, visible, and infrared regions of the electromagnetic spectrum. This phenomenon, known as cathodoluminescence, is strong enough in certain minerals, such as benitoite ($BaTiSi_3O_9$), to be directly observed by the human eye in an electron-beam instrument equipped with an optical microscope. Cathodoluminescence can be understood in terms of the electronic band structure of the solid, shown schematically in Fig. 3.51. Such materials are characterized by a filled valence band, in which all possible electron states are occupied, and an empty conduction band; the valence and conduction bands are separated by a band gap of forbidden states with an energy separation of E_{gap}. When an energetic beam electron scatters inelastically in such a solid, electrons from the filled valence band can be promoted to the conduction band each leaving a hole (i.e., the absence of an electron) in the conduction band which creates an electron–hole pair. In cadmium sulfilde and silicon, the gap energies are 2.4 eV and 3.6 eV, respectively. If no bias exists on the sample to sweep the electron–hole pair apart, the electron and hole

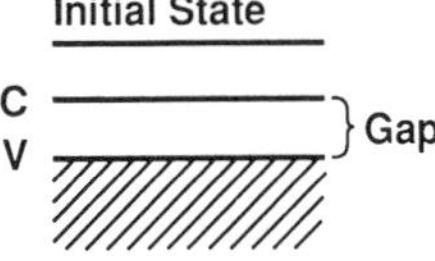

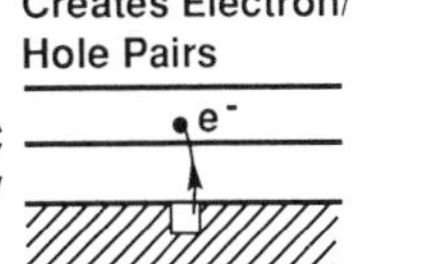

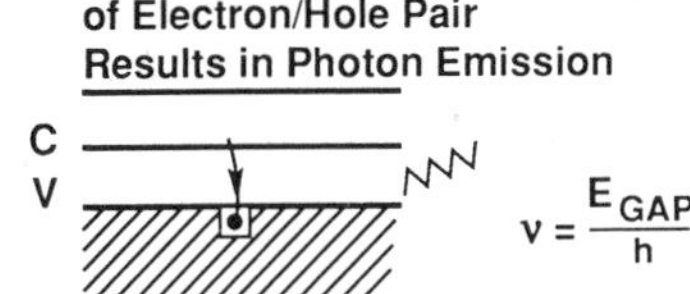

Figure 3.51. Schematic illustration of the process of emission of cathodoluminescent radiation from electron–hole pair formation and recombination.

recombine. The excess energy, equal to the gap energy, is released in the form of a photon of electromagnetic radiation. Since the band gap is well defined, the radiation peaks at certain energies and is characteristic of the emitting compound (note the emphasis on *compound* rather than *elemental* specificity). When impurity atoms are present, additional states become available, and more transitions (photon energies) become possible. Spectra of cathodoluminescence radiation emitted from GaAlAs are shown in Fig. 3.52. The decay of the electron–hole pair can be modified by the presence of impurity atoms or physical crystal defects such as dislocations, leading to shifts in the energy and intensity of the radiation, as well as the time constant of the decay.

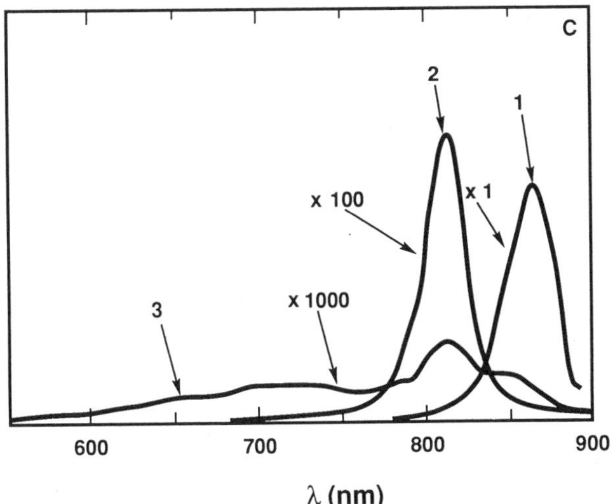

Figure 3.52. Spectra of cathodoluminescence emission from semiconductor GaAlAs: (1) bulk material; (2) thin material; (3) thin material in the presence of lattice defects; temperature = 25° C, E_0 = 150 keV (Petroff *et al.*, 1978).

3.5.5. Specimen Heating

A significant fraction of the energy of the primary beam is eventually transformed into heat in the specimen. For a fixed beam energy and current, the power absorbed by the specimen decreases as the atomic number of the specimen increases, since the higher backscattering coefficient of high atomic number materials means that more of the beam energy is lost with the backscattered electrons. Kanaya and Ono (1984) have calculated the temperature rise in the immediate beam impact area (a cylindrical volume defined by the incident beam area and a specimen depth of 5 nm) as a function of beam energy, current, dwell time, and sample composition and configuration. In general, for a 10 pA/20 keV/5 nm diameter beam incident on a bulk target, the temperature rise is approximately 10°C for carbon, and less than 1°C for metallic targets such as gold and tungsten. At the higher beam currents, >10 nA, necessary for x-ray microanalysis with wavelength diffraction spectrometers, the temperature increase can be between 10 and 100°C in the immediate beam impact area. If the conduction of heat away from the region of maximum energy deposition by the beam is limited because of specimen thermal conductivity or geometric configuration, then the temperature rise may be much greater, leading to melting or evaporation.

3.6. Summary

The concepts of the interaction volume of the primary beam electrons and the sampling volume of the emitted secondary radiation are important both in the interpretation of SEM images and in the proper application of quantitative x-ray microanalysis. These concepts are summarized in Fig. 3.53, which schematically shows the principal

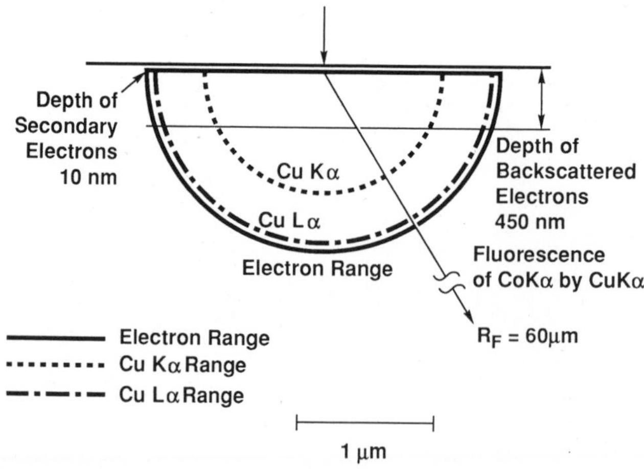

Figure 3.53. Schematic illustration of the electron-beam interaction in a copper–10% cobalt alloy, showing the electron interaction volume, the backscattered-electron sampling depth, the secondary-electron sampling depth, and the x-ray generation range for Cu $K\alpha$, Cu $L\alpha$, and Co $K\alpha$; E_0 = 20 keV.

interactions and products for a target of intermediate atomic number (copper—10% cobalt) and a typical beam energy used for microscopy and microanalysis ($E_0 = 20\,\text{keV}$). The points to note in Fig. 3.53 include the following:

1. The limiting envelope of the interaction volume for the primary beam is approximated as a hemisphere with a radius equal to the Kanaya–Okayama range.
2. The sampling volume for the backscattered electrons is shown as a side view of a disk with a depth of 0.3 of the Kanaya–Okayama range. Note that the diameter of this disk is determined by the lateral extent of the interaction volume.
3. The sampling volume of the secondary electrons is shown as a side view of a disk with a thickness of 10 nm and a diameter equal to that of the backscattered electron disk.
4. The x-ray generation ranges for Cu Kα, Co $K\alpha$, and Cu L are calculated from the Andersen–Hasler x-ray range equation. Since the self-absorption of copper for its own Cu $K\alpha$ x rays is small, the sampling volume is nearly the same size as the x-ray generation range. Cu L radiation is much more strongly absorbed. Even though the x-ray generation range is larger for Cu L than for Cu $K\alpha$, the sampling volume as observed by an external spectrometer will be much smaller for Cu L. The x-ray generation range for Co $K\alpha$ is slightly larger than that for Cu $K\alpha$, since the critical excitation energy is lower for Co.
5. The range of characteristic fluorescence of Co $K\alpha$ ($E_c = 7.71\,\text{keV}$) by Cu $K\alpha$ ($E = 8.04\,\text{keV}$) for 99% of the total fluorescence generation is about 60 μm, a factor approximately 50 times greater in linear dimension and 125,000 times greater in volume than the range of direct electron excitation of Co $K\alpha$. The fluorescence range cannot be directly shown in Fig. 3.53 because of the scale of the drawing.
6. Figure 3.53 can be described as only approximate, since only the limiting spatial envelopes of the various processes are shown. The volume density of each process is not constant with position. Secondary-electron signals decrease exponentially with distance from the surface. Backscattered electrons initially rise with increasing depth, reach a peak, and then decrease to zero. X-ray signals show a similar behavior, described by the $\phi(\rho z)$ curve, increasing to a peak at about 0.3 of the Kanaya–Okayama electron range and then undergoing an exponential decrease in intensity with further increases in depth.

4

Image Formation and Interpretation

4.1. Introduction

This chapter will consider the formation and interpretation of SEM images. One of the most surprising aspects of scanning electron microscopy is the apparent ease with which SEM images of three-dimensional objects can be interpreted by any observer, including young children with no prior knowledge of the instrument. This aspect of the SEM is often taken for granted, and yet it is one of the most important reasons for the great utility and wide acceptance of the instrument. SEM images are routinely presented in textbooks and popular scientific articles with little or no mention of the type of microscopy employed in preparing the image or of the complex way in which the image was constructed. It can safely be assumed that the reader will automatically perceive the true nature of the specimen without any instruction on the origin of the image. For this to be true, the SEM imaging process must in some way mimic the natural experience of human observers in visualizing the world around them. Such a situation is somewhat surprising in view of the unusual way in which the image is formed, which seems to differ greatly from normal human experience with images formed by light and viewed by the eye. In the SEM, high-energy electrons are focused into a fine beam, which is scanned across the surface of the specimen. Complex interactions of the beam electrons with the atoms of the specimen produce a wide variety of radiation products. A mix of this radiation is collected by a detector, most commonly the Everhart–Thornley scintillator–photomultiplier detector, and the resulting signal is amplified and displayed on a cathode ray tube or television screen scanned in synchronism with the scan on the specimen. Despite this complicated and unusual path to the formation of the image, the result is somehow straightforward to interpret, at least for the important class of objects that can be described as topographic (three-dimensional) in nature and are viewed at low to intermediate magnifications (up to 10,000 diameters).

While this "look–see" approach to scanning electron microscopy will suffice for some problems, the experienced microscopist frequently discovers situations in which it is necessary to make use of advanced concepts of image interpretation. Such situations are encountered with special types of specimens and contrast mechanisms or in the case of high-magnification images (>10,000 diameters), where questions of resolution limits must be addressed. The intellectual tools which the microscopist must bring to bear on the imaging problem include knowledge of electron optics, beam–specimen interactions, detector characteristics, and signal/image processing to make a proper interpretation. The first two topics, electron optics and beam–specimen interactions, were covered in detail in Chapters 2 and 3, while the remaining topics will be the subject of this chapter.

Among the questions which constantly arise in microscopy studies are what types of features can be seen in an image (calling for information about the specimen), and how small a region showing a particular characteristic can be detected (calling for information about resolution)? These questions of information and resolution are closely related, and in this chapter equations will be developed that permit the microscopist to make an estimate of what can be seen. That is, the inevitable limitations on the performance of the SEM can be estimated numerically. Recognizing that such limitations exist and understanding how to optimize instrument parameters to achieve the best performance are critical to advanced use of the SEM.

Digital imaging has become a major development in SEM imaging. While it is still true that most images are viewed by the conventional analog display process, the use of "real time" digital frame stores is rapidly increasing. Comments will be made throughout this chapter on the special value of computer-aided digital imaging.

4.2. The Basic SEM Imaging Process

4.2.1. Scanning Action

The detailed description of the electron optics of the SEM given in Chapter 2 can be briefly summarized by considering the degree of control which the microscopist has in choosing the parameters that characterize the beam as focused to form a probe at the specimen plane: probe diameter, d, which ranges from a minimum of 1 nm in the highest-resolution instruments (typically limited to 5 nm in conventional instruments) to a maximum of 1 μm; probe current, i_B(pA to μA); and probe convergence $\alpha(10^{-4}$ to 10^{-2} radians). Of course, these three parameters cannot be selected independently because they are interrelated, as described by the brightness equation (equation 2.3). The brightness equation gives the theoretical limit of the beam

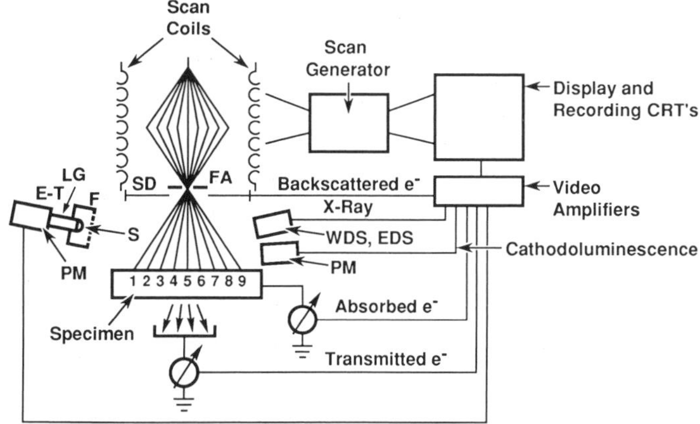

Figure 4.1. Schematic illustration of the scanning system of the SEM. Abbreviations: FA, final aperture; SD, solid-state backscattered-electron detector; EDS, energy-dispersive x-ray spectrometer; WDS, wavelength-dispersive x-ray spectrometer; CRTs, cathode ray tubes, and E–T, Everhart–Thornley secondary/backscattered–electron detector, consisting of F, Faraday cage; S, scintillator; LG, light guide; and PM, photomultiplier. Successive beam positions are indicated by the numbered rays of a scanning sequence.

performance; lens defects (aberrations) result in beams that are either larger than expected for a given beam current and convergence or lower in current for a specified diameter and convergence.

The SEM imaging process is illustrated schematically in Fig. 4.1. The electron beam, defined by the parameters d, i_B, and α, leaves the electron column, enters the specimen chamber, and strikes the specimen at a single location on the optic axis of the column. The beam electrons interact both elastically and inelastically with the specimen, forming the limiting interaction volume from which the various types of radiation emerge, including backscattered, secondary, and absorbed electrons, characteristic and bremsstrahlung x rays, and in some materials, cathodoluminescence radiation (long-wavelength photons in the ultraviolet, visible, and infrared). By measuring the magnitudes of these signals with suitable detectors, a determination of certain properties of the specimen, e.g., local topography, composition, etc., can be made at the single location where the electron beam strikes. In order to study more than a single location and eventually construct an image, the beam must be moved from place to place by means of a scanning system, as illustrated in Fig. 4.1. Scanning action is usually accomplished by energizing electromagnetic coils arranged in sets consisting of two pairs, one pair each for deflection in the X and Y directions. (By convention, the Z direction of the X–Y–Z coordinate system runs parallel to the optic axis of the microscope, while the X and Y axes define a plane perpendicular to the beam.) A typical double-deflection scanning system, as shown in Fig. 4.1, has two sets of electromagnetic scan coils, located in the bore of the final (objective) lens. As controlled by the scan generator, the upper coils act to drive the beam off-axis and the lower coils deflect the beam to cross the optic axis again, with this second crossing of the optic axis taking place in the final (beam-defining) aperture. This system has the advantage that because the scan coils are within the lens, the region below is kept open so that the specimen can be placed close to the lens at a short working distance, minimizing the spherical aberration coefficient and improving the beam characteristics. By locating the beam-defining aperture at the second crossover, large scan angles necessary for low magnifications can be obtained without cutting off the field of view on the aperture (Oatley, 1972).

Scanning action is produced by altering the strength of the current in the scan coils as a function of time, so that the beam is moved through a sequence of positions on the specimen (e.g., locations 1, 2, 3, 4, etc. in Fig. 4.1) and each detector samples the electron–specimen interaction at a defined sequence of points. In an analog scanning system, the beam is moved continuously, with a rapid scan along the X-axis (the line scan), and a slow scan, typically at 1/500 of the line rate, at right angles along the Y-axis (the frame scan), so that a good approximation to an orthogonal scan is produced. In digital scanning systems, only discrete beam locations are allowed. The beam is addressed to a particular location (X, Y) in a matrix, remains there for a fixed time (the dwell time), and is then sent to the next point. The time to shift the beam between points is negligible compared to the dwell time. The image is constructed on a cathode ray tube (CRT) scanned in synchronism with the scan of the specimen, controlled by the same scan generator. The signal derived from one of the detectors is amplified and used to control the brightness of the CRT (*intensity modulation*), often with some form of signal processing applied to enhance the visibility of the features of interest, as will be covered in detail below. At the resolution at which a human observer usually examines or photographs the CRT image, there is no effective difference between images prepared with analog and high-density (512×512, 1024×1024, or higher) digital scans; both types of images appear continuous. An added benefit of the digital scan is that the numerical address of the beam location is accurately and reproducibly known, and therefore the information about the electron interaction can be digitally encoded in the form (X, Y, I), where X and Y give the address and I is the intensity. Such encoding provides the basis for digital image processing, to be discussed later in this chapter.

4.2.2. Image Construction (Mapping)

The information flow from the scanning electron microscope consists of the scan location in X–Y space and a corresponding set of intensities from the set of detectors (backscattered-electron, secondary-electron, transmitted-electron, specimen-current, x-ray, cathodoluminescence, etc.) that monitor the beam–specimen interactions. This information can be displayed to the observer in two principal ways.

4.2.2.1. Line Scans

In the line-scan mode, the beam is scanned along a single vector on the specimen, e.g., in the X or Y direction. The same scan generator signal is used to drive the horizontal scan of the CRT. The resulting synchronous line scan on the specimen and the CRT produces a one-to-one correspondence between a series of points in the specimen space and on the CRT or *display space*. In such a line scan displayed on a CRT, the horizontal position is related to distance along a particular line on the specimen. The effective magnification factor M between the specimen space and the CRT space is given by the ratio of the lengths of the scans:

$$M = \frac{L_{\mathrm{CRT}}}{L_{\mathrm{spec}}}, \tag{4.1}$$

where L denotes the length of the scan.

The intensity measured during the line scan by one of the detectors, e.g., the signal from an x-ray spectrometer or a backscattered-electron detector, can be used to adjust the Y-deflection (Y-modulation) of the CRT, which produces a trace such as that illustrated in Fig. 4.2. For example, if the signal is related to the characteristic x-ray intensity measured by an x-ray spectrometer, the Y-deflection of the CRT is proportional to the amount of a specific element present at each location. Although line scans are the simplest type of scanning action, they are almost never used alone and are typically superimposed on the more familiar two-dimensional SEM image. Line scans are extremely useful for diagnostic work, where the signal profile across a feature is needed. Line scans can be used to display small signal changes that can easily be detected as offsets in

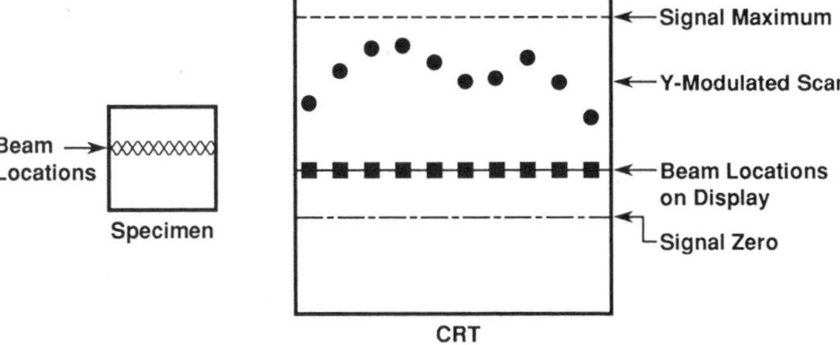

Figure 4.2. Forming a line scan: the beam is scanned along a locus of points on the specimen, and the beam on the CRT is scanned along a similar locus parallel to the horizontal (x-axis). A signal measured from one of the detectors (backscattered electrons, x-rays, etc.) is used to adjust the vertical position (y-axis). For quantitative measurements, the trace corresponding to the locus of points should be recorded, along with traces corresponding to the signal zero and maximum values.

Y-modulation but which would be difficult to discern in a conventional intensity-modulated area image. In recording a line scan, as shown in Fig. 4.2, it is important to record

1. the scan locus (i.e., the location in the raster at which the line scan is taken),
2. the Y-modulated scan of the signal,
3. the signal zero level and
4. the signal maximum.

4.2.2.2. Image (Area) Scanning

To form the SEM image with which we are most familiar, the beam is scanned on the specimen in an $X-Y$ pattern while the CRT is scanned in the same $X-Y$ pattern, as illustrated in Fig. 4.3. Again, a one-to-one correspondence is established between the set of beam locations on the specimen and the points on the CRT, and the linear magnification of the image is given by Eq. (4.1). To display the beam–specimen interaction information, the signal intensity S derived from a detector is used to adjust the brightness of the spot on the CRT (intensity modulation, shown schematically in Fig. 4.4.) Typically, an SEM provides one or more slow-scan CRT visual displays, a photographic CRT, and a television-rate display. On some instruments, one CRT visual display combines both slow-scan and rapid-scan capabilities. The television-rate display operates with a rapid-decay phosphor that can respond to scan speeds of 0.033 s per frame, which the observer perceives as flicker free. The visual-display CRT contains a phosphor with a relatively long persistence (usually one second or more) to permit the observer to view the image with scan speeds in the range 0.5–5 s. Finally, a high-resolution (small spot size) CRT with a short-persistence phosphor is used for photographic recording, almost invariably with rapid-development film.

To create an SEM image, one must construct an intensity map in

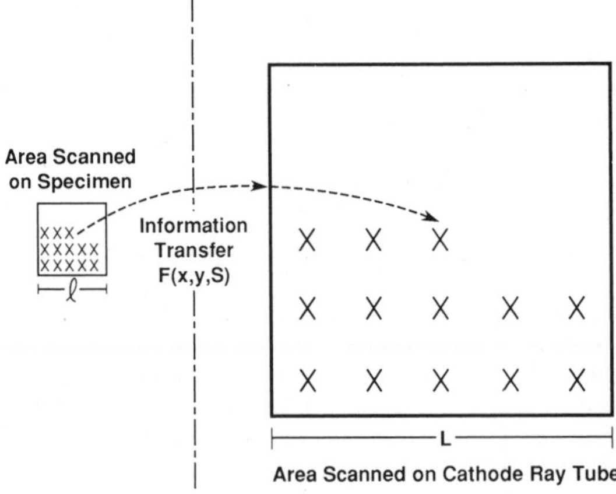

Figure 4.3. The principle of image display by area scanning. A correspondence is established between a set of locations on the specimen and one on the CRT. Magnification = L/l.

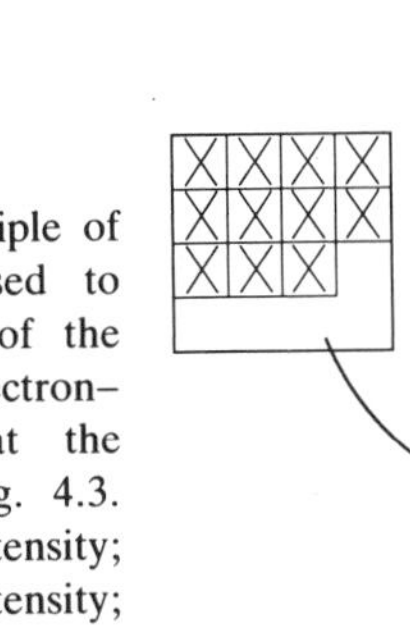

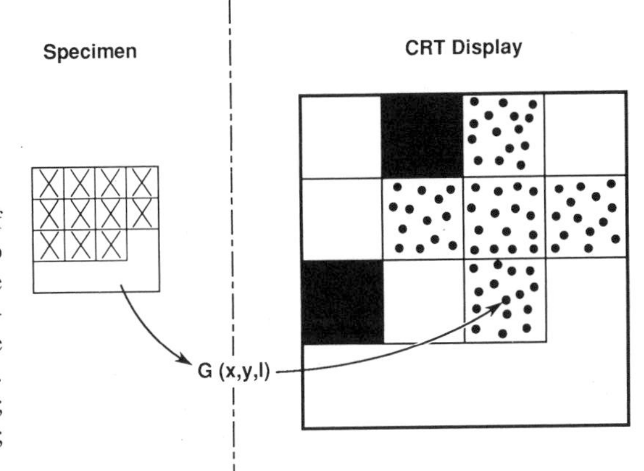

Figure 4.4. The principle of intensity modulation used to display the magnitude of the signal produced by electron–specimen interaction at the locations scanned in Fig. 4.3. Black represents low intensity; stippled, intermediate intensity; white, high intensity.

the analog or digital domain. Unlike an optical microscope or transmission electron microscope, no true image exists in the SEM. In a true image, actual ray paths connect points on the specimen to corresponding points in the image as displayed on a screen or detected by the eye or film. It is not possible to place a sheet of film anywhere in the SEM and record an image, as it is in a light-optical or transmission electron microscope. In the SEM, the image is produced by the mapping operation which transmits information from the specimen space to the display space. That such an abstract process of creating an image can produce readily understandable images of topographic objects is a considerable surprise, fortunately a pleasant one. The information contained in the image can convey the true shape of the object if the specimen and CRT scans are synchronous and are contrived to maintain the geometric relationship of any arbitrarily chosen set of points on the specimen and on the CRT. When this condition is satisfied, as shown in Fig. 4.5, a triangle on the specimen produces a triangle of the same shape on the CRT. The

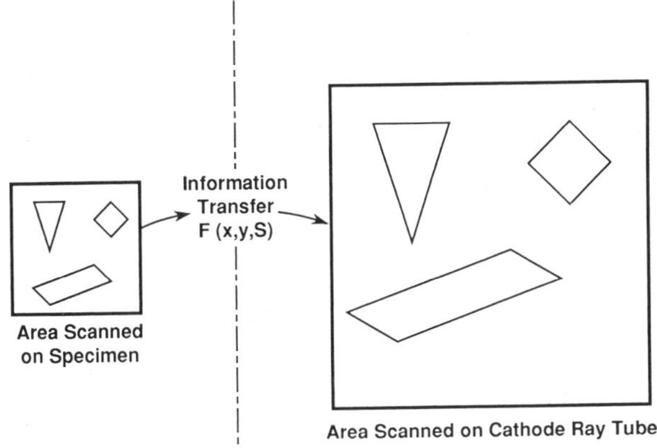

Figure 4.5. Shape correspondence between the scanned field on the specimen and the display on the CRT. In a perfect scan system, the shapes of objects in the plane of the scan in the microscope are transferred without distortion to the CRT display.

influence of projection distortion and various scan defects on the shape of objects in SEM images will be described below, and the nature of the intensity variations which produce the gray-level shading of various objects in the image will be discussed in Section 4.4.

4.2.2.3. Digital Imaging: Collection and Display

A digital image consists of a numerical array (X, Y, I) in a computer memory, each entry of which consists of three values, two for the position and one for the signal intensity. To create a digital image, the X–Y pattern of scanned locations is made to correspond to a matrix array in a computer memory or in a framestore (a dedicated computer memory board consisting of an array of memory registers adequate to record one or more complete images). Each scan position is generated as a digital address and converted by a digital-to-analog converter (DAC) into a voltage to drive the scan circuit. The number of discrete positions is specified as the digital resolution, e.g., 512×512, 1024×1024, etc. The pattern does not have to be square; rectangular or other scan shapes can be generated. When the beam has been addressed on the specimen, the analog signal intensity is measured by the detector(s). The analog signal amplifier is adjusted to bring the range of the signal into the input acceptance range of the analog-to-digital converter (ADC). The voltage signal produced by the signal amplifier is digitized to a value I for the computer and stored as a discrete numerical value in the corresponding register (X, Y, I). Typically, the intensity is digitized into a minimum of 8 bits, which gives $2^n = 2^8 = 256$ discrete levels.

With only 256 levels in the digital representation of the signal, it is possible to recognize changes only at the level of $1/256 = 0.4\%$ between subsequent digital functions, given that the input signal spans the entire range of the DAC. Generally, to avoid the possibility of saturating the response of the DAC, the input signal range is adjusted more conservatively so that the signal does not reach the 0 or 255 digital levels. Note that once the digitization has been performed, it is not possible to recover information not recorded in the original digitization step. For some applications, it is desirable to provide more discrete digital levels for subsequent processing, so the original signal can be digitized to 12 bits (4096 levels) or 16 bits (65,536 levels). This additional signal is gained by digitizing for longer integration times. Generally, we do not wish to record all images at the maximum digital resolution available. The penalty for recording large digital images is the need to provide mass storage to save the images. A $1024 \times 1024 \times 256$-level image requires over 1 megabyte of storage. The choice of the digital X–Y resolution should be made to slightly oversample the image in terms of the picture element (see Section 4.2.4), and the digital resolution of the intensity depends on the need for subsequent contrast manipulation.

The digital image is viewed by converting the numerical values stored in memory back into an analog signal (digital-to-analog conversion) for display on a CRT. The density and speed of computer memory is such that large-image arrays (512 × 512 or larger) can be stored digitally and read out in real time, producing a "flicker-free" image. Such digitally stored images have many advantages. One special advantage is the possibility of minimizing the radiation dose necessary to view the specimen. In order to continuously view an image with an analog SEM system, the beam must be continuously scanned on the specimen in order to refresh the display. The information in each scan is discarded unless recorded by film or videotape. Radiation damage or beam-induced contamination is generally proportional to the total dose, so that analog imaging with repeated scans exposes the specimen to possible degradation. Digital images are stored in memory and can be repeatedly read out to the analog display for inspection of that particular frame. Thus, the beam can be scanned for a single frame on the specimen to accumulate data in a memory and then "blanked" or deflected off the specimen, generally into a Faraday cup for measurement of the current as a stability check, while the digital memory is repeatedly read to refresh a CRT display. As described in detail below, image-processing functions can be applied to this stored image to enhance the visibility of objects of interest. Once the displayed image has been satisfactorily adjusted, the resulting image can be transmitted to the photographic-recording CRT or other form of hard-copy recording device.

4.2.3. Magnification

The magnification of the SEM image is changed by adjusting the length of the scan on the specimen for a constant length of scan on the CRT, as given by Eq. (4.1), which gives the linear magnification of the image. There are several important points about the SEM magnification process:

Numerical Value of Magnification. Since the maximum CRT scan length is fixed to the full dimension L of the tube, e.g., 10 cm, an increase in magnification is obtained by *reducing* the length of the scan on the specimen. Table 4.1 gives the size of the area sampled on the specimen as a function of magnification.

Area Sampled. The capability of the SEM to obtain useful images

Table 4.1. Area Sampled as a Function of Magnification[a]

Magnification	Area on sample
10X	$(1\ \text{cm})^2$
100X	$(1\ \text{mm})^2$
1,000X	$(100\ \mu\text{m})^2$
10,000X	$(10\ \mu\text{m})^2$
100,000X	$(1\ \mu\text{m})^2$
1,000,000X	$(100\ \text{nm})^2$

[a] Assumes CRT screen measures 10 cm × 10 cm.

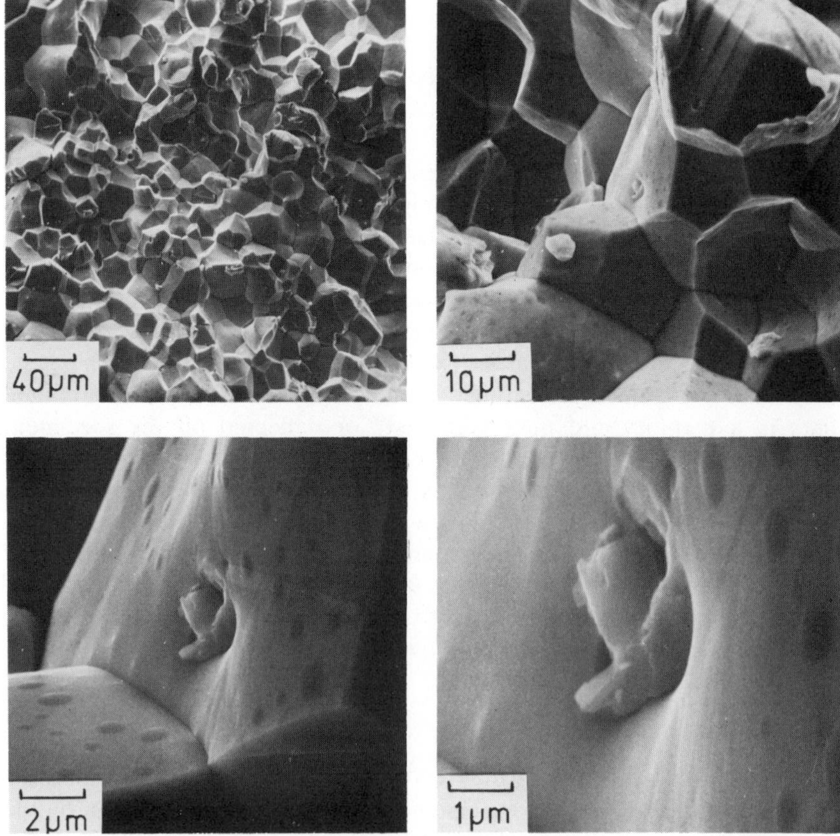

Figure 4.6. Magnification series of a fracture surface of high-purity iron, illustrating the rapid surveying capability of the SEM. The images are recorded at constant objective-lens strength and working distance. Note that focus is maintained throughout the series and there is no image rotation.

at high magnification sometimes leads to a "cult of magnification," the tendency to represent a specimen by a limited number of high-magnification images. The practical importance of Table 4.1 becomes clear when we consider the problem of surveying a specimen to determine its significant features. Clearly, when only one or two images are taken at high magnification, the microscopist samples only a few square micrometers, which may represent only a few parts per million or parts per billion of the possible specimen area when the specimen has overall dimensions of square millimeters to square centimeters. Such a small sample may not be sufficient to obtain an adquate impression of the specimen's characteristics. For survey work, a combination of both low-magnification and high-magnification imaging should be used, as shown in Fig. 4.6, and an adequate number of areas must be recorded to gain a valid description of the specimen.

Zoom Capability. Magnification in the SEM depends only on the excitation of the scan coils and not on the excitation of the objective lens, which determines the focus of the beam. Thus, once the objective lens is adjusted in strength to focus the image at high magnification, lower magnifications of the same region remain in focus as the scan strength is increased to scan a larger area. This *zoom magnification* feature is very useful for rapid surveying of the specimen, as shown for a fracture surface in Fig. 4.6.

Lack of Image Rotation. The image does not rotate as the magnification is changed, as can be seen in Fig. 4.6, provided the objective lens excitation is constant. As discussed in Chapter 2, a relative rotation of the image occurs if the working distance, i.e., the pole-piece-to-specimen distance, is changed, and the objective lens strength must be altered to focus the beam at the new working distance. Image rotation with a change in working distance is illustrated in Fig. 4.7.

Absolute Value of the Magnification. Can the magnification printed on the SEM screen as a numerical value or a *micrometer marker* be trusted? If accurate measurements are to be made, the magnification should be verified by means of an external standard. Calibrated gratings with known spacings provide suitable standards. Standard Reference Material (SRM) 484, available from the National Institute of Standards and Technology (NIST), is a stage micrometer consisting of electrodeposited layers (nominal spacings 0.5, 1, 2, 5, 10, 30, 50 μm) of nickel and gold (McKenzie, 1990). This SRM permits image magnification calibration to an accuracy of 5% at the micrometer scale. The limitation in accuracy arises because of the difficulty in exactly locating the edges of the calibration structures. Extension of the stage micrometer approach to smaller structures is the subject of ongoing research.

4.2.4. Picture Element (Pixel) Size

The *picture element* or *picture point,* related to scanning action and magnification, is critical to the interpretation of SEM images. The picture element can be understood with the aid of Fig. 4.3. As described above, the scan consists of a series of discrete locations to which the beam is addressed. The picture element is the size of the area *on the specimen* from which information is transferred to the CRT image or computer memory. Usually the picture element is considered as a circle or square, described by a linear measure of diameter or edge length. Considering the rectilinear scan to be divided into equal-sized square boxes filling all of the scan frame, the linear dimension D_{PE} of the picture element is given by the expression

$$D_{PE} = \frac{L_{spec}}{N_{PE}}, \tag{4.2}$$

where L_{spec} is the length of the scan on the specimen from Eq. (4.1) and N_{PE} is the number of discrete locations (picture elements) along the scan line (often listed as the number of lines per frame on the SEM

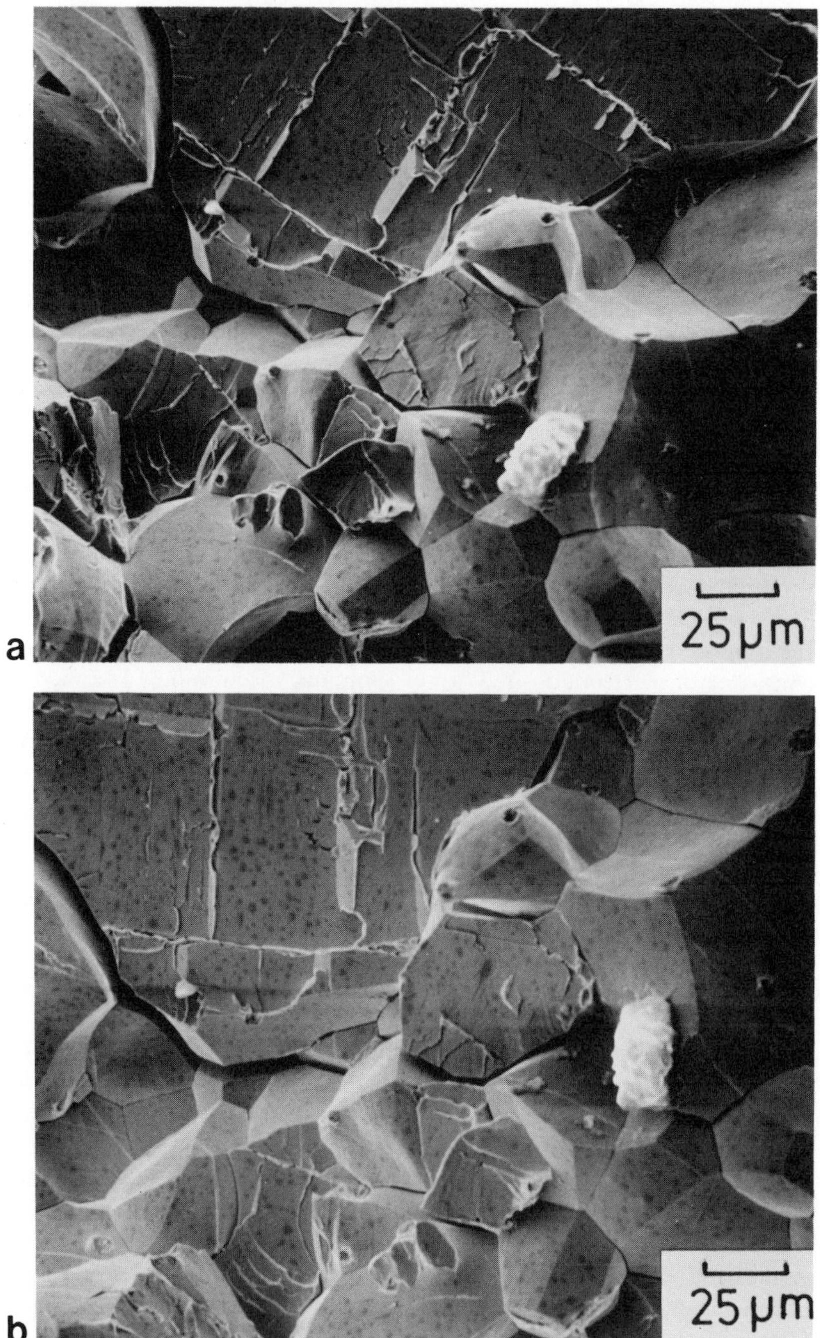

Figure 4.7. Images of a fracture at constant magnification but with different working distances. The change in lens excitation causes a pronounced image rotation.

Table 4.2. Size of Picture Element as a Function of Magnification[a]

Magnification	Edge of picture element
10X	10 μm
100X	1 μm
1,000X	0.1 μm (100 nm)
10,000X	0.01 μm (10 nm)
100,000X	1 nm

[a] 1000 × 1000 scan matrix; 10 cm × 10 cm display on CRT.

scan-control module). Table 4.2 shows the length of the picture element's edge as a function of magnification for the case of a high-resolution (1000 × 1000) image.

For a given choice of magnification, images are considered to be in sharpest focus if the signal measured when the beam is addressed to a given picture element comes only from that picture element. From Table 4.2, the picture element decreases to nanometers at high magnification. Overlap of information from adjacent picture elements must eventually occur as the magnification is increased, because of the finite size of the interaction volume. This overlap of pixel information is manifested as blurring in the image, as illustrated in Fig. 4.8.

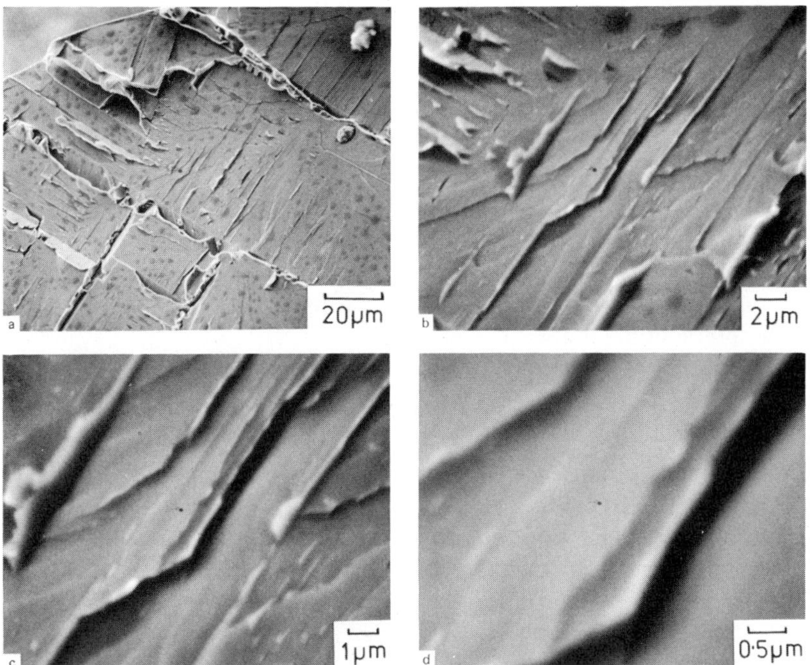

Figure 4.8. Images of a fracture surface illustrating the effects of pixel overlap (hollow magnification). The finest details in the image at low magnification appear sharp, while at the highest magnification, blurring can be observed.

Increasing the magnification beyond this point results in "hollow magnification." To determine the limiting magnification for a given beam size, energy, and specimen composition, we must calculate the signal-producing area and compare it to the pixel size. How can the signal "leak out" to overlap adjacent picture elements? First, the beam diameter at the specimen plane is controlled by the strength of the final lens. As the lens strength is adjusted, the crossover moves along the optic axis (Z-axis) and the effective beam size at the specimen is varied. On the CRT display, we observe this focusing operation as a maximization of fine detail when the beam is focused at the specimen plane containing the feature of interest. A second and frequently more important source of signal leakage occurs as a result of electron scattering and the finite size of the interaction volume from which the imaging signals emanate.

How can the size of the signal-producing area be estimated from these two contributions? Consider the case of backscattered electrons. The dimension of the signal-producing area contains contributions from the finite size of the beam as well as the projected surface area of backscatter emission due to electron scattering. Table 3.4b provides a description of the radius of the emission area (normalized to the Kanaya–Okayama range) that contains various percentages of the backscattered electrons. The diameter of the effective signal-producing area is found by adding in quadrature the beam diameter d_B and the diameter of the signal-producing area d_{BSE}:

$$d_{eff} = (d_B^2 + d_{BSE}^2)^{1/2}. \tag{4.3}$$

Consider a 10-keV beam focused to 50 nm that is incident on specimens of aluminum and of gold; the specimens have flat surfaces set normal to the incident beam. From Table 3.4b, the diameter of the region emitting 90% of the backscattered electrons is 0.98 R_{KO} or 1.3 μm (1300 nm) for aluminum and 0.50 R_{KO} or 0.13 μm (130 nm) for gold. From Eq. (4.3), the total lateral spreading of the backscatter signal, considering both the incident beam and the BSE spread due to the interaction volume, is

$$\text{Aluminum} \qquad d_{eff} = 1300\text{ nm} = 1.3\ \mu\text{m}$$
$$\text{Gold} \qquad d_{eff} = 140\text{ nm} = 0.14\ \mu\text{m}.$$

(Note that, because of the addition in quadrature and the large difference between the beam size and the interaction volume, the beam size makes no significant contribution to the spreading for aluminum, and only a small contribution for gold.) When these values of d_{eff} are compared to the pixel size from Table 4.2, we see that even at a low magnification of 100X, the adjacent picture elements are slightly overlapped for aluminum; for gold, this situation is encountered above 700X. Thus, at surprisingly low magnifications, adjacent pixels begin to overlap. The consequences of this situation will be considered in detail later.

4.2.5. Low-Magnification Operation

A large fraction of the images recorded with the SEM are prepared with modest magnifications, less than 5000X. What is the proper strategy for choosing instrument parameters for low-magnification operation? Should we, for example, choose to operate with a very finely focused beam in the 10-nm range or less, which is appropriate for high-magnification operation? We know from the discussions in Chapter 3 that for a conventional thermal-source SEM, a severe penalty in beam current must be paid if we choose a very small probe. In arguments to be developed later in this chapter (Section 4.5), we shall learn that the beam current is a critical parameter in establishing the visibility of an object: the more current in the beam, the lower the contrast level that can be seen. From the example presented in Section 4.2.4 for estimating the size of the effective signal-sampling region, we found that the beam size could be a small contributor to the overall spread of the BSE signal. We could therefore safely increase the size of the focused probe to 100 nm or more by weakening the first condenser lens, directly increasing the current in the beam approximately as the square of the beam's diameter. Increasing the beam's diameter from 10 to 100 nm would increase the beam current by a factor of 100 while causing no increase in d_{eff} for aluminum and increasing d_{eff} for gold only from 140 to 170 nm.

The key point in low-magnification operation is to use a weakly excited first condenser lens and a large final aperture to obtain a large beam which carries a much larger current than a finely focused, high-resolution probe.

4.2.6. Depth of Field (Focus)

The large depth of field that can be obtained in SEM images is one of their most striking capabilities, and rather than high resolution, is often the more valuable feature to users. How can we optimize the depth of focus? The concept of picture elements can be used to understand more fully the concept of depth of focus and to develop an imaging strategy for depth-of-focus microscopy. (Strictly speaking, there is no image field in the SEM, since the image is constructed by scanning, but the term *field* is conveniently carried over from conventional optical microscopy.)

As explained in Chapter 2 and illustrated in Fig. 4.9, the action of focusing the beam to a crossover ("plane of optimum focus" in Fig. 4.9) inevitably produces angular divergence of the rays that constitute the focused beam. This divergence causes the beam to broaden above and below the plane of optimum focus. If we consider a rough specimen with features at various distances from the final lens, then beams of different sizes strike the features from place to place on the specimen, depending on their altitude or depth relative to the plane of

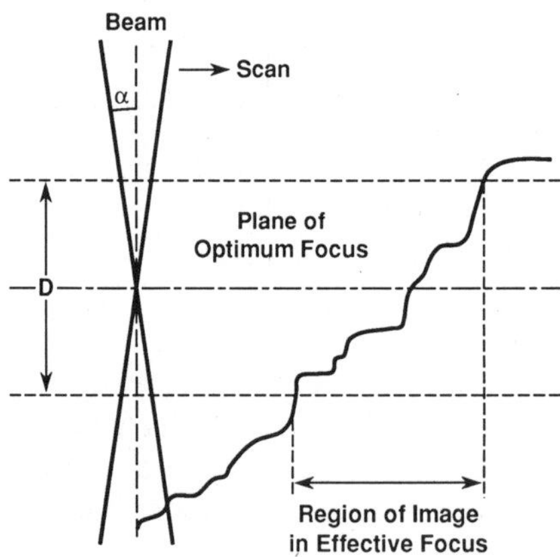

Figure 4.9. Schematic illustration of the depth of focus (field) in a SEM image.

optimum focus. For a sufficiently rough specimen, it is clear from Fig. 4.9 that the beam intersecting some features will become so large that some portion of the specimen will not appear in focus.

To calculate the depth of focus, we must know at what distance above and below the plane of optimum focus the beam has broadened to a noticeable size. The depth-of-focus criterion then depends on where the beam reaches a condition of overlapping adjacent pixels. The geometrical argument in Fig. 4.9 indicates that, to a first approximation, the vertical distance $D/2$ required to broaden the beam of minimum size r_0 to a radius r is given by

$$\tan \alpha = \frac{r}{D/2}. \tag{4.4a}$$

For small angles, $\tan \alpha \approx \alpha$,

$$D/2 \approx r/\alpha \tag{4.4b}$$

$$D \approx 2r/\alpha, \tag{4.4c}$$

where α is the beam divergence, as defined by the semicone angle. Consider that r equals the pixel size of the image. On a high-resolution CRT (spot size = 0.1 mm = 100 μm), most observers will find that defocusing becomes objectionable when two pixels are fully overlapped. The pixel size on the spcimen is then given by $0.1/M$ mm, where M is the magnification. Substituting this expression into Eq. (4.4b) gives a practical expression for the depth of focus:

$$D \approx \frac{0.2}{\alpha M}\ \text{mm}. \tag{4.4d}$$

Equation (4.4) indicates that to increase the depth of focus D, the operator can choose to reduce either the magnification M or the divergence α. Changing the magnification is not generally an option, since the magnification is chosen to fill the image with the details of interest. This leaves the divergence as the adjustable parameter. The divergence is adjusted by the selection of the final aperture radius, R_{Ap}

Table 4.3. Depth of Focus (Field) in μm

Magnification	α (rad)		
	5×10^{-3}	1×10^{-2}	3×10^{-2}
10X	4,000	2,000	670
50X	800	400	133
100X	400	200	67
500X	80	40	13
1,000X	40	20	6.7
10,000X	4	2	0.67
100,000X	0.4	0.2	0.067

and the working distance D_W:

$$\alpha = R_{\mathrm{Ap}}/D_W \tag{4.5}$$

A typical set of final aperture sizes, specified by the diameter, are 100 μm, 200 μm, and 600 μm, and a typical working distance is 10 mm, with a possible increase to 50 mm or more in some instruments, depending on the sample stage. The depth of focus calculated from Eq. (4.4) is given in Table 4.3 for several combinations of the operating parameters. Figure 4.10 gives examples of the appearance of

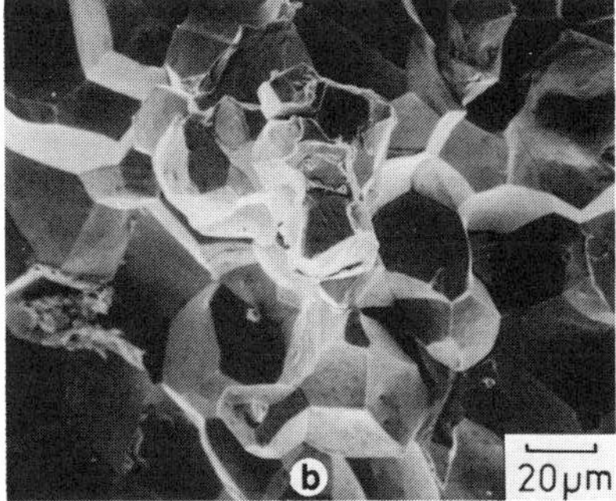

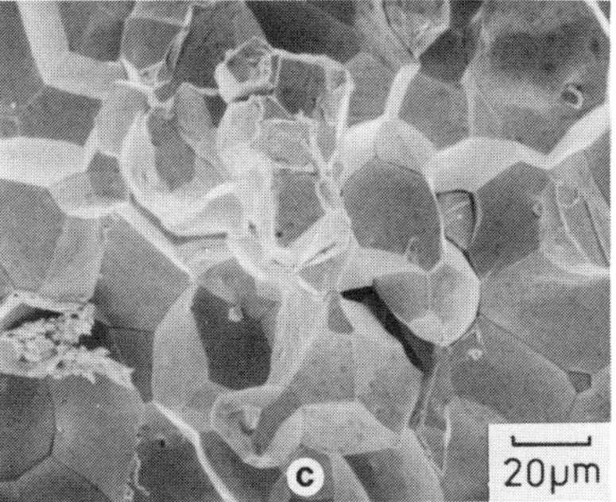

Figure 4.10. Appearance of a fracture surface with different depths of focus obtained by varying the aperture size and the working distance. (a) Small depth of focus, most of the field of view is out of focus (15-mm working distance, 600-μm aperture), (b) Intermediate depth of focus; more of the surface appears in focus (15-mm working distance, 100-μm aperture), (c) maximum depth of focus, entire field in focus (45-mm working distance, 100-μm aperture). Beam energy 20 keV.

a rough object, a fracture surface, under various conditions of depth of focus. Considering the values of the depth of focus compared to the width of the image field in Table 4.3, it is possible to choose operating parameters to achieve a condition in which all objects within a specified cubical volume of space are in focus.

Choosing the SEM operating parameters to optimize the depth of focus is of special importance when stereomicroscopy is performed. Stereomicroscopy, discussed in detail in Section 4.9.2, permits visualization of the three-dimensional structure of an object. The perceived stereo effect is generally much stronger when the entire field of view of an object appears in focus. Conversely, when only a single image of a three-dimensional object is viewed, it is sometimes difficult to judge the roughness if the image is prepared with high depth of focus. In Fig. 4.10a, which was prepared with poor depth of focus, it is actually easier to judge which part of the specimen is nearest to the polepiece than in Fig. 4.10c, prepared with a high depth of focus.

4.2.7. Image Distortions

4.2.7.1. Projection Distortion: Gnomonic Projection

The SEM image is a two-dimensional representation of three-dimensional objects. The SEM image is geometrically equivalent to a projection obtained by extending the beam vector to intersect a plane perpendicular to the optic axis of the instrument. The nature of this projection is shown schematically in Fig. 4.11. (Note that this drawing must not be construed to imply that an actual image exists in the instrument below the specimen.) The scan can be considered to originate as a series of ray vectors with their point of origin in the final aperture. The geometrical equivalent of the image actually created on the CRT is the intersection of these vectors extended to meet a plane perpendicular to the beam axis.

As the magnification is lowered, the angle ϕ of the scan ray increases from the optic axis. At low magnifications, the angle of the scan cone can be substantial. For example, at 10X, the scanned field is 10 mm wide, and if the working distance is 10 mm, the semicone angle of the scan is 26.5°. The position S_x of the scan on the specimen is given by

$$S_x = WD \tan \phi, \tag{4.6}$$

where WD is the working distance. This type of projection is known as a gnomonic projection. The magnitude of the distortion of the gnomonic projection increases with distance from the optic axis (centerline) of the system. As the magnification is reduced and the scan excursion increases, the position of the scan deviates from linearity. For the 10X, 10-mm-working-distance example, the scan position at the 25° location is 6.5% greater than that predicted by linear extrapolation of the length of the scan at 5°. These nonlinear effects must be considered when accurate measurements are to be made.

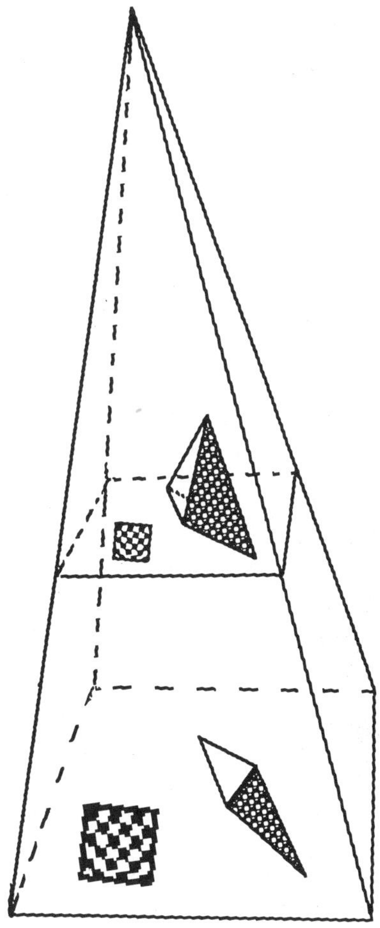

Figure 4.11. Construction of a SEM image as a projection onto a plane perpendicular to the beam axis.

At higher magnifications (>500X), the total scan cone angle is reduced to 0.5° or less, and the beam changes angle only slightly across the scan field. The distortion caused by the gnomonic projection can be neglected because the beam is effectively perpendicular to the projection plane at every point in the scan field and the projection is effectively parallel at all points.

4.2.7.2. Projection Distortion: Image Foreshortening of Tilted Objects

When the specimen is a flat surface set normal to the beam and a square raster is scanned synchronously on the specimen and the CRT, the magnification is identical along any direction in the image; the shape of objects in the field is reproduced accurately in the SEM image, as depicted schematically in Fig. 4.5. An example is shown in Fig. 4.12b, which shows a grid with square openings and a spherical particle. As expected, the grid, which lies in the plane perpendicular to the beam, is seen with openings reproduced as squares. The sphere is a three-

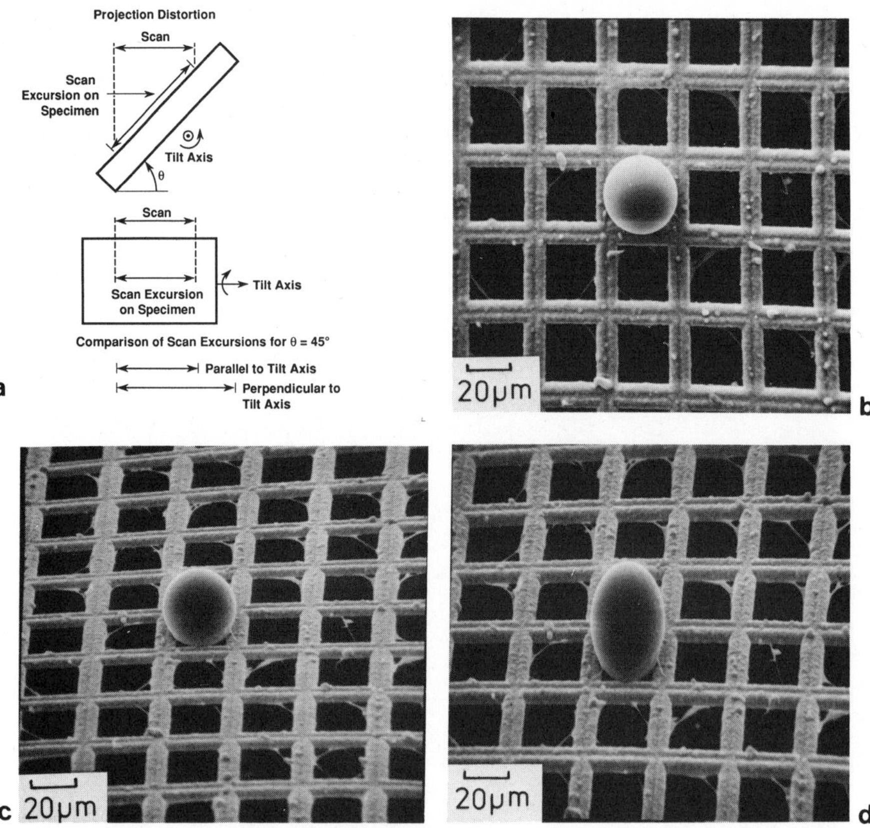

Figure 4.12. (a) Projection distortion (image foreshortening), which results from the tilt of a specimen for a tilt angle of 45°. The scan has the correct length parallel to the tilt axis (horizontal in the displayed image), but is lengthened perpendicular to the tilt axis (vertical in the displayed image), thus reducing the magnification. (b) Latex sphere on a copper grid, specimen plane normal to the beam (0° tilt). (c) Grid-sphere specimen tilted to 55°. (d) Grid-sphere specimen tilted to 55°, tilt correction applied, note restoration of square grid openings, but distortion of sphere.

dimensional object, which is projected onto the plane perpendicular to the beam as a circle.

When the planar specimen is given a simple tilt around an axis parallel to the image horizontal, the projection situation changes, as illustrated in Figure 4.12a. The projection of the scan line on the projection plane intercepts the expected length of specimen parallel to the tilt axis, but perpendicular to the tilt axis, a greater length of the specimen is projected into the same length of scan. The magnification is therefore *lower* perpendicular to the tilt axis compared to its value parallel to the tilt axis. This effect is referred to as *foreshortening*. The length of the scan excursion perpendicular to the tilt axis $S_\perp$, in terms of the scan length parallel to the tilt axis $S_\parallel$, is given by

$$S_\perp = \frac{S_\parallel}{\cos\theta} = S_\parallel \sec\theta, \tag{4.7}$$

where θ is the tilt angle.

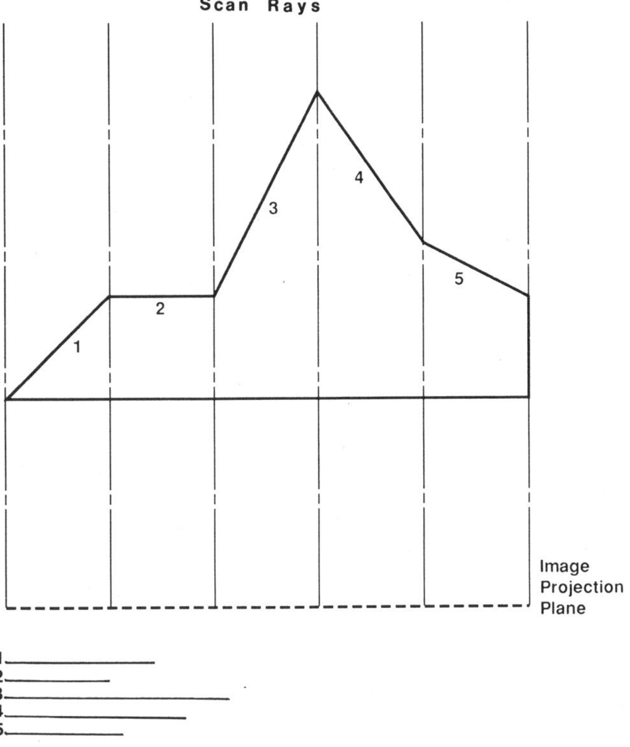

Figure 4.13. Projection distortion of facets with different inclinations to the beam. The magnification is sufficiently high that the scan angle is negligible and the scan rays are nearly parallel to the optic axis. Although the numbered facets have different lengths, their images in projection all have the same apparent length. Below the image projection plane are shown the relative true lengths of the numbered facets.

Consider a specimen consisting of randomly tilted, intersecting flat surfaces or facets imaged at high magnification, where the scan angle is small so that the rays are nearly parallel to the optic axis, as shown in Fig. 4.13. The projection of these surfaces to form the image produces a final image in which the effective magnification varies with the local tilt angle according to Eq. (4.7). The more highly tilted a facet is, the more foreshortened its image will be. Although each of the numbered facets in Fig. 4.13 is of different length, in the image each appears to have the same length. Thus, linear measurements made within the image of a single facet will be in error by the secant of the tilt angle of that facet as compared to a facet normal (at 0° tilt) to the beam. For example, if a facet is tilted 60° from the horizontal, then measurements of the length of features on that surface will be *underestimated* by a factor of 2, compared to measurements within a facet at 0° tilt in the same image. Unfortunately, it is not possible to obtain information on the local tilt from a single SEM image, so calculating a correction is not feasible. For such situations, the technique of stereomicroscopy must be applied to correct for the local tilt and achieve accurate measurements.

4.2.7.3. Corrections for Tilted Flat Surfaces

Tilt Correction. If the tilt angle is accurately known, the effect of foreshortening can be calculated by means of Eq. (4.7), and a quantitative correction can be applied to the distance measured in the image. *Note that such a correction applies only to a flat specimen at a known tilt angle and only when the tilt axis is parallel to the scan line.* Alternatively, the correction can also be applied by means of electronically reducing the strength of the scan in the direction perpendicular to the tilt axis by the amount predicted by Eq. (4.7). The scan generator control for this operation is usually called "tilt correction." It must be emphasized that electronic tilt correction is valid only for the special class of specimens consisting of a flat plane tilted to a known angle. *Tilt correction is an invalid operation for any three-dimensional specimen* and in fact would introduce unnecessary and misleading distortions into the image. To appreciate these distortions, consider the images in Fig. 4.12c and d. In Fig. 4.12c, the specimen is tilted, as evidenced by the foreshortening of the square openings of the grid, while the projection of the three-dimensional sphere is still a circle, as expected. When tilt correction is applied, Fig. 4.12d, the grid openings are restored to their proper square shape, but the image of the sphere is now highly distorted, because the sphere is a three-dimensional object and the normal gnomonic projection of the SEM image has been disrupted by the modification of the scan strength in orthogonal directions.

Dynamic Focus Correction. An additional feature found on some SEMs is *dynamic focus,* which should not be confused with tilt correction. Dynamic focusing, which is illustrated in Fig. 4.14a, is achieved by adjusting the strength of the final condenser lens as a function of the scan position perpendicular to the tilt axis so as to cause the plane of optimum focus to coincide with the surface at all working distances. When a flat surface is highly tilted, the lens is strengthened when the scan is at the top of the field and progressively weakened as the scan moves down the slope of the tilted object, thus keeping the specimen in optimum focus at all points in the image, as illustrated in Fig. 4.14b. Dynamic focus depends on satisfying a simple (and known) relationship between the scan position and the working distance. This relationship is established only if the specimen is flat and tilted to a known angle, with the tilt axis parallel to the scan line. *The same restrictions apply to dynamic focus as apply to tilt correction, namely, that it is invalid to apply this operation to the image of any three-dimensional object.*

4.2.7.4. Scan Distortion: Pathological

The SEM image may contain unexpected distortions due to pathological defects in the scan. Such distortions can arise from improper adjustment of the scan, typically because the two orthogonal scans are produced by different circuits which may not be adjusted to exactly the same value of deflection. The scan pattern may also become nonlinear, particularly at the extremes of the field, even at high magnifications where the nonlinearity of the gnomonic projection is not significant. Scan distortions may also increase with the scan speed, particularly at television rates, due to hysteresis in the scan coils.

Distortion that can occur at the edge of the image field is illustrated in Fig. 4.15, which shows the change in the appearance of a spherical object as it is translated from the edge to the center of the field. Such scan distortions may not be noticed when images are of irregular objects that lack any obvious points of reference. To determine the existence of scan distortions, it is necessary to make careful measurements on known regular objects. Carefully fabricated grids make excellent test specimens. Another good choice is polystyrene spheres. NIST Standard Reference Materials 1960 (9.89 μm) and 1961 (26.64 μm) consist of polystyrene spheres fabricated in microgravity in earth orbit aboard the NASA Space Shuttle (McKenzie, 1990). These objects exhibit an extremely high degree of sphericity and uniformity of diameter. As such, the spheres can serve as measures of both the absolute magnification and the uniformity of scan field.

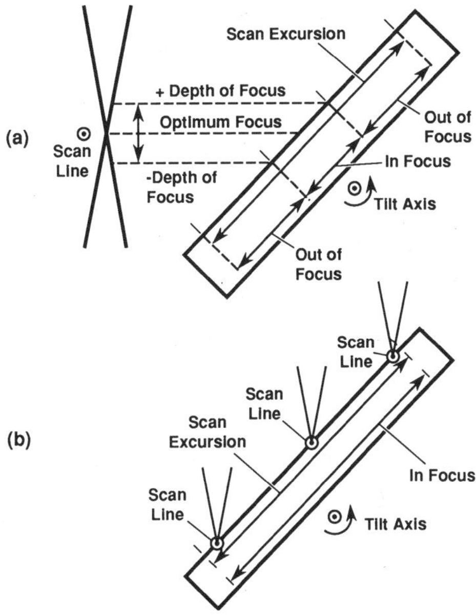

Figure 4.14. Schematic illustration of the technique of dynamic focus. (a) Normal focusing situation obtained with a constant third-lens strength. There is a plane of optimum focus, a region of the sample above and below this plane which remains in effective focus because of the depth of focus, and a region of the specimen which is out of focus because the working distance exceeds the depth of focus. (b) Dynamic focusing is achieved by adjusting the strength of the final lens as a function of the scan position perpendicular to the tilt axis so as always to bring the plane of optimum focus to coincide with the specimen surface. (c) Image of a grid corresponding to the situation illustrated in (a). (d) Image corresponding to (c), with dynamic focusing applied.

Figure 4.14. (Continued).

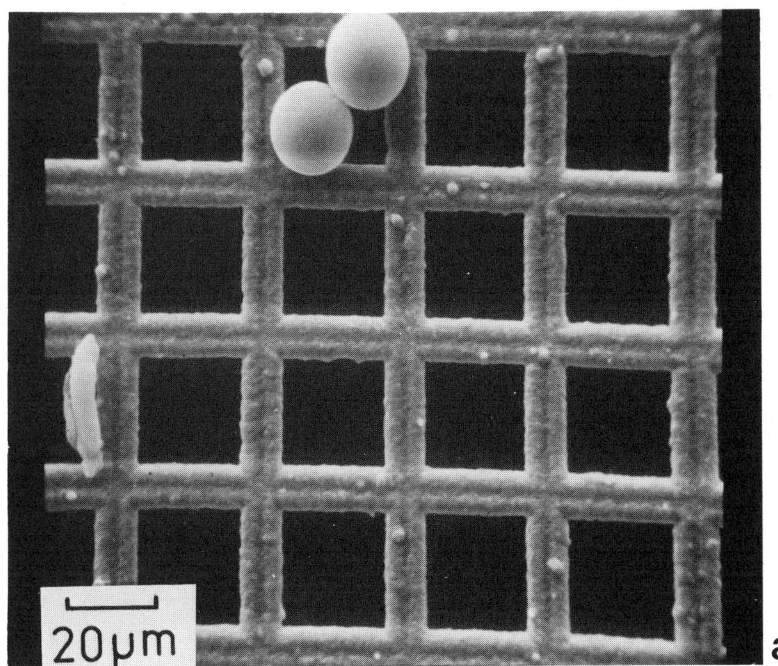

Figure 4.15. Illustration of image distortion that can occur at the edge of the image field. (a) Note the pronounced distortion of the image of the spheres to an elliptical shape when placed at the top of the field of view. (b) The expected round shape is obtained when the spheres are translated to the center of the field.

4.2.7.5. Moiré Effects

Although the recorded SEM image appears to be continuous to the human eye, it is composed of a grid of picture points with a spatial periodicity. We are thus always looking at the specimen through a grating, although the image appears continuous to the unaided eye. Viewing a typical specimen with irregular features through this grating introduces no deleterious effects. However, when the specimen itself has features with a periodic structure, as does a grid, the superposition of the periodicities of the specimen and the image can lead to the formation of moiré fringes. Moiré fringes are the interference patterns between gratings of similar period. These patterns may appear in SEM images and should not be confused with real features of the specimen. Figure 4.16 shows the appearance of moiré fringes in low-magnification images of a copper grid. Moiré fringes can be identified because the pattern changes rapidly with specimen rotation relative to the fixed scan.

4.3. Detectors

In order to form an image in the SEM, an appropriate detector must be employed to convert the radiation of interest that leaves the specimen into an electrical signal for manipulation and display by the signal processing electronics. From Chapter 3, we know that a rich variety of radiation leaves the specimen, including backscattered electrons, secondary electrons, characteristic and bremsstrahlung x rays, cathodoluminescence, absorbed current (specimen current), phonons, currents and potentials induced in certain types of semiconductors, etc., with each specific type of radiation potentially carrying information on different characteristics of the specimen. In this chapter, we will consider detectors for the electron signals and cathodoluminescence. X-ray detectors will be discussed in Chapter 5.

When considering a detector, we need to ask four questions. The answers to the first two of these questions are illustrated in Fig. 4.17.

Where is the Detector Relative to the Beam and the Specimen? The position of the detector is described by the take-off angle ψ which is defined as the angle from the surface of the specimen to the line connecting that point on the specimen to the center of the detector face.

How Large is the Detector? The size of the detector is described by the solid angle Ω, the ratio of the area of the face of the detector to the square of the radial distance to the beam impact point, $\Omega = A/r^2$. The unit of measure of solid angle is the steradian (sr).

How Efficient is the Detector for Converting Radiation which Strikes the Collection Area into a Useful Signal? The fraction of single incoming radiation events which produce a measurable response from the detector is called ε. Often, ε will change for different radiation energies, e.g., the response may not be uniform over the broad energy range of an incoming signal such as backscattered electrons.

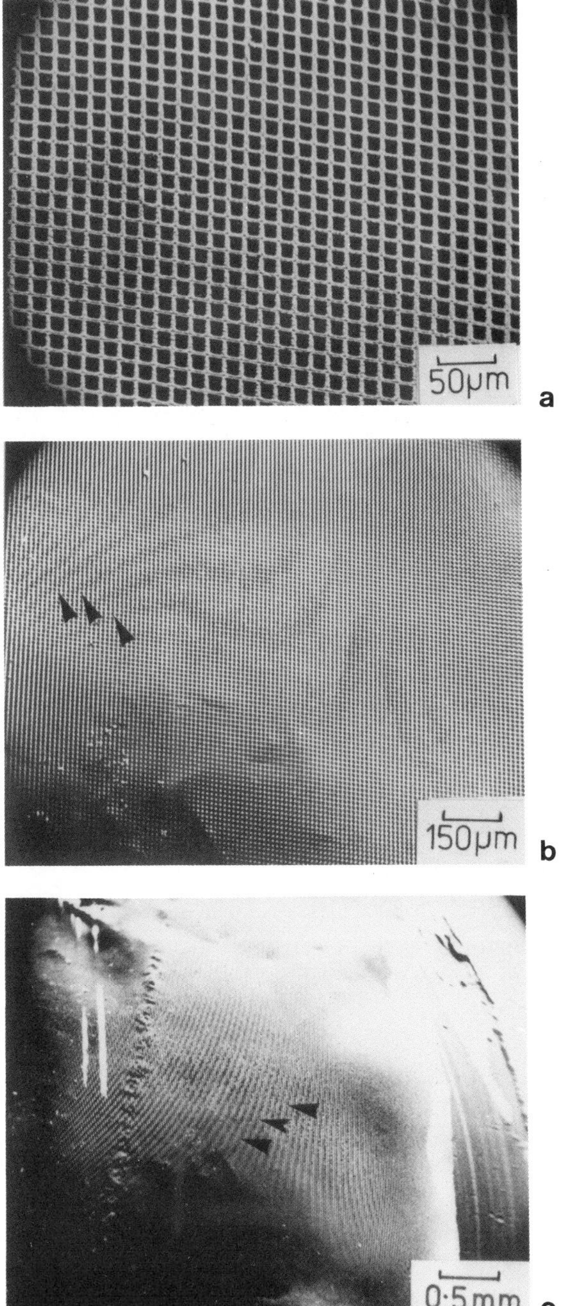

Figure 4.16. SEM images of a grid showing the appearing of moiré fringes as a function of magnification. The moiré fringes appear as the magnification is decreased and the periodicity of the scanning raster approaches the periodicity of the grid. The dark lines are moiré fringes and are artifacts; these features do not actually exist on the specimen. (a) Regular grid pattern at high magnification. (b), (c) Lower magnifications showing moiré fringes (arrows). The fringes appear as wavy lines due to misalignment between the specimen pattern and the scanning pattern.

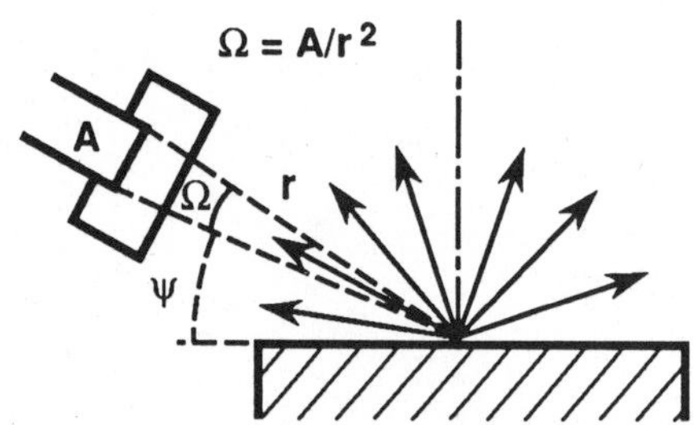

Figure 4.17. General characteristics of detectors. The position of the detector relative to the beam and the specimen is described by the take-off angle, ψ. The size of the detector is given by the solid angle Ω, which is equal to the area of the detector divided by the square of the radial distance to the beam impact point, $\Omega = A/r^2$.

What is the Bandwidth of the Detector/Amplification System? Bandwidth refers to the range of signal frequencies that the detector can process. Since the SEM image is scanned, the signal appears as a function of time, and the spatial detail in the image is converted into the domain of time. For example, as the beam is scanned across fine scale features, the signal changes rapidly. Fine-scale spatial features are converted to high-frequency signals. For comparison, the general shading of coarse features varies much more slowly with the scan, so that these features are converted to low-frequency signals. A detector–amplification system typically has a high-frequency cutoff beyond which information is lost.

4.3.1. Electron Detectors

From Chapter 3, the electrons that escape the specimen under bombardment by keV electrons fall into two classes with widely differing properties:

1. *Backscattered electrons* are beam electrons which escape the specimen as a result of multiple elastic scattering and have an energy distribution $0 \leq E_{BSE} \leq E_0$, with the energy distribution peaking in the range $0.8–0.9E_0$ for targets of intermediate and high atomic number.
2. *Secondary electrons* are specimen electrons given a small amount of kinetic energy by inelastic collisions with beam electrons and are emitted with energies in the range $0 \leq E_{SE} \leq 50$ eV, with a most probable energy of 3–5 eV.

These widely differing properties present a considerable challenge to the design of a detector to make use of both signals; alternatively, the differences permit the design of detectors that are selective for one of the signals.

4.3.1.1. Everhart–Thornley Detector

The electron detector most commonly used in scanning electron microscopy is the combined secondary/backscattered-electron detector developed by Everhart and Thornley (1960), based upon a sequence of earlier developments by the Cambridge group under Prof. Sir Charles Oatley (Oatley, 1972). The Everhart–Thornley (E–T) detector is so popular that it is extremely rare to find a conventional SEM not equipped with it. The development of this detector provided the first efficient use of the rich secondary/backscattered electron signal with higher amplifier gain, low noise, and robust, low-maintenance performance. The rise of the SEM as a tool of broad use to the scientific community is due in considerable measure to the utility of this detector. Because of its efficient collection of secondary electrons, the E–T detector is often mistakenly considered only a secondary-electron detector. As the following discussion will demonstrate, backscattered electrons form an important part of the E–T signal.

The E–T detector, illustrated in Fig. 4.18, operates in the following manner: An energetic electron (keV energy) strikes the scintillator material, which may be a doped plastic or glass target or a crystalline compound such as CaF_2 doped with europium (for a review of scintillators, see Pawley, 1974). The interaction of the energetic electron with the scintillator produces photons, which are conducted by total internal reflection in a light guide (a solid plastic or glass rod) to a photomultiplier. Since it is now in the form of light, the signal can pass through a quartz glass window, which forms a vacuum seal, to the first electrode of a photomultiplier. At this photocathode, the photon flux is converted back into an electron current, and the electrons are accelerated onto the successive electrodes of the photomultiplier, producing a cascade of electrons. The typical gain of the photomultiplier, 10^5 to 10^6, is adjustable by selecting the voltage on the electrodes. The photomultiplication process provides high gain with small noise degradation and high bandwidth. The response is sufficiently fast to operate at television scan rates.

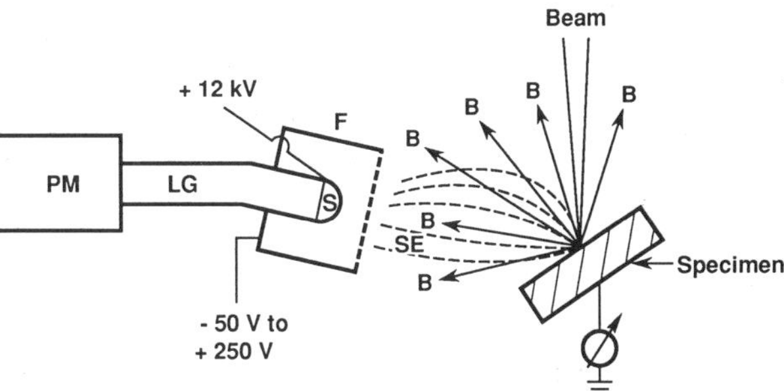

Figure 4.18. Schematic diagram of the Everhart–Thornley detector: B, backscattered-electron trajectories; SE, secondary-electron trajectories; F, Faraday cage (bias range −50 V to +250 V); S, scintillator, with thin metallic coating; high bias (+12 kV) supply to the scintillator coating; LG, light guide; PM, photomultiplier.

Backscattered electrons originating from incident beams with energies from 10–30 keV carry sufficient energy to excite the scintillator directly, even in the absence of post-specimen acceleration. By applying a large positive potential (+10 to +12 kV) to the face of the scintillator, the low-energy secondary electrons can be accelerated to a sufficient energy to generate light in the scintillator. To protect the beam from unwanted deflection by this large potential, the scintillator is surrounded by a Faraday cage, which is insulated from the scintillator bias. In order to collect the low-energy secondary electrons with higher efficiency than the fraction which would be collected by the simple geometric efficiency given by the solid angle, a separate bias potential can be applied to the Faraday cage, typically selectable in the range −50 to +250 V. Note that this range from negative to positive provides the possibility of completely rejecting secondary electrons (−50 V) or efficiently collecting secondary electrons (+250 V). Let us consider the action of the detector for two bias conditions:

Negative Bias. When the E–T detector is biased negatively, only backscattered electrons are detected. All secondary electrons are rejected, including those emitted from the specimen in the direction of the E–T detector within the solid angle for geometric collection. The take-off angle and solid angle of collection for the E–T detector for the direct collection of backscattered electrons are illustrated in Fig. 4.17. Those high-energy backscattered electrons which leave the specimen with motion directly toward the face of the scintillator (that is, along a "line of sight") are collected; all other backscattered electrons emitted from the specimen are lost. For a specimen at 0° tilt (normal beam incidence), the E–T detector usually views the specimens from a take-off angle of approximately 30°. For a 1-cm-diameter scintillator placed at a distance of 4 cm from the beam impact point on the specimen, the solid angle of the E–T detector is about 0.05 sr. As a fraction of the 2π-steradian space above a flat specimen, this solid angle gives a geometric efficiency of only 0.8% for the negatively biased case. The collection efficiency also depends on the angular distribution of the emitted signal. Consider the cosine angular distribution over which the backscattered electrons are emitted at normal beam incidence. The portion of the distribution collected at different take-off angles is illustrated in Fig. 4.19. Because of the nonuniform

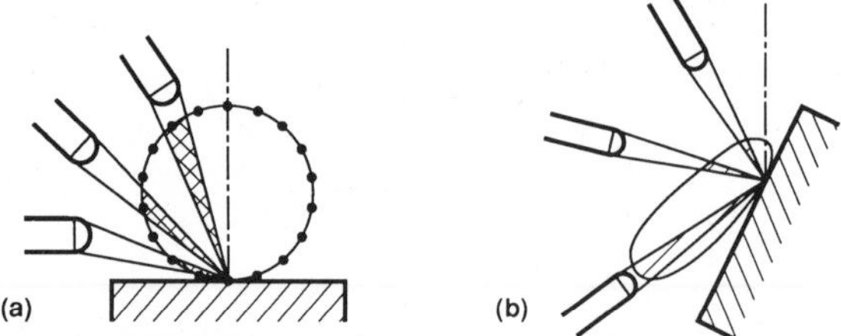

Figure 4.19. (a) Relative collection of backscattered electrons emitted in a cosine distribution by negatively biased E–T detectors placed at various take-off angles. (b) Relative collection of backscattered electrons emitted from a highly tilted surface.

angular distribution, at a low detector take-off angle, the fraction of the backscattered electrons collected is even less than that given by the geometric efficiency.

The rigid nature of the light pipe of the E–T detector means that it occupies a fixed position in the specimen chamber. The take-off angle and collection situation thus also depends on the specimen tilt. When the specimen is highly tilted, the angular distribution of backscattered electrons becomes skewed in the forward direction. Figure 4.19b shows how the collection situation changes with detector position. Comparing Fig. 4.19a and b, the detector position which is most favored for collection at 0° specimen tilt is least favored at high specimen tilt.

The energy response of the scintillator of the E–T detector depends on the exact nature of the material used, but it generally increases with increasing electron energy up to a saturation point. There is a threshold energy below which the detector produces no response.

Positive Bias. The postively biased E–T detector behaves in a profoundly different manner. The direct effect of the positive bias is to permit secondary electrons to enter the Faraday cage for subsequent acceleration by the bias on the scintillator. In addition to those secondaries emitted from the specimen into the solid angle of collection of the E–T detector, the attractive positive bias acts to deflect the trajectories of secondaries emitted from the specimen over a much wider range of solid angle into the detector, as shown in Fig. 4.20. Calculations of the trajectories of secondaries under the field of the E–T detector reveal that secondary electrons can be collected even if there is not direct line of sight from the specimen to the detector. From a flat surface, the collection efficiency can approach 100%.

The positive bias on the Faraday cage does not affect the directly collected backscattered electrons emitted from the specimen into the solid angle of the detector. They are collected regardless of the cage bias. The positive bias does, however, have an unexpected effect on the indirect collection of backscattered electrons, as illustrated in Fig. 4.21. The vast majority of backscattered electrons follow trajectories which miss direct collection by the E–T detector. These trajectories do cause the backscattered electrons to strike the polepiece and the specimen chamber's walls, where they cause the emission of secondary electrons, the SE_{III} component. The backscattered electrons may

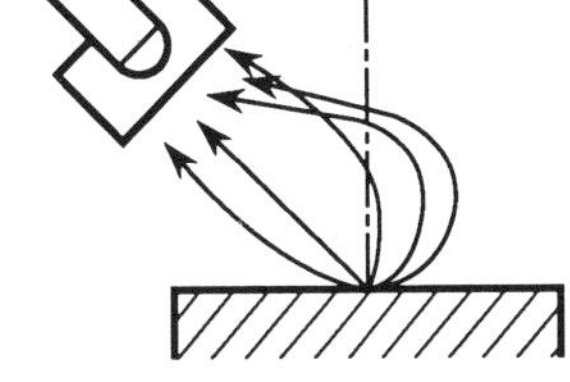

Figure 4.20. Schematic illustration of deflection of trajectories of secondary electrons of various energies by positive potential on the Faraday cage.

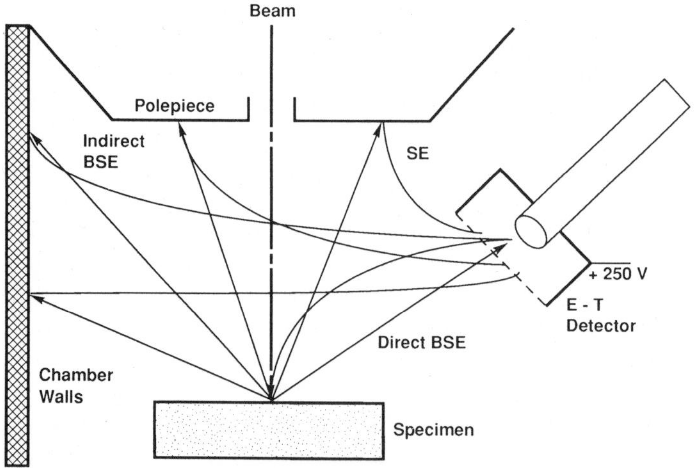

Figure 4.21. Schematic illustration of the indirect collection of backscattered electrons by a positively biased E–T detector. The backscattered electrons strike the polepiece and chamber walls, where they create secondary electrons. These secondaries are collected by the E–T detector with high efficiency. Although nominally a contribution to the secondary signal, they really represent the backscattered-electron signal component.

actually bounce repeatedly in the chamber before losing all energy, and at each collision, more secondary electrons are released. These secondary electrons are collected with high efficiency by the positive bias on the E–T detector. Although nominally part of the secondary-electron signal, this signal component actually represents the detection of the remote backscattered electron signal, since the magnitude of the SE_{III} signal scales with changes in the backscattered electron signal as it leaves the specimen. Some idea of the efficiency of collection for these indirect backscattered events is given by the measurements of Peters (1984). As listed in Table 4.4, for a material such as gold with a high backscatter coefficient, over 60% of the total signal of a positively biased E–T detector arises from indirect collection of the backscattered electrons. While the negatively biased E–T detector was highly directional in its acceptance of backscattered electrons, the positively biased E–T detector accepts both this direct component and the indirect component, which is effectively collected over a much larger solid angle, approaching 2π sr. The positively biased E–T detector is

Table 4.4. Collection of Secondary-Electron Signals from a Gold Specimen by the Everhart–Thornley Detector[a]

Signal	Source	Percentage of total
SE_I	Beam-produced secondaries	9%
SE_{II}	Backscatter-produced secondaries	28%
SE_{III}	Remote backscatter-produced secondaries	61%
SE_{IV}	Beam-produced secondaries from aperture	2%

[a] Peters (1984).

thus properly thought of as a *combined secondary and backscattered electron detector*. Since the backscatter coefficient is typically larger than the secondary-electron component, in many situations the backscattered electron signal is the most significant component of the total signal.

4.3.1.2. Dedicated Backscattered-Electron Detectors

There are usually two to ten times more backscattered electrons than secondary electrons emitted from the sample. Backscattered electrons carry much useful information on specimen composition, topography, crystallinity, etc. Because of the large differences in energy between backscattered and secondary electrons, it is relatively easy to develop detectors which are sensitive only to high-energy backscattered electrons. Dedicated backscattered-electron detectors are frequently included in the basic specification of an SEM.

The negatively biased E–T detector provides a signal composed exclusively of backscattered electrons, but it has a very low geometric efficiency and is highly directional in its collection. Dedicated backscattered-electron detectors are designed to greatly increase the solid angle of collection.

Scintillator Backscatter Detectors. Scintillator detectors operate on the principle that, with an incident beam energy of 10 keV or more, the backscattered electrons carry sufficient energy to excite the scintillator even in the absence of post-specimen acceleration. Without such acceleration, the secondary electrons have no effect on the scintillator, so the signal from an unbiased scintillator detector consists only of contributions from backscattered electrons. The elimination of biasing also has the benefit that the detector potential does not disturb the beam, so the detector can be placed close to the specimen for more efficient collection. Wells (1974a) and Robinson (1975) have described passive scintillator detectors with large collection angles. Figure 4.22a shows a large passive scintillator placed in proximity to a tilted specimen; the detector's take-off angle has been optimized to permit collection of forward-scattered electrons from a tilted surface (Wells, 1974a).

By making the scintillator of the same material as the light guide, designs that collect over much larger solid angles are possible (Robinson, 1975). Figure 4.22b shows a combined scintillator–light guide placed above the specimen. A hole drilled through the material permits access for the beam, and the detector occupies nearly 2π sr of solid angle. For a specimen placed at normal beam incidence, only electrons scattered out of the specimen at low angles to the surface are lost, as well as a small fraction at very high take-off angles near the incident beam that are scattered into the beam-access hole. In this configuration, the detector surrounds the specimen nearly symmetrically, so that the signal is integrated in all directions, nearly eliminating sensitivity to trajectory effects. Note that if the specimen is tilted,

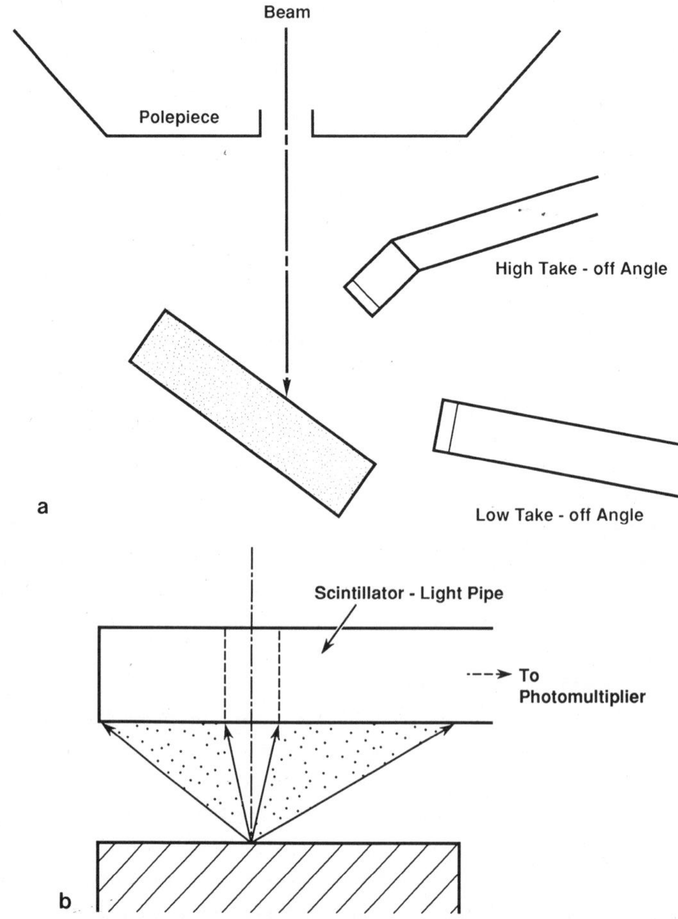

Figure 4.22. Scintillator backscattered electron detectors. (a) Large scintillator–light guide placed in proximity to the specimen but placed asymmetrically (Wells, 1974a). (b) Combined scintillator–light pipe designed as an inverted bowl placed symmetrically above the specimen and collecting a solid angle of nearly 2π sr (Robinson, 1975).

backscattered electrons are directed away from such a detector placed above the specimen, the efficiency is significantly decreased, and sensitivity to trajectory effects develops.

BSE-to-SE Conversion Detector. The backscattered-to-secondary-electron conversion detector operates on the same principle that provides the sensitivity of the positively biased E–T detector for indirect collection of backscattered electrons through collection of the remotely generated secondary electrons. The BSE-to-SE conversion detector (Moll *et al.*, 1978; Reimer and Volbert, 1979) seeks to optimize this conversion while excluding the secondary electrons produced at the surface of the specimen by the incoming beam electrons and the exiting backscattered electrons. As shown in Fig. 4.23, the separation of the two classes of secondaries is accomplished by surrounding the specimen with a grid biased to -50 V. Secondary

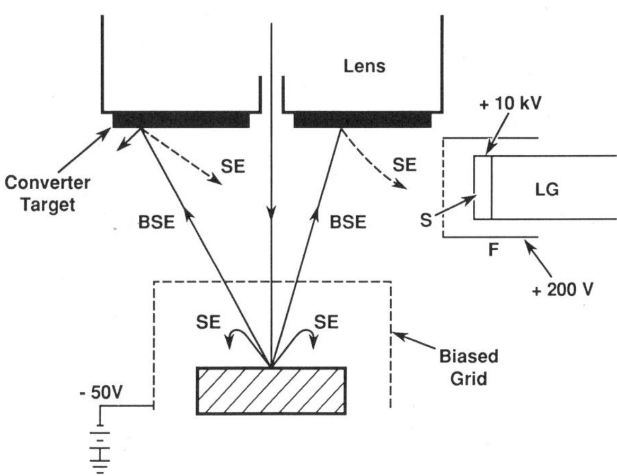

Figure 4.23. Schematic diagram of a backscatter-to-secondary-electron conversion detector.

electrons cannot pass through this grid, while the high-energy backscattered electrons pass through the grid openings and are not significantly affected by the grid potential. To improve the yield of secondary electrons produced by the backscattered electrons, a converter target covered with a thin layer of a material such as MgO with a high secondary yield is applied to the bottom of the objective polepiece. These remote secondary electrons are then collected by the conventional E–T detector operated with a positive bias on the Faraday cage.

By collecting backscattered electrons striking over a large fraction of the polepiece and chamber walls, the BSE-to-SE conversion detector achieves a large solid angle of collection and greatly reduced sensitivity to trajectory effects. Because the yield of secondary electrons generally increases as the energy of the exciting electron decreases, the BSE-to-SE conversion detector is unique in that its gain per backscattered electron actually increases at low incident-beam energy. The BSE-to-SE detector can therefore be used for low-voltage microscopy.

Solid-State Diode Detector. The solid-state diode detector operates on the principle of electron–hole pair production induced in a semiconductor by energetic electrons (Kimoto and Hashimoto, 1966; Gedcke *et al.*, 1978). The electronic structure of a semiconductor consists of an empty conduction band separated by a band gap of forbidden energy states from the filled valence band. When energetic electrons scatter inelastically in the semiconductor, some electrons are promoted to the conduction band, each leaving the absence of an electron, or hole, in the valence band. For silicon, the energy to create an electron–hole pair is about 3.6 eV, so when a single 10-keV backscattered electron strikes the detector, about 2800 electron–hole pairs are produced. If not separated, the electron and the hole recombine. As shown in Fig. 4.24, if an external bias is applied or the internal self-bias of certain electronic structures is used, the free

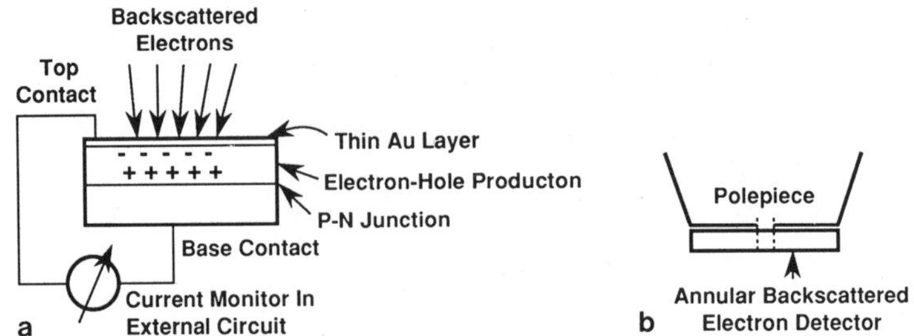

Figure 4.24. (a) Schematic diagram of a solid-state backscattered-electron detector, showing separation of electron–hole pairs by an applied external potential. (b) Typical installation of a solid–state detector on the polepiece of the objective lens.

electron and the hole move in opposite directions, separating the charges and preventing recombination. The charge collected on the external electrodes is used as the input to a current amplifier. The initial energy to electron–hole pair conversion gives a gain of about 1000, so the current amplification problem is not as great as directly amplifying the specimen current (see Section 4.3.1.3).

A number of important features of solid-state detectors can be noted:

1. The solid-state detector has the form of a flat, thin wafer (typically several millimeters thick), which can be obtained in a variety of sizes, from small squares to large annular detectors. Annular detectors or arrays of smaller discrete detectors are typically placed on the polepiece of the objective lens, as shown as in Fig. 4.24.

2. The size of the detector permits it to be placed in close proximity to the specimen, which provides a large solid angle for high geometric efficiency.

3. Multiple detector arrays can be readily created by using separated solid-state chips, so that the discrete signals can be viewed separately or combined. The advantages of detector arrays will be described below.

4. The detector is sensitive to high-energy backscattered electrons only and not to secondary electrons. The detector is also affected by x rays, but these are rare compared to backscattered electrons.

5. A typical energy response for a solid state detector is shown in Fig. 4.25. (Such a response curve can be readily measured by placing the detector directly under the beam and using a Faraday cup to quantify the beam current to normalize the results.) Above a threshold, the response of the detector is linear with energy. The energy threshold is a result of the loss of energy suffered by the energetic electrons during penetration of the surface electrode ($\sim$ 20 nm) and silicon dead layer ($\sim$200 nm) to reach the active volume of the detector. Typically, the threshold is in the range 2–5 keV. As shown in Fig. 4.25, the position of the threshold depends on the thickness of the front-surface electrode. This threshold has both

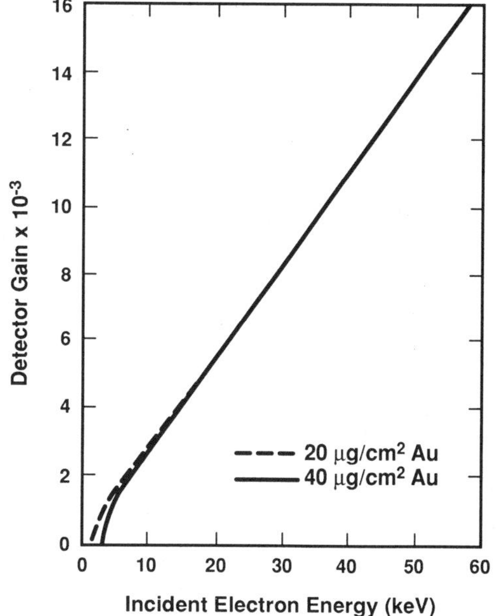

Figure 4.25. Response of an advanced solid-state detector with increasing electron energy (Gedcke *et al.*, 1978).

favorable and unfavorable impacts on performance. On the positive side, the loss of the low-energy backscattered electrons actually improves contrast and resolution in most situations, since the low-energy backscattered electrons are likely to have undergone more elastic and inelastic scattering and are therefore likely to emerge far from the beam impact point and with little specific information on specimen characteristics. On the other hand, the threshold obviously eliminates the use of the solid-state detector for low-voltage microscopy.

6. Through the mechanism of electron–hole production, the solid-state detector acts to boost the signal by about three orders of magnitude prior to the current amplification stage. A current amplifier is required, preferably of the operational amplifier type; such an amplifier can also be used to amplify the direct specimen current signal (Fiori *et al.*, 1974).

7. Because of the capacitance of the solid-state device, a relatively narrow bandwidth is obtained, which compromises the performance at TV scan rates. High capacitance produces a large time constant which has the effect that the detector signal does not accurately follow a large, rapid change in signal such as occurs across a sharp edge, resulting in apparent blurring at TV rates. When the scan speed in reduced to a slow visual scan, the available detector bandwidth becomes adequate and the image quality improves. Photographic scan recording is sufficiently slow that no bandwidth limitations are encountered. "Premium" detector designs can overcome this limitation, but even for these detectors, the high-speed performance is usually noticeably poorer when compared to that of the E–T detector.

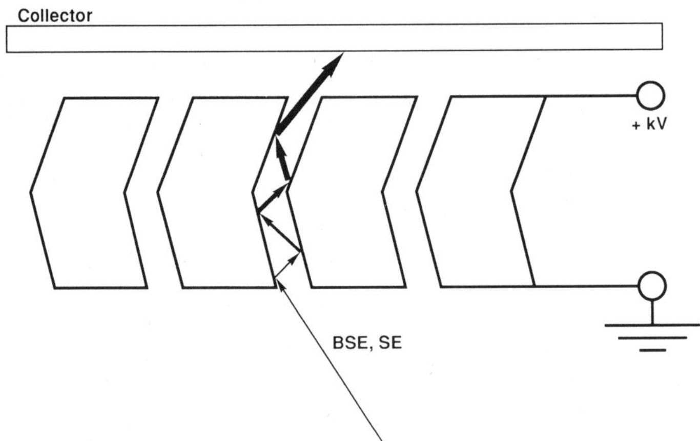

Figure 4.26. Schematic illustration of the channel-plate electron multiplier.

Channel Plate Detector. The interest in low-voltage microscopy has stimulated the development of electron detectors that can operate with high performance for electron energies in the range of low keV to hundreds of eV (Joy, 1989). The channel plate electron multiplier detector, illustrated in Fig. 4.26, operates on the principle of backscatter-to-secondary-electron conversion, but in this case the conversion occurs within the detector. The channel plate consists of at least one array of short (mm) glass capillaries with an accelerating potential applied at the exit face. When an energetic electron strikes the entrance of a capillary, it liberates secondary electrons, which are accelerated down the capillary in response to the applied potential. Each capillary acts as a miniature electron multiplier, so that every time the electrons strike the walls, more secondary electrons are emitted, building into a cascade. To increase the gain, two channel plates can be stacked to produce a chevron-shaped array. Because the secondary-electron yield actually increases as the energy of the incoming electron decreases into the keV range, the channel plate detector is ideal for low-voltage operation. Note that without post-specimen acceleration, the channel plate detector is not sensitive to secondary electrons.

4.3.1.3. Specimen Current (The Specimen As Detector)

Consider the interaction of the beam electrons to produce backscattered and secondary electrons. We normally consider discrete events involving single particles. For a 20-keV beam on copper, about 30 out of 100 beam electrons are backscattered ($\eta = 0.3$). The remaining 70 beam electrons lose all their energy in the solid, are reduced to thermal energies, and are captured by the solid. About 10 units of charge are ejected as secondary electrons ($\delta = 0.1$). This leaves 60 excess charges in

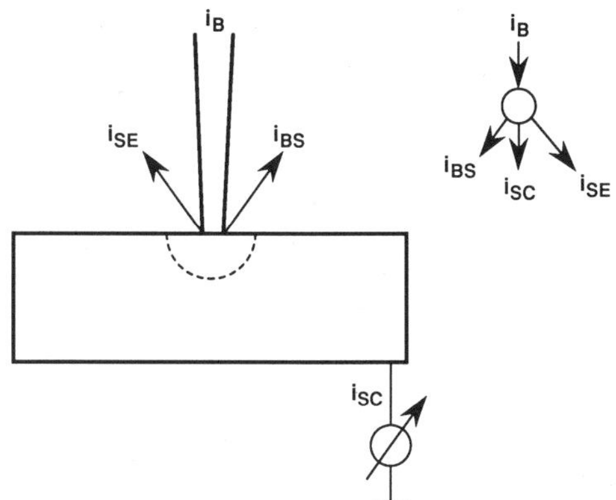

Figure 4.27. Illustration of currents which flow in and out of specimen: i_B, beam current; i_{BS}, backscattered electron current; i_{SE}, secondary-electron current; i_{SC}, specimen current. The junction equivalent of the specimen is also shown.

the target. What is the fate of these electrons? To understand this, an alternative view is to consider the currents, defined as charge per unit time, which flow in and out of the specimen. Viewed in this fashion, the specimen can be treated as an electrical junction, as illustrated schematically in Fig. 4.27, and is subject to the fundamental rules which govern junctions in circuits. By Thevinin's junction theorem, the currents flowing in and out of the junction must exactly balance, or else there will be net accumulation or loss of electrical charge and the specimen will charge on a macroscopic scale. If the specimen is a conductor or semiconductor and if there is a path to ground from the specimen, then electrical neutrality is maintained by the flow of a current, designated the *specimen current* (also *target current* or *absorbed current*), either to or from ground, depending on the exact conditions of beam energy and specimen composition. What is the magnitude of the specimen current?

Considering the specimen as a junction, the current flowing into the junction is the beam current i_B, and the currents flowing out of the junction are the backscattered electron current i_{BS} and the secondary electron current i_{SE}. For charge balance to occur, the specimen current i_{SC}, is given by

$$i_{SC} = i_B - i_{BS} - i_{SE}. \tag{4.8}$$

For the copper target, the backscatter current is $i_{BS} = \eta i_B = 0.3 i_B$ and the secondary electron current is $i_{SE} = \delta i_B = 0.1 i_B$. Substituting these values in Eq. (4.8) gives the result that the specimen current will be $i_{SC} = 0.6 i_B$, double the largest of the conventional imaging currents, the backscattered-electron signal. If a path to ground is not provided so that the specimen current can flow, the specimen charges rapidly.

Note that in formulating Eq. (4.8) no consideration is given to the large difference in energy carried by the backscattered and secondary electrons. Since current is the passage of charge per unit time, the ejection of a 1-eV secondary electron from the specimen carries the same weight as that of a 10-keV backscattered electron.

The specimen obviously serves as its own collector for the specimen current. As such, the specimen current signal is readily available just by insulating the specimen from ground and attaching a wire to ground to collect the specimen current. Does the specimen current signal actually convey useful information? As described below under contrast formation, the specimen current signal contains exactly the same information as that carried by the backscattered- and secondary-electron *currents*. Since our detectors measure a convolution of backscattered and/or secondary-electron current with other characteristics such as energy or directionality, the specimen current signal can give a unique view of the specimen (Newbury, 1976).

To make use of the specimen current signal, the current must be routed through an amplifier on its way to ground. The difficulty is that we must be able to work with a current similar in magnitude to the beam current, without any high-gain physical amplification process such as electron–hole pair production in a solid-state detector or the electron cascade in an electron multiplier. To achieve acceptable bandwidth at the high gains necessary, most current amplifiers take the form of a low-input-impedance operational amplifier (Fiori *et al.*, 1974). Such amplifiers can operate with currents as low as 10 pA and still provide adequate bandwidth to view acceptable images at slow visual scan rates (1500-line frame/s).

4.3.2. Cathodoluminescence Detector

Cathodoluminescence, the emission of electromagnetic radiation in the ultraviolet, visible, or infrared wavelengths, occurs for materials with a band gap structure and is useful for the characterization of semiconductors, certain minerals, and biological specimens. The radiation is detected with an appropriate photomultiplier for the photon-energy range of interest and may be dispersed through an optical spectrometer prior to detection to permit the recording of cathodoluminescence spectra.

The critical element in the design of a detection system is establishing an efficient coupling with the specimen, since cathodoluminescence is often weak. In the simplest case, the coupling consists of a lens and a light pipe to channel the light to the photomultiplier. To create a high geometric collection efficiency, an ellipsoidal mirror is placed so that the specimen occupies one of the foci and the entrance to the light guide is at the other focus (Horl and Mugschl, 1972; van Essen, 1974), as shown in Fig. 4.28. A hole drilled through the mirror permits the beam to reach the specimen. Light emitted from the specimen is focused from the ellipsoid to the second focus with nearly 2π-sr collection efficiency.

The ellipsoidal mirror surrounds the specimen, obviously precluding the efficient collection of backscattered or secondary electrons with any of the conventional detectors. In this situation, the specimen current signal is useful for electron imaging, since it only depends on the backscattered and secondary electrons leaving the specimen. The fact that they are intercepted by the mirror does not influence the specimen current, so electron-image information is still available with the specimen current.

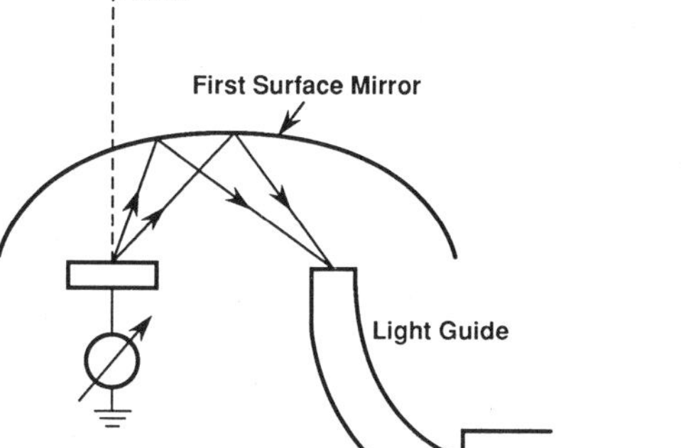

Figure 4.28. Cathodoluminescence radiation collection with an ellipsoidal mirror. The specimen is placed at one focus, and the light is reflected to the other focus, where it enters a light pipe for transmission to the photomultiplier.

4.4. Image Contrast at Low Magnification (<10,000x)

An SEM image is produced by recording as an intensity level on a CRT or film the product(s) of the interaction of the electron beam with the specimen at a series of points selected by the location of the beam. The task of image interpretation is to relate such gray-level images to the characteristics of the specimen so as to derive useful information. Two critical concepts form the basis of image interpretation:

1. the origin of contrast mechanisms and
2. the relationship of the size of the beam-interaction volume/signal-emitting region to the size of the picture element in the recording.

The understanding of contrast formation in the SEM has developed progressively throughout the history of the field and is the subject of a large literature (e.g., Oatley *et al.*, 1965; Oatley, 1972; Wells, 1974a, b; Goldstein *et al.*, 1975; Peters, 1982; 1984; Joy, 1991)

The following discussion considers the interpretation of image contrast at low magnification, $M < 10{,}000\text{X}$. For these low magnifications, the beam size and the spread of the signal due to the finite size of the interaction volume are such that the signal for each pixel is generated only from that pixel, so that neighboring pixels are essentially independent. The overlap of signals emitted from adjacent pixels is at a minimum. We should recognize that the selection of $M < 10{,}000\text{X}$ for the low-magnification regime of imaging is a "gray" number. Depending on the composition of the specimen, the spread of the beam can vary by more than a factor of five. Moreover, the shape of the surface-distribution function is also important. Higher-atomic-number targets produce more sharply peaked distributions of backscattered electrons.

4.4.1. Contrast

Contrast is defined according to the equation

$$C = \frac{(S_2 - S_1)}{S_2} \qquad S_2 > S_1, \tag{4.9}$$

where S_1 and S_2 represent the signals detected at any two arbitrarily chosen points in the scan raster that defines the image field. By this definition, C is always positive and is restricted to the range $0 \leq C \leq 1$. The concept of contrast and its numerical meaning form one of the important basic factors in scanning electron microscopy. Contrast is a measure of the real information in the signal related to the properties of the specimen, which we wish to determine. As discussed in detail later in this chapter, when the contrast available between two features of interest has a low value, establishing the visibility of those features requires a special strategy in operating the SEM, and even with proper application of the imaging strategy, image resolution is significantly poorer than for optimum situations of high contrast.

The signals in Eq. (4.9) are those which leave the detector, prior to any amplification in the signal-processing chain that eventually produces the displayed or recorded image. In discussing contrast, we must consider the specimen and detector as a complete system. Thus, the contrast carried in the signal leaving the specimen is a result of events within the specimen (e.g., scattering from different kinds or densities of atoms) or in its immediate vicinity (e.g., by electric or magnetic fields just above its surface). This contrast can be subsequently modified by the response of the detector; for example, the detector may be more sensitive to high-energy backscattered electrons. The signal leaving the detector represents the maximum amount of information available for the particular set of operating conditions employed. Subsequent amplification and signal processing, described in later sections, can serve only to control the way in which the information is displayed. The information in the signal cannot be increased after the detector. All subsequent processing, while useful for improving the match of the image display to the characteristics of the observer's visual system, cannot recover information not already present in the signal as it leaves the detector.

Contrast Components. Contrast can be influenced by a complex mix of the characteristics of the beam–specimen interaction, the properties of the specimen, the nature of the signal carriers, and the position, size, and response of the detector. We can identify three different *contrast components*:

Number Component. The number component refers to contrast which arises simply as a result of different numbers of electrons leaving the specimen at different beam locations in response to changes in the specimen characteristics at those locations.

Trajectory Component. The trajectory component refers to contrast effects resulting from the paths the electrons travel after leaving the specimen.

Energy Component. The energy component arises when the contrast is carried by a certain portion of the backscattered-electron energy distribution. Typically, the high-energy backscattered electrons are the most useful for imaging contrast mechanisms.

4.4.2. Compositional (Atomic Number) Contrast

4.4.2.1. Compositional Contrast with Backscattered Electrons

Consider a simple specimen, shown in Fig. 4.29a, consisting of a flat, atomically smooth block of an amorphous (i.e., noncrystalline) pure element, which is sufficiently thick relative to the electron range to be effectively infinite in thickness to the beam electrons at the incident energy. In forming an SEM image of this object, a sufficiently high magnification is chosen so that the area scanned is much smaller than the lateral area of the specimen, ensuring that the beam and the interaction volume never approach the edge of the specimen. Under such conditions, the backscattered-, secondary-, and absorbed-electron signals emitted from all beam locations in the scan are identical, except for statistical fluctuations (an extremely important factor to be considered subsequently in Section 4.5). From Eq. (4.9), no contrast is formed, and therefore no information is obtained about the specimen. Now consider a slightly more complicated specimen formed by placing two such amorphous regions adjacent, but with differing atomic numbers and an atomically sharp interface, as shown in Fig. 4.29b. Why might the signals differ from the two regions? If $Z_2 > Z_1$, then based upon our knowledge of electron-beam–specimen interactions, we can predict with a high degree of confidence that the behavior of backscattering is such that $\eta_2 > \eta_1$. The secondary-electron coefficients δ_2 and δ_1 from the two regions may also differ, but in a less predictable manner. Assume that, to at least a first approximation, the secondary coefficients are similar in the two materials. Considering then only the backscattered electrons collected for beam locations

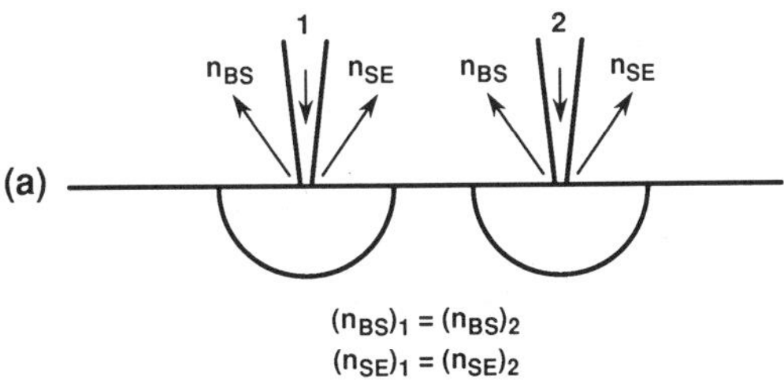

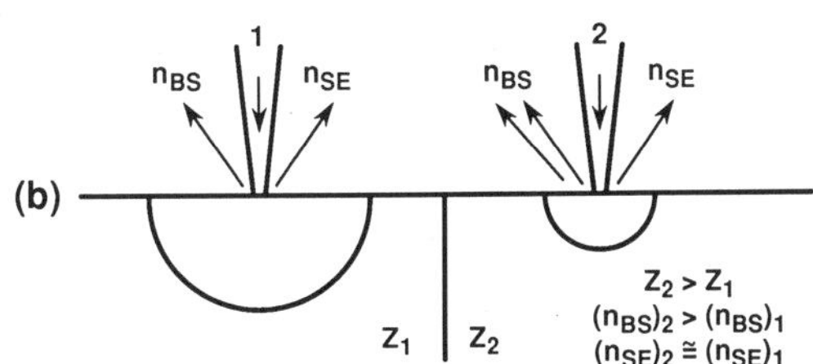

Figure 4.29. Schematic illustration of electron currents emitted from a sample which is (a) the same composition at two beam locations, and (b) different in composition. n_{BS} is the number of backscattered electrons, and n_{SE} is the number of secondary electrons.

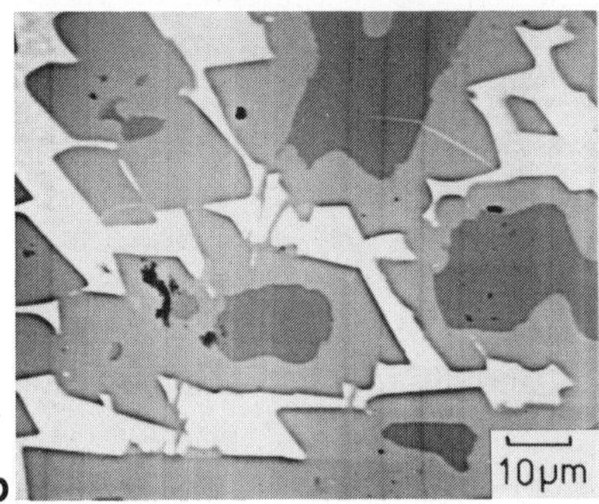

Figure 4.30. Atomic number (compositional contrast) observed in an aluminum–nickel alloy (Raney nickel). (a) Backscattered-electron image derived from a negatively biased Everhart–Thornley detector. (b) Direct specimen-current image of the same region; note the contrast reversal compared to (a).

sufficiently far from the interface so that the interaction volume is not affected by the interface region, contrast should exist between two regions. This contrast mechanism is referred to as "atomic number," "compositional," or "material" contrast. An example of this contrast mechanism is shown in Fig. 4.30, which depicts a multiphase aluminum–nickel alloy. Note the reversal of the sense of the contrast between two emitted electron and absorbed electron images, as predicted by Eq. (4.9).

Atomic number contrast has the following characteristics:

1. Because of the nearly monotonic increase of η with Z (or weight-fraction-averaged Z for compound targets), regions of high average atomic number appear bright relative to regions of low atomic number. Thus, the gray levels in Fig. 4.30a can be interpreted as regions of different composition with increasing average atomic number corresponding to increasing brightness.

2. The magnitude of atomic-number contrast can be predicted. If we make use of a detector sensitive to the number of backscattered electrons, e.g., a negatively biased Everhart–Thornley detector, a solid-state diode, or the specimen current, then the signal from the detector is proportional to the backscatter coefficient, neglecting any energy response characteristics of the detector:

$$S_{\text{detector}} = e_{\text{BS}}\eta, \tag{4.10}$$

where e is the efficiency with which the backscattered electrons are detected. The efficiency involves both the geometric effects i.e., the angular size of the detector and the energy response of the detector. From Eq. (4.9), the contrast can be calculated as

$$C = \frac{(S_2 - S_1)}{S_2} = \frac{(e_{BS}\eta_2 - e_{BS}\eta_1)}{e_{BS}\eta_2} = \frac{(\eta_2 - \eta_1)}{\eta_2}. \quad (4.11)$$

In Eq. (4.11), e_{BS} can be cancelled only if the same detector response per electron is obtained from the backscattered electrons emitted from both materials. This condition holds exactly only for the specimen-current signal. A detector such as the solid-state diode has a response which, above a threshold, increases proportionally to the energy of the backscattered electrons. Thus, if element 2 has a high atomic number (e.g., gold) while element 1 has a low atomic number (e.g., aluminum) the greater fraction of high-energy backscattered electrons from the high-atomic-number target has the effect of making e_{BS} an increasing function of Z. The energy-distribution effect on the contrast can be calculated by convolving the energy distribution from Fig. 3.21 with the energy response of the backscatter detector, Fig. 4.25. In an extreme case, e.g., gold versus carbon, the contrast may be increased by as much as a factor of two by the energy dependence of the detector.

Equations (4.09)–(4.11) introduce the concept of a contrast calculation, that is, relating a property of the specimen, e.g., composition, to the signal which it modifies, e.g., backscattering, to produce contrast, which provides the information in the measured signal. A contrast calculation is the first step in placing scanning electron microscopy on a quantitative basis.

3. The contrast calculated from Eq. (4.11) and the Reuter (1972) equation fitted to η vs Z (Eq. 3.19) for several pairings of elements is listed in Table 4.5. Elements separated by one unit of atomic number produce low contrast; e.g., Al and Si yield a contrast of only 0.067 (6.7%). For pairs widely separated in atomic number, the contrast is much larger; e.g., Al and Au produce a contrast of 0.69 (69%).

4. When pairs of elements separated by one unit of atomic number are considered, the predicted contrast decreases as the atomic number increases, as shown in Table 4.5 and Fig. 4.31. This behavior occurs because the slope of the η-vs-Z curve decreases with increasing Z. The decrease in contrast with Z is quite pronounced, dropping from 0.14 (14%) for B–C to 0.067 (6.7%) for Al–Si and 0.0041 (0.41%) for Au–Pt. As discussed later in this chapter, contrast above 10% is relatively easy to image in the SEM, contrast in the range 1–10% requires a careful strategy, and contrast <1% requires extreme measures to successfully image it. As noted in Chapter 3, the curve of η vs Z is in fact not perfectly monotonic, and in some cases the backscatter coefficient may actually decrease with a unit increase in Z, resulting in an unexpected reversal in the atomic number contrast.

5. Specimen tilt influences atomic number contrast. As shown in Fig. 3.17, the backscatter coefficients from highly tilted specimens tend

Table 4.5. Atomic Number Contrast (E_0 = 20 keV)

Z_A	Z_B	η_A	η_B	C
13 (Al)	14 (Si)	0.153	0.164	0.067
13 (Al)	26 (Fe)	0.153	0.279	0.451
13 (Al)	79 (Au)	0.153	0.487	0.686
		Adjacent elements		
5 (B)	6 (C)	0.055	0.067	0.14
13 (Al)	14 (Si)	0.153	0.164	0.067
26 (Fe)	27 (Co)	0.279	0.287	0.028
41 (Nb)	42 (Mo)	0.375	0.379	0.013
57 (La)	58 (Ce)	0.436	0.439	0.0068
78 (Pt)	79 (Au)	0.485	0.487	0.0041

toward the same value of unity at grazing incidence, so the atomic-number contrast between any two elements decreases as tilt increases.

6. The directionality of backscattering also has an influence on atomic number contrast. For normal beam incidence, backscattering follows a cosine distribution, so the most favorable detector placement to maximize atomic number contrast is at a high take-off angle directly above the specimen. A backscatter detector placed at a low take-off angle intercepts a smaller fraction of the backscattered electrons; these electrons tend to be those that have lost the most energy and are less sensitive to atomic number effects. At high tilt angles, backscattering is strongly peaked in the forward direction, so that the most favorable detector location for atomic number contrast is below the beam impact point on the specimen. For a highly tilted specimen, the usual location of the solid state backscattered electron detector on the polepiece of the final lens is not favorable for atomic number contrast.

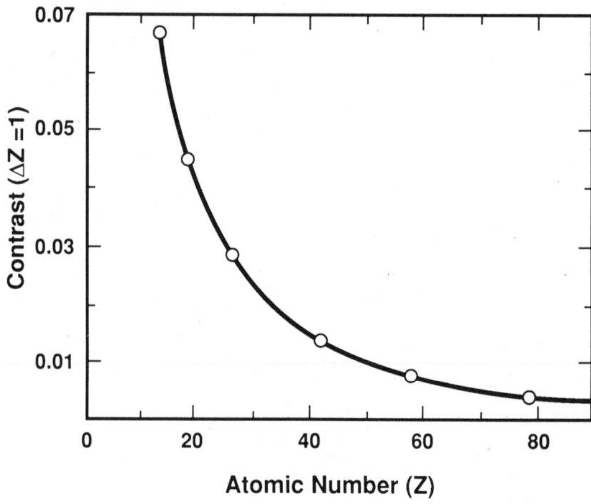

Figure 4.31. Atomic number contrast from adjacent pairs of elements ($\Delta Z = 1$) as predicted by a fit to the η-Z-curve.

4.4.2.2. Compositional Contrast with Secondary Electrons

When atomic number contrast is imaged with the Everhart–Thornley detector operated in its conventional positively biased mode, a secondary-electron component and an additional backscattered-electron component are added to the signal and the contrast calculation becomes more complicated:

$$C = \frac{(S_2 - S_1)}{S_2} = \frac{[(e_{\mathrm{BS}}\eta_2 + e_{\mathrm{SE}}\delta_2) - (e_{\mathrm{BS}}\eta_1 + e_{\mathrm{SE}}\delta_1)]}{(e_{\mathrm{BS}}\eta_2 + e_{\mathrm{SE}}\delta_2)}, \quad (4.12)$$

where the subscripts represent the efficiency with which backscattered electrons and secondary electrons are collected. As discussed above, the Everhart–Thornley detector in the positively biased mode has a complex response to secondary and backscattered electrons. The efficiency for backscattered electrons actually increases markedly when the bias is changed from negative (backscatter only) to positive (backscatter plus secondary) through the indirect collection of backscattered electrons by means of the remotely generated secondary electrons, increasing the total signal. Although the total signal increases, the contrast from the backscattered electron contribution is expected to be unchanged, because e_{BS} is similar for the two materials. At high beam energies, 10–30 keV, the contribution from secondary electrons in Eq. (4.12) is generally similar for two materials, because δ is not a strong function of Z. At beam energies below 10 keV, the secondary coefficient increases and can influence the observed atomic number contrast. The lack of reliable data on secondary-electron coefficients, especially at low beam energy, makes interpretation of atomic number contrast difficult in this energy range.

While the secondary-electron contribution to compositional contrast is generally small and unpredictable, there are some noteworthy exceptions in semiconductor materials, where secondary electron signals can vary strongly with very small changes in composition (Lifshin and DeVries, 1972; Sawyer and Page, 1978). An example of this effect in reaction-bonded silicon carbide is given in Fig. 4.32, which shows images of the same field of view when the Everhart–Thornley detector is biased positively (backscattered plus secondary electrons) and negatively (backscattered electrons only). Reaction-bonded silicon carbide consists of grains of silicon carbide partially melted in an excess of molten silicon. As the liquid-plus-solid mixture cools, the dissolved silicon carbide deposits epitaxially on the unmelted grains, forming crystalline material of the same orientation as each original grain. The excess silicon is the last material to freeze out between the silicon carbide grains. In Fig. 4.32b, the image prepared with backscattered electrons shows only the expected atomic number contrast between the silicon (bright) and the silicon carbide (dark). Figure 4.32a, the corresponding secondary-plus-backscattered-electron image, shows reversed contrast between the silicon (black) and silicon carbide (bright), but more striking is the extraordinary contrast within the silicon carbide grains, which generally appear to have a dark center surrounded by a band of bright material.

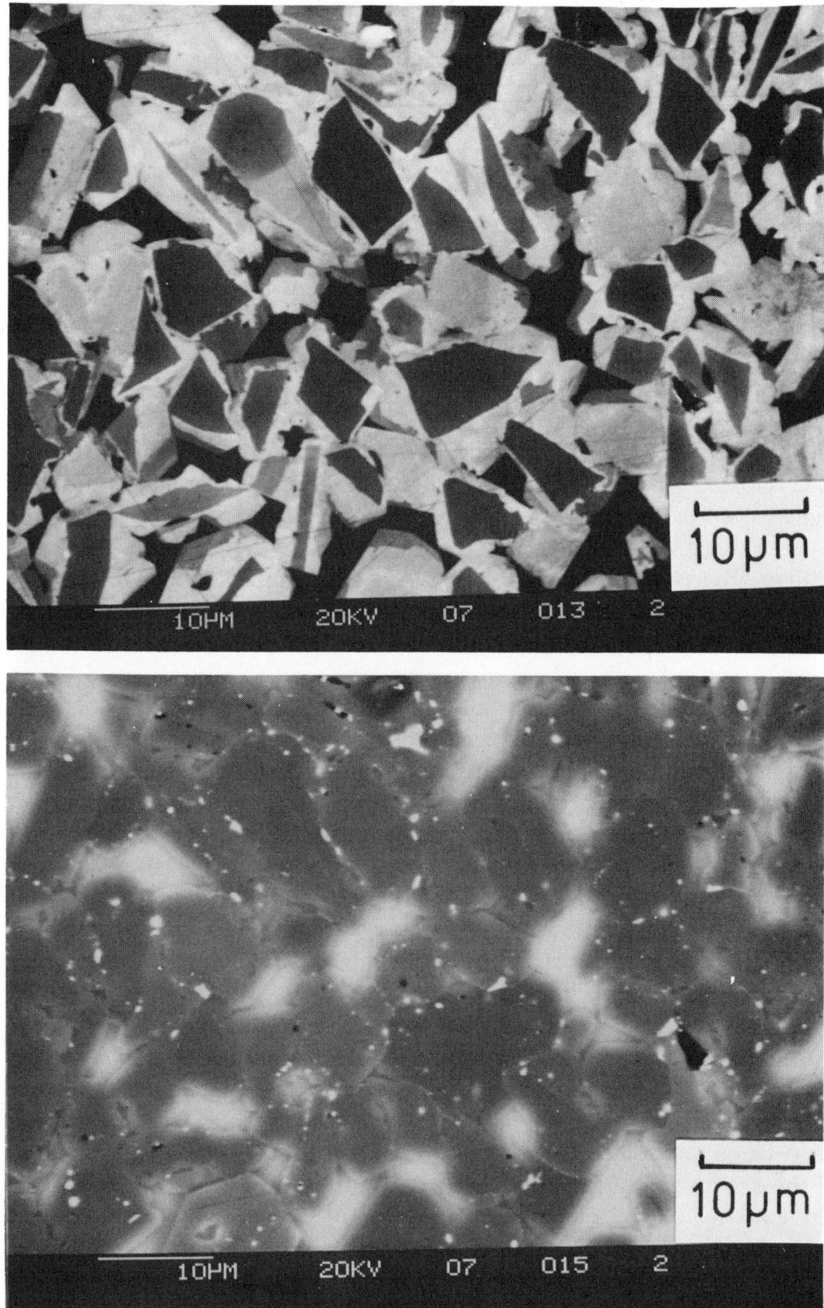

Figure 4.32. Compositional contrast observed from a sample of reaction-bonded silicon carbide. (a) Secondary- plus backscattered-electron signal derived from a positively biased Everhart–Thornley detector. (b) The same area viewed with backscattered electrons only from a negatively biased Everhart–Thornley detector. (Courtesy of Dr. Trevor Page.)

Sawyer and Page (1978) proposed that this zoned contrast is a result of impurity doping effects in the semiconductor. Secondary electrons are generated by scattering with weakly bound conduction-band electrons. In a high-purity semiconductor, the valence band is filled and is separated by an energy gap from the empty conduction band (see Fig. 3.51). When atoms of an impurity are present, additional energy levels in the band gap may become occupied, providing a source of electrons for thermal promotion to the conduction band. The zoned contrast within the grains is seen as a result of the differing impurity content between the unmelted seed and the new, epitaxial growth from the melt. The difference in impurity levels between the seed and epitaxial material is so slight that it cannot be detected in the pure backscattered electron image or by wavelength-dispersive x-ray spectrometry. Newbury *et al.* (1986) demonstrated by means of secondary ion mass spectrometry and ion microscopy that the spatial distribution of the trace aluminum constituent in the silicon carbide, present at a level of approximately 100 ppm, matched the observed secondary-electron contrast.

Compositional contrast in the secondary-electron signal is very sensitive to the condition of the sample surface. An evaporated carbon layer applied to the specimen or a thin contamination layer produced during electron bombardment can completely suppress secondary-electron effects such as those seen in Fig. 4.32a. Compositional contrast effects in the secondary-electron signal would probably be seen more frequently if it were not for the contamination that often builds up on specimens due to inadequate cleaning of the surface as well as residual hydrocarbon gases in SEMs equipped with conventional oil-pumped vacuum systems, conventional trapping, and polymer vacuum seals.

4.4.2.3. Compositional Contrast with Specimen Current

To understand the appearance of the specimen in an image prepared with the specimen-current signal, consider how the signals change between any two locations. Using Eq. (4.8) as a starting point, the difference in signals between any two locations can be calculated. To simplify the argument, consider that the backscattered- and secondary-electron signals are combined into a emissive signal

$$i_E = i_{\mathrm{BS}} + i_{\mathrm{SE}}. \tag{4.13}$$

With this substitution, Eq. (4.8) becomes

$$i_B = i_E + i_{\mathrm{SC}}. \tag{4.14}$$

The difference in the signals between any two pixels is found by taking differences for each term:

$$\Delta i_B = 0 = \Delta i_E + \Delta i_{\mathrm{SC}}. \tag{4.15}$$

Because the electron-optical column is carefully constructed to maintain a constant beam current, the difference in the beam current between any two points in the image is zero, except for statistical fluctuations.

Equation (4.15) can thus be rearranged to give the relationship between the emissive and the specimen current signals:

$$\Delta i_{\mathrm{SC}} = -\Delta i_E. \tag{4.16}$$

The sense of the contrast is thus opposite in the specimen-current image compared to the image recorded with a detector of emissive-mode signals. This contrast reversal can be seen in the specimen-current image of Fig. 4.30b, where regions appear bright which are dark in the emissive-mode image of Fig. 4.30a.

While this contrast reversal may seem to be a trivial change, a more subtle difference exists between the specimen-current and emissive-mode signals. Specimen current is sensitive only to the numbers of electrons leaving the specimen and is completely insensitive to their trajectories. As long as a secondary or backscattered electron leaves the specimen, it contributes information to the specimen-current signal, regardless of its fate (to be collected or to be lost). Of all possible detectors, only specimen current is sensitive to number effects only. Trajectory and energy effects are completely eliminated. If the specimen is biased to suppress secondary emission, the specimen-current signal can be rendered sensitive to backscattered-electron effects only.

Because the direct specimen-current image gives the reversed sense to atomic number contrast, it is common practice to apply a signal processing operation to artificially reverse the sense of the signal (reversed specimen-current image) so that a bright area corresponds to higher atomic number.

4.4.3. Topographic Contrast

4.4.3.1. Origin

Topographic contrast includes all of those effects by which the topography (shape) of the specimen can be imaged. Since the vast majority of applications of the SEM involve studying the shapes of specimens, topographic contrast is the most important of all contrast mechanisms. As discussed in detail below and in Chapter 3, topographic contrast arises because the number and trajectories of backscattered electrons and the number of secondary electrons depend on the angle of incidence between the beam and the specimen's surface. Except for the lowest magnifications, the incident beam can be thought of as effectively parallel over the scanned field. The angle of incidence varies because of the local inclination of the specimen. At each point the beam strikes, the numbers of backscattered and secondary electrons measured by the Everhart–Thornley detector give direct information on the inclination of the specimen. Surprisingly, an observer, even an untrained observer or a child, can interpret such topographic images and understand the true shape of the specimen without any prior knowledge of electron interactions or detector characteristics. In this section we will examine in detail the process of forming the topographic image to understand this convenient but unexpected property of SEM images.

From the information presented in Chapter 3, the following effects can be expected to contribute to the formation of topographic contrast:

1. The backscatter coefficient increases as a monotonic function of the specimen tilt. The more highly inclined the local surface is to the incident beam, the higher the backscatter coefficient. This effect produces a *number component* contribution to topographic contrast in backscattered electrons.

2. The angular distribution of backscattered electrons is strongly dependent on the local surface tilt. At normal beam incidence (local tilt angle $\theta = 0°$), the angular distribution relative to the local surface normal follows a cosine function. As the local tilt angle increases, the angular distribution gradually deviates from the cosine law and becomes distorted, with the peak in the distribution away from the incident beam. For $\theta > 50°$, the angular distribution becomes highly asymmetric, with the peak in the forward scattering direction. The maximum likelihood for scattering occurs in a plane defined by the beam vector and the local surface-normal vector. The directionality of backscattering from tilted surfaces contributes a *trajectory component* to the backscattered-electron signal.

3. The secondary-electron coefficient varies monotonically with the specimen's tilt angle, varying approximately as a secant function. Tilted surfaces thus yield more secondary electrons than surfaces normal to the beam. This effect introduces a *number component* to topographic contrast in the secondary-electron signal. The angular distribution of secondary-electron emission does not vary significantly with tilt angle.

The topographic contrast actually observed depends on the detector used and its placement relative to the specimen and on the exact mix of backscattered and secondary electrons detected. The Everhart–Thornley detector is most commonly used for imaging topographic contrast, but depending on the application, a dedicated backscattered-electron detector may be used as an additional source of information. To understand how the number and trajectory components combine in the signal from the Everhart–Thornley detector to produce the topographic images with which we are familiar, we will construct in a stepwise fashion the image of a specimen which presents only topographic contrast. The specimen is a fracture surface from a piece of high-purity iron so that there is no contribution from atomic number contrast. Such a contrast-component separation experiment can be performed because the E–T detector permits separate collection of a portion of the backscattered electrons while excluding secondary electrons.

4.4.3.2. Topographic Contrast with the Everhart–Thornley Detector

Backscattered Electron Component. We shall first consider the appearance of the fracture surface with the portion of the backscattered-electron signal collected by the E–T detector along the line of sight to the specimen. This backscattered-electron signal is selected by placing a negative bias (usually in the range −300 to −50 V) on the Faraday cage which surrounds the scintillator so as to exclude *all* secondary electrons, as shown schematically in Fig. 4.33a. Under these conditions, three properties of the E–T detector affect the contrast which is observed.

1. The detector is located on one side of the specimen, so it has an anisotropic view of the specimen.
2. The solid angle of collection of the detector is small, so that only a small fraction of the backscattered electrons can be collected.
3. The detector is at a low take-off angle relative to the horizontal plane (i.e., the detector's line of sight is at a high angle relative to the beam).

These three factors combine to give the detector a highly directional view of the specimen. The actual appearance of the

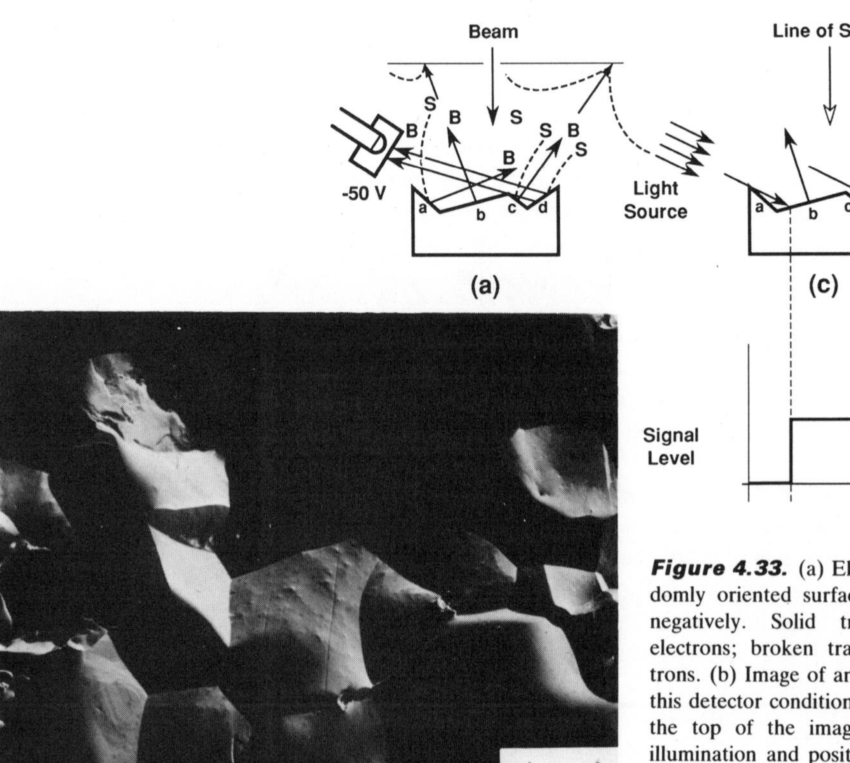

Figure 4.33. (a) Electron collection from randomly oriented surfaces. E–T detector, biased negatively. Solid trajectories, backscattered electrons; broken trajectories, secondary electrons. (b) Image of an iron fracture surface with this detector condition; the detector is located at the top of the image. (c) Equivalent optical illumination and position of the line of sight to produce a similar image. (d) Schematic of signal as a function of beam position.

fracture surface under this condition, shown in Fig. 4.33b, is characterized by bright regions from which very high signals are collected and dark regions from which essentially no signal is collected. As a result, the contrast in the image is harsh, with few intermediate gray levels represented. This appearance can be explained as follows, with reference to the schematic diagram, Fig. 4.33a: Although all surfaces struck by the primary electron beam produce backscattered electrons, only those surfaces which face the scintillator direct at least some of their backscattered electrons toward it, resulting in a detectable signal. Secondary electrons are rejected from all surfaces. Because backscattering from tilted surfaces is increasingly directional with tilt angle, any surface tilted away from the detector sends very few backscattered electrons toward the detector, resulting in a low signal, so that the surface appears dark. The signals are so low that the only information conveyed from these surfaces is the fact that the tilt must be *away* from the E–T detector. Shadowing effects of topography are also found. In Fig. 4.33a, a portion of face *d* is screened from the detector by the ridge formed by surfaces *b* and *c*.

As observers of SEM images, we subconsciously tend to interpret images based on our most familiar imaging situation, namely, light interacting with everyday objects. Oatley *et al.* (1965) demonstrated that a highly effective light-optical analogy can be constructed for the SEM imaging process of topographic contrast from rough, three-dimensional objects in images prepared with the E–T detector. The light-optical analogy to Fig. 4.33a and b is illustrated in Fig. 4.33c. If we wanted to view the rough object with light and produce the same sense of contrast as seen in the SEM image, it would be necessary to place a directional light source such as a flashlight at the position of the E–T detector and place our eye as detector looking down on the specimen from the position of the electron beam. As in Fig. 4.33a the surfaces tilted toward the flashlight appear bright, and those tilted away appear dark. Shadowing effects from local topography are also found; ridges which occur between the detector and the object shadow the object. The light strength received by the viewer along the line of sight is shown in Fig. 4.33d.

Secondary + Backscattered Electron Component. Let us now consider the case where the Faraday cage is biased positively (usually in the range +50 V to +300 V), as illustrated in Fig. 4.34a, so that secondary electrons are collected. The signal is now a complex mixture of three components:

Direct Backscattered Electrons. The schematic diagram of Fig. 4.34a has been drawn to indicate that the backscattered electrons directly collected or lost in the positive bias case behave in virtually the same way as for negative bias. A positive bias that is small (relative to the incident beam energy) does not significantly deflect the trajectories of high-energy backscattered electrons to alter their *direct* collection.

Secondary Electrons. The small positive bias does profoundly

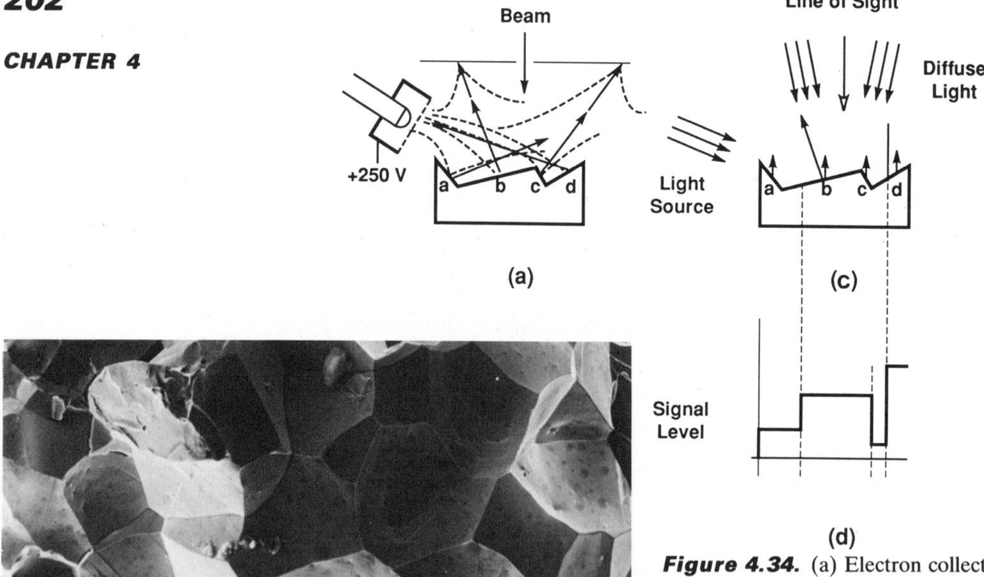

Figure 4.34. (a) Electron collection from randomly oriented surfaces. E–T detector, biased positively. Solid trajectories, backscattered electrons; broken trajectories, secondary electrons. (b) Image of an iron fracture surface with this detector condition; the detector is located at the top of the image; same field as in Fig. 4.33b. (c) Equivalent optical illumination and position of the line of sight to produce a similar image. (d) Schematic of signal as a function of beam position.

alter the collection of secondary electrons. Secondary electrons are emitted from all surfaces struck by the primary beam, with the number increasing sharply as the secant of the local tilt angle. The contrast in the secondary-electron signal produced by this number effect can be estimated as follows: The secondary-electron coefficient δ given off by a surface with tilt θ is given by

$$\delta = \delta_0 \sec \theta, \tag{4.17}$$

where δ_0 is the secondary-electron coefficient at 0° tilt. The difference in signal $d\delta$ between two surfaces with a difference in tilt angle $d\theta$ is found by differentiating Eq. (4.13):

$$d\delta = \delta_0 \sec \theta \tan \theta \, d\theta. \tag{4.18}$$

The contrast C is found by forming the ratio of Eqs. (4.18) and (4.17):

$$C = d\delta/\delta = \frac{\delta_0 \sec \theta \tan \theta \, d\theta}{\delta_0 \sec \theta} = \tan \theta \, d\theta \, . \tag{4.19}$$

In addition to the number component of contrast given by Eq. (4.19), there is a strong trajectory effect on secondary-electron collection. Secondary electrons are emitted from all surfaces according to a cosine distribution relative to the local surface normal. However, because secondary electrons are low in energy (most are emitted with less than 10 eV), the positive collection field on the Faraday cage of the E–T detector can cause wide deflections of the secondary-electron trajectories, resulting in extremely high collection efficiency from surfaces which face the detector and at least partial collection from surfaces tilted away from the detector. Some secondary electrons are collected from all surfaces which the primary beam strikes.

Indirect Backscattered Electrons. In Figs. 4.33a and 4.34a, the backscattered electrons not intercepted by the detector for direct collection strike the polepiece of the final lens and the walls of the specimen chamber. These backscattered electrons retain sufficient energy to cause secondary electrons to be emitted wherever they strike interior surfaces of the microscope, and due to elastic scattering, they may re-emerge and travel to strike other microscope surfaces, producing still more secondary electrons. All of this scattering and secondary-electron production takes place within the dwell time of the beam on the pixel to which it has been addressed. The positive bias on the E–T detector results in such an efficient secondary electron "sponge" that a significant fraction of these remotely produced secondary electrons from the chamber walls are collected as part of the total signal. What do they contribute to the signal? Surprisingly, these secondaries do not represent the true secondary-electron component, which is related to the specimen location to which the beam is addressed. These remote secondary electrons actually represent *indirect* collection of backscattered electrons. The number of remote secondaries rises and falls with the number of backscattered electrons emitted from the sample.

The image obtained with the positively biased E–T detector is shown in Fig. 4.34b, which depicts the same area as Fig. 4.33b. With positive bias on the E–T detector, the harsh trajectory component of backscattered-electron topographic contrast is reduced, since backscattered electrons are collected with at least some efficiency no matter how well screened a feature is from direct sight to the detector due to the mechanism of indirect collection of remote secondary electrons. Note that those features which face the detector and appear bright in Fig. 4.33b remain the brightest features in Fig. 4.34b because of the unchanged situation for the direct collection of backscattered electrons. Features facing the detector appear as highlights in the image. The equivalent lighting situation for the positively biased E–T detector is illustrated in Fig. 4.34c, showing the same situation as Fig. 4.33 but augmented with a general diffuse illumination. Now from the point of view of an observer looking down on the scene, all surfaces are illuminated, and those tilted toward the directional source of light are highlighted.

4.4.3.3. Light–Optical Analogy

The useful fact that SEM images of rough, three-dimensional objects prepared with a positively biased E–T detector can be readily interpreted, even by viewers with no prior knowledge of the nature of the SEM Image-formation process, is a direct result of the existence of a familiar light-optical equivalent to the SEM imaging process. As observers of our environment, we are used to viewing scenes lighted from above by a highly directional source, the sun, supplemented by diffuse illumination due to light scattered by the atmosphere. The light-optical analogy shown in Fig. 4.34c is equivalent to this situation. To properly interpret the SEM image, we must match the components of the two systems that have similar characteristics. Thus, the electron beam is a highly directional component, since only those positions of the specimen that can be directly struck by the beam are accessible in the final SEM image. The human eye is highly directional; with unaided vision we cannot see the person standing behind a tree. To establish the light-optical analogy, we must place the viewer at the position of the electron beam, i.e., viewing the specimen from above as if looking through the final aperture with our eyes. The positively biased E–T detector has a highly directional signal component, the direct BSE component, and a diffuse component, the true SE and remote BSEs collected as the SE_{III} component. The E–T detector thus has the characteristics of the sun in our everyday illumination situation, with direct sunbeams and atmospherically scattered light. To interpret the SEM image, we therefore must imagine that the direct illumination seems to come from the E–T detector, as if we place a flashlight at its position in the specimen chamber and we view that specimen by looking along the electron beam. To complete the recipe for the light-optical analogy, since the everyday illumination is from above (sun high in the sky), the E–T detector must be located at the top of our SEM frame, looking down on the specimen.

The light-optical analogy is a powerful effect, virtually automatic in its influence on an observer, but to ensure proper interpretation of the topography of surfaces, it is vital that we always establish this condition of top lighting in our SEM images. When the lighting source comes from an unexpected direction, e.g., below or from the side, our ability to recognize features is compromised. (Consider the disorienting view of a human face lit from below with a flashlight in a dark room!) We are so conditioned to expect top lighting that failure to provide top lighting can cause the apparent sense of the topography to invert, particularly where the topography is shallow and there are no clues from experience or prior knowledge of a specimen. *It is important to note that the apparent lighting may not be correct in an arbitrarily selected SEM.* (In a multiuser facility, always assume the worst and check!). The apparent lighting is controlled by the position of the E–T detector relative to the scanned field. The E–T detector is rigidly fixed in the specimen chamber. To establish the proper illumination for the light-optical analogy to function, the scan must be oriented to place the apparent position of the E–T detector at the top of the image, both on the visual CRT and on the final photographs. The SEM usually has an electronic scan rotation feature that permits the microscopist to orient the image on the screen to any position through a rotation of 360°. This can be a dangerous operation, since the apparent position of the E–T detector travels around the image field as the scan is rotated. If the apparent position of the E–T detector is placed at the bottom of the image field, the sense of topography may invert. To avoid this error, the following procedure can be used to establish the proper setting for scan rotation: The microscopist should first determine in what way the mechanical translation of the specimen stage moves the specimen toward the E–T detector. That is, does the X- or Y-drive (or a combination) move the stage toward the E–T detector? With the SEM operating in the normal fashion, operate this mechanical drive (or drivers) to move the specimen toward the E–T detector, and observe the apparent motion on the CRT screen. The apparent motion should be from the bottom to the top of the frame. If not, the scan rotation should be adjusted to produce this result. (Mark this important position on the control knob). This will ensure that top lighting is obtained.

A comparison of proper and improper lighting caused by the choice of the E–T detector position is shown in Fig. 4.35. For some (but not all) observers, the perceived sense of the topography reverses for the bottom-lighting situation.

Establishing a top-lighting condition produces a situation where the sense of the topography can be properly interpreted in most cases. Occasionally images will be encountered where the sense of the topography is not obvious, perhaps because of the shallowness of the relief or some local peculiarity in the mix of secondary and backscattered electrons collected. In such cases, tilting or rotating the specimen to view that particular field from a different angle may be necessary, or the preparation of stereo pairs may be needed to properly assess the nature of the topography.

4.4.3.4. Topographic Contrast with Other Detectors

A typical arrangement of detectors surrounding a topographic specimen is shown in Fig. 4.36. The appearance of a rough surface varies significantly with the type, solid angle, and take-off angle of the detector used to collect the signal. The appearance of topography may differ sharply from the familiar image obtained with the positively biased E–T detector because the light-optical analogy may not be satisfied for other types of detectors.

Dedicated Backscattered-Electron Detector. Consider a dedicated backscattered-electron detector having a large solid angle and placed at a high take-off angle, as shown in Fig. 4.36 (Robinson, 1975). The detector consists of two segments, A located at the top of the image field and B located at the bottom. Figure 4.37a shows the same field of view as Figs. 4.34 and 4.35 with the sum of the signals from the detectors, A + B. The image is distinctly different in many respects from the conventional positively biased E–T image. With a

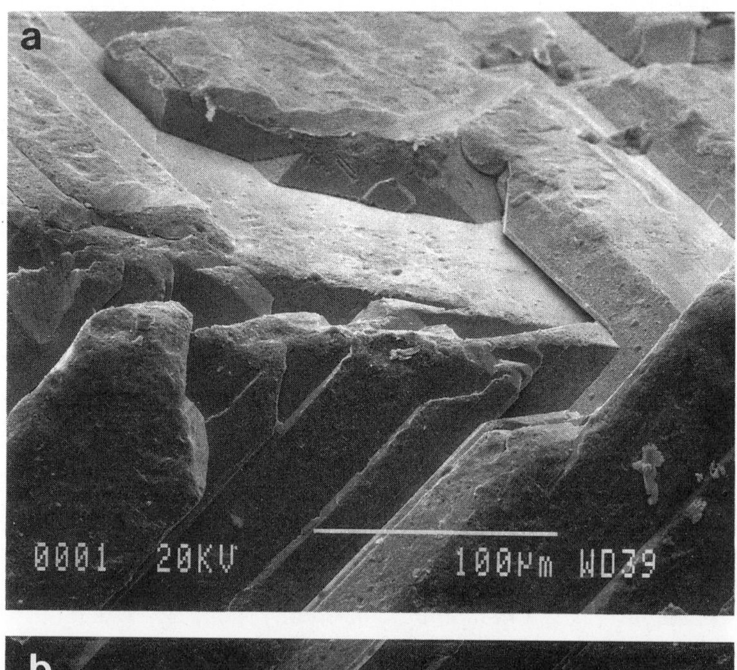

Figure 4.35. Comparison of topographic (positively biased E–T detector) images taken (a) with the proper sense of lighting by placing the E–T detector at the top of the field of view and (b) with the improper sense of lighting by placing the E–T detector at the bottom of the field.

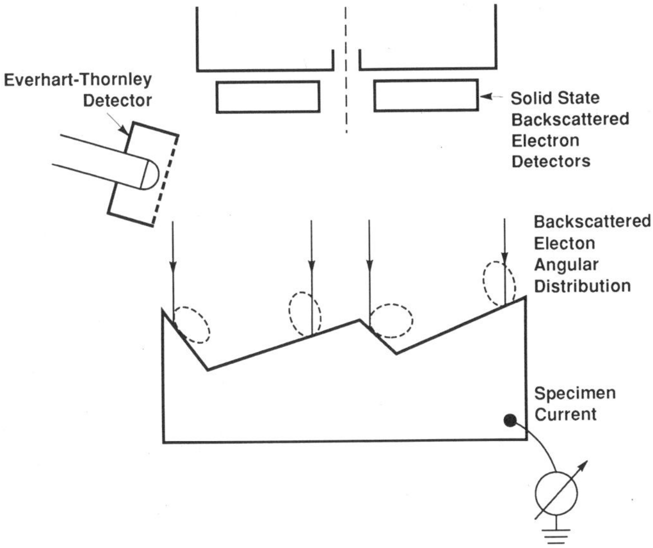

Figure 4.36. Typical arrangement of detectors in a SEM chamber relative to a rough surface with random facets.

large-solid-angle detector placed near the beam at a high take-off angle, a signal is obtained preferentially from surfaces nearly perpendicular to the beam, while highly tilted surfaces appear dark because electrons backscattered from these surfaces are directed away from the detector. The topographic contrast is somewhat reduced, since electrons emitted from surfaces tilted in opposite directions are still likely to be collected because of the large solid angle of the detector. However, the sense of the topographic contrast is reversed, so the shapes appear "inside out" (compare Figs. 4.35b and 4.37a). Another striking difference is the greater sensitivity to the small dimple-like structures in the dedicated backscattered-electron detector image.

When the image is formed as the difference between the two detectors, A–B, the appearance of the topography again changes and becomes very similar to the conventional positively biased E–T image as far as the general shading of the facets is concerned. Note, however, that the fine-scale dimpling is suppressed in the difference image.

Specimen-Current Signal. To understand the appearance of the specimen-current image of the topography, two properties of specimen current must be recalled. First, from Eq. (4.16), the sense of topographic contrast is expected to be reversed in the specimen-current image as compared to the emissive-mode image. Second, specimen current is sensitive only to number effects and completely insensitive to trajectory effects, which contribute strongly to emissive-mode images prepared with the E–T detector, and to effects of the energy distribution of backscattered electrons, to which solid state detectors are sensitive.

Figure 4.37. The same field of view as in Figs. 4.34 and 4.35, but viewed with a dedicated backscattered-electron detector having a large solid angle and placed at a high take-off angle. The detector consists of two segments, A and B. E_0 = 15 keV. Segment A is located at the top of the field and segment B at the bottom. (a) Image obtained from the sum (A + B) of the detector signals. (b) Image obtained from the difference (A − B) of the two signals. (c) The same field of view as Figs. 4.34 and 4.35, but viewed with the direct specimen-current signal; (d) reversed-contrast specimen-current image.

The direct specimen–current image is shown in Fig. 4.37c. The sense of the topography appears reversed relative to the emissive-mode image, Fig. 4.35b, as expected from Eq. (4.16). The fine-scale dimples on the grain surfaces appear prominently in the specimen current image. An unexpected effect is the uniform appearance of facets at approximately equal tilt to the beam. The facets of the grain boundary triple junction in the lower left of the image appear uniform in the specimen-current

Figure 4.37. (Continued).

image, but these facets have different brightness in the positively biased E–T image and in the dedicated backscattered-electron image because of trajectory effects, which are completely absent in the specimen-current image. Since both backscattering- and secondary-electron emission increase monotonically with tilt, the magnitude of the specimen current can be used to quantitatively assess the local tilt angle.

The distraction of the reversal of topography encountered in the direct specimen-current image can be eliminated by using contrast reversal in the signal-processing chain to produce the image shown in Fig. 4.37d, which displays the proper sense of the topography as expected from the point of view of the emissive-mode detector. Unless this contrast reversal is applied, the sense of topography obtained from a direct specimen-current image is incorrect.

4.4.3.5. Separation of Contrast Components

When the character of the specimen is such that multiple contrast mechanisms can operate, the resulting SEM image will contain information from each source. The visibility of each contrast component in the final image depends upon the relative strength of the signal modulation that each causes. A strong contrast mechanism dominates weak mechanisms in terms of the relative visibility. Because contrast mechanisms have different characteristics in terms of their number and trajectory components, separation of competing contrast mechanisms is sometimes possible through the use of multiple detectors. The classic case is the separation of atomic number and topographic contrast (Kimoto and Hashimoto, 1966). Atomic number (compositional) contrast has almost completely a number component, while topographic contrast has both a number component and a very strong topographic component. Two detector strategies, backscattered detector arrays and comparisons of emissive and specimen current, can separate number and trajectory components.

A solid-state backscatter detector with separated sectors can function to separate number and trajectory components when operated in the sum and difference modes. In the sum mode, number effects dominate because the detector has such a large solid angle that no matter in which direction electrons leave the specimen, they are collected by the detector, eliminating trajectory effects. In the difference mode, trajectory effects dominate, because each detector views the specimen from a different location, and therefore those features of the specimen which face a particular detector dominate the signal from that detector. When differences are taken between the signals from various detectors, the illumination appears to come from the detector signal with the positive multiplier (e.g., A–B is accomplished by scaling B with a negative multiplier and adding). Since each detector in a difference signal receives the same number components of contrast from the specimen, these number effects are eliminated in the difference image. An example of sum- and difference-mode imaging is shown in Fig. 4.38. The specimen is Raney nickel, which contains multiple phases that produce atomic number contrast and surface defects that produce topographic contrast. As a point of familiar reference, the specimen is viewed in Fig. 4.38a with a positively biased E–T detector, showing predominantly topographic contrast. In Fig. 4.38b, the summation signal from all four components of a four-quadrant solid-state BSE detector is shown, and the atomic number contrast dominates the image. In Fig. 4.38c the difference mode for a detector at the top of the field of view minus a detector at the bottom is shown. Topographic contrast dominates, with apparent top illumination. When the difference is reversed, Fig. 4.38d, the sense of the illumination reverses. In Fig. 4.38e and f, other difference combinations are shown. In general, the sense of the illumination appears to come from the detector with the positive coefficient. These difference-mode images can be misleading if care is not taken to ensure that top

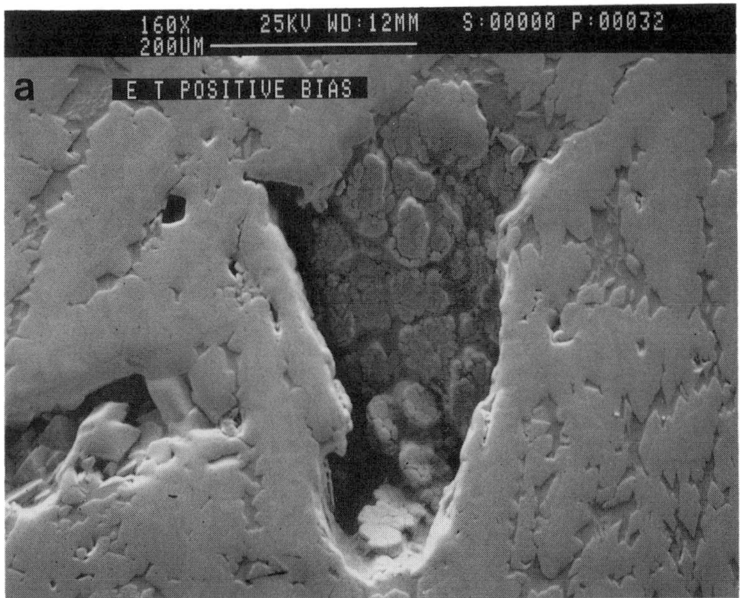

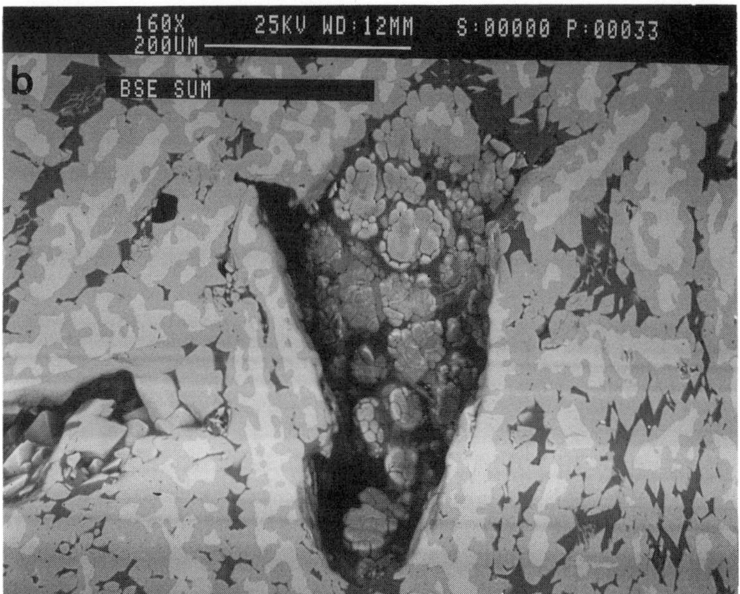

Figure 4.38. Separation of contrast components. The specimen is Raney nickel, which contains multiple phases that produce atomic number contrast and surface defects that produce topographic contrast. (a) The specimen is viewed with a positively biased E–T detector, showing predominantly topographic contrast. (b) Four-quadrant solid-state BSE detector, sum mode: atomic number contrast dominates the image. (c) Difference mode, TOP − BOTTOM: topographic contrast dominates, with apparent top illumination. (d) Difference mode, BOTTOM − TOP: topographic contrast dominates, with apparent bottom illumination. (e) Difference mode, RIGHT − LEFT: topographic contrast dominates, with apparent illumination from the right. (f) Difference mode, LEFT − RIGHT: topographic contrast dominates, with apparent illumination from the left.

illumination is achieved. Otherwise, the sense of the perceived topography may appear to reverse. The sensitivity to topography is not uniform. Linear features that run parallel to a line connecting the detector components produce very low contrast, while features running perpendicular to that line are seen in high contrast. The topographic contrast also appears much stronger than that which we obtain from the conventional E–T detector. There is also evidence that artifacts can appear in difference images at phase boundaries (Reimer, 1976).

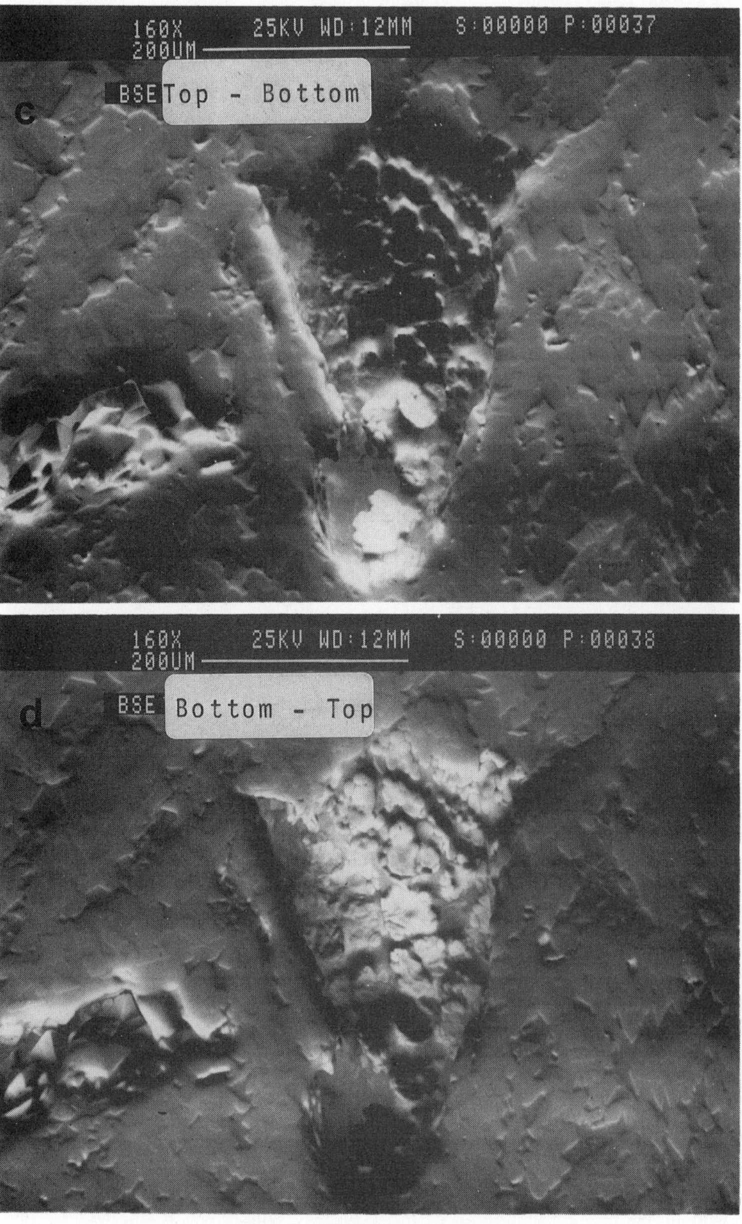

Figure 4.38. (Continued)

Comparison of emissive (backscattered- and secondary-electron) images with specimen-current images can also separate contrast mechanisms (Newbury, 1976). The specimen-current signal is totally dominated by number effects and is completely independent of trajectory effects, while an asymmetric backscattered-electron deflection such as the negatively biased E–T detector is extremely sensitive to trajectory effects. Figure 4.39 shows a comparison of images of a two-phase alloy with

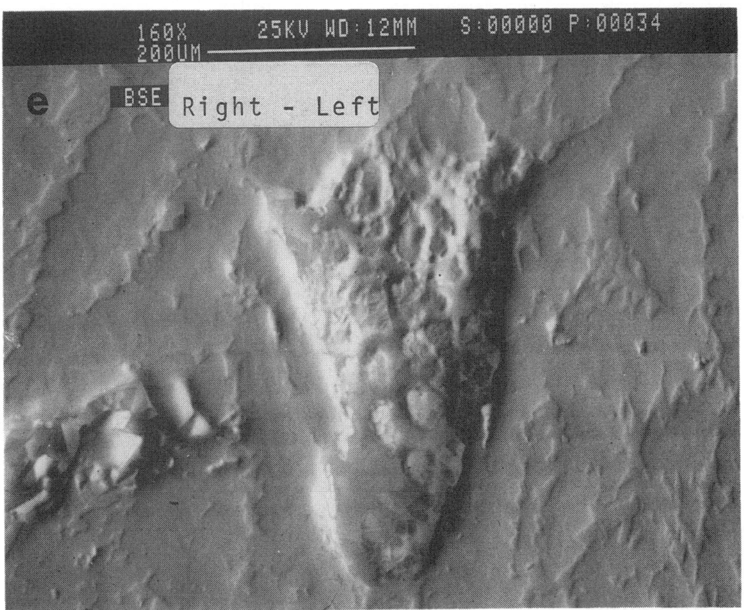

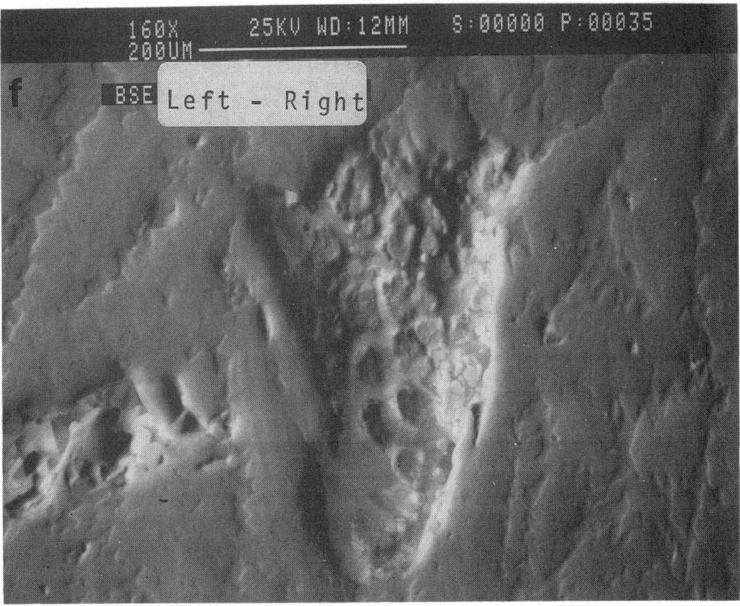

Figure 4.38. (Continued).

surface topography made with (a) a negatively biased E–T detector and (b) specimen current, with the contrast reversed so that the phase with the higher average atomic number appears bright. In Fig. 4.39a, the ridges of the topography dominate the image. The negatively biased E–T detector is located at a low take-off angle above the surface, which increases the effect of the apparent oblique illumination and increases the sensitivity of the image to the trajectory effects inherent in the topographic contrast. In Fig. 4.39b, the atomic number contrast dominates the image obtained with the specimen current signal, and the topographic contrast is greatly diminished.

4.4.3.6. Other Contrast Mechanisms

In addition to compositional and topographic contrast, other contrast mechanisms exist that permit the microscopist to characterize a wide variety of specimens. These contrast mechanisms include crystallographic contrast (electron channeling), contrast from magnetic domains (Type I, secondary-electron deflection; Type II, backscattered-electron deflection; Type III, secondary-electron polarization), contrast from specimen surface potential (voltage contrast), electron beam–induced conductivity contrast (EBIC, charge collection microscopy), and thermal-wave contrast. The mechanisms are described in detail in Newbury *et al.* (1986). A point to note is that many of these contrast mechanisms are weak. Electron channeling contrast is typically 2–5%, while Type II magnetic contrast is only 0.1–0.3%.

a
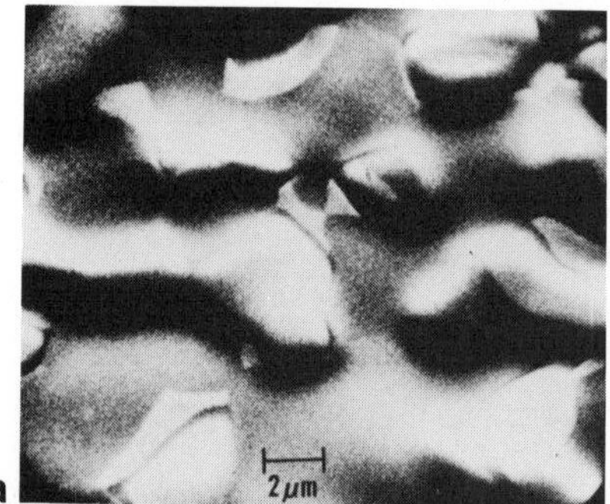

b
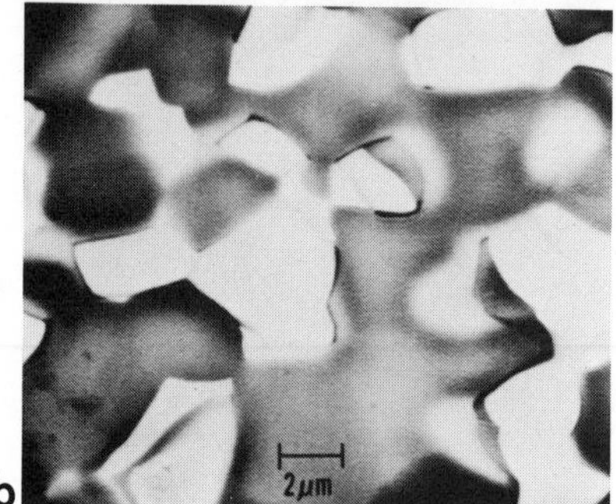

Figure 4.39. Separation of contrast components with emissive mode and specimen-current mode images. The specimen is eutectic lead–tin alloy, which has two phases, with surface topography in the form of ripples due to polishing. (a) Negatively biased E–T detector; the detector is located at the top of the image. (b) Specimen-current, contrast reversed so that the phase with the higher average atomic number appears bright.

If a specimen has characteristics (compositional differences, topography, etc.) that can be imaged with the SEM, what criteria must be satisfied to produce a final image on the CRT that conveys this contrast information to the observer? These criteria can be separated into two categories:

1. the relationship of the contrast potentially available to the quality of the signal, which is degraded by the presence of random fluctuations referred to as noise; and
2. the techniques of signal processing that must be applied to actually render the contrast information in the amplified signal visible to the observer viewing the display.

Signal Quality and Contrast Formation. We are all familiar with the everyday problem of attempting to tune in a distant television station on a home receiver. If the station's signal is weak, we find that the visibility of detail in the picture is obscured by the presence of noise, that is, random fluctuations in the brightness of the image points, which are superimposed on the true changes in signal (contrast) which convey the image information. We know from our experience that features with strong contrast in the original image are the easiest to see in a noisy image, while a feature which is low in contrast, small in size as a fraction of the image field, or rapidly changing in position, is difficult to see. The presence of randomness or noise sets a limit to the information available in an image; this is a common theme in all imaging processes.

As shown in Fig. 4.40, we can study the nature of the signal in the SEM by making a line scan across the specimen and plotting the strength of the measured signal (from an E–T detector, solid-state detector, specimen current, etc.) as a vertical deflection on an oscilloscope whose horizontal axis is scanned in synchronism with the scan on the specimen. The SEM imaging process involves the

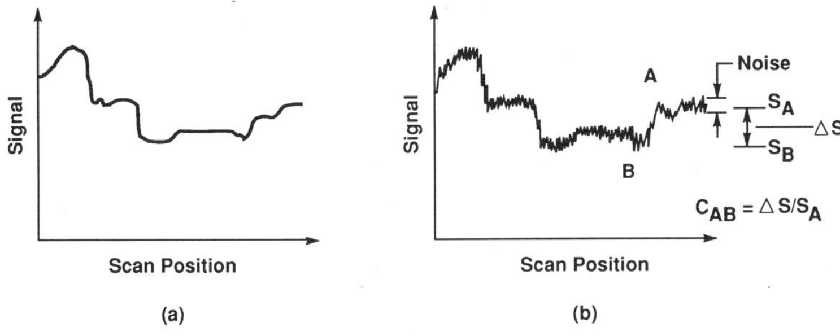

Figure 4.40. (a) Schematic illustration of a single line scan across a field of view in the SEM, as displayed on an oscilloscope. (b) Schematic illustration of multiple line scans along the scame scan vector on the specimen superimposed on the oscilloscope. A and B represent two arbitrarily chosen points of interest.

measurement of discrete events, e.g., the secondary and/or backscattered electrons collected at the detector that arrive with a random distribution in the time of a sampling period, which is the dwell time of the beam at a pixel. Measuring the signal S consists of counting a number of discrete events n at the detector. Because of the random distribution of the events in time, repeated counts with the beam at the same location on the specimen will vary about the mean value $\bar{n}$ with a distribution whose standard deviation is given by $\bar{n}^{1/2}$. Since the presence of noise degrades the signal, the signal quality is expressed as the signal-to-noise ratio, S/N:

$$\frac{S}{N} = \frac{\bar{n}}{\bar{n}^{1/2}} = \bar{n}^{1/2}. \tag{4.20}$$

As the mean number of counts increases, the signal quality improves because the random fluctuations become a less significant fraction of the total signal. In Fig. 4.40b, the noise can be estimated as the thickness of the line trace, and the signal at a point of interest can be measured; the ratio gives S/N.

Rose (1948) made an extensive study of the ability of observers to detect the contrast between objects of different size and the background in a scanned TV image in the presence of various levels of noise. He found that for the average observer to distinguish small objects against the background, the change in signal ΔS due to the contrast had to exceed the noise N by a factor of 5:

$$\Delta S > 5N \qquad \text{(Rose criterion)}. \tag{4.21}$$

The Rose visibility criterion can be used to develop the relation between the threshold contrast, that is, the minimum level of contrast potentially available in the signal, and the beam current. The noise can be considered in terms of the number of signal events:

$$\Delta S > 5\bar{n}^{1/2} \tag{4.22}$$

Equation (4.22) can be expressed in terms of contrast ($C = \Delta S/S$) by dividing through by the signal:

$$\frac{\Delta S}{S} = C > \frac{5\bar{n}^{1/2}}{S} = \frac{5\bar{n}^{1/2}}{\bar{n}} \tag{4.23}$$

$$C > \frac{5}{\bar{n}^{1/2}} \tag{4.24}$$

$$\bar{n} > \left(\frac{5}{C}\right)^2. \tag{4.25}$$

Equation (4.25) indicates that in order to observe a specific level of contrast C a mean number of signal carriers, given by $(5/C)^2$, must be collected per picture element. Considering electrons as signal carriers, the number of electrons which must be collected per picture element in the dwell time τ can be converted into a signal current i_s:

$$i_s = \frac{\bar{n}e}{\tau} \tag{4.26}$$

where e is the electron charge (1.6×10^{-19} coulombs). Substituting Eq. (4.25) into Eq. (4.26):

$$i_s > \frac{25e}{C^2\tau}. \tag{4.27}$$

The signal current i_s differs from the beam current i_B by the efficiency of signal collection ε, which depends on the generation of signal carriers by the beam–specimen interaction and the geometric size and response of the detector:

$$i_s = i_B\varepsilon. \tag{4.28}$$

Combining Eqs. (4.27) and (4.28):

$$i_B > \frac{25(1.6 \times 10^{-19}\ \text{coulombs})}{\varepsilon C^2 \tau}. \tag{4.29}$$

The picture–element dwell time τ can be replaced by the time to scan a full frame, t_f, from the relation:

$$\tau = \frac{t_f}{n_{\text{PE}}}, \tag{4.30}$$

where n_{PE} is the number of picture elements in the whole image. Substituting Eq. (4.30) into Eq. (4.29):

$$i_B > \frac{(4 \times 10^{-18}) n_{\text{PE}}}{\varepsilon C^2 t_f}\ \text{A}. \tag{4.31}$$

For a high-quality image, there are typically 1000×1000 picture elements, so Eq. (4.31) can be stated as

$$i_B > \frac{4 \times 10^{-12}}{\varepsilon C^2 t_f}\ \text{A}. \tag{4.32}$$

Equation (4.31), referred to as the *threshold equation,* defines the condition for the minimum beam current, the *threshold current,* necessary to observe a certain level of contrast C with a signal production collection efficiency ε (Oatley, 1972). Alternatively, if we specify the beam current which is available and the signal production collection efficiency ε, then we can calculate the minimum contrast, the *threshold contrast,* which can be observed in an image. Objects in the field of view that do not produce this threshold contrast cannot be distinguished from the noise of random background fluctuations.

A useful way to understand the relationships of the parameters in the threshold equation is the graphical plot shown in Fig. 4.41. This plot has been derived from Eq. (4.32) with the assumption that the

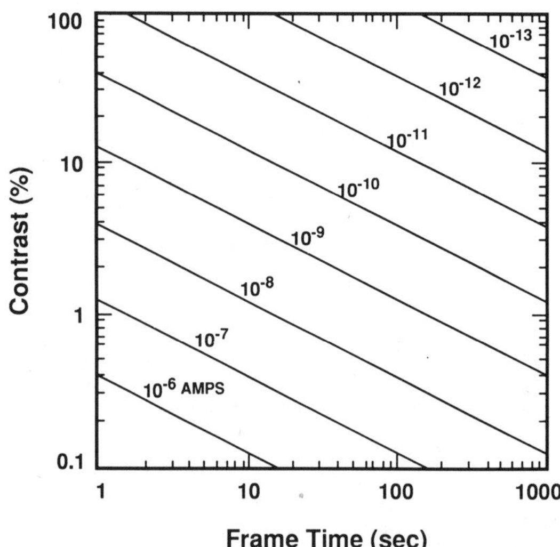

Figure 4.41. Graphical representation of the relationship of parameters of contrast, frame time, and beam current in the threshold equation. Assumptions: the signal production/collection efficiency $\varepsilon = 0.25$, and the scan is 1000×1000 picture elements.

signal generation–collection efficiency is 0.25, that is, one signal-carrying electron (backscattered or secondary) is collected for every four beam electrons that strike the specimen. This collection efficiency is a reasonable assumption for a target such as gold, with high backscattering and secondary-electron coefficients, when the electrons are detected with a positively biased E–T detector. From Fig. 4.41, we can immediately see that to image a contrast level of 0.10 (10%) with a frame time of 100 s, a beam current in excess of 10^{-11} A must be supplied. If the specimen produces a contrast level of only 0.005 (0.5%), a beam current of 10^{-8} A must be used. Conversely, if we define the current in the beam, Fig. 4.41 demonstrates that there will always be a level of contrast below which nothing will be visible. For example, if a beam current of 10^{-10} A is used for a 100-s frame time, all objects producing contrast less than 0.05 (5%) against the background will be lost. Once we know the current required to image a specific contrast level, we can calculate the probe size that can be obtained with this current, considering all of the aberrations that degrade electron-optical performance, using the equations in Chapter 2. Table 4.6 shows the Gaussian (brightness limited) probe that can image various levels of natural contrast. A severe penalty in minimum probe size is incurred when the contrast is low. The effects of aberrations would be to make this situation worse.

The Rose criterion is actually a conservative estimate of the threshold conditions. For an object which forms a large fraction of the image or which has an extended linear nature, such as an edge or a fiber, the ability of an observer's visual process to effectively combine information extended over many contiguous pixels actually relaxes the

Table 4.6. Gaussian Probe Size for Various Contrast Values (tungsten filament, 20 keV)

Contrast	d_{min} (nm)	Contrast	d_{min} (nm)
1.0	2.3	0.025	91
0.5	4.6	0.01	230
0.25	9.1	0.005	460
0.10	23	0.0025	911
0.05	46	0.001	2,300

strict visibility criterion (Bright *et al.*, 1991). However, it must be recognized that for objects producing contrast near the threshold of visibility, different observers may have substantially different success in detecting features in images.

4.6. High-Resolution Microscopy: Intermediate (10,000–100,000x) and High Magnification (>100,000x)

To achieve high-resolution microscopy, three conditions must be satisfied

1. As described in Chapter 2, the electron-optical performance must be optimized to produce a finely focused probe at least as small as the size of the finest features to be examined.
2. The imaging signal must originate in the immediate vicinity of the beam-impact area on the specimen if it is to be sensitive to fine-scale details on the scale of the probe.
3. From the previous discussions, the probe must carry sufficient current to establish sufficient signal quality in view of the contrast produced to satisfy the threshold current criterion.

The development of LaB_6 and field-emission electron sources for the SEM has made high-resolution microscopy a practical reality. However, as the magnification is increased above 10,000 diameters, the nature of the contrast which we view changes (Joy, 1984; 1991; Peters, 1982; 1984). The simple light-optical analogy that proves so useful in the interpretation of low-magnification images is not necessarily valid at high magnifications, because the lateral and depth ranges of the imaging signals begin to have dominant effects on the appearance of the image in comparison to the signals related to the local inclination of the specimen's surface toward the E–T detector. The success of the light-optical analogy depends on the existence of a robust rule such that bright features are those which face the E–T detector. When the range of the imaging signals begins to dominate images, the brightness of features may have nothing at all to do with their relative orientation to the detector. Proper image interpretation at high magnification depends therefore on the

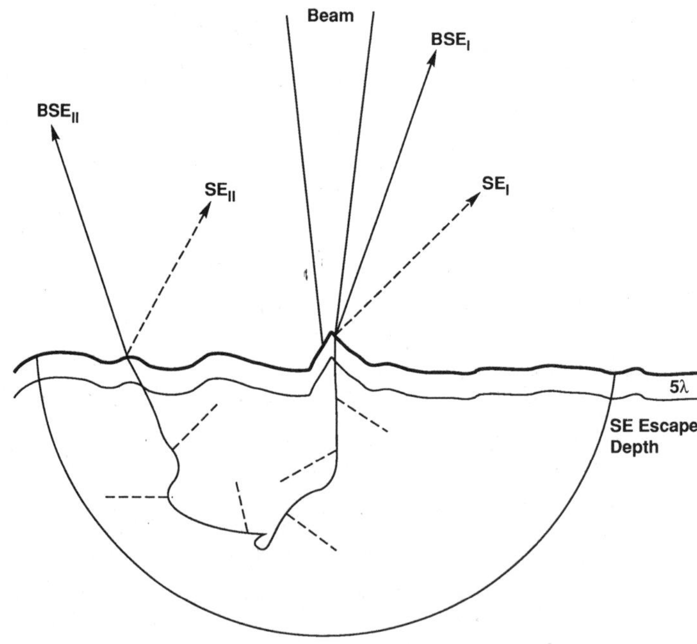

Figure 4.42. Backscattered- and secondary-electron signals emitted with a finely focused high-resolution beam.

microscopist's knowledge of electron–specimen interactions. As shown in Fig. 4.42, the beam–specimen interactions create a complex distribution of backscattered and secondary-electron signals that are available for imaging.

4.6.1. Electron–Specimen Interactions in High-Resolution Microscopy

4.6.1.1. Backscattered Electrons

As revealed by Monte Carlo electron-trajectory calculations, the process of electron backscattering has two distinct regimes. As the beam electrons enter the specimen through the beam impact area, elastic scattering begins. The average elastic scattering angle is small, generally less than 5°, so that the dominant tendency is for the beam electrons to continue into the specimen. However, elastic scattering can deflect the electron trajectory by any amount up to 180° in a single event. From Eq. (3.5), we can estimate that elastic events which produce scattering angles in the range 90°–180° occur with a frequency of 1 : 10,000 compared to events over the entire range from 1°–180°. Beam electrons which suffer high elastic-scattering angles promptly leave the specimen in the immediate vicinity of the beam impact area and therefore carry information specific to that area. This class of backscattered electrons is designated BSE_I. If the beam is focused to a small diameter, the BSE_I signal constitutes a source of high-resolution information about the specimen. Beam electrons that continue to scatter suffer multiple elastic interactions whose cumulative effect brings a significant fraction back to the surface to emerge as backscattered electrons. These backscattered electrons, designated BSE_{II}, are the dominant part of the total backscattering coefficient η and are emitted relatively uniformly over a much larger area whose diameter is similar to the Kanaya–Okayama range of the beam below the surface. The BSE_{II}-signal is spatially disconnected from a finely focused beam and does not carry high-resolution information.

The BSE_I and BSE_{II} electrons can be distinguished by the energy which they carry. Since inelastic scattering occurs throughout the path of the electron in the specimen at a rate of approximately 1–50 eV/nm, depending on electron energy and specimen composition, the farther an electron travels through the specimen, the more energy it loses. BSE_I electrons are promptly scattered out of the specimen with only a few nanometers to tens of nanometers of travel subject to inelastic scattering, so they lose only a few eV to hundreds of eV and retain virtually all of their incident energy. BSE_{II} electrons escape after traveling much greater distances, so their energy is substantially reduced, forming a continuum down to low energies (by definition, $50\,\mathrm{eV} < E_{BSE} < E_0$), as shown in Fig. 3.21. BSE detection with energy selection thus permits separation of the BSE_I and BSE_{II} signals.

Wells (1974a,b) has developed a high-resolution imaging procedure based on collecting only the BSE_I signal, which he designates the "low loss backscattered electrons." As shown in Fig. 4.43, to increase the fraction of the BSE_I signal and impart a strong directionality to the emission of the BSE_I signal that can improve collection, the specimen is tilted to a high angle. An energy-filtering collector is placed below at a low take-off angle to the surface. With the surface highly tilted, the electrons can escape the specimen as a result of a much smaller and much more probable elastic scattering angle. The highly tilted surface tends to produce preferential scattering in the forward direction, so by placing the detector as shown, the BSE_I signal can be collected with much greater efficiency than would be obtained with the normal beam-incidence situation shown in Fig. 4.42. Since the BSE_I signal represents such a small fraction of the total backscattered signal, such an arrangement is critical if adequate signal is to be obtained to overcome limitations imposed by the statistics of signal measurement, as described in Section 4.5.

The BSE_I signal has been used to image topographic contrast with a spatial resolution approaching 1–3 nm. Figure 4.43b and c show examples of BSE_I images prepared with a first-generation high resolution SEM based upon a LaB_6 source (Broers, 1974; Broers, 1975), revealing fine-scale detail with dimensions close to that of the focused probe. Because of the high tilt and the narrow acceptance angle of the energy-filtering detector, the contrast of topographic features is high,

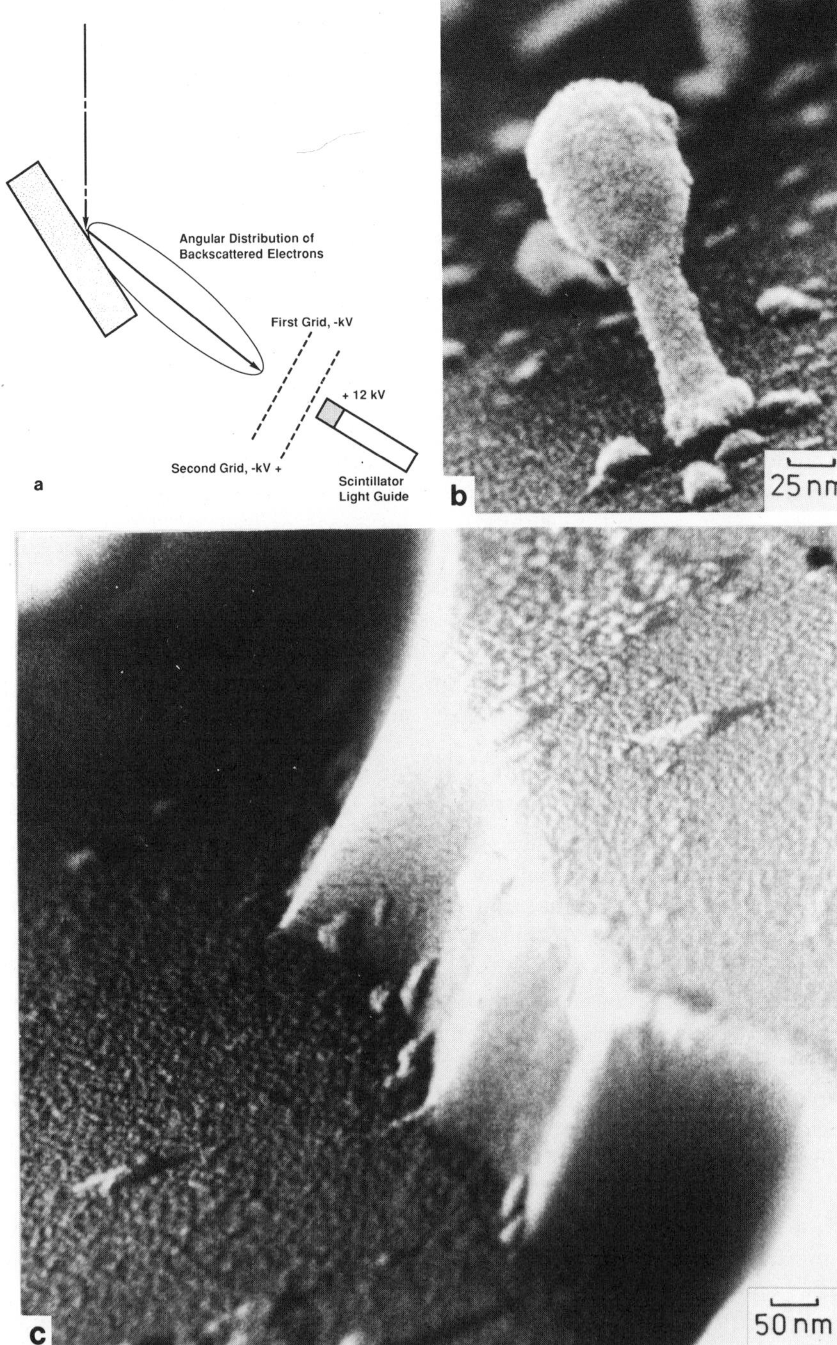

Figure 4.43. (a) Energy filtering detector and tilted specimen for high-resolution imaging utilizing the BSE_1 low-loss backscattered electrons (Wells, 1974a; b). High resolution images prepared with the BSE_I (low-loss backscattered electron) signal. (b) T4 coliphage (Broers, 1975; courtesy of A. Broers); Sample coated with 20 nm of gold–palladium (c) Etched SiO_2 step on silicon wafer (Broers, 1974; courtesy of A. Broers).

since scattering is lost from surfaces tilted away from the detector direction.

In addition to high-resolution lateral detail, the BSE_I image is also profoundly sensitive to surface structure. The condition imposed by the energy selection on the distance traveled within the specimen not only constrains the BSE_I electrons laterally to the vicinity of the beam impact area but also restricts the depth to which they can penetrate prior to backscattering. Wells (1974a) has demonstrated that the BSE_I image of a subsurface hole in an aluminum specimen covered by a thin (4 nm) oxide layer reveals details only of the thin oxide layer, whereas the conventional, positively biased E–T image, which is dominated by the low-resolution and highly penetrating BSE_{II} signal, is insensitive to the thin oxide layer and reveals the underlying hole.

4.6.1.2. Secondary Electrons

The situation for secondary electrons is more complicated. As shown in Fig. 4.42, as the incident beam penetrates down through the first 5–10 nm of the specimen, the secondary electrons produced by inelastic scattering have a significant probability of escape, which decreases exponentially with depth. These secondaries, designated SE_I, are sensitive to conditions in the beam impact area and are therefore capable of carrying high-spatial-resolution information from a finely focused primary beam (Peters, 1982; Pawley, 1984). Monte Carlo electron-trajectory simulations of a point beam reveal that the SE_I signal from a target such as aluminum leaves the surface with a Gaussian intensity profile characterized by a full width at half maximum (FWHM) of approximately 2 nm (Joy, 1984).

As the beam electrons penetrate deeper than 10 nm below the surface, secondaries continue to be produced along the trajectory because of inelastic scattering, but the low-energy secondaries are restricted in range, and they cannot escape and be detected. However, as the BSE_{II} electrons leave through the surface of the specimen, the secondaries generated along the trajectories within 5–10 nm of the surface are also able to escape, forming the SE_{II} component of the total secondary-electron signal. The SE_{II} component has a Gaussian distribution as well, but with a FWHM of 0.2–0.5 of the Kanaya–Okayama range, and therefore the SE_{II} component does not contribute to the high-resolution signal.

Unfortunately, the SE_I and SE_{II} secondaries have the same energy and angular distributions and cannot be separated on any physical basis. As noted previously in the discussion of the E–T detector, the backscattered electrons strike the walls of the specimen chamber and polepiece, forming a third class of secondaries, SE_{III}, which when collected really represent the BSE_{II} component of the signal. Peters (1982) has demonstrated that the SE_{III} component can be reduced as a component of the total secondary-electron signal by covering the polepiece and chamber walls with a material having a low secondary-electron yield, such as a carbon plate. The SE_{III} component can be virtually eliminated through the use of a suppression grid near the polepiece. A small secondary-electron component entirely unrelated to the specimen, designated SE_{IV}, originates as beam electrons pass through the final aperture and strike its edge. The suppression grid can also eliminate the SE_{IV} component.

Given that the SE_I and SE_{II} secondaries cannot be separated and that the low-resolution BSE_{II} component is also collected with great efficiency by the positively biased E–T detector through the mechanism of SE_{III} collection, the total signal is dominated by these low-resolution components in the beam energy range typically chosen for scanning electron microscopy, 10–20 keV. For this energy range, the SE_{II} and BSE_{II} signal components are emitted over a sufficiently large area to ensure that any response to fine-scale details in the immediate vicinity of the beam is overwhelmed by longer-range effects. The only type of specimen for which high-resolution information is achieved with a 10–20 keV incident beam is an isolated thin edge, as described below in Section 4.6.5. Typical SEM operating conditions in the 10–20 keV range are therefore the least suitable to achieving high resolution microscopy.

4.6.2. High-Resolution Imaging at High Voltage

Some separation of the SE_I/BSE_I and SE_{II}/BSE_{II} signals can be achieved when a high beam energy is used to image a low atomic number specimen. As shown in Fig. 4.44, for a 30-keV probe focused to 1 nm on a silicon target, the SE_I (and BSE_I) component has a spatial distribution of about 2 nm, but the range of the primary electrons $R_{KO} = 9.5\ \mu m$ is so

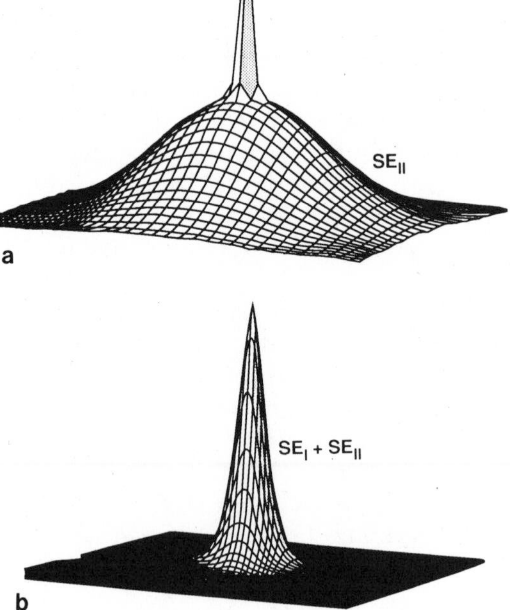

Figure 4.44. (a) Secondary-electron emission for a 1-nm, 30-keV probe focused onto a silicon target. The SE_I has a FWHM of 2 nm, while the SE_{II} component has a FWHM of approximately 10 μm. (b) Situation at 1 keV, where FWHM is similar for both.

great that the SE_{II} (and BSE_{II}) signals emerge over an area approximately 10,000 times larger in radius. If the sample is scanned at a low magnification, <10,000X, then the field of view is 10 μm across or larger. As the beam moves from pixel to pixel, the SE_{II} component changes in response to the coarse features of the field of view, and the signal at an individual pixel contains low-resolution components from many pixels around it. However, at high magnification, >100,000 diameters, the field of view is 1 μm or less, and as the beam is scanned within this area, the effective movement of the large SE_{II} distribution is negligible, so the absolute magnitude of the SE_{II} component remains relatively constant. Although the SE_{II} component has not been removed from the signal, by remaining relatively constant, its effect on image contrast is minimized. The changes in the total signal are almost all due to the SE_{I} component, so that high-resolution information is preserved in the image. The random noise contributed by the SE_{II} component is significant and contributes to resolution limitations imposed by signal quality. Figure 4.45a shows a high-resolution image, prepared with a field emission SEM

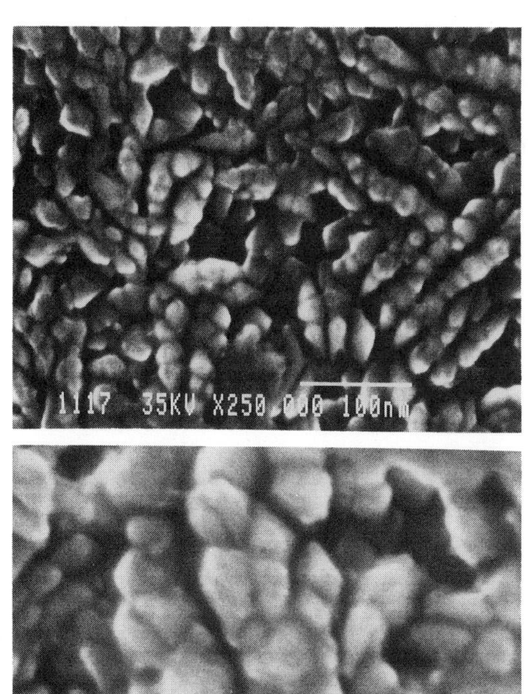

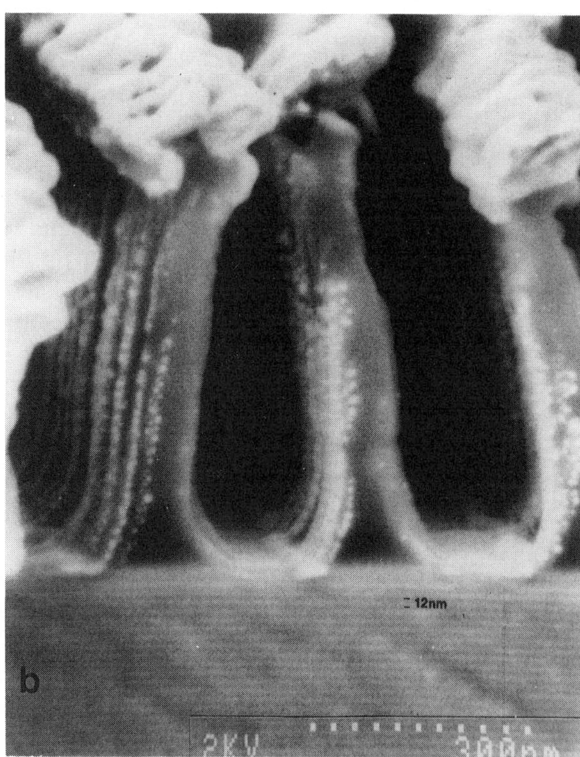

Figure 4.45. (a) (left-side) High-resolution secondary-electron images of the surface of the magnetic storage of a computer hard disc. Image recorded at 35 keV on a field-emission SEM with signal collection from an E–T detector. Two magnifications of the same area are shown. (b) (right-side) High-resolution secondary-electron image of a complex structure formed from a polymethylmethacrylate grating on a tungsten carbide–layered x-ray mirror. Beam energy = 2 keV on a field-emission SEM. (Image courtesy T. Reilley, Hitachi.)

operated at 35 keV, of the surface of the magnetic storage medium of a computer hard disc.

When low-atomic-number targets, e.g., the typical matrix of biological or polymer samples, are to be imaged, the low secondary-electron yield typical of such compositions creates limitations to high-resolution imaging which arise from problems of the poor signal-to-noise ratio. The signal-to-noise ratio can be improved by enhancing the secondary yield by the application of an ultrathin (<5 nm) coating of a metal which has a high secondary coefficient but which is sufficiently thin to cause negligible elastic scattering of the beam electrons. Gold and gold–palladium alloys have been the traditional choices for coating due to their high secondary yields, but films of these metals often develop fine structures which act to mask the true fine details of the specimen. To avoid artifacts arising from such structure in the coatings, advanced thin-film deposition techniques such as ion beam sputtering and alternative coating metals such as chromium must be used (Peters, 1982, 1985). Coatings of chromium as thin as 1 nm have proven to achieve the desired improvements in secondary-electron emission and lack of added structure.

Surprisingly, when the specimen is a bulk intermediate- or high-atomic-number material imaged under the same conditions, the spatial resolution that can be attained is actually decreased, despite the fact that the interaction volume from such targets has dimensions smaller by a factor of 2 to 10 than that for silicon. The reason for this behavior can be found in the relative contributions and spatial distributions of the SE_I and SE_{II} components. First, for intermediate- and high-atomic-number targets, the ratio SE_{II}/SE_I is greater (see Table 3.8), so the absolute effect of the SE_{II} component on degrading the signal quality is larger. Second, when the FWHM of the SE_{II} distribution is reduced while the FWHM of the SE_I component remains constant, the effect of the unwanted SE_{II} modulation of the signal during scanning actually increases at any magnification, so that the "low resolution" regime persists to higher magnifications.

4.6.3. High-Resolution Imaging at Low Voltage

An alternative approach to achieving high-resolution microscopy is to operate at low beam energies, less than 5 keV (Pawley, 1984; Joy, 1984). The dimensions of the interaction volume decrease rapidly with beam energy, approximately as the 1.7 power. Lowering E_0 from 20 keV to 2 keV results in a decrease in the range of a factor of 50, so that the FWHM dimensions of the SE_{II} and BSE_{II} components begin to approach that for SE_I. This is a desirable situation, since the useful high-resolution component of the secondary electron signal now dominates the total signal and the backscattered electrons contribute high-resolution information as well. The total secondary-electron signal rises at low beam energies, further improving the signal-to-noise ratio, a distinct benefit for satisfying the threshold equation. Finally, the situation greatly improves for the intermediate- and high-atomic-number specimens, since the spatial range of signals decreases to a greater extent for strongly scattering materials. Figure 4.45b shows a complex structure formed from a polymethylmethacrylate grating on a tungsten carbide–layered x-ray mirror imaged with a high-performance field-emission SEM operated at 2 keV. Layers of 12-nm spacing are readily resolved, and a small contrast between the layers due to their compositional differences is also visible.

There are drawbacks to operation at low beam energy (Pawley, 1984). The source brightness falls proportionally to the beam energy, so the electron-optical performance degrades the probe size, and the available beam current may fall below the threshold current for any useful contrast. Low-energy beams are also much more susceptible to deflection or degradation by stray electromagnetic fields. Contamination can also be a serious limitation to low-voltage microscopy, since the range of the signals is so short that the contamination may dominate the image as it builds up. The field-emission source on a well-shielded, clean-vacuum, high-performance SEM is a workable answer to these problems, but great care must be taken to achieve reproducible, high resolution performance.

4.6.4. Resolution Improvements: the Secondary-Electron Signal

The secondary-electron signal can be manipulated to a certain extent to improve the signal quality. Although the various sources of secondary electrons produce electrons which cannot be distinguished on the basis of their characteristics, there is an important spatial separation of signals that can be utilized. The SE_{III} signal is formed by the interaction of the backscattered electrons with the polepiece and chamber walls. Since this source is physically separated by several centimeters from the beam impact on the specimen, where the SE_I and SE_{II} signals are formed, these components can be separated. Peters (1982, 1984) has demonstrated a significant improvement in contrast by removing the SE_{III} component. The SE_{III}'s can be by suppressed with a low-secondary-coefficient shield such as carbon on the polepiece or by using a positively biased grid near the polepiece. The effect on a high-resolution image is shown in Fig. 4.46, where fine details can be observed after suppression of the SE_{III} component. In terms of the signal quality, the threshold current for these features is lowered by eliminating a noise source, SE_{III}.

Another approach to manipulating the secondary signal is to alter the surface properties to improve the secondary yield. The secondary yield is low for carbonaceous specimens such as biological and polymer specimens. Traditionally, a relatively thick layer of gold or gold–palladium has been sputtered on such specimens, both to eliminate charge and to improve the secondary yield. However, the relatively thick layers of heavy metals also increase the backscattering, which acts to degrade resolution. More recently, ultrathin films (1–2 nm) of chromium have proved more effective (see Chapter 13), by increasing the secondary yield without significantly altering the scattering of the high energy beam electrons.

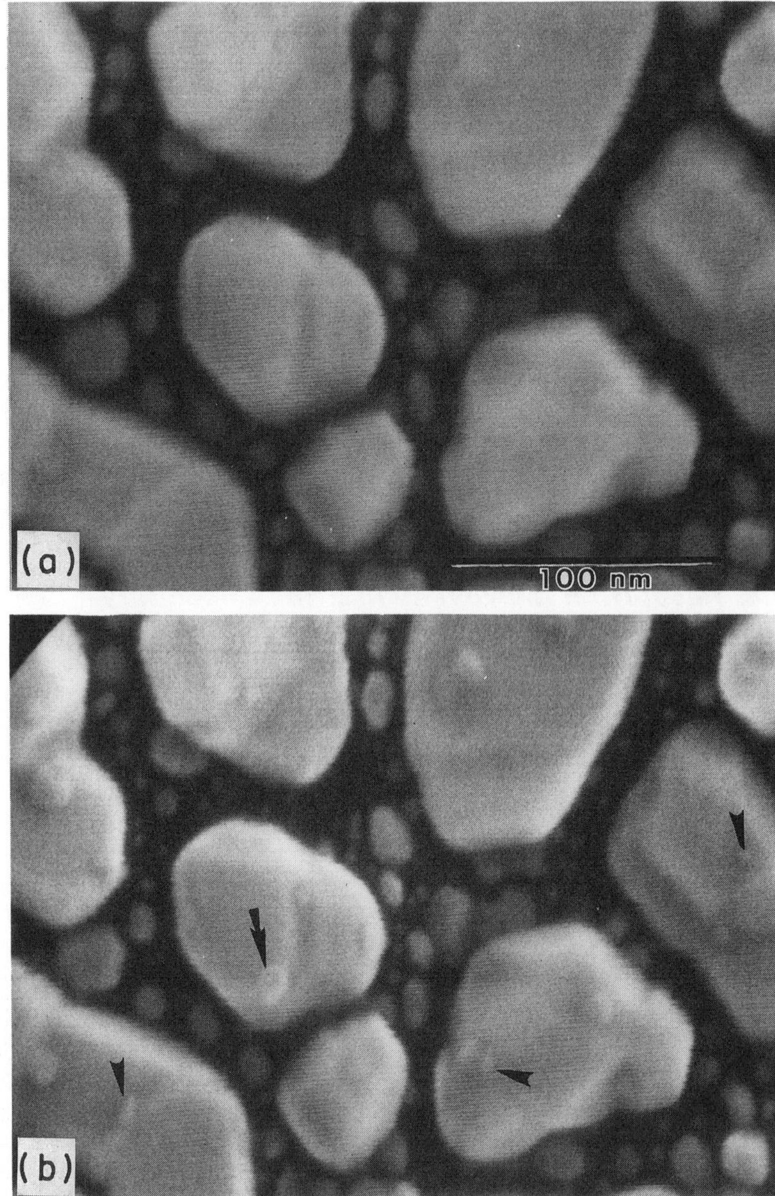

Figure 4.46. Improved contrast by separation of SE_{III} from the $SE_I + SE_{II}$. (a) Image of gold islands on graphite with $SE_I + SE_{II} + SE_{III}$ obtained with conventional E–T operation. (b) $SE_I + SE_{II}$ only, with SE_{III} suppressed by the use of a low secondary-yield shield. Note improved contrast of small features indicated by arrows (Peters, 1984; micrographs courtesy of K.-R. Peters).

4.6.5. Image Interpretation at High Resolution

Given that high-resolution images can be obtained in the intermediate (10,000–100,000X) and high (>100,000X) magnification ranges, special care must be taken when interpreting the contrast observed. The simple topographic contrast situation of the light-optical analogy may not be valid because there no longer exists the unambiguous, direct relationship between image brightness and the orientation of a feature relative to the E–T detector that exists at low magnification, when the signals are constrained by their range to originate from the specific pixel addressed by the beam. Peters (1984) and Joy (1984) have noted several effects for which the range of the imaging signals plays a dominant role in determining the contrast observed in the image:

1. Mass-thickness contrast arises when the specimen is inhomogeneous in the vertical direction on the scale of the sampling depth of the backscattered electrons so that backscattering changes in response to the amount of material present, modifying the backscattered- and secondary-electron signals. Mass-thickness contrast can have a purely atomic number character when compositional interfaces or gradients are present.
2. If the thickness of a structure such as an edge is less than the sampling depth of the backscattered electrons, and has a uniform composition but varying thickness, contrast can arise in the BSE_{II} and SE_{II} signals from the differences in thickness alone.
3. If a structure such as an edge is thinner than the range of the incident beam electrons, then the energetic beam electrons may penetrate through the structure, generating additional BSE_{II} and SE_{II} electrons at the exit surface. The energetic electrons which pass through the thin edge may continue to scatter, striking other specimen or microscope surfaces and generating SE_{III} signals. All of these signals may be collected by the E–T detector.
4. Lateral range effects occur when the beam is placed at a pixel location, but because of the range of the backscattered and secondary electrons, signals are generated over a distance encompassing neighboring pixels, or even quite remote parts of the specimen or sample chamber.

Because of the basic nature of the SEM imaging process, all of the signal collected when the beam is addressed to a given pixel is assigned to that pixel when the image is constructed, no matter how far from the pixel the signal is actually generated and collected.

The mass-thickness and lateral-range effects manifest themselves in various ways in high-resolution images, but one of the most common is the appearance of bright edges which stand out in strong contrast to the apparently smooth interiors of the features, as illustrated in Fig. 4.47a. The interaction situation, illustrated in Fig. 4.47b, is seen to be a combination of mass-thickness effects and lateral-range effects. The backscattered electrons, the SE_{II} secondary electrons formed by backscattered electrons leaving the beam impact area, and the remotely generated SE_{III} secondary electrons all represent delocalized signals which contain no information about the beam impact area. These signals form a large fraction of the total signal when the beam strikes a sufficiently thick portion of the specimen to completely contain the interaction volume ("bulk" position in Fig. 4.47a). Fine-scale features on the scale of the beam tend to be lost because the delocalization of the signal causes the contrast of such features to fall below the threshold contrast. These "bulk" regions of the specimen may appear very smooth, but this apparent smoothness is potentially highly misleading. When the beam approaches a thin edge in such a specimen, the signal-generation situation changes radically. At an edge, the interaction volume is greatly reduced in size compared to the interior because the beam penetrates the edge. Because of the increased surface-to-volume ratio, more secondary electrons can be created to augment those made by the beam's electrons entering the specimen, as a result of backscattered electrons escaping though the entrance surface and side of the specimen, as well as from those transmitted through the edge. These transmitted electrons continue to scatter off other surfaces of the specimen and the chamber walls, producing SE_{III} electrons from all collisions. The excellent collection properties of the positively biased E–T detector for secondary electrons results in a large signal collection in this situation. The total signal

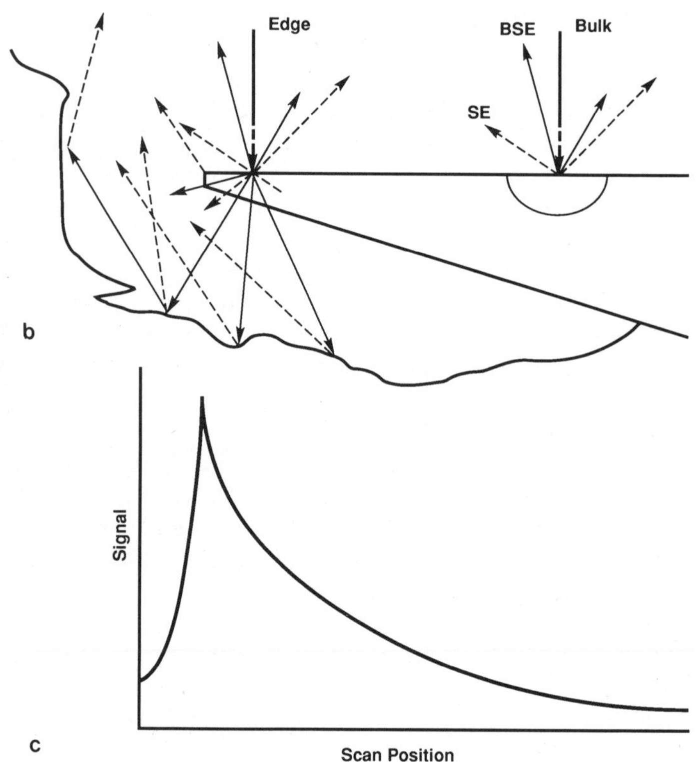

Figure 4.47. (a) High-magnification image of copy paper showing particles with sharp edges but little interior structure. (b) Schematic illustration of the signal-generation situation that takes place at edges, producing an image such as that seen in (a).

collected from all these surfaces hit by beam and backscattered electrons is assigned to the pixel corresponding to the beam location at the edge and is therefore very high compared to the bulk signal behavior of the interior, as illustrated in Fig. 4.47c. The edges of the specimen are thus observed in high contrast compared to the interior. The contrast between the edge and the interior exceeds the threshold contrast, and the threshold equation allows us to image these details with a small beam current and consequently a small beam diameter. When these sharp-edge details, which obviously demonstrate high spatial resolution, are combined with the smooth interiors of bulk objects in the same field of view, the overall appearance to an observer can be highly misleading. The high spatial resolution obtained for thin edges simply does not ensure that similar fine-scale details can be observed in the interior of a sample. Away from the edges, the contrast produced by fine details against the general background is greatly reduced, owing to enlarged interaction and sampling volumes. The contrast of such details is below the threshold contrast for the same value of the beam current that can be successfully used to observe the thin edges. This argument suggests that there will always be some threshold level of contrast below which objects cannot be distinguished above the background signal level. For high-spatial-resolution imaging conditions, the threshold contrast is in fact quite high, and many specimen features will be lost, since only a few types of features produce high contrast. *Thus, the apparent absence of detail in an SEM image is unfortunately no guarantee that fine-scale features are absent.*

Other contrast features dominated by range effects are observed in metal-coated organic and polymer specimens. Detailed treatments of the interpretation of such features have been published by Peters (1982, 1985) and by Joy (1984, 1991).

4.7. Image Processing for the Display of Contrast Information

The time-serial nature of the SEM signal permits the real-time application of processing operations to modify the signal actually displayed on the viewing or photographic CRT. With proper signal processing, we can readily observe very-low-contrast features by creating a bright image which is comfortable to view in a modestly darkened room. The great utility of this use of signal processing can be best appreciated when we compare the situation in light or transmission electron microscopy where the image is viewed by the human eye and no enhancement of low-contrast situations is possible with conventional image displays in real time. Low-contrast features strain the eye of the unaided viewer. Only by the addition of television microscopy accessories, which directly mimic the image manipulation process of the SEM by converting the continuous optical-microscope or TEM image into a time-serial image can real-time enhancement be achieved.

Scanning electron microscopy is at a technical crossroads with regard to the processing of images. Most equipment is still equipped with analog signal processing, in which special electronic circuits modify the signal at various stages in its amplification from the detector for direct presentation on the final CRT display. However, digital image processing is rapidly becoming a standard accessory. Many new SEMs are equipped with digital framestores that permit on-line digital image processing in which modification of the image is performed numerically in the digital domain. A variety of powerful off-line digital-image-processing programs are available for personal computers (e.g., Rasband, 1991). Digital image processing offers new features and increased flexibility over purely analog signal processing. Both analog signal processing and digital image processing will be considered in this discussion, but it should be recognized that all signal processing functions applied by analog methods can readily be mimicked by digital image processing.

4.7.1. The Visibility Problem

If the conditions imposed by the threshold equation are satisfied and the specimen contains the features of interest, then the SEM signal should contain information in the form of contrast. Can an observer be expected to see that contrast? The answer may be "no" if the contrast displayed on the CRT is less than the contrast sensitivity of the eye of the observer.

The display of the SEM image takes place on a CRT that is intensity modulated to produce a series of gray levels, with the signal adjusted to produce a gray level on the display which increases in brightness as the signal from the detector increases. If linear amplification is used to modify the signal to a level suitable for display, the natural contrast, that is, the changes in the signal level caused by the characteristics of the specimen, is equal to the display contrast in the CRT image. That is, if a linear amplification factor K is applied to the signal S to produce a display intensity I,

$$I = KS \tag{4.33}$$

and

$$C_{\text{natural}} = \frac{S_{\max} - S_{\min}}{S_{\max}} = \frac{KS_{\max} - KS_{\min}}{KS_{\max}} = \frac{I_{\max} - I_{\min}}{I_{\max}} = C_{\text{image}} \tag{4.34}$$

Since the human eye is relatively insensitive to intensity modulation, we normally adjust the signal processing by examining a line-scan trace on an oscilloscope where the x-axis scan is synchronized with the x-scan of the SEM and the signal intensity is plotted as a y-deflection. In such an oscilloscope trace, an observer's sensitivity to changes in the signal is enhanced to the point of recognizing changes of 1% or even less. The y-deflection of the oscilloscope trace can be calibrated in voltage and in the gray level of the final photograph.

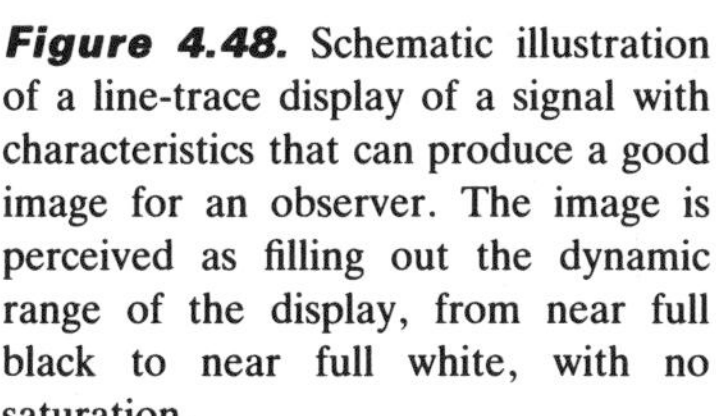

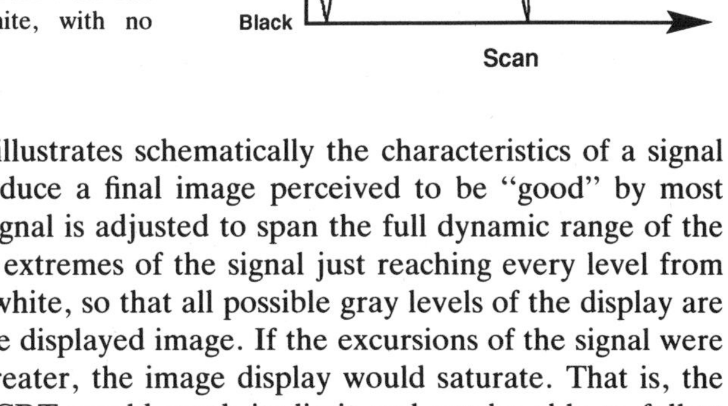

Figure 4.48. Schematic illustration of a line-trace display of a signal with characteristics that can produce a good image for an observer. The image is perceived as filling out the dynamic range of the display, from near full black to near full white, with no saturation.

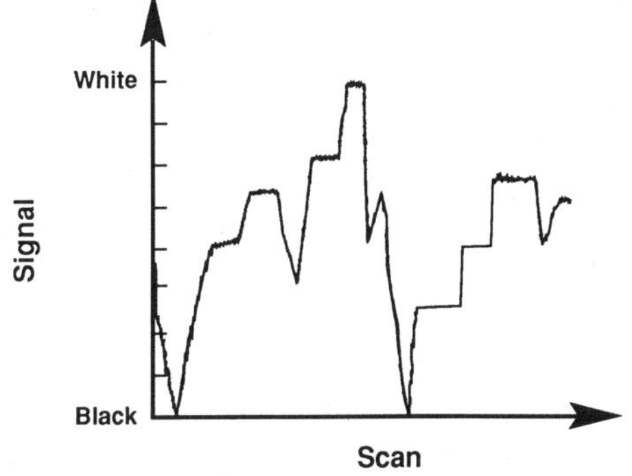

Figure 4.48 illustrates schematically the characteristics of a signal which would produce a final image perceived to be "good" by most observers. The signal is adjusted to span the full dynamic range of the display, with the extremes of the signal just reaching every level from full black to full white, so that all possible gray levels of the display are represented in the displayed image. If the excursions of the signal were adjusted to be greater, the image display would saturate. That is, the response of the CRT would reach its limit and not be able to follow further increases or decreases in the signal. Regions of saturation in the image would be seen as featureless areas of full white or full black.

Once the oscilloscope signal has been established with characteristics such as those in Fig. 4.48, there remains the problem of properly adjusting the display CRT. Separate CRT controls for brightness and contrast permit the manipulation of the final display to produce a satisfactory range of gray levels. These CRT brightness and contrast controls are not to be confused with similarly named controls, described below, in the signal-processing chain. (Note that CRT brightness is often turned down to a low value to protect the CRT phosphor from unnecessary degradation when the instrument is left in a standby condition or when certain long-dwell operations, such as single-pixel selection, are used.)

If the natural contrast is high, 70–100%, as it typically is in the case of topographic contrast, then linear amplification will produce a satisfactory final CRT image in which the natural contrast is converted into image contrast that spans most or all of the gray-scale range from nearly black to nearly white. Details of features in such an image are easily perceived by an observer. Such an image is shown in Fig. 4.49. The average observer can detect about 12–15 levels of gray in such an image, so that a change in gray level corresponding to a minimum signal change of about 5–7% can be seen in the image.

4.7.2. Analog Signal Processing

For many imaging situations in the SEM where the contrast is weak or has other deficiencies, linear amplification does not produce

Figure 4.49. Example of linear amplification applied to a sample which produces high natural contrast by the topographic contrast mechanism. Positively biased E–T detector. (Specimen: pollen grains; courtesy JEOL.)

satisfactory characteristics for the CRT display. In such cases, appropriate signal-processing functions must be applied to modify the displayed signal.

4.7.2.1. Display of Weak Contrast (Differential Amplification)

The most frequently encountered display problem in SEM imaging is the situation in which the natural contrast in the signal is too weak to be readily visible in the final display following linear amplification. Such a situation occurs when the modulation of the signal caused by the natural contrast is less than 5%. Weak natural contrast occurs for topographic contrast when the differences in local inclination among features are small or for atomic number contrast when the difference in atomic number is small, as shown in Fig. 4.31. Some contrast mechanisms are so inherently weak that they never produce a natural contrast above 0.05 (5%). Examples include electron channeling contrast, which is a result of differences in the orientation of crystalline samples, and magnetic contrast, which results from differences in the local magnetic field (Newbury *et al.*, 1986). If analog signal and/or digital image processing were not available in the SEM, these contrast mechanisms could not be made visible.

The signal processing technique most frequently applied to display weak contrast is differential amplification, commonly designated as "contrast expansion," "black level," or "dark level" on the control panel of the SEM. For the Everhart–Thornley detector, two separate adjustments control differential amplification, the black level and the gain. Figure 4.50 shows schematically the changes in the signal profile in the successive stages of differential amplification to enhance weak contrast in an image. Initially, the line trace depicts linear amplification of low natural contrast (Fig. 4.50a). The modulations of the signal (contrast) due to specimen features are confined to one or two gray levels and are therefore poorly visible. An example of a gray-scale image corresponding to this type of signal situation is shown in Fig. 4.51a. The first operation of differential amplification is to subtract a constant (DC) level from the signal at every point in the image. This action can be viewed on an oscilloscope; as a potentiometer control for the Everhart–Thornley detector—usually labelled "black level," "dark level," or "contrast"—is increased, the effect is to lower the overall signal (Fig. 4.50b). On the CRT display, the image becomes dark. The second operation consists of increasing the gain, labelled "photomultiplier gain" or "brightness," to amplify the difference signal, which expands the range of gray levels over which the specimen contrast is displayed (Fig. 4.50c). The weak contrast is now easily visible in the final display, as shown in Fig. 4.51b. For the Everhart–Thornley detector, this amplification is accomplished with the "gain" or "brightness" control.

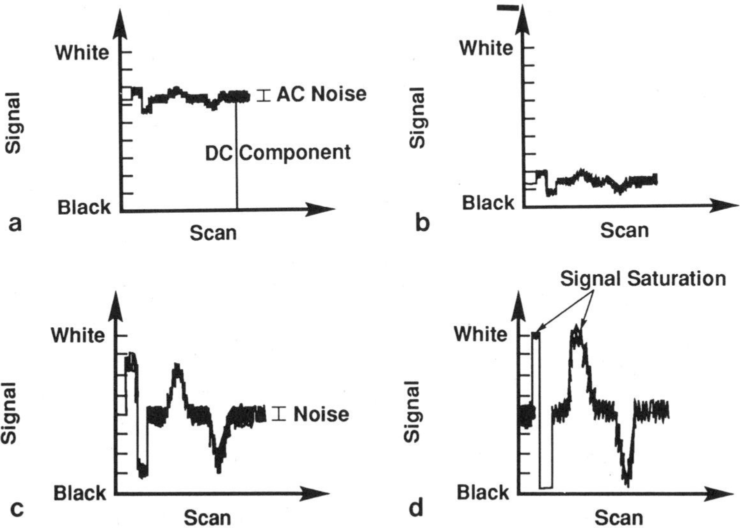

Figure 4.50. Schematic illustration of the steps required to apply differential amplification to enhance weak contrast in an image. (a) Line trace of a specimen which produces low natural contrast. The modulations of the signal (contrast) due to specimen features, confined to one or two gray levels, are poorly visible. (b) A constant (DC) level is subtracted from the signal at every point. (c) The difference signal is amplified, expanding the range of gray levels over which the specimen contrast is displayed. (d) Excessive application of differential amplification leads to signal saturation.

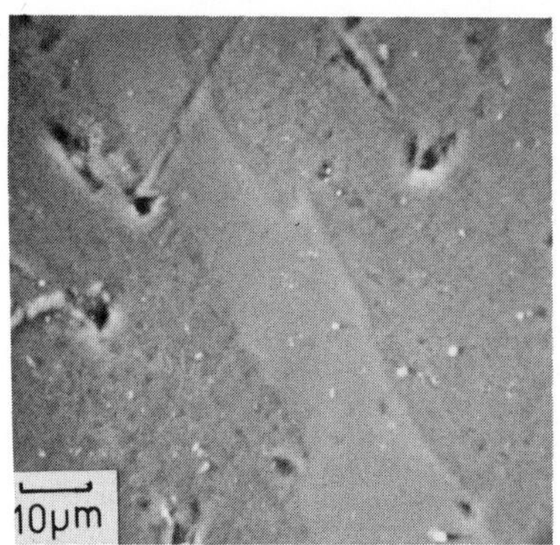

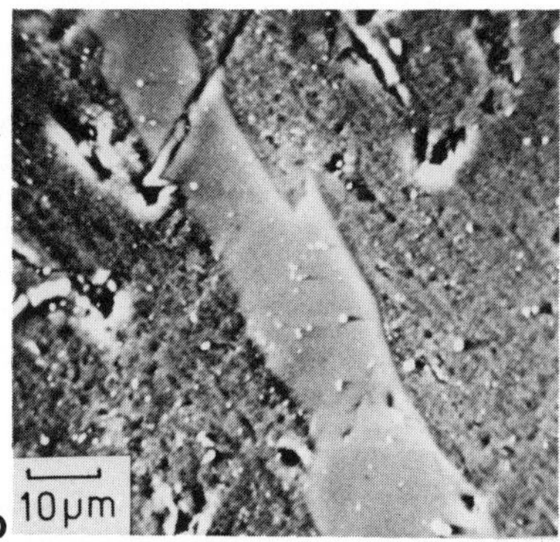

Figure 4.51. Weak atomic number contrast in an aluminum–silicon eutectic enhanced by differential amplification. (a) Natural contrast approximately 7% displayed with linear amplification. The brighter silicon band is barely visible against the darker surrounding aluminum. (b) Same image as (a), but with differential amplification applied. Note the bright areas of saturation around some of the voids in the specimen.

The signal equation for differential amplification has the following form:

$$S_{\text{out}} = K(S_{\text{in}} - S_{\text{constant}}), \tag{4.35}$$

where S_{in} is the input signal, S_{constant} is the constant signal subtracted from every value of S_{in}, K is the linear amplification factor, and S_{out} is the output signal. With careful adjustment of the differential amplification, the SEM display can be used to observe natural contrast levels as low as 0.1% displayed in gray scale images.

It is possible to apply excessive differential amplification. Figure 4.50d shows the effect on the signal trace of applying further differential amplification. The excursions due to the specimen contrast are now so large that in some regions the signal reaches a condition of saturation in which the displayed gray level can increase or decrease no further to follow the changes in the signal. Saturated regions appear in full white or full black, and the observer must realize that information is potentially lost in such regions. In Fig. 4.51b, the application of differential amplification was successful in ehancing the weak atomic number contrast, but note that in the regions of the holes in the specimen, the display is saturated in the white at the edges of the holes and in the black in the center, and we lose information in those regions. Saturation has other possible drawbacks. Not only is information lost in the saturated area, but for saturation at the white end of the display, the phenomenon of "blooming" can occur, where the excessive brightness can actually excite neighboring areas on the phosphor or film. In addition, a speckled, "grainy" texture may be observed in the image. This speckling is the direct visualization of the noise or randomness in the imaging signal, which was discussed in Section 4.5.1. In Fig. 4.50a, the noise, which can be estimated as the thickness of the line trace, is so small that it does not cause a sufficient excursion of the signal to excite a neighboring gray level. Since the noise is confined to a single gray level, it is not visible in the display. After differential amplification, the true contrast spans more gray levels and is more readily visible, but the noise has been expanded as well, and it now causes sufficient excursion in the signal to excite neighboring gray levels, becoming visible as speckle.

With the control over the display of contrast provided by differential amplification, we might be tempted to think that we have transcended the limitations imposed by the threshold equation. However, this is unfortunately not correct. A critical parameter in the threshold equation is the contrast, which enters as a squared term and therefore has a strong influence on the threshold current condition. It must be recognized that the threshold equation makes use of the natural contrast in the signal, that is, the modulations of the signal as it leaves the specimen–detector system, and not any subsequently altered contrast. Although we can increase the display contrast when the natural contrast is weak, we actually do not improve the signal-to-noise ratio. The image contrast is increased, but so also is the noise, so the threshold situation is not altered by differential amplification.

4.7.2.2. Enhancement of a Selected Contrast Range (Gamma Processing)

In second type of contrast problem which arises frequently, the specimen produces contrast spanning the whole dynamic gray-scale range of the display, but the features of interest occupy only a limited range of gray levels near the white or black end of the scale, resulting in poor visibility despite the full use of the gray scale. Such a situation is illustrated schematically in Fig. 4.52. If differential amplification is applied to enhance contrast at the dark end of the gray scale, large areas of the image are saturated at the white end, an unsatisfactory result.

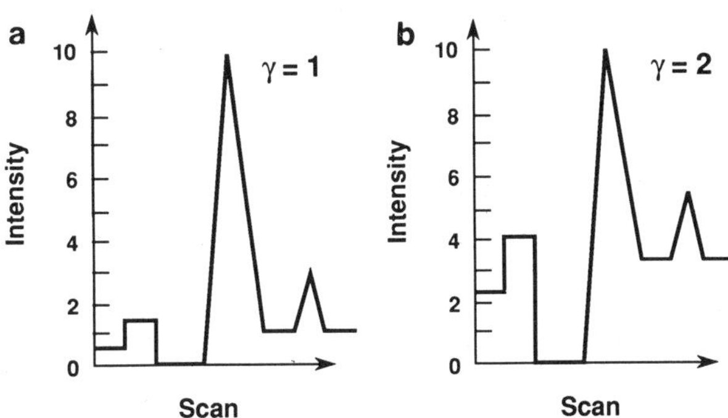

Figure 4.52. (a) Schematic illustration of a line trace for which the signal spans the full dynamic range of the display, but the information of interest lies in a narrow gray-scale range near the zero of signal (dark regions). (b) Same signal trace, but with nonlinear amplification applied ($\gamma = 2$).

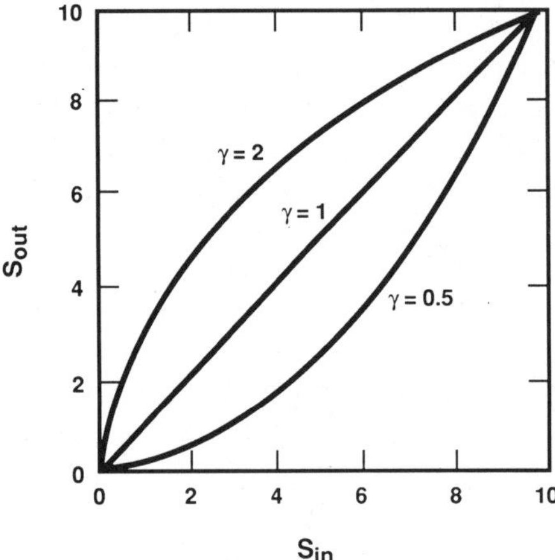

Figure 4.53. Signal response functions for nonlinear amplification (gamma processing). For $\gamma = 2$, the contrast at the dark end of the gray scale is expanded, improving the visibility of dark objects. For $\gamma = 0.5$, the expansion of contrast occurs at the bright end of the gray scale.

Nonlinear amplification, also known as *gamma processing,* provides selective contrast expansion at either the black or white end of the gray scale while preventing saturation or clipping of the displayed image. The signal formula for nonlinear amplification has the form

$$S_{out} = KS_{in}^{-\gamma}, \tag{4.36}$$

where γ is an integer (typically 2, 3, 4) or a fraction ($\frac{1}{2}$, $\frac{1}{3}$, $\frac{1}{4}$) and K is a linear amplification constant. The linear amplification is needed following the nonlinear operation to expand the signal range to cover the full dynamic range. The signal response produced by Eq. (4.36) is shown in Fig. 4.53. For $\gamma = 2$, a small range of input signals near the dark end of the display is distributed over a larger range of output signal, covering more gray levels and making the contrast more easily visible. The effect of $\gamma = 2$ on the line trace of Fig. 4.52a is shown in Fig. 4.52b. The signals at the white end are compressed into fewer gray levels. Note, however, that there is no difference in the signal saturation following gamma processing. For $\gamma < 1$, the expansion is obtained at the white end of the display, with compression at the dark end. An application of nonlinear amplification is illustrated in Fig. 4.54. The features in the hole are not visible in the linear image, Fig. 4.54a, but after nonlinear amplification with $\gamma = 2$, the features are readily visible, while the saturation in the bright regions of the image is controlled.

4.7.2.3. Enhancement of Selected Spatial Frequencies (Derivative Processing)

Because of the time-resolved nature of the SEM image, we can consider the image to consist of components of various frequencies in a time or spatial domain. When the beam scans across the edge of a feature, such as the fracture surface shown in Fig. 4.55a, the signal changes rapidly, producing high-frequency components of the image. When the beam is scanning across one of the grain faces visible in Fig.

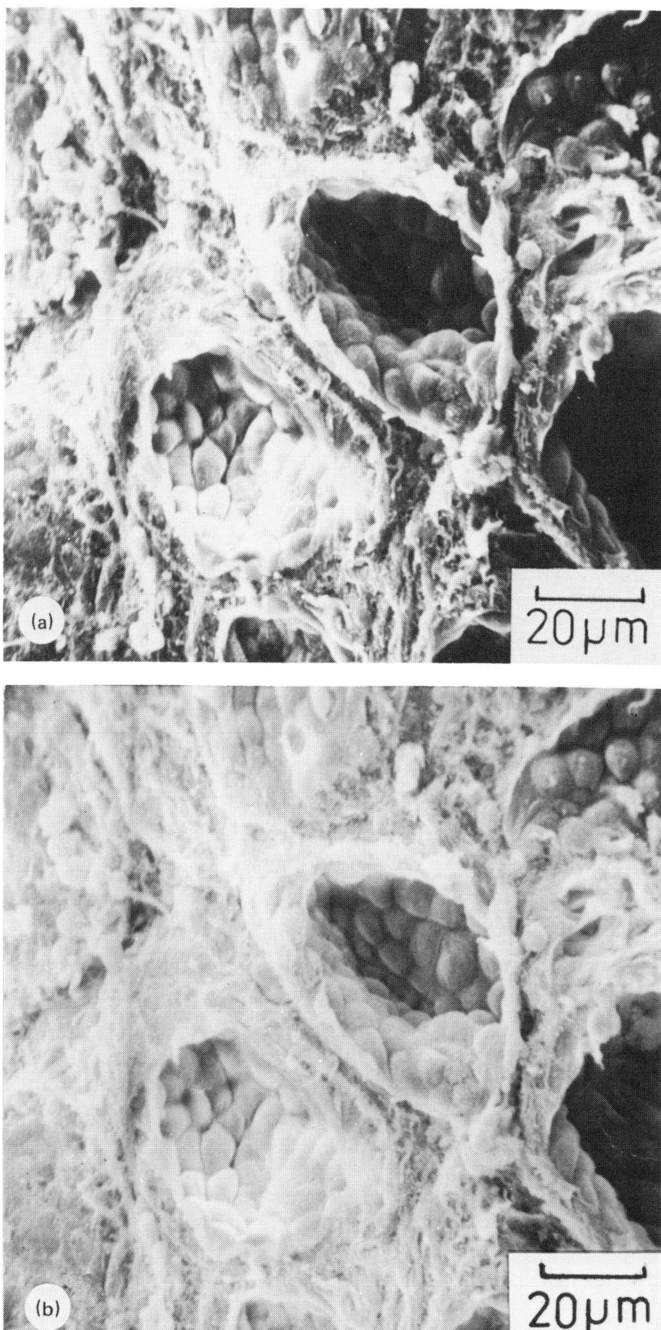

Figure 4.54. Application of nonlinear amplification (gamma processing) to improve the visibility of detail in a hole in mouse thyroid tissue. (a) Linear image. (b) Gamma processing, with $\gamma = 2$.

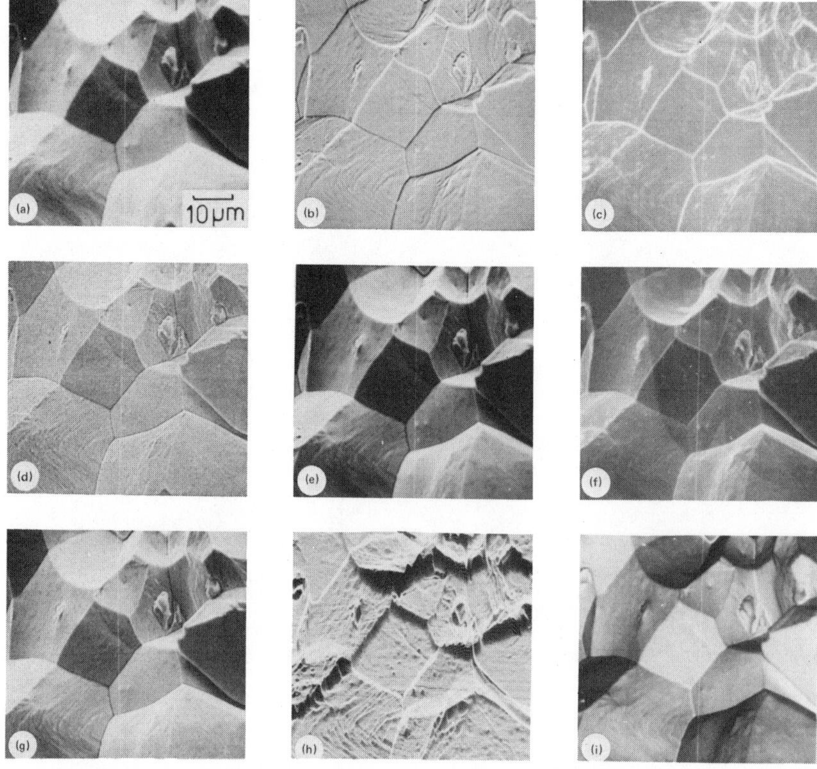

Figure 4.55. Images of an iron fracture surface with various types of signal processing applied: (a) direct + E–T image; (b) first time derivative; (c) absolute value of the first time derivative; (d) second time derivative; (e) 50% direct image plus 50% first time derivative; (f) 50% direct image plus 50% absolute first time derivative; (g) 50% direct image plus 50% second time derivative; (h) Y-modulation image; (i) reversed contrast. Beam energy 20 keV.

4.55a, the signal changes very slowly and consists of low-frequency components. When we consider an image, the edges of objects are often of the most interest to us in defining their position, size, and shape. Edges can be selectively enhanced by the application of time-derivative processing. Signal differentiation consists of the following mathematical transformation of the input signal:

$$S_{\text{out}} = \frac{dS_{\text{in}}}{dt} \tag{4.37}$$

The modification of the signal caused by the operation in Eq. (4.37) is illustrated schematically in Fig. 4.56. This transformation acts as a frequency filter, passing the high-frequency components of the image and blocking the low-frequency components. Edges in the original signal are replaced by sharp spikes in the first-time-derivative signal. Note that the sign of S_{out} depends on the sense of the change of S_{in}, and both positive- and negative-tending derivatives are possible depending on the slope of S_{in}, as shown in Fig. 4.56b. To detect both senses of change, S_{out} is further adjusted by linear amplification so that values of dS_{in}/dt near zero produce the middle gray level; positive-valued derivatives appear toward white and negative-valued derivatives appear toward black. The magnitude of the derivative signal depends not on the magnitude of the original signal but on its rate of change. The derivative is equally sensitive to a signal change no matter what the level the original signal is. Thus, if an image consists of bright and dark features with similar rates of signal change at the edges, the derivative will produce similar values for both types of objects. Other useful time-derivative operators, illustrated in Fig. 4.57, include the absolute value of the first derivative, in which negative-tending values are recorded with unchanged magnitude but the opposite sign, and the second time derivative, which has a sharper response.

Derivative functions applied by analog circuits suffer from a major artifact. Although the beam dwells for the same time on each picture element, analog scanning actually consists of a rapid scan along a line, typically 1–10 ms, and a much slower frame scan perpendicular to that line, typically 1–100 s, three orders of magnitude slower. This arrangement of scans has the consequence that when an analog time derivative is applied, it produces large derivative signals for objects with edges perpendicular to the line scan, but very low values for edges parallel to

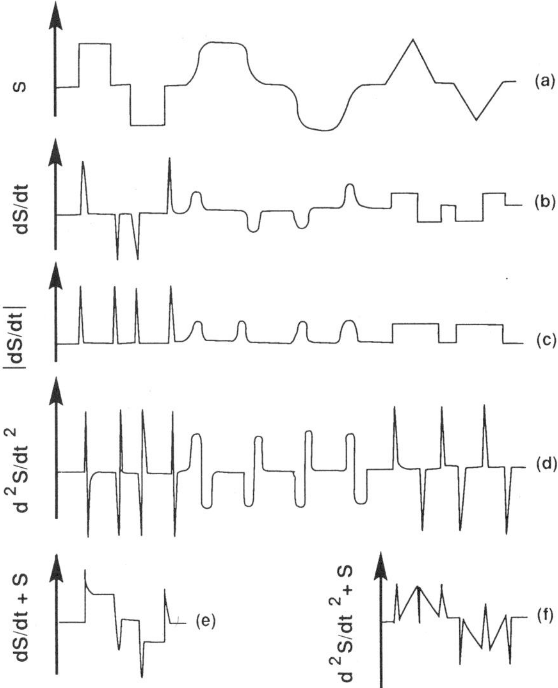

Figure 4.56. Signal traces of time-derivative processing operations: (a) direct signal; (b) first time derivative; (c) absolute value of the first derivative; (d) second derivative; (e) signal mixing, direct signal plus first derivative; (f) signal mixing, direct signal plus second derivative.

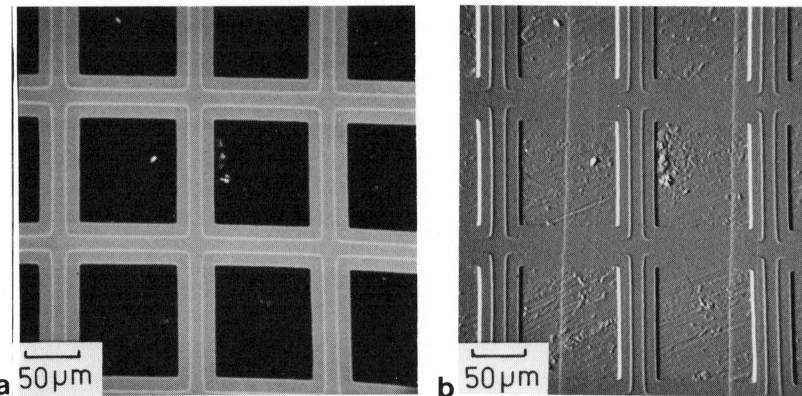

Figure 4.57. Images of a grid illustrating the anisotropy of unidirectional time-derivative processing. (a) Specimen-current image of a copper grid on a carbon planchet, reversed contrast. (b) First time-derivative image, with the scan line horizontal. Note the enhancement of the vertical edges of the grid openings which run perpendicular to the scan line, and the loss of the horizontal edges of the grid, which are parallel to the scan line.

the line scan. The effect of this anisotropy on an image is illustrated in Fig. 4.57. Figure 4.57a shows the reversed-contrast specimen-current image of a copper grid on a carbon planchet. The first-time-derivative image, Fig. 4.57b, with the scan line horizontal, shows strong enhancement of the vertical edges of the grid openings, which run perpendicular to the scan line, but loss of the horizontal edges of the grid, which are parallel to the scan line.

Time-derivative processing suffers from a second serious limitation, its high sensitivity to noise. Noise, which changes from pixel to pixel, resides in the highest-frequency component of the image. The action of a time derivative on the image is to selectively enhance the high frequencies, so the noise is preferentially enhanced. Time-derivative processing is effective only in situations of high signal-to-noise ratio.

Another caution with respect to time-derivative processing arises because of the possibility of improper interpretation of the apparent spatial resolution of the derivative image (Fiori *et al.,* 1974). We normally recognize objects by their edges, and we have a tendency to prefer sharp edges (Goldmark and Hollywood, 1951). Since the derivative produces a maximum value corresponding to the position of the maximum rate of change of the original signal, the sharpening effect may lead to an apparent improvement in the spatial resolution as the width of the image of a boundary may become narrower in the derivative image. This improvement may be misleading, since the maximum rate of change in the signal may not coincide with the physical edge of the boundary. Time-derivative processing at high magnifications must be applied with caution.

4.7.2.4. Signal Mixing

The pure-derivative images of a rough object have an unfamiliar appearance, as shown in Fig. 4.55b, c, and d, since the action of the derivative blocking the low-frequency components of the image removes the general shading of the topography, which is critical to establishing the light-optical analogy. While the derivative images have better edge definition, the images look flat, and the three-dimensional nature of the surface cannot be discerned.

This loss of topographic contrast can be overcome through the method of signal mixing (Heinrich *et al.*, 1970). Because all of the signals are simultaneously available, the output signal for the display can be assembled as a sum of signal components:

$$S_{\text{out}} = aS_{\text{out}} + b\frac{dS_{\text{in}}}{dt}, \tag{4.38}$$

where a and b are linear scaling constants individually selectable over the range $0 \leq a, b \leq 1$. Examples of signal mixing for the time-derivative signals mixed with the original signal are shown in Fig. 4.55e, f, and g. In all cases the topographic contrast of the original image is restored while retaining the sharpening of the time derivative.

Signal mixing can be used with other combinations of the wide range of signals produced by the SEM, such as mixing signals from different detectors. An example of signal mixing with various combinations of the direct and derivative signals is shown in Fig. 4.56, which depicts an iron fracture surface.

4.7.2.5. Contrast Reversal

Situations sometimes arise in which it is desirable to reverse the contrast which naturally appears in an image. An example is the direct specimen-current image, which has the opposite sense of contrast to the corresponding emissive-mode image. Since we intuitively expect to interpret images according to the characteristics of the emissive mode, direct specimen-current images are often confusing, both for topographic contrast, where the sense of topography appears reversed, and for atomic number contrast, where light elements are unexpectedly bright compared to heavy elements. It is therefore useful to artificially reverse the contrast during signal processing. This reversal is accomplished by the following signal transformation:

$$S_{\text{out}} = S_{\text{max}} - S_{\text{in}}. \tag{4.39}$$

An example of this contrast-reversal transformation applied to an image of a rough surface is shown in Fig. 4.37.

4.7.2.6. Y-Modulation

An observer is much more sensitive to low values of contrast if the modulation due to that contrast is displayed as a deflection of a line trace rather than as an intensity-modulated image. Normally, we examine only a single Y-modulated line trace at a time. A Y-modulation image can be composed by moving the locus of the line scan and superimposing

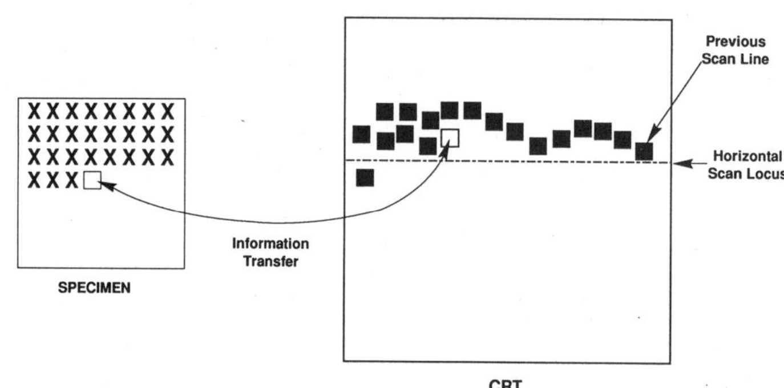

Figure 4.58. Schematic illustration of the generation of a Y-modulation image.

repeated traces on the image, as shown in Fig. 4.58. A conventional raster scan is made on the specimen. On the CRT, the horizontal position of the corresponding point to a pixel is determined as usual, but the vertical position is plotted according to the strength of the signal detected at that point. The intensity of all points plotted is maintained constant. An example of the resulting Y-modulation image is shown in Fig. 4.55h. The Y-modulation image is sensitive to small changes in the signal such as those produced by slight changes in topography. However, the image is subject to serious distortions, since the position of a point depends on a complex mixture of the actual position and the magnitude of the measured signal.

4.7.3. Digital Image Processing

Digital image processing offers great advantages over analog signal processing (see Newbury *et al.*, 1986a, Chapter 5 for a more detailed treatment). In digital image processing, the image scan is digitized and recorded in a computer memory as a matrix of triple values (x, y, I), where x and y represent the location coordinates and I is the signal intensity. The digitization of the image is typically carried out in a "frame grabber," which acquires individual television-rate scan frames. The contents of the frame grabber can be read out at television rates to give the effect of a real-time image.

The digital arrays of signals can be processed by an ever increasing array of computer-aided imaging functions. Digital image processing can be divided into two categories: *Digital image enhancement* seeks to render information contained in the image visible to an observer and includes those functions described above under analog signal processing as well as numerous processes available only in the digital domain. *Digital image interpretation* includes measurements of features, such as the number of particles in a field of view, their size distribution, etc.

4.7.3.1. Real Time Digital Imaging

Imaging with a frame grabber has several advantages. To improve the signal-to-noise of the image, successive scan frames can be effectively superimposed by summing and averaging, and the constantly updated "result" image can be continuously viewed by the observer in "real time" instead of the noisy individual frames. The effect of summing and averaging is to provide photographic-quality, reduced-noise images for direct viewing but with the images accumulated during a rapid scan. When a satisfactory image has been accumulated, it can be stored by the computer, and the memory can be slow-scanned to construct a conventional photographic CRT image or a hard-copy print-out. If the specimen has an insulating character and tends to charge, rapid scanning can often overcome or reduce the charging to an acceptable level. However, with a conventional analog photographic system, the scan must be slowed by a factor of 10 to 100 to improve the signal-to-noise ratio in the recorded image. Unfortunately, the slow scan often causes the charging to become severe, since the beam dwells for a longer time on each pixel and deposits proportionally more charge.

4.7.3.2. Off-Line Digital Image Processing

All of the mathematical operations contained in the circuits of analog signal processing and described by the equations given above can be readily expressed as mathematical operations in the digital domain, and many additional operations are possible (for a more detailed discussion, see Chapter 5, of Newburty *et al.*, 1986a). For example, differential amplification is carried out directly, as given by Eq. (4.35), except that the mathematical operations are directly implemented within a computer program. A numerical value is subtracted from each pixel, and the difference value is scaled to span the full gray-scale range of the output on a digital-to-analog converter.

Digital image processing makes the implementation of complicated mathematical transformations much more straightforward. The time derivative operators described above, with repeated overlaid scans, are extremely difficult to accomplish in the analog mode with exact registration of the successive scans. In digital image processing, the successive scans of the time derivatives are replaced by the formal mathematics of the transformations of functions such as the space gradient and the space Laplacian, which provide the edge enhancement. The Laplacian provides a formally isotropic image transformation. Such operations are implemented digitally through the application of a processing "kernel." The operation is considered for a specific pixel in an image, and the kernel consists of a grid of pixels containing and surrounding that pixel. For example, the simplest kernel consists of the central pixel and the four immediately adjacent pixels, ± 1 digital address in the x and y directions. A digital process is expressed as a multiplier for each of the pixels in the kernel that scales the intensity in that pixel. After the kernel operation is applied, the result is stored in the specific pixel address under process but in a new digital matrix array that stores the result as it is calculated. Thus, the original data is preserved, and a new result image is available for inspection and further processing. An example of digital image processing with a computer-aided imaging system implemented on a personal computer is shown in Fig. 4.59.

4.7.3.3. Digital Imaging for Minimum-Dose Microscopy

Digital image processing also is ideally suited for "minimum-dose microscopy," where the goal is to obtain the maximum information with the least damage to the specimen. Once a digital image with a satisfactory signal-to-noise level is recorded and preserved, the image can be viewed indefinitely with the beam blanked off the specimen. Various digital image processing functions can be performed on the stored image to derive the maximum information before exposing the specimen to any further possible damage. Since damage depends on both the total dose and the dose rate, digital image acquisition with a frame grabber operating at TV rates can be an effective solution.

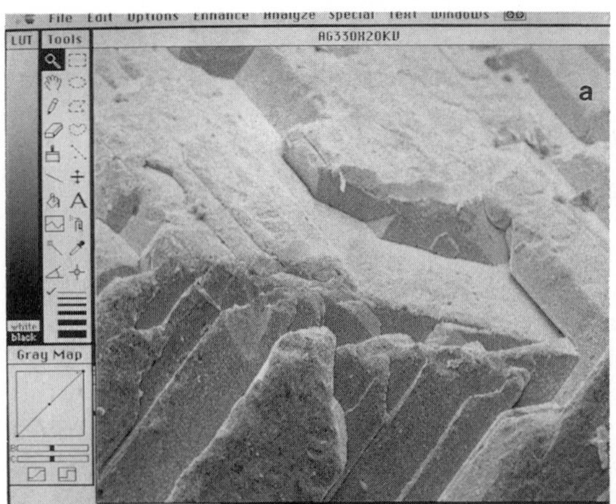

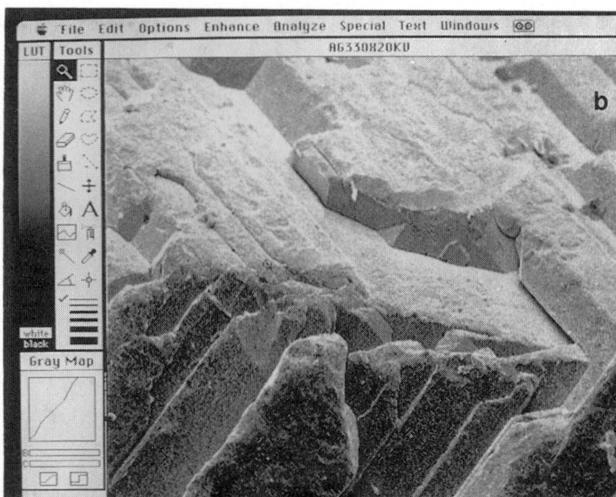

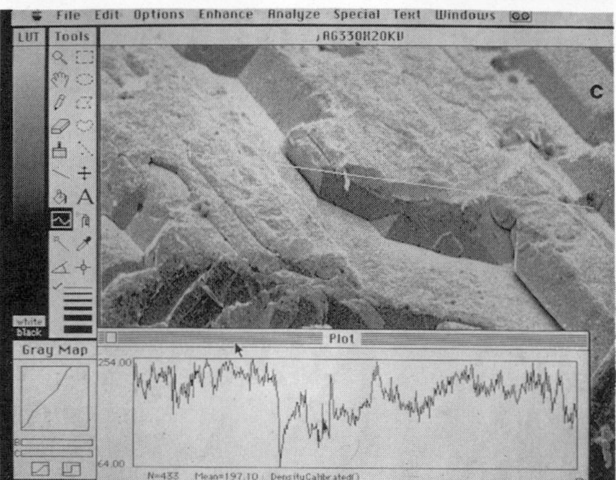

Figure 4.59. Digital image processing using a personal computer. (a) Original image of a silver crystal. (b) The displayed contrast in the image of a silver crystal is modified by the use of histogram equalization; note change in input–output relation in "Gray Map" window. (c) An intensity profile is taken through the image. The computer display has been photographed to show intentionally some of the processing tools available to the user. Many others are located on pull-down menus. (Software: NIH Image, by Wayne Rasband of the National Institutes of Health, Bethesda, Maryland.)

4.8. Defects of the SEM Imaging Process

The SEM imaging process, powerful though it is, is nevertheless subject to several defects. A number of these defects have been discussed at the appropriate places in previous chapters and sections: resolution limitations imposed by electron optics, beam–specimen interactions, and signal-to-noise considerations; distortions of the image inherent in the scanning process due to projection effects or arising as a result of pathological failures of the scan pattern or misapplication of special scanning functions such as tilt corrections; moiré effects due to the scan grid; and deviations from the light-optical analogy critical for proper interpretation of topographic contrast. In this section we will consider other major defects of the imaging process.

4.8.1. Contamination

The term *contamination* describes the collective phenomena by which the surface of a specimen undergoes a deposition of a foregin substance, generally assumed to be a carbonaceous material derived from the breakdown of a hydrocarbon. The effect of this surface deposit is generally observed as a "scan square" when the magnification is reduced while centered on the previous higher magnification field, as shown in Fig. 4.60. Contrast arises in such an image because of the change in the secondary-electron coefficient δ caused by the deposition of the foreign material. Such a phenomenon is a nuisance for low-magnification work, but its impact can be reduced by taking images required over a range of magnifications as a function of increasing magnification rather than decreasing magnification. Decreased image contrast inevitably results from the accumulated contamination, but when an expedient practical solution is required, this technique will often suffice.

Contamination can pose severe limitations to high-resolution operation. An example of the deposition of a halo of contamination around fine Al–W dendrites is shown in Fig. 4.61. Such contamination may be mistaken for true specimen details, or it may simply obscure true features. In severe situations, the local build-up of contamination can be quite extraordinary. Hydrocarbon molecules seem to be attracted to the beam location, migrating along the specimen surface. High-resolution imaging in the high-voltage analytical electron microscope has revealed "fingers" of contamination that build up to heights of several hundreds of nanometers during fixed-beam analysis (Joy *et al.*, 1986).

Clearly it is most desirable to avoid contamination in the first place. In a modern, well-maintained pumping system operated with efficient liquid nitrogen trapping of volatiles from the diffusion pump vapor, the source of contamination is almost always the specimen

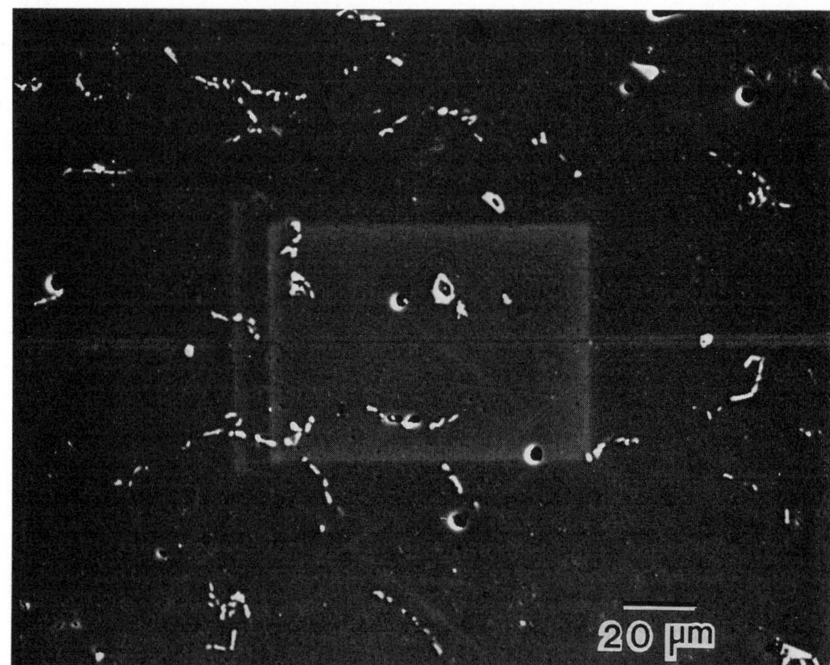

Figure 4.60. Typical appearance of contamination. After scanning at an elevated magnification both in image and line-scan modes, when the magnification is reduced, the area scanned at higher magnification is visible as a "scan square." Specimen: aluminum alloy; beam energy = 15 keV.

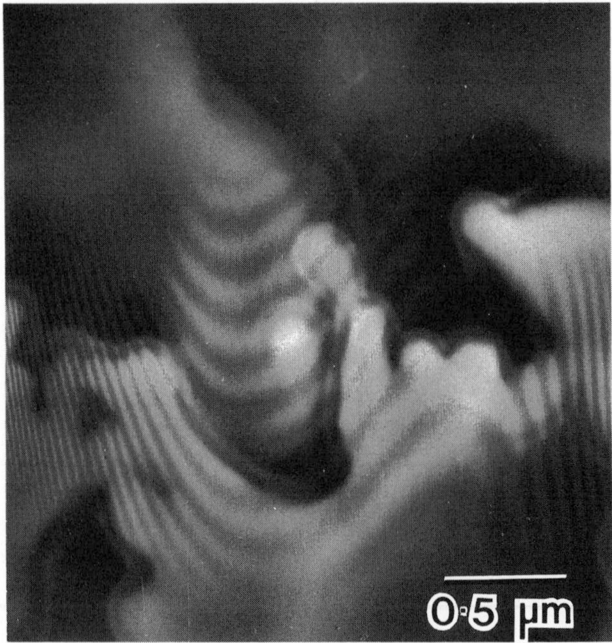

Figure 4.61. Contamination at high magnification: After scanning for an extended period, a halo or envelope of contamination has formed around the aluminum–tungsten dendrites.

itself. Procedures to properly clean and handle specimens are described in Chapter 11. Several additional techniques are worth consideration to control contamination:

1. Anticontamination plates, liquid nitrogen–cooled surfaces placed in the vicinity of the specimen, can reduce the hydrocarbon background in the area where it matters most.
2. Migration of hydrocarbon molecules can be reduced by cooling the specimen. Alternatively, mildly heating the specimen can drive off hydrocarbons or lower the sticking coefficient to prevent further deposition.
3. Exposure of the specimen to an intense ultraviolet light source prior to SEM imaging can cross-link a surface contamination layer and fix it in place, so that molecular migration to the electron beam is greatly reduced.

4.8.2. Charging

When the beam interacts with the specimen, some of the charge injected by the beam is emitted in the form of the backscattered and secondary electrons. For many situations, a greater fraction of the injected charge remains in the specimen as the beam electrons lose all of their initial energy and are captured by the specimen. This charge flows to ground if the specimen is a conductor and a suitable connnection exists. If the ground path is broken, even a conducting specimen quickly accumulates charge and its surface potential rises. It may actually become high enough to act as an electron mirror, although a distorted one, so the scanning beam is actually reflected off the specimen to scan on the walls of the chamber. Figure 4.62 shows the highly distorted "fisheye lens" view of the specimen chamber that results.

Charging is more often observed whenever a specimen or a portion of it is an insulator. When the electron beam strikes an insulator, the charges injected by the beam cannot readily flow to ground. The resulting accumulation of charge is a complex, dynamic phenomenon. The specimen is in a continually changing state of surface potential due to the accumulation and discharge of electrons.

Charging manifests itself in images in a variety of ways. The collection of secondary electrons by the Everhart–Thornley detector depends sensitively on the electric field lines around the specimen established by the collection field of the Faraday cage on the E–T detector and the local specimen field, which is grounded in the case of a conducting sample connected to the stage. When local charging alters the surface potential, the field lines to the collector are disrupted, and the collection of secondaries can be greatly altered. A contrast mechanism known as *voltage contrast* develops in which the potential distribution across the surface is imaged. Some areas appear extremely bright because they are negative relative to the E–T

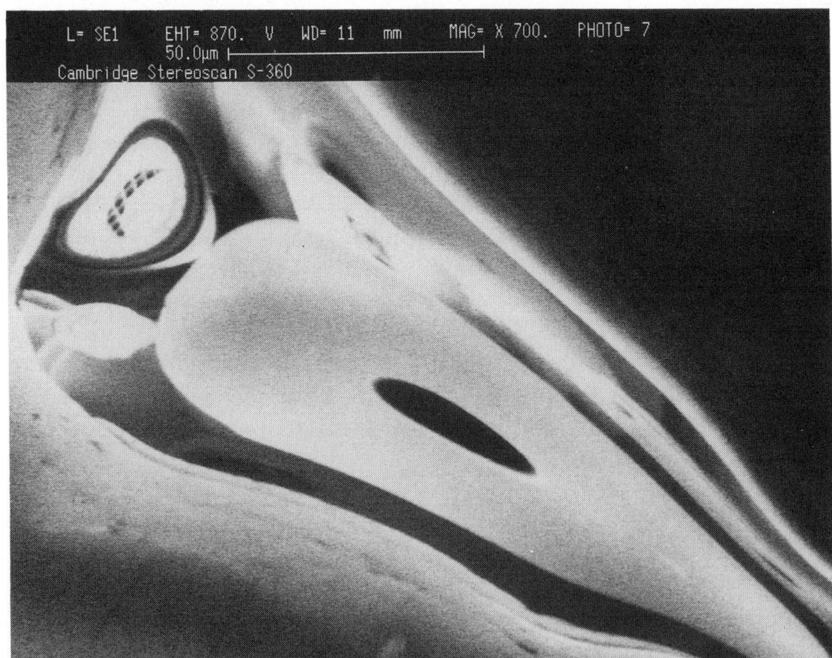

Figure 4.62. Extreme specimen charging, in which the beam is reflected off the specimen and scans the specimen chamber, producing a highly distorted "fisheye lens" view of the polepiece, the Faraday cage of the E–T detector, and the chamber walls.

detector and enhance secondary collection, while others appear black because they charge positively and suppress the collection of secondaries. The difficulty is that the contrast due to surface potential becomes so large that it overwhelms the contrast from the true features of the specimen. In more extreme cases, charging can reach such high values that the beam is deflected and becomes positionally unstable. The charged regions may undergo catastrophic breakdown in which a sudden discharge occurs, sharply changing the beam position or electron collection situation. Such a high level of charging manifests itself as discontinuities in the scan, such as streaks, or distortions of objects in the image. Figure 4.63 shows examples of charging observed from a calcite crystal examined in an uncoated condition. At 15 keV, discontinuities in the scan and harsh black–white voltage contrast are observed. At 5 keV, the scan is stable, but the harsh voltage contrast overwhelms the true topographic contrast.

An example is shown in Fig. 4.64 of a more insidious charging artifact where the charging phenomena appear to be a feature of the specimen. In Fig. 4.64a, which was recorded at a beam energy of 2.4 keV, there is no evidence of significant charging in the image, except that each of the spheres has a curious structure in its center. These structures are remarkably similar on all the spheres, and could easily be mistaken for a real feature of the specimens. In Fig. 4.64b, when the beam energy is lowered to a value of 1.53 keV, the features

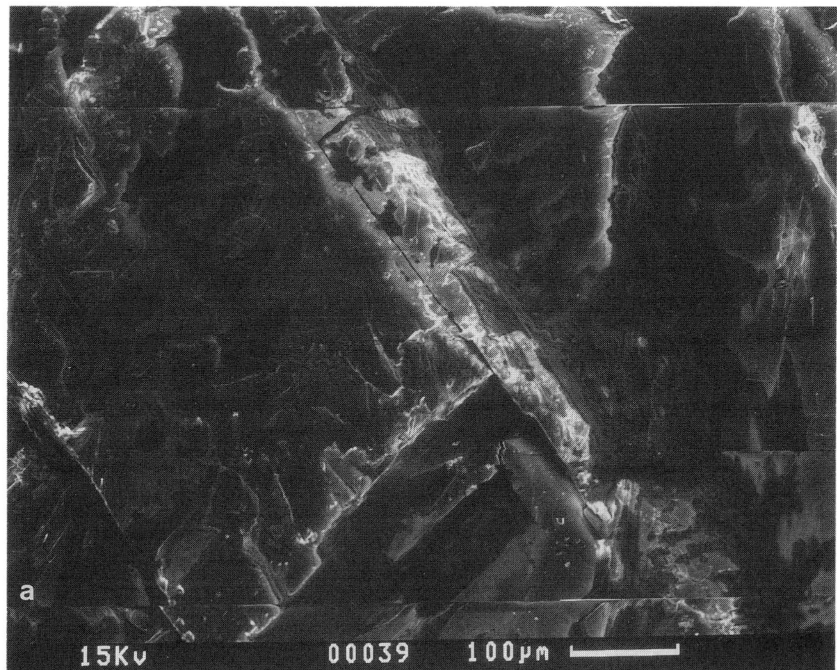

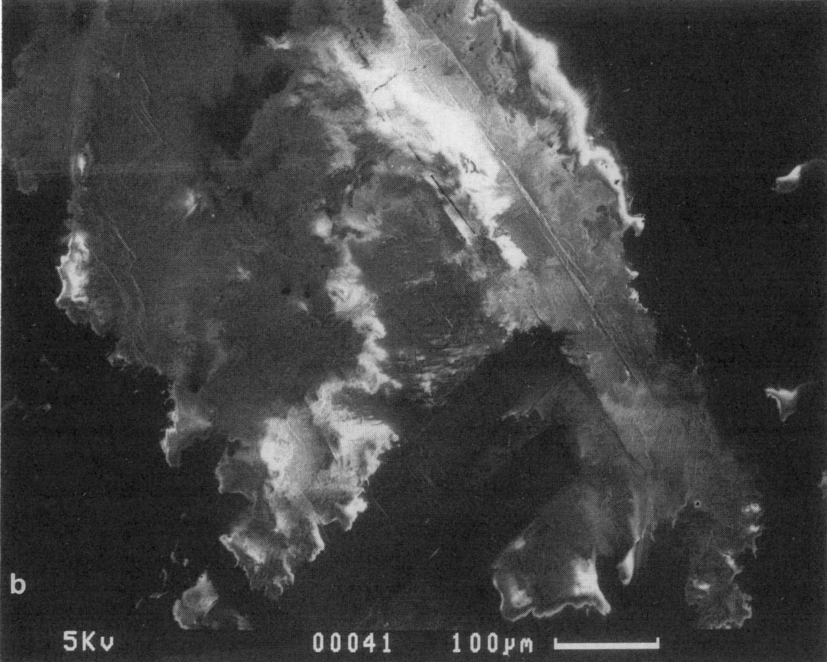

Figure 4.63. Charging observed during SEM imaging of an uncoated calcite crystal. (a) $E_0 = 15\,\mathrm{keV}$; note the bright/dark regions and the scan discontinuities. (b) $E_0 = 5\,\mathrm{keV}$; the scanned image is now stable, but the bright/dark voltage contrast dominates the true features of the image. (c) $E_0 = 1.5\,\mathrm{keV}$; operation at the E_2 crossover point produces a stable image in which topographic contrast showing the true features of the specimen is evident. The bright area on the tilted top surface shows that not all regions of charging can be simultaneously controlled.

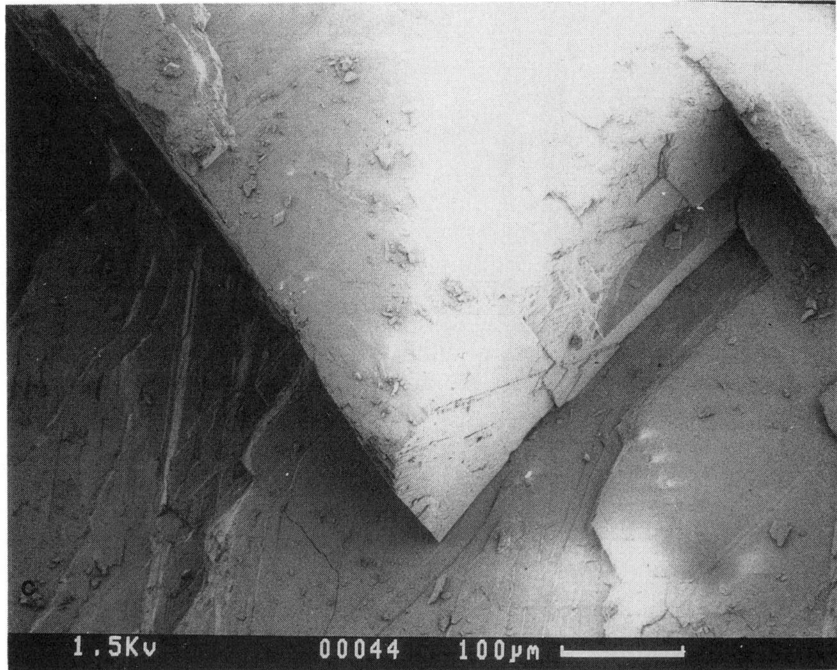

Figure 4.63. (Continued).

change in form, giving each sphere an apparent "eye." This eye is a local charged region, and each sphere is acting as a tiny mirror optic. In fact, each of the "eyes" in Fig. 4.64 is a microscopic view of the specimen chamber similar to that seen in Fig. 4.62. Finally, when the beam energy is lowered to 0.87 keV (870 eV) and the magnification is lowered by a factor of two to reduce the electron dose per unit area, the spheres are observed without any interior artifacts.

Charging in SEM images is a dynamic phenomenon because the SEM is a dynamic imaging system. The beam must be constantly moved to produce an image. From the point of view of a pixel in the image field, the beam dwells for a certain time and injects charge. The surface potential of the pixel changes to a level that depends on the injected charge and the local capacitance, and then the beam moves on to the next pixel, which at low magnification can be micrometers or more away. That charged pixel then begins to discharge with a time constant which depends on the capacitance and the local resistance. By the time the beam returns in the next scan frame, the potential of the pixel will have altered. If the scan is very rapid, as in TV-rate scanning, the discharge may not be significant, and the specimen may reach a dynamic equilibrium. Such a dynamic situation can particularly frustrate the microscopist who sets up an image in a rapid TV-scan mode and then tries to photograph it with a long-dwell time scan.

Charging can be minimized or avoided by a number of practical techniques. These methods will be described according to progressively increasing charging.

4.8.2.1. Incipient Charging

Rapid Scanning. E–T detector images are extremely sensitive to charging; only a few volts will create black–white contrast excursions. Often when charging is minor, appearing only as a few regions of deviant brightness or darkness at slow visual or photographic scan rates, satisfactory images can still be obtained with the E–T detector by TV-rate scanning. In this regard, the development of "real time" digital imaging with framestores, as described in the previous section, is especially welcome, because framestores permit high quality images to be accumulated at TV rates.

BSE Imaging. Although the secondary component of the E–T detector image is sensitive to low levels of charging, backscattered electrons are not. A surface potential of a few volts does not substantially alter the trajectories of high-energy backscattered electrons. Thus, if the BSE-only mode of the E–T detector (negative bias)

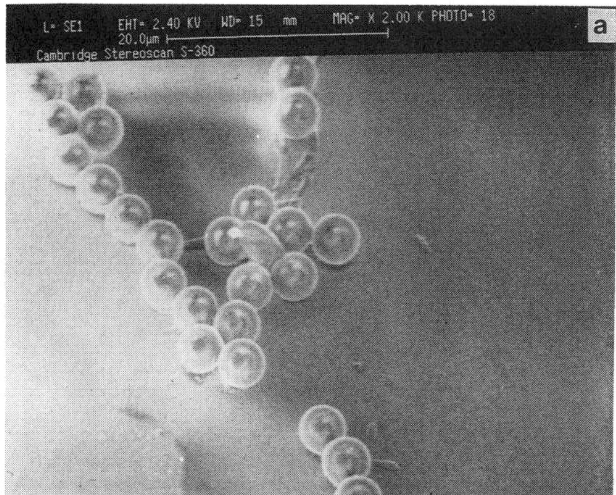

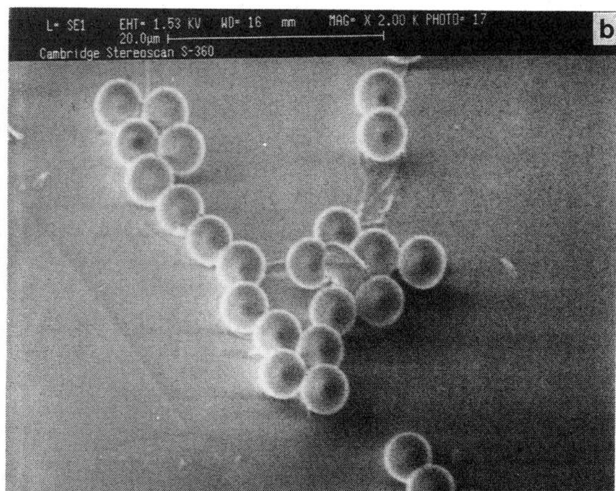

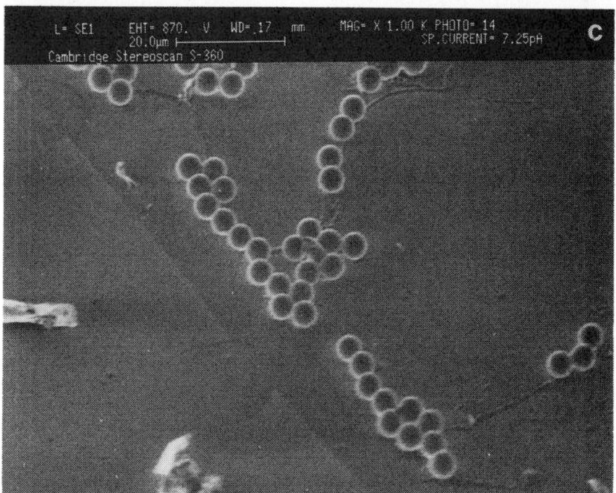

Figure 4.64. Charging that could be mistaken as a specimen feature. (a) The polymer spheres seem to show a complicated interior structure; $E_0 = 2.4$ keV. (b) At $E_0 = 1.53$ keV, the structure has changed into an "pupil", making each sphere look like an "eye". (c) At $E_0 = 0.87$ keV and a reduction in magnification of a factor of two to reduce the dose per unit area, the charging artifact has been eliminated.

or a dedicated backscattered-electron detector is chosen, charging can be eliminated from the image.

4.8.2.2. Severe Charging

When severe charging is observed, with its deflection of the beam and dynamic instability, more elaborate measures are needed.

Specimen Coating. If it is necessary to operate at high beam energy for x-ray microanalysis with accompanying SEM imaging, then a conductive coating must be applied to the specimen. Chapter 13 describes in detail the many coating techniques available. The coating provides a path along which the excess charge injected by the beam can flow to ground. Generally the coating is much thinner than the electron range, so that charge is still injected below the surface. The presence of the grounded, conducting coating within a micrometer of the injected charge provides a large voltage gradient, which leads to breakdown and discharge. As a cautionary note, applying a conductive surface coating may not be sufficient. There must still be a path to ground from the coating. If the specimen has sides which do not conduct, the coating may not reach these areas, and the conducting path may be broken. A track of conducting paint should be applied along these insulating sides between the top coating and the conducting stub.

Low-Voltage Microscopy. The behavior of the secondary and backscattered electron coefficients as a function of beam energy, shown in Fig. 3.27, forms the basis for a solution to charging by means of operation at low-beam energy. If the beam energy is significantly above the second crossover point E_2, then $\eta + \delta < 1$ and the specimen will charge negatively. This is an unstable situation. If the beam energy is selected to be exactly at the second crossover point E_2, then since $\eta + \delta = 1$, the charges injected by the beam are exactly cancelled by the charge leaving the specimen as secondary and backscattered electrons, and a condition of dynamic equilibrium is achieved. If the beam energy is less than E_2 but greater than E_1, then $\eta + \delta > 1$, more electrons leave the specimen than arrive from the beam, and the specimen charges positively. This causes the potential difference between the specimen and the filament to increase, so that the incident energy for the next beam electron increases. Since E_0 effectively increases, the specimen automatically "hunts" the E_2 point, and stable operation is again achieved. Successful operation in this low-voltage regime is illustrated in Fig. 4.64c, where the "eyes" due to charging have been eliminated.

This approach is not without difficulty. If a specimen has strong topography, then the surface inclination effects on η and δ can locally alter whether the dynamic equilibrium at the E_2 point is achieved throughout the image. Figure 4.63 illustrates this point with a series of images of an uncoated calcite crystal, where different portions of the image become stable and free from charging artifacts as the beam energy is decreased. True topographic features are seen at low voltage in Fig. 4.63, but part of the crystals is still charging.

4.9. Special Topics in SEM Imaging

4.9.1. SEM at Elevated Pressures (Environmental SEM)

4.9.1.1. The Vacuum Environment

The SEM is typically operated with a pressure in the sample chamber below 10^{-4} Pa ($\sim 10^{-6}$ torr), a condition determined by the need to satisfy four key instrument operating conditions:

1. The pressure in the electron gun must be lower than 10^{-4} Pa ($\sim 10^{-6}$ torr) for a conventional thermal-emission tungsten filament and below 10^{-5} Pa ($\sim 10^{-7}$ torr) for the highly reactive LaB_6 source. Although a separate pumping system is typically devoted to the electron source, a high specimen-chamber pressure leads to unacceptably high pressure in the electron source.
2. To preserve the integrity of the focused primary beam, the chamber pressure must be minimized to reduce the number of collisions between the beam electrons and the molecules of the residual gas. Such elastic-scattering events will not only remove intensity from the focused beam, but since the gas scattering can occur at any point along the centimeter-long path from the final lens to the specimen, the scattered electrons can strike the specimen at distances of millimeters from the focused beam.
3. The Everhart–Thornley detector is operated with a bias of +12,000 volts or more applied to the face of the scintillator. If the chamber pressure exceeds approximately 1 Pa ($\sim 10^{-2}$ torr), electrical breakdown may occur between the scintillator (+10,000 V) and the Faraday cage (+250 V), which is located in close proximity.
4. A major source of specimen contamination during examination arises from the cracking of hydrocarbons by the electron beam. A critical factor in determining contamination rates is the availability of hydrocarbon molecules for the beam electrons to hit. To achieve a low-contamination environment, the pumping system must be capable of achieving low ultimate operating pressures augmented with careful cold trapping of gases backstreaming from the pump to minimize the partial pressure of hydrocarbons. So successful is this strategy that in a well maintained modern instrument, beam-induced contamination almost always results from residual hydrocarbons on the specimen which remain from incomplete cleaning.

A significant price is paid to operate the SEM with such a "clean" vacuum. The specimen must be prepared in a condition so as not to evolve gases in the vacuum environment. Many important materials, such as biological tissues, contain liquid water, which will rapidly evaporate at reduced pressure, distorting the specimen and disturbing the stable operating conditions of the microscope. This water, and other volatile substances, must be removed during sample preparation to examine the specimen in a "dry" state, or the water must be immobilized by freezing to low temperatures ("frozen, hydrated samples"). Such specimen preparation is both time-consuming and prone to artifacts, including the redistribution of "diffusable" elements such as salts.

Scanning electron microscopy at elevated pressures in the region of 100–1000 Pa (~1–10 torr), known as "environmental scanning electron microscopy," seeks to obtain the special benefits of operating at higher pressures while maintaining a reasonable level of SEM performance (Danilatos, 1988). The elevated-pressure SEM uses the concept of differential pumping to obtain the desired elevated pressure in the specimen chamber while simultaneously maintaining a satisfactory pressure in the electron gun. Differential pumping consists of establishing a series of regions of successively lower pressure, each region with its own dedicated pumping system and separated by a small aperture. The probability of gas molecules moving from one region to the next is limited by the area of the aperture. In the SEM, these differential pumping apertures are the actual electron-beam apertures, and the sequence of vacuum regions consists of the specimen chamber, one region for each lens, and finally the electron gun. Such a vacuum system can maintain a six-decade pressure differential between the specimen chamber and the electron gun.

A wide variety of gases can be used in the elevated pressure sample chamber, including oxygen, nitrogen, argon, and water vapor. Because the image is extremely sensitive to the chamber pressure of the sample, a means of carefully regulating the pressure is required.

4.9.1.2. Detectors for Elevated-Pressure Microscopy

Backscattered Electrons (Scintillator Detector). As noted above, the Everhart–Thornley detector, or any other detector which employs a high post-specimen accelerating voltage, such as the channel plate multiplier, cannot be used at elevated pressures due to electrical breakdown. The passive backscattered electron detectors, such as the large-area scintillator detectors, are suitable for elevated pressures, since the backscattered electrons suffer negligible energy loss in the gas phase and retain sufficient energy to activate the scintillator without post-specimen acceleration. Figure 4.65a shows a BSE image obtained from Raney nickel in an environment of water vapor at a pressure of 3.8 torr. In fact, an added advantage of elevated-pressure operation is the automatic discharging of the surface of insulators (see below), which means that the bare scintillator can be used without the metallic coating required for conventional reduced-pressure operation.

Secondary Electrons (Gas Amplification Detector). To use the low-energy secondary electrons, a special gas-phase detector has been

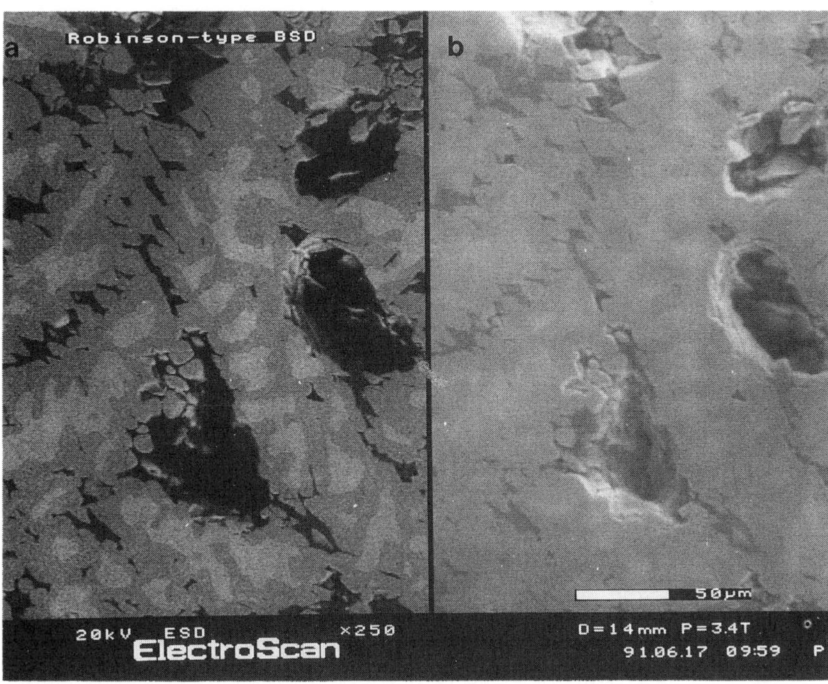

Figure 4.65. SEM at elevated pressure: comparison of images of Raney nickel obtained with (a) conventional large solid angle scintillator BSE detector; (b) gas amplification secondary-electron detector. Beam energy 20 keV; chamber pressure 3.8 torr, water vapor. (Example courtesy of the Electroscan Corp.)

developed whose operation is analogous to the flow-proportional gas detector used in the wavelength-dispersive x-ray spectrometer described in Chapter 5 (Danilatos, 1990). A wire is maintained at a modest accelerating voltage, approximately 1 kV positive, which is less than the breakdown voltage for pressures below 1000 Pa (~10 torr), in close proximity to the grounded specimen. The secondary electrons emitted from the specimen are accelerated toward this wire and undergo collisions with the gas molecules, ionizing them and creating more free electrons. These ejected electrons are also accelerated toward the wire, resulting in a cascade of charge, amplifying the current collected at the wire compared to that emitted originally from the specimen. While backscattered electrons can also contribute to the total signal collected at the wire by collisions with gas molecules, the mean free path for collisions increases rapidly with increasing energy, so the contribution of the high-energy backscattered electrons is much less than that of the secondary electrons. An example of an image of Raney nickel obtained with the gas-amplification detector is shown in Fig. 4.65b.

4.9.1.3. Contrast in Elevated-Pressure Microscopy

An image containing both compositional and topographical features viewed at elevated pressure is shown in Fig. 4.65. Both the dedicated backscattered-electron image from the scintillator detector and the secondary-electron image from the gas-amplification detector reveal the atomic number contrast from the multiphase specimen. The atomic number contrast in the dedicated backscattered-electron image arises in the same fashion as previously discussed for low operating pressures. Atomic number contrast in the gas amplification secondary-electron image arises from the SE_{II} type secondaries, which are produced by the backscattered electrons and scale in a similar fashion with composition.

Topographic contrast created by the holes in the structure in Fig. 4.65 show strong contrast in the scintillator backscattered-electron image, but the secondary-electron image obtained with the gas-amplification detector shows details much deeper in the holes (note the hole near the arrow). These images can be understood with careful analysis. The holes appear dark in the backscattered electron image because, although the primary beam strikes the walls of the hole, the backscattered electrons are strongly reabsorbed by the walls or scattered out of the line-of-sight collection of the scintillator backscattered-electron detector because of the curvature of the crater walls. Because the environmental gas penetrates into the holes, as long as the primary beam can strike a surface and cause it to emit secondary electrons, the electron cascade can develop in the gas and generate a measurable signal at the collection wire. Thus, an image can be obtained from greater depths, as deep as the beam can reach.

4.9.1.4. Resolution

The electron beam profile is inevitably degraded by elastic collisions in the high-pressure environment. However, the result is an extreme case of the situation discussed in Fig. 4.44 for high-resolution imaging at high beam energy. The effect of the gas molecule scattering is to create an extremely wide skirt around the focused beam. Although the focused beam loses intensity, a sufficient fraction of the electrons remain within the central peak of the distribution to perform useful imaging. When this combined focused beam + skirt is scanned across an object, the change in the signal due to the wide skirt is negligible compared to the change in signal due to the focused beam. Thus, although the total available current is lowered by the gas scattering, the image retains a significant level of resolution, comparable to that in low-pressure operation. Note, however, that the existence of this wide skirt significantly broadens the area on the specimen from which x rays are obtained in fixed-beam microanalysis.

4.9.1.5. Benefits of SEM at Elevated Pressures

There are several special benefits of performing scanning electron microscopy at elevated pressures:

1. The abundance of free electrons and positive ions in the immediate vicinity of the specimen due to beam–molecule, backscattered

electron–molecule, and secondary electron–molecule collisions provides for charge dissipation through direct neutralization. Charged areas attract oppositely charged species in appropriate numbers for neutralization. Insulating specimens can therefore be examined without any coating.

2. The partial pressure of gas, combined with a temperature-controlled stage, means that even liquid water can be maintained under operating conditions. Many specimens, including delicate biological specimens, which could not be viewed directly in a low-pressure SEM, can be immediately examined in the elevated-pressure SEM. Sample preparation is minimized, reducing or eliminating important types of artifacts.
3. By choosing the gas–liquid species, pressure, and temperature, many different "environmental" experiments can be conducted, including those on crystal growth, corrosion, chemical reactions, etc.

Figure 4.66 illustrates the power of this approach to study complex reactions in extreme environments. The images depict two stages in a heating experiment involving lead and potassium chloride heated in water vapor. As various processes occurred, including potassium chloride evaporation, condensation, and crystal growth and melting of the lead, the processes could be followed dynamically with recording on video tape. Figure 4.66a shows the growth of potassium chloride crystals at 650°C, while Fig. 4.66b shows a lead film on molten lead at 890°C.

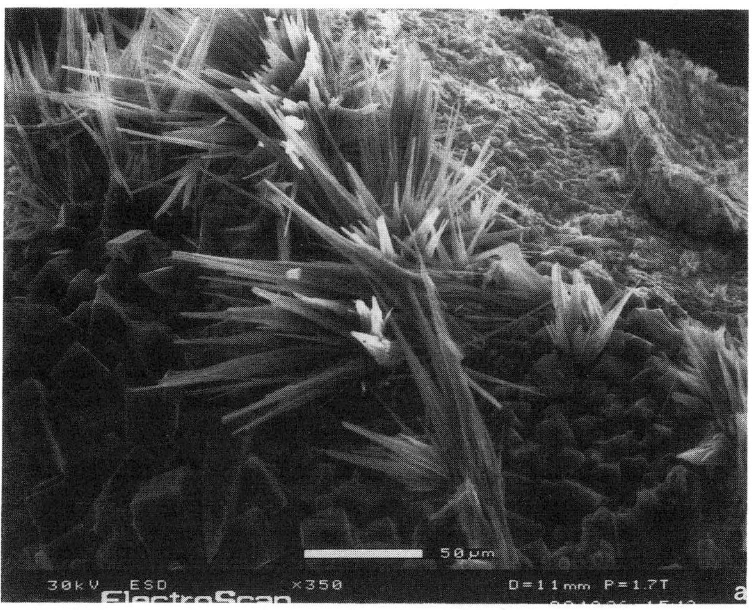

Figure 4.66. Application of the environmental SEM to a dynamic crystallization experiment. Lead and potassium chloride were heated together in water vapor, and the process was followed with continuous recording on videotape. (a) 650 °C, crystals of potassium chloride grown during the heating cycle. (b) 890 °C, lead oxide film on surface of molten lead, with only traces of crystals of potassium chloride remaining after vaporization. (Example courtesy of the Electroscan Corp.)

Figure 4.66. (Continued).

4.9.2. Stereo Microscopy

The large depth of field, typically equal to the scanned width of the image or more, and the good resolution obtainable in the SEM make possible the effective use of stereo techniques in the examination of rough surfaces (Wells, 1960). The stereo effect is obtained by viewing two images of the same area taken with some angular difference between them. The perception of depth arises from the parallax, i.e., the slightly differing images presented to the brain by our two eyes. The greater the difference in the angles the greater the apparent depth, up to a limit where the brain can no longer fuse the two images together. Stereo techniques fall into two categories. Qualitative stereo microscopy makes use of the stereo effect to aid the microscopist in understanding the true shape of an object. Often an object's shape is difficult to ascertain when only images which are two-dimensional projections of the three-dimensional object are available. In fact, the great depth of field of the SEM can be a liability, because it is often difficult to tell those areas of the specimen that are higher or lower when everything is in focus. Quantitative stereo microscopy is used to make measurements on the vertical separation of features of interest.

4.9.2.1. Qualitative Stereo Microscopy

Two methods can be employed to produce stereo pairs with an SEM. The simplest is to translate the sample between exposures; the other is to tilt the specimen between exposures. Both methods produce the slightly differing views needed to produce stereoscopic vision. The translation or shift method is similar to that used by aerial photographers in that the specimen is moved across the field of view so that any given feature is imaged in different parts of the scan raster. The displacement between micrographs is obtained by using the stage shift controls. The area of overlap, common to both micrographs, can be viewed stereoscopically. In order to provide a stereo image containing a sufficient depth effect, the displacement needed is fairly large, typically half the screen width at 20X magnification. This technique is not therefore suitable for high-magnification use, as there would be negligible overlap region.

The tilting method is based upon the use of two images of the same area obtained from two different tilt angles to provide the separate points of view. The tilting method has two advantages. First, it can be applied at all magnifications available on the microscope; second, the whole image frame area is now usable. The optimum tilt angle for electron micrograph stereo pairs depends on the magnification as well as on the actual amount of depth information in the image (Hudson and Markin, 1970). As a guide we can write

$$\sin\frac{\theta}{2} = \frac{\text{maximum parallax}}{(2\Delta HM)}, \qquad (4.40)$$

where θ is the tilt angle, the maximum parallax (i.e., the largest image separation that the brain will tolerate and still fuse the images) is about 5 mm for images at a viewing distance of 25 cm, ΔH is the vertical separation of features in the object, and M is the overall magnification of the micrograph. For objects with a depth of from $\frac{1}{8}$ to $\frac{1}{4}$ of the width of the field of view, the optimum angle is between 5 and 10 degrees. This change in angle can be accomplished either by mechanically tilting the specimen in a goniometer stage or by keeping the specimen fixed and tilting the incident beam. Because beam tilting can be accomplished very rapidly, real-time TV stereo imaging can be performed in this way (Pawley, 1988).

The images forming the stereo pair must contain as much overlap area as possible, and this can be assured by marking the position of some prominent feature on the first image on the visual CRT screen with a wax pencil and then aligning the same feature, after tilting, to the identical position. Alternatively, on microscopes equipped with a framestore, the first image can be held in memory and used to help align the second image of the pair. In general, tilting also changes the vertical position of the sample, and the second image must be restored to a correct focus. Focusing must not be achieved by altering the excitation of the final lens of the SEM because this will vary the working distance and hence change the magnification of the second image relative to the first, as well as causing an image rotation. Instead, the Z motion of the stage should be used to bring the specimen back into focus.

While some observers are able to see the stereo effect by simply holding the stereo pair of micrographs at a slightly closer than normal viewing distance and "crossing" their eyes, most microscopists need optical assistance. Usually the stereo pair is examined with a stereo viewer which, by an arrangement of lenses or mirrors, presents an image of one micrograph to each eye. The integrated image from the two photographs gives an appearance of depth through the brain's interpretation of the parallax effects observed. Other ways of presenting stereo images are also possible. One suitable for use with digital image storage systems is the "anaglyph" technique in which the images, one colored red and the other colored green, are overlaid in the correct registration as a single picture (Judge, 1950). When viewed through appropriately colored glasses the stereo image is seen in a muddy brown color. Polarized illuminating light and crossed filters for each eye can also be used to separate the images prior to their fusing in the brain.

An example of a stereo pair obtained from an irregular surface is shown in Fig. 4.67. When this figure is examined with a stereo viewer, the portions of the specimen that are elevated and those that are depressed can be easily discerned. Such information is invaluable for determining the elevation character of an object, i.e., whether it is above or below adjacent features. Because of the great depth of field of the SEM, such interpretation is difficult from a single image since the eye and brain lack the necessary out-of-focus detail which allows this interpretation in the optical case (Howell, 1975). It must be realized, however, that the depth effect is an illusion since the third-dimension plane normal to the plane of the photographs is formed by the observer's visual process. This may readily be illustrated by rotating the image pair through 180° such that the left micrograph is

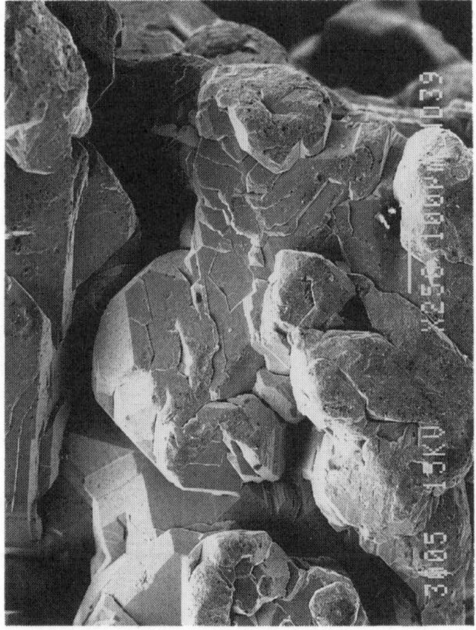

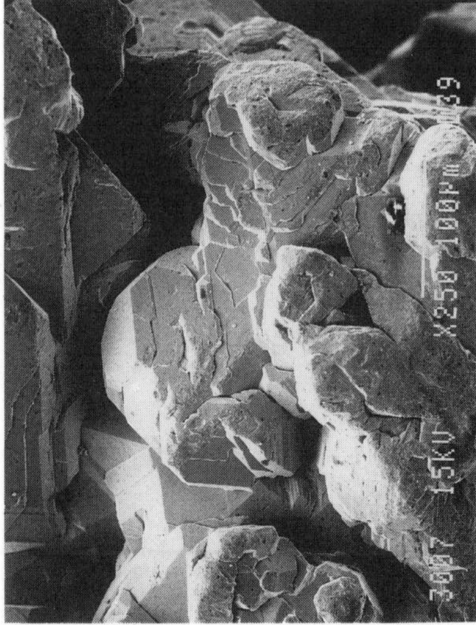

Figure 4.67. Stereo pair of native silver. A typical application of qualitative stereo imaging to view a complicated three-dimensional structure. Illumination is from the left.

interchanged with the right. In this case, regions that originally appeared elevated are now depressed, and *vice versa.*

A standardized convention for orienting the micrographs is thus necessary if correct and unambiguous interpretations are to be made. The following routine, due to Howell (1975), is suitable for all cases. The lower, less tilted (where increasing tilt is defined as tilting toward the E–T detector) micrograph is placed for viewing by the left eye; the more tilted micrograph is placed for viewing by the right eye. The pictures must be placed so that the axis about which they were tilted is parallel to the interocular plane, i.e., when the micrographs are correctly oriented, the tilt axis on the left- and right-hand prints is parallel to the direction of the observer's nose. On many instruments the tilt axis of the stage lies parallel to the bottom edge of the CRT screen, and the micrographs must therefore be rotated 90° counterclockwise to orient them correctly. This then has the disadvantage that the image appears to be illuminated from the side (i.e., the apparent direction of the electron detector) rather than from the top. However, many modern SEMs do have a tilt axis parallel to the specimen–detector line, so as to allow the effective illumination to be from the top of the stereo pair. Even when this desirable tilt axis is not provided, the same effect can be achieved by using a combination of rotation and tilt (Lane, 1970). If the sample is held at some tilt θ, then the stage rotation R and the desired parallax angle P are related by the equation

$$\sin R = \frac{\sin P}{\sin \theta}. \tag{4.41}$$

As an example, for a tilt $\theta = 34°$ and a required parallax of $P = 6°$, the rotation R must be 11°. With a stage giving clockwise motion for an increasing reading on the dial of the rotation control, the left member of the stereo pairs is the image taken at the higher rotation reading. Also, the left-hand micrograph must be rotated clockwise relative to the right-hand micrograph in order to match their respective fields of view. This difficulty can be avoided if electrical image rotation is used to compensate for the mechanical stage rotation.

4.9.2.2. Quantitative Stereo Microscopy

Quantitation of the topography of features in SEM micrographs can be carried out by measurements with stereo pairs (Boyde, 1973; 1974a, b; Wells, 1974a). This can be done even if the operator is not able to perceive the stereo effect (as is the case for about 15% of the population). The only measurements that can be made from the micrograph are the X and Y coordinates of the same feature of interest in both micrographs. These coordinates can be referred to any point that can be identified on both micrographs, either an actual feature or some convenient artifact such as a corner of the scanned area. This point will then be arbitrarily assigned the X, Y, Z coordinates $(0, 0, 0)$ and all subsequent height measurements will be made with respect to this point.

The only parameters required are the X, Y coordinates of a feature (where the X-axis is normal to, and the Y-axis is parallel to, the tilt axis), and the tilt angle difference α between the two halves of the stereo pair. The magnification M must also be known accurately if absolute values are needed. At normal working magnifications in an SEM, it can usually be assumed that the scan is effectively moving parallel to the optic axis; very simple formulas can then be used for quantification. With reference to a fixed point, such as the optic axis in the two photographs, the three-dimensional coordinates X, Y, Z of the chosen feature are given by

$$Z = \frac{P}{2M \sin(\alpha/2)} \tag{4.42a}$$

$$MX = X_L - \frac{P}{2} = X_R + \frac{P}{2} \tag{4.42b}$$

$$MY = Y_L = Y_R \tag{4.42c}$$

where the parallax $P = X_L - X_R$. The subscripts L, R refer to the measured coordinates in the left-hand and right-hand micrographs, respectively. With this convention, points lying above the tilt axis have positive parallax values P. Note that if the measured coordinates Y_L and Y_R are not the same, this implies that the tilt axis is not accurately parallel to Y and the axes must then be rotated to correct this error.

The principles of obtaining quantitative data from a stereo pair can be illustrated from the example in Fig. 4.68. This shows a computer-generated stereo-pair image of an NH_3 molecule. As discussed above, the images are shown so that the one with the lower of the two tilt angles is at the left, and the one with the higher angle is at the right. The angle between them is 6°, and the tilt axis is vertical. A cross marks the center

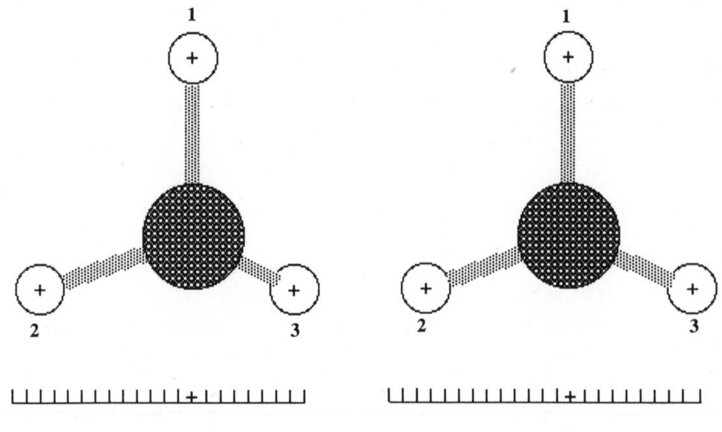

Figure 4.68. Computer-simulated stereo pair of an NH_3 molecule. The tilt between the two images is 6°.

Table 4.7. Positions of Hydrogen Atoms in Stereo-Pair Image Shown in Fig. 4.68 and Calculated Parallaxes

Atom	X_{left}	X_{right}	Parallax
1	0	0	0
2	−11.0	−10.2	−0.8
3	+7.2	+8.6	−1.4

of each hydrogen atom. The task is to determine the relative vertical positions of atoms 1, 2, and 3. For convenience in measurement, a ruler scale has been placed along the X-axis below each half of the stereo pair. The tilt axis of the stereo pair is parallel to the Y-axis. Any point on the pictures can be used as the origin of coordinates provided that the same point is used for each of the pair of micrographs. Here the origin was arbitrarily placed at the point where the trace of the bond from atom number 1 to the nitrogen atom intersects the ruler scales. For convenience the origin is marked with a cross. To apply Eqs. (4.42a–c), it is then necessary only to measure the parallax P, using the ruler to measure the X-coordinates of each of the atoms in turn. Following convention, X-coordinates measured to the right of the origin are positive, while those measured to the left are negative. The data is shown in Table 4.7, with distances being measured in the units of the ruler attached to the images. Putting $\alpha = 6°$ gives $2\sin(\alpha/2) = 0.105$ and hence of the atoms, number one is the highest, lying actually on the tilt axis, number two is next at $(0.8/0.105) = 7.62$ units below, while atom number three is at $(1.4/0.105) = 13.33$ units below the tilt axis. The procedure for computing vertical distances on an actual photomicrograph is exactly similar, although in that case it is sometimes a little more difficult to identify the exact position of the tilt axis and to find a point on each feature of interest suitable for measurement on both halves of the stereo pair.

To apply this procedure to an SEM stereo pair, e.g., to determine the difference in height between faces A and B of the galena crystal in Fig. 4.69, the left (low tilt) and right (high tilt, $+10°$) images are prepared according to the convention described above and oriented so that the parallax axis is vertical. It is a good idea to inspect the stereo pair with a stereo viewer to ensure that the stereo pair is properly arranged, and to qualitatively assess the nature of the topography, i.e., determine how features are arranged relative to one another. Establish a set of X (horizontal) and Y (vertical) axes. With this arrangement, the parallax axis is parallel to the Y-axis, so all parallax measurements will be made along the X-axis. To measure the height difference between faces A and B, as shown in Fig. 4.69, we first locate a distinctive feature on each face, the end of the flake on A and the particle on B noted on Fig. 4.69. Define one feature (arbitrarily) as the reference point $(0, 0)$; the particle on B is chosen. In each photograph, a set of X–Y axes is established.

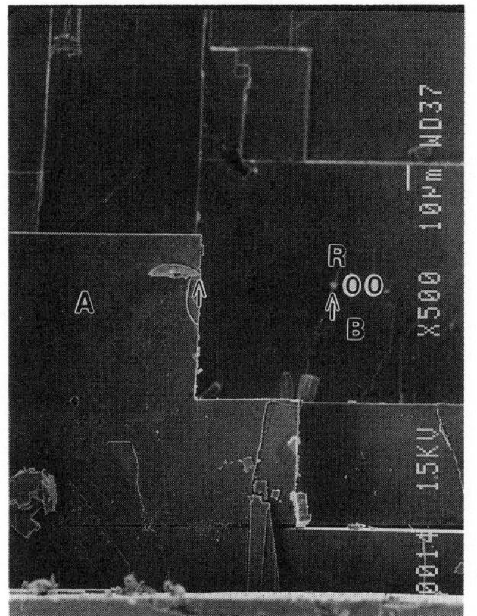

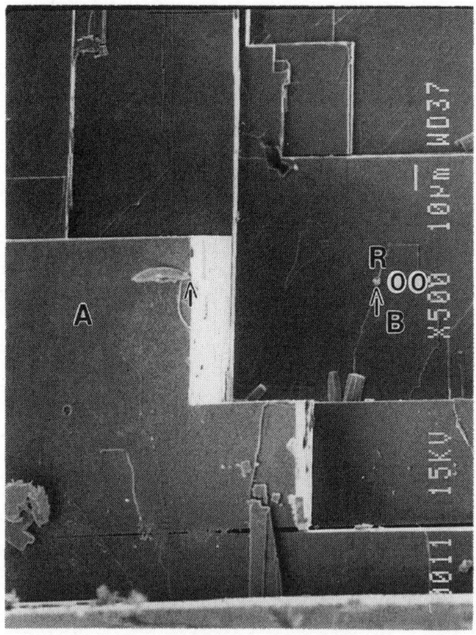

Figure 4.69. Stereo pair of a crystal of galena with surface steps. The height of the fragment that marks the edge of crystal face A relative to the point (0, 0) on crystal face B is calculated from measurements as shown. The tilt difference between the two images is 10 degrees. (Note that because of the arrangement of the tilt axis and E–T detector in this instrument, the lighting appears to come from the right side after rotating the images to place the tilt axis vertical.)

The X-coordinates of the end of the flake are then found as

$$X_L = -26\,\text{mm}$$

$$X_R = -35\,\text{mm}.$$

The parallax is then $P = X_L - X_R = -26\,\text{mm} - (-35\,\text{mm}) = +9\,\text{mm}$.

From Eq. (4.42a), the Z coordinate of the end of the flake is

$$Z = \frac{P}{2M\sin(\alpha/2)} = \frac{9\,\text{mm}}{2 \times 500 \times \sin(10/2)} = 0.103\,\text{mm} = 103\,\mu\text{m}. \qquad (4.43)$$

Since the sign of the parallax is positive, face A is 103 μm *above* face B. This sense of the topography is confirmed by the qualitative stereo view.

If the tilt angle used is very large (i.e., more than 10°), or if the depth variations are large compared with the field of view of the micrograph, then the parallel projection approximation assumed for Eqs. (4.42a)–(c) does not apply, and more general equations must be employed. For details of these the reader is referred to (Wells, 1960; Lane, 1970; Boyde, 1973; 1974a;b).

It does not take much extra effort to take a stereo pair as compared with producing a single micrograph, but the additional information gained is often invaluable, particularly on specimens whose topography is complex and where the high depth of field of the SEM may lead to a quite mistaken view of the actual relative height of various features. When x-ray microanalysis is to be performed from an area with significant surface roughness, it is always good practice to take a stereo pair first to ascertain that the take-off path to the detector is not obstructed by surrounding topography.

4.9.3. STEM in SEM

The examination of submicrometer-sized particles with the SEM often leads to disappointing results when a conventional approach is used. Typically, particles are collected from air samples deposited directly on polycarbonate filters or are extracted from suspensions in liquids by filtering. A typical preparation involves attaching the filter onto a bulk substrate and applying a conductive coating. Such an approach is generally satisfactory for supermicrometer-sized particles, because most of the scattering of the beam electrons takes place in the particle. For submicrometer particles, particularly those composed of low- to intermediate-atomic-number elements, the primary beam penetrates the particle and into the substrate. Although the substrate has a low atomic number and density, the extraneous scattering from the substrate acts as noise and thus affects the signal quality, raising the necessary threshold current and diminishing the quality of the image. Eventually a particle size is reached where the substrate scattering dominates the signal, and the particle contrast drops below the threshold contrast for the beam current available in a small probe. The transition situation is especially difficult. Situations are often encountered where a satisfactory photograph can be prepared with a slow scan or a long-accumulation digital image can be taken of a small particle, but as soon as the scan speed is increased to a rapid visual or television rate to locate another particle while surveying the specimen, the particles effectively disappear because of the higher threshold current at the rapid scan.

To improve the detection and visualization of particles in SEM images, a powerful alternative is to employ the technique of scanning transmission electron microscopy (STEM) in the SEM. If the specimen and its substrate are sufficiently thin, then beam electrons are transmitted. By placing a suitable detector below the specimen, as shown in Fig. 4.70, this transmitted electron signal can be collected and used to form a STEM image. The specimen stage must be altered to allow access to the region below the specimen, and the specimen substrate must be changed from a bulk material to a thin film, such as 20-nm carbon, which is supported on a grid. Procedures for transferring particles from a filter to a carbon film on a grid are given in Chapter 11. The STEM detector can be constructed from a scintillator–light pipe or solid state detector. A simple, inexpensive detector can be made from a high-atomic-number scattering surface placed below the specimen and tilted so that the transmitted electrons are scattered toward the conventional E–T detector in the specimen chamber.

An example is shown in Fig. 4.71 of a field of clay mineral particles viewed simultaneously in the STEM in SEM mode and in the conventional positively biased E–T detector mode and recorded for the same

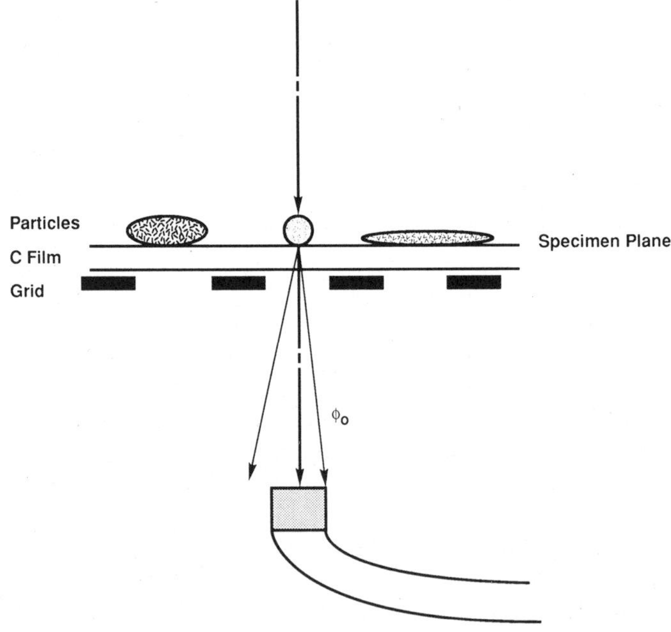

Figure 4.70. Schematic illustration of a detector for STEM in SEM.

time. The images show a similar array of particles and are complementary, in that bright areas in the conventional E–T image, which show areas of strong scattering, appear dark in the STEM image because these scattered electrons are lost to the transmitted signal. The striking thing about the images in Fig. 4.71 is that the conventional E–T image is so very much more noisy than the STEM in SEM mode image. Figure 4.71c shows an E–T detector image similar in quality to the STEM in SEM image, but this image required approximately a factor of 25 longer

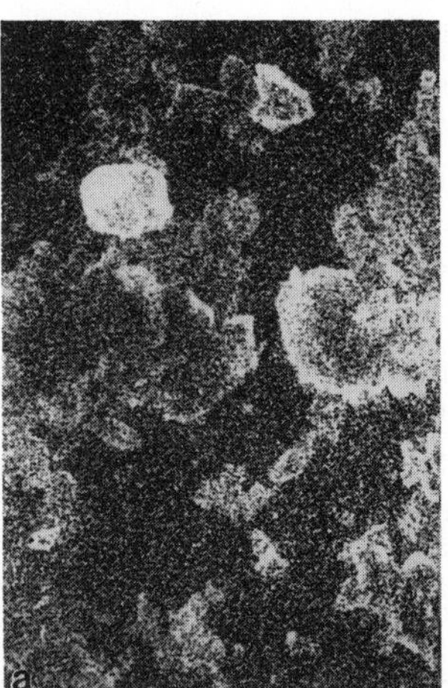

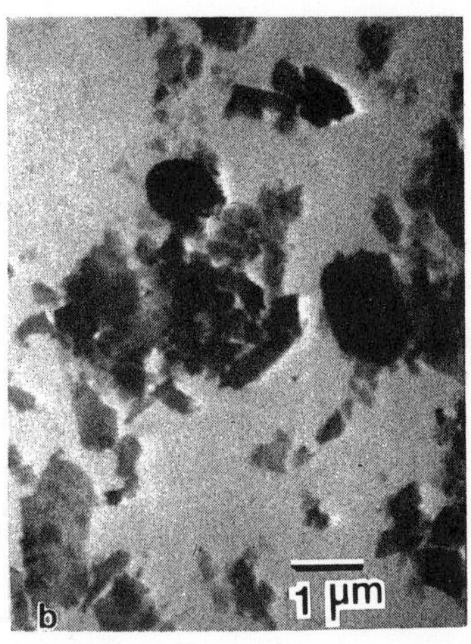

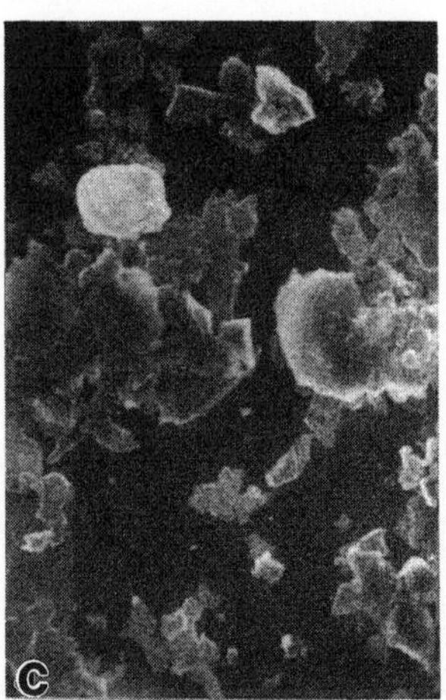

Figure 4.71. Clay mineral particles dispersed on a thin (20 nm) carbon film and viewed by (a) a conventional positively biased E–T detector located above the specimen and (b) a passive scintillator STEM detector placed 4 cm below the specimen. The images in (a) and (b) were recorded simultaneously for the same integration time. (c) Same as (a), but recorded for a longer (by a factor of 25) time. Beam energy = 25 keV.

recording time to establish a similar signal-to-noise level and hence threshold current. Small, low-mass objects have very low scattering power, so that the particle produces an SE_I signal, which is desirable, but very little backscattering occurs. Even modest scattering produces large contrast in the STEM detector, which measures the predominantly forward scattering component of elastic scattering. When operating in the STEM in SEM mode, it is best to use the maximum beam energy available in the SEM to improve both the electron penetration through the particles and the brightness of the source for optimum electron-optical performance.

The magnitude of the contrast in the STEM in SEM imaging mode available from small particles can be estimated. Equation (3.5) gives the probability for scattering through angles greater than a specified value. From Fig. 4.70, the angle of interest is the STEM detector angle. Electrons elastically scattered through an angle greater than the detector angle ϕ_0 miss the detector and contribute to the contrast. From the cross-section, the mean free path $\lambda(>\phi_0)$ for scattering is given by

$$\lambda(>\phi_0) = \frac{A}{Q(>\phi_0)N_A\rho} \text{ cm/event } >\phi_0. \tag{4.44a}$$

Consider that a condition of single scattering exists; that is, each electron suffers no more than one scattering event while passing through the specimen. As the specimen thickness approaches λ, each electron suffers at least one scattering event $>\phi_0$. For thinner samples, the contrast is then estimated as the fraction of the mean free path

$$C = \frac{t}{\lambda} \tag{4.44b}$$

The thickness for a specified contrast is then

$$t = C\lambda = \frac{CA}{Q(>\phi_0)N_A\rho}. \tag{4.44c}$$

Table 4.8 gives the thicknesses of various materials required to produce STEM in SEM contrast levels of 0.10 (10%) and 0.25 (25%) for beam

Table 4.8. Contrast in STEM in SEM Images[a]

		Particle thickness (nm) to produce specified contrast			
		20 keV		30 keV	
Target	Z	10%	25%	10%	25%
C	6	46	110	100	260
Al	13	19	46	42	100
Fe	26	3.3	8.2	7.4	18
Au	79	0.5	1.3	1.1	3

[a] STEM detector angle $\phi_0 = 10°$.

energies of 20 keV and 30 keV and a STEM detector angle of $\phi_0 = 10°$, the angle subtended by a detector 1.8 cm in diameter placed 5 cm below the specimen.

For intermediate- and high-atomic-number particles, the contrast sensitivity of the STEM signal is very high, and very small particles produce high contrast.

4.10. Developing a Comprehensive Imaging Strategy

The advanced microscopist should develop a flexible approach that uses the full range of operating modes of the SEM. The object of an imaging strategy is to organize and apply the large amount of information necessary to optimize the electron-optics, detector selection, signal processing, and image interpretation to solve the problem at hand. Two examples will illustrate this point:

The threshold equation [Eq. (4.31)] and the brightness equation [Eq. 2.3 in Chapter 2] form the basis for developing strategies for obtaining optimal SEM images. Developing a strategy begins with considerations of the specimen. We must first determine what type of contrast mechanism is operating and what numerical level of contrast can be expected. The threshold equation is used to relate this contrast, produced by the specimen–detector system as a result of the beam interaction, to the critical parameters of the microscope, beam current, and frame time. Once the threshold beam current is established, the brightness equation is used to relate this beam current to the other electron-optical parameters, the probe size and divergence. From such a series of calculations, we can select the beam, detector, and scanning conditions to maximize the possibility of detecting the features of interest. Understanding the contrast produced by the specimen is also vital to selecting the proper analog signal processing or digital image processing techniques to ensure that the features of interest will be visible in the final displayed or reconstructed image.

To illustrate the strategy in planning an SEM imaging experiment, let us consider the following situation: We wish to image a flat, polished sample consisting of alternating layers of aluminum and iron of various thicknesses viewed edge-on with backscattered electrons. We want to know the minimum beam current required to photographically image this structure, and the finest detail we could hope to observe. On the basis of the discussion of atomic number contrast, we can calculate Al–Fe contrast as $C = 0.45$ (45%). Assuming an average detection efficiency of 0.20, the threshold beam current needed to image this contrast in a 100-s frame [Eq. (4.32)] is 1×10^{-12} A, or 1 pA. The probe size needed to carry this current can be calculated from the brightness equation. A conventional tungsten filament produces a brightness of 5×10^4 A/cm^2/sr at $E_0 = 20$ keV and a divergence of 10^{-2} radians (200 μm aperture, 10 mm working distance); the minimum probe size is 2.8 nm. Because of aberrations, the actual probe size would be substantially

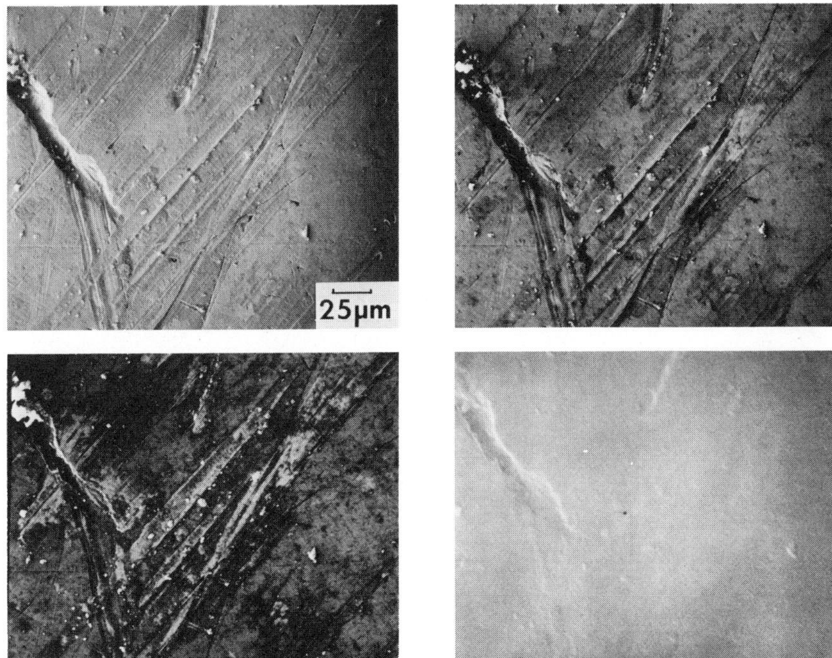

Figure 4.72. Beam energy series of the surface of a platinum aperture. Positively biased E–T detector. (a) 20 keV; (b) 10 keV; (c) 5 keV; (d) 1 keV.

larger, perhaps 5–10 nm. Now consider the same experiment as it is repeated for layers of platinum and gold. The atomic number contrast for this pair of materials is only 0.0041 (0.41%). The threshold beam current for this contrast level is 9.5×10^{-9} amperes (9.5 nA). The minimum probe size to contain this current increases to 280 nm (0.28 μm). Such calculations demonstrate the distinct differences in required operating conditions depending on the nature of the specimen. The microscopist must be prepared to make proper use of the wide range of SEM operating conditions to successfully solve challenging problems.

A second example illustrates the idea of developing imaging strategy to make as much use as possible of the extraordinary range of capabilities of the SEM. Often microscopists tend to get stuck in the rut of applying "standardized" operating conditions, i.e., one beam energy, one beam current, one detector, etc., that produce useful results on an important class of specimens. While those standardized operating conditions may be ideal for some specimens, other problems may require a much different approach. As an example, consider the problem of imaging a layer deposited on a sample. If only a high beam energy, greater than 20 keV, is used, then the beam penetration through a surface layer may be so great that insignificant contrast is generated at its features. If a series of images is obtained over a range of beam energies, often a different view is obtained. Figure 4.72 shows such a series for the surface of a platinum disk. At high beam energies, only a small amount of surface detail is observed, while at low beam energies, strong contrast is obtained from surface features.

5

X-Ray Spectral Measurement: WDS and EDS

5.1. Introduction

Chemical analysis in the scanning electron microscope and electron microprobe is performed by measuring the energy and intensity distribution of the x-ray signal generated by a focused electron beam. The subject of x-ray production has already been introduced in Chapter 3, on electron-beam–specimen interactions, which describes the mechanisms for both characteristic and continuum x-ray production. This chapter is concerned with the methods for detecting and measuring these x rays as well as converting them into a useful form for qualitative and quantitative analysis.

5.2. Wavelength-Dispersive Spectrometer

5.2.1. Basic Design

Until 1968, when energy-dispersive spectrometers (EDS) were first interfaced with microanalyzers, the wavelength-dispersive spectrometer (WDS) was used almost exclusively for x-ray spectral characterization. The basic components of the WDS are illustrated in Fig. 5.1.

A small portion of the x-ray signal generated from the specimen impinges on an analyzing crystal, which consists of a regular array of atoms. The crystal is oriented such that a selected crystallographic plane of these atoms is parallel to its surface. For a particular spacing d and wavelength λ, there exists an angle θ at which the x rays are strongly scattered. The following relationship is known as Bragg's law:

$$n\lambda = 2d \sin\theta \tag{5.1}$$

where n is an integer, $1, 2, 3, \ldots, \lambda$ is the x-ray wavelength, d is the interplanar spacing of the crystal, and θ is the angle of incidence of the x-ray beam on the crystal. The x rays are diffracted (strongly scattered to

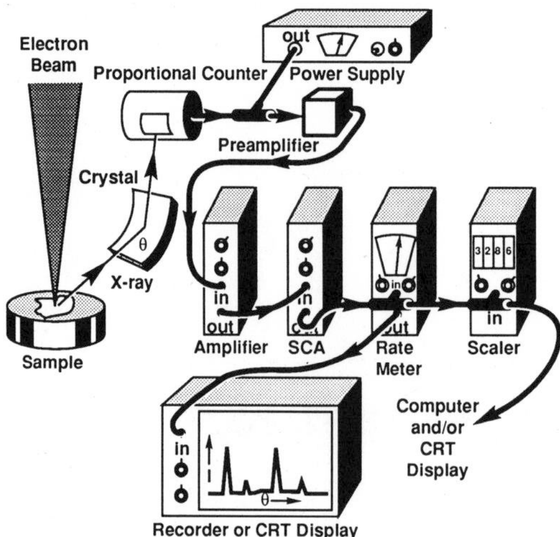

Figure 5.1. Schematic representation of a wavelength-dispersive spectrometer and associated electronics.

a particular angle) and detected by a gas-filled proportional counter. The signal from the detector is amplified, converted to a standard pulse size by a single-channel analyzer (SCA), and then either counted with a scaler or displayed as a rate-meter output on a recorder. As described in Section 3.5.2, each element in the sample can emit one or more series of unique spectral lines. A typical qualitative analysis therefore involves obtaining a recording of x-ray intensity as a function of crystal angle. Each peak observed in the spectrum corresponds to each detectable line, but may have its shape or position altered as a result of the attributes of the spectrometer itself, such as alignment and crystal quality. In the case of longer-wavelength x rays, the shape and position may also be altered by changes in the local chemical environment. Taking these factors into consideration it is, nevertheless, possible to convert peak positions to wavelengths through Bragg's law and then use the Moseley relationship [Eq. (3.33)] to identify the elemental constituents. In practice, crystal-spectrometer readings are either proportional to wavelength or calibrated directly in wavelength. Standard tables, usually available as a computer data base, can then be used for elemental identification. It is also possible to record the digital signal directly into computer memory to allow for a greater variety of display options such as peak identification, intensity measurement, background subtraction, correction of spectral overlap, and multiple spectrometer display.

Bragg's law can be derived with the aid of Fig. 5.2. In this simplified drawing, the x-ray beam approaching the crystal is viewed as a wavefront rather than as an individual photon. The wavefront is then specularly reflected from parallel crystal planes spaced d units apart. Of the two x-ray paths shown in Fig. 5.2, the lower one is longer by an amount $ABC = 2d \sin \theta$. If this distance equals an integral number of wavelengths n, then the reflected beams will combine in phase and an intensity maximum is detected by the proportional counter. If the intensity distribution is measured as a function of θ with a high-quality analyzing

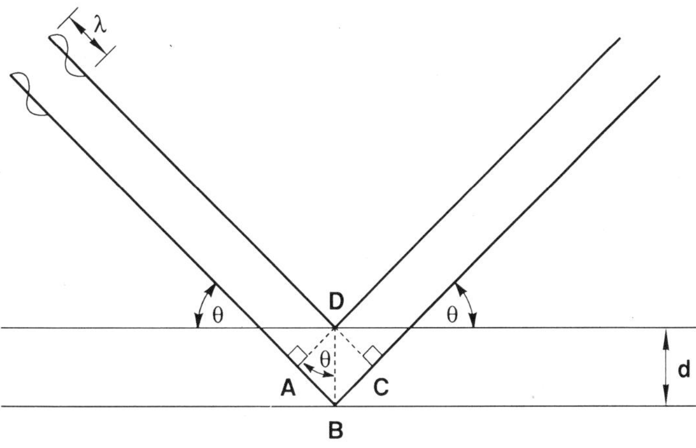

Figure 5.2. Diffraction according to Bragg's law. Strong scattering of x rays of wavelength $n\lambda$ occurs only at angle θ. At all other angles, scattering is very weak.

crystal, the effect of combined reflections from a large number of planes will result in relatively narrow peaks. For example, the measured full width, half-maximum (FWHM) for Mn $K\alpha$ is about 10 eV, compared to the natural value of 2 eV. X rays of wavelengths which do not satisfy the Bragg equation are absorbed by the crystal or pass through it into the crystal holder.

The x-ray signal in focused-electron-beam instruments is fairly weak and can be thought of as originating from a point source; therefore, to maximize the signal at the detector, curved-crystal, fully focusing x-ray spectrometers are used in preference to flat-crystal spectrometers of the type normally associated with tube-excited x-ray emission analysis. In a fully focusing spectrometer of the Johansson type, illustrated in Fig. 5.3a, the x-ray point source, the specimen, the analyzing crystal, and the detector are all constrained to move on the same circle with radius R, called the *focusing circle.* Furthermore, the crystal planes are bent to a radius of curvature of $2R$ and the surface of the crystal itself is ground to a radius of curvature of R. As a result of this geometry, all x rays originating from the point source have the same incident angle θ on the crystal and are brought to a focus at the same point on the detector, thereby maximizing the overall collection efficiency of the spectrometer without sacrificing good wavelength resolution. Clearly, if a flat crystal were used, the angle of incidence of the x-ray beam would vary across the length of the crystal, causing a considerable loss in count rate. This is because only a narrow band of the crystal would satisfy the Bragg condition for the desired wavelength.

Figure 5.3 also illustrates the additional geometric requirement of a constant x-ray take-off angle ψ imposed by having a small region available about the sample and beneath the final lens. The fully focusing requirement is maintained by moving the analyzing crystal along a

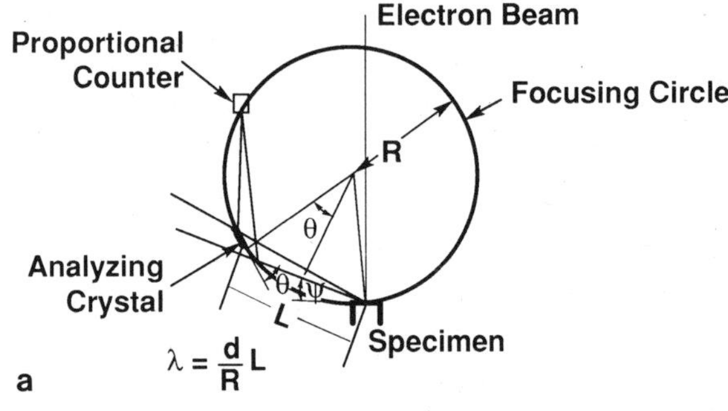

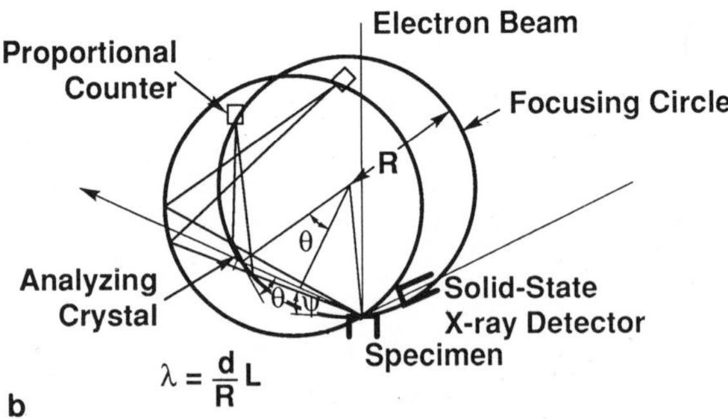

Figure 5.3. (a) Fully focusing wavelength-dispersive spectrometer. R is the radius of the focusing circle, ψ is the take-off angle, and θ is the diffraction angle. (b) Movement of the focusing circle to change the diffraction angle. Note position of EDS detector at the same take-off angle.

straight line away from the sample while rotating the crystal, as illustrated in Fig. 5.3b, and moving the detector through a complex path so as to make the focusing circle rotate about the point source. An interesting consequence of this arrangement is that the crystal-to-source distance L is directly proportional to the wavelength. This can be shown with the aid of Fig. 5.3: First we write

$$\frac{L}{2} = R \sin \theta \quad \text{or} \quad \frac{L}{2R} = \sin \theta. \tag{5.2a}$$

Combining this with Bragg's law, Eq. (5.1) gives

$$n\lambda = 2d \sin \theta = \frac{2dL}{2R} \tag{5.2b}$$

or, for first-order reflections,

$$\lambda = \frac{d}{R} L, \tag{5.2c}$$

with higher-order reflections occurring simply at multiples of the L value for the first-order reflections. Most spectrometers read out directly in L (mm or cm). Actually, most fully focusing spectrometers use crystals that are only bent to a radius of curvature of $2R$ but not ground to be completely tangent to the focusing circle, since grinding a crystal tends to degrade its resolution by causing defects and a broader mosaic spread. This compromise, known as Johann optics, results in some defocusing of the image at the detector but does not seriously impair resolution.

A variation of the use of Johann optics is incorporated in the semifocusing spectrometer, which uses a fixed, nonchangeable source-to-crystal distance. In this arrangement, several crystals bent to various fixed radii of curvature are mounted on a carousel so that each can be switched into position, rather than using a single bending crystal. Since the focusing condition is strictly obeyed for only one wavelength per crystal, a certain degree of defocusing and consequent loss of resolution and peak intensity will be experienced at wavelengths other than the optimum value. The primary use of this type of spectrometer is to measure peak and background intensities for one specific element. Another advantage of this approach is that placement of the x-ray source on the focusing circle is less critical, so that when x-ray images are obtained by scanning the electron beam over the surface of the sample, they are less susceptible to defocusing effects since the entire image is, in fact, defocused.

In instruments with fully focusing crystal spectrometers, only a finite volume of the sample can be considered to be in focus. The shape and dimensions of this region depend on a number of factors, such as the design of the spectrometer, the choice and size of crystal, the wavelength being measured, and the criterion for focus (i.e., 99% or 95% of the maximum intensity). Generally, the region of spectrometer focus is an extremely elongated ellipsoid, whose major axis is proportional to the width of the diffracting crystal and has a length of millimeters. For a diffraction crystal that is only bent to the Rowland circle and not ground, the minor axes are controlled by the sharpness of the diffracted peak. Because of the perfection of the crystal, the diffraction process is very sharply peaked, which translates into short minor axes for the focus ellipsoid, on the order of micrometers. The three possible spectrometer arrangements are illustrated in Fig. 5.4. In most typical spectrometer arrangement in an electron microprobe, the so-called vertical spectrometer, the greatest sensitivity to spectrometer defocusing (i.e., shortest ellipsoid axis) occurs along the Z direction of the sample (i.e., the electron-optical axis). The maximum area that can be scanned, consistent with the criterion for focus, can be experimentally determined by scanning a well-polished (flat) pure elemental sample and measuring the variation of x-ray intensity with position. The criterion for focus is important when comparing capabilities required for quantitative analysis with those for x-ray mapping (Chapter 10). For qualitative analysis and x-ray mapping these values can be doubled or even tripled. If qualitative analysis or maps are needed for irregular surfaces, inclined or horizontal rather than vertical spectrometers are often used because of their reduced defocusing effects with variations in the Z direction of the sample. The horizontal spectrometer, Fig. 5.4b, places the long axis of the focus ellipsoid parallel to the Z-axis, but at the expense of a large angle to the electron beam, which necessitates tilting the specimen to obtain a useful take-off angle. The inclined spectrometer, Fig. 5.4c, tilts the whole spectrometer to simultaneously achieve a high take-off angle and project a large fraction of the long axis of the focus ellipsoid along the Z-axis. To accommodate the physical size of the inclined spectrometer, we are generally limited to two such units on an electron column. Because of possible defocusing effects quantitative analysis is best performed by step-scanning the sample under a centered static beam rather than by beam scanning. Normally, in instruments with a coaxial light microscope, the light and x-ray optics are prealigned. Thus, if the sample is positioned in optical focus with the stage controls, it should also be in x-ray focus. The most critical direction is along the short axis. Since for a vertical spectrometer, the short axis of the defocus elipsoid is parallel to the Z-axis, the extremely shallow depth of focus of the optical microscope is a great advantage in accurately and precisely positioning the specimen at the desired focus point. It should be noted that in a SEM equipped with a WDS system, the absence of an optical microscope with a shallow depth of focus to aid in locating the spectrometer focus can lead to severe problems in quantitative analysis. In this case, the excellent depth of focus of the SEM is a liability, since it is difficult to observe changes in working distance of a few micrometers, which is critical to the x-ray measurement.

Most electron microprobes and SEMs can be equipped with more than one crystal spectrometer. Multiple spectrometers, each containing several crystals, are necessary not only for analyzing more than one element at a time, but also to include the variety of crystals required for optimizing performance in different wavelength ranges. Table 5.1 lists

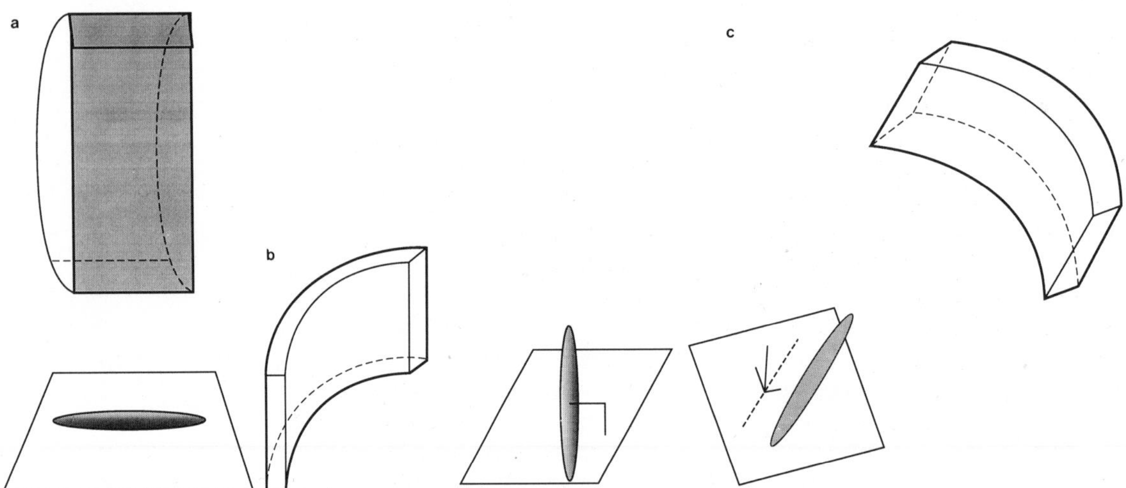

Figure 5.4. Position of the focus ellipsoid for a WDS spectrometer in the (a) vertical, (b) horizontal, and (c) inclined geometries.

Table 5.1. Crystals Used in Diffraction

Name	$2d$(Å)	Lowest atomic number diffracted	Resolution	Reflectivity
α-Quartz(10$\bar{1}$1)	6.687	$K\alpha_1$ 15 P $L\alpha_1$ 40 Zr	High	High
KAP(10$\bar{1}$0)	26.632	$K\alpha_1$ 8 O $L\alpha_1$ 23 V	Medium	Medium
LiF(200)	4.028	$K\alpha_1$ 19 K $L\alpha_1$ 49 In	High	High
PbSt	100.4	$K\alpha_1$ 5 B	Medium	Medium
PET	8.742	$K\alpha_1$ 13 Al $L\alpha_1$ 36 Kr	Low	High
RAP	26.121	$K\alpha_1$ 8 O $L\alpha_1$ 23 V	Medium	Medium

some of the most commonly used analyzing crystals, showing their comparative $2d$ spacings and range of elements covered. Since $\sin\theta$ cannot exceed unity, Bragg's law establishes an upper limit of $2d$ for the maximum wavelength diffracted by any given crystal. More practical limits are imposed by the spectrometer design itself, since it is obvious from Fig. 5.3 that for $\sin\theta = 1$, i.e., $\theta = 90°$, the detector would have to be at the x-ray source point inside the electron-optical column. A lower wavelength limit is imposed by Eq. (5.2) since it becomes impractical to physically move the analyzing crystal closer to the specimen.

It should be noted that most naturally occurring crystals such as LiF are limited to shorter wavelengths because their maximum interplanar spacings are small. Synthetic crystals, such as TAP, provide a very useful extension of the measurable wavelength range, but still do not have sufficiently large d spacings to cover the longest wavelengths encountered, e.g., Be, B, C and N $K\alpha$. Until recently, the measurement of these wavelengths required the use of psuedocrystals grown by the Langmuir–Blodgett technique (Blodgett and Langmuir 1937), illustrated in Fig. 5.5a. In this method a suitable substrate such as mica or glass is dipped into a trough containing a monolayer film of lead stearate or some similar compound floating on the surface of water. The lead stearate is characterized by long chain molecules with a heavy atom on one end. The initial dipping results in the attachment of the heavy metal with its organic tail perpendicular to the substrate. Removal of the substrate from the trough results in a second layer deposited in which the hydrocarbon tails are attached and a row of heavy metal atoms form the external surface. The pseudocrystal is then built up by repeated dippings. The organic tails thus serve as light-element spacers between the reflecting planes of heavy atoms. The interplanar spacing can therefore be established by using organic molecules with varying chain lengths.

An important recent development in the measurement of long-wavelength x rays is the layered synthetic microstructure (LSM). Figure

5.5b is a transmission electron micrograph of a cross-section of an LSM. These diffraction structures are built up by the physical vapor deposition of alternating layers of heavy and light elements (in this case, carbon and tungsten). This fabrication technique permits deposition of layers with any value of the spacing required and with material choices optimized to scatter x rays efficiently. LSM diffractors are typically an order of magnitude more efficient than pseudocrystal diffractors. As shown in Fig. 5.6, the LSMs generally offer considerably improved count rates, under the same instrument operating conditions, relative to the equivalent pseudocrystal with an acceptable loss of resolution. (See also Section 10 of Chapter 9 for a discussion of the LSM crystals used for light-element detection.) This loss of resolution is often an advantage, since the effects of chemical shifts in wavelength are minimized.

5.2.2. The X-Ray Detector

The most commonly used detector with microanalyzer crystal spectrometer systems is the gas-proportional counter, shown in Fig. 5.7. This detector has an excellent dynamic range (0 to 50,000 counts per second or more), covers a wide range of energies, and has a generally high collection efficiency. It consists of a gas-filled tube with a thin wire,

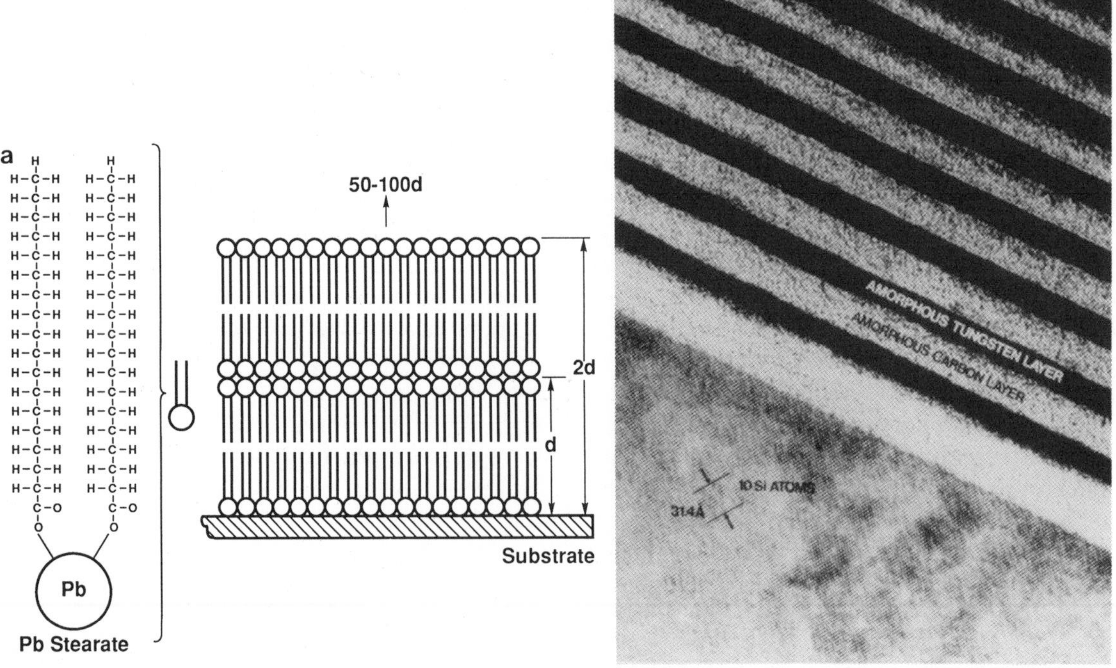

Figure 5.5. (a) Pseudocrystal grown by Langmuir–Blodgett Technique. (b) Transmission electron micrograph of layered synthetic microstructure. (Courtesy of Ovonics.)

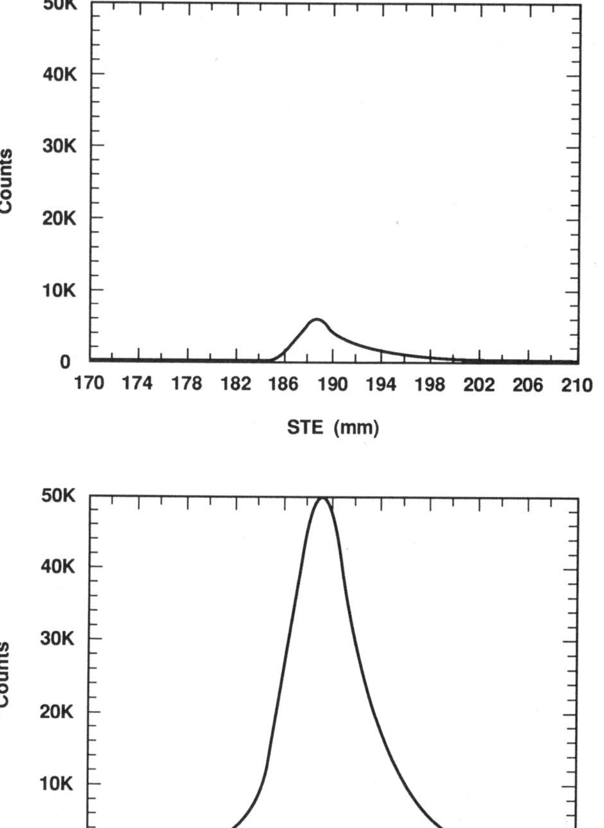

Figure 5.6. Intensity comparison of the B spectrum between stearate (STE) and layered synthetic microstructure (LSM). (Courtesy of C. Nielsen, JEOL.)

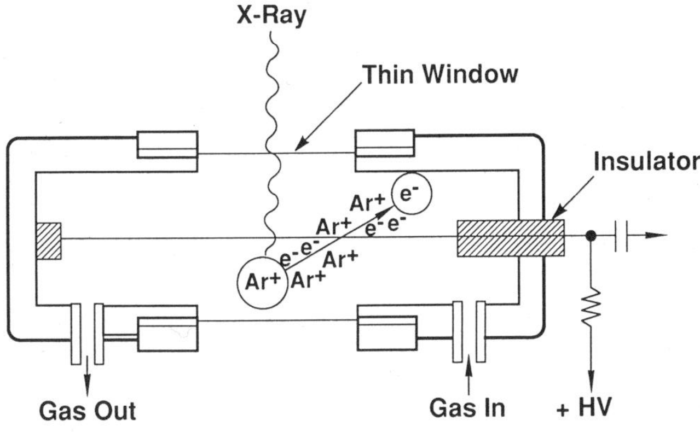

Figure 5.7. Schematic drawing of a gas flow proportional counter.

usually tungsten, held at a 1–3 kV potential, running down the center. When an x-ray photon enters the tube through a thin window on the side and is absorbed by an atom of the gas, it causes a photoelectron to be ejected, which then loses its energy by ionizing other gas atoms. The electrons thus released are then attracted to the central wire, giving rise to a charge pulse. If the gas fill used is P10 (90% argon–10% methane), approximately 28 eV is absorbed per electron-ion pair created. For Mn $K\alpha$, whch has an energy of 5.895 keV, about 210 electrons are directly created by the absorption of a single photon. This would be an extremely small amount of charge to detect without a low noise preamplifier system. However, if the positive potential of the anode wire is high enough, the electrons ejected in the primary events are accelerated sufficiently to ionize gas atoms, and secondary ionizations occur, which can increase the total charge collected by several orders of magnitude. Figure 5.8 shows the effect on the gas amplification factor of increasing the bias voltage applied to a tube. The initial increase corresponds to increasing primary charge collection until it is all collected (a gas amplification factor of 1), and then the curve levels off in the "ionization" region. Increasing the potential beyond this point initiates secondary ionization, the total charge collected increases drastically, and the counter tube enters what is termed the "proportional" region, because the collected charge remains proportional to the energy of the incident photon. Further increasing the voltage causes the tube to enter the Geiger region, where each photon causes a discharge giving rise to a pulse of a fixed size independent of its initial energy, thereby losing the information necessary for pulse-height analysis. A further disadvantage is that the counter "dead time," the time required for the tube to recover sufficiently to accept the next pulse, increases from a few to several hundred microseconds. Any increase in applied voltage beyond the Geiger region results in permanent damage to the tube. In practice, operation in the lower part of the proportional

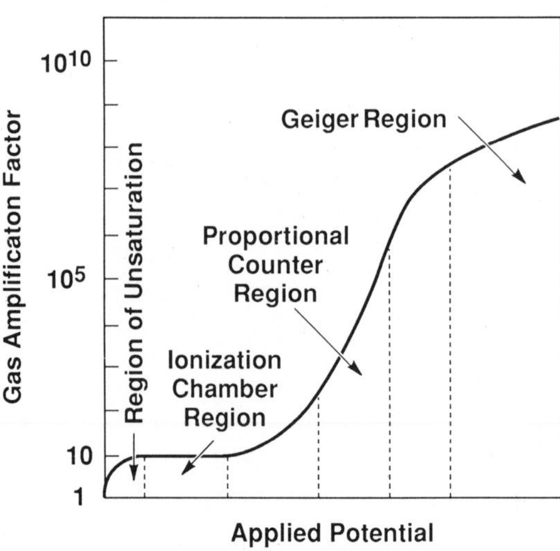

Figure 5.8. The effect of applied counter-tube bias on the gas amplification factor.

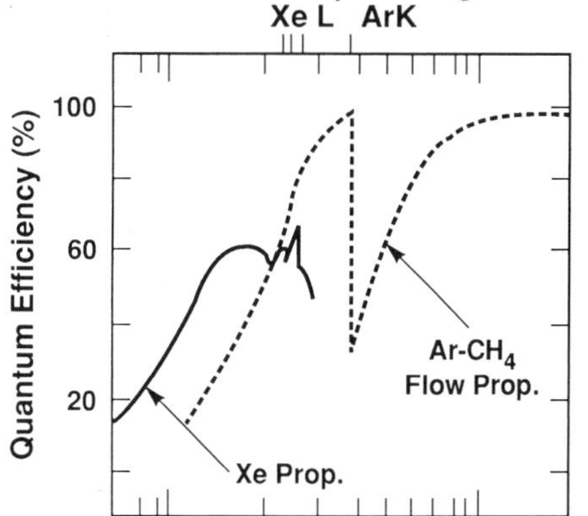

Figure 5.9. Representative collection efficiencies for proportional counters filled with Xe and Ar–CH_4 gases. The Ar–CH_4 counter has a 25 nm (250 Å) Formvar window.

region is preferred to minimize the effect of gain shifts with counting rate and to minimize dead time.

The proportional counter shown in Fig. 5.7 is of the gas-flow type, normally used for detecting soft x rays ($\lambda > 3$ Å). A flowing gas, usually P10, is chosen because it is difficult to permanently seal the thin entrance windows necessary to reduce absorption losses. Since crystal spectrometers are kept under vacuum to eliminate absorption of the x-ray beam in air, it is necessary to support ultrathin windows such as those of Formvar and cellulose nitrate on fine wire screens in order to withstand a pressure differential of 1 atm; however, this causes an additional decrease in detector collection efficiency. Recently, unsupported stretched polypropylene films have come into use with considerable success. Sealed counters, with windows of Be or other materials and containing krypton or xenon are used for shorter x-ray wavelengths since, as shown in Fig. 5.9, they have a higher quantum counter efficiency (the percentage of input pulses detected) than argon-filled detectors at 1 atm. The efficiency of argon-filled detectors for shorter x-ray wavelengths can, however, be increased by increasing the gas pressure to 2 or 3 atm.

5.2.3. Detector Electronics

The role of the detector electronics is to integrate the total charge produced by each x-ray photon and convert it into a single voltage pulse to be further processed for counting or display purposes. On the average, the number of electrons per incident x-ray photon entering the detector is given by

$$n = \left(\frac{E}{\varepsilon}\right)A \tag{5.3a}$$

where E is the incident x-ray energy, ε is the average energy absorbed per electron–ion pair created (about 28 eV for argon), and A is the gas amplification factor, determined by the potential applied to the counter tube. The charge collected by the preamplifier is given by

$$q = en = \left(\frac{E}{\varepsilon}\right)eA \tag{5.3b}$$

where e is the charge of an electron (1.6×10^{-19} C). Because this quantity is very small, the preamplifier is normally located close to the counter tube to avoid stray electronic pickup. The output of the preamplifier is a voltage pulse V_p, where

$$V_p = \frac{q}{C} = \left(\frac{G_p eA}{\varepsilon C}\right)E \tag{5.3c}$$

C is the effective capacitance of the preamplifier, and G_p is a gain factor, usually about unity. A typical premplifier output pulse shape is given in Fig. 5.10a, which shows a rapidly falling negative pulse with a tail a few microseconds long. The signal from the preamplifier is now suitable for transmission several meters over a coaxial cable to the main amplifier, where it is usually inverted, shaped further and amplified to give a Gaussian pulse of the type shown in Fig. 5.10b. The magnitude of this pulse is given by

$$V_A = G_A V_p = \left(\frac{G_A G_p Ae}{\varepsilon C}\right)E, \tag{5.3d}$$

where G_A is the main amplifier gain. The value of V_A for a given x-ray energy is typically set within the range of 2–10 V by operator adjustment of either the G_A or the bias voltage applied to the detector. Since during an analysis all of the quantities shown in the parentheses in Eq. (5.3d)

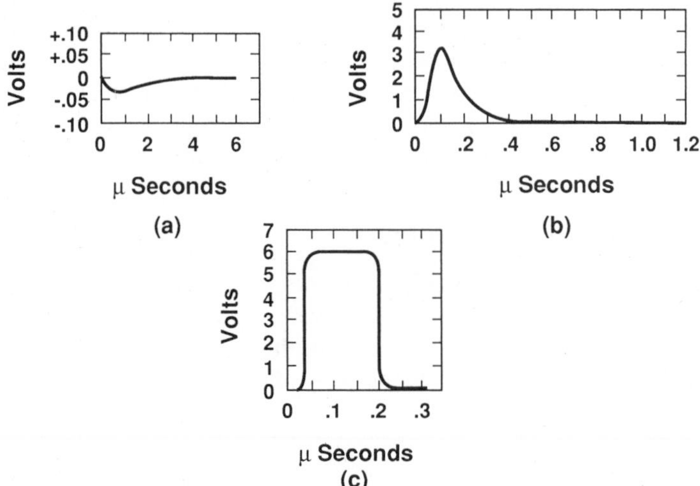

Figure 5.10. Typical WDS x-ray detection pulse shapes: (a) preamplifier, (b) main amplifier, (c) single-channel analyzer output.

are held fixed, it follows that

$$V_A = KE, \tag{5.4}$$

where $K = G_A G_p Ae/\varepsilon C$ is a constant. In other words, the average pulse size for a series of fixed-energy x-ray pulses is directly proportional to the x-ray energy entering the proportional counter, providing the counter-tube voltage is set for the proportional range and the main amplifier is run in a linear region.

Pulse shapes of the type shown in Fig. 5.10 are easily measured with most general-purpose laboratory oscilloscopes (having a 0.1 μs/cm sweep rate or better). Periodic monitoring of amplifier output pulses is highly recommended because it provides a convenient way to observe the manner in which typical pulses are processed by the detector electronics. Undesirable effects such as peak clipping, base line instabilities, noise, and pulse overshoots, indicative of faulty electronics or improper control settings, become readily apparent and can often be corrected easily. The use of an oscilloscope to observe amplifier output pulses is, furthermore, the best method for correctly adjusting amplifier gain and counter-tube bias. Information about the actual voltage distribution of pulses processed for a preselected period can be readily obtained by means of a single-channel analyzer (SCA). Basically the SCA serves two functions. As a discriminator, it is used to select and transmit pulses within a predetermined voltage range for further processing. As an output driver it causes the selected pulses to trigger rectangular pulses of a fixed voltage and duration compatible with scalar and rate-meter input requirements. A typical SCA output pulse shape is shown in Fig. 5.10c. In this example the output pulse is 6 V high and lasts for 0.2 μs.

Pulse-height analysis with an SCA is illustrated schematically in Fig. 5.11. By means of potentiometer controls in the SCA, the operator sets a

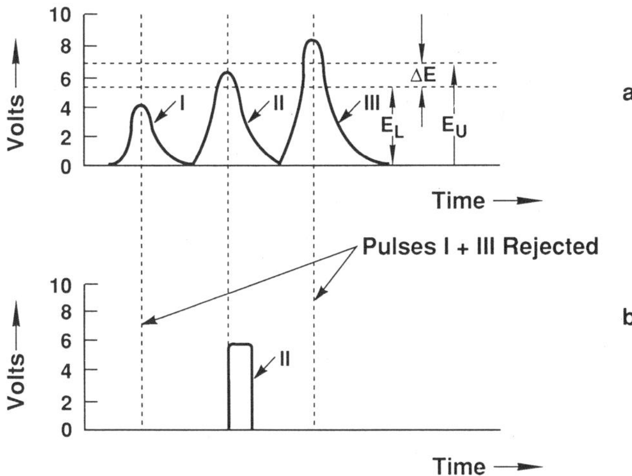

Figure 5.11. Schematic representation of pulse-height analyzer behavior. (a) Main amplifier output; (b) Single-channel analyzer output. $E_L = 5\,V$; $\Delta E = 2\,V$; $E_U = 7\,V$.

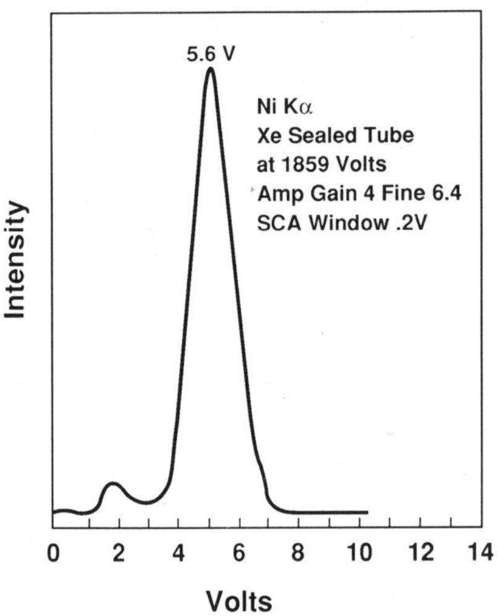

Figure 5.12. Pulse distribution of Ni $K\alpha$ determined by a single-channel analyzer.

combination either of a base line voltage E_L and a window voltage ΔE or of a base line voltage E_L and an upper window voltage E_U. Either of these arrangements is referred to as a *differential setting*. If only the base line is used, then the setting is referred to as *integral*. However, in fact, an upper limit is set by the main amplifier's electronic saturation, usually about 10 V. In the example shown, only pulses from 5 to 7 V (pulse II, 6 V) are acccepted. Pulses larger (pulse III, 8 V) or smaller (pulse I, 4 V) are rejected. In practice pulse-voltage distribution curves with an SCA can be obtained by tuning the WDS to a specific characteristic line, selecting an SCA window voltage ΔE of a few tenths of a volt, and recording the pulse intensity as a function of the base line setting, E_L. This can be done by either scanning the base line at a fixed rate from 0 to 10 V and recording the SCA output with a rate meter (see Fig. 5.12) or manually stepping through a range of base line values E_c, counting the pulses with a scaler, and plotting the results.

Figure 5.13 shows a Cr $K\alpha$ pulse-amplitude distribution from the flow-proportional counter. The principal features to note are two

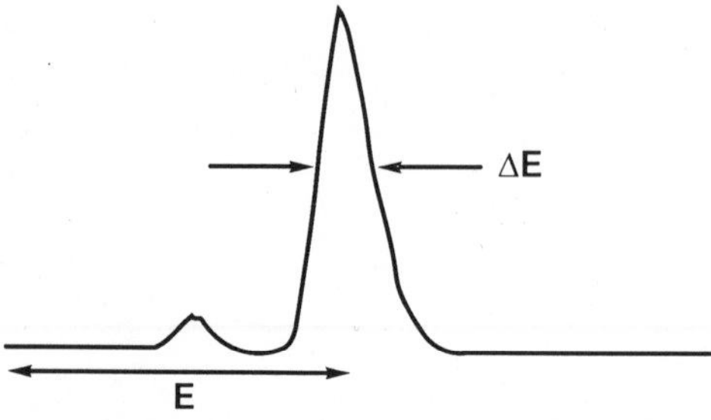

Figure 5.13. Cr $K\alpha$ pulse amplitude distribution from a flow proportional counter on a wavelength-dispersive spectrometer.

relatively broad peaks. The larger of the two corresponds to the main Cr $K\alpha$ pulse distribution. The smaller peak is known as an "escape peak." Its center occurs at a pulse voltage corresponding to an x-ray energy equal to the Cr $K\alpha$ energy minus the characteristic line energy of the element used as the counter-tube gas, which in this case is argon (argon $K\alpha = 2.96\,\text{keV}$) for the example shown in Fig. 5.13. The escape peak corresponds to a special type of proportional-detector response during which either the incoming x-ray photon or the primary photoelectron ionizes a core rather than outer-shell electron in the counter-tube gas. This process is then followed by the emission of a characteristic x-ray photon, which has a high probability of escaping from the counter tube entirely. If it does, the remaining energy for electron–ion pair production will be correspondingly diminished. For these events the amplifier output, Eq. (5.4), must be modified to the form

$$V_A(\text{escape peak}) = K(E - E_{CT}) \tag{5.5}$$

where E_{CT} is the characteristic line energy of the counter-tube gas.

The natural spread of both peaks arises from the fact that each monoenergetic photon entering the detector does not give rise to the same number of electron–ion pairs. Several competing processes occur by which the initial photoelectron may dissipate its energy. The percentage resolution of a detector is defined as 100 times the width of the pulse distribution curve at half-maximum divided by the mean peak voltage. The resolution of a properly functioning counter tube is about 15–20%. The pulse distribution should be approximately Gaussian and free of large asymmetric tails. It is desirable to periodically check this distribution since failure of the electronics or degradation of the counter tube can lead to a shift in the peak position, width, or symmetry, making any prior SCA settings improper.

There has always been some confusion about the precise role of the energy discrimination capability of the SCA. First of all, it cannot improve the energy selection of the spectrometer system for wavelengths close to that of the characteristic line being measured, for this has already been done by the diffraction process itself, which clearly has much higher energy resolution than the flow counter. The SCA can, however, eliminate both low- and high-energy noise as well as higher-order reflections ($n > 1$ in Bragg's law) arising from the diffraction of higher-energy characteristic or continuum x rays with the same $n\lambda$ value, since such events occur at the same λ setting as the line of actual interest. Although photons of various energies may be diffracted at the same spectrometer setting, the pulse distribution from the detector is different from each type of photon. (For examples, see Figs. 9.36–9.38, where the Ni $L\alpha_1$ third-order interference on the C $K\alpha$ is eliminated by the use of a differential window.) Balancing the benefits against the problems introduced in eliminating these effects often leads to the simple solution of restricting the discriminator settings to a high-enough base-line value to remove low-energy noise and adjusting the window or upper level to a value that allows the acceptance of all other pulses. One reason for this

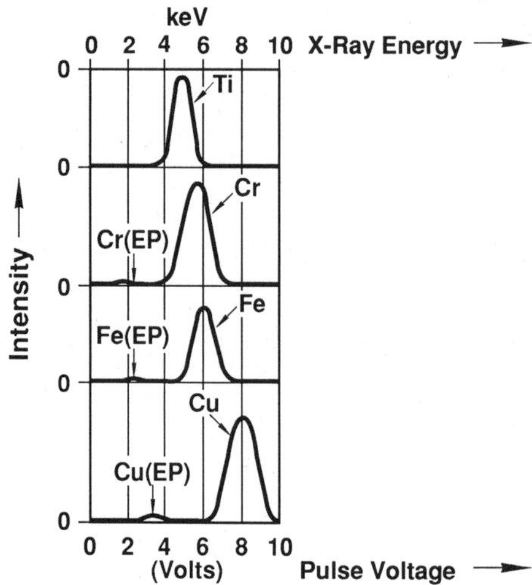

Figure 5.14. Four elemental pulse distributions determined from a proportional counter.

approach is dramatized with the aid of Fig. 5.14, which shows a series of pulse distributions for crystal-spectrometer settings corresponding to several elements readily accessible to a single LiF crystal during a typical scan. It can be seen in scanning from Ti to Cu, that in accordance with Eq. (5.3) the main pulse distribution shifts from a mean value of about 4.5 to 8.0 V. Note that the constant K was intentionally set by means of the amplifier gain to give 1000 eV/V in order to make the voltage and energy scales numerically the same. It is evident that if the SCA window is set from 4 to 6 V for the main Ti pulse distribution, all of the pulses would be excluded when the spectrometer reached the copper setting. Therefore, the use of such a window is totally incompatible with qualitative line scanning and should be used only for fixed-spectrometer-setting quantitative analysis. It should further be pointed out that even when narrow-band SCA settings are used for fixed-spectrometer positions, as in the case of a quantitative analysis, the analyst is not free from problems. It has been shown that under certain conditions the entire pulse distribution may shift to lower voltages with increasing count rate. The exact origin of this phenomenon is not understood, but it is probably related to either main-amplifier baseline stability or a decrease in the effective potential across the tube caused by positive ion buildup around the counter-tube walls. In summary, because the mean value of the pulse-height distribution strongly depends on several parameters, including the counter-tube voltage, the main amplifier gain, and the condition of the counter tube including the gas fill (pressure and type), it is necessary to check that the settings are correct for the intended x-ray lines and the range of elements measured. The latest, fully automated spectrometer systems have greatly simplified the setup of the single-channel analyzer.

A consequence of the finite time for an x-ray pulse to pass through the detector electronics is the possibility that during this time, another x-ray photon may enter the proportional counter. If so, this second photon is lost, because the detector electronics have not yet returned to the baseline for measuring the next pulse. As the rate of photons reaching the detector increases, the probability for these coincident events also increases. The length of time that the detector is unavailable because of pulse processing is termed the *dead time,* and loss of events due to pulse coincidence is termed *coincidence losses.* For accurate measurement of x-ray intensities as the input to quantitative analysis routines, the dead time must be corrected, or errors of several percent will be introduced at higher spectrometer count rates.

In the wavelength-dispersive spectrometer, the dead time correction is applied after the intensity is measured. The dead time relationship is

$$N = \frac{N'}{(1 - \tau N')} \tag{5.6}$$

where N' is the measured count rate, N is the true count rate which we wish to calculate, and τ is the dead time in seconds. To apply Eq. (5.6), the dead time τ must be known. One method for determining τ consists of plotting N' versus the measured beam current, which is directly proportional to N. The deviation of this plot from linearity can be fitted to Eq. (5.6) to determine τ. For the proportional counter used in the WDS, Heinrich *et al.* (1966) showed that the dead-time relation [Eq. (5.6)] could be employed up to count rates of at least $5 \times 10^4\,s^{-1}$.

Any intensity measured with the WDS must be corrected for dead time, by the use of Eq. (5.6), before that intensity is used for quantitative analysis. In a typical WDS, τ is approximately 2 μs, so that the correction for dead time is no more than 2% for count rates below 1×10^4 counts/s. It is possible to electronically "fix" the dead time to a value slightly larger than the longest dead time in the electronic chain. This method allows direct application of Eq. (5.6).

Typical crystal-spectrometer scans of rate-meter output vs wavelength for a nickel-base superalloy are illustrated in Fig. 5.15a for LiF and 5.15b for TAP. The $K\alpha_1$ and $K\alpha_2$ peaks for vanadium, which are separated by only 6 eV, can be separated with a crystal spectrometer. The shape of the diffracted peak is a Lorentzian, rather than the Gaussian found with the energy-dispersive spectrometer. Two other capabilities, namely, light-element detection and peak-shift measurements, are illustrated in Fig. 5.16. This figure shows superimposed computer-controlled scans for boron $K\alpha$ in pure boron, cubic boron nitride, and hexagonal boron nitride. The peak shifts and satellite lines are due to shifts in the outer electron energy states associated with differences in the chemical bonding. Measurements of this type can also be used to fingerprint various cation oxidation states in metal oxides (Holliday, 1963). A more complete description of this effect is given in Section 9.10, where light-element analysis is discussed.

In addition to performing chemical analysis at a fixed point, it is often desirable to analyze variations in the x-ray intensity of one or more

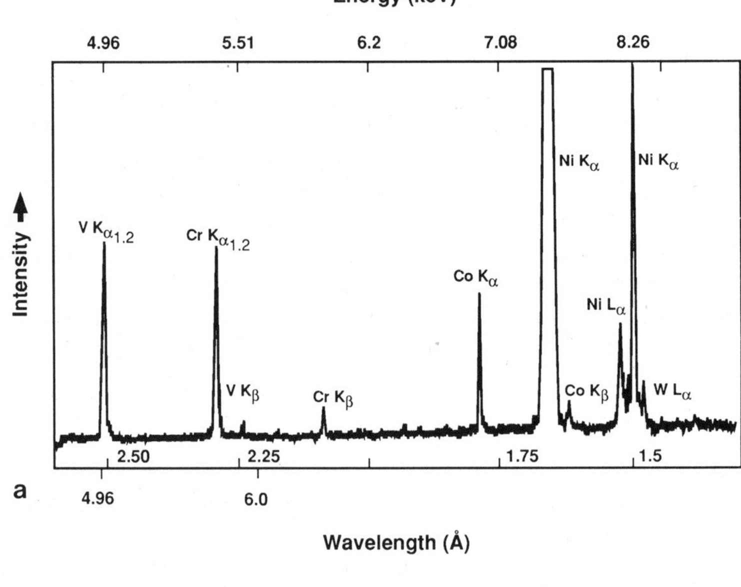

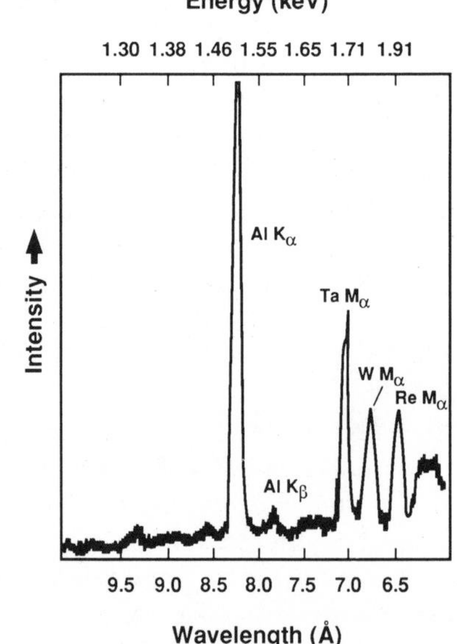

Figure 5.15. Wavelength-dispersive spectrometer scans of a nickel-base superalloy; (a) scan using a LiF crystal; (b) scan using a TAP crystal.

selected elements across a line on a specimen or even over a two-dimensional field of view. In the line-scan mode illustrated in Fig. 5.17, the rate-meter output corresponding to a fixed spectrometer setting is used to modulate the vertical deflection of the CRT as the electron beam is scanned across the sample. Multiple exposures of the CRT are used to superimpose the x-ray and beam position information on an electron

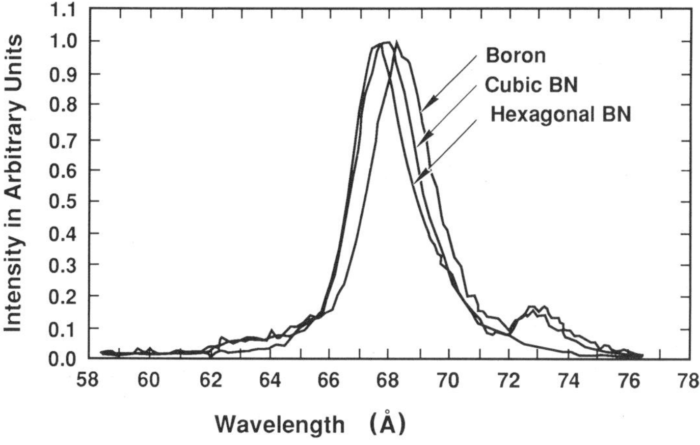

Figure 5.16. Boron $K\alpha$ scans obtained from pure boron, cubic boron nitride, and hexagonal boron nitride.

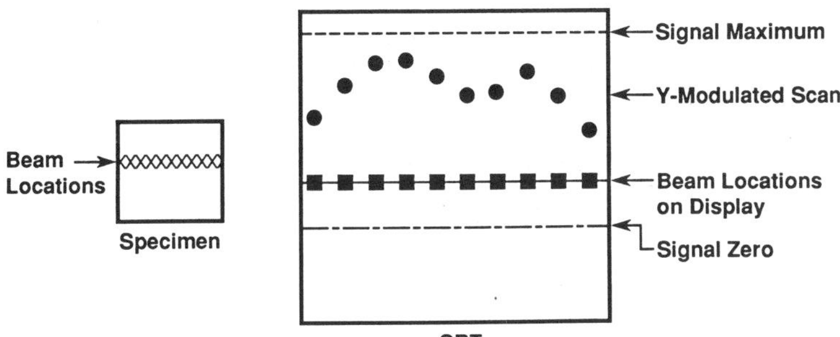

Figure 5.17. Schematic diagram showing how line scans are made.

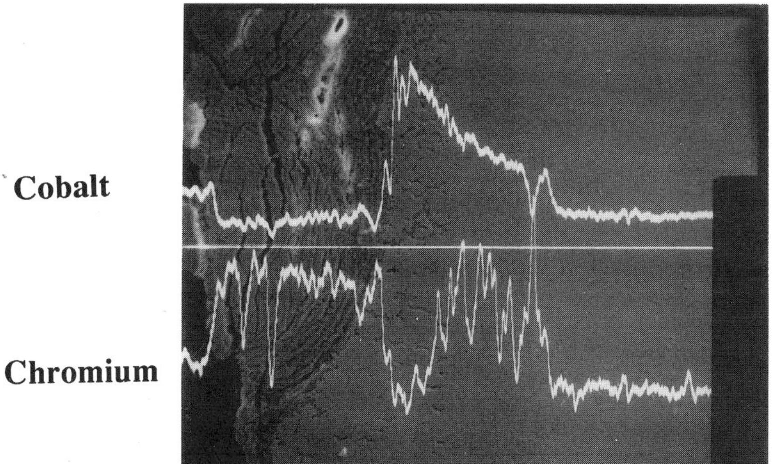

Figure 5.18. Co and Cr line scans across an oxidized high-temperature alloy. Straight line indicates the position of the scan on the secondary-electron image. (Top portion represents cobalt; bottom, chromium.)

image so that the analyst can readily interpret the results. Two examples of line scans are given in Fig. 5.18. Data presented in this manner give qualitative information since a complete quantitative evaluation requires conversion of the intensity data to chemical composition by one of the mathematical methods described in Chapter 9. Furthermore, since deflection of the beam can lead to defocusing of the x-ray spectrometers, quantitative line scans can be performed by stepping the sample under a static beam. Two-dimensional scanning, known as *x-ray dot mapping,* described in detail in Chapter 10, involves taking the output of the SCA and using it to modulate the brightness of the CRT during normal secondary-electron raster scanning. Each x-ray photon detected appears as a dot on the CRT, with regions of high concentration characterized by a high dot density.

5.3. Energy-Dispersive X-Ray Spectrometer

5.3.1. Operating Principles

Fitzgerald, Keil, and Heinrich (1968) first described the use of a lithium-drifted silicon Si(Li) solid-state x-ray detector mounted on an electron probe microanalyzer. Although their system was barely capable of resolving adjacent elements, it did demonstrate the feasibility of interfacing the x-ray detector with an electron-beam instrument. The next few years saw a period of rapid improvement in detector resolution from 600 eV to less than 150 eV, thereby making the technique much more suitable for microanalysis requirements. Today the idea of using solid-state detectors for mid-energy x-ray spectroscopy (1–12 keV) is no longer a novelty, for they can be found on a large percentage of scanning and transmission electron microscopes as well as electron probe microanalyzers. Because of its ease of operation, the Si(Li) detector is often the detector of choice for both qualitative and quantitative analysis. Its use is described in detail in this chapter; however many of the concepts described also apply to intrinsic Ge, which is beginning to find a niche, as well as mercuric iodide, which offers some promise as a room-temperature detector.

In order to understand the operation of the Si(Li) detector, it is first useful to consider a few basic points about semiconductors. Figure 5.19 is a simplified energy diagram of silicon. It illustrates the fact that in undoped silicon (which has four outer-shell electrons) all of the valence band is full, while the next-highest available energy levels in the conduction band are empty. The filled valence band reflects the fact that each silicon atom shares its outer electrons, one each, with its four nearest neighbors. Consequently, there are no excess electrons available for charge transport in the conduction band, nor are there any holes in the valence band. Such a material is called an *intrinsic semiconductor*. It will not conduct current in an applied electric field unless it absorbs energy causing electrons to be promoted into the

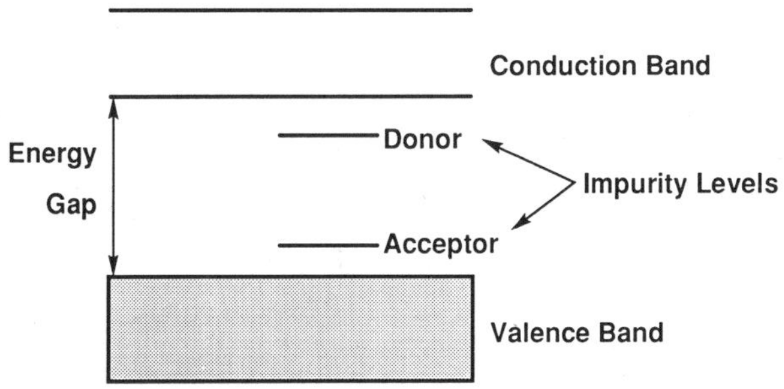

Figure 5.19. Schematic representation of energy levels in silicon.

conduction band, leaving holes behind. For this reason intrinsic semiconductors can make good radiation detectors.

Most electronic devices consist of junctions of doped *p*- and *n*-type silicon. Doping silicon with boron makes it a *p*-type semiconductor (holes are the majority carriers). Doping silicon with phosphorus makes it *n*-type (electrons are the majority carriers). Dopants create new energy levels in the band gap. Since a boron atom has three, rather than four, electrons in its outer shell, when boron is substituted for a silicon atom it removes an electron from the valence band, leaving a hole behind. On the other hand, phosphorus, with five electrons in its outer shell, can provide an excess electron for the conduction band when substituted for a silicon atom. In each case, the net effect of having mobile charge carriers (excess holes or electrons) is to allow current to flow across such a device in the presence of an applied field. Therefore, doped silicon is unsuitable for a detector, because any change in conductivity with the absorption of radiation would be superimposed on an existing background level caused by the presence of mobile charge carriers.

In actual practice, silicon sufficiently pure to achieve the "intrinsic" condition is hard to obtain. Instead, detector crystals are made to behave like intrinsic silicon by a process developed by Pell (1960). Lithium (an *n*-type dopant) is applied to the surface of *p*-type silicon and caused to diffuse into the crystal, forming a small *p–n* junction. While the *p–n* junction boundary is an intrinsic semiconductor, it is only a few micrometers thick. However, when a reverse bias is applied to the *p–n* junction at elevated temperature, the intrinsic region can be enlarged to a few millimeters, making it suitable for a detector material after most of the *p*-type material is removed. It is important to note that both steps are performed above room temperature, but even at room temperature the Li is mobile in the presence of an applied field. For this reason detectors can never be allowed to operate under bias except near liquid-nitrogen temperatures.

The operating principles of a solid-state detector system are illustrated in Fig. 5.20. X-ray photons from the sample pass through a thin window isolating the environment of the specimen chamber from

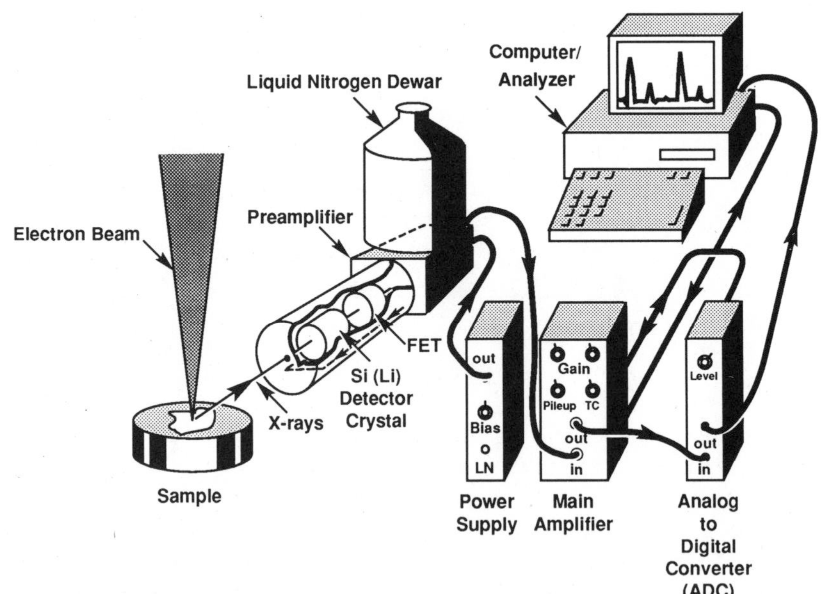

Figure 5.20. Schematic representation of an energy-dispersive spectrometer and its associated electronics.

the detector, into a cooled, reverse-bias *p–i–n* (*p*-type, intrinsic, *n* type) Si(Li) crystal. Absorption of each individual x-ray photon leads to the ejection of a photoelectron, which gives up most of its energy to the formation of electron–hole pairs. They in turn are swept away by the applied bias to form a charge pulse, which is then converted to a voltage pulse by a charge-to-voltage converter (preamplifier). The signal is further amplified and shaped by a linear amplifier and finally passed to a computer x-ray analyzer (CXA), where the data is displayed as a histogram of intensity by voltage. The contents of the CXA memory in all recent instruments either reside directly in a computer or can be transmitted to a computer for further processing such as peak identification or quantification. The key to understanding how an energy-dispersive spectrometer (EDS) works is to recognize that each voltage pulse is proportional to the energy of the incoming x-ray photon. The role of the CXA is to establish this relationship and to present it in a form understandable to the operator.

The typical physical appearance of a detector can be seen in more detail in Fig. 5.21. The lithium-drifted silicon crystal is mounted on a cold finger connected to a liquid-nitrogen reservoir stored in the Dewar. Since the detecting crystal is light sensitive, it is essential to block visible radiation, preferably through the use of an opaque window. Windowless and ultrathin-window EDS can be used if the specimen chamber is light tight and the specimen is not cathodoluminescent. The detector chamber is also sealed under vacuum both to prevent contamination from the specimen region (especially when the

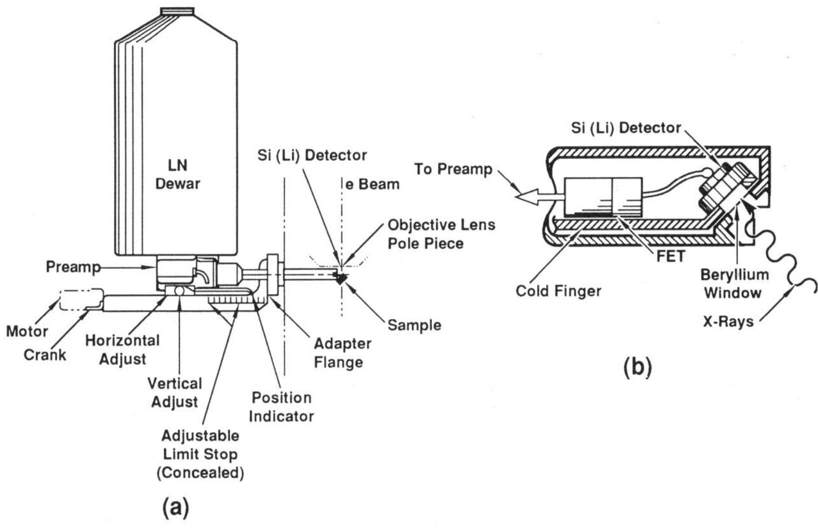

Figure 5.21. (a) Physical appearance of a retractable detector and associated preamplifier electronics. (b) Detail of Si(Li) mounting assembly.

specimen chamber is brought to air pressure) and to more easily maintain the low temperature essential for reducing noise. As mentioned previously, low temperture is also needed to limit the mobility of the lithium ions initially introduced in the silicon crystal to neutralize recombination centers. Again, under no conditions should the bias be applied to a noncooled detector. Many systems, in fact, incorporate safety features to turn the bias off in the event of either the crystal warming up or a vacuum failure. Dewars presently available are capable of maintaining cryogenic conditions for several days without refilling. Peltier-cooled detectors, which do not require liquid nitrogen, are also now becoming available. Note that the crystal and supporting cold finger are well separated from the housing assembly to prevent condensation on the latter and to provide electrical isolation. The feature of being able to mechanically move the detector housing in and out relative to the specimen without breaking the microscope vacuum is quite useful. As will be discussed later, the situation may arise where, for a fixed beam current, the x-ray signal has to be increased to obtain better counting statistics or decreased for better energy resolution. In many cases the desired count rate can be obtained by simply varying the crystal-to-sample distance, which changes the solid angle of detector acceptance, with the solid angle decreasing as the square of the specimen-to-detector distance. Some new instruments incorporate an adjustable diaphragm for this purpose.

The development of the energy-dispersive x-ray spectrometer has made “easy” x-ray spectrometry available to virtually all types of electron-beam instruments. However, one must recognize that, because of the nature of the EDS technique, distortions are introduced into the ideal x-ray spectrum ("spectral artifacts") during the measurement process which must be dealt with in practical analytical spectrometry. In the discussion which follows, we will consider these artifacts at each stage of the detection and amplification process.

5.3.2. The Detection Process

The basic detection process by which this proportional conversion of photon energy into an electrical signal is accomplished is illustrated in Fig. 5.22. The active portion of the detector consists of intrinsic silicon, with a thin layer of *p*-type material, called the "dead layer," on the front surface, coated with a thin gold electrical contact. When an energetic photon is captured, electrons are promoted into the conduction band, leaving holes in the valence band. Under an applied bias, these electrons and holes are swept apart and collected on the electrodes on the faces of the crystal. The process of x-ray capture is photoelectric absorption, and the x-ray photon is annihilated in the process (see Chapter 3). The incident x-ray photon is first absorbed by a silicon atom, and an inner-shell electron is ejected with an energy $h_{\nu} - E_c$, where E_c for silicon is 1.84 keV. This photoelectron then creates electron–hole pairs as it travels in the detector silicon and scatters inelastically. The silicon atom is left in an energetic condition because of the energy required to eject the photoelectron. This energy is subsequently released in the form of either an Auger electron or a silicon x-ray. The Auger electron scatters inelastically and also creates electron–hole pairs. The silicon x-ray can be reabsorbed, which initiates the process again, or it can be scattered inelastically. Thus, a sequence of events takes place leading to the deposition of all of the energy of the original photon in the detector, unless radiation

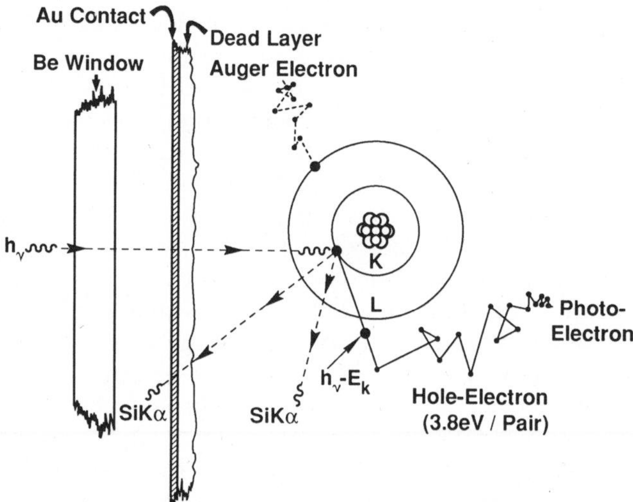

Figure 5.22. The x-ray detection process in the Si(Li) detector.

generated during the sequence, such as a silicon $K\alpha$ photon, escapes the detector, giving rise to the artifact known as the "escape peak," which will be discussed later.

The ideal number of charges n created in the detector per incident photon with energy E (eV) is given by

$$n = \frac{E}{\varepsilon}, \tag{5.6}$$

where $\varepsilon = 3.8\,\text{eV}$ for silicon. For example, if the detector captures one photon having an energy of 5 keV, then from Eq. (5.6) the total number of electrons swept from the detector is approximately 1300, which represents a charge of $2 \times 10^{-16}\,\text{C}$. This is an extraordinarily small charge. To measure charge accurately, noise minimization is essential, hence the need for keeping the detector crystal close to liquid-nitrogen temperature.

5.3.3. Charge-to-Voltage Conversion

Figure 5.23a is a representation of the detector, charge-to-voltage converter (often referred to as a preamplifier), and pulse-shaping linear amplifier from an electronic perspective. Once created, the charge from the detector is rapidly swept into a virtual ground at the input of an operational amplifier, where it is deposited onto the capacitor, C. The first stage of the preamplifier is a carefully selected low-noise field effect transistor (FET), which is also cooled and placed immediately behind the

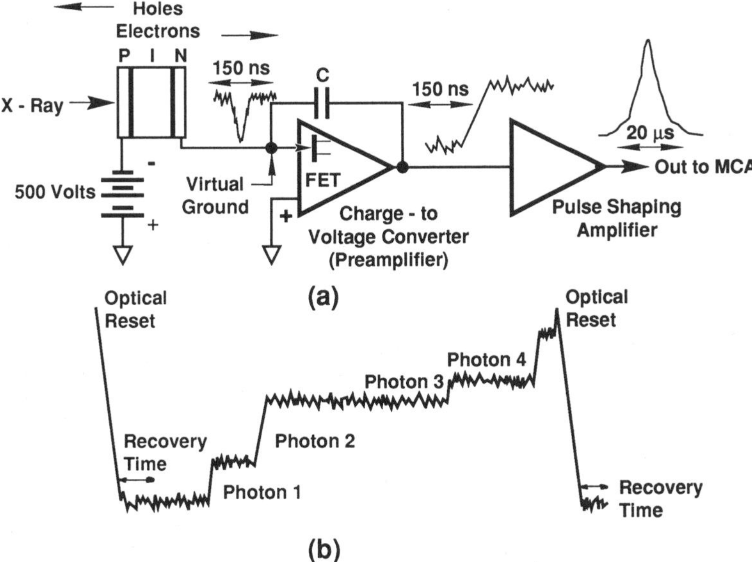

Figure 5.23. Charge-to-voltage conversion process. (a) Representation of the detector charge-to-voltage converter, and pulse-shaping linear amplifier from an electronic perspective. (b) Output of the charge-to-voltage converter after the detection of a series of x-ray photons.

detector (as shown in Fig. 5.21b). The preamplifier output for each event is a voltage step proportional to the incident photon energy. The height of each step is on the order of a few millivolts and occurs over a period of about 150 ns. It is important to recognize that the effect of electronic noise is the introduction of some fluctuation in the output voltage level on both sides of each step. This will be discussed in further detail in the section on factors determining energy resolution. Figure 5.23b shows the output of the charge-to-voltage converter after the detection of a series of x-ray photons. The difference in the step sizes indicates that they have different energies. An increase in the overall voltage level is also observed, and a point could eventually be reached when the amplifier would be saturated. To prevent this from happening it is necessary to discharge the capacitor C to return the operational amplifier output voltage back to zero. Note that only four photons are shown in the figure; however, in an actual system several hundred may be processed before reseting. One method to accomplish this is to employ a light-emitting diode, which when turned on will cause a leakage current to flow in the FET. This technique is called pulsed optical feedback (POF). Since considerable noise is generated when the POF is turned on, it is necessary to gate the main amplifier for about 1 ms. POF preamplifiers are now in common use. Recently, a new type of 5-electrode FET obviates the use of POF and shows considerable promise in noise reduction. It should be noted, however, that the present state of the art of detector manufacture is very near the theoretical resolution limit for silicon (about 100 eV at 5.9 keV).

5.3.4. Pulse-Shaping Linear Amplifier and Pileup Rejection Circuitry

It is the principal function of the pulse-shaping linear amplifier to transform each individual voltage step from the charge-to-voltage converter into a voltage pulse suitable for presentation to the CXA. As shown in Fig. 5.23a, the linear amplifier output pulse starts at zero and rises to a maximum value, typically several volts, and then returns to zero. The pulse height retains proportionality to the energy of the originating x-ray photon. The overall measurement process is particularly challenging because, as mentioned previously, the steps are on the order of millivolts and the voltage level on which they occur is noisy. This noise is dependent not only on the quality of the detector crystal, its environment, and associated preamplifier electronics, but also on the operating characteristics of the linear amplifier. In a Si(Li) detector system, noise reduction is particularly critical because, unlike what occurs in the WDS, all of the spectral dispersion is done electronically. Special circuity must be used to ensure maximum linearity, low noise, rapid overload recovery, and stable high-count-rate performance. Most commercial amplifiers use pole-zero cancellation networks to compensate for pulse overshoot when internal AC coupling is used and DC restoration circuits to clamp the pulse base line to a stable reference voltage. Clearly any loss of linearity leads to a breakdown in Eq. (5.6) and the incorrect assignment of x-ray pulse energy.

For the linear amplifier to convert a step to a pulse requires, conceptually, averaging the noise on either side of the step. The noise intensity, as a function of frequency, is not uniform but has a distinct minimum at a particular frequency. The pulse-shaping circuitry is designed in such a way as to take maximum advantage of the noise minimum. For more information on this topic see Goulding, (1972) or Reed (1975). Up to a point, increasing the time that the noise level is analyzed improves the accuracy of the pulse height as a measure of the energy of the originating x-ray photon. In order to accomplish this, however, voltage-step measuring times have to be extended to several or even tens of microseconds. This can cause a problem, called "pulse pileup," if the number of photons entering the detector exceeds a few thousand per second. Pulse pileup occurs if a photon arrives at the detector before the linear amplifier has finished processing the preceding photon. It appears as an increased output pulse height for the second photon because it is riding on the tail of the first, as shown in Fig. 5.24 for the 6- and 10-μs cases. Pulse pileup can also appear as a single large pulse representing the combined voltages of two pulses, if the second photon arrives before the pulse from the first has reached its maximum value. In the most extreme case, two photons arrive at the detector almost simultaneously, and the output is a single combined pulse corresponding to the sum of the two photon energies. This phenomenon gives rise to what are called "sum peaks," such as the $2K_\alpha$ and the $K_{\alpha+\beta}$ peaks.

Pulse pileup can be reduced by decreasing the processing time for each photon, since shorter linear-amplifier output pulses are less likely to interfere with each other. This is accomplished electronically through user selection of the linear-amplifier time constant (T.C.), which determined the output pulse width (sometimes designated as fast, medium, or slow). It should be recognized, however, that the T.C. is only an indication of pulse width, since the acceptable time between pulses may be many times the value of the T.C. This point is illustrated in Fig. 5.24. It can be seen that pulse II, arriving 20 μs after pulse I, is correctly evaluated as 4 V for the 1-μs T.C. but as 4.5 V for the 6-μs T.C. and 6.5 V for the 10-μs T.C. Pulse pileup situations will lead to the incorrect photon energy assignment because of inaccurate pulse-height measurement.

The electronic elimination of such events is accomplished by means of a pulse-pileup rejection circuit of the type illustrated in Fig. 5.25a, which shows both the principal electronic components and signals at key points (Fig. 5.25b). A second amplifier, called a "fast amplifier," is placed in parallel with the slower pulse-shaping linear amplifier. It has a much shorter T.C., and thus can give rise to narrower output pulses at the expense of pulse-height precision. This is acceptable because the fast amplifier is used only to sense the presence of each pulse arriving at the pulse-shaping linear amplifier (point 2), even if this amplifier is busy

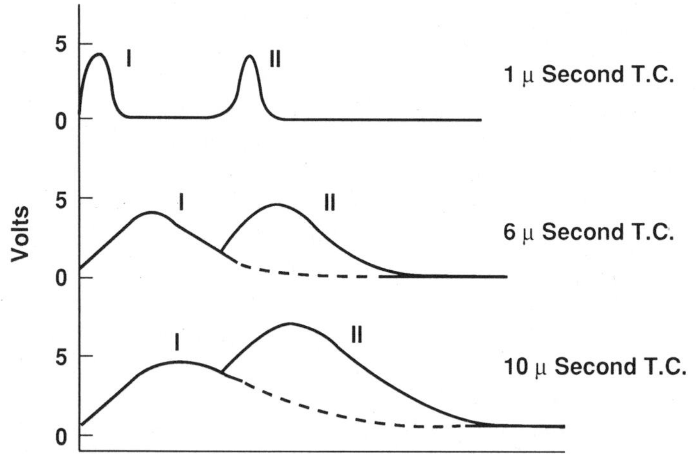

Figure 5.24. EDS main amplifier pulse shapes for different time constants (T.C.).

processing a previous pulse. If the fast amplifier pulse exceeds a predetermined discriminator setting (point 4), then a signal is sent to the pulse pileup inspector which can prevent the output of the main amplifier from going to the CXA. The decision can be made to block both pulses (if the second one arrives before the first one reaches its peak value) or to block only the second one (if the first has peaked and been processed by the CXA but the signal has not returned to the base line).

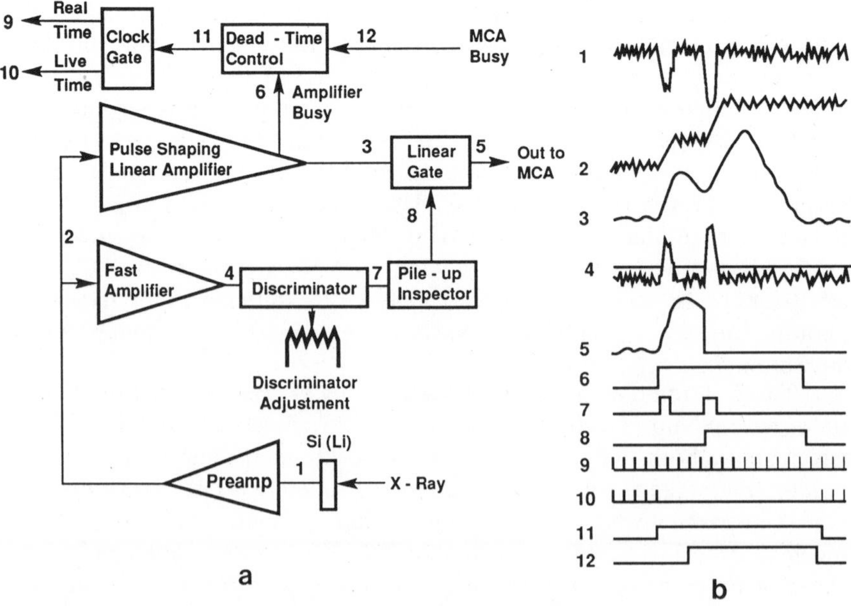

Figure 5.25. (a) Schematic representation of the principal components used with pulse-pileup rejection. (b) Waveforms at various key locations.

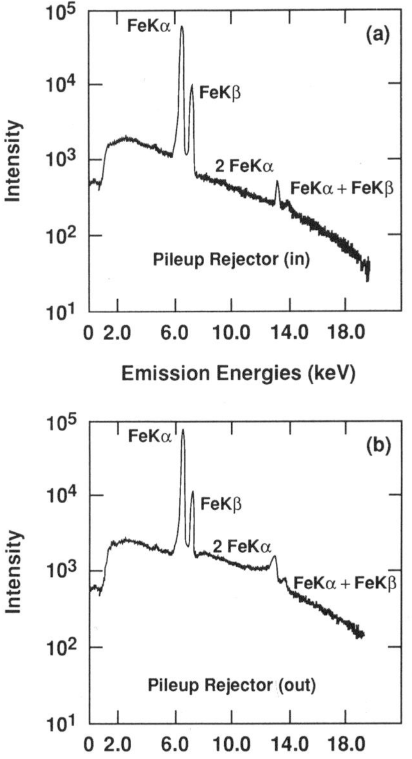

Figure 5.26. The use of pileup rejection to improve an iron spectrum (from Lifshin and Ciccarelli, 1973). (a) Pileup rejector in (b) Pileup rejector out; note more pronounced pileup peaks.

Figure 5.26 shows a comparison between two iron spectra taken with and without pulse-pileup rejection. Pulse pileup causes distortion in the background at energies above the Fe $K\alpha$ peak. Note the presence of the 2 Fe $K\alpha$ and Fe $K\alpha$ + Fe $K\beta$ peaks in both spectra. These peaks are due to the simultaneous arrival of two Fe K_α photons or an Fe K_α and an Fe K_β photon in the detector crystal creating electron–hole pairs in numbers corresponding to a single photon with their total energy. They cannot be eliminated by the pulse-pileup rejector, but the fraction of such peaks relative to the main characteristic peak can be reduced to a very low level by keeping the total count rate below a few thousand counts per second when using a longer T.C. When a quantitative analysis is performed, the number of $K\alpha + K\beta$ and twice the number of $2K\alpha$ pulses should be added to the $K\alpha$ peak intensity to correct for their incorrect energy assignment.

Because of the shorter T.C. of the fast amplifier, its output pulses ride on a higher noise level than those from the pulse-shaping linear amplifier (see Fig. 5.27). Proper setting of the discriminator is crucial, because if it is too low, noise will be accepted as pulses, causing unnecessary pulse rejection; however, if it is too high, low-energy pulses might be missed. Pileup rejection is therefore more complicated for

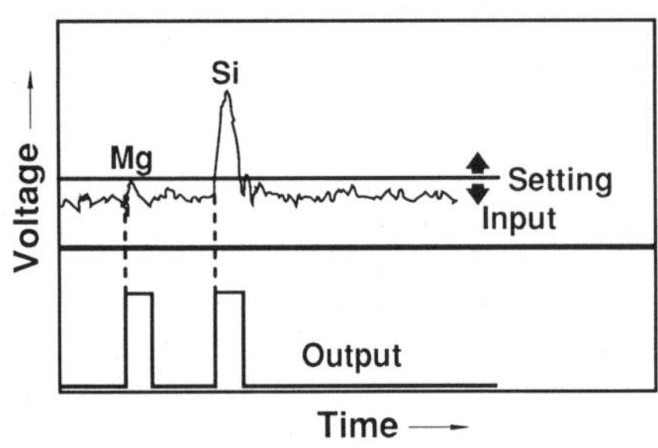

Figure 5.27. Effect of window setting on the discrimination of Mg and Si characteristic peaks.

low-energy x rays which may be hard to separate from the background noise. Even in an optimally adjusted discriminator, there is an energy below which the desired pulses are so close to the noise that the discrimination becomes inadequate. This situation is illustrated schematically for magnesium and silicon in Fig. 5.27. The magnesium pulses are poorly defined above the noise, and therefore the proper setting of the discriminator for magnesium is difficult. Moreover, because of drift in the electronics which results from temperature changes, humidity, time, etc., the discriminator level may change. Thus, although adequate pulse-coincidence discrimination may be achieved for silicon, Fig. 5.28a, the discrimination is almost negligible for magnesium, Fig. 5.28b, where a pileup continuum, double and triple energy peaks are observed.

Operation of the pulse-pileup rejector results in "dead time" during a given measurement, when pulses are not measured. Dead time is often expressed as a percentage of real or clock time. Conversely, the time during which pulses are measured is referred to as "live time." It is often necessary to compare spectral information taken under identical operating conditions, including live times. In a quantitative analysis, for example, live time must be used when ratios of x-ray intensities between samples and standards, taken under identical operating conditions, are required as inputs to the quantitative correction model. In most systems available today, the operator sets the desired live time, and the counting interval is automatically extended to compensate for the dead time. Figure 5.25a shows how this is done. The evenly spaced clock pulses at point 9 (Fig. 5.25b) corresponding to actual elapsed time (real time) are passed through a clock gate and then are output (point 10) to a counter (not shown). Data collection is terminated when a preset number of clock pulses is collected. The decision to temporarily block the flow of clock pulses is determined by the dead-time control circuit inhibit pulse (point 11), activated by a combination of the amplifier busy (point 6) and CXA busy signals (as well as POF busy, also not shown). In the example shown, the clock gate has allowed only 8 of 21 clock pulses to be counted indicating significant dead time (compare the pulse trains at points 9 and 10).

The effect of this loss of pulses is illustrated in Fig. 5.29, which shows an example of the relationship between the CXA input count rate and the linear-amplifier input-count rate for different T.C.s. It can be seen that only at low count rates (less than about 2000 counts per second) are

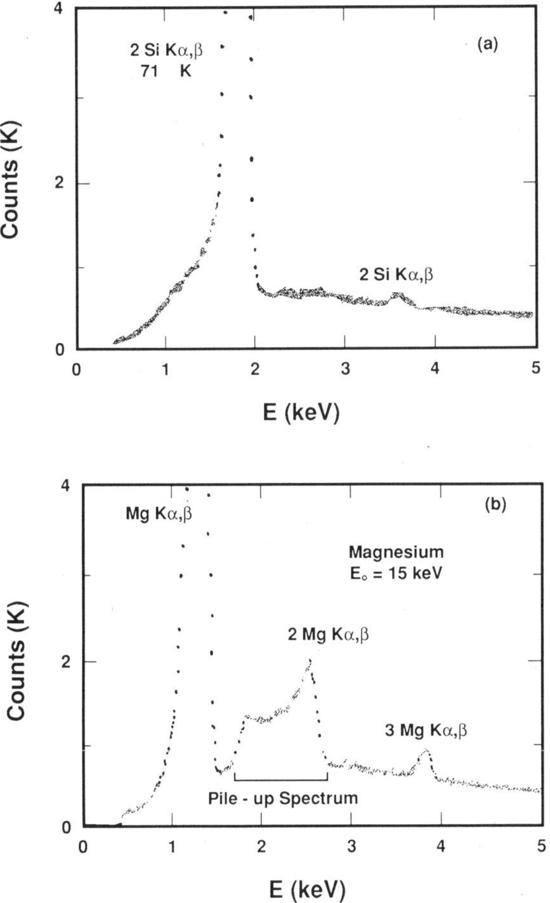

Figure 5.28. (a) Electron-excited EDS spectrum of silicon showing sum peak. (b) Electron-excited spectrum of magnesium showing pileup continuum and sum peaks.

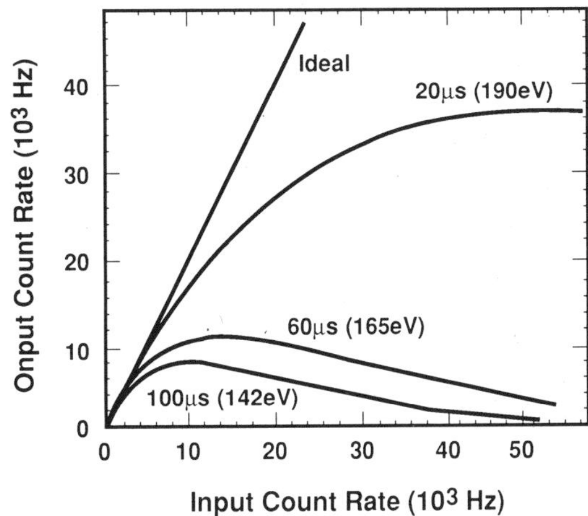

Figure 5.29. Relationship of output and input count rates for three different values of pulse width and resolution.

the two count rates equal. For reasons already discussed, as the amplifier-input count rate increases, pulse-pileup effects become more serious, particularly with the large amplifier time constants. In a quantitative analysis it may, therefore, be necessary to count for a longer time than anticipated on the basis of real time to achieve a desired level of precision based on counting statistics. Figure 5.29 also shows that increasing the amplifier-input count rate by changing the probe current or moving the detector closer to the sample initially results in a linear increase in CXA-input count rate (output count rate in Fig. 5.29) followed by a nonlinear region in which the rate of increase in CXA input is less than the rate of increase of linear amplifier input pulses. Eventually a point of diminishing returns is reached when increasing the main-amplifier input count rate actually results in a decrease in the count rate seen by the CXA. Further increases beyond this point lead to essentially 100% dead time and consequently a total locking up of the system. Figure 5.29 also shows that the onset of the various regions described is determined by a choice of operating curves based on acceptable resolution criteria. It is important to note that Fig. 5.29 is not a universal curve, and similar curves must be established for each system. Indeed, some of the latest pulse processors have almost a factor of two better throughput for a given resolution, compared to that indicated in Fig. 5.29.

The maximum input count rate is not simply that of the element of interest but rather that for the total of all pulses measured across the full range of the spectrum. Therefore, system count-rate performance is determined by the major elements and not by the minor constituents that may be under investigation. Clearly, achieving maximum count rate is critical to optimizing precision. Since pulse pile-up is the principal limiting factor, research continues on how to increase throughput without paying a major premium in resolution. One approach is to truncate the linear-amplifier pulses immediately after they reach their maximum values so that the likelihood of one pulse riding on the tail of the next is reduced. This approach, developed by Kandiah (1971), is now implemented in many available systems.

5.3.5. The Computer X-Ray Analyzer

The preceding section described the output of the pulse-shaping amplifier as a series of well-separated pulses whose magnitudes remain proportional to the energy of each corresponding photon. The role of the CXA is to present this information in a form understandable to the operator. In earlier times, the CXA was referred to as a multichannel analyzer or MCA, a separate piece of hardware. The evolution of computer-based systems has now integrated most functions into software commands and computer displays. In recognition of this change, the term "computer x-ray analyzer" (CXA), is used in this book.

The first step is to determine accurately the peak height of each pulse and to convert it into a digital value. This is done by an analog-to-digital converter (ADC). Two approaches are now in common use. The first is called the "ramp" or Wilkinson type, and the second is known as the "successive approximation" type. Figure 5.30 illustrates the ramp approach. As shown in Fig. 5.30a, the linear-amplifier output is directly coupled to a capacitor through a closed gate. The capacitor is charged to the peak value of the pulse and then the gate is opened, leaving the capacitor holding the maximum peak value. (This type of circuit is called a "sample and hold" circuit.) As shown in Fig. 5.30b, the capacitor is then connected to a constant current source which causes it to discharge linearly with time until a zero detection circuit senses that it is fully discharged. The time from the initiation of the discharge at T_1 until the zero crossing at T_2 is measured by counting clock pulses with a scaler from T_1 to T_2. In this simplified sketch, four clock pulses were counted. In Fig. 5.30c, we see a comparison of the run-down curve for the first pulse with a second pulse of twice the size. Since the pulse is twice as large, the run-down time is twice as long and the number of clock pulses is eight. In general, the number of clock pulses is proportional to the height of each pulse and the number of pulses therefore serves as a digital measure of pulse size. In the example shown, the number of clock pulses is small and the time duration between each pair is not specified. In an actual system, each voltage pulse may give rise to hundreds or thousands of clock pulses and the duration of each may be as short as 0.01 μs (in the case of a 100-MHz ADC). Large numbers of clock pulses are needed to accurately distinguish between voltage pulses of different sizes, while very short pulses are required to reduce any dead time associated with the analog-to-digital conversion process. Note, for example, that even with a 100-MHz ADC, in order to distinguish voltage differences of one part in a thousand, 1000 clock pulses must be counted. This would require 10 μs, which is comparable to pulse-shaping times.

Although it has been commonly used for some time, the ramp ADC suffers from the problem that the conversion time is proportional to the pulse size. The difficulty can be overcome by the use of a successive-approximation ADC. The approach is relatively straightforward, and the time required for the conversion does not depend on the size of the voltage pulse. The concept of successive approximation involves making a series of fixed comparisons of the voltage pulse with a preselected series of values differing by factors of two. For example, in order to determine the size of a pulse between 0 and 8 volts it can first be compared to a 4-V reference, thus reducing the field of possible values by a factor of 2. Once it is determined to be, for example, between 0 and 4 V, it can be compared to a 2-V reference, and so on. Thus each comparison improves the estimate. A 12-bit ADC, for example, can determine a pulse voltage to 1 part in 4096 of the voltage range.

In addition to an ADC, a computer x-ray analyzer system includes

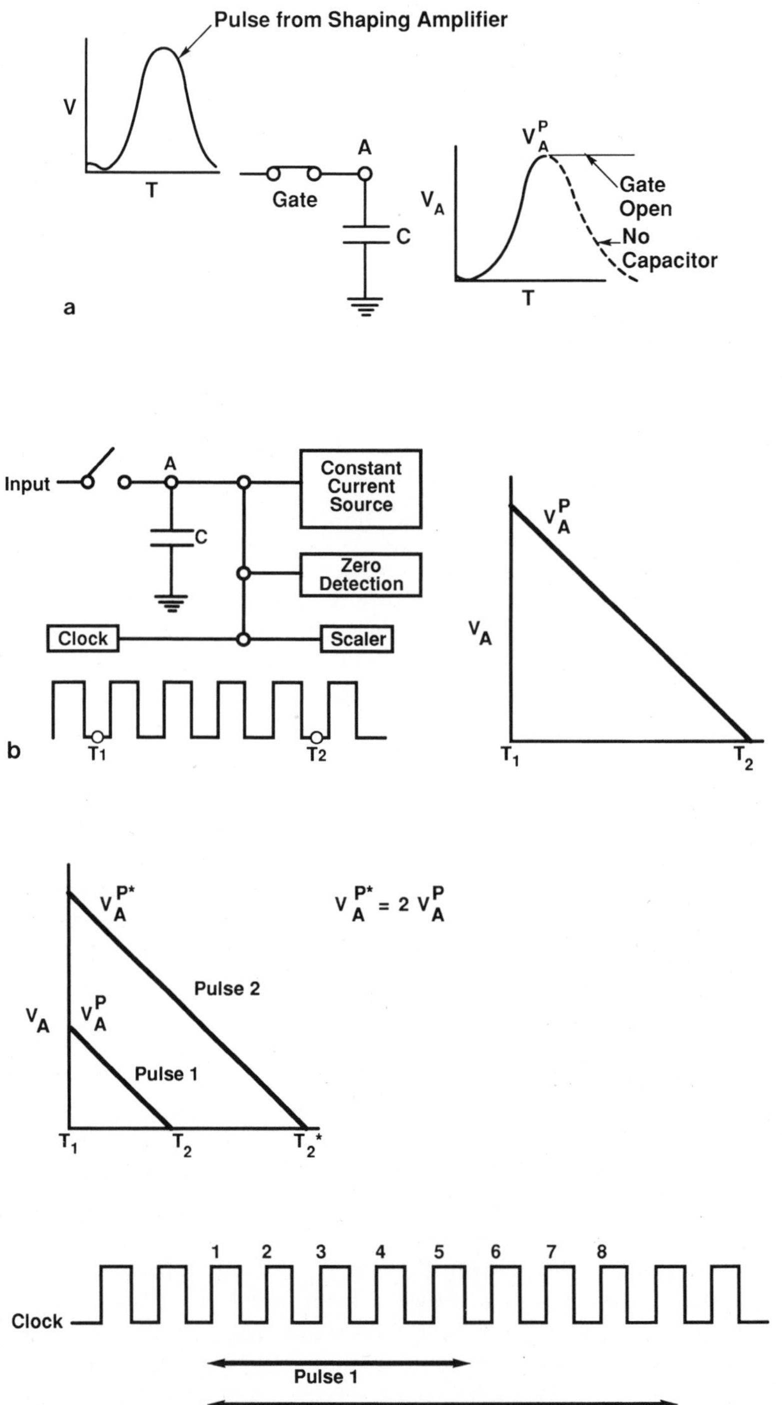

Figure 5.30. Operating principles of a Wilkinson ramp analog to digital converter: (a) Linear amplifier pulse profiles, (b) Discharging cycle of capacitor, (c) Voltage as a function of time and clock pulses.

an acquisition processor, a memory, and operator controls including key pads and analog input devices. It may also include peripherals such as data- and program-storage devices, printers, and network connectivity devices. Since many of these capabilities are available in modern personal computers and work stations, the latter are often modified to meet the needs of energy dispersive spectrometry. This evolution, which has been taking place over the past two decades, represents a significant departure from the stand-alone hardware MCAs first used.

Figure 5.31a shows a simplified picture of the sorting process. For purposes of illustration, ten locations in a computer memory have been assigned to store the numbers of pulses corresponding to specific voltage sizes; e.g., address 1001 contains the count of pulses with energies between 0 and 1 V, address 1002 contains the count of pulses with energies between 1 and 2 V etc. Assume that at the beginning of data collection all of the memory contents are 0. The first pulse from the main amplifier is 5.2 volts high, and the ADC sends a digital signal to the acquisition processor that contains this value. The acquisition processor then causes 1 to be added to the contents of address 1006, because that is where pulses between 5 and 6 V are counted. The display is a CRT which provides a convenient way for the operator to follow what is happening. The horizontal sweep occurs rapidly from 0 to 10 V full scale. It would appear as just a bright line, but is electronically chopped into a series of dots corresponding to integer voltage values. The vertical deflection of each dot is then modulated to reflect the number of counts in each memory address, which initially was zero, but now shows one count corresponding to 5.2 volts.

Figure 5.31b shows a sequence of seven more pulses of different sizes, and the assignment of these pulses to the computer memory by the ADC and the acquisition processor. Figure 5.31c shows a ninth pulse (the last before the data-acquisition time runs out) and a listing of the contents of the computer memory and what the contents of that memory look like on the display. In this simple example, something looking like a peak is beginning to emerge. In actual systems, the total number of memory locations involved may be 1024, 2048, or more and the number of counts stored in each location may be in the tens of thousands or even considerably higher.

As described thus far, pulses have been sorted and displayed by voltage, which, if everything has been done right, should be proportional to the energy of each corresponding x-ray photon in accordance with Eq. (5.6). Calibration is required to make this happen on the display of the CXA. While the exact procedure may vary between systems, the operator first determines the energy range of interest, e.g. 0 to 10 keV. An arbitrary decision is made that 5-V pulses correspond to 5-keV x rays (as shown in Fig. 5.31). At this point it would be totally fortuitous to find that 5-keV x rays produce 5-V pulses. However, if a source is available emitting x-rays of known energies, (e.g., Cu $K\alpha$ at 8.04 keV), then the gain of the linear-amplifier can be adjusted until Cu $K\alpha$ produces pulses 8.04 V high. The process could stop here, except that there is the

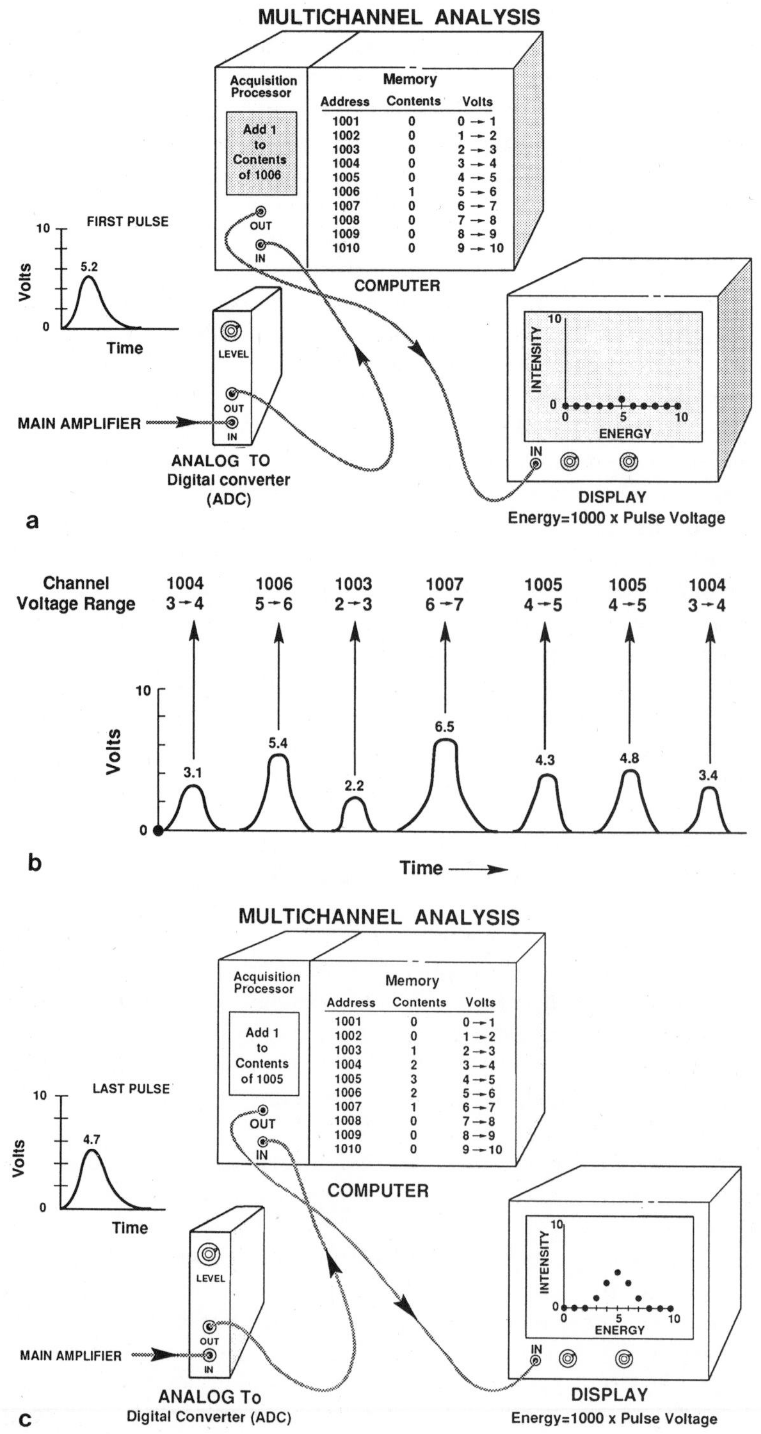

Figure 5.31. Schematic representation of a computer x-ray analyzer. (a) Components of the pulse processor, (b) Pulse sequence and channel destinations of next seven pulses, (c) Final display showing display of processed pulses.

possibility that Cu pulses are not referenced to zero volts. To make this correction required adjusting the amplifier's zero level, and now there are two variables, gain and zero, which requires a second peak at low energy (such as Al $K\alpha$) to uniquely determine their values. This procedure requires an interactive process adjusting the gain and zero level until both peaks appear at the proper positions. Some systems can now do this in a totally automated mode.

Figure 5.32 is an EDS spectrum from the same alloy used to illustrate WDS capabilities in Fig. 5.15. Note that all of the elements are shown in a single spectrum, but the peak resolution is not as good as the WDS. More detailed comparisons of the two techniques will be given later.

Currently available EDS systems are extremely flexible. As shown in Fig. 5.33, peak energies can easily be determined by a dial- or mouse-controllable feature known as a cursor. At the same time its position, in energy, as well as the corresponding number of counts can be read directly from a numerical display. Scale information and labels are also displayed directly on the CRT. Other features include region-of-interest (ROI) displays, line markers, and automatic peak-identification routines. The ROI approach allows the user, often with the help of the cursor, to define a series of energy bands in which a running count of the integrated intensities is recorded. In this manner either integrated peak intensities can be monitored as a prerequisite for quantitative analysis or pulses corresponding to specific elements can be coupled to SEM displays for line scans or area dot maps. Hard-copy outputs from presently available systems include a variety of standard output devices such as dot matrix and laser printers capable of combining both graphics and text for numerous types of qualitative and quantitative reports.

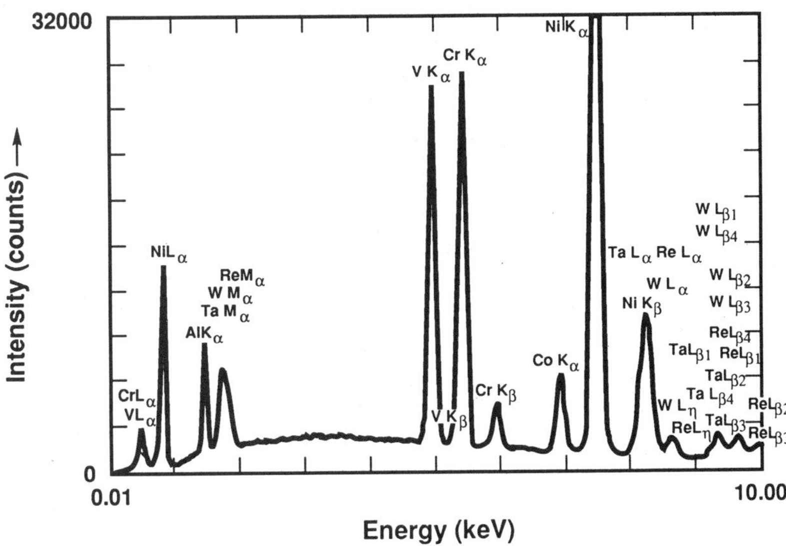

Figure 5.32. EDS spectrum of a high-temperature superalloy.

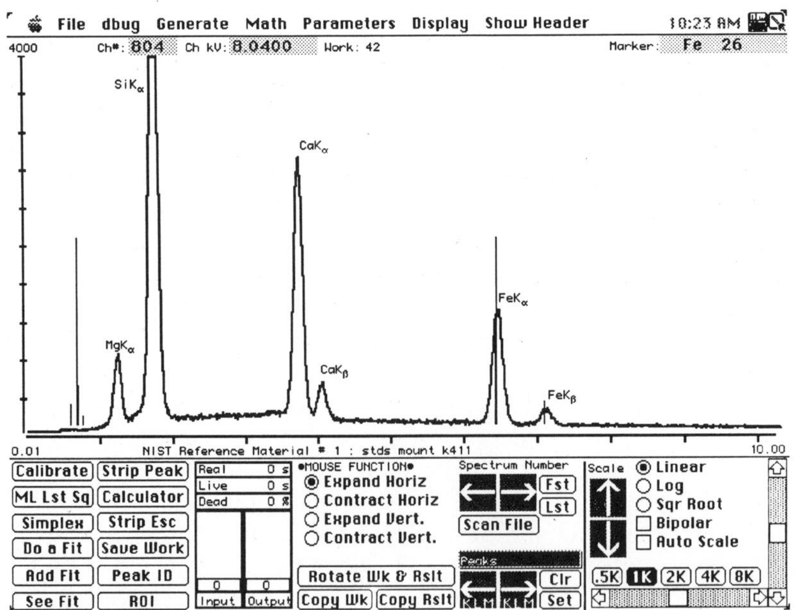

Figure 5.33. Modern EDS CRT display illustrating use of line markers and other features. (Courtesy of NIST.)

5.3.6. Artifacts of the Detection Process

Deviations from the ideal detector process result in the appearance of artifacts, principally peak broadening, peak distortion, silicon x-ray escape peaks, sum peaks, silicon and gold absorption edges, and the silicon internal fluorescence peak.

5.3.6.1. Peak Broadening

The natural width of an x-ray peak is energy dependent and is on the order of 2–10 eV measured at half the maximum of the peak intensity (FWHM). The use of the FWHM as a measure of peak width is traditional in spectroscopy because it is easily measured. The concept of a natural line width arises from the fact that a statistical distribution of x-ray photon energies is associated with nominally identical transitions from an ensemble of atoms of the same type. This variation is a direct consequence of the Heisenberg Uncertainty Principle. For example, for manganese $K\alpha_1$ radiation (5.898 keV), the natural FWHM is approximately 2.3 eV, which makes the natural width about 0.039% of the peak energy. The measured peak width from the Si(Li) spectrometer is degraded by the convolution of the detector-system response function with the natural line width to a typical value of 150 eV for Mn $K\alpha_{1,2}$, or 2.5% of the peak energy. The significant broadening of observed EDS peaks relative to their natural line widths is the result of two factors. First, when a sequence of

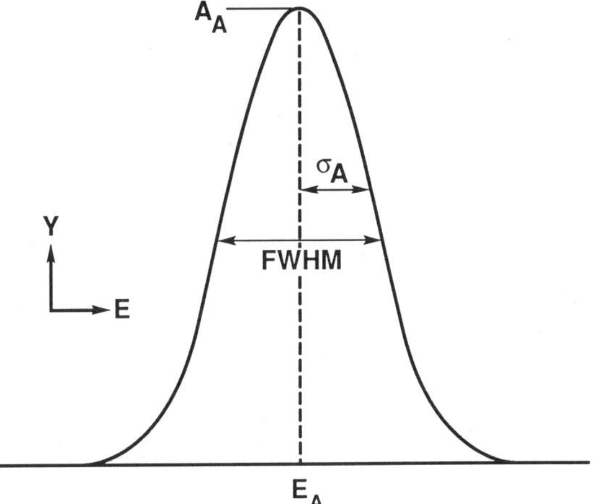

Figure 5.34. Gaussian peak shape representation for x-ray peaks above 1 keV.

monoenergetic photons enters a detector, each photon does not necessarily create exactly the same number of electron–hole pairs given by Eq. (5.6), but rather there is a statistical distribution in the final number of charge carriers created and the value 3.8 eV per electron–hole pair represents only the average. Second, an uncertainty is introduced by the thermal noise of the amplification process, primarily from the FET. The distribution of numbers of charge carriers for a single photon energy Y is reasonably well described by a Gaussian distribution for x-ray peaks above 1 keV, as shown schematically in Fig. 5.34 and in the following expression:

$$Y = A_A \exp\left[-\frac{1}{2}\left(\frac{E_A - E}{\sigma}\right)^2\right], \quad (5.7)$$

where A_A is the maximum x-ray intensity, E_A is the peak energy, and E is the x-ray energy. The relationship between the standard deviation σ, used in statistics, as a measure of broadening, and FWHM, used in spectroscopy, is FWHM $= 2.355\sigma$.

The FWHM of this distribution can be calculated from the two sources of noise by quadrature addition according to the equation

$$\text{FWHM} \propto (C^2E + N^2)^{1/2} \quad (5.8)$$

where C is the uncertainty in the formation of charge carriers, E is the x-ray energy, and N is the FWHM of the electronic noise of the amplification process. C is given by

$$C = 2.35(F\varepsilon)^{1/2}, \quad (5.9)$$

where F is a constant known as the Fano factor (about 0.1 for Si) and $\varepsilon = 3.8$ eV for Si. Figure 5.35a shows values calculated by Woldseth (1973) of the observed FWHM as a function of energy for different conditions of electronic noise. It can readily be seen that even if the noise contribution were totally eliminated, the theoretical energy

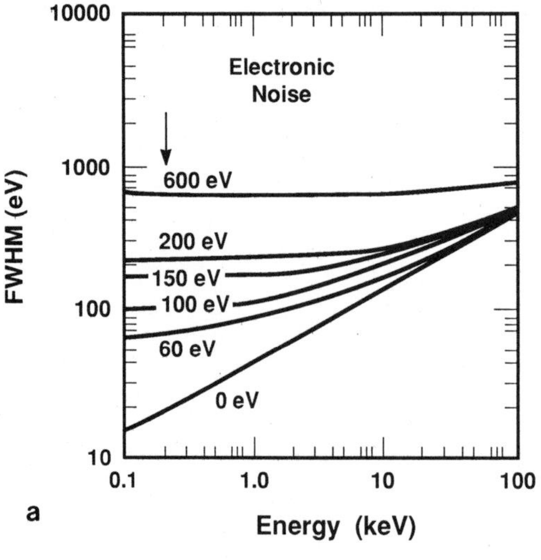

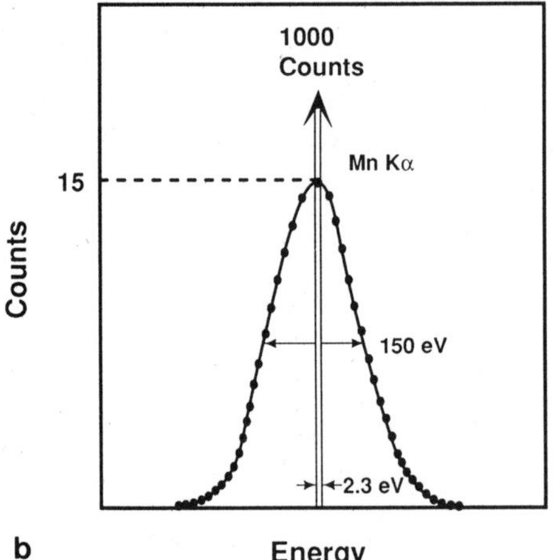

Figure 5.35. (a) Silicon energy resolution including intrinsic and electronic noise effects. (adapted from Woldseth, 1973). (b) Redistribution of peak counts for Mn $K\alpha$ with 150-eV resolution at FWHM.

resolution limit would still be greater than 100 eV for Fe $K\alpha$ at 6.4 keV.

Fiori and Newbury (1978) derived an equation which takes advantage of the fact that the noise term is a constant for a given detector, operated under a fixed set of conditions. Consequently, Eq. (5.8) can be restated in a form which is useful for comparing the width of a peak at an energy of interest to the width of a peak located at any other place in a spectrum:

$$\mathrm{FWHM} = [2.5(E - E_{\mathrm{ref}}) + \mathrm{FWHM}^2_{\mathrm{ref}}]^{1/2}. \tag{5.10}$$

All units in this expression are electron volts.

The immediate consequence of the peak broadening associated with the detection process is a reduction in the height of the peak (counts per energy interval), as compared to the natural peak, and an accompanying decrease in the peak-to-background ratio as measured

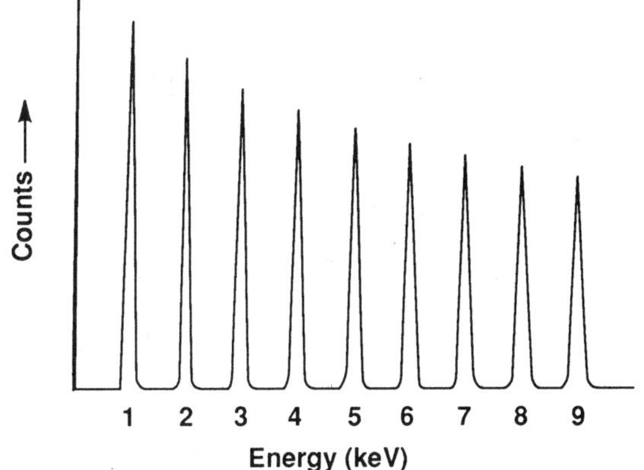

Figure 5.36. Resolution and peak intensity as a function of energy; Mn $K\alpha$ has a resolution of 150 eV.

at a given energy. Figure 5.35b shows the effect of detector broadening on the natural peak width of Mn $K\alpha$. In this example, a natural peak 2.3 eV wide and 1000 counts high is broadened into a peak 150 eV wide and 15 counts high.

A related effect is shown in Fig. 5.36, which is a computer simulation of a series of peaks containing equal numbers of counts measured at different energies. In this case, the variation of FWHM described by Eq. (5.8) results in a variation of peak heights. This is an additional factor, like the overvoltage effects described by Eq. (3.34), which points out the potential danger of estimating relative elemental concentrations by comparing peak heights between elements.

The value of FWHM is useful in estimating the extent of the overlap of peaks which are close in energy. An estimate of the extent of overlap is vital when considering peak interferences in qualitative analysis, where the identification of a low-intensity peak near a high-intensity peak may be difficult, and in quantitative analysis, where the removal of the interference is necessary for accurate determination of the composition.

Peak overlap can also be a problem even for peaks in the same family for a given element. An example is given in the case of KCl, as shown in Fig. 5.37. With a detector of 170 eV resolution (Mn $K\alpha$), the potassium $K\alpha$ and $K\beta$ peaks are nearly resolved, while the chlorine $K\alpha$ and $K\beta$ peaks are not.

5.3.6.2. Peak Distortion

Two different artifacts cause distortion, i.e., deviation from a Gaussian shape on the low-energy side of a peak:

1. The collection of charge carriers created in certain regions of the detector near the faces and sides is imperfect due to trapping and recombination of the electron–hole pairs, leading to a reduction in the value of n predicted by Eq. (5.6) for the incident photon. The resulting distortion of the low-energy side of the peak is known as incomplete

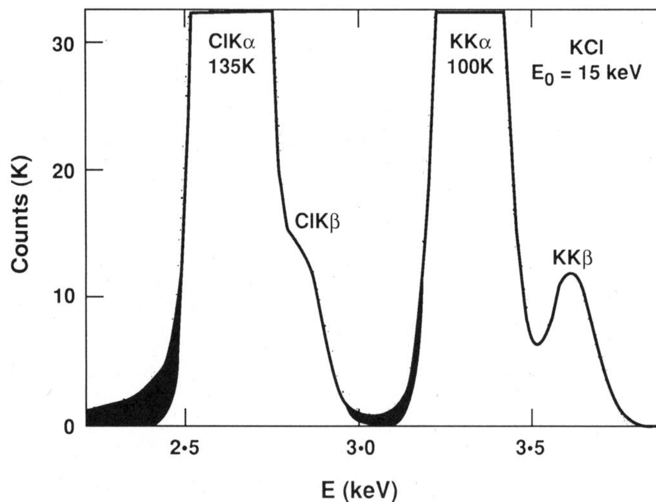

Figure 5.37. EDS spectrum of KCl illustrating peak overlap: K $K\alpha$, K $K\beta$, are nearly resolved. The solid line is a Gaussian fit to the data points. The shaded area represents the deviation caused by incomplete charge collection.

charge collection (Freund *et al.*, 1972; Elad *et al.*, 1973) and its effect is illustrated in Fig. 5.37 for the chlorine $K\alpha$ and potassium $K\alpha$ peaks. The deviation from a Gaussian distribution (shown as a solid line) is a function of energy. For example, the magnitude of the effect is significantly different for chlorine and potassium, which are separated by an atomic number difference of only 2.

2. The background shelf, Fig. 5.38, is a phenomenon in which the presence of a peak increases the background at all energies below the peak value. This effect is more easily seen by observing the spectrum

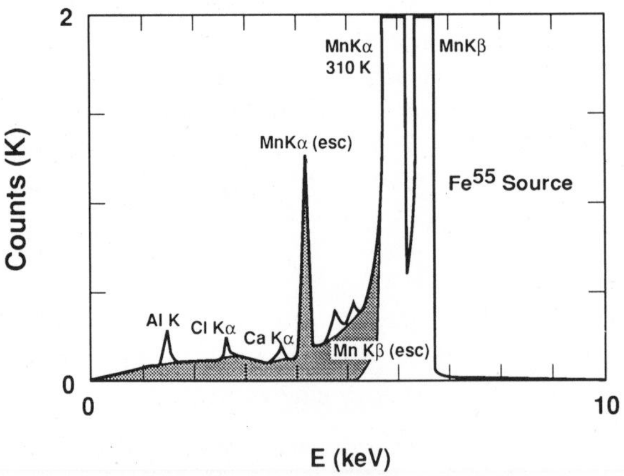

Figure 5.38. EDS spectrum derived from ^{55}Fe radioactive source (emits Mn K radiation). The background shelf is easily observable at energies below Mn $K\alpha$ down to the threshold at 300 eV. The Mn $K\alpha$ and $K\beta$ silicon escape peaks are noted. Extraneous characteristic peaks from the source holder are observed.

obtained from the radioactive source, ^{55}Fe, which emits Mn K lines. This source is chosen because it is readily available and is relatively background free. Additional counts above the expected background result both from the incomplete charge-collection phenomenon extending to low energy and from the escape from the detector of some of the continuum x rays generated by the photoelectron as it scatters inelastically in the silicon. Any radiation lost from the detector reduces the number of charge carriers created. Both the incomplete charge collection and the loss of continuum x radiation lead to a transfer of counts from the peak to the entire energy region, down to zero. Typically, the background shelf at one-half the energy of a peak has a relative intensity of about 0.1% of the parent peak. The total number of counts lost to the full peak due to this effect is approximately 1%. No Si(Li) detector, therefore, is 100% efficient. Although efficiency values exceeding 99% are common, poorly functioning detectors can have a 20% loss for peaks below a few keV, which can result in problems for both qualitative and quantitative analysis.

5.3.6.3. Silicon X-Ray Escape Peaks

The generation of a photoelectron leaves the silicon atom in the detector in an ionized state. In the case of K-shell ionization, relaxation of an L-shell electron results in the emission of either a Si K-series x-ray photon or an Auger electron, as shown in Fig. 5.22. The range of the Auger electron is only a fraction of a micrometer, and hence it is highly probable that this electron will be reabsorbed in the detector. In this case, the total energy deposited leads to creation of the number of electron–hole pairs given by Eq. (5.6). If a Si K x ray is produced, one of two events will occur. In the first case, one of the L shells of another Si atom is ionized and the ejected photoelectron leads to electron–hole pair production with the total number again given by Eq. 5.6. In the second case, the Si K x-ray photon escapes from the detector because an x-ray can travel much further than an Auger electron of the same energy. This is more likely to happen if the Si K x-ray photon is produced near the surface. In this case, the total number of electron–hole pairs will be given by Eq. (5.6), providing the energy term E, is replaced by $(E - E_{\mathrm{Si}\,K\alpha}) = (E - 1.74\,\mathrm{keV})$. The reduction in the number of electron–hole pairs produced when an escape event takes place leads to the creation of an artifact peak called an "escape peak." It appears at an energy equal to the energy of the parent line minus that of the silicon $K\alpha$ x-ray, 1.74 keV. In principle, both Si $K\alpha$ and Si $K\beta$ escape peaks are formed, but the probability for $K\beta$ formation is about 2% of the $K\alpha$; hence only one escape peak is usually observed per parent peak. Escape peaks are illustrated in Fig. 5.38 and 5.39. In Fig. 5.38 the Mn $K\alpha$ and Mn $K\beta$ escape peaks are observed. In Fig. 5.39 the parent peaks are Ti $K\alpha$ (4.51 keV) and Ti $K\beta$ (4.93 keV), and escape peaks are found at 2.77 keV (Ti $K\alpha$–Si $K\alpha$) and 3.19 keV (Ti $K\beta$–Si $K\alpha$). The magnitude of the escape peak

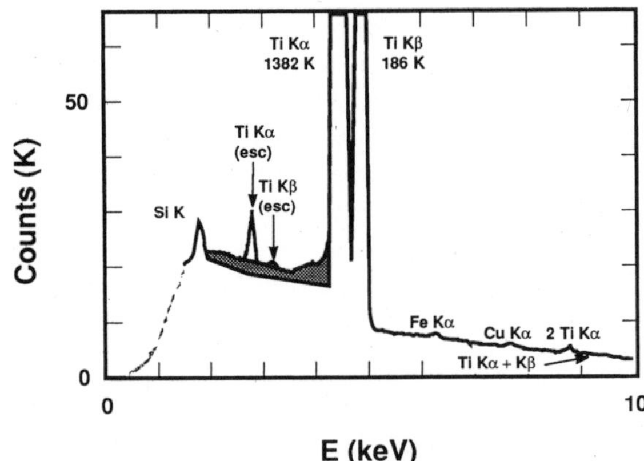

Figure 5.39. Electron-excited EDS spectrum of titanium. The Ti $K\alpha$ and $K\beta$ silicon x-ray escape peaks and the 2 $K\alpha$ and ($K\alpha$ + $K\beta$) sum peaks are noted. Extraneous peaks from the specimen chamber are also observed. Hatched area is contribution due to incomplete charge collection.

relative to the parent peak varies from about 1.8% for phosphorus to 0.01% for zinc K x-rays (Fig. 5.40). Silicon x-ray escape peaks cannot occur for radiation below the excitation energy of the silicon K shell (1.838 keV).

5.3.6.4. Absorption Edges

The radiation emitted from the specimen penetrates several layers of window material before it arrives in the active part of the detector. These materials are either functional or artifacts related to detector fabrication, and their effects must be understood. The nominal purpose of the first window material is to protect the cooled detector chip from the relatively poor vacuum in the specimen chamber, which can be high in water vapor

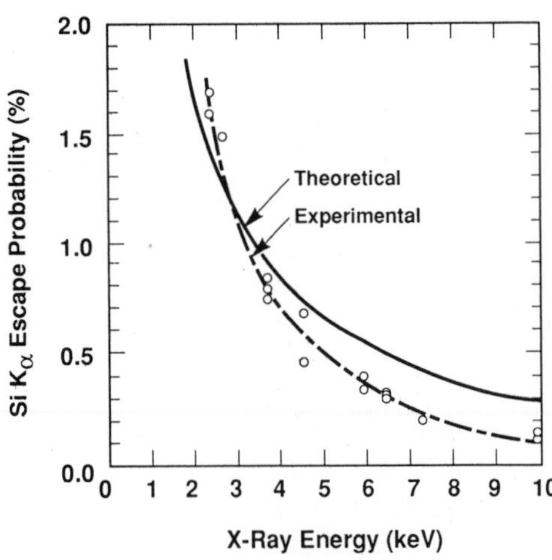

Figure 5.40. The escape probability of Si $K\alpha$ as a function of incident photon energy (adapted from Woldseth, 1973).

and organic components because of frequent exchanges of the specimen. It is very important that these components do not condense on the cooled detector. Even with a protective window, ice will build up, over a period of time, which will absorb lower energy x rays. Modern detectors have special heaters to remove this ice buildup, some of which actually originates from the detector assembly and cryostat at the time of manufacture. This problem can be overcome by the use of high-vacuum construction practices.

Protective windows are made from a variety of material, each having advantages and disadvantages. Historically, beryllium, about 7.6 μm (0.3 mil) thick, has been used. This thickness has the advantage that it can withstand the full-atmosphere pressure differential associated with specimen changes. Furthermore, the detector can be removed from the microscope without the need for isolation valves. It is also opaque to optical photons, which can easily saturate the detector during operation. Recently, protective window materials with considerably less mass-thickness, but still able to withstand a full atmosphere of pressure differential, have gained wide popularity. They are constructed of thin films on a support grid having approximately 85% open space. Low-energy x rays, such as those from oxygen or carbon, pass through only the unbacked portions of the film, while more energetic x rays (>8 keV) can pass through both the film and the support grid. The grid is usually pure silicon or a metal such as nickel, and the window material is either boron nitride, silicon nitride, diamond, or a polymer. Examples of transmission curves for various window materials currently in use are given in Fig. 5.41a and b. Since thin films of these materials are transparent and the detector is also sensitive to visible light, it is necessary either to eliminate possible sources of light or to apply an optically opaque layer to the protective window. Light sources include microscope illumination systems, viewing ports, and light leaks. Specimen cathodoluminescence is rarely intense enough to cause a serious problem although it is encountered with some minerals and semiconductors.

The second window layer is an electrical contact (about 10–20 nm and usually gold). Gold has been observed to form "islands" during the evaporation process and hence is not uniform in thickness. The third window layer consists of inactive *p*-type silicon extending 100–200 nm or less into the detector. This layer is also not uniform and is often called the "dead layer" or "silicon dead zone." The combined effect of these Au and Si layers is often overlooked. In reality, their effect is as great as or greater than that of any of the new "thin" window materials. It is clear from Fig. 5.41a that there really is no such thing as a "windowless" detector.

During the passage of x rays through all of these window layers, absorption occurs. It is important to recall that photoelectric absorption refers to a process in which x rays are diminished in number but do not lose energy; thus, the energies of the observed spectral lines are not altered while passing through the windows. In the case of 7.6 μm

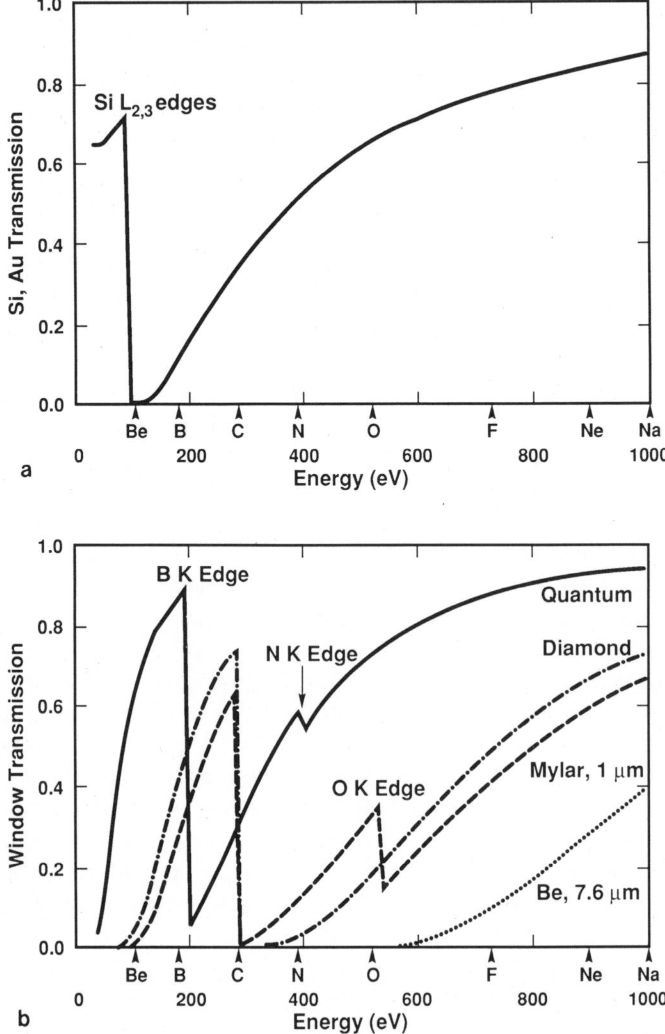

Figure 5.41. (a) Transmission curve for a "windowless" detector with a 0.01-μm Au contact and a 0.1-μm silicon dead layer. The energies of the light element K lines are shown along the energy axis. (b) Transmission curve for several commercially available window materials. The energies of the light-element K lines are shown along the energy axis. The actual transmission of x rays into the active part of the detector is a combination of the transmission characteristics shown in parts (a) and (b) of this figure. Quantum window = 0.25 μm; diamond window = 0.4 μm.

beryllium protective windows, nearly all x rays below about 600 eV are eliminated due to absorption effects. Above 2 keV, virtually all x rays are transmitted. Between these limits, the absorption increases with decreasing energy such that at 1.5 keV about 70% of the x rays are transmitted, while for an energy of 1 keV, the transmission is 45%. Absorption by the gold and silicon layers is much less significant because of the small mass-thickness of these layers. However, a noticeable change in the x-ray

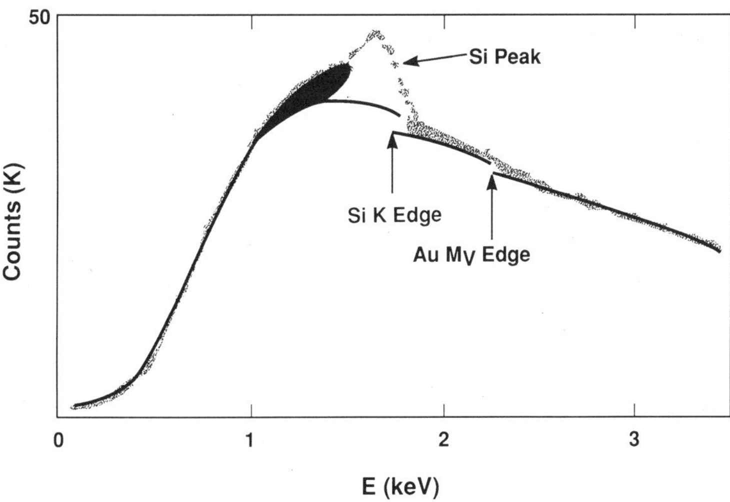

Figure 5.42. Electron-excited EDS spectrum from carbon. The silicon and gold absorption edges are illustrated. The solid line represents a theoretical fit to the continuum. $E_0 = 10\,\mathrm{keV}$. The silicon peak arises from internal fluorescence.

continuum is observed at the absorption edge of silicon and to a lesser degree at that of gold (Fig. 5.42). Just above the energy of the absorption edge, the mass absorption coefficient increases abruptly, resulting in a decrease in the measured continuum x radiation. The height of the resulting step is an indication of the thickness of the layer. Note that the action of the broadening effect of the detection process causes the absorption edge, which in reality is a sharp change of absorption over a range of about 1 eV, to be smeared over a much broader range, typically 100 eV for the silicon absorption edge.

5.3.6.5. Internal Fluorescence Peak of Silicon

The photoelectric absorption of x rays by the silicon dead layer results in the emission of Si *K* x rays from this layer into the active volume of the detector. These silicon x rays, which do not originate in the sample, appear in the spectrum as a small silicon peak, the so-called *silicon internal fluorescence* peak. An example of this effect is shown in the spectrum of pure carbon, Fig. 5.42, which also contains a significant silicon absorption edge. For many quantitative-analysis situations, this fluorescence peak corresponds to an apparent concentration of approximately 0.2 wt % or less silicon in the specimen.

5.3.7. Artifacts from the Detector Environment

The user of an EDS system is usually responsible for the installation of the spectrometer and its associated electronics on the electron-beam instrument. In order to obtain an optimum spectrum, the user may have to overcome a number of artifacts which result from interactions between the EDS system and its environment. These include microphony, ground loops, accumulation of contaminants, and the entry of spurious radiation, including electrons, into the detector. A suggested approach to the setup of a detector to avoid the effects described below is found in the Appendix to this chapter.

5.3.7.1. Microphony

The Si(Li) spectrometer contains a detector and electronic circuitry of extraordinary sensitivity which can respond to radiation of energies other than x rays. In particular, stray electromagnetic and acoustic radiation can affect the recorded x-ray spectrum. The coaxial cable through which the detector–preamplifier communicates with the main amplifier must be carefully routed to prevent it from becoming an antenna. The detector must be shielded against mechanical and acoustic vibration to which the detector acts as a sensitive microphone. The analyst can move a poorly routed cable to eliminate electromag-

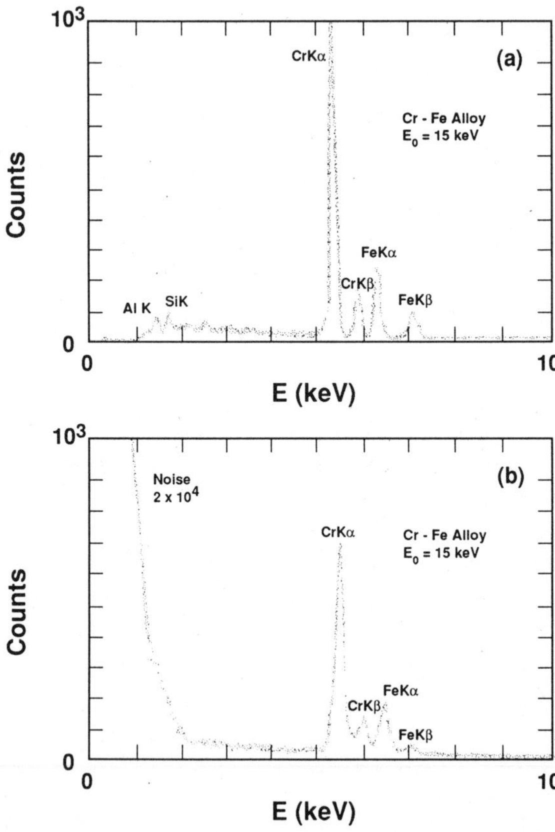

Figure 5.43. (a) Electron-excited spectrum of chromium–iron alloy. (b) Spectrum obtained under same beam conditions as (a), but with acoustic interference.

netic interference but is generally powerless to do anything about the mechanical isolation of the detector. Thus, it is important when evaluating a new detector prior to acceptance to check for microphonic and antenna effects. In Fig. 5.43a, a spectrum obtained under nonmicrophonic conditions contains characteristic peaks and a continuum spectrum with a typical shape showing cutoff at low energies due to absorption in the beryllium window. In Fig. 5.43b, the same spectrum was recorded with several sources of mechanical and acoustic vibrations in the vicinity of the detector—the operation of a wavelength-dispersive spectrometer motor, conversation, etc. The detector responded to these sources, producing an extremely high background in the region from 2 keV to 0 keV. The characteristic peaks are broadened in Fig. 5.43b as compared to Fig. 5.43a due to the noise. While virtually every detector has some microphonic response to high-intensity noise, the detector should be isolated well enough to be insensitive to normal laboratory environment noise and vibration. The response in Figure 5.34b is quite unacceptable. The main amplifier should also be carefully positioned. In general, it should be kept isolated from transformers and devices, such as computers, containing extensive logic circuits.

5.3.7.2. Ground Loops

One of the most insidious artifacts associated with the installation of a detector on an electron-beam instrument is the occurrence of ground loops. We might normally assume that the metal components of the microscope–spectrometer system are all at ground potential with no current flowing between them. In fact, small differences in potential, on the order of millivolts to volts, can exist between the components. These potential differences can cause currents to flow which range from microamperes to amperes. These extraneous currents are commonly referred to as "ground loops" or "ground currents" since they flow in components of the system that are nominally at ground potential, such as the chassis or the outer shields of coaxial cables. Since alternating-current ground loops produce electromagnetic radiation, such currents flowing in coaxial-cable shielding can modulate low-level signals passing through the central conductor. In EDS systems, the signals being processed are at extremely low levels, particularly in the region of the detector and preamplifier; hence, ground loops must be carefully avoided if signal integrity is to be maintained. The interference from a ground loop can manifest itself as degraded spectrometer resolution, peak shape distortion, background shape distortion, or dead-time correction malfunction.

It must be emphasized that the direct influence of the ground loop may not manifest itself in the spectrum as displayed in the multichannel analyzer, but there may still be deleterious effects on other important analytical functions, especially the dead-time correction. The EDS user should not presume that the dead-time correction

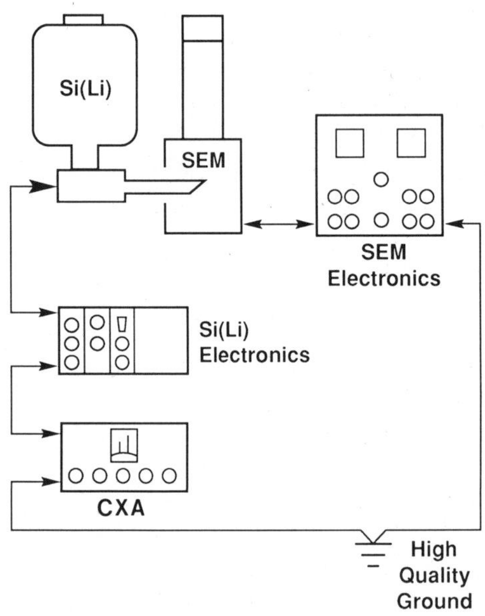

Figure 5.44. Schematic illustration of ideal ground paths in SEM/EDS system. The individual ground paths of the SEM and EDS should not be linked except at a high-quality ground.

circuitry will always be working correctly. After its initial installation and periodically thereafter, a check should be made of the accuracy of the dead-time correction, calibration, etc.

Ground loops are particularly insidious because of the many diverse ways in which they can affect the signal chain. Ground loops can enter the signal chain at any point between the detector and the computer x-ray analyzer. Moreover, a ground loop may occur intermittently. Because of this complexity, it is not possible to describe all of the manifestations of ground loops or provide a sufficiently general procedure to locate and cure them. In dealing with ground loops, prevention is preferable to seeking a cure. Proper attention to eliminating possible ground-loop paths during installation of the system usually minimizes difficulties. Figure 5.44 shows the major components of an EDS system and an electron-beam instrument. The grounding path connecting the components of each system should be created in a logical fashion avoiding cross-connections between microscope components and spectrometer components. An important ground-loop path to avoid is that between the cryostat housing and the microscope. The resistance between the cryostat assembly (disconnected from its cabling) and the microscope column should typically exceed $10^6\ \Omega$. Cross-connections can be inadvertently introduced when the EDS main amplifier or computer x-ray analyzer units share common racks with microscope components such as scan generators, video amplifiers, power supplies, etc. Ideally, it is best to keep the two systems electrically isolated through the use of separate racks. The ground path should be established separately in each system, terminating in a high-quality ground (Fig. 5.45). Note that a high-quality ground is not typically available at the wall power plug. In some

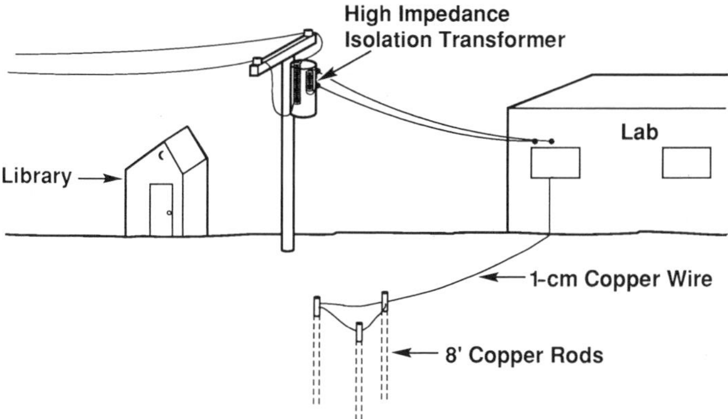

Figure 5.45. Schematic illustration of an ideal-case high-quality ground installation and power feed.

extreme cases it may be necessary to construct a high-quality ground consisting of a copper wire, 1 cm in diameter or greater, leading through the shortest distance possible to an external assembly consisting of three separate 3-m or longer copper rods driven vertically into the ground and reaching into the water table. The EDS user should note that modifications to establish a high-quality ground necessarily involve altering the electrical distribution network of the microscope–EDS system, and as a result, such modifications must be carried out under supervision of a qualified electrician.

5.3.7.3. Ice–Oil Accumulation

The accumulation of contamination in the EDS detector system during long-term operation can lead to degraded performance. Ice can accumulate in two locations:

1. Moisture condensing in the liquid-nitrogen cryostat can form small fragments of ice which "dance" within the cryostat as the liquid nitrogen boils. This vibration can be transmitted to the detector and the sensitive field-effect transistor. An accumulation of ice at the bottom of the cryostat reduces the thermal conduction between the liquid-nitrogen reservoir and the detector assembly, raising the detector temperature above the desired operating value. Some accumulation of ice in the cryostat is inevitable over a long period of time, but the rate of accumulation can be minimized if simple precautions are followed. The liquid-nitrogen supply for the cryostat is typically placed in a transfer Dewar from a sealed main tank. It is important that the liquid nitrogen in the transfer Dewar not be exposed to humid atmosphere since ice crystals would quickly accumulate and then be inadvertently poured into the cryostat. When an accumulation of ice does develop in the cryostat, it is possible to remove the 500-V bias

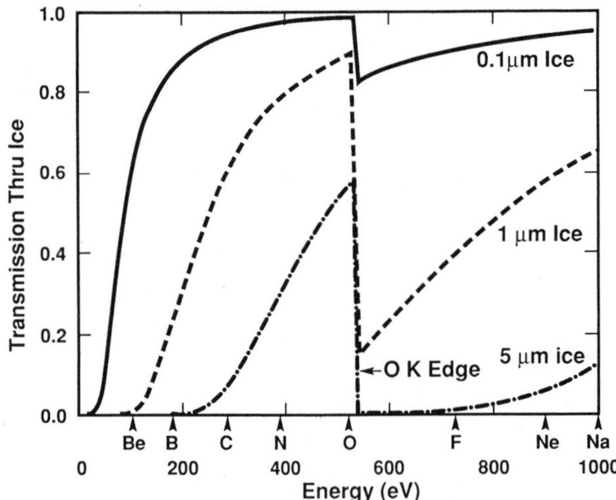

Figure 5.46. Transmission through three thicknesses of ice, which can easily form on the front surface of many Si (Li) detectors.

from the detector and allow it to warm to room temperature. The liquid water which collects in the bottom of the Dewar can be poured, or aspirated, out. Note: Such a recovery procedure should follow the guidelines provided by the manufacturer of the EDS system.

2. Ice can accumulate on the detector surfaces if the vacuum integrity of the assembly is compromised by a pinhole leak, which might be located in the beryllium window. Also, in a windowless or thin-window detector system, the detector acts as a cold finger to condense residual water from the sample chamber or water left in the detector at the time of manufacture. The consequence of having ice on the detector is decreased surface resistance, which introduces resistor noise leading to degraded resolution. The transmission of x rays through the ice is also severely affected, as shown in Fig. 5.46. Many recent detectors have a special heater which sublimes away the ice. It is often necessary to run this heater every few weeks or so to maintain good light-element performance.

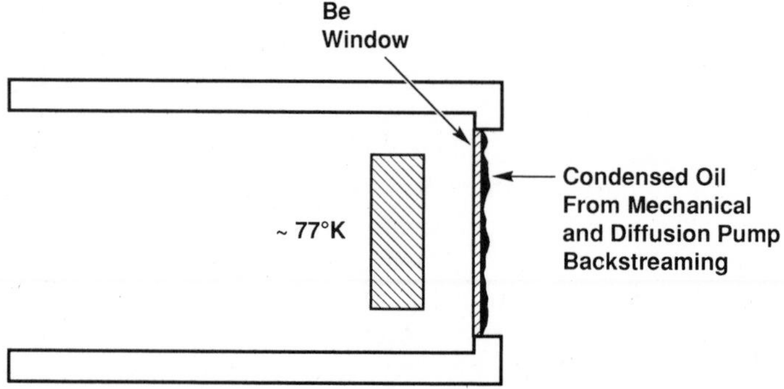

Figure 5.47. Oil accumulation from vacuum system on the window of the EDS system.

The beryllium window is usually several degrees below ambient temperature because of its proximity to the cooled detector chip. As a result, in microscopes with poor vacuum, residual oil and water vapor in the vacuum of the specimen chamber can condense on the window (Fig. 5.47), leading to increased x-ray absorption and loss of sensitivity at low x-ray energies. In some cases, oil can be removed from the window in the field, but only with extreme care. The manufacturer should be contacted for details.

5.3.7.4. Sensitivity to Stray Radiation

Origin. One of the features of the Si(Li) detector which is normally considered a great advantage is its relatively large solid angle of collection as compared to the focusing wavelength-dispersive spectrometer. The solid angle of collection is usually considered from the point of view of the electron-excited source in the specimen, with the apex of the cone placed at the beam-impact point (Fig. 5.48). To appreciate the complete collection situation, however, we must also consider the solid angle of collection from the point of view of the detector, also shown in Fig. 5.48. It is obvious from Fig. 5.48 that the true solid angle of collection can be very large indeed, including not only the entire specimen but often a large portion of the sample stage and chamber walls. Even if a more restrictive collimator is used than that shown in the figure, the area of the sample seen by the detector is still typically several square millimeters or more. The difference in the collection angle between the points of view represented in Fig. 5.48 would be immaterial if the excitation were really confined to the volume directly excited by the focused electron beam. Unfortunately, excitation can occur at a considerable distance from the region of

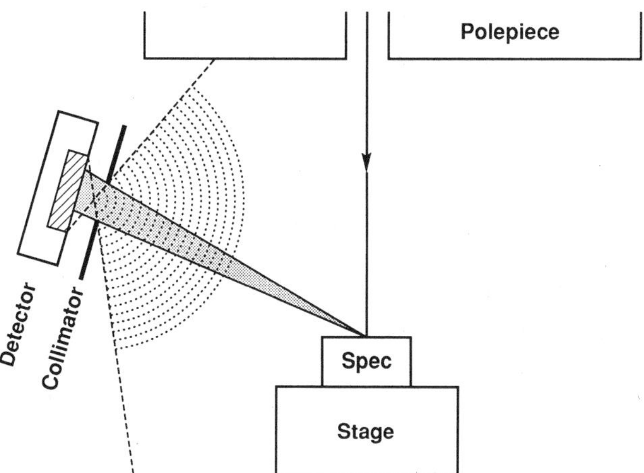

Figure 5.48. Solid angle of x-ray collection from the point of view of the specimen (dark shading) and the detector (light shading).

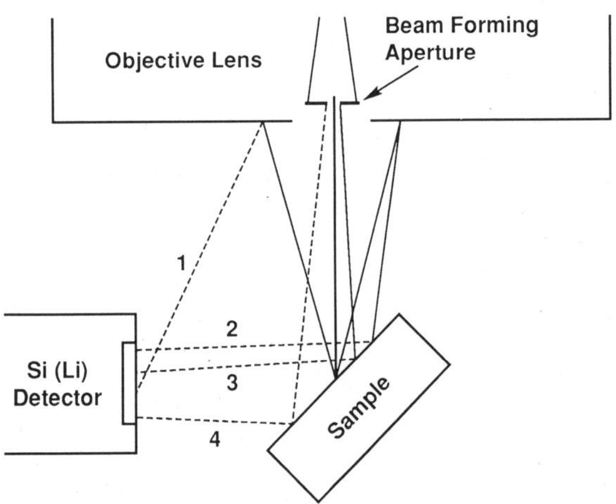

Figure 5.49. Possible sources of remote x-ray excitation in an electron-beam–EDS system. Solid lines, electron paths; dashed lines, x-ray paths: (1) Excitation of polepiece; (2) Remote excitation of sample by backscattered electron; (3) Remote excitation of sample by electron scattered by aperture; (4) Remote excitation of sample by x rays produced at aperture.

impact of the focused beam. A schematic diagram of some typical sources of this remote excitation is shown in Fig. 5.49. Electron-induced remote sources include scattering from apertures, backscattering from the specimen, and rescattering from the pole piece. In this regard it should be noted that a significant fraction of the backscattered electrons from heavy elements retain 50% or more of the incident energy and are thus capable of exciting x-rays from the specimen environment (walls, stage, polepiece). Interaction with several surfaces is possible before an electron comes to rest. X-ray-induced remote sources originate principally from characteristic and continuum x-rays generated by electrons that strike the upper surface of the final beam-defining aperture. These x rays can propagate through the aperture and illuminate a large portion of the sample chamber as an x-ray fluorescence source. The significance of this source of remote excitation depends on the material of the aperture, its thickness, and the beam energy. Thin-film "self-cleaning" apertures of molybdenum produce a strong source of remote excitation, since self-absorption of the Mo $K\alpha$ radiation (critical excitation energy, 20 keV) is low and the thin film (12.5 μm) transmits about 78% of the radiation. Considering that this final aperture may intercept 80% of the total beam, it is obvious that the x-ray fluorescence effect can be extremely deleterious to accurate analysis (Bolon and McConnell, 1976).

Recognition. It is not always obvious that a remote source exists, but its effects can usually be recognized by employing the following procedure: A Faraday cup is fabricated from a block of metal (iron,

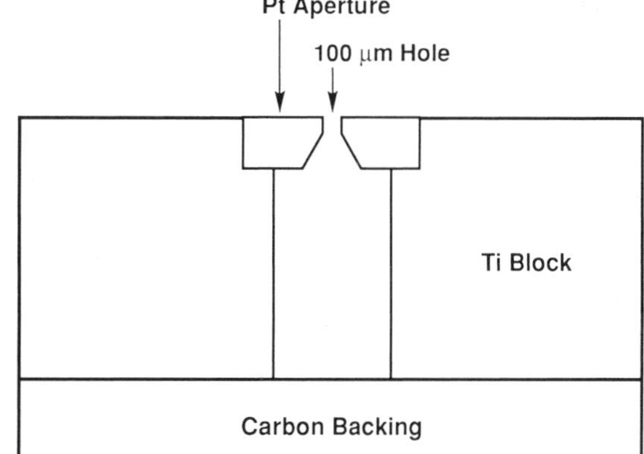

Figure 5.50. Schematic illustration of a cross-section through a Faraday cup suitable for detecting remote sources of x-ray excitation in an electron-beam instrument.

brass, titanium, etc.) e.g., by drilling a blind hole of 3 mm diameter and a few millimeters in depth and press-fitting or carbon-dagging a microscope aperture (20 to 100 μm in diameter) into this hole (Fig. 5.50). The aperture material should be different from that of the block. Spectra are then recorded with the focused beam alternately in the hole, on the aperture material, and on the block. The results of such a test of the system are shown in Fig. 5.51. If no remote sources exist, then with the beam falling into the aperture no spectrum should be obtained. If this "in-hole" spectrum contains x-ray signals of the Faraday cup materials, then the ratios of the intensities of the characteristic lines to the values obtained from the spectra of the aperture and block recorded for the same dose (current × live time) are an indication of the magnitude of the problem. For example, a

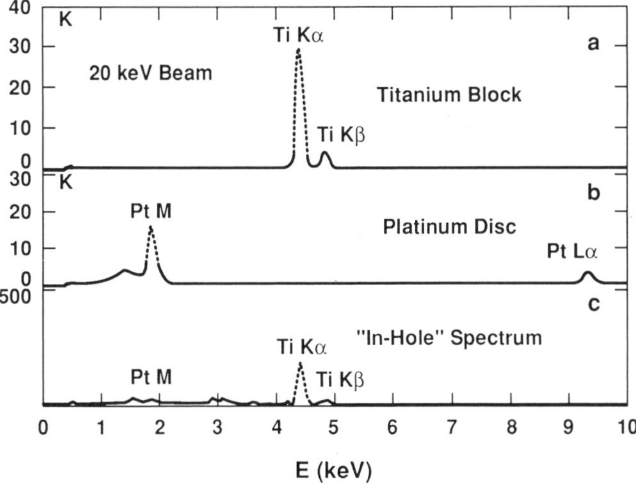

Figure 5.51. Spectra obtained in an electron probe microanalyzer from the components of a Faraday cup: (a) titanium block directly excited by electrons; (b) platinum aperture; (c) "in-hole" spectrum.

strong signal for the aperture material in the in-hole spectrum indicates a source of radiation in the vicinity of about 1 mm of the beam, while a high signal for the block material indicates a more distant source. The extraneous radiation has been observed to be as high as 20% and is typically between 0.01% and 1% (Bolon and McConnell, 1976) of that obtained with the beam directly on the aperture. Examination of the peak-to-background ratio of characteristic peaks observed in the in-hole spectrum can also indicate the type of remote excitation, electron or x-ray. Excitation by x rays produces a much lower continuum. If the peak-to-background ratio is higher in the in-hole spectrum than in the directly excited spectrum of the aperture or block containing the element, the remote excitation is most likely by x-rays.

Observation of characteristic lines of the chamber wall or stage materials indicates excitation at great distances from the beam impact on the specimen. In this respect, it should be noted that a flat, nontilted specimen and a specimen of rough surface may behave differently. A rough specimen is more likely to scatter electrons in all directions to the surroundings, whereas the backscattered electrons from a flat specimen normal to the beam follow a cosine distribution peaked about the normal, with little scattering in the horizontal plane.

Correction of Stray-Radiation Problem. Eliminating stray radiation following its recognition is very much dependent on the particular instrumental configuration. Only general guidelines can be given here.

X-ray fluorescence problems originating in the final aperture can be minimized by

1. use of thick apertures and
2. choosing aperture materials of higher atomic number, such as platinum or tantalum.

Typical beam energies (30 keV or lower) are insufficient to generate K x rays of these materials, and the L lines of both materials have energies of less than 10 keV. For apertures of 12.5 μm thickness, the transmission of $L\alpha$ x-rays of the aperture material is less than 3% for Pt and 4% for Ta. Transmission of the 20 keV continuum is 6% for Pt and 15% for Ta.

Electron-scattering problems are more difficult to deal with. Apertures should be kept clean, because debris particles on the circumference can produce unwanted scattering. Electrons often scatter in the column and pass around apertures if passages exist. Aperture alignment should be optimized. Double apertures can be employed with some success, although their alignment is difficult.

Scattering from the specimen is difficult to control, especially if the sample is rough, such as a fracture surface. Adjacent stage, polepiece, and chamber-wall surfaces can be coated with carbon or beryllium sheets to prevent generation of characteristic x rays in the energy region of interest by the scattered electrons. After all obvious sources of remote excitation have been minimized, a remnant in-hole spectrum may still exist. This in-hole spectrum can be subtracted from that of an unknown, but the procedure is risky, since the background spectrum may depend on the exact circumstances of scattering from the specimen and the surroundings of the specimen and the standard.

Direct Entrance of Electrons into the Detector. The Si(Li) detector is capable of responding to an energetic electron which enters the active region of the detector. A pulse is developed, the height of which is a measure of the energy of the electron. When the beam electrons strike the sample, a significant fraction, about 30% for copper with a beam normally incident, are backscattered with a wide energy range (see Chapter 3). Many of these backscattered electrons retain a substantial fraction of the incident energy. It is inevitable that some of those electrons are scattered in the direction of the detector. The beryllium window with a typical thickness of 7.6 μm (0.3 mil) is capable of stopping electrons with an energy below about 25 keV. Above energies of 25 keV, electrons begin to penetrate the window and activate the detector, although with a loss of energy due to inelastic scattering in the beryllium. When higher-energy (>25 keV) beams are employed, the electrons which enter the detector can have a substantial effect on the background. An example is shown in Fig. 5.45 for a 40-keV beam incident on arsenic. The background below 20 keV is greatly distorted by the electron contribution added to the expected x-ray continuum shown by the solid curve. Above 25 keV, the background is mostly due to the normal x-ray continuum; virtually no electrons are able to penetrate the beryllium window and retain energies in this range. Note that the analyst usually examines the region between 0 and 10 keV or 0 and 20 keV. In the example of Fig. 5.52, it may not be obvious that the background is anomalous unless the entire spectrum is examined.

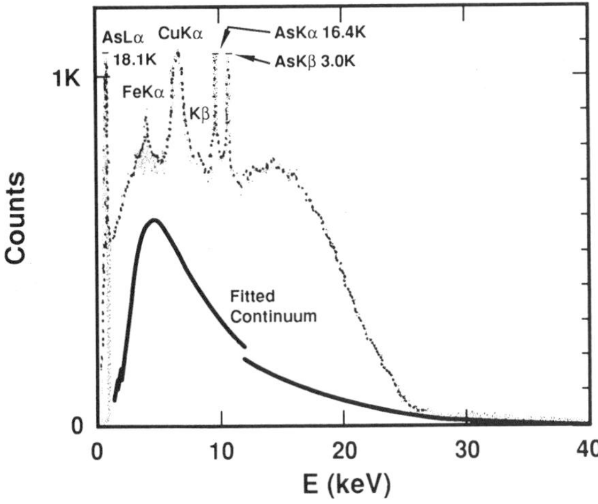

Figure 5.52. EDS spectrum of arsenic excited by a 40-keV electron beam showing an anomalous background below 25 keV due to direct entry of electrons scattered from the specimen into the detector. The x-ray continuum was fitted from the high-energy end of the spectrum.

Artifacts arising from scattered electrons entering the detector can be eliminated with magnetic shielding in front of the detector entrance. In the windowless or thin-window variety of Si(Li) spectrometers, such shielding is an absolute necessity. The artifact may become more pronounced for samples with a high atomic number or samples with surfaces tilted toward the detector. These two conditions will produce the greatest number of high-energy backscattered electrons. Light-element targets which are flat produce a relatively minor effect. Operation with beam energies below 20 keV will also minimize the effect.

5.3.8. Summary of EDS Operation and Artifacts

The process by which the characteristic lines and continuum distribution are modified in going from the generated spectrum in a Mn sample to the observed spectrum in the EDS are summarized in Fig. 5.53. Point a and the associated spectrum corresponds to what the distribution of characteristic lines and continuum might look like if the measurement could be made by an ideal detector located within the sample. Point b shows the spectral distribution as it would exist in the emitted x rays. Note that the principal change is due to absorption, which sharply truncates the low-energy portion of the continuum, introduces a discontinuity associated with continuum absorption at the Mn K absorption edge, and slightly diminishes the characteristic peak intensities. At point c the signal has passed through the Be window, Au surface contact, and silicon dead layers, causing a further reduction in intensity at low energies and introducing some fine structure caused by absorption and fluorescence effects. Finally, at point d the signal has been processed by the Si(Li) diode and

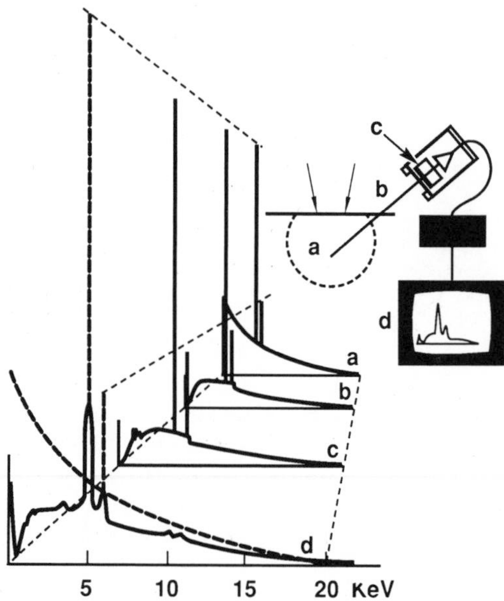

Figure 5.53. Schematic drawing of the relationship between the measured and generated spectral distributions of manganese. (Courtesy of R. Bolon.)

associated electronics. The processing results in a smearing effect which distributes most of the information corresponding to each sharply defined energy increment in the CXA encompassing the FWHM. The effect is to noticeably broaden peaks and absorption edges well beyond their natural width, which leads to reduced peak-to-background ratios and the high probability of peak overlap. In addition, it is at this point that the electronic artifacts of escape peaks and peak-on-peak pileup make their appearance. Although the EDS does introduce a large number of modifications to the generated spectrum, the resultant signal is very useful. A comparison of the EDS and WDS spectrometers is given in the next section.

5.4. Comparison of WDS and EDS

An ideal x-ray detector would be small, inexpensive, and easy to operate, collect most of the x rays emitted from a sample, have a resolution of the natural x-ray line width being measured (a few electron volts), and be capable of collecting spectral data rapidly without losing information. Neither wavelength-dispersive spectrometers nor Si(Li) detectors individually have all of these characteristics, but when used together the two techniques do, in fact, complement each other. Table 5.2 summarizes a comparison of the major features of both modes of detection. An item-by-item analysis of Table 5.2 follows:

5.4.1. Geometrical Collection Efficiency

Geometrical collection efficiency refers to the solid angle of spectrometer acceptance $[(\Omega/4\pi)100\%]$. As was illustrated in Fig. 5.3, the angle

Table 5.2. Comparison between Types of X-Ray Spectrometers

Operating characteristic	WDS Crystal diffraction	EDS Silicon, energy dispersive
Geometrical collection efficiency	Variable, <0.2%	<2%
Overall quantum efficiency	Variable, <30%	≈100% for 2–16 keV
	Detects $Z \geq 4$	Detects $Z \geq 10$ (Be window)
		Detects $Z \geq 4$ (windowless or thin-window)
Resolution	Crystal dependent (5 eV)	Energy dependent (140 eV at 5.9 keV)
Instantaneous acceptance range	≈The spectrometer resolution	The entire useful energy range
Maximum count rate	50,000 cps on an x-ray line	Resolution-dependent, <2000 cps over full spectrum for best resolution
Minimum useful probe size	≈2000 Å	≈50 Å
Typical data-collection time	Tens of minutes	Minutes
Spectral artifacts	Rare	Major ones include: escape peaks, pulse pileup, electron-beam scattering, peak overlap, and window absorption effects

subtended in the plane of the focusing circle of a WDS does not change with λ. However, orthogonal divergence perpendicular to that plane leads to a decreased collection efficiency with increasing λ for a given crystal. In the EDS case the higher geometric collection efficiency is a result of the greater ease in positioning the detector close to the sample (often less than a centimeter). Furthermore, although the solid angle can be varied for retractable detectors (see Fig. 5.16a) the nature of the detection process does not require that the detector be physically moved to encompass its entire energy range as is the case for a WDS.

5.4.2. Quantum Efficiency

Overall quantum efficiency is a measure of the percentage of the x rays entering the spectrometer which are counted. At low beam currents, EDS systems generally have higher count-rate capability than WDS systems per unit beam current, due partially to a higher geometric collection efficiency and partially to higher inherent detector quantum efficiency. With a detector using a Be window, close to 100% of the x rays in the 2.5 and 15 keV energy range striking the detector are collected. At higher energies, a certain percentage of x-ray photons are transmitted through the silicon crystal, while at low energies a certain percentage of photons are absorbed in the window. Significant absorption of soft x-rays can also occur in the surface dead layer or gold contact layer on the detector crystal. Absorption and detector noise generally limit meaningful light-element analysis to $Z \geq 4$.

By comparison, the quantum counting efficiency of crystal spectrometers is generally less than 30%, partially due to transmission losses in the proportional counter tube (see Fig. 5.6) and partially due to losses in the diffraction crystal. Large-d-spacing crystals and thin-window proportional counters make it possible to detect elements down to beryllium. Consideration of the overall quantum counting efficiency leads to the conclusion that the useful energy range for WDS systems is usually about 0.1–15 keV while for EDS (windowless or thin-window) it more typically spans 0.28 to 20 keV.

5.4.3. Resolution

Resolution for both WDS and EDS systems is usually measured as the FWHM. As already described in Eq. (5.6) and illustrated in Fig. 5.35a, EDS resolution is energy dependent in a predictable way, with values normally determined for Mn $K\alpha$ as measured with a ^{55}Fe source. Even at 140 eV it is still about 30 times poorer than that obtainable with WDS using a good quartz or lithium fluoride crystal. The principal effect of reduced resolution is to lower the peak-to-background ratio P/B at a given energy and hence the sensitivity or minimum detectability limit of a given element (see Chapter 9 for a discussion of trace-element detection). The reduction in P/B occurs because it is necessary to sample a wider energy interval containing more background counts to obtain a major

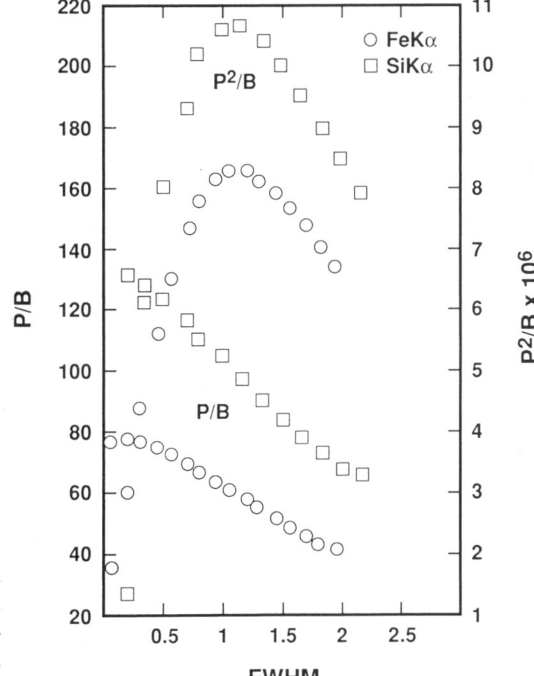

Figure 5.54. Integration limits and their effects on P/B and P^2/B for Fe and Si measured with an EDS (Geller, 1977).

fraction of the peak counts *P*. Figure 5.54, taken from the work of Geller (1977), illustrates this point for the case of a 160-eV detector used to obtain *P* and *P*/*B* for Fe $K\alpha$ and Si $K\beta$ on pure iron and silicon, respectively. The term P^2/B is often used for comparison of sensitivities for the detection of trace amounts of a particular element. The value of P^2/B goes through a maximum when the energy band chosen for peak integration is approximately equal to the FWHM. Note that the peak-to-background ratio must be obtained by integrating both the peak and background counts in the region of interest and taking the ratio of the sums. This quantity is known as the integrated peak-to-background ratio. The situation is somewhat different in evaluating the performance of a WDS system since optimum performance is obtained by tuning the spectrometer to a diffraction angle corresonding to the maximum peak intensity and then detuning it to obtain corresponding background levels.

Integrating WDS peaks by scanning through them is basically an exercise in seeing the system's response to a varying of the spectrometer efficiency. Such scanning in the vicinity of a peak may have to be used, however, as a means of determining the position of a maximum for those quantitative measurements in which the spectrometer is detuned between sample and standard readings and in analysis of the light elements where the characteristic peaks are broadened (See Section 9.10). Integrating WDS peaks does not add to the precision but is a means of avoiding the introduction of systematic errors.

The superior energy resolution of the crystal spectrometer results in significantly higher peak-to-background ratios and better spectral dispersion, thereby minimizing the possibility of peak overlap. This can be readily seen by comparing spectra from the same superalloy standard obtained using a crystal spectrometer (Figs 5.5a and b) with that obtained from a state-of-the-art (130 eV) Si(Li) detector (Fig. 5.32). Figure 5.15a shows clearly distinguished Ta $L\alpha$, Ni $K\beta$, and W $L\alpha$ lines, while these same lines are hardly discernible in Fig. 5.32. Similarly, the Ta, W, and Re $M\alpha$ lines are easily separated using a TAP crystal but remain unresolved with the Si(Li) detector. The inferior resolution of the solid-state detector often makes it necessary to establish the presence of a series of spectral lines for a given element when the identification of a particular peak is ambiguous or the peak of a suspected line is obscured by a line of another element. In such cases it is common practice to use the line markers or collect one or more pure elemental spectra and display them simultaneously with that of the unknown in order to make a direct comparison.

5.4.4. Spectral Acceptance Range

The term "instantaneous acceptance range" refers to that portion of the spectrum that can be measured at any instant of time. For the WDS, we have a serial detection system where only those pulses very close to the selected Bragg angle can be measured, while all others are essentially ignored. The EDS, on the other hand, is a parallel detection system with a large acceptance range.

5.4.5. Maximum Count Rate

As already shown in Fig. 5.29, the maximum useful count rate for systems operating at optimum resolution is about 2000 to 3000 c s^{-1} over the entire energy range of the excited x rays (0 to E_0). If the composition of a minor component is sought, this number may drop well below a hundred counts per second for the element of interest and therefore necessitate long counting times to achieve needed levels of precision. In the WDS, with the spectrometer tuned to a specific element, count rates in excess of 50,000 s^{-1} at the peak energy are possible without loss of energy resolution.

5.4.6. Minimum Probe Size

The previous point leads directly to a discussion of the minimum useful probe size for x-ray analysis. As described in detail in Chapter 2, for each type of source and voltage there exists a maximum probe current which can be associated with any given probe size. For conventional tungsten sources, the probe current varies with the 8/3 power of the beam diameter, with typical values at 20 kV ranging from 10^{-10} A for 20 nm (200 Å), 10^{-8} A for 100 nm (1000 Å), and 10^{-6} A for 1000 nm (10,000 Å). For an EDS system utilizing a 4-mm-diameter detector placed 1 cm from a sample of pure nickel, a count rate of about $10^4\ s^{-1}$ is obtained for a 35° takeoff angle, a 20-nm (10^{-10} Å) probe, and 100% quantum efficiency. According to Fig. 5.29 the count rate of 10^4 would be too high to be consistent with maximum energy resolution, so the operator would have to retract the detector, decrease the EDS time constant, or decrease the probe current by going to a smaller spot. On the other hand the corresponding WDS count rate would probably be more like $100\ s^{-1}$, which would be too low for practical use. For bulk samples (more than a few micrometers thick), spatial resolution for chemical analysis does not improve for probes much less than 0.5 μm in diameter, since the volume of x-ray production is determined by electron-beam scattering and penetration rather than by the probe size. This point is dramatically demonstrated in Fig. 5.55, which shows a series of Monte Carlo simulations of electron-beam scattering and x-ray production for a 0.2-μm-diameter probe with a hypothetical 1-μm TaC inclusion in a Ni–Cr matrix. It can readily be seen that the electron trajectories and consequently the region of x-ray production, particularly at the high voltage, can easily exceed 1 μm, or five times the beam diameter. Since probes of under a few hundred nanometers are of limited value in the examination of such samples, a more complete analysis can be accomplished by boosting the probe current to 10 nA and using the WDS systems, where the benefit of high count-rate capability combined with high energy resolution can be realized. The situation is quite different, however, in the examination of thin foils and biological sections where spatial resolution equal to or even smaller than the foil thickness is possible. In this case, the low x-ray yields necessitate having a detector

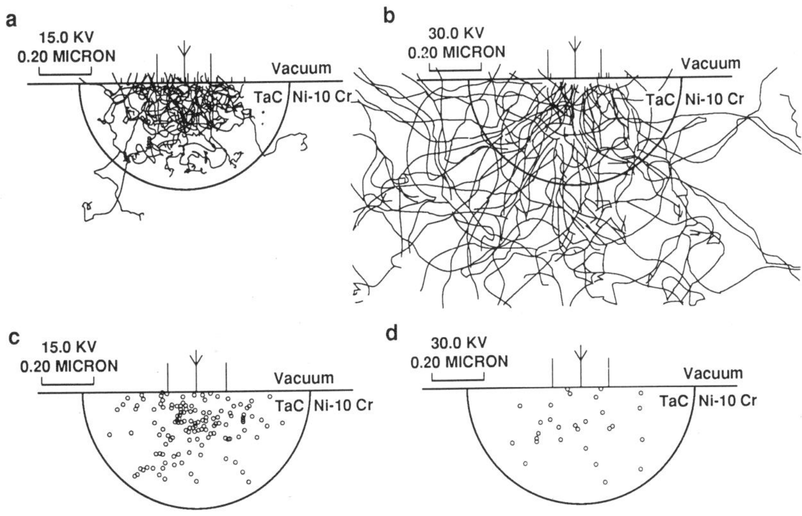

Figure 5.55. Monte Carlo simulated interaction of a 0.2-μm-diam beam with a hypothetical 1-μm-diameter hemispherical TaC inclusion in a NiCr matrix. (a) Electron trajectories, 15 keV, (b) Electron trajectories, 30 keV, (c) Ta $M\alpha$ x rays at 15 keV, (d) Ta $M\alpha$ x rays at 30 keV.

with both the high geometrical collection and overall quantum efficiency of the EDS system. It is for this reason that the EDS has been so successful when coupled to both scanning and analytical electron microscopes.

5.4.7. Speed of Analysis

From a practical point of view, one of the best features of the EDS system is the speed with which data can be collected and interpreted. The continuous acceptance of a large energy range has distinct advantages for qualitative analysis, which partially offset some of the disadvantages previously described. When a WDS system is mechanically scanned, it dwells at each resolvable wavelength for only a short part of the total scan duration. Therefore, while looking at one element or even a section of the background, it is throwing away information about all of the other elements. Unless the WDS is specifically programmed to go to peak positions, anywhere from only 1/100 to 1/1000 of the total data-collection time may be associated with the measurement of each individual peak. In the EDS case a 100-s counting time coupled with a count rate of 2000 s^{-1} leads to a spectrum containing 200,000 counts. Even if half of those counts are background, most of the measurable constituents present in amounts greater than a few tenths of a percent will probably be detected. Furthermore, use of the line markers and other interpretive aids can result in a qualitative analysis in a matter of minutes. In the WDS case, several crystals covering different wavelength ranges would be required, with typical data collection and interpretation times being 5–30 min.

5.4.8. Spectral Artifacts

The WDS system is relatively free of spectral artifacts of the type that cause peak position shifts or incorrect elemental assignments. The EDS, on the other hand, is subject to a number of difficulties which can lead the unsuspecting analyst into trouble. Artifacts arise at each stage of the spectral measurement process. Detection artifacts include peak broadening, peak distortion, silicon escape peaks, absorption, and the silicon internal fluorescence peak. Pulse-processing artifacts include pulse pileup, sum peaks, and sensitivity to errors in dead-time correction. Additional artifacts arise from the EDS–microscope environment, including microphonics, ground loops, and oil and ice contamination of the detector components. Both WDS and EDS can be affected by stray radiation (x rays and electrons) in the sample environment, but because of its much larger solid angle of collection, the EDS is much more readily affected by stray radiation. However, because of the large collection angle, the EDS is less sensitive to spectrometer defocusing effects with sample position.

In summary, this comparison suggests that the strengths of the EDS and WDS systems compensate for the weaknesses of each. The two types of spectrometers are thus seen to be complementary rather than competitive. At the present stage of development of x-ray microanalysis, it is clear that the optimum spectrometry system for analysis with maximum capabilities is a combination of the EDS and WDS systems. As a result of the ongoing revolution in the development of laboratory computers, automated systems which efficiently combine WDS and EDS spectrometers are available from several manufacturers.

Appendix: Initial Detector Setup and Testing

The wide variety of Si(Li) spectrometer–CXA systems precludes a specific description for proper setup and testing which is applicable to all instruments. The manufacturer provides specific instructions on the proper installation and adjustment of the instrument. It is usually left to the analyst in the field to make the actual installation. In this section, some general guidelines and suggestions to supplement the manufacturers' procedures are given in order to highlight the critical areas in the operation of these systems.

1. Before the Si(Li) detector is installed on the instrument, it is very useful to test the system separately by activating the detector with a radioactive source, preferably ^{55}Fe. The ^{55}Fe source emits Mn $K\alpha$ and Mn$K\beta$ x-radiation with negligible continuum. The x-ray spectrum of this source should be recorded with the detector placed on a vibration-free surface such as plastic foam at least 5 cm from the source. The amplifier should be electrically isolated from other electronics. The manufacturer's instructions for setup for optimum resolution should be followed. The total spectrum count rate should be 1000 s^{-1} or less. A total of 100,000 or more counts should be collected. It is best if this spectrum can be retained in digital form by storage on magnetic media. If digital recording is unavailable, the spectrum should be written on graph paper in several scale expansions centered about Mn $K\alpha$. The merits of this source spectrum are the following:

a. The source spectrum serves as a permanent reference to the as-received condition of the system. In the event of a question arising in the future about the quality of the system, a benchmark is available.
b. The resolution in the as-delivered condition can be measured at the usual value, the FWHM for Mn $K\alpha$.
c. To measure the degree of incomplete charge collection, the peak-to-tail ratio and the asymmetry of the Mn $K\alpha$ peak should be determined. The peak-to-tail ratio is typically measured by taking the ratio of the counts in the peak channel of Mn $K\alpha$ to the counts in the background channel at half the energy of Mn $K\alpha$. Because of the low background counts, an average over at least 10 channels should be used. For a typical detector, this ratio should exceed 1000:1. The asymmetry of the Mn $K\alpha$ peak is typically determined by measuring the full width of the peak at one-tenth the maximum amplitude (FWTM). The FWTM should be no worse than 1.9 FWHM.

2. After installing the spectrometer on the SEM or EPMA, a spectrum should be obtained from a manganese target excited with a 15- to 20-keV electron beam at the same count rate and total Mn $K\alpha$ counts as the reference spectrum. The resolution measured on this spectrum should not be degraded by more than a few electron volts over the reference spectrum. If the resolution has deteriorated significantly and/or large numbers of counts are observed in the low-energy region, e.g., Fig. 5.43b, there are several possible sources of trouble: microphony, lack of ground isolation between circuit components (ground loops), and stray radiation from transformers, computers, etc. Remedies for these problems depend on the local situation. Ground loops can usually be eliminated by connecting all components of the system to a single high-quality ground and not interconnecting them.

3. An electron-excited spectrum should be obtained from spectrographically pure carbon with a scanning beam to minimize possible contamination. This specimen should provide a continuum spectrum devoid of characteristic peaks. Such a spectrum from a typical detector, Fig. 5.42, has the following characteristics:

a. The intensity of the lowest-energy channels should nearly reach the base line, within about 3% of the intensity maximum in the continuum hump. If not, this failure may be indicative of a poorly set slow channel discriminator, microphony, stray radiation, or ground loops.
b. A silicon K absorption edge and a silicon internal fluorescence peak are always observed, as shown in the spectrum obtained with a normal detector, Fig. 5.42. In the abnormal detector, Fig. 5.56, the silicon absorption edge and the silicon fluorescence peak are much more pronounced and the magnitude of the absorption edge is a function of the bias voltage on the silicon detector.

4. At this point, the analyst should examine the entire carbon continuum spectrum up to the beam energy for other artifacts. Specifically, the presence of any characteristic peaks, e.g., Ti $K\alpha$ in Fig. 5.56, is an indication of stray radiation and a clue to its source.

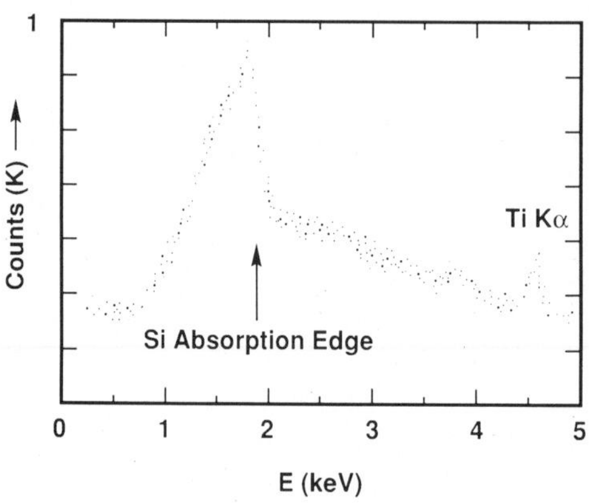

Figure 5.56. Electron-excited spectrum of carbon from an unsatisfactory detector with a thick silicon dead layer producing a silicon absorption edge of a very significant size.

5. Spectra should be obtained from elements which give characteristic lines througout the energy range of interest to check the energy linearity of the system.

6. The performance of the pulse-pileup rejector should be examined as a function of x-ray energy and count rate. It can be expected that the pileup rejector should work well for radiation as energetic as silicon $K\alpha$ and greater. Below silicon $K\alpha$, the pileup rejector frequently becomes progressively less satisfactory. Spectra illustrating the pileup performance from an optimally adjusted system (maximum resolution and maximum allowed count rate), Fig. 5.28a and b, show adequate pulse-pileup rejection for silicon but an almost total failure for magnesium. A pileup continuum below the double energy peak and even a triple energy peak are observed in the magnesium spectrum. In the silicon spectrum, only the double energy peak, which corresponds to coincidence within the time resolution of the pulse rejector, is observed. As a figure of merit, the area of the silicon $K\alpha$ double energy peak should be less than 1/200 of the parent peak at allowed system count rates.

7. For accurate quantitative analysis, the performance of the dead time correction circuit should be tested. The x-ray production in the specimen at any beam energy is proportional to the beam current striking the sample. This fact provides a way to vary the input count rate to the spectrometer system in a controlled fashion. A flat, pure-element target such as iron should be scanned with a beam energy of 15–20 keV. The current should be set initially to give a total spectrum count rate of about 500. The beam current is measured in a Faraday cup. The integrated counts across the peak (peak plus background) are then plotted as a function of beam current for a fixed live time (e.g., 100 s). The plot of counts versus current should be linear over that count rate where the dead-time correction mechanism is functioning properly.

6

Qualitative X-Ray Analysis

6.1. Introduction

The first stage in the analysis of an unknown is the identification of the elements present, i.e., the qualitative analysis. Qualitative x-ray analysis is often regarded as straightforward, meriting little attention. The reader will find far more references to quantitative analysis than to qualitative analysis, which has been relatively neglected in the literature, with a few exceptions (e.g., Fiori and Newbury, 1978). It is clear that the accuracy of the final quantitative analysis is meaningless if the elemental constituents of a sample have been misidentified. As a general observation, the major constituents of a sample can usually be identified with a high degree of confidence, but when minor or trace level elements are considered, errors can arise unless careful attention is paid to the problems of spectral interferences, artifacts, and the multiplicity of spectral lines observed for each element. Because of the differences in approach to qualitative EDS and WDS analysis, these techniques will be treated separately.

The terms "major", "minor", and "trace" as applied to the constituents of a sample in this discussion are not strictly defined and are therefore somewhat subjective. Each analytical method defines "trace" differently. Because the EDS limit of detection in bulk materials is about 0.1 weight %, the following arbitrary working definitions will be used: major 10 wt % or more; minor, 1–10 wt %; trace, less than 1 wt %. Generally, WDS can achieve limits of detection of 100 ppm in favorable cases, with 10 ppm in ideal situations where there are no peak interferences and negligible matrix absorption (See Section 9.9).

In performing qualitative x-ray analysis, we have to make use of several different types of information. Foremost is the specific energy of the characteristic x-ray peaks for each element. This information is available in the form of tabulations, which may be given as an "energy slide rule" (provided by several manufacturers of EDS systems), a graph, or, in the case of a computer-based x-ray spectrum analyzer (CXA), as a visual display marker ("KLM markers"). In employing these aids, the analyst should be aware of some of their potential shortcomings. The energy slide rules and even some computer "KLM" markers may not list all x-ray peaks observed in practice. Frequently, the Ll, $L\eta$, $M\zeta$, and $M_{II}N_{IV}$ lines are neglected, but these lines are frequently detected in practice. Without knowledge of their existence, errors may be made in ascribing these lines to other elements which are not actually present. A comprehensive table of x-ray lines, such as Bearden's compilation (Bearden, 1967a, b), is the ultimate reference, but is often too detailed. For example, this compilation lists 25 L-family lines. Even when the integrated spectral count is as high as 5,000,000, only nine of these will usually be observed in high-quality EDS spectra, because of the low relative intensity of the other 16 or

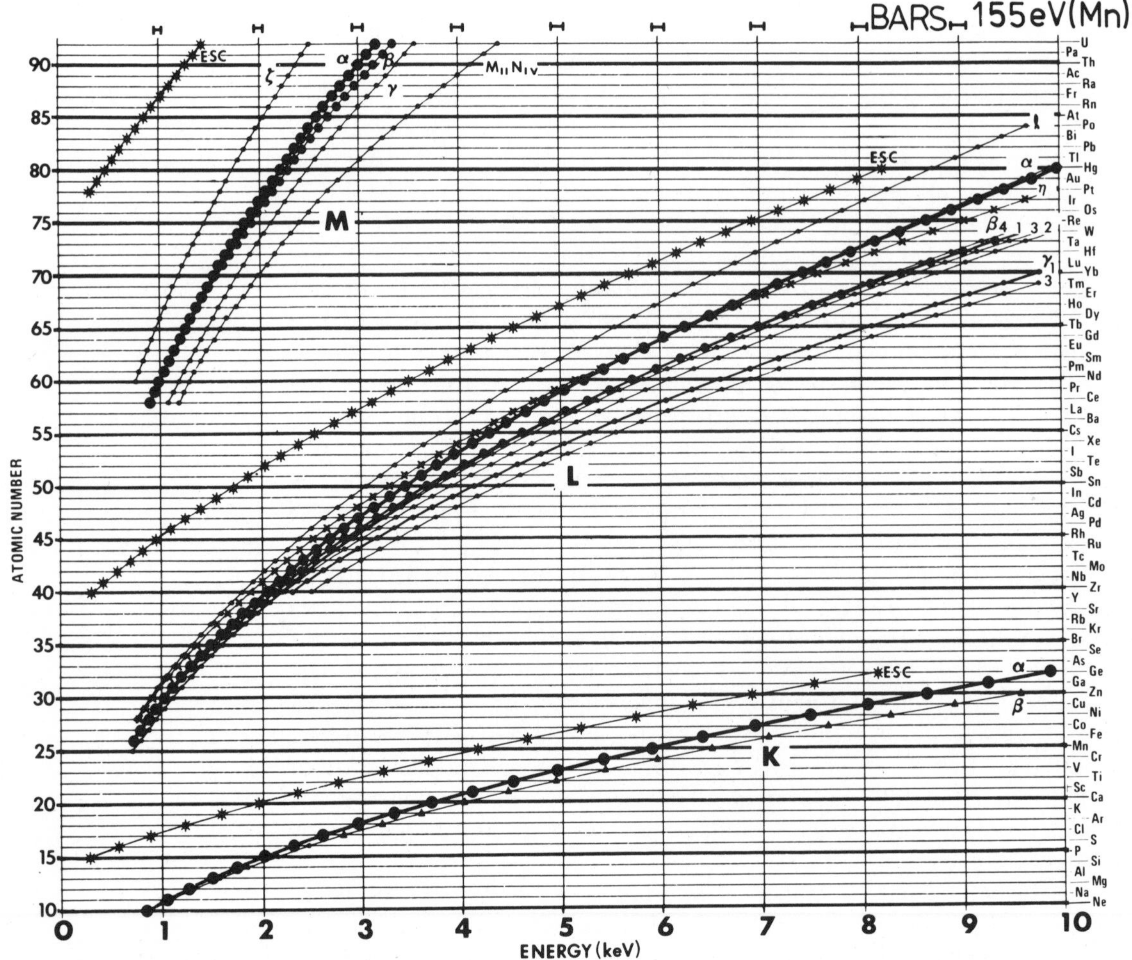

Figure 6.1. Plot of the energy of the x-ray emission lines observed in the range 0.75–10 keV by energy-dispersive x-ray spectrometry (Fiori and Newbury, 1978).

the overlap of other members of the family. Note, however, that the higher resolution of WDS systems increases the number of lines which can be detected, requiring reference to a detailed x-ray compilation.

As an aid to qualitative EDS analysis, Fiori and Newbury (1978) published a graphical representation of all x-ray lines observed in high-quality (5,000,000 counts integrated intensity) EDS spectra in the 0.70–10-keV range, shown in Fig. 6.1. This plot provides a convenient compilation of x-ray energies and also allows for a rapid evaluation of possible interferences. The spectral broadening effect for a 155-eV EDS spectrometer is also plotted, allowing for estimation of peak overlaps. Figure 6.1 is intended as an aid to qualitative analysis in conjunction with a table (or KLM markers) of x-ray energies. Tables 14.6, 14.7 and 14.8 provide $K\alpha$, $L\alpha$, and $M\alpha$ x-ray energies. Values accurate to within ± 10 eV of the x-ray energies are needed for proper peak identification.

The second important piece of information with which the analyst must be familiar is the concept of a family of x-ray peaks. When the beam energy exceeds the critical x-ray ionization energy for a shell or subshell so that it is ionized, all possible transitions involving that ionized shell may take place, producing a family of peaks, which will become more complicated as the electronic structure of the atom increases in complexity. With a beam energy of 15 keV or more, all possible lines of an element in the range 0.1–10 keV will be efficiently excited. The presence in the spectrum of all possible members of a family of lines increases the confidence that can be placed in the identification of that element. Since the family members must all exist, the absence of a particular line should immediately raise suspicion in the analyst's mind that a misidentification may have been made, and other possible elements in that energy range should be considered.

6.2. EDS Qualitative Analysis

6.2.1. X-Ray Lines

The energy-dispersive x-ray spectrometer is an attractive tool for qualitative x-ray microanalysis. The fact that the total spectrum of interest, from 0.1 keV to the beam energy (e.g., 20 keV) can be acquired in a short time (10–100 s) allows for a rapid evaluation of the specimen. Since the EDS detector has virtually constant efficiency (near 100%) in the range 3 to 10 keV, the relative peak heights observed for the families of x-ray lines are close to the values expected for the signal as it is emitted from the sample. On the negative side, the relatively poor energy resolution of the EDS compared to the WDS leads to frequent spectral interference problems as well as the inability to separate the members of the x-ray families which occur at low energy (<3 keV). Also, the existence of spectral artifacts such as escape peaks or sum peaks increases the complexity of the spectrum,

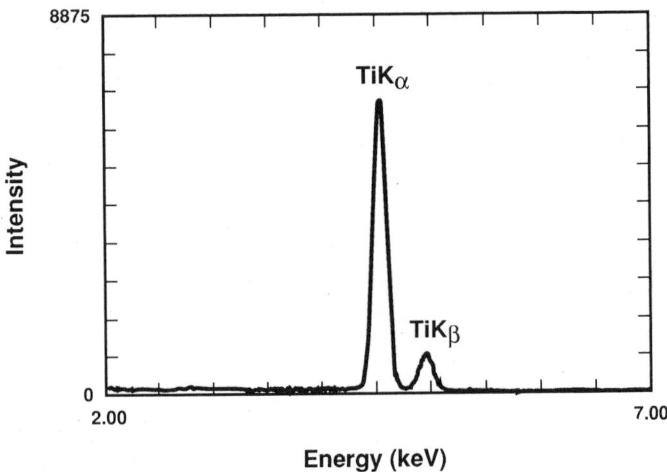

Figure 6.2. Titanium $K\alpha$ and $K\beta$ peaks. $K\alpha = 4.51$ keV. [Ultrathin window (diamond) Si(Li) EDS, 145 eV FWHM at Mn $K\alpha$.]

particularly when low relative-intensity peaks are considered. To aid in the identification of unknowns, it is useful to consider the appearance of the K, L, and M families in EDS spectra as a function of position in the energy range 0.7–10 keV. For x-ray lines located above approximately 3 keV, the energy separation of the members of a family of x-ray lines is large enough so that, despite the line broadening introduced by a typical 145-eV (Mn $K\alpha$) EDS system, it is possible to recognize more than one line. The appearance of typical K, L, and M family lines in the 2.5–10 keV range is shown in Figs. 6.2–6.4.

The approximate weights of lines in a family provide important information in identifying elements. The K family consists of two

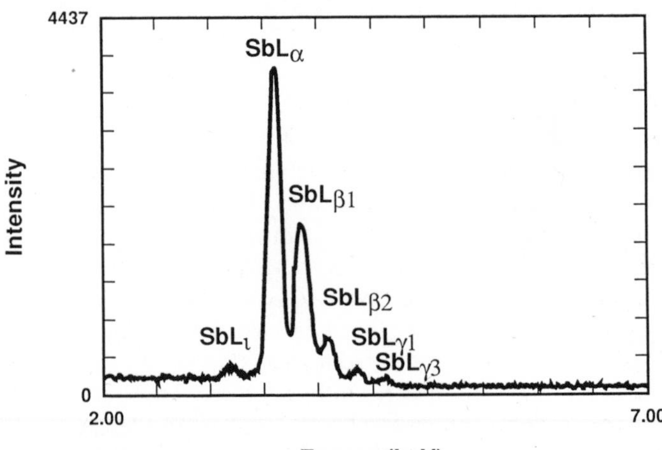

Figure 6.3. Antimony L-family x-ray peaks. $L\alpha_1 = 3.61$ keV. [Ultrathin window (diamond) Si(Li) EDS, 145 eV FWHM at Mn $K\alpha$.]

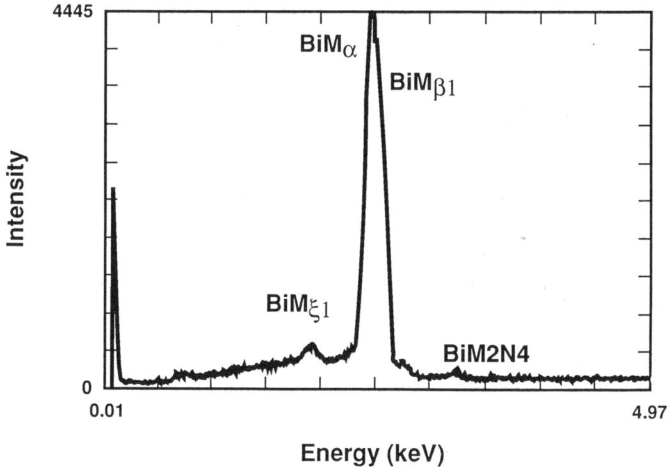

Figure 6.4. Bismuth M-family x-ray peaks. $M\alpha = 2.42$ keV. [Ultrathin window (diamond) Si(Li) EDS, 145 eV FWHM at Mn $K\alpha$.]

recognizable lines, $K\alpha$ and $K\beta$. The ratio of the intensities of the $K\alpha$ peak to the $K\beta$ peak is approximately 10 : 1 when these peaks are resolved, and this ratio should be apparent in the identification of an element. Any substantial deviation from this ratio should be viewed with suspicion as originating from a misidentification or the presence of a second element. The L series as observed by EDS consists of $L\alpha(1)$, $L\beta_1(0.7)$, $L\beta_2(0.2)$, $L\beta_3(0.08)$, $L\beta_4(0.05)$, $L\gamma_1(0.08)$, $L\gamma_3(0.03)$, $L1(0.04)$, and $L\eta(0.01)$. The observable M series consists of $M\alpha(1)$, $M\beta(0.6)$, $M\gamma(0.05)$, $M\zeta(0.06)$, and $M_{II}N_{IV}(0.01)$. The values in parentheses give approximate relative intensities, since these intensities vary with the element in question and with the overvoltage.

Below 3 keV, the separation of the members of the K, L, or M families becomes so small that the peaks are not resolved with an EDS system. The appearance of these families below 2 keV is illustrated in Figs. 6.5 (Si K, 1.74 keV), 6.6 (Y L, 1.92 keV), and 6.7 (Ta M, 1.71 keV). Note that the unresolved low-energy $K\alpha$ and $K\beta$ peaks appear to be nearly Gaussian (because of the decrease in the relative height of the $K\beta$ peak to about 0.01 of the height of the $K\alpha$), while the L and M lines are asymmetric because of the presence of several unresolved peaks of significant weight near the main peak.

All x-ray lines for which the critical excitation energy is exceeded will be observed. Therefore, in a qualitative analysis, all lines for each element should be located. Considering the 0.7–15 keV range (i.e., for a Be window detector), if a high-energy K line is observed [6.4 keV (iron) and above], then a low-energy L line should also be present for the element. Figure 6.8 shows this situation for copper K and L. Similarly, if a high-energy L line is observed [4.8 keV (cerium) or above], then a low-energy M line must also be present. Figure 6.9 illustrates this situation for dysprosium L and M. Because of large

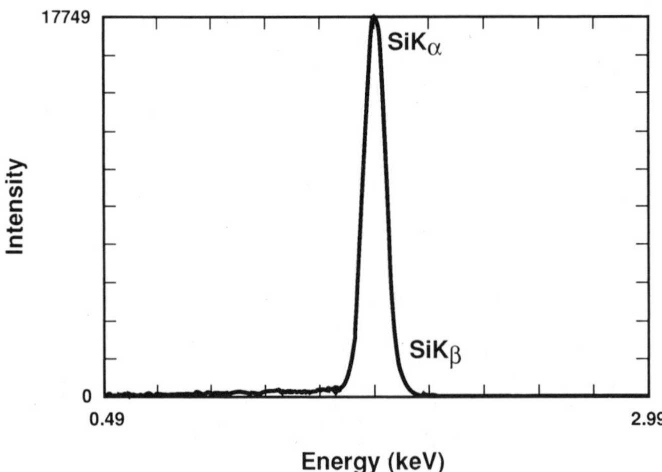

Figure 6.5. Silicon K-family x-ray peaks. $K\alpha = 1.74\,\text{eV}$. [Ultrathin window (diamond) Si(Li) EDS, 145 eV FWHM at Mn $K\alpha$.]

differences in the generation and absorption of low- and high-energy x rays, it is not possible to make use of relative peak heights among K, L, or M families in qualitative analysis.

Thin-window or windowless EDS provides access to the x-ray energy range below 1 keV, where the K-lines of the light elements beryllium, boron, carbon, nitrogen, oxygen, fluorine, and neon occur. Figure 6.10a–c shows the peaks for several of these elements. As the x-ray energy decreases below 1 keV, peak shapes tend to deviate somewhat from the Gaussian shape characteristic of higher energies. A major consideration in energy-dispersive spectrometry below 1 keV

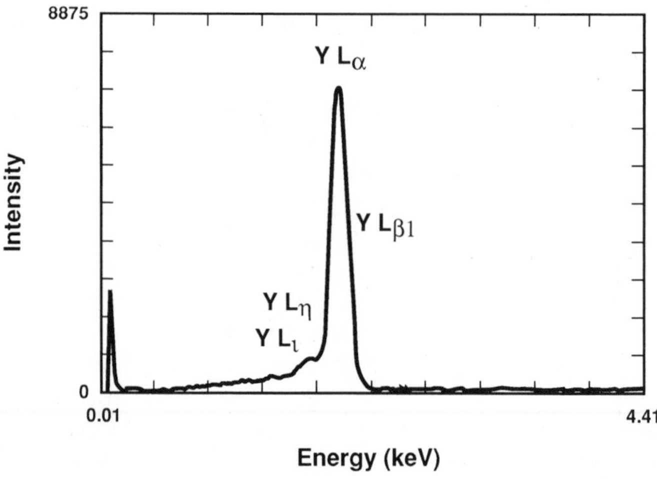

Figure 6.6. Yttrium L-family x-ray peaks. $L\alpha = 1.92\,\text{keV}$. [Ultrathin window (diamond) Si(Li) EDS, 145 eV FWHM at Mn $K\alpha$.]

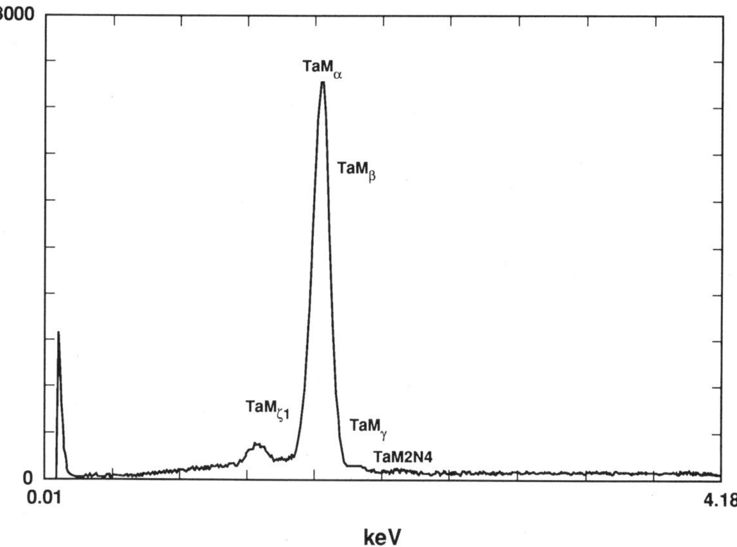

Figure 6.7. Tantalum M-family x-ray peaks. $M\alpha = 1.71$ keV. [Ultrathin window (diamond) Si(Li) EDS, 145 eV FWHM at Mn $K\alpha$.]

is the problem of interferences from L-, M-, and N-family x rays of heavier elements. Figure 6.10d shows the situation for the oxygen K-peak (0.523 keV) in an oxide of an iron alloy containing significant amounts of chromium (Cr $L = 0.571$ keV). The oxygen is barely discernible on the shoulder of the chromium peak. If vanadium (V $L = 0.510$ keV) is added to the oxide, as in Fig. 6.10e, the interference situation becomes untenable without quantitative spectral deconvolution.

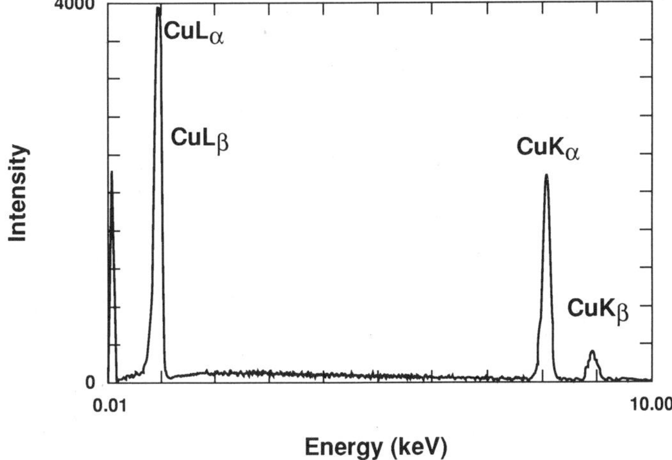

Figure 6.8. Copper K- and L-family x-ray peaks. $K\alpha = 8.04$ keV; $L\alpha = 0.92$ keV. [Ultrathin window (diamond) Si(Li) EDS, 145 eV FWHM at Mn $K\alpha$.]

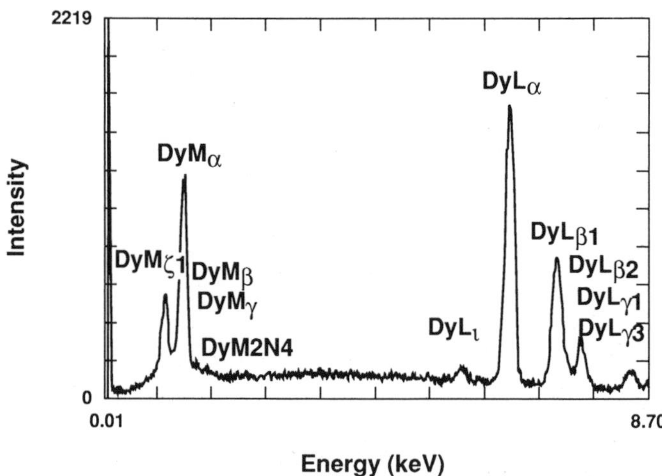

Figure 6.9. Dysprosium L- and M-family x-ray peaks. $L\alpha_1 = 6.495$ keV, $M\alpha = 1.293$ keV. [Ultrathin window (diamond) Si(Li) EDS, 145 eV FWHM at Mn $K\alpha$.]

6.2.2. Guidelines for EDS Qualitative Analysis

From the previous discussion, we can construct a set of general and specific guidelines for qualitative analysis:

6.2.2.1. General Guidelines for EDS Qualitative Analysis

a. Only peaks which are statistically significant should be considered for identification. The minimum size of the peak P after background subtraction should be three times the standard deviation of the background at the peak position, i.e., $P > 3(N_B)^{1/2}$. This peak height can be approximately estimated directly on the EDS display from the statistical scatter in the background on either side of the peak. The thickness of the background trace due to statistical fluctuations in the counts of each energy channel is a measure of $(N_B)^{1/2}$. The peak above the background should be at least three times this thickness. If it is difficult because of statistical fluctuations in the count to decide whether a peak exists above the continuum, then more counts should be accumulated in the spectrum to "develop" the peak. (For more information on detectability limits, see Section 9.10.)

b. In order to satisfy (a) and obtain adequate counts in the spectrum, it is tempting to use a high count rate. However, EDS systems become increasingly susceptible to the introduction of artifacts, such as sum peaks, as the count rate increases. For an EDS system operating at its best energy resolution, the maximum total spectrum input count rate should be kept below $3000\ s^{-1}$. An alternative criterion is that the dead time should be kept below 30%.

c. The EDS spectrometer should be calibrated so that the peak positions are found within 10 eV of the tabulated values. Note that, because of amplifier drift, the calibration should be checked frequently.

d. Suitable x-ray lines to identify the elemental range from beryllium to uranium are found in the energy range from 0.1 keV to 14 keV. To provide an adequate overvoltage to excite x-ray lines in the upper half of this range, a beam energy in the range 20–30 keV should be used. A beam energy of 20 keV is a good compromise between the need for adequate overvoltage and the need to minimize absorption in the sample, which increases as the beam energy and depth of penetration increase. However, such a beam energy may mask the presence of light elements that produce only x-ray energies below 2 keV, since the very high overvoltage, $U > 10$, will lead to deep penetration of the specimen by the electron beam and consequently very high absorption, with 50–99% of the low-energy x-rays absorbed in the specimen. To avoid missing possible light elements, the spectrum accumulation should be repeated, if possible, with a beam energy in the range 5–10 keV. Occasionally, because of interferences, e.g., Mo $L\alpha$ and S $K\alpha,\beta$, it is necessary to confirm the presence of the heavy element by using x-ray lines greater than 14 keV in energy. If the beam energy can be increased to give at least $U > 1.5$ and long spectrum accumulation times are used, then these high-energy x-ray lines can prove valuable.

e. In carrying out accurate qualitative analysis, a conscientious "bookkeeping" method must be followed. When an element is identified, all x-ray lines in the possible families excited must be marked off, particularly low-relative-intensity members. In this way, one can avoid later misidentification of those low-intensity family members as belonging to some other element at a minor or trace concentration. Artifacts such as escape peaks and sum peaks, mainly associated with the high-intensity peaks, should be marked off as each element is identified. Peaks arising from stray excitation in the sample chamber ("system peaks"), previously described in Chapter 5, should also be marked off.

6.2.2.2. Specific Guidelines for EDS Qualitative Analysis

a. Set the vertical gain of the CXA so that all peaks are contained in the display field. Begin at the high-energy end of the spectrum and work downwards in energy since the members of the *K*, *L*, and *M* families are more widely separated at high energies and likely to be resolved.

b. Determine the energy of a large peak. If it corresponds closely to a $K\alpha$ line of an element, immediately look for a $K\beta$ line with about 10% of the $K\alpha$ peak height. $K\alpha$ and $K\beta$ lines of elements starting at sulfur (2.31 keV) can be resolved with a typical EDS spectrometer.

c. If the $K\alpha$ and $K\beta$ pair does not fit the unknown, try the *L* series, noting the multiplicity of high-intensity *L* lines: [$L\alpha(1)$, $L\beta_1(0.7)$, $L\beta_2(0.2)$, $L\beta_3(0.08)$, $L\beta_4(0.05)$, $L\gamma_1(0.08)$, $L\gamma_3(0.03)$,

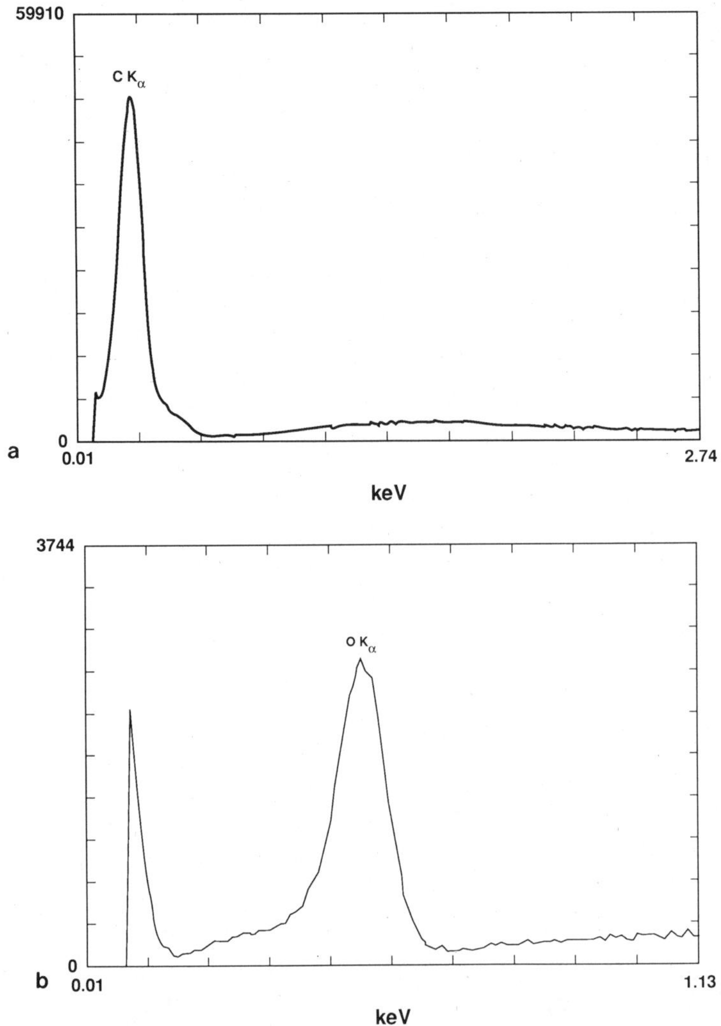

Figure 6.10. Low-energy EDS: K peaks with an ultrathin window (diamond) Si(Li) EDS, 145 eV FWHM at Mn $K\alpha$. (a) Carbon, (b) Oxygen, (c) Gadolinium fluoride with carbon coating, showing C K-, F K-, and Gd M-families. Interferences in low-energy EDS: (d) Oxygen peak region in an oxidized iron–chromium alloy. (e) Oxygen peak region in an oxidized iron–chromium–vanadium alloy.

$Ll(0.04)$, and $L\eta(0.01)$] which must be found to confirm the identification.

d. M family lines can be observed for elements starting at cerium. While $M\alpha(1)$ and $M\beta(0.6)$ are poorly resolved, the lower-intensity $M\zeta(0.06)$, $M\gamma(0.05)$, and $M_{II}M_{IV}(0.01)$ are separated from the main peak and usually visible. These low-intensity M-family members can be very easily misidentified. Note that, even for uranium, the M lines occur at an energy of 3.5 keV or less.

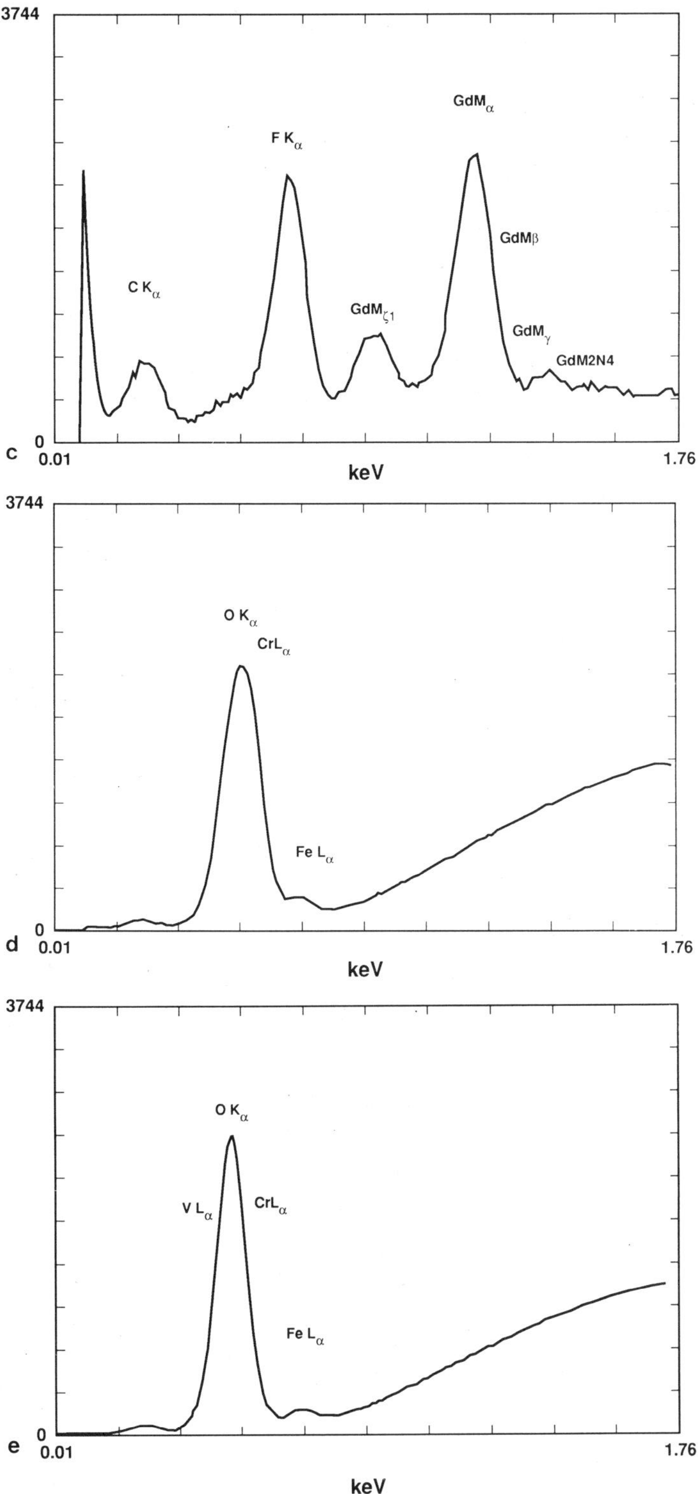

Figure 6.10. (Continued).

e. When an element is identified, all lines of all families (*K, L* or *L, M*) of that element should be marked off before proceeding. Next, escape peaks and sum peaks associated with the major peaks of the element should be located and marked. The magnitude of the escape peak is a constant fraction of that of the parent peak, ranging from about 1% for P $K\alpha$ to 0.01% for Zn $K\alpha$. The sum peak magnitude depends on the count rate.

f. At low x-ray energy (less than 3 keV), peak separation and the limited resolution of the EDS spectrometer will likely restrict element identification to only one peak. Note that low-energy *L* or *M* lines will be accompanied by high-energy *K* or *L* lines in the 4–20 keV range, which can aid identification.

g. When all of the high-intensity peaks in the spectrum have been identified, and all family members and spectral artifacts have been located, the analyst is ready to proceed to the low-intensity peaks. Any low-intensity peaks which are unidentified after the above procedure should belong to elements at low concentrations. Note that for such minor or trace elements, only the α line in a family may be visible. The lines at lower relative intensity in a family will probably be lost in the statistical fluctuations of the background. As a result, the confidence with which minor or trace elements can be identified is necessarily poorer than for major elements. If the positive identification of minor and trace elements is important, a spectrum containing a greater number of counts is needed. In many cases it may be necessary to resort to WDS qualitative analysis for greater confidence in identifying minor or trace elements.

h. Below 1 keV, qualitative EDS analysis is difficult. If the specimen contains elements of intermediate and high atomic number, interferences from the *L*-, *M*-, and *N*-family x rays of these elements may affect the *K* lines of the light elements (see Fig. 6.10d,e). If the qualitative-analysis procedure has been carefully followed to this point, any *L*-, *M*-, and *N*-family x rays in the region below 1 keV and escape peaks originating from parent peaks in the range from 2–3.75 keV have been identified, so that possible interferences have been noted. The analyst can therefore identify the remaining peaks in this region as low-energy *K* x rays of the light elements. In reality, if any *L*-, *M*-, and *N*-family x rays have been found below 1 keV, it is quite likely that the interference situation with the light-element *K*-lines is severe. Unambiguous identification of light elements should then be performed with the WDS, with its much higher spectral resolution. Alternatively, peak deconvolution can be performed on the EDS spectra, but the poor counting statistics often found in the low-energy region limit the deconvolution of peaks when the light-element peak is less than 10% of the overlapping *L*-, *M*-, or *N*-peak.

i. As a final step, by using Fig. 6.1 and Tables 14.6–14.8, the analyst should consider what peaks may be hidden by interference. If it is important to know of the presence of those elements, it will be necessary to resort to WDS analysis.

6.2.3. Pathological Overlaps in EDS Qualitative Analysis

The limited energy resolution of the EDS frequently causes the analyst to be confronted with serious peak overlap problems. In many cases, the overlaps are so severe that an analysis for an element of interest can not be carried out with the EDS. Problems with overlaps fall into two general classes: the misidentification of peaks and the impossibility of separating two overlapping peaks even if the analyst knows both are present. It is difficult to define a rigorous overlap criterion, owing to considerations of statistics. In general, however, it is very difficult to unravel two peaks separated by less than 50 eV no matter what peak-stripping method is used. The analyst should check for the possibility of overlaps within 100 eV of a peak of interest. When the problem involves identifying and measuring a peak of a minor constituent in the neighborhood of a main peak of a major constituent, the problem is further exacerbated, and overlaps may be significant even with 200-eV separation in the case of major versus minor constituents. When peaks are only partially resolved, the overlap can actually cause the peak channels for both peaks to shift by as much as 10–20 eV from the expected value. This phenomenon is encountered in the copper L-family spectrum, Fig. 6.8, where the unresolved Cu $L\alpha$ and Cu $L\beta$ peaks form a single peak whose peak channel is 10 eV higher than the Cu $L\alpha$ energy. Since copper K and L lines are often used as a set of calibration lines for EDS, the peak overlap effect in the L series can lead to a 10-eV calibration error at the low-energy end of the spectrum if the Cu $L\alpha$ energy is used directly. It is better to calibrate the system on a low-energy and a high-energy K line, e.g., pure Mg and pure Zn.

Because of the multiplicity of x-ray lines and the number of elements, it is not possible to list all significant interferences. Figure 6.1 can be conveniently used to assess possible interferences at any energy of interest by striking a vertical line and noting all lines within a given energy width on either side.

It is valuable to list elemental interferences which arise from peak overlaps in frequently encountered compositional systems. Examples in biological and materials science EDS analysis follow:

Biological EDS Analysis. Many interferences encountered in biological x-ray microanalysis arise from the use of heavy metal fixatives and stains in sample preparation. The heavy metals have L- and M-family x-ray lines in the 1 to 5 keV range that can interfere with the K lines of low-atomic-number materials, Na to Ca, which are important biologically. Table 6.1 lists the typical heavy metal stains and the elements for which significant peak overlap occurs.

Materials Science EDS Analysis. Because of the large number of elements encountered in materials science analysis, the number of possible interferences is much greater than in biological analysis, and so the analyst must constantly check his or her work to avoid errors.

Table 6.1. Common Interference Sources in Biological X-Ray Microanalysis

Element in stain or fixative	Interfering x-ray line	Interferes with	X-ray line interfered with
U	M	K, Cu, Ti	$K\alpha$
		Cd, In, Sn, Sb, Ba	$L\alpha$
Os	M	Al, P, S, Cl	$K\alpha$
		Sr	$L\alpha$
Pb	M	S, Cl	$K\alpha$
		Mo	$L\alpha$
	L	As, Se	$K\alpha$
Ru	L	S, Cl, K	$K\alpha$
Ag	L	Cl, K	$K\alpha$
As	L	Na, Mg, Al	$K\alpha$
Cu (grid)	L	Na	$K\alpha$
	Biological elements		
K	$K\beta$	Ca	$K\alpha$
Zn	$L\alpha$	Na	$K\alpha$

Particularly insidious is the interference in the first transition metal series, in which the $K\beta$ line of an element interferes with the $K\alpha$ line of the element with the next-higher atomic number, as indicated in Table 6.2. Quantitative EDS analytical systems can correct for these interferences (see Chapter 7). In qualitative analysis, however, when a minor element is interfered with by a major element x-ray line, it is often very difficult to detect the presence of the minor element.

An example of a frequently encountered problem is the mutual interference of the S K, Mo L, and Pb M lines, which lie within an energy range of only 50 eV. Spectral interference to this degree makes it very difficult to accurately identify the elements present in the sample without applying advanced spectral processing or recourse to wavelength dispersive spectrometry.

Table 6.2. Common Interference Sources in Materials Science X-Ray Microanalysis

Element	Interfering x-ray line	Interferes with	X-ray line interfered with
Ti	$K\beta$	V	$K\alpha$
V	$K\beta$	Cr	$K\alpha$
Cr	$K\beta$	Mn	$K\alpha$
Mn	$K\beta$	Fe	$K\alpha$
Fe	$K\beta$	Co	$K\alpha$
Pb	$M\alpha$	S	$K\alpha$
		Mo	$L\alpha$
Si	$K\alpha$	Ta	$M\alpha$
Ba	$L\alpha$	Ti	$K\alpha$

Incisor of Rat. The spectrum shown in Fig. 6.11 was obtained from the region of mineralized tissue formation in a rat incisor. Following the above procedure, we first find that the peaks at 3.69 and 4.01 keV correspond to Ca $K\alpha$ and Ca $K\beta$, and also locate Ca L in Fig. 6.11a. At lower energy, the peaks at 2.015 and 2.310 keV appear to also be a $K\alpha$, $K\beta$ pair, but while P $K\alpha$ is found to correspond to the 2.015 keV peak, P $K\beta$ is unresolved in Fig. 6.11a. With expansion of the intensity scale, further study reveals that the 2.310 keV peak is S

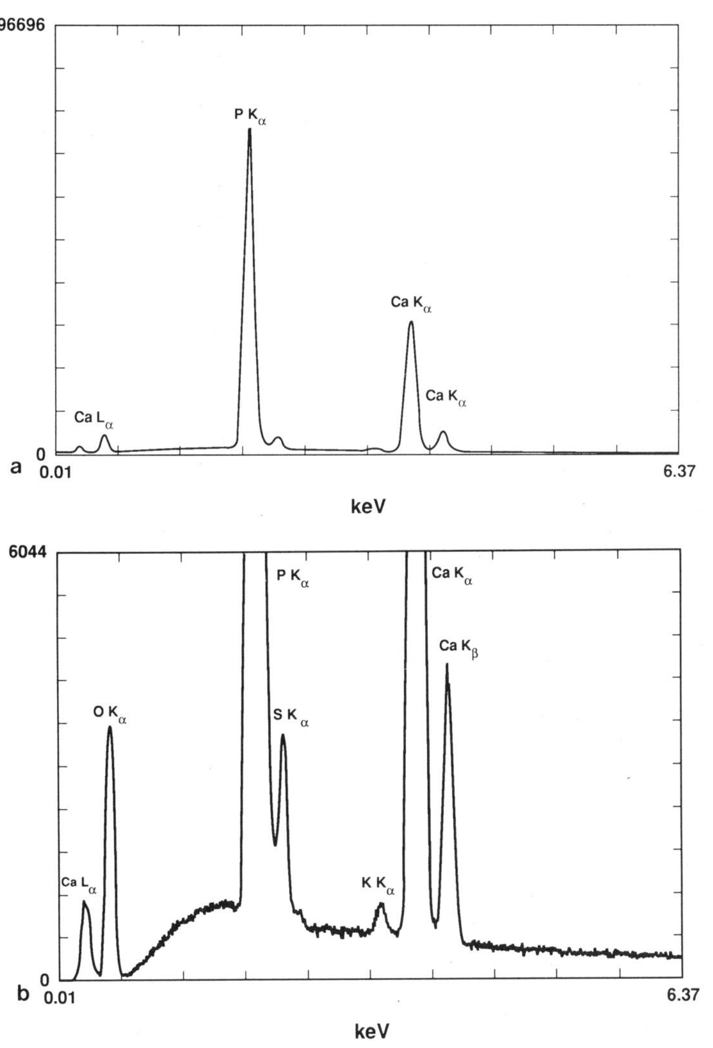

Figure 6.11. Energy-dispersive spectrum [Ultrathin window (diamond) Si(Li) EDS, 145 eV FWHM at Mn $K\alpha$] of enamel-forming region of rat incisor. 20 keV. (a) Identification of the major constituents, calcium (including Ca L) and phosphorus. (b) Further accumulation and expansion for identification of sulfur, oxygen, and potassium.

$K\alpha$ (Fig. 6.11b). The remaining well-resolved low-energy peak is the O K at 0.523 keV. There is an interesting bump in the background just below Ca $K\alpha$ which may be a peak but which is close to the $3(N_B)^{1/2}$ limit. Neither the escape peak of Ca $K\alpha$ nor the sum peak of P $K\alpha$ occurs in this energy region. Further accumulation of the spectrum reveals that the peak is K $K\alpha$, with K $K\beta$ lost under Ca $K\alpha$. The possible interferences of the main peaks (Ca $K\alpha$: Sb, Te; P $K\alpha$: Y, Zr) do not correspond to elements likely to exist in this system.

Multielement Glass. The EDS spectrum of a complicated, multielement glass is shown in Fig. 6.12. Beginning at the high-energy end, the peaks at 6.40 and 7.06 keV correspond to Fe $K\alpha$ and Fe $K\beta$. Note that Fe $L\alpha$ at 0.704 keV is not observed, owing to high absorption. The next series of four peaks near 4.5 keV might at first be thought to be Ti $K\alpha$, $K\beta$ and V $K\alpha$, $K\beta$. However, the peak positions are not proper, nor are the relative peak heights. This series is the Ba L family: Ba $L\alpha$(4.47 keV), $L\beta_1$(4.83), $L\beta_2$(5.16), $L\gamma$(5.53), with others in the family unresolved. The peaks at 3.69 and 4.01 keV are Ca $K\alpha$ and Ca $K\beta$. At low energy, the peak at 1.74 keV is Si $K\alpha$, and at 1.49 keV the peak is Al $K\alpha$. The O K peak is found at 0.523 keV. Expansion of the vertical gain reveals no low-intensity peaks except those associated with artifacts. Note that in this example, if it were important to detect titanium or vanadium, the interference of the barium L family is so severe that EDS analysis will not resolve the question and WDS analysis would certainly be needed.

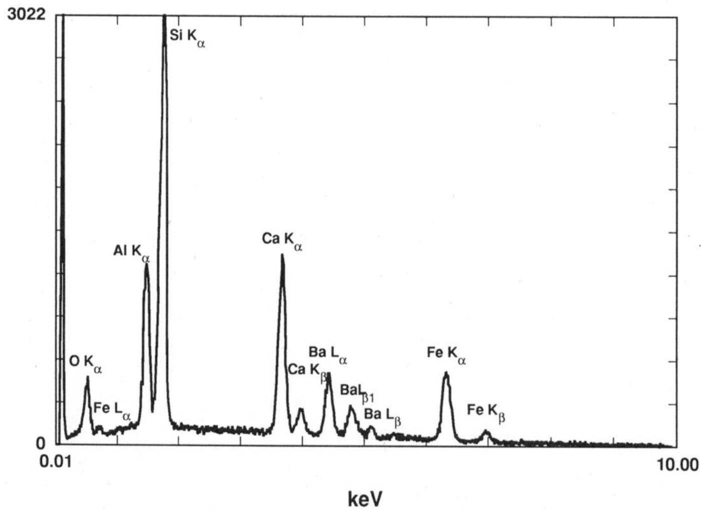

Figure 6.12. Spectrum of multielement glass (NIST K309: Al = 7.94 wt%, Si = 18.7 wt%, Ca = 10.7 wt%, Fe = 10.5 wt%, Ba = 13.4 wt%, O = 38.7 wt%), 20 keV. Identification of Fe, Ba, and Ca in upper energy region of spectrum; identification of silicon, aluminum, iron-L, and oxygen in lower energy region. (Note: oxygen is a major constituent, but its low energy peak is highly absorbed in the specimen.)

6.3.1. Measurement of X-Ray Lines

For qualitative analysis with wavelength-dispersive spectrometers, one seeks to identify the elements in the sample by determining the angles at which Bragg's law is satisfied as the spectrometer is scanned through a range of angles. Peaks are observed at those angles 2θ at which the condition

$$n\lambda = 2d \sin\theta \tag{6.1}$$

is satisfied. The strategy for qualitative WDS analysis is distinctly different from qualitative EDS analysis. The resolution of the WDS spectrometer is much better, typically <10 eV compared to 150 eV for EDS, which leads to a peak-to-background ratio at least 10 times higher for WDS. As a result of the better resolution and peak-to-background ratio, more members of the family of x-ray lines for a given element can be detected and must therefore be accounted for in a logical fashion to avoid subsequent misidentification of minor peaks. This is illustrated in Fig. 6.13, where the EDS and WDS spectra for the cadmium L family are compared. Many of the peaks observed in the WDS spectrum are not detected in the EDS spectrum.

The analyst must recognize the consequences of all of the terms in Bragg's law, Eq. (6.1). For a particular value of λ, as n varies from $n = 1, 2, 3 \ldots$, i.e., the order of the reflection changes, the value of θ_B changes. If the parent line ($n = 1$) is obtained on a given crystal, there

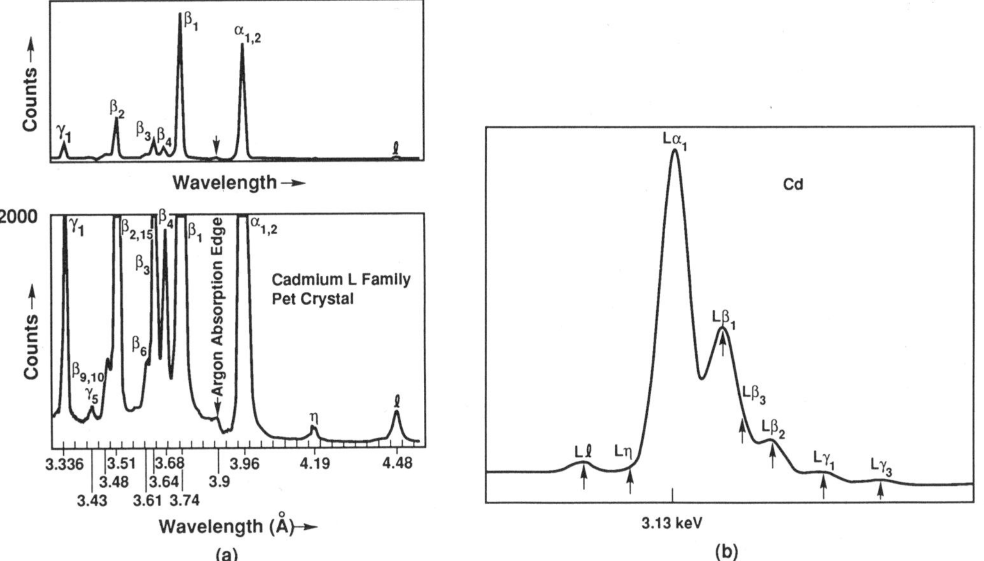

Figure 6.13. (a) WDS spectrum of the Cd L-family lines. Data were taken with a PET crystal. (b) EDS spectrum of the Cd L-family lines.

may be several other values of θ_B at which peaks corresponding to this same λ will be found for $n = 2, 3$, etc. If other crystals, i.e., other d spacings are considered, additional reflections may be detected, e.g., corresponding to $n = 4, 5$, etc. Thus, the complete collection of lines for an element which must be considered includes not only all of the family of x-ray lines but also the higher-order reflections associated with each of the parent lines.

A second consequence is that for a given setting of θ_B on a particular crystal with spacing d, x rays of different energies which satisfy the same value of the product $n\lambda$ will all be diffracted. Thus, if diffraction occurs for $n = 1$ and $\lambda = \lambda_1$, diffraction will also occur for $n = 2$ and $\lambda = (\lambda_1/2)$ and so on. As an example, consider sulfur in an iron–cobalt alloy. The S $K\alpha$ ($n = 1$) line occurs at 5.372 Å, while Co $K\alpha$ ($n = 1$) occurs at 1.789 Å. The third-order Co $K\alpha$ line falls at 3×1.789 Å $= 5.367$ Å. These lines are so close in wavelength that they would not be adequately separated by an ordinary WDS spectrometer. While such interferences are unusual, the analyst must be aware of their possible occurrence. Although the interference cannot be resolved spectroscopically, the S $K\alpha$ and Co $K\alpha$ x rays are of different energies and can be separated electronically. This method takes advantage of the fact that the x-rays of different energies create voltage pulses of different magnitude. In the sulfur–cobalt example, the voltage pulses produced by the Co $K\alpha$ are about three times larger than those of the S $K\alpha$. By using the pulse-height discriminator previously described (Chapter 5), a voltage window can be set around the S $K\alpha$ pulse distribution which excludes the Co $K\alpha$ pulse distribution. It should be noted that once a discriminator window has been established, the spectrometer should not be tuned to

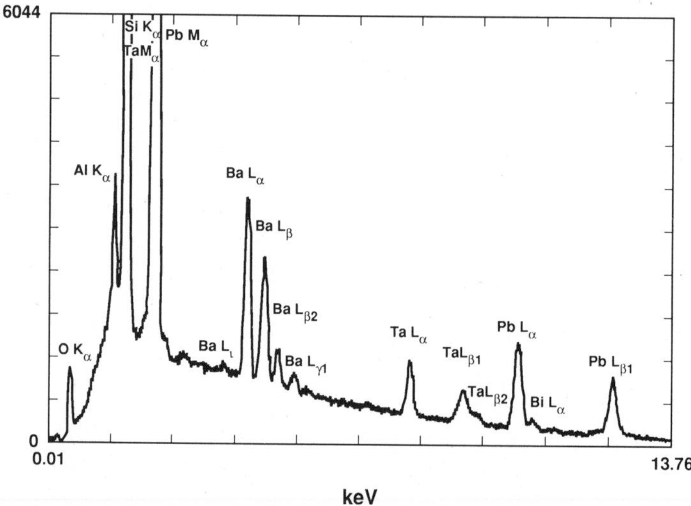

Figure 6.14. EDS spectrum of heavy-element glass. X-ray energy 0–20.0 keV; expanded vertical scale. Identification of Pb, Ta, Ba, and Si as major constituents; identification of minor peaks for Bi, Al, and O. (Note: oxygen is a major constituent, but its low energy peak is highly absorbed in the specimen.)

Table 6.3. Composition of NIST K-251

Element	wt %
O	23.0
Al	1.3
Si	14.0
Ba	9.0
Ta	4.1
Pb	44.1
Bi	4.5

another peak without removal of the discriminator window or resetting to the appropriate value for another peak.

From these considerations, it is obvious that WDS qualitative analysis is not as straightforward as EDS qualitative analysis. Nevertheless, WDS qualitative analysis offers a number of valuable capabilities which are not possessed by EDS, as illustrated in Figs. 6.14 and 6.15, which give a comparison of EDS and WDS spectra on a heavy-element glass (NIST K-251, the composition of which is given in Table 6.3).

1. The high resolution of the WDS allows the separation of almost all overlaps which plague the EDS spectrum. Thus, in Fig. 6.14, bismuth is extremely difficult to detect in the presence of lead in the EDS spectrum but is readily apparent in the WDS spectrum (Fig. 6.15a). Further, the multiplicity of lines which can be observed aids in a positive identification of each element. However, this multiplicity makes the bookkeeping that much more difficult and important in order to avoid misassignment of low-intensity peaks.
2. The improved peak-to-background ratio of the WDS spectrometer provides the capability of detection and identification at a concentration which is about a factor of 10 lower than the EDS (roughly 100 ppm for the WDS compared to 1000 ppm for EDS). (See Section 9.10 for further explanation.)
3. The WDS spectrometer can detect low-Z elements with atomic numbers in the range $4 \leq Z \leq 11$. Windowless or thin-window EDS spectrometers can also detect elements as low as $Z = 4$. However, the detection limits of EDS are much poorer than WDS for these low-Z elements.

The principal drawbacks of WDS qualitative analysis in addition to the multiplicity of lines and higher-order reflections noted above are the following.

1. Qualitative analysis is much slower than in EDS, with scan times of 30 min per crystal required for a full wavelength scan at high sensitivity. With automated WDS, the process can be speeded up by scanning just those portions of the spectrum where peaks occur, and more rapidly

stepping between peak positions. A full qualitative analysis requires the use of four crystals. If multiple spectrometers are available, these crystals can be scanned in parallel. The multiplicity of lines requires much more of the analyst's time for a complete solution of the spectrum, especially if it is done manually (utilizing a full x-ray table, e.g., Bearden, 1967).

2. Much higher beam currents are required for use in the WDS compared to EDS because of the considerably lower geometrical and quantum efficiencies of the WDS (see Chapter 5). For fragile specimens such as biological targets, these high currents may be unacceptable.

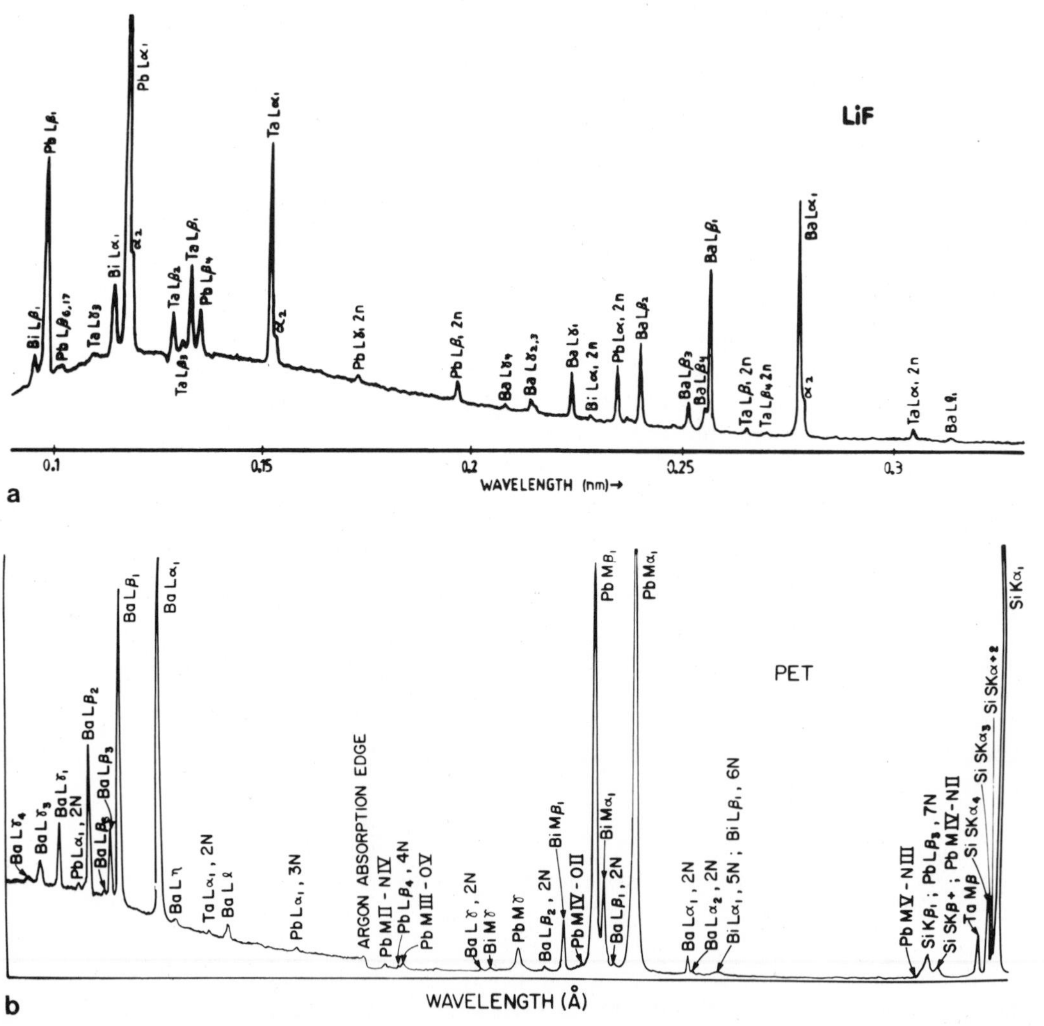

Figure 6.15. WDS spectra of the same heavy-element glass as in Fig. 6.14. (a) LiF spectrum. (b) PET spectrum. (c) TAP spectrum. (d) lead sterate (OdPb) spectrum.

Given a sample of the complexity of that shown in Fig. 6.15a–d, the following procedure should be used:

1. Because of the possibility of a peak originating from a high-order reflection, it is best to start at the highest-energy, i.e., shortest-wavelength end of the spectrum for each crystal where the probability is highest for finding first order, $n = 1$ peaks.

2. The highest-intensity peak should be selected at the short wavelength end of the LiF crystal scan and its wavelength determined. From a complete x-ray reference such as that of Bearden (1967), the

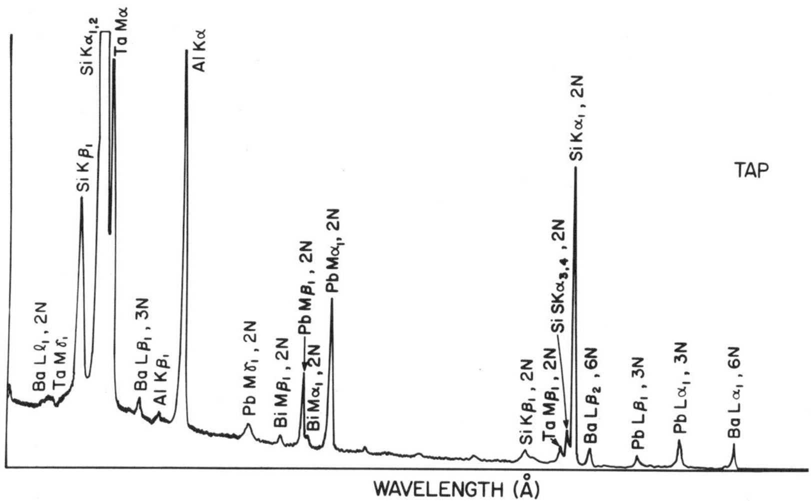

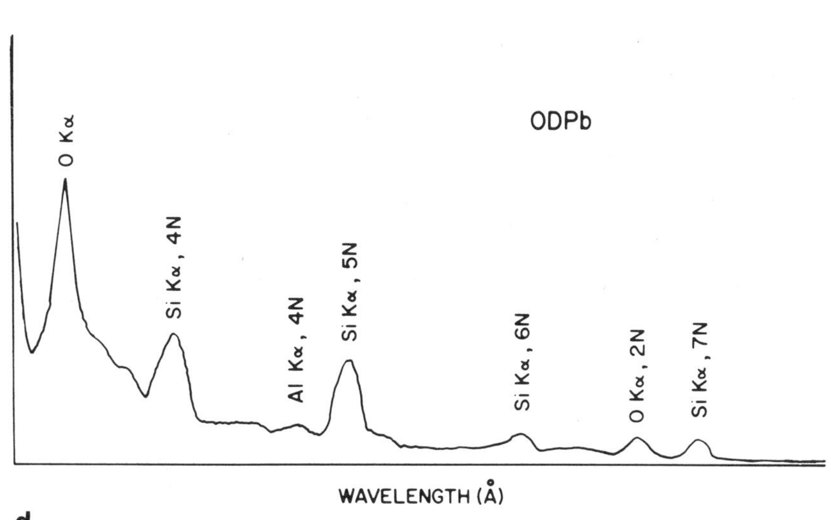

Figure 6.15. (Continued).

elements which could produce the peak in question, such as a $K\alpha_{1,2}$ or $L\alpha_{1,2}$, should be noted. In parallel with the concept of a family of lines introduced in the description of EDS qualitative analysis, when a candidate element is tentatively associated with a peak designated $K\alpha_{1,2}$ ($n = 1$), the analyst should immediately locate the associated $K\beta_1$, peak. Again, the ratio of $K\alpha$ to $K\beta$ should be roughly 10:1. However, because of changes in crystal and detector efficiency, the expected ratios may not always be found. For example, in the Cd spectrum in Fig. 6.13, the efficiency of the detector approximately doubles on the low-wavelength side of the argon K absorption edge. Hence the $L\beta_1$ peak, which should be approximately 60% as big as the $L\alpha$ peak, is actually larger. The efficiency doubles at the argon K edge because the flow-proportional x-ray detector of this spectrometer utilizes P-10 gas (90% Ar–10% methane). The dimensions of the detector and the pressure of the P-10 gas permit a certain fraction of x-rays of wavelength longer than the edge wavelength to pass through the gas without interaction. For those x-rays with wavelengths shorter than the edge wavelength, a greater fraction (approximately twice as many) will interact with the gas and, hence, be detected. Note also that the resolution of the WDS with some crystals such as LiF and quartz is sufficient to show at least some separation of $K\alpha$ into $K\alpha_1$ and $K\alpha_2$, which have a ratio $K\alpha 1$:$K\alpha 2$ of 2:1. Similarly, if an $L\alpha$ peak is suspected, the full L family should be sought. Note that L lines in addition to those listed in Fig. 6.1 (i.e., $L\alpha_{1,2}$, $L\beta_1$, $L\beta_2$, $L\beta_3$, $L\beta_4$, $L\gamma_1$, $L\gamma_3$, $L\iota$, $L\eta$) may be found because of the excellent resolution and peak-to-background ratio. In identifying families of lines, it is possible that, because of wavelength limitations of the crystals, only the main peak may be found (e.g., Ge $K\alpha$ on LiF, where Ge $K\beta$ lies outside the crystal range). Applying this approach to the LiF spectrum in Fig. 6.15a, the major peak at 0.118 nm could correspond to As $K\alpha$, or Pb $L\alpha$. Inspection for other family members reveals the presence of the Pb L family but not the As $K\beta$. Thus Pb is identified with certainty and As is excluded.

3. Once the element has been positively identified and all possible first-order members of the family of x-ray lines have been located, which may require examining spectra from more than one crystal (e.g., Zn $K\alpha$, $K\beta$ on LiF, Zn L family on PET), the analyst should locate all possible higher-order peaks associated with each first-order peak throughout the set of crystals. Continuing the heavy-element-glass example, the higher orders for the Pb L family are noted in the remainder of the LiF spectrum as well as on the PET and TAP spectra in Fig. 6.15b and c. No higher-order Pb L lines are found in the lead stearate spectrum (Fig. 6.15d). Note that the orders can extend to values as high as 7, as shown in Fig. 6.15d for silicon.

4. Only after all possible family members of the element and all higher-order reflections of each member have been identified should the analyst proceed to the next unidentified high-intensity, low-wavelength peak. The procedure is then repeated.

5. Wavelength-dispersive spectrometry is sufficiently complicated

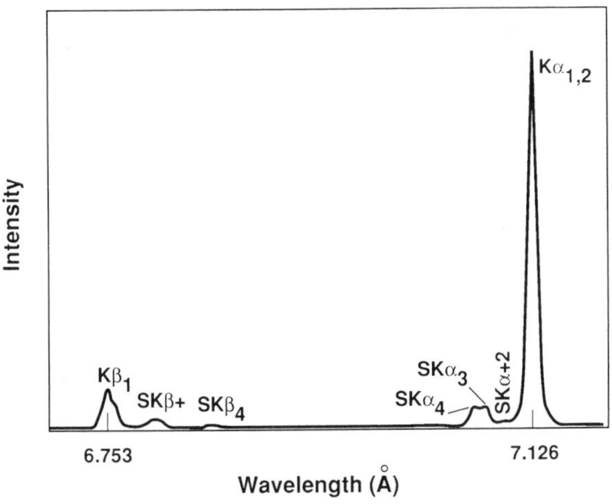

Figure 6.16. WDS spectrum of pure Si taken with a PET crystal at a beam energy of 20 keV. Note presence of satellite lines labeled *S*.

that a number of special considerations arise: (a) For some elements, the parent $K\alpha$ peak ($n = 1$) may lie outside the range of the low-wavelength crystal (usually LiF), whereas higher-order reflections of that $K\alpha$ peak may still be found. Example: Zirconium $K\alpha$ at 0.078 nm (0.785 Å). (b) Satellite lines may also be observed as low-intensity peaks on the high-energy shoulder of a high-intensity peak. Satellite lines arise from doubly ionized atoms and are illustrated in Fig. 6.16. Note that several of the satellite lines of the Si $K\alpha$ peak are nearly as high as the Si $K\beta$ peak.

6.4. Automatic Qualitative EDS Analysis

Most modern computer-based analytical systems for energy-dispersive x-ray spectrometry include a routine for automatic qualitative analysis. Such a routine represents an attempt to create an expert system in which the guidelines described in the preceding sections for manual qualitative analysis are expressed as a series of conditional tests to recognize and classify peaks. Several approaches to automatic qualitative analysis are possible, but in general, the basic steps are:

1. background removal, by spectral filtering or background modeling;
2. peak searching, in which peak positions and amplitudes are determined.
3. peak recognition, in which possible candidate elements for assignment to the peaks are catalogued from look-up tables;
4. peak stripping, in which the intensity for a recognized element, including minor and artifact peaks such as the Si escape peak, is stripped from a candidate peak to determine if any other elements are present in that energy region.

The success with which which such an expert system operates depends on several factors:

1. Has the analyst accumulated a statistically valid spectrum prior to applying the automated qualitative analysis procedure?
2. Are the complete x-ray families included in the look-up tables, such as the Ll and $L\eta$ members of the L-family and the $M\zeta$ and $M_{II}N_{IV}$ members of the M-family?
3. Have x-ray artifacts such as the escape peaks and sum peaks been properly accounted for?

Automatic qualitative analysis systems range from the naive to the sophisticated. In the naive approach, all possible elemental peaks within a specified energy range, e.g., ±25 eV, of a candidate peak are reported without application of any rules such as families of lines and weighting of lines to sort out the invalid assignments. Peak stripping is not usually incorporated into such a system. The analyst is left with the task of applying the guidelines for manual qualitative analysis. In the most sophisticated systems, a true expert system with a rigorous codification of the qualitative analysis guidelines is used to reject misidentifications. The identified constituents are often listed in order of decreasing confidence.

Can the results reported by an automatic qualitative analysis system be trusted? Generally it is difficult to assign a quantitative measure to the degree of confidence with which a qualitative identification is made. The statistics of the peak relative to the local background is a measure of the confidence that a peak actually exists, but there is no numerical value for the likelihood of an identification being correct. For major constituents, the automatically determined answer is virtually always correct, but for minor and trace constituents, the best advice is "*caveat emptor*." The "report table" generated by a sophisticated computer-assisted analytical system may look extremely impressive, but that should not be taken as an excuse to accept its elemental identifications without question. Just as in manual qualitative analysis, the analyst must contribute common sense to solving the problem. Common sense is one of the most difficult concepts to incorporate in an expert system. It is therefore really the responsibility of the analyst to examine each putative identification, major constituents included, and determine if it is reasonable when other possibilities are considered. As always, an excellent procedure in learning the limitations of an automatic system is to test it against known standards of increasing complexity. Even after successful performance has been demonstrated on selected complex standards, the careful analyst habitually checks the suggested results on unknowns; suggestions of the presence of scandium, promethium, or technetium should always be treated with healthy skepticism.

7

X-Ray Peak and Background Measurements

7.1 General Considerations for X-Ray Data Handling

As discussed in Chapter 6, qualitative analysis is based on the ability of a spectrometer system to measure characteristic line energies and relate those energies to the presence of specific elements. This process is relatively straightforward if.

1. the spectrometer system is properly calibrated.
2. the operating conditions are adequate to give sufficient x-ray counts so that a given peak can be easily distinguished from the underlying background level, and
3. no serious peak overlaps are present.

Quantitative analysis, on the other hand, involves.

1. measuring the intensity of spectral peaks corresponding to preselected elements for both samples and standards under known operating conditions,
2. calculating intensity ratios (k values), and
3. converting these k values into chemical concentration by the methods described in Chapters 8 and 9.

Since quantitative analysis can now be performed with relative accuracy approaching 1%, great care must be taken to ensure that the basic measurement of the characteristic x-ray intensity is accurate to at least the 1% level, and preferably better. Such a level of accuracy can be achieved only if the measured response of the x-ray detector system is linear over a wide range of counting rates and if the characteristic x-ray intensity can be accurately extracted from the background. As discussed in Chapter 5, dead-time corrections for WDS proportional counters and live-time corrections for EDS detectors must be performed accurately as part of the overall spectrum-processing procedure. As would be expected, accurate background measurements become increasingly important at lower concentrations as peak-to-background ratios get smaller. For example, a 100% error in a background measurement of a peak 100 times larger than the background introduces a 1% error in the measured peak intensity, whereas the same error in the case of a peak twice background introduces a 50% error.

As discussed in Chapter 5, the characteristics of EDS and WDS x-ray spectra are markedly different because of the great difference in spectral resolution. The WDS generally gives better resolution than EDS by a factor of 20 for the same x-ray energy. The practical consequences of this difference in resolution are that peak-to-background ratios are much lower for EDS, making accurate background correction critical to any quantitative analysis, and peak overlaps are much more common with the EDS, necessitating use of peak deconvolution methods for many analysis situations. Despite the much higher resolution of WDS, peak interferences can still be encountered. Accurate WDS background correction is needed to approach 1% overall quantitative accuracy and to extend useful trace analysis to the levels of hundreds or even tens of parts per million. In general, errors associated with the extraction of characteristic intensities from EDS and WDS spectra are larger than the uncertainties introduced by the quantitative matrix correction procedures discussed in Chapters 8 and 9.

7.2. Background Correction

7.2.1. Background Correction for EDS

As a starting point to perform an accurate background correction, we need to view the characteristic peak and the adjacent background. Because the EDS peaks are so broad, the tails of the Gaussian peak extend over a substantial energy range, interfering with our view of the adjacent background. Background measurements with the EDS are therefore made difficult because of the problem of finding suitable background areas adjacent to the peak being measured. One possible method for background correction is linear interpolation from the adjacent background locations. The accuracy of such a method depends on the background locations selected. Figure 7.1 shows a portion of a pure chromium spectrum. The background value at the peak channel determined by interpolation is about 100 counts if points B and D are used or about 130 counts if points B and C are used. The latter pair is the better because the chromium absorption edge at 5.989 keV produces an abrupt drop in the background at higher energies, which is obscured by the characteristic Cr peaks. If the spectrum had been collected with a detector system with poorer resolution than that used in obtaining Fig. 7.1 (160 eV at 5.89 keV), then point C could not be used. In this case, determination of the background by extrapolation of the low-energy

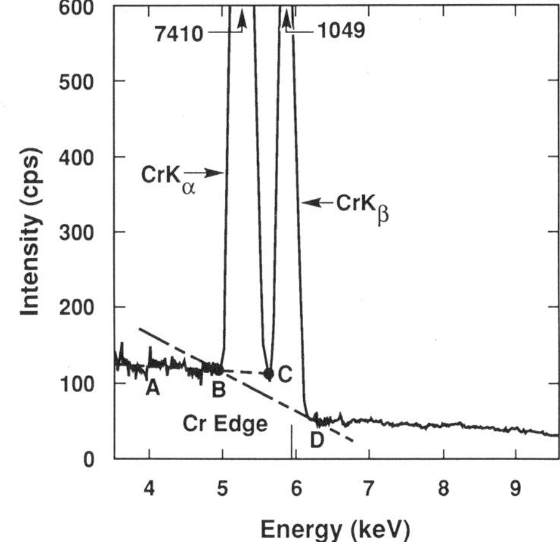

Figure 7.1. EDS background fitting by linear interpolation (points B–C or B–D) or extrapolation (points A–B) in a simple spectrum of chromium (Lifshin, 1975).

points A and B to point C could be used in preference to interpolation of points B and D. The error associated with use of this approach would affect the measurement of the pure chromium standard by less than 1%. If the spectrum had been that of a copper 1-wt % chromium alloy, the Cr portion of the spectrum would look substantially the same with a correspondingly lower peak-to-background ratio. The error introduced by this background extrapolation would rise to 100% for the 1%-Cr constituent.

For a mixture of elements, the spectrum becomes more complex, and interpolation is consequently less accurate. Figure 7.2 shows a stainless steel spectrum, taken under the same operating conditions as Fig. 7.1. There are a sufficient number of $K\alpha$ and $K\beta$ peaks such that the use of extrapolation is questionable for Fe and Ni while interpolation between points A and B must be performed over a 3-keV range. Since

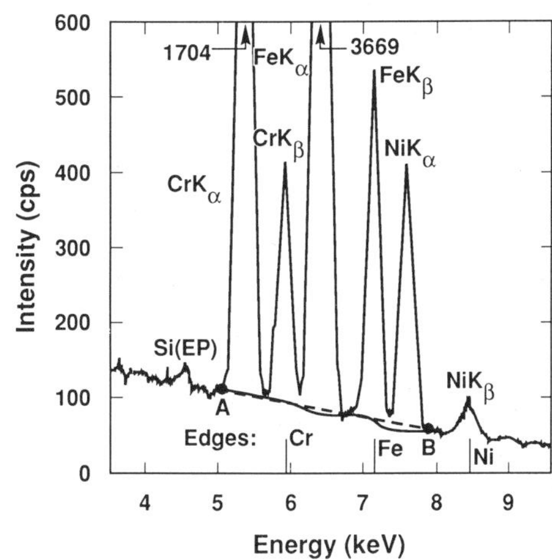

Figure 7.2. EDS background fitting by linear interpolation in a complex spectrum of stainless steel.

the real background is irregular owing to the presence of Cr and Fe absorption edges, an error of about 3% would occur in the Ni intensity measurement in this specimen with background extrapolation by linear interpolation.

Although these errors of a few percent of the amount present in the major constituents are larger than what can be expected with higher-resolution WDS measurements, a much greater difficulty in the example cited would be encountered in the determination of a few percent Mn (Mn $K\alpha = 5.898$ keV) or Co (Co $K\alpha = 6.924$ keV) in the stainless steel. Not only might the signals of Mn and Co be of the same order of magnitude as the background, but they would be obscured by spectral lines of the major constituents (Cr $K\beta$ and Fe $K\beta$).

Compensation for the background, by subtraction or other means, is critical to all EDS analysis. Basically there are two approaches to this problem. In the first approach, a continuum energy-distribution function is either calculated or measured and combined with a mathematical description of the detector response function. The resulting function is then used to calculate a background spectrum, which can be subtracted from the observed spectral distribution. This method can be called *background modeling.* In the second approach, the physics of x-ray production and emission is generally ignored and the background is viewed as an undesirable signal, the effect of which can be removed by mathematical filtering or modification of the frequency distribution of the spectrum. Examples of the latter technique include digital filtering and Fourier analysis. This method can be called *background filtering.* It must be remembered here that a real x-ray spectrum consists of characteristic and continuum intensities both modulated by the effects of counting statistics. When background is removed from a spectrum, by any means, the remaining characteristic intensities are still modulated by both uncertainties. We can subtract away the average effect of the background, but the effects of counting statistics cannot be subtracted away. This last point is extremely important and we will go into more detail in Section 9.9 about the composite statistical considerations when one removes both the average effects of the background and the average effects of mutually interfering peaks in the same energy region of a spectrum.

In practice, both background filtering and background modeling have proved successful.

7.2.1.1. Background Modeling

Background modeling has been investigated by a number of authors (Ware and Reed, 1973; Fiori *et al.*, 1976; Rao-Sahib and Wittry, 1972; Smith *et al.*, 1975; and Small *et al.*, 1987). We will describe here only one approach (Fiori *et al.*, 1976) to communicate the essence of the method. Early theoretical work by Kramers (1923) demonstrated that the intensity distribution of the continuum, as a function of the energy of the emitted photons, is

$$I_E \Delta E = k_E \bar{Z}(E_0 - E)\,\Delta E. \tag{7.1}$$

In this equation $I_E \Delta E$ is the average energy of the continuous radiation produced by one electron in the energy range from E to $E + \Delta E$. E_0 is the incident electron energy in keV, E is the x-ray photon energy in keV, Z is the specimen's average atomic number, and k_E is a constant, often called Kramers' constant, which is supposedly independent of Z, E_0, and $(E_0 - E)$. The "intensity" term I_E in Eq. (7.1) refers to the x-ray energy. The number of photons N_E in the energy interval from E to $E + \Delta E$ per incident electron is given by

$$N_E \Delta E = k_E \bar{Z}\left(\frac{E_0 - E}{E}\right)\Delta E \tag{7.2}$$

This point has been a source of considerable confusion in the literature, since Eq. (7.1) is often used incorrectly to describe the shape of the spectral distribution of the background generated within a bulk specimen. Equation (7.2) for the intensity of the continuum was discussed in Chapter 3. This ambiguity in the definition of intensity was undoubtedly complicated by the fact that the first ionization detectors integrated total radiant energy collected rather than counting individual x-ray photons. Such an ionization detector would produce the same output for one photon of energy $2E$ as it would for two photons, each of energy E, arriving simultaneously in the detector. The difference in the shape of the two curves described by Eqs. (7.1) and (7.2) is illustrated in Fig. 7.3.

The measured number of photons $N(E)$, within the energy range E to $E + \Delta E$, produced in time t with a beam current i (electrons per second), observed by a detector of efficiency P_E and subtending a solid

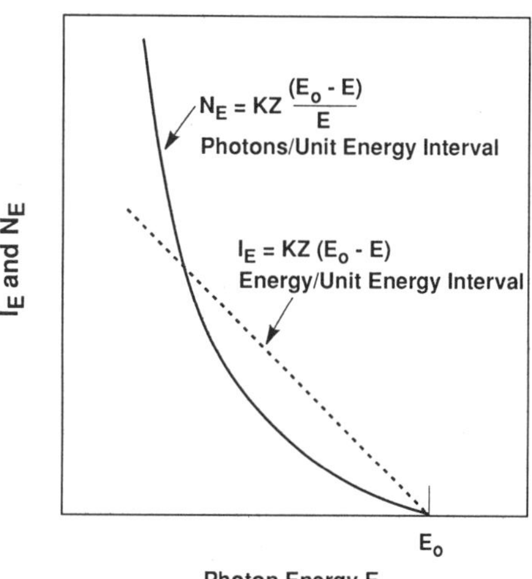

Figure 7.3. The theoretical shape of the x-ray continuum according to Kramers' equation (Green, 1962).

angle Ω, is equal to

$$N(E) = \frac{\Omega}{4\pi} itf_E P_E N_E \,\Delta E$$

$$N(E) = \frac{\Omega}{4\pi} itf_E P_E k_E \bar{Z}\left(\frac{E_0 - E}{E}\right) \Delta E. \tag{7.3}$$

Here, the term f_E (absorption factor for the continuum) denotes the probability of absorption of a photon of energy E within the target. In Eq. (7.3), the term $\Omega/4\pi$, the fraction of a sphere which the detector subtends, is equal to the fraction of photons emitted toward the detector only if the generation of the continuum is isotropic. This assumption of isotropy is a reasonable approximation for a thick target, although a small degree of anisotropy does exist. If we define a equal to $(\Omega/4\pi)it\,\Delta E$, then we obtain

$$N(E) = af_E P_E k_E \bar{Z}\left(\frac{E_0 - E}{E}\right). \tag{7.4}$$

The energy and atomic number dependence of $N(E)$, expressed by Eq. (7.4), is only approximate. The residual correction has been carried out in at least two ways: (1) applying more refined theoretical calculations and modifying the exponents from unity on the energy term and atomic number Z in Kramers' equation; or (2) using an additive energy-dependent correction term in Eq. (7.4), as proposed by Ware and Reed (1973):

$$N(E) = a\left[f_E P_E \bar{Z}\left(\frac{E_0 - E}{E}\right) + F(E)\right], \tag{7.5}$$

where $F(E)$ is taken from an empirically determined table; the authors indicated that for continuum photon energies above 3 keV, the additive energy term was unnecessary. Lifshin (1974) empirically developed and tested a more general relation for $N(E)$:

$$N(E) = f_E P_E \bar{Z}\left[a\left(\frac{E_0 - E}{E}\right) + b\frac{(E_0 - E)^2}{E}\right], \tag{7.6}$$

in which a and b are fitting factors and the remaining terms are as defined for Eq. (7.4).

Fiori *et al.* (1976) found that it is possible to determine a and b for any target by measuring $N(E)$ at two or more separate photon energies and solving the resulting equations in n unknowns. This procedure presumes that both f_E and P_E in Eq. (7.6) are known. Since the determination of $N(E)$ is made from measurements on the target of interest only, the inaccuracies of Kramers' law with respect to atomic number do not affect the method. The points chosen for measurement of $N(E)$ must be free of peak interference and detector artifacts (incomplete charge collection, pulse pileup, double-energy peaks, and escape peaks).

Since a and b are determined for each spectrum being fitted, the quadratic term adds great robustness to the quality of the fit.

Absorption Factor, f_E. For characteristic primary x-ray photons, the absorption within the target is taken into account by a factor called $f(\chi)$. (See Section 9.2) The factor $f(\chi)$ can be calculated from Eq. (9.18) (PDH equation), or a simpler form due to Yakowitz *et al.* (1973) can be used:

$$1/f(\chi) = 1 + a_1\gamma\chi + a_2\gamma^2\chi^2, \tag{7.7}$$

where γ is $E_0^{1.65} - E_C^{1.65}$, E_C is the excitation potential in keV for the shell of interest, and $\chi = (\mu/\rho)^i \csc\psi$. The term $(\mu/\rho)^i$ is the mass absorption coefficient of the element of interest in the target, and ψ is the x-ray take-off angle. The currently used values for the coefficients a_1 and a_2 are $a_1 = 2.4 \times 10^{-6}\,\text{gcm}^{-2}\text{keV}^{-1.65}$, $a_2 = 1.44 \times 10^{-12}\,\text{g}^2\text{cm}^{-4}\text{keV}^{-3.3}$.

If it is assumed that the generation of the continuum has the same depth distribution to that of characteristic radiation and if $E \approx E_C$, then Eq. (7.7) can be modified to

$$1/f_E = 1 + a_1(E_0^{1.65} - E^{1.65})\chi_c + a_2(E_0^{1.65} - E^{1.65})^2\chi_c^2, \tag{7.8}$$

where $\chi_C = (\mu/\rho)^E \csc\psi$. The term $(\mu/\rho)^E$ is the mass absorption coefficient for continuum photons of energy E in the target, and f_E is the absorption factor for these photons. Similar assumptions and expressions have been used by other authors.

As pointed out by Ware and Reed (1973), the absorption term f_E depends on the composition of the target. Consequently, when the composition of the target is unknown, it is necessary to include Eq. (7.6) in an iteration loop when the ZAF method for x-ray microanalysis (see Chapter 9) is being applied.

Detector Efficiency, P_E. As discussed in Chapter 5, the radiation emitted from the target toward the detector penetrates several layers of "window" material before it arrives in the "active" part of the detector. The nominal purpose of the first "window" is to protect the cooled detector chip from the relatively poor vacuum in the specimen chamber.

The first window can be made from a variety of materials. Historically, a beryllium window, typically about 7.6 μm thick, has been used. During the last several years window materials with considerably less mass-thickness have been gaining wide popularity. These window materials are either boron or silicon nitride or diamond or are organic.

The second window material is a surface-barrier contact (about 20 nm thick and usually made of gold). The purpose of this window is to provide an electrical contact to the diode. The gold is not uniform in thickness and tends to form islands.

The third window is an inactive layer of silicon extending 200 nm or less into the detector. This layer is also not uniform and is often called the "dead layer" or "silicon dead zone." The radiation then enters the active (intrinsic) region of the detector which has a thickness typically between 2 and 5 mm.

As an example of how we would model the transmission through these various windows, we will provide an expression for the traditional beryllium window detector. The absorption losses in the beryllium window, the gold, the silicon dead layer, and the transmission through the active silicon zone can be calculated from the linear combination of Beer's Law applied to each material separately:

$$P_E = \exp - \left[\left(\frac{\mu}{\rho}\right)^E_{\text{Be}} t_{\text{Be}} + \left(\frac{\mu}{\rho}\right)^E_{\text{Au}} t_{\text{Au}} + \left(\frac{\mu}{\rho}\right)^E_{\text{Si}} t^D_{\text{Si}}\right]\left\{1 - \exp\left[\left(\frac{\mu}{\rho}\right)^E_{\text{Si}} t_{\text{Si}}\right]\right\} \quad (7.9)$$

In (7.9), t^D_{Si} and t_{Si} are the mass thicknesses (g/cm^2) of the silicon dead layer and of the active detector region, respectively. The mass attenuation coefficients of beryllium, gold, and silicon at the energy E, $(\mu/\rho)^E_{\text{Be}}$, $(\mu/\rho)^E_{\text{Au}}$, and $(\mu/\rho)^E_{\text{Si}}$ are calculated as described by Myklebust *et al.* (1979). Since sufficiently accurate values are not usually available from the manufacturer, estimates of the mass thicknesses can be adjusted to optimize the fit between the calculated and the experimental values.

An example of applying this technique to fitting the continuum from a complex sample is presented in Fig. 7.4. The sample is a mineral,

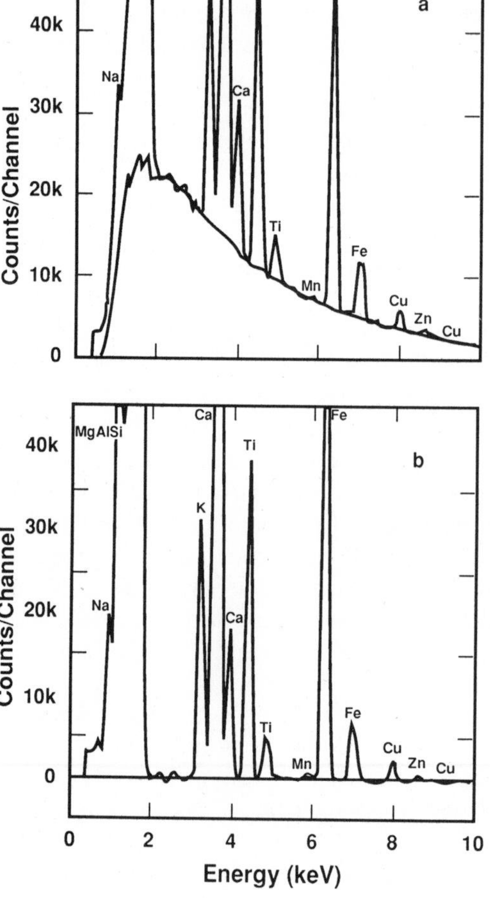

Figure 7.4. Background correction in Kakanui hornblende. (a) Fitted continuum curve calculated by Eq. (7.7); Observed spectrum is superposed. Note the presence of Mn $K\alpha$ peak at 5.9 keV. The concentration of Mn is less than 700 ppm. (b) Background subtracted.

Kakanui hornblende. It is possible to detect even the Mn concentration of less than 700 ppm.

7.2.1.2. Background Filtering

In this method of background calculation, the continuum component of an x-ray spectrum is viewed as an undesirable signal whose effect can be removed by mathematical filtering or by modification of the frequency distribution of the spectrum. Knowledge of the physics of x-ray production, emission, and detection is not required.

Filtering techniques take advantage of the fact that the continuum component of an x-ray spectrum is smooth and slowly varying, as a function of energy, relative to the characteristic x-ray peaks, except at absorption edges. If we mathematically transform (Fourier transform) an x-ray spectrum from energy space into frequency space, the result would be as shown in Fig. 7.5. The horizontal axis gives the frequency of an equivalent sine-wave component. For example, a sine wave with a full period of 0–10 keV would be plotted at channel 1, while a sine wave with a full period in 10 eV would be plotted at channel 1000. The vertical axis $|F|$ gives the population of each sine wave. In this representation the continuum background which has a long-period variation is found at the low-frequency end. The characteristic peaks vary more rapidly than the background and are found at higher frequency. The noise is found at all frequencies.

We can now mathematically suppress the frequency components responsible for the slowly varying continuum and additionally suppress a portion of the frequency components responsible for the statistical noise. By then performing a reverse Fourier transform, we go back to energy space with a spectrum which should now be composed only of charac-

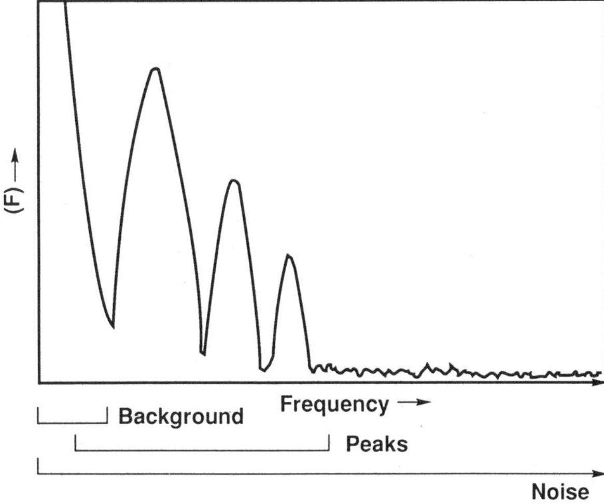

Figure 7.5. Schematic illustration of frequency space (Fourier transform) representation of electron-excited x-ray spectrum. (See text for explanation.)

teristic peaks devoid of continuum and which is, furthermore, smoothed for statistical variation. Unfortunately, close examination of Fig. 7.5 reveals that the three main components of the transform overlap. Consequently, it is not possible to suppress all the undesirable components without sacrificing part of the peak components. Similarly, it is not possible to keep all of the peak components without including part of the undesirable components. The result in either case will be a spectrum with undesirable distortions. Although the Fourier transform approach has been successfully applied to a number of specific systems, the question of what happens to the instrumentally broadened absorption edges and common frequency components between the background and peaks has not been studied in sufficient detail to ensure accurate analysis for all systems. The frequency representation is useful mainly for purposes of instruction.

Another filtering technique now widely used was developed by Schamber (1978). It has been successfully applied to a wide variety of systems and uses what is known as the "top hat" digital filter. The top hat filter is a simple and elegant algorithm—a fact which can easily be obscured by the mathematical formalism required to describe it. Simply stated, the top hat filter is a special way of averaging a group of adjacent channels of a spectrum, assigning the "average" to the center "channel" of the filter, and placing this value in a channel in a new spectrum which we will call the filtered spectrum. The filter is then moved one channel and a new "average" is obtained. The process is repeated until the entire spectrum has been stepped through. The filter in no way modifies the original spectrum; data are taken from the original to create a new spectrum. The "averaging" is done in the following manner. The filter (see Fig. 7.6) is divided into three sections: a central section, or positive lobe, and two side sections, or negative lobes. The central lobe is a group of adjacent channels in the original spectrum from which the contents are summed together and the sum divided by the number of channels in the central lobe. The side lobes, similarly, are two groups of adjacent channels from which the contents are summed together and the sum divided by the total number of channels in both lobes. The "average" of the side lobes is then subtracted from the "average" of the upper lobe.

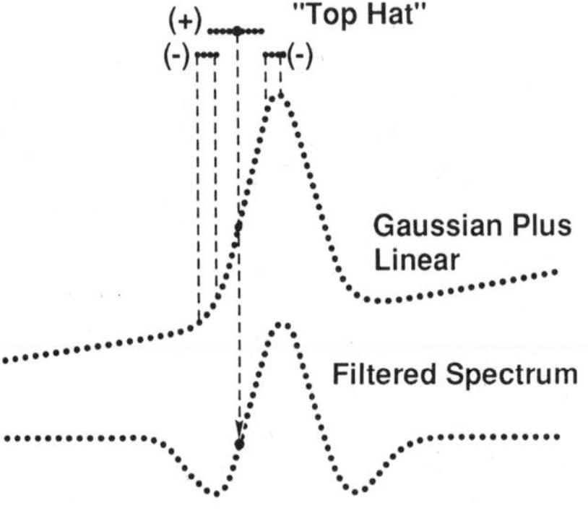

Figure 7.6. Effect of "top hat" digital filter on a spectrum comprised of a Gaussian peak plus a sloped linear background. The filtered spectrum is plotted immediately below the actual spectrum. The channel correspondence for one calculation of the top hat filter is shown.

This quantity is then placed in a new spectrum into a channel which corresponds to the center channel of the filter.

The effect of this particular averaging procedure is as follows. If the original spectrum is straight, across the width of the filter, then the "average" is zero. If the original spectrum is concave upward, across the width of the filter, the "average" is negative. Similarly, if the spectrum is convex upward, the "average" is positive. The greater the curvature, the greater the "average." The above effects can be observed, for a Gaussian superposed on a linear background, in Fig. 7.6. In order for the filter to respond as much as possible to the curvature found in spectral peaks and as little as possible to the curvature found in the spectral background, the width of the filter must be carefully chosen. For a detailed treatment of the subject see Schamber (1978) and Statham (1977). In general, the width of the filter for any given spectrometer system is chosen to be twice the full width at half the maximum amplitude (FWHM) of the Mn $K\alpha$ peak, with the number of channels in the upper lobe equal to or slightly more than the combined number of channels in the side lobes.

Because the top hat filter "averages" a number of adjacent channels, the effects of counting statistics in any one channel are strongly suppressed. Consequently, in addition to suppressing the background under spectral peaks, the digital filter also "smooths" a spectrum. Note that in Fig. 7.6, the top hat filter converts the sloped background into a flat background.

In summary, the effects of passing a top hat digital filter through an x-ray spectrum as recorded by a Si(Li) spectrometer system are to strongly suppress the background and statistical scatter and significantly alter the shape of the spectral peaks. The result strongly resembles the smoothed second derivative; however, this distortion has no adverse statistical or mathematical effects of any consequence. Clear advantages of the method are simplicity and the fact that an explicit model of the continuum is not required. However, since the continuum has been

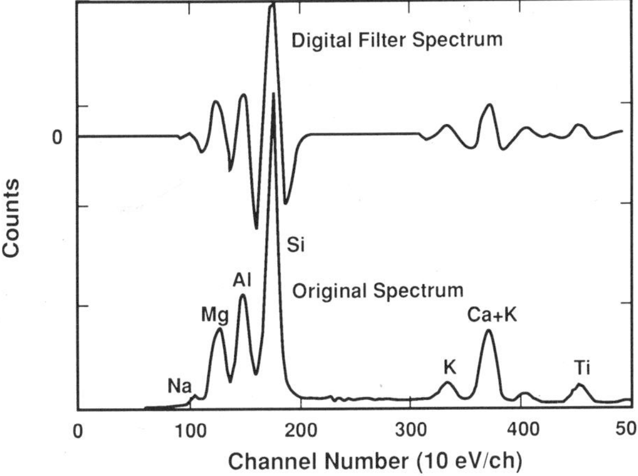

Figure 7.7. Spectrum of Kakanui hornblende (lower plot) and the digital filter of this spectrum (upper plot).

suppressed, the information it carried (i.e., average atomic number, mass-thickness, etc.) is no longer available. The filter method is currently the most widely used procedure to remove background effects. The result of digitally filtering an experimentally generated spectrum of a complex geological specimen, Kakanui hornblende, is shown in Fig. 7.7.

7.2.2. Background Correction for WDS

7.2.2.1. Interpolation

The high spectral resolution of the WDS results in well separated peaks for most analytical situations, even with complex mixtures of elements. Because the peaks are well separated, the background between the peaks is accessible by simply moving the spectrometer off the peak position, as illustrated in Fig. 7.8. The most commonly used method of background measurement with a wavelength-dispersive spectrometer is to detune to wavelengths slightly above and below the tails of the characteristic peak and then establish the value of the background at the peak setting by linear interpolation.

The interpolation method is not foolproof when performing a trace analysis, since the background is not a truly straight line but has curvature. This curvature results not only from the shape predicted by Kramers' law but also from the presence of any nearby large peaks. As described in Chapter 5, WDS peaks have a Lorentzian profile rather than the Gaussian profiles seen in EDS. Lorentzian peak profiles have tails which extend out a larger distance than in those of Gaussian peaks. Thus, when minor or trace constituents are to be measured by WDS, it is good practice to actually scan the spectrometer through the peak to inspect the nearby background regions. Unexpected deviations from the expected smooth background due to higher-order lines, absorption edges, Lorentzian tails, etc. can be recognized, and the background correction can be adjusted appropriately.

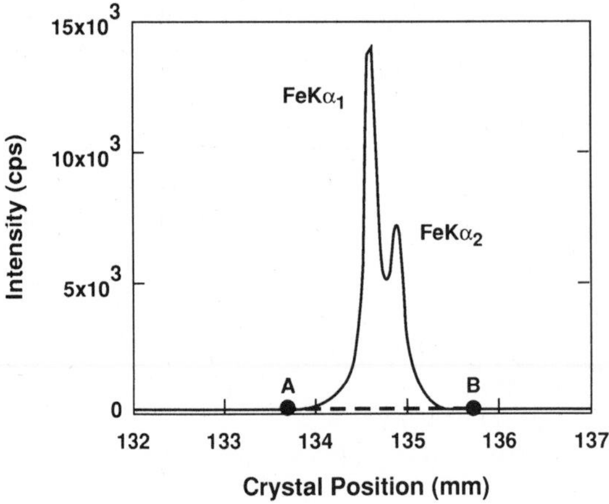

Figure 7.8. Commonly used method of background measurement with a WDS using linear interpolation between measured background positions at A and B. Crystal position is equivalent to wavelength in this figure.

7.2.2.2. Substitute Material Method

In older instruments it is generally not desirable to detune the spectrometer between sample and standard measurements because the spectrometer position cannot be accurately reproduced. Mechanical backlash normally prevents accurate repositioning to the previous peak value. It is also difficult to obtain 1% reproducibility in the intensity by determining maximum signal on a rate meter while manually adjusting the spectrometer control. If we wish to avoid moving the spectrometer when measuring background, it is sometimes possible to locate a substitute material which has approximately the same average atomic number as the material of interest but does not contain the element being analyzed. Since the background generated at any energy is approximately proportional to average atomic number, other conditions being equal, it is then possible to obtain an estimate of the background under the characteristic x-ray peak in the specimen of interest. As an example, when measuring the background for Ni $K\alpha$, a copper standard could be used. This substitute material method must be used with caution. The background region of the spectrum of the candidate substitute material should be inspected by scanning to ensure that no unexpected peaks are present. It is sometimes not possible to find a suitable material with an average atomic number sufficiently close to that of an unknown specimen. When the peak-to-background ratio is low, as for the analysis of concentrations below 1%, the errors introduced by shifting to a substitute material can become unacceptably large. Fortunately, in modern WDS units it is possible to accurately determine a peak position, detune to obtain a background reading, and return to the peak position by simply dialing in the wavelength setting. In modern WDS instruments a computer-controlled step scan in wavelength can be used for repeaking, although some loss of data-collection time will result.

7.3. Peak Overlap Correction

7.3.1. EDS Peak Overlap Correction

To measure the intensity of an x-ray line in a spectrum, we must separate the line from other lines and from the continuum background. The preceding section discussed how to remove the average effect of the background. This section will discuss various ways to isolate an x-ray line from the average effect of other lines when spectral overlap occurs. We will assume in the following discussion that our spectra have been background corrected.

Resolution is the capability of a spectrometer to separate peaks close in energy or wavelength. It is conventionally characterized by the FWHM. As discussed in Chapter 5, the resolution of an energy-dispersive detector is usually specified at the energy of Mn $K\alpha$ (5.89 keV). Typical detectors have a resolution between 130 and 155 eV.

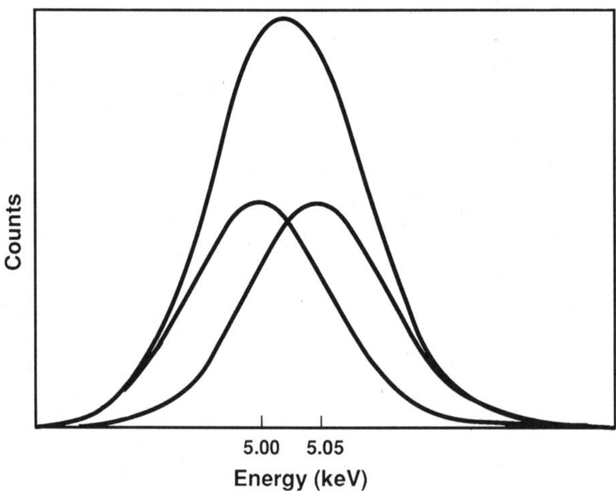

Figure 7.9. Computer generation of two spectral peaks and their convolution.

Due to the complexity of x-ray spectra from multielement specimens, this level of resolution results in frequent situations of peak overlap. It is then necessary to isolate a characteristic x-ray peak from the average effects of other peaks by some type of mathematical procedure.

The computer-generated continuous curves in Fig. 7.9 illustrate the problem. The upper curve, which is a composite of the two underlying curves, is the sort of spectrum we would see from an energy-dispersive detector if there were no background or statistical modulation and the horizontal (energy) axis was essentially continuous rather than discrete. The underlying curves are what we must reconstruct given only the information contained in the upper curve. The area under each of the lower curves represents the number of characteristic x-ray photons recorded by the energy-dispersive detector system and is one of the quantities we require in any quantitative data-reduction scheme (see Chapters 8 and 9).

Procedures which address themselves to the solution of this problem have a variety of names, such as curve-fitting, deconvolution, peak stripping, peak unraveling, multiple linear least-squares curve fitting, etc. We will attempt to make the distinctions clear.

All of the procedures just mentioned start with a data set usually obtained from a computer x-ray analyzer (CXA). The data set obtained exists as a one-dimensional array in a computer memory. Consequently, the data are discrete. Consider now a second but corresponding spectrum also existing in a one-dimensional array. We will call this a "calculated" spectrum. This spectrum can be generated in several ways. One way is to utilize a mathematical model to separately describe each peak in the spectrum. At each channel in the calculated spectrum, the effects of all the peaks at that channel are summed together. This process is called *convolution,* and it is by this process that we can generate a spectrum. The mathematical model used to describe the shape of each peak usually has a minimum of three parameters, one to specify the amplitude of the peak, one to describe its width, and one to describe its position (energy). The Gaussian (normal) profile is the most-often-used model for the peaks.

An alternative method of generating a spectrum is to add together simpler spectra (corrected for the average effects of the background) obtained by actually measuring specimens of known composition under known conditions. Each of these spectra is referred to as a "reference" spectrum. For example, if the unknown spectrum was obtained from a copper–zinc alloy, the two reference spectra would be obtained by measuring pure copper and pure zinc. In this case the only parameters we have to adjust are the amplitudes of each of the two reference spectra. A common error is to confuse a reference with a standard. It is important to note that a reference is merely a shape function. The analytical conditions required for a good standard are often not those required to make a good reference. In general, a reference spectrum is acquired from pure elements or simple compounds which do not have to be stable under the electron beam. As an example, NaCl makes an excellent reference for Na or Cl, but as a standard for either of those elements it is an extremely poor choice because it suffers extreme beam damage, so that peak heights change, but not peak shapes.

Whether we obtain a calculated spectrum by using a mathematical model for each constituent peak and then adding these, or whether we record several simpler spectra, which are then scaled and added together, a criterion is needed to determine if the calculated spectrum matches the unknown as closely as possible. We also require a procedure to adjust the various parameters to provide this "best" fit. It is the procedure by which the parameters are adjusted which serves mainly to distinguish the various methods used to account for spectral overlaps (Fiori and Myklebust, 1979; Fiori *et al.*, 1984; Schamber, 1973; 1977; Statham, 1976). The following sections describe what we mean by linearity and goodness of fit and provide an overview of the important methods for dealing with peak overlaps.

7.3.1.1. Linearity

All of the procedures which address the problem of isolating one peak from the interfering effects of other peaks can be classified into one of two categories: linear procedures and nonlinear procedures. We require a definition of linearity as it applies to the above problem. Linearity (or nonlinearity) is a property of the fitting parameters. If all of the parameters being adjusted to provide the best fit in a given procedure are used in a simple multiplicative or additive fashion, then the procedure is linear. This definition applies only to the parameters actually being adjusted. There can be other parameters which are used in a nonlinear manner, but as long as they remain fixed during the actual peak fitting, the procedure remains linear. An example will help to clarify these points.

The profile of a characteristic x-ray peak obtained from an EDS is closely approximated by a Gaussian (normal) probability distribution function. That is, the contents, Y_i, of any given channel comprising a given Gaussian profile can be calculated from

$$Y_l = A_p \exp\left[-\frac{1}{2}\frac{(E_p - E_i)^2}{\sigma^2} \right], \tag{7.10}$$

where E_i is the energy (in appropriate energy units, e.g., electron volts) of the ith channel, E_p is the energy (same energy units as E_i) of the profile center, and A_p is the amplitude of the profile at its center. It should be noted that the center of the profile does not have to coincide in energy with the mean energy of any given channel. The parameter σ specifies the width of the profile (again, in the same units as E). Of the three parameters, A_p, E_p, and σ, only A_p is linear since it is present with a first-power exponent. Consequently, any procedure which uses Gaussian profiles and adjusts either the width or the profile center during the fitting is a nonlinear procedure. There is one exception worth mentioning. If only one peak is being fitted, it is possible to take the natural logarithm both of the unknown spectrum and of Eq. (7.10) and, after some appropriate algebraic manipulation, derive a new set of parameters that are linear. These parameters can be transformed back into the original parameters after the fitting has been accomplished.

7.3.1.2. Goodness of Fit

We require a criterion with which to evaluate the goodness of fit for a given peak-fitting procedure in a particular application. That is to say, is there a quantity which will tell us how closely our calculated peaks match the peaks in an unknown spectrum? A widely used criterion is the chi-squared (χ^2) criterion. The set of parameters used in a given fitting procedure which cause χ^2 to reach a minimum value will be the set with the greatest likelihood of being the correct parameters. The χ^2 criterion can be reasonably approximated by the following functional form:

$$\chi^2 \approx \sum_i \frac{(Y_i - X_i)^2}{X_i}, \tag{7.11}$$

where Y_i is the content of the ith channel of the unknown spectrum and X_i is the content of the ith channel of the calculated spectrum. For the equation as written to be a proper χ^2 fit it is required that repeated measurements of Y_i for a given channel i be normally distributed with a true variance of X_i and that the equation we have chosen to describe a peak (or the reference spectrum we have chosen) is a correct representation in all aspects (e.g., a mathematical function that would include any peak distortions such as incomplete charge, etc.). Under many conditions of peak fitting, the above conditions are approached with sufficient accuracy that a minimum value of Eq. (7.11) means that the set of fitting parameters which produced the minimum will be the "best choice" parameters. Note that because we square the differences between the two spectra, χ^2 is more sensitive to larger differences than it is to smaller differences.

A useful variation of Eq. (7.11) is the normalized χ^2:

$$\chi_N^2 = \sum_{i=1}^{M} \frac{1}{X_i} \frac{(Y_i - X_i)^2}{M - f}, \tag{7.12}$$

where M is the number of channels used in the fit and f is the number of parameters used in the fit (i.e., A_p, σ, E_P). The utility of Eq. (7.12) lies in the range of χ^2 values it will produce. When $\chi^2 \approx 1$, the fit is essentially perfect, while for values of $\chi^2 \gg 1$, the fit is poor. For a high-quality, low-noise spectrum, a χ^2 value greater than 100 would indicate a bad fit.

7.3.1.3. The Linear Methods

As stated previously, linearity requires that only those parameters that are used in a multiplicative or additive manner may be adjusted during the fitting procedure. A simple additive parameter has the effect of moving a peak or spectrum with respect to the vertical (amplitude) axis, and consequently is useful mainly for accommodating the average effect of the background if background were to be included in the fitting procedure. While it is possible to include background along with x-ray peaks, it unnecessarily complicates the mathematics, can cause convergence problems, and substantially increases the computational time. Consequently, as mentioned earlier, we will assume that the average effect of the background is first removed from a spectrum before we attempt to unravel the spectral overlaps.

Besides addition, the other possible method by which a parameter can enter an expression, and still retain the property of linearity, is multiplication. The practical consequence of multiplication is that only the amplitudes of peaks can be determined in a linear-fitting procedure. At first glance this seems to be a severe restriction because it requires us to know beforehand the exact energy calibration of our spectrometer system (that is, the precise position of each peak in the spectrum), and it requires us to know the exact width of each peak. Assuming values for either of these parameters, especially peak position, which differ from corresponding values in our "calculated" spectrum can cause the amplitudes of some, or all, of the peaks to be incorrectly determined by the fitting procedure. Fortunately, the stability, linearity, and count-rate performance of modern spectrometer systems are generally adequate when certain precautions are observed (see Chapter 5).

The most popular linear method is the multiple linear least-squares technique. Assume that we have a number of channels which comprise a background-corrected, measured spectrum of N overlapped peaks consisting of Y_i counts in each channel. Further assume a calculated spectrum of N corresponding peaks. Each peak is described by a Gaussian profile [Eq. (7.10)] for which we must estimate as best we can the width and energy. The calculated spectrum can be mathematically described by

$$Y_i = \sum_N A_N \exp\left[-\frac{1}{2}\left(\frac{E_N - E_i}{\sigma_N}\right)^2\right], \tag{7.13}$$

where A_N is the amplitude, E_N is the energy, σ_N is the width parameter for each peak N, and E_i is the energy of each channel in the calculated spectrum. E_N, E_i, and σ_N each have dimensions of energy (usually eV or keV). We need to find a set of amplitudes A_N, that will cause the calculated spectrum to match the measured spectrum as closely as possible. The contents Y_i of each channel in the measured spectrum have an associated error, ΔY_i, due to Poisson counting statistics. This error is proportional to $Y_i^{1/2}$. Since channels which are near peak centers can contain many more counts than those which are in the tail of a peak, it is common practice to weight the effect of each channel such that it has approximately the same effect in the fitting procedure as any other channel. The weighting factor W_i is typically chosen to be proportional to $1/(\Delta Y)^2$, which is equal to $1/Y_i$. The principle of least squares states that the best set of A_N will be those which minimize

$$\chi^2 = \sum_i \left(Y_i - \sum_N A_N G_{Ni}\right)^2 W_i, \tag{7.14}$$

where G_{Ni} is the Gaussian profile [the exponential portion of Eq. (7.10)]. The desired set of A_N can be found by the following procedure. Successively differentiate Eq. (7.14) with respect to each of the A_N. The result is a set of N equations of the form

$$2\sum_i \left[\left(Y_i - \sum_N A_N G_{Ni}\right)(-G_{Ni})W_i\right] \tag{7.15}$$

Equating each of these to zero serves to couple the equations (which can now be called simultaneous equations) and permits us to find the desired minimum. One way to solve the set of equations is by matrix inversion. Equation (7.15) can be written in matrix notation as

$$[\mathbf{B}] \times [\mathbf{A}] = [\mathbf{C}], \tag{7.16}$$

where

$$B_{mn} = \sum_i G_{mi} G_{ni} W_i$$

and

$$C_m = \sum_i G_{mi} Y_i W_i.$$

The subscripts m and n refer to the rows and columns of the matrix. Consequently, we can find our best set of amplitudes A_N by

$$[\mathbf{A}] = [\mathbf{B}^{-1}] \times [\mathbf{C}]. \tag{7.17}$$

We note that the set A_N has been found analytically. That is to say, we have determined a set of equations which, when solved, give a unique answer which is most likely (in the statistical sense) to be correct. The analytical property is a major distinction between the linear and nonlinear methods of peak fitting. In the nonlinear methods we do not have a closed-form set of equations which must be solved. We must resort, for example, to iteration techniques or search procedures to give us the answer we need. We must always keep in mind, however, how we achieved the luxury of linearity. We have made the rather bold assumption that we know very well the width and location of every peak in our observed spectrum. The determination of these parameters, to the required accuracy, is quite feasible but is far from being a trivial exercise.

In this description a calculated spectrum was constructed from simple Gaussian profiles. Other functions could have been used, such as a slightly modified Gaussian, to accommodate the small deviation of observed x-ray peaks on the low-energy side from a true Gaussian (described later). Reference spectra (measured on pure elements or simple compounds) are usually used rather than functions to create the "calculated" spectrum. Finally, digitally filtered spectra both for the measured and the calculated spectra could have been used. The original purpose in applying the digital filter was to remove the average effects of the background. The application of a digital filter to a spectrum modifies the x-ray peaks from nearly Gaussian profiles to approximations of smoothed second derivatives of Gaussian profiles (see Fig. 7.7). Neither the application of a digital filter nor the use of other than Gaussian functions affects the linearity of the method just discussed. In all cases the parameter being determined is peak amplitude, which is a linear parameter even in the case of digitally filtered spectra. This latter procedure of using measured references and the application of the digital filter to the spectrum being fit and to the references is now the most widely used method to extract the peak areas required for quantitative analysis. The procedure has been called the "Filter Fit" method.

7.3.1.4. The Nonlinear Methods

We have seen in the preceding section that a major advantage of the linear technique is that it is analytic. We have only to solve a set of equations to get the information we require. The amplitudes returned by a linear peak-fitting procedure are unique and are "most probably" the correct amplitudes. As with everything, very little comes without price, and the price paid to gain linearity is that the user must provide the fitting procedure with information about the width and position of all peaks, rather than the other way around. The linear procedure then uses this information without change, since it is, by definition, correct. There is a school of thought which suggests that the user cannot always be so sure about his or her knowledge of peak width and location and that these should be included, with amplitude, as quantities to be determined by the fitting procedure. Both linear and nonlinear peak fitting work well in general, and each has areas where it is superior. While the nonlinear techniques are capable of determining fitting parameters other than amplitudes, the price we must pay for this luxury is dealing with a procedure which is nonanalytic and quite capable of returning an answer which is wrong if certain precautions are not observed. In the following paragraphs we will describe in some detail one of the methods presently in use, the sequential simplex.

A sequential simplex procedure is a technique which can also be used for selection of the set of independent variables in a mathematical expression which cause the expression to be a "best" fit (in a statistical sense) to a set of data points (Nelder and Mead, 1962; Deming and Morgan, 1973).

In this procedure, each of the n independent variables in the function to be fitted is assigned an axis in an n-dimensional coordinate system. A simplex, in this coordinate system, is defined to be a geometric figure consisting of $n + 1$ vectors [in this discussion we will use the purely mathematical definition of a vector, i.e., an ordered n-tuple $(X_1, X_2, \ldots, X_n)$]. In one dimension a simplex is a line segment; in two dimensions, a triangle; and in three or more dimensions a polyhedron, the vertices of which are the above mentioned $n + 1$ vectors. The simplex is moved toward the set of independent variables which optimize the fit according to a set of specific rules. The function used to determine the quality of the fit for any set of independent variables is called the "response function."

We will use a simple example to demonstrate the essence of the method. Consider the problem of finding the "best" straight line through a set of data points. A straight line is defined by the two quantities m, the slope of the line and b, the y-axis intercept of the line. The functional form is

$$y = mx + b. \tag{7.18}$$

The quantities we wish to optimize by a simplex search are m and b. We start the process by estimating, as best we can, three lines, each defined by its own slope m_i and y-axis intercept b_i (Fig. 7.10). Each of these three lines can be characterized by a vector (m_i, b_i) on an (m, b) plane. The three vectors form a simplex—in this case a triangle—on the plane (Fig. 7.11). If we now define a third orthogonal axis to represent the "response function," the simplex search can be visualized. The search proceeds by calculating the response above each vertex of the simplex, discarding the vertex having the highest (worst) response, and reflecting this point across the line connecting the remaining two vertices. This defines a new simplex, and the process is repeated until the global minimum on the response surface is reached.

The problem of fitting a straight line to a set of data points is, of course, a linear problem. However, the method of sequential simplex is an organized procedure to search for a minimum on a response surface and is usable whether the function describing the data set is linear in its coefficients or not. If the coefficients are linear, or can be made linear by a suitable transformation, the least-squares determination can be accom-

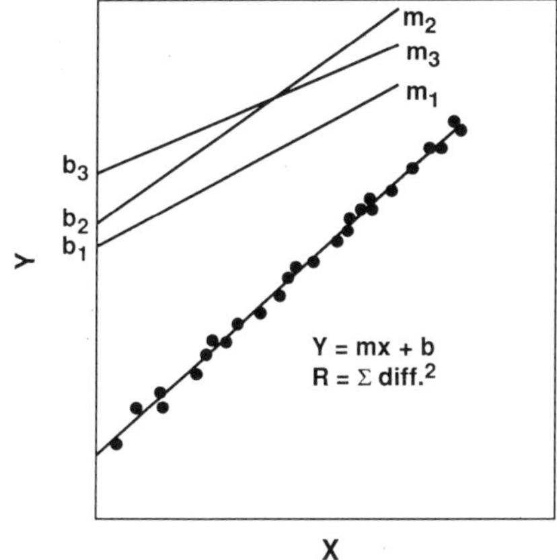

Figure 7.10. Illustration of a simplex procedure. A "true" straight line and three estimates for this line are plotted. Each line is characterized by its own slope *m* and *y*-intercept *b*.

plished analytically. In this case a search procedure would not be used, and one is used here only for the purpose of explanation.

We now require a functional description of an x-ray peak, just as required in the linear procedure. Again, we can use an additive combination of *N* Gaussian profiles [see Eq. (7.13)]. The fitting procedure involves the calculation of the function, which is then compared with the experimental data points. The purpose of the response function is to provide a measure of the goodness of fit to the data points. A response function which can be used is the normalized chi-square function [see Eq. (7.12)].

From Eq. (7.10) we see that the independent variables A_p, E_p, and σ_p must be determined for each peak. Therefore the total number of coefficients n is three times the number of peaks N. Here, values for $n + 1$ different sets of coefficients are chosen to create the initial simplex, and the response R (Fig. 7.11) is determined for each set.

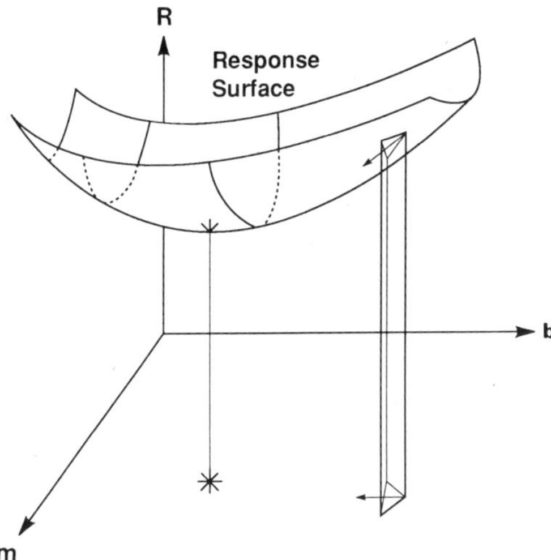

Figure 7.11. Response surface, showing simplex and projection onto factor space. The global minimum is shown by an asterisk.

Fiori and Myklebust (1979) introduced several simplifications which reduce the number of coefficients to be fitted. Since the energies of x-ray lines are well known, they are entered as known quantities, and it is unnecessary to include all of the peak centroids as coefficients. The energy of the principal peak only is used as a coefficient to correct for shifts in energy due to miscalibration of the electronics of the EDS. The stability and linearity of present-day amplifiers permits this assumption. In addition, the width σ_p of the principal peak must be included, since the widths of the other peaks σ_i are related to σ_p [for a Si(Li) detector, Fiori and Newbury (1978)]:

$$\sigma_i = \frac{[2500(E_i - E_P) + (2355\sigma_P)^2]^{1/2}}{2.355} \text{ (keV)} \tag{7.19}$$

The number of coefficients required is, therefore, reduced from $3N$ to $N + 2$. Consequently, only one more coefficient (amplitude) must be included for each additional peak used in a fit. Since the energies and widths of small unresolved peaks are determined as functions of the principal peak, opportunities for obtaining false minima are considerably reduced. Significant savings in computation time are also realized.

A difficulty common to all minimization methods is the possibility that a local minimum may be found before the global minimum. As the number of peaks to be fitted increases so does the possibility of finding false minima. Consequently, the simplex must be started closer to the unique set of coefficients that determine the global minimum than to a set which determine a local (false) minimum.

7.3.1.5. Error Estimation

It is the purpose of this section to discuss the various errors which occur when corrections are applied to account for spectral overlap. By "error" we mean the deviation between the true amplitudes and the calculated amplitudes. We have to consider two types of error. First, there are the unavoidable errors due to fundamental statistical limitations and inadequacies in our algorithms. These errors, in general, cannot be avoided but can often be quantified, or at least estimated. The second type of error is experimental. This type of error can often be avoided but, in general, cannot be quantified. Examples of the second type include imperfect subtraction of the average effect of the background, distortion of spectral peaks due to electrical ground loops, direct entrance of backscattered electrons into the detector, etc.

The amplitude information we receive is rarely an end product. This information is fed to another procedure which accounts for specimen and standard matrix effects (for example, the ZAF data reduction program for x-ray microanalysis). Chapters 8 and 9 show that the matrix-correction programs themselves have associated errors which add in quadrature to the peak-amplitude errors. Correcting for spectral overlap always contributes an error in peak amplitude greater than what would be expected from consideration of the counting statistics alone. This additional error due only to the unraveling process can often dwarf all other errors in quantitative x-ray microanalysis. It can categorically be stated that quantitative analysis utilizing spectra without spectral overlap is always superior to those analyses in which overlap occurs, all other things being equal. For a theoretical treatment of the error due to unraveling see Ryder (1977).

In the discussion of multiple linear least squares, we noted that one way to find the "best" set of amplitudes A was to solve the matrix equation

$$[\mathbf{B}] \times [\mathbf{A}] = [\mathbf{C}] \tag{7.20}$$

by the method of matrx inversion

$$[\mathbf{A}] = [\mathbf{B}^{-1}] \times [\mathbf{C}]. \tag{7.21}$$

The matrix $[\mathbf{B}^{-1}]$ is often called the error matrix because its diagonal, designated B_{ii}^{-1}, has the useful property that

$$\sigma_{Ai}^2 \approx B_{ii}^{-1}, \tag{7.22}$$

where σ_{Ai}^2 is the variance associated with the ith amplitude. Consequently, we have a simple means by which to estimate the quality of our fit in terms of each peak. If we have N peaks we have N estimates of uncertainty, one for each amplitude returned by the fitting procedure. In general, the uncertainty associated with unraveling overlapped spectral peaks increases rapidly as the number of counts under any given peak decreases or as the separation between peaks decreases. At a certain point the uncertainty becomes so great that spectral unraveling becomes dangerous.

Consider the problem of determining the true peak heights for two closely spaced, overlapped peaks such as those shown in Fig. 7.9. The user employing one of the unraveling methods obtains as results a set of calculated amplitudes and an associated χ^2 value. Since this χ^2 value is supposedly the minimum value in the set of χ^2, it indicates that the peak amplitudes must be the best choice. We can calculate χ^2 for a range of peak amplitudes and then examine the so-called χ^2 surface to determine how sensitive the χ^2 parameter is to the selection of the true peak amplitudes. Figures 7.12 and 7.13 show such χ^2 surfaces calculated for the case of two peaks separated by 50 eV and centered at 5.0 and 5.05 keV, obtained from a detector with a resolution of 150 eV at Mn $K\alpha$. R is the χ^2 value obtained for each selected value of peak amplitudes A and B. In Fig. 7.12, the true peak amplitudes were equal at 1000 counts, while in Fig. 7.13 the true amplitude ratio is 5:1 (1000:200 counts).

The χ^2 value plotted on the R axis in these figures is obtained by comparing the values of the hypothetical composite curve generated for each selection of the amplitudes of A and B. It can be seen from both plots that a sharply defined minimum in the response surface does not exist. A long, flat-bottomed valley exists near the true solution. The flat

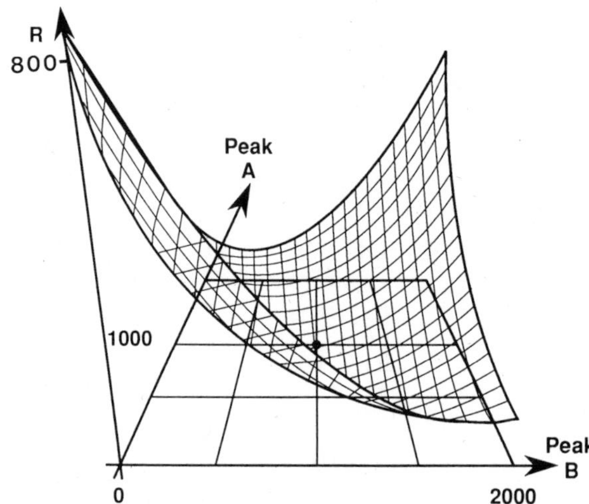

Figure 7.12. Pseudo-three-dimensional plot of χ^2 response surface for deconvolution of two peaks of equal amplitude separated by 50 eV. True solution is indicated by heavy point. Calculation was made by varying peak amplitude while holding peak σ and energy constant.

bottom results from one peak increasing in amplitude while the other decreases the same amount. The steep slopes result when both peaks increase or decrease together. Thus, it is not possible to determine the true solution with high confidence using the χ^2 (or any other) minimization criterion. If the uncertainty due to counting statistics is included, the χ^2 surfaces would be "pot-holed" with local false minima, making the selection of the true solution even more unlikely.

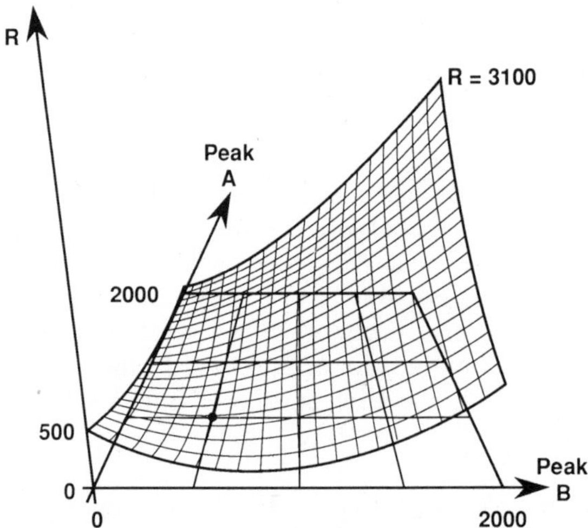

Figure 7.13. Pseudo-three-dimensional plot of χ^2 response surface for deconvolution of two peaks of 5:1 amplitude ratio separated by 50 eV. True solution is indicated by heavy point. Calculation was made by varying peak amplitude while holding peak σ and energy constant.

This discussion demonstrates that spectral unraveling techniques have some limitations. The problem of separating an overlap such as S $K\alpha$ (2.308 keV) and Pb $M\alpha$ (2.346 keV) is difficult and in general will produce values for S and Pb that are less certain than if we determined them in a situation were there was no overlap.

To demonstrate that extra error results when overlap is present, we show that the spread (variance) in area determinations from fitting 25 spectra each of sulphur and lead individually and then when they are combined as PbS produces different results. Figure 7.14 is one spectrum of PbS from the set of 25 as it would be seen by a 10-mm^2 detector mounted 52 mm from a specimen of PbS irradiated by a 20-keV, 0.1-nA electron beam for 100 s. Figure 7.15 is similar but expanded horizontally for detail. It also includes the individual contributions of S and Pb, showing the severe amount of spectral overlap for this material. The spectra were all generated from first principles. The simplex procedure, described above, was used to determine the areas of the S peaks and the Pb peaks from the individual contributions before combining and after combining. This was done for 25 spectra identical except for the statistical variation due to counting noise. The results are given in Table 7.1.

The results are shown as area counts $\pm 3\sigma$. The smaller S peak has 60% greater uncertainty in the case of overlap than if no overlap existed. This type of analysis, where N spectra are used to determine the statistical uncertainty, is called a Monte Carlo method.

The linear methods, in general, each provide a simple means to estimate the fitting error for each peak through the error matrix. The nonlinear methods, on the other hand, do not provide a convenient technique for the error estimates of the individual peaks. It should be noted that both linear and nonlinear methods supply a single measure of the error associated with the entire fit over all of the peaks. This measure is simply the normalized chi-squared, χ^2. However, χ^2 does not tell us

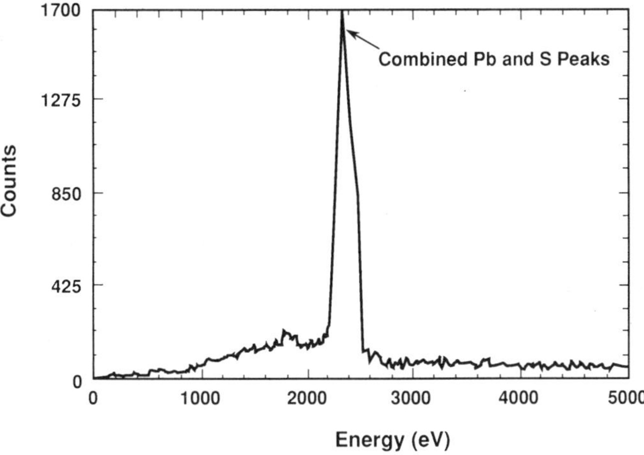

Figure 7.14. Spectrum of PbS showing the severe overlap between the Pb M family and the S K family (E_0 = 20 keV; 0.1-nA, 100-s, 10-mm^2 detector, 40° take-off angle).

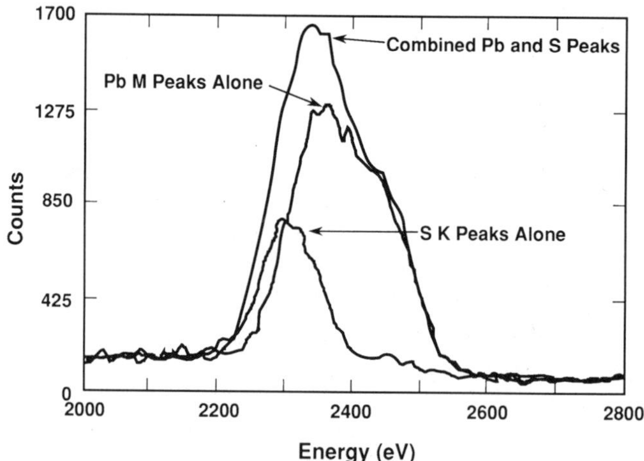

Figure 7.15. Same as Fig. 7.14 but expanded horizontally for greater detail. The individual contributions of Pb and S are also shown.

very much about the errors associated with the individual peaks which comprise the overlapped spectrum. We must note, however, that the error matrix of the linear methods is valid only when the conditions which permit a linear assumption are satisfied. That is, we accurately know each peak width and position in our spectra. If we make an error in either or both of these quantities, then the amplitudes determined by the fitting procedure will be in error and the error matrix will not manifest this deviation.

In conclusion, this section has discussed several of the errors associated with unraveling spectral overlap. It is, unfortunately, a common belief that the techniques which we have been discussing have unlimited ability to provide accurate determinations of peak amplitudes no matter how bad the overlap or how few counts are in the peaks. It is for this reason alone that some measure of expected error (such as the error matrix) for individual peaks is so valuable. The several procedures in use today are powerful tools when intelligently used but all have limits to their capabilities. *Caveat emptor.*

Table 7.1. Area of S and Pb Peaks, Uncombined and in PbS[a]

	S $K\alpha$ area	Pb $M\alpha$ area
Pb	0	12,051 ± 943.8
S	6942 ± 1032	0
PbS	7319.9 ± 1560	11,585 ± 1098

[a] From 25 spectra differing only in statistical variation due to counting noise.

7.3.2. WDS Peak-Overlap Correction

The natural energy distribution of characteristic x rays of a single line is well described by the Lorentzian probability distribution:

$$Y_i = \frac{A_C}{1 + \left[\frac{(E_i - E_{CN})}{\gamma}\right]^2}, \tag{7.23}$$

where Y_i is the amplitude in the ith channel, γ = FWHM/2, A_C is the amplitude at the center energy of the peak, E_{CN} is the center energy, and E_i is the energy of the ith channel away from the center energy. For comparison to this distribution, the peak shape of a characteristic x-ray line, even if it assumed to be a perfect step function (i.e., infinitely narrow), as observed by an energy-dispersive system is well described by a Gaussian curve:

$$Y_i = A_C \exp\left[-\ln(2)\left(\frac{E_i - E_{CN}}{\gamma}\right)^2\right], \tag{7.24}$$

where the terms have the same meaning as those in Eq. (7.23). The ln(2) term is a scaling constant to permit the use of FWHM/2 in the Gaussian function, rather than the usual description, Eq. (5.7).

The relative shape of these two distributions, for the same amplitude, energy and width, is shown in Fig. 7.16. Notice the relatively long tails of the Lorentzian distribution. These tails can extend a considerable distance, up to hundreds of eV, from the parent peak. The existence of these long tails is the reason that background correction is important in WDS when performing a trace-element analysis using an x-ray line in the vicinity of a large line from another element in the specimen.

The instrumental broadening of the energy-dispersive detector is large. Typically the FWHM at the energy of Mn $K\alpha$ is 135–165 eV, while the natural width at Mn $K\alpha_{1,2}$ is just a few eV. Consequently, the Gaussian shape of the energy-dispersive detector dominates the Lorentzian shape of the natural x-ray line. For the case of a WDS, with its

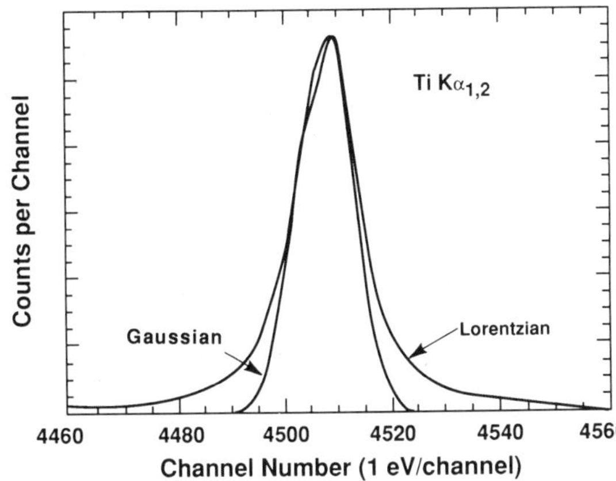

Figure 7.16. Representation of an x-ray peak with FWHM of 10 eV as described by a Gaussian and a Lorentzian function. Note the long tails of the Lorentzian curve.

considerably better energy resolution, the shape of the natural line width cannot be neglected. Often, the peak shapes as observed by a WDS have strong Lorentzian contributions (Remond *et al.*, 1989; Remond, 1990). Stated mathematically, the observed peak shape is the convolution of the natural line profile and the instrumental response function. In the case of the WDS, the instrumental response curve itself has a Lorentzian character due to the deliberately induced crystal imperfections (polygonization) in the structure of several types of diffracting crystals. Another implicating factor is the effect of chemical bonding on the natural x-ray line distributions, which distort the Lorentzian profile. Chemical bonding effects are most pronounced for the $K\alpha$ x-ray lines for the light elements, Be to Si (see Section 9.10).

In general, the instrument response curve for a WDS is not predictable and can change dramatically depending on the choice of diffracting crystal, slit width, etc. The shape can vary from near Gaussian, for the new synthetic multilayer diffraction structures, to strongly Lorentzian, for a lithium fluoride crystal. It is possible, however, to develop a peak description as a convolution of a given quantity of a Lorentzian function with a given quantity of a Gaussian function. This approach provides an instrumental response curve which very closely matches the observed distribution from a given WDS making it possible to "fit" WDS spectra.

There are two situations when we might wish to "fit" a WDS peak. While it is a rare occurrence, it is possible to experience peak interference in WDS similar to that experienced in EDS. There is now the possibility of applying the same type of peak fitting employed for EDS in Section 7.3.1 to the WDS case, since modern WDS spectrometers

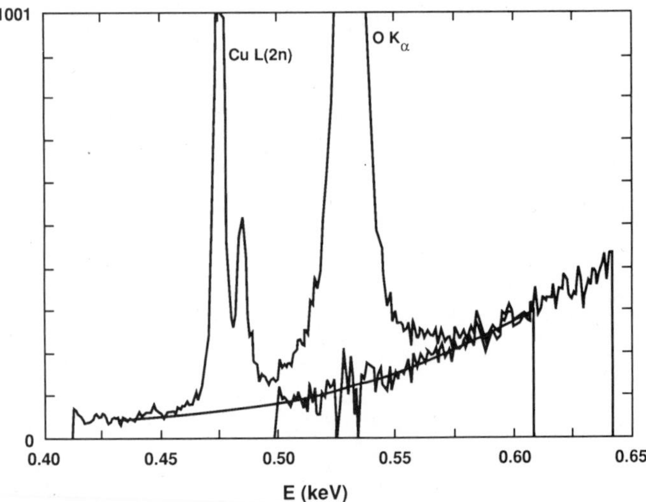

Figure 7.17. Example of a WDS peak deconvolution for an oxygen K peak with an interference from the second-order copper $L\alpha$ peak. The residuals show the background under the stripped peak. Specimen: Cu_2O: $E_0 = 5$ keV.

are controlled by, and record data into, a computer x-ray analyzer. Under computer control, the WDS is scanned over a peak or peaks and the adjacent background region. The recorded "spectrum" is similar to that observed by an EDS, but over a more restricted range in energy. With the WDS spectrum in this form, we may then use either linear or nonlinear methods to fit the peaks, just as in the EDS case.

The second reason for wishing to "fit" a WDS peak is for the case of light-element *K*-line analysis, for elements from Be to Si, or for the low energy (<1 keV) *L* and *M* lines of the heavier elements, if low incident-beam energy is being used. Chemical bonding effects can cause the peak profiles for x-ray lines of these elements to shift in energy or change in shape. Since it is the area of peaks, rather than their amplitudes, that we seek to find for quantitative analysis (see Section 9.10), a method which can extract this quantity from the background is important. Figure 7.17 demonstrates the modeling of the oxygen *K* peak with an interference from a nearby Cu $L\alpha$ second-order peak. Following modeling, the subtraction of the peak from the spectrum leaves just the background distribution, as shown by the fit of the residuals.

8

Quantitative X-Ray Analysis: The Basics

8.1. Introduction

As discussed in Chapters 5 and 6, the x rays emitted from a specimen bombarded with the finely focused electron beam of the scanning electron microscope (SEM) or electron-probe microanalyzer (EPMA) can be used to identify which elements are present (qualitative analysis). With the proper experimental setup and data-reduction procedures, the measured x rays can also be used to quantitatively analyze chemical composition with an accuracy and precision approaching 1%. This chapter provides an *overview* of the basic principles and techniques used for determining chemical composition, on the micrometer scale, with the SEM and EPMA. Our intention is to provide the conceptual basis for an understanding of the x-ray microanalytical data-reduction procedures that today are almost always incorporated into black-box computer-based, integrated analysis systems with which the analyst interacts as a user. As a user, the analyst depends on the knowledge and skill of programmers to have devised an accurate, robust analytical procedure from the diverse approaches available in the literature. Despite the apparent disconnection which has arisen between the analyst as user and the underlying physics as incorporated into the algorithms of the software, it is nevertheless extremely important to grasp the underlying physical principles to become a sophisticated analyst rather than a mere user. The analyst must understand the limitations of the technique as well as its strengths to avoid becoming a slave to a canned system. Moreover, the x-ray microanalysis software often presents choices of data-correction procedures for the user to make, and an optimal choice obviously depends on proper knowledge of the relative merits of each approach. This chapter underlies the material in Chapter 9, which gives the detailed theories and equations for x-ray generation, emission, and the ZAF and $\phi(\rho z)$ techniques for quantitative analysis of flat polished specimens. In addition, Chapter 9 presents material on nonideal sample geometries and light-element analysis.

8.2. Advantages of Quantitative X-ray Microanalysis in the SEM/EPMA

The major advantage of the x-ray quantitative technique in the SEM and EPMA is that the analysis is obtained from a small volume of material. As shown in Chapter 3, x rays can be generated, depending on the initial electron-beam energy and atomic number, from volumes with linear dimensions as small as 1 micrometer. This means that, typically, a volume as small as $10^{-12}\ \mathrm{cm}^3$ can be analyzed. Assuming a typical density of $7\ \mathrm{g/cm}^3$ for a transition metal, the composition of 7×10^{-12} g of material can be determined. From this small mass of the sample selected by the electron–x ray interaction volume, elemental constituents can be determined to concentrations ranging as low as 0.01% (100 ppm), which corresponds to limits of detection in terms of mass of 10^{-16} to 10^{-15} g. For reference, a single atom of iron weighs about 10^{-22} g, so our limit of detection corresponds to only a few million atoms. Why would we wish to analyze such a small volume or mass of the sample to such sensitivities? The answer is that most natural or artificial solid substances are chemically heterogeneous on the microscopic scale. In a wide variety of cases, whether the examples are drawn from fields as diverse as biology or materials science, the macroscopic properties and behavior of the substance are often controlled by chemical structures and processes that take place on the spatial scale of micrometers to nanometers. Quantitative x-ray microanalysis provides the tool to characterize the microstructure.

The ability to detect 10^{-16} to 10^{-15} g of an analyte suggests that use of the SEM or EPMA is a trace-analytical technique. However, the term trace, as used by the analytical community, refers to the capability to detect a minute fraction of a constituent in the analyzed volume. What constitutes a trace-level measurement is not strictly defined. Generally, trace is taken to mean a concentration level near the limits of detection, and therefore what is considered the trace level varies from one technique to another. Thus, some analysts consider that the trace level is parts per million (ppm), while others working with a different technique might insist on parts per trillion (ppt). For the SEM and EPMA, the following arbitrary definitions of concentration levels will be used, based on the practical observation that energy-dispersive x-ray spectrometry in the SEM or EPMA has limits of approximately 0.1 wt %:

Major: $>$ 10 wt %

Minor: 1–10 wt %

Trace: $<$ 1 wt %

Since the focused electron probe of the SEM or EPMA is largely nondestructive with regard to the specimen, the specimen can be reexamined using optical microscopy or other techniques. This capability is extremely important when microanalysis or microstructural information must be complemented by other data from the same sample. Virtually all types of solid materials can be analyzed using the SEM or EPMA. Among these materials are metals and their alloys; minerals and nonmetals such as nitrides, carbides, and oxides; polymers; and biological materials. The last two often contain water, which may boil off in the vacuum during analysis and are poor conductors. In such cases procedures are needed to remove or stabilize the water as ice. For all samples, it is critical to mount the materials in a conducting medium or to coat the materials so that the electrons from the focused probe find a path to ground. For a coated surface, the sample must be connected to a good ground path on the sample holder or stage. Such a connection is often made with conducting carbon paint, which must be thoroughly dried to avoid degrading the instrument's vacuum.

8.3. Quantitative Analysis Procedures

The general procedure for quantitative analysis in the SEM or EPMA can be outlined as follows below. Many of these steps, or even the complete procedure, may be performed automatically by the computer-aided (or-dominated) analytical system. Before attempting a quantitative analysis, the analyst must be sure to perform a *qualitative* analysis to identify the constituents present with a high degree of confidence. This topic has been discussed in detail in Chapter 6. No quantitative analysis, no matter how carefully performed, is useful if the spectral peaks being quantified actually belong to another element!

1. *Obtain the x-ray spectrum of the specimen and standards under defined and reproducible conditions.* Quantitative analysis makes the assumption that all measurements for a given element, in both the specimen and the standards, are performed with identical spectrometer response (efficiency, spectrometer take-off angle, calibration, and resolution), at the same beam energy, and under the same electron dose (beam current multiplied by the spectral accumulation time). The experimental measurement procedure is shown in Fig. 8.1. The x-ray intensity from element i is measured by the same x-ray detector from both the sample and the standard in which the initial electron beam energy E_0, the electron probe current, and the angle (take-off angle ψ) between the specimen surface and the direction of the measured x-rays are held constant. If possible, E_0 should be chosen to obtain an optimum overvoltage U equal to 2.0. By using the ratio of the measured intensities ($I_i/I_{(i)}$) from sample and standards, instrumental factors and other constant factors found both in the specimen and standard will drop out. While simple normalization can

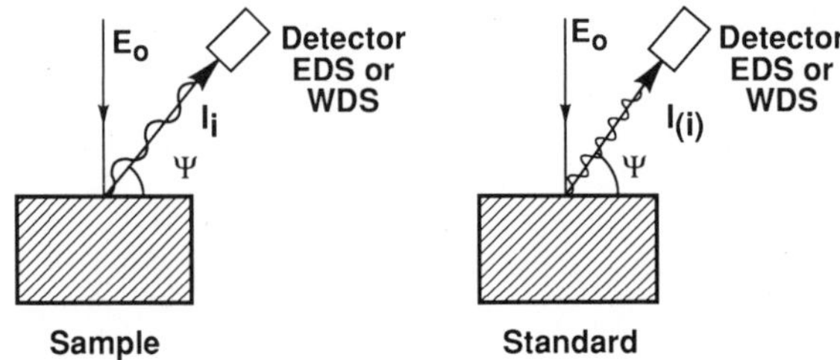

Figure 8.1. General experimental set up for quantitative x-ray microanalysis.

be applied to adjust for differences in dose if spectra are recorded for different times or at different beam currents, such corrections can be made only to the accuracy with which the beam current has been measured and to which the spectrum accumulation time has been controlled. Hence, an accurate beam-current measuring system (specimen stage, Faraday cup, and calibrated picoammeter) and accurate dead-time correction by the EDS or WDS system must be assured.

2. *Measure standards containing the elements that have been identified in the specimen.* A standard for x-ray microanalysis is a substance which has not only a *known* composition but the *same* composition everywhere throughout; that is, the standard must be homogeneous at the microscopic level. Since most substances are not homogeneous, they cannot serve as x-ray microanalysis standards. Thus, a steel may be available that has been carefully and accurately characterized by bulk analytical chemistry procedures, but since steels almost invariably have a complex chemical microstructure, such a material is not suitable as a microanalysis standard. Fortunately, and this is one of the great strengths of electron-excited x-ray microanalysis, accurate quantitative analysis can be performed with simple standards consisting of pure elements or—for those elements whose form is incompatible with the requirements of measurement in a vacuum (e.g., phosphorus)—a simple stoichiometric compound such as GaP.

3. *Process the spectra of the specimen and standards to remove the background, which results from the x-ray continuum, from the x-ray peaks, so that the measured intensities consist only of the characteristic signal.* The background formed by the x-ray bremsstrahlung or continuum is a significant fraction of the characteristic peaks for minor and trace constituents. Incomplete or inaccurate removal adversely affects the accuracy of quantitation. Background removal has been discussed extensively in Chapter 7; different procedures are used for WDS and EDS spectra. Since the x-ray continuum actually depends on the composition, background removal may be incorporated into the quantitative-analysis procedure where it is repeated iteratively as a progressively better estimate for the composition of the specimen is obtained.

4. *Develop the x-ray intensity ratios using the specimen intensity* I_i

and the standard intensity $I_{(i)}$ for each element present in the sample (see Fig. 8.1) and carry out matrix corrections to obtain quantitative concentration values. This critical step is discussed in the next section.

8.4. The Approach to X-Ray Quantitation: The Need for Matrix Corrections

Upon initial examination, it would seem that quantitative analysis should be extremely simple. Just form the ratio of the characteristic x-ray intensity measured from the specimen to that measured from the standard, and that ratio should be equal to the ratio of concentrations between the specimen and the standard. As was first noted by Castaing (1951), the primary generated intensities are roughly proportional to the respective mass fractions of the emitting element. If the fluorescence and absorption contributions are very small, the measured intensity ratios between specimen and standard are roughly equal to the ratios of the mass or weight fractions of the emitting element. This assumption is often applied to x-ray quantitation; it is called the first approximation to quantitative analysis and given by

$$\frac{C_i}{C_{(i)}} = \frac{I_i}{I_{(i)}} = k. \tag{8.1}$$

The terms C_i and $C_{(i)}$ are the composition in weight (mass) concentration of element i in the unknown and in the standard, respectively. The ratio of the measured unknown-to-standard intensities, $I_i/I_{(i)}$, is the basic experimental measurement that underlies all quantitative x-ray microanalysis and is called the "k-value."

Careful measurements performed on homogeneous substances of known multielement composition compared to pure element standards reveal that there are significant systematic deviations between the ratio of measured intensities and the ratio of concentrations. An example of these deviations is shown in Fig. 8.2, which depicts the deviations of measured x-ray intensities in the iron–nickel binary system from the linear behavior predicted by the first approximation to quantitative analysis, Eq. (8.1). Figure 8.2 shows the measurement of $I_i/I_{(i)}$ for Ni $K\alpha$ and Fe $K\alpha$ in nine well-characterized homogeneous Fe–Ni standards (Goldstein *et al.*, 1965). The data were taken at an initial electron-beam energy of 30 keV and a take-off angle ψ of 52.5°. The intensity ratio k_{Ni} or k_{Fe} is the $I_i/I_{(i)}$ measurement for Ni or Fe, respectively. The straight lines plotted between pure Fe and pure Ni indicate the relationship between composition and intensity ratio given in Eq. 8.1. For Ni $K\alpha$, the actual data fall below the first approximation and indicate that there is an absorption effect taking place, that is, more absorption in the sample than in the standard. For Fe $K\alpha$, the measured data fall above the first approximation and indicate that a fluorescence effect is taking place in the sample. In this alloy the Ni

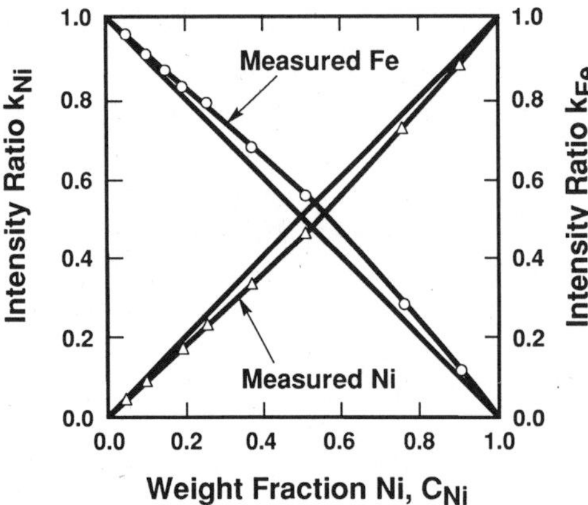

Figure 8.2. Measured Fe–Ni intensity ratio, k-value, versus weight fraction Ni; curves are k-ratios, straight lines represent ideal behavior. 30 keV.

$K\alpha$ radiation is heavily absorbed by the iron and the Fe $K\alpha$ radiation is increased over that generated by the bombarding electrons due to x-ray fluorescence by the Ni $K\alpha$ radiation.

In most quantitative chemical analyses, the measured intensities from specimen and standard need to be corrected for differences in electron backscatter, density, x-ray cross-section, and energy loss as well as absorption within the solid in order to arrive at the ratio of generated intensities and hence the value of C_i. These effects are referred to as matrix or interelement effects. The magnitude of the matrix effects can be quite large, exceeding factors of ten or more in certain systems. Recognition of the complexity of the problem of the analysis of solid samples has led numerous investigators to develop the theoretical treatment of the quantitative analysis scheme first proposed by Castaing (1951).

8.5. The Physical Origin of Matrix Effects

What is the origin of these matrix effects? As discussed in Chapter 3, the x-ray intensity *generated* for each element in the specimen is proportional to the concentration of that element, the probability of x-ray production or the ionization cross-section for that element, the path length of the electrons, and the fraction of incident electrons that remain in the specimen and are not backscattered. It is very difficult to calculate directly the absolute generated intensity for the elements present in a specimen. However, the intensity that the analyst must deal with is the *measured* intensity. The measured intensity is even more difficult to calculate, particularly because absorption and fluorescence of the generated x rays may occur in the specimen, thus further modifying the measured x-ray intensity from that predicted on the basis of the ionization cross-section alone. Instrumental factors such as differing spectrometer efficiency as a function of x-ray energy must also be considered. Many of these factors are dependent on the atomic species involved. Thus, in mixtures of elements, matrix effects arise because of differences in elastic and inelastic scattering processes and in the propagation of x rays through the specimen to reach the detector. For conceptual as well as calculational reasons, it is convenient to divide the matrix effects into those due to atomic number, Z_i; x-ray absorption, A_i; and x-ray fluorescence, F_i.

Using these matrix effects, the most common form of the correction equation is

$$\frac{C_i}{C_{(i)}} = [ZAF]_i \frac{I_i}{I_{(i)}} = [ZAF]_i k_i, \tag{8.2}$$

where C_i is the weight fraction of the element i of interest in the sample and $C_{(i)}$ is the weight fraction of i in the standard. This equation must be applied separately for *each* element present in the sample. The Z, A, and F effects must therefore be calculated separately for each element. Equation 8.2 is used to express the matrix effects and is the common basis for x-ray microanalysis in the SEM or EPMA.

It is important for the analyst to develop a good idea of the origin and the importance of each of the three major nonlinear effects on x-ray measurement for quantitative analysis of a large range of specimens. But before we can examine the terms Z_i, A_i, and F_i, we need to explore the process of x-ray production in more detail.

8.6. X-Ray Production

8.6.1. Effect of Atomic Number

As shown in Chapter 3, the paths of beam electrons within the specimen are well represented by Monte Carlo simulations of electron trajectories. In the Monte Carlo simulation technique, the detailed history of an electron trajectory is calculated in a stepwise manner. At each point along the trajectory, both elastic and inelastic scattering events can occur. The production of characteristic x-rays, an inelastic scattering process, can occur along the path of an electron as long as the energy E of the electron is above the critical excitation energy E_c of the characteristic x ray of interest.

Figure 8.3 displays Monte Carlo simulations of the positions where K-shell x-ray interactions occur for three elements, Al, Ti, and Cu, using an initial electron energy E_0 of 15 keV. The incoming electron beam is assumed to have a zero width and to impact normal to the sample surface. X-ray generation occurs in the lateral direction x and in the depth dimension z. The micron marker gives the distance in both the x and the z dimension. Each dot indicates the generation of an x ray; the black regions indicate that a large number of x rays are generated. This figure shows that the x-ray generation volume decreases with increasing atomic number (Al, $Z = 13$ to Cu, $Z = 29$) for the same initial electron energy. The decrease in x-ray generation volume is due to an increase in elastic scattering with atomic number, which deviates the electron path from the initial beam direction, and an increase in critical excitation energy E_c, with a corresponding decrease in overvoltage U ($U = E_0/E_c$) with atomic number. The decrease in U decreases the fraction of the initial electron energy available for the production of characteristic x rays and the energy range over which x rays can be produced.

One can observe from Fig. 8.3 that there is an uneven distribution of x-ray generation with depth z for specimens with various atomic numbers and initial electron-beam energies. This variation is illustrated by the histograms on the left side of the Monte Carlo simulations. These histograms plot the number of x rays generated with depth into the specimen. In detail, the x-ray generation for most specimens is somewhat higher just below the surface of the specimen and decreases to zero when the electron energy E falls below the critical excitation energy E_c of the characteristic x ray of interest.

As illustrated from the Monte Carlo simulations, the atomic number of specimen strongly affects the distribution of x-rays generated in specimens. These effects are even more complex when considering the more interesting multielement samples as well as in the

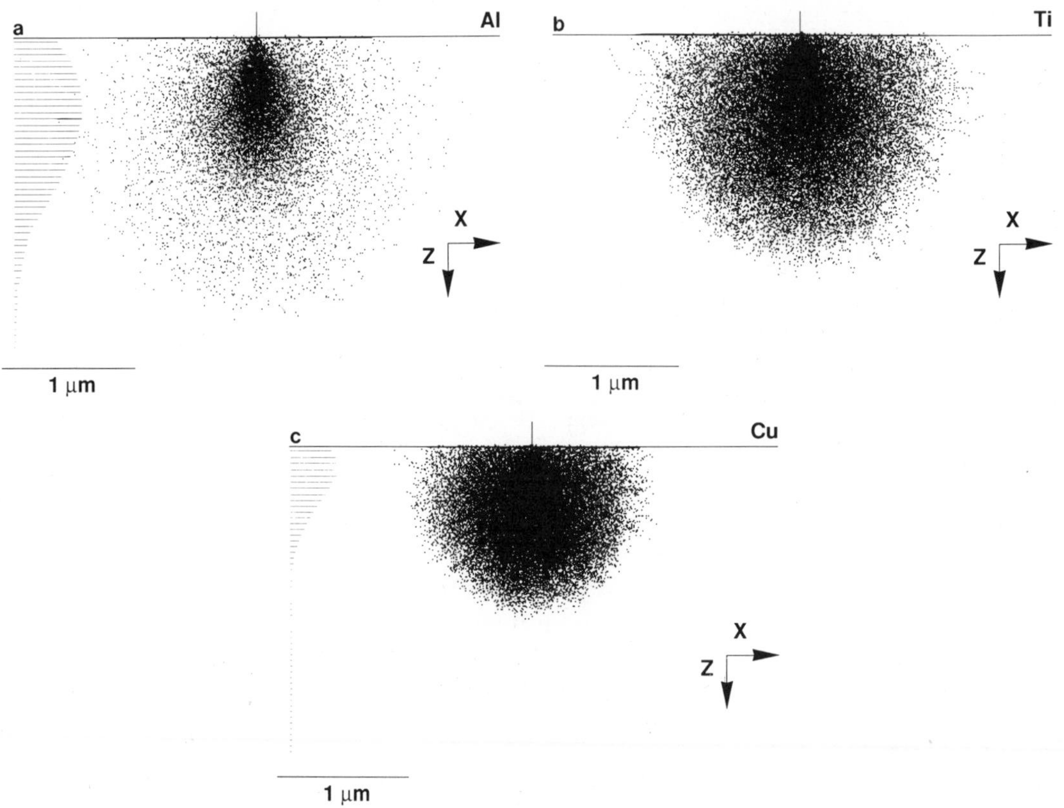

Figure 8.3. Monte Carlo calculations of X-ray generation at 15 keV for (a) Al, (b) Ti and (c) Cu.

generation of *L*- and *M*-shell x-ray radiation. In the next section, we will consider x-ray generation in depth in more detail.

8.6.2. X-Ray Generation with Depth, $\phi(\rho z)$

Figure 8.3 clearly shows that x-ray generation varies with depth as well as with the atomic number of the specimen. In practice it is very difficult to measure or calculate an absolute value for the x-ray intensity generated with depth. Therefore, we follow the practice first suggested by Castaing (1951) of using a relative or a normalized generated intensity which varies with depth, called $\phi(\rho z)$. The term ρz is called the *mass depth* and is the product of the density ρ of the sample and the depth dimension z; ρz is usually given in units of g/cm^2. The mass depth ρz is more commonly used than the depth term z. The use of the mass depth masks the effect of specimen depth when comparing specimens of different atomic number. Therefore it is important to recognize the difference between the two terms as the discussion of x-ray generation proceeds. The $\phi(\rho z)$ curve is normalized by the intensity generated in a free standing layer so thin that no elastic scattering occurs (e.g., <10 nm).

The general shape of the depth distribution of the generated x rays, the $\phi(\rho z)$-versus-ρz curve, is shown in Fig. 8.4. The amount of x-ray production is related to the amount of elastic scattering, the initial electron-beam energy, and the energy of the x-ray line of interest. As the incident beam penetrates the layers of material in depth, the length of the trajectory in each successive layer increases because elastic scattering deflects the electron out of the normal

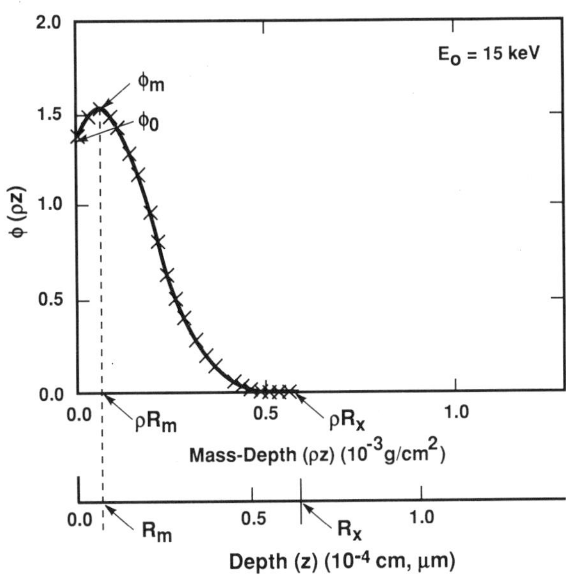

Figure 8.4. Schematic for the measurement of a $\phi(\rho z)$ curve.

straight-line path requiring a longer path to cross the layer and because backscattering results in electrons scattered deeper in the specimen crossing the layer in the opposite direction. Due to these factors, x-ray production increases with depth from the surface, $\rho z = 0$, and goes through a peak ϕ_m at a certain depth ρR_m (see Fig. 8.4). Another consequence of backscattering is that surface layer production ϕ_0 is larger than 1.0 in solid samples because the backscattered electrons excite x rays as they leave the sample. After the depth ρR_m, x-ray production begins to decrease with depth because the backscattering of the beam's electrons reduces the number of electrons available at increasing depth ρz, and the remaining electrons lose energy and therefore ionizing power as they scatter at increasing depths. Finally, x-ray production goes to zero at $\rho z = \rho R_x$, where the energy of the beam electrons no longer exceeds E_c.

Now that we have discussed and described the depth distribution of the production of x rays using the $\phi(\rho z)$-versus-ρz curves, it is important to understand how these curves differ with the type of specimen analyzed and the operating conditions of the instrument. The specimen and operating conditions that are most important in this regard are the average atomic number Z of the specimen and the initial electron-beam energy E_0 chosen for the analysis. Calculations of $\phi(\rho z)$-versus-ρz curves have been made for this chapter using the PROZA program (Bastin and Heijligers, 1990a). In Fig. 8.5, the $\phi(\rho z)$ curves for the $K\alpha$ lines of pure Al, Ti, and Cu specimens at 15 keV are displayed. The shapes of the $\phi(\rho z)$ curves are quite different. The ϕ_0 values increase from Al to Ti due to increased backscattering, which produces additional x-ray radiation. The ϕ_0 for Cu is smaller than the value for Ti because the overvoltage U for the Cu $K\alpha$ x ray lie at an E_0 of 15 keV is low (1.67) and the energy of many of the backscattered electrons is not sufficient to excite Cu $K\alpha$ x rays near the surface. The values of ρR_m and ρR_x decrease with increasing Z, and a smaller x-ray

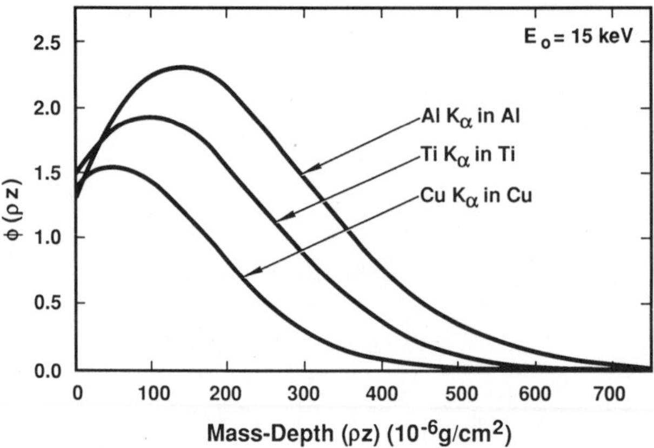

Figure 8.5. Calculated $\phi(\rho z)$ curves for Al $K\alpha$ in Al, Ti $K\alpha$ in Ti, and Cu $K\alpha$ in Cu at 15 keV.

excitation volume is produced. This decrease would be much more evident if we plotted $\phi(\rho z)$-versus-z, the depth of x-ray excitation, since the use of mass depth includes the density.

As discussed above, the x-ray intensity for each x-ray line in the specimen I_{gen}, can be obtained by taking the area under the $\phi(\rho z)$-versus-ρz curve, that is by summing the values of $\phi(\rho z)$ for all the layers $\Delta(\rho z)$ in mass-thickness within the specimen for the x-ray line of interest and then multiplying that area by the x-ray intensity I $(\Delta \rho z)$ from the isolated thin film. With this background on x-ray production, we are now in a position to discuss the three matrix effects (Z_i, A_i, and F_i) which appear in the correction equation Eq. (8.3).

8.7. ZAF Factors in Microanalysis

The matrix effects Z, A, and F all contribute to the correction for x-ray analysis as given in Eq. (8.2). This section discusses each of the matrix effects individually. The combined effect of ZAF determines the total matrix correction.

8.7.1. Atomic Number Effect

In x-ray microanalysis calculations, the atomic number effect Z_i is equal to the ratio of I_{gen} for the standard to I_{gen} for the sample for each element i. Using appropriate $\phi(\rho z)$ curves, I_{gen} can be obtained for the standard and the sample.

Another approach to the atomic number effect is to consider directly the two factors, backscattering (R) and stopping power (S), which determine the amount of x-ray intensity generated in a sample. Dividing the stopping power S for the sample and the standard by the backscattering term R for the sample and the standard yields the atomic number effect Z_i for each element i in the sample. A discussion of these two factors follows.

1. *R*. The process of elastic scattering leads to backscattering, which results in the premature loss of a significant fraction of the beam electrons from the target before all of the ionizing power of those electrons has been expended generating x-rays of the various elemental constituents. From Fig. 3.13, which depicts the backscattering coefficient as a function of atomic number, this effect is seen to be strong, particularly if the elements involved in the unknown and standard have widely differing atomic numbers. For example, consider the analysis of a minor constituent (e.g., 1%) of aluminum in gold against a pure aluminum standard. In the aluminum standard, the backscattering coefficient is about 15% at a beam energy of 20 keV, while for gold the value is about 50%. When aluminum is measured as a standard, about 85% of the beam electrons expend their energy completely in the target, producing the maximum amount of Al $K\alpha$ x rays. In gold, only 50% are stopped in the target, so by this effect,

aluminum is actually under represented in the x rays generated in the specimen relative to the standard. The energy distribution of backscattered electrons further exacerbates this effect. Not only are more electrons backscattered from high-atomic-number targets, but as shown in Fig. 3.21, the backscattered electrons from high-atomic-number targets carry off a higher fraction of their incident energy, further reducing the energy available for ionization of inner shells. The integrated effects of Fig. 3.13 and 3.21 form the basis of the R factor, the fraction of the maximum x-ray production actually generated in the sample, in the atomic number correction of the ZAF formulation of matrix corrections. R decreases with increasing Z and U.

2. *S*. The rate of energy loss due to inelastic scattering also depends strongly on the atomic number. For quantitative x-ray calculations, the concept of the stopping power of the target, given in Eq. (3.12), is used. Using the Bethe formulation for the rate of energy loss, dE/ds [Eq. (3.12)], reveals that the stopping power is a decreasing function of atomic number, so that low-atomic-number targets actually remove energy from the beam electron with mass distance ρz more rapidly than high-atomic-number targets. The S-factor is the second factor of the atomic number correction. The effects of R and S tend to go in the opposite directions and thus tend to cancel.

An example of the importance of the atomic number effect is shown in Fig. 8.6. This figure shows the measurement of the intensity ratios k_{Au} and k_{Cu} for Au $L\alpha$ and Cu $K\alpha$ for four well-characterized homogeneous Au–Cu standards (Heinrich *et al.*, 1971). The data were taken at an initial electron-beam energy of 15 keV and take-off angle of 52.5° and pure Au and pure Cu were used as standards. The atomic number difference between these two elements is 50, 79 (Au) minus 29 (Cu). The straight lines plotted on Fig. 8.6 between pure Au and pure Cu indicate the relationship between composition and intensity ratio given in Eq. (8.1). For both Au $L\alpha$ and Cu $K\alpha$, the absorption effect A_i is less than 1% and the fluorescence effect F_i is less than 2%. For

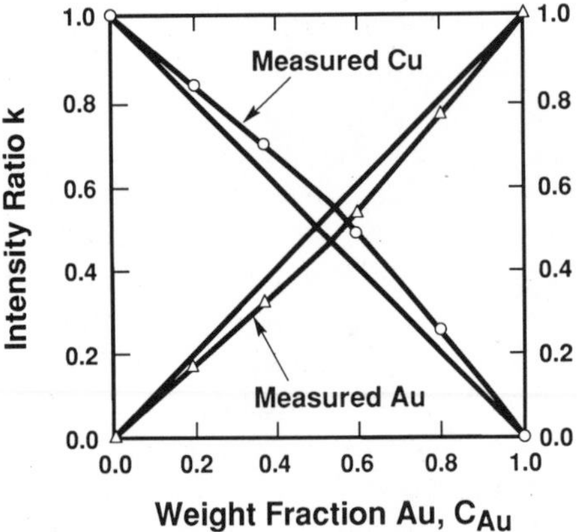

Figure 8.6. Measured Au–Cu intensity ratios, k_{Au} and k_{Cu}, versus weight fraction of Au; curves = k-values, straight lines = ideal behavior, 15 keV.

Cu $K\alpha$, the measured data fall above the first approximation and almost all the deviation is due to the atomic number effect, the difference in atomic number between the Au–Cu alloy and the Cu standard. As an example, for the 40.1 wt % Au specimen, the atomic number effect Z_{Cu} is 0.893, an increase in the Cu $K\alpha$ intensity by 11%. For Au $L\alpha$, the measured data fall below Castaing's first approximation, and almost all the deviation is due to the atomic number effect. As an example, for the 40.1 wt % Au specimen, the atomic number effect Z_{Au} is 1.24, a decrease in the Au $L\alpha$ intensity by 24%. In this example, the S factor is larger than the R factor for the Cu $K\alpha$ x rays leading to a $Z_{Cu} < 1$. Just the opposite is true for the Au $L\alpha$ x rays, leading to a $Z_{Au} > 1$. It is important to note that the $\phi(\rho z)$ curves for multielement samples and elemental standards which can be used for the calculation of the atomic number effect inherently contain the R and S factors.

8.7.2. X-Ray Absorption Effect

Figure 8.7 illustrates the effect of varying the inital electron-beam energy using Monte Carlo simulations of the positions where K-shell

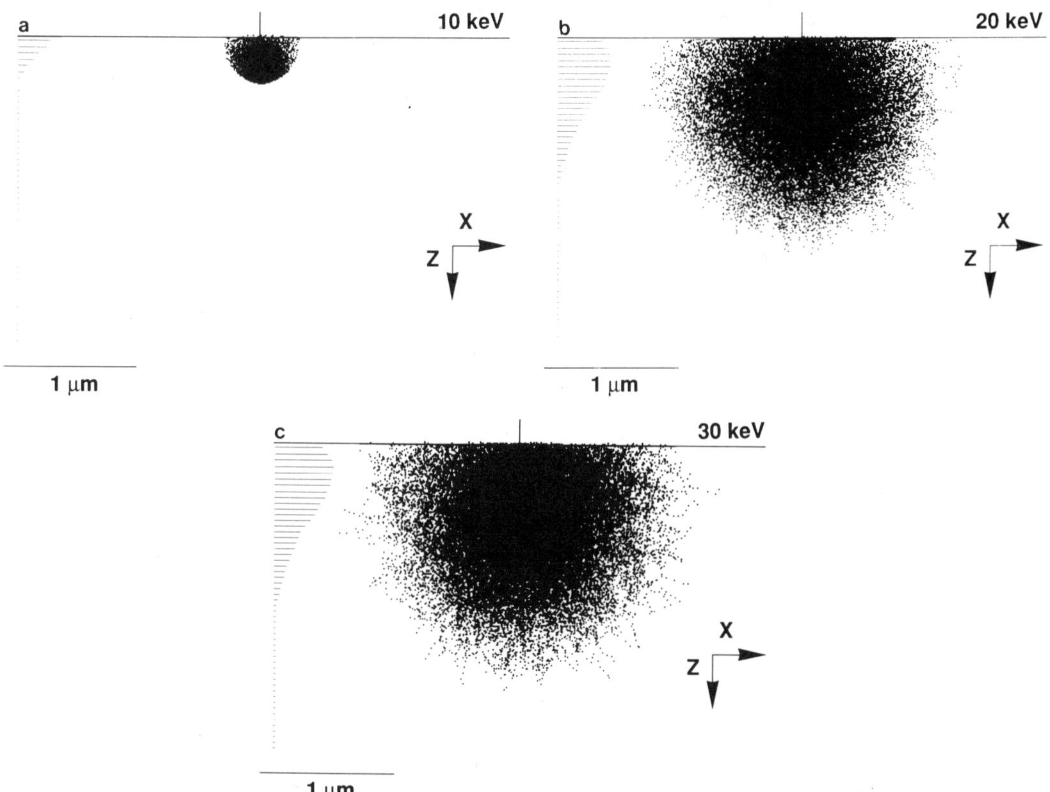

Figure 8.7. Monte Carlo calculations of x-ray generation volume for Cu at (a) 10, (b) 20, and (c) 30 keV.

x-ray interactions occur for Cu at three initial electron energies, 10, 20, and 30 keV. This figure shows that the Cu characteristic x rays are generated deeper in the specimen and the x-ray generation volume is larger as E_0 increases. From these plots, we can see that the sites of inner-shell ionizations, which give rise to characteristic x rays, are created over a range of depth below the surface of the specimen.

To reach the detector, the x-rays have to pass through a certain amount of matter, and as explained in Chapter 3, the photoelectric absorption process lessens the intensity. It is important to realize that the x-ray photons are either absorbed or else pass through the specimen with their original energy unchanged, so that they are still characteristic of the atoms which emitted them. Absorption follows an exponential law, so as x rays are generated deeper in the specimen, a progressively greater fraction is lost to absorption.

From the Monte Carlo plots of Fig. 8.7, the depth distribution of ionization is a complicated function. To quantitatively calculate the effect of x-ray absorption, an accurate description of the x-ray distribution in depth is needed. Fortunately, the complex three-dimensional distribution can be reduced to a one-dimensional problem for the calculation of absorption, since the path out of the specimen towards the x-ray detector depends only on depth. The $\phi(\rho z)$ curves, as discussed previously, give the generated x-ray distribution of x rays in depth (see Figs. 8.4 and 8.5). Figure 8.8 shows calculated $\phi(\rho z)$ curves for Cu $K\alpha$ x rays in pure Cu for 10, 15 and 30 keV. The curves extend deeper (in mass-depth or depth) in the sample with increasing E_0. The ϕ_0 values also increase with increasing initial electron-beam energies since the energy of the backscattered electrons increases with higher values of E_0.

The x rays which escape from any depth can be found by placing the appropriate path length in the x-ray absorption equation for the

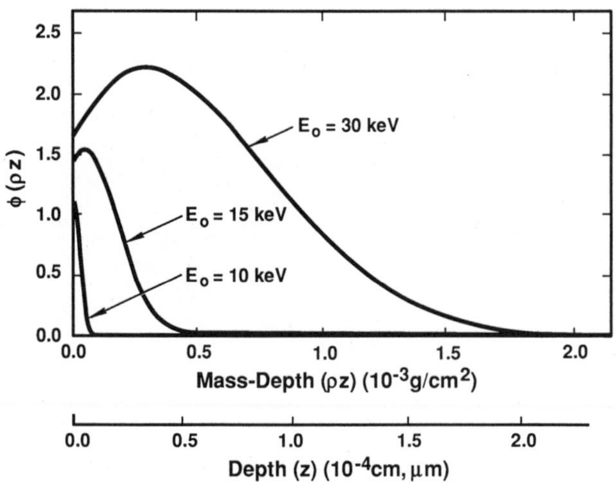

Figure 8.8. Calculated $\phi(\rho z)$ curves for Cu $K\alpha$ in Cu at 10, 15, and 30 keV.

ratio of the measured x-ray intensity I to the generated x-ray intensity I_0 at some position in the sample:

$$\frac{I}{I_0} = \exp\left(-\left(\frac{\mu}{\rho}\right)\rho t\right) \tag{3.46}$$

The terms in the absorption equation are (μ/ρ), the mass absorption coefficient; ρ, the specimen density, and t, the path length PL that the x ray traverses within the specimen before it reaches the surface, $z = \rho z = 0$. For our purpose, I represents the x-ray intensity which leaves the surface of the sample and I_0 represents the x-ray intensity generated at some position within the x-ray generation volume. Since the x-ray spectrometer is usually placed at an acute angle from the specimen surface, the so-called take-off angle ψ, the path length PL from a given depth z is given by $PL = z \csc \psi$, as shown in Fig. 8.9. When this correction for absorption is applied to each layer $\Delta(\rho z)$ in the $\phi(\rho z)$ curve, a new curve results, which gives the depth distribution of emitted x-rays. An example of the generated and emitted depth distribution curves for Al $K\alpha$ at an initial electron-beam energy of 15 keV is shown in Fig. 8.10 for a trace amount of Al in a copper target. The area under the $\phi(\rho z)$ curve respresents the x-ray intensity. The difference in the integrated area between the generated and emitted $\phi(\rho z)$ curves represents the total x-ray loss due to absorption. The absorption correction factor in quantitative matrix corrections is calculated on the basis of the $\phi(\rho z)$ distribution. Figure 8.2, for example, illustrates the large amount of Ni $K\alpha$ absorbed in an Fe–Ni sample.

X-ray absorption is usually the biggest factor that must be considered in the measurement of composition by x-ray microanalysis. For a given x-ray path length, the mass absorption coefficient μ/ρ for each measured x-ray line controls the amount of absorption. The μ/ρ varies greatly from one x-ray line to another and depends on the matrix elements of the specimen (see Chapter 3). For example, the

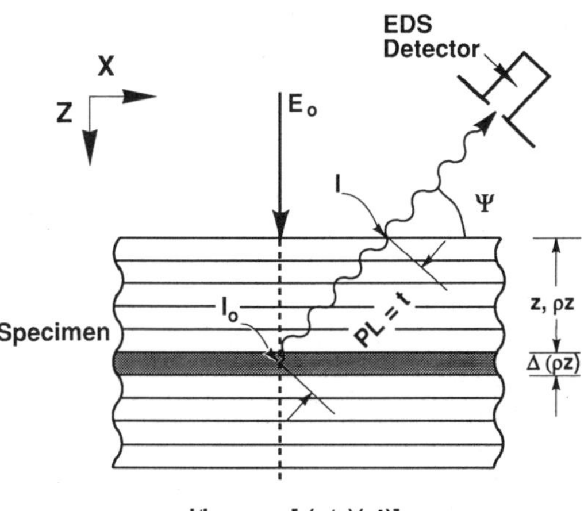

Figure 8.9. Schematic diagram of absorption in the measurement or calculation of $\phi(\rho z)$ emitted. PL=pathlength, ψ = take-off angle.

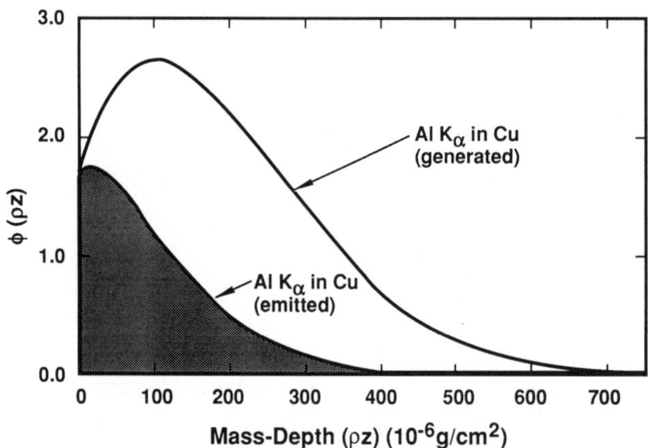

Figure 8.10. Calculated $\phi(\rho z)$ generated and emitted curves for Al $K\alpha$ in a Cu matrix at 20 keV.

mass absorption coefficient for Fe $K\alpha$ radiation in Ni is $90.0\,\text{cm}^2/\text{g}$, while the mass-absorption coefficient for Al $K\alpha$ radiation in Ni is $4837.5\,\text{cm}^2/\text{g}$. Using Eq. (3.46) and a nominal path length of 1 μm in a Ni sample containing small amounts of Fe and Al, I/I_0, the ratio of x rays generated in the sample to the x rays emitted at the sample surface is 0.923 for Fe$K\alpha$ radiation, but only 0.0135 for Al $K\alpha$ radiation. In this example, Al $K\alpha$ radiation is very heavily absorbed with respect to Fe $K\alpha$ radiation in the Ni sample. Such a large amount of absorption must be taken into account in any quantitative x-ray analysis scheme. Even more serious effects of absorption occur in the measurement of the light elements, C, N, O, etc. For example, the mass-absorption coefficient for C $K\alpha$ radiation in Ni is $17{,}270\,\text{cm}^2/\text{g}$, so large that in most practical analyses, no C $K\alpha$ radiation can be measured if the absorption path length is 1 μm. Significant amounts of C $K\alpha$ radiation can be measured in a Ni sample only within 0.1 μm of the surface. In such an analysis situation, the initial electron-beam energy should be held below 10 keV so that the C $K\alpha$ x-ray source is produced close to the sample surface.

As shown in Fig. 8.8 x rays are generated up to several micrometers into the specimen. Therefore the x-ray path length $PL = t$ and the relative amount of x rays available to the x-ray detection system after absotption (I/I_0) vary with the position at which each x ray is generated in the specimen. In addition to the position, ρz or z, at which a given x ray is generated within the specimen, the relation of that position to the x-ray detector is also important, since a combination of both factors determine the x-ray path length for absorption. Figure 8.11 shows the geometrical relationship between the position at which an x ray is generated and the position of the collimator, which allows x rays into the EDS detector. If the specimen is normal to the electron beam (Fig. 8.11), the angle between the

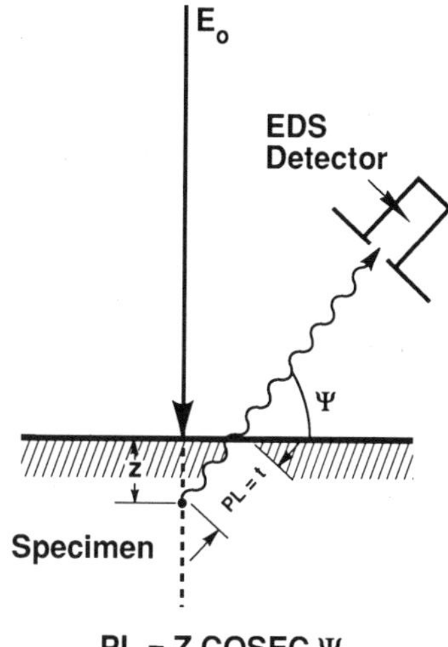

Figure 8.11. Diagram showing the absorption path length in a solid flat polished sample. PL = pathlength, ψ = take-off angle.

specimen surface and the direction of the x rays into the detector is ψ. The angle ψ is called the take-off angle. The path length, t = PL, over which x rays can be absorbed in the sample is calculated by multiplying the depth z in the specimen where the x-ray is generated, by the cosecant (the reciprocal of the sine) of the take-off angle ψ. A larger take-off angle yields a shorter path length in the specimen and minimizes absorption. The path length can be further minimized by decreasing the depth of x-ray generation R_x, that is, by using the minimum electron-beam energy E_0 consistent with the excitation of the x-ray lines used for analysis. Table 8.1 shows the variation of the path length which can occur if one varies the initial electron-beam energy for Al $K\alpha$ x rays in Al from 10 to 30 keV and the take-off angle ψ from 15 to 60 degrees.

The variation in PL is larger than a factor of 20, from 0.35 μm at the lowest keV and highest take-off angle to 7.7 μm at the highest keV and lowest take-off angle. Clearly the initial electron-beam energy and the x-ray take-off angle have a major effect on the path length and therefore on the amount of absorption.

Table 8.1. Path Length PL for Al $K\alpha$ X Rays in Al

E_0	Take-off angle ψ	R_x (μm)	PL (μm)
10	15	0.3	1.16
10	60	0.3	0.35
30	15	2.0	7.7
30	60	2.0	2.3

8.7.3. X-Ray Fluorescence

Photoelectric absorption results in the ionization of inner atomic shells, and those ionizations can also cause the emission of characteristic x rays. For fluorescence to occur, the target must contain a species of atom with a critical excitation energy less than the energy of the characteristic x rays being absorbed. In such a case, the measured x-ray intensity from this second element will include both the direct electron-excited intensity and the additional intensity generated by the fluorescence effect. Generally, the fluorescence effect can be ignored unless the photon energy exceeds the edge energy by less than 5 keV. An example of an analytical system in which a significant fluorescence effect occurs is the Fe–Cr–Ni system. Because of the relative x-ray energies, the iron K-shell is fluoresced by Ni $K\alpha$, and the chromium K-shell is fluoresced by both Fe $K\alpha$ and Ni $K\alpha$.

The significance of the fluorescence correction F_i can be illustrated by considering the binary system Fe–Ni. In this system, the Ni $K\alpha$ characteristic energy at 7.478 keV is greater than the energy for excitation of Fe K radiation, $E_c = 7.11$ keV. Therefore an additional amount of Fe $K\alpha$ radiation is produced. Figure 8.2 shows the effect of fluorescence in the Fe–Ni system at an initial electron-beam energy of 30 keV and a take-off angle ψ of 52.5°. Under these conditions, the atomic number effect Z_{Fe} and the absorption effect A_{Fe} for Fe $K\alpha$ are very close to 1.0. The measured k_{Fe} ratio lies well above the first-approximation straight-line relationship. The additional intensity is given by the effect of fluorescence. As an example, for a 10-wt % Fe–90-wt % Ni alloy, the amount of fluorescence is about 25%.

The quantitative calculation of the fluorescence effect requires a knowledge of the depth distribution over which the absorption of the x rays takes place. The $\phi(\rho z)$ curve of electron-generated x rays is the starting point for the fluorescence calculation, and a new $\phi(\rho z)$ curve for x-ray generated x rays is determined. The electron-generated x rays are emitted isotropically. The calculation considers the propagation of the x-ray intensity generated in each of the layers of the $\phi(\rho z)$ distribution over a spherical volume centered on the depth ρz of that layer. The amount of absorption is calculated based on the radial distance from the starting layer and determining the contributions of absorption to each layer Δz in the x-ray-induced $\phi(\rho z)$ distribution. Because of the longer range of x-rays than of electrons in materials, the x-ray-induced $\phi(\rho z)$ distribution covers a much greater depth, generally an order of magnitude or more, than the electron-induced $\phi(\rho z)$ distribution. Once the x-ray-induced $\phi(\rho z)$-generated distribution is determined, the absorption of the outgoing x-ray-induced fluorescence x rays must be calculated with the absorption path length calculated as above.

The fluorescence factor F_i is usually the least important factor in the calculation of composition by evaluating the $[ZAF]$ term in Eq. (8.2), since secondary fluorescence may not occur or the concentration of the element which causes fluorescence may be small. As we shall see in Chapter 9, of the three effects which control x-ray microanalysis calculations, the fluorescence effect F_i can be calculated with sufficient accuracy so that it rarely limits the development of an accurate analysis.

8.8. Types of Matrix Correction Schemes

Several approaches to the matrix correction problem are currently in use, and even within a given type of correction procedure, there may be several variations. The ZAF method and the closely related $\phi(\rho z)$ method calculate corrections for each of the matrix effects described above. These methods are the most generally applicable and are capable of producing accurate results even in the case in which only pure-element standards are available.

The *ZAF* method relies on experimental data such as the backscattering coefficients and energy distributions and theoretical descriptions such as the Bethe energy loss expression to calculate the atomic number correction factor Z. The x-ray absorption A and fluorescence F factors are calculated on the basis of a limited set of experimentally determined $\phi(\rho z)$ curves. As an illustration of the accuracy of the method, Fig. 8.12 shows error histograms determined from a number of multielement standards treated as unknowns and analyzed against pure-element standards (Bastin *et al.*, 1986). The bins of the horizontal axis of the error histogram are constructed by ratioing the calculated k factor k' to the experimentally measured k factor. A ratio of unity indicates perfect agreement between calculated and measured k values. The vertical axis of the histogram gives the number of analyses within each bin. The error distribution is such that 1-σ (66%) of the analyses fall within relative errors of ±2% and 2-σ (95%) of the analyses fall within ±4%. Virtually all analyses fall within ±10% relative error. In Fig. 8.12b, the tail on the ZAF distribution for $k'/k > 1$ is mainly a result of inaccurate absorption corrections for the light elements.

The $\phi(\rho z)$ method makes use of a data base of several hundred experimentally determined $\phi(\rho z)$ curves to develop generalized equations to predict the generated $\phi(\rho z)$ curve at any beam energy and for any target composition. Because the area under the $\phi(\rho z)$ curve is proportional to the total generated x-ray intensity including the effects of backscattering and stopping power, the $\phi(\rho z)$ expression directly incorporates an atomic number correction Z. Absorption A and fluorescence F are treated directly from the $\phi(\rho z)$ expressions. An example of an error histogram for the $\phi(\rho z)$ method is shown in Fig. 8.12.

As an example, Table 8.2 summarizes the ZAF factors for stainless steel containing 18 wt % Cr, 8 wt % Ni, and 74 wt % Fe that have been calculated using the $\phi(\rho z)$ method (PROZA, Bastin and

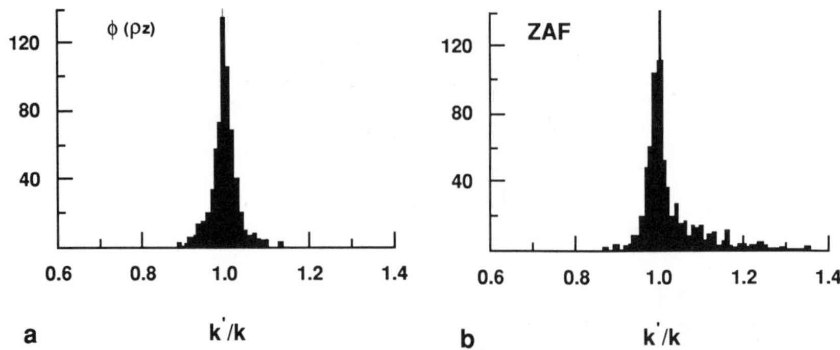

Figure 8.12. Error histogram for (a) $\phi(\rho z)$ and (b) ZAF analysis procedures.

Heijligers, 1990a) for an initial electron-beam energy of 20 keV and a take-off angle of 40°. For Cr $K\alpha$, Ni $K\alpha$, and Fe $K\alpha$, pure-element standards are used. The atomic number effect Z_i for all three elements is very close to 1.00 due to the very small difference in atomic number between specimen and standard. The absorption effect A_i is very small for Cr in the stainless steel. However, A_{Ni} is large, 1.11, indicating that absorption of Ni x rays is about 11% greater in the sample than in the standard. The result of this large absorption effect for Ni is the production of the characteristic fluorescence of Cr $K\alpha$ and Fe $K\alpha$ in the sample. In addition, absorption of Fe $K\alpha$ x rays results in the production of the characteristic fluorescence of Cr $K\alpha$ in the sample. Therefore the fluorescence effect F_i is significant: 0.863, a 14% correction, for Cr and 0.991, a 1% correction, for Fe.

For pure-element standards, applying Eq. (8.2),

$$\begin{aligned} C_{\mathrm{Cr}} &= [ZAF]_{\mathrm{Cr}} \cdot k_{\mathrm{Cr}}, \\ C_{\mathrm{Ni}} &= [ZAF]_{\mathrm{Ni}} \cdot k_{\mathrm{Ni}}, \\ C_{\mathrm{Fe}} &= [ZAF]_{\mathrm{Fe}} \cdot k_{\mathrm{Fe}} \end{aligned} \tag{8.3}$$

From Table 8.2, ZAF_{Cr} is 0.872, ZAF_{Ni} is 1.098, and ZAF_{Fe} is 1.021 for the stainless steel specimen. For this example, measurements of k_{Cr} must be multiplied by 0.872 to obtain C_{Cr}, measurements of k_{Ni} must be multiplied by 1.098 to obtain C_{Ni}, and measurements of k_{Fe} must be multiplied by 1.021 to obtain C_{Fe}.

Table 8.2. X-Ray Microanalysis of a Stainless Steel Alloy[a]

X-ray line	Z_i	A_i	F_i	ZAF_i	k_i
Cr $K\alpha$	1.00	1.01	0.863	0.872	0.207
Ni $K\alpha$	0.988	1.111	1.0	1.098	0.0729
Fe $K\alpha$	1.001	1.029	0.991	1.021	0.724

[a] Containing 18 wt % Cr, 8 wt % Ni, 74 wt % Fe; $E_0 = 20$ keV, $\psi = 40°$.

If standards are available that are similar in composition to the unknown, then a purely empirical method based on interelement coefficients can be used. This empirical method is most frequently applied in EPMA of geological materials, where extensive suites of mineral standards are available very similar in composition to unknowns. With standards close in composition, the error distribution can be extremely narrow.

8.9. Caveats

Several significant requirements must be satisfied if analyses are to be achieved with error distributions similar to those shown in Fig. 8.12.

1. The specimen is assumed to be homogeneous within the volume excited by the electron beam, including the enlarged volume of the x-ray-induced fluoresence and, in addition, homogeneous along the length of the path the x-rays travel through the specimen to the spectrometer. Although the spatial resolution of the technique is approximately 1 μm or less, especially if fluorescence is insignificant, this condition may be difficult to satisfy when the incident beam is positioned near an interface between chemically different phases, if the specimen is a thin film supported on a substrate, or if the specimen contains a submicrometer second-phase distribution. In such cases, advanced quantitative-analysis methods must be applied. Note that this requirement for specimen homogeneity within the analyzed volume eliminates the use of overscanning a field of view containing multiple phases during spectrum accumulation to get an average analysis. Such a procedure can lead to relative errors of hundreds of percent!

2. The specimen must have a flat, highly polished surface placed at known angles to the incident beam and the x-ray spectrometer(s). Quantitative analysis methods all assume that the difference in composition between the unknown and the standard is the only reason that the measured x-ray intensities differ. There exist physical effects such as the specimen's surface roughness, size, shape, and surface tilt which can also influence the interactions of electrons and the propagation of x-rays. Such effects strongly modify the x-ray spectra obtained from microscopic particles and rough objects such as fracture surfaces. Again, analysis of such specimens requires advanced quantitation methods. Application of conventional flat-specimen quantitation methods to such irregular specimens can lead to relative errors of hundreds of percent.

3. The specimen surface should not be chemically polished or etched, since these chemical processes can lead to preferential leaching of constituents from the critical near-surface region of the specimen that x-ray microanalysis characterizes.

This introductory chapter has considered x-ray microanalysis only for flat polished samples, analyzed normal to the direction of the electron beam. It is of interest to obtain chemical analyses from particles and surfaces which are not flat, as well as from tilted samples. Also, x-ray microanalysis of multilayered samples such as those used in microelectronic devices are of specific interest. We will consider the analysis of these types of specimens as well as special materials, such as polymers or biological tissues, in Chapter 9.

9

Quantitative X-Ray Analysis: Theory and Practice

9.1. Introduction

An overview of the basic principles and techniques used to determine chemical composition, on the micrometer scale, with the SEM and EPMA was presented in Chapter 8. We outlined the approach to quantitation, the need for matrix corrections, and the physical origins of the matrix effects. The x-ray production process and the use of $\phi(\rho z)$ curves to describe x-ray production were introduced. Finally, we discussed the three major matrix effects, atomic number (Z), absorption (A), and fluorescence (F) and showed, on a conceptual basis, how they are calculated. This chapter presents the more detailed theory and equations which can be used to determine the three major matrix corrections for quantitative analysis of flat polished specimens. The two commonly used correction schemes, the ZAF and $\phi(\rho z)$ methods, will be described separately although the two schemes are, in many ways, closely related.

The remainder of the chapter addresses itself to quantitative techniques for special geometries and specific types of materials. We will discuss special analysis procedures for biological and polymer samples, as well as for geological samples. We will outline special techniques for nonideal sample geometries such as nonnormal incidence, thin foils, thin films on substrates, and particles and rough surfaces. Finally, we will consider the special case of light-element ($Z < 11$) analysis and present a section on statistics for quantitative microanalysis which considers the precision of the analysis, trace-element analysis, and other topics.

9.2. ZAF Technique

9.2.1. Introduction

In his thesis, Castaing (1951, 1960) outlined a method for obtaining a quantitative x-ray analysis from a micrometer-sized region of solid specimens. Castaing's treatment can be represented by the following considerations: The average number of ionizations n from element i generated in the sample per primary beam electron incident with energy E_0 is

$$n = \left(\frac{N_0\rho C_i}{A_i}\right)\int_{E_0}^{E_c} \frac{Q}{-dE/ds}\,dE, \tag{9.1}$$

where dE/ds is the mean energy change of an electron in traveling a distance ds in the sample, N_0 is Avogadro's number, ρ is the density of the material, A_i is the atomic weight of i, C_i is the concentration of i, E_c is the critical excitation energy for the characteristic x-ray line (K, L, or M) of element i of interest, and Q is the ionization cross-section, defined as the probability per unit path length of an electron of given energy causing ionization of a particular electron shell (K, L, or M) of an atom in the specimen. (See Chapter 3 for a complete discussion of the ionization cross-section.) The term $(N_0\rho C_i/A_i)$ gives the number of atoms of element i per unit volume. Backscattering electrons can be taken into account by introducing a factor R equal to the ratio of x-ray intensity actually generated to that which would have been generated if all of the incident electrons had remained within the specimen. I, the average number of x-rays of element i generated per incident electron is proportional to n:

$$I = (\text{const})C_i R\rho \int_{E_0}^{E_c} \frac{Q}{-dE/ds}\,dE, \tag{9.2}$$

where the constant term contains Avogadro's number and the atomic weight.

In practice it is very difficult to calculate directly the absolute generated intensity I. In fact, the intensity that one must deal with is the measured intensity. The measured intensity is usually even more difficult to obtain because of the need to calculate absorption effects in the specimen and spectrometer components and the physics of the spectrometer detection process. Therefore, as suggested by Castaing (1951), we select a standard for element i and measure the ratio $I_i/I_{(i)}$, where I_i and $I_{(i)}$ are the measured intensities from the sample and standard, respectively. This experimental measurement procedure was discussed in Section 8.3 and shown in Fig. 8.1. The x-ray intensity from element i is measured by the same x-ray detector for both sample and standard. The beam energy, E_0, the electron probe current, i_b and the angle between the specimen surface and the direction of the measured x rays (take-off angle ψ) are held constant. By using the ratio of the measured intensities $I_i/I_{(i)}$, the constant term in Eq. (9.2) drops out. As discussed by Castaing (1951), to a first approximation, A, ρ, Q, and dE/ds may be assumed equal for both the specimen and standard.

In most analyses, the measured intensities from specimen and standard need to be corrected for differences in A, ρ, Q, and dE/ds and for absorption within the solid in order to arrive at the ratio of generated intensities and hence the value of C_i. Recognition of the complexity of the problem of the analysis of solid specimens led numerous early investigators (Wittry, 1963; Philibert, 1963; Duncumb and Shields, 1966) to expand the theoretical treatment of quantitative analysis proposed by Castaing.

All robust, reliable schemes for carrying out the quantitative analysis of solid specimens use a standard of known composition. In many cases, especially for metals, pure elements are suitable. In the case of mineral or petrological samples, homogeneous compound standards close in average atomic number to the unknown are usually chosen. What the analyst measures is the x-ray intensity ratio between the elements of interest in the specimen and the same elements in the standard. Both specimen and standard are examined under identical experimental conditions. The measured intensity ratio, commonly called k_i, must be accurately determined or any quantitative analysis scheme will suffer from the same inaccuracy. In this chapter we will assume that measurements of $I_i/I_{(i)} = k_i$ can be obtained accurately. Measurement techniques to obtain accurate I_i and $I_{(i)}$ values were considered in some detail in Chapter 7. Once the k_i values have been obtained, they must be corrected for several effects including.

1. differences between specimen and standard for electron scattering and retardation, i.e., the so-called atomic number effect expressed by the factor Z_i;
2. absorption A_i of x rays within the specimen and standard;
3. characteristic fluorescence effects F_i and, in a few specific cases, continuum fluorescence in the specimen and standard.

A common form of the matrix equation was presented in Chapter 8 as

$$C_i/C_{(i)} = [ZAF]_i \cdot \frac{I_i}{I_{(i)}} = [ZAF]_i \cdot k_i, \qquad (8.2)$$

where C_i is the weight fraction of the element of interest. The ZAF method attempts to calculate each of the matrix effects using a physics type of approach in which R, Q, dE/ds, and other terms in the Z, A, and F matrix factors are calculated. Because of the difficulty of obtaining specific values for each of these terms, a significant amount of empiricism has necessarily crept into the evaluation of the various ZAF terms of the ZAF method. Each of the matrix effects listed above will be discussed in detail in the following sections.

9.2.2. The Atomic Number Correction Z

The atomic number effect arises from two phenomena, namely, electron backscattering R and electron retardation S, both of which depend upon the average atomic number of the target. (See Chapter 8 for an introductory discussion of the R and S terms.) Therefore, if there is a difference between the average atomic number of the specimen, given by

$$\bar{Z} = \sum_j C_j Z_j, \tag{9.3}$$

and that of the standard, an atomic number correction is required. For example in an Al–3 wt % Cu alloy, the value of $\bar{Z}$ is 13.3, so a somewhat larger atomic number effect for a Cu ($Z = 29$) analysis would be expected than for an Al ($Z = 13$) analysis if pure-element standards are used. In general, unless this effect is corrected for, analyses of heavy elements in a light-element matrix generally yield values which are too low, while analyses of light elements in a heavy-element matrix usually yield values which are too high.

At present, the most accurate formulation of the atomic number correction Z_i for element i appears to be that given by Duncumb and Reed (1968):

$$Z_i = \frac{R_i \int_{E_c}^{E_0} \frac{Q}{S}\, dE}{R_i^* \int_{E_c}^{E_0} \frac{Q}{S^*}\, dE}, \tag{9.4}$$

where R_i and R_i^* are the backscattering correction factors of element i for standard and specimen, respectively:

$$R_i = \frac{\text{total number of photons actually generated in the sample}}{\text{total number of photons generated if there were no backscatter}}$$

Q is the ionization cross-section, defined as the probability per unit path length of an electron with a given energy causing ionization of a particular inner electron shell of an atom in the target (see Chapter 3), and S is the electron stopping power—$(1/\rho)(dE/dX)$ (see Chapter 3) in the region $1 \leq E \leq 50$ keV. The stopping power S_i given by Bethe (1930) is

$$S_i = (\text{const}) \frac{Z}{A} \frac{1}{E} \ln\left(\frac{C_1 E}{J}\right), \tag{9.5}$$

where C_1 is a constant and J is the so-called mean ionization potential. To calculate the term $(C_1 E/J)$, C_1 is equal to 1.166 when E and J are given in kilovolts.

R Factor. The R factor represents the fraction of ionization remaining in a target after the loss due to the backscattering of beam electrons. Values of R lie in the range 0.5–1.0 and approach 1 at low atomic numbers. The backscattering correction factor varies not only with atomic number but also with the overvoltage $U = E_0/E_c$. Figure 9.1 shows this variation. As the overvoltage decreases towards 1, fewer electrons are backscattered from the specimen with energies $> E_c$, and consequently less of a loss of ionization results from such backscattered electrons.

There are several tabulations of R values as a function of the pure

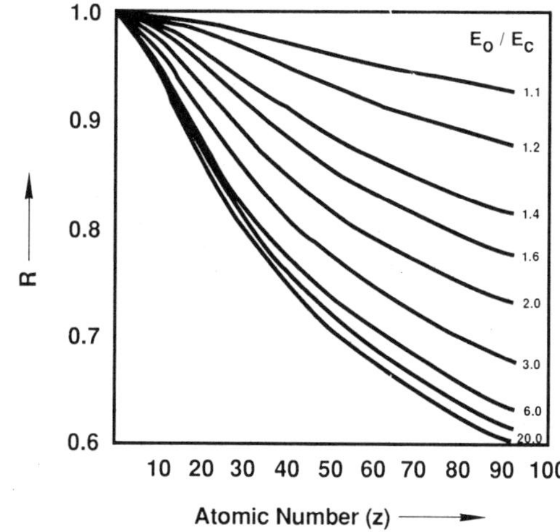

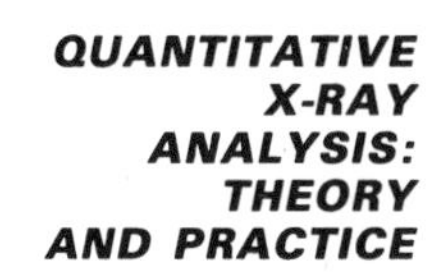

Figure 9.1. Fraction of ionization R remaining in a specimen of atomic number Z after loss of beam electrons due to backscatter (Duncumb and Reed, 1968).

element's atomic number and the overvoltage U (Duncumb and Reed, 1968; Green, 1963; Springer, 1966). The tabulation given by Duncumb and Reed very nearly agrees with experimental determinations where comparisons can be made. The electron backscattering correction factors R from Duncumb and Reed (1968) were fitted by Yakowitz *et al.* (1973) with respect to overvoltage U and atomic number Z as follows:

$$R_{ij} = R_1' - R_2' \ln(R_3' Z_j + 25), \tag{9.6}$$

where

$$R_1' = 8.73 \times 10^{-3} U^3 - 0.1669 U^2 + 0.9662 U + 0.4523$$

$$R_2' = 2.703 \times 10^{-3} U^3 - 5.182 \times 10^{-2} U^2 + 0.302 U - 0.1836$$

$$R_3' = (8.887 U^3 - 3.44 U^2 + 9.33 U - 6.43)/U^3$$

The term i represents the element i, which is being measured, and j represents each of the elements present in the standard or specimen, including element i. The term R_{ij} then gives the backscattering correction for element i as influenced by element j in the specimen.

Duncumb and Reed (1968) have postulated that

$$R_i = \sum_j C_j R_{ij} \tag{9.7a}$$

for the standard, and

$$R_i^* = \sum_j C_j R_{ij} \tag{9.7b}$$

for the sample. To evaluate R_i, one must obtain from the Duncumb–Reed tabulations, as given in the data base (Chapter 14) or in Eq. (9.6), the value of R_{ij} for each element in the specimen or standard.

S Factor. There are several relations for Q, as discussed in Chapter 3. Despite the fact that Q values differ by several percent depending on the values of the constants, Heinrich and Yakowitz (1970) have shown that discrepancies in Q values have only a negligible effect on the final value of the concentration. A simplifying assumption is often made that Q is constant and therefore cancels in the expression for Z_i, Eq. (9.4). With the advent of fast laboratory computers, the integration can be carried out directly.

The value of J to be used to calculate S_i from Eq. (9.5) is a matter of controversy, since J is not measured directly but is derived from experiments done in the MeV range. The original Bethe expression was developed for hydrogen only. Present J values allow the expression to be used for other elements. The most complete discussion of the value of J is given by Berger and Seltzer (1964). These authors postulate, after weighing all available evidence, that a "best" J-vs-Z curve is given by

$$J = 9.76Z + 58.8Z^{-0.19}\,(\text{eV}). \qquad (9.8)$$

The Berger–Seltzer J values as a function of Z are listed in Chapter 14.

In order to avoid the integration in Eq. (9.4), Thomas (1964) proposed that the average energy E may be taken as $0.5(E_0 + E_c)$, where E_c is the critical excitation energy for element i. The average energy is used as E in Eq. (9.5) with little loss in accuracy. Hence

$$S_{ij} = (\text{const})\frac{Z_j}{A_j(E_0 + E_c)}\ln\left[\frac{538(E_0 + E_c)}{J_j}\right], \qquad (9.9)$$

where i represents the element i which is measured, and j represents each of the elements present in the standard or specimen, including element i. The constant in Eq. (9.9) need not be evaluated, since it cancels out when the stopping power for the sample and standard are compared [Eq. (9.11)]. There is experimental evidence that, to evaluate S_i and S_i^*, a weighted average of S_{ij} and S_{ij}^*, respectively, can be used, namely,

$$S_i = \sum_j C_j S_{ij} \qquad (9.10\text{a})$$

for the standard and

$$S_i^* = \sum_j C_j S_{ij}^* \qquad (9.10\text{b})$$

for the specimen. The final values of S_i and S_i^* are obtained using Eq. (9.10).

If the integration in Eq. (9.4) is avoided and Q is constant, we have as the final form of Z_i,

$$Z_i = \frac{R_i}{R_i^*}\frac{S_i^*}{S_i}. \qquad (9.11)$$

The major variable which influences the atomic number factor Z_i is the difference between the average atomic number of the specimen and that of the standard.

A sample calculation of the atomic number correction for a Al–3 wt % Cu alloy is given in Table 9.1. The atomic number corrections Z_{Cu} and Z_{Al} are obtained for an operating energy E_0 of 15 keV. The

Table 9.1. Atomic Number Correction for an Al–3 wt % Cu alloy

a. Input data

	Al	Cu
Z	13	29
A	26.98	63.55
E_c (keV)	1.56	8.98

b. Output data

Operating conditions:	$E_0 = 15$ keV	
For Cu $K\alpha$:	$U_{Cu} = 1.67$	
	$J_{Cu} = 314.05$	$J_{Al} = 163.0$
	$S_{CuCu} = 0.144$	$S_{CuAl} = 0.179$
	$R'_1 = 1.641$	$R'_2 = 0.189$ $R'_3 = 0.792$
	$R_{CuCu} = 0.910$	$R_{CuAl} = 0.968$
	$S^*_i = 0.178$	
	$R^*_i = 0.966$	
using a Cu standard:	$S_i = 0.144$, $R_i = 0.910$	$Z_{Cu} = 1.16$
For Al $K\alpha$:	$U_{Al} = 9.62$	
	$S_{AlAl} = 0.238$	
	$R'_1 = 2.073$	$S_{AlCu} = 0.189$
	$R_{AlAl} = 0.910$	$R'_2 = 0.332$ $R'_3 = 0.623$
	$S^*_i = 0.236$	$R_{AlCu} = 0.823$
	$R^*_i = 0.907$	
using an Al standard:	$S_i = 0.238$, $R_i = 0.910$	$Z_{Al} = 0.997$

values of J_{Cu} and J_{Al} are obtained from Eq. (9.8). The values of S_{Cu} and S_{Al} for either Cu $K\alpha$ radiation ($E_c = 8.98$ keV) or Al $K\alpha$ radiation ($E_c = 1.56$ keV) are calculated using Eq. (9.9). The constant in Eq. (9.9) is set equal to 1 since it eventually drops out in later calculations. The backscattering corrections R_{Cu} and R_{Al} for either Cu $K\alpha$ or Al $K\alpha$ are obtained using Eq. (9.6). The S_i, S^*_i and R_i, R^*_i terms are calculated using Eqs. (9.7) and (9.10). The final corrections Z_{Cu} and Z_{Al}, obtained from Eq. (9.11), are 1.16 and 0.998. For 30 keV, Z_{Cu} and Z_{Al} are 1.11 and 0.999. It is interesting to note that the Z_{Cu} correction is decreased by ~50% from 1.16 to 1.08 at 15 keV when a lower atomic number standard of θ phase, $CuAl_2$ ($Z = 21.6$), is used in favor of pure Cu ($Z = 29$). The Z_{Cu} correction is also decreased from 1.11 to 1.05 when operating at a higher E_0 (30 keV).

Heinrich and Yakowitz (1970) have investigated error propagation in the Z_i term. In general, the magnitude of Z_i decreases as the overvoltage $U = E_0/E_c$ increases, but very slowly (5% for a tenfold increase in U). The uncertainty in Z_i remains remarkably constant as a function of U since the uncertainties in R and S tend to counterbalance one another. Thus, no increase in the error of Z_i is to be expected at low U (>1.5) values, and hence the choice of operating with low U to minimize errors in the absorption correction as discussed in the next section is also useful for obtaining the highest accuracy in the Z_i factor.

9.2.3. The Absorption Correction A

9.2.3.1. Formulation

Since the x rays produced by the primary beam are created at some nonzero depth in the specimen (see Figs 8.3 and 8.8), they must pass through the specimen on their way to the detector. On this journey, some of the x rays undergo photoelectric absorption because of interactions with the atoms of the various elements in the sample, as discussed in Chapter 3. Following the initial formulation of Castaing (1951), the intensity dI_i of characteristic radiation, without absorption, generated from element i in a layer of thickness dz having density ρ at some depth z below the specimen surface is

$$dI_i = I_i(\Delta\rho z)\phi_i(\rho z)\, d(\rho z), \qquad (9.12)$$

where $\phi_i(\rho z)$ is defined as the distribution of characteristic x-ray production of element i with depth and $I_i(\Delta\rho z)$ is the x-ray intensity of an isolated thin film. (See Section 8.6.2 for a description of $\phi(\rho z)$.) In the absence of absorption, the total flux generated for element i, I_{igen}, is

$$I_{igen} = I_i(\Delta\rho z)\int_0^\infty \phi_i(\rho z)\, d(\rho z). \qquad (9.13)$$

Considering absorption of the generated x rays, the total flux I_{iem} emitted is

$$I_{iem} = I_i(\Delta\rho z)\int_0^\infty \phi_i(\rho z) \exp - [\mu/\rho)^i \csc \psi(\rho z)]\, d(\rho z), \qquad (9.14)$$

where $\mu/\rho)^i$ is the x-ray mass attenuation coefficient of the specimen for the characteristic x-ray line of element i and ψ is the take-off angle, the angle between the direction of the measured x-ray and the sample surface (see Fig. 8.12). The path length over which absorption takes place is $z \csc \psi$ (see Fig. 8.12). The quantity $\mu/\rho)^i \csc \psi$ is called χ.

Philibert (1963) referred to the generated intensity I_{igen} as $F(0)$, when χ is zero. He also referred to the emitted intensity I_{iem} as $F(\chi)$. Using these terms, the ratio $F(\chi)/F(0)$ is called $f(\chi)$, which is equivalent to

$$f(\chi) = \frac{\int_0^\infty \phi_i(\rho z) \exp - [\mu/\rho)^i \csc \psi(\rho z)] d(\rho z)}{\int_0^\infty \phi_i(\rho z) d(\rho z)} \qquad (9.15)$$

The ratio $f(\chi)$ is called the *standard absorption term* (Philibert, 1963).

For the determination of the absorption correction A_i for any element i in a composite specimen, we use the equation

$$A_i = \frac{f(\chi)_{\text{std}}}{f(\chi)_{\text{spec}}}, \tag{9.16a}$$

where *std* and *spec* refer to standard and specimen, respectively. An alternative nomenclature found in the literature gives

$$A_i = \frac{f(\chi)}{f(\chi)^*}, \tag{9.16b}$$

where the specimen is noted by an asterisk and the standard is left unmarked.

9.2.3.2. Expressions for f(χ)

The absorption correction factor $f(\chi)$ of a specific characteristic line of element i depends on the respective mass absorption coefficient μ/ρ, the x-ray emergence angle ψ, the initial energy of the electron beam E_0, the critical excitation energy E_c for K, L, or M radiation from element i, and the mean atomic number and mean atomic weight of the specimen. Hence we can write

$$f(\chi) = f[(\mu/\rho)\csc\psi, E_0, E_c, \bar{Z}, \bar{A}]. \tag{9.17}$$

There have been a number of experimental measurements of the $\phi(\rho z)$ for pure elements, starting with those of Castaing (1951). Direct measurements of $\phi(\rho z)$ curves will be discussed in Section 9.3. If measured $\phi(\rho z)$ curves are available, $f(\chi)$ and $f(\chi)^*$ can be obtained directly using Eqs. (9.13) and (9.14).

Another approach to obtaining $f(\chi)$ and $f(\chi)^*$ is the use of equations which describe $\phi(\rho z)$ curves for various elements, x-ray lines, and initial electron-beam energies E_0. This approach, now called the $\phi(\rho z)$ method, is described in Chapter 8 and Section 9.3. The most commonly used absorption correction, that of Philibert–Duncumb–Heinrich, was developed, however, by using an empirical expression for $\phi(\rho z)$. Thus the ZAF and the $\phi(\rho z)$ methods are in many cases interrelated.

Philibert (1963) obtained an equation for the functionality of $\phi(\rho z)$ using a simplified equation to fit the form of experimental $\phi(\rho z)$ curves available at that time. In order to simplify the calculations of $f(\chi)$, he chose a $\phi(\rho z)$ equation in which ϕ_0, the intersection of the $\phi(\rho z)$ curve with the surface $\rho z = 0$, was set equal to 0. As discussed in Section 8.6.2 and shown in Figs. 8.5, 8.6 and 8.9, ϕ_0 is always >1.0 due to the effect of backscattered electrons.

The equation for $f(\chi)$ developed by Philibert (1963) is

$$\frac{1}{f(\chi)} = \left(1 + \frac{\chi}{\sigma}\right)\left(1 + \frac{h}{1+h}\frac{\chi}{\sigma}\right), \tag{9.18}$$

where

$$h = 1.2A/Z^2 \tag{9.19}$$

and A and Z are, respectively, the atomic weight and atomic number of element i. The absorption parameter χ equals $\mu/\rho)$ csc ψ, where μ/ρ is the mass absorption coefficient of element i in itself. The parameter σ (the Lenard coefficient) is a factor which accounts for the voltage dependence of the absorption or loss of the primary electrons.

Duncumb and Shields (1966) proposed that the dependence of E_c should be taken into account in the formulation of σ in Eq. (9.18). Later, Heinrich (1969), after critical examination of existing experimental $f(\chi)$ data, suggested a formula for σ, namely,

$$\sigma = \frac{4.5 \times 10^5}{E_0^{1.65} - E_c^{1.65}}. \tag{9.20}$$

This development of the absorption effect [Eqs. (9.18)–(9.20)] is known as the Philibert–Duncumb–Heinrich (PDH) equation and is currently the most popular expression for $f(\chi)$. Accordingly, the PDH equation is commonly used in microprobe correction schemes to calculate $f(\chi)$.

In using the PDH equation to calculate $f(\chi)$ or $f(\chi)^*$ of element i in multicomponent samples, the effect of other elements in the sample or standard must be considered. These elements have an effect on the values of h, χ, and σ. After h is evaluated for each element in the specimen, an average h value is obtained using

$$h = \sum_j C_j h_j, \tag{9.21}$$

where j represents the various elements present in the sample including element i and C_j is the mass fraction of each element j. To obtain χ, the mass absorption coefficient $\mu/\rho)^i_{\text{spec}}$ for element i in the sample must be calculated. The mass absorption coefficient is given by

$$\mu/\rho)^i_{\text{spec}} = \sum_j \mu/\rho)^i_j C_j, \tag{9.22}$$

where $\mu/\rho)^i_j$ is the mass absorption coefficient for radiation from element i in element j and C_j is the weight fraction of each element in the specimen including i. The σ value is obtained from Eq. (9.20) using the E_c value of element i.

9.2.3.3. Practical Considerations

The effects of the errors in input parameters μ/ρ, ψ, and E_0 have been considered in detail by Yakowitz and Heinrich (1968). The major conclusions of their study are:

1. Serious analytical errors can result from uncertainties in input parameters. Mass absorption coefficients for the light elements are a particular problem.

2. In order to reduce the effects of these uncertainties in input parameters, the value of the absorption function $f(\chi)$ should be 0.7 or greater.
3. To maximize $f(\chi)$, the path length for absorption in the sample should be minimized. Samples should be run at low overvoltage ratios, and instruments should have high x-ray emergence angles. (See Section 8.7.2 and Table 8.1.)

The PDH absorption correction is particularly sensitive to errors when the amount of absorption is high and most of the emitted signal comes from regions close to the sample's surface. Typical examples of this situation include measurements of the light-element x-rays (C, N, O) in metal matrices (Ti, Fe, Cu). In this analytical case, the functionality of the $\phi(\rho z)$ curve assumed by Philibert (1963), close to the sample surface, is particularly in error. In such cases, the calculation of $f(\chi)^*$ may have rather substantial errors.

It is particularly important to discuss the case of quantitative analysis using an energy-dispersive detector or crystal spectrometer mounted on a typical SEM. The most important considerations are the correction for background (as discussed in Chapter 7) and the need to know the exact value of ψ, the x-ray take-off angle. In the SEM the specimen is usually tilted in order to provide a suitable value of the take-off angle. Uncertainties in the take-off angle effect mainly the absorption correction calculation; such uncertainties increase as ψ decreases (Yakowitz and Heinrich, 1968). Thus, take-off angles of 30° or larger are needed. Furthermore, it is of crucial importance to ensure that the specimen and standard are measured at identical x-ray take-off angles.

9.2.3.4. Calculations of the Absorption Factor A

The three major variables that influence the absorption factor A_i are the operating voltage E_0, the take-off angle ψ, and the mass absorption coefficient for the element of interest i in the specimen $\mu/\rho)^i_{\text{spec}}$. Since A_i is determined by the ratio $f(x)/f(x)^*$, both terms should be similar if the absorption factor A_i is to approach 1. As A_i approaches 1, the measured intensity ratio becomes a better approximation for the concentration ratio of element i from sample to standard.

The significance of the absorption factor can be illustrated by considering two binary systems, Fe–Ni and Al–Mg. In both binaries, the atomic numbers of the two elements are so close that no atomic number correction Z_i need be made. We will consider Ni $K\alpha$ absorption in Fe and Al $K\alpha$ absorption in Mg. In both cases secondary fluorescence (Ni $K\alpha$ by Fe or Al $K\alpha$ by Mg) does not occur, so that an F_i correction need not be made. Calculations of $A_i = f(x)/f(x)^*$ were made for both systems using the PDH correction and Eqs (9.16) and (9.18)–(9.21).

Tables 9.2 and 9.3 contain the input data for the absorption calculation and also list the various terms χ, σ, h which are evaluated in the calculation. In the case of Fe–Ni, a 10 wt % Ni alloy was considered (Table 9.2). Calculations were performed for two operating voltages, 30

Table 9.2. Absorption of Ni $K\alpha$ in an Fe–Ni Binary

a. Input data

	Ni	Fe
μ/ρ Ni $K\alpha$ (cm^2/g) in absorber	58.9	379.6
Z	28	26
E_c (keV)	8.332	7.111
A	58.71	55.85

b. Output data

Ni–Fe sample	E_0 (keV)	ψ (deg)	χ	σ	h	$f(\chi)$	$f(\chi)^*$	A_{Ni}
Fe–10% Ni sample	30	52.5	438	1870	0.0982	—	0.794	1.21
Ni standard			74.2	1870	0.0899	0.959	—	
Fe–10% Ni sample	30	15.5	1300	1870	0.0982	—	0.555	1.60
Ni standard			220.4	1870	0.0899	0.886	—	
Fe–10% Ni sample	15	52.5	438	8310	0.0982	—	0.945	1.05
Ni standard			74.2	8310	0.0899	0.990	—	
Fe–10% Ni sample	15	15.5	1300	8310	0.0982	—	0.853	1.14
Ni standard			220.4	8310	0.0899	0.972	—	
Fe–50% Ni sample			820.4	8310	0.0945	—	0.902	1.08
Ni standard			220.4	8310	0.0899	0.972	—	

Table 9.3. Absorption of Al $K\alpha$ in a Al–Mg Binary

	Al	Mg
μ/ρ Al $K\alpha$ (cm^2/g) in absorber	385.7	4376.5
A	26.98	24.305
Z	13	12
E_c (keV)	1.56	1.303

b. Output data

Al–Mg sample	E_0 (keV)	ψ (deg)	χ	σ	h	$f(\chi)$	$f(\chi)^*$	A_{Al}
Mg–10% Al sample	30	52.5	5,013	1,657	0.201		0.165	4.48
Al standard			486	1,657	0.192	0.738		
Mg–10% Al sample	30	15.5	14,884	1,657	0.201		0.04	11.7
Al standard			1,443	1,657	0.192	0.469		
Mg–10% Al sample	15	52.5	5,013	5,286	0.201		0.443	2.04
Al standard			486	5,286	0.192	0.902		
Mg–10% Al sample	15	15.5	14,884	5,286	0.201		0.178	4.23
Al standard			1,443	5,286	0.192	0.753		
Mg–50% Al sample			8,910	5,286	0.197		0.291	2.58
Al standard			1,443	5,286	0.192	0.753		
Mg–10% Al sample	7.5	52.5	5,013	17,506	0.201		0.742	1.306
Al standard			486	17,506	0.192	0.969		

and 15 keV, and two take-off angles, 15.5° and 52.5°. The case of $E_0 = 30$ keV, $\psi = 52.5°$ is illustrated in Fig. 8.2 with experimental data. The smallest $f(\chi)$ and $f(\chi)^*$ factors are calculated at $E_0 = 15$ keV and $\psi = 52.5°$. The amount of absorption is minimized because x rays are generated close to the surface and the absorption path length is smaller at high take-off angles. In this case the A_{Ni} factor is 1.05, requiring only a 5% correction. On the other hand, at 30 keV and a low take-off angle of 15.5°, the absorption factor is 1.60, requiring almost a 60% correction. It is clear that A_i will be minimized at low overvoltage and high ψ angles.

Table 9.3 illustrates the absorption corrections necessary for Al $K\alpha$ in a Mg–10 wt % Al alloy. At $E_0 = 15$ and 30 keV, the $f(x)^*$ values are so low that corrections of over 200% are necessary. The very high $\mu/\rho)$ value of Al $K\alpha$ in Mg (4376.5 cm^2/g) is responsible for the large absorption correction. Since E_c for Al is only 1.56 keV, it is possible to perform an x-ray microanalysis at a lower operating voltage than 15 keV. The excitation region is closer to the surface and the absorption path length is decreased. At an operating voltage of 7.5 keV and a take-off angle of 52.5° (note Table 9.3), the absorption factor is only 1.306. In this case a reasonably small correction can be applied.

It is clear from these calculations that the analyst should be wary of large $\mu/\rho)$ values and operation at high overvoltages and low ψ angles. Clearly a reasoned choice of SEM operating conditions and x-ray lines with small mass absorption coefficients can help minimize the corrections needed in the ZAF procedure. Alternatively, the analyst could turn to the $\phi(\rho z)$ method, which incorporates a more advanced absorption correction specifically devised to consider the high absorption situations, which often cannot be avoided in practical analysis situations.

9.2.4. The Characteristic Fluorescence Correction F

If the energy of a characteristic x-ray peak E from element j in a specimen is greater than E_c of element i, then parasitic fluorescence must be accounted for in the correction procedure for element i. Such a fluorescence correction is necessitated because the energy of the x-ray peak from element j is sufficient to excite x-rays secondarily from element i. Thus, more x-rays from element i are generated than would have been produced by electron excitation alone. The correction becomes negligible, however, if $(E - E_c)$ is greater than 5 keV.

Electrons are attenuated more strongly than x-rays of comparable energy. Thus, fluorescent radiation can originate at greater distances from the point of impact of the electron beam than primary radiation. (Note Fig. 3.49). Hence, the mean depth of production of fluorescent radiation is greater than that of primary radiation. Therefore, the intensity of fluorescent emission that can be measured by the x-ray detector relative to that of primary emission increases with increasing x-ray emergence angle.

Since x-ray fluorescence always adds intensity for element i, an equation of the following form can be used:

$$F_i = \frac{\left(1 + \Sigma_j \frac{I^f_{ij}}{I_i}\right)}{\left(1 + \Sigma_j \frac{I^f_{ij}}{I_i}\right)^*}. \tag{9.23}$$

The correction factor I^f_{ij}/I_j relates the intensity of radiation of element i produced by fluorescence by element j, I^f_{ij}, to the electron-generated intensity of radiation from element i, I_i. The total correction factor is the summation of the fluorescence of element i by all the elements j in the sample. The most popular version of the correction factor I^f_{ij}/I_j was derived by Castaing (1951) and modified by Reed (1965). For element i fluoresced by element j in a specimen containing these or additional elements, we have

$$I^f_{ij}/I_i = C_j Y_0 Y_1 Y_2 Y_3 P_{ij}. \tag{9.24}$$

C_j is the concentration of the element causing the parasitic fluorescence, i.e., the fluorescer; Y_0 is given by

$$Y_0 = 0.5 \frac{r_i - 1}{r_i} \omega_j \frac{A_i}{A_j},$$

where r_i is the absorption-edge jump ratio for element i—for a K line $(r_i - 1)/r_i$ is 0.88, and for an L line $(r_i - 1)/r_i$ is 0.75, although each term has a small atomic number dependence (Armstrong, 1988); ω_j is the fluorescent yield for element j (see Chapter 3 and Table 14.10); A_i is the atomic weight of the element of interest and A_j is the atomic weight of the element causing the parasitic fluorescence; $Y_1 = [(U_j - 1)/(U_i - 1)]^{1.67}$, where $U = E_0/E_c$. The term $Y_2 = \mu/\rho^j_i)/\mu/\rho)^j_{\text{spec}}$, where $\mu/\rho)^j_i$ is the mass absorption coefficient of element i for radiation from element j, and $\mu/\rho)^j_{\text{spec}}$ is the mass absorption coefficient of the specimen for radiation from element j. The values of ω, A, and $\mu/\rho)$ are given in Chapter 14. The other term Y_3 accounts for absorption:

$$Y_3 = \frac{\ln(1 + u)}{u} + \frac{\ln(1 + v)}{v},$$

with

$$u = \frac{\mu/\rho)^i_{\text{spec}}}{\mu/\rho)^j_{\text{spec}}} \cdot \csc \psi,$$

where $\mu/\rho)^i_{\text{spec}}$ is the mass absorption coefficient of the specimen for radiation from element i, Eq. (9.22) and

$$v = \frac{3.3 \times 10^5}{(E_0^{1.65} - E_c^{1.65})\mu/\rho)^j_{\text{spec}}},$$

where E_c is evaluated for element i. Finally, P_{ij} is a factor for the type of fluorescence occurring. If KK (a K line fluoresces a K line) or LL

fluorescence occurs, $P_{ij} = 1$. If LK or KL fluorescence occurs, $P_{ij} = 4.76$ for LK and 0.24 for KL. In the Y_3 factor, an exponential function was used to represent the $\phi(\rho z)$ curve, assuming a point source at the surface.

The Reed relation is used in most computer-based schemes for correction. Heinrich and Yakowitz (1968) tested the Reed model for its response to input parameter uncertainties. They found that ω_j produces the worst uncertainties; the other variables produce negligible errors. The extension of I_{ij}^f/I_i, to fluorescence of element i by more than one element j is given by Eq. (9.23). The term I_{ij}^f/I_i is calculated by Eq. (9.24) for each element j which fluoresces element i. The effects of all these elements are summed, as shown in Eq. (9.23). If the standard is a pure element or element i is not fluoresced by other elements present in a multicomponent standard, Eq. (9.23) for F_i can be written in the more standard form:

$$F_i = \frac{1}{\left(1 + \Sigma_j \dfrac{I_{ij}^f}{I_i}\right)^*}. \tag{9.25}$$

The fluorescence factor F_i is usually the least important factor in the ZAF correction, since secondary fluorescence may not occur or the concentration C_j in Eq. (9.24) may be small.

The significance of the fluorescence correction F_i can be illustrated by considering the binary system Fe–Ni. In this system, the Ni $K\alpha$ characteristic energy, 7.478 keV, is greater than the energy for excitation of Fe K radiation, $E_c = 7.11$ keV. Therefore an additional amount of Fe $K\alpha$ radiation is produced. In this system the atomic number correction Z_i is <1% and can be ignored. Calculations of F_{Fe} in a 10-wt % Fe–90-wt % Ni alloy were made using the expression given in Eqs. (9.24) and (9.25).

Table 9.4 contains the input data used for the F_i calculation. Calculations of F_{Fe} were performed for two operating voltages, 30 and 15 keV, and two take-off angles, 15.5° and 52.5°. The amount of Fe fluorescence given by $I_{Fe\text{–}Ni}^f/I_{Fe}$, listed in Table 9.4, ranges from 16.8% to 34.6%. It increases with increasing take-off angle and operating voltage. To minimize the amount of the fluorescence correction, low-kilovoltage operation is suggested. Low-kilovoltage operation also minimizes the absorption correction A_{Fe} and A_{Ni} (Table 9.4). Although the fluorescence correction is minimized at low ψ angles, the error associated with the term F_i does not increase with the take-off angle (Heinrich and Yakowitz, 1968). Therefore low-E_0, high-ψ operation, recommended so as to minimize A_i, is satisfactory even in cases requiring large fluorescence corrections.

If the concentration of C_j, [Eq. (9.24)] decreases, the amount of fluorescence I_{ij}^f/I_i also decreases. For example, if the Ni content changes in a binary Fe–Ni alloy from 90 wt % to 50 wt %, resulting in a 50-wt % Fe–50-wt % Ni alloy, the amount of fluorescence at $E_0 = 15$ keV, $\psi = 15.5°$ decreases by more than a factor of 2, from 16.8% to 6.5%. Clearly the relative effect of fluorescence increases markedly as C_i

Table 9.4. Fluorescence of Fe $K\alpha$ in a 10-wt % Fe–90 wt% Ni Alloy

a. Input data

	Fe	Ni
μ/ρ Fe $K\alpha$(cm^2/g) in absorber	71.4	90
μ/ρ Ni $K\alpha$ (cm^2/g) in absorber	379.6	58.9
ω	—	0.37
A	55.847	58.71
E_c (keV)	7.11	8.332
C (wt fraction)	0.1	0.9

b. Output data

ψ (deg)	E_0 (keV)	$\frac{I^f_{Fe-Ni}}{I_{Fe}}$	F_{Fe}	A_{Fe}	A_{Ni}
52.5	15	0.263	0.792	1.002	1.005
15.5	15	0.168	0.856	1.008	1.015
52.5	30	0.346	0.743	1.011	1.023
15.5	30	0.271	0.787	1.030	1.065

decreases and C_j increases. Figure 8.2 gives experimental data on the effect of fluorescence at $E_0 = 30$ keV and $\psi = 52.5°$.

Reed (1990) reviewed the approximations used in the derivation of the characteristic fluorescence correction as well as the constants used in the calculations. More recent data for fluorescence yields and absorption-edge ratios are preferred. In addition, it is suggested to replace the $(U - 1)^{1.67}$ term in the equation for Y_1 by $(U \ln U - U + 1)$. The effect of these changes is generally small when the correction is large for high atomic numbers ($Z_i > 20$). For low atomic numbers ($Z_i < 20$) where the absolute size of the correction is small, the relative difference is greater and is caused mainly by using more recent values of the fluorescence yield.

9.2.5. The Continuum Fluorescence Correction

Whenever electrons are used to excite x-ray spectra, a band of continuous radiation, with an energy range from 0 to E_0, always accompanies the characteristic peaks used as analytical lines. This continuum arises because the incoming electrons decelerate in the electrical field of the atoms to produce quanta of different energies as they proceed through the solid.

The continuum band contains quanta of energy sufficient to excite any characteristic radiation that can be directly excited by the beam electrons, since there is always some continuum radiation in the range E_c to E_0. Calculation of the intensity of the continuum-induced radiation is complicated because of the following considerations:

1. The effect must be integrated over an energy range from E_c to E_0. The cross-section for excitation varies with the continuum energy.

2. Knowledge of the functional dependence of the x-ray continuum on specimen atomic number and the beam energy is required.
3. The difference in the continuum fluorescence effect between sample and standard depends greatly on the relative mass absorption coefficients for the line of interest.

Henoc (1968) has derived a functional expression for the intensity of continuum-induced fluorescence I'_c:

$$I'_c = f(\bar{Z}, \omega, r, \mu/\rho), \psi), \tag{9.26}$$

where $\bar{Z}$ is the average atomic number, ω is the fluorescence yield, r is the absorption jump ratio, $\mu/\rho)$ is the mass absorption coefficient, and ψ is the take-off angle.

Accurate computation of I'_c requires extensive program space and computer time. Because the effect can be safely ignored in many cases, most ZAF programs do not include a correction for continuum fluorescence.

Myklebust *et al.* (1979) investigated the continuum correction and came to the following conclusions:

1. The magnitude of the continuum fluorescence correction cannot be neglected in cases where $f(\chi)$ of the material is 0.95 or greater. This corresponds to analysis using x-ray lines from an element in a light matrix, e.g., analysis of the "heavy" component in an oxide.
2. Voltage variation and errors in the take-off angle, fluorescence yield factor, and jump ratio have no significant effect on the continuum fluorescence correction.
3. Continuum fluorescence corrections can be ignored when

$$\begin{gathered} f(x) \leq 0.95 \\ C_i \geq 0.5 \\ \bar{Z}_{\text{std}} \simeq \bar{Z}_{\text{spec}} \end{gathered} \tag{9.27}$$

4. All of the results support the choice of the operating parameters suggested for quantitative analysis, i.e. low overvoltage and high take-off angle.

Perhaps the most important conclusion is that uncertainties may result if one employs hard x-ray lines from heavy elements such as Zn or Hg found in small amounts in matrices of light elements such as carbon. For trace amounts of copper in bulk biological or polymeric specimens, the continuum radiation contribution provides most of the observed intensity. Hence, when analyzing for small amounts of heavy elements in light-element matrices, the lowest-energy characteristic x-ray line should be chosen as the analytical line. If it is necessary to use a hard x-ray line such as Cu $K\alpha$ in a biological matrix, then the standard should consist of a dilute copper constituent in a similar light matrix, such as lithium borate glass.

9.2.6. Summary Discussion of the ZAF Method

We now consider the way in which the individual corrections are utilized to actually calculate the composition of the specimen. In the usual experiment, one determines a set of k values, but the factors Z, A, and F depend upon the true composition of the specimen, which is unknown. This problem is handled by using the measured k values as the first estimate of composition. In this way,

$$C_i = \frac{k_i}{\sum_i k_i} \tag{9.28}$$

for all elements i in the specimen. These first-estimate concentrations are normalized to sum to 100%. The resulting mass fractions are used to compute the initial ZAF factors for each element. The estimated concentrations and ZAF factors for each element are used to calculate k_i values which are then compared to the measured k_i values. Iteration proceeds until results converge.

Before an x-ray microanalysis is performed, it is wise to calculate the magnitude of the ZAF factors. When the corrections for a given element are significant, the analyst can attempt to minimize the corrections by changing the operating conditions before the analysis is undertaken. The analyst can often control E_0 and sometimes ψ and can if necessary employ standards of similar composition to the sample. The following two examples show calculations of ZAF corrections and point out how these corrections can be minimized.

For both Cu $K\alpha$ and Al $K\alpha$ in the Al–3-wt % Cu alloy considered in Section 9.2.2, the F_i factor is 1.0, since neither element is effectively fluoresced by the other. The absorption correction for Cu $K\alpha$ in the Al–Cu alloy A_{Cu} is 1.0 at 15 keV and $\psi = 52.5°$ since the mass absorption coefficients are very small; $\mu/\rho)_{\mathrm{Cu}}^{\mathrm{Cu}\,K\alpha} = 53.7\ \mathrm{cm^2/g}$ and $\mu/\rho)_{\mathrm{Al}}^{\mathrm{Cu}\,K\alpha} = 49.6\ \mathrm{cm^2/g}$. For Cu $K\alpha$ in the Al–Cu alloy, the whole ZAF correction is 1.16 using a pure Cu standard and is due entirely to the atomic number effect. On the other hand, Al $K\alpha$ in the Al–Cu alloy is highly absorbed by Cu, $\mu/\rho)_{\mathrm{Al}}^{\mathrm{Al}\,K\alpha} = 385.7\ \mathrm{cm^2/g}$, and $\mu/\rho)_{\mathrm{Cu}}^{\mathrm{Al}\,K\alpha} = 5377\ \mathrm{cm^2/g}$. By using the PDH absorption correction described in Section 9.2.3, A_{Al} was calculated for the Al–Cu alloy for the operating condition $E_0 = 15$ keV, $\psi = 52.5°$. The calculation gave $f(\chi) = 0.902$, $f(\chi)^* = 0.880$, and $A_{\mathrm{Al}} = 1.025$. The absorption correction is surprisingly small but is understandable when one realizes that only 3 wt % of the highly absorbing element Cu is present in the sample. For Al $K\alpha$ in the Al–3-wt % Cu alloy, the whole ZAF correction is 1.023, primarily consisting of the small absorption correction A_{Al}, 1.025, and the very small atomic number correction Z_{Al}, 0.998.

A second example of the use of the ZAF approach is the microanalysis of a seven-component meteoritic pyroxene of composition 53.5-wt % SiO_2, 1.11-wt % Al_2O_3, 0.62–wt % Cr_2O_5, 9.5-wt % FeO, 14.1-wt % MgO, and 21.2-wt % CaO (Hewins, 1979). The $K\alpha$ lines were used for all elements and the wt % of O was calculated using the assumed

Table 9.5. ZAF Calculation for a Meteoritic Pyroxene at $\psi = 52.5°$ and $E_0 = 15$ and 30 keV

	Concentration (wt %)	Z_i	A_i	F_i	$[ZAF]_i$	k_i
$E_0 = 15$ keV	53.3 SiO_2	0.980	1.11	0.998	1.091	0.489
	1.11 Al_2O_3	0.995	1.19	0.987	1.17	0.00947
	0.62 Cr_2O_3	1.106	1.019	0.98	1.106	0.0056
	9.5 FeO	1.14	1.01	1.0	1.15	0.0828
	14.1 MgO	0.971	1.20	0.993	1.16	0.122
	21.2 CaO	1.028	1.033	0.996	1.06	0.200
$E_0 = 30$ keV	53.3 SiO_2	0.985	1.302	0.997	1.278	0.417
	1.11 Al_2O_3	1.0	1.484	0.983	1.460	0.0076
	0.62 Cr_2O_3	1.086	1.07	0.974	1.13	0.0055
	9.5 FeO	1.11	1.033	1.0	1.14	0.0832
	14.1 MgO	0.977	1.47	0.991	1.43	0.0989
	21.2 CaO	1.018	1.11	0.994	1.124	0.189

stoichiometry of the cation–anion ratio of the oxides. Pure oxide standards SiO_2, Al_2O_3, Cr_2O_3, FeO, MgO, and CaO were used for each of the metallic elements. Calculations of ZAF and the k ratios were made for a take-off angle of 52.5% and two operating voltages, 15 and 30 keV (Table 9.5).

The calculations show that there is a significant atomic number effect, ~10%, for Cr and Fe, the two elements which have a significantly higher Z than the average atomic number of the sample. The atomic number correction is somewhat less at $E_0 = 30$ keV. The fluorescence effect is small, with a maximum correction of 2% to 3% for Cr. Although Fe $K\alpha$ strongly excites Cr $K\alpha$, the Fe content is relatively small, only ~7.5 wt %, minimizing the fluorescence effect. The absorption effect is clearly the most important of the three corrections. Corrections up to 20% at $E_0 = 15$ keV and up to 48% at $E_0 = 30$ keV are necessary. If the take-off angle ψ were less than 52.5°, the absorption correction would be much larger. Clearly, operation at low E_0 is desirable, since the A_i correction is thereby minimized. Therefore, for almost all silicate analyses, high take-off angle, low-kilovoltage operation is recommended.

The ZAF method can often be used to obtain quantitative results from measured relative-intensity data. Fortunately, the absorption correction A_i, usually the most important correction factor, is rather insensitive to the shape of the $\phi(\rho z)$ curve assumed. As a result, even crude approximations to the $\phi(\rho z)$ distribution (Philibert, 1963) have given acceptable accuracy, at least in limited applications. The ZAF method with the equations suggested for Z_i, A_i, and F_i in this chapter may not be applicable to all types of specimens. The procedure may not be successful in an analysis using x-ray lines of energies less than 1 keV, not only because of the lack of knowledge of required input parameters but because of the approximations used in the correction models, particularly for the absorption correction (see Section 9.10). For these reasons, if the analyst has a choice of possible analytical lines for an element (e.g., Cu $K\alpha$ at 8.04 keV and Cu $L\alpha$ at 0.92 keV), the analysis of low-energy x-ray lines is likely to be less accurate than analysis utilizing higher-energy x-ray lines.

An example of the accuracy that can be expected when good specimens are available is shown in Fig. 8.13. This figure is a histogram showing the relative error distribution in a group of homogeneous, well-analyzed samples. These alloys were examined under a variety of operating conditions in a commercial EPMA, and crystal spectrometers were used. One way to sum up the implications of the results in Fig. 8.13 is in gambling terms: the odds are 2 to 1 against an analysis being within 1% relative (1% of amount present), even money on the analysis being within 2.5% relative, and 7 to 1 for the analysis being within 5% relative. These odds are accuracy estimates based on nearly optimum specimens, i.e., those properly prepared and placed normal to the electron beam in a commercial EPMA.

9.3. $\phi(\rho z)$ Technique

9.3.1. Introduction

The $\phi(\rho z)$ method uses calculated $\phi(\rho z)$ curves for the determination of the atomic number Z and absorption A in the microchemical analysis of specimens in the SEM–EPMA using Eq. (8.2). The concept of $\phi(\rho z)$ curves was introduced in Chapter 8 along with a description of how the curves can be used to calculate the atomic number and absorption corrections. The atomic number correction for a given element i is obtained by calculating the area under the $\phi(\rho z)$ curve, representing the total number of x rays of element i generated in the standard and dividing it, by the area under the $\phi(\rho z)$ curve for element i in a specimen for the same operating conditions (Fig. 8.1). Using Eq. (9.13), Z_i can be expressed as

$$Z_i = \frac{\int_0^\infty \phi_i(\rho z)\, d(\rho z)}{\int_0^\infty \phi_i^*(\rho z)\, d(\rho z)}, \tag{9.29}$$

where the specimen is noted by an asterisk and the standard is left unmarked.

The absorption correction A_i can be expressed in terms of $\phi(\rho z)$, using Eq. (9.15) for both $f(\chi)$ and $f(\chi)^*$. In this case,

$$A_i = \frac{\left(\dfrac{\int_0^\infty \phi_i(\rho z) \exp - [\mu/\rho)^i \csc \psi(\rho z)]\, d\rho z}{\int_0^\infty \phi_i(\rho z)\, d\rho z}\right)}{\left(\dfrac{\int_0^\infty \phi_i^*(\rho z) \exp - [\mu/\rho)^i \csc \psi(\rho z)]\, d\rho z}{\int_0^\infty \phi_i^*(\rho z)\, d\rho z}\right)}. \tag{9.30}$$

The contribution of Z_i and A_i is given by the equation

$$Z_iA_i = \frac{\int_0^\infty \phi_i(\rho z) \exp - [\mu/\rho)^i \csc \psi(\rho z)]\, d\rho z}{\int_0^\infty \phi_i^*(\rho z) \exp - [\mu/\rho)^i \csc \psi(\rho z)]\, d\rho z}, \quad (9.31)$$

and these terms are often considered together in the correction procedure.

Combining the characteristic fluorescence correction F_i, as given in Section 9.2.4, we have the complete correction, ZAF_i, which can be used along with the measured intensity ratio k_i to calculate the concentration C_i of element i [(see Eq. 8.2)].

The following sections describe the process for determining $\phi(\rho z)$ for multielement specimens and give examples of how the $\phi(\rho z)$ correction is used to calculate the atomic number, absorption and full quantitative correction.

9.3.2. The $\phi(\rho z)$ Curves

9.3.2.1. Definition

Figure 9.2a and b shows the scheme for defining the depth distribution of the generated x rays, $\phi(\rho z)$. The x-ray generation volume in the specimen is divided into 10 to 20 or more thin layers of equal mass thickness, $\Delta(\rho z)$. We can calculate or measure the number of x rays generated for a given x-ray line of interest in each layer, $I(\rho z)$, ρz in mass-depth from the specimen surface and $\Delta(\rho z)$ in mass-thickness, for a given number of beam electrons. For normalization purposes, we can also calculate or measure the number of x rays generated for the same x-ray line in a thin film of the specimen (same composition) with the same mass thickness $\Delta(\rho z)$, isolated in space (see Fig. 9.2b), again for

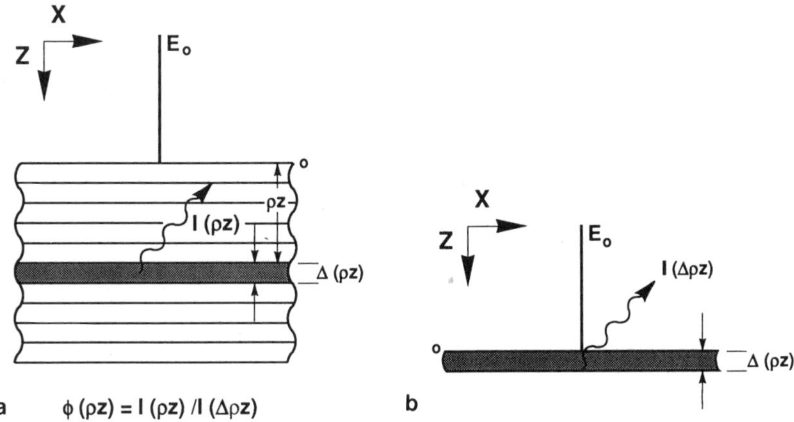

Figure 9.2. Schematic definition of $\phi(\rho z)$ curve. (a) X-ray generation in a solid sample. (b) X-ray generation in a thin foil.

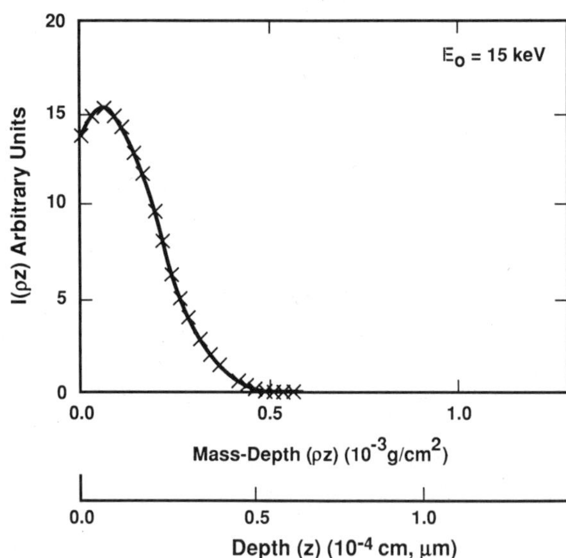

Figure 9.3. Initial measurement of intensity versus mass-depth to derive a $\phi(\rho z)$ curve for a solid.

the same number of beam electrons. We define the x-ray intensity in the isolated thin film as $I(\Delta\rho z)$. The depth distribution $\phi(\rho z)$ of generated x rays at a mass-depth ρz is then $I(\rho z)$ divided by $I(\Delta\rho z)$. The $\phi(\rho z)$ term varies with mass-depth ρz, and with depth z from the specimen surface $\rho z = z = 0$, to the position where x rays are no longer generated, $\rho z = \rho R_x$, $z = R_x$.

Figure 9.3 shows, schematically, the variation of $I(\rho z)$ with mass-depth ρz using over 20 layers of equal mass-thickness $\Delta(\rho z)$. The depth distribution of generated x rays, $\phi(\rho z)$ is obtained by dividing the intensities $I(\rho z)$ in Fig. 9.2a by $I(\Delta\rho z)$ in Fig. 9.2b, as discussed in Chapter 8 and Fig. 8.4. The total generated intensity, I_{gen}, can be obtained by adding together the contributions of $\phi(\rho z)$ from each layer $\Delta(\rho z)$ in mass-thickness; that is, taking the area under the $\phi(\rho z)$ curve and multiplying that area by the x-ray intensity of the isolated thin film $I(\Delta\rho z)$. The total generated intensity, I_{gen}, can be determined for a specific x-ray line, specimen, and initial electron-beam energy and, as we shall see later, can also be used to calculate the atomic number effect.

The general shape of the depth distribution of the generated x rays, the $\phi(\rho z)$-vs-ρz curve (Fig. 8.4), provides information on the effect of Z and of E_0. For convenience, we define the intersection of the $\phi(\rho z)$ curve with the surface, $\rho z = 0$, as ϕ_0 and the maximum in the $\phi(\rho z)$ curve at $\rho z = \rho R_m$ as ϕ_m. The ultimate depth of x-ray generation where $\phi(\rho z) = 0$ occurs where ρz equals ρR_x (see Fig. 8.4). The major reason that ϕ_0 is larger than 1.0 in solid samples is the effect of the backscattered electrons. The backscattered electrons excite x rays in a solid sample as they leave it. In a thin-film specimen of mass-thickness $\Delta(\rho z)$, beam electrons are not able to backscatter as in the solid specimen. Therefore

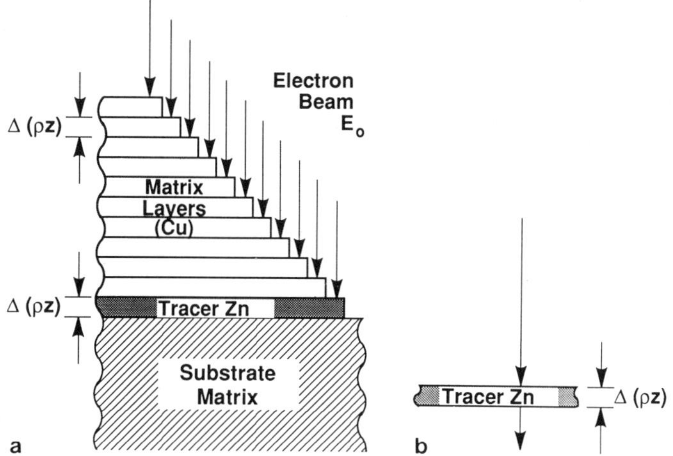

Figure 9.4. Experimental setup for the measurement of $\phi(\rho z)$ curves by the tracer technique. (a) Tracer. (b) Isolated thin film.

the ratio ϕ_0 of the intensities from the top layer in the solid sample to the isolated thin film is ≥ 1.0. The ϕ_0 value increases as backscattering increases; that is, it increases with the atomic number of the specimen.

9.3.2.2. Measurement of $\phi(\rho z)$ Curves

Many $\phi(\rho z)$ curves have been measured experimentally. The experimental setup first proposed and used by Castaing (1951) to obtain $\phi(\rho z)$ curves, called the tracer technique, is shown in Fig. 9.4. We illustrate the method by the measurement of $\phi(\rho z)$ for Cu $K\alpha$. A thin film of Zn, $\Delta(\rho z)$ in thickness, the tracer, is deposited on a substrate and coated by a number of successive layers of Cu (matrix), each $\Delta(\rho z)$ in mass-thickness. The emitted intensity of Zn $K\alpha$ radiation from the tracer is measured by placing the beam on each successive deposit above the tracer (Fig. 9.4a). The intensity from the thin-film Zn tracer itself serves as the isolated film in space (Fig. 9.4b). The generated $\phi(\rho z)$ curve can be calculated after correction for the absorption of Zn $K\alpha$ from the tracer in the overlying matrix layers. Zn was selected as the tracer in this case because it is of similar atomic number to Cu and the Zn $K\alpha$ x-ray line has an energy similar to but higher than that of Cu $K\alpha$, so that it is not fluoresced by Cu $K\alpha$. The measured Cu $K\alpha$ $\phi(\rho z)$ curve, using Zn $K\alpha$ as a tracer, at 25 keV (Brown and Parobek, 1972) is shown in Fig. 9.5 and illustrates the type of $\phi(\rho z)$-vs-ρz curve that can be measured. The general shape of the $\phi(\rho z)$ curve and the variation of the curve with atomic number and initial electron-beam energy were discussed in some detail in Sections 8.6.2 and 8.7.2.

During the last 30 yrs, considerable effort has been made to increase the number of measured $\phi(\rho z)$ curves so that generalized expressions can be obtained for $\phi(\rho z)$. Once these generalized expressions became

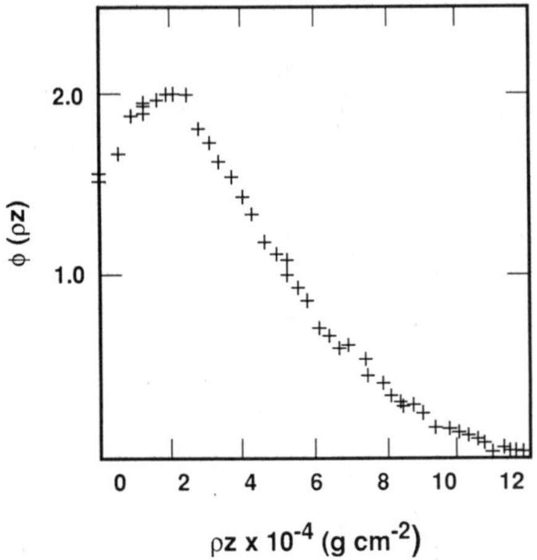

Figure 9.5. Measured Cu $K\alpha$ $\phi(\rho z)$ curve using a Zn $K\alpha$ tracer for 25 keV (Brown and Parobek, 1972).

available, the quantitative-analysis scheme using the $\phi(\rho z)$ method was developed. The next section considers the development of generalized expressions for the x-ray generation function.

9.3.3. Calculation of $\phi(\rho z)$ Curves

Numerous researchers have attempted to accurately model $\phi(\rho z)$ curves, for example Wittry (1957), Kyser (1972), and Parobek and Brown (1978). A key observation was made by Packwood and Brown (1981) with regard to the functionality of the $\phi(\rho z)$ curves. They showed that by plotting ln $\phi(\rho z)$ versus $(\rho z)^2$, straight line behavior was obtained (Fig. 9.6), at least beyond the maxima ϕ_m in the $\phi(\rho z)$ curves. This behavior implies that the $\phi(\rho z)$ curves are Gaussian in character, centered at the surface of the sample. However, there is also a loss of intensity near the surface. In general, the $\phi(\rho z)$ curves can be represented by the following

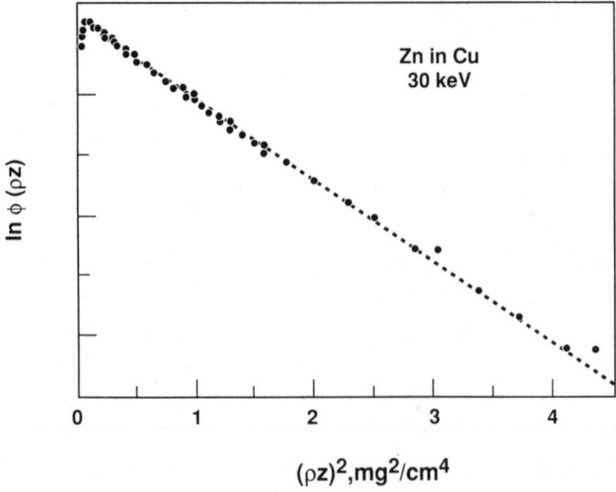

Figure 9.6. Plot of ln $\phi(\rho z)$ versus $(\rho z)^2$ showing that the $\phi(\rho z)$ curves are Gaussian in character [Packwood and Brown (1981)].

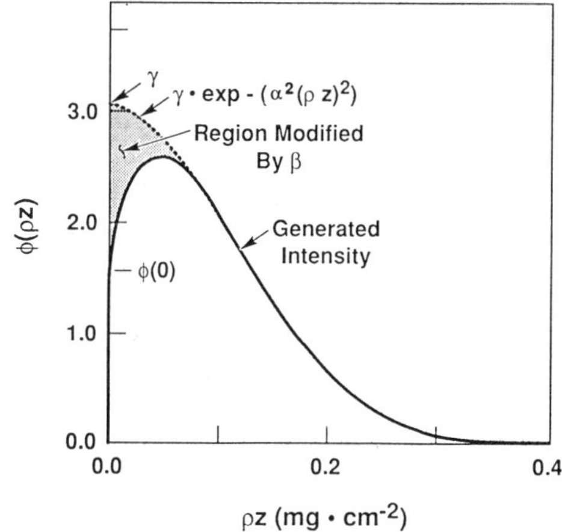

Figure 9.7. Example of a $\phi(\rho z)$ curve showing the functional behavior of the four Gaussian parameters, α, β, γ, and $\phi(0)$ in the Packwood–Brown $\phi(\rho z)$ equation.

equation as given by Packwood and Brown (1981):

$$\phi(\rho z) = \gamma \exp - \alpha^2(\rho z)^2\left[1 - \frac{(\gamma - \phi(0)) \exp - \beta\rho z}{\gamma}\right]. \quad (9.32)$$

The four parameters α, β, γ, and $\phi(0)$ describe the modified Gaussian curve and are a function of E_0, absorption-edge energy, and matrix atomic number and atomic weight. The functional behavior of the terms in the Packwood–Brown $\phi(\rho z)$ equation is shown in Fig. 9.7.

In Eq. (9.32) the Gaussian expression for the $\phi(\rho z)$ curve as a function of mass-depth ρz is given by

$$\gamma \exp - \alpha^2(\rho z)^2, \quad (9.33)$$

in which

1. γ can be regarded as a scaling factor or surface intensity for the basic surface-centered Gaussian, and
2. α represents the decay rate of the Gaussian and describes the electron-beam penetration to E_c for a given x-ray line.

The basic surface-centered Gaussian function is modified in the near-surface regions by a transient function

$$-(\gamma - \phi(0)) \exp(-\alpha^2(\rho z)^2 - \beta\rho z), \quad (9.34)$$

which reduces the x-ray intensity at the surface. The term $\phi(0)$, the relative x-ray intensity generated at the sample surface, is the same as ϕ_0 for the $\phi(\rho z)$ curve. The β term describes the transition process by which the collimated electron probe which enters the specimen changes to scattering in random directions in the specimen. Once the four parameters are determined for a sample and standard for each element i present in the sample, Z_i and A_i can be determined by integration of Eqs. (9.29)–(9.31). One advantage of the Packwood–Brown equation is that it is possible to directly integrate the equation in order to obtain the corrections for atomic number and absorption.

The performance of Eq. (9.32) in the $\phi(\rho z)$ approach for matrix correction is, of course, largely dependent upon the successful parameterization of the α, β, γ, and $\phi(0)$ terms. Packwood and Brown (1981) fit Eq. (9.32) to an extensive array of experimental $\phi(\rho z)$ data using an optimization method. Improvements were also made by considering, theoretically, electron interactions with the specimen. Since the initial Packwood–Brown method was proposed, numerous investigators, for example Brown and Packwood (1982), Bastin *et al.* (1984a;b), and Bastin and Heijligers (1986a), have proposed newer values for the four terms in the $\phi(\rho z)$ equation. Most of these improvements were due to the availability of new experimental $\phi(\rho z)$ data, and new k ratio measurements on well-characterized specimens, particularly samples containing the light elements.

Another approach to the calculation of $\phi(\rho z)$ curves has been developed by Pouchou and Pichoir (1984) and commonly called the PAP approach. They use a polynomial expression for $\phi(\rho z)$, the coefficients of which are given by four parameters; the surface ionization $\phi(0)$ or ϕ_0, the depth of maximum ionization R_m, the maximum depth of ionization R_0, and the integral F of the $\phi(\rho z)$ distribution. The ϕ_0, R_m, and R_x parameters are shown in Fig. 8.4.

The integral F is equal to

$$\int_0^\infty \phi_i(\rho z)\, d\rho z, \tag{9.35}$$

which is equivalent to the generated intensity $I_{i\text{gen}}$ divided by $I_i(\Delta\rho z)$ (see Eq. (9.13)). The integral F of $\phi(\rho z)$ is calculated theoretically by an evaluation of the atomic number correction R_i/S_i. The atomic number correction of Pouchou and Pichoir (1987; 1991) differs in detail from that of Duncumb and Reed (1968), which is discussed in Section 9.2.2.

Bastin and Heijligers (1990a,b, 1991a) have adopted the Pouchou and Pichoir method of calculating the integral F of $\phi(\rho z)$. They use the calculation of F to obtain a more correct value of the parameter β in the Packwood–Brown equation, (9.32). Using this procedure, the values of ϕ_m and R_m in the $\phi(\rho z)$ curve, Fig. 8.4, are more accurately described. This improvement in the $\phi(\rho z)$ procedure has been incorporated in a new calculation scheme called PROZA. The PROZA method is used to calculate the examples discussed later in this section and the $\phi(\rho z)$ curves shown in Chapter 8. The detailed equations for F, and the parameters α, β, γ, and $\phi(0)$ in the Packwood–Brown equation by Bastin and Heijligers (1990a,b; 1991a) are given in Appendix 9.2.

The correct calculation of $\phi(\rho z)$ curves is particularly important for application in other quantitative analysis schemes; thin films on substrates, particles and the analysis of tilted samples (see later sections of this Chapter). The $\phi(\rho z)$ method is less mature than the ZAF method and we expect further developments in this area for application to quantitative microchemical analysis.

9.3.4. Atomic Number Correction Z_i

Equation (9.29) gives the formulation of the atomic number correction Z_i. The Packwood–Brown equation for $\phi(\rho z)$, Eq. (9.32), can be integrated for specimen and standard in closed form as indicated by Brown and Packwood (1982). According to Bastin *et al.* (1984), the atomic number correction Z_i is

$$Z_i = \frac{\{\gamma - [\gamma - \phi(0)]R(\beta/2\alpha)\}\alpha^{-1}}{\{\gamma - [\gamma - \phi(0)]R(\beta/2\alpha)\}^* \alpha^{*-1}} \tag{9.36}$$

where R represents the value of the fifth-order polynomial in the approximation for the complementary error function for the argument in parentheses. The equations describing α, β, γ, and $\phi(0)$ are given in Appendix 9.1. The equation describing Z_i and $R(\beta/2\alpha)$ is given in Appendix 9.2. In the PAP and PROZA approaches to the calculation of $\phi(\rho z)$ discussed in Section 9.3.3, the integral F of $\phi(\rho z)$ is calculated theoretically and

$$Z_i = \frac{F}{F^*}. \tag{9.37}$$

In the following paragraphs, we will illustrate the calculation of the atomic number correction Z_i with a specific example.

Figure 9.8 shows $\phi(\rho z)$-versus-ρz curves calculated using the PROZA program (Bastin and Heijligers, 1990a,b, 1991a) at an initial beam energy of 15 keV for Al $K\alpha$ and Cu $K\alpha$ radiation for the pure elements Al and Cu. These curves are compared in Fig. 9.8 with calculated $\phi(\rho z)$-versus-ρz curves at 15 keV for Al $K\alpha$ and Cu $K\alpha$ in a binary sample containing Al with 3 wt % Cu. The ϕ_0 or $\phi(0)$ value of the Cu $K\alpha$ curve in the alloy is smaller than that of pure Cu because the average atomic number of the Al–3-wt % Cu sample is so much lower, almost the same as pure Al. In this case, less backscattering of the primary high-energy electron beam

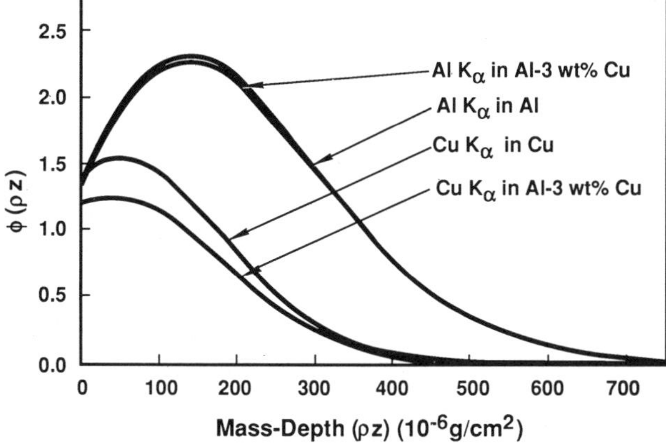

Figure 9.8. Curves of $\phi(\rho z)$ versus ρz for Al $K\alpha$ and Cu $K\alpha$ radiation in Al, Cu and Al–3 wt% Cu at 15 keV, calculated using the PROZA program.

Table 9.6. Generated X-Ray Intensities in Cu, Al, and Al–3 wt % Cu alloy at E_0 = 15 keV, calculated by the PROZA Program

Sample	X-ray line	$\phi(\rho z)_{gen}$ Area (g/cm^2)	Z_i	α	β	γ	$\phi(0)$
Cu	Cu $K\alpha$	3.34×10^{-4}	1.0	4,737	5,235	2.39	1.39
Al	Al $K\alpha$	7.85×10^{-4}	1.0	3,158	5,160	4.169	1.33
Al–3 wt % Cu	Cu $K\alpha$	2.76×10^{-4}	1.21	4,719	2,141	2.464	1.20
Al–3 wt % Cu	Al $K\alpha$	7.89×10^{-4}	0.995	3,159	5,242	4.160	1.342

occurs, and fewer Cu $K\alpha$ x rays are generated. On the other hand, the Al $K\alpha$ $\phi(\rho z)$ curves for the alloy and the pure element are essentially the same, since the average atomic number of the specimen is so close to that of pure Al. Although the variation of $\phi(\rho z)$ curves with atomic number and initial operating energy is complex, a knowledge of the pertinent x-ray generation curves is critical to understanding what is happening in the specimen and the standard for the element of interest.

As discussed before with reference to Eq. (9.13), the x-ray intensity for each element's x-ray line i in the specimen $I_{i\text{gen}}$ can be obtained by first taking the area under the $\phi(\rho z)$-versus-ρz curve, that is, by summing the values of $\phi(\rho z)$ for all the layers $\Delta(\rho z)$ in mass-thickness within the specimen for the x-ray line i of interest. We will call this area $\phi(\rho z)_{i\text{gen}}$ Area, and then multiplying that area by the x-ray intensity from the isolated thin film $I_i(\Delta\rho z)$. Table 9.6 lists the values calculated using the PROZA program, of the $\phi(\rho z)_{\text{gen}}$ Area for the 15 keV $\phi(\rho z)$ curves shown in Fig. 9.8 (Cu $K\alpha$ and Al $K\alpha$ in the pure elements and Cu $K\alpha$ and Al $K\alpha$ in an alloy of Al–3-wt % Cu) and the corresponding values of α, β, γ, and $\phi(0)$. A comparison of the $\phi(\rho z)$ Area values for Al $K\alpha$ in Al and in the A1–3-wt % Cu alloy shows very similar values while a comparison of the $\phi(\rho z)$ Area values for Cu $K\alpha$ in pure Cu and in the Al–3-wt % Cu alloy shows that about 17% fewer Cu $K\alpha$ x rays are generated in the alloy. The latter variation is due to the different atomic numbers of the pure Cu and the Al–3-wt % Cu alloy specimen. The different atomic number matrices cause a change in ϕ_0 (see Table 9.6) and in the height of the $\phi(\rho z)$ curves.

The atomic number correction Z_i can be calculated by taking the ratio of I_{gen} for the standard to I_{gen} for the specimen where pure Cu and pure Al are the standards for Cu $K\alpha$ and Al $K\alpha$ respectively. The values of the calculated ratios of generated x-ray intensities, pure-element standard to specimen (atomic number effect, Z_{Al}, Z_{Cu}) are also given in Table 9.6. As discussed in Sections 9.2.2 and 8.7.1, it is expected that the atomic number correction for a heavy element (Cu) in a light element matrix (Al–3-wt % Cu) is greater than 1.0 and the atomic number correction for a light element (Al) in a heavy element matrix (Al–3-wt % Cu) is less than 1.0. The calculated data in Table 9.6 also shows this relationship. It is interesting to compare the atomic number correction Z_{Cu} (Cu $K\alpha$ in Al–3-wt % Cu, E_0 of 15 keV) for the ZAF and $\phi(\rho z)$

procedures. The correction using the Duncumb–Reed atomic number procedure (Table 9.1) is 1.16, while the $\phi(\rho z)$ procedure (Table 9.6) yields a value of 1.21. Such differences are significant and can result in a 5 percent relative error in the final composition.

9.3.5. Absorption Correction A_i

Equation (9.31) gives a formulation for the atomic number and absorption correction Z_iA_i. The absorption correction, Eq. (9.30), is obtained by dividing the Z_iA_i correction by the atomic number correction Z_i, as discussed in the previous section. The Packwood–Brown equation for $\phi(\rho z)$, Eq. (9.32), can be integrated for specimen and standard in closed form, as indicated by Brown and Packwood (1982). According to Bastin *et al.* (1984a,b), the combined Z_iA_i correction is

$$Z_iA_i = \frac{[\gamma R(\chi/2\alpha) - (\gamma - \phi(0))R((\beta + \chi)/2\alpha)]\alpha^{-1}}{[\gamma R(\chi/2\alpha) - (\gamma - \phi(0))R((\beta + \chi)/2\alpha)]^*\alpha^{*-1}} \tag{9.38}$$

where R represents the value of the fifth-order polynomial in the approximation for the error function, which comes in when the integrals are solved in closed form. The term χ is $\mu/\rho)^i \csc \psi$, as defined in Section 9.2.3 for the absorption correction. The equations describing α, β, γ and $\phi(0)$ are given in Appendix 9.1. An equation to describe Z_iA_i, $R(\chi/2a)$, and $R((\beta + \chi)/2\alpha)$ is given Appendix 9.2. In the following paragraphs we will examine the calculation of the absorption correction A_i with some specific examples.

In a several-step process, we can calculate the emitted intensity I_{em} for element i, which results when the generated intensity I_{gen} for element i is absorbed within the specimen. In the first step, the value of $\phi(\rho z)$ for each layer ρz in mass-depth from the surface of the specimen and $\Delta\rho z$ in mass-thickness is multiplied by $\exp - [\mu/\rho)^i \csc \psi(\rho z)]$. Figure 8.9 shows the geometrical relationships within the specimen that are considered for the calculation of the x-ray intensity for one x-ray line and for one layer $\Delta(\rho z)$ in mass-thickness, ρz in mass-depth from the surface. These calculations yield a $\phi(\rho z)$-versus-ρz curve called "$\phi(\rho z)$ emitted" (see Fig. 9.9). In the second step, the area under the $\phi(\rho z)$-versus-ρz emitted curve is obtained. We call this area $\phi(\rho z)_{em}$ Area. Finally, the area under the $\phi(\rho z)$-versus-ρz emitted curve ($\phi(\rho z)_{em}$ Area) for element i is multiplied by the x-ray intensity from element i in the isolated thin film $I(\Delta\rho z)$. The value of I_{em} obtained contains the combined effects of atomic number, through the $\phi(\rho z)$-versus-ρz curve, and absorption.

Figure 9.9 shows the $\phi(\rho z)$-versus-ρz (emitted) curves for Cu $K\alpha$ in pure Cu and Al $K\alpha$ in pure Al, calculated at an initial electron-beam energy of 15 keV (see also Fig. 9.8). A 40° take-off angle was assumed, and the calculation was made using the PROZA program (Bastin and Heijligers, 1990a,b, 1991a). The Cu $K\alpha$ $\phi(\rho z)$ emitted curve falls essentially on the $\phi(\rho z)$ generated curve since the mass absorption

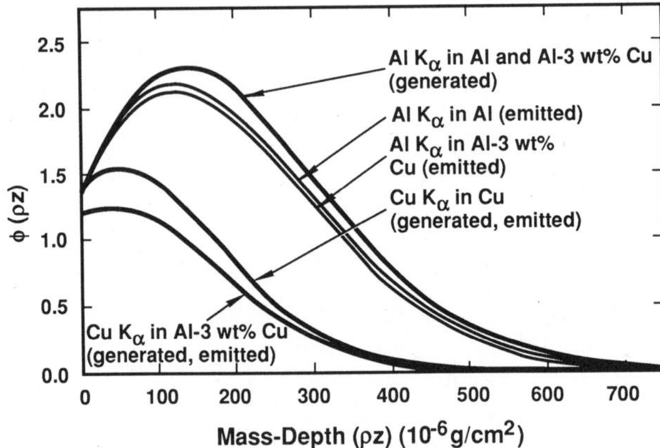

Figure 9.9. Calculated $\phi(\rho z)$, generated and emitted, versus ρz curves for Al $K\alpha$ and Cu $K\alpha$ radiation in Al, Cu, and Al–3 wt% Cu at 15 keV. The take-off angle is 40°.

coefficient is so small, 52 cm^2/g. The Al $K\alpha$ $\phi(\rho z)$ emitted curve for pure Al falls below that of the $\phi(\rho z)$ generated curve and the position of the maximum x-ray production, ρR_m, is somewhat closer to the surface. Both of these changes to the $\phi(\rho z)$ generated curves are due to the much higher Al $K\alpha$ mass absorption coefficient and the increasing amount of absorption with mass-depth. The values of ϕ_0 are not changed, because absorption at the surface is limited.

Figure 9.9 also shows the Cu $K\alpha$ and Al $K\alpha$ $\phi(\rho z)$ generated and emitted curves in an Al–3-wt % Cu alloy. As previously shown (Fig. 9.8), the generated $\phi(\rho z)$ curve for Cu $K\alpha$ in the Al–3-wt % Cu alloy has a lower ϕ_0 value and a lower ρR_m value due to the smaller amount of electron backscattering in the lower-atomic-number specimen. As in the pure-element case, the Cu $K\alpha$ emitted curve (Fig. 9.9) shows little effect of absorption and parallels the generated curve. The Al $K\alpha$ $\phi(\rho z)$ generated and emitted curves for the alloy parallel those of pure Al since the atomic number of the Al–3-wt % Cu specimen is almost the same as that of pure Al. The Al $K\alpha$ emitted curve in the alloy is slightly lower than the Al $K\alpha$ emitted curve in pure Al because the presence of Cu increases the Al $K\alpha$ mass absorption coefficient in the alloy.

Figure 9.10 shows the $\phi(\rho z)$-versus-ρz generated and emitted curves (calculated using the PROZA program) for the case of a trace amount (0.1 wt %) of Al in a pure Cu matrix. A 40° take-off angle was assumed. In this example, the Al $K\alpha$ radiation is highly absorbed by the Cu matrix, and the Al $K\alpha$ radiation that is measured is coming from a restricted region close to the specimen's surface. Although the Al $K\alpha$ radiation is generated from the same mass-depth as in the Al–3-wt % Cu alloy, ϕ_0 is increased from 1.34 in the Al-rich alloy (Fig. 9.9) to 1.76 in the Cu-rich alloy (Fig. 9.10) due to increased electron backscattering.

The effects of increasing the initial electron-beam energy are to increase the depth and lateral size of the x-ray source region and to increase the path length for the absorption of x rays. For x-ray lines that

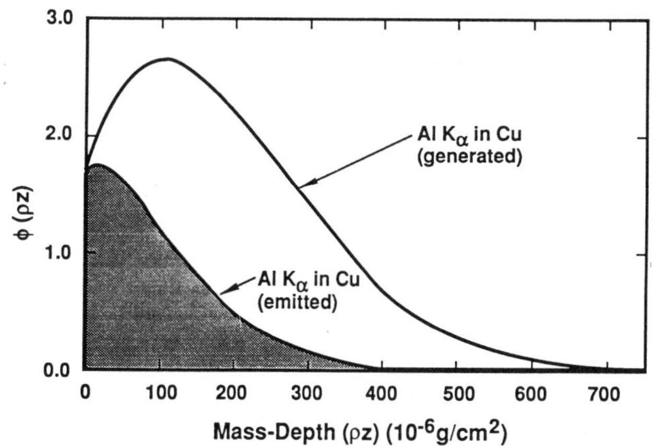

Figure 9.10. Calculated $\phi(\rho z)$ versus ρz generated and emitted curves for Al $K\alpha$ (0.1 wt % Al) in Cu at 15 keV. The take-off angle is 40°.

are heavily absorbed, such as Al $K\alpha$ in a Cu matrix, the amount of absorption is very large and the x rays which are emitted lie much closer to the surface than would be expected from examining the generated $\phi(\rho z)$ curves and the ρR_x value. The analyst should be aware that in these cases the region from which the x-ray microanalysis is being conducted is much smaller than the calculated ρR_x values would lead one to believe. As discussed in Chapter 3, in some cases of multielement analysis the x-ray microanalysis regions for the various characteristic x-ray lines measured are quite different in size. Therefore, to obtain an accurate x-ray microanalysis, a knowledge of x-ray generation with depth for the x-ray lines of interest and the corresponding x-ray emission with depth is critical.

Table 9.7 lists the calculated values (using the PROZA program) of the area under the $\phi(\rho z)_{\text{gen}}$ and $\phi(\rho z)_{\text{em}}$ curves at 15 keV and a 40° take-off angle, shown in Figs. 9.9 and 9.10. The $\phi(\rho z)_{\text{em}}$ Area values for Al $K\alpha$ in pure Al and the Al–3-wt % Cu alloy are quite similar (within 3%). However, the $\phi(\rho z)_{\text{em}}$ Area values for Cu $K\alpha$ in pure Cu and in the Al–3-wt % Cu alloy show a difference of 21%. The difference in Cu $K\alpha$ intensity is *not* due to x-ray absorption but reflects the difference in the atomic number of the two specimens. Note that the areas under the

Table 9.7. X-Ray Absorption Effect in Al–Cu Alloys, E_0 = 15 keV, ψ = 40°

Sample	X-ray line	$\phi(\rho z)_{\text{gen}}$ Area (g/cm²)	$\phi(\rho z)_{\text{em}}$ Area (g/cm²)	Z	ZA	A
Cu	Cu $K\alpha$	3.34×10^{-4}	3.30×10^{-4}	1.0	1.0	1.0
Al	Al $K\alpha$	7.85×10^{-4}	6.91×10^{-4}	1.0	1.0	1.0
Al–3 wt % Cu	Cu $K\alpha$	2.76×10^{-4}	2.73×10^{-4}	1.21	1.21	1.0
Al–3 wt % Cu	Al $K\alpha$	7.89×10^{-4}	6.67×10^{-4}	0.995	1.04	1.05
Cu–0.1 wt % Al	Cu $K\alpha$	3.33×10^{-4}	3.30×10^{-4}	1.003	1.0	0.998
Cu–0.1 wt % Al	Al $K\alpha$	9.16×10^{-4}	3.07×10^{-4}	0.857	2.25	2.63

$\phi(\rho z)_{gen}$ and $\phi(\rho z)_{em}$ curves are almost the same for the Cu $K\alpha$ x-ray line in pure Cu and in the Al–3-wt % Cu alloy. When one compares the $\phi(\rho z)_{em}$ Area values for Al $K\alpha$ between pure Al and a small amount of Al (0.1 wt %) in pure Cu, one observes a very large difference (55%). This large difference in the $\phi(\rho z)_{em}$ Area is due to the combined effect of absorption and atomic number.

Calculations of the ratio of I_{em} values, standard to sample, of the same x-ray line, reflect the combined effects of atomic number Z and absorption A and are labeled ZA in Table 9.7. To obtain the effect of absorption A alone, ZA is divided by Z. Values of Z, ZA, and A are listed in Table 9.7 for the Al–Cu alloy discussed above. Values of the atomic number effect Z and the absorption effect A which are close to 1.0 indicate that the x-ray analysis, using the combination of x-ray line, specimen and standard, electron-beam energy and take-off angle, needs only a small amount of correction when calculations of composition from measured x-ray intensities are made. On the other hand, in the case of the measurement of 0.1-wt % Al in Cu, the Al $K\alpha$ has a large absorption effect (over 60%) and the value of A_{Al} is 2.63. Calculations using the ZAF approach, for the same alloy with a 40° take-off angle and an E_0 of 15 keV give a value of A_{Al} of 2.92. There is a difference of over 10% when significant amounts of absorption are considered, as in this case.

9.3.6. Summary Discussion of the $\phi(\rho z)$ Method

We now consider the way in which the individual corrections are used to calculate the composition of the specimen. In the usual experiment, one determines a set of k values. The Z, A, and F corrections depend on the actual composition of the sample, which is unknown. In the $\phi(\rho z)$ correction, the $\phi(\rho z)$ curves and the values of I_{gen} and I_{em} are also a function of the composition of the sample. As with the ZAF method, this problem is handled by using the measured k values as the first estimate of composition. These first-estimate concentrations are normalized to sum to 100%. In this way,

$$C_i = \frac{k_i}{\sum_i k_i} \tag{9.39}$$

for all the elements i in the specimen. The resulting mass fractions are used to compute the parameters $\phi(0)$, α, γ, and β according to Eqs. (9.A1)–(9.A14) and Table 9.A1 in Appendix 9.1. Using these parameters, the atomic number and absorption correction can be calculated. The fluorescence correction is obtained using Eqs. (9.23)–(9.25). The estimated concentrations and calculated ZAF factors for each element are used to calculate predicted k_i values. These predicted k_i values are compared to the measured k_i values, and new estimates of C_i are computed. Since the $\phi(\rho z)$ parameters $\phi(0)$, α, γ, and β are all functions of composition C_i [see Eqs. (9.A1), (9.A3)–(9.A5), (9.A8)–(9.A10), and (9.A12) in Appendix 9.1], an iterative process is pursued

Table 9.8. X-Ray Microanalysis of Two Al–Cu Alloys, $E_0 = 15\,\text{keV}$, $\psi = 40°$, Pure Element Standards

Sample	X-ray line	Z_i	A_i	F_i	ZAF_i	k_i
Al–3 wt % Cu	Cu $K\alpha$	1.21	1.0	1.0	1.21	0.0248
	Al $K\alpha$	0.995	1.05	1.0	1.045	0.928
Cu–0.1 wt % Al	Cu $K\alpha$	1.003	0.998	1.0	1.001	0.998
	Al $K\alpha$	0.857	2.63	1.0	2.254	0.00444

until results converge. In the calculation of these parameters for $\phi(\rho z)$, the concentrations of the elements enter as weight-averaged fractions (see Appendix 9.1). The following two examples show calculations of ZAF factors by the $\phi(\rho z)$ method and point out how these corrections can be minimized.

Table 9.8 summarizes the ZAF factors calculated using the $\phi(\rho z)$-versus-ρz generated and emitted curves for the Al–3-wt % Cu and the Cu–0.1-wt % Al alloys discussed in Sections 9.3.4 and 9.3.5. For Cu $K\alpha$ and Al $K\alpha$, pure-element standards are used. Even though the Cu $K\alpha$ x ray has more than sufficient energy to excite Al $K\alpha$ x rays, that energy is more than 5 keV greater than E_c for Al $K\alpha$ radiation. The amount of fluorescence is minimal, and in these alloys F_{Cu} and F_{Al} are 1.0 and therefore no fluorescence correction is needed. The absorption effect A_i is also 1.0 for Cu in the Al–3-wt % Cu alloy (Table 9.8), indicating again that no correction is needed. The absorption effect A_i for Al in the Al–3-wt % Cu alloy is small, 1.05, indicating only a 5% correction must be applied. The atomic number effect Z_i for Al is 0.995. The small atomic number correction for Al is due to the very small difference in atomic number between specimen and standard. The atomic number effect Z_i for Cu, however, is much more significant, 1.21. The Cu $K\alpha$ radiation in the Al–3-wt % Cu alloy is decreased primarily due to the increased stopping power of the light Al matrix compared to the heavier copper.

For pure-element standards, and applying Eq. (8.2) as repeated in Section 9.2.1:

$$C_{\text{Cu}} = [ZAF]_{\text{Cu}} \cdot k_{\text{Cu}}$$
$$C_{\text{Al}} = [ZAF]_{\text{Al}} \cdot k_{\text{Al}} \tag{9.40}$$

From Table 9.8, ZAF_{Cu} is 1.21 and ZAF_{Al} is 1.045 for the Al–3-wt % Cu specimen. For this example, measurements of k_{Cu} must be multiplied by 1.21 to obtain C_{Cu} and measurements of k_{Al} must be multiplied by 1.045 to obtain C_{Al}.

In the Cu–0.1-wt % Al alloy, the ZAF factor for Cu $K\alpha$ is very close to 1.0, at 1.001. However the Al $K\alpha$ radiation shows a significant atomic number correction Z_{Al} of 0.857 and an even more significant absorption correction A_{Al} of 2.63 (Table 9.8). For this sample, measurements of k_{Cu} must be multiplied by 1.001 to obtain C_{Cu}, but measurements of k_{Al} must be multiplied by 2.254 to obtain C_{Al}.

Table 9.9. Microanalysis Calculation for Pyroxene (E_0 = 15 keV, ψ = 52.5°)[a]

a. Pure element standards, calculation of element concentration

Element	(wt %)	Oxide	(wt %)	Z_i	A_i	F_i	ZAF_i	k_i
Si	24.9	SiO_2	53.3	1.023	1.215	0.999	1.241	0.201
Al	0.59	Al_2O_3	1.11	1.047	1.360	0.992	1.412	0.0042
Cr	0.42	Cr_2O_3	0.62	1.201	1.016	0.985	1.201	0.0035
Fe	7.38	FeO	9.5	1.212	1.006	1.00	1.220	0.0605
Mg	8.5	MgO	14.1	1.010	1.475	0.996	1.482	0.0574
Ca	15.2	CaO	21.2	1.077	1.032	0.997	1.108	0.137
O[b]	43.0			1.118	1.804	1.00	2.02	0.2129

b. Pure oxide standards, calculation of oxide concentration

Element	(wt %)	Oxide	(wt %)	Z_i	A_i	F_i	ZAF_i	k_i
Si	24.9	SiO_2	53.3	0.973	1.106	0.999	1.075	0.496
Al	0.59	Al_2O_3	1.11	0.989	1.180	0.992	1.158	0.00959
Cr	0.42	Cr_2O_3	0.62	1.101	1.019	0.985	1.105	0.00561
Fe	7.38	FeO	9.5	1.136	1.008	1.00	1.145	0.0830
Mg	8.5	MgO	14.1	0.977	1.208	0.996	1.175	0.120
Ca	15.2	CaO	21.2	1.034	1.033	0.997	1.065	0.199
O[c]	43.0							

[a] Calculations made using the PROZA x-ray quantitation program (Bastin and Heijligers, 1990a,b; 1991a).
[b] FeO used as the O $K\alpha$ standard.
[c] Measured by stoichiometry.

Most specimens of interest are multielement, and much more complex than the simple binaries discussed so far in this chapter. Tables 9.9 and 9.10 summarize calculated *ZAF* factors for the seven-component pyroxene sample considered in the discussion of the composition measurement using the ZAF correction. A take-off angle ψ of 52.5° is assumed. Two values of E_0, 15 keV and 30 keV, are chosen in order to compare the values of the ZAF terms. Tables 9.9 and 9.10 contain *ZAF* factors calculated when pure elements are used as standards and when simple oxides are used as standards. For oxygen analysis, a FeO standard is used since a pure-element standard is not available.

For the pyroxene microanalysis, using elemental standards and an initial electron-beam energy of 15 keV (Table 9.9a), the atomic number effect Z_i for all the elements is greater than 1.0. The atomic number correction is needed, since the average atomic number of the specimen is less than that of each elemental standard and of FeO, the standard for oxygen. The absorption effect A_i is high, 20 to 50%, for Si, Al, and Mg. The absorption effect for oxygen is quite large, more than 80%, primarily due to the high O $K\alpha$ mass absorption coefficients in the metallic elements. The fluorescence effect F_i is small, very close to 1.0, except for Cr. X-rays of Cr $K\alpha$ are excited by Fe $K\alpha$ radiation, and F_{Cr} is 0.985, a 1.5% effect. Since Z_{Cr} is 1.201, a 20% effect, the fluorescence effect is relatively unimportant. The ZAF calculation varies from a minimum of 1.1 for Ca to 1.5 for Mg, 10 to 50% corrections, for the metallic elements

Table 9.10. Microanalysis Calculation for Pyroxene (E_0 = 30 keV, ψ = 52.5°)[a]

a. Pure element standards, calculation of element concentration

Element	(wt %)	Oxide	(wt %)	Z_i	A_i	F_i	ZAF_i	k_i
Si	24.9	SiO_2	53.3	1.019	1.714	0.998	1.74	0.143
Al	0.59	Al_2O_3	1.11	1.045	2.215	0.989	2.29	0.0026
Cr	0.42	Cr_2O_3	0.62	1.167	1.059	0.98	1.21	0.0035
Fe	7.38	FeO	9.5	1.168	1.026	1.00	1.20	0.0616
Mg	8.5	MgO	14.1	1.010	2.585	0.994	2.60	0.0328
Ca	15.2	CaO	21.2	1.058	1.111	0.995	1.17	0.130
O[b]	43.0			1.10	2.448	1.00	2.70	0.160

b. Pure oxide standards, calculation of oxide concentration

Element	wt %	Oxide	wt %	Z_i	A_i	F_i	ZAF_i	k_i
Si	24.9	SiO_2	53.3	0.977	1.304	0.998	1.272	0.421
Al	0.59	Al_2O_3	1.11	0.994	1.488	0.989	1.463	0.00759
Cr	0.42	Cr_2O_3	0.62	1.087	1.070	0.98	1.140	0.0054
Fe	7.38	FeO	9.5	1.111	1.033	1.00	1.148	0.0828
Mg	8.5	MgO	14.1	0.982	1.521	0.994	1.484	0.0950
Ca	15.2	CaO	21.2	1.024	1.114	0.995	1.135	0.187
O[c]	43.0							

[a] Calculations made using the PROZA x-ray quantitation program (Bastin and Heijligers, 1990a,b, 1991a).
[b] FeO used as the O $K\alpha$ standard.
[c] Measured by stoichiometry.

in the pyroxene. The ZAF calculation for oxygen is a factor of 2.02, over 100%. Even a small error in calculating ZAF using the $\phi(\rho z)$ technique causes a significant error in the chemical analysis of the mineral.

Geologists usually take advantage of pure oxide standards for x-ray microanalysis. In this way the measurement of oxygen is unnecessary. Unfortunately, one must assume stoichiometry in the oxides to take advantage of these standards. Because the atomic number of oxide standards is much closer to the average atomic number of the specimen, the magnitude of the atomic number effect is closer to 1.0 and a smaller correction is needed. In this example (compare Table 9.9a and b), Z_{Cr} is reduced from 1.20 to 1.10 and Z_{Al} from 1.047 to 0.989. The amount of the absorption effect is also decreased, since the amount of absorption of a characteristic x-ray line in the standard is much closer to the amount of absorption in the specimen. For Si $K\alpha$, Al $K\alpha$, and Mg $K\alpha$, which are most heavily absorbed, the value of the absorption effect A_i is halved (compare Table 9.9a and b). The fluorescence effect F_i is unaffected because fluorescence does not occur either in the pure oxide or in the metal standards. In this analysis situation, the ZAF calculation is reduced from a minimum of 1.07 for Ca to a maximum correction of ~1.18 for Mg. Small errors in the ZAF calculation do not affect the analysis in a major way, and oxygen is calculated by stoichiometry. The analysis of ceramic oxides as well as carbides and nitrides can be accomplished more accurately by using compound standards, as long as stoichiometry is preserved. If this is not the case, carbon, oxygen, or nitrogen must be measured separately.

One can compare the calculation of the composition of pyroxene at an initial beam energy of 15 keV with that using a higher initial beam energy of 30 keV. Since the path length for absorption is longer, there is a larger absorption correction. In comparing A_i values at 15 and 30 keV (Tables 9.9a, 9.10a) for pure-element standards, A_i for Si, Al, and Mg increase from a 20-to-50% correction at 15 keV to a 70-to-150% correction at 30 keV. The oxygen absorption correction effect increases as well from 80 to 140%. At 30 keV, the atomic number effect Z_i is decreased as much as 4%, but this is not nearly enough to compensate for the increase in the absorption effect. Therefore, as discussed earlier, a lower electron beam energy is preferable for microanalysis measurements.

The use of pure-oxide standards at 30 keV (Table 9.10b) can minimize the effect of Z_i and A_i with regard to pure-element standards. However, the absorption effect for Si, Al, and Mg is still twice that obtained at an initial electron beam energy of 15 keV. As in almost all samples, x-ray microanalysis should be performed at as low an initial electron-beam energy as possible and using a standard for each element with an atomic number as close to that of the average atomic number of the specimen as possible.

It is of interest to compare the ZAF calculations using the ZAF and $\phi(\rho z)$ techniques for the pyroxene sample run under similar conditions, 15 and 30 keV, $\psi = 40°$ using oxide standards (Tables 9.5, 9.9 and 9.10). The calculated ZAF_i values are very similar, within 1% relative at 15 keV and 30 keV, except for Mg $K\alpha$. Mg is the lowest-atomic-number element analyzed and the atomic number correction, calculated by the $\phi(\rho z)$ method, is 5% higher than that calculated by the ZAF method.

In general, the ZAF and $\phi(\rho z)$ techniques yield quite similar results, within a few percent relative, for analysis of $K\alpha$ lines of atomic number 12 (Mg) and above when analyses are made at E_0 values between 15 and 30 keV. For the light elements (O, N, C, etc.) and low energy (<1 keV) $L\alpha$ lines, as well as for analyses at 10 keV and below, the ZAF correction scheme yields less accurate values. In the ZAF correction scheme, the R and S factors in the Z correction were optimized for higher-analyses. The $\phi(\rho z)$ curve used for the A correction is seriously in error near the surface of the sample, where most of the emitted radiation for low-energy x rays comes from. The $\phi(\rho z)$ calculation scheme, on the other hand, has been developed with explicit consideration being given to operation at low keV and with low-energy x-rays (Bastin *et al.*, 1984a,b; 1990a). The added advantage of using the $\phi(\rho z)$ technique is its versatility. With detailed knowledge of the $\phi(\rho z)$ curves of the elements producing radiation in a specimen, one can more easily handle problems when the specimen is tilted away from 90° to the electron probe and where consideration of geometries, such as those found in thin films and substrates and in particulate material, must be given (see Sections 9.4 and 9.8).

9.4. Quantitative Analysis with Nonnormal Electron-Beam Incidence

In most EPMAs, the beam impinges at right angles to a flat, polished specimen, the configuration assumed for the ZAF and $\phi(\rho z)$ methods discussed in Sections 9.2 and 9.3. In most modern SEMs, the EDS x-ray detector is placed at an appropriate take-off angle, >30°, so that x-ray analysis with a nontilted specimen is possible. However, in many older SEMs, the EDS detector is at a low take-off angle, perhaps as low as 0°, and the specimen must then be tilted through at an angle θ away from the normal to the electron beam in order to provide a suitable value of the take-off angle. Even with EDS detectors placed at nonzero take-off angles, specimens are often tilted toward the EDS detector to obtain an even higher x-ray take-off angle. The geometry for a tilted specimen in the SEM is shown in Fig. 9.11. The tilt angle is θ and the total take-off angle is ψ. In this type of geometry, a smaller mean depth of x-ray ionization is obtained, and for low-energy x-rays (the most strongly absorbed) the measured intensity can rise by a significant amount.

The effect of specimen tilt on the accuracy of quantitative analysis was investigated by Monte Carlo calculations of electron–solid interactions some years ago (Curgenven and Duncumb, 1971; Duncumb, 1971). These early studies clearly indicated that changes occur in x-ray distribution and emission as a function of specimen tilt. An example of the effect of tilt is given by a series of Monte Carlo calculations for angles 45° and above, as shown in Fig. 9.12. Experience with tilted specimens indicates that if specimen and standard are flat polished and set to the same tilt angle and the tilt angle is not extreme ($\theta < 60°$), large quantitative analysis errors do not occur (Bolon and Lifshin, 1973; Colby *et al.*, 1969).

The chief difficulty in handling tilted specimens is developing the effect of tilt on the ZAF or $\phi(\rho z)$ corrections, which are all based on

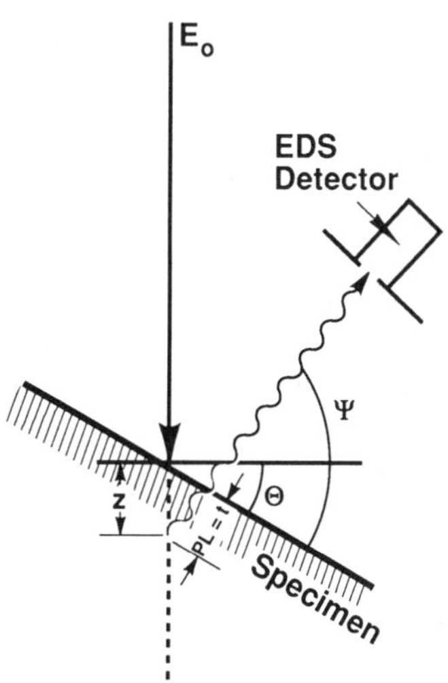

Figure 9.11. Geometry for a tilted specimen. The tilt angle θ and take-off angle ψ are indicated.

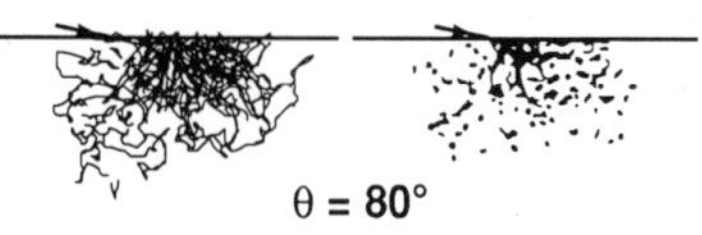

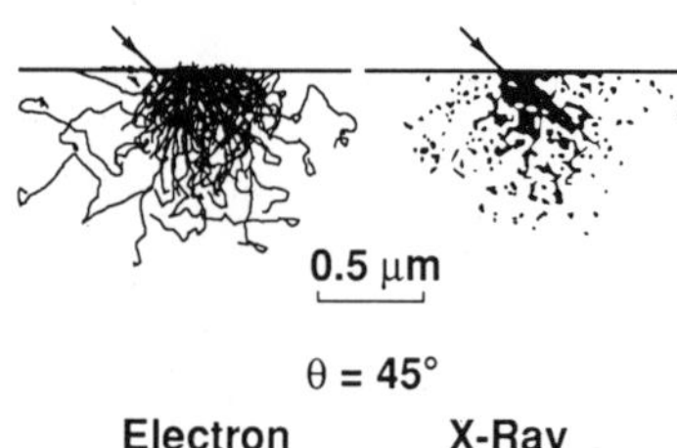

Figure 9.12. Monte Carlo electron-trajectory simulations and inner-shell ionization event sites for copper at 20 keV, tilted to $\theta = 45°$ and $\theta = 80°$ (Curgenven and Duncumb, 1971).

normal beam incidence. Reed (1971) studied the effect of the backscattering factor R, which changes considerably when the sample is tilted. When sample and standard are measured at the same tilt angle, the effect of tilt on R is similar in both. Even with large differences in atomic number between sample and standard, Reed estimates that the maximum effect of tilt is only 7%.

The absorption correction, on the other hand, is more likely to be affected for the tilted sample. As shown in Fig. 9.12, the x-ray generation volume is closer to the surface in a highly tilted sample. Since absorption is exponentially dependent on path length (see Fig. 9.11), the tilt effect can lead to a sharp reduction in absorption. There is no agreement in the literature on a method to compensate for tilt effects in the absorption correction. The simplest correction involves modifying $\chi = \mu/\rho)^i_{\text{spec}} \csc \psi$ by a sin $(90° - \theta)$ term, where θ is the tilt angle:

$$\chi(\theta) = \mu/\rho)^i_{\text{spec}} \csc \psi \sin(90° - \theta) \qquad (9.41)$$

The new value of χ can be used in the PDH correction for absorption [Eqs (9.18)–(9.22)].

In the last decade, major progress has been made in the development of $\phi(\rho z)$ approaches to handle tilted, flat polished specimens. Monte Carlo calculations have shown how x-ray production varies with depth in tilted specimens and suggested modifications in the form of the $\phi(\rho z)$ curves to allow calculations of the Z and A factors for variations in tilt angle θ. Figure 9.13 shows Monte Carlo calculations for Cu $K\alpha$ radiation in Cu for tilt angles of 0 and 30° and $E_0 = 20$ keV (Packwood, 1991). In these calculations the reference layer for the $\phi(\rho z)$ calculation has the same tilt angle as the specimen. The area I_{gen} under the $\phi(\rho z)$ curve is clearly less at the 30° tilt angle. Analysis of these and other $\phi(\rho z)$ curves calculated by Monte Carlo techniques as well as $\phi(\rho z)$ measured by Packwood (1991) has allowed an analysis of how tilt angle θ influences the four parameters α, β, γ, and $\phi(0)$ in the Packwood–Brown $\phi(\rho z)$ formulation [Eq. 9.32]. The maximum beam penetration at R_x described by the α parameter does not vary with tilt angle. Therefore α at a tilt angle of θ is equal to α at normal incidence, $\theta = 0°$. The surface ionization term $\phi(\theta)$ is relatively insensitive to tilt angle, so that

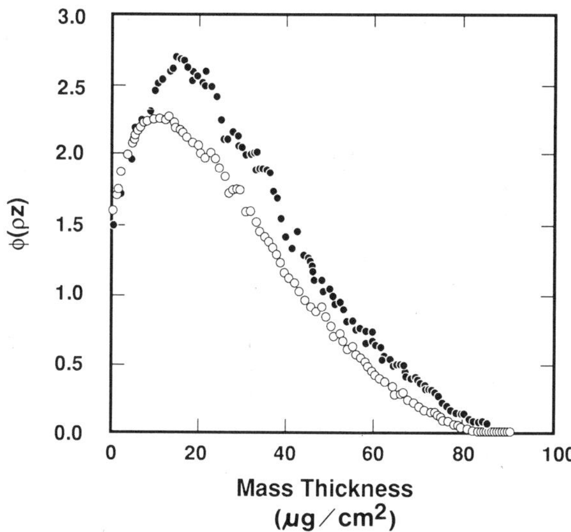

Figure 9.13. Monte Carlo calculations for Cu $K\alpha$ in copper at 0° (closed circles) and 30° tilt (open circles), reference layer at same orientation, $E_0 = 20$ keV (Packwood, 1991).

$\phi(\theta) = \phi(0)$. The value of γ, which gives the intercept with $\rho z = 0$ at the surface of the specimen, Fig. 9.7, does vary with tilt angle as given by

$$\gamma(\theta) \sim \gamma \cos^2 \theta. \tag{9.42}$$

This change in γ is mainly caused by the increase by factors of $1/\cos\theta$ in the projected thickness of the reference and trace layers of the specimen. Finally, the β parameter, which describes the transition process by which the collimated electron probe that enters the specimen changes to a random motion in the specimen is observed to vary in a way given by

$$\beta(\theta) = \frac{\beta[\gamma - \phi(0)]}{[\gamma(\theta) - \phi(\theta)]}. \tag{9.43}$$

The α, β, γ, $\phi(0)$ terms as a function of tilt angle can be put into the expressions for $\phi(\rho z)$. With this change, values of I_{gen} and I_{em} can be calculated, along with values of the corrections for atomic number Z and absorption A (see Section 9.3).

Pouchou *et al.* (1989, 1990) have also used Monte Carlo techniques to suggest changes in their formulations of $\phi(\rho z)$ by introducing the XPP model for tilted specimens. They agree generally with Packwood (1990, 1991) in that they observe little or no change in R_x and $\phi(\theta)$ with tilt angle θ. Their correction is greatly influenced by the observation that the mean depth of x-ray production varies noticeably with tilt angle, approximately with $\cos\theta,^{3/2}$ and that this variation has a major effect on the absorption correction. Their changes in the basic $\phi(\rho z)$ model, also yields promising improvements in the treatment of tilted specimens. Therefore, it appears that the use of the $\phi(\rho z)$ method for tilted specimens will, in the next few years, yield acceptable compositional

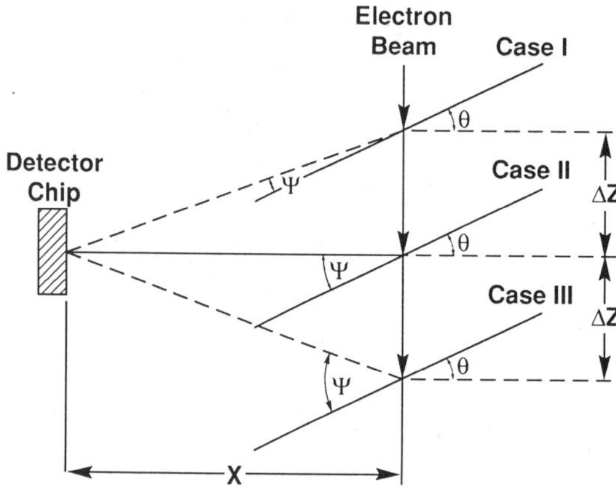

Figure 9.14. Geometry for determining x-ray emergence angle ψ for a tilted specimen and fixed EDS position.

values even for low-energy x rays with high values of the absorption factor A.

A most important factor in analyzing tilted specimens is the proper measurement of the take-off angle ψ. In some older SEMs, the EDS detector is placed normal to the electron beam, as shown in Fig. 9.14. For the cases illustrated in this figure, the specimen is tilted at an angle θ, as indicated on the SEM stage control; the center of the detector chip in the EDS is at a distance X from the impact point of the electron beam on the specimen. The distance of the impact point above (Case I) or below (Case III) the detector chip's centerline is ΔZ. Then

Case I:

$$\psi = \theta - \arctan(\Delta Z/X)$$

Case II:

$$\psi = \theta \tag{9.44}$$

Case III:

$$\psi = \theta + \arctan(\Delta Z/X)$$

It is usually advantageous to try to adjust the height of the specimen so that Case II is obtained. The closer the detector chip to the EDS detector (smaller X), the greater the uncertainty in the take-off angle. Furthermore, since the solid angle accepted by the detector increases as the distance X decreases, using the position of the center of the detector as a basis for computing ψ becomes only an approximation.

9.5. Standardless Analysis

Virtually all modern energy-dispersive x-ray analytical systems include in their quantitative analysis procedures an option referred to as "standardless analysis." Although the standardless procedure is gaining popularity, there are many potential drawbacks to the use of this technique. The use of the standardless technique without some independent confirmation of the composition in test specimens can lead to very large errors, which may go unrecognized. This section will discuss the procedure and its limitations.

Standardless analysis is the simplest of the correction procedures. The spectrum of the unknown is recorded without concern for the electron dose as long as the dead time is acceptable. To perform a quantitative analysis, it is necessary only to supply the correct beam energy and the elements to be analyzed. The analysis total will always be exactly 100%.

The underlying assumption of the standardless-analysis approach is that we can calculate the intensity measured from standards with accuracy equal to that of measuring them directly, based on certain properties of the single spectrum obtained from an unknown. There are various routes to develop a standardless-analysis approach. The most general approach makes use of four equations to predict the x-ray intensity that would be obtained from a pure-element standard. The first is the equation that describes the absolute generation of x rays, Eq. (3.36). The second is an expression to correct the loss of x-ray production because of backscattering, the R-factor, Eq. (9.6). The third is a correction for absorption in the putative standard and sample, the absolute absorption factor, Eq. (9.18). The fourth is a calculation of the absolute efficiency of the spectrometer, Eq. (7.9). Normalization is then used to bring the total concentration to 100%.

It is useful to consider how well the factors incorporated into these equations are known:

1. The absolute generation of x rays depends on two critical factors, the ionization cross-section Q_i and the fluorescence yield ω. The product $Q_i\omega$ is reasonably well known for the K-family of elements with atomic numbers from 20 to 32, but the product becomes uncertain for lower atomic numbers. The product is not well known for L- and M-family x rays. When an analysis must involve x-ray measurements involving more than one family, interfamily calculations are particularly difficult. Some approaches attempt to rectify this problem by tying the calculation to specific experimental measures, e.g., Cu K–Cu L or Au L–Au M determined at a typical operating condition, e.g., 20 keV.

2. The absolute value of the R-factor is reasonably well known, probably within 5%.

3. The absolute value of the absorption factor is also reasonably well known, probably within 5% for x-ray lines above 3 keV, and less accurately known below 3 keV. Below 1 keV the absorption factor is poorly known.

4. The greatest problem probably arises when an absolute calculation of the energy-dispersive spectrometer's efficiency is needed over a wide energy range. The absolute value of the efficiency is plotted in Fig. 5.41, and it can be seen that the efficiency is reasonably smooth and predictable above 3 keV. Below 3 keV, the efficiency changes rapidly due to absorption in all of the components. Comparing 0.5 keV with 3 keV,

the efficiency typically drops by a factor of 5 to 10. Moreover, this is the region of the spectrometer response where "spectrometer aging" effects, i.e., the changing of efficiency with time in service, are most strongly manifested. The inevitable buildup of ice, even in detectors with protective windows, causes substantial changes in the detector efficiency in this region. It is possible with most older (a few years old) EDS detectors and many new ones to experience substantial loss of counts from x-ray peaks due to incomplete charge collection. The magnitudes of loss are at best about 1% for peaks at any energy to as much as 20% for peaks in the range of about 1.8–3 keV. Below 1 keV, the effects are even worse. Thus, a standardless analysis that involves soft x-ray lines in the 0.5–2 keV range, where the spectrometer efficiency is likely to change, and hard x-ray lines in the 5–10 keV range, where spectrometer efficiency is likely to remain constant in response, may be in substantial error.

The standardless approach works quite well for the analysis of stainless steel, which typically involves measuring the transition elements Fe, Cr, Ni, Mn, Co, and V. The K-lines of these elements span the range 5.0–7.5 keV where the spectrometer efficiency is virtually constant and close to unity. The product $Q\omega$ is well behaved, and a reasonably accurate analysis can be obtained, especially if the procedure has been "optimized" for steel analyses.

Even with a standardless analysis that appears to work for steels, it is interesting to examine the effect of changing the energy of the beam. If the procedure is robust, it ought to work over a range of beam energies for analysis, e.g., 15–25 keV. This is a reasonable test to see if a particular implementation of standardless analysis is suitable.

When multiple x-ray families must be brought to bear, the situation becomes worse. Figure 9.15 gives the results of the analysis of NIST Standard Reference Material gold–copper alloys, which can be analyzed only by taking combinations of lines that mix families or span a large difference in x-ray energy, e.g., Cu $K\alpha$ (8.04 keV) versus Au $L\alpha$ (9.71 keV) or Au $M\alpha$ (2.12 keV). The errors range as high as 20% relative, which can be compared to results obtained with ZAF correction relative to standards that gave errors of 1–2% relative. Moreover, the errors are not constant with beam energy.

The foundation for achieving accurate quantitative electron probe microanalysis, whether by ZAF, $\phi(\rho z)$, or the Bence–Albee method (Section 9.7), rests upon the standardization step, which consists of measuring an element in the unknown against that same element in a standard under identical conditions of beam energy, dose (current × livetime), spectrometer efficiency, and choice of x-ray line. If the k intensity ratio is formed as a ratio between x-ray lines of the same type of radiation, several important terms in the expression for absolute x-ray intensity [Eq. (3.36)] and, most important, spectrometer efficiency, cancel in the ratio. It is not worth the apparent convenience of the standardless approach to concede the accuracy of the robust methods. Additionally, the loss of meaning in the analytical total of standardless analysis will inevitably hide those situations where an unexpected

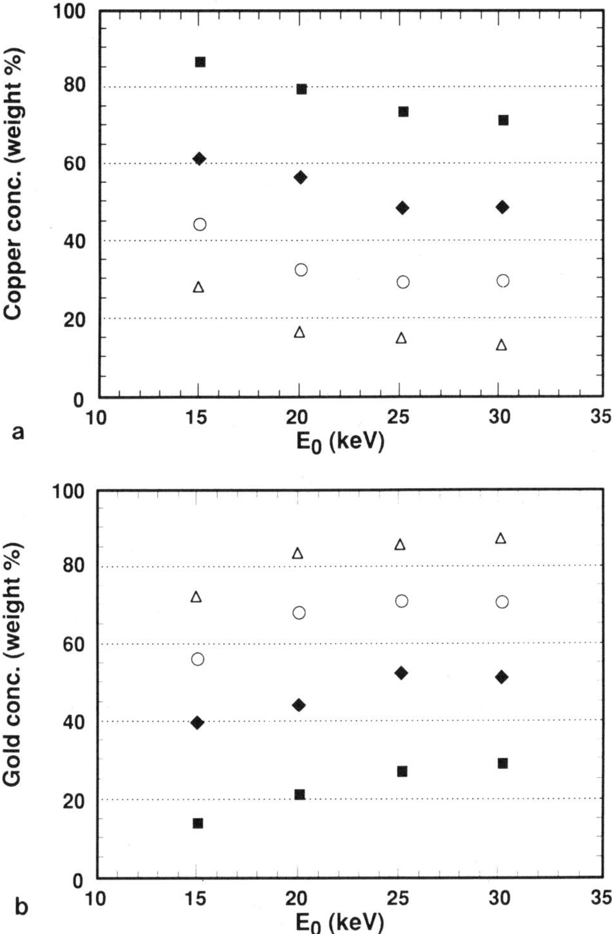

Figure 9.15. Plot of the results of a standardless analysis procedure applied to energy-dispersive x-ray analysis of NIST SRM 482 gold–copper alloys, nominal 20, 40, 60, 80 wt% concentrations. (a) Results for copper, (b) results for gold.

constituent appears at the analyzed location or the specimen differs significantly from ideal bulk material because of voids, surface relief, etc.

The benefit for careful application of the formal methods of quantitative analysis by the ZAF, $\phi(\rho z)$, or Bence–Albee method with unknowns measured against known standards is analytical performance described by the error histograms shown in Fig. 8.12. The penalty for the apparent ease of the standardless approach is the real possibility of very large errors lying far outside the comfortable bounds of Fig. 8.12. A similar histogram of errors for standardless analysis is not available. However, based on the experiences of the authors, we believe that for many samples the levels of the errors for standardless analysis would be embarrassingly large.

We summarize with a few suggestions on how to test a standardless procedure, based on the above observations: Use a material that contains light and heavy components, such as Cu–Au. Use x-ray lines from different families, such as *K* and *M*. Try the analysis at several beam energies, such as 15 keV and 25 keV. Try the analysis at low and high count rates, and try taking and processing the data at several x-ray take-off angles.

9.6. The Biological or Polymer Specimen: Special Procedures

The analysis of biological and polymeric specimens is made difficult, and sometimes impossible, by several forms of radiation damage directly caused by the electron beam. At the energies found in the SEM and the analytical electron microscope (10–300 keV), it is possible for the kinetic energy of individual beam electrons to break or rearrange chemical bonds. At the energies and currents found in the SEM and EPMA (5–50 keV, 1–1000 nA), it is also possible to cause highly damaging temperature elevations. Indeed, when analyzing this class of specimen it should be assumed that significant mass loss will occur during the measurement at each point of the specimen. If all constituents were lost at the same rate, then simply normalizing the result would compensate. Unfortunately, the matrix constituents (principally carbon compounds and water) can be selectively lost, while the heavy elements of interest in biological microanalysis (e.g., P, S, Ca) remain in the bombarded specimen and appear to be at effectively higher concentration than existed in the original specimen. What is then required of any analytical procedure for biological and polymeric specimens is a mechanism to provide a meaningful analysis under these conditions of a specimen that undergoes continuous change. Marshall and Hall (1966) made the original suggestion that the x-ray continuum could serve as an internal standard to monitor specimen changes.

Relatively few inelastic collisions suffered by the incident-beam electrons in any target result in x-ray production. Most inelastic collisions result in the loss of only very small amounts of energy through mechanisms such as ionization of an atom by the ejection of an outer-shell electron or collective excitation of outer-shell electrons. Only a few electrons are multiply scattered in a thin target. Therefore, most scattered-beam electrons lose very little energy in total in traversing a thin target, and almost none are absorbed or backscattered. Typically, the rate of energy loss in a low-atomic-number target is about 0.1 eV/nm, so that only 10–100 eV is lost through a 100-nm section. That this is the case in fact defines a target as thin. We may therefore conveniently assume that the energy of the electrons involved in collisions producing x rays from a thin target is approximately the incident beam energy. This assumption permitted Marshall and Hall (1966) and Hall *et al.* (1966) to develop the well-known procedure for thin specimens, to be discussed below, which is used extensively in the biological community and is applicable in many types of polymer analysis. The technique was developed for applications in the EPMA and works well in this environment. However, most applications today are found in the analytical electron microscope (AEM) operating at energies of 80–120 keV. This is because of the considerably improved spatial resolution found when analyzing thin (<100 nm) specimens at the higher beam energies found in the AEM. For those situations where the analyst is using a SEM and can tolerate the larger analysis volume, the Hall method is a powerful tool.

We will consider the derivation of the Marshall–Hall method from first principles. More information on the physics of characteristic and continuum x-ray generation can be found in Chapter 3.

9.6.1. The Characteristic Signal

We can predict the number of characteristic x rays generated into a full sphere (4π sr) from the following relation:

$$I_{\text{Ch}} = \left(N_0 \rho \frac{C_{\text{A}}}{A_{\text{A}}}\right) Q_A \omega_A F_A N_e \, dz, \qquad (9.45)$$

where N_0 is Avogadro's number, ρ is the density of the target in the analyzed volume, C_A is the weight fraction of the analyte in the volume, A_A is the atomic weight of element A, Q_A is the ionization cross-section for the shell of interest and has dimensions of area; the fluorescent yield ω_A is the probability that an x ray will be emitted due to the ionization of a given shell, F_A is the probability of emission of the x-ray line of interest relative to all the lines that can be emitted because of ionization of the same shell; N_e is the number of electrons that have irradiated the target during the measurement time, and dz is the target's thickness in the same units of length used in Q and ρ.

The quantity in parentheses in Eq. (9.45) is the number of atoms of analyte per unit volume of target and is obtained by the following rationale. By definition, the weight fraction of an element A is the mass of A divided by the total mass. The mass of A per cm^3 is the weight fraction of A times the density ρ, given in g/cm^3, in the analytical volume. We emphasize that density is a measured quantity and refers to the mass of all the atoms per unit of volume. To convert mass to a number of atoms we use Avogadro's number, $N_0 = 6.02 \times 10^{23}$, the number of atoms in a mole of an element. Therefore, the number of atoms of element A per cm^3 is $(C_A\rho)(N_0/A_A)$, where A_A is the gram-atomic weight of a mole of element A. The total number of atoms N, in a volume of 1 cm^3 is then the sum over all the elements i in the volume:

$$N = N_0 \rho \sum_i \frac{C_i}{A_i}. \qquad (9.46)$$

The basic functional form of the characteristic cross-section is due to

Bethe (1930):

$$Q = 6.51 \times 10^{-20} \frac{n_s b_s}{U E_C^2} \ln(c_s U), \tag{9.47}$$

where the constant is the product πe^4 (in $\text{keV}^2\ \text{cm}^2$; e is the charge of an electron), n_s is the number of electrons that populate the sth shell or subshell of interest ($n_k = 2$), E_c is the energy required to remove an electron from a given shell or subshell (the critical excitation energy), U is the overvoltage ratio E_0/E_c (where E_0 is the energy of the impinging beam electron), and b_s and c_s are constants for the sth shell or subshell. The area in cm^2 of Q is essentially the size of the K, L, or M shell target that a beam electron must hit to produce an ionization of that shell. In general this area is about 100 times smaller than the area of the entire atom.

9.6.2. The Continuum Signal

The interaction of a large number of beam electrons with a thin foil of a given element produces an emitted continuum spectrum having a distribution shaped approximately as $1/E$, where E is the photon energy. The magnitude (height) of this distribution is proportional to the number of atoms composing the foil. Other factors such as beam current and measurement time scale the magnitude linearly. The beam energy E_0 and observation angle have a second-order effect on the shape of the distribution and will not be discussed here. The observation angle is the angle between the x-ray detector and the direction of travel of the electron beam and is not to be confused with the x-ray take-off angle, usually designated by ψ. We must make this distinction for thin targets because the only effect of tilting a specimen away from the normal is to thicken the specimen, whereas changing the observation angle causes the detector to intercept a different part of the nonuniform (anisotropic) distribution of the continuum distribution.

We can predict the number of continuum X-rays generated into a unit steradian ($1/4\pi$ of a full sphere) from the following relation:

$$I_{\text{CO}} = \sum_i (N_i Q_i) N_e, \tag{9.48}$$

where N_i is the number of atoms of element i, Q_i is the continuum cross-section for that element and has dimensions of area, and N_e is the number of electrons that have irradiated the N_i target atoms during the measurement time. Q_i is usually assumed to be differential in photon energy and in observation angle ϕ. Consequently, it is not necessary to include the differential terms in Eq. (9.48). We note that continuum cross-sections usually are defined in units of 1 sr, while characteristic cross-sections are usually specified as radiating into the full sphere of 4π sr. The reasons for this are beyond the scope of this book and the reader is referred to Fiori *et al.* (1986) for a detailed treatment and further references. Using the fact that the total number of atoms in 1 cm^3 is:

$$N_0\rho \sum_i \frac{C_i}{A_i}, \tag{9.49}$$

Eq. (9.48) can be rewritten for a thin film as:

$$I_{\mathrm{CO}} = N_0\rho \sum_i C_i \frac{Q_i}{A_i} N_e\, dz, \tag{9.50}$$

where dz is the target thickness in the same units of length as used for Q_i and ρ.

9.6.3. Derivation of the Hall Procedure in Terms of X-Ray Cross-Sections

The density of the target within the analytical volume of the electron probe is generally not known unless the material being irradiated is a standard reference material. We also note that the thickness of a target at the site of impact of the electron beam is rarely a known quantity. However, it is obvious that a change in either local density or thickness causes a proportionate change in the number of atoms with which a beam electron is likely to interact in passing through a thin target. Characteristic and continuum generation are equally affected. If mass loss is occurring during the measurement period, then we must assume that atoms are being lost proportionally. To treat these changes more easily, density and thickness are frequently combined into the single quantity mass-thickness denoted by $(\rho\, dz)$ with dimensions of g/cm^2. We must stress that the concept of mass thickness is useful only if the beam energy does not change significantly in traversing the target thickness dz. A typical biological or polymer target should be less than several hundred nanometers in thickness for the above assumption to be accurate. With this caveat, we form the ratio of Eqs. (9.45) and (9.50):

$$\frac{I_{\mathrm{Ch}}}{I_{\mathrm{CO}}} = \frac{N_0 \dfrac{C_A}{A_A} Q_A \omega_A F_A N_e \rho\, dz}{N_0 \sum_i C_i \dfrac{Q_i}{A_i} N_e \rho\, dz}. \tag{9.51}$$

We note that $(\rho\, dz)$, the mass-thickness, Avogadro's number, and the number of electrons incident on the film during the measurement period all cancel in the ratio. To further simplify the formula, we take advantage of the fact that, for the analysis of a given element, the fluorescent yield and associated relative transition probability are constants that can be gathered into one grand constant h. Thus we have

$$\frac{I_{\mathrm{Ch}}}{I_{\mathrm{CO}}} = h \frac{C_A \dfrac{Q_A}{A_A}}{\sum_i C_i \dfrac{Q_i}{A_i}}, \tag{9.52}$$

where Q_A is the characteristic cross-section for a particular element A and Q_i is the continuum cross-section for the ith element; the subscript i must span *all* the elements present in the electron beam–target interaction volume. Equation (9.52) is the most general formula for the method proposed by Hall.

We now put into the numerator of Eq. (9.52) the characteristic cross-section used by Marshall and Hall [the basic Bethe Eq. (3.9)] and into the denominator their simplified choice of a continuum cross-section (for a derivation, see Fiori *et al.* 1986). We collect all numerical constants into h to obtain

$$\frac{I_{\text{Ch}}}{I_{\text{CO}}} = h \frac{\dfrac{C_A}{(A_A U E_C^2)} \ln(c_s U)}{\sum_i \left(\dfrac{\dfrac{C_i Z_i^2}{A_i}}{E_0 E} \right) \ln 1.166 \dfrac{E_0}{J_i}} . \tag{9.53}$$

We can further simplify this formula by taking advantage of the fact that the electron beam energy E_0 is held constant during the analysis and the energy band dE of continuum measurement, centered at the energy E, is usually not changed during the analysis. Consequently, dE, E, and E_0 (and hence U) can be absorbed into k, resulting in the simpler form

$$\frac{I_{\text{Ch}}}{I_{\text{CO}}} = h \frac{C_A}{A_A \sum_i \left(C_i \dfrac{Z_i^2}{A_i} \ln 1.166 \dfrac{E_0}{J_i} \right)}, \tag{9.54}$$

where J_i is a function of atomic number such as is given in Eq. (3.8).

This is the usual representation of the Marshall–Hall equation, which is a more rigorous expression of the continuum correction concept proposed by Hall *et al.* (1966) and often referred to as the Hall method or correction.

The equation can be used in the following manner: A measured characteristic-to-continuum ratio from a well-known material is set equal to the right side of Eq. (9.54). Since the target being irradiated is a reference standard, we know the atomic numbers Z_i, atomic weights A_i, and weight fractions C_i. Since all other terms except the constant h are known, h can be calculated for the given set of experimental conditions. Next, we hold these conditions constant and measure the characteristic and continuum intensities from the specimen. To calculate a weight fraction C_A of the analyte from the specimen we must know the weight fraction of each and every one of the elements that compose that part of the specimen irradiated by the electron beam. Crucial to the Hall procedure is the following assumption: The analytes of interest are minor or trace constituents such that the quantity $\sum (C_i Z_i^2 / A_i)$ for the biological or polymeric specimen to be analyzed is dominated by the matrix. In this case the unknown contribution of the analyte C_A, to the sum may be neglected. Furthermore, the value of the sum is known or can be estimated from other information about the detail being analyzed (the actual, localized, material being irradiated by the electrons, and *not* some bulk property). Some representative values of $\sum (C_i Z_i^2 / A_i)$ are 3.67 (water), 3.08 (polycarbonate), and 3.28 (protein with S). Typically the range is between 2.8 and 3.8 for most biological and many polymeric materials. This basic theme can be extended, and several analytes may be analyzed simultaneously and at higher concentrations. The details of this extension are beyond the scope of this book, but a full description and derivation can be found in Kitazawa *et al.* (1983). Most modern commercial computer x-ray analyzer systems have the full Hall procedure included in their suite of analysis tools.

The Hall procedure works well for thin specimens in the energy regime of the SEM, EPMA, and AEM. The method does not work for specimens in which the average atomic number is expected to vary significantly from one analysis point to another, or from that of the standard. Consequently, the procedure can not be used for most applications in materials science.

9.6.4. Error Analysis

It is usual practice during the development of any matrix correction procedure to undergo a rigorous testing phase where materials of well-known concentrations, which are homogeneous at the microlevel and stable under the electron beam, are used as unknowns to uncover systematic errors in the procedure. However, in the case of biological and polymeric analysis, this technique poses more problems than it does in materials analysis. This class of specimens often has none of the above properties. Consequently, it has been difficult to develop new algorithms and to determine if there has been improvement over existing procedures.

9.6.5. Bulk Targets and Analysis of a Minor Element

Since the density and average atomic number of biological and many polymeric specimens is low, the excitation volume for x rays is large. Indeed, this volume can easily exceed 10 μm in extent. Consequently, one of the major advantages of the method is diminished, the ability to analyze very small volumes.

For those cases in which the increased excitation volume is acceptable, it is possible to perform a classical ZAF analysis. If the same assumptions used in the Hall procedure above can be made, then the analysis is quite tractable. If the concentrations of all the major elements are known and it can be assumed that they will remain constant for all of the analyzed points, then these concentrations can be fed to the ZAF procedure directly. This method is known as "analysis by fixed concentration" and several, but not all, ZAF or $\phi(\rho z)$ procedures permit this type of analysis.

For more detail on the subjects discussed in this section, we refer the interested reader to the literature and recommend Hall (1979a), Hall and

Gupta (1984), Hall (1979b), Kitazawa *et al.* (1983), Halloran and Kirk (1979), Leapman *et al.* (1984a;b), Wall (1979), Kopf *et al.* (1986), Dorge *et al.* (1978), Somlyo *et al.* (1985), and Somlyo *et al.* (1977).

9.7. Special Procedures for Geological Analysis

9.7.1. Introduction

In the early 1960s, the ZAF method was not well developed and analysis errors were often quite significant. Furthermore, computer data reduction was much less generally available. In response to this state of affairs, Ziebold and Ogilvie (1964) developed a hyperbolic or empirical correction method for quantitative analysis of binary metal alloys. They expressed this relationship for element 1 in the sample in the form

$$\frac{(1 - k_1)}{k_1} = a_{12}\frac{(1 - C_1)}{C_1}, \tag{9.55}$$

or

$$\frac{C_1}{k_1} = a_{12} + (1 - a_{12})C_1, \tag{9.56}$$

where k_1 is the intensity ratio, sample to standard, of element 1 [Eq. (8.1)] and C_1 is the concentration of element 1 in the sample. The term a_{12} is the a-factor, which is a constant for element 1 in a binary system of elements 1 and 2 for a given value of electron-beam energy E_0 and take-off angle ψ. If this relationship is correct, a plot of C_1/k_1 versus C_1 is a straight line with a slope of $(1 - a_{12})$. Such a hyperbolic relationship between C_1 and k_1 has been shown to be correct for several alloy and oxide systems. However, it is difficult to obtain standards to determine a-factors for a large variety of binary systems.

Over twenty years ago, Bence and Albee (1968) proposed the use of empirical a-factors, similar to procedures proposed by Ziebold and Ogilvie (1964), to correct electron microprobe analyses of multielement oxide and silicate minerals. As discussed by Armstrong (1988), this paper has been one of the most-cited articles in the geological sciences, and the Bence–Albee analytical technique has played an important role in microbeam analysis in geology. In the intervening years since the original Bence–Albee paper, new empirical a-factors have been developed, and other studies have indicated that the original a-factors produced some systematic errors in processing geological data. Although ZAF and $\phi(\rho z)$ procedures have been developed which yield accurate compositional values in oxide and silicate systems (see Sections 9.2 and 9.3), most analysts continue to use the Bence–Albee procedure for geological samples. The following sections discuss the method in some detail and present the application of the method for quantitative analysis.

9.7.2. Formulation of the Bence–Albee Procedure

Ziebold and Ogilvie (1964) also developed an analysis procedure for a ternary system containing elements 1, 2, 3 with compositions C_1, C_2, C_3 and intensity ratios k_1, k_2, and k_3. For element 1

$$\frac{(1 - k_1)}{k_1} = \frac{a_{123}(1 - C_1)}{C_1}, \tag{9.57}$$

where

$$a_{123} = \frac{(a_{12}C_2 + a_{13}C_3)}{(C_2 + C_3)}. \tag{9.58}$$

Alternatively, Eq. (9.57) can be given as

$$\frac{C_1}{k_1} = a_{123} + (1 - a_{123})C_1. \tag{9.59}$$

Similar relationships can be written for elements 2 and 3. In this procedure, the ternary a-factors are determined from each of the individual binary a factors a_{12}, a_{13}, and a_{23}. In such an analysis, a-factors must be measured or calculated for three binary systems.

Extending this formulation, Bence and Albee (1968) showed that, in a system of n components, for the nth component

$$\frac{C_n}{k_n} = \beta_n,$$

where

$$\beta_n = \frac{k_1 a_{n1} + k_2 a_{n2} + k_3 a_{n3} + \cdots + k_n a_{nn}}{k_1 + k_2 + k_3 + \cdots + k_n} \tag{9.60}$$

In this equation, a_{n1} is the a-factor for the $n1$ binary, i.e., for the determination of element n in a binary consisting of element n and element 1. Similarly, a_{n2} is the a-factor for a binary consisting of element n and element 2, and so on. The value of a_{nn}, which is the a-factor for n in n, is unity. Similarly, the notation a_{1n} represents the a value for element 1 in a binary of elements 1 and n. The value of k_1 is the relative intensity ratio found for element 1 in the specimen referred to the standard used for element 1. This notation can be used for all the elements in the specimen and, for example, k_n is the relative intensity ratio for element n referred to the standard used for element n. It is important to note that the a-factor values must be known for the experimental conditions, E_0 and ψ, used for analysis.

To solve completely a system of n components, one needs $n(n - 1)$ values for a. The a value for each of the n pure elements is unity, as noted above. Thus $n(n - 1)$ standards are required. Bence and Albee (1968) obtained a number of suitable standards of geologic interest and determined their respective a values at various initial electron-beam energies and at one take-off angle. The standards used were binary oxides. As an example, a_{NiFe} represents the a-factor for Ni in the NiO–FeO binary and is used for the determination of the element Ni in a binary consisting of NiO and FeO.

Given the necessary matrix of a values, the analyst chooses his analytical standards and determines $k_1, k_2, \ldots, k_n$. Then $\beta_1, \beta_2, \ldots, \beta_n$ can be calculated. From these values, a first approximation to $C_1, C_2, \ldots, C_n$ is found. We can call these values C_1', C_2', etc. At this point, a new set of β values, using the first-approximation concentrations $C_1', C_2', \ldots$ are calculated as $\beta_1', \beta_2', \ldots, \beta_n'$:

$$\beta_n' = \frac{C_1' a_{n1} + C_2' a_{n2} + C_3' a_{n3} + \cdots + C_n' a_{nn}}{C_1' + C_2' + C_3' + \cdots + C_n'} \tag{9.61}$$

Iteration continues until the differences between successive calculated values of β are arbitrarily small. In practice, the result really depends on how good the standard is; in geologic work, standards of nearly the same average atomic number as the specimen are usually chosen. Hence, atomic number effects are reduced. Fluorescence effects are usually small in silicate minerals and oxides. Hence, most of the final correction is for absorption.

9.7.3. Application of Bence–Albee Procedure

Tests of the validity of this procedure indicate that it yields results comparable to those obtainable with the ZAF method. The chief obstacle to its general use is the need for many homogeneous, well-characterized standard materials. For this reason, the a-factor method is sometimes combined with the ZAF or the $\phi(\rho z)$ method. In this case, one calculates the necessary a-factor matrix for n elements by assuming a concentration value C and using the ZAF or $\phi(\rho z)$ method to compute the corresponding k values. The value of the a-factor can be calculated from Eq. (9.55). For example, Albee and Ray (1970) have calculated correction factors for 36 elements relative to their simple oxides at $E_0 = 15$ and 20 keV and at several take-off angles, using the ZAF technique. Nevertheless, a calculated a-factor is subject to all of the uncertainties associated with the ZAF or $\phi(\rho z)$ methods, as discussed in Sections 9.2 and 9.3. The use of such corrections, however, permits x-ray microanalysis using a small suite of simple crystalline oxides as standards, and the calculation procedure is much simpler than the ZAF or $\phi(\rho z)$ corrections.

As an example of the use of the empirical method, one can consider the microanalysis of the seven-component pyroxene sample discussed previously for the ZAF and $\phi(\rho z)$ analysis (Tables 9.5, 9.9, and 9.10). The $K\alpha$ lines for Si, Al, Cr, Fe, Mg, and Ca are used, and the analysis is calculated for $E_0 = 15$ keV and $\psi = 52.5°$. The appropriate a-factors $[n(n-1) = 30]$ for these operating conditions, taken from Albee and Ray (1970), are listed in Table 9.11. The β_i factors and intensity ratios k_i calculated for each element i from Eq. (9.60) for the meteoritic pyroxene are given in Table 9.12. In addition the calculated ZAF_i factors and k_i values from the ZAF and $\phi(\rho z)$ procedures are listed in Table 9.12. The

Table 9.11. Empirical a Factors for Analysis of a Meteoritic Pyroxene ($\psi = 52.5°$, $E_0 = 15$ keV)[a]

Element	MgO	Al_2O_3	SiO_2	CaO	Cr_2O_3	FeO
Mg	1.0	1.03	1.09	1.26	1.50	1.76
Al	1.62	1.0	1.01[b]	1.14	1.28	1.37[b]
Si	1.29[b]	1.34[b]	1.0	1.05	1.12	1.19[b]
Ca	1.08	1.06	1.08	1.0	0.91	0.92
Cr	1.13	1.11	1.12	1.14	1.0	0.88
Fe	1.15	1.13	1.15	1.14	1.08	1.0

[a] Albee and Ray (1970).
[b] T. Bence, (1978).

β and ZAF factors differ by no more than 2% relative, and all three techniques give comparable results.

Other investigators have proposed that more accurate results can be obtained if the constant a-factor were replaced with a polynomial function of concentration (Bence and Holzwarth, 1977; Laguitton *et al.*, 1975). Armstrong (1988) has calculated new a-factors using ZAF, $\phi(\rho z)$, and Monte Carlo techniques. He has found that many of the binary systems show significant deviations from the constant a-factor assumption. In many cases, the deviations exceed 2% (including geologically important binaries such as $MgO-Al_2O_3$, MgO–FeO) for $E_0 = 15$ keV and $\psi = 40°$. Armstrong (1988) found that if the constant a-factor is replaced by a second-order polynomial function of concentration, the Bence–Albee procedure employing the concentration-dependent a-factors duplicates the results of the correction procedure on which it is based. A modification of the conventional Bence–Albee procedure is suggested in which the constant a-factor in Eq. (9.60) is replaced with the polynomial equation

$$a_{12} = c + d\left[\frac{C_1}{(C_1 + C_2)}\right] + e\left[\frac{C_1}{(C_1 + C_2)}\right]^2, \tag{9.62}$$

where C_1 and C_2 are the oxide weight fractions of the elements in each calculated binary and the constants c, d, and e are used for each binary

Table 9.12. Calculation Using the a Factor for a Meteoritic Pyroxene ($\psi = 52.5°$, $E_0 = 15$ keV)

Concentration (wt %)	a factor		ZAF		$\phi(\rho z)$	
	β_i	k_i	ZAF_i	k_i	ZAF_i	k_i
53.3 SiO_2	1.074	0.496	1.091	0.489	1.075	0.496
1.11 Al_2O_3	1.16	0.00957	1.17	0.00947	1.158	0.00959
0.62 Cr_2O_3	1.102	0.00563	1.106	0.0056	1.105	0.00561
9.5 FeO	1.133	0.0838	1.15	0.0828	1.145	0.0830
14.1 MgO	1.179	0.120	1.16	0.122	1.175	0.120
21.2 CaO	1.047	0.203	1.06	0.200	1.065	0.199

oxide pair. Values of the c, d, and e coefficients at $E_0 = 15$ keV and $\psi = 40°$ are given in Armstrong (1988). Using these coefficients, the Bence–Albee correction appears to yield compositions of the same accuracy as the ZAF or $\phi(\rho z)$ method. Under these circumstances, the Bence–Albee method continues to be a viable technique for the x-ray microanalysis of oxide and silicate materials of geologic interest. The Bence–Albee method has the special advantage of working directly with the oxide standards that are most appropriate for the analysis of geological unknowns; additionally, because the calculations involve only simple arithmetic, final results are obtained in the shortest time of any of the correction methods.

9.7.4. Specimen Conductivity

Almost all specimens for geological analysis are coated with a thin layer of C or metal to prevent surface-charging effects. However, x rays are generated in depth in the specimen, and the surface conducting layer has no influence on electrons that are absorbed micrometers into the specimen. Electrical fields may be built up deep in the specimen, which can ultimately affect the shape of the $\phi(\rho z)$ curves.

These potential problems have been discussed in some detail by Bastin and Heijligers (1991b). The first problem to consider is the potential of electron energy loss in the conducting layer. Such a loss can happen due to the quality of the conducting layer and the current in the electron beam. Such an effect can be observed by measuring the short-wavelength limit of the continuum (see Chapter 3). If the value of the highest energy continuum x-rays is lower than E_0, the coating is a major problem. If it is not corrected, high energy x-ray lines may not be measured, or their intensity may be diminished by the loss of the overvoltage for x-ray production.

The second problem lies in the possibility that the nonconducting specimen beneath the surface coating may undergo an electrical field buildup, which in some way may influence x-ray production so that existing ZAF or $\phi(\rho z)$ corrections do not describe the x-ray generation and absorption in the specimen. This effect has been observed by Bastin and Heijligers (1991b) in the oxygen analysis of conducting and nonconducting oxides. The x-ray distribution in nonconducting oxides for O $K\alpha$ is not the same as in conducting oxides. Although no simple solution has as yet been offered, the potential problem must be considered.

Most geologists don't measure oxygen directly, but get their oxygen values by stoichiometry. If oxygen analysis is desired, then C coatings are not optimal because they are highly absorbing for both O and N $K\alpha$ x rays (see Section 9.10 for an in-depth discussion of light-element analysis). Nevertheless, it is important to be aware of the subtle differences in metallic-element analysis which may occur in conducting and nonconducting silicate minerals, oxides, sulfides, etc.

9.8.1. Introduction: The Analytical Total

A fundamental assumption of quantitative EPMA is that the measured x-ray intensity differs between the specimen and the standards only because of differences in the respective compositions. The ZAF, $\phi(\rho z)$, and Bence–Albee matrix correction procedures described in the previous sections are applicable only to specimens with ideal geometry, that is, polished flat and sufficiently thick to contain the interaction volume of the primary electrons *and* the range of secondary fluorescence, defined by the absorption of the characteristic and continuum x rays. The electron range is generally contained within a linear distance of 1 to 10 μm below the surface, while the absorption range for x rays can extend to 100 μm or more for energetic photons.

There exist important classes of specimens whose geometry differs markedly from the ideal flat, semi-infinite solid, including freestanding foils, films supported on substrates, particles, and rough surfaces. For such specimens, specimen geometry factors, such as size and shape, significantly influence the measured x-ray intensities. Quantitative x-ray microanalysis of these special samples requires modifications of the conventional bulk specimen procedures. In the following discussion, the general principles of these special sample analysis procedures will be covered. The specific details of the corrections vary from one analytical system to another, and the analyst should be especially careful to understand the limits of application of the correction procedures in the particular system in use.

When must these special correction procedures be applied? The difference between bulk and thin materials depends on how the specimen's dimensions compare to the range of the electrons. The size and shape become important when the depth or lateral dimension of the specimen, such as the thickness of a film, approaches the value calculated for the electron range in a flat bulk target of the same composition. For dimensions smaller than the electron range, electrons escape from the specimen or penetrate into the substrate, removing a portion of the energy which would have gone into the production of x rays if the specimen had been a bulk sample.

Is there any clue when such a geometrical factor situation is encountered, particularly if the analyst has no prior knowledge of the nature of the specimen? The analytical total in a conventional ZAF or $\phi(\rho z)$ quantitative procedure provides an important indication when the electron–x ray interaction in the specimen differs significantly from the ideal bulk case. When all elemental constituents are independently measured against standards or, as in the case of oxygen, included through stoichiometric calculations, then the *nonnormalized* analytical total conveys significant information. Considering the systematic errors in the matrix corrections, the analytical total should fall within 99–101% if the specimen corresponds to the ideal bulk case and the analysis is conducted with careful attention to instrument operating conditions. If the analytical total exceeds 102%, the most likely explanation is a mistake in establishing consistent conditions of electron dose and spectrometer efficiency under which the specimen and standards are measured. If the analyst is confident that the measurement conditions are consistent, then the deviation is likely to be due to an effect of the geometry of the specimen as described below. Similarly, if the analytical total is less than 98%, then, after eliminating the possibility of a mistake in the dose/spectrometer conditions, the possibility exists that a constituent has been overlooked. The qualitative analysis should be repeated to search for the missing element(s). If none is found, then the low total is likely to be due to an effect of the geometry of the specimen.

9.8.2. Films on Substrates

Films on substrates present a case in which one dimension, the specimen thickness along the beam direction, becomes an important factor in the analysis. The $\phi(\rho z)$ description of the depth distribution of x-ray production provides the basis for understanding the origin of the effect of the geometry of the specimen and the modified analytical procedures to correct for them. Consider the $\phi(\rho z)$ curve for a flat bulk target of pure element A, generated by Monte Carlo simulation and shown in Fig. 9.16. If the specimen is now modified to consist of the same pure element A in the form of a film supported on a substrate of an element B with a similar but lower atomic number, e.g., nickel (A) on iron (B), how does the $\phi(\rho z)$ curve change? The elastic and inelastic scattering of the beam electrons is only slightly modified by the change in material as the electrons penetrate through the interface between elements A and B, so the range and backscattering are not significantly

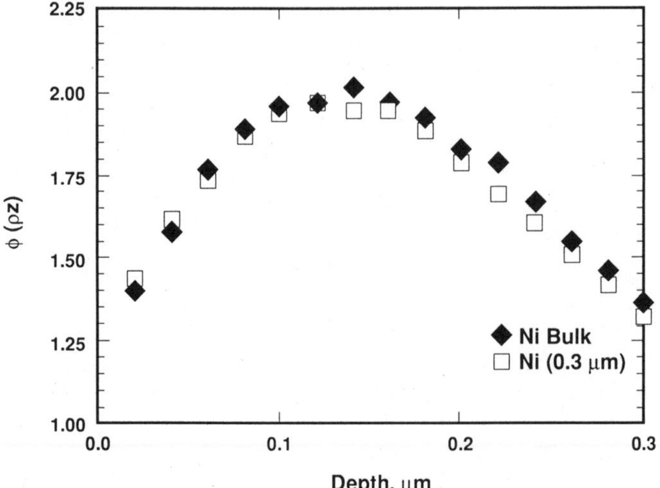

Figure 9.16. X-ray depth-distribution function for Ni $K\alpha$ in bulk nickel and an 0.3-μm thick Ni film on iron.

affected. Since the characteristic x rays for B are lower in energy than for A, there can be no secondary fluorescence of A by B x rays. Since the average atomic numbers of the pure specimen and the film-on-substrate specimen are nearly identical, the continuum-induced fluorescence is also nearly identical. Thus, as shown in Fig. 9.16, the $\phi(\rho z)$ curve for the film on substrate is nearly identical to the equivalent segment of the $\phi(\rho z)$ curve for the bulk pure-element specimen. Because of the similarity of the $\phi(\rho z)$ curves for the pure bulk A and the A-film on B-substrate, the intensity of element A generated in films of various thicknesses can be found with good accuracy by integrating the $\phi(\rho z)$ curve of pure A to the depth which corresponds to the thickness of the film. The emitted intensity of A can be found by performing an absorption correction from the $\phi(\rho z)$ curve. Finally, since the experimental measurement consists of forming the k-value (the ratio of the x-ray intensity measured from the A–B film sample to pure bulk A), then the measured k-value can be predicted mathematically:

$$k = \frac{i_{A-B}}{i_A} = \frac{\int_0^t \phi(\rho z)\exp\left[-\left(\frac{\mu}{\rho}\right)\rho z \csc\psi\right] dz}{\int_0^R \phi(\rho z)\exp\left[-\left(\frac{\mu}{\rho}\right)\rho z \csc\psi\right] dz}. \tag{9.63}$$

For the film, the integration is performed from the surface ($z = 0$) to the film thickness ($z = t$), while for the bulk, the integration extends to the x-ray range ($z = R$).

Figure 9.17a and b show $\phi(\rho z)$ and Ni k-values for various thicknesses of a film of nickel on different substrates, as calculated by Monte Carlo electron-trajectory simulation. Such plots are known as "working curves." If the analyst is presented with a new specimen of an A film on a B substrate but of unknown thickness, the thickness can be readily determined by measuring the k-value and directly reading the corresponding thickness from the appropriate working curve. In practice, standard films of specified thickness and composition are unlikely to be available, so the analyst is most likely to calculate the working curve for the film by means of Eq. (9.16) or a Monte Carlo electron-trajectory simulation rather than to determine the curve by actual experiment.

If films of element A are placed upon substrates of radically different atomic number, then the resulting $\phi(\rho z)$ curves differ significantly from the pure-element case because the electron scattering (and characteristic and continuum-induced fluorescence) changes markedly when beam electrons cross the interface between the film and the substrate. The calculated curves for nickel on aluminum and nickel on gold compared to nickel on iron are also shown in Fig. 9.17b. Only electron effects are considered in these calculations; secondary fluorescence is not included.

In its most general form, the thin-film analysis problem consists of determining the thickness of the film and the concentrations of multiple constituents. In more complex situations, the film may consist of several layers with different compositions, with certain elements present in more

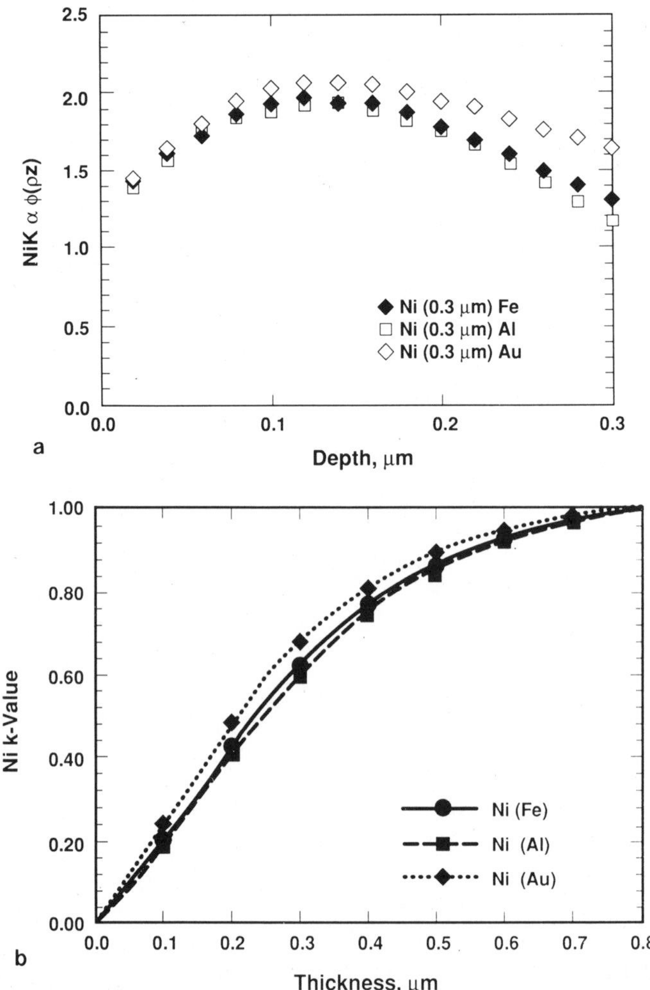

Figure 9.17. (a) X-ray depth distribution function for Ni $K\alpha$ on bulk aluminum, iron, and gold. (b) Corresponding plots of the k-values for Ni films of various thicknesses on Al, Fe, and Au substrates.

than one layer or in the substrate. Two approaches are available for mathematically predicting the k-values from films for comparison to experimental measurements of k. The Monte Carlo method directly simulates the hypothetical geometry of the specimen and calculates the electron interaction and x-ray absorption as each electron trajectory is followed on a point-by-point basis (Kyser and Murata, 1974). The $\phi(\rho z)$ thin-film method consists of modifications to the conventional bulk $\phi(\rho z)$ method to consider the changes in the $\phi(\rho z)$ function caused by differences in scattering and energy loss based upon the assumed sturcture of the film on substrate (Yakowitz and Newbury, 1976; Bastin and Heijligers, 1990c; Pouchou and Pichoir, 1991). As shown in Fig. 9.18a, various electron and x-ray paths must be considered. Considering these effects, the form of the $\phi(\rho z)$ function for a film compared to a

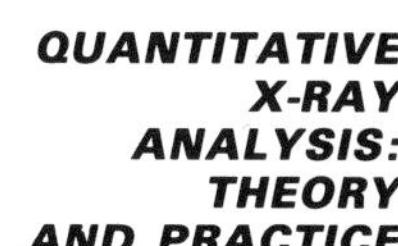

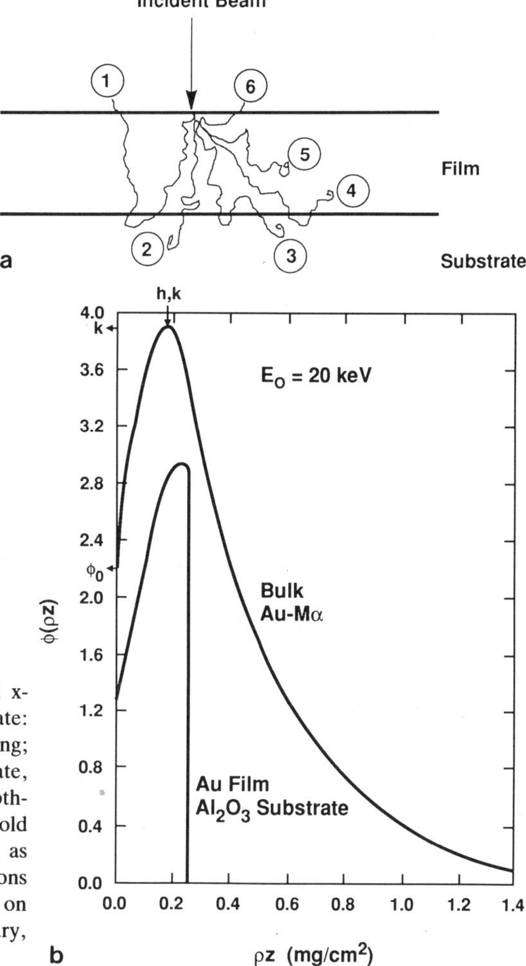

Figure 9.18. (a) Electron and x-ray paths for a film on substrate: Paths 1 and 6 lead to backscattering; paths 2–5 remain in film or substrate, (b) comparison of the x-ray depth-distribution function for bulk gold and for a gold film on alumina, as calculated from analytic expressions for bulk targets and for films on substrates (Yakowitz and Newbury, 1976).

bulk target is shown in Fig. 9.18b. The procedure to generate the $\phi(\rho z)$ curve for the film(s) on substrate must be altered to take into account the differing scattering situations: electron scatters only in film; electron traverses film and scatters in substrate; electron re-enters film; etc. Yakowitz and Newbury (1976) and Pouchou and Pichoir (1991) have given detailed analytic treatments for adapting a $\phi(\rho z)$ equation to the film-on-substrate case. Examples are shown in Fig. 9.19a and b of the $\phi(\rho z)$ functions for Al $K\alpha$ emitted from a constant-thickness layer on various substrates and for various Al thicknesses on a tungsten substrate, as calculated with the PAP model. The accuracy of this approach can be judged from a comparison of theory with experimental measurements of x-ray emission from films as a function of beam energy, as shown in Fig. 9.20a and b.

To apply the method to solve the composition and structure of a multicomponent film, the model is used to calculate curves of k versus thickness and composition based upon an assumed structure. Since the

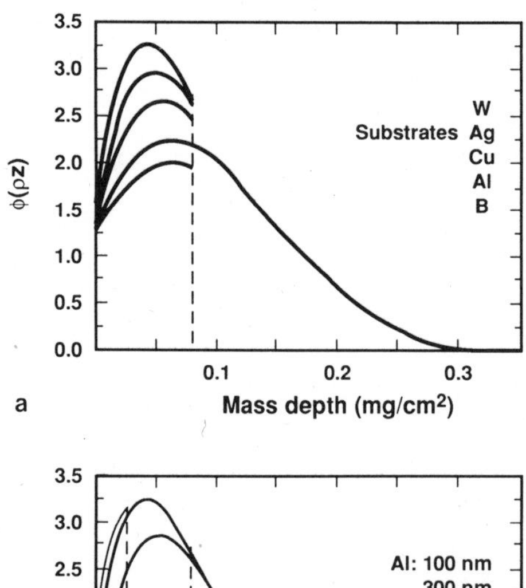

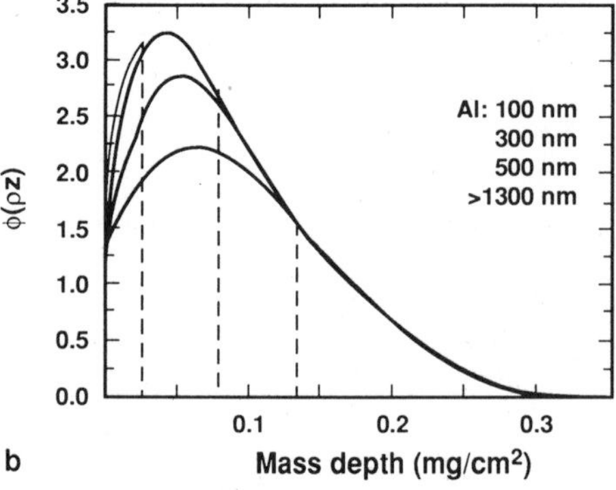

Figure 9.19. Description of $\phi(\rho z)$ functions for films by the method of Pouchou and Pichoir. (a) 300 nm of Al on W, Ag, Cu, Al, and B; $E_0 = 10\,\text{keV}$. (b) Various thicknesses of Al on W; $E_0 = 10\,\text{keV}$ [adopted from Pouchou and Pichoir (1991)].

$\phi(\rho z)$ formalism is expressed in terms of analytic functions, it is possible to calculate readily any postulated film (Bastin and Heijligers, 1990c; Pouchou and Pichoir, 1991). The analysis is thus performed interactively on a graphics terminal to compare measured results with theoretical predictions. For complex layered films, a useful strategy is to progressively increase the beam energy so as to vary the range of beam penetration. The changes in the x-ray spectrum as a function of beam energy can provide vital information to distinguish the layers which would not be possible from a spectrum measured at a single energy.

The Monte Carlo approach can also be used to simulate any film-on-substrate geometry, but in order to generate k-versus-thickness curves, large numbers of trajectory simulations must be performed, and the heavy demand on computation has limited its applicability. The continued development of computers and decreased cost of computing may make the Monte Carlo approach viable for real-time analysis (Romig *et al.*, 1990).

9.8.3. Foils

Foils, which are self-supporting thin layers and may be regarded as unsupported films, represent a simplified case of the film-on-substrate situation, since the substrate is absent, and the same methods described above can be applied with appropriate adjustments. Eliminating the

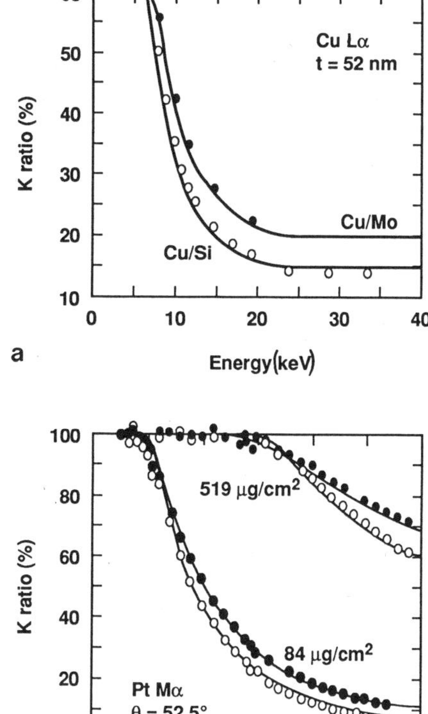

Figure 9.20. Comparison of theoretical predictions by the PAP $\phi(\rho z)$ model of emitted x-ray intensities. (a) Cu on Mo and Cu on Si. (b) Overlayer of Pt on Si (○) and Au (●).

substrate, however, opens another avenue of analytical attack, which is based upon the techniques of quantitative analysis used in analytical electron microscopy (Joy *et al.* 1986). X-ray microanalysis in the AEM is based on the observation that at high beam energies (typically 100–400 keV), elastic scattering and energy loss are minimized. The x-ray absorption path length through a foil is small, which can effectively eliminate the need for an absorption correction, and as a consequence, the fluorescence correction also vanishes. With the matrix effects eliminated, to at least the first order, a simple sensitivity-factor approach to quantitative analysis can be used.

If multielement standards in the form of thin foils are available, analysis can proceed by a simple interelement sensitivity-factor approach (Cliff and Lorimer, 1975):

$$k_{AB} = \left[\frac{C_A}{C_B}\right]\left[\frac{I_B}{I_A}\right], \qquad (9.64)$$

where C is the weight (mass) concentration of the element in the standard, I is the measured x-ray intensity, and A and B represent any two elements. B is the reference element, generally a major constituent of the standard whose concentration is known with good accuracy. Once values of k_{AB} are measured experimentally or calculated from first principles, they can be applied to the analysis of unknowns. Any measured interelement x-ray intensity ratio I_A/I_B can be converted to a concentration ratio C_A/C_B by multiplying the intensity ratio by k_{AB}. The concentrations of the individual constituents in terms of the concentration of the reference element B can be obtained by rearranging Eq. (9.64):

$$C_A = C_B k_{AB} \frac{I_A}{I_B} \tag{9.65}$$

If all elemental constituents are measured, the absolute concentrations can be calculated from the following relationship:

$$C_B + \sum_i C_i = 1, \tag{9.66}$$

where C_B is the concentration of the reference element and C_i represents all other constituents. Since Eq. (9.65) can be substituted into Eq. (9.66) for each element $A \cdots i$, Eq. (9.66) is reduced to an equation in one unknown, C_B. When Eq. (9.66) is solved to give an absolute value of the concentration for the reference element in the unknown, C_B, this value can be substituted into Eq. (9.65) to give absolute values of the concentrations of all other constituents. Usually analytical errors are in the range 5–10% relative, typically limited by the accuracy with which the standard concentrations are known. Goldstein *et al.* (1977) have described a criterion for the limit of specimen thickness for which the thin-film approximation is valid for any pair of elements, based upon the differential x-ray absorption:

$$\left[\left(\frac{\mu}{\rho}\right)_{\text{spec}}^{B} - \left(\frac{\mu}{\rho}\right)_{\text{spec}}^{A}\right] \csc \psi \rho t < 0.2, \tag{9.67}$$

where μ/ρ is the mass absorption coefficient for the respective element in the specimen, ψ is the spectrometer take-off angle, ρ is the density, and t is the thickness.

The great power of the ratio method incorporated in the sensitivity factor is that it directly incorporates the instrumental response, including spectrometer efficiency, peculiar to the instrument used. The method also effectively eliminates specimen-thickness effects that would dominate absolute x-ray intensity measurements or measurements against an external standard.

The maximum beam energy in an SEM is typically 30–50 keV. While this energy is significantly below the energy range for AEM, typically 100–400 keV, the thin-foil approach can still be effective in the SEM, particularly if the SEM is operated in the scanning transmission electron microscopy (STEM) imaging mode, which gives much more sensitive imaging of the structure of the foil (see Chapter 4). The combination of the high-resolution and imaging and analysis obtained in the STEM-in-SEM mode substantially exceeds the performance obtained with the same specimens mounted for conventional bulk SEM methods.

For films on substrates and foils, only the depth dimension is comparable to the electron range, and hence the modifications to conventional analytical procedures need only be parameterized in that one dimension. The lateral dimensions of films and foils are comparable to conventional bulk specimens and are effectively infinite to the beam. When microscopic particles and rough surfaces are considered, however, all three dimensions of the specimens are important since any or all of them may be comparable to the electron range. An example of the complexity of the situation is illustrated in Fig. 9.21. As the beam energy is increased, the electrons, which are well contained within the particle at 5 keV, spread to completely fill the particle and escape from all surfaces at 17 keV, and penetrate mostly through the bottom of the particle at 30 keV. This complex situation gives rise to three geometry effects on the x-ray intensities which have parallel physical origins to the ZAF factors: the mass effect, the absorption effect, and the fluorescence effect (Small *et al.*, 1979). In the following discussion, we will consider both particles and rough bulk samples together, because of their similarities.

9.8.4.1. Mass Effect

When the dimensions of a particle approach those of the interaction volume in a bulk solid of the same composition, electrons escape from the sides and bottom of the particle, as illustrated in Fig. 9.21. These escaping electrons carry away energy that in the bulk case would have contributed to the generation of x rays. The generation of x rays in particles thus decreases relative to a bulk standard because the mass of the target decreases. The intensity response, normalized to a bulk standard, as a function of particle diameter for spherical particles is shown in Fig. 9.22 for two different characteristic x rays, Fe $K\alpha$ (6.40 keV) and Si $K\alpha$ (1.74keV). The mass effect *always* leads to a lower intensity measured in the particle compared to the standard. The effect becomes significant for particle diameters below 5 μm, although the onset size depends on the average atomic number and the beam energy. Since the electron range increases as the 1.7 power of the beam energy (Chapter 3), a decrease in the beam energy can significantly reduce the particle size at which the mass effect becomes significant.

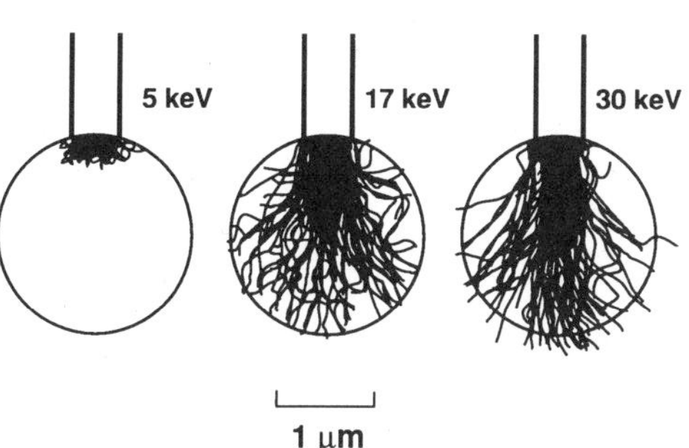

Figure 9.21. Monte Carlo electron-trajectory calculations for the interaction of electrons in aluminum particles as a function of energy (5, 17, and 30 keV). The beam diameter is 0.5 μm, and the particle diameter is 2 μm (Newbury *et al.*, 1980).

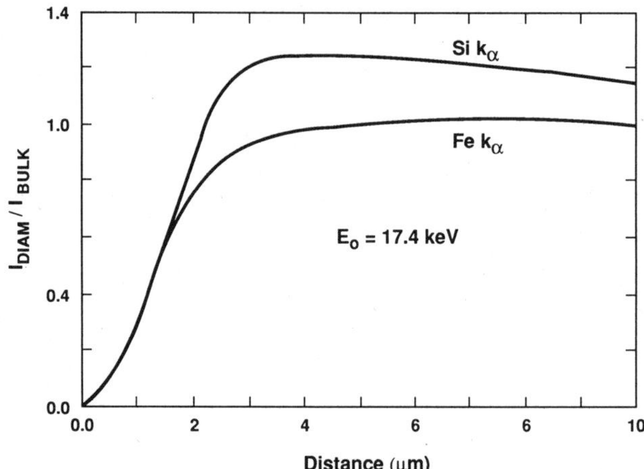

Figure 9.22. Monte Carlo electron-trajectory calculations of the normalized intensity (I_{diam}/I_{bulk}) as a function of diameter for spherical particles of NIST glass *K*-309(15% Al_2O_3, 40% SiO_2, 15% CaO, 15% Fe_2O_3, 15% BaO).

The mass effect for rough samples occurs because of the dependence of the excited volume of the specimen on the local inclination of the surface. As shown in Section 3.3.2.3, the effect of tilting a surface is to decrease the electron range and increase the backscattering, which decreases the production of x rays by removing energy from the specimen. Figure 9.23 shows the x-ray production from tilted surfaces, normalized to 0° tilt. Near 0° tilt, the deviation with increasing tilt is small, but above 20° tilt, the decrease in x-ray production becomes severe.

9.8.4.2. Absorption Effect

As shown in Fig. 9.24 for a flat bulk specimen placed at defined angles to the incident beam and the x-ray detector, there is a well-defined

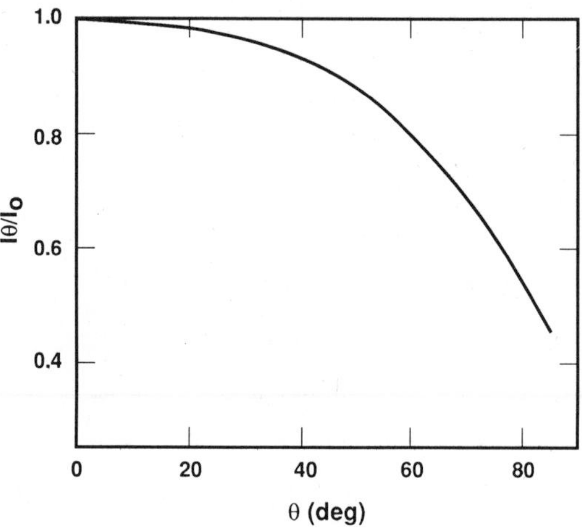

Figure 9.23. Monte Carlo electron-trajectory calculation of the emission of x-rays from a tilted specimen of iron, with the intensity normalized to 0° tilt. Beam energy = 20 keV.

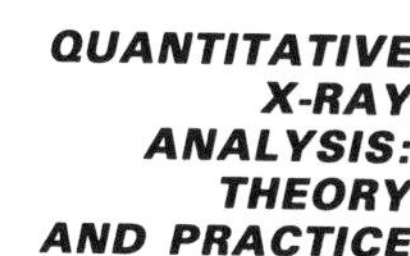

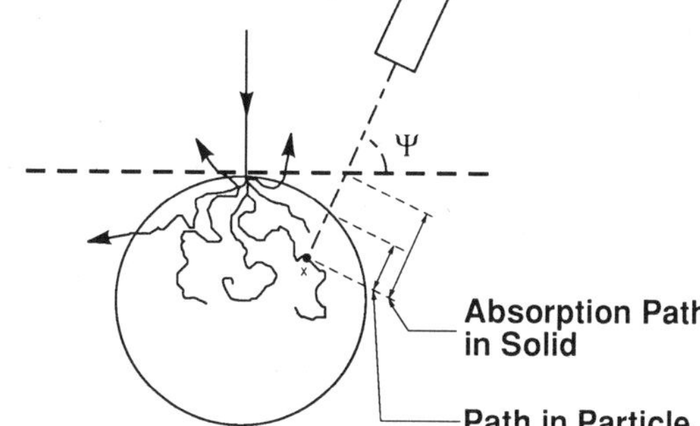

Figure 9.24. Schematic diagram comparing the absorption path in a particle with the equivalent path in a bulk target.

and calculable path along which the x rays travel through the specimen to reach the detector. Because x-ray absorption follows an exponential relation, any changes in this path length can have a strong effect on the fraction of x rays which escape the specimen. As shown in Fig. 9.24, the curved surface of a spherical particle can substantially shorten the x-ray absorption path for a significant portion of the x rays produced on the side of the particle facing the detector. The absorption effect can act to increase or decrease the emitted x-ray intensity, depending on beam placement on the particle relative to the x-ray detector and the particle size. Beam placement and particle size are the critical parameters:

Beam Placement. For particles slightly larger than the interaction volume, the location of the beam impact point relative to the x-ray detector can profoundly affect the measured spectrum, as shown in Fig. 9.25. When the beam is placed on the side of the particle facing the detector (Fig. 9.25, position 1), the absorption path is a minimum and the complete x-ray spectrum, including low-energy x rays below 3 keV (except for loss due to absorption in the spectrometer components) is detected efficiently. If the beam is placed symmetrically at the top of the particle, and the detector is at a high take-off angle > 30° (Fig. 9.25, position 2), a situation similar to that of a bulk target is obtained, and the spectrum is still reasonable. However, if the beam is placed on the back side of the particle away from the detector (Fig. 9.25, position 3), the extra absorption path through the particle severely attenuates the low-energy x-rays. Examples of spectra that illustrate these beam

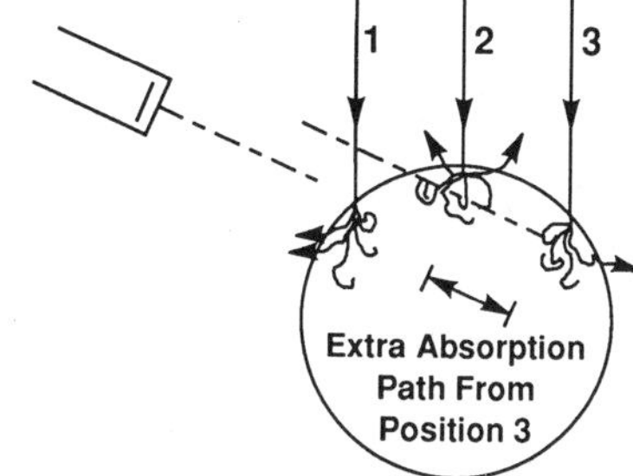

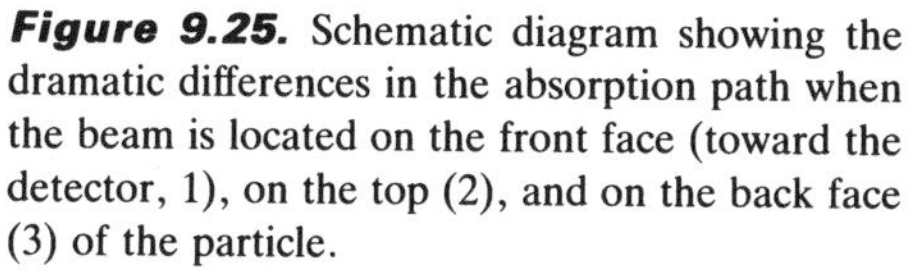
Figure 9.25. Schematic diagram showing the dramatic differences in the absorption path when the beam is located on the front face (toward the detector, 1), on the top (2), and on the back face (3) of the particle.

locations on a particle are shown in Fig. 9.26. Note that the ratio Si/Fe changes drastically in Fig. 9.26a and b between the front and back surfaces. The Si $K\alpha$ radiation at 1.74 keV is strongly absorbed, whereas the Fe $K\alpha$ at 6.40 keV is virtually unabsorbed. In the analysis of unknowns, the relative heights of characteristic lines are not a useful guide. A cursory inspection of the spectra in Fig. 9.26a and b would

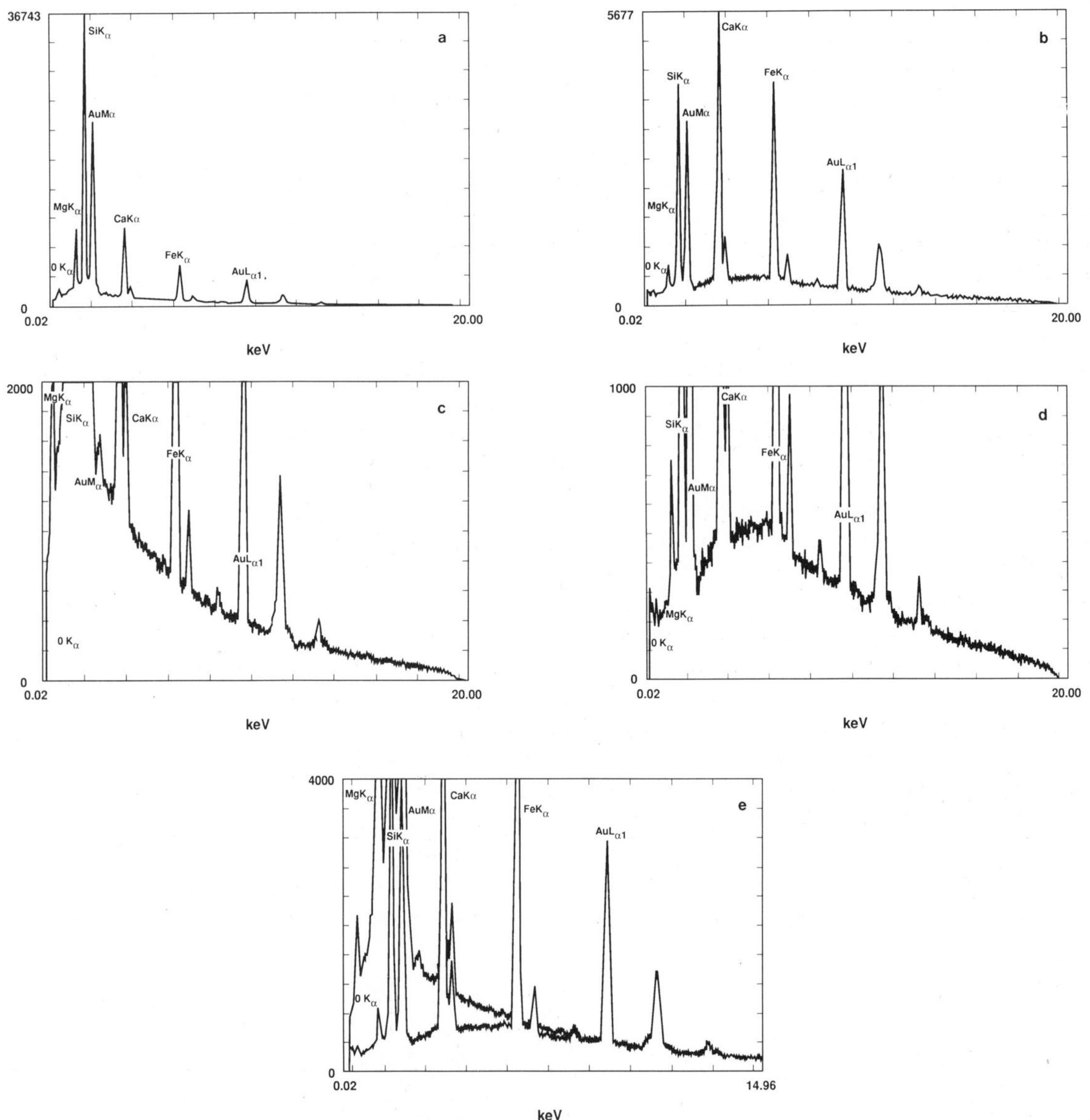

Figure 9.26. Energy-dispersive x-ray spectra obtained from a 20-μm-diameter spherical particle of NIST SRM glass K-411 (15% MgO, 55% SiO_2, 15% CaO, 15% FeO). The gold peak is an artifact from the sputtered coating. (a), (c) Front face, toward the detector. (b), (d) Back face, away from the detector. (e) Comparison of the overlaid spectra normalized at Fe $K\alpha$ showing large attenuation of the low-energy portion below 5 keV.

suggest a markedly different composition between these two locations, which are in fact identical. The best diagnostic tool for recognizing anomalously high absorption is the x-ray continuum background, which is readily observable in an EDS spectrum. The spectrum measured with a high take-off angle detector from a flat surface normal to the beam with an incident energy of 20 keV shows an increasing background with decreasing energy down to approximately 2 keV. Below 2 keV, the specimen's self-absorption and the absorption in the Be spectrometer window cause the intensity to fall despite the continued rise in intensity of the continuum generated in the specimen (see Chapter 3). When an anomalous absorption situation is encountered (beam placement on back side of particle, in a surface pit, etc.) the roll-off in the background occurs at a higher x-ray energy, typically in the 3–7 keV range. This situation is illustrated in the overlay plot of the particle spectra shown in Fig. 9.26e, where the background from the back-side spectrum rolls over at 5.5 keV. To minimize problems in beam location during small-particle analysis, the particle can be bracketed by a rapid scan raster so that all particle locations are excited by the beam. While useful, this approach obviously precludes any study of particle inhomogeneity, since the spectrum is averaged over the entire particle during the scan.

The effect of the extra path length on attenuating the low-energy end of the spectrum is exacerbated if the detector axis is placed at a right angle to the electron-beam axis. While this arrangement has been replaced in virtually all modern SEMs, many earlier instruments used this geometry. When the detector axis is at a right angle to the beam, the take-off angle, which is the angle above the surface to the detector axis, is zero when the specimen surface is normal to the beam. A zero take-off angle causes complete absorption of the x-rays for a bulk specimen, so to achieve an adequate take-off angle, the specimen is tilted to a high angle, typically in the range 30°–50°, as shown in Fig. 9.27a. If the specimen is at the same elevation as the detector, the angle of tilt equals the take-off angle. Consider now the analysis of spherical particles placed on a flat substrate. If the beam is placed at the top center of the particle, then for a high take-off angle detector, a low absorption situation is obtained, as shown in Fig. 9.27a. If a low take-off angle detector is used, then even when the flat surface containing the particles is tilted toward the detector, a low take-off angle is obtained for particle analysis when the beam is placed in the top center of the particle.

Particle Size. The absorption effect is further complicated by the particle size, as illustrated in Fig. 9.27c and d. For small particles completely excited by the beam (Fig. 9.27c), a low absorption situation is obtained with virtually any detector angle, while for a large particle in the size range greater than 5 μm, a high-absorption situation is encountered if the detector is at a right angle to the beam. The x rays must escape through the side of the particle, and they consequently suffer additional absorption because of the extended path length.

Even for high take-off angle detectors, the absorption effect can produce a complicated response. Over a wide range of particle diameters,

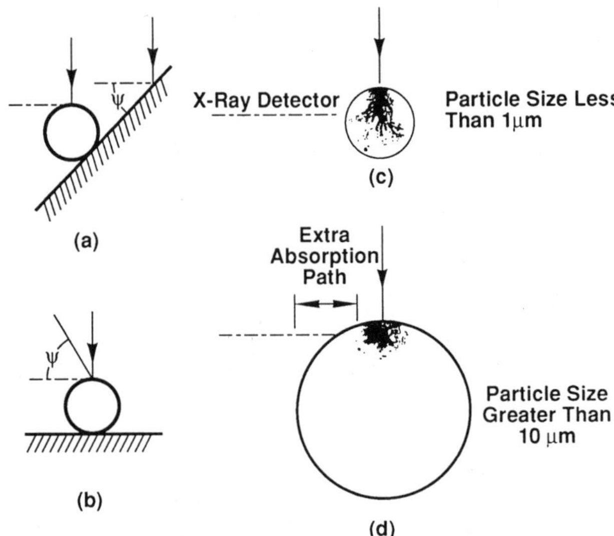

Figure 9.27. Schematic diagram illustrating the effect of particle size and detector position on x-ray path length. (a) Flat specimen tilted to produce a high take-off angle: low absorption. A particle on this same surface gives a low take-off angle if the beam is placed at the top center: high absorption. (b) High take-off angle detector, beam placed on top center of particle: low absorption (c) Low take-off angle detector, small particle fully excited by beam: low absorption. (d) Low take-off angle detector, particle much larger than interaction volume: high absorption.

the interaction volume is effectively contained within the particle but the surface curvature can still influence the measured x-ray spectrum. The particle absorption effect can actually increase the emitted intensity compared to that of a flat bulk standard. This effect is illustrated by the plot for Si $K\alpha$ radiation in Fig. 9.22, where the k-value exceeds unity over a substantial range of particle size. This occurs because Si $K\alpha$ suffers substantial absorption in the bulk standard due to its relatively low energy, so the decreased absorption path due to the particle curvature has a marked effect on the emitted x rays, and the Si $K\alpha$ k-value exceeds unity over a substantial size range. For an energetic line such as Fe $K\alpha$ shown in Fig. 9.22, there is very little absorption in the flat bulk case, so the tendency of particle curvature to reduce the absorption has a negligible effect, and the Fe $K\alpha$ k-value does not exceed unity for any particle size.

The absorption effect is much more complicated as a function of particle size than is the mass effect. Depending on the exact conditions of beam and detector placement and particle size, the absorption effect can lead to *increased or decreased* x-ray emission relative to a flat bulk standard.

For rough bulk specimens the absorption path effects can be even more profound. If the local surface inclination is random, then the absorption path can vary enormously from nearly complete absorption to a nearly ideal high take-off angle–low absorption situation. The spectra

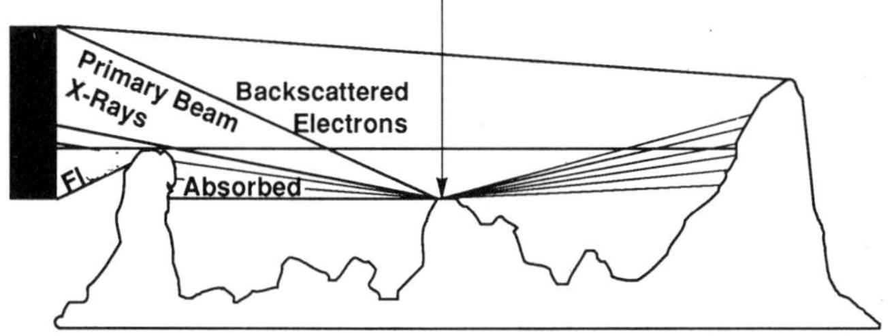

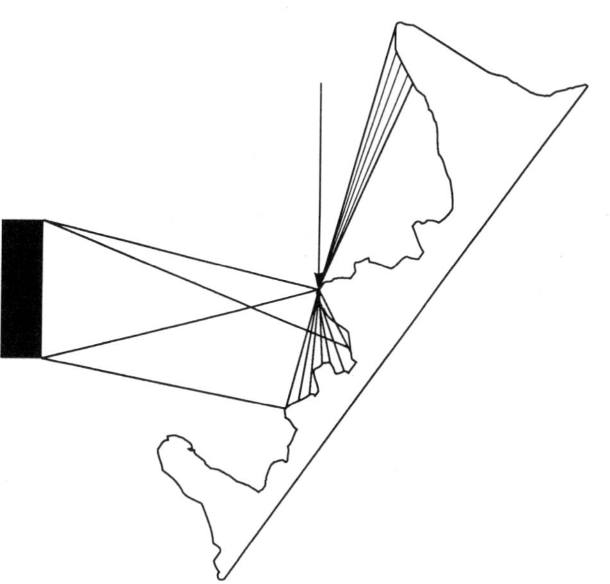

Figure 9.28. Schematic representation of the scattering processes which can occur when rough surfaces are examined. Fl = fluorescence.

from rough surfaces can vary even more dramatically than those shown for the particle in Fig. 9.27. Additionally, it is often difficult to judge from a single image of a rough surface what the local surface inclination is relative to the x-ray detector. Stereo pair imaging can be of some help in this regard, but it may not be possible to achieve a low absorption situation without considerable tilting or rotation of the specimen relative to the detector. As illustrated in Fig. 9.28, the electron scattering situation from a rough surface can be so complicated that remote excitation from other parts of the surface may contribute significantly to the spectrum. This remote excitation situation is especially true for the EDS detector, which is nonfocusing and generally collimated to no better than an area a few millimeters in diameter around the beam impact point on the specimen. The WDS detector, with its relatively small acceptance ellipsoid, having dimensions of 10–100 μm, may be a better choice to minimize remote scattering effects in spectra from rough surfaces.

9.8.4.3. Fluorescence Effect

Fluorescence of characteristic x rays of element A by other characteristic or continuum x rays with energies above the critical energy E_A takes place over a distance of 10–100 μm in a bulk target, as illustrated in Fig. 3.49. Even when a particle is sufficiently large to just contain the complete electron-interaction volume so that virtually all direct electron-initiated ionizations that occur in a flat bulk target also occur in the particle, most of the fluorescence radiation observed in the bulk case is lost in the particle case. For example, consider an interaction volume approximated as a hemisphere with a radius of $R_e = 1\ \mu$m. If the range for x-ray absorption, and subsequent fluorescence, is $R_x = 10\ \mu$m, then the volume ratio of x-ray to electron production is $R_x^3/R_e^3 = 1000{:}1$, so that only 0.1% of the bulk fluorescence is obtained from the particle. The fluorescence effect tends to decrease the emission of characteristic x rays compared to a bulk target.

The fluorescence effect actually can be more significant in particle analysis than in bulk analysis, where it usually cancels out and is ignored. As discussed in Chapter 3, characteristic fluorescence is significant only when the higher-energy characteristic x-rays have energies no greater than 5 keV above the edge in question, because the efficiency of ionization falls off rapidly as the photon energy increases above the ionization energy. Fluorescence by the continuum is normally ignored in conventional bulk specimen analysis because the effect is similar in the bulk specimen and the bulk standards and therefore cancels in the k-ratio and need not be calculated as part of the matrix correction. The major exception in bulk analysis is the case of hard lines of heavy-element minor constituents in a light matrix, e.g., Au–Lα in Al, where the continuum-induced fluorescence may reach levels of 20% or more. For particle analysis against bulk standards, the fluorescence correction becomes significant in all cases, because the fluorescence radiation is almost completely lost from small particles. The absolute magnitude of the indirect fluorescence radiation is given in Fig. 3.48, where it reaches a level of 10–15% of the total x rays produced for x-ray energies in the 8–12 keV range, e.g., Zn $K\alpha$ or Au $L\alpha$. Thus, if all other particle effects were calculated with exact accuracy, the effect of the loss of continuum fluorescence would be to lower the analytical total depending on the line energy of the elements.

9.8.4.4. Compensating for Geometric Effects in Quantitative Analysis

From the previous discussion, geometric factors can exert a complex effect on quantitative analysis, ranging from underestimates to overestimates of the composition. Several techniques are available for dealing with the geometric effects; only the most widely used will be described in this text. More exhaustive overviews of particle analysis has been given by Small (1981) and Armstrong (1991). These approaches include

1. ignoring the geometric effects,
2. simple normalization,
3. analytic solutions for special particle shapes, and
4. the peak-to-background method.

Energy-dispersive x-ray spectrometry has a particular advantage in particle analysis over an array of wavelength-dispersive spectrometers because the EDS can collect the entire spectrum along a single direction and therefore along a single x-ray absorption path. The individual spectrometers of a WDS array view the specimen along markedly different directions and therefore different absorption paths. These differences in absorption path introduce additional uncontrolled variables when the specimen has a complex shape or inhomogeneous nature, adding a further layer of complexity to the already complex problem of geometrical correction.

Ignoring the Geometric Effects. Some analysts choose to report the non-normalized results of analyses of particles or rough surfaces measured against bulk standards *along with the analytical total.* While the raw analytical total gives a measure of the deviation of the unknown from an ideal flat bulk target, this procedure is not recommended. Depending on the particle's mass-thickness, the raw concentrations obtained from particle analyses can deviate by a factor of 10 or more from the true concentrations of the elements in the particle. Such particle analyses cannot really be compared because the magnitude of the geometric effects may be so large that even the relative order of the apparent concentrations may not be correct. Although the raw analytical total can be used at any time to calculate a normalization factor, as described in the next section, there exists a significant potential for misinterpretation of the non-normalized concentration data if the accompanying analytical total is not carried along in any subsequent evaluation of the data.

Normalization. A procedure that is less prone to inadvertent mistaken use of the non-normalized particle concentrations requires the analyst to perform a simple normalization and to report the normalization factor as a measure of the severity of the geometric effects. Normalization is performed by dividing each raw concentration by the sum of the raw concentrations, including any constituents calculated by assumed stoichiometry, such as that for oxygen:

$$C_{n,i} = \frac{C_i}{\sum_j C_j}, \tag{9.68}$$

where the normalization factor is $1/\sum C_j$. Normalization places the concentration data on a sensible basis for comparison to other measurements. If the normalization factor is separated from the normalized concentration data during subsequent data interpretation, the concentration data still remain on a reasonable basis. However, careful procedure requires that both the normalized concentrations and the normalization factor be reported, since the normalization factor directly indicates the

Table 9.13. Errors Observed in Analysis of Particles by Conventional Methods with Normalization and by an Analytic Method[a]

Constituent	Actual concentration	Conventional analysis, normalized	Percentage relative error	Analytic particle analysis	Percentage relative error
Pyrite[b]					
S	53.45	54.8	+2.5	53.5	+0.1
Fe	46.55	45.2	−2.9	46.5	−0.1
Olivine[b]					
MgO	43.94	50.7	+15	43.8	−0.3
SiO_2	38.72	39.3	+1.5	38.9	+0.5
FeO	17.11	9.7	−43	17.1	0

[a] Armstrong (1978).
[b] Average of 140 particles.

magnitude of the geometrical effects. The greater the normalization factor, the more likely the existence of significant errors in the normalized concentrations.

Several caveats of the normalization procedure must be noted. The analysis must be performed by a correction procedure such as ZAF or $\phi(\rho z)$ that does not automatically force a normalization of the results, as "standardless" analysis does. Normalization is successful only when all constituents of the unknown are measured or, as in the case of oxygen, included by means of stoichiometric calculations. Normalization is most effective for the analysis of particles in the size range below 3 μm for a 20-keV beam. For such particles, the mass effect dominates and all x-ray energies are affected similarly. For example, in Fig. 9.22 the initial portion of the I_{diam}/I_{bulk} curve is identical for both Si $K\alpha$ and Fe $K\alpha$, despite the substantial difference in energy. The analysis of pyrite (FeS_2) in Table 9.13 shows an example where normalization of the results was highly successful. Normalization produces inadequate, and possibly misleading, results when the particle constituents and size are such that the mass effect influences the hard x-ray energies, while the absorption effect dominates the low-energy lines. Such a case is illustrated in the analysis of olivine in Table 9.13. The normalized magnesium concentration overestimates the true concentration, while the normalized iron concentration is a severe underestimate. A histogram of the absolute values of the relative errors of normalized concentrations from a large number of analyses of particle standards is shown in Fig. 9.29.

Analytic Shape Factors. Armstrong and Buseck (1975) have developed an analytic approach to calculate correction factors for particle geometric effects. Their method is based on bracketing the particle and overscanning during collection of the x-ray spectrum. Correction factors are developed by rigorously calculating the modification of the generation of x rays by analytical modeling of the penetration of electrons into and propagation of x rays through specific particle shapes. These shapes include the sphere, hemisphere, squared pyramid, and rectangular, tetragonal, cylindrical, and right triangular prisms. The correction factors

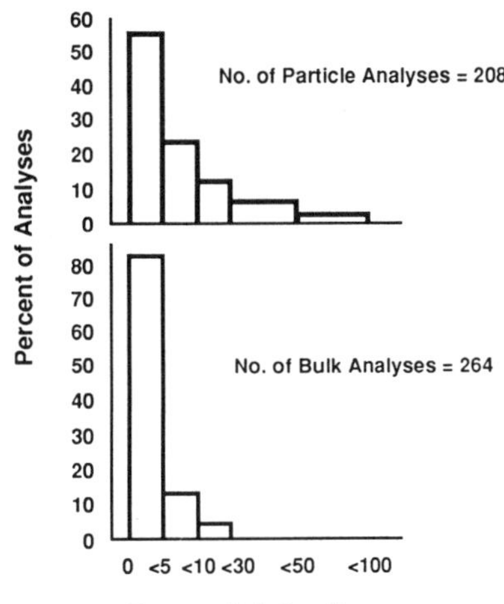

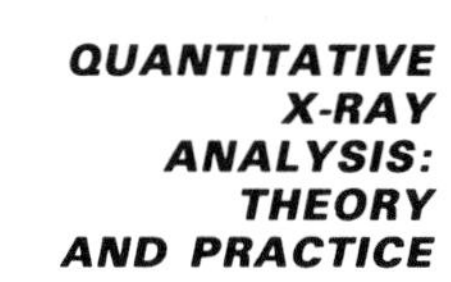

Figure 9.29. Histogram of relative errors (absolute values) of the normalized values from analyses of standard particles compared to the histogram for ZAF analysis of bulk multielement standards (Small, 1981).

R_{AB} are calculated in terms of the ratio of predicted k-factors for two particular elements, k_A/k_B, versus particle thickness along the electron beam. An example of a set of calculated factors for several particle shapes and for calcium and silicon is shown in Fig. 9.30. In practice, the analyst must judge which shape most closely corresponds to the unknown particle, and what its thickness is along the beam. The correction factors for all of the constituents are then determined from the appropriate curve for the shape most similar to the unknown. As listed in Table 9.13 and shown in the error distributions of Fig. 9.31a and b, this method can produce highly accurate results. The method has two main caveats. The use of overscanning obviously precludes the recognition of inhomogeneous particles, and the analyst's judgement is important in deciding on the appropriate shape/correction factor.

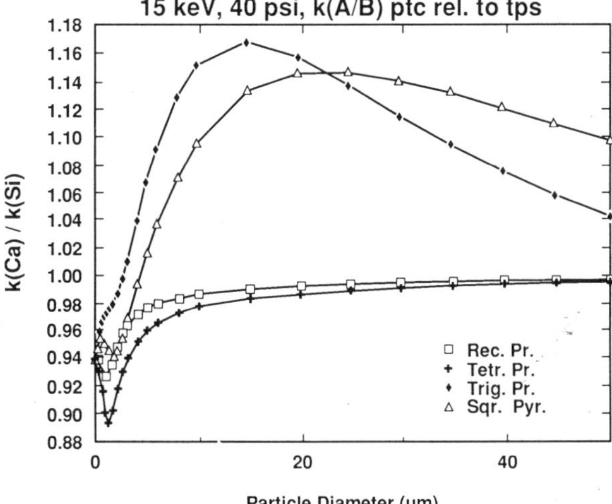

Figure 9.30. The theoretical k-factor ratio for calcium and silicon versus particle thickness for several particle shapes (Armstrong, 1991).

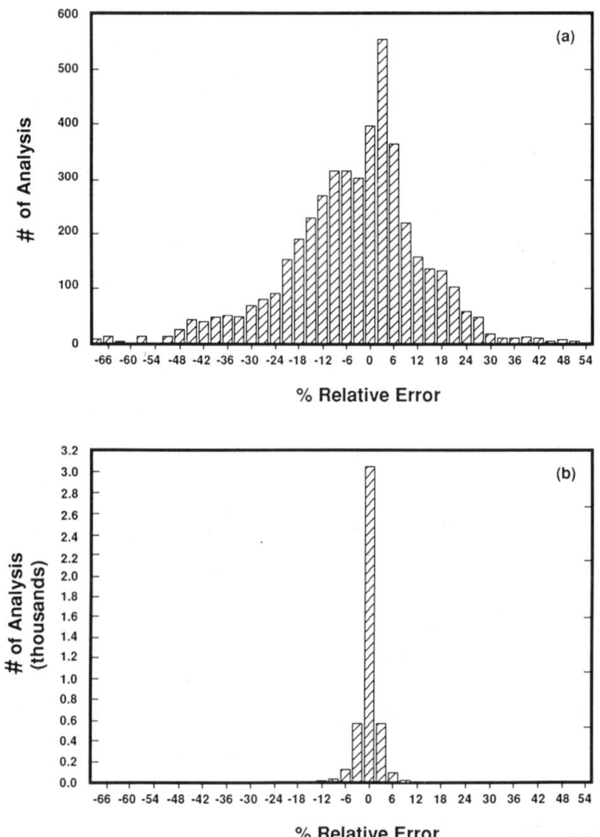

Figure 9.31. Comparison of error histograms of particle analyses with (a) conventional bulk ZAF procedures and (b) analytic modeling correction factors (Armstrong, 1991).

Peak-to-Background Method. The peak-to-background method (Small *et al.*, 1978; 1979; Statham and Pawley, 1978; Statham, 1979) is based on the observation that, although the intensity of characteristic x rays emitted from a particle or rough surface is highly dependent on geometric effects, the peak-to-background ratio taken between the characteristic x rays and the continuum x rays *of the same energy* is much less sensitive to those effects. Although the two types of radiation are produced by different processes (inner-shell ionization versus deceleration in a Coulombic field), the two types of radiation are produced in nearly the same volume. Both forms of radiation scale similarly with the geometric mass effect. Both have a similar, although not identical, depth distribution so the absorption paths are similar. Since the energies of both forms of radiation are identical by definition, the geometric absorption effect also scales for both. Therefore, the continuum intensity I_B can be used as an automatic internal normalization factor for the geometric effects. Thus, although $k = I_{\text{particle}}/I_{\text{bulk}}$ (where "bulk" refers to the same composition as the particle but in flat bulk form) is a strong function of particle size, the ratio $(I_{\text{particle}}/I_{B,\text{particle}})/(I_{\text{bulk}}/I_{B,\text{bulk}})$ is nearly independent of particle size, except for very small particles where anisotropy of the continuum emission becomes significant (Newbury *et al.*, 1980). This experimental observation, which has been confirmed by theoretical calculations, can be employed in several ways (Small *et al.*, 1978; Statham, 1979). One useful approach is to incorporate the following correction scheme into a conventional ZAF method (Small *et al.*, 1978; 1979). Given that

$$\frac{I_{\text{particle}}}{I_{B,\text{particle}}} = \frac{I_{\text{bulk}}}{I_{B,\text{bulk}}}, \tag{9.69}$$

a modified particle intensity, P^*_{particle}, can be calculated, which is equivalent to the intensity that would be measured from a flat bulk target of the same composition as the particle:

$$P^*_{\text{particle}} = I_{\text{bulk}} = I_{\text{particle}} \times \frac{I_{B,\text{bulk}}}{I_{B,\text{particle}}}. \tag{9.70}$$

To apply Eq. (9.70) for the analysis of an unknown particle or rough surface, the quantities I_{particle} and $I_{B,\text{particle}}$ can be measured directly from the energy dispersive x-ray spectrum. $I_{B,\text{bulk}}$ is in general not measurable, since a standard identical in composition is generally not available. However, as described in Section 9.2, an *estimate* of the concentrations of elements in the unknown is always available in the ZAF procedure, including the first step, where $C_i = k_i/\sum k$. The value of $I_{B,\text{bulk}}$ can therefore be estimated from the background measured on pure-element standards:

$$I_{B,\text{bulk}} = \sum_i C_i I_{i,B,\text{standard}} \tag{9.71}$$

where $I_{i,B,\text{standard}}$ is the pure-element bremsstrahlung at the energy of interest and C_i is the concentration of that element.

Examples of particle analyses with the peak-to-background modification to the normal ZAF procedure are given in Table 9.14. In general, the errors are equal to or substantially less than those produced

Table 9.14. Errors Observed in the Analysis of Particles by the Peak-to-Background Method[a]

Constituent	Actual concentration	Conventional ZAF	Percentage error	P/B ZAF	Percentage error
Pyrite					
S	53.4	39.9	−25	52.9	−1.0
Fe	46.6	35.8	−23	46.4	−0.5
Talc					
Mg	19.3	10.3	−47	18.5	−4
Si	29.8	15.8	−47	29.0	−3

[a] Small *et al.* (1979); single analyses on known standards.

Table 9.15. Errors Observed in the Analysis of Rough Objects (Fracture Surfaces of Homogeneous Au–Cu alloys) by Peak-to-Background Modified ZAF Methods[a]

Constituent	Actual conc.	Conventional ZAF	Relative error, %	Normalized ZAF	Relative error, %	P/B ZAF	Relative error, %
Analysis 1							
Au	60.3	28.2	−53	49.6	−18	58.0	−4
Cu	39.6	28.7	+28	50.4	+27	44.0	+11
Analysis 2							
Au	60.3	1.08	−98	29.1	−52	52.0	−14
Cu	39.6	2.63	−93	70.8	+79	41.0	+3
Analysis 3							
Au	80.1	95.8	+20	73.8	−8	76.9	−4
Cu	19.8	34.0	+72	26.2	+32	19.1	−3

[a] Small *et al.* (1979).

by simple normalization, especially when the particle is in the size range where the absorption effect is significant. Because the P/B (peak-to-background) method can deal with the absorption effect, the technique offers considerable promise for dealing with the especially difficult analytical problems posed by rough, irregular surfaces. Examples of analyses of rough surfaces of microscopically homogeneous alloys are given in Table 9.15. Particularly interesting are those analyses where the geometrical effects are so serious that simple normalization produces unacceptably large errors, but P/B corrections result in reasonable values.

The peak-to-background method has the particular advantages that the procedure applies to fixed-beam analysis, so that inhomogeneous particles can be analyzed, and the analyst does not need to make a judgment on the specifics of particle size and shape. The main disadvantage arises from the impact of the counting statistics of the background on the corrected k-value in Eq. (9.70). The background from the particle spectrum in Eq. (9.70) exerts a particularly large influence on the analytical precision, since it may be a much smaller number than the background from the standard, and therefore have a larger relative standard deviation. Moreover, it is especially important that the particle spectrum does not contain large contributions from the substrate. Even if beryllium or carbon is chosen for a bulk substrate, there will be a particle size for which the beam penetration through the particle becomes sufficiently large that the spectral background is dominated by the contribution from the substrate. Techniques for mounting particles on thin films for SEM and EPMA are discussed in Chapter 11.

Limitations. No matter what correction technique is applied for geometric effects, the analyst must be aware that significant pitfalls exist in the analysis of inhomogeneous particles and rough surfaces, especially where questions involving minor or trace constituents are involved. Figure 9.28 shows the effects of remote excitation of the sample by electrons scattered from the fixed beam position on a rough surface.

Since the solid angle of acceptance of the EDS typically encompasses millimeter dimensions at the specimen location, it is difficult to avoid contributions to the spectrum from such remote scattering. These remote sources may contribute only a small percentage of the total spectrum, but in cases where the minor and trace constituents are important, such remote scattering may introduce significant artifacts.

Similarly, when inhomogeneous particles are examined, the penetration of the primary beam makes it difficult to confine the analysis to a single phase. For small, inhomogeneous particles, with thicknesses less than 0.2 μm a useful alternative strategy is to operate in the STEM mode at maximum beam energy to minimize lateral scattering and analyze the particle as a thin foil, as described above.

9.9. Precision and Sensitivity in X-Ray Analysis

Up to this point we have discussed the corrections for quantitative x-ray microanalysis and the errors associated with the calculation procedure. In addition, one must consider the precision or sensitivity of a given analysis. By the precision or sensitivity of an analysis we mean the scatter of the results due to the very nature of the x-ray measurement process. We can take advantage of the statistical equations which describe precision or sensitivity to determine the chemical homogeneity of a sample, the variation of composition from one analysis point to another, and the minimum concentration of a given element which can be detected.

9.9.1. Statistical Basis for Calculating Precision and Sensitivity

X-ray production is statistical in nature; the number of x-rays that are produced from a given sample and interact with radiation detectors is completely random in time but has a fixed mean value. The distribution or histogram of the number of measurements of x-ray counts for a fixed time interval from one point on a sample versus the number of x-ray counts may be closely approximated by the continuous normal (Gaussian) distribution. Individual x-ray count results from each sampling lie upon a unique Gaussian curve for which the standard deviation is the square root of the mean ($\sigma_c = \bar{N}^{1/2}$). Figure 9.32 shows such a Gaussian curve for x-ray emission spectrography and the standard deviation $\sigma_c = \bar{N}^{1/2}$ obtained under ideal conditions. Here, $\bar{N}$ is considered to be the most probable value of N, the total number of counts obtained in a given time t. Inasmuch as σ_c results from fluctuations that cannot be eliminated as long as quanta are counted, this standard deviation σ_c is the irreducible minimum for x-ray emission spectrography. It is not only a minimum, but fortunately it is a predictable minimum. The variation in percentage of total counts can be given as $(\sigma_c/\bar{N})100$. For example, to obtain a result with a maximum of 1% deviation in N, at least 10,000

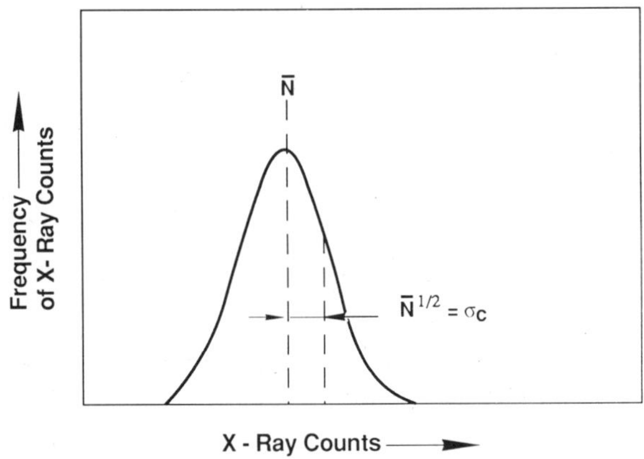

Figure 9.32. Gaussian curve for x-ray emission spectrography. The standard deviation σ_c is obtained under ideal conditions (Liebhafsky *et al.*, 1955).

counts must be accumulated. A common error is to misapply this simple fact to counting samples which have been background corrected. When a constant (the background) is subtracted from a counting sample, the result no longer obeys Poisson statistics, and other, more complicated, relations must be used (Newbury *et al.* 1986; Fiori *et al.*, 1984). As the magnitude of the peak diminishes and becomes numerically comparable to the background on which it rides, it is extremely important not to make this mistake, since large errors can result.

As Liebhafsky *et al.* (1955) have pointed out, the real standard deviation S_c of the experiment is given by

$$S_c = \left[\sum_{i=1}^{n} \frac{(N_i - \bar{N}_i)^2}{n - 1} \right]^{1/2}, \tag{9.72}$$

where N_i is the number of x-ray counts for each determination i and

$$\bar{N}_i = \frac{\left(\sum_{i=1}^{n} N_i \right)}{n} \tag{9.73}$$

where n is the number of measurements of N_i. The standard deviation S_c equals σ_c only when operating conditions have been optimized. In most SEM instruments, drift of electronic components and of specimen position (mechanical stage shifts) create operating conditions which are not necessarily ideal. The high-voltage filament supply, the lens supplies, and other associated electronic equipment may drift with time. After a specimen is repositioned under the electron beam, a change in measured x-ray intensity may occur if the specimen is off the focusing circle of a WDS x-ray spectrometer or if the take-off angle ψ of the specimen varies when an EDS is being used. In practice, for typical counting times of 10 to 100 s/point, the actual standard deviation S_c is often about twice σ_c. If longer counting times are used, S_c/σ_c increases due to instrument drift.

Only when counting times are short and the instrument is electronically stable does S_c approach σ_c. Besides the sample signal, sources of variation may also occur if data from reference standards or background standards are required (Ziebold, 1967). These, as well as faulty specimen preparation, may also affect the precision of an analysis. Therefore, both instrumental factors and signal variations must be considered when the precision of an analysis is determined.

Before proceeding with the following sections it is important to review what we mean by Peak P and Background B in x-ray microanalysis and how we measure P and B (see Chapter 7). For the WDS, we measure the intensity at an energy at the center of the peak. To measure the rest of the peak profile it is necessary to scan the peak by moving the spectrometer through a range of wavelengths or energies encompassing the peak. We further note that such a WDS spectrometer tuned to the center of a peak is incapable of determining whether an x ray is a peak or background x ray. Consequently, what is measured is Peak + Background (P + B) and there is no way to directly tell one from the other. What is required is to estimate the background that is actually under the peak by one of several methods discussed in Chapter 7 and subtract this value from the measured $P + B$. Consequently, when we write $P - B$ what we really mean is $(P + B) - B$, where the second B is an estimate of the first B. The purpose of the parentheses is to emphasize the fact that the terms $(P + B)$ are really one quantity. With a WDS spectrometer, it is often an easy matter to get a very good estimate of the background under a peak by detuning the spectrometer on each side of the peak and making measurements (see Chapter 7), keeping track of how far each measurement is from the peak and how long each measurement took. Since the spectral resolving power of the WDS is $\leq$10 eV, the effects of interfering peaks are *usually* not a problem.

When an EDS is used to make measurements of P and B, the process is very much more complicated, and in some cases, impossible. As discussed in Chapter 7, it can be very difficult to get a good estimate of the background under a peak where the background is strongly curved (1 to 3 keV) or where absorption edges occur. Indeed, in cases where a digital filter is used to remove background, it is important to note that the filter averages each channel in a spectrum with its neighbors in such a way that complicates or prevents the use of many of the methods about to be discussed (the channels are no longer independent variates and the ratio of the magnitude of the peaks to the magnitude of the background is altered in the filtering process). Peak overlap is often the rule in EDS, and we must now contend with the statistical effects of the peaks of one element interfering with the background determination of the peaks of another element.

In the following sections it is safe to assume that the methods discussed will work well with data obtained with WDS and will occasionally work with data obtained with EDS. In the last part of this section we will discuss newer procedures to handle the more complex EDS cases.

9.9.2. Sample Homogeneity

An analyst is often asked if a sample or a phase is *homogeneous*. In order to answer this question, the x-ray data must be obtained so that it can be treated statistically. Either one can set up criteria for homogeneity and apply them, or one can measure the range of variation in composition of a sample, at a certain confidence level, and report that number. Either method allows a more quantitative statement to be made than just a simple "yes" or "no" to questions concerning homogeneity. The following discussion presents homogeneity criteria that one can use and explains how to calculate the range and level of homogeneity.

A simplified criterion that has been used to establish the homogeneity of a phase or a sample states that for a homogeneous sample all the data points n, each taken at a different position on the sample, fall within the $\bar{N} \pm 3\bar{N}^{1/2}$ limits, where N is the raw, uncorrected intensity from the specimen (Yakowitz *et al.*, 1965). If this criterion is satisfied, one assumes that the sample is homogeneous. The variation

$$\frac{\pm 3\bar{N}^{1/2}}{\bar{N}} 100\,(\%) \tag{9.74}$$

for the element of interest in the sample represents the level (in percent) of homogeneity that is measured for the sample, remembering that there must be an irreducible minimum level, because x-ray production is statistical. If 100,000 counts are accumulated at each point in a sample and all these points fall within the limits $\bar{N} \pm 3\bar{N}^{1/2}$, the sample is homogeneous and the level of homogeneity is ±0.95%, according to Eq. (9.74). A level of homogeneity of $\leq \pm 1\%$ is often desired. If the concentration in the sample C is 10 wt %, the range of homogeneity, that is the minimum variation of concentration that can be validly measured, is ±0.1 wt %. This equation is valid only when the P/B ratio is high enough to use $\bar{N}$ directly without a background correction.

A more exacting determination of the range (in wt %) and level (in %) of homogeneity involves the use of the standard deviation S_c of the measured values and the degree of statistical confidence in the determination of $\bar{N}$. The standard deviation includes effects arising from the variability of the experiment, e.g., instrument drift, x-ray focusing errors, and x-ray production. The degree of confidence used in the measurement states that we wish to avoid a risk α of rejecting a good result a large percentage (say 95% or 99%) of the time. The degree of confidence is given as $1 - \alpha$ and is usually chosen as 0.95 or 0.99, that is, 95% or 99%. The use of a degree of confidence means that we can define a range of homogeneity, in wt %, for which we expect, on the average, only 5% or 1% of the repeated random points to be outside this range. The range of homogeneity in wt % for a degree of confidence $1 - \alpha$ is

$$W_{1-\alpha} = \pm C\left(\frac{t_{n-1}^{1-\alpha}}{n^{1/2}}\right)\frac{S_c}{\bar{N}}, \tag{9.75}$$

where C is the true weight fraction of the element of interest, n is the

Table 9.16. Values of Student t Distribution for 95% and 99% Degrees of Confidence[a]

n	$n-1$	t_{n-1}^{95}	t_{n-1}^{99}
3	2	4.304	9.92
4	3	3.182	5.841
8	7	2.365	3.499
12	11	2.201	3.106
16	15	2.131	2.947
30	29	2.042	2.750
∞	∞	1.960	2.576

[a] Bauer (1971).

number of measurements, $\bar{N}$ is the average number of counts accumulated at each measurement, and $t_{n-1}^{1-\alpha}$ is the Student t value for a $1-\alpha$ confidence level and for $n-1$ degrees of freedom. Student t values for t_{n-1}^{95} and t_{n-1}^{99} for various degrees of freedom, $n-1$, are given in Table 9.16 (Bauer, 1971). It is clear from Table 9.16 that at least four individual measurements, $n = 4$, should be made to establish the range of homogeneity. If fewer than four measurements are made, the value of $W_{1-\alpha}$ is too large.

The level of homogeneity, or homogeneity level, for a given confidence level $1-\alpha$, in percent is given by

$$\pm\frac{W_{1-\alpha}}{C} = \pm\frac{(t_{n-1}^{1-\alpha})S_c(100)}{n^{1/2}\bar{N}} \quad (\%). \tag{9.76}$$

It is more difficult to measure the same level of homogeneity as the concentration, present in the sample, decreases. Although $W_{1-\alpha}$ is directly proportional to C, the value of S_c/N increases as C and the number of x-ray counts per point decreases. To obtain the same number of x-ray counts per point, the time of the analysis must be increased.

More sophisticated studies of homogeneity can be performed by making several measurements on each of the n analysis points on the sample. This type of analysis includes the difference between analysis points and the error due to counting statistics. In addition, differences in homogeneity between individual samples can be considered using a statistical analysis. The appropriate analysis procedures for these studies are given by Marinenko *et al.* (1979; 1981).

9.9.3. Analytical Sensitivity

Analytical sensitivity is the ability to distinguish, for a given element, between two concentrations C and C' that are nearly equal. X-ray counts N and N' for both concentrations therefore have a similar statistical variation. If one determines two concentrations C and C' by n repetitions of each measurement, taken for the same fixed time interval, then these two values are significantly different at a certain degree of confidence,

$1 - \alpha$, if

$$\bar{N} - \bar{N}' \geq \frac{2^{1/2}(t_{n-1}^{1-\alpha})S_c}{n^{1/2}} \tag{9.77}$$

and

$$\Delta C = C - C' \geq \frac{2^{1/2}C(t_{n-1}^{1-\alpha})S_c}{n^{1/2}(\bar{N} - \bar{N}_B)}, \tag{9.78}$$

in which C is the concentration of one element in the sample, $\bar{N}$ and $\bar{N}_B$ are the average number of x-ray counts of the element of interest for the sample and for the continuum background on the sample, respectively, $t_{n-1}^{1-\alpha}$ is the "Student factor" dependent on the confidence level $1 - \alpha$ (Table 9.16), and n is the number of measurements. Ziebold (1967) has shown that the analytical sensitivity for a 95% degree of confidence can be approximated by

$$\Delta C = C - C' \geq \frac{2.33}{n^{1/2}} \frac{C\sigma_c}{(\bar{N} - \bar{N}_B)}. \tag{9.79}$$

The above equation represents an estimate of the maximum sensitivity that can be achieved when signals from both concentrations have their own errors but instrumental errors are disregarded. Since the actual standard deviation S_C is usually about two times larger than σ_C, ΔC is in practice approximately twice that given in Eq. (9.79).

If $\bar{N}$ is much larger than $\bar{N}_B$, Eq. (9.79), can be rewritten as

$$\Delta C = C - C' \geq \frac{2.33C}{(n\bar{N})^{1/2}}, \tag{9.80}$$

and the analytical sensitivity in percent that can be achieved is given as

$$\frac{\Delta C}{C}(\%) = \frac{2.33(100)}{(n\bar{N})^{1/2}}. \tag{9.81}$$

For an analytical sensitivity of 1%, ≥54,290 accumulated counts, $n\bar{N}$ from Eq. (9.81), must be obtained from the sample. If the concentration C is 25 wt %, $\Delta C = 0.25$ wt %, and if the concentration C is 5 wt %, $\Delta C = 0.05$ wt %. Although the analytical sensitivity improves with decreasing concentration, it should be pointed out that the x-ray intensity decreases directly with the reduced concentration. Therefore, longer counting times become necessary to maintain the 1% sensitivity level.

Equation (9.81) is particularly useful for predicting the procedures necessary to obtain the sensitivity desired in a given analysis. If a concentration gradient is to be monitored over a given distance in a sample, it is important to predict how many data points should be taken and how many x-ray counts should be obtained at each point. For example, if a gradient from 5 to 4 wt % occurs over a 25-μm region and 25 1-μm steps are taken across the gradient, the change in concentration per step is 0.04 wt %. Therefore ΔC, the analytical sensitivity at a 95% degree of confidence, must be ≤0.04 wt %. Using Eq. (9.80), since $\bar{N}$ is much larger than $\bar{N}_B$, $n\bar{N}$ must be at least 85,000 accumulated counts per step. If only ten 2.5-μm steps are used across the gradient, the change in concentration per step is 0.1 wt %, and now $n\bar{N}$ need be only $\geq$13,600 accumulated counts per step. By measuring 10 as opposed to 25 steps, the analysis time is cut down much more than the obvious factor of 2.5 since the number of required accumulated counts per step, due to sensitivity requirements, also decreases. Furthermore, if the data is plotted, the eye "integrates" neighboring points and permits the unambiguous determination of even smaller differences.

9.9.4. Trace-Element Analysis

As the elemental concentration C approaches the order of 1.0 wt % in x-ray microanalysis, $\bar{N}$ is no longer much larger than $\bar{N}_B$. This concentration range, below 1.0 wt %, is often referred to as the trace-element analysis range for x-ray microanalysis. The requirement in trace-element analysis is to detect significant differences between the sample and the continuum background generated from the sample.

To develop a useful procedure for trace detection, we need a criterion that will guarantee that a given element is present in a sample. This criterion can be called the "detectability limit" DL. The detectability limit is governed by the minimum value of the difference $\bar{N} - \bar{N}_B$, which can be measured with statistical significance. Leibhafsky *et al.* (1960) have discussed the calculation of detectability limits. They suggest that an element can be considered present if the value of $\bar{N}$ exceeds the background $\bar{N}_B$ by $3(\bar{N}_B)^{1/2}$. A more sophisticated analysis must consider the measured standard deviation, the number of analyses, and the confidence level desired. By analogy with Eq. (9.77) we can also define the detectability limit DL as $(\bar{N} - \bar{N}_B)_{\text{DL}}$ for trace analysis as

$$(\bar{N} - \bar{N}_B)_{\text{DL}} \geq \frac{2^{1/2}(t_{n-1}^{1-\alpha})S_c}{n^{1/2}}, \tag{9.82}$$

where S_c is essentially the same for both the sample and background measurement. In this case we can define the detectability limit at a confidence level $1 - \alpha$ (Table 9.16) that the analyst chooses. The 95% or 99% confidence level is usually chosen in practice. If we assume for trace analysis that the x-ray calibration curve of intensity versus composition is expressed as a linear function, then C, the unknown composition, can be related to $\bar{N}$ by the equation

$$C = \frac{\bar{N} - \bar{N}_B}{\bar{N}_S - \bar{N}_{SB}} C_S, \tag{9.83}$$

where $\bar{N}_S$ and $\bar{N}_{SB}$ are the mean counts for the standard and the continuum background for the standard, respectively, and C_S is the concentration in wt % of the element of interest in the standard. The detectability limit C_{DL}, that is, the minimum concentration which can be measured, can be calculated by combining Eqs. (9.82) and (9.83) to yield

$$C_{\rm DL} = \frac{C_S}{\bar{N}_S - \bar{N}_{SB}} \frac{2^{1/2} t_{n-1}^{1-\alpha} S_c}{n^{1/2}} \tag{9.84}$$

The relative error or precision in a trace-element analysis is equal to $C_{\rm DL}/C$ (100%) and approaches ±100% as C approaches $C_{\rm DL}$.

The background intensity $\bar{N}_{\rm B}$ must be obtained accurately so that the unknown concentration C can be measured, as shown by Eq. (9.83). It is usually best to measure the continuum background intensity directly on the sample of interest. Other background standards may have different alloying elements or a different composition, which will create changes in absorption with respect to the actual sample. Also, such background standards may have different amounts of residual contamination on the surface, which is particularly bad when measuring x-ray lines with energies≤1 keV. The background intensity using a WDS is obtained after a careful wavelength scan is made of the major peak to establish precisely the intensity of the continuum on either side of the peak. Spectrometer scans must be made to establish that these background wavelengths are free of interference from other peaks in all samples to be analyzed. It is difficult to measure backgrounds using the EDS with the accuracy needed to do trace-element analysis. (Measurements of continuum background with the EDS are discussed in Chapter 7). If no overlapping peaks or detector artifacts appear in the energy range of interest and the continuum background can be determined, trace-element analysis can be accomplished using the EDS detector.

Ziebold (1967) has shown that trace-element sensitivity or the detectability limit DL can be expressed as

$$C_{\rm DL} \geq \frac{3.29a}{(n\tau P \cdot P/B)^{1/2}}, \tag{9.85}$$

where τ is the time of each measurement, n is the number of repetitions of each measurement, P is the pure-element counting rate, P/B is the peak-to-background ratio of the pure element (i.e., the ratio of the counting rate of the pure element to the background counting rate of the pure element), and a relates composition and intensity of the element of interest through the Ziebold and Ogilvie (1964) empirical relation (see Section 9.7). To illustrate the use of this relation, the following values were used for calculating the detectability limit for Ge in iron meteorites (Goldstein, 1967) using a WDS system. The operating conditions were

beam energy, 35 keV, $\tau = 100$s

specimen current, 0.2 μA, n = 16

$P = 150{,}000$ counts, $\alpha = 1.0$

$P/B = 200$

With these numbers, Eq. (9.85) gives $C_{\rm DL} > 15$ ppm (0.0015 wt %);

the actual detectability limit obtained after calculating S_C and solving Eq. (9.84) was 20 ppm (0.0020 wt %) Goldstein (1967). Counting times on the order of 30 min were necessary to achieve this detectability limit. In measuring the carbon content in steels, detectability limits of the order of 300 ppm are more typical with a WDS system if one uses counting times on the order of 30 min, with the instrument operating at 10 keV and 0.05 μA specimen current. If more specimen current can be used without increasing specimen contamination, the detectability limit will be lower.

A criterion often used to compare the sensitivity of wavelength and energy-dispersive detectors is the product $P * P/B$ in the Ziebold relation, Eq. (9.85). Geller (1977) has compared WDS and EDS detectors in which each detector was optimized for maximum P^2/B. The specimen used in the analysis was a synthetic mineral D1-JD-35 containing Na, Mg, Si, and Ca as the oxides. The EDS was operated at a specimen current of 1.75 nA at 15 keV. An optimum count rate of 2000 cps on the specimen was obtained so that summed peaks would not be significant in the analysis. The WDS was operated at a specimen current of 30 nA at 15 keV. An optimum count rate was set at 13,000 cps on Si so that the maximum count rate caused less than 1% dead tme in the proportional counter. Table 9.17 lists the C_{DL} for the elements present in the sample. For optimized operating conditions, the minimum detectability limit using the WDS is almost ten times better than the EDS for all elements. Therefore, if the solid specimen is not affected by the higher specimen current, the WDS can offer a factor-of-10 increase in sensitivity over the EDS. Typical C_{DL} for the WDS and EDS detectors are 0.1 and 0.01 wt %, respectively.

Table 9.17. Comparison of the Minimum Detectability Limit of Various Elements Using an EDS and WDS Detection System on the Basis of Optimized Operating Conditions[a]

Analysis	Element	P (cps)	B (cps)	P/B	Wet chem. (wt %)	C_{DL} (wt %)
EDS	Na $K\alpha$	32.2	11.5	2.8	3.97	0.195
	Mg $K\alpha$	111.6	17.3	6.4	7.30	0.102
	Al $K\alpha$	103.9	18.2	5.7	4.67	0.069
	Si $K\alpha$	623.5	27.3	22.8	26.69	0.072
	Ca $K\alpha$	169.5	19.9	8.5	12.03	0.085
WDS	Na $K\alpha$	549	6.6	83	3.97	0.021
	Mg $K\alpha$	2183	8.9	135	7.30	0.012
	Al $K\alpha$	2063	16.1	128	4.67	0.008
	Si $K\alpha$	13390	37.0	362	26.69	0.009
	Ca $K\alpha$	2415	8.2	295	12.03	0.009

[a] Geller (1977). Analysis of D1-JD-35: EDS data collected at 2000 cps for 180 s, dead-time corrected (25%) (1.75 nA probe current at 15 keV); WDS data collected for 30 s for each element; 180 s total analysis time (30 nA probe current at 15 keV).

9.9.5. Variance under Peak Overlap Conditions in the EDS

The microprobe assay of a specimen must provide both a mean and a variance about that mean. Since we are concerned here with microanalysis, the mean refers to the estimate of the weight concentration at a single analytical point, or some local grouping of points, from a homogeneous region of the specimen. The variance about this mean then represents the uncertainty due to counting statistics plus those aspects of the data reduction procedure that will contribute uncertainty, such as peak unraveling and continuum suppression. The accuracy of the estimate is a measure of the closeness of our estimate to the true value of the concentration. The task of predicting the variance about this estimated concentration can range from easy to quite difficult. As with the WDS, a spectrum observed with an EDS consists of characteristic and the continuum x rays. The x-ray peaks we wish to determine may overlap with the peaks from other elemental constituents of the specimen. Furthermore, the peaks are always superposed onto a smoothly varying spectrum of x rays arising from the continuous process, and both the characteristic and continuum signals are modulated by the effects of counting statistics.

The DL and variance about a measured concentration depend on the magnitude of the peak and background intensities, the degree of peak overlap, and the algorithms used to extract the required peak intensity and background intensity values below the peak. In general there is no straightforward way of estimating the quantities required for a standard statistical treatment for EDS such as was discussed for WDS in the previous sections. Therefore, some analysts, when faced with the problem of providing good error estimates, resort to the time-consuming but reliable technique of direct measurement. In this method the specimen is sampled n times each at a number of representative locations. For each of the n replicate measurements at each location, one will go through all the EDS spectral processing (removal of background and peak overlap effects) and data reduction steps (such as ZAF) required to arrive at an elemental concentration. From the n results at each location, the analyst can then predict, by conventional statistical methods, the expected variance for each of the elemental concentrations at the various, presumably representative locations. Knowing then the expected variances, the analyst can proceed with a strategy of single measurements at each analytical point in the specimen. For specimens with many phases or a wide range of compositions, this procedure can be quite time consuming.

There is an ever growing body of knowledge concerning the physics of electron–specimen interaction and of the energy-dispersive x-ray spectrometer used to detect the resulting x rays. Fiori and Swyt (1989) have suggested that it is possible to generate from first principles an x-ray spectrum which will be more than sufficiently accurate in all of the germane physical and statistical properties to represent an actual spectrum from a real specimen. From a generated spectrum one can then deduce accurate estimates of variance about the mean compositional values. One may also accurately estimate the DL of an element in a matrix without the need to produce a set of calibration standards. Furthermore, one may adjust the experimental parameters to determine the optimum ones, which produce the lowest DL. All of this can be done before presenting a specimen to the electron beam.

In this spectrum-generation process, one can enter a hypothetical set of concentrations and analytical conditions and generate a spectrum that has the correct number of counts for specific operating conditions. This spectrum does not have counting noise on it and represents the true mean spectrum. One can add counting noise, an application of the Monte Carlo method in statistics, and then go through all the procedures involved in extracting net peak areas and conversion of these to elemental concentrations. By repeatedly adding new counting noise to the same mean spectrum and repeating all the steps to get a concentration for each of the resulting spectra, one can obtain a set of measurements from which it is possible to determine with a high degree of certainty the DL and variance about a concentration mean for an element in a given chemical matrix. The variance term, is, of course, what is required to derive a sensitivity as described in the preceding section. In summary, there is a good prospect that more advanced statistical methods may allow one to obtain precision estimates, even from complex EDS spectra.

9.10. Light-Element Analysis

9.10.1. Introduction

Quantitative x-ray analysis of the low-energy (<0.7 keV) $K\alpha$ lines of the light elements is difficult in the SEM or EPMA. Table 9.18 lists the low-atomic-number elements considered in this section along with the energies and wavelengths of the $K\alpha$ lines. X-ray analysis in this energy range is a real challenge for the correction models developed for quantitative analysis since a large absorption correction is usually necessary. Unfortunately, the mass absorption coefficients for low-energy x rays are very large and the values of many of these coefficients are still

Table 9.18. Characteristics of the $K\alpha$ lines of the Light Elements

Element	Symbol	Z	λ (Å)	E (keV)
Beryllium	Be	4	114.0	0.109
Boron	B	5	67.6	0.183
Carbon	C	6	44.7	0.277
Nitrogen	N	7	31.6	0.392
Oxygen	O	8	23.62	0.525
Fluorine	F	9	18.32	0.677

not well known. The low-energy x rays are measured using large d spacing crystals in a WDS system or thin-window or windowless EDS detectors.

Although special problems, some of which are referred to in the previous paragraph, make the analysis of the light elements difficult, much progress in light element analysis has been made in the past decade. The following sections discuss the special problems encountered in light-element analysis and suggest procedures that should be followed for successful quantitative light-element analysis.

9.10.2. Operating Conditions for Light-Element Analysis

One can reduce the effect of x-ray absorption by analyzing at high take-off angles ψ and low electron-beam energies E_0. The higher the take-off angle, the shorter the path length for absorption within the specimen. The penetration of the electron beam is decreased when lower operating energies are used, and x-rays are produced closer to the surface (see Chapter 3.) Figure 9.33 shows the variation of boron $K\alpha$ intensity with electron-beam energy E_0, at a constant take-off angle for boron and several of its compounds (Shiraiwa *et al.*, 1972). A maximum in the boron $K\alpha$ intensity occurs when E_0 is 5 to 15 keV, depending on the sample. This maximum is caused by two opposing factors.

1. an increase in x-ray intensity due to increasing E_0 and overvoltage U, and
2. a decrease in intensity due to the fact that x rays are produced deeper in the sample and are more highly absorbed as the incident electron-beam energy increases.

The absorption effect clearly dominates at high values of E_0. Light-element x-ray analysis should be carried out at an electron-beam energy

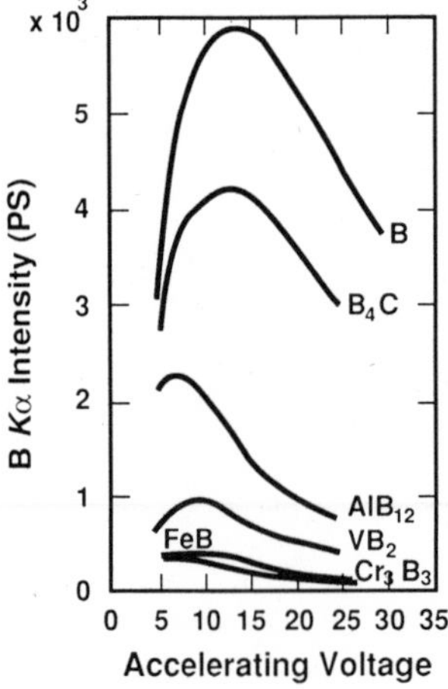

Figure 9.33. Boron $K\alpha$ intensity versus electron-beam energy E_0 for boron and several of its compounds (Shiraiwa *et al.*, 1972).

where the maximum in the intensity of the light-element x ray of interest is obtained, usually between 8 and 15 keV. In this energy range, one easily exceeds the critical excitation potential. Additionally, if the electron-beam energy is set much lower, the x rays generated may be significantly affected by surface contamination developed during specimen preparation. The operating conditions may have to be changed from optimum conditions if E_0 is so low that the characteristic x rays of the remaining elements in the sample can not be excited ($E_0 < E_c$). However, even if one uses these procedures, selecting the optimum E_0 and optimizing ψ, the effect of absorption within the sample for the low-energy x rays is still significant.

In order to increase the count rate for the light elements, one might be inclined to increase the beam current to much higher levels (~100 to 300 nA) than those used for x-ray microanalysis with the WDS (10 to 30 nA). As pointed out by Bastin and Heijligers (1990), this increase in beam current may lead to dead-time problems in the WDS system for the metal x-ray lines since these intensities are usually measured simultaneously with the light-element x-ray lines. In addition, one may run into appreciable pulse-height-shift problems for the metal x-ray lines. The dead-time problem can be avoided by either measuring the metals and light elements separately, both under optimum conditions, with obvious experimental disadvantages, or by using higher orders $n\lambda$ of the metal lines.

Pulse-shift problems can become prominent when a narrow window is applied in the pulse-height analyzer (see Chapter 5) and the differences in count rates between standard and specimen are very large. In such cases, the pulse may move out of the energy window when going from standard to specimen. These problems can be minimized if more efficient diffraction crystals, which allow for higher intensities for the light elements, are used. For the EDS system, the maximum count rate for the light elements is limited by the total x-ray intensity, all characteristic x-ray peaks plus continuum background (see Chapter 5). Therefore, raising the beam current only increases the EDS dead time beyond the useful range (20 to 30%).

9.10.3. X-Ray Spectrometers

9.10.3.1. WDS

Crystals with large d spacings are needed to diffract the long-wavelength x rays of the light elements. Traditional crystals which have been used in x-ray microanalysis are TAP, with a $2d$ spacing of 25.8 Å, which can detect O and F; lead stearate (STE), with a $2d$ spacing of 100.4 Å, which can detect B to O; and laurate, with a $2d$ spacing of 70.0 Å, which can detect C to O. Recently, layered synthetic microstructures have been developed that have been optimized for specific light elements. These large-d-spacing crystals are discussed in more detail in Chapter 5.

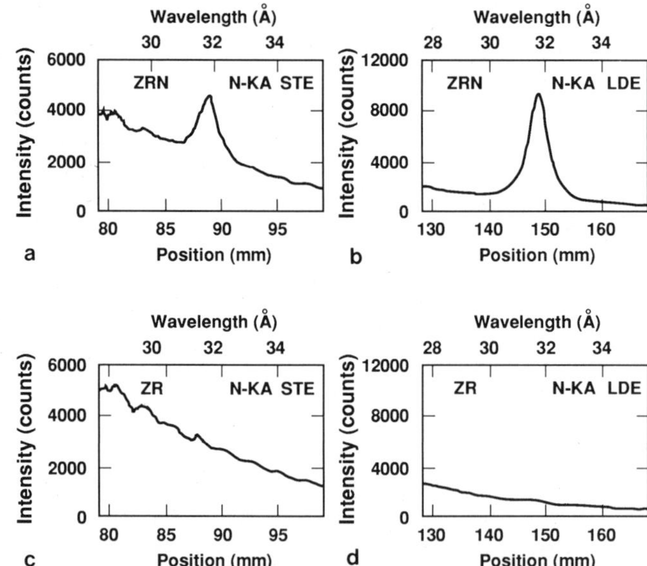

Figure 9.34. (a), (b) Nitrogen spectra (intensity versus crystal position in mm or wavelength, λ) recorded from ZrN with a conventional STE crystal and a LDE crystal, respectively (c), (d) Backgrounds measured for nitrogen from elemental zirconium with a conventional STE crystal and a LDE crystal. Experimental condiitons, 10 keV, 300 nA (Bastin and Heijligers, 1991).

Figure 9.34 shows a comparison of nitrogen spectra recorded from ZrN with a conventional STE crystal and a W/Si (LDE) synthetic multilayer crystal (Bastin and Heijligers, 1991a). The data were taken at 10 keV with a beam current of 300 nA. The LDE crystal consists of 200 pairs of alternating W (a few Å) and Si (~60 Å) layers with a $2d$ spacing of 59.8 Å. This crystal has been designed specifically for the analysis of N, O, and F. There is a large increase in peak count rate with a much higher peak-to-background ratio. In addition, the LDE crystal is not sensitive to the higher-order $(n\lambda)$Zr reflections observed with the STE crystal. Such low-intensity and smooth-background-intensity spectra from LDE makes the measurement of peak and background intensity more accurate. For measurements of B and Be, an OVH synthetic multilayer crystal consisting of a Mo/B_4C multilayer with a $2d$ spacing of 149.8 Å has been developed. Bastin and Heijligers (1991) have shown a dramatic increase of B intensity over that measured by STE of a factor of about 50. By using the OVH crystal, it may be possible to do light-element analysis for B at the same beam currents now used for the metallic elements. In this case, excessive dead-time corrections can be overcome for the metallic elements. Synthetic multilayer crystals such as Ni–C have been developed for C. The STE crystal has also been used successfully for C analysis by a number of investigators.

The proportional counter in the WDS spectrometer has been optimized for soft x rays (see Chapter 5). A flow-proportional counter is used with normal counter voltages of 1400 to 1700 V along with high

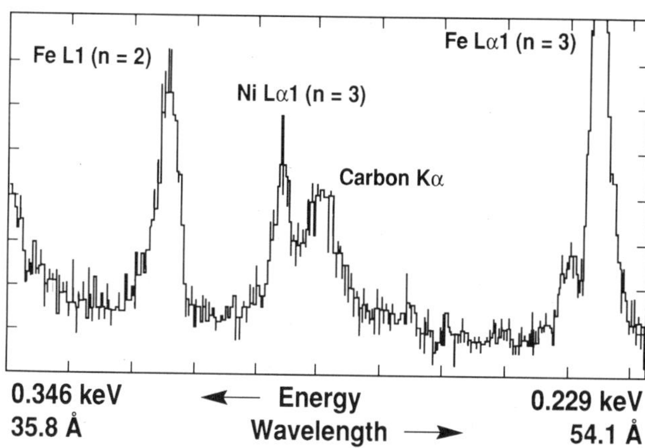

Figure 9.35. Wavelength dispersive spectrometer scan across the carbon peak of a Fe–9.9 wt% Ni–1.04 wt% C alloy with the pulse height analyzer open. Operating conditions: beam energy = 10 keV, take-off angle = 40°, beam current = 110 nA. Maximum intensity full scale = 64.

amplifier gain. The flow gases used are P–10 gas (90% argon–10% methane) or neon–methane mixtures. A thin organic film is used for an entrance window with high transmission of the light elements. The detector window, which must sustain a vacuum on one side of the membrane, is usually composed of polypropylene, formvar, or collodion.

In the WDS there is the potential of interference by higher-order metal lines with the light-element lines, especially with a lead stearate crystal, which is very efficient in transmitting higher-order reflections. Difficult elements in this respect are metals such as Cr, Mn, Fe, Ni, and the group Zr, Nb, and Mo. As pointed out by Bastin and Heijligers (1990), due to the proximity of the M_5 absorption edge, the latter group causes problems with most of the ultralight elements. As an example, one can consider the case of the third-order Ni $L\alpha$ peak, which has a $n\lambda$ (n = 3) value of 43.68 Å. The C $K\alpha$ peak falls at λ = 44.7 Å, and there is a clear overlap of the two peaks. Figure 9.35 shows a WDS wavelength scan across the carbon $K\alpha$ peak in a Fe-9.9 wt %–Ni-1.04 wt % C alloy (Lyman *et al.*, 1990). In this scan, the pulse-height analyzer was not used and was completely open. The Ni $L\alpha$ third-order peak clearly interferes with the C $K\alpha$ peak. Also observed are two $n\lambda$ L-shell peaks from the matrix Fe. Problems involving $n\lambda$ interference are usually handled by using a rather narrow window in the pulse-height analyzer of the detection system (see Chapter 5). Alternatively, WDS peak fitting can be used (see Section 7.3.2). The higher energy $L\alpha$ line from Ni and Fe is not counted by the proportional detector. Figure 9.36 gives a schematic of the pulse height distribution curves for the C $K\alpha$ peak and the third-order Ni $L\alpha$ peak. With the pulse-height analyzer window set to reject the higher-energy Ni $L\alpha$ peak, only the C $K\alpha$ peak is measured by the detection system. Figure 9.37 shows a WDS wavelength scan across the carbon $K\alpha$ peak in the same Fe–Ni–C alloy with the pulse-height

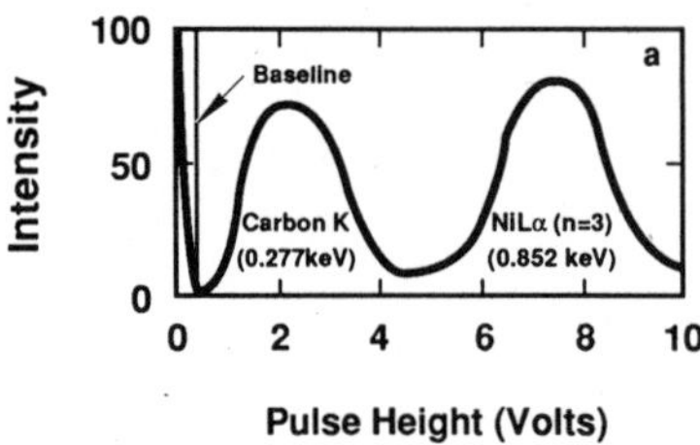

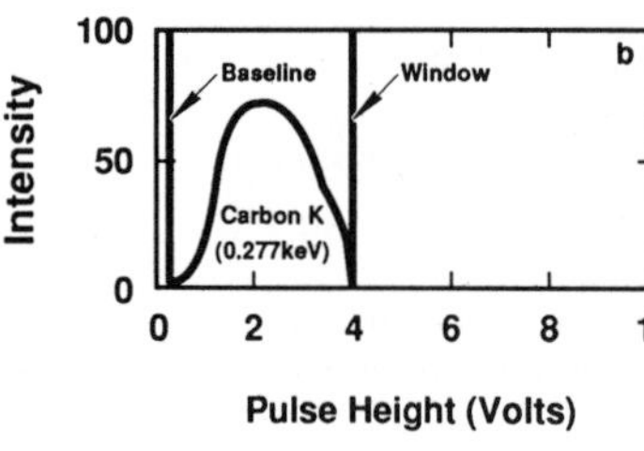

Figure 9.36. Schematic pulse height distribution curves. (a) Pulse height distribution curve for Fig. 9.35, showing acceptance of overlapping carbon and nickel signals. (b) Pulse height distribution curve for Fig. 9.37 showing rejection of iron and nickel signals by restriction of the energy of the pulses accepted by the counting system.

analyzer optimized to reject the iron and nickel signals. The Ni third-order $L\alpha$ peak is removed. However, the Fe L peaks from the matrix Fe are not completely removed, making background measurements difficult. Using the pulse-height analyzer to remove interfering peaks also involves the risk of increasing the effects of small pulse shifts in the detector as a result of large differences in count rates between standard and sample. Serious errors in the measured k ratios can result because of the large differences in count rates between elemental standards and the sample.

Complications in light-element analysis using the WDS may also arise because of the direct overlap of L spectra from the heavier metals.

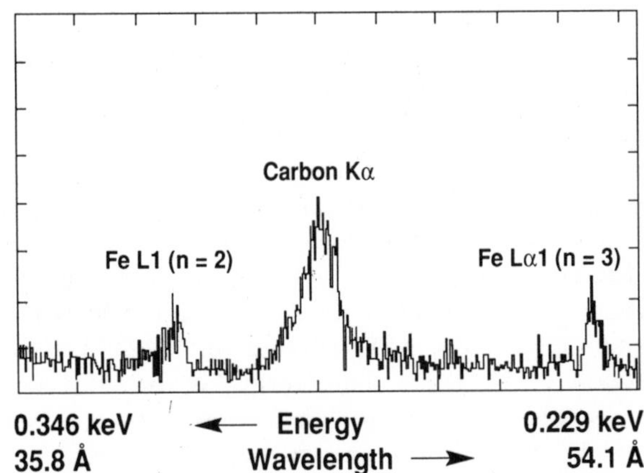

Figure 9.37. WDS scan across the carbon peak of a Fe–9.9 wt% Ni–1.04 wt% Cu alloy with the pulse height analyzer adjusted to allow only the carbon peak to be accepted. Operating conditions: beam energy = 10 keV, take-off angle = 40°, beam current = 110 nA. Maximum intensity full scale = 64.

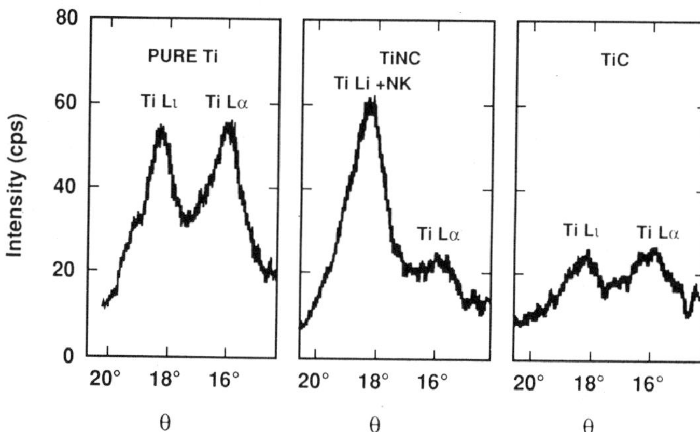

Figure 9.38. Comparison of the Ti L spectra (intensity versus diffraction angle θ) obtained from pure Ti (left), TiNC (center), and TiC (right), measured at 10 keV (Duncumb and Melford, 1966).

One example is the case of titanium carbonitride inclusions (TiNC) in steels. Figure 9.38 shows a comparison of the Ti L spectra obtained from TiNC, TiC, and pure Ti (Duncumb and Melford, 1966). The titanium carbonitride phase gives a more intense peak at the Ti Ll wavelength than that of pure Ti. This peak contains mainly Ti Ll at 31.4 Å together with a small amount of nitrogen $K\alpha$ emission indistinguishable from it at 31.6 Å. The Ti $L\alpha$ line at 27.4 Å is heavily absorbed by nitrogen and is about one-third as intense as that from pure Ti. Quantitative analysis of Ti, C, and N in these inclusions is therefore very complex. However, Duncumb and Melford (1966) were able to determine the Ti and C contents of the inclusion directly by measuring the Ti $K\alpha$ and C $K\alpha$ intensities directly and employing a TiC standard to obtain appropriate k ratios.

9.10.3.2. EDS

The EDS can be used to measure the characteristic peaks of the light elements (see Chapter 5). Using the EDS, one does not have to worry about multiple-wavelength interference ($n\lambda$). However, the method suffers from inherently poor spectral resolution (about 100 eV) and a number of spectrum-processing problems (Love and Scott, 1985). With spectral resolution of about 100 eV, there can be overlap of the light-element peaks with each other (see Table 9.18) and with the L or M lines of metals in the samples of interest. Figure 9.39 shows the interference of the Fe L peaks with the N $K\alpha$ peak in a steel containing ~6 wt % nitrogen in a nitride layer (Love and Scott, 1985). A background can be established under the N $K\alpha$ peak, and a peak intensity can be measured. However, in general, the accuracy of the measurement and the minimum detectability limit of the nitrogen in the metal matrix is not as good as that which can be obtained by the use of the wavelength-dispersive detector. The EDS has a useful role to play in light-element

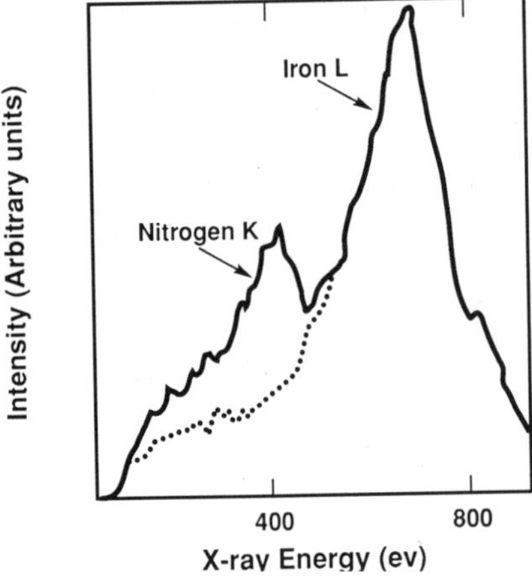

Figure 9.39. Nitrogen $K\alpha$ peak from a nitrided steel measured with an EDS, showing interferences, from Fe L peak: Operating conditions 7 keV, 10 nA (Love and Scott, 1985).

analysis, especially when one is dealing with beam-sensitive materials where low beam currents are necessary. In the cases for which suitable spectrum-processing methods are available, light-element peak x-ray intensities can often be obtained for quantitative analysis. Section 7.3 gives a detailed analysis of the peak overlap correction for EDS. These methods are directly applicable to EDS peaks in the light-element energy range.

9.10.4. Chemical Bonding Shifts

9.10.4.1. WDS

For the light elements, the x-ray emission spectra consist mainly of a single band produced by the transition of a valence electron to a vacancy in the K shell. As pointed out by Fisher and Baun (1967), the valence electrons are the ones most affected by chemical combination and the emission band can and does reflect the often large effects of changes in chemical bonding between atoms. These changes are signified by wavelength shifts, by increases or decreases in the relative intensities of various lines or bands, and by alteration of shape. Such shifts may cause problems when quantitative light-element analysis is desired. Figure 9.40 shows the C K band from carbon deposited by the electron beam as well as the C K band of electrode-grade graphite and various carbides (Holliday, 1967). The wavelength shift for the carbides relative to graphite is significant and can easily be observed with the WDS. A similar wavelength shift has been observed for B $K\alpha$, as shown in Fig. 5.16.

Large errors in k-ratio measurements can be made when peak shape alterations occur, either in the sample, the standard, or both. As discussed by Bastin and Heijligers (1991) in the case of boron, these errors can even be made when the same compound is used as a standard; the influence of the crystallographic orientation can cause variations in

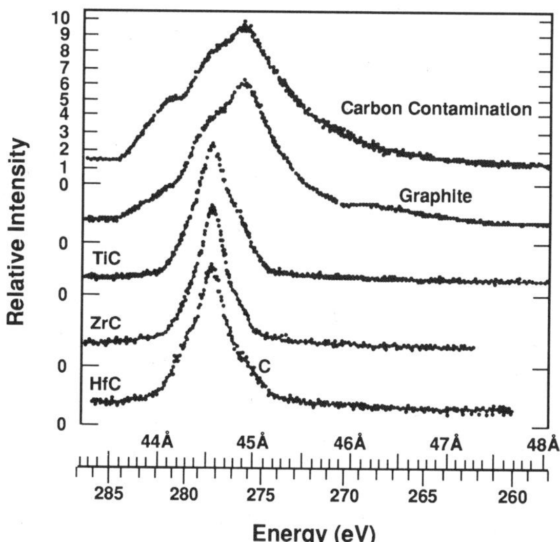

Figure 9.40. C K spectra from various C-containing samples obtained at an energy of 4 keV. The carbon contamination was deposited from the electron beam (Holliday, 1967).

peak intensity on the order of tens of percent. Such peak shape changes are worse for the lightest element, boron, and slightly less for carbon, where crystallographic effects and polarization phenomena are absent. For nitrogen and oxygen, the effects of shape alterations rapidly decrease. For oxygen, the effect is hardly noticeable, at least for the LDE crystal (Bastin and Heijligers, 1991).

To measure the intensity of the light elements in both the sample and standards when wavelength shifts occur, it is critical to measure the integral intensity under the characteristic x-ray peaks. Integral measurements are performed in WDS by scanning the spectrometer in small steps (<10 eV) across the characteristic peaks. As noted above, these wavelength shifts are more significant as the atomic number of the light element decreases. It is very time consuming to measure integral intensities of the characteristic x-ray peaks for every analysis. Bastin and Heijligers (1984, 1986) have introduced the area peak factor analysis method that has led to a considerable reduction in time and effort to measure the integral intensity of x-ray peaks. Section 7.3.2 discusses accurate peak fitting methods for WDS spectrometry of low-energy peaks.

The area peak factor (APF) is the ratio between the k-ratio measured by using the integral intensity from sample and standard and the k-ratio measured by using the peak intensity from sample and standard. The APF is valid only for a given compound with respect to a given standard and for a given spectrometer with its unique light-element crystal, but on a given instrument it is independent of E_0. Once an APF has been determined, future measurements of that compound can be carried out by measuring the peak intensity from sample and standard. Multiplication of the k-ratio obtained from peak intensity measurements

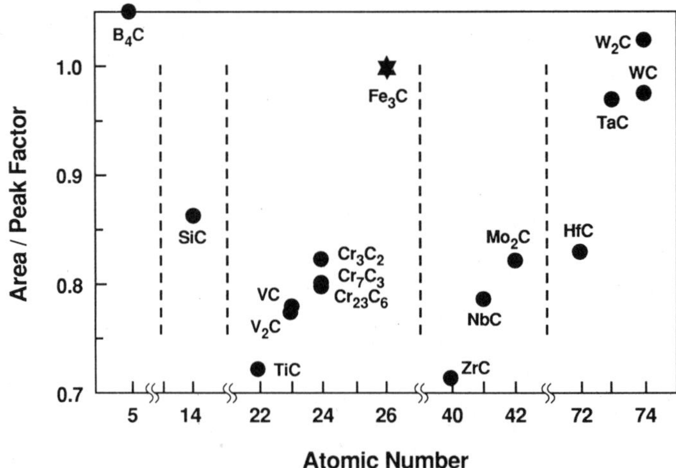

Figure 9.41. Area peak factors for C $K\alpha$ radiation emitted from binary carbides, relative to Fe_3C, as a function of the atomic number of the metal in the carbide. The value for Fe_3C is equal to 1.0. STE crystal, 10 keV, 300 nA. Dashed vertical lines mark the transition from one period to another in the periodic table (Bastin and Heijligers, 1991).

by the appropriate APF will yield the correct k-ratio. Figure 9.41 shows the area peak factor measured by Bastin and Heijligers (1990) for C $K\alpha$ in binary carbides. The standard is Fe_3C, and the APF for that carbide is 1.0. Values of APF between 0.7 and 1.1 are measured. When the APF is plotted versus the atomic number of the metal in the carbide as in Fig. 9.41, a saw-tooth like variation is shown throughout the periodic system. The minimum values of APF are held by very strong carbide formers, such as Ti, and Zr, while weaker carbide formers such as Fe have values much closer to unity.

It is important to use the area peak factor method to measure peak intensities when peak shape alterations are observed, for a given light element, between sample and standard. Fortunately, multilayer crystals appear to be less sensitive to peak shape alterations than STE or other crystals used for the light elements. The APF values measured on one specific WDS spectrometer cannot be transferred to another one because the APF is strongly related to the resolution of the crystal and spectrometer. However, there is evidence that for instruments comparable in terms of crystal and resolution, the same APF values apply within experimental error. This apparent agreement should, however, be checked before tabulated values of the APF values (See Bastin and Heijligers, 1990a,b, 1991a) are used.

9.10.4.2. EDS

Because of the poorer energy resolution of the EDS, peak shifts due to chemical bonding cannot be observed. Therefore the analyst need not be concerned about energy shifts in measuring intensity ratios for the light elements. For EDS, the peak is automatically measured as an integral and there is no need to be concerned with measuring area peak factors. In EDS analysis of the light elements, procedures for dealing with peak overlaps and the measurement of background are of much more importance than considerations of peak shifts.

9.10.5. Standards for Light-Element Analysis

A primary standard for light-element analysis must give a strong, stable, and reproducible peak intensity. For example, in carbon analysis, neither spectrographic nor pyrolitic graphite provides reproducible carbon x-ray intensities. However, various metallic carbides give reproducible carbon intensities and are adequate carbon standards; for example, Fe_3C for steels. In addition, the expected count rate for carbon is of the same order of magnitude as that in most binary carbides. This favorable situation would rule out problems of pulse shifts in cases of pulse-height discrimination. Furthermore, Fe_3C can easily be prepared and has a fixed composition since it is a line compound.

Reliable quantitative results can be obtained by comparison with standards whose composition is close to that of the specimen. Figure 9.42 shows the calibration curves for the x-ray analysis of carbon $K\alpha$ at 10 keV using a STE crystal in well-characterized alloys of Fe, Fe–10-wt % Ni and Fe–20-wt % Ni containing specific concentrations of carbon (Fisher and Farningham, 1972). The carbon intensity ratio is the carbon $K\alpha$ line intensity from a given alloy standard less its background, divided by the line intensity from a Cr_3C_2 carbon standard less *its* background. At a given carbon level, the addition of Ni to the steel standard decreases the C intensity, since Ni increases the mass absorption coefficient for C $K\alpha$.

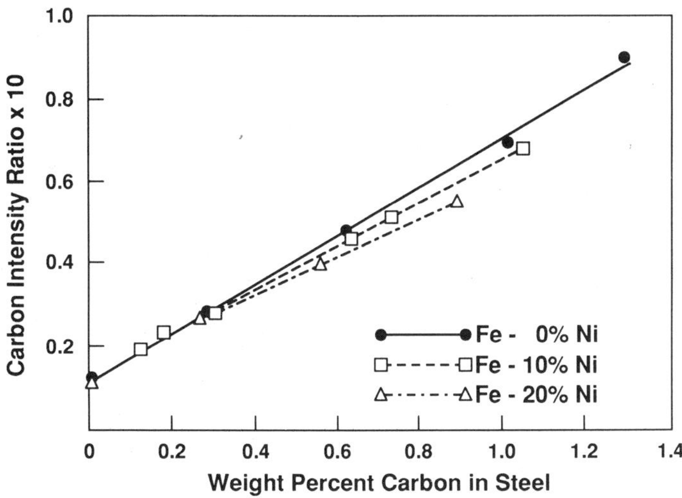

Figure 9.42. Calibration curves for EMPA of carbon in nickel steels (Fisher and Farningham, 1972).

9.10.6. Surface Contamination

A sample subjected to electron bombardment in a diffusion-pumped vacuum gradually becomes covered with a contamination layer due to polymerization, under the action of the beam, of organic matter adsorbed on the surface (Castaing, 1960). The organic molecules come from the oil vapors of the vacuum pumps and the outgassing of any organic material present in the instrument. The effect is not very troublesome unless the deposited layer absorbs the emitted x rays to a significant extent. For low-energy x-rays (<0.7 keV), the absorption of the x-ray lines by the contamination layer can be severe. The problem, in the case of carbon analysis, is increased because the contamination layer contains carbon. This situation leads to the observation of an increasing carbon $K\alpha$ count rate as a function of analysis time.

Two methods have been used to avoid a contamination layer. Castaing and Descamps (1954) showed that directing a low-pressure jet of gas onto the specimen at the region bombarded by the electron beam suppressed contamination. When air is introduced into the vicinity of the sample, oxygen oxidizes the hot carbon deposit and the high-energy electron beam acts to produce an ion bombardment–sputtering condition. Air jets have been installed on various SEM instruments and can be build in any laboratory for a nominal cost (Duerr and Ogilvie, 1972; Bastin and Heijligers, 1990). Another method to avoid a contamination layer is to provide a surface within the SEM–EPMA that is cold relative to the surface of the specimen. Organic molecules tend to collect on the colder surface rather than on the specimen. The cold surface or cold finger must, however, be placed very close to the specimen. Cold fingers have been installed on various instruments and have reduced the contamination rate. As shown by Bastin and Heijligers (1990), it appears

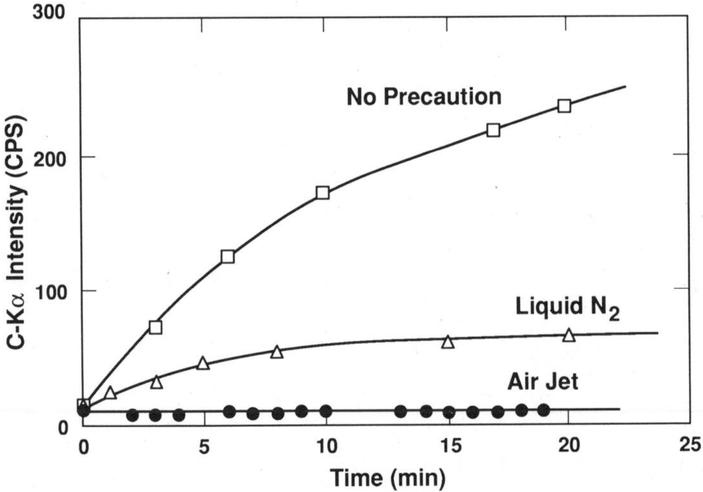

Figure 9.43. Carbon contamination rate on polished copper without anti contamination device, with liquid nitrogen cold finger, and with air jet. Conditions: 10 keV, 100 nA. Oil diffusion pumped vacuum system (Bastin and Heijligers, 1990).

that the maximum effect can be obtained by using an air jet (see Fig. 9.43). Experience has shown that one must minimize the effect of the contamination layer (air jet or liquid nitrogen or both) in light-element quantitative analysis for all the light elements.

When examining Fig. 9.42 for the C $K\alpha$ calibration curves in Fe–Ni–C alloys, one can observe that the C $K\alpha$ intensity ratio (carbon background subtracted) does not go to zero as expected when the carbon content is zero, even though no contamination layer is observed. It is probable that a thin carbon film had been present on the Fe standard prior to the analysis. This specimen-borne contamination is caused by numerous factors such as specimen preparation and exposure to the laboratory atmosphere. Even in ion-pumped instruments, contamination layers are observed by Auger electron spectrometry on the specimens unless the specimen surfaces in the instrument are cleaned. These contamination layers are not caused by the instrument but are due to the mobilization of specimen-borne contamination. The presence of an oxygen surface layer on gold, measured by Bastin and Heijligers (1991) with an LDE crystal, where the solubility of oxygen is negligible, shows the presence of a specimen-borne (oxygen rich) surface layer on the sample when it is placed in the SEM–EPMA. Effects of the presence of a specimen-borne contamination layer can be minimized if samples and standards are prepared as close to the time of analysis as possible and precautions are taken in cleaning the surface of the sample to minimize leaving a residue on it that can affect the x-ray analysis. Detailed specimen-preparation procedures to minimize the development of surface-contamination layers are discussed by Bastin and Heijligers (1990). As long as specimen-borne contamination can be characterized and is reproducible, its contribution to the analysis can be subtracted from the measured data at an appropriate time.

Conducting surface layers are often evaporated or sputtered onto specimens. Although it is necessary to eliminate charge buildup in the specimen, the effect of these surface layers in light-element analysis can be very significant. If the specimen is a conductor, such as a metallic sample or a conducting oxide, it is important to avoid using any conducting surface layer. A conducting sample can be mounted in a conducting medium such as bakelite containing a fine dispersion of copper particles. If a coating is necessary, it is important to choose its composition carefully. For analysis of carbon, a carbon coating cannot be used, and an Al coating is suggested. For oxygen analysis, carbon is a strong O $K\alpha$ absorber, and copper has been used as a surface coating. The copper coating often leads to deviations in the measured k-ratio over time. As reported by Bastin and Heijligers (1991b), the electron-beam energy decreases across the copper film and the overvoltage for the production of characteristic x-rays in the specimen drops. It is suggested that the analysis be performed without a coating at a higher electron-beam energy or by burning a hole in a carbon coating by using an airjet. The effect of nonconducting samples on the generation of light-element x-rays has not been explored in detail. Therefore the analyst must be aware of potential problems with this type of specimen.

9.10.7. Measurement of Background Intensities

9.10.7.1. WDS

The usual procedure for measuring the background intensity for analysis with the WDS is to measure the background at equal distances (energy or θ angle) on either side of the peak and interpolate between these values (see Chapter 7). This type of extrapolation can be very dangerous in light-element analysis as a result of the complexity of the backgrounds, which are rarely flat and straight (see Figs. 9.34 and 9.37). Fortunately, the synthetic multilayer crystals are less sensitive to higher-order peaks, and the continuum background is smooth (see Fig. 9.34). The principal causes of the complex backgrounds are the presence of a residual carbon contamination peak and the presence of remnants of interfering higher-order metal x-ray lines. As discussed by Bastin and Heijligers (1990), higher-order metal x-ray lines can not be completely removed by the pulse-height analyzer. The most difficult elements appear to be Cr, Zr, Nb, Mo, and the second-order O $K\alpha$ peak that occurs at 47.24 Å.

It is clear therefore that the measurement of the background for the light elements is by no means simple and straightforward. The commonly used procedure of linear background interpolation yields only small errors when the count rates are high and the peak-to-background ratios are also high. This is true, of course, if there are no peaks in the continuum region used for interpolation. Therefore, it is always necessary to measure the background intensity versus the diffraction angle θ around the light-element peak to make sure that no extra peaks are present. If such peaks are present in the background, then more careful attention must paid to the background.

Measurement of the background is very important if one attempts to determine low levels of carbon or oxygen (<1 wt %) or if interfering peaks are present. In those cases, ignoring the structure of the background and of contamination effects may lead to errors of a factor of 2 or more. It is often useful to obtain samples, called "blanks", which do not contain the light element of interest. For example, to obtain measurements of C levels below 1 wt %, a measurement of residual contamination level (blank or background level) at the position of the characteristic peak can be made on a decarburized piece of high-purity iron. This type of measurement was made during the analysis of the Fe–Ni–C standards shown in Fig. 9.42. As discussed briefly in Section 9.10.7, the residual carbon level is not zero and a small C intensity is measured on the high-purity iron sample. This residual carbon level represents the background for the measurement. Therefore, for low-concentration measurements of the light elements, suitable blank samples must be developed for the analysis or the measured background around the peak must be relatively smooth so that a background intensity can be measured near the peak position.

9.10.7.2. EDS

A detailed discussion of the measurement of background intensities for EDS is given in Section 7.2.1 and will not be repeated here. In general, two approaches are available for background subtraction: calculated or measured continuum distribution (background modeling), and mathematical modification of the continuum spectrum (background filtering). Both approaches have been used successfully. The specific problems encountered in light-element continuum background measurement are the close proximity of the noise peak to the energy range of interest, the major effect of absorption of the thin window, Au layer, and Si dead layer on continuum intensity (see Chapter 5), and the effect of the tails of overlapping and other low-energy x-ray peaks.

9.10.8. Quantitation Procedures for the Light Elements

The k-ratios for a given light element are difficult to obtain, as discussed in the previous sections. We need to optimize the beam energy and take-off angle to minimize absorption. Furthermore, for the WDS we must choose x-ray diffraction crystals to maximize peak and peak-to-background ratios and we must eliminate peak overlaps in sample and standard, using pulse-height analysis if necessary. For the WDS, one must also use integral peak intensities if wavelength shifts occur. For the EDS, considerations of peak energy shifts and overlapping $n\lambda$ peaks are unnecessary. We also have to choose a standard which is not only well characterized, but is conducting (if possible), with minimum peak shifts. The surface contamination must be minimized by using an air jet, liquid nitrogen cold finger or both, and conducting surface layers must be avoided if possible. Finally we must investigate the nature of the background intensity around the light element peak position in sample and standard and apply suitable techniques for the WDS or EDS systems to measure the background accurately.

We will now consider the quantification process, which enables the analyst to take the measured k-ratios and calculate the concentration of the light element(s) and the other components of the sample. The ZAF and $\phi(\rho z)$ calculation schemes have been discussed in some detail in Sections 9.2 and 9.3. However these calculation schemes are very severely tested when used for light-element analysis. The main reason for concern is illustrated in Fig. 9.44 which shows calculated $\phi(\rho z)$ curves for B $K\alpha$ radiation in AlB_2 with a beam energy of 10 keV (Bastin and Heijligers, 1986). Both the $\phi(\rho z)_{gen}$ and $\phi(\rho z)_{em}$ curves are shown. The extremely high absorption of the B $K\alpha$ is illustrated. In this case the emitted intensity comes from only one quarter of the generated x-ray depth with the majority of the emitted intensity coming from within 100 nm (1000 Å) of the surface. The real test of the calculation schemes is that the $\phi(\rho z)$ curve must be well known close to the surface of the sample and that, because of the large amount of absorption, the mass absorption coefficient $\mu/\rho)^i$, must be very well known.

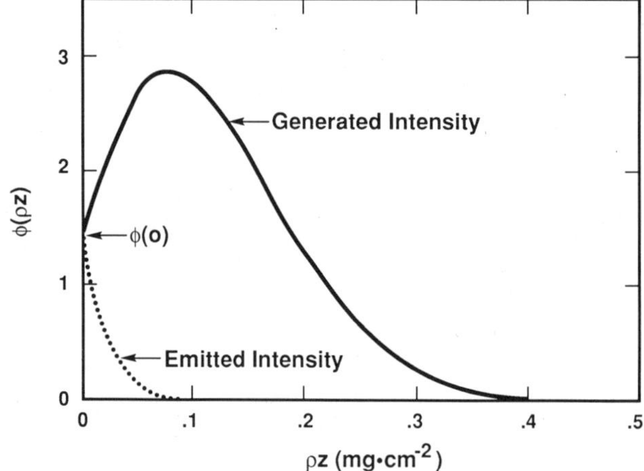

Figure 9.44. Calculated $\phi(\rho z)$ curves for B $K\alpha$ radiation in AlB_2 for a beam energy of 10 keV (from Bastin and Heijligers, 1986).

Table 9.19 illustrates the variation of mass absorption coefficients available in the literature for C $K\alpha$ x rays. Differences of more than 100% are not uncommon. In general, for the light elements, a variation of 1% in the mass absorption coefficients leads to a variation of 1% in the calculated ZAF factors. The large uncertainty in the mass absorption coefficients introduces a large uncertainty into the results no matter which matrix correction program is used to calculate the ZAF factors.

Bastin and Heijligers (1991) discuss methods by which more consistent mass absorption coefficients for the light elements can be obtained. Basically, the approach is to determine a large data base for analyses of B, C, N, and O in well-characterized standards under a wide variety of beam energies (4 to 30 keV). Almost 1000 measurements have now been made for these elements. Using these data, the appropriate mass absorption coefficients can be back calculated. If correct values are obtained, the correction scheme should give consistent values at different beam energies. Bastin and Heijligers (1991) show that mass absorption coefficients obtained in this way give improvements in the results

Table 9.19. Mass Absorption Coefficients for Carbon $K\alpha$ X-Rays According to Various Sources (g/cm^2)

Absorber	Barkalow *et al* 1972	Henke *et al.* 1982	Bastin and Heijligers 1990	Pouchou and Pichoir 1991
Fe	15,000	13,900	13,500	13,500
W	18,000	18,800	18,600	18,000
Mo	19,000	16,400	20,600	20,500
Cr	10,000	10,600	10,700	10,700
V	15,000	8,840	9,700	8,850
C	2,300	2,350	2,373	2,170

calculated for *all* the matrix correction programs normally applied to the light elements, not just those of the authors. Pouchou and Pichoir (1991) have also used the data of Bastin and Heijligers (1986; 1990; 1991) to obtain a consistent set of mass absorption coefficients for B, C, and N. The values for the C $K\alpha$ mass absorption coefficients in selected elements, from Bastin and Heijligers (1990) and Pouchou and Pichoir (1991), shown in Table 9.19 are remarkably consistent. Nevertheless, the methods used to obtain more consistent mass absorption coefficients cannot yield values of mass absorption coefficients with better than 1 to 2% accuracy. Although improvement in the accuracy of the mass absorption coefficients for the light elements has been made, it is important for the user to be sure to know the value of the light-element mass absorption coefficient as well as the source of the mass absorption coefficient selected for the calculation of the ZAF factors. Because of the major variations reported in mass absorption coefficients for the light elements, the analyst should check the values of the mass absorption coefficients being used in the matrix correction program against the available coefficients on a case-by-case basis. The light-element mass absorption coefficients listed in Table 14.3 were taken from various compilations of Heinrich (1986).

Sections 9.2 and 9.3 discuss the various ZAF and $\phi(\rho z)$ correction procedures used with flat polished samples analyzed with the electron beam normal to the sample. Because the light-element x rays are emitted so close to the sample's surface, the x-ray generation function used in either correction method must be very accurate. The Philibert–Duncumb–Heinrich absorption correction commonly used in the ZAF correction method assumes a ϕ_0 value of zero. Figure 9.44 shows that ϕ_0 is greater than 1.0. While this absorption correction can be used successfully in many analysis situations, it is not applicable to low electron-beam energies and condiitons where x rays are emitted close to the surface. The PDH absorption correction has been successfully used for the light elements only in cases where the correction is similar for sample and standard and where the errors in the correction essentially cancel out when the ratio $f(\chi)/f(\chi)^*$ is calculated.

The recently developed $\phi(\rho z)$ methods have been increasingly successful in calculating accurate light-element analyses. The details of the $\phi(\rho z)$ method have been given in Section 9.3, and the $\phi(\rho z)$ method can be applied to carefully measured k-ratios for the light elements as well as the other elements present in the sample. In recent review papers, Pouchou and Pichoir (1991) and Bastin and Heijligers (1991) show that, with sufficient care in the measurements and a good correction program in combination with consistent mass absorption coefficients, it is possible to obtain very good quantitative results (better than 2 to 4% accuracy) for light elements, especially in the electron-beam energy range of 8 to 20 keV. Although specific authors favor their own approaches to quantitation using the $\phi(\rho z)$ method, it appears that several schemes produce acceptable results. One can expect even further progress in developing correct $\phi(\rho z)$ curves for the light elements as more $\phi(\rho z)$ curves are measured and more light-element measurements, under correct measuring conditions, are made.

Appendix 9.1: Equations for the α, β, γ, and $\phi(0)$ Terms of the Packwood–Brown $\phi(\rho z)$ Equation

The following equations use the terms given by Bastin and Heijligers (1990a,b; 1991):

9.A1.1. $\phi(0)$

The equation for $\phi(0)$ is given by Pouchou and Pichoir (1987):

$$\phi(0) = 1 + 3.3(1 - 1/U_0^{\gamma})\bar{\eta}^{1.2}, \qquad (9.\text{A1})$$

where

$$U_0 = E_0/E_c \qquad (9.\text{A2})$$

for the x ray of interest and

$$\gamma = 2 - 2.3\bar{\eta}. \qquad (9.\text{A3})$$

(Note this is not the same γ referred to in the Packwood–Brown expression.) The mean backscatter coefficient $\bar{\eta}$ is

$$\bar{\eta} = 1.75 \times 10^{-3}\bar{Z}_p + 0.37[1 - \exp(-0.015\bar{Z}_p^{1.3})], \qquad (9.\text{A4})$$

where $\bar{Z}_p$ is a partial weight-fraction average atomic number,

$$\bar{Z}_p = \left(\sum_i C_i Z_i^{0.5}\right)^2 \qquad (9.\text{A5})$$

and i represents the elements present in the sample.

9.A1.2. α

The equation for α is

$$\alpha_{ij} = \frac{2.1614 \times 10^5 Z_j^{1.163}}{(U_{0i} - 1)^{0.5} E_{0i}^{1.25} A_j}\left[\frac{\ln(1.166E_{0i}/J_j)}{E_{ci}}\right]^{0.5}, \qquad (9.\text{A6})$$

where Z, A, and J are atomic number, atomic weight, and ionization potential of the matrix element j and E_0, E_c, and U_0 are the beam energy, critical excitation energy, and overvoltage ratio for the x-ray line i of interest.

The ionization potential of the matrix element j is given by

$$J = Z[10.04 + 8.25\exp(-Z/11.22)]\ (\text{eV}). \qquad (9.\text{A7})$$

For a compound target, a matrix of α_{ij} values (α for the x-ray line of element i interacting with element j of the matrix) is calculated. The α_i

value for the x-ray line of element i in a multielement target containing j elements is

$$\frac{1}{\alpha_i} = \frac{\Sigma_j C_j \dfrac{Z_j}{A_j} \dfrac{1}{\alpha_{ij}}}{\Sigma_j C_j \dfrac{Z_j}{A_j}}. \tag{9.A8}$$

9.A1.3. γ

The equation for γ is,
for U_0 of element $i \leq 6$,

$$\gamma = 3.98352 U_{0i}^{-0.0516861} \times (1.276233 - U_{0i}^{-1.25558Z^{-0.1424549}}). \tag{9.A9}$$

For U_0 of element $i > 6$,

$$\gamma = 2.814333 U_{0i}^{0.262702Z^{-0.1614454}} \tag{9.A10}$$

For a compound target, the weight-fraction averaged atomic number is substituted for Z in both cases.

In order to accommodate the change in ionization cross-section with atomic number for light-element x rays (E_c of element $i < 0.7$ keV), it is necessary to multiply α [Eq. (9.A9) or (9.A10)] by the term

$$\frac{E_c}{-4.1878 \times 10^{-2} + 1.05975 E_c}. \tag{9.A11}$$

9.A1.4. β

The calculation of β is more complex (!) and involves an evaluation of the integral F:

$$F = \int_0^\infty \phi_i(\rho z)\, d\rho z. \tag{9.35}$$

As discussed in Section 9.3.4 and by using Eqs. (9.36) and (9.37), the total intensity generated in a specimen (or standard) is

$$F = \frac{[\gamma - (\gamma - \phi(0)) \cdot R(\beta/2\alpha)]\sqrt{\pi}}{2\alpha}, \tag{9.A12}$$

where $R(\beta/2\alpha)$ represents the value of the fifth-order polynomial in the approximation for the complementary error function for the argument $(\beta/2\alpha)$ in parentheses. Equation (9.A.12) is the formal solution of the Gaussian integral of $\phi(\rho z)$ between 0 and infinity in closed form. After rearranging, we can give $R(\beta/2\alpha)$ as

$$R(\beta/2\alpha) = \frac{\gamma - 2\alpha F/\sqrt{\pi}}{\gamma - \phi(0)} \tag{9.A13}$$

If we know the value of F, we can calculate the value of β from Eq. (9.A13) using known values of α, γ, and $\phi(0)$ and Eqs. (9.A1) and

Table 9.A1. Values of $R(\beta/2\alpha) = x$ and $\beta/2\alpha$[a]

Values of x	$\beta/2\alpha$
$0.9 < x < 1$	$0.9628832 - 0.9642440x$
$0.8 < x \leq 0.9$	$1.122405 - 1.141942x$
$0.7 < x \leq 0.8$	$13.43810 \exp(-5.180503x)$
$0.57 < x \leq 0.7$	$5.909606 \exp(-4.015891x)$
$0.306 < x \leq 0.57$	$4.852357 \exp(-3.680818x)$
$0.102 < x \leq 0.306$	$(1 - 0.5379956x)/(1.685638x)$
$0.056 < x \leq 0.102$	$(1 - 1.043744x)/(1.604820x)$
$0.03165 < x \leq 0.056$	$(1 - 2.749786x)/(1.447465x)$
$0 < x \leq 0.03165$	$(1 - 4.894396x)/(1.341313x)$

[a] Bastin and Heijligers (1990a,b; 1991a).

(9.A8)–(9.A11). First one must evaluate F using the atomic number correction of Pouchou and Pichoir (1987). The total intensity generated F under the $\phi(\rho z)$ curve is

$$F = \text{P.I.}/Q_i(E_0), \tag{9.A14}$$

where P.I. is the primary generated intensity I_{igen}, from the sample equal to R/S [Eq. (9.11)] and $Q_i(E_0)$ is the ionization cross-section of element i, which determines the intensity $I_i(\Delta\rho z)$ for the thin-film tracer. The equations for η (backscatter coefficient), $\bar{W}$ (averaged reduced energy of backscattered electrons), R (backscattering factor), and J (ionization potential) are those used by Pouchou and Pichoir (1987), and the reader is referred to the original paper for the detailed equations.

Once the value of F is obtained, we can use Eq. (9.A13) to calculate β. To determine β, we first calculate $R(\beta/2\alpha)$ from Eq. (9.A13) and then determine $\beta/2\alpha$. Bastin and Heijligers (1990a,b; 1991a) developed a procedure to solve for $\beta/2\alpha$ by substituting $R(\beta/2\alpha) = x$, obtained from Eq. (9.A13) into a table which divided $R(\beta/2\alpha) = x$ into nine regions. If we substitute $x = R(\beta/2\alpha)$, we obtain the best fit. Table 9.A1 gives the appropriate values.

Appendix 9.2: Solutions for the Atomic Number and Absorption Corrections

Once α, β, γ, and $\phi(0)$ are calculated (Bastin and Heijligers, 1990a,b, 1991a) according to the method in Appendix 9.1, the Z_i and Z_iA_i corrections can be obtained. The essential feature of this procedure is that the time-consuming process of numerical integration of the Gaussian functions can be avoided by writing

$$Z_i = \frac{\{\gamma - [\gamma - \phi(0)]R(\beta/2\alpha)\}\,\alpha^{-1}}{\{\gamma - [\gamma - \phi(0)]R(\beta/2\alpha)\}^*\,\alpha^{*-1}}, \tag{9.36}$$

in which Z_i represents the ratio between the integrals of the generated intensities in standard and specimen for element i. Likewise, the ratio between the emitted intensities can be written as

$$Z_iA_i = \frac{[\gamma R(\chi/2\alpha) - (\gamma - \phi(0))R((\beta + \chi)/2\alpha)]\alpha^{-1}}{\gamma R(\chi/2\alpha) - (\gamma - \phi(0))R((\beta + \chi)/2\alpha]^* \alpha^{*-1}}, \qquad (9.38)$$

in which R is a fifth-degree polynomial used in the approximation of the error function, which comes in when the integrals are solved in closed form through a Laplace transformation.

The polynomial R is described by

$$R = a_1 t + a_2 t^2 + a_3 t^3 + a_4 t^4 + a_5 t^5, \qquad (9.A15)$$

with

$$t = 1/(1 + px),$$

where

$$p = 0.3275911$$
$$a_1 = 0.254829592$$
$$a_2 = -0.284496736$$
$$a_3 = 1.421413741$$
$$a_4 = -1.453152027$$
$$a_5 = 1.061405429.$$

The input value for x is either $\beta/2\alpha$, $\chi/2\alpha$, or $(\beta + \chi)/2\alpha$, as required from Eq. 9.36 or Eq. 9.38.

10

Compositional Imaging

10.1. Introduction

Scanning electron micrographs prepared with the backscattered-electron signal provide direct information on compositional heterogeneity through the mechanism of atomic number contrast. Such images are quite useful for characterizing microstructures, but the compositional information that they contain is nonspecific. That is, while the sequence of the average atomic numbers of the phases present can be recognized and classified by the relative gray levels in the image, that image conveys no specific information to identify the elements present or the concentration levels.

The use of x-ray signals derived from spectrometers to prepare scanning images containing element-specific information was developed very early in the history of the EPMA (Cosslett and Duncumb, 1956; Duncumb, 1957). Despite the difficulties in obtaining satisfactory images because of the considerable time penalty imposed by this mode of operation, the method of "x-ray area scanning" or "dot mapping" rapidly became one of the most popular modes of x-ray analysis. The reasons for this popularity arise from the elemental specificity of the elemental information and from the power of visual presentation of information. The human visual process is able to rapidly assimilate information presented in the form of images, and often a problem can be solved merely by determining where in a microstructure a particular constituent is localized.

With proper care in operating procedure, dot maps can be an extremely useful mode of operation to produce valuable qualitative information. However, the dot-mapping procedure is subject to artifacts which become significant when minor or trace elements are to be mapped. To overcome these limitations, the technique of quantitative compositional mapping has been developed, in which a complete compositional analysis is performed at each location in a matrix scan (Newburty *et al.*, 1990a;b). When displayed on an analog device such as a CRT, the gray or color scale of a particular pixel in a compositional map is related to the actual concentration of each constituent.

10.2. Analog X-Ray Area Scanning (Dot Mapping)

10.2.1. Procedure

X-ray area scanning is performed by the same technique as conventional SEM imaging: the beam scans the specimen in a raster pattern in the usual fashion, and the synchronously scanned CRT is intensity modulated by the signal derived from the detector, in this case either the energy-dispersive or wavelength-dispersive x-ray spectrometer. The difficulty with this approach is the low intensity of the x-ray signal. Compared to the backscattered-electron signal, the x-ray signal from an element is lower by a factor of 10,000 to 100,000 or more. The x-ray signal is generally too weak, except in the case of a very-high-concentration (>25%) constituent for which there exists an efficiently excited x-ray line, to allow the depiction of a continuous gray scale (such as we could derive from the x-ray ratemeter signal). During the scanning process in the dot-mapping mode, whenever an x-ray is detected in a preselected energy range by the x-ray spectrometer, the corresponding beam location on the CRT is marked by raising the intensity of the CRT beam to full brightness (white). A dot is thus created on the display to note the detection and mark the location of the x-ray event, and this display is recorded photographically. As shown in Fig. 10.1, the image thus created has the appearance of a field of fine white dots, hence the term "dot mapping"; the procedure is also referred to as "area scanning."

To produce acceptable results such as those illustrated in Fig. 10.1, the dot-mapping procedure requires careful attention to the instrumental operating conditions:

X-Ray Excitation. We wish to maximize the count rate consistent with the spectrometer characteristics, so we choose a beam energy sufficiently high to achieve an overvoltage of at least $U = 2$ for the x-ray line(s) of interest, and a beam current sufficient to produce the maximum acceptable deadtime, e.g., 30–40% with an EDS or 100,000 cps for a WDS on a bulk pure-element standard. The elements of interest in the unknown are at lower concentrations and produce proportionally lower x-ray count rates. For WDS, we could choose to increase the beam current on the unknown still further, since all other x-ray peaks from the specimen are excluded by the diffraction process. For EDS, this option is not available, because the limiting dead time is determined by the entire x-ray spectrum rather than just the peak of interest for mapping.

Number of Dots Required for High-Quality Maps. In conventional SEM imaging with electron signals, a photograph is normally recorded in a single sweep of the scan with a frame time of 50 to 100 s. Because of the low x-ray signal rate, the image accumulation must be made for substantially longer times. Experience has shown that to obtain high-quality dot maps, 10^5–10^6 x-ray pulses must be accumulated over the entire image field, with the optimum number highly dependent on the uniformity of the spatial distribution of the constituents being mapped (Heinrich, 1975). A good procedure is to collect for total counts rather than total time, scanning continuously over the frame and using a running integration on the x-ray spectrometer counts to determine the cutoff point. An added advantage of this approach is that any changes in operating conditions during the long time necessary to accumulate the dot map will be integrated evenly over the entire recorded image, rather than producing a noticeable drift in the dot map, which would be the case for a single frame exposure.

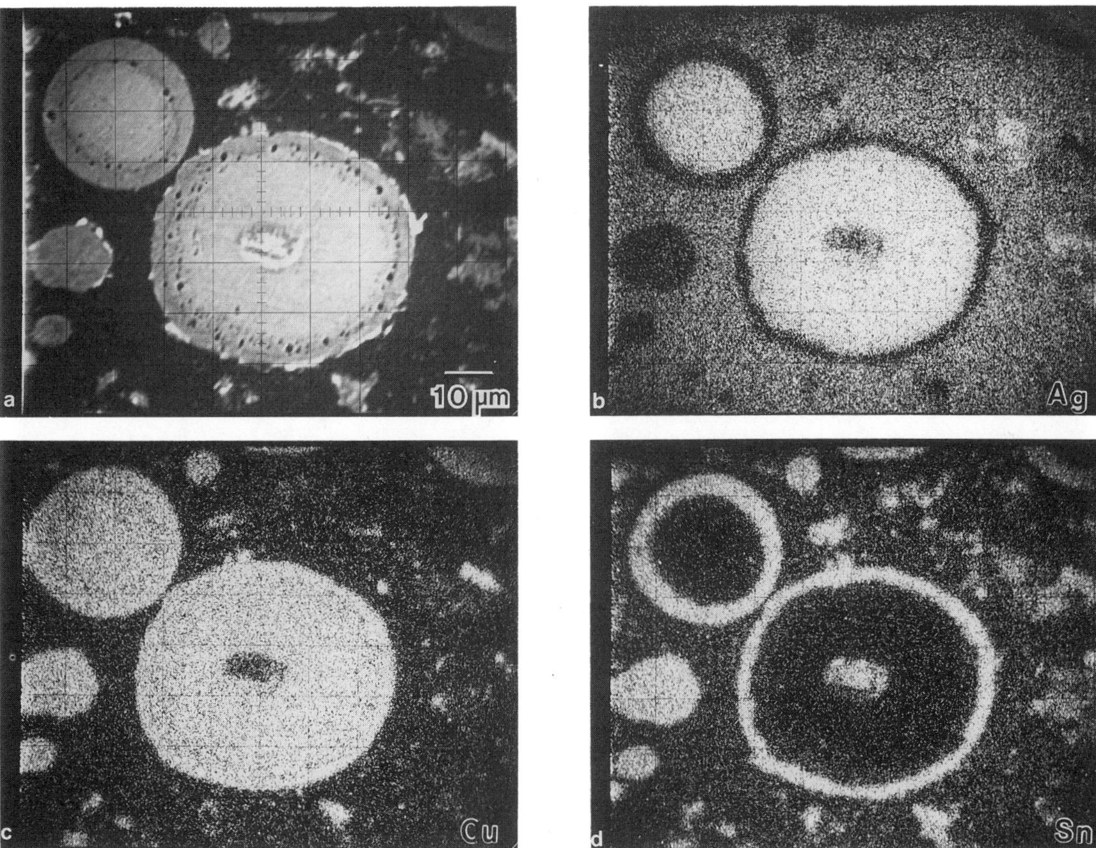

Figure 10.1. X-ray dot map (area scan) of a ternary alloy: (a) Specimen-current (inverted contrast) image showing atomic number contrast; (b) Dot map of silver ($L\alpha$); (c) Dot map of copper ($K\alpha$); (d) Dot map of tin ($L\alpha$). The dot maps have been prepared with the characteristic x-ray signal derived from a wavelength-dispersive x-ray spectrometer. (Courtesy of Robert Myklebust, NIST.)

Achieving satisfactory results with dot mapping often requires trial and error. Figure 10.2 illustrates a typical dilemma. In Fig. 10.2a, accumulation of 10,000 dots gives good contrast in the higher-concentration areas, but when 40,000 dots are accumulated to develop contrast in lower-concentration areas (Fig. 10.2b), the higher-concentration areas are completely saturated. (Note: The calcium in Fig. 10.2 is localized over a small portion of the image. If this constituent were more evenly dispersed over the image, proportionally more counts would have been needed to produce the contrast shown.)

Accumulation Time for Dot Mapping. A considerable time penalty must be paid to achieve high-quality dot maps. For the optimum case of a wavelength-dispersive x-ray spectrometer and a high probe current of 100 nA or more, the count rate from a pure standard can reach 5×10^4–1×10^5 cps. If the element of interest is present in the unknown at a level of 10 wt %, the count rate is reduced

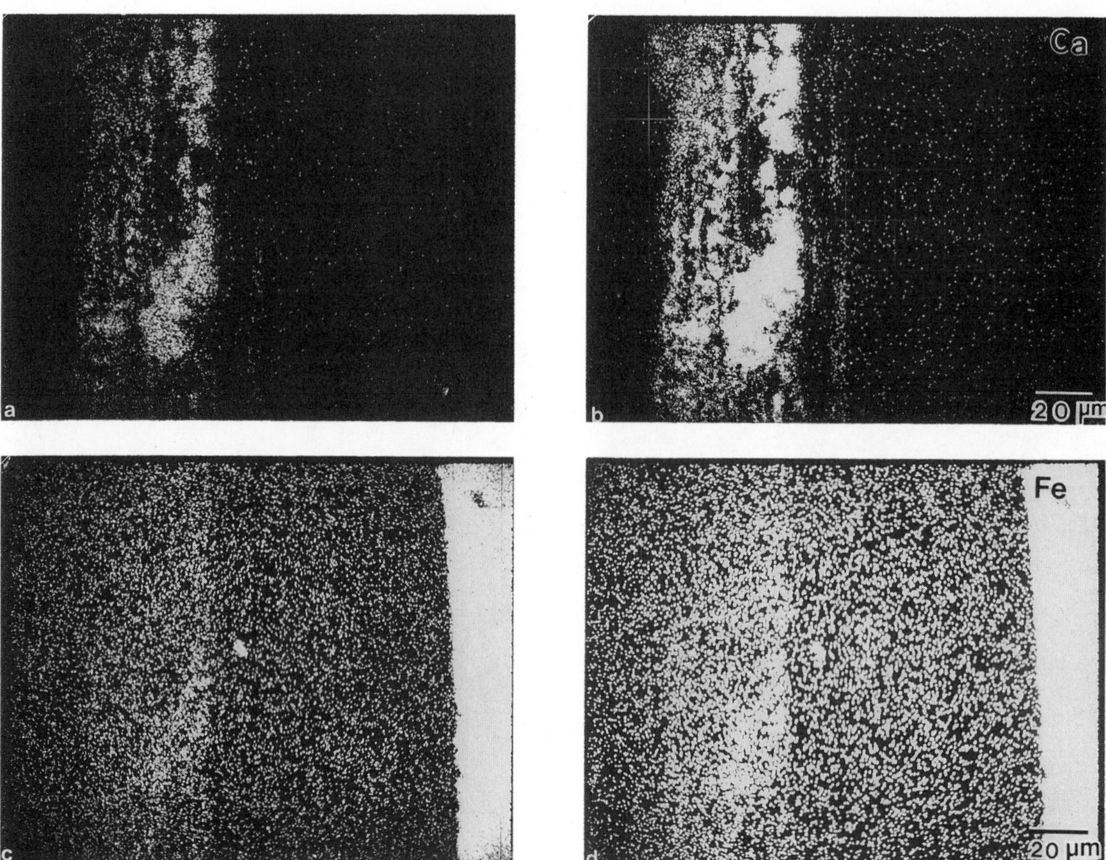

Figure 10.2. Choice of recording parameters in x-ray dot mapping. Effects of number of dots: (a) Good contrast is found in high-concentration region with 10,000 dots recorded; (b) further accumulation to 40,000 dots leads to improved contrast in lower-concentration area at the expense of saturation in higher-concentration region. Choice of dot brightness: (c) Properly adjusted CRT dot; (d) dot adjusted to be too bright (note loss of contrast in image.) Specimen: coating on steel; $E_0 = 20$ keV. (a) and (b) Ca $K\alpha$; (c) and (d) Fe $K\alpha$. (Courtesy of Ryna Marinenko, NIST.)

to 5×10^3 cps, neglecting matrix effects, and an integrated count of 10^6 x rays can be accumulated in 200 s. For a constituent present at the 1 wt % level, the scanning time would have to be increased to 2000 s. If the x-ray line is less efficiently excited or the beam current is reduced, the accumulation times must be increased accordingly.

For energy-dispersive spectrometry, it is important to choose carefully the proper EDS pulse-processing time (or amplifier time constant, see Chapter 5). If the peaks of interest are well separated in the EDS, then we can accept a poorer resolution to obtain a higher limiting count rate. The typical limiting count rate, even with the poorest resolution (shortest amplifier time constant), is about 10,000 cps, if we wish to avoid possible count-rate artifacts, described below. For EDS, this count rate applies to the entire energy range of the incident beam, since photons from all excited peaks and the x-ray continuum all contribute to the dead time. The count rate from a constituent present at the level of 10 wt % may be only 100–500 cps, depending on the excitation of other elements in the specimen, so the accumulation time for 10^6 x rays is 2000 to 10,000 s. We therefore tend to accept lower-quality maps with EDS because of these count-rate limitations. For concentrations lower than 10 wt %, the accumulation times can become prohibitive. If the spectrum is complex, so that higher resolution and longer pulse-processing times are used, the limiting count rates will be lower, and accumulation times longer.

Adjustment of the CRT Conditions. To achieve the best dot maps, the analyst must carefully adjust the recording conditions of the CRT, usually labelled "brightness" and "contrast." The ideal dot just reaches the level of saturating the CRT phosphor. Such a dot is very fine and near the resolution limit of the human eye, and when adequate counts are accumulated, the resulting dot maps show good spatial resolution. However, if the constituent is present at a low concentration, it may not be practical to spend the time to accumulate sufficient counts, and the dot map prepared with such fine dots will appear dark. Under such circumstances, the CRT brightness and contrast are often increased so that the dot "blooms" and enlarges, becoming more readily visible. Such an image appears brighter, but the spatial resolution is degraded. These effects are illustrated in Fig. 10.2c and d, where a consequence of brighter, larger dots is seen to be a loss of resolution and contrast.

10.2.2. Limitations and Artifacts

While dot mapping is a useful technique, it is subject to several significant limitations:

Qualitative, Not Quantitative Information. The dot map is only qualitative in nature, conveying the spatial location of constituents but not the amount present. The critical information for quantitative analysis, namely, the count rate at each point in the image, is lost in the dot map recording process. If any particular pixel is examined, the analytical information at that pixel consists of either "present" (white dot) or "absent" (no dot). Since no gray-scale information exists, quantitiatve information cannot be represented. The area density of dots, as seen in Fig 10.1 and 10.2, obviously suggests the variation in composition, but the information is useful only when comparing areas within an image which are much larger than single pixels. When a feature covering only a small area is examined, the noise in the map overwhelms changes due to specimen composition.

Sequential Recording. Dot maps are usually recorded for one constituent at a time directly on film media. Although parallel recording is feasible, most systems contain only one photo-record CRT. Considering the time required to produce a dot map of a single constituent, sequentially mapping for multiple constituents can impose a prohibitive time investment.

Lack of Flexibility for Subsequent Processing. Recording on film media greatly reduces the flexibility for subsequent processing of the information. For example, registration of multiple images for color superposition on film is difficult.

WDS Defocusing, EDS Decollimation. To prepare a dot map, the beam is typically scanned in a raster pattern which carries it off the coincident point of the optic axis of the microscope and the WDS spectrometer focusing or the EDS collimation axes. For WDS, as

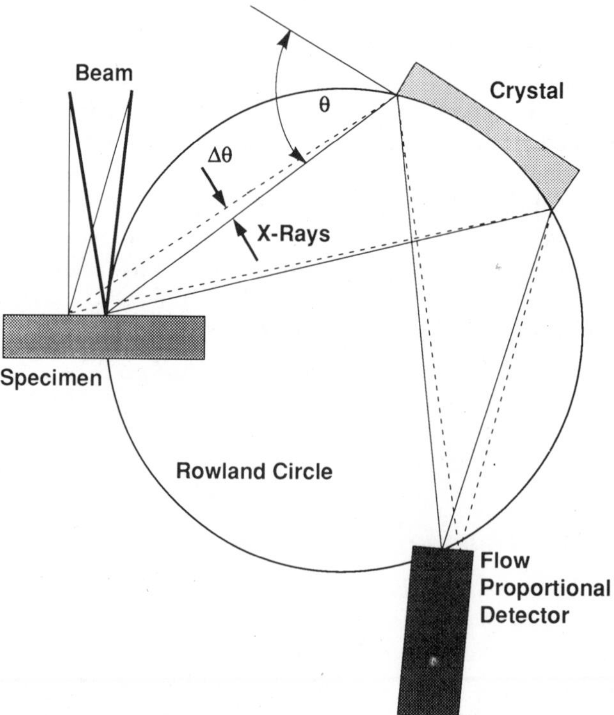

Figure 10.3. WDS defocusing during mapping. As the beam is scanned off the optic axis of the WDS, the displacement in the specimen plane is equivalent to a change in the angle away from the Bragg angle θ_B of the x rays approaching the crystal by an angle $\Delta\theta$.

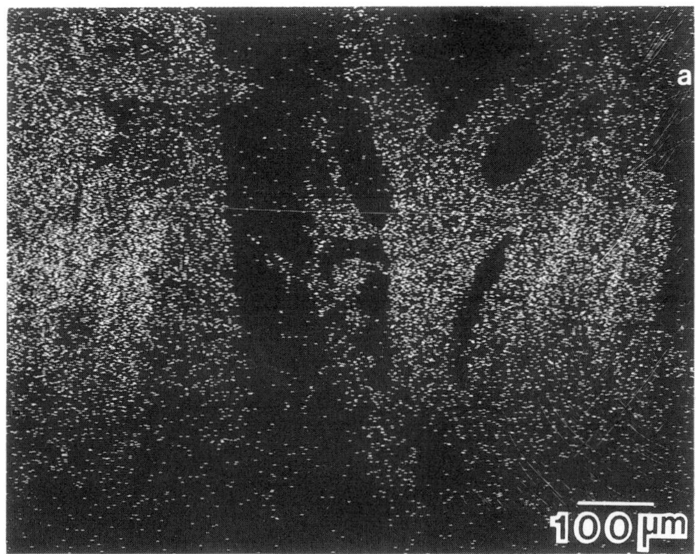

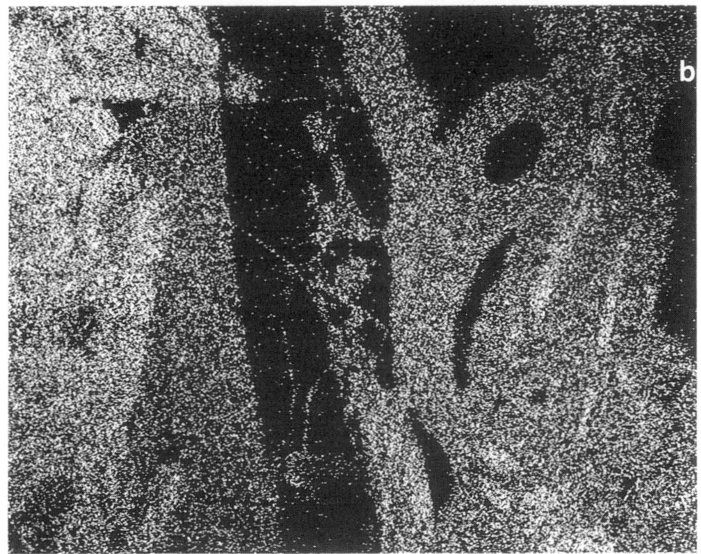

Figure 10.4. (a) Example of a dot map with defocusing evident. The band of intensity corresponds to the exact focus. The band axis is parallel to the width dimension of the diffraction crystal. Aluminum $K\alpha$ from a geological sample. (b) Correction of the defocusing by dynamic crystal rocking. The diffraction crystal was rocked during the scan to move the band of intensity from top to bottom in synchronism. (Cameca Instruments, 1974.)

shown in Fig. 10.3, a consequence of this beam motion is that the x-ray source at the specimen moves off the Rowland circle, so the Bragg condition for diffraction is not maintained during the scan. Because x-ray diffraction peaks in the WDS are quite narrow in angular range, the detected intensity falls sharply as the Bragg condition is lost, a situation called defocusing, shown in Fig. 10.4a for

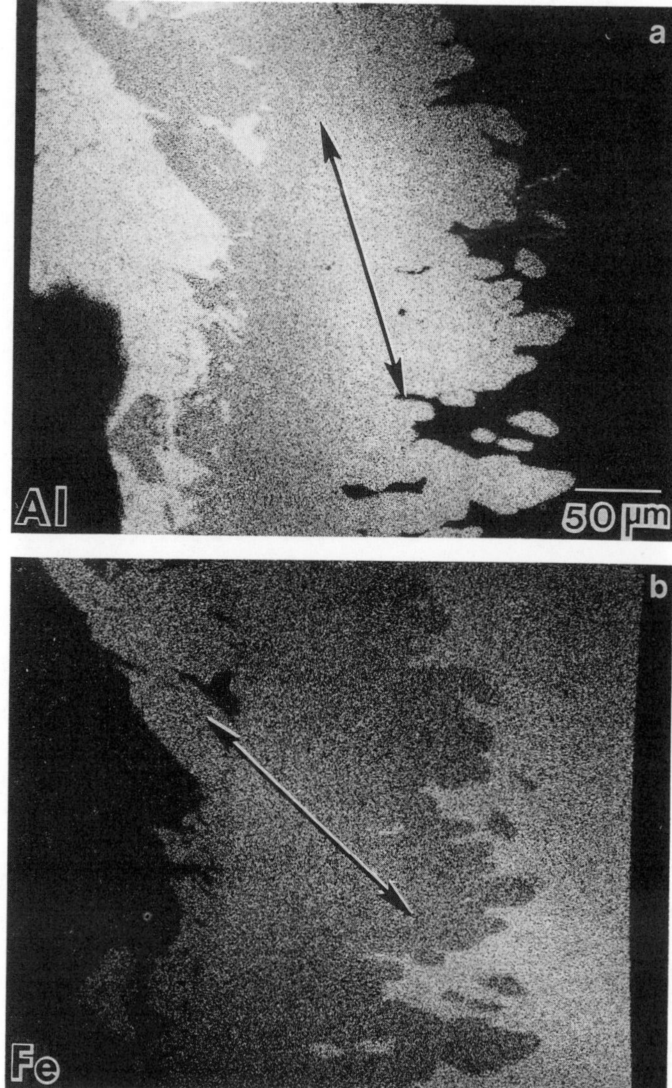

Figure 10.5. Example of the WDS defocusing artifact as seen in dot maps of an iron–aluminum electrical connection interface. The arrow marks the axis of the defocus band. (a) Aluminum $K\alpha$ map. (b) Iron $K\alpha$ map. Note the nonuniformity of the intensity and the loss in contrast near the bottom of the aluminum map; note also the different orientation of the focus bands in the two maps because of the different positions of the wavelength spectrometers relative to the specimen.

a vertical spectrometer. However, the Bragg condition is maintained for beam positions on the specimen that are parallel to the width of the diffracting crystal. A band of constant transmission is thus seen in a dot map, with the intensity falling rapidly on either side. This defocusing effect becomes more pronounced as the magnification is reduced and the scan excursion becomes greater, becoming extremely noticeable below approximately 500X. The defocusing artifact can dominate images and in some cases actually overwhelm the compositional contrast in the dot map, as shown in Fig. 10.5. The aluminum and iron dot maps show an uneven intensity distribution that interferes with the true contrast, even for these major constituents. In the iron map, the spectrometer transmission is a maximum along the diagonal indicated. The two iron-containing phases each have nearly uniform composition, but because of defocusing the contrast between the phases is satisfactory to show fine details of the interface between the phases only in the lower middle portion of the image, and it is nearly lost at the top and bottom.

A practical solution to defocusing is to move the band of maximum transmission to coincide with the scan position. In practice, the scan is rotated to bring the scan line parallel to the spectrometer's focusing band, and the band is moved synchronously with the frame scan by rocking the spectrometer crystal (Cameca Instruments, 1974). The resulting map is uniform over much greater distances (lower magnifications), as shown in Fig. 10.4b.

EDS spectrometers operate on the line of sight and are not focusing devices, so the defocusing artifact does not exist. However, the EDS system is commonly equipped with a collimator to restrict unwanted collection of remotely generated x rays. This collimator typically limits the view of the specimen to a few millimeters laterally from the optic axis. If the beam is moved sufficiently far off axis to achieve low magnifications, <50X, the solid angle of collection of the spectrometer diminishes relative to that obtained when the beam is located on the optic axis and the geometric collection efficiency decreases, producing a drop in intensity in an x-ray dot map.

Count Rate Effects. The long pulse-processing time of the EDS leads to significant dead time effects. As shown in Fig. 5.29, the EDS output is paralyzable due to dead time. That is, as the rate of photons arriving at the detector increases, the output of discrete pulses from the detector or amplifier first increases linearly, but eventually a maximum output count rate is reached beyond which the output actually decreases with further increases in the input photon rate. If a sample consists of two phases with quite different concentration levels, then the x-ray count rates from areas with the different concentrations can be thought of as occupying two linearly scaled locations along the input pulse rate axis. However, because of dead time, if the beam current is so high that the input count rate from the high-concentration phase is "over the hump" of the dead-time response curve, then the high concentration phase produces fewer output pulses than the low-concentration phase, reversing the apparent chemical contrast in the dot map. In practice, even more complicated effects may be encountered. Because the EDS detector's dead time is created by photons of all energies reaching the detector, the observed output pulse rate from a preselected energy window is actually susceptible to count-rate effects from the whole spectrum, including both characteristic peaks and continuum background.

Lack of Background Correction: Detection Limits. In the conventional analog dot-mapping procedure, no background correction is possible since any x ray in the acceptance window of the spectrometer must be counted. The limit of detection in the dot mapping mode is poor. Characteristic and continuum x rays which occur in the energy acceptance window of the WDS or EDS spectrometer are counted with equal weight. With WDS, detection limits in the dot mapping mode are approximately 0.5–1 wt %, while for EDS systems, which have a much poorer peak-to-background ratio, the limit is approximately 2–5 wt %.

Lack of Background Correction: Continuum Image Artifact. A second consequence of not performing a background correction is the susceptibility of dot mapping to an image artifact in which compositional contrast may seem to exist when an element is not actually present in the sample. This artifact can occur in maps where a background correction is not applied, because of the dependence of the x-ray continuum on the average atomic number of the target [Kramers' expression, Eq. (3.31)]. Thus, if a specimen consists of phases of different composition, the background at any x-ray energy varies proportionally with the local average atomic number. This effect can be ignored when mapping major constituents (>10 wt %), but the continuum artifact can dominate minor- or trace-level maps. Because of the poorer peak-to-background ratio, this artifact is more significant for EDS spectrometry than for WDS. A typical EDS P/B ratio for the FWHM measurement window is in the range 30/1 to 50/1. At mass concentration levels of 1–3 wt %, the background contributes as much intensity as the characteristic peak. If the analyte is present at the same concentration in two phases whose average atomic numbers differ by a factor of two, there is an apparent doubling in the concentration of the analyte due to the change in continuum between the two phases. An example of such false contrast due to the continuum artifact is shown in Fig. 10.6. The specimen is aluminum–copper (eutectic). The maps have been prepared with the energy-dispersive x-ray spectrometer, with an energy window at the position of aluminum, copper, and scandium. The contrast in the scandium image suggests that scandium is present and is preferentially segregated to the copper-rich phase. In reality, no scandium is present and the false contrast is entirely due to the atomic number dependence of the continuum. When a proper background correction is applied, the apparent scandium contrast is eliminated. (Myklebust *et al.*, 1989). For WDS with its superior peak-to-background ratio, the continuum artifact becomes significant below 1 wt %.

Poor Contrast Sensitivity at High Concentration Levels. The dot mapping technique has poor contrast sensitivity, particularly at high concentration. The quality of a dot map depends on the concentration level of a constituent. With sufficient time expenditure, it is possible to prepare a dot map of a region containing a constituent at 5 wt % against a background which does not contain that con-

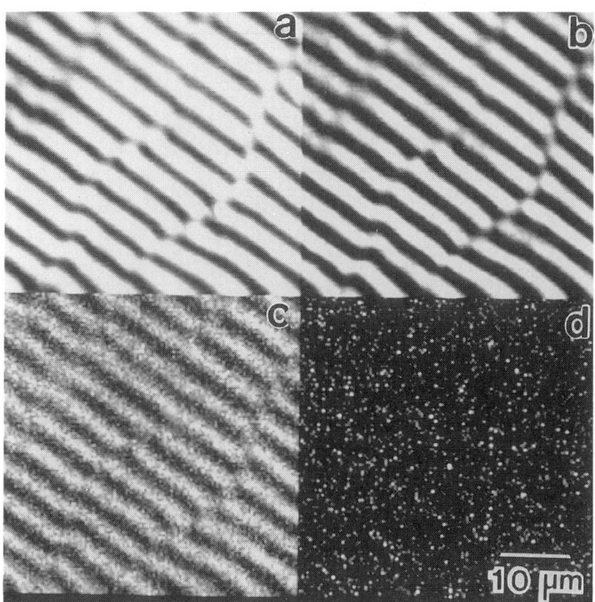

Figure 10.6. Example of false compositional contrast, a mapping artifact which occurs because of the dependence of the x-ray bremstrahhlung on the composition of the specimen. The specimen is aluminum–copper (eutectic). A map has been prepared with the energy-dispersive x-ray spectrometer, with an energy window at the position of (a) aluminum, (b) copper, and (c) scandium. The contrast in image (c) suggests that scandium is present and is preferentially segregated in the copper-rich phase. But no scandium is present. (d) When a proper background correction is applied, the apparent scandium contrast is eliminated (Myklebust *et al.*, 1989).

stituent. However, it is virtually impossible to visualize 5-wt % increases in concentration above a generally high concentration level, for example, 50 wt %. A high density of dots is produced by the high concentration level of the analyzed element, reducing the visibility of the contrast resulting from any small changes in concentration, as illustrated in Fig. 10.2a and b.

10.3. Digital Compositional Mapping

10.3.1. Principles

Digital compositional mapping consists of collecting count-rate data and carrying out a complete quantitative analysis at every pixel in a scan. Corrections are performed for every instrumental and matrix effect, so that a set of data matrices of concentration values is produced. These digital data matrices are reconstituted into analog images for viewing by creating a gray or color scale in which the specific gray level or color value is related to the numerical concentration(s). The following steps must be performed

1. data collection,
2. dead-time correction,
3. defocusing correction (WDS) or decollimation correction (EDS),
4. background correction,
5. standardization (k-value calculation),
6. matrix correction, and
7. quantitative display.

10.3.1.1. Data Collection

As in the conventional quantitative x-ray microanalysis of single beam locations, the collection of the actual count rates at each pixel is the starting point for quantitative compositional mapping. The beam is addressed under digital control to a specific location on the specimen, and the count rate is measured over a specified interval ranging from milliseconds to several seconds. If multiple-wavelength dispersive spectrometers are available, the measurement is performed simultaneously, with each spectrometer measuring a different element. If an EDS is used, the complete spectrum may be collected for subsequent processing, which gives the most flexible analysis approach, or if mass storage is limited, the data collection may be restricted to specific energy windows covering the constituents of interest and background regions.

10.3.1.2. Dead-time Correction

Dead-time corrections are applied in the same manner as in conventional fixed-beam analysis. WDS data are corrected for dead-time through the use of Eq. (5.6). EDS data are collected for a specified livetime, so that the EDS dead-time correction circuit automatically compensates for local differences in count rate.

10.3.1.3. Defocusing or Decollimation Correction

As in the case of analog dot mapping, the WDS defocusing and EDS decollimation effects present a significant instrumental artifact, which must be corrected if meaningful compositional maps are to be obtained.

Four approaches have been demonstrated to correct WDS defocusing (Newbury *et al.*, 1990a; b):

1. Stage Scanning. The beam can be fixed and the specimen stage scanned mechanically in the x–y plane (and, if necessary, in the z-direction as well with automatic focusing of the optical microscope) so that all measurements are made on the optic axis of the system. With optically encoded stages, the reproducibility of stage motions is better than 1 μm. For low-magnification maps, recorded at magnifications below 100X, stage scanning is the preferred mode of operation to minimize spectrometer artifacts in the final map.

2. Crystal Rocking. The spectrometer crystal can be rocked in synchronism with the x–y scan to maintain the focus line of the spectrometer to be coincident with the beam position. With full digital computer control, multiple WDS spectrometers can be positioned simultaneously. This is the same procedure used for analog dot mapping, as shown in Fig. 10.4b.

Methods (1) and (2) have the advantage that all measurements are made at the same statistical variance for the same concentration.

3. Standard Mapping. The specimen and standard can be scanned under the same conditions of magnification so the defocusing is identical in both maps. When the k-value map is calculated with intensity measurements from corresponding pixels in the two maps, the spectrometer transmission factor cancels in the k-ratio, and defocusing is eliminated.

4. Peak Modelling. An exact geometric correspondence exists between the shape of the defocusing response along a line perpendicular to the line of maximum spectrometer transmission and the shape of an x-ray peak generated by scanning the WDS spectrometer. This correspondence forms the basis for calculating a correction factor for any beam position from a scan of the x-ray peak.

Methods (3) and (4) require no mechanical motions, but are applicable only above approximately 100X. Below 100X, the defocusing becomes so severe that the spectrometer is completely detuned off the peak. A second defect in these methods arises at low magnification because the count rate depends on the position in the scan field. The statistical variance of the count from individual pixels differs across the image even if the concentration is constant. The variance in the background count becomes important and can influence the visibility of true contrast from trace constituents at low magnifications.

EDS decollimation effects at low magnification can be corrected by determining the intensity response function on a large, flat pure-element standard such as a silicon wafer. Since the correction is purely geometric and doesn't depend on the energy of the x-rays, the same correction function applies for all subsequent measurements.

10.3.1.4. Background Correction

The problem of background correction in compositional mapping is critical if meaningful images of minor- and trace-element distributions are to be obtained free from the artifacts of the continuum dependence on composition.

For EDS, strategies similar to those used in conventional fixed-beam analysis can be applied in mapping (see Chapter 7). For the background filtering techniques, the complete spectrum collected at a pixel is processed to yield the peak intensities. For the background-modeling procedure, energy windows located at appropriate background positions in the spectrum must be collected at each pixel, along with the energy windows selected to collect the characteristic peaks. A caution must be observed if the specimen components vary locally so that the peak of an unexpected element might appear at the energy position of one of the background windows. This problem can be avoided if the complete EDS spectrum is collected at every pixel and fully processed, including qualitative analysis, so that all constituents are recognized. Background correction by the modelling method was used in Fig. 10.6d to eliminate the false contrast due to the continuum.

For WDS, the conventional fixed-beam background-correction procedure of detuning the spectrometers to measure the background imposes too great a time penalty in the mapping mode (see Chapter 7). A more time-efficient strategy makes use of the composition dependence of the background described by Kramers' equation. The method relies on the major points, that the background is controlled by the contributions of the major constituents and that the compositional maps can be calculated iteratively. The background is first measured with each WDS on an element Z_{ref} not present in the specimen. This first estimate of background is used to correct the measured characteristic intensities for each pixel for the first calculation. The major element concentrations calculated in the first loop are then used to calculate the average atomic number at each pixel, Z_{av}. The background is recalculated for the next cycle with the scaling factor $Z_{\mathrm{av}}/Z_{\mathrm{ref}}$. After three iterations, an accurate background correction is obtained which permits minor- and trace-element determinations to the level of a few hundred parts per million.

10.3.1.5. Standardization (k-Value)

Standardization (k-value calculation) in compositional mapping is identical to the procedures used for conventional analysis (see Chapters 8 and 9). Pure-element, simple-compound, or multiple-element standards can be employed. A single-point determination of the standard is sufficient for all methods except for standard mapping. Since the specimen and standard must be scanned under identical conditions, a homogeneous, featureless standard as large as the area to be scanned on the unknown is required. Any defects in the standard such as pits or scratches appear in the k-value map.

10.3.1.6. Matrix Correction

As in conventional analysis, any matrix correction procedure, including ZAF, $\phi(\rho z)$, and Bence–Albee, can be applied in quantitative compositional mapping (see Chapter 9). Because of the large number of data points to be processed, the emphasis is on the speed of the calculation, although the speed of computation now available makes even the most elaborate calculations practical with a reasonable time penalty (<10 min) incurred to fully calculate corrections at all pixels in a 256×256 image.

10.3.1.7. Combined EDS–WDS Strategy

The availability of both EDS and WDS spectrometers on many analytical SEMs and EPMAs makes possible a combined EDS–WDS approach to compositional mapping in which the best features of each spectrometer are optimized. The EDS system is used for total spectrum imaging at each pixel, making possible a complete qualitative analysis to recognize unexpected constituents which may be locally segregated. The EDS spectrometer is used for the measurement of major constituents whose peaks are sufficiently separated to avoid problems with peak overlap. WDS spectrometers are employed for trace and minor constituents, and especially for those elements where a problem of peak interference exists. With a combined EDS–WDS mapping strategy, ten constituents or more can be simultaneously mapped (Newbury *et al.*, 1988).

10.3.1.8. Statistics in Compositional Mapping

The statistics of a single pixel in a compositional map are always inferior to the statistics of conventional fixed-beam analysis because of the disparity in integration time. Typically, the beam dwell per pixel is selected in the range 0.1–5 s, depending on the number of pixels in the image. Nevertheless, even with modest dwell time per pixel, statistically meaningful measurements can be obtained. For the case of WDS measurements, assume the beam current is limited to 500 nA by the requirement that a submicrometer beam is needed to retain adequate spatial resolution in maps recorded at magnifications of 1000X or lower. With a 500-nA beam current, a count rate of 100,000 cps can be obtained from a pure-element standard. Such a count rate is within the limit imposed by the deadtime of the WDS, typically 1–2 μs. If a pixel dwell time of 2 s is chosen, which corresponds to a 9.1 hr accumulation time for a 128 × 128 map, then the counting statistics given in Table 10.1 can be obtained at the indicated concentration levels, considering only the counts due to the characteristic peak and ignoring any loss of signal due to absorption. Note that the existence of background due to the x-ray continuum and the need to subtract it from the characteristic peak intensity will serve to worsen these counting statistics at the lowest concentrations (see Section 9.9). Alternatively, if we wish to accumulate to a specified counting statistic of, for example, a relative standard deviation of 3%, then the time to achieve that statistic as a function of the mass concentration is shown in Table 10.2.

10.3.2. Advantages

A comparison of a digital compositional map with a conventional dot map is shown in Fig. 10.7 (Marinenko *et al.*, 1987). Digital compositional

Table 10.1. Counting Statistics with 9.1 Hour Accumulation Time

Mass concentration	Relative standard deviation
0.1 (10%)	0.7%
0.01 (1%)	2.2%
0.001 (0.1%)	7.1%

Table 10.2. Accumulation Time for 3% Relative Standard Deviation

Mass concentration	Time (hr)
0.1 (10%)	0.51
0.01 (1%)	5.1
0.001 (0.1%)	51

mapping provides many significant advantages over the conventional dot map:

1. The finished compositional map viewed on an analog display consists of a gray or color-scale image where the intensity depends on the actual concentration. Recognition of particular numerical values of concentration can be aided through the choice of the display scale. The thermal color scale, which is the sequence of colors of a heated black body, is an example of a logical color scale where the attention value of the color sequence progresses in a monotonic fashion. The thermal scale can render visible a concentration scale covering two decades. As illustrated in Fig. 10.7d, the presentation of the concentration data as a

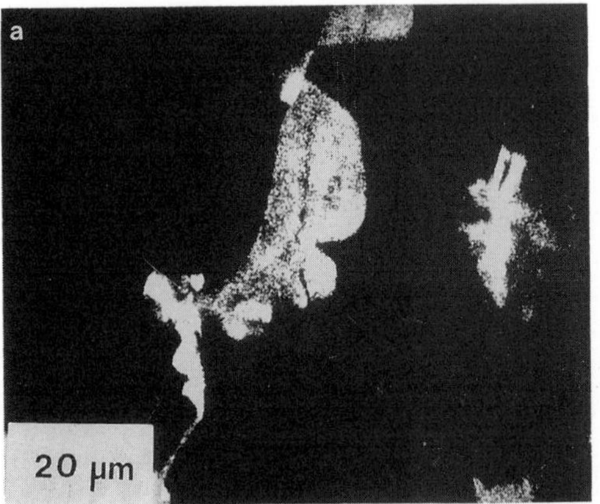

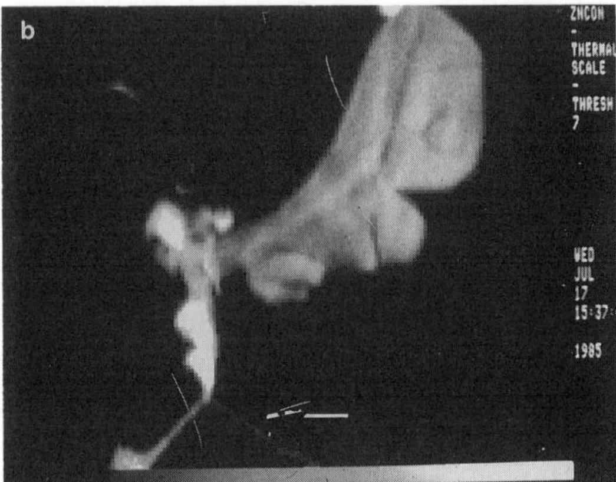

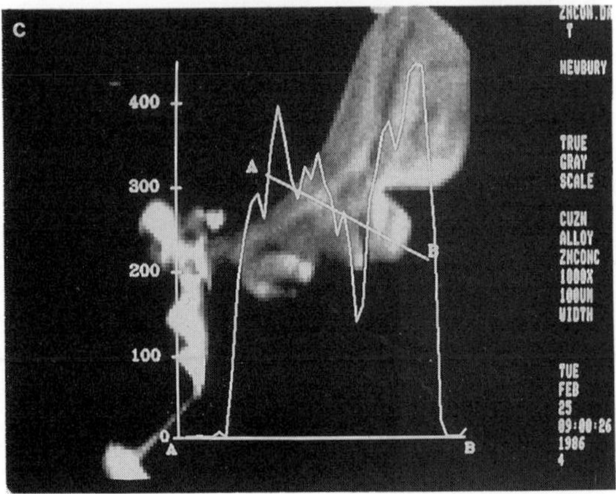

Figure 10.7. Comparison of (a) analog dot map and (b) digital compositional map of zinc at the grain boundaries of polycrystalline copper. The compositional map reveals concentrations as low as 0.1 wt % as shown by the narrow boundary indicated by the arrow. (c) Superimposed readout of the composition along the vector AB. (d) Thermal scale presentation of the data, where the range is 0.1–10 wt % (see color plate facing p. 558). (Sample courtesy of Daniel Butrymowicz, NIST.)

thermal-scale image extends the visibility to concentrations spanning the range from 0.2 to 10 wt %.

2. Each pixel in the image is supported by full numerical concentration data, including the statistics of the measurement, so that quantitative concentrations can be recovered from any pixel or group of pixels. Figure 10.7c shows a plot of concentration along a selected vector superimposed on the compositional map.

3. Because an accurate background correction is made, mapping of minor and trace levels can be carried out successfully. Although the

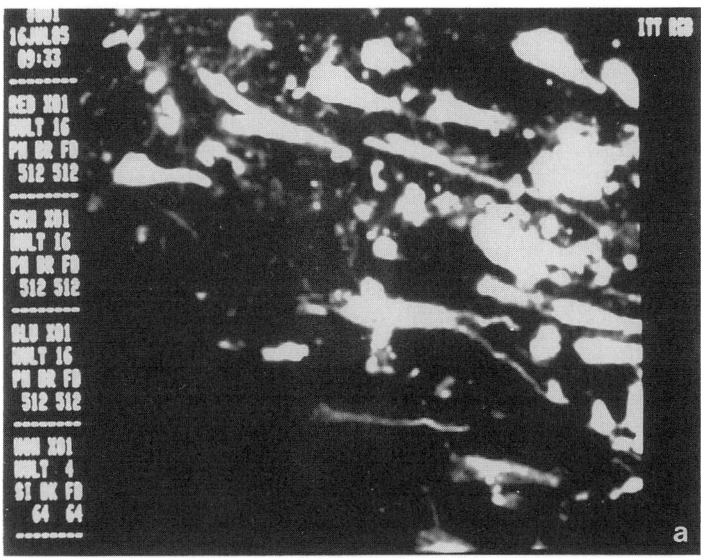

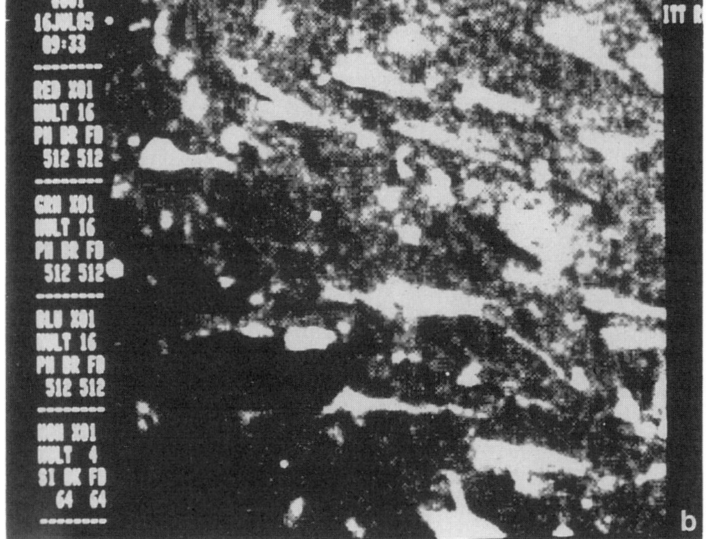

Figure 10.8. Trace-level imaging of trace constituents in human brain tissue: (a) Calcium, white = 7000 parts per million, (b) Aluminum, white = 500 parts per million (Garruto *et al.*, 1984).

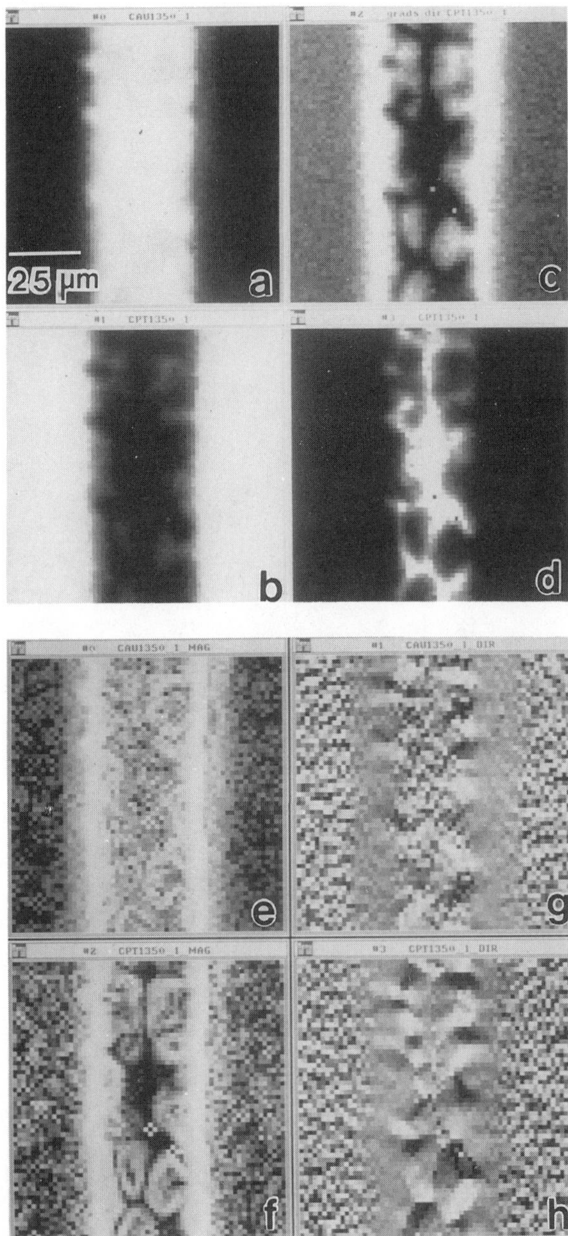

Figure 10.9. Compositional maps of a gold–platinum diffusion couple, illustrating the application of image-processing functions: (a) Au map; (b) Pt map, linear gray scale depiction; (c) Au × Pt; (d) Au/Pt; (e) Au and (f) Pt, magnitude of image gradient; (g) Au and (h) Pt, direction of image gradient. (Sample courtesy of C. Handwerker, NIST. Images courtesy of R. Marinenko, D. Bright, NIST.)

single-pixel statistics may be modest by comparison to data from fixed beam analysis, an important feature of mapping at trace levels is that the observer is very successful at recognizing extended low-contrast features in an image under circumstances were the individual pixels that make up the image would be impossible to recognize against the statistical fluctuations in the background. An example of mapping at concentration levels below 100 ppm is illustrated in Fig. 10.8. Often, the significance of such structures would not be decipherable from a collection of fixed-beam analyses, since the relationship of pixels defines a meaningful structure.

4. The statistical quality of the data is sufficiently great to permit application of many image-processing functions to increase the visibility of features in the maps. As an example of the transformations that can be applied, Fig. 10.9 shows an image sequence of compositional maps of a gold–platinum diffusion couple in which image product (Pt × Au), image ratio (Au/Pt), and image gradient operations are applied. The image product shows where both constituents are simultaneously maximized, the ratio shows where Au is maximized relative to platinum, and the image gradient depicts where the concentrations are changing most rapidly.

5. If the quantitative analysis is performed relative to standards, the analytical total has significance. A compositional map prepared with the analytical total can reveal features such as voids or regions where an unanalyzed element appears. Figure 10.10 illustrates the use of the analytical total, where oxygen has been included on the basis of assumed

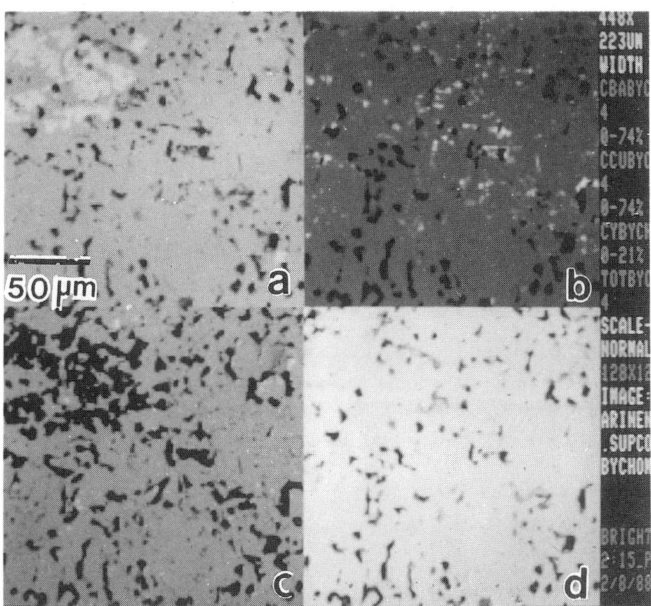

Figure 10.10. Compositional maps of a $YBa_2Cu_3O_7$ high-T_c superconductor: (a) Ba; (b) Cu; (c) Y; (d) analytical total, including oxygen by assumed stoichiometry. (Sample courtesy of J. Blendell, NIST.)

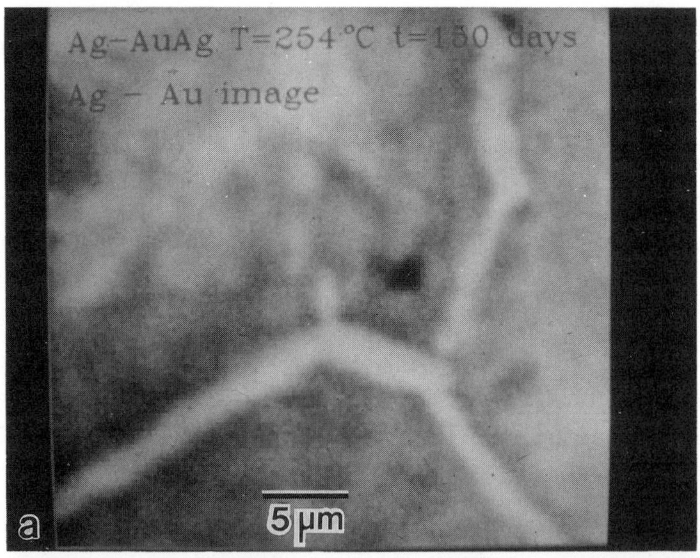

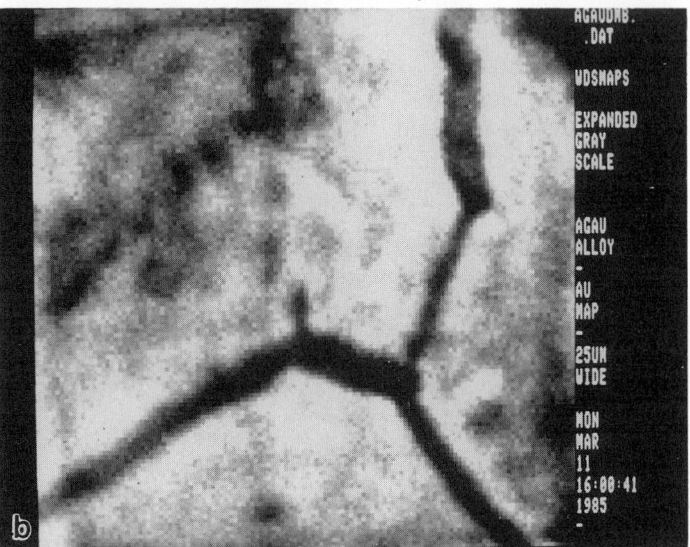

Figure 10.11. Enhancement of compositional contrast at high concentrations. Increase in the silver concentration at grain boundaries of a silver–gold alloy: (a) Compositional map of silver; the bright area represents a 5% increase in silver against a general background of 65% silver; (b) The corresponding gold image, showing a deficit in concentration. (Sample courtesy of Daniel Butrymowicz, NIST.)

stoichiometry, to show the location of voids in a $YBa_2Cu_3O_7$ high T_c superconductor.

6. Compositional contrast occurring at any point in the concentration range can be imaged, because digital image-enhancement techniques can be applied to the concentration data to improve visibility. Figure 10.11 shows an example of an increase in concentration of silver at the

grain boundaries of a silver–gold alloy. In Fig. 10.11a, the bright area at the boundary represents a 5% increase in silver against a general background of 65% silver–35% gold. The corresponding deficit in gold is shown in Fig. 10.11b.

7. By manipulating the maps with the computer, color superposition becomes straightforward. Figure 10.12 shows an example of a multielement map for a polyphase ceramic.

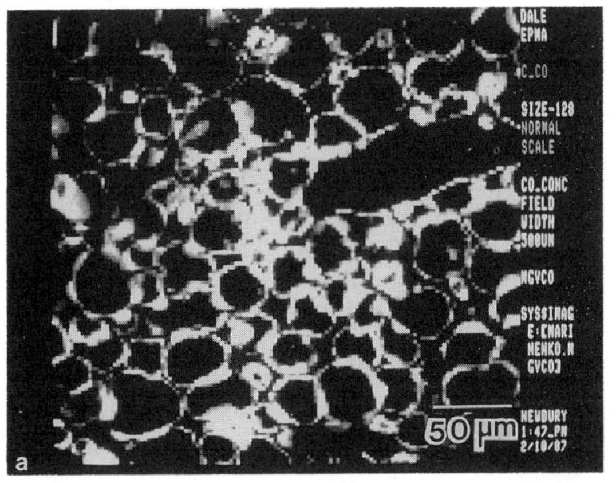

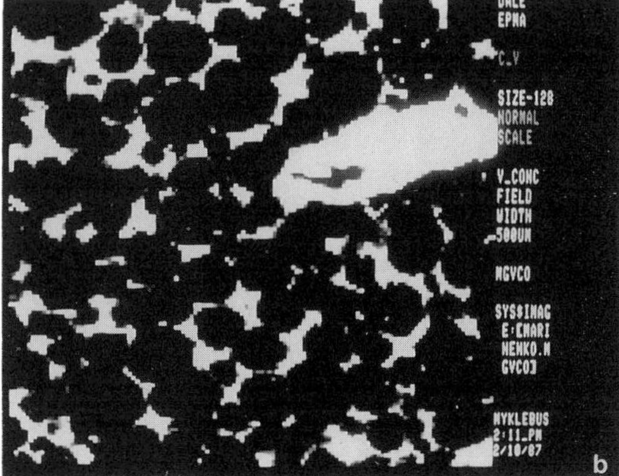

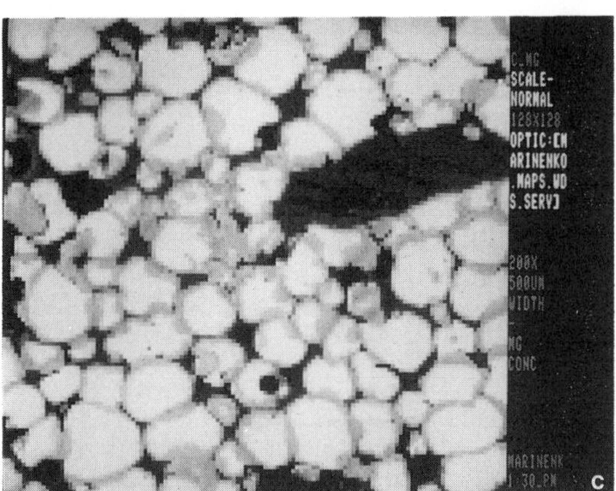

Figure 10.12. Color superposition of compositional maps: (a) cobalt; (b) vanadium; (c) magnesium; (d) color overlay with red = magnesium, green = cobalt, blue = vanadium (see color plate facing p. 558). (Sample courtesy of J. Blendell and C. Handwerker, NIST. Images courtesy R. Marinenko, NIST.)

11

Specimen Preparation for Inorganic Materials: Microstructural and Microchemical Analysis

Specimen preparation for SEM and EPMA is in many ways still as much of an art as a science. This chapter *outlines* a variety of procedures used to prepare inorganic materials for SEM and EPMA, and supplemented by laboratory exercises (Lyman *et al.*, 1991), will provide the microscopist and analyst with the starting point for most specimen-preparation problems for inorganic materials. This chapter specifically addresses the preparation of metals; ceramic and geological materials; sands, soils, and clays; electronic devices and packages; semiconductor materials; and particles and fibers. In addition, specimen preparation for both microstructural and microchemical characterization are discussed.

The important class of polymeric materials presents special problems both for specimen preparation and for subsequent examination in the SEM–EPMA. Because of the similarity of specimen preparation techniques for polymeric and biological materials, these topics are considered separately in Chapter 12.

11.1. Metals

Metallic specimens must be prepared according to the type of information the microscopist or analyst is attempting to derive from the specimen. We will discuss specimen preparation for topographic analysis of rough specimens (e.g., fracture surfaces), microstructural analysis of fine features using secondary and backscattered electrons, and microchemical analysis of specimens using x rays. Topographic and microstructural analysis are usually carried out in the SEM. Qualitative elemental analysis is routinely performed in the SEM equipped with an energy-dispersive x-ray spectrometer, while quantitative microchemical analysis is typically reserved for the analytical SEM or the EPMA equipped with both energy-dispersive and wavelength-dispersive spectrometers.

One of the great strengths of scanning electron microscopy is the fact that many specimens can be examined with virtually no specimen preparation. Specimen thickness is not a consideration, as is the case in transmission electron microscopy. Therefore, bulk specimens can be examined in the SEM with a size limited only by considerations of accommodation in the specimen stage. For the examination of images of topographic contrast from metal and ceramic specimens, the only specimen preparation necessary is to ensure that the specimen is thoroughly degreased so as to avoid hydrocarbon contamination and, in the case of insulators, to provide a conductive coating. If low-voltage microscopy is to be used, the cleaned surface can be examined without the need for the conductive coating. If we wish to examine the surface as received, without any alteration, an environmental SEM may be used. Techniques for cleaning surfaces include solvent cleaning and degreasing in an ultrasonic cleaner, mechanical brushing, replica stripping, and chemical etching. These techniques should be applied starting with the least damaging one and employing only the minimum cleaning necessary. Usually the first step is to use a solvent wash such as acetone, toluene, or alcohol in an ultrasonic cleaner. Several specific techniques for cleaning metal surfaces are described by Dahlberg (1976).

Since we wish to examine the surface of the material, it is important to remove contaminants that may have an adverse effect on secondary-electron emission. The electron beam can cause cracking of hydrocarbons, resulting in the deposition of carbon and other breakdown products on the specimen during examination. Contamination during operation frequently can be detected by making a magnification series, from high magnification (small scanned area) to low magnification (large scanned area). The deposit forms quickly at high magnification because of the increased exposure rate. When the high-magnification area is observed at low magnification, a "scan square" of contamination is observed, as shown in Fig. 4.60. It is therefore important to avoid introducing volatile compounds into the SEM. Residual hydrocarbons from the diffusion pump's oil can also produce contamination under the influence of the beam. This problem can be avoided for the most part by using traps cooled with liquid nitrogen to condense hydrocarbon vapors. In a modern, well-maintained pumping system with liquid nitrogen cold trapping, contamination problems are almost always due to a dirty specimen rather than a dirty vacuum system, unless the system has been contaminated by a previous dirty specimen. *Care should always be taken to handle specimens and specimen holders with gloves to avoid introducing volatile compounds from fingerprints into the vacuum system.*

Specimen preparation for metals does, however, become an important consideration under certain circumstances. As explained in Chapter 4, a weak-contrast mechanism such as electron channeling is frequently impossible to detect in the presence of a strong-contrast mechanism such as topographic contrast. The signal excursion is dominated by the strong contrast mechanism, with the result that the weak-contrast mechanism is compressed into one or two gray levels in the display and is lost to view. It is therefore necessary to eliminate specimen topography when we desire to work with weak-contrast mechanisms, including electron-channeling contrast, types I and II magnetic contrast, and even atomic number (compositional) contrast when the difference in average atomic number between the phases is small. Quantitative x-ray microanalysis also requires a perfectly smooth surface. Chemical polishing or electropolishing can produce a mirror surface nearly free from topography in metal specimens. A large amount of literature describing such polishing techniques exists for most metals and alloys (Kehl, 1949; McCall and Mueller, 1974; Samuels, 1982; van der Voort, 1984; ASM Metals Handbook, 1985; Tegart, 1959). Metallographic mechanical polishing also removes topography and gives a high-quality mirror surface, but such mechanical polishing results in the formation of a shallow layer (~ 100 nm) of intense damage in most metals. Such a layer completely eliminates electron-channeling contrast, and in magnetic materials the residual stresses in the layer result in the formation of surface magnetic domains characteristic of that particular stress state. If we are interested in domains characteristic of the bulk state of the material, such a residual stress layer must be avoided. Mechanical polishing to produce a flat surface followed by brief electropolishing or a chemical treatment to remove the damaged layer often gives optimum results. Electropolishing alone can occasionally result in a polished but wavy surface. The presence of a damaged layer is usually of no consequence for microchemical analysis and need not be removed by chemical or electrochemical polishing. However, if chemical polishing or electropolishing attacks phase boundaries, the interface chemistry may be modified. In general, specimen preparation remains an art, with each material presenting a different problem to the investigator.

11.1.1. Specimen Preparation for Surface Topography

The preparation procedure for surface topography examination in the SEM is relatively straightforward. The key consideration is to ensure that the sample surface is clean and undamaged. The first step is to cut large specimens to fit the specimen holder as needed. Next the specimen is degreased in a solvent such as clean acetone, using an ultrasonic cleaner if the specimen can withstand ultrasonic vibration without losing important surface material. A final wash with methanol removes any remaining surface film. It is important to ensure that the solvent does not compromise the integrity of the surface. (**Warning:** *Flammable solvents in an ultrasonic cleaner may be hazardous. Read all safety and disposal information pertaining to the solvent. Note: In many places it is illegal to dispose of used solvents directly down the drain. Contact your local environmental group to make sure you are using proper disposal procedures.*) The cleaned specimen can then be mounted onto a specimen stub either mechanically with a clamp or with glue, conductive paint, or conductive double-sticky tape. If a nonconductive glue is used, a track of conductive paint should be applied from the specimen to the stub to ensure good electrical contact. The specimen should then be dried in a clean, low-temperature (75 °C) oven. The sample should never be "pumped dry" in the SEM chamber or airlock. It is often good practice to coat the specimen with a conductive layer prior to examination in the SEM. Even for metallic samples, nonmetallic inclusions and surface oxide deposits can charge. When the electron beam strikes an insulating material such as a silicate, oxide, or inclusion in a metal specimen, the absorbed electrons accumulate on the surface, since no conducting path to ground exists. The accumulation of electrons builds up a space charge region. The problem of charging and methods for avoidance, including coating, are discussed in detail in Chapter 13. The specimen is now ready for examination in the SEM.

11.1.2. Specimen Preparation for Microstructural and Microchemical Analysis

The SEM may be used as an extension of the light-optical microscope for typical metallographic specimens. Such a specimen can also be used to determine the composition of individual phases and inclusions in the specimen by quantitative x-ray microanalysis. Often the same specimens prepared for light metallography may be examined directly in the SEM for microstructural analysis. Similar specimens, but without the final etch, can be microchemically examined in the SEM or EPMA.

The first step is to slice the specimen into pieces small enough to be placed on an appropriate mount. If the cut surface is the one to be ultimately polished and examined, the cut should be made with a slow-speed diamond saw or slurry-wire saw to minimize the depth of heat- or mechanically damaged material. The specimen should then be mounted following standard metallographic practice. Either an epoxy, cold mount, bakelite, or in some cases a fusible metal alloy should be employed. Although a specimen itself may be conducting (e.g., an alloy steel), it may be useful to mount the specimen in conductive epoxy or bakelite infiltrated with small copper particles to prevent charging during examination.

11.1.2.1. Final Specimen Preparation for Microstructural Analysis

After mounting, specimens are polished using standard metallographic practice and appropriate polishing compounds. A typical grinding and polishing sequence might be 320, 400, 600 silicon carbide papers followed by 3 μm, 1 μm, and 0.25 μm alumina or diamond polishing compounds. This work may be done manually or by using automated rotating wheels. After each grinding or polishing step, one must be careful to clean the entire mount in an appropriate solvent, such as soap and water, with an ultrasonic cleaner to avoid contaminating the next finer stage. Electropolishing may be needed to remove surface damage and potential smearing of one phase over the next. After polishing, the surface is etched to bring out the phase structure, using standard chemical, electrochemical, or ion-etching procedures. The detail in these samples can be as small as a few tens of nanometers in the secondary-electron mode, depending on the material. Often the degree of etching required for SEM observation is more than for light metallography. Heavy etching, however, may generate surface artifacts that could be confused with the true microstructure. The resolution in the BSE imaging mode (50–300 nm) is generally inferior to that obtainable with the conventional SE–BSE imaging mode on etched samples, but still better than that obtained by light-optical metallography (500 nm–1 μm). Etched samples should never be used for x-ray microanalysis (see below). Rinse and dry the specimen carefully before placing it in the SEM. (**Warning:** *The solutions used for chemical and electrochemical polishing and etching may be dangerous. They may be highly corrosive, flammable, or explosive. Use only in a hood, with appropriate safety glasses and after proper training. Contact your local safety personnel for detailed instructions in handling these materials. Also take note of local environmental laws regarding disposal of used polishing and etching solutions.*) Even without etching, the phase structure may be apparent in backscattered-electron images due to atomic number contrast. Figure 11.1 shows atomic number contrast of second-phase particles in a polished cross-section of an aluminum alloy (ASTM 6351). Note the large second-phase particles, the fine dispersion of precipitates with submicrometer sizes, and the precipitate-free zone around the large particles, where the fine particles

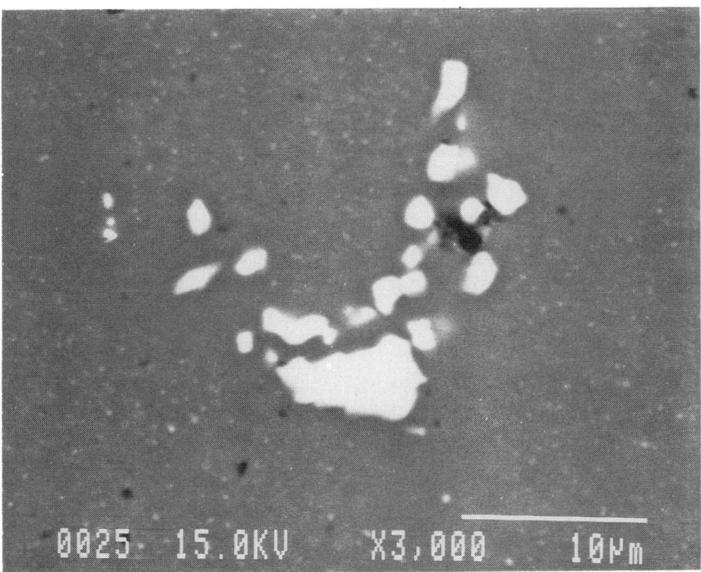

Figure 11.1. Atomic number contrast of a polished cross-section of aluminum alloy ASTM 6351. Note the large second-phase particles and the fine dispersion of submicrometer precipitates. Note also the precipitate-free zone around the large particles. Beam energy, 15 keV; solid-state backscattered-electron detector.

have been depleted. The fine dispersion of particles is well below the resolution limit of an optical microscope.

The finished specimen must be attached to the SEM stub either mechanically or by an adhesive cement, conductive paint, or tape. A track of conductive paint must be placed on the stub and from the stub to the specimen to ensure good electrical contact between the specimen and ground. It is prudent practice to coat the surface of the specimen with a thin conductive film to prevent localized charging around nonmetallic inclusions or other nonconducting areas. If the specimen is mounted in a nonconducting medium such as plastic, it is absolutely necessary to apply a conductive coating to the surface, as described in Chapter 13. A common mistake to avoid is the assumption that the plastic mount can be left uncoated if a conducting specimen is large enough (for example, a 1-cm-diameter metal specimen held in a 2.5-cm-diameter plastic mount) and a conducting paint pathway is applied from the metallic specimen to the stage. The microscopist or analyst may think that the specimen can be carefully located with the calibrated stage motions or with an ancillary optical microscope before the electron beam is turned on, so that only the metal specimen is directly hit by the beam. In fact, the bare plastic located millimeters to centimeters from the beam still charges. The reason is that backscattered electrons bounce off the specimen and scatter throughout the specimen chamber. Many backscattered electrons bounce a second time off the polepiece of the instrument. Some of these doubly scattered electrons strike the bare plastic, which builds up a charge, disturbing secondary-electron collection and even beam stability.

11.1.2.2. Final Specimen Preparation for Microchemical Analysis

For qualitative EDS or WDS x-ray analysis, the specimen-preparation procedure is similar to that described above for microstructural analysis. However, to eliminate systems peaks of Al and Cu, which may occur from electron scattering off the mounting stub, it is useful to use carbon planchets or stubs for specimen mounting.

For quantitative analysis, the specimen-preparation procedure is the same as described above for microstructural analysis, except that these samples must be "perfectly" flat. Since the x-ray analysis performed is essentially an analysis of a prepared surface, it is very important that the prepared surface be truly representative of the specimen. Over the years, a number of qualitative criteria for a properly prepared surface have evolved. These criteria are

1. the specimen should be polished as flat and scratch-free as possible and
2. the specimen should be analyzed in the unetched condition so as not to alter its topography or surface chemistry.

Such criteria were set forth primarily for metallurgical specimens; they can also be applied directly to ceramic and petrographic–geological specimens. Flatness of both specimen and standard is a prime requisite.

For pure elements and homogeneous materials, it is feasible to prepare relatively flat surfaces, since their hardness does not vary greatly over the specimen. This results in fairly uniform material removal during grinding and polishing. Unfortunately, most specimens have complex microstructures with multiple phases. In cases where phases of different hardnesses coexist, sharp steps may occur at the phase boundaries after polishing. These steps may cause anomalous absorption and must be taken into account if it is necessary to carry out analysis of regions near the boundary. One way to minimize the effects of relief at phase boundaries is to rotate the specimen 180° and remeasure the intensities to confirm the presence of a suspected absorption effect. If the absorption effect exists, it can be minimized by rotating the specimen so the x-rays are detected in a direction parallel to the phase boundary.

When the specimen's topography deviates from a flat surface, the differences in local inclinations affect the result of the x-ray microanalysis and are superimposed on the x-ray statistical uncertainty. Yakowitz and Heinrich (1968) showed that one should use high values of the take-off angle ψ in order to minimize effects of local surface inclinations. In addition they showed that effects of differences in surface preparation between specimen and standard will contribute to errors in the measured intensity ratio k.

A somewhat more insidious and more general problem than sharp steps is that of relief due to polishing, etching, and repolishing. This relief problem results in variable surface contours over the specimen's face, that is, a hilly surface. These contours can occur in the same phase as well as across phase boundaries. Although grain boundaries show the effect most markedly, variation within the grain may often be as large as or larger than at grain boundaries. Relief polishing is often minimized by using lower-nap polishing cloths and higher wheel speeds. If a polish–etch–polish procedure is being used, an etchant giving the lowest relief possible is the most desirable. Quantitative analysis of such specimens may be difficult since the possibility of errors due to topography cannot be precluded. If etching is needed to allow one to find regions of interest in the sample, the specimen should be etched, marked with fiduciary microhardness indentations, photographed, and repolished. The region of interest can then be relocated between the microhardness marks by imaging the indents in the SEM; additionally, the chemical microstructural features between the indents can be imaged with atomic number contrast from the BSE detector. It is best not to use conductive coatings, but rather to use conductive mounting materials when possible. If a conductive coating is used to allow analysis of nonconducting inclusions or to avoid charging of the mounting medium, the coating should be of a light element to reduce absorption effects. Carbon and aluminum are good choices. Care is required to coat the unknown sample and the microanalysis standards in an identical manner. The best way to do this is to mount the unknown and the standard in the same mount, if possible. Co-mounting unknowns and standards also ensures that the same tilt and take-off angles are obtained in the SEM.

Another problem in specimen preparation is to obtain a flat surface in samples containing inclusions in various matrices which requires that we polish two phases of variable hardness that are adjacent to one another. Unfortunately, the inclusion is often small and harder than the matrix, and hence may be pulled out during specimen preparation. The standard technique to retain inclusions is to use as little lubricant as possible during mechanical polishing or to electropolish the specimen. The entire junction between matrix and inclusion usually can be examined at high magnification with an optical microscope. If the specimen is properly prepared, both matrix and inclusion should be in sharp optical focus simultaneously. A detailed treatment of the subject of inclusion polishing and identification is given by Kiessling and Lange (1964).

An associated problem is the introduction of polishing abrasive into the material of interest. One should be suspicious of any inclusions containing elements used in the polishing preparation. This is particularly true when specimens containing cracks or porosity are examined. In such cases a different preparation technique is often warranted in order to determine whether or not such inclusions are artifacts. In extremely soft materials such as lead- or indium-based alloys, where local smearing of constituents could lead to erroneous EPMA results, electropolishing may be considered. Again, care and patience should be exercised in order to obtain the flattest surface possible. Techniques for the preparation of metallographic and petrographic sections of specific materials can be found in the literature (ASTM, 1960; Anderson, 1961; Cadwell and Weiblen, 1965; Kehl, 1949; Taylor and Radtke, 1965; Tegart, 1959).

The need for truly flat specimens is clear when a specimen is mechanically driven under a fixed electron beam for line- or step-scanning operations. In such cases, the x-ray flux can vary both with specimen height and surface roughness because of defocusing of the x-ray spectrometer. This variation in x-ray intensity is particularly severe when the vertical WDS geometry (see Chapter 5) is used and may render the x-ray data unusable for the determination of specimen composition.

11.1.3. Preparation of Standards for X-Ray Microanalysis

As discussed in Chapters 8 and 9, quantitative x-ray microanalysis is based on measuring the x-ray intensity emitted by an unknown specimen relative to that of a suitable standard. The most easily obtainable standards are those of the pure elements. Data correction procedures can then be applied to reduce the measured intensity ratio to composition. As discussed in Chapter 9, the matrix correction techniques may not yield sufficiently accurate ZAF factors for the most demanding levels of accuracy when pure-element standards are used. Furthermore, even if accurate correction procedures were available for pure-element standards, some elements such as sulfur, chlorine, potassium, and gallium cannot be obtained in solid, stable form for use in x-ray microanalysis. For these reasons, the use of intermediate compositional standards similar to the unknown in composition has become widespread. There are three basic approaches to obtaining intermediate standards:

1. to prepare an entire series of standards so that an empirical calibration curve can be established for the system of interest,
2. to obtain a single standard in order to characterize a particular constituent, and
3. to obtain single standards to monitor instrumental performance or as "bench marks" for correction procedures to be applied to similar systems.

The basic requirements for all such standards are that they be homogeneous at micrometer spatial resolution, carefully analyzed by independent techniques, stable with respect to time, and properly prepared for use in the SEM or EPMA. Usually, the most stringent requirement is that of homogeneity. This factor should be carefully checked by means of line scans, area scans, and point counting (see Section 9.9). In choosing compounds for use as EPMA standards, it is important to select covalently bonded oxides, particularly ones that are conducting, since these materials are less likely to suffer beam damage. In the case of the alkali metals and the halogens, the analyst must usually resort to salts such as NaCl or KI. These materials are nonconductive, are susceptible to beam damage, and should only be used at low current densities. It is also necessary to prepare polished salts in the absence of water and to protect the polished standards from humid atmospheres.

Preparation of entire sets of standards for a single system is usually confined to metallurgical applications. It is warranted when a system is to be studied in great detail, for example, in the case of phase diagram determination. With proper standards closely matched to the compositions of the unknowns, the accuracy of analysis can be made to approach the limits imposed by the x-ray counting statistics for such systems. Figure 8.2 shows a measured calibration curve used for the analysis of Fe and Ni in binary Fe–Ni alloys. The calibration curve was established with the aid of nine well-characterized, homogeneous standards (Goldstein *et al.*, 1965). The Ni $K\alpha$ radiation is heavily absorbed by the iron, and the Fe $K\alpha$ radiation is increased due to x-ray fluorescence by the Ni $K\alpha$ radiation.

The method of obtaining a single standard to characterize a particular constituent may be adopted when no suitable elemental standard is available. Before use, the material (for example FeS, GaP, etc.) should be characterized for homogeneity and composition. Usually, data taken on specimens using such a standard must be corrected for matrix effects in the same way as for those specimens analyzed using elemental standards.

The National Institute of Standards and Technology (NIST, formerly NBS) has prepared one ternary and several binary metal alloy systems suitable for use as standards (NIST Standard Reference Materials Catalog, McKenzie, 1990). At present, the Standard Reference Materials available are a low alloy steel (Michaelis *et al.*, 1964); Au–Ag and Au–Cu 20-, 40-, 60-, and 80-wt % (nominal) alloys (Heinrich *et al.*, 1971); W–20% Mo alloy (Yakowitz *et al.*, 1969); Fe–3.22% Si alloy (Yakowitz *et al.*, 1971); two cartridge brasses (Yakowitz *et al.*, 1965); and an Fe–Cu–Ni alloy (Yakowitz *et al.*, 1972). The principal value of these standards is to provide test specimens for the development, testing, and refinement of analysis techniques. In addition, multielement glasses have now been issued by NIST which can serve as standards for the analysis of typical rock-forming minerals (basalt, granite) and a series of special glass compositions (e.g., 20% SiO_2–80% PbO) is available (Marinenko *et al.*, 1979).

11.2. Ceramics and Geological Specimens

The preparation of these bulk inorganic materials is very similar to the specimen preparation methods described previously for metals. Therefore, this section will highlight the differences in specimen preparation. For further general information on the prepration of ceramic and geological specimens, several excellent references are available (Buehler, 1973; Smith, 1976).

11.2.1. Initial Specimen Preparation

In some cases, it will be necessary to examine an as-fractured sample of a ceramic or glass or a piece of a mineral. The information contained in these specimens is topographic, and such specimens are treated exactly as described above for rough metal specimens. However, in this case, since ceramics and minerals are not conductors, coating of the specimen is usually necessary. Examination at very low voltage or in an environmental SEM may negate the need for coating.

If bulk specimens are to be analyzed for microstructural details, the same rough preparation procedure outlined for metals is used. Sectioning with low- and high-speed diamond saws is the most common way of obtaining specimens of the appropriate size. Since many minerals cleave, it is also possible to prepare the initial specimen in that manner.

In ceramics and geological materials, some phases are transparent at optical wavelengths. Thin sections, typically 25 to 50 μm thick, allow one to examine these materials with transmitted light in the SEM or microprobe equipped with a light-optical microscope and simultaneously to take x-ray data. The samples are each initially prepared by slicing a wafer of the appropriate thickness from a larger parent specimen with a low-speed diamond saw or slurry saw.

11.2.2. Mounting

If bulk specimens are to be analyzed for microstructural details, the same mounting procedure outlined previously for metals is used. The use of conductive mounting media is encouraged to prevent charging during examination under the electron beam. Thin sections of ceramics and geological specimens are often used for microanalysis. The samples are prepared by slicing a wafer to the appropriate thickness, polishing one side, and attaching that side to a glass slide with a glue which is stable in a vacuum. The free surface of the sample is then polished.

11.2.3. Polishing

If bulk specimens or thin sections are to be analyzed for microstructural details, the same polishing procedure outlined above for metals is used. Care may be required in selecting the abrasive media, since some mineral phases may be harder than the polishing compound. Diamond paste is often used since it is harder than any phase in the specimen, except for any diamond phases. Diamonds can be prepared for microscopy and analysis, as encountered in recent work on man-made diamond thin films, by polishing slowly with diamond paste.

11.2.4. Final Specimen Preparation

The final step in bulk specimen preparation may be etching to help discriminate among phases. If etching is required after polishing, chemicals are usually ineffective (except for those containing HF). Usually, the natural spectral reflectance of each phase allows one to discriminate among phases using transmitted-light techniques, especially polarized-light methods. Similarly, differences in average atomic number of the various phases allow one to discriminate among phases by imaging with backscattered electrons. If etching is required, plasma-ion etching is generally successful. Since ceramics and geological materials are usually nonconductors, coating is required. Thin films of carbon or evaporated metal are most suitable for microstructural analysis (see Chapter 13).

As with metals, qualitative EDS and WDS chemical analyses can be performed on ceramic and geological specimens prepared for microstructural analysis. One must be careful not to use metallic specimen stubs or conductive metal coatings because extraneous x-ray lines from coating elements may be added to those of the specimen. In some cases, a metallic coating can be used if one is certain that a given metal is not in the specimen and that the metal does not have interfering x-ray lines. Metals tend to give a more even and uniform coating and may suppress charging more effectively than a carbon coating.

11.3. Electronic Devices and Packages

The preparation of electronic devices and packages utilizes the same techniques described above for metals and ceramics. In this case, we are defining the device to be the integrated circuit (IC). The examination of a specific junction in the IC is accomplished either by the SEM using voltage-contrast or charge-collection microscopy (Newbury *et al.*, 1986) or by cross-sectional transmission electron microscopy (TEM). Classical SEM and EPMA techniques (topographic contrast, atomic number contrast, x-ray mapping, and point analyses) are used to examine:

1. the integrated circuit (e.g., metallization, passivation glass, etc.),
2. the chip carrier (e.g., die bonding, often using an intermediate-temperature solder, or wire bonding, often by thermocompression or ultrasonic bonding), or
3. the attachment of the chip carrier, or individual IC, to a ceramic substrate using thick- or thin-film technology and soldering, or to a printed wiring board using a solder.

In order to examine the specimen using secondary and backscattered electrons as well as x rays, the specimen must be properly prepared. Since these devices and packages are typically constructed from ceramics and metals, most of the techniques described above apply. Some chip carriers (commercial, nonhermetic) are made from polymers. Procedures for specimen preparation of polymers are described in Chapter 12.

11.3.1. Initial Specimen Preparation

It is often desirable to examine an IC or package directly. Many current-generation ICs now present submicrometer features. General observation of metallization lines on an IC can not be achieved optically. At low beam energy, or at high beam energy after coating with carbon, it is possible to observe the metallization and passivation patterns directly. Such procedures allow us to examine the lines and passivation to inspect for metallization coverage at steps and for the examination of voids in the metal, as shown in Fig. 11.2. For larger structures, such as a die placed in a chip carrier, it is possible to examine features such as wire bonds directly with optical microscopy. If higher-resolution details of a wire bond, for example, are required, direct observation in the SEM with secondary electrons is often used. X-ray microanalysis, even qualitative, is often difficult for ICs and packages. Features are often small, surface topography is relatively large, and material layers are often very thin ($<1\ \mu m$), making it difficult to obtain a spectrum from only a specific region of interest in the specimen. For examination of fine chemical details, it is recommended that the IC, chip carrier, etc. be cross-sectioned and polished before examination. A detailed description of SEM and EPMA examination of devices and packages is given by Romig, *et al.* (1991).

The IC, chip carrier or circuit assembly is a composite of metals, ceramics and sometimes polymers. If the sample to be cross-sectioned is large enough to handle, it can be sliced with a low-speed diamond saw. If the sample is small, such as an IC, it may be necessary to embed it in epoxy before sectioning. In some cases, the diamond saw sectioning is done dry to prevent contamination and corrosion problems which might

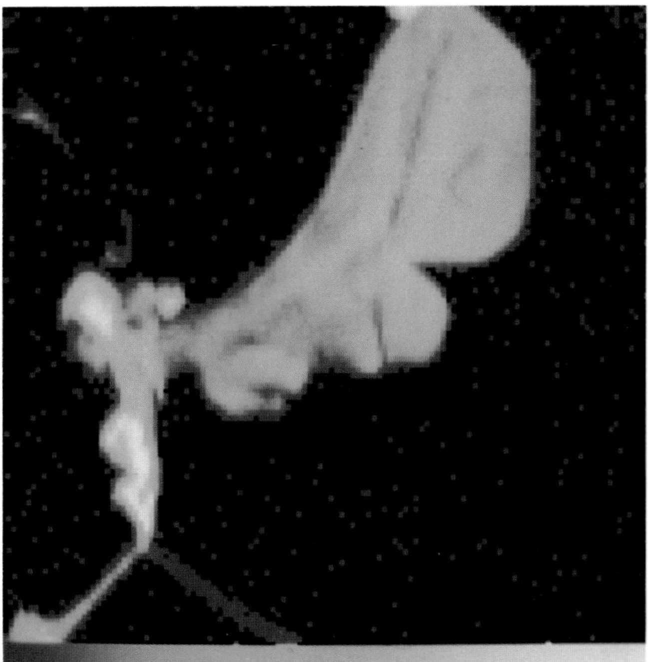

Figure 10.7d. Thermal scale presentation of digital compositional map of zinc at the grain boundaries of polycrystalline copper. The composition range is from 0.1 to 10 weight percent.

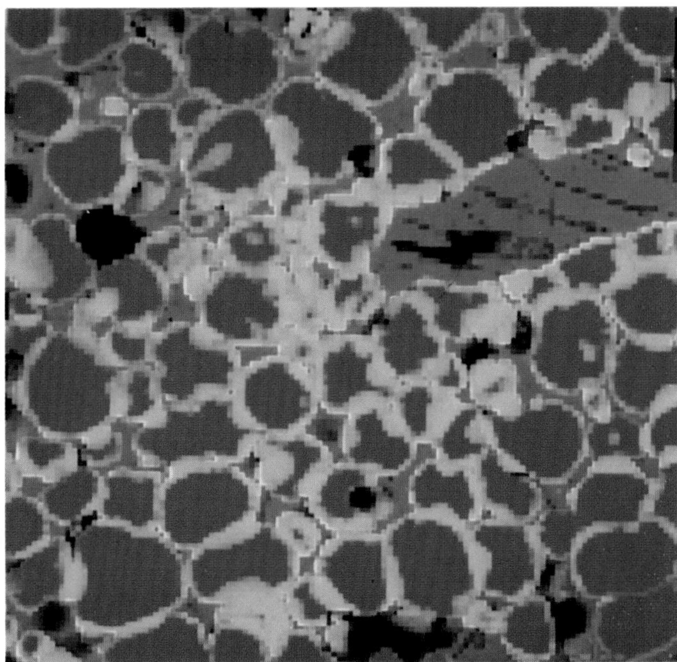

Figure 10.12d. Primary color overlay of compositional maps for a polyphase ceramic: red, magnesium; green, cobalt; blue, vanadium. Note: yellow phase = magnesium + cobalt; purple phase = magnesium + vanadium.

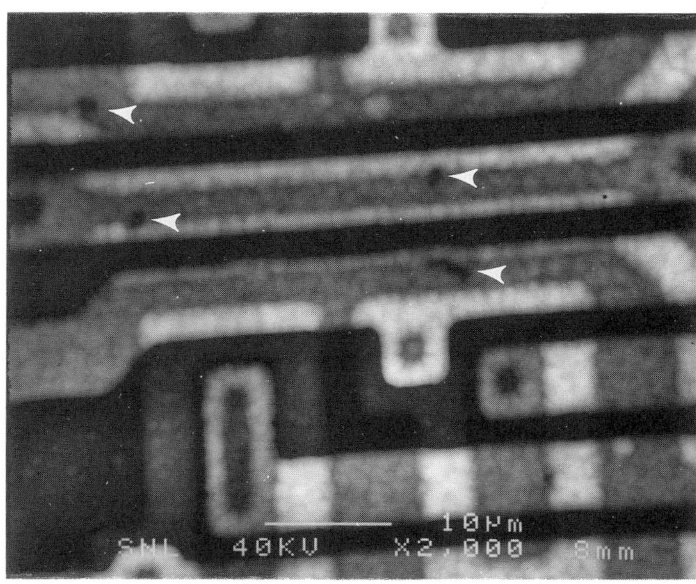

Figure 11.2. Aluminum metallization line covered by a passivation glass layer. The metal line width is approximately 3 μm and the passivation glass layer is approximately 1 μm thick. The image was taken at 40 keV in the backscatter mode. Note that the voids in the aluminum metallization caused by stress voiding are apparent (marked by arrows). Micrograph courtesy of J. A. Van den Avyle.

compromise the analysis. To bring out structural details, the cross sectioned device can be selectively ion milled.

11.3.2. Mounting

To examine very small features, such as the metal and passivation on an IC, the die attachment joint, the internal details of a wire-bond or surface-mount solder joint, it is necessary to mount the specimen. Typically, the cross-sectioning cut described above comes close to the region of interest. If the sample is small and can not be cross-sectioned prior to mounting, it is mounted as one piece. An epoxy or other liquid mounting medium is required. The samples have a great deal of relief and topography and readily trap air in the form of bubbles during mounting. If possible, the liquid mounting medium should be poured and cured in a vacuum. Agitation during curing is also sometimes helpful. The use of a vacuum and agitation should eliminate, or at least minimize, the presence of bubbles. Bubbles in the mount can be very detrimental, since they lead to edge rounding during polishing and may trap abrasives and polishing or etching chemicals which would destroy the final specimen.

The mounted specimen is then sawed or ground to the plane of interest, depending on the amount of material to be removed and the relative location of the region of interest. A clear mounting medium may help one to find the proper plane of polish more easily. Coarse grinding is accomplished through a series of abrasive papers as described above. In order to minimize edge rounding, the entire specimen may be electroplated with nickel prior to polishing, or hardened ceramic or steel balls may be placed in the mount with the specimen.

11.3.3. Polishing

Polishing is done in the same way as described for metals and ceramics. Special difficulties may arise because of the extreme hardness differences between, for example, a ceramic substrate and a soft Pb–Sn solder. Careful polishing with very hard abrasives (e.g., diamond) and very light loads usually produces suitable results. However, one must always be careful of abrasive pickup in the softer constituents such as solder.

11.3.4. Final Preparation

Very often, simple differences in average atomic number between the various constituents in the specimen allow us to see the features of interest. If additional details are desired, etching may be required. Chemical etches work for the metallic constituents, while plasma-ion etching may be required for the ceramics.

The as-polished specimen should be suitable for microchemical examination. For qualitative microanalysis, the specimen used for microstructural characterization (i.e., etched) may be used again. For quantitative microanalysis, the specimen must be polished flat. As with metals and ceramics, the specimen can be etched and photographed and then repolished to remove any surface topography. If needed, fiduciary hardness marks can be made in the etched specimen to locate a specific area in the repolished specimen. Figure 11.3 shows a typical cross-section of a package as imaged in the secondary, backscatter, and x-ray mapping modes.

11.4. Semiconductors

Most semiconductor specimens can be prepared using the general methods described in previous sections. However, special surface preparation is often needed to see certain contrast mechanisms, such as voltage-contrast, electron-channeling, and charge-collection microscopy. For example, to perform charge-collection microscopy of semiconductor materials, such as examining dislocations in silicon, it is necessary to place a metal layer called a Schottky barrier on the specimen. This requires specialized chemical handling systems and an UHV metal deposition system. For a discussion of these SEM microcharacterization techniques, see Newbury *et al.* (1986).

11.4.1. Voltage Contrast

Active circuits are observed with voltage contrast. The specimen must be electrically isolated from the specimen stage and activated by

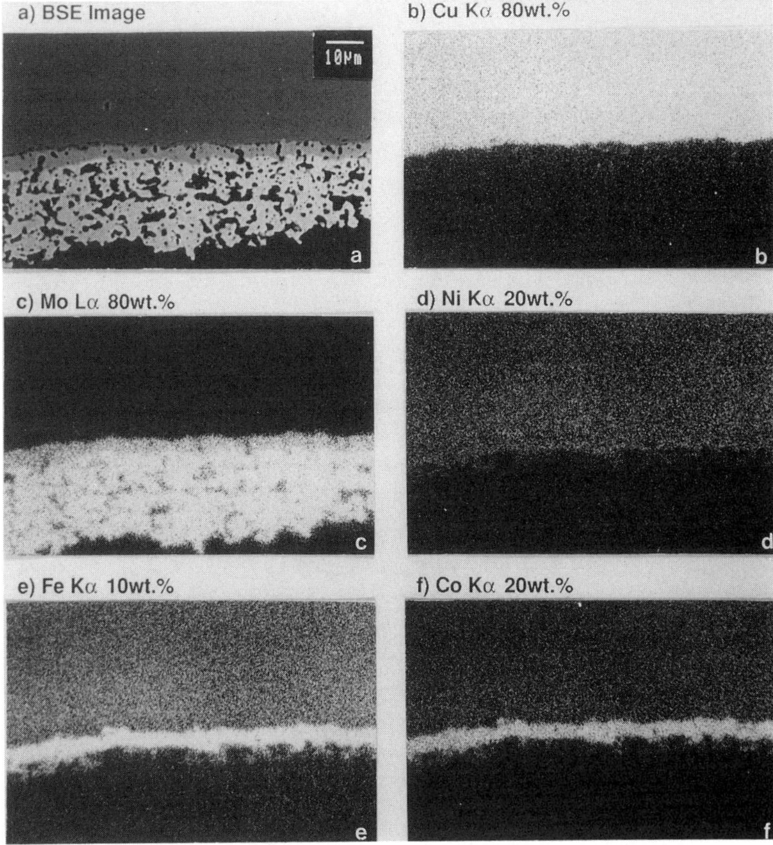

Figure 11.3. Wavelength-dispersive elemental x-ray maps showing detail of the Cu braze/Mo–Mn metallization interface. This sample was brazed for 10 minutes with a peak temperature of 1109°C. (a) Backscattered-electron image. (b) Cu $K\alpha$ 80 wt% map. (c) Mo $L\alpha$ 80 wt% map. (d) Ni $K\alpha$ 20 wt% map. (e) Fe $K\alpha$ 10 wt% map. (f) Co $K\alpha$ 20 wt% map. Photographs courtesy of J. J. Stephens.

leads reaching the specimen from the electrical feedthrough. The activation of the circuit may be static (conventional voltage contrast) or dynamic (stroboscopy) by connection to an external signal generator. The image is observed in the secondary-electron mode and no specimen preparation is needed except that the specimen surface must be free of dust, lint, and grease. Cleaning in alcohol and deionized water, followed by thorough air drying, is often the only preparation required.

11.4.2. Charge Collection

Charge-collection microscopy, including the most common form, called EBIC (electron beam induced conductivity), is a technique used to examine inactive p–n junctions as well as defects in semiconductor materials. There are two principal reasons to examine inactive devices: to examine p–n junctions with EBIC and to examine the packaging details of the circuit (bond pads, interconnection metallization lines, passivation layer integrity, etc.). Loose dust and grease must be removed from the surface of the sample. If the passivation layer is to be examined, a conductive coating may be required; however, it may be possible to examine the specimen at low beam energies (about 1 keV) without a coating, although the beam penetration is very shallow. For analysis of *p–n* junctions, metallizations, lines, etc., it is necessary to remove the passivation layer. The removal technique depends on the nature of the passivation layer. Hydrofluoric acid is often used for glass, while organic solvents are used for polymer-based passivation layers. Plasma etching is also occasionally used. The sample must be properly attached to the stage and grounded. In some cases it is possible to examine metallization lines through the passivation glass by using the high-energy backscattered-electron signal. A beam energy in excess of 25 keV is required. Resolution of such images is poor, but is generally sufficient to allow identification of voids in metals.

If EBIC is to be used to examine a semiconductor material that is not in a device with a depletion region, a Schottky barrier may be used to create a depletion region near the top surface. The deposition of high-quality Schottky barriers requires great care. The surface must be properly cleaned and polished, similar to that required for electron channeling, and the barrier metal must be deposited in a very clean evaporator, free from any trace of hydrocarbon contamination. One should use a liquid nitrogen–trapped diffusion pump or a turbo pump. The barrier must be continuous, but thin (20–40 nm). For intrinsic or *n*-type material, Au seems to be the best barrier material, while Ti seems to work best for *p*-type material. A significant amount of experimentation may be required to produce a suitable Schottky barrier for a given situation. A detailed description of the procedure used to deposit good Schottky barriers is described in Newbury *et al.* (1986).

11.4.3. Electron Channeling

A flat, strain-free surface is required for electron channeling. To achieve flatness, it is often necessary to polish the sample metallographically with 0.05-μm alumina (or other abrasive), as described in Section 11.1. The mechanical polishing does, however, leave a strained surface layer (the so-called Bilby layer) which destroys the channeling effect. This layer must be removed either chemically or by electropolishing. The solution required depends on the exact nature of the material. Dilute solutions of hydrofluoric acid can often be used for Si.

11.5. Sands, Soils, and Clays

Sands, soils, and clays can be examined from two perspectives:

1. the inorganic constituent or
2. the organic constituent.

Sands, soils, and clays are simply mixtures of inorganic particles of different size ranges mixed with water, a variety of organic compounds, and often living organisms (Bear, 1964). Sands are characterized on the basis that the largest average particle is in the range 0.1 to 2 mm in diameter. The mineral constituent is usually quartz, basalt, or gypsum, although other more exotic minerals, e.g., monazite, are possible. The inorganic components are often moistened with salt brine or water, except in desert regions. Organics, such as hydrocarbons, are often mixed with sands in oil-bearing formations. In addition, a variety of microorganisms may exist in the sand. Soils are simply very fine sands made of clay minerals often mixed with water, organic compounds, and microorganisms. The inorganic particle size typically ranges from 100 μm to 0.1 mm. If the particle size become smaller than 100 μm, the material is referred to as a *clay* rather than a *soil.*

For characterization of an as-received specimen containing significant volatile components, the sand, soil, or clay may be placed directly in an environmental SEM, which allows for direct observation of all constituents in the specimen. The organics are often the constituents of primary importance in sands, soils, and clays. Specimen preparation of the organic constituents makes use of the techniques described in Chapter 12. For examination of the inorganic components, the material must be thoroughly washed in water and the appropriate solvent applied to remove the organic component. Often acetone is a good first choice. If the specimen is to be examined for water-soluble materials and the specimen is initially dry, all rinsing must be done with nonhygroscopic solvents. (**Warning:** *Flammable solvents in an ultrasonic cleaner may be hazardous. Read all safety and disposal information pertaining to the solvent. Note: In many locations it is illegal to dispose of used solvents directly down the drain. Contact your local environmental group to make sure you are using proper disposal procedures.*) After being thoroughly washed, the inorganic particulate must be dried. The particulate should be very widely and evenly spread during drying to prevent agglomeration of the particles. The loose particles are then prepared for SEM and EPMA analysis as outlined in Section 11.6.

The study of soil fabric with the SEM requires that the fluid, which is an aqueous solution, be removed from the specimen before it is placed in the instrument. If the soil specimen has a high moisture content or tendency to shrink upon moisture loss, it is difficult to dry the specimen without disturbing the original structure (Lohnes and Demirel, 1978). Six techniques have been applied for removal of pore fluid (Tovey and Wong, 1973). These techniques are

1. oven drying,
2. air drying,
3. humidity drying,
4. substitution drying,
5. freeze drying, and
6. critical-point drying.

The first two techniques are straightforward and self-explanatory. Humidity drying is the process of drying the specimen under controlled humidity. Substitution drying involves the replacement of the pore fluid in the soil by a liquid of low surface tension such as methanol, acetone, or isopentane prior to drying (Lohnes and Demirel, 1978). The last two techniques are the same as those used by biologists for organic material and are described in Chapter 12. In general, air drying is the most widespread technique for low moisture and stiff soils, whereas soils having a fragile fabric can be dried by rapid freeze drying (Lohnes and Demirel, 1978).

After drying the soil specimen, it is necessary to expose an undisturbed surface for study by fracturing or peeling. In order to fracture a specimen, the dried material is scored and then bent or pulled to create a tensile failure surface. Peeling involves applying an adhesive material to the specimen surface and then stripping it off to remove disturbed surface particles of the specimen with the adhesive. The stripping process is repeated several times. For a more complete description of SEM specimen preparation for soils, the reader is referred to a review article by Lohnes and Demirel (1978). Dengler (1978) has also described an ion-milling technique used to expose the matrix microstructure. To examine both the inorganic and organic constituents of a sand, soil, or clay, the sample is examined in an environmental SEM. Mounting of the inorganic constituent is described in Section 11.6, which discusses the mounting of particles and fibers.

11.6. Particles and Fibers

The preparation of particles and fibers for microscopy and microanalysis raises special challenges for several reasons:

1. Most particles and fibers are nonconductive. Environmental particles are typically composed of crustal minerals such as silicates. Because of the enormous surface-to-volume ratio of particles compared to bulk material, even technological products such as metals that typically exist in metallic form under ambient conditions often oxidize when prepared in the form of fine particles and exposed to humid atmosphere. Nonconducting particles that charge under the beam may be mechanically unstable during SEM observation and may be lost due to electrostatic repulsion.

2. The techniques used for collection of particles and fibers often have a major influence on the choice of subsequent preparation techniques. Entrainment of particles and fibers in filter materials may result in a situation which the particles and fibers are obscured from direct observation, necessitating at least partial removal of the filter or even complete separation and transfer of the particles. Separation techniques must avoid chemical and morphological modifiction of the particles during such preparation.

3. Electron beam penetration will become significant for particles with thicknesses less than 1 μm. Subsequent scattering from the substrate will cause a reduction in image contrast and an increase in extraneous contributions to the x-ray spectrum. Several of the most general cases of particle and fiber preparation are described below. For a more comprehensive treatment, see McCrone and Delly (1973) and Brown and Teetsov (1976).

11.6.1. Particle Substrates

The ideal particle substrate should provide adequate support and mechanical stability for particles, should not contribute to the electron and x-ray signals, and should not have surface topographical features of its own. Bulk materials as well as supported thin films are used as particle substrates.

11.6.1.1. Bulk Substrates

Bulk materials are generally chosen as substrates for large particles, with thicknesses $>5\,\mu$m, and for those collection situations such as impaction where the velocity of the collection fluid, typically air, may be too great for thin materials to survive. Typical bulk substrates in use include low-atomic-number bulk materials such as graphite, silicon, and beryllium. These materials give robust support, but each bulk substrate has specific deficiencies:

Beryllium. The lowest-atomic-number material of practical use, beryllium takes a high polish, which can minimize surface features. (**Warning:** *Beware of the highly toxic nature of beryllium oxide if polishing is attempted*!) "Commercially pure" beryllium often has minor levels of impurities such as iron and calcium. Although these contaminants are at trace levels (<1 wt %), the low electron and x-ray absorption of beryllium allows for the excitation and detection of intermediate- and high-atomic-number elements. Also, all substrates contribute a continuum background to the x-ray spectrum, with the background intensity proportional to the average atomic number.

Silicon. Highly polished, extremely pure silicon single-crystal wafers are readily available at low cost from semiconductor manufacturing suppliers. These silicon wafers are essentially featureless at the highest practical SEM resolution. They provide a good substrate for imaging particles of intermediate and heavy atomic numbers. The principal drawback of silicon arises in x-ray microanalysis because of the Si $K\alpha,\beta$ peak, which is excited with high efficiency, even at low beam energies.

Graphite. Graphite is generally the best choice for a bulk particle substrate because of its low atomic number and ready availability at high purity. The principal drawback is the softness of graphite and its susceptibility to scratches. Scratches and pits are often found on even the best polished surfaces, which may contribute confusing features that tend to obscure particle images. This drawback is particularly unfortunate

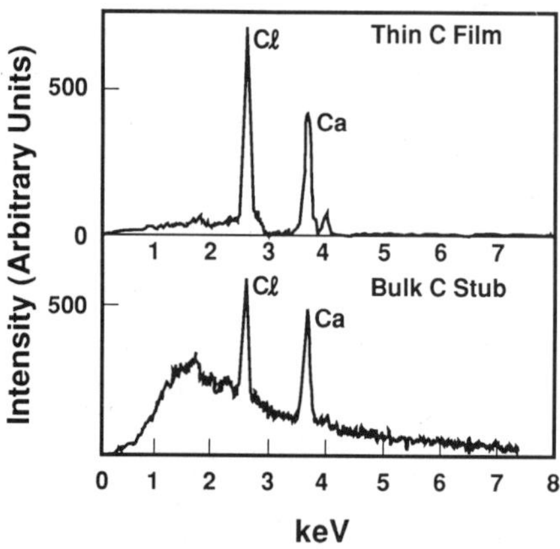

Figure 11.4. Comparison of the x-ray spectra of calcium chloride particles on a thin carbon and a bulk carbon substrate (Barbi and Skinner, 1976).

when searching many fields of view to locate rare particles. Pyrolitic graphite is a highly purified form of carbon that takes a high polish and is recommended as a substrate.

11.6.1.2. Thin-Foil Substrate

Optimum contrast and minimum spectral background for particle analysis is obtained with a thin-foil substrate. The foil is typically carbon supported by a metal grid. To eliminate scattering from electrons that pass through the specimen, the grid is placed over a several-millimeter-deep blind hole in a block of low-atomic-number material such as graphite. Alternatively, the grid specimen is mounted for STEM, as described in Chapter 4. Figure 4.71 shows SEM and STEM images from such a particle preparation. The x-ray spectrum obtained from a particle mounted on a thin foil is shown in Fig. 11.4, compared to the equivalent situation for a particle on a bulk substrate. A significant reduction in the spectral background can be seen in Fig. 11.4 for the thin-foil substrate. This reduction results in an improvement in the limit of detection.

Self-supporting carbon foils as thin as 10–20 nm can be prepared by thermal evaporation of carbon to form a film on a glass slide, flotation of the film onto a water surface, and deposition of the carbon as a foil on grids of metal (beryllium, copper, nickel, or gold) or carbon. The grid openings are typically squares 50–200 μm on a side. This grid-supported carbon foil has sufficient strength to permit particle deposition by a variety of methods. Particles as large as 10 μm in diameter or more can be supported, although the usual application is for particles below 2 μm in diameter.

11.6.2. Transfer and Attachment of Particles to Substrates

Various particle transfer and attachment techniques are available for bulk substrates. In general, we desire to have the particles quantitatively transferred to the substrate without selective loss (fractionation), well dispersed across the substrate without aggregation, and with sufficient mechanical stability to avoid loss during subsequent handling or examination in the SEM. Selection of the optimum technique depends on the nature of the particle specimen.

11.6.2.1. Abundant, Loose Particles

Dry Deposition. From the point of view of particle integrity, dry deposition is the preferred method of particle transfer, since it minimizes the possibility that chemical modification of the particles may occur, as is often the case when particles are exposed to solvents such as water, alcohol, etc.

When the specimen is received in the form of abundant, loose particles, transfer to the substrate can be made by dry deposition onto a sticky surface. A conductive carbon tape that is sticky on both sides can be attached to a bulk substrate to provide a suitable attachment surface. After the particles are sprinkled across the sticky tape, any loose excess is removed by inverting the substrate, gently shaking, and knocking on the backside of the substrate with a tool. After removal of excess particles, a final thin (10–20 nm) carbon coating can be applied if the particles are nonconducting. The analyst should be aware that the particle distribution obtained with this preparation may not be uniform across the surface or even representative of the original distribution if particle settling and size segregation occur in the source or during transfer and attachment.

Dry deposition is also useful with carbon foils supported on metal grids. The particles can be gently spread across the grid, and some will remain attached by electrostatic attraction. The grid can then be given a thin (10–20 nm) carbon coating by thermal evaporation to improve the mechanical stability of the particles under the electron beam. Again, particle fractionation can occur if one size or composition class is selectively lost due to poor attachment.

Dispersal from a Liquid Suspension. Particle dispersal from a liquid with ultrasonic vibration to break up particle aggregates is generally an effective means to improve the uniformity of the final particle area density (DeNee, 1978). A surfactant can be added to water dispersions to aid the breakup of aggregates. Several methods for liquid transfer are possible:

1. An aliquot of the ultrasonified particle-bearing liquid can be pipetted directly onto the bulk or thin-foil substrate.
2. The bulk or thin-foil substrate can be dipped into the particle-bearing liquid.
3. The bulk substrate can be incorporated directly into the base of a glass chimney to receive the charge of particle-bearing liquid, and the liquid can then be evaporated to dryness to deposit the particles on the substrate.

Some uneven deposition may occur because of particle trapping in the last droplets of liquid to evaporate, and in the chimney technique, some particles will be lost to the glass walls. Particle samples, especially environmental samples, tend to incorporate an organic chemical fraction that becomes a naturally sticky residue during liquid dispersal and provides a stable attachment to the substrate. A final thin carbon coating can be applied for insulating particles.

Careful attention must be paid to obtaining a liquid medium that does not alter the particle composition. This can be a difficult problem, since some constituents, such as alkali and alkaline earth elements, are vulnerable to leaching in a wide variety of liquids. Chlorofluorocarbon compounds (Freons) are relatively inert and generally do not cause selective leaching of constituents, but their recognized environmental hazard make them unacceptable.

11.6.2.2. Particle Transfer from a Filter

Filters are often encountered in particle-analysis problems. The most uniform particle deposition with area is obtained by carefully filtering a particle-bearing liquid. Many particle samples, such as environmental collections, are collected as a deposit on a filter. Many types of filters are available for various collection applications, and the appropriate preparation depends on the nature of the filter and the particles. The following examples are of particular interest.

1. The nuclepore polycarbonate filter collects particles by impaction onto a smooth filter surface. The nuclepore filter is sufficiently flat and featureless to permit direct examination of particles in the SEM without removal from the filter, after a conductive coating is first applied to the filter to eliminate charging by both the filter and the particles.

2. Tortuous-path filters entrain many of the particles in the interior of the filter medium, necessitating their exposure or removal, as illustrated in Fig. 11.5. The entrained particles can be exposed by partial etching of the filter with exposure from one side to vapor from a suitable solvent such as chloroform. After coating, the etched filter can be directly examined in the SEM. If transfer of the particles to a thin-foil carbon substrate is desired, the partially exposed particles can be overcoated with a carbon film by evaporation. The carbon-coated filter is then turned over, and the solvent is applied from the backside to complete the removal of the filter, with the particles remaining trapped in the carbon layer for transfer to the SEM.

3. If the particles are not degraded by oxidation, the filter can be removed by low-temperature ashing in an oxygen plasma. The filter residue containing the particles can be deposited directly on a resistant substrate such as silicon and then given a final carbon coating for mechanical stability and charge dissipation.

11.6.2.3. Particles in a Solid Matrix

Second-phase particles and inclusions in a metallic or nonmetallic matrix may be so small that attempts to analyze them with SEM or

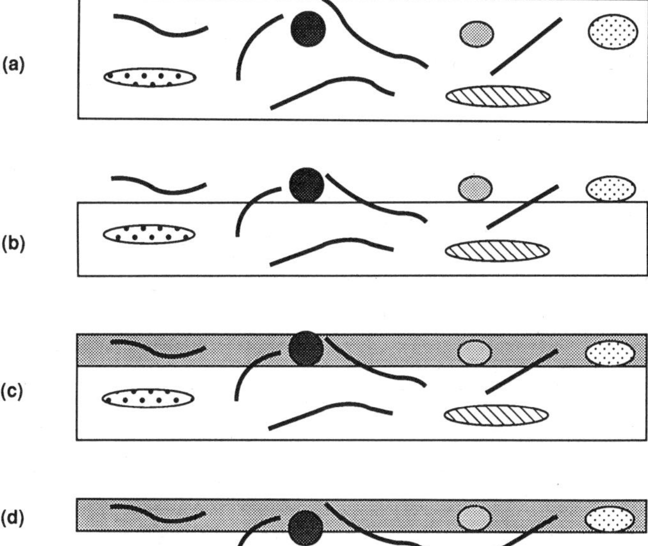

Figure 11.5. Release of particles entrained in a filter by a two-stage etching procedure. (a) Particles entrained in a filter. (b) Particles are partially exposed in the initial etch. (c) A carbon film is evaporated on the etched surface, entraining the particles again. (d) The filter is re-etched from the back side to remove the remaining filter material, leaving the particles in the carbon foil.

EPMA in polished cross-sections may lead to unwanted excitation of the surrounding matrix. If it is important to ascertain whether a constituent of the matrix is also contained within the particle, it is necessary to remove the particle from the matrix for separate analysis. The two-stage etching method described above for exposing the particles trapped in a filter can be applied in this case as well. The specimen is prepared as a thin sheet, and one side is given a flat, metallographically polished surface. This surface is etched in a suitable chemical that attacks the matrix without disturbing the particles. After etching has partially exposed the particles, a carbon coating is applied to envelop the particles. Carbon is generally highly resistant to chemical attack, permitting further etching. The specimen is again etched from the backside to remove the remaining sheet, leaving the particles in the carbon film. The SEM can be used to monitor intermediate steps in the etching process to ensure that proper exposure of the particles is achieved. The images can be used to monitor any possible changes in the particles caused by the matrix etching chemicals.

11.6.3. Transfer of Individual Particles

The most difficult preparation problem in particle analysis is the transfer of *selected,* individual microscopic particles to the SEM. Generally, such samples must be extremely rare to warrant the difficult procedures necessary to manipulate individual particles. Often, these particles may be located by optical or other imaging or analytical

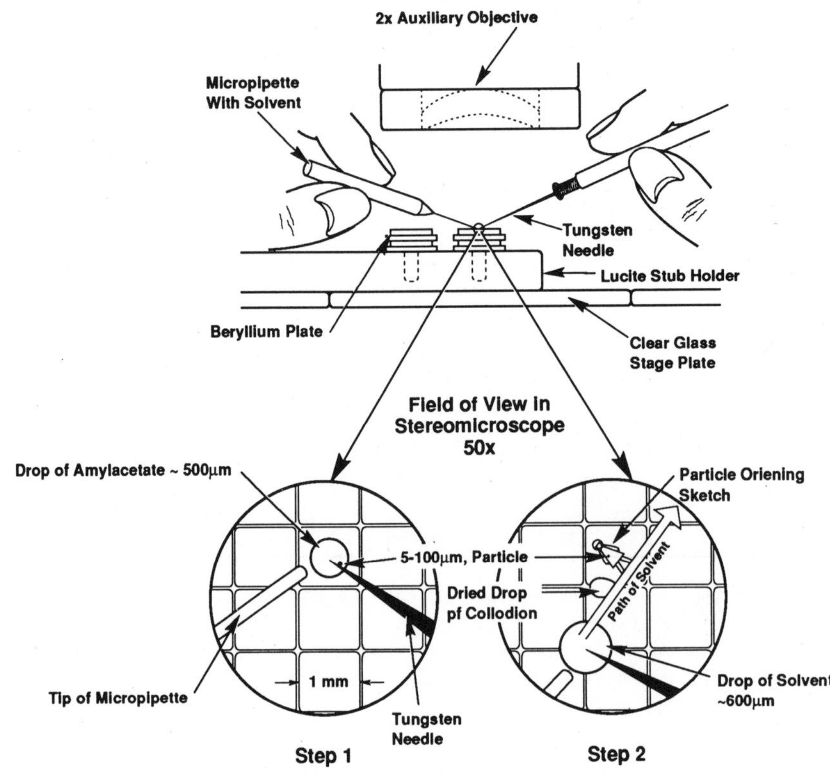

Figure 11.6. Stages in manipulating a selected microscopic particle to a prepared substrate (Brown and Teetsov, 1976).

techniques. Reference grids are used to ensure accurate particle transfer between techniques.

As illustrated in Fig. 11.6, particles greater than approximately 1 μm in size can be moved directly by picking them up dry on fine needles made of tungsten while observing under a stereo optical microscope or macroscope (Brown and Teetsov, 1976). A highly polished specimen substrate ruled with a finder grid is prepared to receive the individual particle by first drying a microdrop of collodion and then placing a microdrop of a solvent such as amyl acetate nearby. The particle is transferred from the needle to the drop of solvent, and with the tungsten probe, the drop is pushed to the collodion. The solvent dissolves the collodion, and the mixture containing the particle is spread across a small region. When the solvent evaporates and the collodion dries, a 20–50-nm-thick layer binds the particle to the substrate. A thin carbon coating ensures conductivity to the conductive substrate.

12

Sample Preparation for Biological, Organic, Polymeric, and Hydrated Materials

12.1. Introduction

Success or failure in obtaining comprehensive structural or chemical information from organic, polymeric, biological, and hydrated specimens examined by electron-beam instrumentation depends critically on the way the samples are prepared. This general statement is nowhere more important than in scanning electron microscopy and x-ray microanalysis.

There are considerable technical difficulties in achieving even adequate, let alone perfect, sample preparation. On the one hand we have an electron-beam instrument, and on the other hand we have, for example, living organisms or a piece of plastic. The interaction of the beam of electrons with the specimen can readily cause both thermal and radiation damage, and the highest signal-to-noise ratios are obtained from flat or thin thermodynamically stable specimens of high atomic number. Let us now consider organic samples and the fairly typical materials derived as a product of biological activity. The samples are invariably soft, wet, and three-dimensional, composed of low-atomic-number elements, with an overall low mass-density. They tend to be thermodynamically unstable, needing a continuous influx of energy in order to maintain their form and functional activity. Also, they have low thermal and electrical conductivity and are very sensitive to radiation damage.

Quite obviously, the biological and organic material and the electron-beam instrument are at odds, and some sort of compromise must be made to bring these two extreme situations closer to each other. There are two alternatives. One may compromise either the instrument or the material. Both alternatives will be considered, one briefly and the other in more detail.

12.2. Compromising the Electron-Beam Instrument

We can do three things with the instrument (and this applies mainly to the SEM) to enable it to be used with unprepared biological and hydrated organic material: add an environmental stage to the system, use a microscope system (called an "environmental microscope") specially designed to deal with such materials, or run the system under conditions which help preserve specimens even at the cost of nonoptimal performance.

12.2.1. Environmental Stages

We can provide within the microscope column a microenvironment similar to the natural environment of the specimen (Valdre, 1979). An environmental cell can be made by placing the specimen in a small chamber that is either provided with electron-transparent windows or closed off from the rest of the column except for small apertures through which there is a constant but small flow of gas. The interior of the chamber is either flooded with an aqueous medium or continuously flushed with an inert carrier gas at a pressure of 30–50 kPa saturated with water vapor. In either situation the electron beam travels through a few millimeters of a less-than-optimal vacuum before interacting with the surface of the specimen itself.

The mean free path of a high-energy electron in liquid water or water vapor is extremely short, and the inevitable scattering of the electrons may result in both specimen damage and image degradation. A critical look at the results which have been obtained using environmental stages shows that very little, if any, new information has accrued. The real advantage of this approach would be in the ability to observe dynamic cell processes such as cell division. This has not been achieved, and the evidence presented by Reimer (1975), Glaeser and Taylor (1978), and Glaeser (1980) convincingly demonstrates that the radiation damage which occurs while observing even the fastest-multiplying bacterium would invalidate the biological significance of any observations.

12.2.2. Environmental Microscopes

Electrical insulators, polymers, and wet specimens can be examined in environmental microscopes in which pressures in the 0.1–1 kPa range are maintained in the specimen chamber. Interactions between the electron beam and the gas molecules can be used to neutralize the surface charge on insulators, and specific gas-specimen interactions can be used to amplify the emitted electron signal. In addition, the gas composition may be altered for investigations of chemical reactions. The environmental microscopes work on the same principle as environmental chambers, except that the whole specimen chamber is maintained at circa 1 kPa while the electron column remains at high vacuum. This is achieved by a series of pressure-limiting apertures within the probe-forming lens. The interactions of the probe electrons, the signal electrons and the sample are complex and closely interconnected with each other by gas composition and pressure, working distance, signal-collection field and accelerating voltage. Secondary electrons formed by gas ionization are collected by a positively biased gaseous detector. Recent papers by Danilatos (1991) and Farley and Shah (1991) provide a good review of environmental scanning microscopy. It is a relatively new form of instrumentation, the potential and limitations of which have yet to be fully realized. It has been used to image thin layers of water, uncoated polymers, and biological samples and to follow dynamic environmental processes such as the crystallization of salts from aqueous solutions (Fig. 12.1).

12.2.3. Nonoptimal Microscope Performance

It is possible to run the SEM under less than optimal conditions and still obtain useful information about untreated biological and organic samples. The less than optimal conditions include low-voltage operation, low beam current, short record times, decreased signal-to-noise ratios, and poorer microscope vacuum. Such conditions lessen the energy input

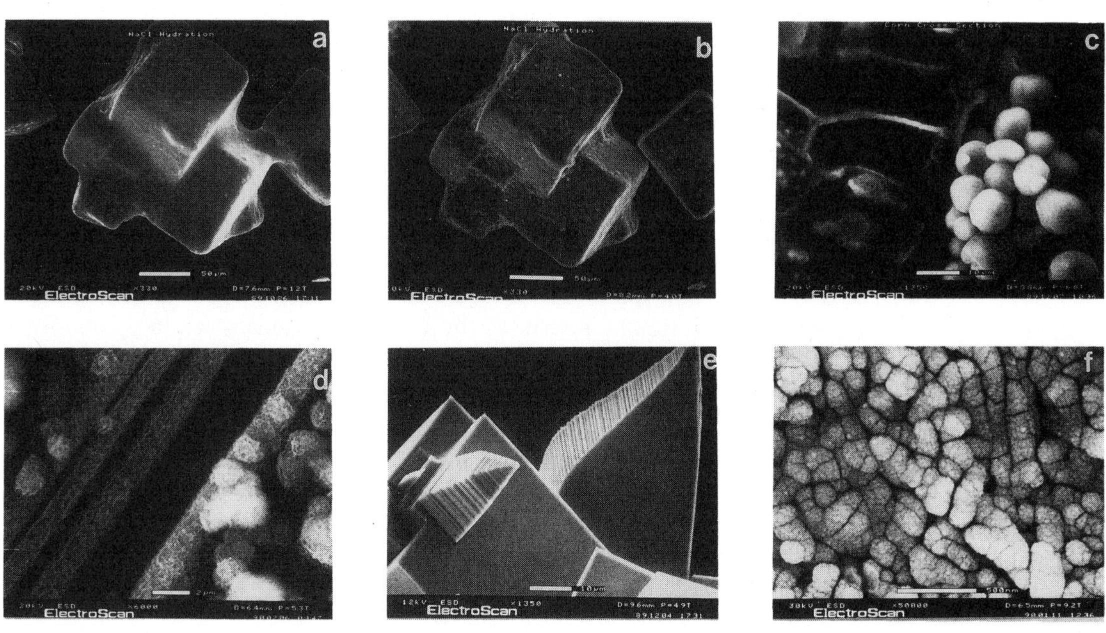

Figure 12.1. Images taken using an environmental microscope. (a) Sodium chloride crystals in wet state: 20 keV, 1600 Pa, magnification bar = 50 μm. (b) Sodium chloride crystals in dry state: 20 keV, 533 Pa, 10 °C, magnification bar = 50 μm. (c) Corn (*Zea mais*) cross-section: 12 keV, 1040 Pa, 10 °C, magnification bar = 20 μm. (d) Solvent-etched polymer: 20 keV, 706 Pa, magnification bar = 2 μm. (e) Potassium chloride crystals grown at 600 °C: 12 keV, 653 Pa, magnification bar = 10 μm. (f) Resolution test specimen of gold-coated oxide particles: SE, 30 keV, 1226 Pa, magnification bar = 500 nm. [Pictures courtesy of Danilatos (1991).]

into the specimens and enable the specimens to be examined untreated and in a near-natural state, although usually at much reduced resolution. [This is not the case with dedicated low-voltage microscopes which have been specifically designed to operate in the range 500 → 5000 eV (see Chapter 2)]. This approach of nonoptimal operation has been particularly useful in the examination of plant material, the cells of which, although containing a large proportion of water, are invariably bound by a cellulose wall, which is frequently covered by a waterproofing or suberized layer. The net effect is to reduce the rate of water loss from the plant organ, permitting it to be observed for several minutes before obvious dehydration effects become apparent. By the same token, one may observe hard tissues, such as teeth, bone, chitin, seeds, spores, wood, many polymers, and pollen grains, all of which have a low water content. Examination without preparation is much less successful with most animal tissues, aquatic plants, microorganisms, and hydrated organic material, all of which shrivel up during drying. However, even this approach should not be dismissed out of hand because useful information can be obtained even from grossly distorted specimens.

It should by now be apparent that attempts to compromise the instrument are not going to achieve very much in terms of obtaining maximum information from biological and organic specimens, and we should look at the ways in which we may compromise the specimen to enable it to survive in the microscope column.

12.3. Compromising the Sample

This is the principal matter of this chapter, and our discussion will center around the conversion of a normally wet and/or organic sample to a dry state while still preserving its three-dimensional shape and chemical constitution to allow it to be examined and analyzed in the high vacuum of the microscope. This conversion—which, by and large, involves changing the organic matrix of the cells and tissues to an inorganic form with a concomitant loss of water—can be separated into three distinct phases:

1. fixation, which involves stabilizing, either by cross-linking, denaturation, or precipitation, the labile organic molecules present in the cells and tissues;
2. dehydration, the removal of water; and
3. coating, which involves the application of a thin layer of conducting material to an otherwise nonconducting surface.

This latter phase is more or less unique to SEM and x-ray microanalysis and is considered separately in Chapter 13 of this book. The procedures of fixation and dehydration are common to all methods used to prepare hydrated organic materials for examination in electron-beam instruments. There is, however, an important distinction between those procedures used in connection with *structural* studies and those used for *in situ chemical* analysis.

In the former, the emphasis lies in preserving the structural integrity of macromolecules at the expense of the low-molecular-weight species which are lost invariably during processing. The opposite is generally true in chemical analysis, where the substances of interest are either electrolytes or low-molecular-weight compounds, which are generally confined to the aqueous phase of the sample. The procedures designed to preserve these soluble components generally do not adequately preserve the structural macromolecules. This dichotomy of experimental approach places severe constraints on what we should and should not do to the sample. With the possible exception of cryofixation, no one single preparative technique will suffice.

The first prerequisite for determining which preparative technique is applicable is to clearly identify the problem. With this in mind, it is recommended that the experimentalist assemble all the information known about the sample, i.e., its identity, structure, chemical constitution, lability, and, if appropriate, its metabolism. A similar list should be made about the technology and instrumentation available (or what will be needed) to provide solutions to the problem. Finally, and this is sometimes the most difficult part, the specific information it is hoped to obtain from the sample should be set out. Armed with information about the sample and instrumentation, one is in a better position to design a good preparative protocol.

This chapter will consider sample preparation under six main headings:

1. Ambient-temperature preparation for structural studies of hydrated organic specimens: Biological structure.
2. Ambient-temperature preparation for analytical studies of hydrated organic specimens: Biological analysis.
3. Ambient-temperature preparation for structural and analytical studies of anhydrous organic materials: Plastics and polymers.
4. Low-temperature preparative methods.
5. Preparative damage, artifacts, and radiation effects.
6. Preparative procedures for specific groups of specimens.

In addition to the general approach provided here, we will provide literature citations which give the nuts and bolts of sample preparation, i.e. how to make up buffers, fixatives, and stains; how to embed, section, and fracture; how to excise the unsullied sample; and how to introduce its ravished counterpart into the instrument.

There is not, unfortunately, one single book which provides all the information needed. The books by Glauert (1974) (a thoroughly revised version will appear in 1992) and by Hayat (1982; and 1989), although general texts for electron microscopy, contain much useful information on specimen handling procedures for SEM and x-ray microanalysis. The compendiums edited by Revel *et al.* (1984) and Muller *et al.* (1986) also contain many useful papers that provide a rigorous explanation of the assumptions made in choosing the most appropriate technique for a given hydrated, organic, or biological sample.

12.4. Correlative Microscopy

Any detailed investigation should attempt to correlate the information obtained from a wide range of instruments. In many ways the SEM is a useful starting point because its magnification range ranges from that available with a good hand lens to that of a high-resolution TEM. The SEM also presents us with a familiar image, albeit from samples maintained under high vacuum. Correlative studies are relatively easy to carry out either by processing the specimen for TEM or light microscopy (LM) after examining the specimen in the SEM. An example of a correlative study is given in Fig. 12.2, and further details may be found in the papers by Barber (1976), Wetzel *et al.* (1974), and Lytton *et al.* (1979) and in the book edited by Echlin and Galle (1976). Albrecht and Wetzel (1979) give details of methods which may be used to correlate all three types of images with histochemical techniques, and Junger and Bachmann (1980) give details of correlative studies with light microscopy, autoradiography, and SEM. Wherever possible, the catalogue of correlative studies should include examining the untreated samples, by eye, by hand lens, and by binocular microscope.

The other aspect of the preparative procedure that requires discussion is the realization that the biological and organic objects we see and record in the SEM are artifacts. Preparative procedures have been described as the art of producing creative artifacts, and to a certain extent this is true. Broadly speaking, we seek to convert the unstable organic specimen into a stable and largely inorganic form. The result is a series of amplitude-contrast images in which it is very difficult to relate accurately a given signal level to a feature in the untreated sample. It is important to realize that one observes only images, which although representative of the material, are in actuality far removed from it. This aspect of electron microscopy has been forcibly brought home in the studies of unfixed and unstained frozen-hydrated and frozen-dried specimens. In such samples, one can no longer see the familiar images obtained by the more conventional techniques, and it may well be that we are going to have to re-educate ourselves to recognize structures in frozen-hydrated samples. Acceptance of the fact that everything we observe in the SEM is more or less an artifact in no way lessens the value of these instruments in research.

12.5. Techniques for Structural Studies

These are the methods used to provide most of the information we obtain using SEMs. The preparative procedures follow an orderly

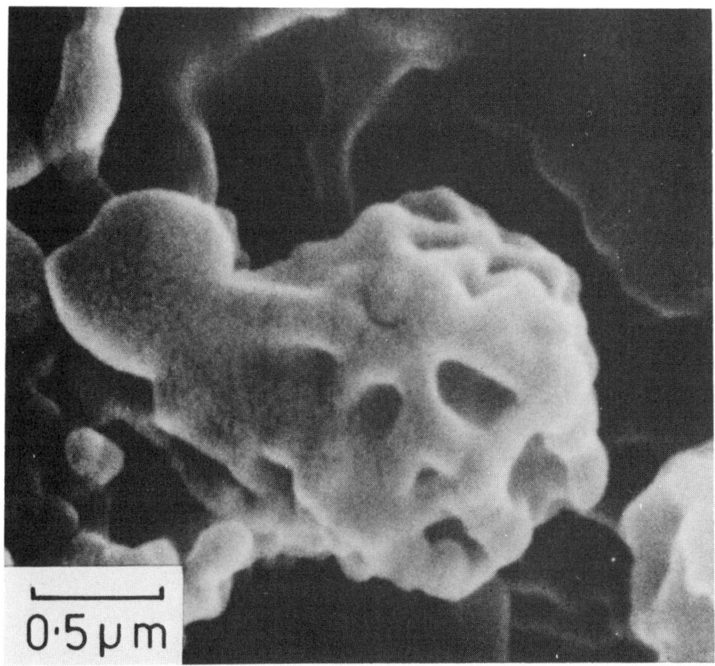

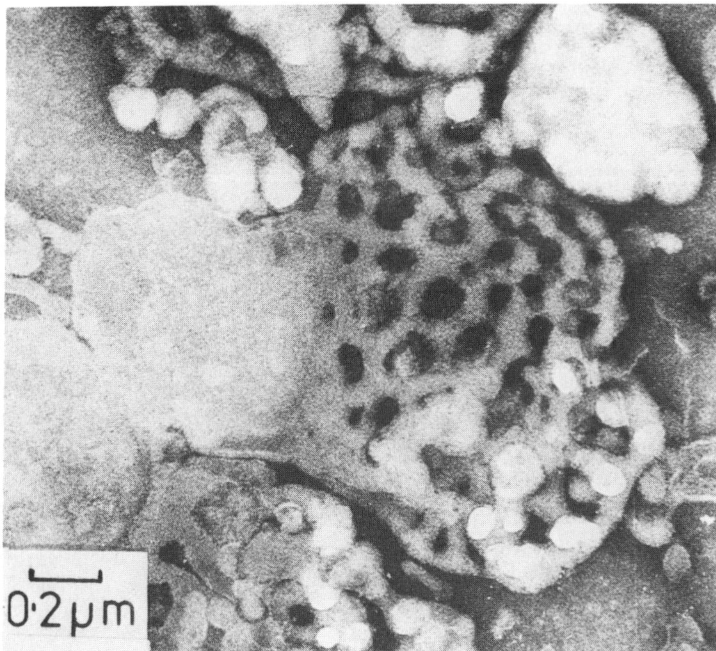

Figure 12.2. Correlation of SEM and TEM images. (a) SEM image of an isolated Golgi complex from rat liver (Sturgess and Moscarello, 1976). (b) TEM image of a negatively stained Golgi complex from rat liver (Sturgess and Moscarello, 1976).

progression of selection, cleaning, stabilization, dehydration, and coating, with exposure of regions of interest coming before or after dehydration.

12.5.1. Specimen Selection

Care must be exercised over the way the specimen is sampled. The sample should be the smallest compatible with the features to be examined. Small specimens are more easily handled and processed during fixation and dehydration. Small specimens can be more adequately stabilized and dehydrated, lessening the chance of volatile components being released inside the microscope. Pieces of tissue up to 1–2 mm^3 may be surgically excised from multicellular organisms; somewhat smaller pieces of plant material would be desirable.

The preparations of microbial, hematological, and unicellular aquatic organisms present their own set of problems and Watson *et al.* (1980), Wouters *et al.* (1984), and Maugel *et al.* (1980) provide a number of useful suggestions on how these types of specimens may best be preserved. Paradoxically, the main problem is their small size. Although the fixation and dehydration times can be very short owing to the small specimen size, problems are created in handling the specimens. The usual procedure is either to collect and handle the samples by centrifugation from an aqueous suspension (Wouters *et al.*, 1984) or to fix the specimens firmly to a large, solid support such as a filter pad, glass substrate, or a freshly cleaned mica- or tissue-compatible metal disk with or without a plastic coat. Walz *et al.* (1984) have used polylysine-coated polyacrylamide beads as a means for attaching intact cells. The cells remain attached throughout processing and can be released at the end of preparation. Similar problems are experienced with cultured mammalian cells, and the papers by Allen (1983) and Evers *et al.* (1983) provide a number of useful methods which can be applied to these specimens. Alternatively, microorganisms may be studied *in situ* on the surface—naturally or artificially exposed—of the tissue it is inhabiting, using techniques described by Garland *et al.* (1979). The attached specimens may then be taken through the preparative procedures. Glass and mica substrates have the added advantage that correlative studies can be carried out using the light microscope, as shown in Fig. 12.3.

Plastic tissue-culture dishes should be used only if one is quite certain that they are not soluble in the organic fluids used in the tissue preparation. Many mammalian tissue-culture cells are best handled in this way, and it is relatively easy to attach such cells to glass cover slips which have been coated with a monolayer of polylysine (Sanders *et al.* 1975). It should be appreciated that very few of the procedures devised to attach samples to the specimen stub are 100% effective. Varying amounts of the sample are lost during processing. The sample support coating methods can also be used for other cells, such as microorganisms and unicellular algae, as well as particulate matter. Small particulate matter may also be harvested on plastic-covered microscope grids, fine metal

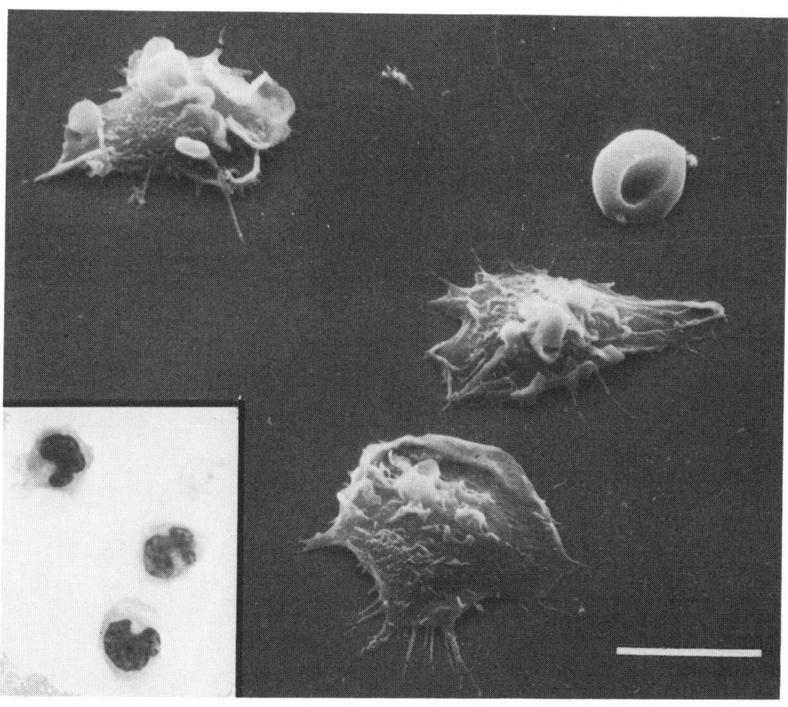

Figure 12.3. SEM image of human monocytes on a glass support. Magnification bar = 10 μm (Wetzel *et al.*, 1974).

gauzes, or, as shown in Fig. 12.4, on one of the many different commercially available filter membranes (Millipore, Nuclepore, Flotronic, etc). If filter membranes are to be used, it is important to check their solubility in the various processing fluids and to realize that a membrane which is smooth to the naked eye is frequently very rough and corrugated when observed in the SEM. Such highly irregular surfaces are unsuitable for specimens such as cell organelles and viruses. McKee (1977) and Drier and Thurston (1978) published a useful review of the preparative techniques which can be used with microorganisms such as bacteria and fungi. DeNee (1978) and Chatfield and Dillon (1978) give details of the ways in which particles may be collected, handled, and mounted for SEM.

12.5.2. Specimen Cleaning

It is important to make sure that the surface to be examined is clean and free of extracellular debris. Some extracellular debris such as mucus, blood, or other body fluids may obscure the whole surface to be examined; others substances such as dust and cell and tissue fragments may obscure parts of the surface and result in an incomplete image. The cleaning of excised tissue pieces should, within reason, be carried out as rapidly as possible to lessen the chance of changes induced *post mortem*

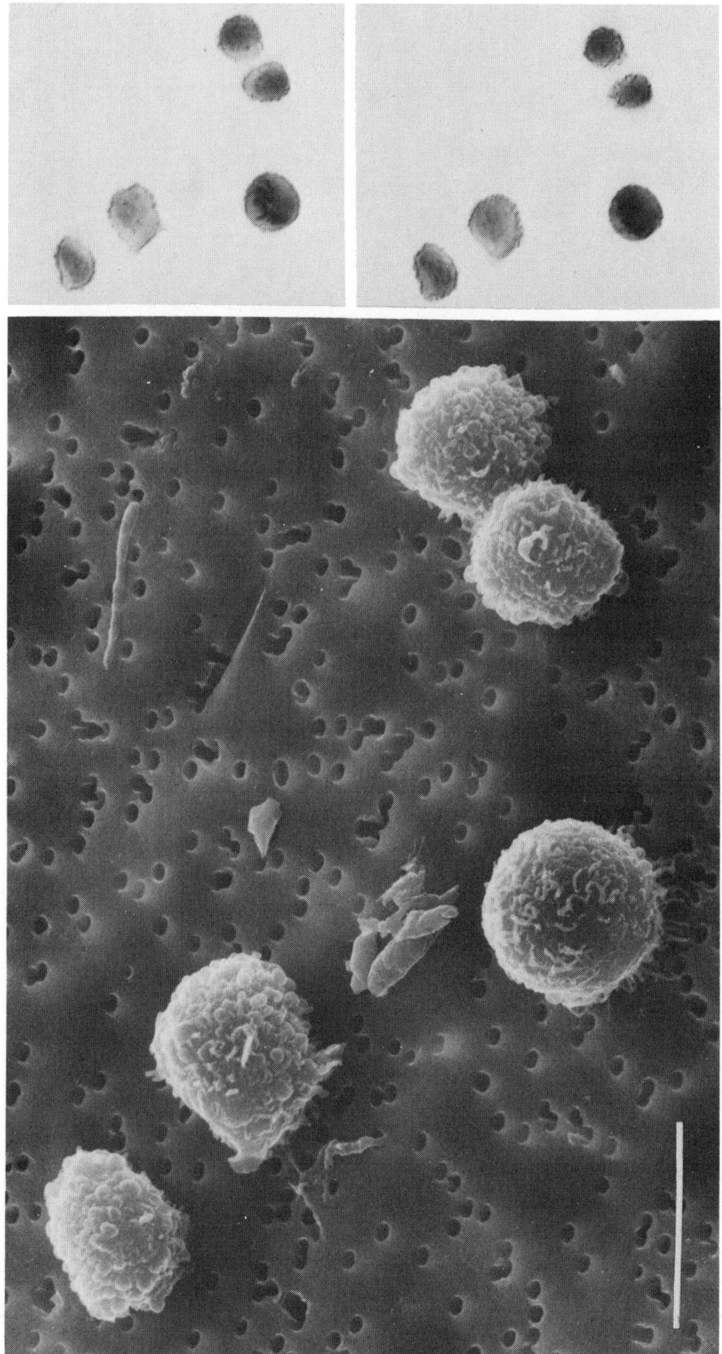

Figure 12.4. SEM image of CHO tissue culture cells on filter support. Magnification bar = 10 μm (Wetzel *et al.*, 1974).

and those caused by oxygen-starvation (anoxia), which are particularly damaging to mammalian tissues.

The aqueous contaminants usually found associated with multicellular animal tissue can often be removed by gently washing or irrigating the specimen with a buffer of the same strength and composition as the cell fluids and maintained at a temperature compatible with the natural environment of the specimen. The internal spaces may be washed clean by continual internal flushing. Should gentle washing not suffice, then more turbulent washing can be tried with or without a small amount of a surfactant. If the biochemical nature of the contaminant is known, then enzymatic digestion can be tried. The types of enzymes which have been used are mainly proteolytic, saccharolytic, or mucolytic and include collagenase, hyaluronidase, neuraminidase, elastase, chymotrypsin, pronase, and some of the amylases. Once the surface has been wetted it should not be allowed to dry before passing on to the chemical fixation. In the case of some hard tissues, only the inorganic matrix is of interest. The organic material may be conveniently removed by gently washing with a 2–5% solution of sodium hypochlorite at room temperature. Organisms with a characteristic siliceous or calcareous skeleton such as diatoms, foraminifera, and globigerina may be cleaned with oxidizing agents such as potassium permanganate or potassium persulfate. In all cases, great care must be exercised to ensure that only the contaminants are removed and that the surfaces remain undamaged. Samples from large multicellular specimens are probably best washed before, during, and after excision to avoid contamination with extracellular fluids.

Aquatic specimens such as unicellular algae, protozoa, and the smaller invertebrates, unless maintained in pure culture, are frequently mixed with a wide range of organisms and other plant and animal debris. Differential centrifugation and repeated washing in an appropriate isotonic buffer usually precipitate the organisms from the debris (or *vice versa*) to give a relatively clean suspension.

Dry specimens, such as calcified and chitinaceous tissue, seeds, wood, spores, pollen grains, and the aerial parts of plants, usually need only dusting, preferably with clean air, although a Freon duster may also be used. Such procedures should not disturb the microflora and phylloplane fungi which are an integral feature of many plant surfaces. Walker (1978) has reviewed some of the procedures which may be used in the preparation of geological samples, including fossilized plant and animal specimens.

Robust specimens such as hard plant or animal tissue can be cleaned in a sputter coater or in a cold-plasma gas discharge. With some sputter coaters, it is possible to reverse the polarity of the specimen and the target, allowing the plasma to etch and presumably clean the surface. Similarly, a cold-plasma discharge of a gas such as oxygen can quickly remove organic material from the surface of an inorganic specimen. Such techniques should be used with extreme caution on suitably stabilized tissue at an erosion rate that does not alter the specimen surface.

Surface cleaning should not be confined to the initial stages of specimen preparation, but should be carried out wherever it appears necessary. For example, fixation and dehydration may also change the nature of a surface contaminant, so that it is more easily removed after these procedures than before. Many of the fracturing techniques (see Section 12.5.5.2), particularly those carried out on dry tissues, result in small fragments that can obscure the specimen. Such fragments should be removed before proceeding to the next stage in the preparation process. The cleaned sample is now ready for the next stage in the process—cell and tissue stabilization.

12.5.3. Specimen Stabilization

Specimen stabilization or fixation is the process whereby the highly mobile solubilized state of the cytoplasm is immobilized by processes of precipitation, denaturation, and cross linking. During structural fixation, most of the ions and electrolytes and many of the low-molecular-weight molecules such as sugars and amino acids are irretrievably lost from the cell, and only the large macromolecular complexes remain.

Many of the fixatives used in SEM have been derived from TEM studies. There are, however, a number of important principles which must be borne in mind when making up a fixative. If one is going to examine either the natural surface of a specimen or tissue or a surface which has been exposed and cleaned prior to fixation, then it is important that the fixative solution should be approximately isotonic with the cell or tissue fluids. In this context, the term "fixative solution" refers to all the components in the fixative other than the fixative itself. This includes the components of the aqueous buffer, balancing ions, and electrolytes and nonelectrolytes such as sucrose. The osmolarity of the fixative solution is as important in obtaining satisfactory fixation as the actual concentration of the fixative. If a specimen or tissue block is to be sectioned or fractured after fixation, then the fixative solution should be slightly hypertonic. The fixation time in the former case can usually be quite short, but in the latter case it must be long enough to allow the fixative to penetrate to the center of the specimen. Longer fixation times have the added advantage of hardening the tissue, making it somewhat easier to fracture at a later stage in the preparative procedure.

The other factors that must be considered when making up a fixative are pH, balancing ions, and temperature. The first two should be as close as possible to the natural environment of the cell and tissue fluids. Fixation of mammalian tissues is usually carried out near 277 K. Care should also be taken with the choice of buffer, as it is important that its effective capacity is within the range of the pH of the specimen. Fixation may take place in several ways. The specimen may be exposed to the vapor phase of the fixative, immersed in or floated on a solution of the fixative, and in the case of large animals, perfused *in vivo* with a fixative solution. Vapor fixation is useful for preserving delicate aerial structures, while immersion and perfusion fixation ensure that the fixative reaches the appropriate parts of the specimen as rapidly as possible. Flotation fixation has proved effective for leaves, petals, and skin.

Chemical fixation of cells and tissues has moved away from morphological preservation alone to a situation where the chemical constitution of the sample is of equal importance. This more informed approach of combining good morphology and chemical activity has led to the development of new fixatives as well as to a better understanding of the existing methods. In a seminal paper on the subject, Bullock (1984) has taken a careful look at the current state of fixation for electron microscopy. While not offering a panacea for all tissue types, the suggestions in the paper do allow a more intelligent approach to fixation. A number of examples will suffice to demonstrate this new approach.

Glutaraldehyde still remains the fixative of choice for the structural and enzymatic component of cells. Despite its slow rate of entry, it is still widely used, though at much lower concentrations than originally suggested. A study by Coetzee and van der Merwe (1984) has shown that although fixation in glutaraldehyde invariably leads to more or less extraction of a variety of elements, compounds, and macromolecules, the amount of material actually extracted is also very dependent on the buffer used to make up the fixative. Least extraction was found in phosphate-buffered 2–3% glutaraldehyde. An alternative to glutaraldehyde would be one of the bifunctional cross-linking agents such as the dimethyladipimidates, which are known to cross-link proteins at low concentrations although Tzaphidou and Matthopoulos (1988) have shown that the penetration is quite slow.

The move to lower concentrations of fixative also applies to osmium tetroxide where data (Luftig and McMillan, 1981; Aoki and Tavassoli, 1981; Bendayan and Zollinger, 1983) indicate that brief exposure to a low concentration of osmium tetroxide both preserves protein structure and retains antigenicity. The work of Aoki and Tavassoli (1981) is of particular interest because these workers found that brief exposure to osmium tetroxide, followed by thiocarbohydrazide, and then a second exposure to osmium tetroxide, adequately preserved actin filaments. Such a technique would have an immediate appeal for high-resolution SEM. If one is concerned solely with structural preservation, then a good starting point would be an organic aldehyde followed by osmium tetroxide, both in identical fixative solutions. Glutaraldehyde (2–5%) is the most commonly used aldehyde, but it can be supplemented with 1–2% formaldehyde and/or acrolein, which penetrate tissues faster than the dialdehyde. Care should be exercised when using acrolein as it is a volatile, inflammable, and toxic liquid. The final concentration of the fixative chemical is dictated by the specimen and the constitution of the fixative solution, but it should rarely exceed 5% of the total fixative. Most tissues are conveniently fixed overnight at room temperature, and it is useful to keep the fixative agitated during this time.

Following the initial fixation, the tissue should be washed in isotonic buffer and postfixed for up to 4 h in 1–2% osmium tetroxide at the same temperature used for aldehyde fixation. The samples should be washed free of any excess osmium and briefly rinsed in buffer. Most tissues are fixed by immersion, either by placing a small piece of excised tissue into a large volume of fixative, or by initially allowing a large volume of fixative to continuously bathe the specimen *in situ* before it is excised and immersed in the fixative. The alternative approach of perfusion is best suited to mammalian tissues, and details of the procedure are found in one or other of the general texts of microscopy mentioned earlier.

The lipid component of cells and the hydrocarbon components of organic materials are notoriously difficult to fix and retain *in situ*. Here the aldehydes are less useful, and it is necessary to use osmium tetroxide at a concentration somewhat higher than would give optimal preservation of proteins. Wolf-Ingo *et al.* (1982) have found that a complex of palladium–tungsten, lead–iron, and cobalt–molybdenum salts helps to preserve phospholipids, cholesterol, and lipoprotein particles for subsequent examination in the SEM. The other two major chemical components of cells, nucleic acids and carbohydrates, appear best preserved after glutaraldehyde and formaldehyde fixation.

Bullock (1984) discusses at some length the various additives which may be included in a fixative. These include calcium and lithium ions, ferri- and ferrocyanides, and tannic and picric acids. While each in its own way enhances fixation in a particular case, there does not seem to be any reason to use them in all fixatives. Tannic acid should be of interest to scanning microscopists, as its mordanting action allows an increased loading of metal salts to biological tissue (see Section 13.4).

Bullock raises the question of the newer fixatives, such as the imidoesters and the carbodiimides which have been suggested as replacements for some of the aldehydes. The advantage of these new materials appears to be not so much in morphological preservation but rather, in combination with glutaraldehyde, of retaining chemical activity for subsequent identification by immunocytochemical methods. Because of the increased interest in retaining biological activity as well as structure, there has been a move away from the traditional phosphate and cacodylate buffers to organic buffers such as PIPES [piperazine-*N,N′*-bis-2-ethanesulfonic acid)]. Such buffers avoid the precipitation of inorganic ions and effectively contribute to the osmalility and pH balance of the fixative. But for structural studies alone, the traditional phosphate buffer is to be preferred, as it may cause less extraction of tissue components.

In recent years there has been much interest in microwave-enhanced processing of biological tissue (Boon and Kok, 1987; Armati *et al.*, 1988; Wild *et al.*, 1989, and Argall and Armati, 1990). A brief pulse of microwaves (4–8 s at 2450 MHz) can be used either on native tissue or in buffered saline on samples immersed in fixative. The quality of fixation is equal to that obtained by immersion fixation and has the added advantage that fixation is completed very quickly.

It should be emphasized that each specimen is unique and requires its own peculiar fixation. It is important to experiment with the fixation of a new tissue, varying the specimen size, time and temperature of fixation, type, concentration, and pH of the final fixative and correlating the findings with observations in the light microscope and the TEM. Finally, it should be remembered that tissue stabilization is one of a number of processes in tissue preparation best carried out as a continuous sequence of events. It is important to follow fixation either by dehydration or by staining followed by dehydration. The process may be temporarily stopped either when the specimen is in 70% ethanol or when the specimen is completely dehydrated and stored in a desiccator before coating.

Once fixation is complete, a number of different options are available for further treatment of the specimen. One may proceed direct to dehydration (Section 12.5.4), one may attempt to reveal the internal parts of the specimen (Section 12.5.5), or one may attempt to localize areas of specific chemical or physiological function (Section 12.6).

12.5.4. Specimen Dehydration

Sample dehydration is carried out in two steps. In the first step, the liquid water is either dissolved in and replaced by organic fluids (chemical dehydration) or solidified (freezing). In the second step the organic fluids are removed from the specimen by either critical-point drying or ambient-temperature sublimation and the solidified water removed by low-temperature sublimation (freeze drying). The processes of

1. chemical dehydration,
2. freeze drying,
3. critical point drying, and
4. ambient-temperature sublimation

will be discussed in turn, together with variations of these four closely interrelated dehydration procedures.

12.5.4.1. Chemical Dehydration

The chemical removal of water from samples involves passage through increasing concentrations of organic liquids, accompanied by tissue extraction and shrinkage, and some changes in tissue volume must inevitably take place. Boyde and coworkers (1977) have made a series of careful studies on the volume changes in a variety of plant and animal tissues after various dehydration regimens. They found that critical-point-dried material may shrink by up to 60%, freeze-dried material by 15%, and material air-dried from volatile liquids by 80% of the original volume. Although plant material usually shrinks less than animal tissue, each specimen must be considered separately. Provided the measured change in volume is uniform in all directions and the same in all parts of the specimen (and there is little evidence that this is ever the case), corrections may be made to any measurements made on the sample.

Before embarking on a discussion of chemical dehydration, it might be useful to consider alternative methods of tissue drying. Air drying is not a satisfactory procedure because it results in such gross distortion of the tissue due to the high surface tension of water, which may set up stresses as high as 20–100 MPa during the final stages of evaporation.

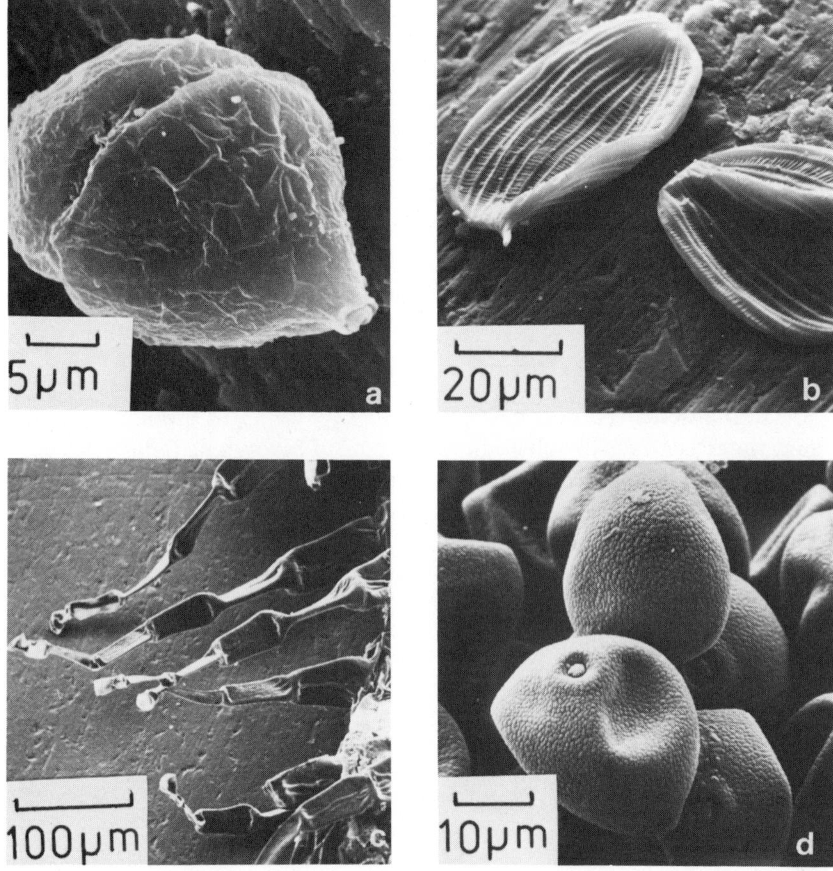

Figure 12.5. Examples of distortion due to air drying specimens. (a) Dinoflagellate alga *Prorocentrum micrans*. (b) Euglenoid flagellate alga *Phacus sp*. (c) Leaf hairs of *Solanum sarrachoides*. (d) Pollen grains of the grass *Poa pratense*.

These forces become increasingly larger as the structures being dried become smaller (Fig. 12.5). Rapid drying of samples from the aqueous phase by placing them in a vacuum has the added disadvantage of ice crystal formation due to the evaporation of the water at reduced pressure. The formation of the ice crystals can severely distort the specimen. Drying tissues from volatile organic liquids with a low surface tension, such as diethyl ether or 1:2 epoxy propane, diminishes but does not negate the problem of surface tension.

Chemical dehydration can be achieved by passing the fixed tissues through a graded series of methyl or ethyl alcohol or acetone solutions. This may be done either by complete replacement of one concentration with the next or by slowly but continuously dripping a large volume of 100% alcohol or acetone into a small container holding the specimen which has been partially dehydrated and suspended in circa 20% alcohol or acetone. This latter technique has the advantage in that it does not subject the specimen to sudden changes in concentration. Care should be taken not to use denatured alcohols, as their use can give rise to small, needlelike contaminants on the specimen's surface. The temperature of dehydration does not appear important, although it is usual to carry out dehydration at the same temperature used for fixation. It is important to remember that 100% alcohol or acetone is very hygroscopic, and it may be necessary to carry out the final stages of dehydration either in a dry atmosphere or in the presence of a suitable drying agent. Rapid chemical dehydration can be achieved using 2-2-dimethoxypropane.

Tissue samples are placed into acidulated solutions of 2-2-dimethoxypropane, which rapidly converts the tissue water to acetone and methanol. Once the endothermic reaction has ceased, the tissues are transferred through three changes of ethanol, methanol, or acetone. If rapid preparation is needed, then this is an effective dehydration procedure. The slower procedures are preferred because there may be a danger of tissue distortion by the more rapid procedure, particularly with larger samples. Kahn *et al.* (1977) have made a careful comparison of ethanol and dimethoxypropane (DMP) dehydration methods on the preservation of tissue cultures, and as Fig. 12.6 shows there is little to

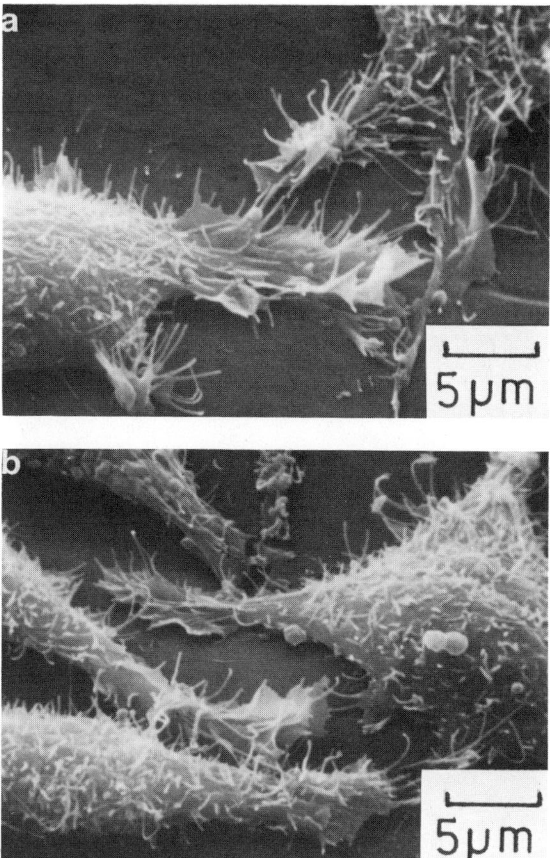

Figure 12.6. Cultured mouse fibroblast cells. (a) Dehydrated with ethanol. (b) Dehydrated with 2-2dimethoxypropane (Kahn *et al.*, 1977).

choose between the two techniques. Tissue distortion during dehydration may be diminished if in the early stages the dehydrant (ethanol, acetone, etc) is diluted with the buffer used as the fixative vehicle.

Liepins and de Harven (1978) have shown that it is possible to dry single-tissue culture cells in a desiccator from Freon 113 at room temperature. Cells attached to cover slips are fixed and dehydrated to

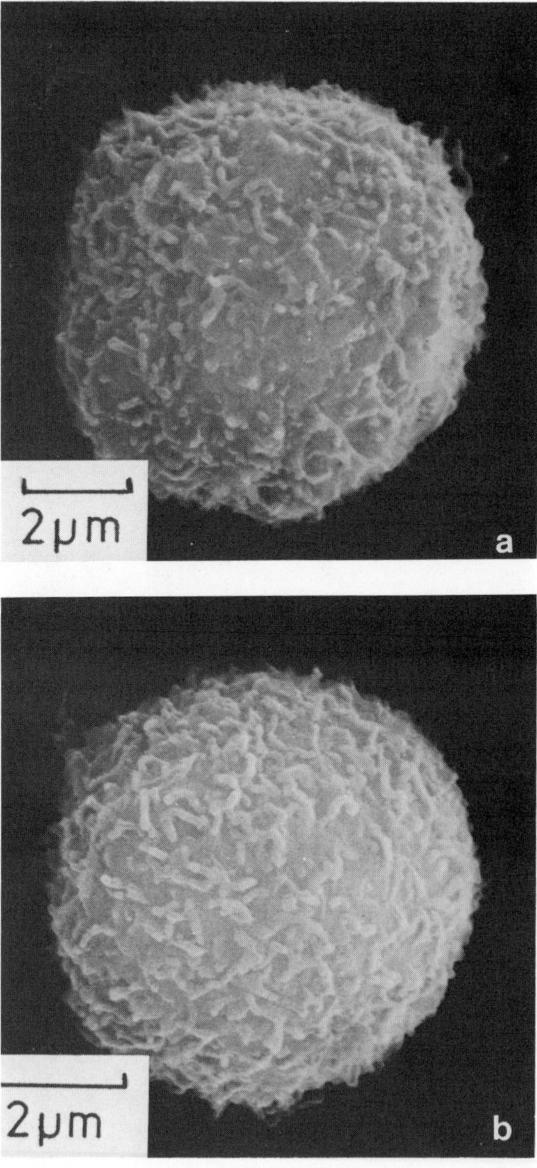

Figure 12.7. A comparison of critical point drying and direct drying using Freon 113. (a), (b) Mastocytoma cells showing surface villi: (a) prepared by critical point drying; (b) prepared by direct drying. (c), (d) Surface of a mastocytoma cell examined in field emission SEM: (c) prepared by critical point drying—note the porosity of the cell surface membrane; (d) prepared by direct drying method—note that the cell surface membrane does not show perforations. From Liepins and de Harven (1978).

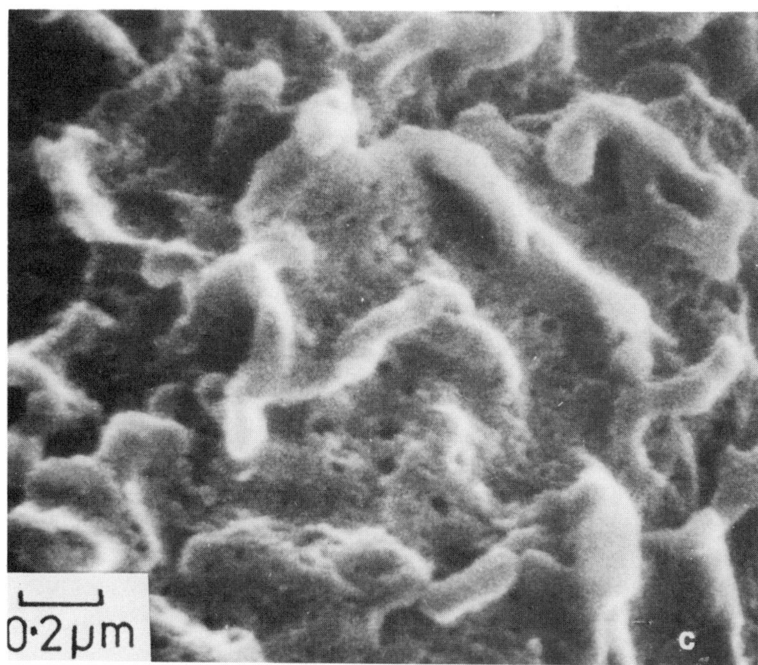

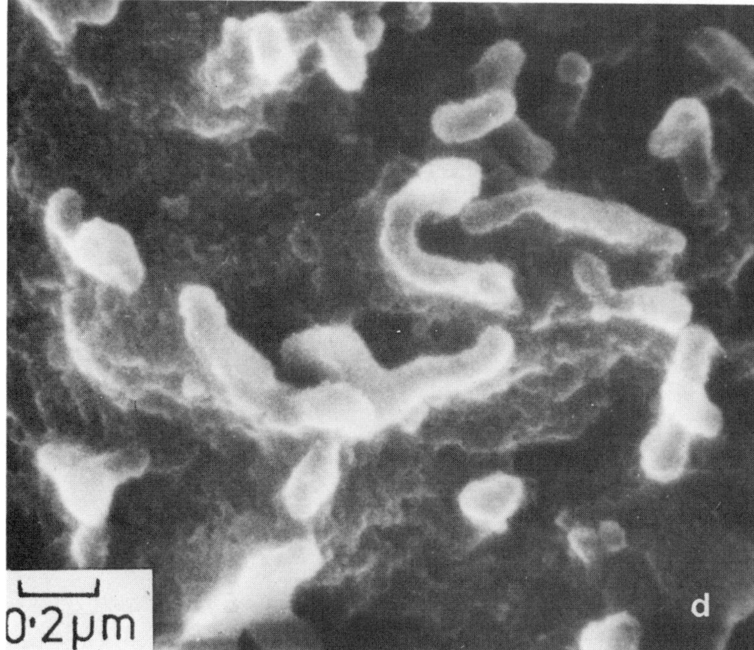

Figure 12.7. (Continued).

100% ethanol and then passed through gradual changes to 100% Freon 113. The specimens immersed in Freon 113 are transferred to a desiccator containing Drierite and evacuated to a pressure of 3 Pa (3×10^{-2} torr). The liquid phase of the Freon 113 evaporates in a few minutes, leaving cells whose ultrastructure is comparable to that of cells dried by critical-point drying (Fig. 12.7). Further studies are necessary to show whether this technique can be effectively applied to pieces of tissues. Similar experiments have been carried out on hard tissues using hexamethyldisilazane (Nation, 1983). Lamb and Ingram (1979) have devised a method which allows specimens to be dried directly from ethanol under a stream of dry argon without the use of the critical-point drier. The technique gives results as good as those obtained by other methods, although with sensitive samples such as ciliated epithelium the critical-point drying method gives better results. Doucet and Bradley (1989) describe a technique of spin drying for drying micrometer-size particles. The particles are chemically dehydrated and suspended in hexane or ethanol, and a droplet is placed on a glass cover slip or a well-polished carbon stub which is spun at 1000 rpm in a vacuum. The volatile liquid evaporates rapidly, and the fine particles remain on the substrate without agglomeration.

It is difficult to give a set time for effective dehydration because, like fixation, dehydration is very dependent on specimen size, porosity, and whether the internal or the external features of the cell are to be examined. As a general guide, most tissue blocks of 1–2 mm^3 are effectively dehydrated after 15 min each in 15%, 30% , 50%, 70%, 95%, and 100% ethanol or acetone followed by three 10-min changes in anhydrous solvent, the whole procedure being completed in about 2 h. Rostgaard and Tranum-Jensen (1980) have shown that there is better structural preservation if the dehydration is continuous rather than being carried out in a series of steps.

The timing of the preparative procedures is important to avoid any unplanned delays during the process. Thus it is usually convenient to select and clean tissue samples on the afternoon of day 1, to fix them in aldehyde overnight; to wash them, postfix them in osmium, dehydrate, and mount them on specimen holders during day 2; and to coat the samples early on day 3 and spend the remainder of the day examining the specimen in the microscope. The final stage of dehydration, unless one is going to embed the material for subsequent sectioning or fracturing, is the critical-point drying procedure. If the samples are to be critical-point dried to remove the organic liquids, they must remain submerged in the liquid throughout the procedures.

12.5.4.2. Critical-Point Drying

Scanning microscopists have rediscovered the critical-point drying method first developed years before by Anderson (1951) for studies on bacterial flagella. The rationale behind the procedure is that the two-phase stage (vapor and liquid) of the majority of volatile liquids disappears at a certain temperature and pressure, the so-called critical point. At the critical point the two phases are in equilibrium, the phase boundary disappears, and there is, therefore, no surface tension. Much has already been written about the theory and practice of critical-point drying, and reference may be made to the papers by Barlett and Burstyn (1975) and Cohen (1979) for the details the technique.

The practical aspects of the method involve passing the tissue from the dehydrating fluids (methanol, ethanol, acetone) through the intermediate fluids (amyl acetate, Freon TF) into the transition fluid (liquid CO_2 or Freon 13) in the high-pressure critical-point drying apparatus (or "bomb"). Ideally, one would like to critical-point dry from water, but the critical temperature of water (648 K) is so high that one would effectively cook the specimen. The transition fluids are the chemicals from which critical-point drying is carried out. Pawley and Dole (1976) have shown the importance of complete removal of the intermediate fluid before the drying is carried out.

Although the critical-point drying procedure is roughly the same for all pieces of equipment, important differences do exist between the various commercially available critical-point drying bombs, and it is important to follow the procedures which the particular manufacturer recommends. As long as the operator remembers that the process goes on inside a bomb, with its attendant high pressures, and uses the technique with care, then critical-point drying can be carried out safely and effectively. It is important to check regularly the valves, seals, and, where appropriate, the viewing windows. Further safety precautions are given in the papers by Cohen (1979) and Humphreys (1977). Most critical-point drying is carried out using liquid CO_2. Once the critical point drying has been completed, the specimens should be immediately transferred to a desiccator, where they may be stored prior to mounting and coating.

In spite of the convenience and widespread use of critical-point drying, there is compelling evidence that liquid dehydration followed by critical-point drying should be discarded in favor of freeze-drying (see Section 12.5.4.3). Nevertheless, it is accepted that liquid dehydration in one of the primary, secondary, or polyhydric alcohols will continue to remain a method of choice. This convenience must be weighed against the chemical extraction, distortion, and shrinkage which accompany these methods. For high-resolution SEM studies, and for investigations in which one hopes to obtain some measure of the chemical activity of the specimen, these techniques should be dispensed with and the water removed by freeze-drying.

The evidence against the use of critical-point drying as a final step in the dehydration procedure is impressive. Following the first detailed study on this aspect of specimen preparation by Boyde *et al.* (1977), a number of papers have appeared which confirm the realization that critical-point drying causes gross (up to 70%) and spatially unequal dimensional changes in most specimens.

These dimensional changes have been measured either directly on the sample during drying by morphometric analysis using an image analyzer (Boyde *et al.*, 1977) or more simply by mapping the changes in relative position of externally applied fluorescent markers (Campbell and Roach, 1981). Other studies of shrinkage have been carried out by Boyde and Maconnachie (1979) and by Eskelinen and Saukko (1983). There is every reason to believe that critical-point drying may cause selective solubilization of chemical components of cells. After all, supercritical CO_2 is used to "naturally" decaffinate coffee.

In the absence of freeze-drying equipment, it is possible to diminish the deleterious effects of critical-point drying by the inclusion of various additives to the stabilization and dehydrating fluids. Post-osmication or treatment with Ca^{2+} and Cu^{2+} ions (Boyde and Maconnachie, 1979); lithium and uranyl salts, and cetylpyridinium chloride; successive treatment with glutaraldehyde, osmium, tannic acid, and uranyl salts (Wollweber *et al.*, 1981); and the use of tannic acid and guanidine hydrochloride to enhance osmium binding (Gamliel *et al.*, 1983) all reduce shrinkage, in some cases to as low as 5%. The selective tissue erosions and perturbation of surface morphology still remain a problem.

It would be wrong, however, to think that all critical-point drying gives rise to a distorted sample. In a careful study on rat hepatocytes, Nordestgaard and Rostgaard (1985) found that both freeze-drying and critical-point drying caused specimen shrinkage. At low magnification, i.e., up to 3000X, there was little to choose between freeze-drying and critical-point drying, but at high resolution, critical-point drying was superior to freeze-drying.

12.5.4.3. Low-Temperature Drying

Although this is strictly a low-temperature preparation method, it is appropriate to discuss it here as it is an effective alternative to critical-point drying in spite of the danger of ice crystal damage. Infiltrating the specimens with a penetrating cryoprotectant, such as glycerol or dimethyl sulfoxide, does to some extent alleviate volume shrinkage but in turn creates other problems. The cryoprotectants are retained in the specimen after drying and slowly outgas and obscure the specimen and contaminate the microscope column. This outgassing can be reduced by examining specimens at low temperature or by using high-molecular-weight nonpenetrating polymeric cryoprotectants such as polyvinylpyrrolidone or hydroxyethyl starch. Practical hints on how best to carry out the procedure are given in Section 12.8.7 and in the review by Boyde (1978), but an outline of the procedure is as follows: Small pieces of the specimen are quench-frozen in a suitable cryogen such as melting ethane or propane and transferred under liquid nitrogen to the precooled cold stage of the freeze dryer. The chamber of the freeze dryer is evacuated to its working vacuum of 20–100 mPa, at which point the ice sublimes from the sample. Ice sublimation is a function of temperature and pressure—the lower the temperature, the lower the pressure required to enable the water to leave the sample. Most freeze drying is carried out at about 200 K and at a pressure of 20–100 mPa. Lower temperatures diminish ice recrystallization but require much lower pressure or longer drying times. It is important to provide a condensing surface for the water molecules removed from the sample. A liquid-nitrogen-cooled trap a few millimeters from the specimen is the most effective way to achieve this, but chemical desiccants such as phosphorus pentoxide or one of the zeolites also suffice. The actual time taken for freeze drying depends very much on the properties of the sample such as size, shape, the amount of free and bound water, and the relative proportions of fresh to dry weight. The drying procedure should be worked out for each specimen, and it is important to allow the specimen to gradually reach ambient temperature while under vacuum before it is removed from the freeze dryer. There is still considerable debate regarding the merits of freeze-drying versus critical-point drying. On balance, it would appear that the freeze-drying method is better for preserving cellular detail. The critical-point drying method provides more information about the whole specimen. A good demonstration of the results obtained by critical-point drying and freeze-drying is shown in Fig. 12.8.

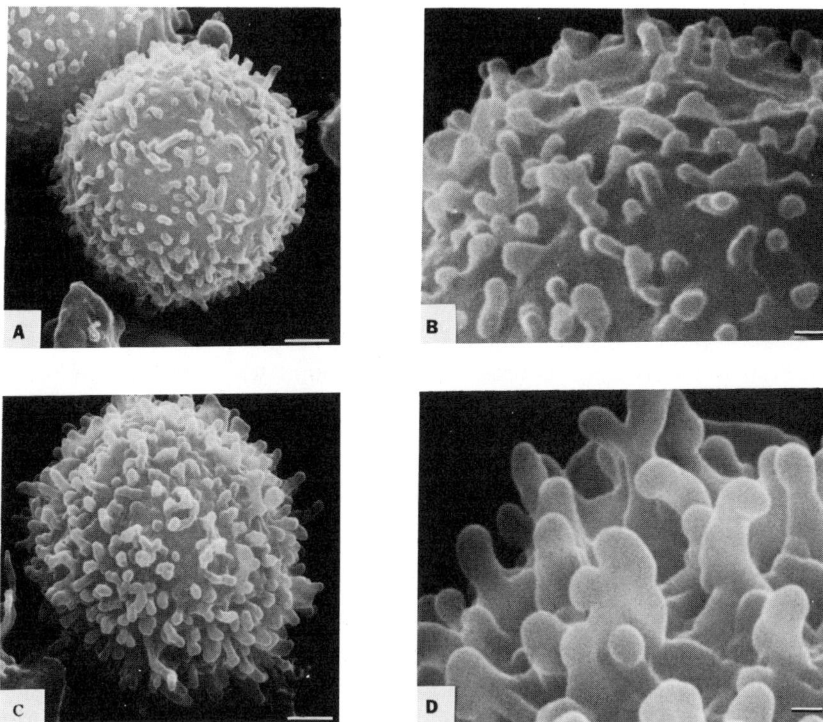

Figure 12.8. A comparison of the results obtained after critical-point drying and freeze drying. (a) and (b) are of critical-point dried lymphocytes showing narrow microvilli and a relatively smooth underlying surface. Magnification bar in (a) = 1.0 μm, in (b) = 0.2 μm. (c) and (d) are of the same specimens, freeze dried. The microvilli are much wider and the cell surface is more complex. Magnification bar in (c) = 1.0 μm, in (d) = 0.2 μm (Schneider *et al.*, 1978).

It is also possible to freeze dry tissues which have been dehydrated in ethanol or acetone and then quenched to about 163 K. The cold-solvent-dehydrated samples are dried at 243–253 K at a pressure of between 1.0 and 0.1 Pa. Katoh (1978) gives details of a similar scheme in which specimens are fixed, dehydrated in ethanol, and then gradually infiltrated with Freon TF. The specimen is then placed on a large block of aluminum and the two are quench-frozen in liquid nitrogen. The cold specimen and metal block are then placed in a vacuum desiccator and the specimen is dried under a vacuum of about 1 Pa while the metal block slowly warms up. Because it is possible to freeze-dry unfixed tissue, this procedure is doubly useful in that it provides samples for morphological study and analysis. In a similar procedure, devised by Akahori *et al.* (1988) and Kaneko *et al.* (1990), delicate biological samples have been freeze dried from *t*-butyl alcohol which forms a solid at 301 K. The samples are fixed and chemically dehydrated in an ethanol series and then gradually infiltrated with *t*-butanol. The samples were slowly cooled to (273 K) and freeze-dried at between 273 and 283 K. A metal block in the vacuum chamber, precooled with LN_2, acts as an effective condensor. The same procedure may be used with partially dehydrated specimens (50% water) because *t*-butanol is miscible with water. The results are impressive and are as good as may be achieved with critical-point drying. As with critical-point-dried material, samples should be stored in a desiccator following drying as they can readily absorb water.

12.5.4.4. Ambient-Temperature Sublimation

A number of organic compounds are miscible with the commonly used chemical dehydrants and form solids which sublime at ambient temperatures. Camphor has been used (Cleaton-Jones, 1976), but because its maximum sublimation rate is at 477 K it has been discarded in favor of materials which sublime at much lower temperatures. One such material is Peldri II, a fluorocarbon compound which is solid at room temperature and melts above 298 K. Once solidified it sublimes with or without vacuum to dry samples without producing surface-tension effects. Details of the method are given by Kennedy *et al.* (1989) and Kan (1990), and some comparative results are shown in Fig. 12.9.

12.5.5. Exposure of Internal Surfaces

One of the prime uses of a scanning microscope is to examine and analyze surfaces. Many of the surfaces we wish to examine are at the natural interface of the object and its environment. Many more are hidden from view, and we must seek ways of exposing them for examination. A number of techniques have been devised which enable these internal surfaces to be examined. The techniques fall into four main classes: *sectioning,* in which there is an orderly procession through a hardened sample by means of tissue slices; *fracturing,* in which the

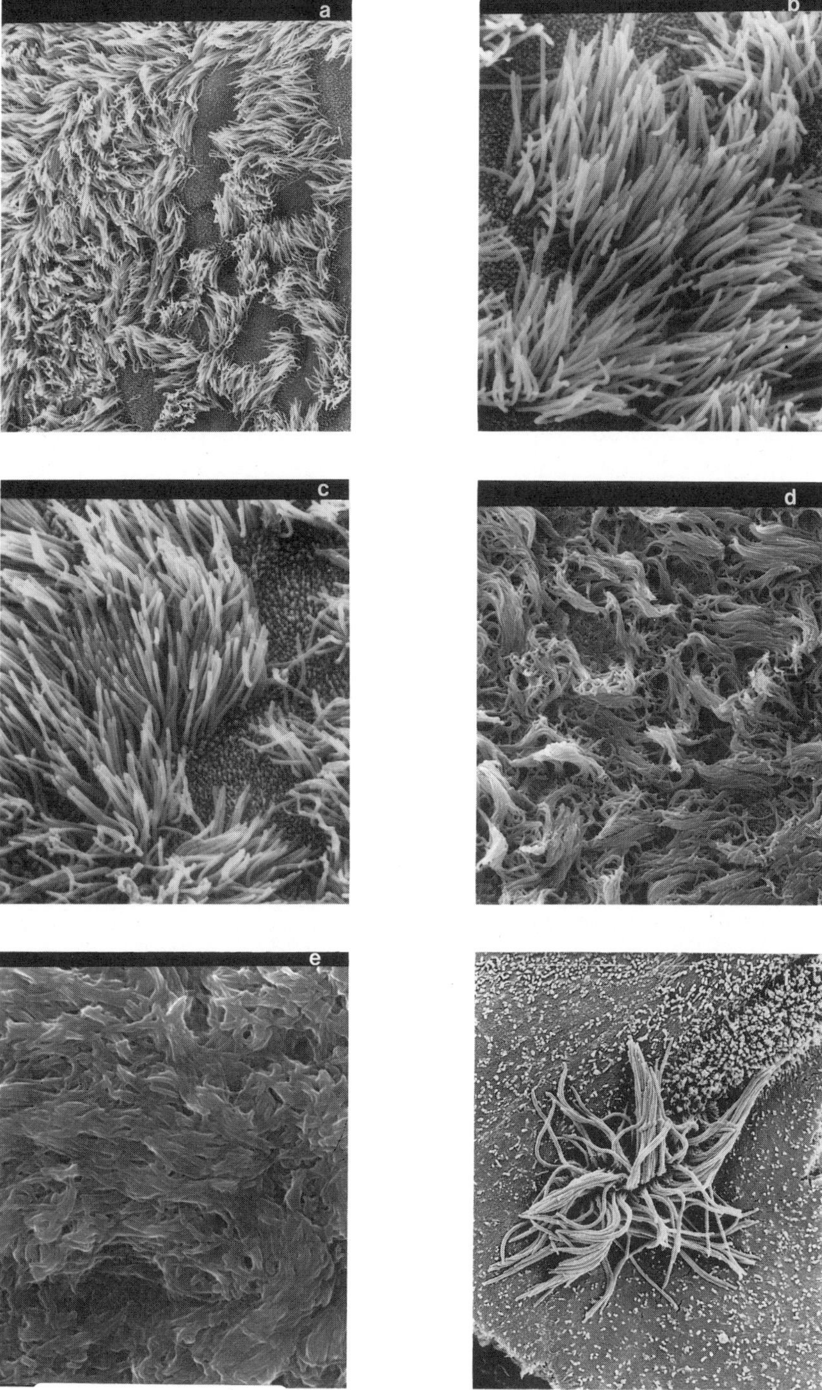

Figure 12.9. Samples of rabbit tracheal epithelium (a), (b), and (f) dried by sublimation from a solidified fluorocarbon, Peldri II; (c) critical-point dried; (d) hexamethyldisilizane dried; and (e) vacuum dried from alcohol. The Peldri II sublimation results are as good as may be obtained using critical-point drying (Kennedy *et al.*, 1989).

specimen is broken open, hopefully along a surface of structural continuity but frequently at some random point, to reveal two highly contoured surfaces; *replication* in which a plastic cast is made of the structure concerned and revealed either after the replica is peeled off or after the organic material is removed by corrosion techniques; and *etching,* in which the surface is progressively eroded with a beam of energetic ions.

The internal contents may be revealed more or less at any stage during the preparative procedure, even in the microscope, by means of microdissection. However, most biological tissues and hydrated organic samples are very soft and are more conveniently sectioned or fractured after stabilization and dehydration.

12.5.5.1. Sectioning

Tissue sections may be prepared using conventional light and TEM techniques and viewed in the SEM in either the transmission or the reflective mode. One of the advantages of the SEM is that, at a given accelerating voltage, there is an up to twenty-fold increase in the section thickness which may be examined in the transmission mode, although at somewhat reduced resolution. Sections examined in the reflective mode show more detail if the embedding medium is removed before examination, leaving the material behind. For thin sections (20–200 nm) the material is fixed, dehydrated, and embedded in one of the epoxy resins. Sections are cut on an ultramicrotome and the resin removed with organic solvents, sodium methoxide, or controlled erosion in an ion beam or cold-plasma gas discharge (Steffens, 1978). Alternatively, the thin sections may be mounted directly onto a solid substrate and examined in the SEM without further treatment. A remarkable amount of detail may be seen in seemingly flat sections, particularly from frozen-dried sections, using the BSE signal.

Thicker sections (0.2–5.0 μm) can also be cut from resin-embedded material and the resin removed using the technique described by Thurley and Mouel (1974). Sections 2–10 μm thick may be cut on ordinary microtomes from material embedded in resin or paraffin wax and then deparaffinized with xylene or the resin removed by one of the aforementioned methods. The sections are either attached to small glass cover slips using for example Haupt's adhesive for correlative light microscopy–SEM studies or picked up on electron transparent supports for correlative TEM–SEM studies. Humphreys and Henk (1979) have used an oxygen plasma to surface-etch thick (0.1 μm) resin sections of kidney tissue prepared using standard TEM techniques and revealed ultrastructural details of subcellular organelles (Fig. 12.10).

Warner (1984) considers gallium to be a useful alternative to plastic and paraffin embedding. Its low melting point (303 K), low toxicity, and high conductivity make it potentially very useful as an embedding medium for plastics and polymers. The gallium-embedded samples can be sectioned and fractured. Although gallium wets surfaces relatively easily,

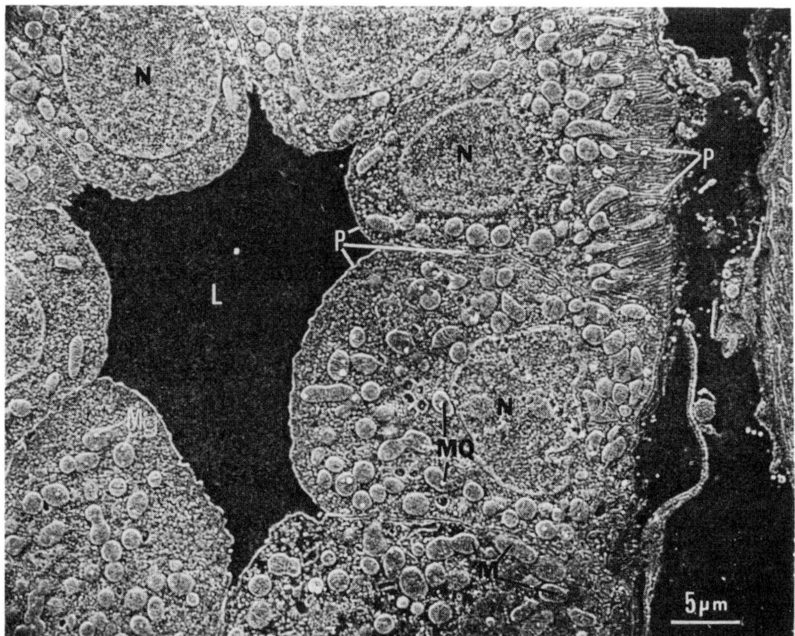

Figure 12.10. Scanning electron micrograph of a 1-μm-thick section of a convoluted tubule of a mouse kidney embedded in an Araldite–Epon mixture. The section was surface etched with an oxygen plasma, metal shadowed with platinum–palladium, and sputter coated with gold. Subcellular structures are identified by the etch-resistant residues forming patterns on the surface of the section after etching. Plasma membranes (*P*) are recognizable at cell surfaces lining the lumen (*L*), between adjacent cells, and at the upper right of the micrograph as infoldings from the basal lamina. *N*, nuclei; *M*, mitochondria; *MQ*, mitochondrial matrix granules (Humphreys and Henk, 1979).

it does not penetrate samples and so cannot be used to infiltrate soft biological tissue.

A different approach to the study of internal structures is given by Nagele *et al.* (1984) and Hawes and Horne (1985). The material is stabilized by conventional fixation and embedded in polyethylene glycol. The embedded tissue is sectioned to the desired region, immersed in solvent to remove the polyethylene glycol, critical-point dried, and examined in the SEM. As Nagele *et al.* have shown, Fig. 12.11, this procedure allows clean fractures to be made through tissues with a minimum of distortion. Bauer and Butler (1983) achieved essentially the same result using polyvinyl alcohol (PVA). The tissues are gradually infiltrated with increasing concentrations of PVA, and the internal surfaces are exposed by cryofracturing. The fracture faces may be etched with ethanol. The internal structures may also be exposed by fracturing paraffin wax (Shennaway *et al.*, 1983) or sectioning polyester wax embedded material (Norenburg and Barrett, 1987). The embedding materal is easily removed by solvents from either the block face or the sections. Such samples can be used for light and scanning microscopy.

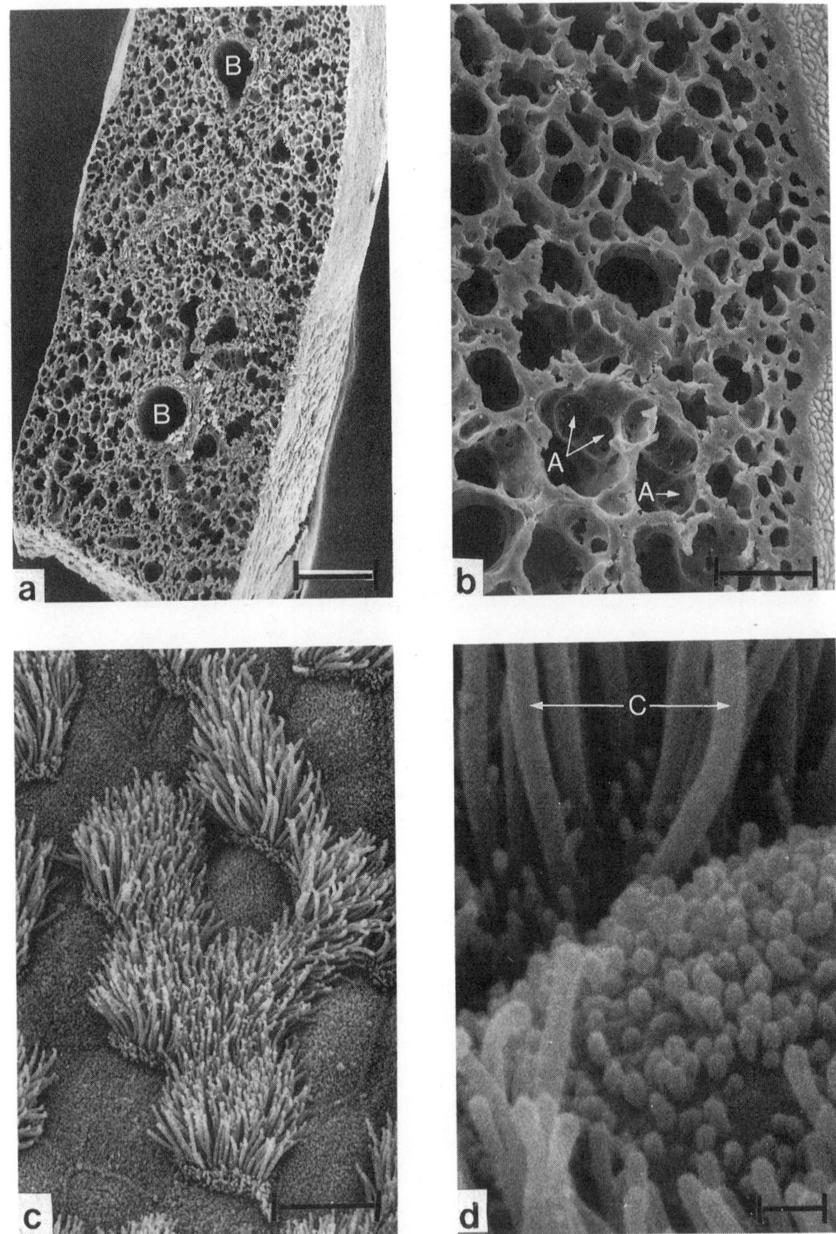

Figure 12.11. Fracture faces from mouse lung tissue embedded in polyethylene glycol. (a) Low-magnification SEM photograph showing a lobe of a mouse lung, fractured to reveal internal detail; B = bronchioles (magnification, ×40; scale bar = 250 μm). (b) Enlarged portion of (a), showing alveoli (A) (magnification, ×128; scale bar = 100 μm). (c) Ciliated epithelial surface of a bronchiole (magnification, ×2640; scale bar = 5.0 μm). (d) Enlarged portion of (c), showing good ultrastructural preservation of fine surface detail (magnification, ×16,000; scale bar = 0.5 μm). (Nagele *et al.*, 1984.)

12.5.5.2. Fracturing

It is sometimes more satisfactory to expose the interior of the specimen using fracturing techniques. Hard tissues can be broken before fixation, and clean, cut surfaces of wood samples may be readily obtained by using a razor blade. Exley *et al.* (1974; 1977) recommend that low-density woods are best cut fresh or after soaking in cold water, whereas denser woods may need softening in hot water before they cut cleanly. Cytoplasmic debris may be removed with a 20% solution of sodium hypochlorite, and following dehydration and coating, the small blocks of wood are mounted in such a way that the edge between two prepared faces points in the direction of the collector, as shown in Fig. 12.12. Somewhat softer tissues can be teased apart with dissecting needles before or after fixation, but the most satisfactory fractures are obtained from brittle material.

Flood (1975) and Watson *et al.* (1975) proposed a dry fracturing technique in which a piece of adhesive tape is gently pressed against

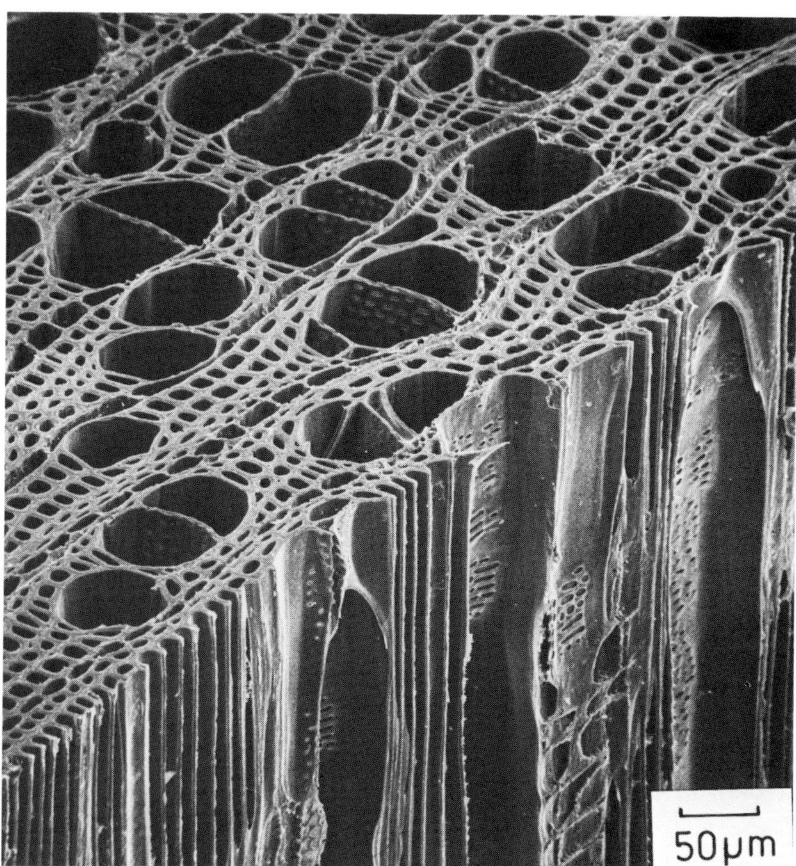

Figure 12.12. Transverse and radial longitudinal surfaces of the wood of *Notofagus fusca* (Exley *et al.*, 1974).

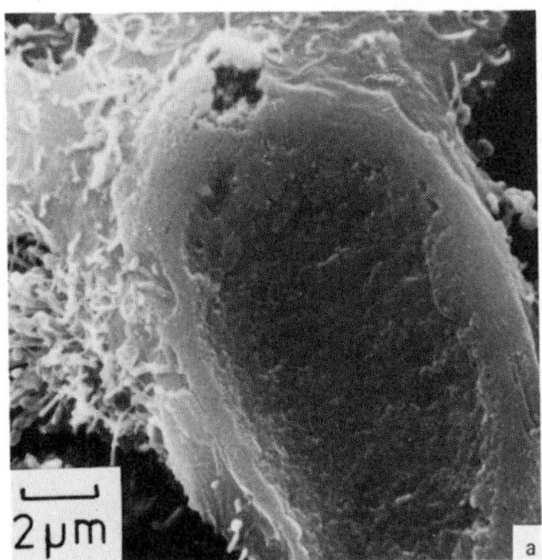

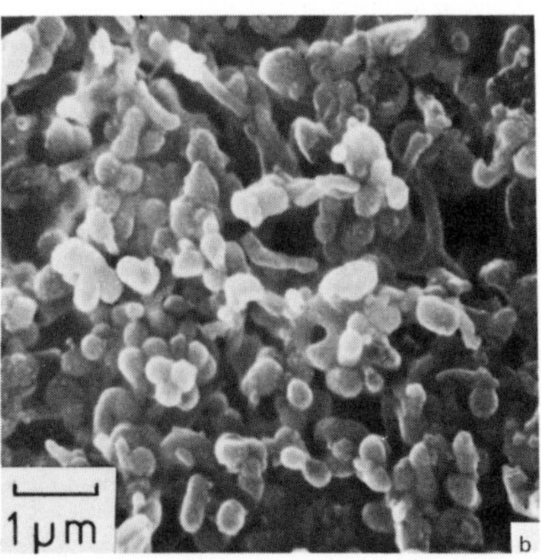

Figure 12.13. (a) Flattened Hela epitheloid cell, dry stripped by adhesive-tape method, and showing the undersurface of the raised margins. (b) Surface of a Hela cell stripped by the adhesive-tape method, revealing rounded organelles (Watson *et al.*, 1975).

dehydrated and critical-point tissues. The piece of tape is removed and in the process tears away some of the tissue, revealing the internal contents. Complementary fractures may be obtained by placing a small piece of tissue between two pieces of adhesive, which are pulled apart and attached side by side on the specimen stub. Larger pieces of tissue may be dry-fractured either by cutting and teasing apart with a small scalpel or by initiating a cut in the specimen, the two edges of which are then held by fine forceps and pulled apart. Examples of such samples prepared by this technique are shown in Fig. 12.13. The sticky type method can be used equally well on single cells attached to a glass substrate with polylysine, revealing organelles and cellular substructure.

Although most resin-embedded material is sectioned to reveal the internal contents, some may be fractured. Westphal and Frosch (1985) have found that cells and tissues, when embedded in a hydrophilic melamine resin, are as brittle as glass and can easily be fractured.

An alternative approach is to use a cryofracturing technique in which unfixed, fixed (Haggis *et al.*, 1976), ethanol-infiltrated (Humphreys *et al.*, 1977), paraffin-infiltrated (Dalen *et al.*, 1983), or resin monomer–infiltrated (Tanaka, 1974) material is immersed in liquid nitrogen to convert the liquid to a brittle solid mass. This brittle material is then fractured under liquid nitrogen using a blunt scalpel, the fractured pieces returned to the appropriate place in the sequence of tissue processing, and the preparation completed. Examples of the results of these techniques are shown in Fig. 12.14. Haggis and Bond (1979) have used the cryofracturing technique with great effect in a study of chick erythrocyte chromatin by quench-freezing glycerinated specimens and fracturing them under liquid nitrogen. After freeze fracture, the cells or tissues are thawed into fixative. Fixation is very rapid, as the fixative quickly diffuses into the fractured cells, but soluble constituents have time to diffuse out, to give a deep view into the cell as shown in Fig. 12.15.

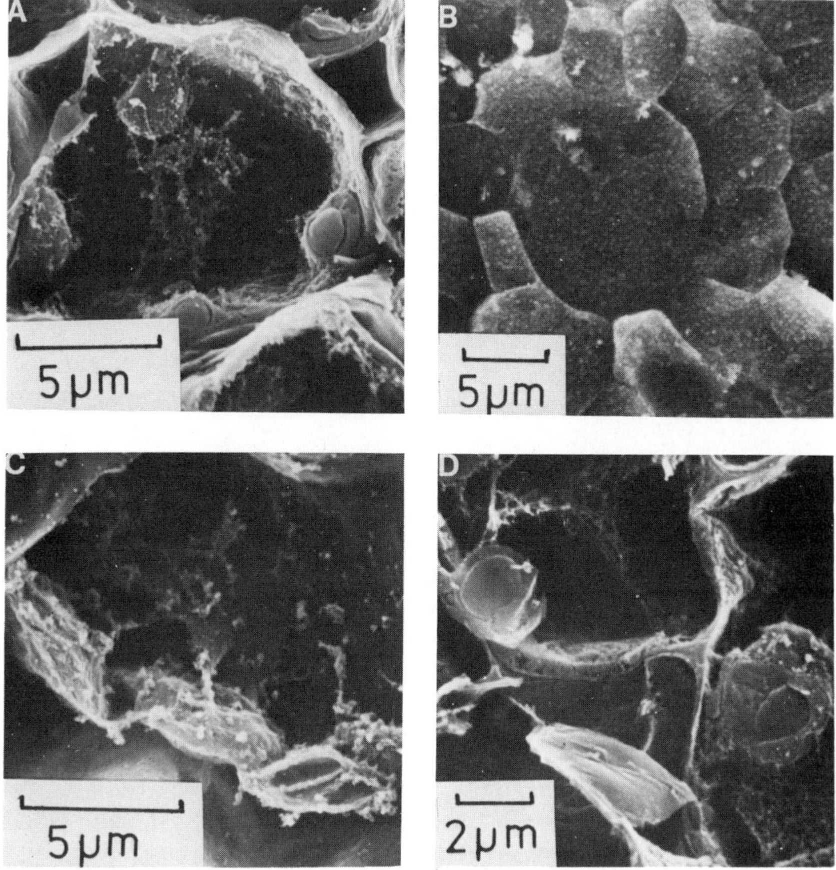

Figure 12.14. Low-temperature SEM. (a) Ethanol cryofracture cells from root of *Lemna minor*. (b) Unfixed, frozen-hydrated fracture surface of cells from root of *Zea mais*. (c) Glycerol cryofracture of cells from root of *Lemna minor*. (d) Resin cryofracture of cells from root of *Lemna minor*.

Haggis (1983) has extended the freeze–fracture–fix thaw technique to high-resolution SEM. Cryoprotected cells are rapidly frozen and fractured under liquid nitrogen. The fractured cells are thawed in a fixative solution which, with or without the aid of detergents, has the effect of removing a large proportion of the soluble material in the cell, exposing components of the cytoskeleton and the nuclear chromatin. These structures may be preserved and observed after critical-point drying and rotary coating with platinum. The results are quite impressive, as the technique allows visualization of microtubules and intermediate filaments with the added advantage of seeing them dispersed within the three-dimensional space of the cell. The results obtained using high-resolution SEM are directly comparable with the information obtained by freeze-etch replicas. A more general approach to the use of freeze-fracture techniques for viewing internal cell structures by SEM is given in an earlier paper by Haggis (1982). The different methods for the

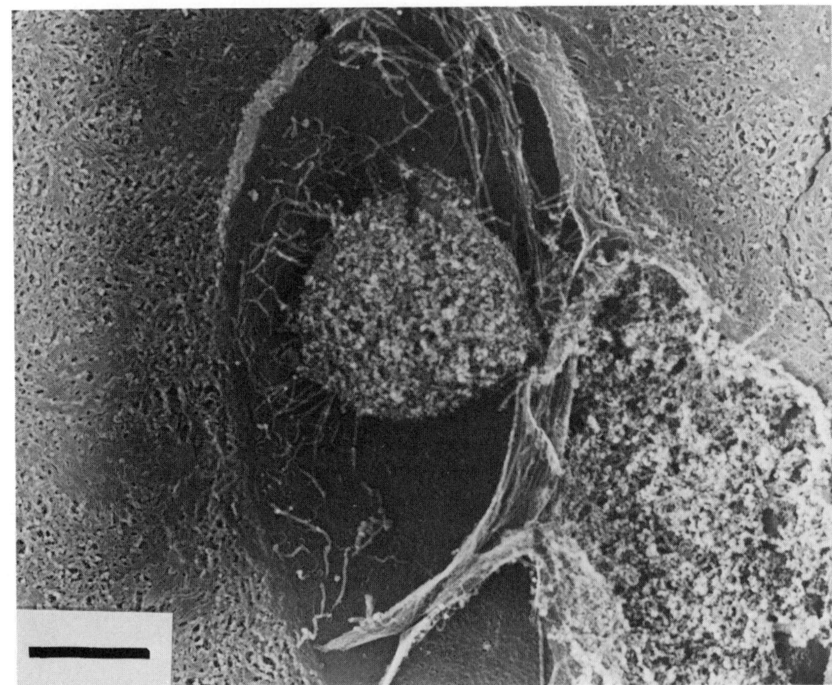

Figure 12.15. A whole chicken erythrocyte with a prominent fracture nucleus, cytoplasmic strands to the left and microtubles to the right. Magnification bar = 1 μm (Haggis and Bond, 1979).

selective removal of soluble cell components are discussed in more detail and include osmium digestion, glycerol extraction, delayed fixation, and treatment with detergents. A similar method has been devised by Inoue (1983). Tissues are fixed with osmium tetroxide, rinsed with distilled water, and infiltrated with 5% DMSO before being quench-cooled. The tissue pieces are then cryofractured and the fracture pieces rinsed in distilled water to remove cytoplasmic debris. Such methods allow three-dimensional visualization of mitochondria and endoplasmic reticulum in the SEM. Barnes and Blackmore (1984, 1988) and Barnes (1990) have successfully applied this technique to plant material, Fig. 12.16 shows details of subcellular ultrastructure.

The cryofracturing methods have a number of advantages over the dry fracturing techniques, not the least of which is a diminution in the amount of debris left on the surface of the specimen, which can either obscure the surface or charge up in the electron beam. Generally speaking, cryofracturing is more easily carried out on animal than on plant material because of the more homogeneous density of soft animal material and the presence of cellulosic and lignified cell walls in plant material, which causes the plant material to splinter more easily and give very irregular fracture faces. We will return to cryofracturing of chemically untreated material in Section 12.8.6.

Jones (1981) has devised a method of ultraplaning to reveal the

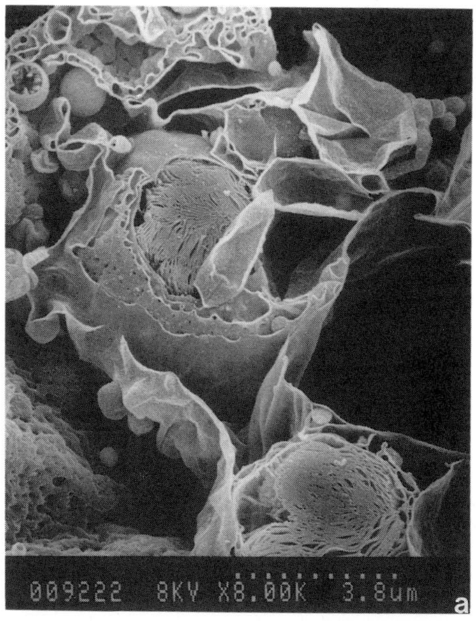

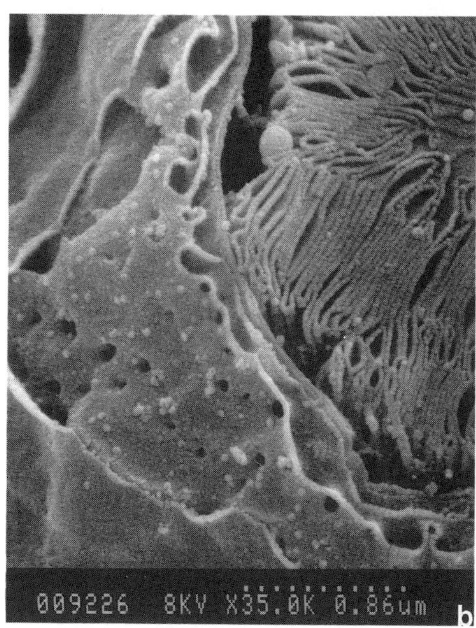

Figure 12.16. (a) SEM images of cryofractured cells. (a) Part of a mesophyll cell of *Aucuba japonica* showing chloroplasts, endoplasmic reticulum, and mitochrondria. Magnification bar = 3.8 μm. (b) Detail of (a) showing stacks of chloroplast lamellae and endoplasmic reticulum. Magnification bar = 860 nm (Barnes, 1990).

internal contents of tissues. Tissues are impregnated with osmium and uranyl acetate and embedded in a high-molecular-weight polymer of polyethylene glycol to produce a hard block. The block face is then planed on an ultramicrotome to give a very smooth surface. The polyethylene glycol is removed from the block face by repeated washes in water, and the tissue, after staining with lead salts, is taken through a dehydration–critical-point drying regime before being examined in the SEM. An alternative procedure involves a celloidin-paraffin embedding, with the embedding medium being removed by xylene. The use of heavy metal impregnation lessens the need for coating. The advantage of the procedure is that it permits a controlled and accurately oriented plane of section to be made and allows one to view a large surface area at low magnification. In addition, the quality of the preparation also allows high-resolution studies to be made on the same surface. This type of preparation would also permit the use of backscattered imaging.

12.5.5.3. Replication

Papers by Pameijer (1979) and Ohtani *et al.* (1982) give details of a wide range of replica techniques which may be applied to natural external surfaces, as shown in Fig. 12.17, which compares the original surface and a replica. Replicas can also be made of internal surfaces by infiltrating an organ with a low-viscosity plastic and allowing the plastic to polymerize, after which the tissue is digested away, leaving a cleaned solid cast of the surfaces. The procedures for making corrosion casts have been reviewed recently by Hodde *et al.* (1990) and Lametschwandtner *et al.* (1990). Such replicas are particularly useful for the interpretation of areas with large internal surfaces such as the lung or kidney, and for the investigation of the microvasculature of a wide range of organs. Figure 12.18 shows details of the villous capillary network produced by using the latex method described by Nowell and Lohse (1974). Methyl methacrylate (Batson's casting medium) has been used by Kardon and Kessel (1979) to make detailed casts of the microvasculature of rat ovary. Nopanitaya *et al.* (1979) have introduced a modification of the methyl methacrylate medium by the addition of Sevriton, a low-viscosity plastic, to give a mixture about ten times less viscous than the Batson's casting medium. Attempts to use corrosion-casting replica techniques on plant tissue have not been successful, presumably because plants do not have a mechanism to pump fluids throughout their structure. However, Green and Linstead (1990) have made excellent replicas of complex living shoot structures. A dental-impression plastic is applied to the surface of the sample and allowed to set. The resulting complex mold is everted and filled with liquid epoxy resin. The hardened cast faithfully reflects the surface features.

A novel modification of the replica method is given by Grocki and Dermietzel (1984). These workers found that simultaneous perfusion of fibrinogen and thrombin combined to form fibrin inside cerebral blood vessels which could then be more easily excised from the brain tissue. Morrison and Buskirk (1984) have ingeniously combined sequential *in situ* microdissection in the SEM of methacrylate casts of the ciliary microvasculature of monkey eyes. Scott (1982) provides a useful compendium of materials which may be used to make replicas of specimens in field settings away from the laboratory.

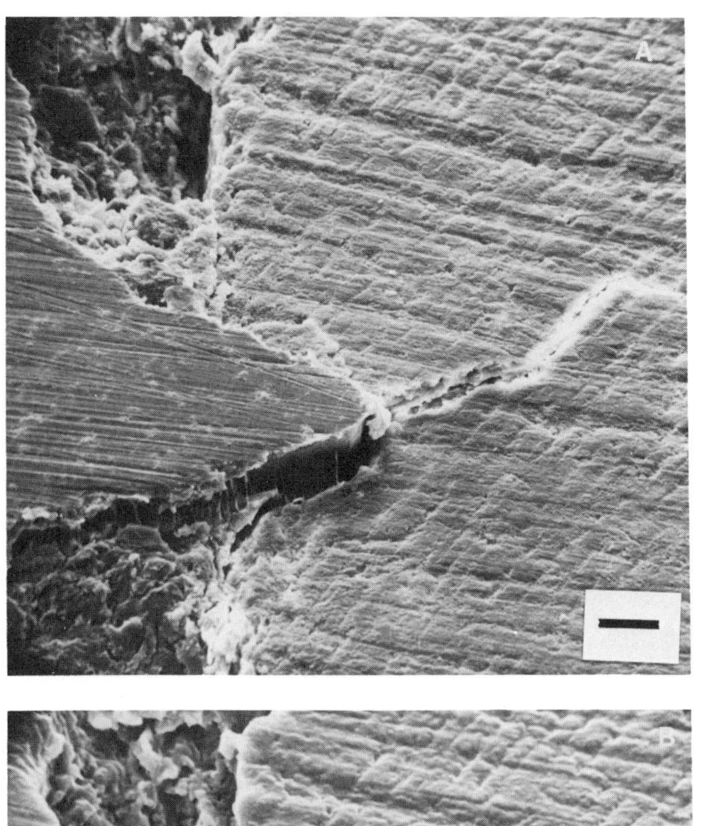

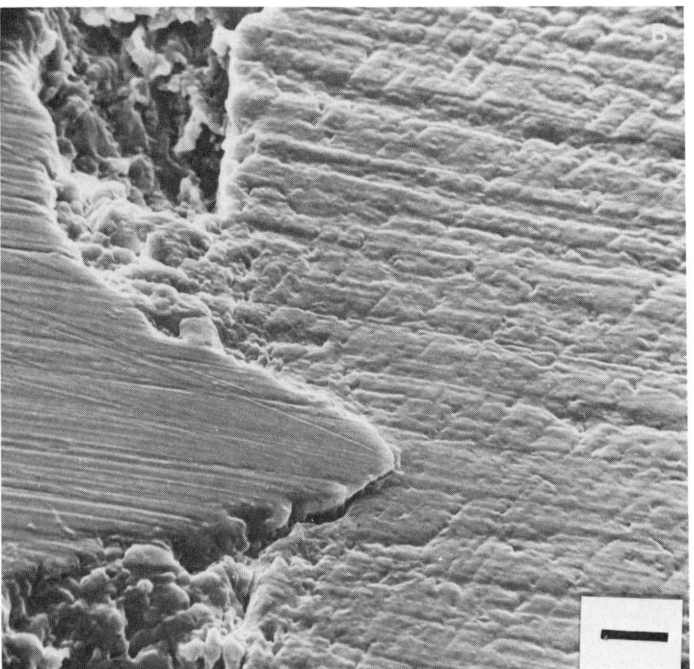

Figure 12.17. (a) SEM image of original of a fracture face of a human tooth containing a metal pin. Magnification bar = 2 μm. (b) SEM image of an acetate tape and silicone rubber replica of a fracture face of a human tooth containing a metal pin. Magnification bar = 2 μm (Pameijer, 1979).

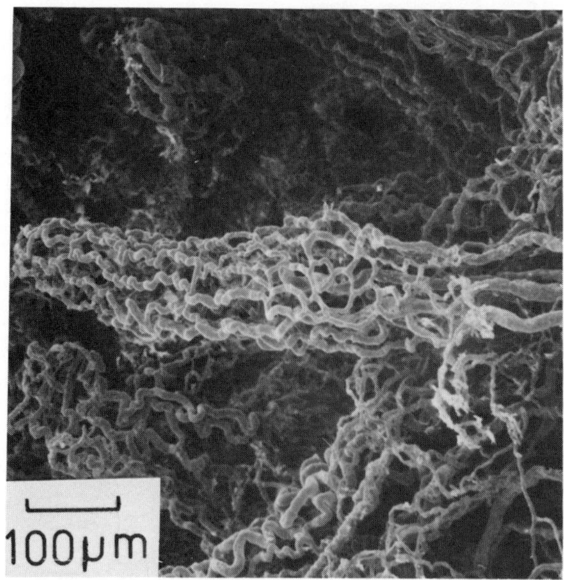

Figure 12.18. Latex-injection replica of a villous capillary network from a dog (Nowell and Lohse, 1974).

12.5.5.4. Surface Etching

Linton *et al.* (1984) and Claugher (1986) have found that both ion-beam etching and radiofrequency etching are useful ways to selectively etch biological tissue and thus expose internal structures. The etching role for ion-beam sputtering appears to be related to the physical sputtering mechanism rather than the chemical activity of the incident ion. Yonehara *et al.* (1989) have made a high-resolution study of biological material which had been progressively etched (Fig. 12.19). The best results were obtained from material which had been treated with osmium tetroxide–tannic acid conductive staining method (see Section 13.4 for details of the method). Earlier work by Kuzurian and Leighton (1983) had shown that the oxygen plasma etching technique could be used on the entire block face of resin-embedded material. There was an increased visualization of tissue components when examined in the SEM. Kuzurian and Leighton have coupled the etcher with a small microtome and sputter coated inside the microscope chamber. This ingenious arrangement allows a continual process of cutting a smooth block face, etching the surface, and coating the etched surface, thus enabling a continuous three-dimensional analysis to be made of a piece of tissue.

12.5.6. Specimen Supports

One of the prime considerations in choosing a support is that it should provide some form of conducting pathway, for even the most

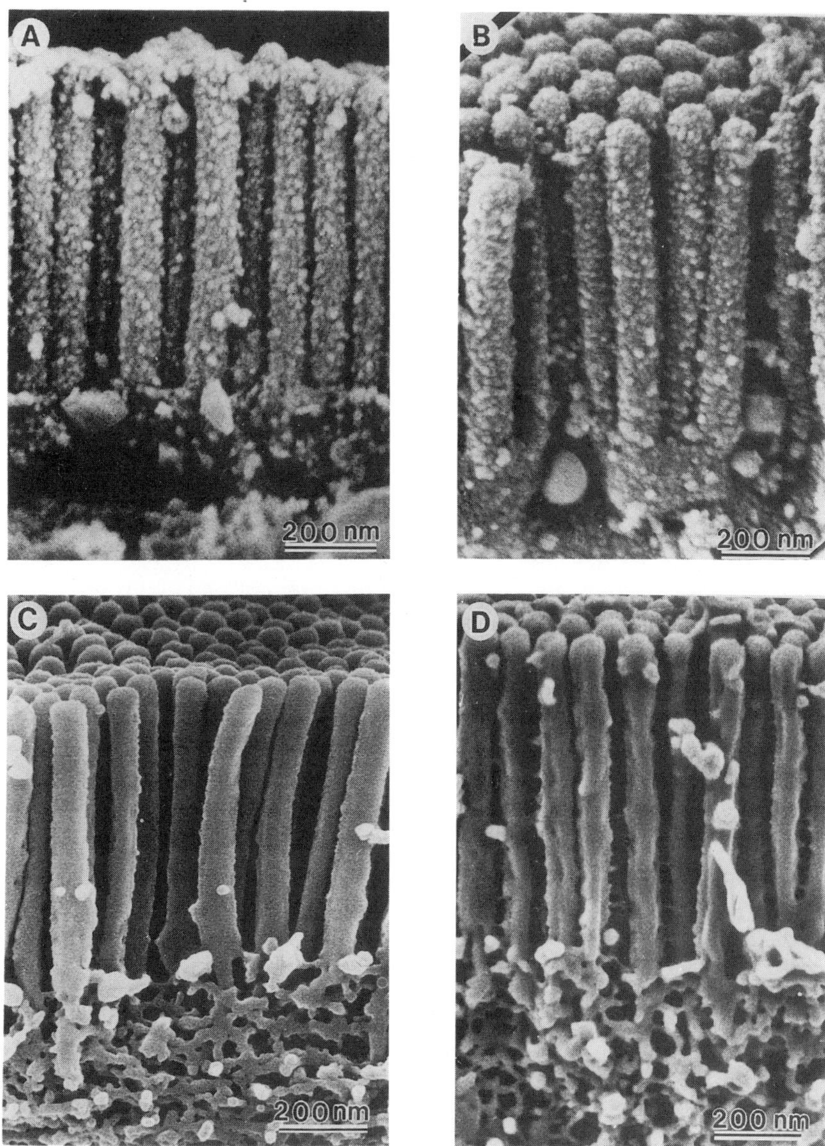

Figure 12.19. SEM images showing the progressive removal of material from microvilli and apical cytoplasm of absorptive cells using ion-beam etching techniques. (a), (b), (c), and (d) show etching times of 0, 5, 10, and 15 min. Photographs taken using a field-emission SEM at 2.0 keV (Yonehara *et al.*, 1989).

suitably metal-coated sample will quickly charge up if electrically insulated from the microscope stage.

In some cases, it is more convenient to place the specimen on a substrate which in turn is mounted on the stub. There are many different substrates, ranging from glass and plastic cover slips, metal and crystalline disks, plastics, waxes, and a whole range of membrane filters. These specimens and supports may be processed either singly or in batches using the type of multiple specimen holder described by Evers *et al.* (1983). Johansen *et al.* (1983) find that mica is a convenient support of biological samples, being more "charge resistant" than glass. Some progress has been made toward providing biologically active substrates. Tsuji and Karasek (1985) have found that cultured endothelial cells will reorganize on collagen disks, which can be easily processed for SEM. Kumon *et al.* (1983) found that gelatin-covered glass cover slips, impregnated with osmium vapor and coated with polylysine, eliminated charging artifacts. The choice of substrate will obviously be influenced by the specimen, and such factors as transparency to photons and electrons, solvent solubility, porosity, topography, composition, and conductivity must be considered. Once the sample has been processed, it is necessary only to attach the support to the specimen stub using some form of conductive paint such as silver dag or colloidal carbon. It is important to paint a small area on the specimen support and to continue this paint through over the edge and onto the specimen stub. The specimen should then be placed in a 313 K oven or low-vacuum desiccator for several hours to ensure that the solvent in the conductive paint has been completely removed before the specimen is coated. Care should be exercised when mounting membrane filters because the conductive paint can infiltrate the filter by capillary action and obscure the specimen, and the paint solvents may dissolve plastic specimen supports. Because specimens emerge dry from the critical-point dryer or freeze dryer they may be attached directly to the metal stub by a wide variety of methods. One of the simplest ways is to use double-sided tape. The specimens are dusted or carefully placed on the adhesive and, in the case of large specimens, a small dab of conductive paint is applied from the base of the specimen across the adhesive and onto the metal specimen stub. Double-sided adhesive tape is a poor conductor, and it is important to establish a conductive pathway between the specimen and the metallic support.

The specimens can be attached to the stubs by a whole range of adhesives, glues, and conductive paints. Rampley (1976) and Whitcomb (1981) have made a careful study of the various glues that are available and have recommended several useful ones. Murphey (1982) has written an encyclopedic review of the materials and procedures for specimen mounting. Whichever adhesive is used, it is important that it does not contaminate the specimens, does not remain plastic and allow the specimen to sink in, and contains only volatile solvents which can easily be removed. It should also be resistant to moderate heat and electron-beam damage.

The metal specimen support should be clean and have good surface finish to allow the specimens to be attached more easily. With a little care, it is possible to load several specimens on the same stub. Identifying marks can be scratched on the surface, and the specimens should be firmly attached to prevent cross-contamination and misinterpretation. Dvorachek and Rosenfeld (1990) describe a simple method of relocating minute samples on a sample stub. If one is concerned about cross-

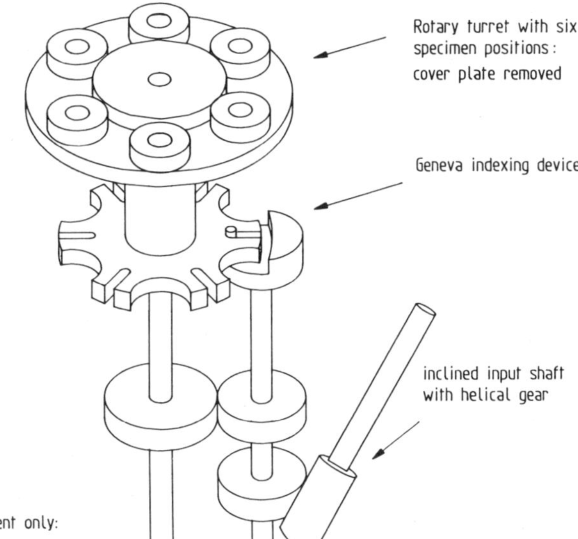

Figure 12.20. A rotary multiple-specimen holder for scanning electron microscopy.

contamination, then each sample should be loaded on its own stub and several stubs loaded on the type of multiple specimen holder described by Watt and Warrack (1989), shown in Fig. 12.20.

12.5.7. High-Resolution Scanning Microscopy

To obtain the ultimate resolution from an SEM requires considerable effort. Many of the methods used in preparing samples for high-resolution studies are no more than modifications of existing techniques discussed elsewhere in this chapter. A recent paper by Scala *et al.* (1991) reviews the methods which may be used on biological material. An examination of the methods used in high-resolution SEM reveals a number of principles of specimen preparation. It is useful to briefly consider these principles which can be adapted to a specific experimental situation.

12.5.7.1. Isolation of Object of Interest

A cell is packed with structural features all bathed in a low-molecular-weight soup. Much of the effort toward optimizing preparative methods in electron microscopy has been centered on preserving everything *in situ.* This has worked well for TEM, which essentially provides a projected two-dimensional planar view of the object. Such procedures are usually satisfactory for routine SEM, but the presence of all the overlapping structures can limit the three-dimensional view which the SEM can provide.

Thus, we must seek ways of isolating the object we wish to study, not by removing it from its natural environment, but by keeping it *in situ* and removing the obscuring structures. If we assume that no useful information can be obtained from the electron microscopy of small organic and inorganic molecules and water, it is permissible to remove these from the cell. One of the most successful ways of achieving this is to use the freeze–fracture–thaw technique described in Section 12.5.5.2. This method removes much of the cytoplasmic matrix by placing the freeze-fractured tissue samples in dilute fixative. On the assumption that one wishes to view the organelles and macromolecular components of the cytoskeleton, this method, which has been improved and extended by Weiser *et al.* (1991) and Kendall *et al.* (1991), provides an excellent view of these structures. Figure 12.21 shows what can be achieved by these methods.

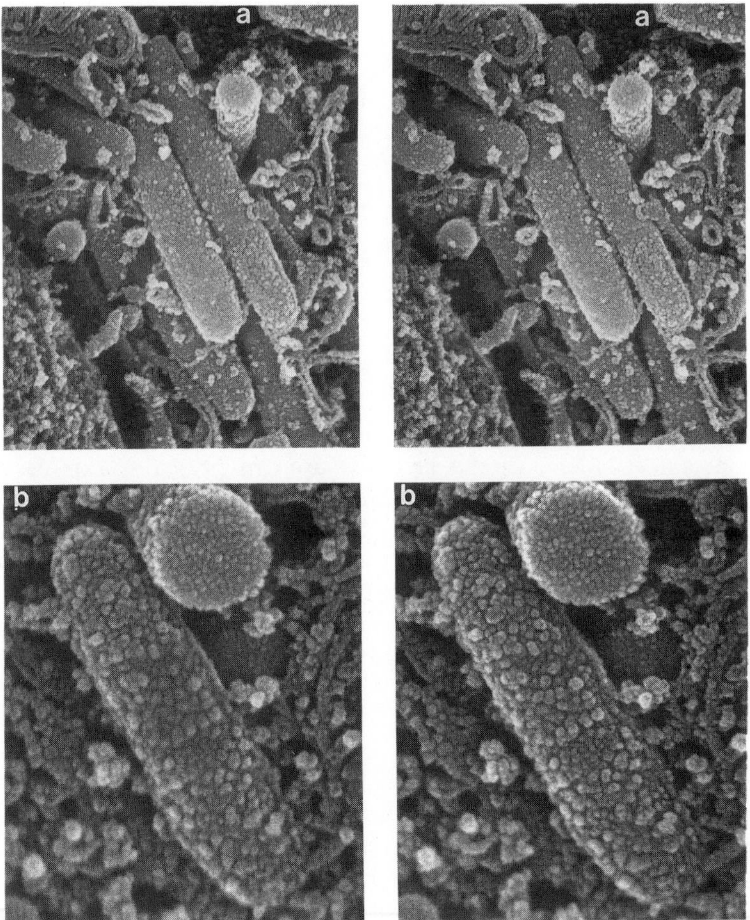

Figure 12.21. High-resolution SEM images of chick retinal pigment epithelium. (a) Stereo pair image showing the fingerlike melanosomes. Picture width = 2.2 μm. (b) Stereo pair showing pigmented granules on melanosome surface. Picture width = 960 nm (Weiser *et al.*, 1991).

A similar approach has been recommended by Yamada *et al.* (1983). Specimens were fixed in osmium tetroxide, conductively stained with uranyl acetate, dehydrated in ethanol, and then frozen and fractured. The critical point-dried material was rotary shadowed with a very thin (~2 nm) layer of platinum–carbon and showed the three-dimensional structure of *Curcurbita* prolamellar bodies.

12.5.7.2. Stabilization and Conductive Staining

Nearly all high-resolution studies have employed some form of conductive staining method whereby the sample is infiltrated with heavy metal ions using the procedures described in Section 13.4. Care must be taken not to allow any of the insoluble deposits, which frequently form during conductive staining, to become deposited on the structures of interest.

12.5.7.3. Specimen Coating

A thin layer of metal on a specimen whose electrical properties have been enhanced by conductive staining improves the quality of the image. Full details of the coating procedures that could be used are given in Chapter 13, but they center around the application of thin layers of elements such as niobium or chromium or very thin layers of noble metals. Irrespective of what preparative method is employed and what instrument is used, contamination in the microscope may still conspire against obtaining high resolution images. The nature of contamination is discussed in detail in a review by Reimer (1984). Although ultrahigh vacuum (10^{-9} to 10^{-11} torr) and cryostages provide a considerable reduction in contamination, the high current densities of field emission instruments can still give contamination rates as high as $10\,nm/s^{-1}$. Kanaya *et al.* (1988) describe a method using prebombardment of the specimen surface with argon ions which reduces the contamination to more acceptable levels.

12.6. Specimen Preparation for Localization of Metabolic Activity and Chemical Specificity

12.6.1. Introduction

Most modern scanning electron microscopes have multiple detector systems that permit the *in situ* detection and analysis of elements by their x rays, molecules by the electrons backscattered from specifically localized heavy metals, and macromolecules by the emissivity of gold-labelled antibodies bound to specific cellular epitopes. The impressive analytical capacity of the SEM (and x-ray microanalyser) presents an imposing challenge to sample preparation. The chemical entities of interest may be dissolved salts and organic molecules, the end products of enzymatic reactions, the enzymes themselves, and the macromolecules which make up the structural framework of the sample. With the exception of low-temperature techniques, no one single preparative method is likely to provide all the answers, although certain general approaches do seem more likely to succeed than others. Elemental x-ray microanalysis is best achieved by using preparative techniques which avoid any chemical intervention. Analytical methods based on a back-scattered electron signal center on histochemical and staining techniques that rely on precipitating local concentrations of elements as heavy metal deposits. Immunocytochemical methods depend on retaining the identity of a stabilized molecule or macromolecule and providing some sort of emissive signal that may be detected. Most of these methods are interchangeable; all are dependent on retaining the structural integrity of the sample to ensure *in situ* localization of the chemical species being analyzed.

The best approach to the multifaceted problem would be first to consider the general strategies of chemical analysis and then consider the tactics which might be employed to provide specific answers. It is again recommended that one should first garner information about the specimen and the precise nature of the chemical analysis one hopes to carry out. These factors will have a major influence on the design of any experimental protocol. We will begin by first considering the nature of the problem and then proceed to discuss the approaches one may wish to adopt to preserve the chemicals of interest.

12.6.2. The Nature of the Problem

This section is included to provide a reminder of the type of information we need to assemble about the specimen and about instrumentation.

12.6.2.1. The Form of the Substance Being Analyzed

Elements can exist in a number of forms; as amorphous or crystalline deposits, covalently bound to molecules or macromolecules, ionically bound to molecules or macromolecules, or as the unbound ion or soluble element freely diffusible in cell and tissue fluids. Molecules and macromolecules may be similarly disposed. Many macromolecules, although less soluble, are nevertheless very labile and dependent on water molecules to maintain their three-dimensional structural and functional integrity. Yet others are as tough as old boots (protein) or as sturdy as a tree (cellulose).

12.6.2.2. Precision of Analytical Investigation

The investigations can be carried out at two levels of sophistication. In *qualitative analysis* one attempts to find whether a particular element or molecule is present in a cell or tissue (see Chapter 6). Figure 12.22

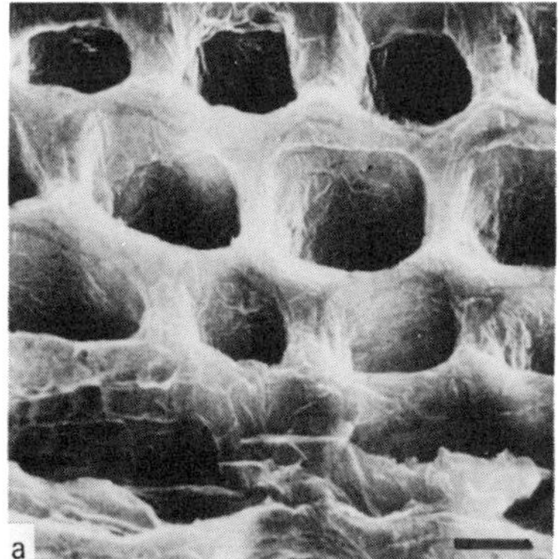

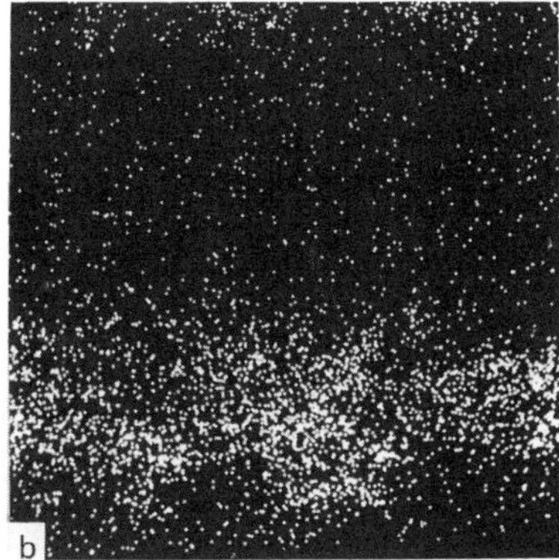

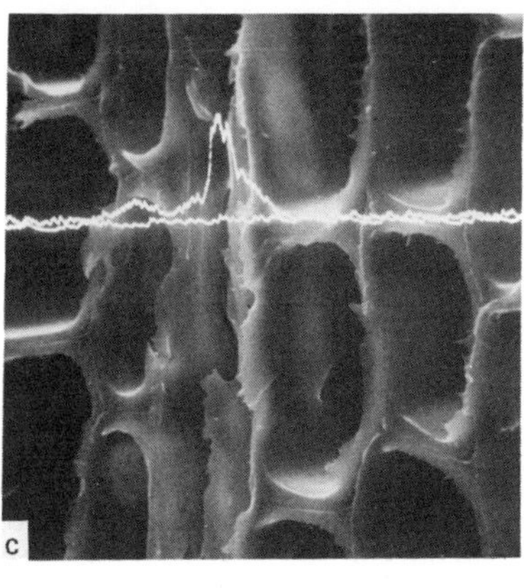

Figure 12.22. A piece of pine wood impregnated with a copper-containing preservative. (a) Secondary-electron image. Magnification bar = 40 μm. (b) X-ray map for Cu $K\alpha$ of same area as (a). (c) Line scan across the piece of wood, showing Cu $K\alpha$ peak in the vascular ray region in both (a) and (b).

shows an example of qualitative-x-ray microanalysis of a piece of wood impregnated with a copper-containing preservative. In *quantitative analysis* one is trying to measure whether one part of a sample contains more or less of a particular chemical than another part (see Chapter 8). The ultimate aim of quantitative analysis is to find out, as accurately as possible, how much of a particular element or molecule is present in a given volume of tissue. Most published work on *in situ* analysis falls into the first category, and provided adequate care is taken during specimen preparation, the technique can give valid information. The second category of accurate quantitative analysis is more difficult to achieve and involves far more than just the optimal preparation of the specimen. It is necessary, for example, to have accurate standards, high instrumental stability, and computational accessories capable of analyzing a multiplicity of repetitive data (see Chapter 9).

It is also important to know what level of structural spatial resolution is required in a given investigation. Thin sections viewed by transmitted electrons and surfaces viewed by a fine reflected probe give the best spatial resolution. Many investigations do not require such high resolution and as a consequence do not need quite so much attention to the preparative techniques. A good rule of thumb would be to aim for sample preservation that would give a structural spatial resolution ten times better than the analytical spatial resolution. Thus in an x-ray analtyical study of fracture faces of cells in which only the major compartments are being analyzed, the analyzed microvolume would be on the order of 2–5 μm, the structural spatial resolution need be no better than 100–500 nm. An immunocytochemical study of membranes, using the BSE signal from a 20-nm gold probe as a means of localizing an active site, needs a spatial resolution of 2–10 nm, bearing in mind the increased microvolume from which the BSE signal is being emitted.

12.6.2.3. Types of Specimens

It is convenient to categorize specimens for analysis into six types. Each category shows varying degrees of analytical spatial resolution, depending on the volume of the sample contributing to the analysis. This volume is a function of instrumentation (i.e., beam diameter and depth of electron penetration) and the nature of the analytic technique (i.e., the fact that immunocytochemical methods are surface-analysis methods and x-ray microanalysis and backscattered imaging sample a subsurface microvolume).

Bulk Samples. These are defined as specimens too thick to allow transmission of either photons or electrons and for which morphological information can be obtained only with a reflected signal. Bulk biological or polymer specimens are less useful for x-ray microanalysis and backscattered imaging because of the rather poor x-ray spatial resolution (i.e., 2–5 μm), but are a good type of specimen in immunocytochemical methods. Hard tissues such as bone, shells, woods, and fossils may be fractured, and such specimens are useful for qualitative analysis, but their surfaces are so irregular as to make accurate quantitative analysis practically impossible. This problem can be partly overcome by polishing the surface, but it is important to use pure abrasives to avoid contamination. The fracture faces of soft material can also be analyzed, but only after the tissue has gone through stabilization, dehydration, and hardening or has been frozen and fractured at low temperature and examined and analyzed in the frozen-dried or frozen hydrated state. It is possible to obtain large areas of smooth fracture surfaces on frozen-hydrated bulk specimens, and providing the specimen-detector geometry is optimized, the technique can give useful analytical information on specimens which are difficult to section at low temperatures.

Thick Sections. Thick sections (i.e., between 0.2 and 2.0 μm) are a useful compromise to bulk samples as reasonably good morphological information can be obtained using scanning transmission images and the specimens are sufficiently thick to contain enough material to be analyzed with good x-ray and BSE spatial resolution. Because most plant and animal tissues are very soft, they must first be stabilized and strengthened before any sections can be cut. As will be discussed later, most of these preparative procedures can cause serious losses of soluble substances from tissues and should be used with great care. Alternatively, one may analyze freeze-dried or frozen-hydrated sections.

Thin Sections. Thin sections, i.e., less than 200 nm, usually show the greatest amount of morphological detail and potentially provide the highest x-ray spatial resolution, but their thickness may reduce the amount of material to very low levels because of the small microvolume of the section being analyzed. This is not a problem with immunocytochemical methods, which are primarily directed to surface antigens. However, one is faced with the same problems of preparation found with thick sections, and the only sensible recourse is to use thin sections of freeze-dried or freeze-substituted material.

Microorganisms, Isolated Cells, and Organelles. There are now many reliable biochemical and manipulative procedures that allow the isolation of single cells and organelles. It is not proposed to discuss these methods, but to consider how such cells and tissue fragments may be prepared for *in situ* chemical analysis. In some respects, isolated cells and organelles only a few micrometers thick can be treated as particulate matter, but their very softness and high water content place severe restrictions on the preparative procedures that may be used. Depending on their size, isolate cells and organelles—and, for that matter, inorganic particulate matter—can for analytical purposes be treated either as bulk samples or sections. Some examples of specimen preparation for isolated cells are given in the papers by Masters *et al.* (1979) and Cameron and Smith (1980).

Particles and Fibers. It is sometimes necessary to carry out analysis on particulate matter which has been eluted from tissues with physiological saline, collected as a dry powder by some form of filtration, or individually isolated from the tissue or organic matrix by micromanipulation. Henderson and Griffiths (1972) have devised a simple extraction–replication technique by which a plastic replica is made of the natural and fractured surface. Particles that were in the specimen's surface are removed with the replica and may be analyzed. Dry particles can be analyzed *in situ* on a filter as long as they are sufficiently well dispersed or embedded in a resin which is then either fractured and polished or cut into sections. Alternatively, the particles can be suspended in a nonaqueous solvent and spread as a thin layer on a suitable substrate. Particles also occur in a matrix of organic material, and this may be removed by washing in sodium hypochlorite solution, boiling in strong solutions of KOH, or cold-plasma ashing. An example of this type of analysis of foreign bodies in tissue is given in the paper by Champness *et al.* (1976), who studied the presence of asbestos in lung tissue.

Liquid samples. Much of the interest in biological x-ray microanalysis is centered on the study of physiologically active fluids, which may be derived from single cells or from spaces surrounded by a few cells. Methods have been developed to extract such liquid samples by micropuncture and to determine the elemental composition and concentration of liquids with volumes of only a few picoliters. The techniques developed by Ingram and Hogben (1967), Morel and Roinel (1969), and Lechene (1978) involve placing small drops of the extracted fluids on a highly polished beryllium surface immersed in a bath of paraffin oil. The aqueous drops do not mix with the hydrophobic paraffin oil and remain as separate drops on the beryllium surface. The paraffin oil is removed by xylene or chloroform, and the droplets are frozen-dried, leaving the nonvolatile material in a microcrystalline form. Hundreds of samples can be processed at a time, as the surface of the beryllium block may be marked with a grid to facilitate sample identification. Lechene and his colleagues have automated many of the analytical processes, and the method is now both accurate and reproducible. Roinel (1988) provides a comprehensive review of these procedures.

A slightly different procedure, described by Quinton (1978) and Van Eekelen *et al.* (1980) involves placing the microdroplets on a thin supported film of parlodion and desiccating the samples by flash evaporation rather than freeze-drying. Beeuwkes (1979) has extended the microdroplet technique to the analysis of organic samples by precipitating very small amounts of urea with thioxanthen-9-ol and then carrying out analysis for the sulphur in the precipitate. A recent paper by Bodson *et al.* (1991) has shown that it is posible to collect picoliter samples from selected compartments of plant cells. This procedure would provide an additional way of carrying out x-ray microanalysis of diffusible elements in plant cells.

12.6.2.4. Types of Instrumentation

The differences among the three main forms of electron-beam instrumentation used in the laboratory for structural and analytical studies have become blurred. Transmission microscopes can be fitted with x-ray spectrometers and scanning attachments; scanning microscopes can be fitted with transmitted-electron detectors, and x-ray microanalysers come with both transmitted- and reflected-signal detectors. Each type of instrument and their hybrids have advantages and disadvantages, and it is important for the experimenter to appreciate the limitations the instrumentation at his or her disposal may put on the preparative technique. Thus a TEM with an EDS detector could be used only on thin sections and an SEM using the BSE signal would be limited to bulk samples or, at best, thick sections. The instrumentation determines the type of preparation.

12.6.2.5. Types of Analytical Applications

Although the SEM and x-ray microanalyser are powerful tools for *in situ* analysis of chemical content, they are not the panacea for all problems. Listed below are the types of investigations one might expect to find in the life sciences and in studies of organic material, and suggestions as to which analytical approach might be adopted.

1. Localization of ions and electrolytes of known physiological function (x-rays, BSE, histochemistry, staining).
2. Study of systems which had been deliberately or naturally perturbed,

i.e., drugs, disease and precipitating agents (x-rays, BSE, histochemistry, staining).
3. Localizing enzymes and enzyme activity (BSE, immunocytochemistry, staining).
4. Analysis of very small volumes of expressed or extracted physiological fluids (x rays).
5. Relationship between elemental composition and a particular structural feature (x rays, immunocytochemistry).
6. Distribution of heavy elements in an organic matrix (x rays and staining).
7. Analysis of particulate matter (x rays, histochemistry).

12.6.3. General Preparative Procedures

The following discussion revolves around the premise that the chemical species being analyzed are either very soluble (electrolytes, small molecules) or very labile (molecules and macromolecules). Accordingly, the preparative procedures are designed to keep soluble components at their natural location in the specimen, to stabilize labile components, and finally, if necessary, to link these soluble and labile components to other chemical species that can be detected and analyzed in the SEM and x-ray microanalyser. Many of the methods used are no different from the procedures used for structural studies, and the important differences will be emphasized. The methods we will discuss are far from ideal, and at the end of this section we will consider the criteria we should adopt to indicate satisfactory specimen preparation.

12.6.3.1. Before Fixation

In general, there should be minimal delay in obtaining the piece of tissue or cells from the experimental organism. Enough is known about the morphological effects of anoxia, stress, and *post-mortem* changes to suggest that local chemical concentrations are also likely to change. Care must also be taken to remove any contaminating body fluids such as blood, mucus, or extracellular fluids which might infiltrate the specimen and give rise to spurious x-ray signals.

12.6.3.2. Fixation

It is of paramount importance to choose a fixative which does not cause loss, redistribution, or masking of the elements and molecules to be analyzed. This is one of the first things which must be checked by carrying out, for example, flame spectrophotometry on the fixation fluids before and after fixation. Very few systematic studies have been undertaken to examine the effect of sample preparation on the elemental concentration of tissues. Holbrook *et al.* (1976) have examined the effects of a number of fixatives on human skin; the details of their findings are shown in Fig. 12.23. These studies and others provide a convincing

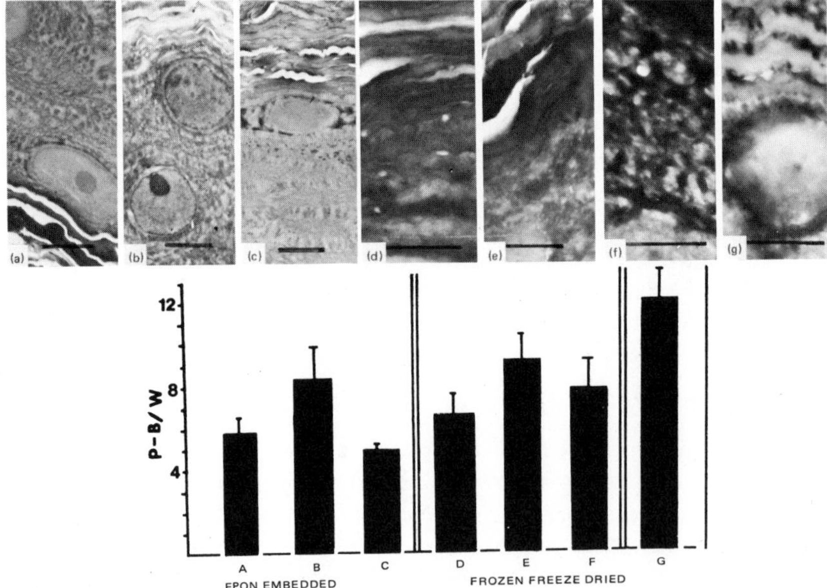

Figure 12.23. Electron micrographs and histograms of peak-to-background ratios comparing the amount of sulfur retained in human stratum corneum cells prepared using various experimental protocols. (a)–(e): magnification bar = 5 μm (f)–(g): magnification bar = 2 μm. The histograms labeled A–G correspond to the micrographs labeled (a)–(g). (a) Osmium fixation, Epon embedding; (b) glutaraldehyde fixation, Epon embedding; (c) glutaraldehyde and osmium fixation, Epon embedding; (d) osmium fixation, freeze dried; (e) glutaraldehyde, freeze dried; (f) glutaraldehyde and osmium fixation, freeze dried; (g) unfixed, freeze dried. Note that the better elemental retention is associated with poorer morphological preservation (Holbrook *et al.*, 1976).

demonstration that *all* chemical fixation techniques have some effect on the natural distributions of elements in tissues. It would be naive to think that only electolytes are lost from specimens processed by wet chemical fixation. Coetzee and van der Merwe (1984; 1989) have shown that a whole range of low-molecular-weight compounds are lost or redistributed during fixation. Their work provides useful recommendations for preparative protocols. A review article by Morgan *et al.* (1978) contains a summary of the elemental losses recorded during various fixative procedures from a wide range of tissue types. It paints a somber picture and should be sufficient to discourage the use of fixatives, particularly when diffusible elements are the subject of analysis. Commonly used fixatives such as the organic aldehydes have been shown to cause a dramatic loss of soluble material from cells. Brief fixation results in gross ionic changes, and sectioning with a trough liquid leads to extraction of elements. If aldehyde fixatives are used, it is probably better to use the freshly distilled material. The loss of soluble elements is exacerbated by postfixation in osmium and permanganate, which have the added disadvantage of masking some of the elements being analyzed. If it is quite clear that some sort of tissue stabilization is necessary, and if fixatives have to be used, the fixation time should be as brief as possible. It is preferable to use fixatives in the vapor phase rather than the more usual liquid state in order to lessen the redistribution of soluble elements. Osmium tetroxide vapor and formaldehyde vapor have been used, but the depth of penetration of the fixative is limited and only small specimens can be successfully treated in this way. Ingram and Hogben (1967) have applied osmium vapor treatment to freeze-dried blocks of tissue, and Somlyo *et al.* (1977) have treated freeze-dried sections. The principal advantage of the osmium treatment is that it improves tissue contrast, but it should be used only if the presence of osmium is not going to interfere with the analysis.

Organic buffers such as bicarbonate, piperzine-*N*-*N'*-bis-2-ethanol sulfonic acid, veronal acetate, and collidine are preferred to the more commonly used inorganic buffers because there is always a danger that the latter may contribute unwanted elements to the cells (for example, cacodylate buffer contains arsenic). It is also important to check that the pH of the preparative fluids does not cause changes in elemental solubility. For those investigators who are interested in "wet-chemical" preparative techniques for microanalysis, it is recommended they read the book by Morgan (1985), and the paper by Chandler (1985).

12.6.3.3. Histochemical Techniques

It has been long known to light and electron microscopists that it is possible to localize areas of specific physiological or metabolic activity by the use of highly specific dyes and histochemical agents. The same idea can, in principle, be applied to x-ray microanalysis. Figure 12.24, from the work of Ryder and Bowen (1974), shows electron-dense deposits resulting from a modified Gomori technique for acid phosphatase; the x-ray spectra confirm the presence of lead. Unlike the situation in electron microscopy, the end product need not necessarily be a heavy metal as

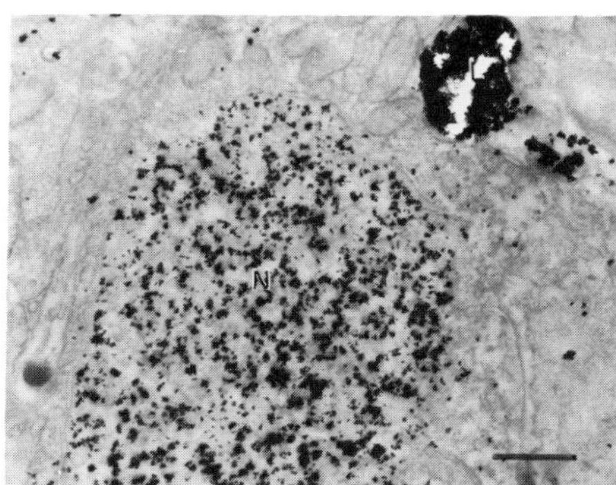

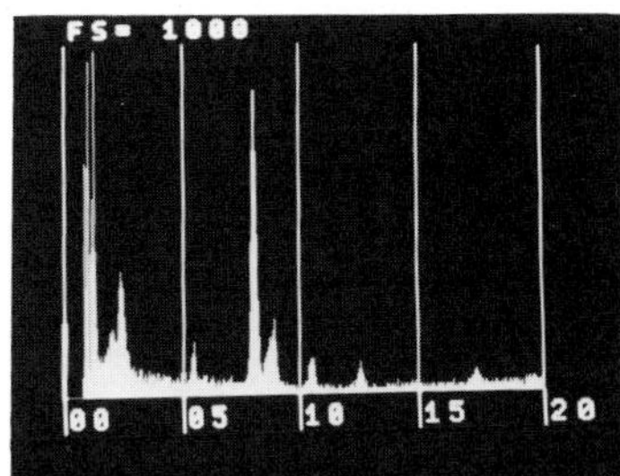

Figure 12.24. (a) TEM photograph of a section of planarian acidophil cells exposed to lead after a modified Gomori test for acid phosphatase. Note deposit of lead. Magnification bar = 1 μm. (b) Energy-dispersive x-ray spectrum from stained nucleus seen in (a), showing the $K\alpha$ peak for phosphorus. Note the $M\alpha$, $L\alpha$, and $L\beta$ peaks for lead and the K peaks for copper (Ryder and Bowen, 1974).

long as it is sufficiently distinct from the bulk of the sample matrix. For example the Alcian blue stain for mucin contains copper, sulfur, and chlorine, all of which are readily detected by x-ray microanalysis. Ryder and Bowen (1974) have localized phosphatase activity using an azo-dye which has three covalently linked chlorine "marker" atoms. Van Steveninck and Van Steveninck (1991) have recently published a comprehensive review of the histochemical–analytical techniques that may be applied to plant material. An earlier review article dealing principally with animal tissue has been written by Bowen and Ryder (1978).

12.6.3.4. Precipitation Techniques

A number of workers have attempted to precipitate diffusible elements *in situ* before fixation takes place. The technique is based on the reaction of an ion with a heavy metal compound to produce an electron-dense precipitate that either is observable in the TEM or can be detected in the x-ray microanalyzer. Table 12.1 shows some of the precipitating reagents which have been used on biological tissue. Alternatively, if the ion itself is a heavy element, organic reagents may be used as the precipitating agents. The same principle can be applied to the intracellular localization of certain enzymes whose reaction products are inorganic ions.

Van Steveninck and Van Steveninck (1991) have published a critical review of the precipitation techniques for diffusible substances, and Fig. 12.25, from the work of Ellisman *et al.* (1980), shows ways in which the method can be applied in biological tissues. It is important that the specificity of the reaction be adquately tested. Thus the organic silver salt precipitation technique for chloride also gives a precipitate with bromide, and the widely used pyroantimonate technique for sodium also gives a precipitate with potassium, magnesium, calcium, and manganese. Simpson *et al.* (1979) have published a review of the pyroantimonate precipitation methods and provide a number of criteria which may be

Table 12.1. Precipitation Reactions for *in Situ* Demonstration of Inorganic Ions and ATPases in Plant and Animal Tissues

Animal specimens		Plant specimens	
Ion or ATPases	Precipitating reagent	Ion or ATPases	Precipitating reagent
Cl^-	Ag acetate or Ag lactate	Cl^-	Ag acetate or Ag lactate
Na^+	K-pyroantimonate	Na^+	K pyroantimonate
Na^+, Mg^{2+}, Ca^{2+}	K-pyroantimonate	Na^+, Mg^+,Ca^{2+}	K pyroantimonate
Na^+, K^+, Mg^{2+}	K-pyroantimonate	Na^+, Ca^{2+}	Benzamide
Ca^{2+}, Mn^{2+}		Ca^{2+}	NH_4 oxalate
Ca^{2+}	K-pryoantimonate	PO_4^-	Pb acetate
	K pyroantimonate or K oxalate	ATPases	$Pb(NO_3)_2$
PO_4^-	Pb acetate		
ATPases	$Pb(NO_3)_2$ or Ca^{2+}		

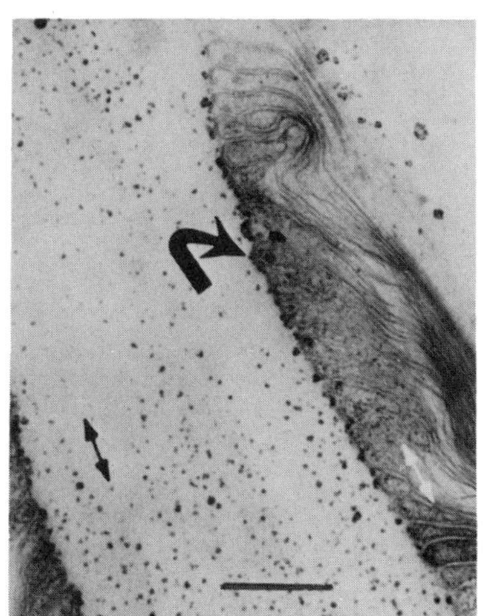

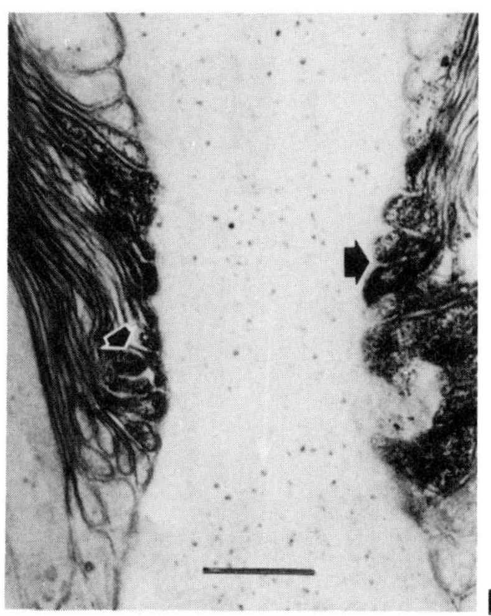

Figure 12.25. Localization of sodium and calcium in nerve tissue using the osmium–pryoantimonate precipitating technique. (a) TEM image of an unstained node of Ranvier showing the distribution of pyroantimonate precipitates. Large grains occur in axoplasm (white arrow), and smaller grains are restricted to the cytoplasm of the paranodal loops (black arrows). Magnification bar = 0.5 μm. (b) TEM image of a portion of the paranodal region of the node of Ranvier showing pyroantimonate precipitates. Small-grain precipitates are seen in the cytoplasm of the paranodal loops (white arrow) and large grain precipitates are within the axoplasm (black arrow). Magnification bar = 0.5 μm. X-ray microanalysis by means of wavelength-dispersive spectrometers show that the granular deposits contain both Na^{+} and Ca^{++} (Ellisman *et al.*, 1980).

used when applying this method to the analysis of cations. One of the criticisms of the pyroantimonate method and precipitation methods generally is that the precipitating agent may act as a focal point and cause diffusible elements to move from their natural location in the cells and tissues to the region where precipitation occurs.

It is important to know the extent of loss from the specimen during sample preparation. Läuchli *et al.* (1978) were able to show that after preloading the tissue with ^{36}Cl only about 4% of the chloride was lost during the silver precipitation technique. Localization artifacts, occurring either before, during, or after the precipitation technique is carried out, must be considered. Yarom *et al.* (1975) demonstrated that calcium precipitated with pyroantimonate was lost from the tissue during alcohol dehydration. Van Stevenick *et al.* (1976) were able to demonstrate by x-ray microanalysis that silver-precipitated chloride deposits are relatively unaffected by postfixation in osmium.

12.6.3.5. Dehydration

There is no doubt that chemical dehydration contributes to the loss of diffusible substances from cells and tissue initiated by chemical fixation. Although no critical comparison studies have been done, there appears to be little difference in the effect of ethanol, methanol, or acetone as dehydrating agents. However, Thorpe and Harvey (1979) have found that plant material dehydrated in dimethoxypropane showed considerably improved ion retention (Na^+, K^+, Cl^-) compared with dehydration in acetone. The classical dehydration procedures may be circumvented by using the inert dehydration water-soluble resins and the glutaraldehyde–urea embedding technique of Pease and Peterson (1972), or by passing glutaraldehyde-fixed material through increasing concentrations of glutaraldehyde up to 50%, after which the tissue is passed directly to Epon 812 (Yarom *et al.* 1975). An alternative procedure involves infiltrating fixed specimens with increasing concentrations of polyvinyl alcohol solution (MW 14,000) up to a final concentration of 20%. Water is then eliminated by dialysis and the resultant hard gel cross-linked with glutaraldehyde. However, these procedures do not appear to significantly reduce the loss of soluble materials from the specimens rested. Simply air drying the specimen also causes elemental redistribution. However, the critical-point drying procedure, which usually comes at the end of fixation and dehydration, is unlikely to make any difference to the level of soluble substances which have long since been removed during dehydration. Critical-point drying may have a severe effect on bound elements, particularly when fluorocarbons are used as the transition fluid. There are, unfortunately, few definitive data on the solubilization effects of the critical-point drying method, even though it is known from industrial studies that supercritical fluids have quite different solubility constants than do those at ambient pressures and temperatures. The solvent evaporation drying procedure described by Boyde and Maconnachie (1979) may also be useful in preparing specimens if it can be demonstrated that minimal elemental extraction occurs during processing. In this technique, fixed and chemically dehydrated specimens are infiltrated with Freon 113 and then rapdily exposed to a vacuum pressure of 1–10 Pa. The initial high rate of Freon 113 evaporation cools the remaining fluorocarbon, which solidifies and sublimes at a slower rate. The absence of any water avoids the problem of ice crystal damage and the morphological results are as good as those produced by critical point drying.

12.6.3.6. Embedding

Most fixed and dehydrated biological tissues need infiltrating with either resin or paraffin wax to enable sections to be cut, but the resin effectively dilutes constituents by increasing specimen mass. This step in the preparative procedure may be omitted if very thin specimens or fine suspensions are to be examined or if the specimens are to be examined only by reflected photons or electrons.

The resins can extract and redistribute elements, and there is no doubt that the embedding procedure can contribute elements to the materials to be analyzed. Some of the epoxy resins can contain relatively high amounts of sulfur, while epoxy resins based on epichlorohydrin contain small amounts of chlorine. The low level of chlorine (0.73%) in

Spurr's low-viscosity resin is still too high where critical studies for chlorine are to be carried out. It is advisable to carry out an elemental analysis of all embedding chemicals before they are used for specimen preparation. Methacrylate resins are theoretically free of mineral elements, but can cause shrinkage during polymerization and are unstable in the electron beam.

12.6.3.7. Sectioning and Fracturing

It may be difficult to see how sectioning and fracturing could influence the analytical results, but a number of effects should be considered. Knife edges do not stay sharp indefinitely and are progressively worn away by integration with the specimen. One should consider where the material from the knife edge disappears to during sectioning. Glass typically contains boron, sodium, and potassium, together with traces of other elements; diamond knives are pure carbon and unlikely to be of any significance in specimen contamination. Steel knives are often used to cut thick sections and could quite possibly contaminate specimens. It is important to clean steel knives thoroughly after they have been sharpened to remove all traces of the honing compound. This cleaning is best achieved using a soft cloth and organic fluids such as acetone followed by methanol. Attention should be paid to cleanliness of the trough liquid (if this is being used) and to section treatment after cutting. For example, if resin sections need flattening, it is probably better to avoid chloroform vapors, which can be readily absorbed by resin sections, giving rise to high levels of chlorine contamination.

12.6.3.8. Specimen Supports

Ideally, the specimen support should be a good conductor and made of material that will make no contribution to the x-ray signal from the specimen. For bulk specimens or sections to be examined by secondary electrons, it is usual to place the specimens on highly polished ultrahigh-purity carbon, aluminum, or beryllium disks. Single crystals of silicon doped with boron are also useful. These materials are reasonably good conductors and make only a small contribution to the x-ray background. Materials to be examined using light optics should be placed on quartz or clear plastic slides which can be thinly coated, ~5–7 nm, with aluminum to provide a conductive layer. For sectioned material a whole range of specimen supports is available, mostly based on the standard 3.08-mm TEM grid. It is possible to buy grids made of copper, titanium, nickel, aluminum, beryllium, gold, carbon, and nylon. These may be used with and without a plastic support film. There is a tendency to use grids made of low-atomic-number elements such as aluminum, carbon, or beryllium because these materials make significantly less contribution to the x-ray background. Aluminized or carbon-coated nylon films have been used as specimen supports and have the advantage of being strong and transparent to electrons and photons although they make a contribution to the background. Bulk specimens should be fixed to the specimen support using a high-purity conductive paint such as colloidal carbon, and it is important to check that the adhesive does not have any interfering x-ray lines.

12.6.3.9. Specimen Staining

The cautionary remarks directed toward the use of fixatives are equally applicable to staining. In view of the danger of extracting soluble constituents and the likelihood of introducing heavy elements whose x-ray peaks can mask or interfere with elements of interest, it is probably best to avoid any staining. If the contrast of the specimen image is unacceptably low, then some compromise has to be made or the specimen has to be examined by scanning transmission electron microscopy or electron energy-loss spectroscopy, which gives higher contrast images, or by using secondary-electron imaging, which gives surprisingly good information from thin sections.

12.6.4. Localizing Regions of Biological Activity and Chemical Specificity

The preceding section was concerned mainly with the preparative procedures used in connection with x-ray microanalysis, in which the chemical concerned is localized by production of its characteristic x-rays. In this section we will consider three analytical procedures that rely on the emission of electrons that can be used to localize the chemicals of interest. The procedures described are primarily used for the *in situ* localization of molecules and macromolecules.

12.6.4.1. Backscattered-Electron Cytochemical Methods

This analytical method, peculiar to the SEM, has been developed to take advantage of the BSE signals. The flux and nature of backscattered electrons are discussed in Chapters 3 and 4. The production of backscattred electrons increases roughly with mean atomic number so that ten times more of the incident electron beam is backscattered by gold than by carbon. It is these differences in backscattering coefficient which form the basis of the qualitative analytical procedure. A high-atomic-number inclusion in a low-atomic-number background gives a stronger backscattered signal, which may be used to form contrast in an image to spatially locate the inclusion. Such inclusions could, for example, be a heavy-metal dust particle inhaled into pulmonary tissue, a heavy-metal ligand attached to the end product of an enzymatic reaction, or a heavy-metal stain with a specificity for a given chemical grouping within a cell such as that achieved by Spicer *et al.* (1983) for carbohydrate moieties. The backscattered signal usually appears bright against a dark background. Biologists have frequently found it convenient to reverse this contrast so that regions of strong backscattering appear dark against a light background. As backscattered imaging is usually carried on specimens also examined in the light microscope, the correlative microscopy is easier to interpret as the areas of interest appear dark in both cases.

De Harven (1987) has proposed a number of conditions which should be met in order to make full and proper use of this analytical technique.

1. There should be maximal specific localization of high-atomic-number elements within the sample.
2. There should be minimal nonspecific distribution of high-atomic-number elements in and around the sample.
3. The cytochemical methods must result in an adequate level of preservation of cell structure.
4. The backscattered detector must have a high efficiency and should not interfere with the secondary mode of operation.
5. The accelerating voltages used should be selected to maximize the backscattered signal.

Before discussing details of the various cytochemical procedures which can result in the deposition of a heavy metal at a specific site in the tissue, it is useful to briefly discuss the types of detectors which can be used to measure these signals (see also Chapter 4). The backscattered image is usually a mixture of specimen composition and surface topography. The BSE annular detectors based on silicon surface barrier diodes, which are simple segmented disks with electrical leads, can be mounted just below the final lens concentric to the beam, looking down on the specimen. By electrically adding or subtracting the signals from one or more segments of the semiconductor disk, one may separate the backscattered signal, to a first approximation, into its compositional and topographic images (see Chapter 4). Adding the signals enhances the compositional image and gives reasonable atomic number contrast, while subtracting them produces good topographic contrast.

In order to obtain maximum topographic contrast using secondary electrons with the conventional ET detectors, it is useful to be able to suppress the backscattered-electron-to-secondary-electron conversion above the specimen. This may be simply achieved using a device described by Volbert and Reimer (1980), and can also be used to advantage in enhancing the backscattered image coming from the specimen. These detectors produce a higher resolution than the ET detectors at much lower incident beam currents, thus reducing the problem of beam damage. There is more than adequate contrast available for most biological samples. In addition, depending on the beam energy, the method does allow us to look into the interior of biological specimens and obtain a backscattered-electron signal from an inclusion 1–2 μm below the surface of the specimen.

A number of light-microscope cytochemical methods have been

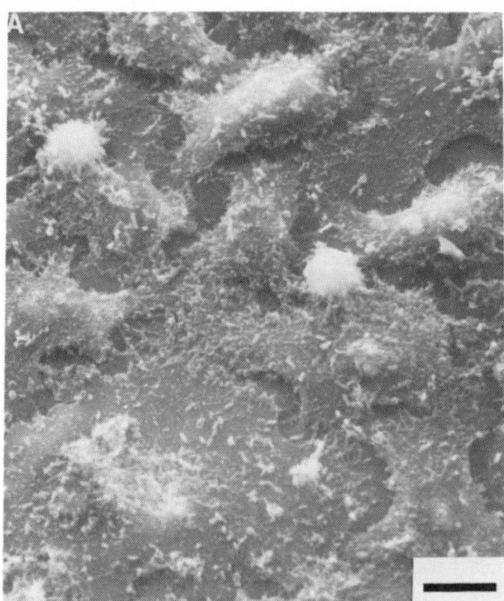

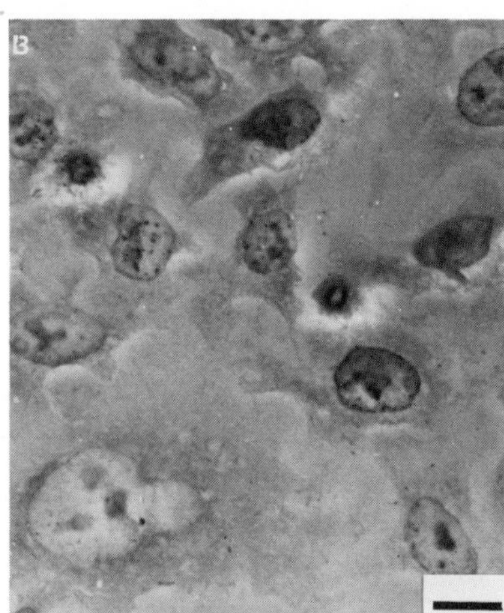

Figure 12.26. (a) Secondary-electron image and (b) backscattered-electron image of Hela tissue culture cells stained with Wilder's silver stain. Carbon coated. Magnification bar = 10 μm (Abraham, 1977).

adapted for use in the SEM. This has the advantage that the techniques come from a well-tested pedigree and form a useful basis for correlative microscopy. Becker and Geoffrey (1981) provide an excellent review of the application of backscattered imaging to histochemical studies. We mention only a few examples here, and readers should refer to standard histochemistry texts for further details and methods of application.

Silver Methenamine Staining, Z = 47. Aldehyde-fixed tissues are exposed to solutions of silver methenamine, which is an effective, although not specific, stain for chromatin. Figure 12.26 shows how Wilders stain can be used on Hela tissue culture cells.

Periodic Acid–Thiocarbohydrazide–Osmium Staining, Z = 76. This is a derivative of the periodic acid–Schiff technique used to reveal the presence of polysaccharides and mucopolysaccharides by oxidizing 1–2 glycol groups to aldehydes, which react with the Schiff reagents. Thiocarbohydrazide is a multidentate ligand used as a bridging agent in cytochemical studies. As we shall see in Chapter 13, on coating techniques, thiocarbohydrazide is also able to bind osmium to other metals, thus enhancing both their overall electron density and their electrical conductivity.

Diaminobenzidine–Osmium Staining, Z = 76. Aldehyde-fixed tissues are treated with a solution of diaminobenzidine, thoroughly washed, and briefly exposed to aqueous osmium tetroxide. This method is widely used to demonstrate the presence of oxidases and peroxidases in both sectioned (Fig. 12.27) and bulk (Fig. 12.28) samples. The diaminobenzidine reaction can also be used in conjunction with nickel and cobalt salts.

Modified Lead–Gomori Method, Z = 82. Tissues are incubated in a phosphate-containing substrate, i.e., β-glycerophosphate, together with lead acetate at an adjusted pH. The acid and alkaline phosphatases split

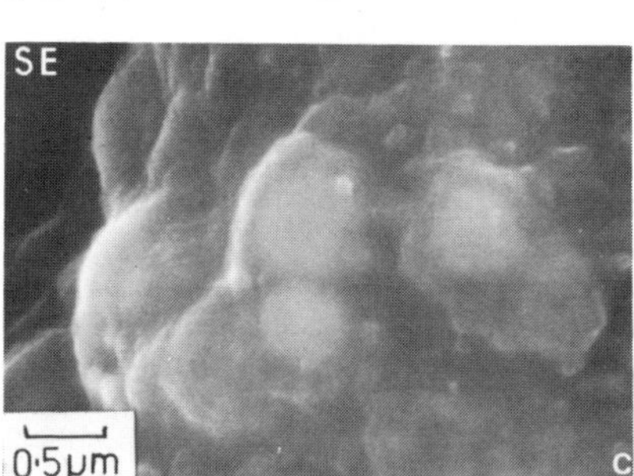

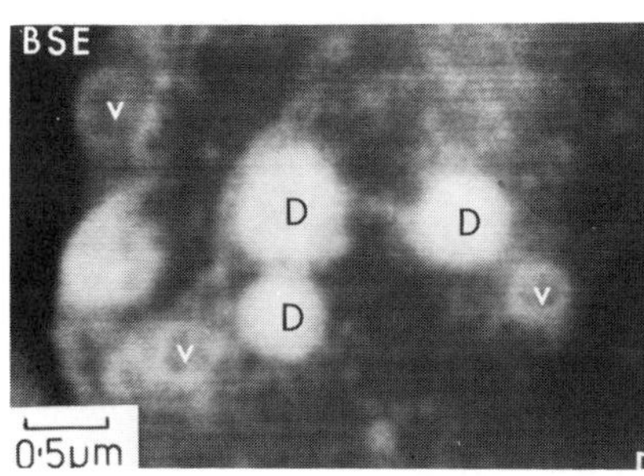

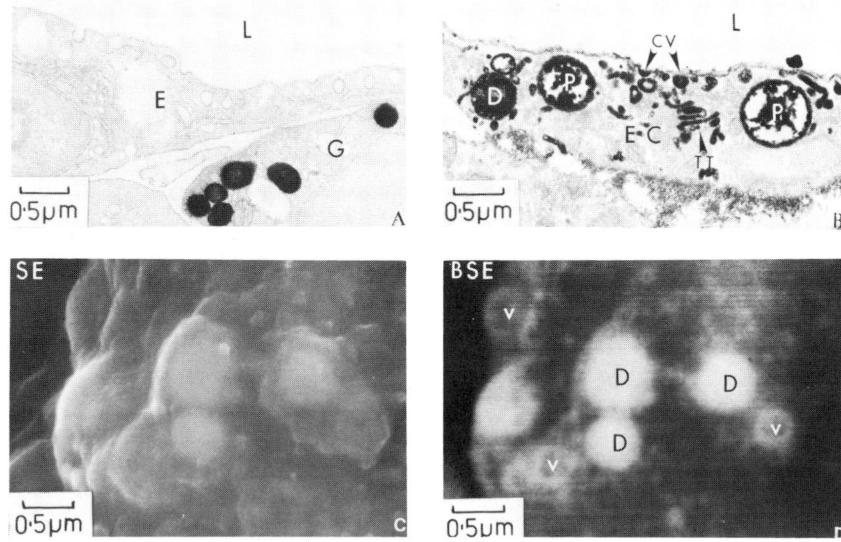

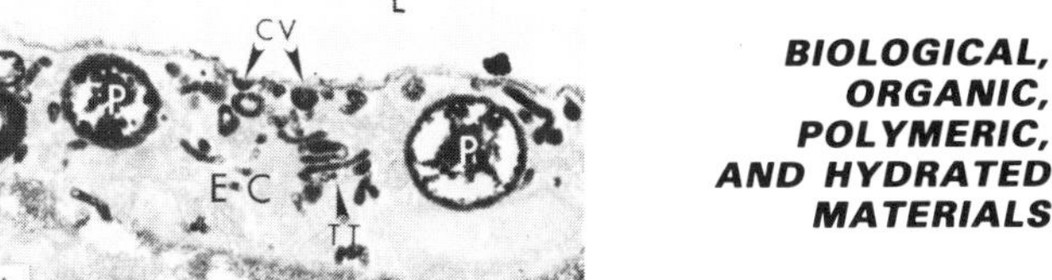

Figure 12.27. (a) TEM image of a control rat bone marrow stained with DAB osmium stain without endogenous peroxidase activity. (b) TEM image of rat bone marrow containing peroxidase activity stained with DAB osmium stain for peroxidase. (c) Secondary-electron image of rat bone marrow containing peroxidase activity stained with DAB osmium. (d) Backscattered-electron image of same specimen shown in (c) (Becker and DeBryun, 1976).

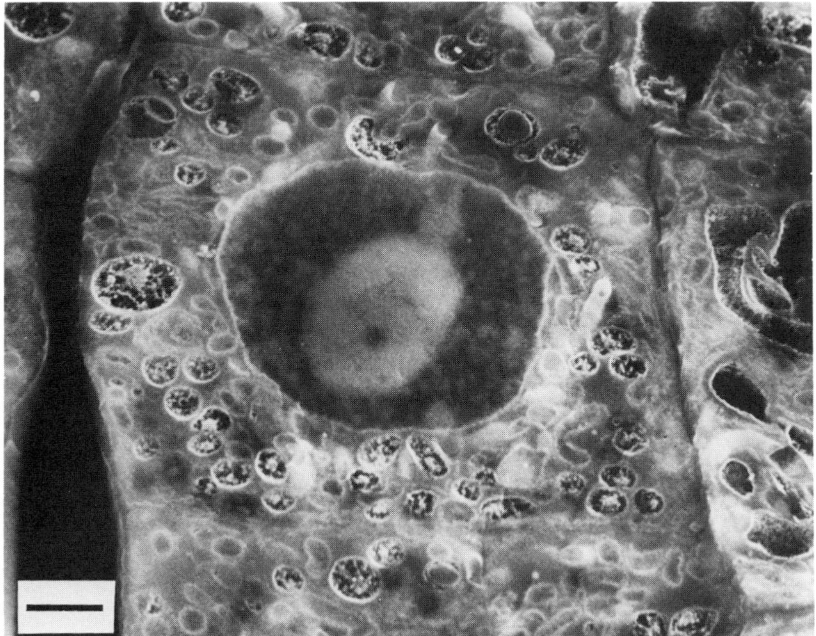

Figure 12.28. SEM photomicrograph of meristematic cell from root tip of *Zea mais*. Specimen processed by osmium ligation technique, dehydrated, and embedded in resin monomer. Fractured in frozen resin, washed in acetone, and critical point dried. Uncoated. Magnification bar = 2 μm (Woods and Ledbetter, 1976).

off the phosphate group which interacts with the lead ion to give an insoluble deposit of the lead phosphate. A recent paper by Campos *et al.* (1991) reviews this method and also provides details of a new technique to localize esterases using osmium as a marker.

In addition to the active methods, where a specific chemical reaction results in the localization of the heavy metal, it is also possible to exploit the situation where cells and tissues absorb or take up heavy metals. Four examples will suffice to illustrate this phenomenon:

Thorotrast Method, Z = 90. Cells take up colloidal thorium by a process of phagocytosis. The thorium is in turn sequestered in particular cells of the tissue.

Iron Carbonyl Method, Z = 26. Soligo *et al.* (1981) have shown that iron carbonyl is a useful marker for phagocytic activity. Within 30 min of application, iron globules can be detected inside cells.

Uranium Staining, Z = 92. Silyn-Roberts (1983) describes a simple method of estimating pore density and the ratio of blocked to unblocked pores in egg shells by staining with a dilute solution of uranyl acetate. Ulrich and McClung (1983) have used staining with uranyl acetate to produce a backscattered image of sufficient intensity to follow the movement of nuclei during cell fusion.

Silver Staining, Z = 47. Traquair (1987) has used silver nitrate to stain calcium oxalate crystals in leaf tissue.

These few examples should serve to illustrate the usefulness of this technique in analytical studies carried out on both thick and thin sections and both intact and fractured bulk material.

12.6.4.2. Radioactive Labelling Methods

Autoradiography is an analytical procedure in which specific radioactive molecules are incorporated into a known synthetic pathway, either to reveal the spatial localization of the pathway in the tissue or to follow the synthesis, turnover, and movement of molecules in the cells. The presence of the radioactive label is revealed by the development of silver grains in a thin layer of photographic emulsion placed on the tissue, into which the radioactive label has been incorporated. A one-to-one spatial juxtaposition is presumed to exist between the site of radioactivity in the tissue and its manifestation as a silver grain in the overlying emulsion.

Autoradiography has been used in light and transmission electron microscopy for a long time. It has only more recently been used in scanning microscopy. The principle of the method is the same, but there are important variations in the way the radioactive label is visualized in the SEM. Whereas the silver grains appear black in light-microscope images and electron-dense in TEM images, they appear bright in SEM images due to the fourfold higher emissivity of the silver against the background biological matrix. The registration of the silver grains is further facilitated in the SEM because of the increased depth of focus. In addition, the presence of the silver grains can be confirmed using x-ray microanalysis.

SEM autoradiography was, for a long time, a somewhat low-resolution analytical technique, but some new methods have been devised which alleviate some of these problems and make it a more useful analytical technique. The problem of autoradiography on rough surfaces cannot be easily resolved, and better results are obtained from either flat sections or ultraplaned surfaces, as previously mentioned. The problem of alternating wet–dry–wet preparative sequences has been overcome by performing the autoradiography on samples maintained in the liquid state, Weiss (1980). Chang and Alexander (1981) describe a method for SEM autoradiography whereby high-resolution cell surface detail is preserved without gross redistribution of silver grains. This technique provides a convenient medium-sized marker for topographic studies of biosynthesized molecules and of cell surface receptors using radiolabelled ligands.

The modified procedures work reasonably well, but it is still difficult to localize silver grains and read the autoradiographs from rounded cells. This can be made easier by examining samples at different focal planes and by using stereo pairs. It is important to combine the techniques of backscattered imaging and the transmitted electron signal to localize the silver grains. A new procedure, autofluorography, described by Chang and Alexander (1981), makes SEM autoradiography even more attractive as an analytical tool. Specimen integrity is maintained by keeping the emulsion and biological material in a liquid phase until the final critical-point drying. Autofluorography differs from autoradiography in that the energy from the decaying atomic nucleus excites a molecule of a scintillator, which in turn emits a photon and exposes the emulsion. The exposure time is shorter than for autoradiography (hours rather than weeks), but the process may turn out to be less efficient than autoradiography.

In all these radiographic techniques, special attention has to be paid to tissue-stabilization procedures to ensure that losses of radioactivity during processing are measured. There is also the problem of the fixation-induced losses of diffusible elements and small molecules.

12.6.4.3. Immunocytochemical Methods

Immunohistochemical localization of antigens by SEM is a sophisticated and precise way of localizing and delimiting sites of specific biological activity in cells and tissues. It successfully combines the analytical and morphological advantages of the SEM and moves the microscope away from being involved solely with the descriptive anatomy of specimens. It is difficult, in fact probably impossible, to identify antibody molecules by themselves; one requires a marker which can be linked to the antibody and is, in turn readily recognizable in the SEM. Such markers must be carefully chosen so as not to change the antibody's activity and specificity, and they should have a net charge similar to that of the natural antibody to avoid any nonspecific ionic interactions.

There is neither space nor detailed expertise on our part to discuss the chemistry and structure of immunoglobulins and antibodies. The book by Sternberger (1986) provides a suitable introduction to the subject. Although the chemical basis for antibody specificity remains uncertain, it is thought that certain residues within the N-terminal parts of the amino acid may form the basis of the antigen-combining site. Other regions of the four amino acid chains provide contact amino acids for bound ligands. It is these regions to which specific marker molecules may be attached. Immunoglobulins are proteins made up of five types, each with its own characteristic properties and molecular weight. The IgM immunoglobulins are of special interest because they have an increased combining power for most antigens and should prove useful for immunocytochemical studies. The most useful introduction and detailed exposition of methodology is the series by Bullock and Petrusz (1981 *et seq.*) and the book by Polak and Varndell (1984).

There are several important prerequisites for a good cell surface marker for SEM. It must have distinct properties of size and shape and be readily visualized against a cell surface background which itself may have a regular or complex surface topography. It must be small enough to allow good target-site localization and should form stable complexes with the particular antibody. Each type of marker requires its own method of attachment to the antibody, and it will be necessary to consult the specialist literature dealing with this aspect of the procedure.

A wide range of objects has been used as markers, including viruses such as T_4 phage, which have an easily recognizable shape; enzymes such as peroxidase, which has in turn a strong affinity for diamino-benzidine-osmium; proteins such as haemocyanin and ferretin; carbohydrates such as chitosan; and microspheres of latex, silica, and sepharose. With a few exceptions, none of these markers are now used, having been displaced by the highly successful colloidal gold marker.

Colloidal gold is proving to be one of the most useful marker molecules for immunocytochemistry. Colloidal gold sols range in color from orange to violet, have a strong secondary- and backscattered-electron coefficient, are electron dense, and have a distinctive x-ray signal. They are ideally suited for correlative microscopy. Colloidal gold has been successfully applied in many TEM studies because it can be detected unambiguously in thin sections. It is more difficult to detect small (10–15 nm) particles using the SE signal because for high-resolution studies it is necessary to coat the sample, and the coating may obscure the SE signal from the very small particles. The BSE signal can readily detect gold particles in the range 10–15 nm. The backscattered-electron signal is considerably increased by the silver enhancement procedure by which many silver atoms are bound to the colloidal gold label (Scopsi *et al.*, 1986; Goode and Maugel, 1986; Namork and Heier, 1989). The gold particles range in size from 5 to 150 nm, are easily prepared, and have a very low nonspecific binding. Gold particles can be rapidly labeled with a variety of macromolecules including polysaccharides, glycoproteins, lectins, and of course antibodies with little apparent change in biological activity (Fig. 12.29).

It is possible to take advantage of the distinctive size range of the

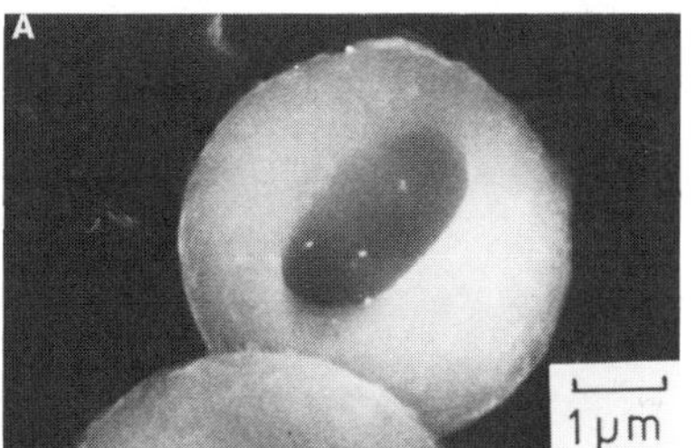

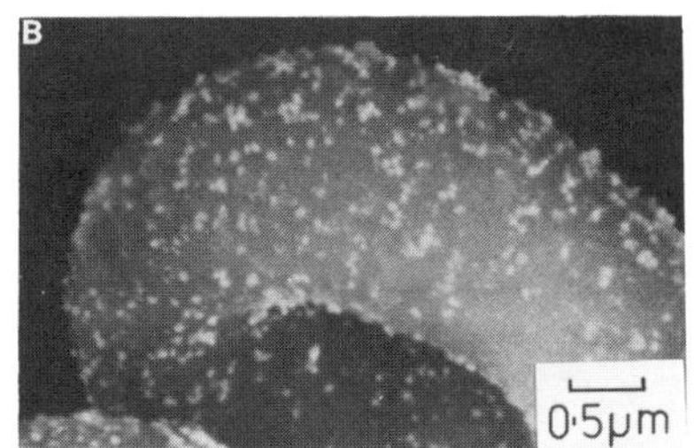

Figure 12.29. (a) Human RBC, control experiment in the presence of an inhibitor to prevent gold granules absorbing on the surface (Horisberger and Rosset, 1977). (b) Human RBC marked with gold granules labeled with soya bean agglutinin and wheat-germ agglutinin (Horisberger and Rosset, 1977).

colloidal gold particles and devise a multiple marking technique. Thus, one antibody could be linked to 10-nm colloidal gold particles while another antibody could be linked to 100-nm particles. Figure 12.30 shows double-labeled red blood cells. The size (~5 nm) of the smallest gold particle makes it particularly useful for studying binding sites and

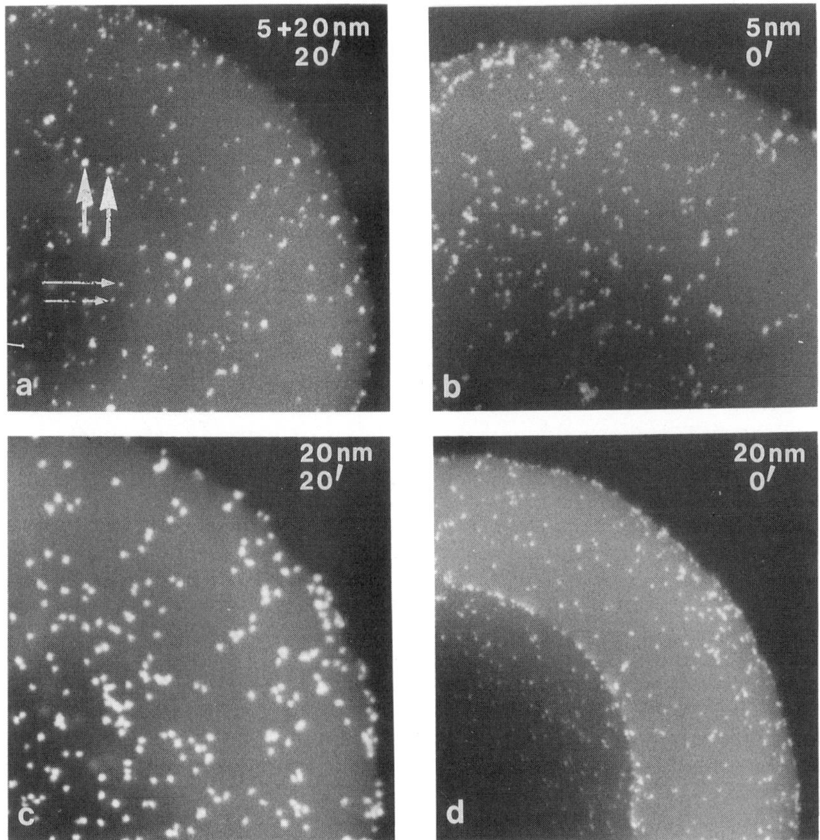

Figure 12.30. Red blood cells labeled with gold probes and silver-enhanced for 20 min. (a) Double labeled with 5-nm (small arrows) and 20-nm probes (large arrows). (b) Single labeled with 5 nm probes. (c) Single labeled with 20 nm probes. (d) Unenhanced, labeled with 20 nm probes. Picture width = 2.75 μm (Namork and Meier, 1989).

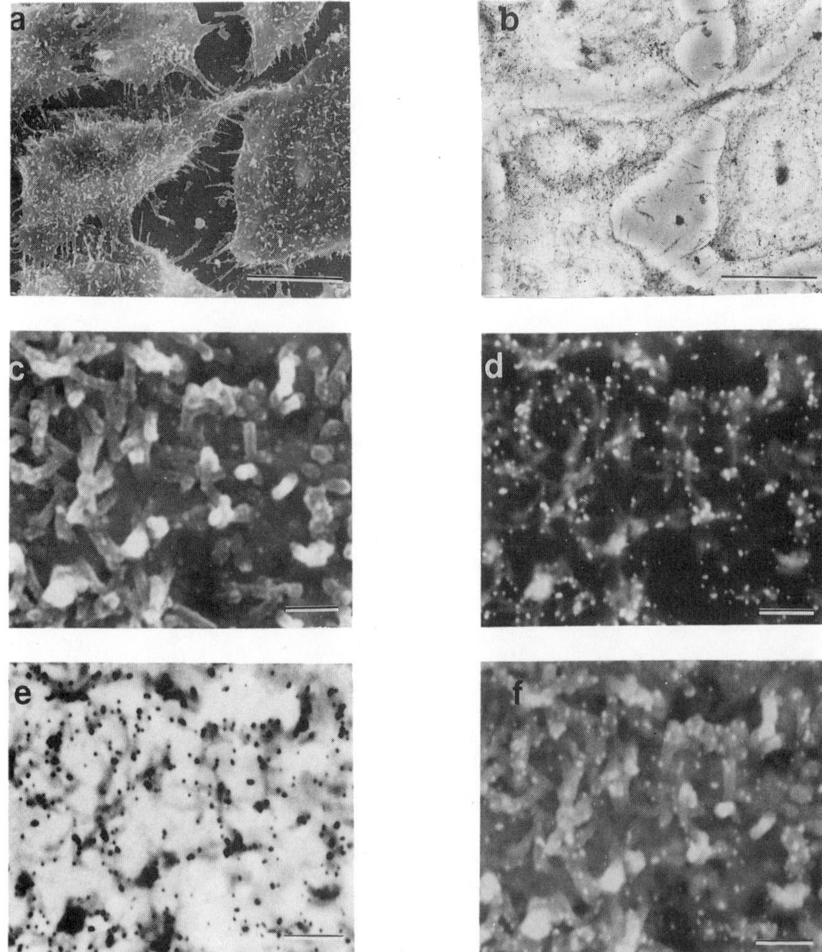

Figure 12.31. SEM multiple display illustrating the distribution of wheat germ agglutinin (WGA) receptors on the surface of human colon adenocarcinoma-derived HT29 cells. The WGA lectin-binding sites are marked by silver-enhanced 10-nm gold particles. Secondary electron (SE), backscattered electron (BSE), and mixed SE + BE signals are used to image the carbon-coated cells in a JEOL 840 SEM at 15 kV accelerating voltage, a working distance of 8 mm, and a probe current of 6×10^{-10} A. The multiple display shows how choice and electronic processing of the SEM signal can significantly enhance image interpretation. (a) Low magnification SE image: no gold labeling identifiable. Magnification bar = 10 μm. (b) The same field as in (a) but viewed in the BSE mode. Gold labelling is clearly seen as individual black dots (reverse polarity). Magnification bar = 10 μm. (c) Higher-magnification SE image: some gold particles can be recognized but are not quantifiable. Magnification bar = 1 μm. (d), (e) BSE image of the same field, gold particles low appear in high contrast. (d) In normal polarity particles appear as white dots, and (e) in reverse polarity as black dots, and they are quantifiable with increased accuracy. (f) Mixed SE + BSE (normal polarity) image of the same field as in (c). The distribution of the labeled receptors can be correlated with the surface morphology. Magnification bar = 1 μm (Hodges and Carr, 1990).

distribution of receptors or membranes where the density may result in separations as small as 5–10 nm. As Fig. 12.31 shows, the colloidal gold–labeled material can be visualized in different modes of SE and BSE imaging. As discussed earlier, the only disadvantage of the use of gold markers is that it is not possible to coat SEM samples with noble metals. The usual procedure in these situations is either to coat with carbon or to use one of the bulk-conductivity methods discussed elsewhere in this chapter. There are several comprehensive reviews on the general use of colloidal gold marker system generally (Beesley, 1989; DeMay and Moremans, 1986; Hayat, 1989), and as applied in SEM (Horisberger, 1985; Eskelinen and Peura 1988; Hodges *et al.*, 1987; Hodges and Carr, 1990).

Considerable care has to be taken over the specimen-stabilization procedures in order to retain the biological activity and have good morphological and spatial resolution. Formaldehyde and glutaraldehyde very effectively immobilize and cross-link proteins. This can be a potential problem because the functionality and specificity of antibodies are dependent on their conformational features. Although some of the deleterious effects of the organic aldehyde may be alleviated by treatment with sucrose, washing, or neutralizing with glycine, it is probably best to avoid these fixatives if possible. Unfortunately, there is no good alternative. Bullock (1984) suggests using carbodiimides or parabenzoquinone which, although shown to be of use in light microscopy, have yet to be applied at the ultrastructural level. Alternative procedures suggested by Bullock include rapid fixation and dilute fixative (both of which would mitigate against good structural preservation). Bendayan and Zollinger (1983) find that treating glutaraldehyde- and osmium-fixed material with sodium metaperiodate enhanced the antigenic staining capacity.

The well-known protein-denaturing effects of dehydrating agents such as ethanol and methanol can to some extent be reversed by rehydration. But these too should be avoided as far as possible. Therefore, one can use some of the low-temperature methods where tissues are mildly and rapidly fixed, dehydrated at low temperature with nonpolar liquids, and finally embedded in water-soluble resins. Under these conditions, most of the antigenicity is retained.

12.6.5. Criteria for Satisfactory Specimen Preparation

Before ending the section on sample preparation for chemical analysis, it is appropriate to consider how we may judge whether a given set of preparative procedures has been successful. This is more difficult to achieve than for structural studies, where correlative microscopical methods can usually provide satisfactory answers. A number of criteria should be fulfilled when applying preparative procedures for *in situ* chemical analysis.

1. The Normal Structural Relationships of the Specimen Should be

Adequately Preserved. It is difficult to set limits for morphological preservation, but as indicated earlier, a good guideline would be to achieve a morphological spatial resolution one order of magnitude better than the hoped-for analytical spatial resolution.

2. The Amount of Material Lost From, or Gained by, the Sample Must be Known. This can be achieved by carrying out a bulk chemical analysis on one-half of an identical sample before preparation and on the other half after preparation. Although this gives only the total elemental or molecular concentration of the sample, it does show whether there have been major changes as a result of a particular preparation protocol.

3. The Chemical Identity of Material Lost From, or Gained by, the Sample Must be Known. It is important to know not only if there are changes in the elemental concentration in the sample, but whether these changes are selective. Elements and many molecules in biological specimens are usually partitioned into different compartments, each of varying composition. Within any one compartment, chemicals may be loosely or strongly bound, and a preparative procedure may selectively affect the concentration of a particular component.

4. The Amount of Chemical Redistribution and Translocation Within the Sample Must be Known. Although a given preparative procedure may not change the total concentration of chemical species in the system, it may well have caused their gross redistribution. Artificial movement of material in the sample is the most difficult phenomenon to assess. The observation of unusually high concentrations of an element, as either crystals or precipitates, gives some indication that redistribution has occurred.

5. The Chemicals Used to Prepare the Samples Should not Mask the Chemicals Being Analyzed. If one examines the characteristic x-ray spectra of the elements, it is possible to see that their *K, L,* and *M* radiations overlap particularly in spectra derived from an energy-dispersive x-ray spectrometer (Chapter 5). Unless care is taken, the *L* or *M* radiation of one element may overlap and mask the *K* radiation of the element being analyzed. The actual process of analysis may remove material from even the most carefully prepared sample. Analysis of biological and organic material is limited by the transformation of elemental composition under the electron beam. The general mass loss of organic compounds and the loss of specific inorganic elements depend on the molecular structure of the sample, the conditions of analysis, the temperature, and the chemical identity of the element. Sodium, potassium, chlorine, and sulphur are at particular risk, as are some bulk samples and thick sections. Freeze-dried samples are at a greater risk than resin-embedded samples, and mass losses are higher at ambient temperatures and high beam currents.

In many instances it is not necessary or practical to go through elaborate pretesting of preparative procedures, as parallel physiological and biochemical studies give an indication whether the preparative procedures are causing changes. Because no other reliable analytical procedure exists to verify the accuracy of the microanalysis on this scale, it is sometimes difficult to obtain an independent assessment of the accuracy of a given analytical technique.

12.7. Preparative Procedures for Organic Samples Such as Polymers, Plastics, and Paints

12.7.1. Introduction

Polymers and plastics share the following common characteristics: They are made up of low-atomic-number elements, which generally exist as low-melting-point homo- or heteropolymers, or mixtures of both and may be in a crystalline, amorphous, or semicrystalline state. In addition, they have low electron emissivity and poor thermal and electrical conductivity, are radiation sensitive, and, unlike biological materials, have relatively few active ligand binding sites which could accommodate specific chemical stains.

It is appropriate to consider these types of samples separately because with a few exceptions they share another common characteristic—an absence of substantial amounts of water. Although this anhydrous condition obviates the need for dehydration and embedding, paradoxically it makes it more difficult to stabilize labile specimens, because water is a good solvent for the chemicals used in this process. The techniques which will be described here are designed primarily for use with SE and BSE imaging and, to a lesser extent, for x-ray microanalysis.

In a general text on SEM and microanalysis one can do little more than present a broad overview of the preparative techniques which have proved useful for studying the structure and chemical composition of polymers. A more detailed treatment of the subject can be found in the excellent text by Sawyer and Grubb (1987), which covers all aspects of the subject, and Hemsley (1984), which is confined to light microscopy, and in the review paper by White and Thomas (1984), which is directed towards SEM. The book by Echlin (1984) provides a variety of analytical methods that may be used to study organic materials.

12.7.2. Examination of the Surface of Polymers and Fibers

The sample is mounted on a specimen stub, coated, and observed in the SE mode in the SEM (Fig. 12.32). It is important to check that the polymer is unaffected by the adhesive solvent and that the coating method does not damage the surface. Low-angle metal shadowing can be used to distinguish small surface features.

12.7.3. Examination of the Interior of Polymers and Fibers

The same procedures are used for polymers as for biological material, i.e., sectioning and fracturing, in addition to methods which are peculiar to these materials.

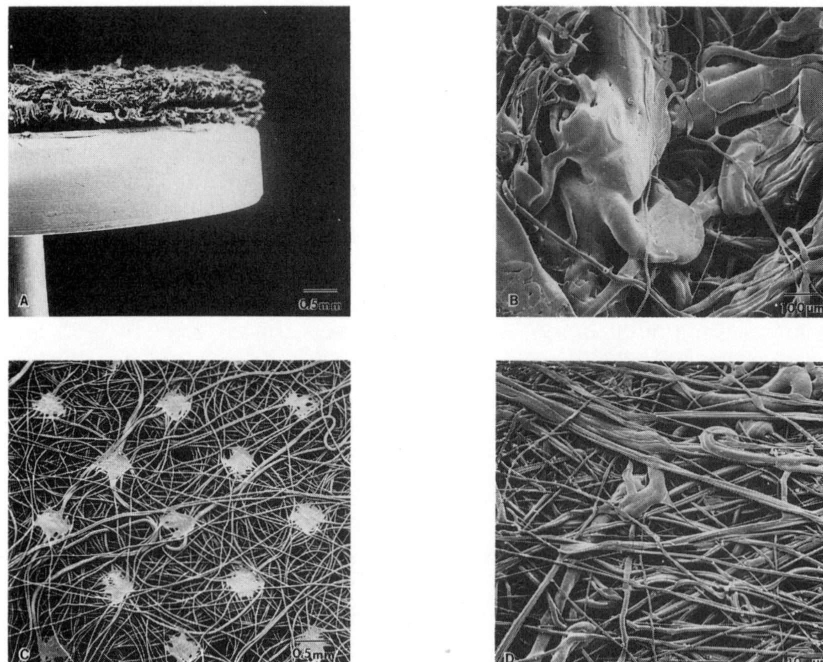

Figure 12.32. SE images of spray-spun nonwoven fabric. A side of the fabric is shown in (a) on a SEM stub. A face view is shown in (b). A calendered nonwoven surface is shown in (c) and (d). The spots on the surface in (c) are regions of local melting that hold the fabric together. At higher magnification, (d) the fibers range from round to deformed shapes (Sawyer and Grubb, 1987).

12.7.3.1. Sectioning

Although most polymers can be cut as thin or thick sections, they have a very wide range of cutting characteristics. Material such as waxes may be soft and brittle; rubbers are soft and tough; nylon is hard and tough; and Perspex (Lucite) is hard and brittle. It may be necessary to embed small samples, although it is difficult to find resins which adhere well to plastics and rubbers. The actual sectioning procedure is no different from that used with biological material and all the visual artifacts are present. Additional problems may arise if the polymer contains a hard additive such as glass or ceramic fibers; the carbon black in rubber types makes curing very difficult. Soft plastics and elastomers are best cut at their glass–rubber transition temperature (T_g), which frequently means using a cryomicrotome.

12.7.3.2. Polished Cut Surfaces

Embedded or unembedded material is cut with a diamond saw to give a flat surface. This surface is polished using standard metallographic methods (Chapter 11), and the surface either examined directly or chemically etched to reveal subsurface detail.

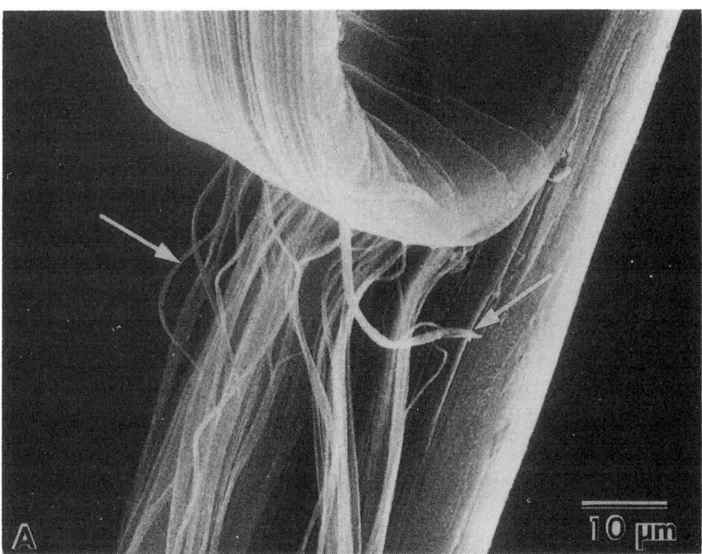

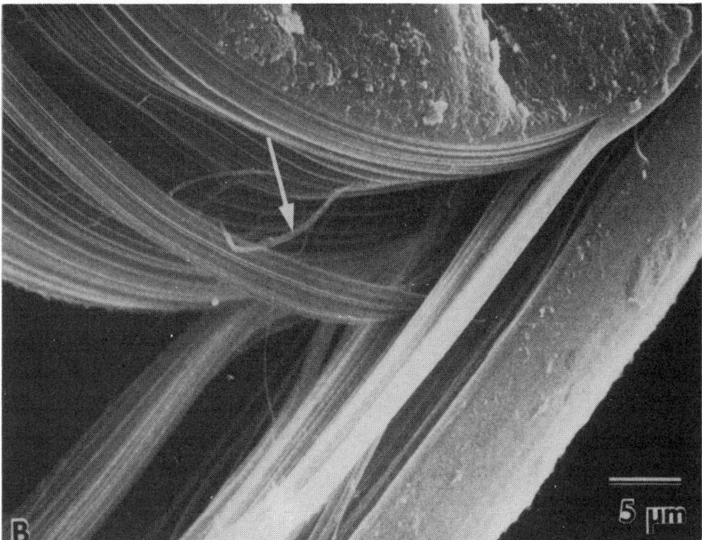

Figure 12.33. Secondary-electron image of a polymer fiber prepared by the peelback method reveals the internal fibrillar texture (arrow) (Sawyer and Grubb, 1987).

12.7.3.3. Peelback Procedure

Peelback is suitable for examining the internal structure of the long axis of polymers. The surface is nicked at an oblique angle and forceps are used to peel away a segment of material at a known orientation (Fig. 12.33).

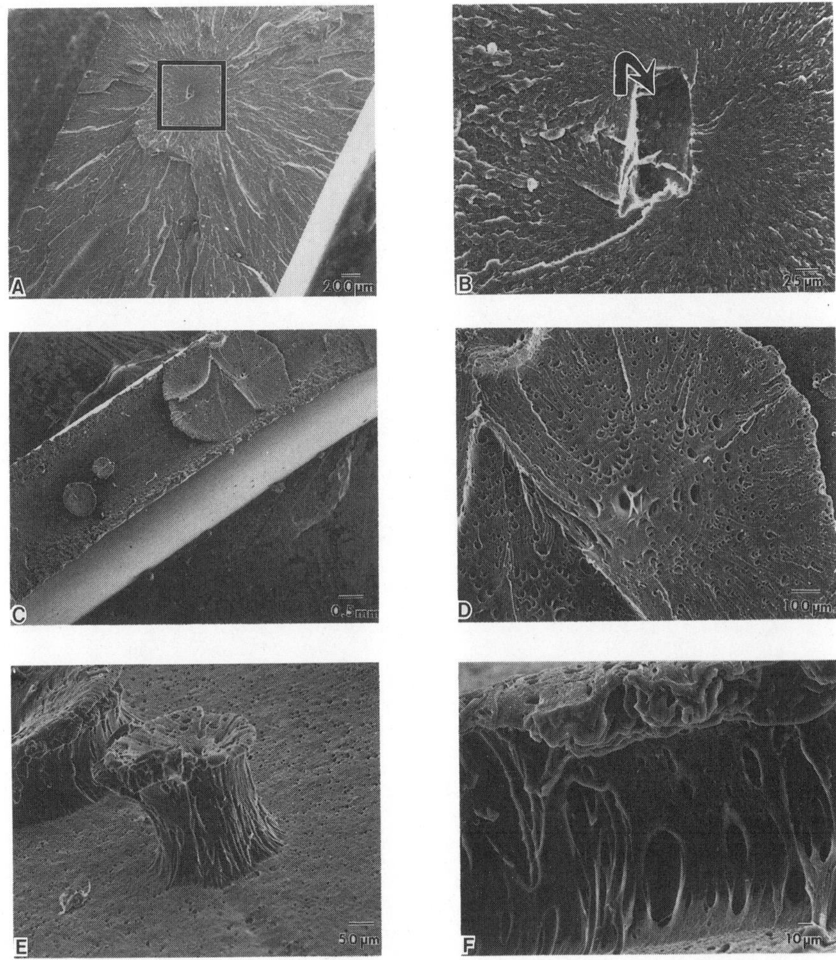

Figure 12.34. SEM images of brittle and ductile fractures. The SEM was used to compare the fracture morphology of two semicrystalline specimens that were processed similarly and yet produced different elongation properties. The nylon specimen with smaller elongation properties (a), (b) has a brittle fracture morphology with a rectangular particle at the locus of failure (b). The specimen with higher elongation properties (c)–(f) exhibited ductile failure (Sawyer and Grubb, 1987).

12.7.3.4. Fracturing

Separation along a line of least resistance can be achieved by either tensile stress or impact fracture with a sharp tool (Fig. 12.34). Soft materials and elastomers must be cooled well below T_g to provide the mechanical strength for a brittle fracture to occur. Fractures are an effective way of revealing the internal structure of composites. For example, Fig. 12.35 shows fracture surfaces of a glass-filled composite.

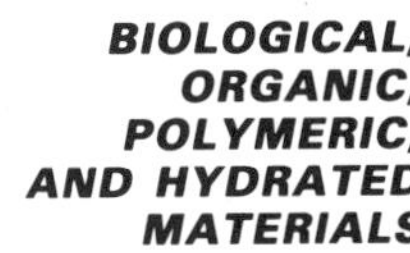

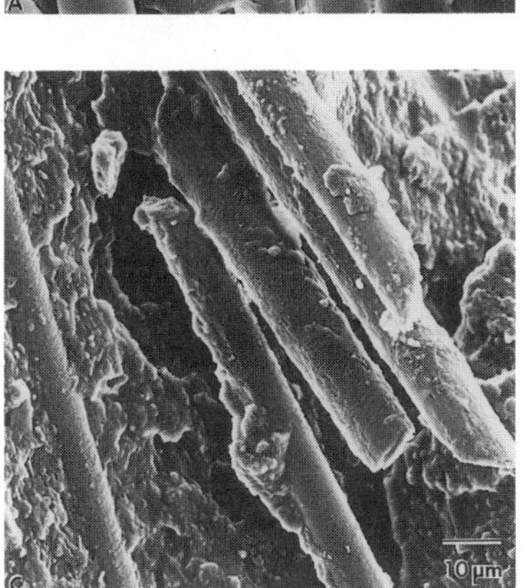

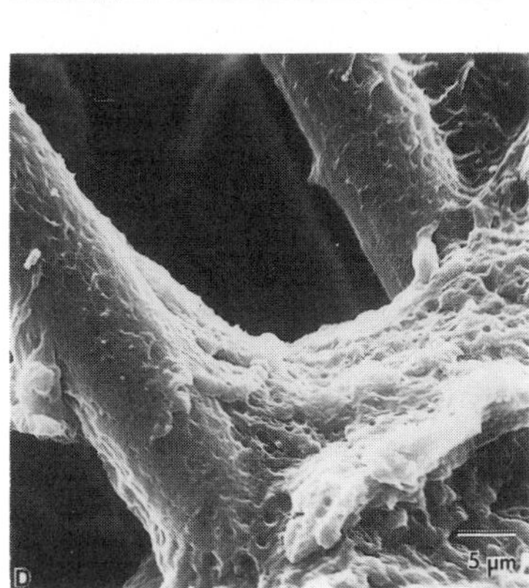

Figure 12.35. Secondary-electron image of a glass-filled composite. Images of two glass-filled composites with different adhesive properties are shown. (a) A composite with poor adhesion, as shown by the clean glass-fiber surfaces resulting from failure at the fiber–resin interface. (b) Holes are seen where fibers with poor adhesion pulled out of the matrix. (c) Failure in the matrix with resin on the fiber surfaces. (d) Higher magnification image of (c) (Sawyer and Grubb, 1987).

12.7.3.5. Replicas

This technique is useful for beam-sensitive material, but considerable attention has to be paid to the chemical composition of the replicating materials. Solvents may affect the polymers; solvent chemical composition may result in a permanent bond with the polymer surface and prevent the release of the replica. Goodhew (1974) is a good primary reference for details of replication methods.

12.7.4. Surface Etching of Polymers

Surface etching may be achieved by either chemical or physical means and is designed to reveal substructural detail of composition by selective removal of one or more components of the polymer (Fig. 12.36). These procedures have to be monitored very carefully as there is frequently incomplete information about the solubility of the polymer and the solvent properties of the chemicals used. Correlative microscopy is essential when using this technique.

12.7.4.1. High-Energy Beam Bombardment

Electron-beam etching can cause bond rupture in plastics such as polymethacrylates, and the resultant low-molecular-weight components may be removed by solvent extraction. The same procedure can be used to cross-link and polymerize other plastic or polymer components.

12.7.4.2. Argon Ion–Beam Etching

Argon ion–beam etching removes materials of varying densities in plastics such as polyethylene terephthalate, but there are problems with charge-up of the specimen, and uneven etching can occur. The sample should be rotated and cooled during the etching process.

12.7.4.3. Oxygen Plasma Etching

Oxygen plasma etching procedure is rather difficult to control, but it has been used to expose inorganic inclusions such as glass and ceramic fibers embeded in the organic polymer matrix.

12.7.4.4. Chemical Dissolution

The procedure takes advantage of the selective solubility of plastics and polymers in various solvents. For example, xylene dissolves polymethylpentene sulphones; carbon tetrachloride dissolves high density polyethylene; and isopropanol vapor selectively removes polystyrene from a rubber matrix.

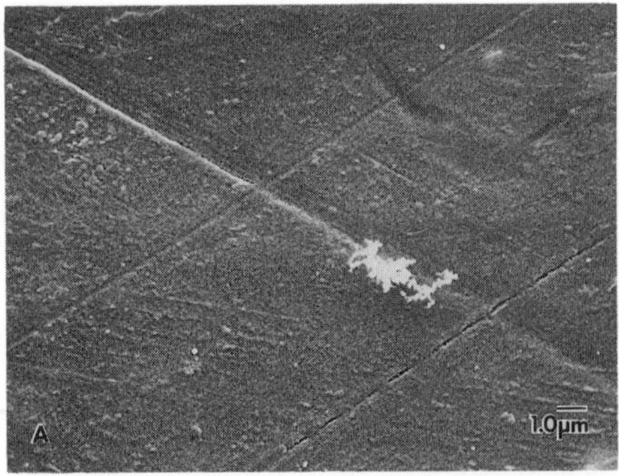

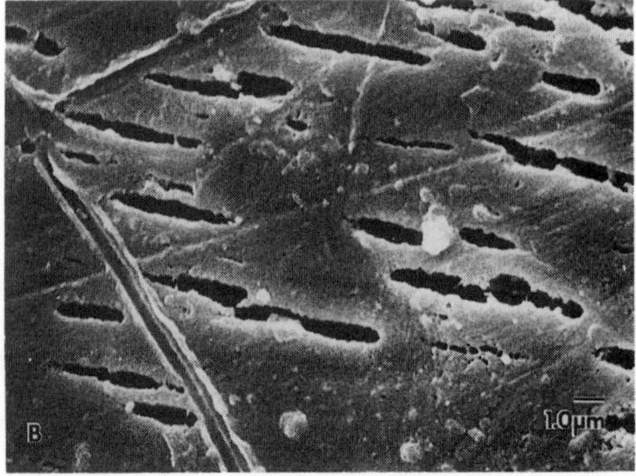

Figure 12.36. Secondary-electron images of unetched and chemically etched polyacetal. (a) Molded polyacetal surface showing a smooth texture with little surface detail. (b) Elongated pits, oriented in the direction of polymer flow after etching for short times (Sawyer and Grubb, 1987).

12.7.4.5. *Chemical Attack*

In this procedure, powerful oxidants are used to remove various components from the material. For example, fuming nitric acid can etch away the amorphous parts of polyethylene; chromic acid etches polypropylene, and sodium methoxide dissolves epoxy resins.

12.7.4.6. *Enzymatic Digestion*

Enzymatic digestion is applicable only to natural polymers and fibers but has been successfully used, for example, to etch hair with papain and wood with different cellulytic and lignolytic enzymes.

12.7.5. Staining of Polymers and Plastics

For the SEM, this technique involves incorporating high-atomic-number elements or their compounds into specific or general sites in order to increase the electron scatter or emissivity. Tables 12.2 and 12.3, from Sawyer and Grubb (1987), list polymer functional groups and stains and specific functional groups and stains, respectively, as a guide to what may be achieved by staining procedures. Most of the stains are positive and are useful in distinguishing lamellar layers and amorphous interlayers in polymers and plastics. Five main groups of stains have been used.

12.7.5.1. *Osmium Tetroxide*

Osmium tetroxide reacts with any unsaturated bonds and can be used as a general stain for a wide range of materials. Some materials may need pretreatment such as alkaline saponification before they will take up the stain.

12.7.5.2. *Chlorosulphonic Acid*

Chlorosulphonic acid cross-links, stabilizes, and stains amorphous materials in semicrystalline olefines. It stains the lamellar regions of

Table 12.2. Polymer Functional Groups and Stains

Polymers	Stains
Unsaturated hydrocarbons, alcohols, ethers, amines	Osmium tetroxide
Acids or esters	(a) Hydrazine (b) Osmium tetroxide
Unsaturated rubber (resorcinol-formaldehyde-latex)	Ebonite
Saturated hydrocarbons (PE and PP)	Chlorosulfonic acid/uranyl acetate
Amides, esters, and PP	Phosphotungstic acid
Ethers, alcohols, aromatics, amines, rubber, bisphenol A, and styrene	Ruthenium tetroxide
Esters, aromatic polyamides	Silver sulfide
Acids, esters	Uranyl acetate

Table 12.3. Specific Functional Groups, Examples, and Stains

Functional group	Examples	Stains
—CH—CH—	Saturated hydrocarbons (PE, PP) HDPE)	Chlorosulfonic acid Phosphotungstic acid Ruthenium tetroxide
—C=C—	Unsaturated hydrocarbons (Polybutadiene, rubber)	Osmium tetroxide Ebonite Ruthenium tetroxide
—OH, —COH	Alcohols, aldehydes (Polyvinyl alcohol)	Osmium tetroxide Ruthenium tetroxide Silver sulfide
—O—	Ethers	Osmium tetroxide Ruthenium tetroxide
—NH_2	Amines	Osmium tetroxide Ruthenium tetroxide
—COOH	Acids	Hydrazine, followed by Osmium tetroxide
—COOR	Esters (butyl acrylate) (polyesters) (ethylene-vinyl acetate)	Hydrazine, followed by Osmium tetroxide Phosphotungstic acid Silver sulfide Methanolic NaOH
—$CONH_2$ —CONH—	Amides (nylon)	Phosphotungstic acid Tin chloride
Aromatics	Aromatics Aromatic polyamides Polyphenylene oxide	Ruthenium tetroxide Silver sulfide Mercury trifluoroacetate
Bisphenol A based epoxies	Epoxy resin	Ruthenium tetroxide

polyethylene (PE), and poststaining with uranyl acetate enhances the staining (Fig. 12.37).

12.7.5.3. Phosphotungstic Acid

Phosphotungstic acid reacts with surface functional groups such as -hydroxyl, -carboxyl, and -amine. It is used to stain nylon and the lamellar regions of polypropylene (PP).

12.7.5.4. Ruthenium Tetroxide

Ruthenium tetroxide is generally considered to oxidize aromatic rings and ether alcohols and has been used to stain rubber inclusions, nylon, polystyrene, and polyethylene. It is a more effective surface stain than osmium tetroxide.

12.7.5.5. Silver Salts

Various silver salts have been used to stain biological fibers and polymers such as hair, wool, and regenerated cellulose.

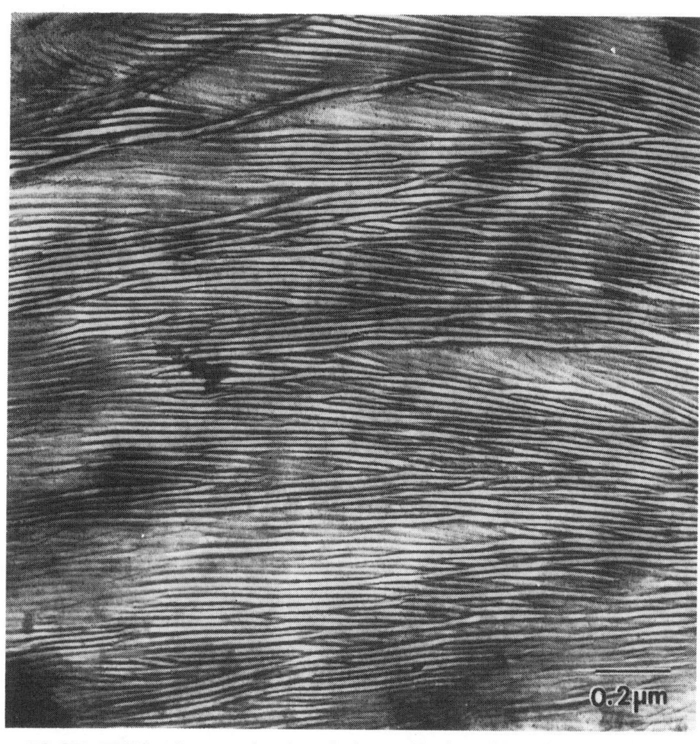

Figure 12.37. TEM micrograph of a cholorosulfonic acid–stained linear polyethylene crystallized isothermally from the melt reveals the electron-dense interlamellar surfaces typical of polyethylene (Sawyer and Grubb, 1987).

All the metal stains have the desired effect of increasing either emissivity or electron scatter and enables the material to be examined by SE or BSE imaging or as thin sections in the transmitted electron mode. Figure 12.38 shows how the BSE signal can be used to image a fractured bulk sample of a mineral-filled polymer composite.

12.7.6. Specialized Preparative Methods

Some polymers contain water or are dispersed in an aqueous phase, i.e., latex suspensions, emulsions and adhesives. The water may be removed by either freeze-drying or critical-point drying. Alternatively, the samples may be studied in the fully hydrated state using a cold stage (Sawyer and Grubb, 1987). Critical-point drying should be used only if there is certain knowledge regarding the solubility (or not) of the polymers in the organic fluids used as chemical dehydrating agents. Wood may require specialized treatment, particularly if the SEM is to be used to provide information about its anatomy. Most wood samples are easily trimmed with a razor blade, although some high-density woods may need softening by boiling in glacial acetic acid and hydrogen peroxide before sectioning. Careful cutting and mounting will enable the various faces to be seen together on the same SEM image. Any cytoplasmic debris can be

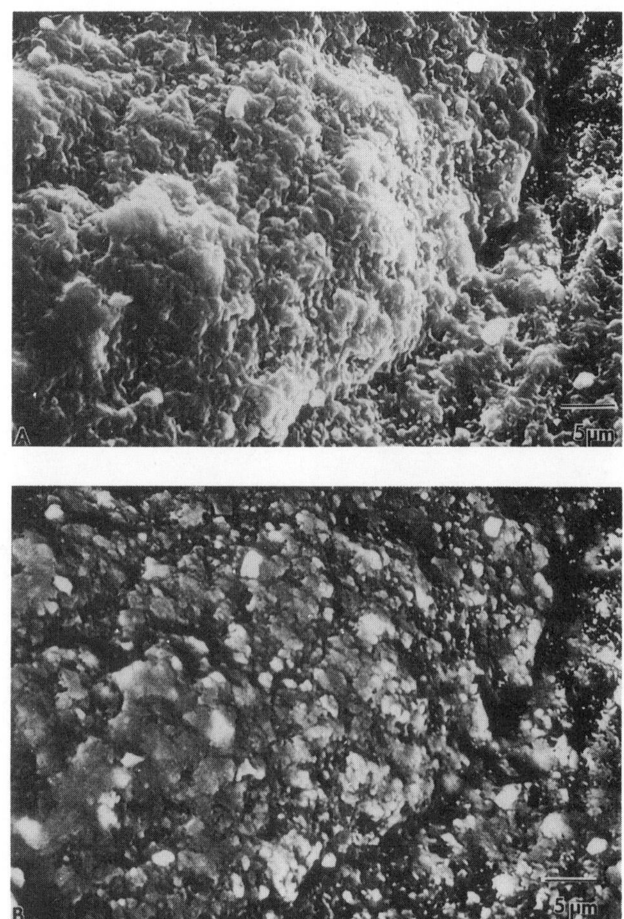

Figure 12.38. (a) Secondary-electron image of a mineral-filled polymer composite. This image does not reveal the nature of the dispersed filler particles. (b) Backscattered-electron image shows the higher-atomic-number mineral filler (Sawyer and Grubb, 1987).

removed by washing in a hypochlorite solution, after which the small wood blocks are dehydrated in methanol, air dried, and sputter coated with Au–Pd. No specialized methods are needed for the *in situ* chemical analysis of polymers other than to make sure the preparative methods do not introduce chemical artifacts into the analytical result. X-ray microanalysis has been used to show chlorine in polyvinylchloride, sulphur in vulcanized rubbers, calcium and aluminum in plastic filter, and titanium in paint.

12.8. Low-Temperature Specimen Preparation for Structural and Analytical Studies

12.8.1. Introduction

Most of the problems associated with ambient-temperature, wet-chemical preparative procedures are considerably diminished by carrying out both the sample preparation and the examination and the analysis at low temperatures. There appear to be six main advantages to using low temperatures for the preparation of organic, hydrated, and biological material:

1. By lowering the temperature, but still maintaining the liquid state, one can slow down chemical reactions and transport processes, such as viscosity and diffusion, and make such processes more amenable to study *in situ*.
2. In the solid state, water and many organic materials show an increase in mechanical strength as their temperature is lowered. This allows these materials to be sectioned, fractured or dissected to reveal subsurface detail.
3. The amount of thermal and ionizing-radiation damage to the specimen, an inevitable consequence of using some of the imaging and analyzing systems, is diminished.
4. There may be no need to use any of the deleterious chemical methods frequently required to stabilize and strengthen organic samples during specimen preparation for examination and analysis at ambient temperatures.
5. It is the only way we may hope to perform the *in situ* examination and analysis of electrolytes, soluble molecules, and hydrophic macromolecules in a form close to their natural state.
6. It is the only way we may hope to study the triphasic state of matter, i.e., soils, food products, emulsions, oil and brine mixtures, gas–liquid exchange systems, etc.

Although low-temperature microscopy and analysis appear to be the panacea for all structural and *in situ* chemical studies, it is not appropriate for all investigations, and like any method has its own set of limitations. The ideal goal of low-temperature methods is the vitrification of the liquid phase in which all the constituents, atoms, molecules, and macromolecular structures remain in their natural position in the sample. In a few instances, we believe this can be achieved; in most cases phase separation occurs and the sample's integrity is compromised by the formation of ice crystals. The phase transformation from a disorganized liquid to a crystalline solid can rarely be avoided although the size of the individual crystallines can be reduced to very small dimensions, if the cooling conditions are optimized. Finally, we can rarely avoid ice crystal artifacts.

The discussion which follows will enumerate the general processes of low-temperature specimen preparation. The interested reader is referred to a text on low-temperature microscopy and analysis (Echlin, 1992).

12.8.2. Water: Properties

Liquid water is vital for living organisms. It is both a reactant and the medium where reactions occur and their products are transported. Water is the most abundant and energetically the least expensive building block of biological material, and when converted to the solid state, it can provide the perfect matrix in which we may study the structure and *in situ* chemistry of hydrated cells and tissues. Water is also an important constituent of nonliving organic materials (polymers) and many inorganic substances (soils, clays, cements, muds). This discussion will consider the nature of the solidfied water matrix and its constituent components in biological, organic and hydrated material, and how this may be formed, manipulated, examined, and analyzed.

Water has several remarkable properties that are a consequence of its molecular structure. The molecule is strongly polarized due to separation of the charges on the oxygen (−) and hydrogen (+) atoms. The constantly shifting weak hydrogen bonds, which can form as a consequence of this charge separation, are responsible for the dynamic structure of liquid water and for its interactions with both dissolved materials and intact macroscopic surfaces. The hydrogen bonding of water influences the properties of the liquid, many of which are intimately associated with the processes of life. The ionic, polar, and apolar associations of water with a wide range of substances indicate that pure liquid water rarely exists, and the different ways in which water associates with materials appear to modify the bulk liquid. This affinity of water molecules for surfaces has given rise to some misunderstanding as to what is meant by the terms "bound" and "free" water. The complex interactions between liquid water and the various components of hydrated systems reveal that some of the water associated with hydrophilic material may exist in a modified form. These variations in the strength and extent of the bonding between water and substrates influence the amount and form of the solid phases to which liquid water may be converted when cooled during many of the preparative methods of cryomicroscopy.

12.8.3. Ice

When water and aqueous solutions are cooled to a sufficiently low temperature, depending on the environmental conditions, they are converted to one of the many forms of ice. In low-temperature microscopy and analysis, we are concerned only with Ice_H, Ice_C, and $Ice_A(Ice_V)$, as the other ice polymorphs exist only under conditions of high pressure. Much is known about Ice_H, the principal form of ice found in most frozen specimens. An appreciation of the thermal, electrical, and mechanical properties of this material is important in understanding how the ice is formed initially and how the frozen material may be subsequently treated during manipulation, examination, and analysis of the specimen.

The conversion of aqueous solutions to the solid state involves three distinct but closely integrated processes. The initial removal of heat is largely specimen dependent. As cooling proceeds, homogeneous and heterogeneous nucleation events intervene to produce the water clusters that grow into ice crystals. The nucleation and the subsequent formation and growth of the ice crystals are kinetic phenomena, which are strongly influenced by temperature. Under exceptional circumstances, the water may be converted to a glasslike state (Ice_V), which preserves the random molecular arrangements characteristic of the liquid state.

The formation of ice and the consequent phase separations within the previously fluid state can severely affect the structural integrity and chemical activity of the systems being examined. This problem is particularly acute in biological systems. Although it is possible to quantify the phase changes in tertiary systems, virtually nothing is known about the complex phase separations that occur in cell and tissue fluids. Such phase separations are best avoided entirely by ensuring that the sample is converted to a vitrified state. For many specimens this is not possible, and it will be necessary to accept that the microcrystalline or partial glass state is an acceptable alternative for most specimens being examined and analyzed at low temperatures.

12.8.4. Rapid Cooling Procedures

The initial act of rapidly cooling the sample is probably the most critical part of all the preparative procedures used in low-temperature sample preparation. Provided the cooling rate is at least $1 \times 10^4\,K\,s^{-1}$, a sample no thicker than 25–50 μm will be preserved with ice crystals smaller than 10 nm (see Table 12.4). There are, however, severe physical limitations to the rate at which heat may be removed from most of the specimens we commonly use in low-temperature microscopy and analysis, and we must accept that only the smallest samples can be preserved without any ice crystal damage. Chemical fixation and cryoprotection are best avoided in morphological studies of all but the larger specimens and should not be used in conjunction with studies involving the analysis of low-molecular-weight, water-soluble materials.

Specimens may be cooled rapidly by a number of different liquid and solid cryogens in a variety of different ways (see Table 12.5). Although most of the methods give cooling rates of about $1 \times 10^4\,K\,s^{-1}$, there is wide variation, and no single method gives good results with all samples

Table 12.4. Comparison of Mean Cooling Rates and Range of Measurements Recorded in the Literature for the Five Main Methods of Sample Cooling

Method	Mean rate of cooling ($10^3\,s^{-1}\,K^{-1}$)	Range of measurements ($10^3\,s^{-1}\,K^{-1}$)	Depth of ice free zone[a] (μm)	Best cryogen (K)
Plunge	10–12	2–500	5–10	Ethane (93)
Spray	50	8–100	20–30	Propane (83)
Jet	20–25	7–90	10–15	Propane (83)
Impact	25	6–55	15–25	Copper (20)
Pressure	0.5	—	500	Nitrogen (77)

[a] Ice crystals cannot be distinguished in TEM, although the spaces once occupied by ice crystals, so-called ice-crystal ghosts, can be seen in chemically prepared samples.

Table 12.5. Mean Cooling Rate and Range of Measured Rates of a Number of Liquid Cryogens

Cryogen	Temperature (K)	Mean cooling rate (10^{-3} K s^{-1})	Range of cooling rates (10^3 K s^{-1})
Ethane	90	13–15	0.8–500.0
Freons	90	6–8	0.2–98.0
Liquid nitrogen	77	0.5[a]	0.03–16.0
Slush nitrogen	63	1–2	0.9–21.0
Propane	83	10–12	0.2–100.00

[a] Primarily due to the "Leiden frost" effect.

(Table 12.6). Immersion fixation into liquid propane or ethane is a general method, impact and jet cooling give satisfactory results for surface layers and thin layers of suspensions, and spray cooling seems best suited for the cryopreparation of microdroplets. The best results for biological tissues have been obtained using high-pressure cooling (Michel *et al.*, 1991; Kiss *et al.*, 1990).

The best approach to adopt when setting up procedures for the rapid cooling of a sample is first to establish the upper limits of ice crystal damage which may be tolerated in the experimental system. Provided sufficient data are available about the thermophysical properties of the specimen, cryogen, and cooling technique, it should be possible to simply calculate the parameters for the rapid cooling procedure that will give the best preservation (Bald, 1987).

Table 12.6. Comparative Immersion Cooling Rates Measured Using Various Procedures, Thermocouples, and Cryogens

Cryogen	Mean cooling rate (K^{-1} s^{-1} × 10^3)					
	Rate A[a]	Rate B[b]	Rate C[c]	Rate D[d]	Rate E[e]	Rate F[f]
Liquid nitrogen	16	0.80	0.70	0.70	0.13	—
Slush nitrogen	21	1.7	—	1.0	—	—
Freon 22	66	9.0	6.3	0.8	2.6	1.5
Propane	98	19.1	8.3	1.8	4.1	1.7
Ethane	—	—	12.0	—	—	2.1

[a] All rates are cited between 273 and 173 K. Rate A recorded with a 70-μm thermocouple with a spring-assisted velocity of 0.5 m s^{-1} (Costello and Corless, 1987).
[b] Rate B recorded with a 300-μm thermocouple with a spring-assisted velocity of 1.4 m s^{-1} (Elder *et al.*, 1982).
[c] Rate C recorded with a 300-μm thermocouple with a spring-assisted velocity of 2.3 m s^{-1} (Ryan *et al.*, 1987).
[d] Rate D recorded with a 50-μm thermocouple rapidly hand dipped (Zierold, 1980).
[e] Rate E recorded with a 360-μm thermocouple allowed to free fall by gravity (Schwabe and Terracio, 1980).
[e] Rate F recorded with a 25-μm thermocouple covered with 415 μm of 20% gelatine at a spring-assisted velocity of 3 m s^{-1} (Ryan *et al.*, 1990). (The rate-F data is probably the most significant, as the recorded rates most closely represent the cooling rates one might expect *inside* a 0.5-mm-thick specimen.)

12.8.5. Cryosectioning

Cryosectioning is one of several procedures used to expose the interior of frozen specimens for examination and analysis. Frozen sections, which may be ultrathin (50–100 nm), thin (100–1000 nm) or thick (1–2 μm), can be either frozen hydrated or frozen dried. Frozen-hydrated sections retain their natural water content throughout preparation and examination; frozen-dried sections are dehydrated after the sections are cut. The actual process of cryosectioning is not completely understood, although several theories claim to explain the way sections are formed (Zierold, 1987). The cryoultramicrotomes and the cryostat microtomes needed for cryosectioning are now more or less standard. The practical procedures include sample preparation, mounting, and trimming, followed by sectioning, section collection, and transfer to the microscope. Cryomicrotomy has its own specific difficulties and problems, and the recognizable and largely explicable artifacts are now understood. Many applications of cryomicrotomy provide a rapid, nonchemical procedure for the close examination of the interior of hydrated material, polymers, and elastomers.

12.8.6. Cryofracturing

Low-temperature fracturing and freeze-fracture replication may be used to expose the interior of a frozen sample. Samples may be fractured either by tensile stress or by the shear forces generated as a knife cleaves the sample. The fracture pathway–unlike sectioning, which follows a more or less predictable pathway—is very much a hit or miss affair and yields unpredictable variations within the exposed surfaces. Although low-temperature fracturing and freeze-fracture replication are similar in many respects, there are important differences. Low-temperature fracturing reveals surfaces which may be examined directly at low temperatures in the SEM (Fig. 12.39), whereas freeze-fracture replication produces high-fidelity copies of surfaces which can be examined at high resolution and ambient temperatures in the TEM and SEM. Recent advances in the instrumentation used for the preparation and examination of fracture surfaces now make these differences less distinct.

The low-temperature fracturing process, which involves fracturing together with optional etching or surface coating, may be carried out either in simple homemade devices or in more sophisticated instrumentation where much attention has been paid to the problems of sample contamination and incipient melting (Fig. 12.40). The equipment is either dedicated to a particular microscope or freestanding and reliant on efficient sample transfer to the cold stage of the microscope. Freeze-fracture replication devices are all freestanding and take the sample through the separate stages of fracturing, optional deep or shallow etching, metal shadowing, and finally the formation of a carbon replica.

Considerable attention has to be paid to problems of contamination and to the various artifacts which occur during the preparative process.

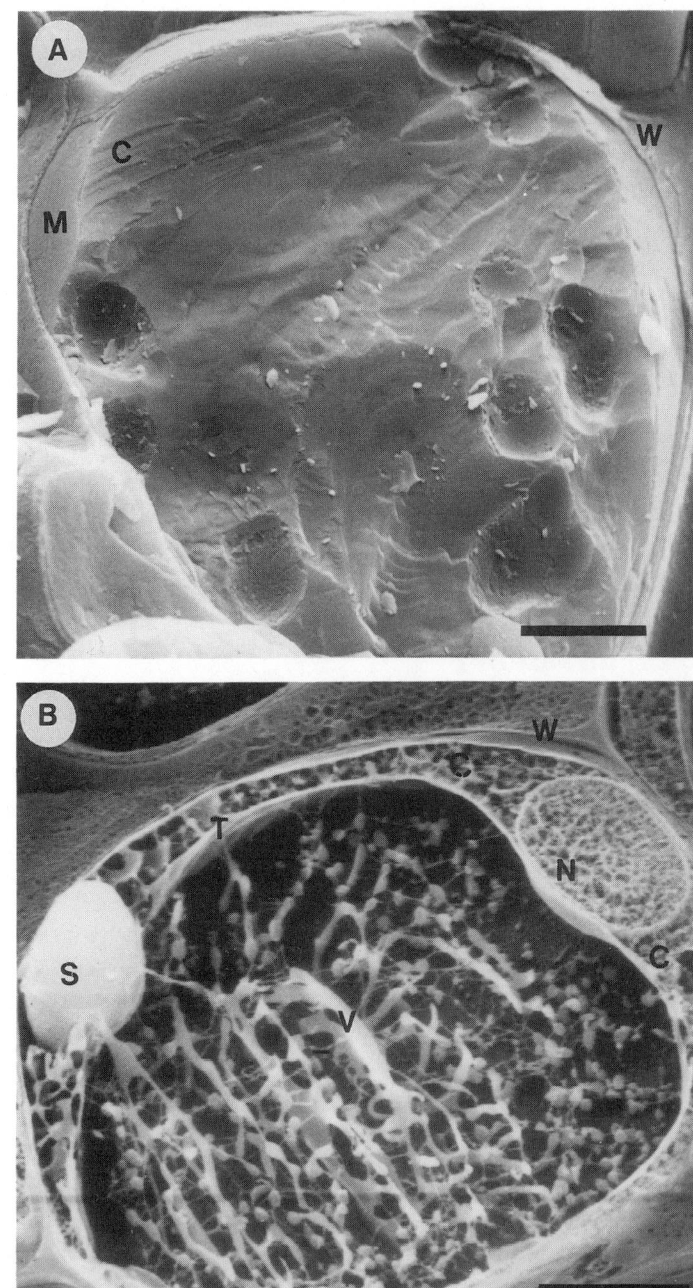

Figure 12.39. Secondary-electron images of (a) a frozen-hydrated and (b) a frozen-dried fracture face of outer cortex cells of roots of duckweed (*Lemna minor* L.). Although the frozen-hydrated fracture face is virtually featureless, it is possible to distinguish the main cellular compartments. The frozen-dried fracture has increased topographic contrast but reveals the large voids once occupied by ice crystals. C, cytoplasm; M, membrane; N, nucleus; S, starch grain; V, vacuole; T, tonoplast, and W, cell wall. Samples coated with 8 nm of gold. SEM operated at 15 keV and 15 pA beam current. Magnification bar = 10 μm.

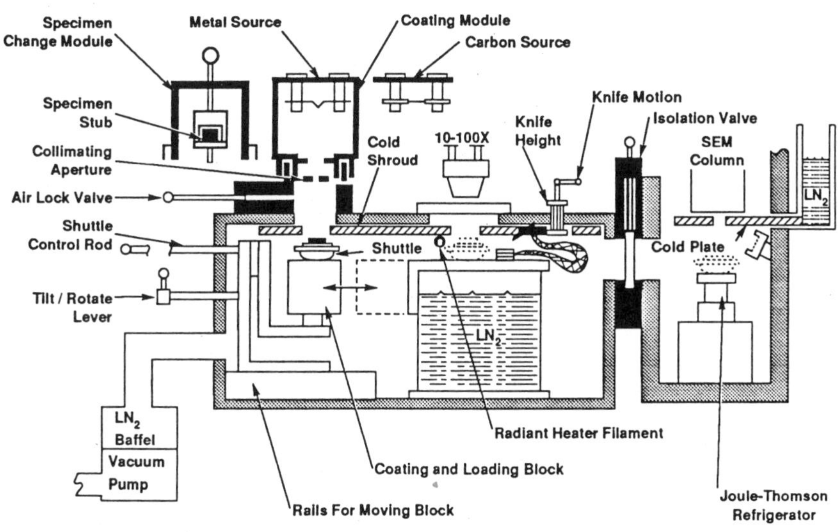

Figure 12.40. Schematic diagram of a cryopreparation system.

Although freeze-fracture replication was originally developed as an ultrastructural technique, it is now being used as an analytical procedure involving autoradiography, immunocytochemistry, and other labeling methods (Boonstra *et al.*, 1991). The two fracturing procedures have broad applications in addition to being used in connection with other low-temperature methods such as freeze solubilization, fracture polishing, and freeze cracking.

12.8.7. Freeze Drying

Freeze drying is a dehydration process in which the water in the sample is first converted to ice, which is then sublimed away at low temperatures and high vacuum. The rate of drying is a function of both the temperature and the vapor pressure of ice (see Table 12.7). The drying process has two distinct phases. The primary drying, which usually takes place at between 173 and 183 K, removes the bulk, unperturbed water from the sample. Secondary drying takes place at more elevated temperatures so as to desorb the unfreezable, perturbed water associated

Table 12.7. Vapor Pressure, Rate of Evaporation, and Evaporation Times of Crystalline Ice at Different Temperatures

Temp (K)	Maximum evaporation rate ($g/cm^2/s$)	Evaporation time per g/cm^2	Vapor pressure of ice
273	8.3×10^{-2}	1.2 s	595 Pa
233	3.4×10^{-3}	4.8 m	13 Pa
213	2.3×10^{-4}	72 m	2.6 Pa
193	8.3×10^{-6}	33.5 h	0.052 Pa
173	5.0×10^{-7}	20 days	1.3×10^{-3} Pa

with hydrophilic surfaces. The different polymorphs of ice sublime at different rates, and this finding has been used to advantage in the design of a freeze dryer which sequentially sublimes the ice as the deep frozen (circa 123 K) samples are warmed through the various phase transitions (Livesey *et al.*, 1991). More conventional freeze drying can be carried out at high (133 μPa) and low (1.33 Pa) vacuum and at various temperatures. Other freeze-drying processes depend solely on the high affinity for water of molecular sieves maintained at liquid-nitrogen temperatures.

Although much is known about the sublimation of crystalline ice, freeze drying of hydrated samples remains an empirical procedure. Each protocol has to be designed around the specimen concerned, and the drying process should be monitored and controlled. Many artifacts are associated with freeze drying, ranging from molecular collapse and aggregation to gross shrinkage of the sample. Freeze drying has a wide range of applications, primarily in high-resolution structural studies of macromolecules and the analysis of sectioned material. It may be used directly to produce specimens for examination and analysis, although the extreme hygroscopic nature of such samples presents its own problems of sample handling. Freeze drying may also be used as part of a more extended process during which the water removed from the sample is replaced by resins, enabling such samples to be sectioned and cryofractured.

12.8.8. Freeze Substitution and Low-Temperature Embedding

Freeze substitution and low-temperature embedding are two distinct processes that are frequently conjoined, first to dehydrate frozen specimens by dissolving the ice in organic liquids and then infiltrating the vacant spaces with liquid resins which are subsequently polymerized. Both procedures are invasive and make extensive use of organic polar and apolar solvents and chemicals given in Table 12.8 (Edelmann, 1991). Freeze substitution is carried out at between 173 to 193 K in order to minimize the disruptive influence of ice recrystallization. Low-

Table 12.8. Organic Liquid Used in Freeze Substitution

Substitution fluid	Temperature (K)	Time (days)
Acetone	200–233	3–4
Acetone + acrolein	190–233	1–2
Butanol	193–233	More than 14
Diethyl ether	200–233	20
Diethyl ether and acrolein	163–233	6–20
Dimethyl ether	158–233	28
Ethanol	200–233	3–4
Methanol	178–243	1
Propanol	193–233	More than 14
Propylene oxide	200–273	10–15

temperature embedding is best done at higher temperatures (223–253 K), which favor higher diffusion rates of the high-viscosity resins (Carlemalm *et al.*, 1985). Both procedures must be carried out under strictly anhydrous conditions.

By careful choice of solvents and resins, and with close control of the times and temperatures of infiltration and embedding, it is possible to dehydrate and embed samples of material without seriously disturbing the supramolecular arrangement and functionality of their constituent hydrophilic macromolecules. The quality of preservation in sections cut from such samples is excellent and may be used for immunocytochemical analysis, autoradiography, and enzyme studies. The ultrastructural preservation is thought to most closely resemble that found in the natural state. Freeze substitution can also be used to prepare samples for SEM (Fig. 12.41). The samples are critical point dried after the solvent substitution procedure (Wharton and Murray, 1990). Although some new techniques are beginning to appear, there is still some uncertainty whether these methods will permit the *in situ* retention of small-molecular-weight compounds and dissolved elements and compounds in the aqueous phases of hydrated specimens. The equipment and techniques needed to carry out all these low-temperatures procedures are neither complicated nor expensive, although the actual processes are time consuming and require great attention to detail. Two additional processes, progressive lowering of temperature (Carlemalm *et al.*, 1982) and isothermal freeze fixation (Hunt, 1984), which are carried out at

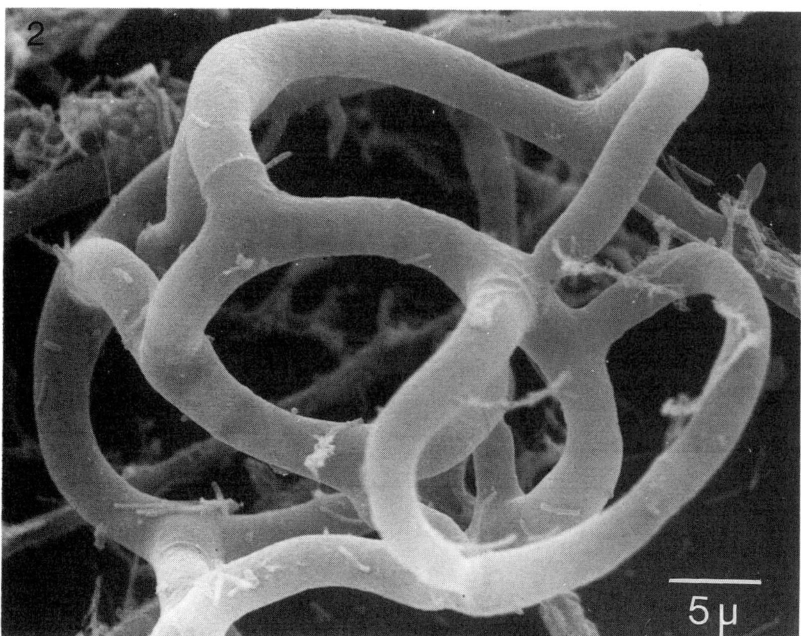

Figure 12.41. A complex nematode-trapping device of the fungus *Arthrobotrys oligospara*. Magnification bar = 5 μm (Wharton and Murray, 1990).

somewhat higher subzero temperatures, are designed to remove water while still retaining the form and location of any ice crystals in near-frozen specimens. These procedures are particularly useful for studying freeze damage in biological specimens.

12.8.9. Low-Temperature SEM

The SEM, with its wide variety of specimen–primary beam interactions, is a particularly versatile electron-beam instrument for studying the structure of frozen specimens (Read and Jeffree, 1991). The finely focused electron probe provides a large depth of field at any one time and allows easy and direct visualization of the three-dimensional structures of an object without recourse to sections or replicas at a spatial resolution approaching that which may be obtained in the TEM (Walther *et al.*, 1990). A wide variety of cold stages and dedicated and nondedicated cryopreparation chambers are available for SEMs and allow frozen samples to be fractured or microdissected, etched and finally coated with a thin layer of evaporated or sputter-coated metal. The initial sample preparation is straightforward and centers on quench cooling well-secured samples in such a way as to ensure that the regions of interest are adequately preserved. Samples may be examined in the fully frozen-hydrated or frozen-dried state and information may be obtained about their elemental, molecular, and structural constitution depending on the nature of the interaction of the primary beam with the specimen.

The interpretation of images obtained from frozen samples of biological and hydrated organic material is strongly influenced by their state of hydration. There is an apparent increase in both the quality and the quantity of morphological information as a fully frozen-hydrated bulk specimen is dehydrated by sublimation of the ice matrix (Fig. 12.42). In sectioned material, dehydration is frequently accompanied by reversals in contrast. Beam damage remains a problem with low-temperature SEM, with mass loss being the principal concern, particularly in high-resolution studies. Low-temperature SEM has a very wide range of applications in both the biological and materials sciences, both from the ability of the technique to provide new information and from its capacity to rapidly provide images of hydrated samples.

12.8.10. Low-Temperature X-Ray Microanalysis

Low-temperature x-ray microanalysis provides one of the very few ways one may hope to obtain *in situ* analytical information about the distribution and local concentration of diffusible ions in biological material (Fig. 12.43). Electron probe x-ray microanalysis is a reasonably noninvasive analytic tool, although the inadequacies of sample preparation and the ever-present problems of beam damage always make this a less than perfect method for high-spatial-resolution studies (Echlin, 1991). The real advantage lies in the fact that it is one of the few techniques that has been developed to the point where a few tens of atoms of an element may be detected and, in turn, directly related to structural features of the sample. Low temperatures and their associated techniques add an enormous potential to such analytical studies. At liquid-nitrogen temperatures and below, the effects of radiation damage are lessened. The solidification of water is an effective way of virtually stopping diffusion of electrolytes and arresting physiological processes, dynamic activities, and most chemical reactions (Echlin and Taylor, 1986; Echlin, 1989).

Techniques and instrumentation are available to analyze frozen-hydrated and frozen-dried bulk samples, single cells, thick and thin sections, and suspended particles with varying degrees of spatial resolu-

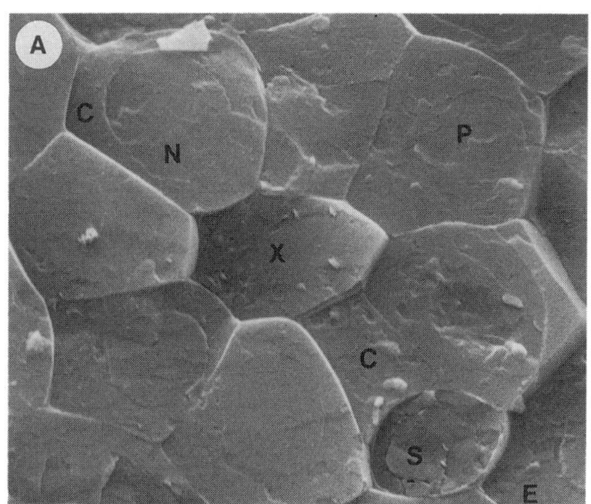

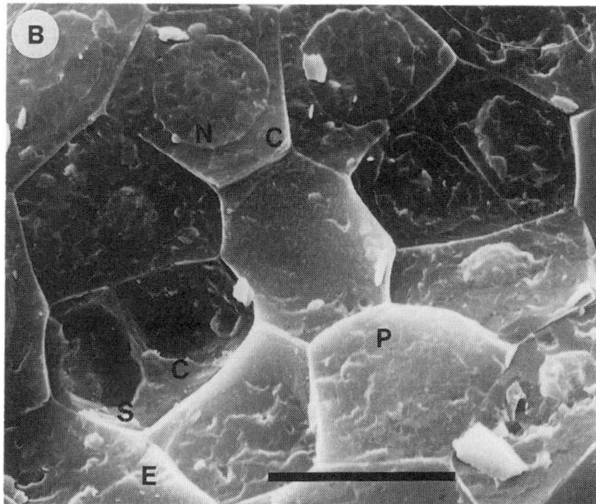

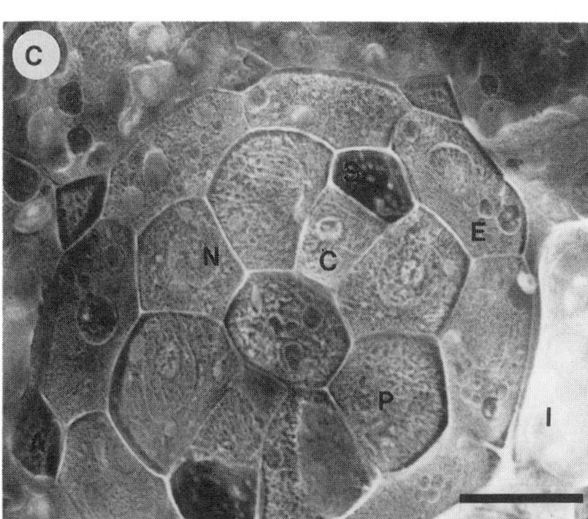

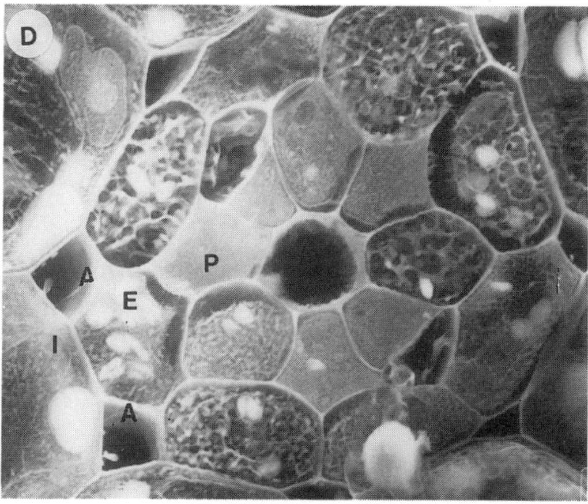

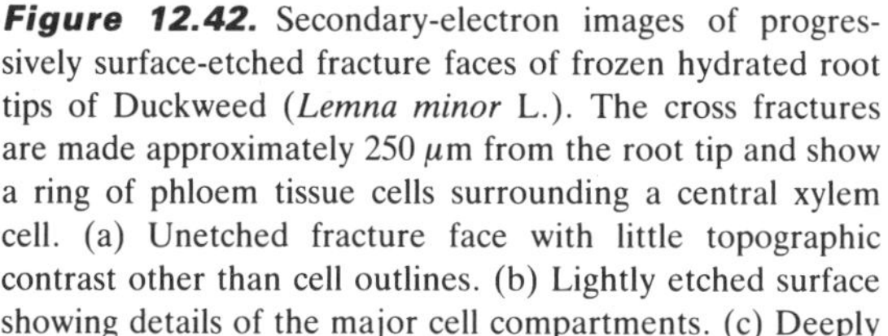

Figure 12.42. Secondary-electron images of progressively surface-etched fracture faces of frozen hydrated root tips of Duckweed (*Lemna minor* L.). The cross fractures are made approximately 250 μm from the root tip and show a ring of phloem tissue cells surrounding a central xylem cell. (a) Unetched fracture face with little topographic contrast other than cell outlines. (b) Lightly etched surface showing details of the major cell compartments. (c) Deeply etched sample showing organelles and ice crystal ghosts. (d) Frozen-dried sample. N, nucleus; C, cytoplasm; X, xylem; P, phloem tissue; CC, companion cell; S, sieve tube; A, air space; E, endodermis; I, inner cortex. Samples (b)–(d) etched at 213 K with an overhead radiant heater and all samples coated with 8 nm of gold. SEM operated at 12 keV and 20 pA beam current. Magnification bar = 10 μm.

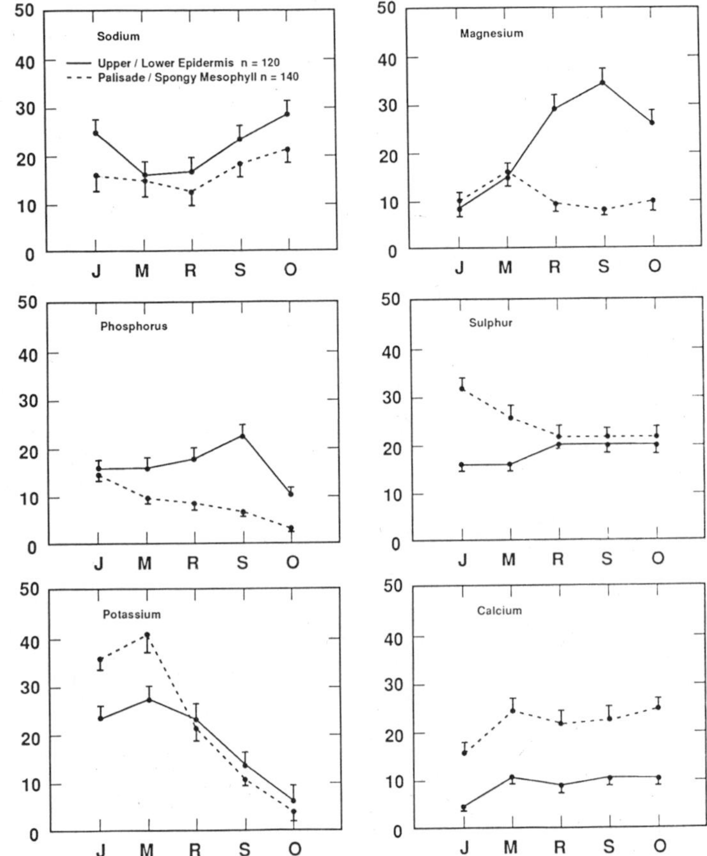

Figure 12.43. Elemental concentrations in cells from the photosynthetic tissues (dotted lines) and nonphotosynthetic tissues (solid lines) of leaves of *Nicotiana tabacum* at five stages of growth. Values are given as mol m^{-3} fresh tissue. J, juvenile; M, mature; R, ripe; S, senescent; O, old. Analysis carried out on frozen hydrated fractured leaves maintained at 110 K (Echlin, 1989).

tion, detection limits, analytic accuracy, and image quality (Table 12.9). The physical processes of x-ray microanalysis are understood and well characterized, and the two types of x-ray detectors, energy-dispersive and wavelength-dispersive spectrometers, can provide precise and accurate analytical data. X-ray microanalysis has benefited from extensive use of computers, which have made it much easier to process the x-ray spectra. A wide range of analytical algorithms is available that, when properly applied, can swiftly and effectively convert crude, uncorrected x-ray spectra into local elemental concentrations. Such local concentrations can be expressed on a numerical weight/weight or weight/volume basis or, when combined with digital imaging processes, can be presented as maps of elemental concentration with a direct, real time association with the structural components of the sample. An added advantage of low-temperature microanalysis is that samples can be examined and analyzed in both the wet and dried state, and measurements can be made of the local water concentration within and outside of multicompartmented

Table 12.9. Characteristics of Various Types of Samples Used in Low-Temperature X-Ray Microanalysis

	Frozen-hydrated bulk specimens	Freeze-dried bulk specimens	Frozen-hydrated 1-μm-thick sections	Freeze-dried 100-nm-thick sections	Freeze-dried ultra thin (50 nm) sections
Lateral analytical resolution	~5 μm	~10 μm	~500 nm	≤50 nm	20 nm
Relative detection limit	~10 mmol/l cell water	~10 mmol/kg dry weight (3 mmol/l cell water)	~10 mmol/l cell water	~5 mmol/kg dry weight (3 mmol/l cell water)	~0.5 mmol/kg dry weight (~0.1 mmol/l cell water)
Absolute detection limit	1.5×10^8 atoms in $1 \times 5 \times 5\ \mu m^3$	5×10^7 atoms in $2 \times 2 \times 2\ \mu m^3$	1.5×10^6 atoms in $1 \times 0.5 \times 0.5\ \mu m^3$	500 atoms in $100 \times 50 \times 50\ nm^3$	40 atoms in $50 \times 20 \times 20\ nm^3$
Ion distribution	Preserved (?) if vitrified	Redistribution possible	Preserved (?) if vitrified	Redistribution possible	Preserved (?) if vitrified
Imaging	With some difficulty	Good, depends on ice-crystal damage	Difficult	Good, depends on ice crystal damage	Good, depends on quality of cryofixation
Problems	Possible radiation damage (mass loss)	Recrystallization, shrinkage, radiation damage (mass loss)	Radiation damage (mass loss)	Recrystallization, shrinkage, some radiation damage (mass loss)	Recrystallization, shrinkage, some radiation damage (mass loss)

specimens. In spite of seemingly insurmountable problems of radiation-induced mass loss from organic and hydrated samples and ice crystallization–induced redistribution of soluble material, low-temperature microanalysis continues to offer the best prospect of accurately localizing the position of chemical elements in a wide range of organic and inorganic samples (Echlin, 1992).

12.9. Damage, Artifact, and Interpretation

12.9.1. Introduction

In spite of the most carefully devised preparation techniques, mistakes do occur and it is important to be able to recognize damage and artifacts in the final image. Damage to the specimen can occur at any stage during preparation or observation, and for this reason correlative studies using alternative imaging systems are very important, as each of the methods allows cross-monitoring of the whole preparative procedure. In spite of the difficulty of identifying exactly when and how damage occurs, it is useful to consider first the damage that can occur during sample preparation and then consider the damage which can occur during (and after) examination and analysis. Before beginning this brief discussion it is important to distinguish between damage and artifact. Damage is manifest when something has gone wrong during preparation, examination, and analysis. Such grave errors are usually easily recognized. Artifacts are more subtle and encompass the deliberately altered state we have to produce in specimens so that they may be examined in an electron-beam instrument.

12.9.2. Sample Damage during Preparation

Sample damage can occur at any point during the preparation process. The most obvious indications of damage arising from faulty preparation are: shrinkage due to extraction and drying; swelling due to faulty fixation; pitting, cracking, and melting due to poor coating procedures; loss, gain, and redistribution of chemical components due to inadequate sample stabilization; and, finally, dehydration and distortion due to ice crystals. Examples of the types of damage which can occur are given in Table 12.10 and in Figs. 12.44 and 12.45. Boyde (1976) and Crang and Klomparens (1988) have each assembled a rogue's gallery of damaged samples; the microscope manufacturer JEOL has assembled a similar macabre exhibition. The best advice one can offer the individual experimentalist is to assemble your own black book of images of damaged specimens together with an explanation of the causes of the damage. This can be kept for future reference. Damage can actually be recognized only by comparison with the unaltered specimen, first at the resolution of the human eye, followed progressively with noninvasive light-optical systems, such as phase and Nomarski optics, video-enhanced microscopy, and confocal microscopy, and then with material treated with vital dyes, histochemical agents, and finally with electron-beam instruments.

Table 12.10. Damage Arising from Faults in Specimen Preparation

Bubbles and blebs on surface—incorrect initial washing, hypertonic fixative
Burst bubbles—hypertonic fixative
Cell surface depressions—incipient air drying
Crystals on surface—buffer electrolytes, medium electrolytes
Detachment of cells—poor freezing
Directional surface ridging—cutting artifact of frozen material
Disruption of cells—poor fixation, poor freezing, incipient air drying
Disruption of cell surface—poor washing, poor fixation
Distortion of cell shape—air drying
Fine particulate matter on surfaces—too much or improper coating
General shrinkage—hypotonic fixatives and washes
Gross damage—poor handling of prepared material, microbial decay
Increased cell dimensions—too much coating
Internal contents clumped and coagulated—poor fixation, dehydration
Large particulate matter—disintegration of specimen, dust, and dirt
Large regular holes—freezing damage
Large surface cracks—movement of coated layer
Microcrystalline deposits—incorrect buffer, fixative, or stain
Obscured surface detail—too much coating
Plastic deformations—incorrect cutting, fracturing
Regular splits and cuts—thermal contraction between cells
Ruptured cell contents—poor critical point drying
Separation of cells—incorrect fixation
Shriveled appendages—hypotonic fixation
Shriveling of cells—overheating during fixation
Slight shrinkage—poor critical point drying
Small holes on surface—incorrect sputter coating
Smooth surface overlaying a featured surface—incorrect drying
Speckled surface—contamination in sputter coater
Stretched cell surface—hypotonic fixation
Strings and sheets of material on surface—incorrect washing
Surface crazing and etching—incorrect sputter coating
Surface folds or ridges—fixative pH is wrong
Surface melting—overheating during coating
Surface sloughing—air drying from organic fluid, slow application of fixative
Swelling of sample—hypertonic buffers and fixatives, poor freeze drying

12.9.3. Specimen Damage during Examination and Analysis

This is primarily a problem with organic, hydrated and biological specimens. In spite of all the tricks we employ during sample preparation to try and convert damage susceptible organic material to damage insensitive inorganic material, the samples may be damaged inside the electron beam instrument. The damage is manifest at two levels, the directly observable, and the non-observable.

12.9.3.1. Observable Damage

The sample may be seen to bubble, blister, move, melt, cavitate, or disappear entirely. These phenomena are due primarily to thermal damage, and such samples would not be used for experimental studies.

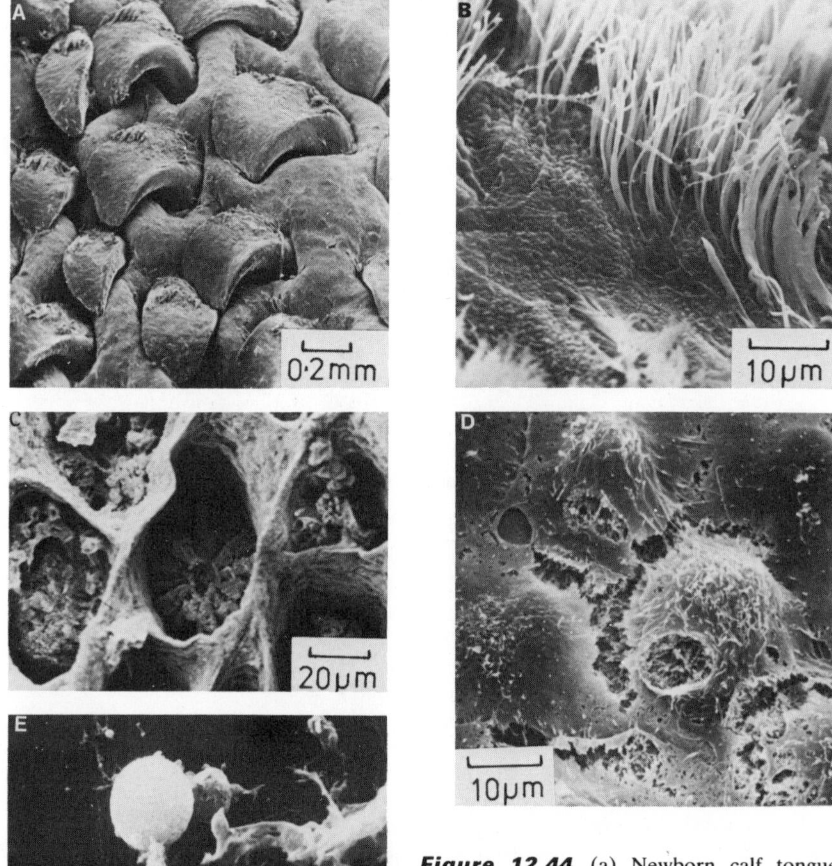

Figure 12.44. (a) Newborn calf tongue fixed with neutral formol saline. Bending due to greater shrinkage of less keratinized side of papillae. (b) Calf trachea with strings of mucus due to insufficient washing before fixatives. Freeze-dried from water. (c) Rhesus monkey cardiac stomach. 5 min treatment with *N*-acetyl, *L*-cystine before fixation causing loss of surface epithelium. Air dried from acetone. (d) K4 cells fixed with hypotonic cacodylate buffer plus 1% formaldehyde +1.9% glutaraldehyde. Note burst blebs. (e) K4 cell rapidly frozen and thawed before fixation in glutealdehyde followed by CPD. Severe blebbing produced (Boyde, 1976).

More subtle yet observable damage may be due to faults in instrumentation such as astigmatism, vibration drift, stray magnetic fields, poor focus, uneven illumination, projection distortion, or image rotation. Strictly speaking, these are faults in imaging in which the sample itself is undamaged. Other faults occur due to the specimen–beam interaction and include charging and contamination. Examples of these types of damage are given in Table 12.11 and Fig. 12.46 and are more extensively discussed in the article by Fleger (1988). Many of the problems associated with observable damage can be overcome by putting less energy into the

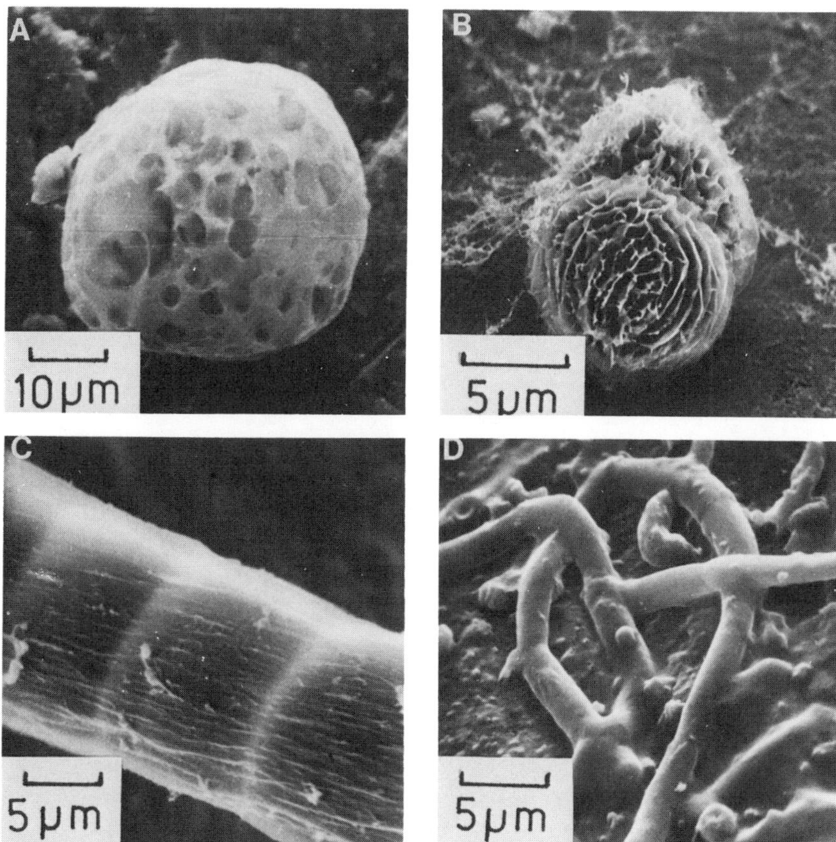

Figure 12.45. Damage during specimen preparation for SEM. (a) Ice-crystal damage on badly freeze-dried pollen of *Clarkia spp.* (b) Incomplete removal of capsule material from surface of freeze-dried alga *Chroococcus turgidus.* (c) Partial collapse of critical-point dried alga *Lyngbia majuscula.* (d) Slime layer obliterating surface of the bacterium *Bacillus megatherium.*

Table 12.11. Damage Arising during Specimen Observation as Seen on the Image

Background distortion—charging
Bright areas—charging
Bright areas in cavities—charging
Bright banding—charging
Bubbles on surface—beam damage
Cracks on surface—beam damage
Dark areas on specimen—charging
Dark raster square—contamination
Image distortion—charging
Image shift—charging
Light and dark raster squares—beam damage
Loss of parts of specimen—beam damage
Raster square on surface—beam damage
Specimen movement—poor attachment, charging
Specimen surface shriveling—beam damage

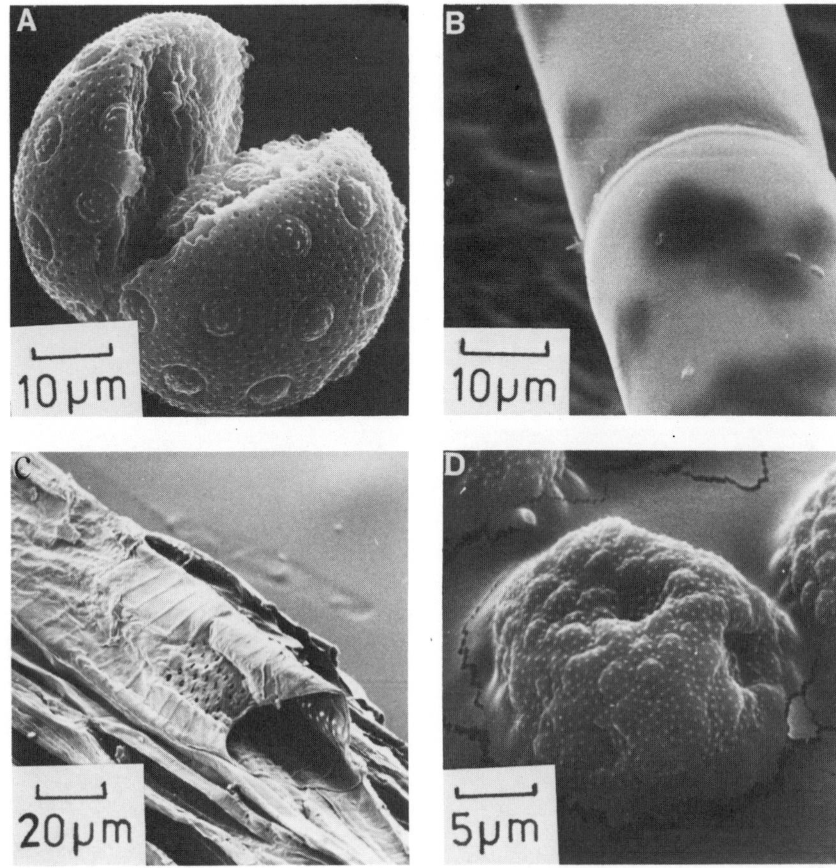

Figure 12.46. Damage seen during the examination of specimens in the SEM. (a) Dry-fractured pollen grain of *Silene zawadski* showing bright edge due to incomplete coating. (b) Fixed and coated filament of the green alga *Oedogonium* showing black contamination spots. (c) Wood vessels of *Tulipa sp.* showing collapse during drying. (d) Pollen grain of *Plantago* media poorly mounted and showing cracking of the coating layer.

sample, optimizing the performance of the instrument, and rechecking the effectiveness of the preparative procedure. The book edited by Crang and Klomparens (1988) contains much useful information on recognizing damage in biological electron microscopy.

12.9.3.2. Nonobservable Damage

This type of damage, which is primarily due to ionizing radiation, is not normally encountered in routine studies by scanning electron microscopy, which utilize beam currents in the picoampere range. Non-observable damage is a problem in x-ray microanalysis and may even be a problem where higher beam currents are used to obtain a good BSE signal for atomic number contrast.

The accuracy with which a structure can be visualized or analyzed in an electron-beam instrument is limited by dehydration and radiation damage. We have sought to show in previous sections in this chapter that the problem of dehydration can be solved either by chemically removing the water or by converting it into its solid phase, in which the nonaqueous parts of the sample are embedded. The problems associated with radiation damage constitute an inevitable and serious consequence of the process of electron beam–specimen interaction which is necessary to produce the image and the analytical signals. Cryoelectron microscopy has added interesting new dimensions to these problems, for although low temperatures are now generally accepted as slowing down radiation damage, the presence of water in a frozen hydrated sample may significantly exacerbate the damage processes. Much has already been written about this radiation damage in electron-beam instruments, and the papers by Glaeser and Taylor (1978), Cosslett (1978), Hobbs (1979), Glaeser (1979), Talmon (1982; 1984, 1987), Dubochet *et al.* (1982), and Jeng and Chui (1984) provide a comprehensive introduction to these problems. The paper by Lamvik (1991) specifically addresses the problems of radiation damage in dry and frozen hydrated organic materials and a paper by Echlin (1991) relates these problems to x-ray microanalysis. In the brief discussion which follows, we will be concerned only with ionizing-radiation damage as there is sufficient evidence to suggest that thermal damage rarely occurs in specimens which are either irradiated at very low beam currents (1–10 pA) or kept at circa 100 K throughout examination and analysis.

The fundamental physical processes that result in damage are complex and incompletely understood. Most of the damage is caused by inelastically scattered electrons which deposit energy as they interact with the sample. The amount of energy deposited depends on the chemical composition and thickness of the sample and the electron-beam energy and current. The energy transferred from the electrons to the specimen is the cause of radiation damage, and these interactions may be categorized according to the physical process involved.

The primary cause of radiation damage is ionization of atoms which occurs so quickly that there is nothing we can do to prevent its happening. In biological, hydrated, and organic material, the secondary reactions of radiation damage are due primarily to the diffusion of free radicals, highly reactive atomic or molecular species each of which possesses an unpaired electron. There is a considerable variation in the number of free radicals formed, and it is unfortunate for cryomicroscopy that water (ice) forms six times as many radicals per 100 eV than aromatic compounds (Henglein, 1984). There is every reason to believe that free radicals are the principal cause of radiation damage in organic and hydrated samples.

The extent to which a sample is damaged by the electron beam is a function of electron exposure (or dose) and temperature. The dose of electrons is most conveniently expressed as the number of electrons which impinge on 1 nm^2 of the sample surface ($e^- \, nm^{-2}$). Dose can also be expressed as the electric charge per unit area, i.e., coulombs (C) per square meter ($C \, m^{-2}$), where C has the units ampere-second. The conversion between the two systems is given by $1\,\mathrm{C\,m^2} = 6.25\,e^-\,\mathrm{nm}^{-2}$, and it is quite simple to calculate the electron dose impinging on the sample during examination and analysis. For example, x-ray microanalysis might be performed using a beam of 100 pA rastered over a 2-μm^2 area for 100 s. The total electron dose is 15.6 ke^- nm^{-2}.

Irrespective of the units, the figures obtained are measures of the charge deposited on the sample rather than the energy deposited. In order to convert charge to energy, it is necessary to know the stopping power (i.e., rate of energy loss) of the incident electrons. At 10 keV, a typical stopping power is of the order of 1 eV/0.1 nm, so the actual energy deposited for a flux of 100 e^- nm^{-2} is on the order of 10^{-5} J/cc, because 1 eV is 1.6×10^{-19} J. Assuming a density of 1.0 for biological material, this would represent a dose of 10^8 J/kg. The conversion from charge deposited to energy deposited is energy dependent because it varies with the stopping power, which in turn decreases as the energy of the incident electrons increases. In this respect it is necessary to have a means of comparing radiation measurements made at different energies. This aspect is considered in some detail by Lamvik (1991) who recommends using the calculation proposed by Isaacson (1977) to scale the dose values by the energy dependence of the inelastic scattering cross-section. The results presented by Lamvik on radiation damage in dry and frozen-hydrated organic material are all scaled to dose values of 100 keV. For example, a dose of 2.5 ke^- nm^{-2} at 30 keV into a solid biological sample corresponds to 6.1 ke^- nm^{-2} at 100 keV.

It should be remembered that all radiation effects are continuous and cumulative and that the simplified calculations given above give only a measure of the amount of energy (or charge) arriving at the surface of the sample; they bear little relationship to the amount of energy being absorbed by the sample. Thus a dose of 15 ke^- nm^{-2} at 100 keV would be expected to deposit substantially more energy into a bulk specimen than into a thin section through which most of the electrons would pass without scattering.

Nonobservable but measurable damage occurs when there are quantifiable changes in electron diffraction patterns, sample mass (primarily C, H, O, and N), contamination and sample elemental composition. It is these latter changes that should concern us here.

High-resolution structural damage, as measured by changes in the diffraction pattern of ordered molecules, is generally not a problem in microanalysis and SEM even though the total dose used for electron-probe analysis is several orders of magnitude greater than that required for low-dose microscopical structural studies. With a few notable exceptions, most x-ray analytical studies are carried out at low spatial resolution, and although the structural changes occur, they are not observed. Mass loss and contamination and to some extent the loss of specific elements such as K, Na, or Cl from the specimen remain a problem. The few systematic studies on the loss of specific elements from samples (Hall, 1986; Rick *et al.*, 1982) show that low temperatures diminish the losses.

Table 12.12. Cumulative Radiation Doses ($ke^- nm^{-2}$) Needed to Obtain Information about Samples Studied by Various Types of Electron-Beam Instrumentation Operating at a Temperature of 300 K

Standard TEM at 100 keV	10–20
Low–dose TEM at 100 keV	0.2–1.0
Secondary-electron SEM at 20 keV	0.1–0.5
Backscattered-electron SEM at 20 keV	1–2
X-ray microanalysis at 20 keV	2–200

Table 12.13. Comparative Radiation Doses ($e^- nm^{-2}$) Associated with Various Phenomena in a TEM Operating at 100 keV and a Temperature of 100 K

Low-dose searching	50
Observation of electron diffraction patterns	100
Low-dose imaging	200
Loss of electron diffraction pattern	1,000
Free-radical formation	1,000
Ice bubbling	2,000
Mass loss	2,000
Removal of 1-nm layer of ice	3,000
Normal viewing	10,000

There is now sufficient information in the literature to enable one to predict the type and amount of damage one may expect to find at given levels of electron exposure. Some of this information is presented in Tables 12.12–12.15. These tables give the range of values we might expect to encounter in various forms of microscopy and electron-beam analysis.

Table 12.14. Beam Damage Caused by a 20-keV, 1-nA Beam Scanning a 2-μm^2 Area of a Frozen-Hydrated Sample Kept at 100 K for 100 s (accumulated dose = 156 $ke^- nm^{-2}$)

1. Obliterates an ultrathin section
2. Removes 20% of a 0.5 μm section
3. Removes 10% of a 1.0 μm section
4. Etches 0.1 μm into a bulk sample

Table 12.15. Electron Doses ($ke^- nm^{-2}$) Used in X-ray Microanalysis of Various Kinds of Biological, Organic, and Hydrated Samples

Frozen-hydrated thin sections (140 K)	50–80
Frozen-dried thin sections (300 K)	50–5,000
Frozen-hydrated thick sections (100 K)	2–3,500
Frozen-dried thick sections (100 K)	70–2,000
Frozen-hydrated bulk samples (100 K)	50–700
Frozen-dried bulk samples (100 K)	3,000

Table 12.12 shows the radiation doses commonly used to search, locate, examine and eventually record information in various types of electron-beam instruments operating at ambient temperature. The values range over three orders of magnitude. Low temperatures ameliorate these effects, but the results are variable and specimen dependent (Lamvik, 1991). The presence of a large area of cold surfaces in the immediate region of the specimen would favor low rates of contamination. But as will be shown later, even traces of water vapor can result in serious damage to the sample. The extent of the protective effect of low temperatures against radiation damage may be seen in Table 12.13, which gives the radiation doses commonly used for different types of operations in the TEM.

A comparison of the data presented in Tables 9.12 and 9.13 shows that, while radiation damage is less of a problem in high-resolution, low-dose structural studies of thin suspensions of biological material, such doses cause serious mass loss in x-ray analytical studies. The severity of this problem is seen in Table 12.14, which shows the mass loss in various types of samples irradiated with an electron dose equivalent to that usually used in x-ray microanalysis. This is an extreme case, and yet very high radiation doses continue to be used in quantitative analytical studies. Table 12.15 shows the range of radiation doses that have been used in x-ray microanalysis of different types of samples. It is worth remembering that a cumulative dose of only 2 $ke^- \, nm^{-2}$ is sufficient to initiate mass loss in the type of samples commonly used in cryomicroscopy at circa 100 K. Studies by David Joy at the University of Tennessee (unpublished) on the mass loss from various fatty acids at 300 K show that at the relatively low exposure of 1.5 $ke^- \, nm^{-2}$, approximately half the mass has disappeared. While these data disregard the cryoprotective effects of low temperature, i.e., an assumed fivefold improvement, they do show that mass losses should be expected in samples irradiated under the conditions commonly used for x-ray microanalysis. Hall and Gupta (1974) found that with a total dose of 2.5 $ke^- \, nm^{-2}$, 30% of a frozen-dried sample was lost at 300 K, whereas only 5% was lost at 90 K.

An examination of the large amount of analytical data obtained using x-ray microanalysis shows that such massive amounts of radiation damage do not always occur. An enquiry into the reasons for these discrepancies reveals that the actual situation may not be as bad as first feared. Joy *et al.* (1989) consider that it is wrong always to assume that the incident electron dose which impinges on the specimen is the same as the electron dose absorbed by the specimen. Provided the specimen is thin enough (i.e., no more than one or two inelastic electron mean free paths at the incident energy) and the electron beam has sufficient velocity, only a very small proportion of the energy of the electron beam is deposited in the sample. As the sample thickness increases and the accelerating voltage decreases, more and more of the beam energy is absorbed by the sample.

While these findings may provide a partial explanation for the declared small mass losses in thin sections analyzed at high accelerating voltages, they provide no comfort for the analysis of 0.5–1.0 thick sections at 20–50 keV. In thick sections and bulk material, although elastic-scattering events and volume diffusion may still play a part, the data of Dubochet *et al.* (1982) suggest that the mass loss, particularly in frozen hydrated samples, is a surface etching effect and as such would be independent of sample thickness. In spite of the significantly reduced spatial resolution, the x-ray microanalysis of frozen-fracture faces of hydrated bulk specimens still has a number of advantages over sectioned material. They are easier to keep at low temperatures, and the necessary thin metal coating layer can leak away a substantial proportion of the electron charge.

A more satisfactory explanation for the increased mass loss in frozen hydrated material may be found in the now well-established fact [see for example Hall (1988) and Lamvick and Davilla (1988)] that frozen-dried samples are much more resistant to beam damage than their frozen hydrated counterparts. The presence of trace amounts of water, either in the sample or in the environment around the sample, provides a rich source of free radicals, which readily attack organic material. This raises the intriguing question as to how low temperatures confer radiation resistance. Do low temperatures act by diminishing the diffusion rate of free radicals, or is the effect due to efficient cryopumping which removes water from the sample and its immediate environments and thus reduces the ionization erosion of the sample? Some recent work by Joy *et al.* (1989) suggests that the latter is the case. The high radiation sensitivity of frozen hydrated thick sections (0.5–1.0 μm) suggests that they be used only under conditions of reduced beam energy.

The only way we may hope to reduce beam damage and mass loss is to reduce the energy input into the sample to the point where one still obtains a statistically significant signal. Such conditions may be found in the newer analytical instruments, which use digital imaging procedures to display and quantify the chemical information. At low magnification, i.e., 1000–3000X, the dose per pixel is on the order of 2–4 ke^- nm^{-2}, but at high resolution (Leapman and Andrews, 1991), doses of 10 Ge^- nm^2 have been used.

The following procedures have been developed to provide valid high-resolution structural and *in situ* information from biological and hydrated organic material. Samples embedded in a thin layer of vitreous ice are imaged at low temperatures and under low dose conditions. The images are then subjected to rigorous spatial averaging techniques and reconstruction methods to extract the structural information.

Frozen hydrated thin sections should not be used for high-resolution studies, and water-content measurements from thick sections should be made only under low-dose conditions. Analysis at the subcellular level should be carried out only on frozen dried sections maintained at 100 K in a scrupulously clean and water-free environment. Frozen hydrated fractured bulk material should be used only for low-resolution analytical studies at the cell and tissue levels.

In addition to the usual information about section thickness, magnification, and bar markers, all studies involving microscopy and analysis should clearly state the conditions under which they were prepared, examined and analyzed. The types of information which should accompany each micrograph and each piece of analytical data include the dose rate, beam energy, sample temperature, sample hydration, and vacuum conditions. Such information is vital for the evaluation of the effectiveness of sample preparation and the validity of the structural and analytical information which the data purports to show (Echlin, 1981; Wergin and Pooley, 1988).

12.9.4. Artifacts

During the course of sample preparation, we deliberately introduce changes into the sample. The reasons for doing this should now be self-evident. These chemically and physically induced changes modify the appearance of the structures we wish to examine and produce the artifacts from which we derive information about our specimens. Some artifacts are consistent and, provided they can be recognized, are useful. The trilamellate appearance of osmium-stained phospholipid bilayers, for example, is a consistent artifact which marks the presence of a biological membrane, although we known that such an image bears little relationship to the living, functioning structure.

While we should try to limit damage to samples, we should aim at producing constructive artifacts. The edge effect we see in nearly all scanning micrographs is a good example of a constructive artifact. The increased electron emission associated with the edge effect arises from beam penetration at the intercept of two plane faces of a solid body (see Chapter 4). The bright line indicates an edge, and we use this artifact in our interpretation of the image. We produce constructive artifacts when we deliberately introduce a heavy metal to mark the site of an enzymatic reaction in a cell, a specific chemical grouping in a polymer, and a colloidal gold particle on an antibody marker. Coating techniques have their own set of artifacts. At high resolution, thin films of noble metals exhibit a fine granularity due not to the structures they are covering, but to the way the metal particles condense and agglomerate on the surface (see Chapter 13). We learn to recognize these artifacts and discount them as not being part of the surface. When metal-coating a planar surface, some parts of the sample may accumulate more metal than others. These so-called decoration artifacts can provide information about the surface because at these regions there is a higher binding coefficient between the substrate and the metal atoms.

Unlike damage, which should be dismissed out of hand, artifacts should be used constructively, because their origin lies in some peculiarity of the specimen and not in some aberration of the preparation technique. However the difficulty comes in separating artifact from damage.

12.9.5. Interpretation

We arrive, at last, at the most critical part of the whole process of microscopy and analysis—what do the pictures mean, what does the analytical data represent. The actual interpretation of SEM images is paradoxically both the easiest and the most difficult part of the whole process. It is easy because we are usually presented with images we think we can recognize. Our inherent stereoscopic vision, and the convention that the photographs are presented as being "lit" from above enables us to appreciate the apparent three-dimensional quality of SEM images. The use of stereo pair images reinforces our interpretation of shape and form. We interpret surface contours on the basis that on an image of a depression, a light area precedes a dark area, whereas with a protrusion, the dark area precedes the light area. But it can also be very misleading, because our brain is using direction of illumination as a code for interpreting topography. If such pictures are turned through 180°, the topography of the surface features is reversed. Additional difficulties arise in relating the SEM image to the original object. The best approach is to try and overlap the lower magnification range of the SEM with the upper magnification range of a stereobinocular light microscope. In addition, correlative microscopy, using sectioned and bulk material prepared by different techniques, can maintain the interpretation process.

The actual interpretation of images is very subjective and depends on previous experience. Our perception of 3 dimensions in two-dimensional images comes from our early experiences and is based on the relative size, shape, and overlap of objects, on shading, contrast and color, and on the motion of objects relative to each other. All this information is continually updated and stored in the brain as visual experience which we use to evaluate new images. This works well in our everyday life, but can lead to misinterpretation when, without any choice in the matter, we apply our learned visual experience to evaluate micrographs. There is very little we can do about it except to appreciate that errors can occur and to be as objective as possible in our interpretation. We should first recognize and reject information that is a result of damage and error, we should look for consistent artifacts and evaluate their positive contribution to the image, and finally we should apply our learned experience of looking at similar objects to interpret the new image which probably no one has ever seen before.

12.10. Specific Preparative Procedures: A Bibliography

This chapter has deliberately taken a general approach to the problem of specimen preparation, offering broadly based strategies and tactics to enable the operator to design the specific procedure for his or her own specimens. This final section provides a brief catalogue of more specific articles and books for preparing the wide range of biological, organic, and hydrated specimens, which are likely to be examined and analyzed in the SEM and x-ray microanalyser. Wherever possible, key references are given, which in turn can provide further sources. In addition to the specific references given below, the following journals contain information about sample preparation procedures for SEM and microanalysis; *Journal of Microscopy, Scanning,* and *Microscopy and Analysis* (U.K.); *Scanning Microscopy* and *Journal of Electron Microscope Technique* (U.S.A.); and *Journal of Electron Microscopy* (Japan).

The comprehensive book by Sawyer and Grubb (1987) is the best source of information on polymers. Techniques for the analysis of paints may be found in Goebel and Stoecklein (1987) and Ward and Carlson (1983). The general text edited by Vaughan (1979) is a useful reference for foods. More undated information may be found in the publication *Food Microstructure* and in a paper by Heathcock (1988). The reviews by Falk (1980) and Barnes and Blackmore (1986) cover a wide range of plant materials. In addition to the papers discussed elsewhere in this chapter, those by Barnett (1988) and Kucera (1986) provide additional details of sample preparation for wood. Nowell and Pawley (1980), while not providing a series of all-embracing recipes have attempted to standardize the preparative procedures used on animal tissue. Additional information may be found in papers by Tsuji (1983) and Waterman (1980); Laschi *et al.* (1987) provide information relating to clinical studies. The preparation of single cells and microrganisms presents its own set of problems. Watson *et al.* (1980), Woulters *et al.* (1984), and Maugel *et al.* (1980) provide a number of useful suggestions on how such specimens are best preserved. Allen (1983) and Evers *et al.* (1983) give details of techniques which can be used on cultured animals cells. In addition to the textbook by Echlin (1992), which covers all aspects of low-temperature microscopy and analysis, the publications of Robards and Sleytr (1985), Steinbrecht and Zierold (1987), and Roos and Morgan (1990) may also be consulted.

13

Coating and Conductivity Techniques for SEM and Microanalysis

13.1. Introduction

Nonconducting samples invariably need some sort of treatment before they can be examined and analyzed under optimal conditions in electron-beam instruments that rely on an emitted signal to provide information. The treatment is necessary to eliminate or reduce the electric charge that builds up rapidly in a nonconducting specimen when it is scanned by a beam of high-energy electrons. Figs. 4.63a–c and 4.64a–c show examples of pronounced and minor charging as observed in the SEM. In addition to charging phenomena, which result in image distortion, the primary beam also causes thermal and radiation damage, which can lead to a significant loss of material from the specimen. In many situations the specimen may acquire a sufficiently high charge to decelerate the primary beam, and the specimen may even ultimately act as an electron mirror (see Fig. 4.62).

The procedures needed to obviate these problems fall into two groups: modification of the specimen and modification of the environment in which the specimen is being studied. The specimen can be modified either by increasing its surface conductivity by coating it with a thin layer of conducting material or by increasing its bulk conductivity by infusing it with conducting compounds. These coating techniques and infusion methods will occupy most of this chapter.

It is not always necessary to treat the sample; indeed, at times this may be undesirable. We will also discuss ways of either modifying the instrumentation or applying noninvasive treatments to nonconducting samples that still allow significant information to be obtained. Most of what will be discussed applies directly to scanning electron microscopy and x-ray microanalysis, and only passing reference will be made to transmission electron microscopy.

The discussion is directed primarily toward biological material and nonmetallic organic samples, simply because these types of specimens are invariably poor conductors and are more readily damaged by the electron beam than are most inorganic materials. However, it is safe to assume that the methods which will be described for organic samples are equally effective for nonconducting inorganic specimens, such as ceramics.

This chapter will concentrate on the practical aspects of the more commonly used techniques, which are now standard procedure in most electron-microscope and analytical laboratories. It is not proposed to enter into a detailed discussion of the theoretical aspects of thin-film technology, but readers interested in this aspect of the subject are referred to the book by Maissel and Glang (1970).

13.2. Specimen Characteristics

13.2.1. Conductivity

The single most important reason for coating or increasing the bulk conductivity is to increase the electrical conductivity of the sample. Materials of high resistivity, i.e., exceeding $10^{10}\ \Omega$ m, charge rapidly under the incident beam and may develop a potential sufficient to cause a dielectric breakdown in regions of the specimen. This leads to variations in the surface potential, giving rise to the complex and dynamic image artifacts commonly referred to as "charging." These artifacts are manifest as deflection of low-energy secondary electrons, increased emission of secondaries within the crevices of a rough specimen, periodic bursts of secondary-electron emission, and deflection of the electron beam, all of which degrade the resolving power and analytical capabilities of the system by introducing astigmatism, instabilities, undue brightness, and spurious x-ray signals (Fig. 13.1). This undesirable situation is frequently made more difficult because many of the adhesives used to attach specimens to the substrate are themselves nonconductors and prevent any electrical charge from leaking away, even from conductive samples. A suitable conducting path may be established with silver or carbon paint, but if one is concerned only about electrical conductivity, then a thin coating layer of gold, silver, or copper will suffice to eliminate the problems associated with charging. Even though metallic samples are usually conductive, there are situations where one may wish to examine nonconducting areas, e.g., inclusions, and in these cases it is necessary to apply a thin coating layer. The conductivity of the thin film should be sufficient to ensure that the specimen current is drained to ground without the development of a significant surface potential.

13.2.2. Thermal Damage

Specimen heating is not usually a problem in most samples examined in the SEM, because the probe current is usually in the

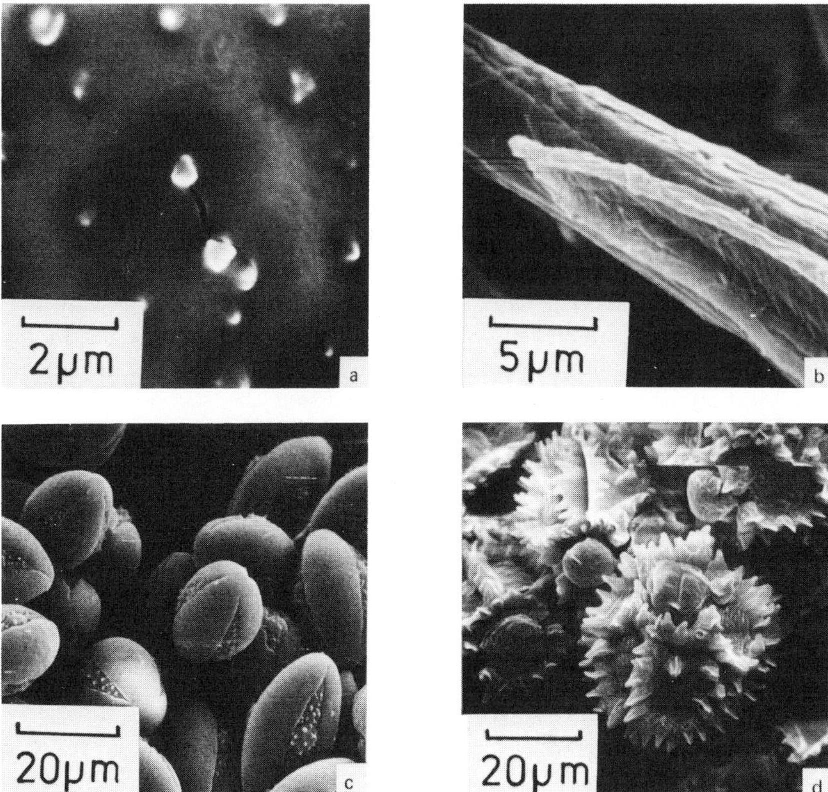

Figure 13.1. Artifacts formed during examination in the SEM. (a) Beam damage causing cracks on the surface of pollen from *Ipomoea purpurea*. (b) Lines across image due to beam instability from charging–discharging. Wood fibers of *Quercus ilex*. (c) Charging causing small flecks on image of *Aesculus hippocastanum* pollen. (d) Charging causing image shift on *Lygodesmia grandiflora* pollen.

picoampere range. Although higher currents are frequently used for TV scanning, these too are unlikely to degrade the specimen seriously. Thermal effects are potentially more serious for cathodoluminescence and x-ray microanalysis, for which the probe currents are in the nanoampere and even the microampere range. Excessive heating in the SEM can lead to specimen movement and instability, and in extreme situations to breakdown and destruction. The phenomenon described as "beam damage" is most certainly a heating effect and is manifest as blisters, cracks, and even holes in and on the surface of the specimen (Fig. 13.1a). In the EPMA, the higher beam currents can cause a rapid loss of organic material from plastics, polymers, and biological samples and can even result in substantial elemental losses.

Thermal damage can be reduced by working at lower beam currents and using a thin specimen in close contact with a good thermal conductor. Alternatively, the specimens may be coated with a thin film of a good heat conductor such as copper, aluminum, silver, or gold. It should be noted that a thin (1–3 nm) continuous layer of metal is not, on its own, sufficiently conductive, and at least double that thickness of metal is needed to conduct away the heat that can readily build up in a specimen.

13.2.3. Secondary- and Backscattered-Electron Emission

The thin metal coating layer or metallic ions used to infuse a porous nonconducting sample in order to make it electrically and thermally conductive is also the source of the bulk of the secondary electrons. As demonstrated in Chapter 4, high topographic spatial resolution is possible on bulk samples using the secondary-electron signal generated by the probe at the site of the incident beam, provided secondary electrons generated in the microscope chamber or column by backscattered electrons or by stray electrons from the probe are adequately suppressed. While it is quite impractical to try to eliminate stray backscattered electrons from the sample, a thin (1–3 nm) metal coating layer enhances the quality of the high-resolution secondary-electron signal. At lower spatial resolution, a 10-nm layer of metal such as gold would certainly improve the coefficient of secondary-electron emission δ for an organic specimen examined at low kilovoltage.

The backscattered-electron signal is widely used in SEM to obtain images which display variations in both atomic number and topography. Backscattered electrons have also been used in conjunction with standard cytological techniques to localize regions of physiological interest in biological tissue. Thus, if one has gone to considerable effort to obtain deposits of lead or silver at specific sites in a piece of tissue, it seems inappropriate to mask the atomic number contrast these locations give either by infusing the sample with heavy-metal salts or by covering them with a layer of high-density metal. The appropriate technique would be to apply a thin layer of a low-atomic-number conductor such as carbon that would not significantly scatter the incident beam and would allow that layer to reach the specimen.

For high-resolution, scanning electron microscopy, where the image is dependent on the scattering of high-energy electrons from the specimen's surface, it is necessary to coat the samples with a thin layer of a heavy metal that shows no structure at the 1-nm resolution level. Experimental evidence suggests that the refractory metals such as tantalum or tungsten yield such a coating.

13.2.4. X-Ray and Cathodoluminescence Emission

X-ray and visible-light photons are characteristic of the material from which they are derived, and any attempts at increasing the conductivity of nonconducting samples will seriously affect the quality of these signals. Because charging can significantly influence the generation of these two signals, it is usually necessary to coat the specimen with a thin layer of a light-element conductor such as carbon or aluminum, which has minimal effect on the quality of the analytical information.

13.2.5. Mechanical Stability

Particulate matter and fragile organic material are more firmly held in position on the specimen stub after coating with a thin layer of carbon. In many cases it is possible to place such material directly onto the specimen support and stabilize it with a very thin layer of carbon applied from two directions. This simple technique avoids the use of adhesives, most of which are highly nonconductive. Metal coatings, particularly those deposited by a sputtering process, are quite strong and contribute to the increased mechanical strength of otherwise fragile material.

13.3. Untreated Specimens

Before discussing the various procedures that can be used to make a sample conductive, it is appropriate to consider how we may examine a sample in an unsullied state. In some instances, the conductivity treatment may interfere with important signals coming from the sample, i.e., x-ray photons and BSE. Similarly, the treatment may prevent other studies from being carried out, for example, studies of the electrical properties of a semiconductor or of the cytochemical localization of an active constituent in a biological sample. In a few, rare instances, the operator may be under strict instructions not to perturb a precious or unique sample in any way—to do nothing other than examine it in the microscope. The methods and techniques to be described are all noninvasive and largely of a physical nature and, as such, do not interfere with any of the chemical properties of the specimen.

Several methods can be used to examine untreated specimens in the SEM, including operation at low beam energy, incorporation of a second beam of electrons or ions to discharge the specimen, and examination of the specimen in the presence of water. Reduction of charging by operating at low accelerating voltages makes use of the special characteristics of electron emission from solids (Oatley, 1972), as described in detail in Chapter 3 (see Fig. 3.27) and Chapter 4 (see Fig. 4.64). This method uses the self-compensation for surface charge obtained by operating with the incident beam energy such that the total electron emission for both backscattered and secondary electrons, $\eta + \delta$, is greater than unity. The development of shielded and stabilized SEMs based on conventional sources, as well as field-emission low-voltage SEMs capable of routine operation in the low-voltage regime, with beam energies in the range 0.5–5 keV, has made this method of charge compensation practical. It is especially useful if the beam energy can be changed in small steps, 0.1 keV

(100 eV) or less, since the incident beam energy must be carefully chosen to be at a value at which dynamic charge balance is obtained. This value is affected by the material being examined, and the microscopist should also be prepared to tilt the specimen, since surface tilt has a strong influence on η and δ. The high-brightness field-emission SEM can also achieve high-resolution images at low accelerating voltages (Nagatani *et al.*, 1987). Joy (1989) discusses how charging may be minimized in low-voltage SEM (LVSEM). In order to obtain maximum information from samples studied at LVSEM, a thin layer of platinum can be applied to nonconducting samples (Holy *et al.*, 1991; Erlandsen *et al.*, 1989).

Low beam energy prohibits x-ray microanalysis since the overvoltage is inadequate except for x-rays of extremely low energies. To avoid this limitation, several attempts have been made to utilize a second beam of low-energy electrons or ions to discharge the sample during bombardment with the high-energy electron beam. Spivak *et al.* (1972) employed a pulsed beam of low-energy electrons to flood the specimen during the scan-line flyback. The principle of the sample discharge is the same as that shown in Fig. 3.27. Crawford (1980) describes a low-energy beam of positive alkali ions which can stabilize the surface potential and allow insulators to be examined without the benefit of a coating layer. A similar approach has been adopted by Kotorman (1980) for the examination of FET wafers. The accelerating voltage was adjusted to below 1 kV in an attempt to eliminate trapped charges. Muranaka *et al.* (1982) found that uncoated biological specimens bombarded with an argon beam (50–100 μA, 1–5 keV) for 30 min prior to examination in the microscope showed no surface charging effects, even when examined at a beam energy of 25 keV. It is clear that such an argon-beam dosage would erode the specimen surface and could remove surface contaminants which might favor charging. However, Muranaka *et al.* consider that the elimination of surface charging is caused by a neutralization between negative charges of electrons and positive charges of ions trapped within the specimen. A similar effect has been observed by Kanaya *et al.* (1982). Prebombardment of biological and organic material with an argon beam allows such samples to be observed at up to 20 keV in an SEM without metal coating. The suppression of the negative charge was thought to be due to positive ions being trapped in the porous surface. An ion-beam dose of between 0.75 and 1.25 mA/min appeared to be the optimal condition for limiting specimen charge-up without any sample etching. Irradiation of the sample with x rays has been used to increase the surface conductivity of nonconductors. It is not entirely clear how this works, although there is an uncomfortable suspicion that the x rays cross-link contaminating hydrocarbons to form a thin (but dirty) carbonaceous conductivity layer.

Lametschwandtner *et al.* (1980) have devised a simple method of preventing highly porous nonconducting specimens from charging. It is usually very difficult to attach such specimens to the specimen stub using conductive paints because the capillary nature of the specimens results in complete infiltration with the paint and consequent disappearance of the surface detail. These workers found that it was possible to attach their vascular corrosion cast specimens to the substrate using small pieces of copper wire, which had previously been attached to the stub with colloidal silver paint. The specimen is pressed onto these conductive projections and may be examined in the SEM without any coating. Panitz *et al.* (1985) have used atomically flat Au(III) facets of gold as substrates for examining biological particles for field-ion microscopy. The same technique has been used by Garcia *et al.* (1989) to image samples for STM. The surfaces are prepared by melting gold wire and forming a molten sphere which is allowed to solidify. Such surfaces form a nearly ideal substrate because they are easy to prepare, highly conductive, highly absorptive for organic material, and extremely flat. The disadvantage is that the underlying gold gives a high backscattered-electron signal which reduces sample contrast. Warner (1984) has used gallium as an entrapping medium for nonconducting organic samples. Samples can be inserted into the low-melting-point (303 K) metal which has good conductivity and is stable under the electron beam.

A paper by Jeszka *et al.* (1981), while not directly concerned with coating, has interesting implications for the reduction in specimen charging of nonconductors. Jeszka *et al.* investigated the properties of conductive polymers made by growing microcrystalline complexes of tetracyanoquinodimethane (TCNQ) and tetrathiotetracene (TTT) in cast films of polycarbonate polymer. At a concentration of 1% TCNQ/TTT, the dark resistivity of 15-μm-thick films is $3 \times 10^{-2}\,\Omega\,\text{cm}$ at room temperature. These so-called organic metals do not contain any of the metallic elements associated with high conductivity. Such polymer films may provide the basis for an inexpensive conductive support to be used in connection with x-ray microanalysis of light elements. A review by Bryce and Murphy (1984) gives details of the unusual properties and technological potential of these interesting materials.

Heating has been investigated as a method to reduce charging effects on nonconductors. Lundquist *et al.* (1988) were able to image, using secondary electrons, sintered ceramics, silicon oxide films, and yttrium aluminum garnet (YAG) crystals by heating them to 573 K. The discharge mechanisms which negate charging include mobile ion migration and thermal ion emission. Finally, nonconductive samples can be tightly wrapped in thin metal foil and a small window cut in the foil to expose the area of interest. This method has proved useful in studying fossil bones.

If the specimen contains water, it will have sufficient conductivity to discharge. Considerable success has been obtained with examining and analyzing uncoated biological material at low temperatures (Echlin, 1992). In conditions where water is retained in the frozen state, so too are ions and electrolytes, which provide conductivity. If specimens can be examined in an environmental cell, water can also be retained in the liquid state to discharge the primary beam (Robinson and Robinson, 1978). The same advantage can be obtained by using the new environmental SEMs (Danilatos, 1991). Some success has been obtained in examining uncoated paints and electrical insulators in such instruments.

13.4. Bulk Conductivity Staining Methods

Unlike the physical procedures described in the previous section, the effectiveness of the methods described here depends on altering the chemical characteristics of the sample either by infiltration with metal salts or metal deposits or by chemically coating the surface with a conductive layer.

The early attempts at these procedures were not very successful, as they were based on extended fixation in heavy-metal fixatives such as salts of osmium, manganese, chromium, silver, and lead. Osmium tetroxide is the most widely used solution, as it is both a fixative and a conductive stain. Chromium trioxide is also a fixative and a stain, but is less effective than osmium. The silver, lead, and manganese salts are not fixatives and do not provide as much conductive staining as osmium. While there was a marginal improvement, there were not enough natural binding sites in the specimens to obtain a sufficiently high loading of the heavy metals to make any substantial change in the conductivity. Similarly, staining during dehydration with sodium, chloroaurate or uranyl nitrate also failed to make any significant change in the sample's conductance.

By using reactive heavy metals such as osmium, lead, and uranyl ions together with bifunctional mordants, some of which may be ligands, such as the thiocarbohydrazides, phenylenediamines, and galloylglucoses, it was possible to make a substantial increase in metal loading on the samples and a significant improvement in conductivity. The process of mordanting is not new to microscopy; it has long been used for enhancing the staining ability of dyes. The bifunctional mordant acts as a molecular bridge between the specimen and the dye (or heavy metal) and thus enhances the reactivity of the molecule of interest. Thiocarbohydrazide is a particularly useful mordant, as it can form a bridge between osmium molecules. One terminal amine group reacts with the osmium previously applied to the tissue, and the other terminal amine group, in turn, binds to further osmium ions or, as Munger (1977) has shown, to uranyl or lead ions. The mordants phenylenediamine (Estable-Puig *et al.* 1975) and tetramethyl ethylenediamine (Wilson et al. 1979) are presumed to act in a similar fashion as molecular bridges. In a detailed study on the galloylglucoses, usually referred to as tannins, Simionescu and Simionescu (1976) found that these substances act as a mordant between osmium and lead. At present, there is no satisfactory explanation of the chemical nature of the binding, but is is thought to involve carboxyl and hydroxyl groups. Of all the mordants and heavy-metal ions which have been tried, osmium–thiocarbonhydrazide and osmium–tannin are the most successful, and examples of typical procedures are given below.

Osmium–Thiocarbohydrazide-Osmium Method (OTO). Specimens are fixed in glutaraldehyde, washed, and postfixed in osmium tetroxide. The osmicated tissue is then thoroughly washed in buffer and incubated in the thiocarbohydrazide solution. After a second thorough wash to remove any unbound thiocarbohydrazide, the tissue is incubated for about 1 hour in osmium tetroxide. Following a final thorough wash, the specimen is dehydrated and dried and is then ready for examination in the microscope. There are a number of variations of the method, the most important one being several alternating passes through thiocarbohydrazide and osmium.

Tannic Acid–Osmium Method (TaO). Aldehyde-fixed tissues are soaked in a guanidine hydrochloride–tannic acid solution for several hours, thoroughly washed, and passed into osmium tetroxide for an equally long period. The samples are then washed, dehydrated and dried. Figure 13.2 shows results of such a preparation.

Tannic Acid–Osmium–Thiocarbohydrazide Method (TaOTh). Aldehyde-fixed material is immersed in a glutaraldehyde–tannic acid mixture. After washing, the tissue is passed into an osmium solution, washed, and then placed in thiocarbohydrazide. After another wash, the tissue passes into a fresh osmium solution and is subsequently washed, dehydrated, and dried. These alternating procedures can be repeated several times in order to increase intensification.

Osmium–Dimethyl Sulphoxide–Osmium Method (ODO). Osmium-fixed material is immersed in a dimethyl sulfoxide solution for half an hour. The samples are quench-frozen and fractured under liquid nitrogen and the fractured pieces thawed in a DMSO solution before being postfixed in an osmium solution. This method has proved particularly useful for examining the internal structure of cells and tissue—see Fig. 13.3. This procedure may be followed by the TaO method described above. In all four procedures of controlled osmication, it is important to avoid osmium black precipitation on any of the surfaces that are to be examined at high magnification. This is achieved by exposing the surface of interest *before* the osmication procedure takes place to ensure that diffusible nonbound reagents and reaction products can be removed by extensive washing.

In addition to these four principal procedures, some success has been obtained with ammoniacal silver and with osmium vapor (Kubotsu and Veda, 1980).

Silver Nitrate Methods. Cells are fixed in glutaraldehyde–DMSO solution, washed to remove excess aldehyde, and treated with ammoniacal silver nitrate. The tissue is then briefly immersed in photographic fixer, dehydrated, and dried.

Osmium–Hydrazine Vapor Method. This method differs from the other methods in that the samples are exposed to the vapor of reactants rather than the solutions. Dried specimens are exposed to osmium tetroxide vapor followed by vaporized hydrazine hydrate, which reduces

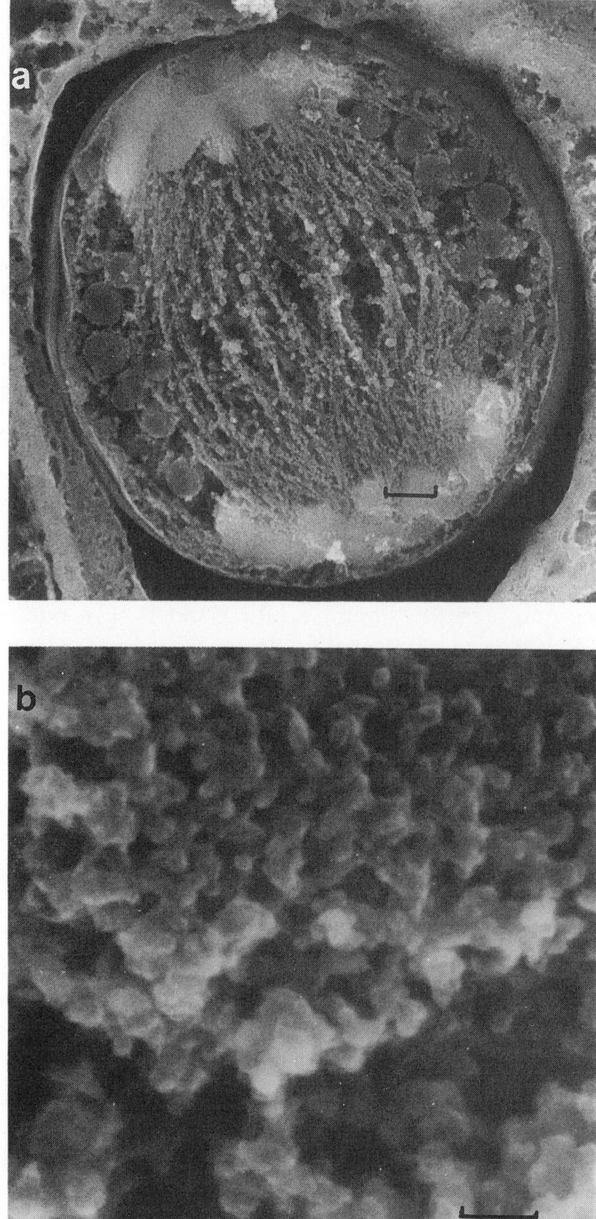

Figure 13.2. Conductive staining methods. (a) Fractured pollen mother cell of *Tradescantia* treated with the TaO conductive staining method. The uncoated samples show details of the chromosomes and spindle fibers. Marker = 2 μm. (b) Detail of (a) showing fine structure of the chromosomes and 15–20-nm spindle fibers. Marker = 50 nm (Naguro *et al.*, 1990).

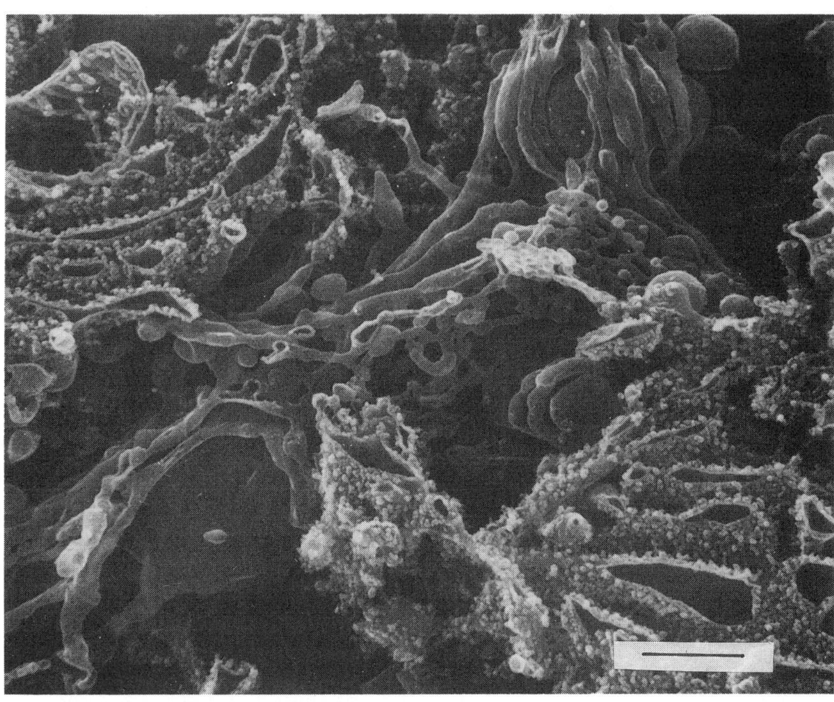

Figure 13.3. Conductive staining methods. A general view of Golgi bodies from a fractured rat lacrimal glandular cell prepared using the O–D–O method. The sample, which is uncoated, shows details of rough endoplasmic reticulum, cis-Golgi cisternae, and mitochondria. Marker = 0.5 μm (Inoué, 1986).

the osmium salt to the metal. Although this method gives satisfactory results with plastics and polymers, it is not recommended for use with labile biological material due to the corrosive natures of the hydrazine vapor. It does, however, give good results with biological material when used in the liquid phases.

Full operational details of these methods are given in the excellent review papers by Gamliel (1985), Murphey (1978; 1980), Murakami and Jones (1980), Murakami *et al.* (1982; 1983), Munger (1977), and Tanaka (1989).

The results of these bulk conductivity preparative techniques are impressive, and high-resolution images have been obtained from completely uncoated samples. There are different degrees of success, the best images at magnifications of up to 225,000X being obtained from the tannic acid–osmium method at beam energies of 20–25 keV and a 100-pA beam current. Tanaka (1989) has used the bulk conductivity method followed by a thin (1–2 nm) coating of Pt to examine the fine structure of mitochondria at high resolution (Fig. 13.4). A recent study by Naguro *et al.* (1990) revealed the substructure of 15-nm fibrils in *Tradescantia* chromosomes in material treated by the tannic acid–osmium procedure but not coated with a thin metal layer (see Fig. 13.2).

The advantages of the bulk conductivity methods over the surface conductivity approach are as follows:

1. Metal-impregnated samples may be repeatedly dissected to reveal internal details without significant charging effects.
2. Very rough-textured or deeply indented surfaces can be rendered conductive.
3. The surface conductivity is more even, and there are fewer edge effects.
4. It makes it possible to obtain sufficient contrast from low-profile specimens.
5. Samples treated by these methods show considerably less shrinkage during dehydration and drying.
6. By substantially increasing the density of the sample by the inclusion of heavy metal ions, beam penetration is diminished and spatial resolution is increased.

A simple comparison of the electrical and thermal properties of the metals used for these procedures shows that they compare favorably with the material used for thin films (Table 13.1). The success of osmium with biological material is no doubt due to its optimal properties of fixation and staining.

The main disadvantages are the increased length of time for specimen preparation, an image with a somewhat lower S–N ratio, and in some cases a resolution lower than that obtained from metal-coated samples. Some of the reagents (i.e. tannic acid, hydrazine, ammoniacal silver nitrate) can cause tissue damage, and prolonged exposure to

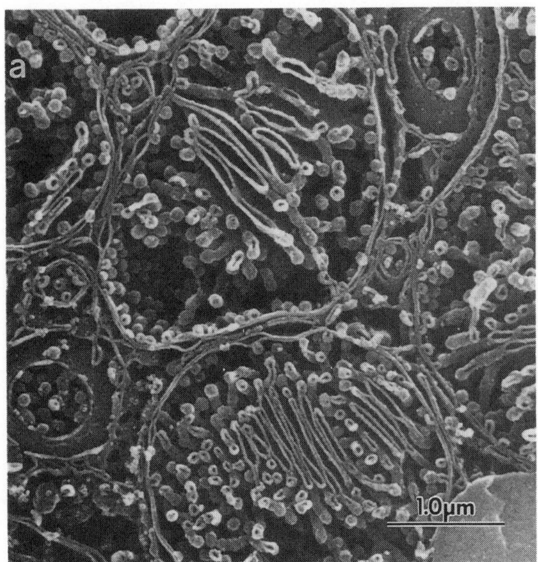

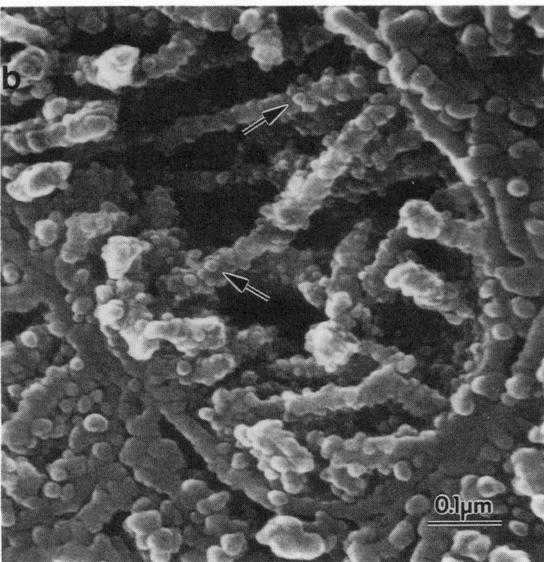

Figure 13.4. Conductive staining methods. (a) Mitochondria of a rat suprarenal gland cell. Lamellar, tubular, and vesicular cristae are present (Hanaki *et al.*, 1985). (b) Inner membrane particles on mitochondrial tubules (arrows) from a rat liver cell (Osatake *et al.*, 1985).

Table 13.1. Thermal Conductivity and Electrical Resistivity of Some Elements and Organic Materials

	Thermal conductivity (W/cm per K at 300 K)	Resistivity (Ohm cm at 300 K)
Gold	3.17	2.40
Palladium	0.72	11.0
Platinum	0.72	10.0
Aluminum	2.37	2.83
Carbon	1.29	3500
Osmium	0.87	9.5
Lead	0.35	22.0
Uranium	0.28	30
Silver	4.29	1.60
Wood	0.05	10^{10}–10^{13}
Bone	0.05	10^{8}
Water	0.006	10^{4}–10^{7}

osmium tetroxide would mitigate against high-resolution SEM at the molecular level. The osmium maceration technique introduced by Tanaka and Naguro (1981), Inoue (1983), and Osatake *et al.* (1985) uses diluted osmium solutions to remove the cytoplasmic matrix in order to more clearly visualize intracellular organelles. The presence of such large amounts of heavy metals precludes any attempt at x-ray microanalysis for elements of natural origin. In some instances, the conductive staining procedures are too successful and result in such high backscattered-electron coefficients that the topographic contrast is reduced. McCarthy *et al.* (1985) have modified the tannic acid procedure so that it may be used for specimens to be examined in both TEM and SEM. As will be shown in a later section of this chapter, the conductive staining techniques in combination with high-resolution coating procedures is the preferred method of obtaining high-resolution SEM images of biological and organic materials. It should be realized that the two methods for increasing specimen conductivity are not mutually exclusive. Some very useful images have been obtained from samples that have been impregnated with metal ions, critical-point dried, and then lightly coated with a thin metal coating (i.e., 3–5 nm). This method has the treble advantage of a substantial increase in bulk conductivity, diminished beam penetration, and high secondary-electron emissivity from the metal layer.

The procedures based on chemically coating the specimen with a conductive layer are less effective in reducing charging than the bulk conductivity methods. Such procedures, which include spraying with organic antistatic agents derived from polyamines, organic suspensions of noble metal colloids, or graphite, or covering the sample with a thin polymer film such as formvar or styrene–vinylpyridine (Pease and Bailey, 1975), are useful only at very low magnification and even then may obscure surface detail.

With the possible exception of the techniques which increase the bulk conductivity of the sample, none of these methods gives anything like the resolution and information content which may be obtained from properly coated samples. Indeed, these alternative methods have diminished in usefulness now that it has been shown that one can adequately coat even the most delicate and thermally sensitive specimens. Further, it can be argued that if a sample cannot survive the vacuum found in a sputter coater, it is unlikely to survive the high vacuum of an electron-beam instrument.

The techniques devised to increase bulk sample conductivity are especially useful when they are used in conjunction with coating techniques. Bulk sample conductivity methods are frequently used in connection with the examination of fractured surfaces of three-dimensional specimens whose internal morphology is being investigated. For example, the substructure of fractured surfaces may be revealed by atomic number contrast from osmium incorporation rather than by topographic contrast derived from surface irregularities.

13.5. Specimen Mounting Procedures

Whether samples are naturally conductive or whether they are made conductive by one means or another, they have to be firmly attached to the specimen support used in the microscope. If samples are to be coated with a thin metal film, it is more convenient to attach the sample to the specimen support *before* the coating procedures. For samples treated to increase their bulk conductivity, it is more convenient to attach the samples *after* these procedures are completed. Murphey (1982), in a monumental paper, describes in great detail the procedures which one may use for mounting specimens. This is frequently a sadly neglected area of specimen preparation, and one suspects that otherwise perfectly prepared specimens are spoiled by lack of attention to this final detail. Murphey, while agreeing that the materials and methods chosen for SEM sample mounting must necessarily be a compromise among the aims of the study, the characteristics of the sample, and the availability of instrumentation, nevertheless provides practical details on how these procedures should be carried out. The specimen stubs are usually made of a conductive material, and care must be taken that the atomic number of the stub material does not interfere with backscattered-electron and x-ray signals coming from the specimen. For these two latter modes of operation, carbon, beryllium, and possibly aluminum are the materials of choice. At least a dozen stub types are available (e.g., flat, grooved) and they should all be cleaned before use and carefully checked to make sure that they fit into the microscope. In some cases, it is more convenient to place the specimen on a substrate which in turn is mounted on the stub. There are many different substrates, ranging from glass and plastic cover slips to metal and crystalline disks, plastics, waxes, and a whole range of membrane filters. Johansen *et al.* (1983) find that mica, although not ideal, can be used as a convenient support for biological samples, being more "charge" resistant than glass. It is usually necessary to coat nonconductive substrates with a thin conductive layer (1–2 nm Pt, 5 nm Au) before the samples are attached to the surface. Tanaka *et al.* (1989) use carbon-coated plates as substrates for high-resolution SEM. The plates may be made hydrophilic by UV irradiation to facilitate attachment of samples. Some progress has been made toward providing biologically active substrates. Tsuji and Karasek (1985) have found that cultured endothelial cells reorganize on collagen disks, which can be easily processed for SEM. Kumon *et al.* (1983) found that gelatin-covered glass and silicon cover slips, impregnated with osmium vapor and coated with polylysine, eliminated charging artifacts. Peters and Carley (1988) used polylysine-coated silicon chips as a support for membrane specimens to be examined by high-resolution SEM. The silicon chips, cut from wafers with a (200) orientation, have a high electrical conductivity and need no metal coating prior to specimen attachment. The polylysine serves to bind the biological sample to the wafer surface. The disadvantage is that the polylysine is not the best adhesive surface, and there is a tendency for the sample to charge. The choice of substrate is obviously influenced by the specimen, and such factors as transparency to photons and electrons, solvent solubility, porosity, topography, composition, and conductivity must be considered.

It is frequently necessary to attach the specimen or the substrate to the stub. A bewildering variety of adhesives is available for this purpose, and the characteristics which must be considered include adhesive viscosity, stickiness, setting time, resistance to heat, solvent and beam damage, physical and chemical composition (many good glues have unacceptably high vapor pressures, which contaminate the electron microscope column), transparency and surface topography, and ease of application. Witcomb (1985) gives a comprehensive review of the available glues together with a list of their properties. Thin adhesive layers are more easily produced using liquid glues, and one of the epoxy glues with or without a conductive additive is a useful starting material for dry samples.

13.6. Thin-Film Methods

One of the most effective ways of increasing the surface conductivity of a nonconductor is to cover it with a thin (1–20 nm) layer of a conducting material, usually a metal. Thin conductive films can be produced in a number of ways, although some methods are impractical for use with the SEM and EPMA. Thus, chemical deposition techniques such as electroplating and anodization are rarely used, as they rely on immersing the sample in solutions at high or low pH. In addition, the two procedures are usually effective only for applying thick (i.e., several micrometers) metal coatings. Carbon films can be made using glow discharge techniques. An evacuated chamber is backfilled with hydrocarbon vapor such as acetylene or benzene at low pressure and in an anoxic environment. A glow discharge is created which causes polymerization of the hydrocarbon vapor to a sooty carbon deposit. This method is too unpredictable to be of any practical use in electron microscopy. The thin films we need for electron microscopy can be produced in a variety of ways (Maissel and Gland, 1970), but of these methods only thermal evaporation and sputtering are useful for coating specimens for SEM and x-ray microanalysis. In discussing these methods below, it is important to consider the properties of an ideal film. Such a film should be thermally and electrically conductive and should not exhibit any structural features above a scale of 3–4 nm resolution to avoid introducing unnecessary image artifacts. For high resolution studies, the structural features of the thin film should be in the 0.5–1.5 nm range. The ideal film should be of uniform thickness and cover all parts of the specimen regardless of the surface topography and should not contribute to the apparent chemical composition of the specimen or significantly modify the x-ray intensity emitted from the sample.

13.7. Thermal Evaporation

Carbon, metals, and some inorganic insulators, when heated by an appropriate method in a vacuum, begin to evaporate rapidly into a monoatomic state when their temperature has been raised sufficiently for their vapor pressure to exceed 1.3 Pa (10^{-2} torr). The high temperatures necessary to permit the evaporation of the materials can be achieved by three different methods.

In the *resistive heating* technique, an electric current is used to heat a container made of a refractory material such as one of the metal oxides or a wire support made of a high-melting-point material such as tungsten, molybdenum, or tantalum (Fig. 13.5). The material to be evaporated is placed in or on the container, which is gradually heated until the substance melts and evaporates. The same method is used to evaporate carbon, except that no container is used. Two thin opposing rods of carbon are heated until the material evaporates. In the *electric arc* method, an arc is struck between two conductors separated by a few millimeters. Rapid evaporation of the conductor's surface occurs. This is the way in which carbon and some of the high-melting-point metals are evaporated. For most high-melting point materials such as tungsten, tantalum, and molybdenum, the most effective way to heat the substance is to use an *electron beam* (Fig. 13.6). In this method the metal evaporant is the anode target, and it is heated by radiation from a cathode maintained at 2–3 keV. This is a very efficient means of heating, as the highest-temperature region is the vapor-emitting surface and not the evaporant source material. An additional ad-

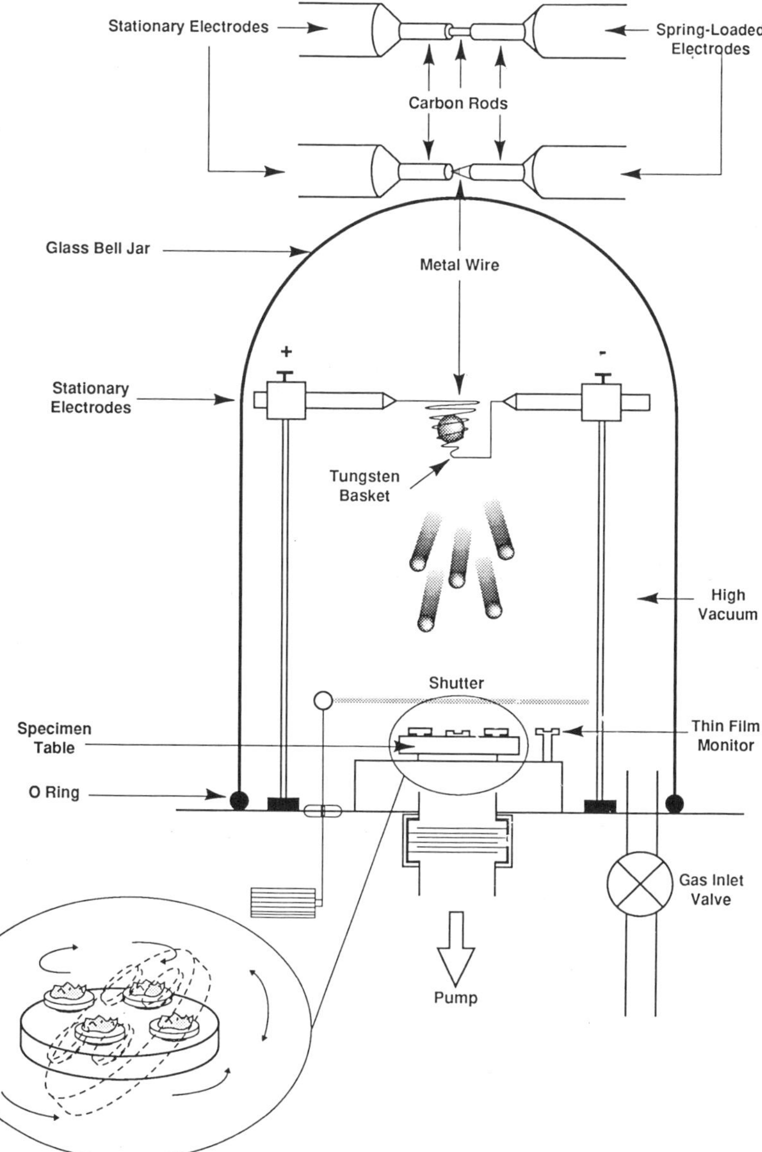

Figure 13.5. Diagram of a high-vacuum thermal evaporation unit.

vantage is that the metal evaporant is deposited with very small grain size. Electron-gun evaporation can also be used for evaporating some of the lower-melting-point metals such as chromium and platinum which also have a very small particle size.

We can conveniently consider evaporation under two headings, high- and low-vacuum techniques, both of which produce thin films suitable for most SEM and x-ray microanalysis investigations. Some specialist coating techniques, including high-resolution methods, are discussed at the end of this chapter.

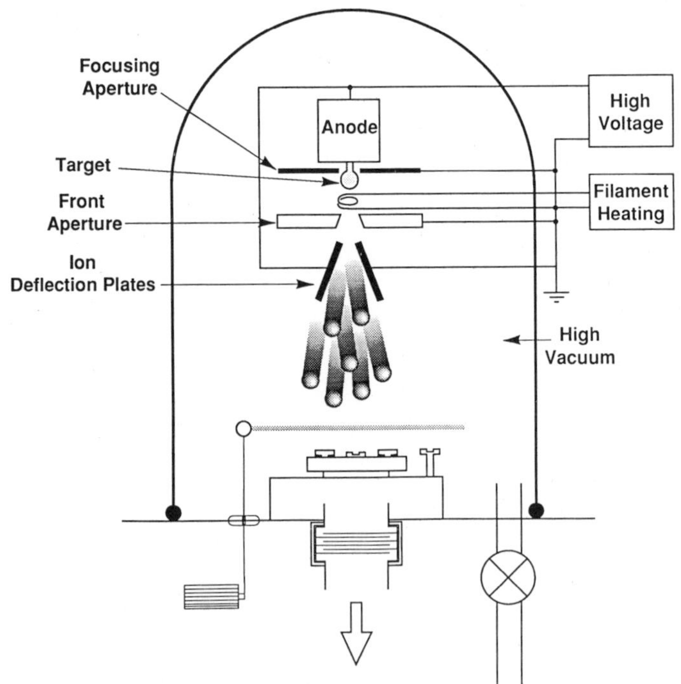

Figure 13.6. Diagram of an electron-beam evaporation assembly.

13.7.1. High-Vacuum Evaporation

In this context "high vacuum" is considered to be the range between 10 μPa and 100 mPa ($\sim 10^{-7}$ to 10^{-3} torr). High-vacuum evaporation techniques are commonly used in electron-microscope laboratories.

The formation of a thin film is complex and proceeds through a series of well-characterized steps: nucleation, migration, and coalescence to form a continuous film. The first atoms arriving at the surface of the specimen stay there only if they can diffuse, collide, and adhere to each other on the surface to form nucleation sites of a critical size. The stronger the binding between the adsorbed atoms and the substrate, the higher the nucleation frequency and the smaller are critical nuclei. Most biological and organic samples are likely to have variable binding energy, which results in the variation of critical nuclei across the surface, producing uneven film deposition. For this reason, precoating with carbon at low vacuum to cover the specimen with a homogeneous layer results in the smaller, more evenly sized critical nuclei when they are subsequently coated with metal. For example, the nucleation density of gold can be substantially increased by pre-coating with 5–10 nm of carbon. As deposition continues, the nucleation centers grow in three dimensions to form islands which gradually coalesce to give a continuous film (Fig. 13.7). During deposition of a metal film, the atoms aggregating at the surface frequently form crystals. Metals with high surface mobility such as silver and gold form

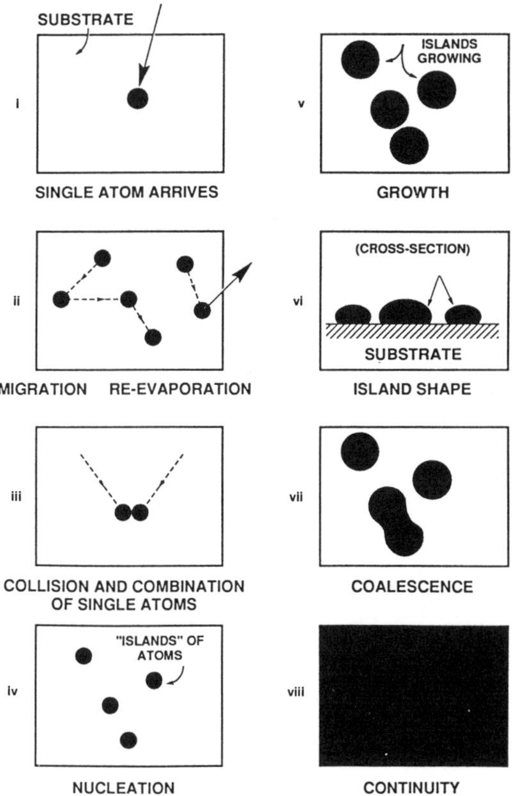

Figure 13.7. Steps in the formation of a thin film.

relatively large crystals, which grow into a continuous film at a thickness of about 5 nm. The steps in the growth of a thin film of silver are shown in Fig. 13.8. Metals of low surface mobility such as tantalum, niobium, platinum, and chromium form very small crystals, and the films become continuous at much smaller total thickness. The condensation of a metal vapor onto discontinuous microcrystalline aggregates rarely produces films with a grain size smaller than 3 nm.

Slayter (1980) describes techniques for producing very thin films in which the grain size is considerably reduced. He finds, for example, that the number of crystallites per unit area for platinum and tungsten is at a maximum in films 0.5 nm thick. High-melting-point metals form more crystallites per unit area, and although the scattering power of such films is diminished, they have sufficient contrast to be used in high-resolution STEM and TEM instruments. The effect is enhanced at low substrate temperatures and in clean environments.

There still remains some uncertainty as to how much damage thermal evaporation can cause an organic sample. The momentum of an atom evaporated at between 2300 and 4800 K is quite considerable, and this energy is dissipated in the sample. Colquhoun (1984) has calculated that the energy of a vaporized platinum atom is about 13 kcal/mol, which would be about enough to disrupt intermolecular

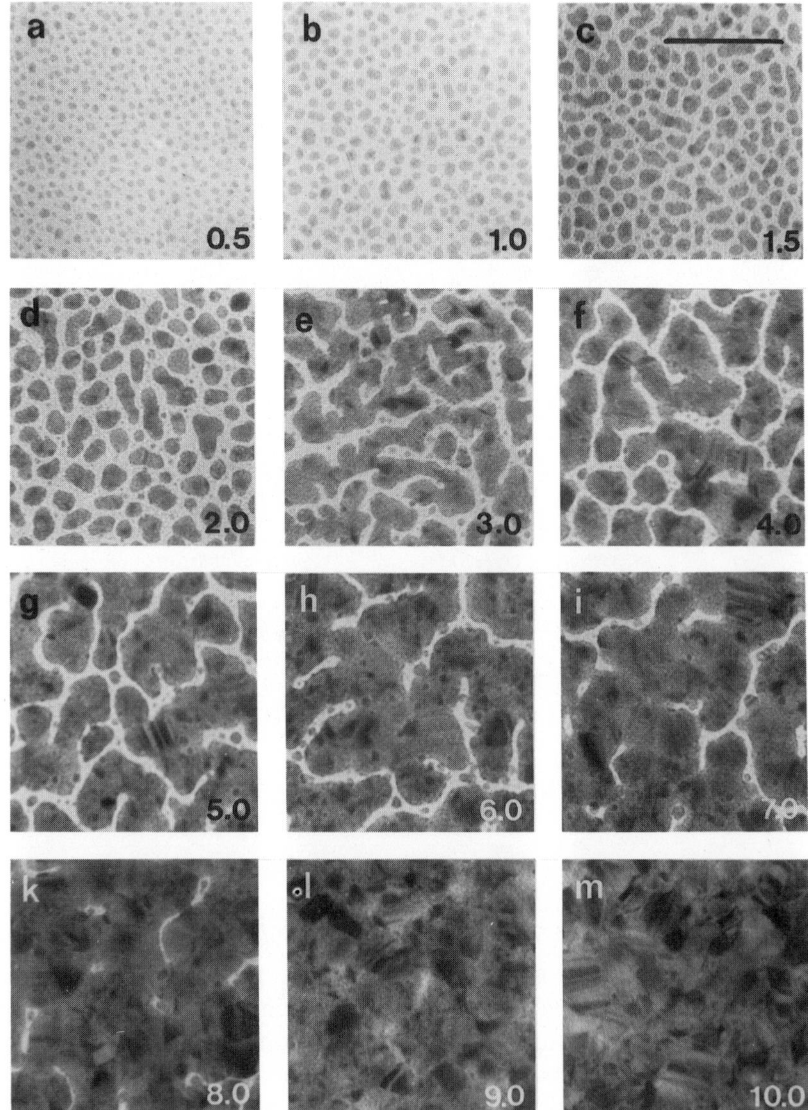

Figure 13.8. Growth of silver film at 293 K. Films of increasing average mass-thickness grow in two phases. First, after nucleation, crystals coalesce (a)–(e), then smaller crystals grow on top of existing crystals (f)–(h) and bridge gaps (i)–(l) until the film is optically continuous (m) (Peters, 1984).

forces but not enough to break bonds. Colquhoun considers that 13 kcal/mol is a median value and that some of the evaporated atoms would be travelling at much higher speeds, with energies approaching 40–50 kcal/mol. Such energies would be sufficient to break bonds in molecules. Colquhoun has calculated that a substantial proportion of the radiant flux from evaporating platinum consists of UV radiation, which would also disrupt chemical bonds. Although some of these problems may be overcome by cooling the specimen, there is still some doubt as to whether metal evaporation methods favorably reproduce the structure of heat-sensitive samples.

A large number of interacting factors influence the formation of small-grain-size films for high resolution SEM. The melting point, crystallographic lattice, purity, and reactivity of the metal evaporant, the surface temperature and energy input into the specimen during coating, and the rate and amount of metal deposited, all influence the final appearance of the thin film. The thinnest films with the finest grain size have been produced either in 0.2-to-2.0 nm films of tungsten thermally evaporated from an electron gun onto a substrate cooled to 77 K or in platinum–iridium–carbon films. A recent study by Wepf *et al.* (1991) showed that 1.5-nm films of platinum–iridium–carbon have extremely fine granularity and are suitable for STM as well as high-resolution SEM (Fig. 13.9). Slayter (1980) uses a platinum deposition rate of 1.2 nm/min and a tungsten deposition rate of 0.9 nm/min and concludes that a 0.5-to-1.0 nm layer of platinum would be the best coating for high-resolution secondary imaging as it combines fine grain size and high crystallite density with good secondary-emission properties. A 1.0-nm layer of platinum would be the best coating for low-loss backscattered imaging, and a 0.5-to-1.0 nm layer of tungsten would be useful for STEM applications where the highest resolution is necessary and secondary-electron emission is less important. Slayter favors a much increased target–specimen

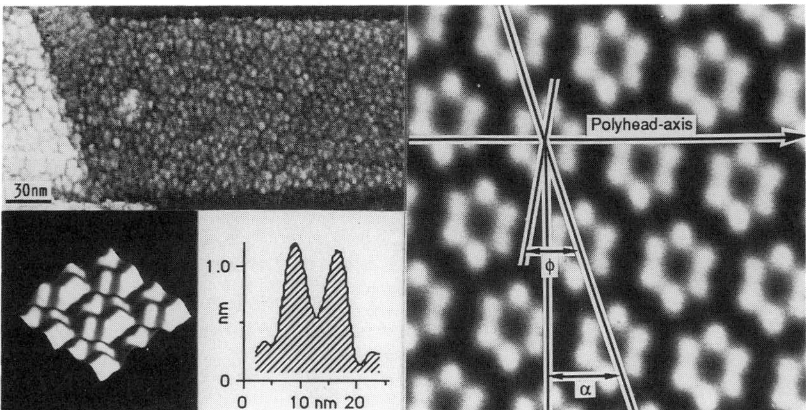

Figure 13.9. Thin films for high resolution microscopy. Raw STM data of a freeze-dried and Pt–Ir–C-coated bacteriophage T4 type III polyhead after removing a least-squares-fitted plane. The topography is shown in a gray-tone representation (top left). It has a spread-flattened diameter of 105 nm, an angular pitch α of 20°, and a lattice constant of 13 nm. Note that the capsomeres can be distinguished as hexamers already in the input data. The averaged capsomere morphology is shown in the right-hand figure in plan view, in the left-hand bottom figure as a surface-view representation, and in the center bottom figure as a section along a line through the center of the capsomere and the tops of two protomeres. The capsomere is composed of six protomeres whose tops lie on a 6-nm-diameter circle; the capsomere has an orieintation angle ϕ of approximately 30° (Amrein *et al.*, 1989).

distance, e.g., 60 cm for optimal evaporation of tungsten, in order to minimize the photon radiation on the sample; Peters (1980) has shown this radiation can cause an increase in crystallite size. Provided a number of precautions is taken, thermal evaporation of tungsten onto relatively flat specimens such as thinly spread macromolecules can provide high-resolution (1–2 nm) information. A thin layer of evaporated carbon applied to the specimen after coating appears to stabilize the coating layer and to diminish recrystallization of the thin-film layer under the influence of the electron beam. Pulker (1980) has observed that other heavy metals are an excellent prenucleation material before applying a layer of gold. This finding is consistent with the well-established procedure of precoating samples with a thin layer of carbon before applying a second coat of one of the noble metals. It remains to be shown whether this initial layer provides a clean surface or sites for high nucleation density.

13.7.2. The Apparatus

There are four basic requirements for a high-vacuum evaporator:

1. It should have high pumping speeds at low gas pressure to ensure that there is a rapid removal of gases liberated from the evaporant source and specimen during coating
2. There should be minimum backstreaming into the coating chamber of vapors from the pumps
3. The system should be easy to take apart for cleaning and maintenance
4. Adequate provision should be made for electrical connections for multiple evaporation devices and sample manipulation.

A diagram and a photograph of a high-vacuum evaporation unit are shown in Figs. 13.5 and 13.10.

Most units are pumped out with a diffusion pump backed by a rotary pump. The larger the throat size of the diffusion pump, e.g., 150 mm versus 75 mm, the shorter the pump–down time of the evaporation chamber, although there may be only a marginal improvement in the ultimate pressure obtained. Some units are equipped with turbomolecular pumps backed by rotary pumps. Such units give shorter pump–down times and cleaner vacuums, although the ultimate pressure is no better than that obtained using a diffusion pump. The rotary pump should be a two-stage pump capable of being ballasted. Gas ballasting is one of the most effective means of preventing vapor condensation in a rotary pump. The method consists of leaking a small quantity of air into the rotary pump during the compression cycle so that the exhausted vapor is mixed with a noncondensable gas. This decreases the compression necessary to raise the exhaust value and prevents vapor condensation. Rotary-pump vapor condensation can also be decreased by raising the pump temperature and by placing a desiccant such as P_2O_5 in the exhaust line. To avoid backstreaming, the system should not be pumped below 10 Pa (10^{-1} torr) on the rotary pump alone, and it should not be exhausted into the laboratory but vented into a fume cupboard or to the outside. There seems little necessity to resort to ion pumps, sublimation pumps, or exotic cryopumping systems as ultrahigh-vacuums (130 nPa to 130 pPa) are not required for the commonly used evaporation methods.

Whatever system is used, it is necessary to ensure that the backing and roughing lines from the pumping unit to the chamber are fitted with baffles or an activated alumina foreline trap to minimize backstreaming of pump oil. Water-cooled baffles are quite effective, although liquid-nitrogen baffles are better. A liquid-nitrogen trap in the roughing line minimizes diffusion of forepump oil into the evaporation chamber during roughing operations and maintains the forepressure of a diffusion pump at approximately 10 mPa. It is also useful to have a liquid-nitrogen trap above the diffusion pump, between it and the coating chamber. Care should be taken to note where the vacuum pressure is read in the system. If the pressure is read near the pumps, it is likely to be ten times better than that found in the evaporation chamber. The evaporation chamber should be made of glass and be as small as is convenient for the work to be carried out. A safety guard should be fitted over the evaporation chamber to minimize danger from implosion fragments. The chamber should contain at least four sets of electrical connections to allow evaporation of two different materials, specimen rotation, and thin-film measurement. Wherever possible, the sample should be placed above the evaporating source. If this is not possible, the sample should be placed not directly below the source, but to one side. This avoids

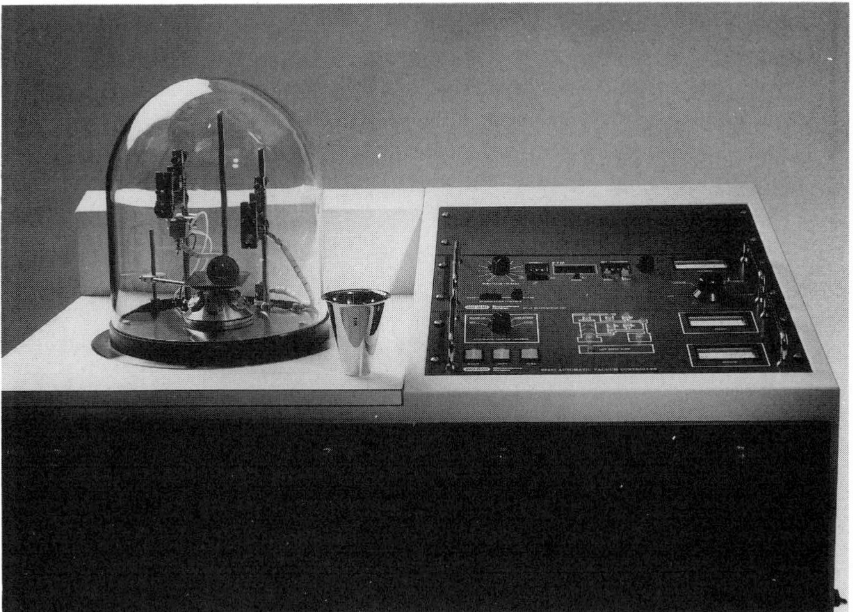

Figure 13.10. Photograph of a high-vacuum evaporation unit. (Courtesy of VG Microstructure-Polaron.)

the embarrassment of pieces of molten metal or carbon sparks accidentally landing on the specimen. The electrical power for the evaporation sources should be variable, and it is useful to have a pushbutton control to allow maximum power to be applied in short bursts. The unit should be brought up to atmospheric pressure by means of a controllable needle valve which can be connected to a dry inert gas such as nitrogen.

13.7.3. Choice of Evaporant

The choice of material to be evaporated and the manner by which it is to be applied is very dependent on the particular application in hand. A tabulation of selected elements and their properties, which are useful for coatings, is listed in Table 14.11. For most SEM work, gold, gold–palladium, or platinum–carbon is used. Silver has a high secondary-electron coefficient and is one of the best substances for faithfully following the surface contours. It also has the advantage that it is easily removed from the sample's surface using a dilute solution of potassium ferricyanide and sodium thiosulphate (Mills, 1988). Unfortunately, silver suffers from the disadvantages that it easily tarnishes and has a larger grain size than other metals, although this does not present a problem at magnifications up to 15,000X. Gold has a high secondary emission, is easily evaporated from tungsten wire, but has a tendency to graininess and agglomerates during coating (Fig. 13.11). It requires a thicker coating layer to ensure a continuous film. A 60 : 40 gold–palladium alloy or palladium alone shows less granularity than gold and yields one of the thinnest continuous films. Unfortunately, both metals easily alloy with the tungsten holder. Platinum–carbon, when evaporated simultaneously, produces a fine grain size but rather low conductivity. Maeki and Benoki (1977) have found that low-angle (~30°) shadowing of specimens with evaporated chromium before carbon and gold–palladium coating improves the image of samples examined in the SEM. The finest granularity is obtained from high-melting-point metals, but they can usually be evaporated only with electron-beam heaters. An exception to this is described by Panitz *et al.* (1985), who were able to evaporate slowly (1 nm min^{-1}) tungsten from an incandescent filament of the metal. This should be attempted only in a good, clean vacuum to ensure that the sample remains uncontaminated during the long coating time. In addition, it is known that specimens which are poor thermal conductors can experience significant temperature increases during coating. Figures 13.12 and 13.13 show the differences in surface features of some of the metal thin films applied to nonconducting samples.

High-vacuum evaporation techniques are also used to evaporate carbon either as a thin precoating layer before a thin metal film is deposited, or in situations where metal films cannot be used, e.g., x-ray microanalysis and selective backscattered-electron imaging. Although very thin, virtually structureless, and relatively easy to make, carbon

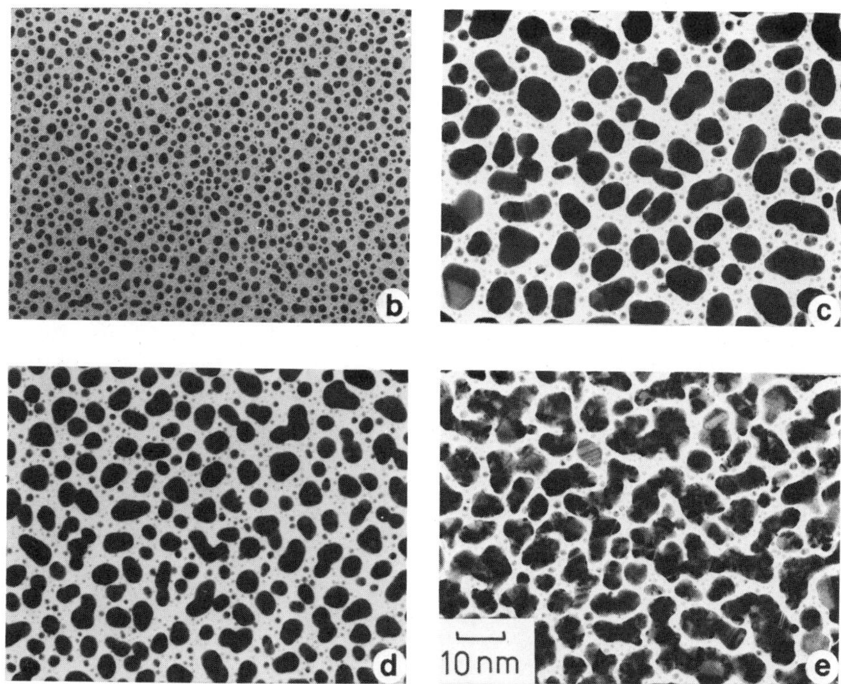

Figure 13.11. Electron micrographs of ultrathin gold films evaporated at 0.05 nm/s. Thicknesses are (b) 1 nm, (c) 4 nm, (d) 6 nm, and (e) 15 nm (Kazmerski and Racine, 1975).

films are not suitable on their own for conventional SEM studies because the yield of secondary electrons is rather low. Carbon suffers from another disadvantage, in that the high evaporation temperature (3000 K) may damage sensitive specimens. Peters (1984b) describes a simple method of obtaining reproducible thin and ultrathin carbon films using a flash evaporation method with carbon fiber as the source material. Films of any thickness up to 20 nm can be produced with a minimum of photon radiation into the sample. In addition, the flash-evaporated films have less background structure than films produced by evaporation from carbon rods. Such films have a wide application in both microscopy and analysis and are particularly useful for heat-sensitive samples. The recent study by Wepf *et al.* (1991), showed that thin films of platinum–iridium–carbon have extremely fine granularity and are suitable for STM as well as high-resolution SEM. The best images for SEM are obtained from samples coated with noble metals and their alloys. One should aim to apply sufficient material to give a maximum signal along with minimal surface charging and distortion of surface features. A 10-nm layer of a noble metal usually gives the optimal values for most routine studies.

Most of the commonly used metal evaporants are available in the form of wire. It is recommended that thick wire, i.e., 0.5–1.0 mm, be used, as short pieces can be easily looped over the appropriate

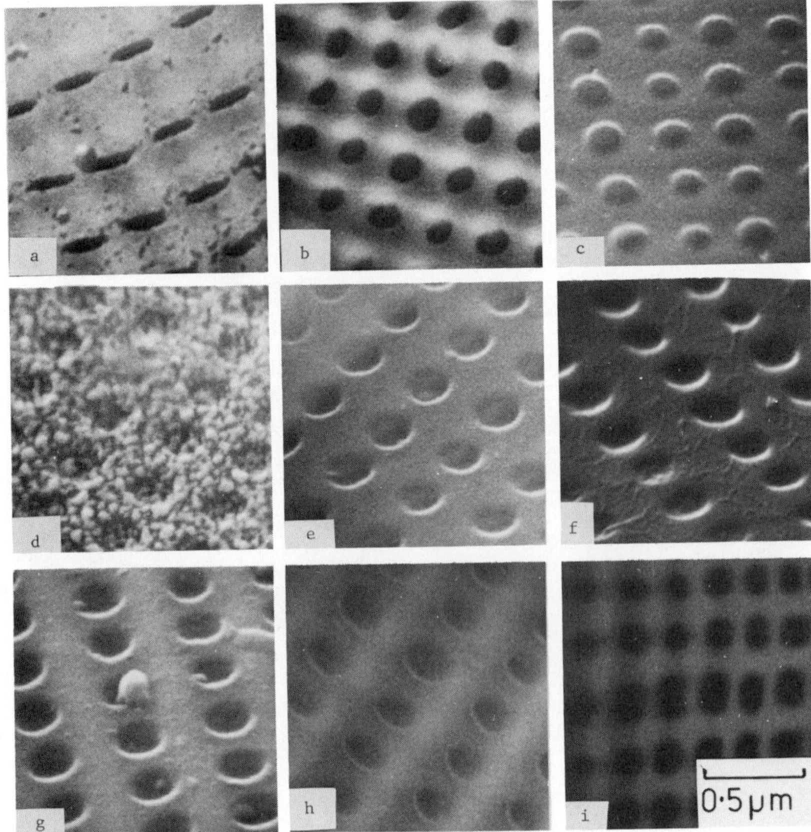

Figure 13.12. Cleaned diatom frustules coated with 10 nm of various metals and examined at 20 keV: (a) gold, (b) aluminum, (c) copper, (d) silver, (e) chromium, (f) gold–palladium, (g) titanium, (h) tin, (i) calcium.

refractory wire. Substances not available in wire form are available as powders or chips and can conveniently be evaporated from refractory crucibles or from boats made of high-melting-point metals. Carbon is most commonly available as rods made from compressed spectroscopically pure graphite. Care should be taken to ensure that the evaporant does not alloy or form compounds with the refractory substance. Most metals with a melting point below 2000 K can be evaporated from a support wire or boat made from a refractory metal such as tungsten, molybdenum, or tantalum. Such a support should be a good electrical conductor, have a very low vapor pressure, and be mechanically stable.

13.7.4. Evaporation Techniques

When a pressure of about 10 mPa has been reached, the refractory support wire and, where appropriate, the carbon rods, can be heated to a dull red color. Gentle heating causes a sharp rise in the

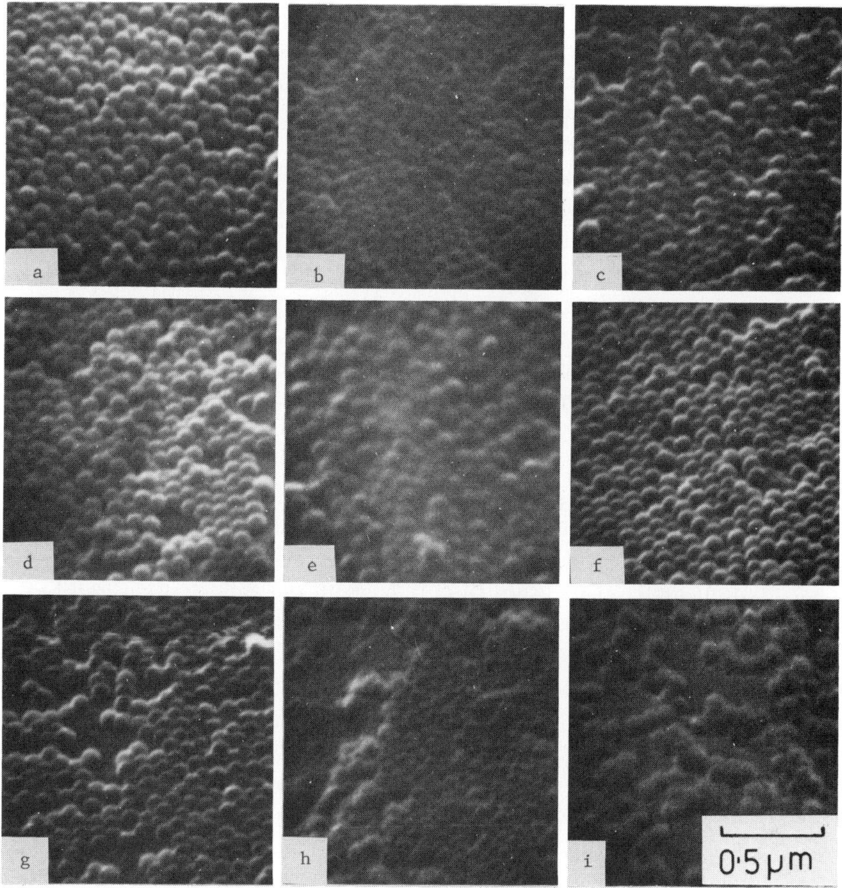

Figure 13.13. Polystyrene latex spheres 0.109 μm in diameter coated with 10 nm of various metals and examined at 20 keV: (a) gold, (b) aluminum, (c) copper, (d) silver, (e) chromium, (f) gold–palladium, (g) titanium, (h) tin, (i) calcium.

vacuum pressure brought about by outgassing and removal of residual contamination. This initial outgassing is particularly important with carbon rods, as it removes any organic binders used in their manufacture. Once the outgassing is complete, the current is turned down and pumping continued. If a carbon coating is to be applied to the specimen before metal coating, this is best done at a pressure of about 10 mPa. During coating, the specimens are rotated and tilted to give an even coating on all surfaces.

Following deposition of the carbon, the vacuum pressure is decreased between 1.0 mPa and 100 μPa. An electric current may now be passed through the tungsten wire holding the metal to be evaporated. This process should be done carefully and should be observed only though the type of dark glasses used by welders. It is best to gradually increase the current until the tungsten wire or the tips of the carbon rods just begin to glow and then back off a small amount. This allows the metal evaporant to heat up slowly and melt as the tungsten wire current is slowly increased.

The metals commonly used in evaporation form a molten sphere in the V of the tungsten wire. The molten sphere should be allowed to remain for a few moments to remove any residual contamination (which may be monitored as a transient rise in pressure). The current should then be further increased until the sphere of molten metal appears to shimmer and rotate. When this point is reached, evaporation has started and the shutter may be opened to expose the now-rotating specimen to the evaporation source. In order to achieve a uniform coating on complex sculptured specimens, it is essential that they be rotated rapidly (6–8 rps) during coating because the evaporating material travels in straight lines. The ideal arrangement is to rotate the specimen in a planetary motion while continuously tilting it through 180°.

This slow heating, melting, and evaporation of the metal evaporant is most important, and most particularly so with aluminum, which alloys with tungsten. If the source is heated too rapidly, the metal evaporant melts at its point of contact with the tungsten wire and falls off. However, the speed of coating is important in obtaining films of good quality, and the higher the speed of evaporation, the finer the structure of the layer. This optimal high speed of coating must be balanced against the higher thermal output from the source and the increased chance of the wire support alloying with the substrate and melting.

The thickness of the evaporated film can be measured by a number of techniques which will be described later, the most convenient of which is a quartz crystal thin-film monitor mounted inside the vacuum chamber. The thickness deposited also depends on the particular specimen being studied and the type of information required from the sample. It has usually been considered necessary to apply sufficient metal to give a continuous film, since it has been assumed that only a continuous film would form a surface conductive layer on a nonconductive sample. Although the physical nature of the charge transport in particulate films is not clearly understood, recent work suggests that particulate layers may be a useful method for SEM specimens because discontinuous metal films exhibit a significant DC conductivity. Paulson and Pierce (1972) have shown that discontinuous films can conduct a limited current by electron tunneling between evaporated island structures and suggest that such discontinuous films may be useful for the examination of nonconductive specimens viewed at very low current. Specimens have been successfully examined at 20 keV that have been coated with only 2.5 nm of gold, which is unlikely to have formed a continuous layer (Fig. 13.14).

The color of the layer on a white card or glass slide can give a quick practical estimate of thickness. For most specimens, a carbon layer visible as a chocolate color and a gold layer which is of a reddish-bronze color by reflected light and blue-green to transmitted

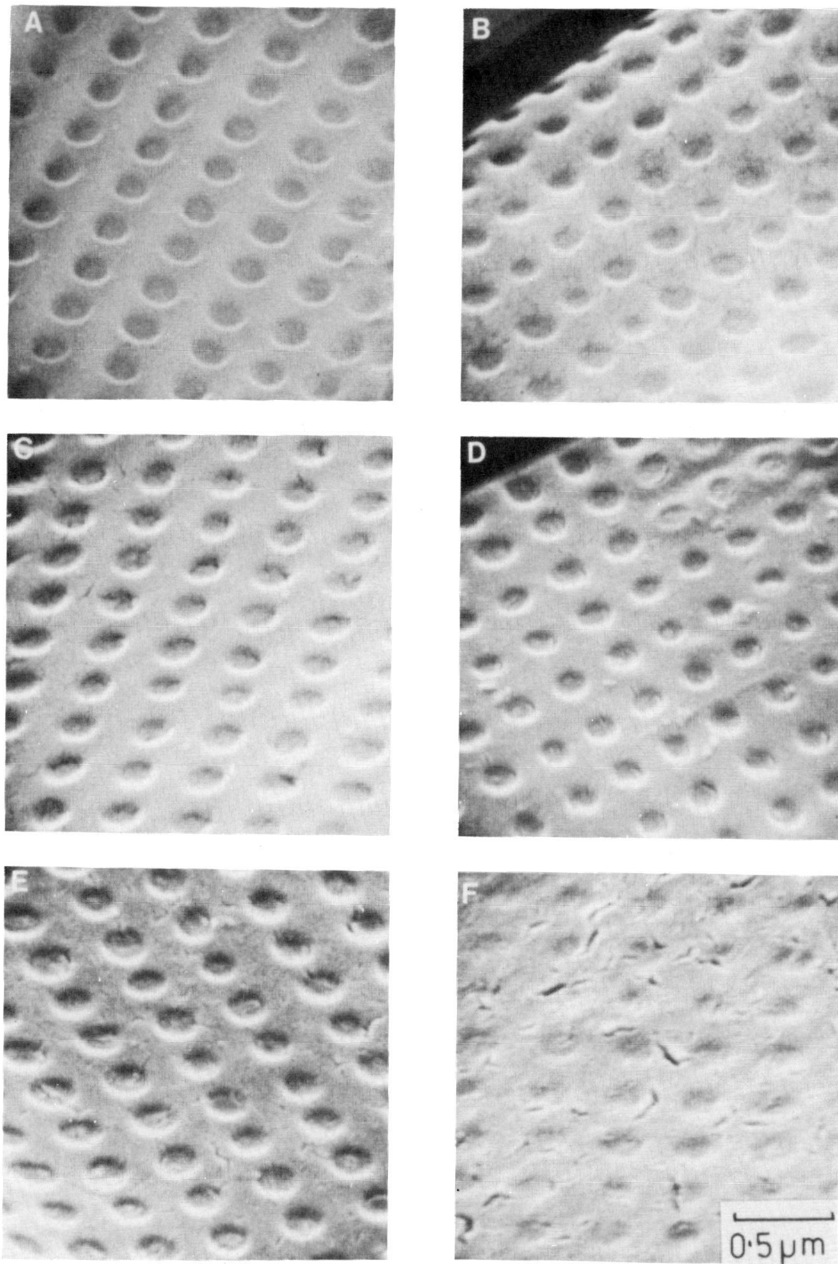

Figure 13.14. Cleaned diatom frustules coated by evaporation, and increasing thicknesses of gold examined using secondary electrons at 30 keV: (a) 2.5 nm, (b) 5.0 nm, (c) 10 nm, (d) 20 nm, (e) 50 nm, (f) 100 nm. Note that the very thin layers give optimal information, whereas the thicker layers obscure the surface of the specimen.

light are sufficient. For aluminum coating, sufficient metal is deposited when the layer is a deep blue to transmitted light. Once these parameters have been worked out for a particular evaporation unit with a fixed geometry of specimen and source it is necessary only to cut off a standard length of evaporant metal wire for each coating.

13.7.5. Artifacts Associated with Evaporative Coating

If the coating procedure has been carried out properly, artifacts are rarely seen. Some of the possible causes of artifacts are discussed below.

Heat of Radiation. The intensity of thermal radiation reaching the specimen depends on the source of heat and the source-to-specimen distance. The heat of radiation can be diminished by using a small source or by moving the specimen farther away. The best practical solution is to use a small source at high temperature, have at least 150 mm between the source and the sample, and employ a mechanical shutter to control the process. Provided the specimen is adequately shielded from the target and the shutter opened only at the working temperature, little damage is likely to occur.

Thermal artifacts appear as minute holes, surface distortions, and smooth, micromelted areas on inclined fracture faces of biological and organic materials that are bombarded vertically. If specimens are being damaged by heat radiation, the damage can be diminished by placing a cold plate with an aperture over the specimen.

Contamination. Contamination is due primarily to dirt and volatile substances in the vacuum system being deposited on the specimen, and it is for this reason that care must be taken to clean the system properly before use. The most effective way to reduce this problem is to surround the specimen with a cold surface. However, this is probably unnecessary except for very-high-resolution studies. Contamination may be recognized as uneven coating and hence charging, as small randomly arranged particulate matter, and in the most extreme situations, as irregular dark areas on the specimen.

Agglomeration Effects. Agglomeration of the evaporated material occurs to some extent with most metal coatings and is a result of uneven deposition of the evaporated metal. Agglomeration occurs if the cohesive forces of the film material are greater than the forces between the film molecules and the substrate. Because of geometrical effects, rough surfaces are particularly difficult to coat evenly, and it is inevitable that parts which protrude receive more coating then crevices and holes.

Film Adhesion. Poor film adhesion is associated with hydrocarbon and water contamination, and, in the case of plastics, with the presence of a thin liquid layer of exuded plasticizer. Discontinuous and poorly adhesive films are recognized by a "crazed" appearance and have a tendency to flake easily. In the microscope there are variations in the image brightness, and charging occurs on isolated "islands" of material not in contact with the rest of the film.

13.7.6. Low-Vacuum Evaporation

In low-vacuum evaporation, carbon is evaporated in an atmosphere of argon at a pressure of about 1 Pa. The carbon atoms undergo multiple collisions and scatter in all directions. This technique is useful for preparing tough carbon films and for coating highly sculptured samples prior to analysis by x-rays, cathodoluminescence, and backscattered electrons. However, its general usefulness for SEM specimens is questionable, particularly as the yield of secondaries from carbon is very low. There is little doubt that the multiple scattering and surface diffusion of the carbon allows it to more effectively cover rough specimens, and for this reason it would be a useful method to use if a sputter coater was not available.

13.8. Sputter Coating

Although sputter coating has been known for a long time, it is only recently that it has become more widely used for producing thin films for electron microscopy. In the process of sputtering, an energetic ion or neutral atom strikes the surface of a target and imparts momentum to the atoms over a range of a few nanometers. Some of the atoms receive enough energy in the collision to break the bonds with their neighbors and are dislodged. If the velocity imparted to them is sufficient, they are carried away from the solid (Wehner and Anderson, 1970).

The glow discharge normally associated with sputter coating is a result of electrons being ejected from the negatively charged target. Under the influence of an applied voltage, the electron accelerates toward the positive anode and may collide with a gas molecule, leaving behind an ion and an extra free electron. The glow discharge is located some distance from the target. The positive ions are then accelerated toward the negatively biased target where they cause sputtering. The sputtered target material leaves the surface in all directions, and herein lies the single advantage of this coating procedure. The multidirectional trajectories of the sputtered target atoms, some of which are further scattered by interaction with other target atoms and the bombarding gas ions, mean the specimen is coated from all angles. The nucleation and growth of thin films deposited on a carbon substrate by sputtering has been studied by Kanaya *et al.* (1988). Nucleation theory predicts that nuclei become stable when they contain a few atoms. Since the smallest clusters observable by microscopy contain at least single atoms or pairs of atoms, some growth must have occurred in any detectable deposits. It was found that monolayer films consisting of single atoms, pairs, or several-atom clusters are obtained at the optimum amount of incident atoms. At high accelerating voltages, many electrons are ejected per

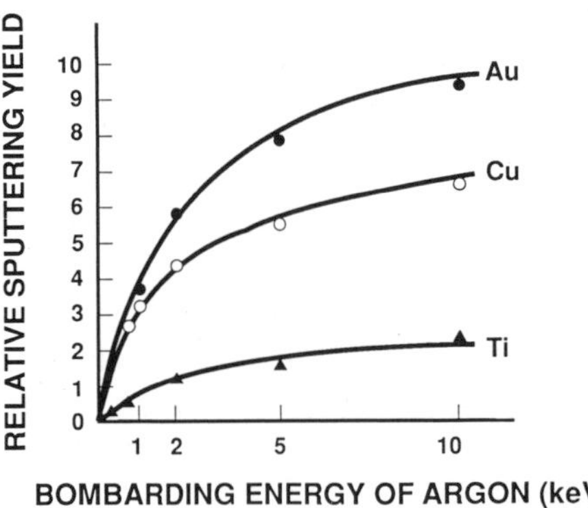

Figure 13.15. Relation between relative sputtering yield for various target materials and the bombarding energy of argon.

impinging ion, and these electrons have sufficient energy to damage delicate targets.

A number of factors affect the deposition rate. The sputtering yield increases slowly with the energy of the bombarding gas ion (Fig. 13.15). The current density has a greater effect than the voltage on the number of ions striking the target and thus is the more important parameter determining the deposition rate (Fig. 13.16). Variations in the power input can have a dramatic effect on the properties and composition of sputtered films; for example, with increasing power levels, aluminum films became smoother, with fewer oxide particles. As the pressure of the sputtering system is increased, the ion density also increases. There is a linear increase in current, and hence sputtering rate, between pressures of 3 and 20 Pa. But because there is an increased tendency at higher pressures for the eroded material to return to the target, a compromise pressure of between 2 and 4 Pa is commonly used.

Impurities in the bombarding gas can appreciably reduce the deposition rate. Gases such as CO_2 and H_2O are decomposed in the glow discharge to form O_2, the presence of which can halve the deposition rate. The presence of oxygen in the sputtering environment can also give rise to free radicals which may damage organic samples. Deposition rate decreases with an increase in the temperature of the specimen, although this phenomenon may not be peculiar to sputter coating. Finally, the

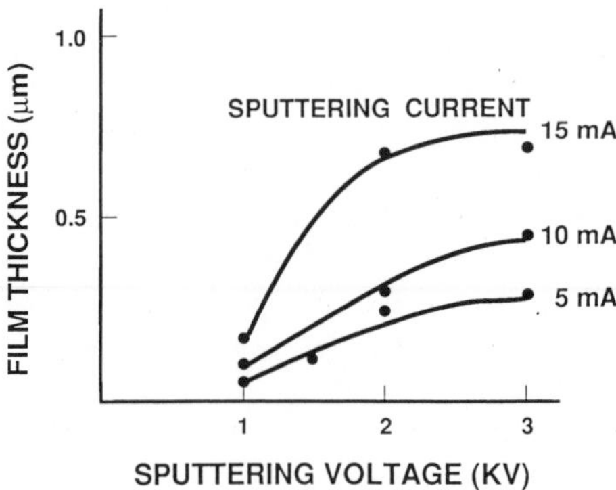

Figure 13.16. Relation between film thickness and sputtering voltage for sputtering currents of 5, 10, and 15 mA.

deposition rate is higher the closer the target is to the specimen, but this also increases the heat load on the specimen. The ejected particles arrive at the substrate surface with high kinetic energy as either atoms or clusters of atoms, but not as vapor. There is some evidence that the sputtered atoms have sufficient energy to penetrate one or two atomic layers into the surface on which they land.

The sputtering process may be carried out in several different ways. These include diode sputtering, plasma–magnetron sputtering, triode sputtering, radio frequency sputtering, ion-beam sputtering and Penning sputtering. Radio frequency sputtering is mostly confined to industrial coating procedures, and triode sputtering, which was devised as a means of lessening damage to samples, is now no longer used as a laboratory coating method. Of the remaining methods, diode and plasma–magnetron sputtering are used for most routine SEM coating procedures and ion-beam and Penning sputtering are used in high-resolution studies. Each of the four methods will be briefly discussed in turn.

13.8.1. Diode or Direct-Current Sputtering

This is the simplest, most reliable, and most economical type of sputtering and is the basis of a number of commercially available dedicated instruments as well as sputtering attachments for use in thermal evaporators. These instruments, which operate between 1 and 3 keV, are sometimes referred to as diode sputtering units as well as DC sputtering units. A DC sputter-coating unit consists of a small bell jar containing the cathode target and a water-cooled specimen-holder anode, which sits on top of a control module containing the vacuum gauge, high-voltage power supply, leak valve, and timer (Fig. 13.17). The unit is attached to an independent rotary pump used to maintain the working vacuum of 2–4 Pa. The detailed mode of operation and application of this type of instrument has been described by Echlin (1978). One of the potential problems of this type of coater is that delicate specimens can be thermally damaged. This problem can be overcome to some extent by intermittent sputtering or by cooling the substrate to circa 263 K with a Peltier stage. There is general agreement that low specimen-surface temperatures favor high nucleation density resulting in finer-grain films and smaller film thickness. We will return to low-temperature coating in Section 13.9.2. Diode sputter coating, with or without substrate cooling, still remains a popular way of applying a coating layer to a nonconducting sample. In addition to applying thin films to samples, diode coaters may be easily modified for glow-discharge activation of carbon support films for use in the TEM of macromolecules (Benada and Pokorny, 1990). Although the basic instrumentation has remained unchanged for the past 15 years, there is a tendency to automate the coating procedure. One can view this only as a mixed blessing. For the novice and for a laboratory with a high throughput of standard samples, automation is a distinct advantage, as it allows relatively inexperienced operators to obtain reasonably good and, more important, reproducible results. However, for someone interested

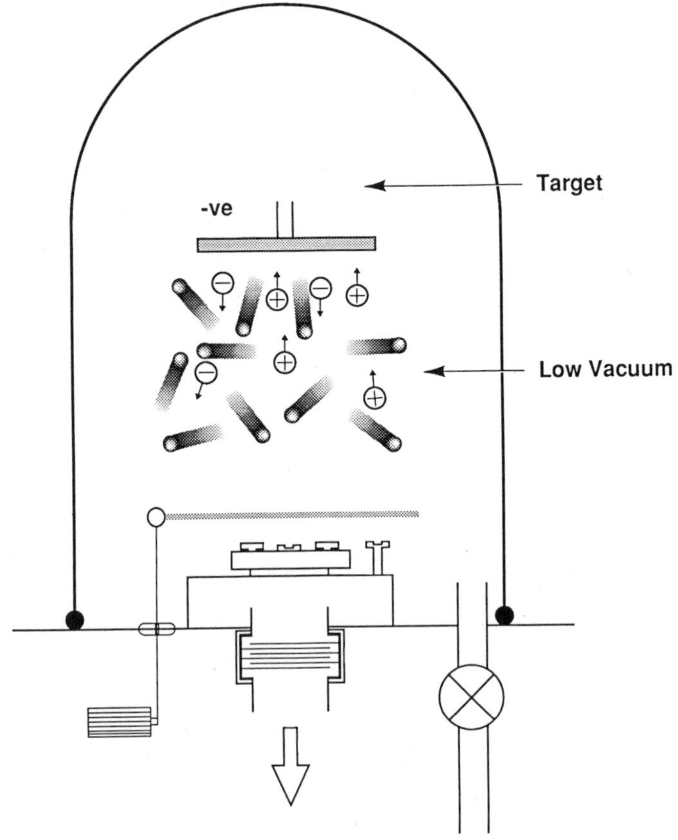

Figure 13.17. Diagram of a diode sputter coater.

in using the full capabilities of the SEM, fully automated coaters do not allow very much leeway in varying the basic coating parameters.

There has been little attempt to diminish the high contamination rates of some sputter coaters although it could be argued that, for routine SEM at low magnifications, a small amount of contamination is of little consequence. Nevertheless, it is important to use pump oils with a low back-streaming coefficient, to use rotary pumps with sufficient capability to enable an activated alumina trap to be placed in the backing line, and to use high-purity (low hydrocarbon) argon as the sputtering gas. Kemmenoe and Bullock (1983) carried out a comparative structural analysis of diode and ion-beam sputter-coated films and concluded that for low-resolution SEM, where thicker films can be tolerated, it is probably best to diode-sputter noble metals at a high deposition rate, provided the specimens are not heat sensitive. They concluded that the small increase in grain size is outweighed by the marked improvement in film structure. Kemmenoe and Bullock found that the average grain size of the thinnest gold and platinum films was more or less independent of the mode or rate of deposition. However, as the film thickness increased, there were significant changes in both the size of the grains and the gross appearance of the film structure.

The heat problem of diode sputtering can also be overcome by redesigning the diode coater to incorporate devices to keep the specimen cool throughout the coating procedure (Panayi *et al.* 1977). The electron bombardment of the specimen is significantly reduced by replacing the disk-shaped target of the diode coater with an annular target. The older triode coaters incorporated a positively charged electrode to attract the electrons although this had little effect on reducing the thermal input into the specimen. The studies carried out by Echlin *et al.* (1980) on heat-sensitive samples have shown that provided attention is paid to minimizing contamination, it is better to work at lower accelerating voltages and sputter for a longer time at a lower sputtering rate, e.g., for the noble metals, about 4 nm/min.

13.8.2. Plasma–Magnetron Sputtering

Thermal damage to the specimen is effectively reduced by fitting a permanent magnet at the center of the target and an annular polepiece around the target. This arrangement deflects the electrons outside the periphery of the specimen holder. This sputter coater is referred to as a plasma–magnetron coater and has largely replaced the diode sputter coater. Plasma–magnetron or "cool" sputter coaters use permanent, high-flux, rare earth ceramic magnets to deflect the high-energy electrons generated in the glow discharge away from the specimen. The magnetic field can also be used to improve the efficiency of sputter coating by increasing the interacting path length of the high-energy electrons in the discharge. Various configurations for the magnets have been tried (Nockolds *et al.*, 1982; Echlin *et al.*, 1982; Robards *et al.*, 1981; and Echlin, 1978) but they all have the same effect. The electrical field of the cathode and the magnetic field combine to trap electrons in a ring close to the surface of the cathode. This maximizes the sputtering efficiency of the target while minimizing the thermal input into the specimen. It also allows the glow discharge to be maintained down to a level of about 0.05 Pa, thus diminishing problems of contamination. A diagram and a photograph of a plasma–magnetron coater is shown in Figs. 13.18 and 13.19. The plasma–magnetron sputter coater is now the most commonly used device for coating specimens for SEM. The coater operates at between 100 and 1000 V and can be used to sputter the noble metals, nickel, and chromium. Aluminum may be sputtered provided pure argon (O_2, ≤ 2 ppm) is used. For a variety of reasons (Echlin *et al.*, 1985), it is not possible to sputter carbon by either diode or plasma–magnetron laboratory coaters. The sputtering rates for noble metals can be as high as 1 nm/s with minimal thermal damage to the sample. Studies by Echlin *et al.* (1985) have shown that it is possible to sputter coat noble metals and refractory metal at voltages as low as 100–250 V. The average particle size is 1–2 nm, significantly smaller than what can be obtained with diode sputter coaters and with plasma–magnetron coaters operating at higher voltage. The advantage of reduced particle size is offset by the longer coating times needed for refractory metals which, in turn, could give rise to sample heating and contamination.

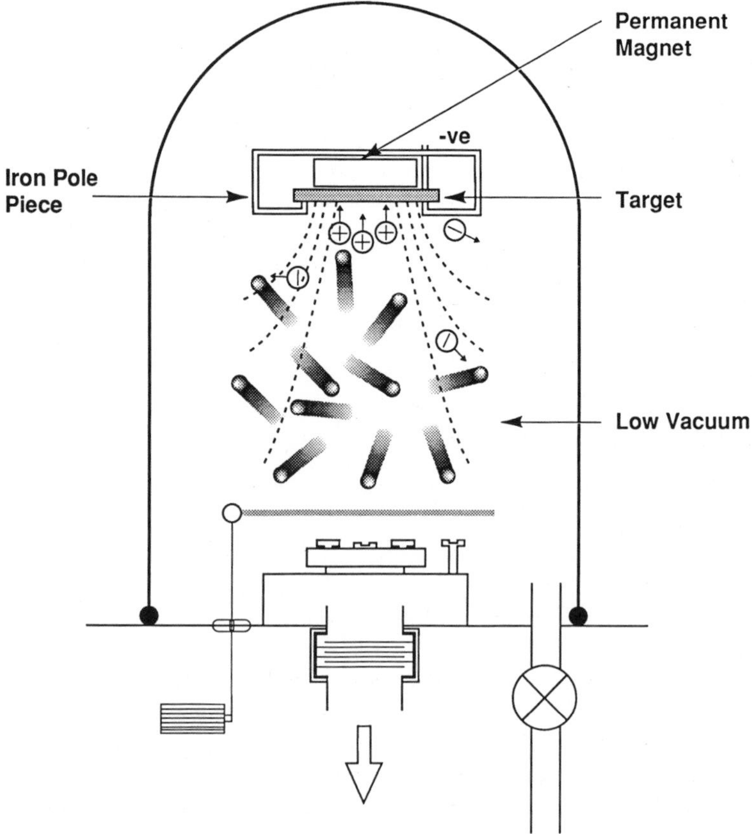

Figure 13.18. Diagram of a plasma–magnetron sputter coater.

Both diode sputtering and plasma–magnetron sputtering can be used in conjuncation with the process of sputter shadowing described by Colquhoun (1984). By placing a horizontal circular shield a few millimeters above the surface of the sample, it is possible to restrict the coating to a cone of vertical deposition. The distance between the shield and the surface and the diameter of the shield with respect to the surface determine the angle of shadowing at each point on the surface of the specimen. These dimensions are near or below the minimum free path of the sputtered atoms. Therefore multiple scattering is minimized. The specimens are thus coated in a manner analogous to rotary shadowing. Colquhoun (1984) has successfully applied this technique to shadow strands of DNA. The results are comparable to those obtained by nondirectional metal evaporation shadowing and rotary shadowing with minimal radiant and metal-deposition heating of the sample. This technique has the added advantage of providing on a single grid a spectrum of contrasting coating thickness with a heavy deposition at the periphery and a light deposition at the center. The simple device is easy to make and may be placed at the anode of any sputter coater.

13.8.3. Ion-Beam Sputtering

Ion beam sputtering is an effective technique for producing high-resolution films. The method employs a collimated or focused argon beam at 5–10 keV and an ion current of 300–500 μA. The ion source is a cold cathode saddle-filled source in which the plasma is contained within the source and only ions and neutrals escape through a hole in the cathode. The energetic ions strike the target atoms and impart momentum in elastic collisions; atoms lying near the surface of the target are ejected with energies in the range of 0–100 eV. These sputtered atoms are then deposited on the sample surfaces which have a line of sight to the sputter target. The vertical target holder, which can be carbon, a noble metal, or one of the refractory metals, is set at an angle of 30° from glancing incidence to the horizontal ion beam, 40 mm from the ion source. The target is rotated normal to the beam to ensure even erosion. The specimen is placed 20–25 mm from the target, and is simultaneously rotated in an axial direction and rocked between 0 and 90° to ensure omnidirectional coating. Unlike the two sputtering procedures already discussed, the bulk of the sputtered atoms travel in a straight line because of the much better vacuum (20–30 mPa) in the coating chamber (Fig. 13.20).

An advantage of this arrangement is that a field-free region exists between the target and substrate so that negative ions and electrons are not accelerated toward the substrate. Multiple coatings can be applied

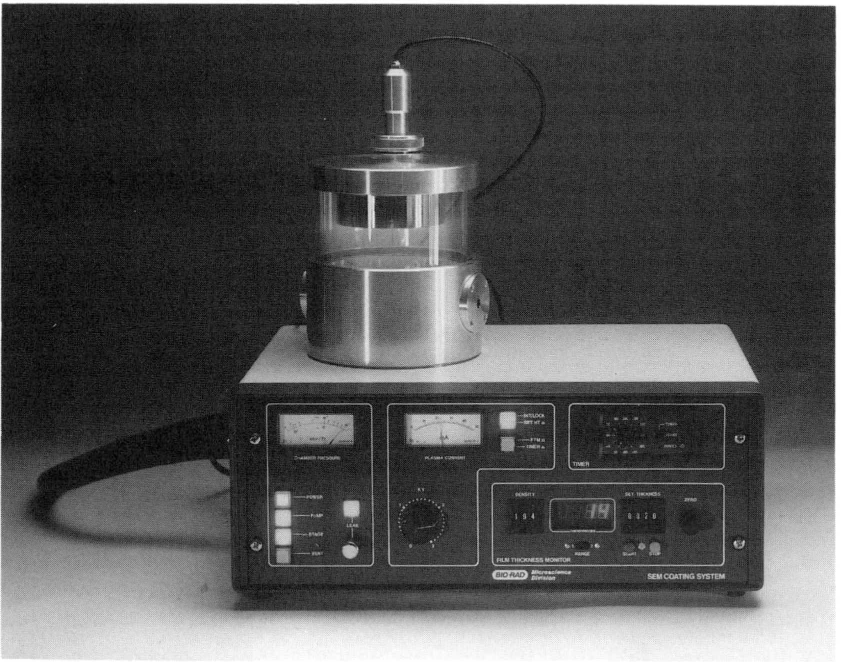

Figure 13.19. A plasma–magnetron sputter coater. (Courtesy VG Microstructure-Polaron.)

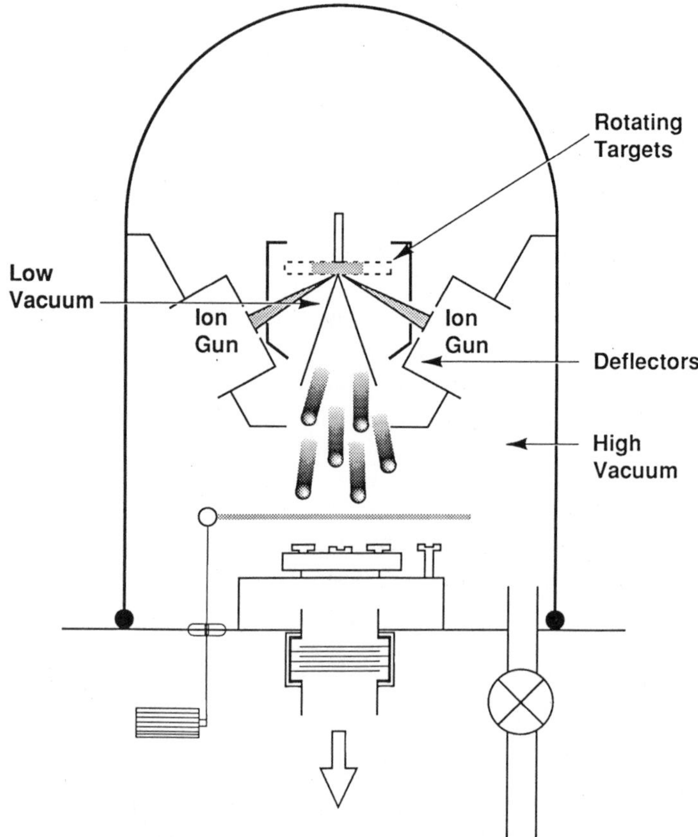

Figure 13.20. Diagram of an ion-beam sputter coater.

from different targets, providing care is taken to prevent cross-contamination of the targets during sputtering. If a nonconducting target is used, the buildup of positive charge can be suppressed by an electron flood gun. The coating rates are comparatively slow although there is some confusion over the actual rates. Franks *et al.* (1980), using optical absorption techniques, quote a figure of between 3 and 7 nm/min for noble metals, and this is within the same order of magnitude as plasma–magnetron sputter coating. The figure for carbon is much slower, between 0.5 and 1.0 nm/min. This rate is approximately 20 times faster than the rate reported by Kemmenoe and Bullock (1983) and Clay and Peace (1981), using the same equipment, who state the typical coating times for gold are between 20 and 60 min. If the Franks *et al.* figure is correct, it would mean that Clay and Peace and Kemmenoe and Bullock have been coating specimens with up to 0.4 μm of gold. An examination of their micrographs suggests that the actual coating layer is much thinner. The answer to this discrepancy probably lies in the inaccuracy of the quartz thin-film monitor used to measure the film thickness. The papers by Flood (1980) and Peters (1980) detail the reasons for this inaccuracy. Most workers are now aware of this inaccuracy, and experiments suggest a five- to tenfold error. Another reasón for this discrepancy lies in the fact that it is necessary to rotate and tilt the specimens during coating to ensure an even coating. As a general principle (assuming minimal sample irradiation with electrons, ions, and photons) high deposition rates and low substrate temperatures are recommended for smooth continuous films of minimum thickness, whereas low deposition rates and low substrate temperatures favor the formation of very thin films, which are unlikely to be continuous. On the available evidence, it appears that ion-beam sputtering is slower than diode sputtering, occurs at about the same rate as Penning sputtering, and takes less time to adequately coat a flat specimen than one with a rough surface. Ion-beam sputtered films have the advantage of being more uniform and of finer grain size than diode sputtered films.

Franks *et al.* (1980), Kemmenoe and Bullock (1983), and Clay and Peace (1981) provide a convincing series of pictures in which they compare specimens coated by diode sputtering, ion-beam sputtering, and thermal evaporation. In all instances, the ion-beam coated material has the finest grain size, which in some cases is below the resolving power of the microscope. In addition, ion-beam sputter coating is better than diode or plasma–magnetron sputter coating for the production of thick films for use in routine SEM. Kemmenoe and Bullock (1983) have examined 15-nm-thick ion-beam sputtered films and find they are continuous and without significant grain size. Films produced by ion-beam sputter deposition are initiated on a high density of nucleation sites as very fine equiaxial crystals, although the actual density of nucleation and crystallite sizes is very dependent on substrate and target materials.

Because the ion source operates at a high vacuum, it can be used to preclean substrates before coating, thus minimizing substrate surface contamination, although it must be appreciated that this precleaning also causes a small amount of surface etching. This surface cleaning can also be used in conjunction wtih thermal evaporation procedures, and it is usual to continue surface cleaning during the early phases of thermal evaporation to ensure the production of fine-grain films with better adhesion characteristics. Boone (1984) has modified a commercial ion-milling machine to deposit fine-grain films for electron microscopy. An ion-milling machine usually consists of two ion guns which simultaneously ion-etch two surfaces of a specimen. In the system described by Boone, only one surface is milled, and the other is coated with a thin film.

Another advantage of the ion-beam sputtering technique is that the specimen is not exposed to high-energy electrons or ions, and thus there are few or no heating artifacts. Because the coating is carried out at a much higher vacuum than that used for diode coating, there is virtually no multiple scattering. This means the specimen must be rotated or tilted during the coating procedure. Care must be taken with measurements of film thickness, as Kemmenoe and Bullock (1983) have shown an eightfold increase in the deposition rate on a stationary compared to a rotating specimen.

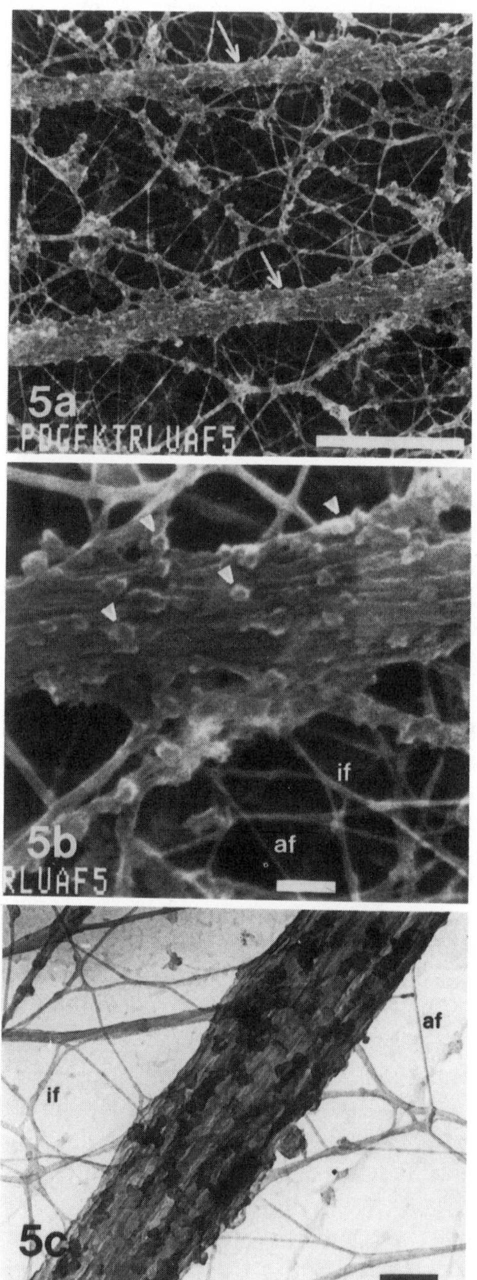

Figure 13.21. Ion-beam sputtered samples of the cytomatrix. All three micrographs are taken from the same specimen (cultured human fibroblasts on a gold grid). The specimen was treated as follows: rinsed in HBSS, extracted in a microtubule-stabilizing buffer containing 0.5% Triton X-100 for 2 × 10 min, fixed in buffered glutaraldehyde, treated with 1% aqueous uranyl acetate, frozen in liquid propane, freeze-dried, and then coated with 1.5 nm of tungsten within the freeze-dryer. Stress fiber bundles of actin are seen connecting various types of filaments (microtubules, intermediate filaments (if) and actin filaments (af). All scale bars correspond to 100 nm. The two upper micrographs (5a, 5b) are taken in the SEM; the lower picture (5c) is a TEM image taken with the JEOL 2000 EX microscope (Lindroth *et al.*, 1991).

In spite of the apparent disadvantage of extended coating times, ion-beam sputtering appears to be the answer to many of the problems associated with diode and plasma–magnetron sputter coating. The ability to use carbon as a target material is most useful, and the high-resolution images which have been obtained, even with gold as a target material, suggest this technique will find a wide range of applications from conventional SEM to high-resolution STEM. The studies of Kemmenoe and Bullock (1983) and Inoué (1983) show that for medium- and high-resolution SEM, ion-beam sputtered films of platinum give the best result. This technique produced very-fine-grain films (1–2 nm) relatively independent of film thickness, at least in the range of thickness one would normally use in SEM. Alternatively, 1-to-5-nm-thick films of ion-beam sputtered tungsten had a slightly smaller grain size, and would also provide films for high-resolution SEM, where contrast is less important than resolution. Figure 13.21 shows the high quality of the thin films which may be obtained using this method of thin-film formation.

The success of ion-beam sputtering can be attributed to a number of factors. When compared to diode and plasma–magnetron sputter coating, the deposition rate is slower, the environment and substrate surfaces are cleaner, and the specimen is not exposed to high-energy electrons and ions. These three factors all favor the formation of thin films with smaller grain size. For high resolution SEM and STEM, ion-beam sputtering and, as we will shortly see, Penning sputtering should be the methods of choice.

13.8.4. Penning Sputtering

Peters (1980; 1986a, b) gives details of Penning sputtering, which can be used to make very thin metal coatings for both high-resolution SEM and TEM. Penning sputtering (or, more correctly, "Penning discharge") is a form of ion-beam sputtering in which the target is placed inside the discharge area where the ions are generated. The specimen is shielded from the target by a small aperture, and the gun assembly is arranged to focus the argon ions on the target in a very narrow spot (Fig. 13.22). The actual vacuum in the sputter head is only a few millipascals, and the specimen is held at approximately 15 μPa during coating. This means that the vacuum pressure during coating is one order of magnitude lower than that used during ion-beam sputtering. Even when an ion current of 10 mA is applied, virtually no argon atoms leave the chamber. The absence of argon ions among the target atoms is an important feature of this form of sputtering. The positive ions and electrons generated in the plasma are deflected by large electrodes immediately outside the chamber, with the result that a virtually pure stream of target atoms is directed onto the specimen. The mean free energy of the target atoms is an order of magnitude greater than can be obtained by thermal evaporation, and unlike the situation in diode and plasma–magnetron sputtering systems, this primary energy is not reduced by scattering with residual gas atoms in the atmosphere. The high primary energy of the target atoms coupled with a low deposition rate means that metal is deposited onto the substrate with little lateral movement and that these anchored atoms act as nucleation centers establishing a horizontal crystal growth with minimum vertical growth. Peters (1984b) has found that the continuity of metal films depends on the mobility of the metal atoms. Increased substrate heating, applied as secondary energy during metal deposition,

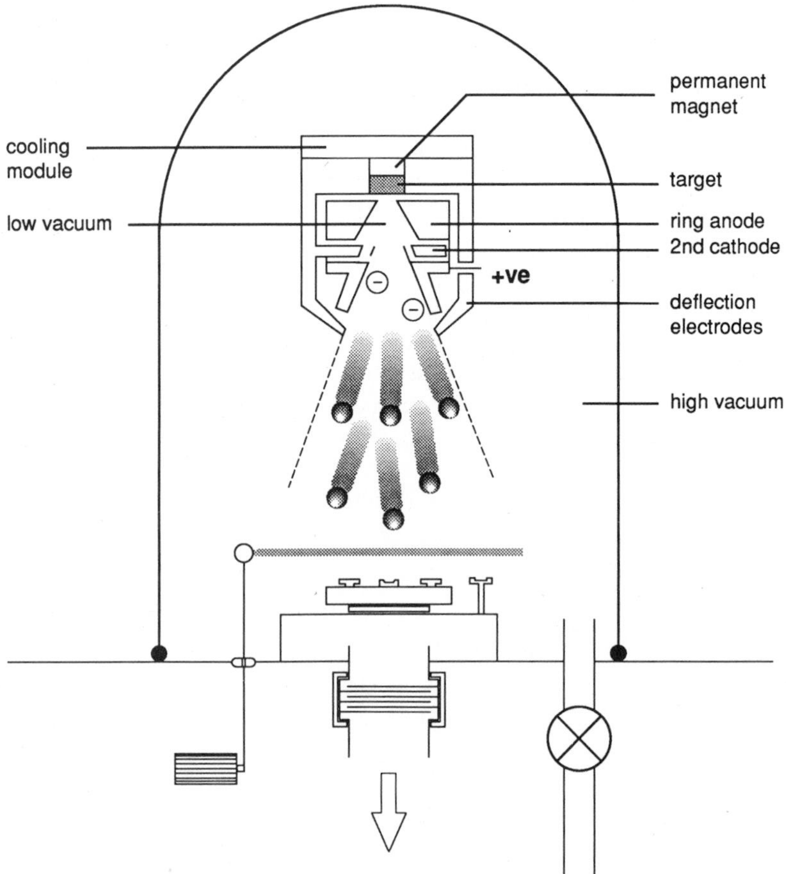

Figure 13.22. Diagram of a Penning sputter coater.

causes increased surface diffusion. Low surface mobility allowed extensive surface covering with coalesced small crystals. Increased surface mobility prevented coalescence and resulted in large crystallites. Peters also found that the thickness required to achieve film continuity increased with higher atom mobility. Penning sputtering can thus produce very thin films well suited for high-resolution SEM (Figs. 13.23 and 13.24). Although the Penning sputtering head works at 1–2 mPa, the specimen may be held at 50 μPa without a cold trap and 5 μPa with a cold trap. The high-vacuum conditions and the presence of a large liquid nitrogen–cooled trap very close to the specimen result in very low contamination rates, another prerequisite for high-resolution film formation.

Because the mean free path of the target atoms is many times larger than the target–substrate distance, the multiple scattering advantage of diode coating is lost, and it is necessary to rotate and tilt complex sculptured specimens during coating. By conventional sputter coating standards, the deposition rates are slow: for the noble metals, about 2 nm/min and for high-melting-point refractory metals, about 0.5 nm/min. But Penning sputtering is not a conventional sputter coating

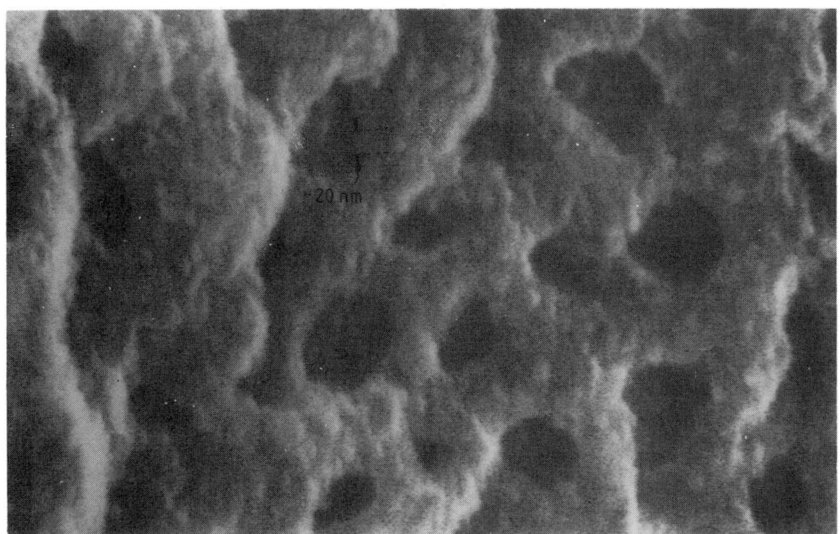

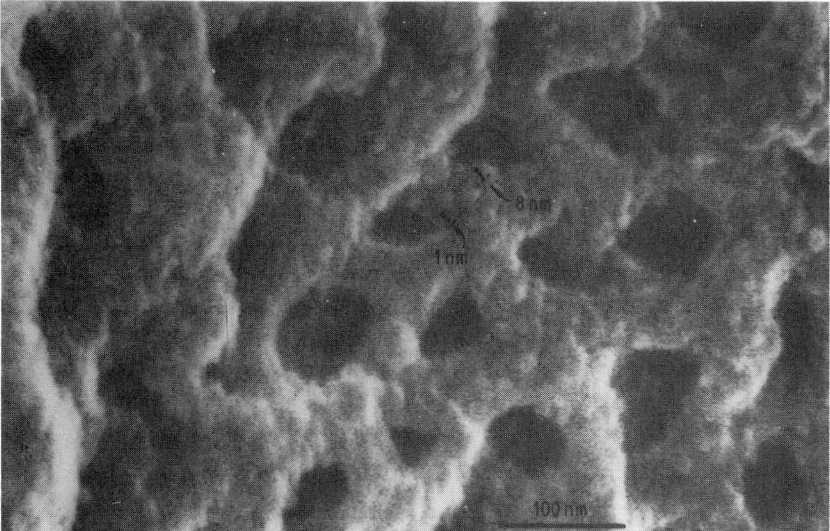

Figure 13.23. Penning sputtering. High-quality, high-magnification stereopair image of a cell surface coated with a 1-nm continuous thin Cr film using Penning sputtering (Peters, 1985).

technique, and there would be little advantage gained in using it for preparing specimens for conventional low-resolution SEM. The technique is designed for high-resolution microscopy and has been successfully used to apply ultrathin (about 1.0 nm) films of molybdenum, tantalum, and tungsten to tubulin filaments and ferritin macromolecules. In a study on retinal rod cells (Peters *et al.*, 1983), 3–5 nm of gold was used for routine survey micrographs, whereas 1–2 nm of niobium or chromium was used for high-resolution observations.

Although individual metal atoms and monoatomic layers may be

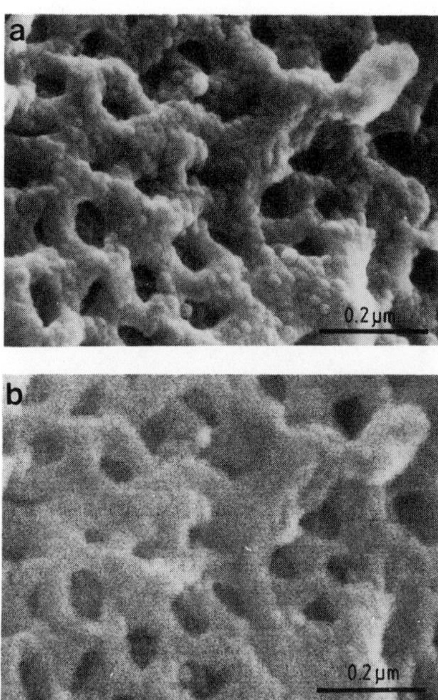

Figure 13.24. High resolution coating procedures: (a) Secondary-electron image and (b) backscattered-electron image of a cell surface coated with a 5-nm continuous thin tantalum layer using Penning sputtering (Peters, 1985).

imaged in STEM instruments, it is necessary to make thin films several atomic layers thick in order for the samples to have sufficient contrast to be imaged by other modes of electron microscopy. Topographic resolution on organic specimens with conventional secondary-electron imaging is rather low, because of the multiple scattering within the specimen. Backscattered electrons (BSE) give rise to secondaries which degrade the resolution of the final image. Peters (1984a, b) showed that these backscattered effects could be virtually eliminated from the secondary-electron signal by shielding the final polepiece with a carbon-coated aluminum plate. Peters considered that since the true secondary electrons are generated in the probe cross-secion at the surface of the specimen, the BSE absorption plate should give an improvement in the topographic resolution. The use of the BSE absorption plate coupled with thin (2.0 nm) Penning sputtered layers of low-atomic-number metals such as niobium or chromium enabled Peters *et al.* (1983) to obtain ultrahigh-resolution images from organic samples.

There is still some discussion regarding the minimum thickness of metal which is necessary for various modes. Peters considers 0.5–1.0 nm to be sufficient for conventional TEM in the bright field mode, 2–4 nm for the low-loss SEM mode, and 1–2 nm for the secondary-emission SEM mode. Although such thin films are discontinuous, they are effective in reducing charging and provide sufficient contrast for imaging. It is doubtful whether such a thin layer would be effective in reducing the thermal effects that may be brought about by irradiation with a high-energy beam of electrons. As a result, it would be useful to have the specimen on a cold stage.

13.8.5. Sputtering Techniques

The general description which follows applies to plasma–magnetron sputter coaters, as they are most frequently used to apply thin films for SEM studies. There will be variations of these procedures depending on the make of sputter coater being used.

It is important to ensure that the coater is fitted to an adequate supply of clean, dry argon or, where appropriate, another noble gas. A small trace of nitrogen in the argon probably does not matter, but it is important that the gas be free of water, carbon dioxide, and oxygen. Traces of water vapor, which are recognizable by a blue tint in the glow discharge, can easily be removed by passing the gas through a column of desiccant. The final traces of volatile material may be removed from the target by bombarding it with a current of 20 mA at 1.5 kV at a pressure of between 2 and 8 pA. The target can be considered clean when there is no degradation of vacuum when the high-voltage discharge is switched on.

Specimens are placed on the specimen table and the unit pumped down to about 10–15 Pa (10^{-1} torr) using a two-stage rotary pump fitted with an activated alumina foreline to prevent oil backstreaming. It is most important not to let the unit pump for a long time at the ultimate pressure that may be obtained by the rotary pump, because this will cause backstreaming, resulting in contamination. Because of this it is advisable to have the argon leak valve slightly open to ensure a continual flow of inert gas through the system, giving a pressure of about 6–7 Pa. If the unit is fitted with a water-cooled or, better still, a Peltier module cold stage, this should be turned on and the specimens cooled to the working temperature.

One of the commonest sources of poor pumping performance in both sputter and evaporative coaters is the continual outgassing of the specimen and the adhesive used to attach it to the support. It is recommended that biological and organic material, after it has been dried and attached to the specimen support, be placed in a 310-K vacuum oven overnight to ensure that all volatile substances are removed before coating. This same procedure should be followed for any sample which has been attached to the specimen stub using a volatile adhesive. It is usually necessary to flush the coating five or six times with an inert gas before proceeding with sputtering.

Once the specimen has been cooled and an adequate pressure has been reached, the sputter coating may proceed. The system is pumped down to about 2 pA and the high voltage turned on. The argon leak valve is opened until a plasma discharge current of 12–15 mA is recorded at a pressure of 2–3 Pa. The timer is set and coating continued until the desired thickness has been deposited on the specimen. The coater should then be back-filled with a dry gas (N_2 or Ar) once atmospheric pressure is reached, and the specimens should be removed and stored in a clean, dry container. The best practice is to use the samples as soon as possible after they have been coated.

It is possible to sputter using nitrogen as an alternative to argon. However, extended discharge times or higher plasma currents are necessary to give the same thickness of coating. Sputter coating in air should be avoided because the presence of water vapor, carbon dioxide, and oxygen gives rise to highly reactive ions which can quickly degrade the specimen.

13.8.6. Choice of Target Material

Platinum or gold–palladium targets have been found to be satisfactory for routine specimen preparation for the SEM (see Table 14.10). Noble-metal alloys are particularly useful, for unlike evaporative methods, alloys sputter congruently, and the mixed coating layer limits the epitaxial growth of crystals on the sample surface. Targets are available in most of the other noble metals and their alloys as well as elements such as nickel, chromium, copper, and aluminum. There are differences in the sputtering yield from different elements, and these must be borne in mind when calculating the coating thickness. The claims that carbon can be sputtered at low kilovoltage in a diode sputter coater are probably erroneous. The deposits of "carbon" achieved are most likely to be hydrocarbon contaminants degraded in the plasma than material eroded from the target. Most plasma–magnetron coaters can be fitted with a small carbon-evaporation head in which carbon films are prepared by low-vacuum preparative methods (see Section 13.7.6). Sputtering with aluminum requires special attention. The oxide layer which rapidly forms on the surface of aluminum is resistant to erosion at low keV, and the rather poor vacuum makes it difficult to deposit the metal. The metal can be sputtered in a plasma–magnetron coater fitted with high-field-strength magnets to concentrate the plasma onto a small area of the target. In addition, it is necessary to have an improved, oxygen-free vacuum environment. For details of other targets, particularly those made by nonconducting materials and bombarding gases, the reader is referred to the book by Maissel and Glang (1970).

13.8.7. Coating Thickness

For a given instrumental setting, there is a linear relationship between the deposition rate and the power input (see Fig. 13.16). In practical terms, this means that the thickness of the coating layer is dependent on the gas pressure, the plasma current, the voltage, and the duration of discharge. It is difficult to accurately deposit thin films below 5 nm, and it is necessary to rely on evaporative or specialized sputtering techniques for such films. However, for most routine studies biological material requires about 15–20 nm of metal coating. Once one has

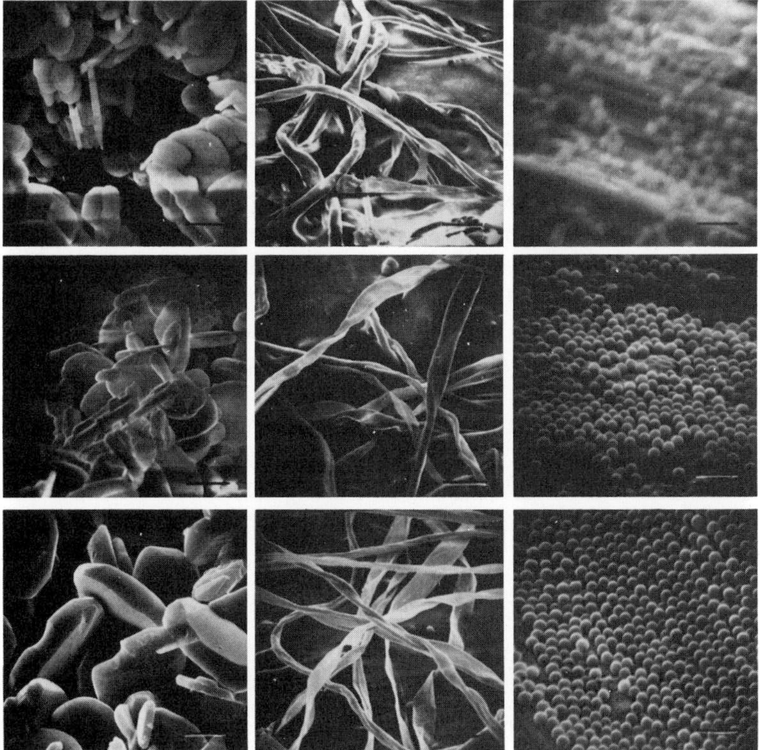

Figure 13.25. Comparison of uncoated, evaporative-coated, and sputter-coated nonconductors. Left, Al_2O_3; center, cotton wool; right, polystyrene latex spheres. Top row, uncoated; center row, evaporative-coated with 10 nm of gold; bottom row, sputter-coated with 10 nm of gold. Magnification bar = 1 μm.

established the ideal thickness for a given specimen, it is easy to accurately reproduce this on other samples.

13.8.8. Advantages of Sputter Coating

One of the main advantages of the technique is that it provides a continuous coating layer even on those parts of the specimen that are not in line of sight of the target. Figure 13.25 compares the major coating processes. The continuous layer is achieved, since sputtering is carried out at relatively high pressures, and the target atoms suffer many collisions and are travelling in all directions as they arrive at the surface of the specimen. Highly sculptured structures or complexly reticulate surfaces are adequately coated. This ability of the target atoms to "go around corners" is particularly important in coating nonconductive biological materials, porous ceramics, and fibers. Although complete coating can be achieved without rotating or tilting the specimen, and using only a single source of coating material, a rotation and rocking movement of the specimen and multiple target heads certainly improves the quality of coating. Providing the accelerating voltage is sufficiently high, it should be possible to sputter coat a number of nonconducting substances such as the alkaline halides and alkaline earth oxides, both of which have a high secondary-electron emission. The difficulty comes when such samples are transferred through air to the microscope, because surface layers rapidly oxidize. Similarly, it should be possible to sputter substances which dissociate during evaporation. Film-thickness control is relatively simple, and sputtering can be accomplished from large-area targets that contain sufficient material for many deposition runs. There are no difficulties with large agglomerations of material landing on the specimen, and the specimens may more conveniently be coated from above. The surfaces of specimens can be easily cleaned before coating either by ion bombardment or by reversing the polarity of the electrodes. The plasma may be manipulated with magnetic fields, providing greater film uniformity and reducing heating of the specimen. As mentioned earlier, it is not possible to sputter-coat carbon in a conventional plasma–magnetron coater. It is possible, however, to adapt a sputter coater to evaporate carbon at a low pressure (Fig. 13.26).

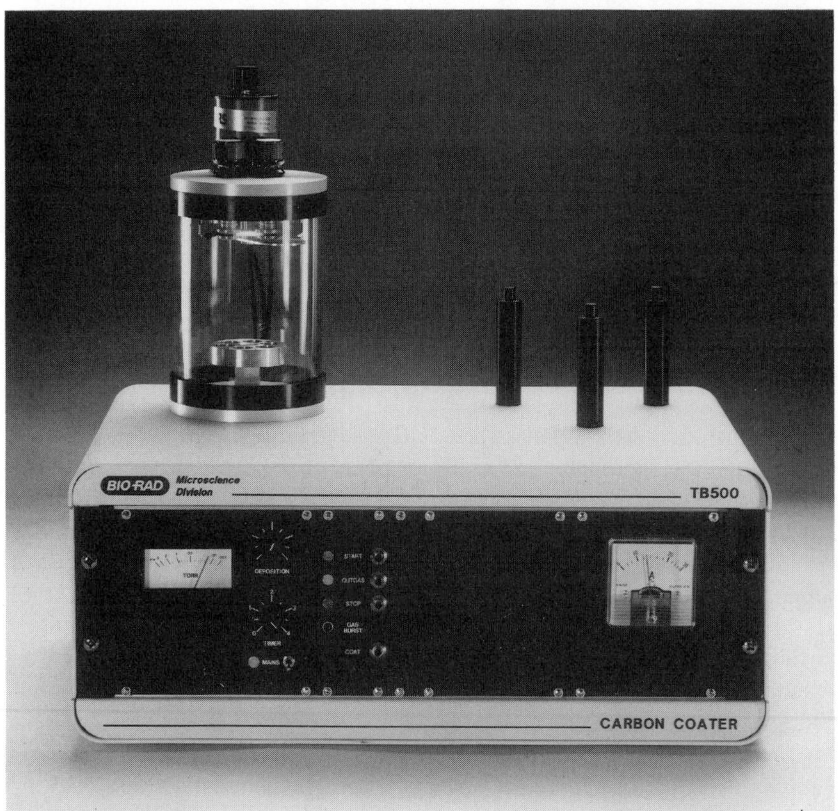

Figure 13.26. Diode sputter coater fitted with an evaporative carbon coater. (Courtesy VG Microstructure-Polaron.)

13.8.9. Artifacts Associated with Sputter Coating

Sputter coating of SEM samples has attracted some unfavorable comments, because in the hands of certain users it has resulted in thermal damage to delicate specimens and to surface decoration artifacts. There is little doubt that the earlier diode sputter coaters could cause damage to heat-sensitive specimens, particularly if coating was carried out for a long time at high plasma currents on uncooled specimens. The thermal damage problem has been greatly reduced with the advent of the plasma–magnetron or "cool" sputter coaters. A careful examination of the examples of decoration artifacts which have been reported in the literature reveals that in most cases scant regard has been paid to cleanliness during coating, and that the artifacts were due to contamination from backstreaming oil vapors or the use of impure or inappropriate bombarding gases. Nevertheless it would be improper to suggest that artifacts never occur with sputter-coated materials, and some of the more common problems are discussed below:

Thermal Damage. A significant rise in the temperature of the specimen can be measured during sputter coating. The sources of heat are radiation from the target and electron bombardment of the specimen. There is an initial rapid rise in temperature, which then begins to level out and, depending on the nature of the material being coated, may cause thermal damage. Depending on the accelerating voltage and the plasma current, the temperature rise can be as much as 40 K above ambient. However, as mentioned earlier, heating effects can be entirely avoided either by using the plasma–magnetron "cool" coater, where the heat input due to electron bombardment is only 200 mW, or by operating a conventional diode coater intermittently at low power input.

When thermal damage occurs, it is manifest as melting, pitting, and in extreme cases complete destruction of the specimen. While accepting that thermal damage can be a problem in sputter coating, in nearly all cases where this has been reported, it is due to the specimen being subjected to inordinately high power fluxes.

Decoration Artifacts. Leaving aside the now-obvious agglomeration of gold particles, this artifact becomes a serious problem only at high resolution. Peters (1985) has made an extensive study of this phenomenon and has shown that decoration artifacts are a problem in high-surface-mobility elements such as Au and Pt, which will form a continuous layer only at thicknesses between 5–10 nm. Peters was able to show that, even when thinner discontinuous layers were used, both gold and platinum particles moved from the sides and aggregated on the top of the surface molecules. This was not a problem with the much thinner continuous layers of Ta and Cr which have a low surface mobility.

Surface Etching. This is a potential hazard in sputter coating and may be caused either by stray bombarding gas ions or more likely, by metal particles hitting the surface with sufficient force to erode it away. It is possible to find very small holes in the surface of sputter-coated specimens examined at high resolution. It is not clear whether these are the result of surface etching or simply thermal damage.

Film Adhesion. Film adhesion is much less of a problem with sputter-coated films than with evaporated films and is probably due to the fact that the metal particles penetrate the surface of the specimen, forming a strong bond. However, sputter-coated samples should not be subject to wide excursions of temperature or humidity, both of which can give rise to expansion and contraction, with a consequent rupture of the surface film.

Contamination. Because of the low vacuum used in most sputter coaters, the problem of backstreaming from the rotary pumps, and the difficulty of placing effective cold traps in the backing lines, contamination can be a potentially serious problem, particularly if no foreline traps have been fitted. Many of the artifacts which have been described are probably due to contamination, and care should be taken in setting up and using the sputter coater.

13.9. Specialized Coating Methods

Although a description has been given of the general principles of evaporative and sputter coating, there are a number of specialized coating methods that should be discussed because they are applicable to both electron microscopy and x-ray microanalysis.

13.9.1. High-Resolution Coating

As the resolution of SEMs increases, greater attention must be given to the resolution limits that may be imposed by the coating layer, although the coating layer alone is not the sole determinant in preparing samples for high-resolution studies. These procedures are discussed elsewhere in this book and in a number of useful review papers (Peters, 1986a; Walther and Hentschel, 1989; Peters, 1986b). The processes of high-resolution microscopy include isolation of the object of interest, stabilization, conductive coating (discussed earlier), and then, finally, thin-film coating, which will be discussed here. For many instruments that can resolve about 10–15 nm, the granularity of the film is of little consequence, with the possible exception of the agglomeration of gold particles. Most SEMs are routinely operating in the 5–10 nm range, and many instruments can give between 1 and 3 nm resolution in the secondary-electron mode. Transmission electron microscopists have long been concerned about high-resolution films, as they are necessary for specimen support, for shadowing, and for making replicas of frozen and ambient-temperature fractured material. Thin films of chromium, gold–palladium, platinum, and zirconium have much finer grain size, and carbon, platinum–carbon, tungsten, and tantalum all have virtually no grain size when examined in the TEM. The graininess of the deposits generally decreases with an increase in the melting temperature of the coating material and with a decrease in the temperature of the substrate being coated. But it is not just simply a case of using coating layers with a fine granularity; the films also have to be thinner than the range of the critical signal electrons and to follow faithfully the surface features of the specimen.

The most convenient means of preparing thin films for medium-high-resolution (5 nm) SEM is to plasma–magnetron sputter coat 3–5-nm-thick films of platinum or gold–palladium onto specimens maintained below ambient temperature. The images generally appear satisfactory up to a magnification of 50,000X; above this magnification, smooth surfaces appear granular. Tanaka *et al.* (1991) obtained excellent high-contrast images up to 80,000X, using 3-nm-thick ion-beam-sputtered platinum.

There appears to be no single high-resolution coating procedure which can satisfactorily coat all samples, although most of the successful methods share the following characteristics:

1. High-melting-pont, low-surface-mobility metals are used.
2. The coating layers are very thin, 1–3 nm.
3. The specimens are maintained at low temperatures.
4. The coating is carried out in a clean environment.

The following few examples will help to illustrate the type of procedures which are being used. Lindroth and Sundgren (1989), Bell *et al.* (1988), and Lindroth *et al.* (1987) have used either a high-vacuum, plasma–magnetron sputter coater (Nockolds *et al.*, 1982) operating at 1 keV, 50 mA plasma current, and a pressure of 400 mPa, or an ion-beam coater operating at 1 keV, 8.5 mA filament current, and a pressure of 13 mPa. The best coating was with ion-beam sputtered tungsten. Peters (1985; 1986a) has used a Penning sputter coater inside a high-vacuum coater (1.3 mPa) and finds the best images at 250,000–400,000X were obtained with a 1–2 nm layer of Ta and to a lesser extent with a 1–2 mm layer of Cr (Peters and Carley, 1988). Discontinuous films of platinum were granular. Tanaka (1989); Tanaka *et al.* (1989), and Tanaka and Fukodone (1991) have used 1–2 nm Pt ion-beam-sputtered material up to a magnification of 250,000X; above this the platinum particles could be discerned. They find that both Ta and W gave poor secondary-electron images. Hermann *et al.* (1988) have deposited 0.9–2.7 nm-thick layers of Cr and Ge using electron-beam evaporation at a pressure of 13 μPa. During coating the sample was rotated and cooled to 188 K and the evaporation surface continuously moved through a predetermined 90° arc. The technique, referred to as double-axis rotary shadowing, gave results comparable to Penning sputter material up to magnifications of 380,000X. The Cr films were better then the germanium films. Probably the most spectacular high-resolution coating has been obtained by Wepf *et al.* (1991). These workers have deposited 1.5-nm-thick films of a platinum–iridium–carbon mixture by ultrahigh-vacuum electron-beam evaporation at 200 nPa onto samples rotated and cooled to 193 K. The composition of the coating material varies, but the best results were obtained with films containing at least 25% C. Quite remarkably, the films were stable at atmospheric pressure, unlike more conventional

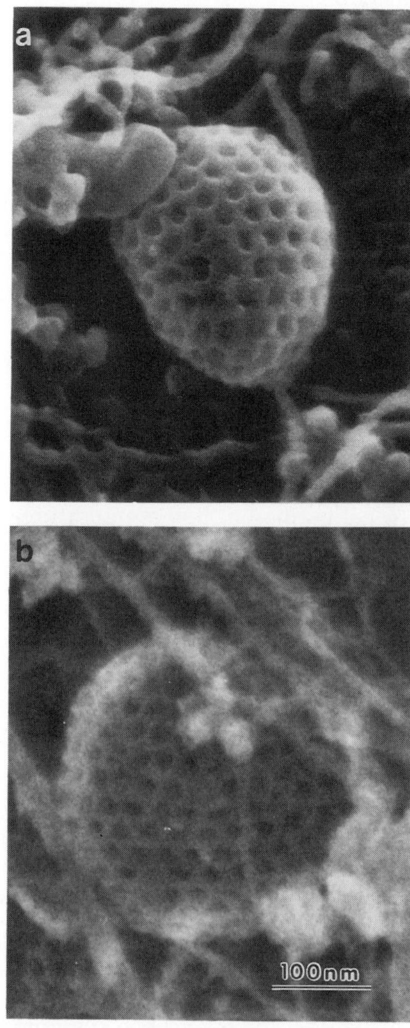

Figure 13.27. Clathrin-coated vesicles in cultured fibroblasts. (a) Metal-coated specimen. (b) Uncoated, conductively stained specimen. The samples were made by the Mitsushima's method (Tanaka *et al.*, 1989).

high-resolution films of Ta–W and Pt–C, which require an additional thin layer of carbon to stabilize the structure. Thse fine-granularity ultrathin films are proving useful for TEM, SEM, and STEM up to 350,000X magnification.

The thin-film coating method used for making samples conductive should be adapted to the size of the specimen. Plasma–magnetron sputter coating with Pt and Au–Pd is suitable for objects larger than cell organelles. Alternatively, as Tanaka (1989) has shown (Fig. 13.27) bulk-conductivity methods can provide images as good as those obtained by medium-resolution coating methods. In the size range between organelles and viruses, Penning sputtering, ion-beam sputtering, and

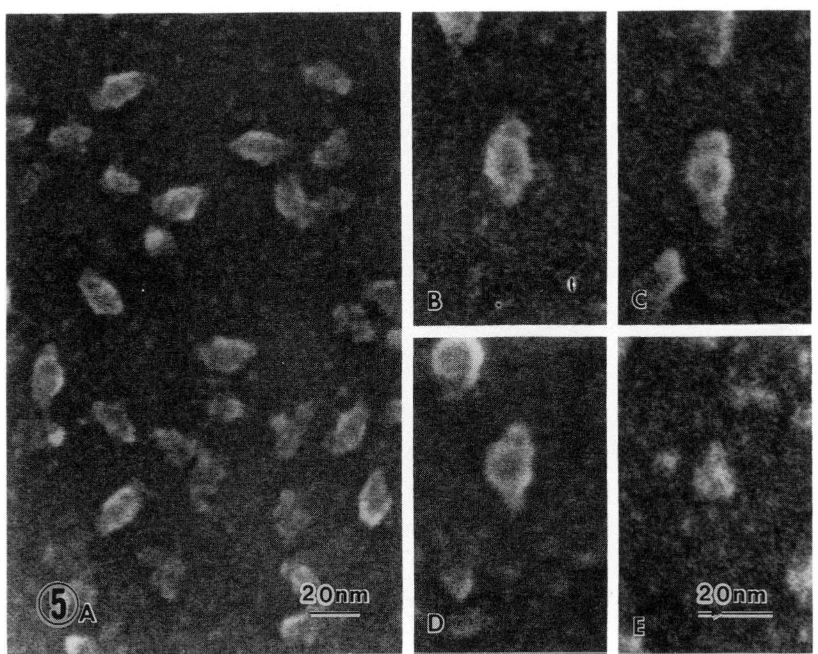

Figure 13.28. High-resolution SEM images of conductive stained bovine thyroglobulin molecules. The molecules appear as roughly spindle-shaped bodies (Nakadera *et al.*, 1991).

electron-beam evaporation of refractory metals such as W, Ta, and noble-metal mixtures should be used. For objects smaller than viruses, the best results have been obtained from uncoated material that had either been conductively stained with great care or carefully maintained on a conductive surface. Yasuda *et al.* (1989) have directly visualized uncoated antibody molecules attached to colloidal gold particles at a magnification of 1,000,000X, and Nakadera *et al.* (1991) have visualized at 800,000X uncoated haemocyanin, ferritin, and apoferretin all of which had been infiltrated with heavy-metal salts and then rapidly frozen and freeze-dried (Fig. 13.28). The samples were supported on conductive, carbon-coated carbon plates. Tanaka *et al.* (1991) have used similar procedures to directly visualize colloidal gold band molecules and cell-surface receptors (Fig. 13.29).

13.9.2. Coating Samples Maintained at Low Specimen Temperatures

There are two reasons for coating specimens at low temperatures: increased spatial resolution of the coating layer and low-temperature SEM, where specimens are studied under deep-frozen, fully hydrated conditions.

Low Temperatures for High Spatial Resolution. There is a general agreement that low specimen temperature favors high nucleation density, resulting in finer-grain films. Low temperatures also favor

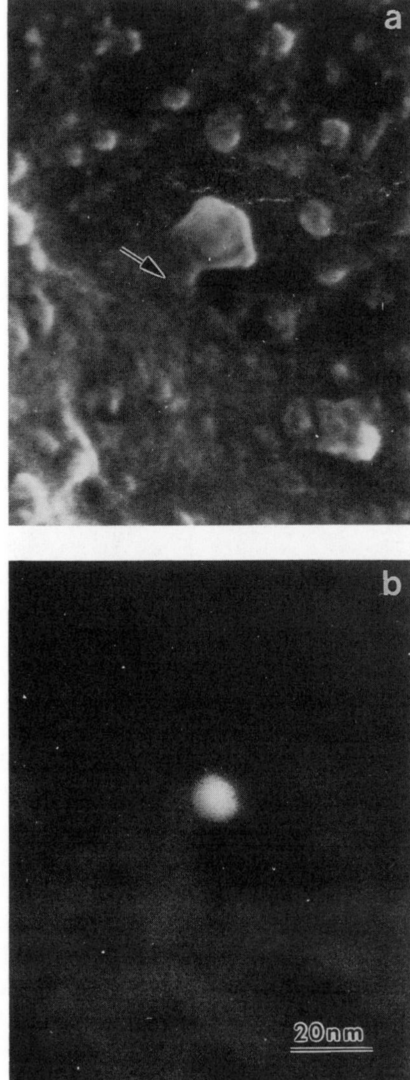

Figure 13.29. High-resolution SEM images of conductively coated cell surface receptors bound to an immunogold particle. (a) Secondary-electron image. (b) Backscattered-electron image of the same area. The bright spot corresponds to the gold in the particle (Tanaka *et al.*, 1991).

thinner films, and smaller dimensional changes on specimens that are being coated from an angle to emphasize small changes in relief. Slayter (1980) showed that the number of nucleation sites is inversely proportional to the substrate temperature and that twice as many crystallites developed per unit area on a sample held at 77 K when coated with platinum than one held at 300 K. This temperature effect is seen only when all other parameters have been optimized, and it is of paramount importance that the surface to be coated is clean and free of contamination. This is particularly important when frozen-hydrated tissue surfaces are being coated, as for example when a replica is made of a frozen-fractured cell. Haggis (1985) has shown that if the specimen temperature is too high, it is impossible to make a good replica because of the stream of water molecules leaving the frozen surface.

The grain size in plasma–magnetron sputtered films generally becomes smaller as the substrate temperature is lowered in the range 303–203 K. In all instances, the particle size was smaller whether examined by SEM, TEM, or low-loss imaging. These studies have been extended to the 203–103 K range by Echlin (1981) and Echlin *et al.* (1985). The coating was carried out in a liquid nitrogen–cooled sputter coating using annular targets of gold–palladium and platinum. Care was taken to minimize contamination, and samples were transferred to and from the specimen chamber only at ambient temperatures. Half-masked carbon film–coated electron-microscope grids were coated at various temperatures with a nominal 15 nm of gold–palladium 80 : 20 and platinum at a rate of 4 nm/min using high-purity argon at 2.0 kV, 15 mA plasma current, and a chamber pressure of 4 Pa.

Gold–Palladium Films. At 303 K, there is evidence of crystal growth and the particle size is quite large. At 273 and 223 K, there is little difference in the appearance of the film, and the average size of the interlocking granules, which form the islands, is 6–7 nm. At 173 K, the islands of metal are less continuous and there are small granules in the clear space between the islands. AT 110 K, the individual islands are somewhat larger, but the particles between the islands are smaller. The smallest particles are 1.5 nm.

Platinum Films. Although there is no evidence of extensive crystal growth at 303 K, the intergranular spaces are large and devoid of fine particles. The films deposited at 273 and 223 K are made up of a series of discontinuous islands similar to the Au–Pd films deposited at the same temperature. The discontinuous islands are still present at 173 K, but are absent from the film deposited at 110 K.

At lower temperatures, there is less surface migration and an increased sticking coefficient for metal to metal than for metal to substrate. This means that at a lower temperature, there are fewer large islands, more small islands, and smaller interisland particles. There seems to be little advantage in depositing 10–15-nm films at really low temperatures, although there are advantages to lowering the temperature to about 220 K. If thin (i.e., 1–5 nm) films are to be deposited, then there are advantages in working at lower temperature. Lindroth *et al.* (1991) maintained specimens at 183 K while they were plasma–magnetron sputter coated with 1.5 nm of W. As has been shown in the earlier section on high-resolution coating, for ultrathin films of 1–3 nm, there is every advantage of working at really low temperatures, i.e., 20 K (Gross *et al.*, 1985) and obtaining very small particle size.

13.9.3. Coating Frozen-Hydrated Material Maintained at Low Temperatures

This subject is considered in detail by Echlin (1992), *Low Temperature Microscopy and Analysis,* but a brief summary of the problems and procedures is included here.

Frozen-hydrated samples have poor bulk and surface conductivity and generally charge in the electron beam. This is not surprising, considering the poor conductivity of ice at low temperatures (Durand *et al.,* 1967; Hobbs, 1974).

The papers by Marshall (1975; 1980), Echlin and Moreton (1973), and Echlin (1978a, b) contain detailed discussions of the charging phenomena in frozen samples. The studies by Brombach (1975) and Fuchs *et al.* (1978) have shed some light on what happens when a high-energy electron beam interacts with frozen specimens. These workers placed electrodes in and around model specimens of pure ice in order to measure the specimen current at various positions in the sample. They concluded that charging is not a serious problem, provided the sample is mounted on a good conductor and that the mean depth of penetration of the electrons is more than half the thickness of the specimen. Thin samples would not charge, a feature borne out by many observations on frozen sections, but bulk samples would charge, even if irradiated at quite low accelerating voltage. Fuchs and colleagues showed that a space charge built up just below the surface of uncoated frozen samples. This space charge had a flat, pancake-like profile, unlike the more usual teardrop shaped electron-interactive volume associated with bulk samples. They found that the application of a coating layer to the surface would effectively eliminate surface charging but would not necessarily reduce the internal space charging. The studies of Fuchs and his colleagues and in particular the work by Brombach (1975) have had a persuasive influence on the way we consider the interaction of a fine probe of electrons with a bulk-frozen hydrated sample. It went some way in explaining the ambiguous charging phenomena occasionally found with frozen samples. Although the electrical conductivity is small, it may be larger than the conductivity of frozen-dried material. Such differences are probably related to the continuity of frozen-hydrated material compared to the spatial discontinuity of the dried material. Marshall and Condon (1985) found that the backscattered-electron yield rises rapidly in charging uncoated frozen-hydrated samples, whereas it remains more or less constant in samples which had been coated. The proposed change in shape of the interactive volume, i.e, pancake shape versus teardrop, remains an enigma. The studies by Brombach (1975) were on pure ice and have never been reported as being extended to frozen biological or hydrated organic samples. All the available evidence would suggest that we do need to coat frozen-hydrated samples with a thin conductive layer if we are going to use the SEM to its full advantage. The technology and procedures for coating have been discussed elsewhere, but extra care is needed when coating frozen samples.

The evaporative and sputter coating techniques which will now be described are those which are specifically designed for use with low-temperature SEM carried out at relatively low magnification and hence low spatial resolution.

Evaporative Coating. The noble metals and their alloys are used as coating materials in low-temperature morphological studies and can be applied using either a conventional vacuum evaporator fitted with an air lock and cold table [see Echlin and Burgess (1977) or Marshall (1980)] or using the ancillary evaporation head associated with the AMRAY Biochamber (Pawley and Norton, 1978). Because the evaporated metal travels only in straight lines, the frozen sample should be rotated and tilted during the coating process to ensure that it receives a complete and even coating. Because the evaporation surface reaches quite high temperatures (circa 3000 K with carbon), the radiant energy may be sufficient to cause incipient ice sublimation on exposed surfaces of the frozen sample. For this reason, the specimen should not be placed close to the heat source, and the input power to the evaporation source should be kept as low as possible.

The geometric arrangement of the evaporation head on the AMRAY Biochamber (see Fig. 12.40) overcomes many of these problems. The evaporant source is sufficiently well collimated to ensure that the cone of evaporation just covers the diameter of the specimen stub. The frozen sample sits on a metal shuttle, which may be moved to different positions on the cold module. Immediately after fracturing, the cold shuttle is moved to the top of a large precooled metal cube, which may be rotated and tilted during evaporation. During the initial warmup and outgassing phase, the specimen remains out of the line of sight between the source and the sample. As the metal evaporation starts (and this may be observed through a porthole on the side of the evaporation head), the cold cube with the shuttle and specimen is moved into position under the evaporant source and immediately rotated and tilted by means of the exterior control mechanism. This ensures that even the most complexly sculptured surface is coated properly and, because the sample is moving rapidly during the coating process, the radiant heat load into the sample is minimized. A different approach has been adopted by Inoué and Koike (1989), who have fitted a small evaporation coater directly to one of the side ports of the SEM. The sample holder may be temporarily disconnected from its cooling block and tilted at 90 degrees toward the small cooling unit. The specimen is coated with metal from a direction perpendicular to the sample surface. A small fenestrated aluminum plate is placed between the evaporator and the sample holder to reduce contamination of the specimen chamber with evaporated metal and to minimize thermal damage to the sample. Presumably the sample can be moved through the tilting arc during coating. The results obtained with a 3–5 nm layer of gold are quite impressive.

Evaporative coating has been used in a relatively unsophisticated manner by a number of workers to coat frozen samples. No attempt is made to measure the coating thickness, but from the appearance of the image, it looks as if most samples are coated with 10–15 nm gold. The noble metals such as gold, gold–palladium, and platinum are relatively easy to apply. Carbon is not a good material, as there is a serious risk of heat damage to the frozen sample. This risk is diminished if carbon fiber is used as the evaporative source (Peters, 1984b). Although evaporative techniques can provide good conductive thin metal films on frozen specimens without any surface damage, most of the more recent SEM studies on frozen samples have relied on a sputter coater to provide the conductive layer.

Sputter Coating. A number of factors affect the deposition rate, and it is possible to list the features which would be desirable in a sputter coater to be used to coat frozen-hydrated samples.

1. The target material should be one of the noble metals or their alloys.
2. The bombarding gas should be pure, dry argon.
3. The target geometry should be designed to give high sputtering rates.
4. The sample should be protected from stray electrons by placing a high-flux, rare-earth permanent magnet behind the target. In addition to sweeping aside any electrons, such magnets localize the energized gas plasma to a small part of the target and ensure high sputtering rates.
5. The sample should be kept cold, i.e. lower than 123 K, and protected from any thermal radiation generated in the target. The grain size of sputter-coated films becomes finer at low specimen temperatures due to facilitated coalescence of the metal film.
6. In general, one should aim at using the lowest possible voltage and current commensurate with obtaining the thinnest satisfactory coating layer within a reasonable period of time in order to minimze the thermal input into the specimen. If sample heating is a problem, it is better to sputter coat intermittently at somewhat higher energy input, rather than coat continuously with a lower-energy input. The periodic coating allows the target heat to cool down between periods of sputtering, whereas there is a slow rise in temperature with continuous sputtering.
7. The sputtering should be carried out in a clean, contamination-free environment. Unlike evaporative coating, sputter coating is carried out at a relatively poor vacuum of 13 Pa (~10 torr). However, this should be a clean environment, easily obtained by repeated flushing with argon.

A number of pieces of equipment are now available which enable frozen-hydrated samples to be sputter coated for low temperature SEM [Robards and Crosby (1979) and Robards *et al.* (1981)]. The frozen sample sits on a cold module and is virtually surrounded by a cold shroud, except for a small circular cutout immediately below the retractable gold target. The target is fitted with a quadrupole magnet, which has the effect of increasing the rate of sputtering while at the same time diminishing the input of stray electrons to the sample. The chamber is pumped down to a good vacuum of 130 mPa dry argon. The high-resolution sputter head operates at a low energy input (20 mA at 1.0 kV to 25 mA at 0.44 kV) and a pressure of 13 Pa (10^{-1} torr). During a typical coating run of 3–4 minutes, the sample temperature rises by no more than 10 K. No attempt has been made to measure the coating thickness, which is estimated to be about 10 nm. For samples maintained at circa 110 K, such a small temperature rise would have no effect on the frozen-hydrated surface. Sputter coating may also be carried out using a planar magnetron sputtering head (Müller *et al.*, 1991). These workers found that a 3-nm layer of platinum was sufficient to prevent charging in a frozen-hydrated sample. In dried materials, it was found necessary to apply an additional 10 nm of C to prevent charging.

Although it is theoretically possible to convert an ambient-temperature plasma–magnetron sputter coater to work at low temperatures, this has not in fact been done. The conversion itself would be quite simple; there is, however, the added complication of providing a suitable airlock to enable frozen-hydrated samples to be transferred without fear of contamination and incipient melting. All sputter coating of frozen samples is carried out using coating modules that are part of a more complex cryopreparation unit. Sputter coating of frozen hydrated samples at low temperature is now a more or less routine procedure, and a wide range of results has been obtained. Its ease of use and avoidance of sample damage makes it the coating technique of choice.

13.9.4. Coating Techniques for X-Ray Microanalysis

The coating procedures used for x-ray microanalysis are no different from those generally employed with nonconductive samples. It is, however, necessary to carefully consider the chemical nature of the coating material.

The usual rule of thumb for x-ray microanalysis is that one should aim for the thinnest coating yielding a stable specimen and a maximum x-ray flux. This is because the thinner the coating, the less x-ray absorption within it, and the less the energy loss of the primary electron beam entering the specimen. Furthermore, the thinner the coating, the smaller the excitation of x-rays from the coating itself. For gold and gold–palladium coatings, which are often used in scanning electron microscopy to provide enough secondary-electron flux, the characteristic or continuum radiation produced could interfere with the x-ray lines of interest if an EDS is being used. Particular problems can occur if the element of interest is present in small or trace amounts. The usual thicknesses used range between 5 and 30 nm. For 5–10-nm films of carbon, aluminum, gold, and gold–palladium, the energy loss of the primary beam appears to be of small consequence, even at low nominal voltages. However, the beam and backscattered electrons obtained from the specimen could excite x-ray radiation from the film. This process may be particularly serious for gold and gold–palladium coatings on specimens with average atomic numbers greater than 10.

Examination of mass attenuation coefficients μ/ρ (see Chapters 3 and 14) shows that for x-ray lines from 0.5 to 1.5 keV, aluminum is lowest, followed in order by carbon, gold, and gold–palladium. In the region above 1.5 keV, the carbon μ/ρ value is lower than those for aluminum, gold, and gold–palladium. However, gold is best when electrical and thermal conductivities are considered, with aluminum about one-third as good, and carbon poor. It would seem that, for general purposes, aluminum is favored by its physical properties for use with x-ray lines of 0.8–4 nm, while carbon is favored outside this region.

In summary, the usual coating materials of noble metals and their alloys are impracticable for most biological samples and light-element materials analyzed using EDS for three reasons:

1. The L and M x-ray lines of these heavy metals (Au, Pd, Pt, etc.) may interfere with the K lines of the light elements.
2. A film of sufficient thickness to overcome the problems of charging (and thermal damage) may attenuate the incoming electron beam and absorb the emitted x-ray photons.
3. X-ray photons from the coating material may make a contribution to the general background associated with the sample. This could compromise the accuracy of analytical procedures that rely on a background measurement as part of the quantitative algorithm.

Only four elements may be used as coating materials in connection with quantitative analytical studies. The important properties of these materials, beryllium, carbon, aluminum, and chromium, are listed in Table 13.2. Each material has both advantages and disadvantages.

Beryllium. This is probably the best material to use. It has medium thermal and electrical properties, negligible x-ray interference and contribution to background, low evaporation temperature, and is easy to apply. Unfortunately it is very toxic when in the finely particulate and thus potentially ingestible form. Marshall and his colleagues have made extensive use of beryllium as a coating material in connection with their analytical work on insect tissue. Any potential users of beryllium should consult their papers [Marshall and Carde (1984); Marshall *et al.* (1985)] for details of the procedures and precautions.

Table 13.2. Properties of Commonly Used Coating Materials for X-Ray Microanalysis

Property	Beryllium	Carbon	Aluminum	Chromium
Density (kg m^{-3})	1800	2300	2700	7200
Thermal conductivity (W cm)	2.00	1.29	2.37	0.93
Resistivity (Ohm cm)	4.57	3500	2.83	13.0
Evaporation temperature (K)	1520	3000	1275	1450
X-ray energy (keV)				
$K\alpha$	0.109	0.227	1.487	5.415
$K\beta$				5.946
$L\alpha$				0.598

Carbon. This is the most popular coating material. It is easy to apply, and has little effect on the x-ray yield from the sample. The electrical properties are rather poor. However, carbon suffers from one grave disadvantage: its high temperature of evaporation, which can easily cause thermal damage to the specimen. Marshall (1977) measured a 40-K temperature rise in samples held at 140 K which had been continuously coated with carbon for 30 s, and Vesely [quoted in Echlin and Taylor (1986)] has measured a 473-K temperature rise in samples during carbon coating. Echlin *et al.* (1980; 1981) found evidence of surface etching to a depth of 150 nm in samples held at 110 K and heavily coated with carbon for 15 s while the sample was being continually rotated and tilted. Echlin *et al.* (1980) placed a small aperture between the evaporation source and the sample in order to reduce the thermal load into the frozen sample. Although a thin layer of carbon appears to diminish charging in bulk samples, it is not the best coating material for such specimens. Carbon coating has been used extensively to coat frozen-dried sections and to provide a thin conducting layer to support such specimens.

Aluminum. In spite of its good electrical properties and low evaporation temperature, aluminum should not be used to coat light-element materials. The real disadvantage of using aluminum as a coating material in connection with x-ray microanalysis is that its x-ray emission at 1.487 kV would create difficulties with the EDS analysis of some of the light elements because of peak overlaps and contribution to the general background. Aluminum has only a minimum effect on the absorption of x rays from light elements. In spite of these problems, aluminum has been used extensively to coat the large-area plastic films which are used to support sections.

Chromium. This material is in many respects a compromise. Its use was first suggested by Marshall (1977) and extended in the more recent work by Echlin and Taylor (1986). The thermal and electrical properties are sufficient to prevent specimen charging and diminish beam-induced thermal damage. The low evaporation temperature does not cause thermal damage to the surface of samples, and the characteristic x-ray lines are sufficiently far away from the characteristic lines of the elements of biological interest to cause any problems with identification.

However, the higher density of chromium could well present problems of beam attenuation and absorption of soft x-rays emitted from samples. Echlin and Taylor (1986) have investigated this and related problems using standards of known composition and have proposed that a small correction should be made for the absorption of x rays from low-atomic-number elements.

The coating procedures for all four elements are relatively easy to carry out, provided due care is taken with beryllium. All four materials are applied using evaporative techniques at high vacuum with the sample rotated or tilted during the actual coating process. Such movements of the sample ensure an even coating on rough surfaces. The relatively low sputtering yield of these materials, primarily due to the formation of an oxide layer on the target, precludes using this method to apply thin films to thermally sensitive samples. The long sputtering times, i.e. minutes at low vacuum, which are used would cause the target material to warm up and thus encourage sample contamination and surface etching.

13.10. Determination of Coating Thickness

A number of techniques can be used to determine coating thickness, and Flood (1980) reviews the available methods. However, all the methods give an average value on a flat surface and it is necessary to briefly consider this type of measurement in relation to the coating thickness one might expect on rough surfaces. If the coating thickness is too thin, it will be ineffective in eliminating the charge; it is it too thick, it will obscure surface detail. A coating layer of a certain minimum thickness is necessary to form a continuous layer, and this thickness varies for different elements and for the relative roughness of the surface. On an absolutely flat surface, a layer of carbon 0.5 nm thick forms a continuous coat. A layer ten times as thick is required to form a coat on an irregular surface. As a general rule, thicker coatings are required for irregular surfaces to ensure a continuous film over all edges, cavities, and protrusions.

13.10.1. Estimation of Coating Thickness

In high-vacuum thermal evaporation, one can assume that all the vapor molecules leave each part of the evaporant surface without preferred direction and pass to the substrate surface without colliding with residual gas molecules. By allowing for the incident angle of the vapor at the specimen, and assuming that all the incident vapor molecules have the same condensation coefficient, the thickness distribution of the coating may be calculated. The formula below may be applied to a flat, nonrotating surface oriented at an angle θ to the source direction and can be used to calculate the thickness from the quantity of material to be evaporated:

$$T = \frac{M}{4\pi R^2 \rho} \cos \theta, \tag{13.1}$$

where T is the thickness in centimeters, M is the mass of the evaporant in grams, ρ is the density in g cm^{-3}, R is the distance from source to specimen in centimeters, and θ is the angle between the normal to the specimen and the evaporation direction. With Eq. (13.1), the thickness can be calculated to within $\pm 50\%$ if the geometry of the evaporation rig and substrate is known and if the material in question evaporates slowly and does not alloy with the heater. It is assumed that evaporation is from a point source and occurs uniformly in all directions. The uncertainty in T arises in part because of uncertainty in the sticking coefficient of the evaporant or the substrate. The efficiency is material dependent and is rarely better than 75%. The thickness calculated by Eq. (13.1) must therefore be reduced by multiplying by the sticking efficiency (approximately 0.75).

The thickness of coating T in nanometers that may be obtained using a sputter coater is given by

$$T = kIVt, \tag{13.2}$$

where I is the plasma discharge in milliamperes, V the accelerating voltage in kilovolts, t the time in minutes, and k a constant depending on the bombarding gas and the target material. The constant k must be calculated for each sputter coater, target material, and bombarding gas from experimental measurements.

13.10.2. Measurement during Coating

Several methods are available, which vary in their sensitivity and accuracy. The devices are placed in the coating chamber close to the specimen and measure the thickness during the coating process. It is possible to measure the density of the evaporant vapor stream either by measuring the ionization that occurs when vapor molecules collide with electrons or by measuring the force impinging particles exert on a surface. Mass-sensing devices may be used for all evaporant materials; these operate by determining the weight of a deposit on a microbalance or by detecting the change in oscillating frequency of a small quartz crystal on which the evaporant is deposited.

The most frequently used measuring device is the quartz crystal thin-film monitor. The thin-film monitor resonates at a frequency dependent on the mass of material deposited on its surface. The frequency of a "loaded" crystal is compared to the frequency of a clean crystal, and the decrease in frequency gives a measure of the film's thickness. A typical sensitivity value for a crystal monitor is a 1-Hz frequency change for an equivalent film of 0.1 nm C, 0.13 nm Al, and 0.9 nm Au. The uncertainty of these devices is on the order of $\pm 0.1\ \mu\mathrm{g\ cm}^{-2}$. By multiplying this value of the uncertainty by the evaporant density, the uncertainty in the film thickness can be determined.

Peters (1980) has calculated that temperature fluctuations of the quartz crystal thin-film monitor must be held to within ±0.06 K for carbon measurements to be within ±0.1 nm and must be within ±0.13 K for metal measurements to be within ±0.02 nm. The influence of temperature is variable and related to the angle at which the quartz disk is cut relative to the main axes of the crystal. Most disks are cut along an axis where the frequency response is stable between 273 and 333 K (Flood, 1980). This is within the temperature range normally experienced in plasma–magnetron, Penning, and ion-beam sputter coaters and in thermal evaporation units. Slayter (1980) carefully cross-checked measurements of thin-film thickness, using quartz thin-film monitors, interferometry, and the direct-weight microbalance and found the accuracy to be within 15% for carbon films in the range 10–70 nm.

It is evident that accurate measurement of thin-film thickness can be made using quartz thin-film monitors provided sophisticated (and expensive!) equipment is used. But as Slayer (1980) is at pains to point out, these monitors do not produce an absolute measurement, because they average out discontinuous thin metal coats. Crystallites within the film could easily have thicknesses greater than the average coat thickness. Quartz crystal monitors measure the *average mass-thickness,* which, if the film is discontinuous, is much less than the *average metric thickness.* There does not appear to be a simple relationship between the two methods of measurement. Kemmenoe and Bullock (1983) have calculated that a gold film of 4-nm average metric thickness would average about 1 nm in mass-thickness. As the film increases in thickness and becomes less discontinuous, the two forms of measurement come closer to giving the same value.

It would be naive to equate a thin film-thickness measurement given on a planar quartz crystal monitor with that on a complexly sculptured specimen being rotated and rocked during the coating process. Peters (1980) has calculated that, using a tooling factor of 50% to correct for differences in location of monitor and specimen, only 12.5% of the coating thickness measured on the thin-film monitor actually lands on a tumbling specimen. Specimen tumbling was introduced to overcome the self-shadowing phenomena described by Konig and Helwig (1950). In a later paper, Peters (1986a) cites a deposition factor of approximately 25% of the measured metal film thickness for sputtering of a tumbling model cylinder. Hermann *et al.* (1988) have applied theoretical calculations to measure the layer thickness of chromium and germanium films deposited by their double-axis rotary shadowing method. Slayter considers that there may be as much as a tenfold increase in film thickness on topographical features presented at normal irradiance to the beam. Clay and Peace (1981) find that sections of embedded ion-beam and diode-sputtered films are 5–10 times thicker than the reading given on the thin-film monitor. The range of inaccuracy is quite large: in a specimen being rotated and tilted, the thickness is underestimated by a factor of 8; in a stationary specimen normal to the beam, the coating thickness may be overestimated by a factor of 10. The accuracy of thin-film measurement is also a function of the method by which the film is deposited. Because of energy transfer to the quartz crystal monitor, measurements in evaporative and diode coaters may be off by ±100%. This is less of a problem with ion-beam and Penning sputter coaters.

What do these discrepancies mean to the practicing microscopist concerned with providing a reasonable coating thickness to ensure maximum specimen–image information transfer? For thick continuous films, i.e., more than 15 nm, the average thickness given by a quartz thin-film monitor is fairly close to the average metric thickness that may be measured on a planar surface. If accurate measurements are required on films less than 10 nm in thickness, the indicated average mass-thickness given by the thin-film monitor should be checked by some independent means [for details see Flood (1980)]. Provided standardized conditions are used for coating, i.e., set target–substrate distance, residual gas pressure, source and substrate temperature, and so on, it should be possible to achieve reproducibility to within ±20% (Flood, 1980). Alternatively, if a standard coating procedure is to be repeatedly used on more or less the same type of specimen, a coating thickness should be applied which results in an optimal image in the microscope. This thickness is measured during coating by a quartz thin-film monitor positioned at a fixed place relative to the specimen, and it is this thickness which is applied each time to the specimen. The actual thickness on the specimen is probably irrelevant because, on a complexly sculptured sample, there may be a 20-fold variation in thickness, depending on the topography of the specimen and whether it is stationary or moving. It is nevertheless necessary to calibrate the thin-film monitor, especially if it is of the less expensive variety, in order to have some idea of its accuracy.

A novel method of calibration, which requires no special equipment, has been described by Johansen and Namork (1984). Using transmission electron micrographs, these workers measured the thickness of thin platinum films sputter coated on 20-nm colloidal gold particles, which protruded from the edge of torn, rolled-up carbon films. Up to a platinum film thickness of 10 nm, there was a 17% increase in metric thickness over mass-thickness. This method of calibration is particularly useful for films less than 10 nm thick, where the inaccuracy of measurement is greatest.

It is also possible to monitor specific film properties such as light absorption, transmittance, reflection, and interference effects by various optical monitors. Film thickness of conductive materials can be measured by *in situ* resistance measurements and the thickness of dielectric materials measured by capacitance monitors. A further sophistication of most of these *in situ* techniques is that they can be used to control the coating process.

13.10.3. Measurement after Coating

The most accurate measurements of film thickness are made on the films themselves. These methods are based on optical techniques, gravimetric measurements, and x-ray absorption and emission. Multiple-beam interferometry is the most precise and, depending on the method used, can be as accurate as to within 1 or 2 nm. The Fizeau method, which can be used to check film thickness, involves placing a reflective coating on top of the deposited film step and measuring a series of interference fringes. The film thickness can also be measured by sectioning flat pieces of resin on which a coating has been deposited and measuring the thickness of the metal deposit in the TEM. The accuracy of this method is dependent on being able to section the resin and to photograph the section at right angles to the metal deposit. A simple method for accurately determining film thickness and grain size has been described by Roli and Flood (1978). They found that linear aggregates of latex spheres accumulate coating material only along their free surface. Transverse to the linear aggregate, the sphere diameter increases in thickness by twice the film thickness, while parallel to the aggregate the increase in thickness corresponds to the film thickness. Using this technique, they have measured the film thickness derived from various coating procedures to within ± 2 nm. Film thickness may be estimated by interference color or, in the case of carbon, by the density of the deposit on a white tile.

Broers and Spiller (1980) have provided some interesting information on measuring the roughness of thin metal films. The roughness measurements are made using variations in the scatter of incident soft x-ray radiation on the film surface. The method has been used on films in the 0.2-to-1.5-nm range, and the results correlate well with measurements made on low-loss image micrographs of the same films. Broers and Spiller find the smoothest surfaces are on thin (0.75 nm) films of tungsten and a rhenium–tungsten alloy (68% Re). The noble metals give a film surface roughness 2–5 times more coarse then the refractory metals. It is hoped that further studies will lead to the provision of standard reference surfaces of known roughness. Such standards would be invaluable in studies which try to relate the film thickness on a quartz crystal to that actually landing on a rough specimen. Wepf *et al.* (1991) used wavelength-dispersive x-ray analysis to measure the composition and hence the average mass thickness of Pt–Ir–C films.

13.10.4. Removing Coating Layers

Having taken all the trouble to deposit a coating layer on a sample, it is sometimes necessary to remove it. Provided care is taken, hard specimens can be restored to their pristine, uncoated state. If the sample is a polished section, the coating can be easily removed by returning to one of the final polishing operations (6 μm diamond or 1 μm Al_2O_3). If the sample is rough or a flat sample cannot be repolished, chemical means must be used to remove the coating.

If it is known that the coating layer is to be removed after examination of a rough sample, it is preferable to use aluminum as the coating material because of its ease of removal. Sylvester-Bradley (1969) removed aluminum from geological samples by immersing them in a freshly prepared solution of dilute alkali for a few minutes. Sela and Boyde (1977) describe a technique for removing gold films from mineralized samples by immersing the specimen in a cyanide solution. Gold–Palladium may be removed using a solution of 10% $FeCl_3$ in ethanol for 1–8 h (Crissman and McCann, 1979).

Carbon layers can be removed by reversing the polarity in the sputter coater and allowing the argon ions to strike the specimen. Care must be taken only to remove the carbon layer and not to damage the underlying specimen. Carbon layers on inorganic samples such as rocks, powders, and particulates can be effectively removed in an oxygen-rich radio frequency plasma.

13.11. Artifacts Related to Coating and Bulk-Conductivity Procedures

Leaving aside the inaccuracies which may occur during the measurement of film thickness, coating artifacts may be considered under two headings, contamination and decoration effects. Contamination can occur during coating by trapping gas atoms or hydrocarbons or after coating as a result of gas absorption and dust deposits on the surface. At a gross level, contamination in coating systems can be recognized as a yellow-brown layer on a white background and as a blue-black tarnish on metal surfaces. Other, more subtle signs of contamination are the large cracks on the surface of thin layers of films deposited over a long period of time. A reexamination of some micrographs of plasma–magnetron sputter-coated tantalum prepared over a 15-min period (Echlin and Kaye, 1979) has revealed such contamination-related cracks. Once the sputter coater was cleaned [for details see Echlin *et al.* (1980)], this artifact was no longer a problem. Provided the few simple housekeeping measures detailed earlier are followed, contamination with the standard plasma–magnetron sputter coater should not be a problem in the production of coating layers for conventional SEM. Contamination may also be kept significantly lower by operating the systems at very low voltages. Peters (1991) has calculated that operating voltages of 100–200 V are insufficient for generating contamination products from hydrocarbons. Under these conditions, much longer coating times are necessary and it is important to diminish the thermal input into the specimen to prevent the accumulation of recrystallization effects. High-resolution thin films (i.e., 1–2 nm) can be produced only under high-vacuum conditions. There is no doubt from the studies made by Peters (1980), Slayter (1980), Gross *et al.* (1985), and Wepf *et al.* (1991) that a high, clean vacuum favors the formation of ultrathin, low-grain-size films.

Decoration effects are more insidious, for they may be either a delicate reflection of a true inhomogeneity on the specimen surface or a result of contamination or uneven coating. A good example of "natural" decoration effect occurs where a very thin layer of metal is deposited onto alkali crystals, i.e., NaCl (Willison and Rowe, 1980). A "natural" decoration effect is a real substructural feature whose appearance is enhanced by a thin metal coating. Such decoration effects must be distinguished from artifacts due to an initial random event that enhances the sticking power of the incoming molecule and in turn gives a decoration artifact. Monomolecular steps between crystal lattices can be demonstrated, provided the step heights are sufficient to catch and hold onto the absorbed atoms. The images appear as long lines of oriented particles that correspond to the crystal planes. This molecular-level effect disappears once the film layer becomes more than a few nanometers thick.

Walzthony *et al.* (1984) found that various heavy metals and carbon displayed distinctive bending patterns on a series of elongated helical macromolecules that had been rotary-shadowed at low elevation angles. They considered that diffusion, nucleation, and coalescence of heavy metals were different among the molecules examined. The differences in grain patterns reflected differences in the physiochemical properties of the underlying molecules, such as surface relief and charge distributions. Thus, Ta and W, which probably carry a positive charge, preferentially decorate clusters of negative charges along the myosin macromolecule. A preliminary coating with C masks these effects, presumably by neutralizing some of the surface charges. Walzthony *et al.* consider that positive "staining" with heavy metals which condense and coalesce at specific sites along macromolecules could be used to provide information about the structure of these specimens. Haggis (1985) has found an analogous phenomenon in platinum-coated actin, with surface temperatures playing an important role in regulating the effect. A similar molecular-decoration artifact can be seen in some high-resolution freeze-etched surfaces contaminated with water vapor. The water vapor shows a preferential affinity for hydrophilic sites and can give rise to some quite misleading images.

Natural decoration effects may also occur at the SEM resolution level. If there are surface features that favor condensation and nucleation, they accrete more of the metal coating and increase the topographic contrast. Thus, Rosowski *et al.* (1984) found that the surfaces of diatom frustules coated with a thin metal film showed a granular matrix not present on the supporting glass surface. The authors consider that minute surface irregularities on the silica frustule favored the formation of fine granules. This same general effect may be seen on contaminated surfaces. The presence of contaminants causes an unequal distribution of surface forces, and the migrating atoms/molecules settle at places with the highest binding energy. This would be considered an unnatural effect, and the difficulty comes in separating these unnatural effects from natural effects. The only sure way of resolving these difficulties is to ensure that the coating is carried out in as clean a vacuum as possible, and the finer the resolution required from the specimen, the higher the vacuum under which the coating should be carried out.

Consideration must be given to the way the thin films are applied, as it is possible to relate some decoration effects to the mode of application. Pashley *et al.* (1964) were able to demonstrate that the radiation from an evaporation source can melt nucleation centers, giving rise to roundish drops and delaying the formation of a continuous film. Peters (1980) has shown that the excessive photon radiation that usually occurs during thermal evaporation may generate crystallites, which in turn give the film a resolvable structure. This is not a feature of ion-beam and Penning sputtering. Flood (1980) has demonstrated that it is possible to suppress the coalescence and growth of islands and diminish crystal growth in films that are sputtered rather than thermally evaporated. One type of decoration artifact, which may be found in specimens coated unidirectionally from an angle, usually referred to as shadowing, is the appearance of lines running normal to the shadowing direction. The self-shadowing occurs when there is diminished grain development in the shadow of a developing grain. A closely related artifact called capping is the exaggeration of specimen dimensions by accumulation of shadowing material. Neugebauer and Zingsheim (1979) show that misapplication of the rotary shadowing method can produce images in which known solid spherical particles appear to be penetrated with holes. These kinds of artifacts are uncommon in SEM preparations, which are usually omnidirectionally coated, but they are a problem in high-resolution coating for SEM–STEM, in freeze-etch replicas, and in macromolecule shadowing techniques; the excellent book by Willison and Rowe (1980) considers these matters in some detail.

It would be wrong to think that, once the coating layer has been deposited, no further effects can occur. Pulker (1980) shows that the number of gold particles decreased and their relative size increased after a thin layer of gold had been heated to 603 K. Peters (1980) showed the same effects at 293 K. These tempeature rises can readily occur in electron-beam instruments. A thin layer of carbon (5 nm) evaporated on top of a carefully applied ultrathin layer of metal (0.5 nm Ta–W) reduced recrystallization in the film. Rosowski *et al.* (1984) have found that an initial thermal evaporation with Au–Pd can protect delicate structures from the thermal damage that can occur with diode sputter coating. Unless high-resolution films are going to be examined immediately after application, it is advisable to coat the sample with an additional layer of carbon. This has not proved to be the case with the thin Pt–Ir–C described by Wepf *et al.* (1991). Provided the films contain at least 25% C, they were quite stable in a normal atmosphere, and when irradiated within a range 1.2–80 K $e^{-}\,\mathrm{nm}^{-2}$. This precaution is probably not necessary for specimens to be examined by conventional SEM.

One coating effect that is now, fortunately, relatively easy to recognize occurs where the coating layer is so thick that it falsifies the surface morphology of the specimen. Blaschke (1980) gives a nice example of this effect together with a good example of the decoration of crystal edges with gold spheres— a result of injudicious coating.

13.12. Conclusions

In concluding this chapter on specimen coating and conductivity, it would be appropriate to make some recommendations on coating methods for practicing microscopists:

1. For the very highest spatial resolution, uncoated samples either in contact with a good conductor or conductively stained, appear to offer the best prospects.
2. For high-resolution SEM and STEM (spatial resolutions 1–3 nm), where one is attempting to push the instrumental capabilities to their limit, Penning and ion-beam sputtering together with ultrahigh-vacuum thermal evaporation must be the methods of choice. The coating material and final thickness of the film must be dictated by the exigencies of the experiment. A mixture of Pt–Ir and C would appear to be suitable for

most morphological, high-resolution problems. For high-resolution STEM studies, heavy metals such as tantalum and tungsten, which have a low specific mobility, appear to be the best choice. For high-resolution SEM, where one can be sure to be collecting only the primary secondary electrons, then platinum, chromium, or niobium is the material of choice. Where this is not the case, the noble metals and their alloys should be used. In all instances, a film thickness of 1–2 nm appears to give the best result. In situations where one wishes to achieve high-resolution spatial decoration, Peters (1984b) recommends fast evaporation of a small amount of metal with high specific surface mobility. The crystallites round up as a result of vertical growth and give the best contrast with the smallest amount of metal. Increased spatial resolution can be obtained with metals with lower mobilities, i.e., platinum, rhodium, and tungsten.

3. For medium-resolution SEM (spatial resolutions 3–5 nm), low-voltage plasma–magnetron sputter coating with either gold–palladium or platinum gives the best result. Alternatively, ion-beam or Penning sputtering with W or Pt gives an equally good coating layer. The final film thickness should be in the range 3–8 nm.

4. For routine scanning microscopy on coated specimens (spatial resolutions greater than 8–12 nm), a plasma–magnetron sputter coater using Au–Pd working at a relatively high deposition rate gives good, even films. However, care must be taken with thermolabile samples. The final film thickness should be in the range 8–12 nm.

14

Data Base

Table 14.1. Atomic Number, Atomic Weight, and Density of Elements

Element		Atomic number	Atomic weight	Density (g/cm^3)
H	Hydrogen	1	1.008	—
He	Helium	2	4.003	—
Li	Lithium	3	6.941	0.534
Be	Beryllium	4	9.012	1.848
B	Boron	5	10.81	2.5
C	Carbon	6	12.01	2.34 (amorphous C) 2.25 (graphite)
N	Nitrogen	7	14.01	—
O	Oxygen	8	16.00	—
F	Fluorine	9	19.00	—
Ne	Neon	10	20.18	—
Na	Sodium	11	22.99	0.97
Mg	Magnesium	12	24.31	1.74
Al	Aluminum	13	26.98	2.7
Si	Silicon	14	28.09	2.34
P	Phosphorus	15	30.97	2.20 (red)
S	Sulfur	16	32.06	2.07
Cl	Chlorine	17	35.45	—
Ar	Argon	18	39.95	—
K	Potassium	19	39.10	0.86
Ca	Calcium	20	40.08	1.54
Sc	Scandium	21	44.96	2.99
Ti	Titanium	22	47.90	4.5
V	Vanadium	23	50.94	6.1
Cr	Chromium	24	52.00	7.1
Mn	Manganese	25	54.94	7.4
Fe	Iron	26	55.85	7.87
Co	Cobalt	27	58.93	8.9
Ni	Nickel	28	58.71	8.9
Cu	Copper	29	63.55	8.96
Zn	Zinc	30	65.37	7.14
Ga	Gallium	31	69.72	5.91
Ge	Germanium	32	72.59	5.32
As	Arsenic	33	74.92	5.72
Se	Selenium	34	78.96	4.79

(*continued*)

Table 14.1. (*Continued*)

Element		Atomic number	Atomic weight	Density (g/cm³)
Br	Bromine	35	79.90	3.12
Kr	Krypton	36	83.80	—
Rb	Rubidium	37	85.47	1.53
Sr	Strontium	38	87.62	2.6
Y	Yttrium	39	88.91	4.472
Zr	Zirconium	40	91.22	6.49
Nb	Niobium	41	92.91	8.6
Mo	Molybdenum	42	95.94	10.2
Tc	Technetium	43	98.91	11.46
Ru	Ruthenium	44	101.1	12.2
Rh	Rhodium	45	102.9	12.4
Pd	Palladium	46	106.4	12.00
Ag	Silver	47	107.9	10.5
Cd	Cadmium	48	112.4	8.64
In	Indium	49	114.8	7.3
Sn	Tin	50	118.7	7.3
Sb	Antimony	51	121.8	6.68
Te	Tellurium	52	127.6	6.24
I	Iodine	53	126.9	4.94
Xe	Xenon	54	131.3	—
Cs	Cesium	55	132.9	1.87
Ba	Barium	56	137.3	3.5
La	Lanthanum	57	138.9	6.189
Ce	Cerium	58	140.1	6.75
Pr	Praseodymium	59	140.9	6.769
Nd	Neodymium	60	144.2	7.00
Pm	Promethium	61	[145]	—
Sm	Samarium	62	150.4	7.49
Eu	Europium	63	152	5.245
Gd	Gadolinium	64	157.3	7.86
Tb	Terbium	65	158.9	8.25
Dy	Dysprosium	66	160.5	8.55
Ho	Holmium	67	164.9	8.79
Er	Erbium	68	167.3	9.15
Tm	Thulium	69	168.9	9.31
Yb	Ytterbium	70	173.0	6.959
Lu	Lutetium	71	175.0	9.849
Hf	Hafnium	72	178.5	13.1
Ta	Tantalum	73	180.9	16.6
W	Tungsten	74	183.9	19.3
Re	Rhenium	75	186.2	21.0
Os	Osmium	76	190.2	22.5
Ir	Iridium	77	192.2	22.4
Pt	Platinum	78	195.1	21.45
Au	Gold	79	197.0	19.3
Hg	Mercury	80	200.6	13.55
Tl	Thallium	81	204.4	11.85
Pb	Lead	82	207.2	11.34
Bi	Bismuth	83	209.0	9.8
Po	Polonium	84	[210]	[9.24]
At	Astatine	85	[210]	—
Rn	Radon	86	[222]	—
Fr	Francium	87	[223]	—
Ra	Radium	88	226.0	5.0
Ac	Actinium	89	[227]	—
Th	Thorium	90	232.0	11.5
Pa	Protactinium	91	231.0	15.4
U	Uranium	92	238.0	19.05(α), 18.9(β)

Table 14.2. Common Oxides of the Elements

Symbol, atomic no.	Oxide	Wt. fraction of element	Wt. fraction of oxygen	Symbol, atomic no.	Oxide	Wt. fraction of element	Wt. fraction of oxygen
Na 11	Na_2O	0.7419	0.2581	Cd 48	CdO	0.8754	0.1246
Mg 12	MgO	0.60317	0.3968	In 49	In_2O_3	0.8271	0.1729
Al 13	Al_2O_3	0.5292	0.4708	Sn 50	SnO	0.8812	0.1188
Si 14	SiO_2	0.4675	0.5326		SnO_2	0.7876	0.2124
	SiO	0.6371	0.3629	Sb 51	SbO_2	0.7918	0.2081
P 15	P_2O_5	0.4363	0.5636		Sb_2O_3	0.8353	0.1647
K 19	K_2O	0.8301	0.1699	Te 52	TeO_2	0.7995	0.2005
Ca 20	CaO	0.7147	0.2853	Ba 56	BaO	0.8110	0.1889
Sc 21	Sc_2O_3	0.6519	0.3481	La 57	La_2O_3	0.8973	0.1027
Ti 22	TiO_2	0.5995	0.4005	Ce 58	CeO_2	0.8141	0.1859
	TiO	0.7496	0.2504	Pr 59	PrO_2	0.8149	0.1851
	Ti_2O_3	0.6519	0.3481	Nd 60	Nd_2O_3	0.8573	0.1427
V 23	V_2O_3	0.6797	0.3203	Sm 62	Sm_2O_3	0.8623	0.1377
	V_2O_5	0.5601	0.4399	Eu 63	Eu_2O_3	0.8636	0.1364
Cr 24	Cr_2O_3	0.6842	0.3158	Gd 64	Gd_2O_3	0.8676	0.1364
	CrO_3	0.5199	0.4800	Tb 65	Tb_4O_7	0.8688	0.1312
Mn 25	Mn_3O_4	0.7203	0.2797	Dy 66	Dy_2O_3	0.8598	0.1402
	MnO_2	0.6319	0.3681	Ho 67	Ho_2O_3	0.8730	0.1270
Fe 26	Fe_2O_3	0.6994	0.3006	Er 68	Er_2O_3	0.8745	0.1255
	FeO	0.7773	0.2227	Tm 69	Tm_2O_3	0.8756	0.1244
Co 27	CoO	0.7865	0.2135	Yb 70	Yb_2O_3	0.8782	0.1218
Ni 28	NiO	0.7858	0.2142	Lu 71	Lu_2O_3	0.8794	0.1206
Cu 29	CuO	0.7988	0.2012	Hf 72	HfO_2	0.8479	0.1520
	Cu_2O	0.8882	0.1118	Ta 73	Ta_2O_5	0.8829	0.1171
Zn 30	ZnO	0.8034	0.1966	W 74	WO_3	0.7930	0.2070
Ga 31	Ga_2O_3	0.7439	0.2561	Re 75	Re_2O_7	0.7688	0.2312
Ge 32	GeO_2	0.8194	0.1806		ReO_3	0.7950	0.2050
As 33	As_2O_5	0.6519	0.3481	Os 76	OsO_4	0.7482	0.2518
	As_2O_3	0.7574	0.2426	Ir 77	IrO_2	0.8573	0.1427
Se 34	SeO_2	0.7116	0.2884		Ir_2O_3	0.8886	0.1114
Rb 37	Rb_2O	0.9144	0.0856	Pt 78	PtO_2	0.8591	0.1409
Sr 38	SrO	0.8456	0.1544	Hg 80	HgO	0.9261	0.0739
Zr 40	ZrO_2	0.7403	0.2597	Tl 81	Tl_2O_3	0.8949	0.1051
Nb 41	Nb_2O_5	0.3495	0.6505	Pb 82	PbO_2	0.8662	0.1338
Mo 42	MoO_3	0.6665	0.3335		PbO	0.9283	0.0717
Ru 44	RuO_4	0.6124	0.3876	Bi 83	Bi_2O_3	0.8970	0.1030
	RuO_2	0.7595	0.2405	U 92	UO_2	0.8815	0.1185
Rh 45	Rh_2O_3	0.8109	0.1819		U_3O_8	0.8480	0.1520
Pd 46	PdO	0.8693	0.1307				
Ag 47	Ag_2O	0.9309	0.0690				
	AgO	0.8708	0.1292				

Table 14.3. Mass Absorption Coefficients for $K\alpha$ Lines (? denotes significant uncertainty due to the presence of an absorption edge)

Absorber \ Emitter		183 eV B	277 eV C	392 eV N	525 eV O	677 eV F	849 eV Ne	1041 eV Na	1254 eV Mg	1487 eV Al	1740 eV Si	2014 eV P	2308 eV S	2622 eV Cl
1	H	1,723	473.7	156.8	60.7	26.2	12.3	6.2	3.3	1.9	1.1	0.7	0.4	0.3
2	He	10,677	3,190	1,136	465.7	210.1	101.8	52.4	28.4	16.1	9.5	5.9	3.7	2.4
3	Li	32,746	10,420	3,917	1,682	789.0	394.9	208.6	115.2	66.4	39.8	24.6	15.7	10.3
4	Be	69,937	23,698	9,372	4,195	2,039	1,052	570.3	322.0	188.9	114.8	71.9	46.4	30.7
5	B	2,861?	39,834	16,610	7,754	3,902	2,074	1,154	666.1	398.6	246.4	156.6	102.3	68.6
6	C	5,945	2,147	23,586	11,575	6,079	3,354	1,929	1,147	704.2	445.4	289.1	192.4	131.1
7	N	10,118	3,757	1,553	16,433	8,823	4,973	2,917	1,766	1,101	705.8	462.9	310.6	213.0
8	O	15,774	5,998	2,529	1,181	11,928	6,844	4,084	2,512	1,590	1,032	684.6	463.7	320.3
9	F	21,803	8,465	3,635	1,720	875.7	8,440	5,110	3,189	2,046	1,344	901.2	616.2	429.1
10	Ne	31,623	12,500	5,458	2,616	1,345	731.8	6,756	4,268	2,771	1,841	1,247	860.8	604.3
11	Na	40,649	16,323	7,235	3,513	1,823	998.4	574.4	5,086	3,337	2,240	1,532	1,067	754.8
12	Mg	54,026	21,998	9,886	4,858	2,545	1,404	811.1	488.0	4,168	2,823	1,948	1,367	975.1
13	Al	66,069	27,234	12,393	6,159	3,258	1,810	1,051	634.2	397.5	3,282	2,282	1,614	1,159
14	Si	83,702	34,881	16,054	8,063	4,304	2,409	1,407	851.9	535.0	347.2	2,774	1,976	1,429
15	P	83,768	41,087	19,107	9,692	5,220	2,943	1,728	1,051	661.7	430.1	287.4	2,230	1,622
16	S	101,883?	50,415	23,667	12,117	6,581	3,737	2,208	1,348	851.6	554.6	370.9	254.6	1,922
17	Cl	6,687	56,799?	26,896	13,888	7,604	4,350	2,584	1,586	1,005	655.8	439.3	301.7	212.0
18	Ar	7,943	52,948	29,456	15,332	8,459	4,872	2,912	1,796	1,142	747.1	501.2	344.6	242.3
19	K	10,616	5,287	36,562?	19,173	10,655	6,179	3,715	2,302	1,469	964.0	648.0	446.1	313.9
20	Ca	13,257	6,520	36,659	22,561	12,624	7,368	4,456	2,774	1,778	1,170	788.2	543.4	382.7
21	Sc	14,826	7,221	3,809	23,946	13,486	7,920	4,817	3,014	1,940	1,281	864.7	597.1	421.0
22	Ti	17,132	8,290	4,324	22,693	14,987	8,854	5,415	3,405	2,200	1,458	986.7	682.7	482.0
23	V	19,490	9,403	4,860	24,847?	16,504	9,805	6,029	3,810	2,472	1,643	1,116	773.4	546.9
24	Cr	22,734	10,978	5,635	3,156	16,086	11,199	6,921	4,395	2,864	1,910	1,301	903.8	640.1
25	Mn	25,257	12,257	6,266	3,481	17,498	12,245	7,606	4,852	3,175	2,126	1,452	1,011	717.4
26	Fe	28,799	14,103	7,202	3,974	2,347	13,799?	8,612	5,518	3,626	2,437	1,669	1,165	828.6
27	Co	31,293	15,527	7,945	4,365	2,559	12,755	9,321	5,999	3,958	2,669	1,834	1,284	914.9
28	Ni	35,672	18,009	9,262	5,077	2,959	1,820	10,607	6,856	4,542	3,074	2,119	1,488	1,062
29	Cu	37,111	19,140	9,925	5,441	3,158	1,930	9,469	7,163	4,764	3,236	2,238	1,575	1,127
30	Zn	40,796	21,586	11,319	6,222	3,602	2,191	8,431?	7,826	5,224	3,561	2,471	1,744	1,251
31	Ga	42,943	23,414	12,450	6,878	3,979	2,411	1,527	7,035	5,494	3,758	2,615	1,851	1,331
32	Ge	45,420?	25,647	13,864	7,715	4,469	2,701	1,704	7,510?	5,885	4,040	2,820	2,002	1,443
33	As	40,611	27,954	15,399	8,650	5,027	3,036	1,908	1,244	5,430	4,359	3,052	2,173	1,570
34	Se	46,930?	29,433	16,564	9,408	5,495	3,320	2,083	1,353	5,690?	4,582	3,219	2,298	1,665
35	Br	56,392?	31,863?	18,364	10,564	6,211	3,761	2,357	1,527	1,023	4,286	3,519	2,519	1,830

36	Kr	49,240	27,206	19,453	11,352	6,727	4,088	2,563	1,658	1,107	4,490?	3,697	2,654	1,933
37	Rb	46,766	31,266	20,986	12,440	7,440	4,543	2,853	1,843	1,228	842.6	3,413	2,863	2,090
38	Sr	37,887	34,433?	22,328	13,461	8,134	4,997	3,146	2,033	1,353	925.8	3,641?	3,062	2,241
39	Y	18,359	30,587	19,613?	14,611	8,929	5,523	3,489	2,258	1,501	1,026	717.7	2,829	2,419
40	Zr	—	31,304	22,346	15,583	9,637	6,007	3,813	2,472	1,644	1,122	783.5	3,003?	2,573
41	Nb	3,822	29,760	25,061?	16,633	10,418	6,548	4,178	2,716	1,808	1,233	860.2	613.6	2,359
42	Mo	4,208	26,159	26,458?	17,407?	11,048	7,007	4,497	2,934	1,956	1,335	930.0	662.4	1,899?
43	Tc	4,712	18,656	23,753	16,961	11,911	7,625	4,926	3,227	2,156	1,472	1,026	729.6	529.5
44	Ru	5,093	4,160?	24,016	18,583	12,396	8,013	5,212	3,431	2,299	1,571	1,095	778.5	564.3
45	Rh	5,609	4,612	24,067	20,608?	13,152	8,587	5,626	3,722	2,502	1,714	1,195	849.3	615.2
46	Pd	6,048	5,017	22,758	17,440	13,682?	9,025	5,956	3,962	2,673	1,835	1,281	910.6	659.3
47	Ag	6,613	5,545	19,417?	18,651	14,138	9,636	6,408	4,286	2,904	1,998	1,397	993.9	719.5
48	Cd	6,988	5,938	4,213	18,814	15,215	9,971	6,681	4,494	3,058	2,111	1,478	1,053	762.3
49	In	7,490	6,469	4,577	18,706	15,401	10,502	7,090	4,798	3,279	2,271	1,594	1,136	823.4
50	Sn	7,870	6,933	4,895	17,404	13,412	10,416	7,407	5,042	3,462	2,405	1,692	1,209	876.4
51	Sb	8,267	7,457	5,260	3,563	14,012	11,334	7,776	5,323	3,672	2,560	1,806	1,292	938.1
52	Te	8,425?	7,818	5,516	3,719	14,034	11,020	7,966	5,485	3,801	2,660	1,882	1,349	980.6
53	I	9,187?	8,599	6,077	4,082	14,394	9,975	7,989	5,939	4,134	2,904	2,061	1,480	1,078
54	Xe	9,701	9,051	6,417	4,296	3,367?	10,405	8,561	6,166	4,311	3,039	2,163	1,557	1,136
55	Cs	10,431	9,690	6,902	4,610	3,094	10,928	8,489	6,527	4,583	3,242	2,314	1,670	1,220
56	Ba	10,915	10,111	7,249	4,835	3,232	11,052	7,357	6,108	4,759	3,379	2,419	1,750	1,281
57	La	11,654	10,726?	7,756	5,171	3,445	11,224?	7,894	6,655	5,039	3,590	2,577	1,869	1,370
58	Ce	12,397	11,467	8,292	5,532	3,676	2,497	8,439	6,542	5,338	3,814	2,745	1,996	1,466
59	Pr	13,230	12,021	8,862	5,923	3,928	2,658	8,992	7,100?	4,995	4,056	2,927	2,133	1,570
60	Nd	13,879	12,418	9,274	6,216	4,117	2,777	9,340	6,157	5,342	4,229	3,059	2,234	1,648
61	Pm	14,509	12,814	9,717	6,541	4,329	2,912	2,324?	6,538	5,724?	4,511	3,252	2,369	1,745
62	Sm	15,042	13,141	10,110	6,841	4,530	3,040	2,095	6,885	5,459	4,585?	3,334	2,446	1,811
63	Eu	15,744	13,638	10,606	7,224	4,789	3,208	2,204	7,299	5,854?	4,279	3,452	2,549	1,897
64	Gd	16,016	13,787	10,838?	7,439	4,941	3,306	2,265	7,515	4,969	4,490	3,487	2,590	1,937
65	Tb	16,532	14,177	11,469	7,826	5,214	3,487	2,382	2,004?	5,295	4,294	3,604	2,692	2,022
66	Dy	16,865	14,443	11,627	8,122	5,431	3,633	2,477	1,731	5,559	4,535	3,417	2,764	2,084
67	Ho	17,240	14,783	11,855	8,477	5,694	3,812	2,596	1,808	5,861	3,898?	3,636	2,857	2,162
68	Er	17,676	15,223	12,174	8,841	5,970	4,003	2,723	1,892	6,156	4,130	3,865?	2,955	2,243
69	Tm	17,968	15,595	12,451	9,244	6,279	4,220	2,869	1,990	1,823?	4,385	3,666	3,070?	2,335
70	Yb	18,049	15,848	12,648	9,519	6,507	4,386	2,983	2,065	1,464	4,581	3,845	2,984	2,397
71	Lu	18,204	16,244	12,978	9,919?	6,827	4,617	3,142	2,173	1,537	4,835	3,287	3,168	2,490
72	Hf	17,904	16,326	13,077	10,336	7,097	4,818	3,283	2,269	1,601	5,037	3,447	2,965	2,564
73	Ta	17,642	16,549	13,312	10,511	7,421	5,059	3,454	2,387	1,682	1,698?	3,632	3,133	2,490
74	W	17,167	16,706	13,522	10,673	7,740	5,301	3,627	2,507	1,764	1,267	3,811	3,300	2,623

(*continued*)

Table 14.3. (*Continued*)

	Absorber \ Emitter	183 eV B	277 eV C	392 eV N	525 eV O	677 eV F	849 eV Ne	1041 eV Na	1254 eV Mg	1487 eV Al	1740 eV Si	2014 eV P	2308 eV S	2622 eV Cl
75	Re	16,540	16,883	13,781	10,884	8,097	5,572	3,822	2,644	1,859	1,333	4,003	2,793	2,469
76	Os	15,557	16,900	13,947	11,032	8,399	5,808	3,996	2,766	1,945	1,392	1,509	2,916	2,583
77	Ir	14,376	16,964	14,198	11,259	8,893	6,119	4,223	2,928	2,058	1,472	1,071	3,073	2,730
78	Pt	12,828	16,945	14,435	11,490	9,073	6,417	4,443	3,086	2,170	1,550	1,126	3,219	2,293
79	Au	10,876	16,874	14,695	11,755	9,285	6,762	4,698	3,269	2,301	1,643	1,192	3,381?	2,418
80	Hg	8,312	16,509	14,777	11,899	9,407	7,062	4,923	3,433	2,418	1,727	1,251	1,499?	2,524
81	Tl	—	15,970	14,791	12,009	9,510	7,371?	5,156	3,604	2,541	1,815	1,314	966.5	2,633
82	Pb	—	15,338	14,828	12,163	9,655	7,646	5,420	3,798	2,681	1,916	1,386	1,018	2,752
83	Bi	—	14,635	14,939	12,410	9,884	7,827	5,721	4,018	2,841	2,031	1,469	1,078	1,398
84	Po	—	13,633	14,926	12,591	10,072	7,981	6,052	4,261	3,017	2,158	1,561	1,145	852.2
85	At	—	12,495	15,006	12,897	10,373	8,228	6,546?	4,531	3,214	2,301	1,665	1,220	907.2
86	Rn	—	10,352	14,109	12,405	10,045	7,981	6,346	4,548	3,231	2,315	1,676	1,228	912.0
87	Fr	—	8,457	13,838	12,510	10,214	8,134	6,467	4,790	3,409	2,445	1,771	1,297	962.8
88	Ra	—	6,089	13,323	12,459	10,274	8,206	6,526	4,985	3,553	2,551	1,848	1,354	1,005
89	Ac	—	—	12,728	12,413	10,359	8,306	6,610	5,276	3,722	2,676	1,939	1,421	1,054
90	Th	—	—	11,748	12,076	10,223	8,235	6,561	5,234	3,814	2,745	1,991	1,460	1,082
91	Pa	—	—	10,953	12,040	10,368	8,400	6,703	5,346	3,992	2,876	2,087	1,531	1,134
92	U	—	—	9,500	11,397	10,017	8,173	6,535	5,212	4,011	2,893	2,101	1,541	1,142
93	Np	—	—	8,185	11,053	9,956	8,190	6,567	5,239	4,200	2,991	2,174	1,595	1,182
94	Pu	—	—	6,449	10,307	9,564	7,945	6,392	5,103	4,088	2,990	2,175	1,597	1,184

	Absorber \ Emitter	2958 eV Ar	3314 eV K	3692 eV Ca	4091 eV Sc	4511 eV Ti	4952 eV V	5415 eV Cr	5899 eV Mn	6403 eV Fe	6930 eV Co	7478 eV Ni	8048 eV Cu	8639 eV Zn
1	H	0.2	0.1	0.1	0.1	0.0	0.0	0.0	0.0	0.0	0.0	0.0	0.0	0.0
2	He	1.6	1.1	0.8	0.5	0.4	0.3	0.2	0.2	0.1	0.1	0.1	0.1	0.0
3	Li	6.9	4.7	3.3	2.4	1.7	1.3	0.9	0.7	0.5	0.4	0.3	0.3	0.2
4	Be	20.8	14.4	10.1	7.2	5.3	3.9	2.9	2.2	1.7	1.3	1.0	0.8	0.6
5	B	46.9	32.8	23.3	16.8	12.4	9.2	6.9	5.3	4.1	3.2	2.5	2.0	1.6
6	C	91.1	64.6	46.5	34.1	25.3	19.1	14.5	11.2	8.7	6.9	5.4	4.4	3.5
7	N	148.6	105.7	76.4	56.0	41.7	31.4	24.0	18.5	14.4	11.4	9.0	7.2	5.8
8	O	224.9	160.6	116.5	85.7	63.9	48.3	36.9	28.5	22.3	17.5	13.9	11.2	9.0
9	F	303.3	217.9	158.6	117.1	87.6	66.3	50.8	39.3	30.7	24.2	19.3	15.5	12.5

10	Ne	430.2	310.9	227.5	168.7	126.6	96.1	73.7	57.2	44.8	35.4	28.2	22.6	18.3
11	Na	541.2	393.6	289.6	215.7	162.5	123.8	95.2	74.0	58.0	45.9	36.6	29.4	23.8
12	Mg	704.0	515.2	381.2	285.3	215.9	165.0	127.3	99.2	78.0	61.8	49.4	39.7	32.2
13	Al	842.6	620.4	461.6	347.2	263.8	202.4	156.7	122.4	96.5	76.6	61.3	49.4	40.1
14	Si	1,045	773.8	578.9	437.5	333.9	257.2	199.8	156.5	123.7	98.4	78.9	63.7	51.8
15	P	1,193	888.6	668.1	507.4	388.9	300.8	234.5	184.3	146.0	116.5	93.6	75.7	61.7
16	S	1,422	1,064	804.1	613.4	472.2	366.6	286.8	226.1	179.7	143.7	115.7	93.8	76.5
17	Cl	1,558	1,172	889.5	681.5	526.8	410.5	322.2	254.8	203.1	162.8	131.5	106.8	87.3
18	Ar	173.5	1,246	950.1	730.9	567.1	443.5	349.3	277.1	221.5	178.1	144.1	117.2	96.0
19	K	224.9	164.2	1,152	890.0	693.1	543.9	429.8	342.0	274.1	220.9	179.1	146.1	119.9
20	Ca	274.4	200.3	148.5	1,022	798.7	628.9	498.5	397.8	319.7	258.3	209.9	171.6	141.0
21	Sc	302.0	220.6	163.6	123.1	832?	657.2	522.4	418.1	336.8	272.8	222.2	182.0	149.9
22	Ti	346.1	252.9	187.6	141.2	107.7	83.1	570.9	458.0	369.9	300.3	245.2	201.2	166.0
23	V	393.1	287.5	213.3	160.6	122.5	94.6	73.8	498.4	403.4	328.2	268.5	220.8	182.5
24	Cr	460.6	337.2	250.4	188.5	143.8	111.1	86.7	68.4	454.8	370.8	303.9	250.4	207.3
25	Mn	517.0	378.8	281.4	212.0	161.8	125.0	97.6	76.9	61.3	401.9	330.1	272.4	225.9
26	Fe	598.0	438.6	326.1	245.8	187.7	145.0	113.2	89.3	71.1	57.1	370.2	306.0	254.1
27	Co	661.4	485.7	361.5	272.6	208.2	160.9	125.7	99.1	79.0	63.5	51.4	329.4	274.0
28	Ni	769.2	565.7	421.4	318.1	243.1	187.9	146.8	115.8	92.3	74.1	60.0	49.0	311.2
29	Cu	817.9	602.4	449.3	339.4	259.5	200.7	156.8	123.8	98.7	79.3	64.2	52.4	43.0
30	Zn	909.6	670.9	501.0	378.9	289.9	224.3	175.4	138.4	110.4	88.7	71.8	58.6	48.2
31	Ga	969.9	716.6	535.9	405.6	310.6	240.5	188.1	148.5	118.4	95.2	77.1	62.9	51.7
32	Ge	1,054	779.9	584.0	442.6	339.2	262.8	205.6	162.5	129.6	104.2	84.4	68.9	56.6
33	As	1,149	852.0	639.0	484.8	372.0	288.4	225.8	178.5	142.4	114.5	92.8	75.7	62.3
34	Se	1,221	907.0	681.3	517.6	397.6	308.5	241.7	191.1	152.5	122.7	99.5	81.2	66.8
35	Br	1,345	1,001	753.3	573.1	440.7	342.3	268.3	212.3	169.5	136.4	110.6	90.3	74.3
36	Kr	1,424	1,062	800.4	609.8	469.5	365.0	286.4	226.7	181.1	145.8	118.2	96.6	79.4
37	Rb	1,543	1,153	870.9	664.6	512.2	398.7	313.0	248.0	198.2	159.6	129.5	105.8	87.1
38	Sr	1,658	1,242	939.4	718.0	554.1	431.8	339.3	269.0	215.2	173.3	140.7	115.0	94.6
39	Y	1,794	1,346	1,020	780.9	603.6	470.8	370.4	293.9	235.2	189.6	153.9	125.9	103.6
40	Zr	1,913	1,438	1,092	837.4	648.1	506.2	398.6	316.6	253.5	204.5	166.1	135.9	111.9
41	Nb	2,048	1,543	1,174	901.5	698.8	546.5	430.8	342.4	274.5	221.5	180.0	147.3	121.3
42	Mo	2,156	1,628	1,241	954.4	740.9	580.2	457.9	364.3	292.2	235.9	191.9	157.1	129.4
43	Tc	1,984	1,749	1,336	1,029	800.2	627.4	495.7	394.8	317.0	256.1	208.4	170.7	140.7
44	Ru	1,564	1,818	1,391	1,074	836.0	656.4	519.3	414.0	332.6	269.0	219.0	179.5	148.0
45	Rh	452.7	1,656	1,479	1,144	892.0	701.4	555.5	443.3	356.6	288.6	235.1	192.8	159.1
46	Pd	484.8	1,302	1,545	1,197	934.7	736.0	583.7	466.3	375.4	304.1	247.9	203.4	167.9
47	Ag	528.7	394.9	1,409	1,274	996.5	785.7	623.9	499.0	402.1	326.0	266.0	218.4	180.4
48	Cd	560.1	418.0	1,087	1,315	1,031	813.9	647.1	518.2	418.0	339.2	277.0	227.5	188.1

(continued)

Table 14.3. *(Continued)*

Absorber	Emitter	2958 eV Ar	3314 eV K	3692 eV Ca	4091 eV Sc	4511 eV Ti	4952 eV V	5415 eV Cr	5899 eV Mn	6403 eV Fe	6930 eV Co	7478 eV Ni	8048 eV Cu	8639 eV Zn
49	In	604.9	451.3	341.1	1,188	1,086	859.1	683.9	548.3	442.8	359.6	293.9	241.6	199.8
50	Sn	644.0	480.4	362.9	916.4	1,128	893.6	712.3	571.8	462.2	375.7	307.3	252.8	209.2
51	Sb	689.7	514.5	388.6	297.3	1,011	934.8	746.2	599.7	485.3	394.9	323.3	266.2	220.4
52	Te	721.5	538.4	406.6	311.0	763.6	955?	763.6	614.4	497.7	405.4	332.2	273.7	226.8
53	I	793.7	592.5	447.5	342.2	264.7	881.1	821.9	662.1	537.0	437.8	359.1	296.1	245.5
54	Xe	837.3	625.5	472.6	361.4	279.5	669.0	728.3	684.6	555.9	453.7	372.4	307.3	255.0
55	Cs	900.7	673.4	509.0	389.3	301.1	235.4	767.3	722.2	587.0	479.6	394.1	325.5	270.3
56	Ba	946.8	708.5	535.9	410.0	317.1	247.8	579.7	639.2	606.2	495.8	407.7	337.1	280.1
57	La	1,015	760.1	575.3	440.3	340.6	266.2	210.0	673?	638.6	522.8	430.4	356.1	296.2
58	Ce	1,087	815.6	617.8	473.1	366.1	286.2	225.8	517.8	577.8	551.9	454.7	376.6	313.4
59	Pr	1,166	875.8	664.0	508.8	393.8	307.9	243.0	193.5	445.3	583.4	481.2	398.8	332.2
60	Nd	1,226	921.6	699.5	536.3	415.4	324.9	256.4	204.2	460.7	518.9	499.3	414.2	345.3
61	Pm	1,298	975.4	740.3	567.6	439.6	343.9	271.4	216.1	173.6	393.2	519.8	431.6	360.1
62	Sm	1,352	1,019	775.3	595.5	461.7	361.4	285.4	227.3	182.7	406.5	461.9	447.4	373.6
63	Eu	1,421	1,075	819.5	630.5	489.5	383.5	303.0	241.5	194.1	157.2	351.6	402?	391.5
64	Gd	1,457	1,105	844.6	651.0	506.1	396.9	313.8	250.2	201.2	162.9	358.8	410.6	400.3
65	Tb	1,526	1,161	889.3	686.6	534.5	419.6	332.0	264.8	213.0	172.5	140.8	311.9	359.0
66	Dy	1,578	1,204	923.8	714.4	556.9	437.6	346.5	276.5	222.6	180.3	147.2	321.6	370.4
67	Ho	1,642	1,255	965.2	747.6	583.5	459.0	363.7	290.4	233.8	189.5	154.7	127.1	279.8
68	Er	1,708	1,308	1,008	781.7	610.9	481.0	381.4	304.8	245.5	199.0	162.5	133.6	290.4
69	Tm	1,782	1,367	1,055	819.5	641.1	505.2	401.0	320.6	258.4	209.5	171.1	140.7	116.4
70	Yb	1,832	1,408	1,088	846.0	662.5	522.6	415.0	332.0	267.7	217.2	177.5	145.9	120.8
71	Lu	1,906	1,467	1,135	883.5	692.5	546.7	434.4	347.8	280.5	227.7	186.1	153.1	126.7
72	Hf	1,965	1,514	1,173	913.7	716.8	566.2	450.3	360.6	291.0	236.3	193.2	159.0	131.6
73	Ta	2,037	1,571	1,218	949.7	745.6	589.4	469.0	375.8	303.4	246.5	201.6	165.9	137.4
74	W	2,105	1,625	1,261	983.9	773.0	611.5	486.8	390.3	315.3	256.2	209.6	172.5	142.9
75	Re	2,181	1,685	1,308	1,022	803.2	635.7	506.4	406.2	328.3	266.9	218.4	179.8	149.0
76	Os	2,090	1,731	1,345	1,051	826.8	654.8	521.9	418.9	338.6	275.4	225.4	185.7	153.9
77	Ir	2,207	1,796	1,396	1,092	859.6	681.2	543.2	436.2	352.7	287.0	235.0	193.6	160.5
78	Pt	2,075	1,853?	1,442	1,129	889.1	705.0	562.5	451.8	365.6	297.5	243.7	200.9	166.6
79	Au	2,192	1,801	1,496	1,172	923.8	732.9	585.1	470.2	380.6	309.8	253.9	209.3	173.6
80	Hg	2,294	1,882	1,538	1,206	951.1	755.0	603.0	484.9	392.6	319.7	262.1	216.1	179.3
81	Tl	2,398?	1,770	1,472	1,240	978.8	777.6	621.4	499.9	404.9	329.9	270.5	223.1	185.2

82	Pb	2,003	1,859	1,544	1,280	1,012	804.3	643.2	517.7	419.5	341.9	280.5	231.4	192.1
83	Bi	2,110	1,962	1,471?	1,328	1,051	836.2	669.1	538.9	436.9	356.2	292.3	241.2	200.3
84	Po	2,228	2,077?	1,557	1,306	1,095	872.6	698.9	563.1	456.8	372.6	305.8	252.5	209.7
85	At	2,358	1,750	1,656	1,386	1,147	915.1	733.5	591.5	480.1	391.7	321.6	265.6	220.6
86	Rn	1,293	1,756	1,665	1,268	1,136	907.9	728.6	588.0	477.5	389.8	320.1	264.5	219.8
87	Fr	722.6	1,853	1,761	1,341	1,131	948.1	761.9	615.4	500.1	408.5	335.6	277.3	230.5
88	Ra	753.2	1,931	1,458	1,406	1,182	981.5	790.1	639.0	519.7	424.7	349.1	288.5	239.8
89	Ac	789.5	1,250	1,537	1,485	1,147	972.8	827.2	669.9	545.3	445.9	366.7	303.2	252.1
90	Th	810.0	613.8	1,591	1,542	1,191	1,007	851.4	690.8	563.0	460.7	379.1	313.5	260.8
91	Pa	848.9	642.6	1,677	1,292	1,268	991.9	899.7	731.8	597.4	489.3	402.8	333.3	277.3
92	U	854.4	646.4	1,151	1,326	1,304	1,020	864.7	749.4	613.0	502.7	414.1	342.8	285.3
93	Np	884.2	668.5	694.7	1,403	1,387	1,086	916.3	794.6	651.9	535.5	441.6	365.7	304.4
94	Pu	884.8	668.6	509.8	1,405	1,128	1,125	889.9	755.4	676.3	556.9	459.8	381.0	317.3

Absorber	Emitter	9252 eV Ga	9886 eV Ge	10,544 eV As	11,222 eV Se	11,924 eV Br	12,649 eV Kr	13,395 eV Rb	14,165 eV Sr	14,958 eV Y	15,775 eV Zr	16,615 eV Nb	17,479 eV Mo	18,367 eV Tc	19,279 eV Ru
1	H	0.0	0.0	0.0	0.0	0.0	0.0	0.0	0.0	0.0	0.0	0.0	0.0	0.0	0.0
2	He	0.0	0.0	0.0	0.0	0.0	0.0	0.0	0.0	0.0	0.0	0.0	0.0	0.0	0.0
3	Li	0.2	0.1	0.1	0.1	0.1	0.1	0.0	0.0	0.0	0.0	0.0	0.0	0.0	0.0
4	Be	0.5	0.4	0.3	0.3	0.2	0.2	0.2	0.1	0.1	0.1	0.1	0.1	0.1	0.0
5	B	1.3	1.0	0.8	0.7	0.6	0.5	0.4	0.3	0.3	0.2	0.2	0.2	0.1	0.1
6	C	2.9	2.3	1.9	1.6	1.3	1.1	0.9	0.8	0.7	0.6	0.5	0.4	0.4	0.3
7	N	4.7	3.9	3.2	2.6	2.2	1.8	1.5	1.3	1.1	0.9	0.8	0.7	0.6	0.5
8	O	7.3	6.0	4.9	4.1	3.4	2.9	2.4	2.0	1.7	1.5	1.3	1.1	0.9	0.8
9	F	10.2	8.3	6.9	5.7	4.8	4.0	3.3	2.8	2.4	2.0	1.8	1.5	1.3	1.1
10	Ne	14.9	12.2	10.1	8.4	7.0	5.9	4.9	4.2	3.5	3.0	2.6	2.2	1.9	1.7
11	Na	19.4	16.0	13.2	10.9	9.1	7.7	6.5	5.5	4.6	4.0	3.4	2.9	2.5	2.2
12	Mg	26.3	21.6	17.9	14.8	12.4	10.4	8.8	7.4	6.3	5.4	4.6	4.0	3.4	3.0
13	Al	32.8	27.0	22.3	18.6	15.5	13.0	11.0	9.3	7.9	6.8	5.8	5.0	4.3	3.8
14	Si	42.4	34.9	28.9	24.1	20.1	16.9	14.3	12.1	10.3	8.8	7.6	6.5	5.7	4.9
15	P	50.5	41.7	34.5	28.8	24.1	20.3	17.1	14.6	12.4	10.6	9.1	7.9	6.8	5.9
16	S	62.8	51.8	43.0	35.9	30.1	25.3	21.4	18.2	15.5	13.3	11.4	9.9	8.5	7.4
17	Cl	71.7	59.3	49.2	41.1	34.5	29.1	24.6	21.0	17.9	15.3	13.2	11.4	9.9	8.6
18	Ar	79.0	65.4	54.4	45.5	38.2	32.3	27.4	23.3	19.9	17.1	14.7	12.7	11.0	9.5
19	K	98.8	82.0	68.3	57.2	48.1	40.6	34.5	29.4	25.1	21.5	18.6	16.0	13.9	12.1
20	Ca	116.5	96.8	80.7	67.7	57.0	48.2	40.9	34.9	29.9	25.6	22.1	19.1	16.6	14.4
21	Sc	124.0	103.2	86.2	72.4	61.0	51.6	43.9	37.5	32.1	27.6	23.8	20.6	17.9	15.6
22	Ti	137.6	114.6	95.9	80.6	68.0	57.7	49.1	41.9	35.9	30.9	26.7	23.1	20.1	17.5
23	V	151.5	126.4	105.9	89.2	75.3	63.9	54.5	46.6	39.9	34.4	29.7	25.8	22.4	19.5

(continued)

Table 14.3. *(Continued)*

Absorber		9252 eV Ga	9886 eV Ge	10,544 eV As	11,222 eV Se	11,924 eV Br	12,649 eV Kr	13,395 eV Rb	14,165 eV Sr	14,958 eV Y	15,775 eV Zr	16,615 eV Nb	17,479 eV Mo	18,367 eV Tc	19,279 eV Ru
24	Cr	172.4	144.1	120.9	101.9	86.2	73.2	62.4	53.4	45.9	39.5	34.2	29.6	25.8	22.5
25	Mn	188.1	157.5	132.3	111.6	94.5	80.4	68.6	58.8	50.5	43.6	37.7	32.7	28.5	24.9
26	Fe	212.0	177.7	149.4	126.3	107.1	91.1	77.9	66.8	57.4	49.6	42.9	37.3	32.5	28.4
27	Co	228.9	192.0	161.7	136.8	116.1	98.9	84.6	72.6	62.5	54.0	46.8	40.7	35.5	31.0
28	Ni	260.3	218.7	184.4	156.1	132.7	113.1	96.9	83.2	71.7	62.0	53.7	46.8	40.8	35.7
29	Cu	271.2	228.1	192.5	163.2	138.8	118.5	101.5	87.3	75.3	65.1	56.5	49.2	42.9	37.6
30	Zn	39.8	249.3	210.6	178.7	152.2	130.0	111.5	95.9	82.8	71.7	62.2	54.2	47.4	41.5
31	Ga	42.8	35.6	221.4	188.0	160.2	137.0	117.6	101.2	87.4	75.7	65.8	57.4	50.1	43.9
32	Ge	46.8	39.0	32.6	201.9	172.1	147.3	126.5	109.0	94.2	81.7	71.0	61.9	54.2	47.5
33	As	51.5	42.9	35.9	30.2	186.0	159.2	136.8	118.0	102.0	88.5	77.0	67.2	58.8	51.6
34	Se	55.2	46.0	38.5	32.4	27.4	23.3	144.4	124.6	107.8	93.5	81.4	71.1	62.3	54.7
35	Br	61.5	51.2	42.8	36.1	30.5	25.9	22.1	136.5	118.2	102.6	89.4	78.1	68.4	60.1
36	Kr	65.8	54.8	45.8	38.6	32.6	27.7	23.7	20.3	124.7	108.3	94.4	82.5	72.3	63.5
37	Rb	72.1	60.0	50.3	42.3	35.8	30.4	26.0	22.3	19.2	117.1	102.1	89.3	78.3	68.8
38	Sr	78.4	65.3	54.7	46.0	38.9	33.1	28.3	24.2	20.8	18.0	109.4	95.7	84.0	73.9
39	Y	85.8	71.5	59.9	50.4	42.7	36.3	31.0	26.6	22.9	19.7	17.1	103.3	90.7	79.8
40	Zr	92.7	77.2	64.7	54.5	46.1	39.2	33.5	28.7	24.7	21.3	18.5	16.1	96.5	85.0
41	Nb	100.5	83.8	70.2	59.2	50.1	42.6	36.4	31.2	26.8	23.2	20.1	17.5	15.3	90.7
42	Mo	107.3	89.5	75.0	63.2	53.5	45.5	38.9	33.3	28.7	24.8	21.5	18.7	16.3	14.3
43	Tc	116.7	97.3	81.6	68.8	58.2	49.5	42.3	36.3	31.2	27.0	23.4	20.4	17.8	15.6
44	Ru	122.8	102.5	85.9	72.4	61.3	52.2	44.6	38.2	32.9	28.5	24.7	21.5	18.8	16.4
45	Rh	132.0	110.2	92.4	77.9	66.0	56.2	48.0	41.2	35.5	30.7	26.6	23.1	20.2	17.7
46	Pd	139.4	116.4	97.7	82.4	69.8	59.4	50.8	43.6	37.5	32.5	28.2	24.5	21.4	18.7
47	Ag	149.8	125.2	105.0	88.6	75.1	63.9	54.7	46.9	40.4	35.0	30.3	26.4	23.1	20.2
48	Cd	156.3	130.6	109.6	92.5	78.4	66.8	57.1	49.0	42.3	36.5	31.7	27.5	24.1	21.1
49	In	166.1	138.9	116.6	98.5	83.5	71.1	60.8	52.2	45.0	38.9	33.8	29.4	25.7	22.5
50	Sn	174.1	145.6	122.3	103.3	87.6	74.7	63.9	54.9	47.3	40.9	35.5	30.9	27.0	23.7
51	Sb	183.5	153.5	129.0	109.0	92.5	78.8	67.5	58.0	50.0	43.2	37.6	32.7	28.6	25.0
52	Te	188.9	158.2	133.0	112.4	95.4	81.3	69.6	59.8	51.6	44.7	38.8	33.8	29.5	25.9
53	I	204.6	171.4	144.2	122.0	103.6	88.3	75.6	65.0	56.1	48.5	42.2	36.7	32.1	28.2
54	Xe	212.7	178.3	150.1	127.0	107.9	92.0	78.8	67.7	58.5	50.6	44.0	38.3	33.5	29.4
55	Cs	225.6	189.2	159.3	134.9	114.6	97.8	83.8	72.1	62.2	53.9	46.8	40.8	35.7	31.3
56	Ba	233.9	196.3	165.4	140.1	119.1	101.7	87.1	75.0	64.7	56.1	48.7	42.5	37.2	32.6
57	La	247.5	207.8	175.2	148.5	126.3	107.8	92.5	79.6	68.7	59.5	51.8	45.2	39.5	34.7

58	Ce	262.1	220.3	185.8	157.5	134.0	114.5	98.2	84.6	73.0	63.3	55.1	48.0	42.0	36.9
59	Pr	278.0	233.8	197.3	167.4	142.5	121.8	104.5	90.0	77.8	67.4	58.7	51.2	44.8	39.3
60	Nd	289.2	243.3	205.5	174.4	148.6	127.0	109.1	93.9	81.2	70.4	61.3	53.5	46.8	41.1
61	Pm	301.8	254.1	214.8	182.4	155.4	132.9	114.2	98.4	85.1	73.8	64.2	56.1	49.1	43.1
62	Sm	313.3	264.0	223.3	189.7	161.7	138.4	118.9	102.5	88.7	77.0	67.0	58.5	51.3	45.0
63	Eu	328.6	277.1	234.5	199.3	170.0	145.6	125.1	107.9	93.4	81.1	70.6	61.7	54.0	47.5
64	Gd	336.3	283.7	240.3	204.4	174.4	149.4	128.5	110.9	96.0	83.3	72.6	63.4	55.6	48.9
65	Tb	351.8	297.0	251.7	214.2	182.9	156.8	134.9	116.4	100.8	87.6	76.3	66.7	58.5	51.4
66	Dy	363.2	306.9	260.2	221.6	189.3	162.4	139.8	120.7	104.6	90.9	79.2	69.3	60.7	53.4
67	Ho	323.7	319.0	270.7	230.7	197.2	169.2	145.7	125.9	109.1	94.8	82.7	72.3	63.4	55.8
68	Er	244.7	331.5	281.4	240.0	205.3	176.2	151.8	131.2	113.8	98.9	86.3	75.5	66.2	58.3
69	Tm	255.0	296.4	293.4	250.3	214.3	184.0	158.6	137.2	119.0	103.5	90.3	79.1	69.4	61.1
70	Yb	261.7	221.7	301.3	257.2	220.3	189.3	163.2	141.2	122.5	106.7	93.1	81.5	71.6	63.0
71	Lu	272?	230.4	268.5	267.3	229.1	196.9	169.9	147.1	127.7	111.2	97.1	85.0	74.7	65.7
72	Hf	109.6	237.1	201.9	236.0	235.9	202.9	175.2	151.7	131.7	114.7	100.2	87.8	77.1	67.9
73	Ta	114.4	245?	209.0	244.2	244.1	210.1	181.5	157.2	136.6	119.0	104.0	91.2	80.1	70.6
74	W	119.1	99.8	215.7	184.6	216.0	216.8	187.3	162.4	141.1	123.0	107.6	94.3	82.9	73.1
75	Re	124.2	104.1	223?	191.0	164.0	224.2	193.8	168.1	146.2	127.5	111.5	97.8	86.0	75.8
76	Os	128.3	107.6	90.6	195.9	168.3	197.0	198.6	172.3	149.9	130.8	114.5	100.4	88.3	77.9
77	Ir	133.8	112.2	94.6	203?	174.4	150.4	176.4	178.4	155.3	135.6	118.7	104.1	91.6	80.9
78	Pt	138.9	116.5	98.2	83.2	179.8	155.1	181.5	183.7	160.0	139.7	122.3	107.4	94.6	83.4
79	Au	144.8	121.5	102.4	86.8	186?	160.7	139.2	163.0	165.5	144.6	126.7	111.3	98.0	86.5
80	Hg	149.6	125.6	105.9	89.7	76.4	164.9	143.0	124.2	169.6	148.2	129.9	114.1	100.6	88.8
81	Tl	154.6	129.7	109.4	92.7	79.0	67.5	146.7	127.6	149.0	151.9	133.1	117.0	103.1	91.1
82	Pb	160.4	134.6	113.5	96.3	82.0	70.1	151.2	131.5	114.7	134.0	137.0	120.5	106.2	93.8
83	Bi	167.2	140.4	118.5	100.5	85.6	73.2	62.9	136.2	118.9	138.5	141.6	124.5	109.8	97.1
84	Po	175.1	147.1	124.1	105.3	89.7	76.7	65.9	141.6	123.7	108.2	126.0	129.2	114.0	100.8
85	At	184.3	154.8	130.7	110.9	94.5	80.8	69.5	59.9	129.2	113.1	99.3	115.5	118.8	105.1
86	Rn	183.6	154.3	130.2	110.5	94.2	80.6	69.3	59.8	127.6	111.8	98.2	113.8	117.1	103.6
87	Fr	192.6	161.9	136.7	116.0	98.9	84.6	72.8	62.8	54.4	116.3	102.2	90.9	104.2	107.5
88	Ra	200.5	168.5	142.3	120.8	103.0	88.1	75.8	65.4	56.7	119.9	105.4	92.9	82.1	110.5
89	Ac	210.8	177.2	149.6	127.0	108.3	92.7	79.8	68.9	59.7	51.9	109.8	96.8	85.5	98.4
90	Th	218.0	183.3	154.8	131.5	112.1	96.0	82.6	71.3	61.8	53.7	112.4	99.1	87.6	77.6
91	Pa	231.9	195.0	164.7	139.9	119.3	102.2	87.9	75.9	65.8	57.2	49.9	104.2	92.1	81.7
92	U	238.6	200.7	169.5	144.0	122.8	105.2	90.5	78.1	67.7	58.9	51.4	105.9	93.7	83.1
93	Np	254.6	214.2	180.9	153.7	131.1	112.3	96.6	83.4	72.3	62.9	54.9	48.1	98.8	87.6
94	Pu	265.5	223.3	188.7	160.2	136.7	117.1	100.8	87.0	75.4	65.6	57.3	50.1	101.6	90.1

[a] Heinrich (1986).

Table 14.4. Mass Absorption Coefficients for $L\alpha$ Lines[a] (? denotes significant uncertainty due to the presence of an absorption edge)

Absorber	Emitter	341 eV Ca	395 eV Sc	452 eV Ti	511 eV V	573 eV Cr	637 eV Mn	705 eV Fe	776 eV Co	852 eV Ni	930 eV Cu	1012 eV Zn	1098 eV Ga	1188 eV Ge
1	H	245.1	153.0	98.9	66.3	45.5	32.1	22.9	16.6	12.2	9.1	6.8	5.2	4.0
2	He	1,724	1,110	737.5	506.2	354.9	254.6	184.7	136.0	100.6	75.8	57.5	44.0	33.9
3	Li	5,818	3,833	2,602	1,821	1,299	947.6	697.9	520.9	390.6	297.4	228.1	176.2	137.1
4	Be	13,642	9,180	6,353	4,524	3,281	2,429	1,814	1,371	1,041	801.6	621.4	484.8	380.6
5	B	23,681	16,289	11,495	8,330	6,138	4,610	3,490	2,672	2,053	1,599	1,253	987.2	782.4
6	C	32,820	23,160	16,728	12,377	9,300	7,109	5,474	4,258	3,323	2,625	2,085	1,664	1,335
7	N	2,227	1,523	23,471	17,534	13,300	10,261	7,974	6,260	4,928	3,926	3,144	2,530	2,046
8	O	3,598	2,480	1,751	1,268	17,766	13,810	10,813	8,552	6,785	5.445	4,391	3,559	2,897
9	F	5,134	3,566	2,534	1,845	1,366	1,031	13,170?	10,482	8,368	6.757	5,483	4,471	3,662
10	Ne	7,657	5,356	3,830	2,804	2,086	1,580	1,207	933.2	724.8	8,870	7,236	5,931	4,884
11	Na	10,088	7,103	5,109	3,761	2,811	2,138	1,638	1,270	989.0	780.6	620.4	7,014	5.801
12	Mg	13,706	9,709	7,022	5,196	3,900	2,978	2,291	1,781	1,391	1,100	875.7	701.7	565.9
13	Al	17,092	12,174	8,852	6,580	4,961	3,803	2,936	2,290	1,793	1,421	1,134	910.1	735.0
14	Si	22,036	15,775	11,525	8,606	6,517	5,015	3,885	3,040	2,387	1,897	1,517	1,219	986.3
15	P	26,112	18,779	13,782	10,335	7,857	6,068	4,718	3,704	2,916	2,324	1,862	1,500	1,215
16	S	32,214	23,267	17,146	12,909	9,851	7,635	5,956	4,691	3,704	2,959	2,377	1,919	1,558
17	Cl	36,475	26,447	19,565	14,784	11,322	8,805	6,891	5,444	4,311	3,453	2,780	2,250	1,830
18	Ar	39,810	28,969	21,508	16,308	12,531	9,777	7,675	6,081	4,830	3,879	3,131	2,539	2,069
19	K	42,264	35,964?	26,790	20,379	15,709	12,293	9,679	7,691	6,125	4,932	3,990	3,243	2,649
20	Ca	4,489?	36,066	31,408?	23,964	18,527	14,540	11,481	9,147	7,305	5,897	4,782	3,896	3,188
21	Sc	4,935	3,756	28,503	25,418?	19,706	15,508	12,279	9,808	7,853	6,355	5,166	4,218	3,459
22	Ti	5,630	4,262	3,295?	24,074	21,810?	17,208	13,659	10,938	8,780	7,122	5,802	4,748	3,902
23	V	6,355	4,789	3,684	2,898?	20,528	18,922?	15,057	12,086	9,725	7,906	6,456	5,294	4,359
24	Cr	7,395	5,552	4,253	3,331	2,650?	18,418	17,120?	13,774	11,108	9,051	7,406	6,086	5,022
25	Mn	8,244	6,172	4,712	3,677	2,913	2,347?	15,993	15,029?	12,146	9,918	8,133	6,697	5,536
26	Fe	9,488	7,093	5,401	4,202	3,317	2,663	2,157?	14,502	13,688?	11,200	9,202	7,592	6,289
27	Co	10,470	7,824	5,949	4,617	3,635	2,910	2,349	1,917?	12,654	12,090?	9,953	8,228	6,828
28	Ni	12,193	9,121	6,933	5,373	4,221	3,370	2,713	2,208	1,806?	11,776	11,320?	9,374	7,794
29	Cu	13,036	9,775	7,436	5,760	4,519	3,602	2,893	2,348	1,916	1,582?	10,099	9,765?	8,133
30	Zn	14,816	11,150	8,498	6,586	5,166	4,112	3,297	2,671	2,174	1,791	1,485	9,128	7,615
31	Ga	16,222	12,266	9,377	7,279	5,712	4,545	3,640	2,945	2,392	1,967	1,627	1,355	7,965
32	Ge	17,963	13,663	10,487	8,162	6,413	5,105	4,088	3,304	2,680	2,200	1,817	1,509	1,262
33	As	19,824	15,181	11,710	9,145	7,202	5,740	4,598	3,716	3,012	2,469	2,036	1,689	1,409
34	Se	21,167	16,335	12,673	9,939	7,851	6,270	5,028	4,065	3,294	2,699	2,223	1,842	1,534

35	Br	23,274	18,117	14,148	11,150	8,840	7,078	5,687	4,602	3,731	3.057	2,517	2,083	1,733
36	Kr	24,433	19,200	15,102	11,968	9,529	7,655	6,165	4,998	4,056	3,324	2,737	2,264	1,883
37	Rb	26,100?	20,724	16,429	13,099	10,481	8,451	6,826	5,545	4,508	3,698	3.046	2,520	2,094
38	Sr	23,101	22,061	17,639	14,156	11,386	9,220	7,473	6,087	4,958	4,073	3,358	2,779	2,310
39	Y	26,314	23,528?	18,985	15,344	12,412	10,096	8,214	6,711	5,480	4,510	3,723	3,085	2,565
40	Zr	28,634?	21,990	20,065	16,340	13,297	10,870	8,880	7,280	5,961	4,917	4,066	3,373	2,807
41	Nb	25,679	24,698?	18,619	17,413	14,262	11,719	9,616	7,912	6,499	5,374	4,453	3,700	3,083
42	Mo	26,680	26,118?	20,828	18,194?	15,001	12,393	10,216	8,439	6,955	5,767	4,790	3,987	3,328
43	Tc	26,935	23,504	23,371?	18,042	16,033	13,321	11,034	9,151	7,569	6,296	5,242	4,373	3,656
44	Ru	24,605	23,846	19,827	19,696	15,220	13,821	11,506	9,583	7,956	6,638	5,542	4,634	3,882
45	Rh	20,976	24,013	21,157	20,664?	17,063	14,619?	12,232	10,232	8,527	7,138	5,977	5,010	4,206
46	Pd	14,680?	22,869	21,811	18,311	18,652?	14,657	12,750	10,713	8,963	7,528	6,322	5,313	4,471
47	Ag	4,594	19,810	21,919	19,431	19,236?	16,313	12,803	11,395	9,571	8,066	6,795	5,726	4,830
48	Cd	4,913	4,175	19,879	19,361	16,559	16,296?	13,837	11,746?	9,906	8,377	7,078	5,981	5,057
49	In	5,348	4,535	16,393?	18,886	17,183	14,418	15,119?	12,037	10,434	8,854	7,504	6,358	5,389
50	Sn	5,729	4,850	4,090	17,012?	17,095	14,964	12,356	13,000?	10,322	9,213	7,832	6,654	5,654
51	Sb	6,165	5,212	4,386	3,703	16,413	15,313	13,039	12,948?	11,235	8,992	8,214	6,998	5,961
52	Te	6,471	5,465	4,590	3,867	3,747?	14,879	13,247	11,096	10,928	9,570	8,408?	7,182	6,132
53	I	7,133	6,021	5,049	4,246	3,581	14,555?	13,876	11,994	9,896	9,777?	8,603	7,746	6,630
54	Xe	7,532	6,356	5,325	4,471	3,763	3,190	13,433	12,211	10,330	8,508	9,204?	7,446	6,873
55	Ca	8,097	6,837	5,725	4,800	4,034	3,413	2,896	12,365	10,861	9,111	9,107?	8,133	6,604
56	Ba	8,495	7,181	6,011	5,036	4,226	3,570	3,024	2,577?	10,999	9,465	7,869	7,861	7,056
57	La	9,072	7,683	6,433	5,386	4,515	3,808	3,220	2,739	11,192?	9,954	8,413	6,951	6,949
58	Ce	9,676	8,215	6,885	5,763	4,827	4,067	3,434	2,916	2,482	10,370	8,954	7,477	7,526?
59	Pr	10,310	8,781	7,370	6,171	5,166	4,349	3,667	3,110	2,642	2,264?	9,487	8,028	6,656
60	Nd	10,747	9,190	7,728	6,476	5,421	4,561	3,842	3,253	2,760	2,361	9,784?	8,421	7,040
61	Pm	11,210?	9,632	8,120	6,813	5,706	4,798	4,039	3,416	2,894	2,472	2,121	8,799	7,442
62	Sm	11,785?	10,023	8,477	7,125	5,971	5,022	4,225	3,570	3,020	2,576	2,207	2,208?	7,791
63	Eu	12,175	10,518	8,929	7,521	6,310	5,309	4,466	3,772	3,187	2,715	2,323	1,995	8,192
64	Gd	12,259	10,751?	9,165	7,741	6,505	5,477	4,608	3,890	3,285	2,795	2,388	2,048	2,088?
65	Tb	12,563	11,410?	9,607	8,139	6,854	5,777	4,862	4,105	3,464	2,945	2,513	2,153	1,852
66	Dy	12,764	11,565	9,930	8,442	7,125	6.015	5,066	4,278	3,609	3,067	2,614	2,237	1,922
67	Ho	13,039	11,791	10,317?	8,805	7,451	6,301	5,314	4,489	3,788	3,217	2,740	2,342	2,010
68	Er	13,410	12,107	10,706?	9,175	7,788	6,600	5,573	4,712	3,977	3,377	2,875	2,456	2,105
69	Tm	13,731	12,382	11,179	9,585	8,163	6,934	5,866	4,964	4,192	3,561	3,030	2,586	2,215
70	Yb	13,958	12,577	11,341	9,862	8,428	7,178	6,083	5,156	4,357	3,702	3,150	2,688	2,300
71	Lu	14,326	12,905	11,626	10,266?	8,806	7,520	6,387	5,421	4,587	3,899	3,319	2,831	2,422
72	Hf	14,432	13,002	11,708	10,577	9,115	7,807	6,646	5,651	4,787	4,073	3,468	2,958	2,530
73	Ta	14,681	13,237	11,917	10,758	9,488?	8,151	6,956	5,926	5,027	4,281	3,648	3,112	2,661

(*continued*)

Table 14.4. (*Continued*)

Absorber	Emitter	341 eV Ca	395 eV Sc	452 eV Ti	511 eV V	573 eV Cr	637 eV Mn	705 eV Fe	776 eV Co	852 eV Ni	930 eV Cu	1012 eV Zn	1098 eV Ga	1188 eV Ge
74	W	14,894	13,446	12,108	10,926	9,877	8,488	7,262	6,199	5,267	4,491	3,830	3,269	2,796
75	Re	15,149	13,704	12,349	11,143	10,067	8,865?	7,604	6,505	5,537	4,727	4,035	3,446	2,948
76	Os	15,291	13,871	12,514	11,295	10,201	9,245	7,895	6,769	5,772	4,935	4,216	3,604	3,085
77	Ir	15,511	14,123	12,763	11,527	10,411	9,431	8,285?	7,118	6,081	5,207	4,454	3,811	3,264
78	Pt	15,698	14,361	13,009	11,761	10,626	9,623	8,717	7,451	6,378	5,469	4,685	4,012	3,439
79	Au	15,890	14,623	13,287	12,031	10,876	9,850	8,920	7,838?	6,722	5,774	4,952	4,246	3,642
80	Hg	15,867	14,709	13,418	12,174	11,016	9,980	9,036	8,195	7,020	6,039	5,187	4,452	3,822
81	Tl	15,746	14,728	13,501	12,282	11,129	10,088	9,135	8,283	7,328?	6,314	5,430	4,667	4,010
82	Pb	15,618	14,771	13,622	12,433	11,286	10,240	9,275	8,409	7,617	6,626	5,707	4,910	4,223
83	Bi	15,531	14,889	13,832	12,676	11,534	10.478	9,496	8,610	7,797	7,084	6,020	5,186	4,465
84	Po	15,270	14,887	13,953	12,852	11,728	10,672	9,680	8,780	7,950	7,221	6,367?	5,490	4,732
85	At	15,046	14,978	14,192	13,151	12,044	10,982	9,973	9,051	8,197	7,443	6,763	5,829	5,029
86	Rn	13,796	14,097	13,534	12,634	11,622	10,624	9,663	8,776	7,951	7,219	6,558	5,841?	5,045
87	Fr	13,104	13,844	13,509	12,722	11,764	10,789	9,832	8,940	8,103	7,358	6,683	6,073	5,310
88	Ra	12,096	13,348	13,287	12,648	11,770	10,835	9,898	9,012	8,175	7,426	6,745	6,128	5,572
89	Ac	10,922	12,777	13,041	12,575	11,791	10,904	9,990	9,112	8,275	7,521	6,831	6,205	5,641
90	Th	9,315	11,823	12,456	12,203	11,546	10,737	9,871	9,023	8,205	7,463	6,780	6,159	5,597
91	Pa	7,720	11,062	12,143	12,129	11,602	10,859	10,025	9,189	8,370	7,620	6,926	6,293	5,718
92	U	5,517	9,641	11,178	11,439	11,088	10,459	9,703	8,922	8,143	7,423	6,752	6,136	5,576
93	Np	—	8,366	10,466	11,043	10,877	10,355	9,664	8,920	8,161	7,451	6,784	6,168	5,605
94	Pu	—	6,670	9,328	10,240	10,286	9,903	9,306	8,628	7,918	7,242	6,601	6,006	5,459

Absorber	Emitter	1282 eV As	1379 eV Se	1480 eV Br	1586 eV Kr	1694 eV Rb	1807 eV Sr	1923 eV Y	2042 eV Zr	2166 eV Nb	2293 eV Mo	2424 eV Tc	2559 eV Ru	2697 eV Rh
1	H	3.1	2.4	1.9	1.5	1.2	1.0	0.8	0.6	0.5	0.4	0.4	0.3	0.3
2	He	26.4	20.7	16.4	13.0	10.4	8.4	6.8	5.6	4.6	3.8	3.1	2.6	2.2
3	Li	107.3	84.8	67.4	53.8	43.4	35.1	28.6	23.5	19.3	16.0	13.3	11.2	9.4
4	Be	300.6	239.4	191.8	154.1	125.0	101.8	83.4	68.8	56.9	47.3	39.6	33.2	28.0
5	B	623.6	500.8	404.4	327.5	267.6	219.3	180.8	150.0	124.8	104.5	87.8	74.0	62.8
6	C	1,077	874.8	713.9	584.1	481.9	398.6	331.6	277.4	232.7	196.2	166.1	141.1	120.4
7	N	1,662	1,359	1,116	918.7	762.0	633.4	529.3	444.7	374.3	316.7	268.8	228.9	195.8
8	O	2,369	1,950	1,611	1,334	1,112	929.2	780.2	658.3	556.5	472.5	402.4	343.8	294.9
9	F	3,012	2,493	2,071	1,724	1,455	1,213	1,024	867.4	736.3	627.6	536.5	459.9	395.8

10	Ne	4,038	3,359	2,804	2,346	1,975	1,666	1,412	1,202	1,024	876.3	751.9	646.8	558.4
11	Na	4,819	4,026	3,376	2,837	2,399	2,032	1,729	1,477	1,264	1,085	934.5	806.6	698.7
12	Mg	459.3	5,009	4,217	3,557	3,019	2,567	2,192	1,880	1,614	1,391	1,202	1,041	904.0
13	Al	597.1	489.0	402.7	4,114	3,505	2,990	2,562	2,204	1,899	1,641	1,422	1,235	1,076
14	Si	802.3	657.7	542.0	448.2	373.9	312.9	3,108	2,681	2,317	2,009	1,746	1,520	1,328
15	P	990.1	812.6	670.3	554.8	463.0	387.6	326.5	276.6	2,607	2,266	1,975	1,724	1,510
16	S	1,271	1,044	862.5	714.5	596.8	499.9	421.3	357.1	303.4	259.2	222.3	2,041	1,792
17	Cl	1,495	1,231	1,018	844.0	705.6	591.5	498.8	422.9	359.5	307.2	263.5	226.8	196.1
18	Ar	1,694	1,397	1,156	960.2	803.5	674.2	568.9	482.6	410.4	350.9	301.0	259.2	224.1
19	K	2,173	1,794	1,488	1,237	1,036	870.4	735.0	624.0	531.0	454.1	389.7	335.7	290.3
20	Ca	2,620	2,168	1,800	1,499	1,257	1,057	893.6	759.2	646.5	553.2	475.0	409.2	354.1
21	Sc	2,848	2,360	1,964	1,637	1,357	1,158	979.6	833.0	709.9	607.8	522.2	450.1	389.6
22	Ti	3,220	2,673	2,227	1,860	1,564	1,319	1,117	950.7	810.9	694.9	597.3	515.1	446.1
23	V	3,604	2,998	2,502	2,093	1,763	1,488	1,262	1,075	917.9	787.2	677.2	584.4	506.3
24	Cr	4,160	3,466	2,898	2,428	2,048	1,731	1,470	1,254	1,072	919.8	791.9	683.8	592.8
25	Mn	4,595	3,836	3,213	2,696	2,278	1,928	1,639	1,400	1,197	1,029	886.5	766.1	664.6
26	Fe	5,230	4,373	3,669	3,084	2,609	2,211	1,883	1,610	1,379	1,186	1,023	884.6	767.9
27	Co	5,688	4,765	4,004	3,372	2,857	2,425	2,067	1,770	1,517	1,306	1,128	976.3	848.3
28	Ni	6,504	5,458	4,595	3,874	3,288	2,795	2,386	2,045	1,756	1,513	1,308	1,133	985.3
29	Cu	6,799	5,714	4,819	4,070	3,459	2,944	2,517	2,160	1,857	1,602	1,386	1,202	1,046
30	Zn	7,432	6,257	5,284	4,470	3,805	3,243	2,776	2,386	2,053	1,774	1,536	1,334	1,162
31	Ga	6,684	6,568	5,556	4,707	4,012	3,425	2,936	2,526	2,177	1,883	1,632	1,418	1,237
32	Ge	7,138	6,028	5,951	5,050	4,310	3,685	3,162	2,725	2,351	2,035	1,767	1,537	1,341
33	As	1,182	6,472?	5,491	5,438	4,648	3,979	3,419	2,950	2,548	2,209	1,919	1,672	1,460
34	Se	1,285	1,084	5,753?	4,895	4,884	4,186	3,602	3,112	2,691	2,336	2,032	1,771	1,549
35	Br	1,450	1,222	1,034	4,230	4,565	4,567	3,935	3,403	2,947	2,560	2,230	1,946	1,704
36	Kr	1,573	1,324	1,120	949.7	3,775?	4,108	4,131?	3,577	3,101	2,697	2,351	2,054	1,800
37	Rb	1,749	1,471	1,242	1,053	898.7	3,464?	3,810	3,303	3,341	2,909	2,539	2,221	1,948
38	Sr	1,929	1,621	1,368	1,158	987.8	844.8	726.5	3,523	3,062	3,111	2,718	2,380	2,090
39	Y	2,142	1,800	1,519	1,285	1,095	935.4	803.6	693.8	3,292?	2,873	2,929	2,567	2,257
40	Zr	2,346	1,971	1,663	1,406	1,198	1,023	877.8	757.3	654.9	2,359	2,670	2,730	2,402
41	Nb	2,578	2,168	1,829	1,547	1,317	1,124	964.3	831.3	718.3	623.6	2,193	2,501	2,203?
42	Mo	2,786	2,334	1,979	1,674	1,425	1,216	1,043	898.7	776.1	673.3	586.0	2,013	2,315
43	Tc	3,065	2,582	2,181	1,846	1,572	1,341	1,150	990.9	855.3	741.7	645.1	563.0	1,892?
44	Ru	3,260	2,749	2,325	1,969	1,678	1,432	1,228	1,058	913.0	791.4	688.1	600.1	525.5
45	Rh	3,538	2,989	2,530	2,145	1,829	1,562	1,340	1,154	996.2	863.3	750.5	654.3	572.7
46	Pd	3,768	3,188	2,703	2,294	1,958	1,673	1,436	1,238	1,068	925.7	804.6	701.3	613.6
47	Ag	4,079	3,457	2,936	2,495	2,132	1,823	1,566	1,350	1,166	1,010	878.2	765.4	669.6
48	Cd	4,280	3,635	3,092	2,631	2,251	1,927	1,656	1,429	1,234	1,070	930.4	811.0	709.5
49	In	4,572	3,890	3,315	2,825	2,420	2,074	1,785	1,541	1,332	1,155	1,005	875.9	766.3

(continued)

Table 14.4. (*Continued*)

Absorber	Emitter	1282 eV As	1379 eV Se	1480 eV Br	1586 eV Kr	1694 eV Rb	1807 eV Sr	1923 eV Y	2042 eV Zr	2166 eV Nb	2293 eV Mo	2424 eV Tc	2559 eV Ru	2697 eV Rh
50	Sn	4,808	4,099	3,499	2,987	2,562	2,199	1,894	1,636	1,415	1,229	1,069	932.2	815.7
51	Sb	5,079	4,339	3,711	3,173	2,726	2,342	2,020	1,747	1,512	1,314	1,143	997.7	873.3
52	Te	5,237	4,483	3,841	3,290	2,830	2,436	2,102	1,820	1,577	1,371	1,194	1,043	913.1
53	I	5,675	4,867	4,177	3,584	3,088	2,661	2,300	1,994	1,729	1,504	1,311	1,146	1,004
54	Xe	5,895	5,065	4,355	3,743	3,230	2,787	2,412	2,093	1,817	1,582	1,381	1,207	1,058
55	Cs	6,243	5,374	4,629	3,985	3,444	2,976	2,578	2,240	1,947	1,697	1,482	1,296	1,137
56	Ba	5,754	5,572	4,807	4,145	3,587	3,104	2,692	2,342	2,038	1,778	1,554	1,360	1,194
57	La	6,273	5,889?	5,089	4,394	3,808	3,299	2,866	2,495	2,174	1,898	1,660	1,455	1,278
58	Ce	6,832?	5,618	5,390	4,661	4,044	3,509	3,051	2,660	2,319	2,027	1,775	1,556	1,368
59	Pr	6,703	6,128	5,060	4,948	4,299	3,734	3,251	2,837	2,476	2,166	1,898	1,666	1,466
60	Nd	5,818	5,886	5,411	5,150?	4,480	3,895	3,394	2,965	2,590	2,268	1,989	1,747	1,538
61	Pm	6,186	6,284	5,798?	4,805	4,783	4,151	3,611	3,151	2,750	2,406	2,109	1,851	1,629
62	Sm	6,527	5,426	5,529	5,120	4,278	4,229	3,694	3,233	2,830	2,483	2,181	1,919	1,692
63	Ev	6,936	5,795	4,821?	4,925	4,604	4,351?	3,814	3,349	2,940	2,587	2,278	2,008	1,774
64	Gd	7,164	6,028	5,031	5,155	4,829?	4,050	3,844	3,386	2,980	2,628	2,320	2,049	1,814
65	Tb	7,523?	6,391	5,359	4,471	4,617	4,342	3,664	3,501	3,090	2,731	2,415	2,138	1,895
66	Dy	1,658	6,667	5,626	4,707	4,874?	4,094	3,876	3,290?	3,165	2,803	2,483	2,201	1,955
67	Ho	1,732	1,860?	5,928	4,981	4,188	4,349	4,123?	3,501	3,265	2,897	2,570	2,282	2,029
68	Er	1,811	1,567	6,224	5,260	4,434	3,733?	3,902	3,722	3,170	2,996	2,662	2,366	2,106
69	Tm	1,904	1,646	1,841?	5,558	4,703	3,966	4,155	3,531	3,379	2,894	2,768	2,463	2,194
70	Yb	1,976	1,706	1,478	5,770?	4,908	4,149	3,514	3,704	3,157	3,038	2,837	2,527	2,253
71	Lu	2,078	1,793	1,552	1,346?	5,171	4,386	3,721	3,930	3,351	3,225	2,774	2,625	2,342
72	Hf	2,170	1,870	1,617	1,401	1,675	4,583	3,898	3,322	3,521	3,018	2,916	2,517	2,412
73	Ta	2,282	1,965	1,698	1,470	1,281	4,802?	4,102	3,501	2,990	3,189	2,744	2,660	2,307
74	W	2,397	2,063	1,782	1,541	1,341	1,170?	4,294	3,676	3,143	3,358?	2,891	2,497	2,430
75	Re	2,527	2,175	1,877	1,623	1,411	1,229	1,568	3,864	3,309	2,842	3,052	2,637	2,564?
76	Os	2,645	2,276	1,964	1,697	1,475	1,284	1,122	4,015?	3,450	2,967	2,557	2,758	2,394
77	Ir	2,799	2,410	2,079	1,795	1,559	1,356	1,185	1,586?	3,627	3,126	2,697	2,914?	2,530
78	Pt	2,951	2,540	2,192	1,892	1,643	1,428	1,247	1,092	1,490	3,274	2,828	2,447	2,660
79	Au	3,127	2,693	2,323	2,006	1,741	1,513	1,320	1,156	1,014	3,437?	2,978	2,580	2,242
80	Hg	3,284	2,830	2,442	2,109	1,830	1,590	1,386	1,213	1,063	935?	3,101	2,692	2,342
81	Tl	3,448	2,973	2,567	2,217	1,924	1,671	1,456	1,274	1,116	980.9	1,435	2,805	2,444
82	Pb	3,634	3,135	2,708	2,340	2,030	1,763	1,537	1,344	1,177	1,034	910.4	1,363	2,559
83	Bi	3,846	3,320	2,869	2,480	2,152	1,869	1,629	1,424	1,246	1,095	963.4	850.1	2,685?

84	Po	4,079	3,524	3,047	2,635	2,288	1,987	1,731	1,514	1,324	1,162	1,023	901.9	1,422?
85	At	4,339	3,751	3,246	2,808	2,439	2,119	1,846	1,614	1,412	1,239	1,090	960.3	849.0
86	Rn	4,356	3,769	3,263	2,824	2,454	2,132	1,858	1,624	1,421	1,247	1,096	965.6	853.3
87	Fr	4,590	3,973	3,442	2,981	2,591	2,252	1,963	1,716	1,501	1,317	1,158	1,020	900.6
88	Ra	4,777	4,139	3,587	3,108	2,703	2,350	2,049	1,792	1,567	1,375	1,208	1,064	939.5
89	Ac	4,997?	4,333	3,758	3,257	2,834	2,465	2,150	1,880	1,645	1,443	1,268	1,116	985.5
90	Th	5,091	4,438	3,851	3,340	2,907	2,530	2,207	1,930	1,689	1,482	1,302	1,146	1,012
91	Pa	5,199	4,641?	4,030	3,497	3,045	2,651	2,313	2,024	1,771	1,554	1,365	1,202	1,061
92	U	5,069	4,616	4,050	3,516	3,062	2,667	2,328	2,037	1,783	1,565	1,375	1,210	1,068
93	Np	5,095	4,638	4,226	3,633	3,166	2,758	2,409	2,108	1,846	1,620	1,423	1,253	1,105
94	Pu	4,962	4,517	4,114	3,631	3,165	2,758	2,409	2,109	1,847	1,621	1,425	1,254	1,106

Absorber \ Emitter		2839 eV Pd	2984 eV Ag	3134 eV Cd	3287 eV In	3444 eV Sn	3605 eV Sb	3769 eV Te	3938 eV I	4110 eV Xe	4287 eV Cs	4466 eV Ba	4651 eV La	4840 eV Ce
1	H	0.2	0.2	0.2	0.1	0.1	0.1	0.1	0.1	0.1	0.1	0.0	0.0	0.0
2	He	1.9	1.6	1.3	1.1	1.0	0.8	0.7	0.6	0.5	0.5	0.4	0.4	0.3
3	Li	7.9	6.7	5.7	4.9	4.2	3.6	3.1	2.7	2.3	2.0	1.8	1.5	1.4
4	Be	23.7	20.2	17.2	14.7	12.7	10.9	9.5	8.2	7.1	6.2	5.4	4.8	4.2
5	B	53.4	45.6	39.1	33.6	29.0	25.1	21.8	19.0	16.6	14.5	12.8	11.2	9.9
6	C	103.1	88.7	76.5	66.2	57.4	50.0	43.7	38.2	33.6	29.5	26.1	23.1	20.4
7	N	168.0	144.8	125.0	108.3	94.2	82.1	71.8	62.9	55.2	48.6	43.0	38.0	33.7
8	O	253.7	219.1	189.6	164.6	143.3	125.0	109.5	96.0	84.5	74.5	65.9	58.3	51.7
9	F	341.5	295.7	256.4	223.1	194.6	170.2	149.3	131.1	115.5	101.9	90.2	80.0	71.0
10	Ne	483.3	419.6	364.9	318.3	278.3	243.8	214.3	188.6	166.4	147.0	130.4	115.7	102.8
11	Na	606.5	528.2	460.5	402.8	353.0	310.0	273.0	240.7	212.8	188.4	167.3	148.7	132.3
12	Mg	787.1	687.4	601.0	527.0	463.0	407.6	359.8	317.9	281.6	249.7	222.1	197.7	176.3
13	Al	939.9	823.1	721.6	634.3	558.7	492.9	436.0	386.1	342.7	304.5	271.4	241.9	216.1
14	Si	1,163	1,021	897.4	790.8	698.2	617.4	547.3	485.7	432.0	384.6	343.3	306.6	274.3
15	P	1,326	1,167	1,028	907.8	803.2	711.8	632.4	562.3	501.1	446.9	399.7	357.6	320.4
16	S	1,577	1,391	1,228	1,087	963.6	855.8	761.8	678.6	605.9	541.4	485.1	434.8	390.2
17	Cl	1,725?	1,525	1,349	1,196	1,063	945.7	843.4	752.8	673.3	602.7	541.0	485.6	436.5
18	Ar	194.4	169.4	147.9	1,272	1,132	1,009	901.6	806.1	722.3	647.6	582.2	523.4	471.2
19	K	251.9	219.5	191.6	167.9	147.5	130?	1,094	980.2	879.6	790.0	711.2	640.4	577.4
20	Ca	307.3	267.8	233.8	204.9	180.1	158.7	140.3	124.2	1,010	908.8	819.4	738.8	667.1
21	Sc	338.3	294.8	257.5	225.7	198.3	174.8	154.5	136.8	121.5	108.1	96.5	770.5	696.6
22	Ti	387.5	337.8	295.1	258.7	227.4	200.4	177.2	156.9	139.4	124.0	110.7	98.9	88.6
23	V	440.0	383.7	335.3	294.1	258.5	227.9	201.5	178.5	158.6	141.1	126.0	112.6	100.8
24	Cr	515.4	449.7	393.1	344.9	303.3	267.4	236.5	209.5	186.1	165.6	147.9	132.2	118.3

(continued)

Table 14.4. *(Continued)*

Absorber	Emitter	2839 eV Pd	2984 eV Ag	3134 eV Cd	3287 eV In	3444 eV Sn	3605 eV Sb	3769 eV Te	3938 eV I	4110 eV Xe	4287 eV Cs	4466 eV Ba	4651 eV La	4840 eV Ce
25	Mn	578.1	504.8	441.4	387.4	340.8	300.5	265.8	235.5	209.3	186.3	166.4	148.7	133.1
26	Fe	668.5	583.9	510.9	448.6	394.8	348.2	308.1	273.1	242.7	216.0	193.0	172.5	154.5
27	Co	738.9	645.9	565.5	496.7	437.3	385.9	341.6	302.8	269.2	239.7	214.1	191.4	171.4
28	Ni	858.9	751.3	658.1	578.4	509.5	449.8	398.3	353.2	314.1	279.7	249.9	223.4	200.2
29	Cu	912.8	799.0	700.4	615.9	542.8	479.4	424.7	376.7	335.1	298.5	266.8	238.6	213.8
30	Zn	1,014	888.6	779.5	685.9	604.8	534.5	473.7	420.4	374.1	333.3	298.0	266.6	238.9
31	Ga	1,081	947.6	831.9	732.5	646.3	571.5	506.8	449.9	400.5	357.0	319.3	285.7	256.1
32	Ge	1,173	1,030	904.6	797.1	703.8	622.7	552.4	490.7	437.1	389.7	348.6	312.1	279.8
33	As	1,279	1,123	987.4	870.7	769.3	681.1	604.6	537.4	478.8	427.2	382.3	342.3	307.0
34	Se	1,358	1,193	1,050	926.7	819.4	725.9	644.8	573.4	511.2	456.3	408.5	365.9	328.4
35	Br	1,494	1,315	1,158	1,023	905.0	802.3	713.1	634.5	566.0	505.5	452.7	405.7	364.2
36	Kr	1,581	1,392	1,227	1,085	960.6	852.2	758.0	674.9	602.3	538.2	482.3	432.4	388.3
37	Rb	1,712	1,509	1,332	1,178	1,044	926.9	825.0	735.0	656.4	586.9	526.2	472.0	424.0
38	Sr	1,839	1,622	1,432	1,268	1,125	999.5	890.2	793.7	709.2	634.4	569.1	510.8	459.1
39	Y	1,987	1,755	1,551	1,374	1,220	1,085	967.0	862.7	771.5	690.5	619.8	556.5	500.5
40	Zr	2,118	1,872	1,656	1,469	1,305	1,161	1,036	924.6	827.3	741.0	665.5	597.9	537.9
41	Nb	2,265	2,004	1,775	1,575	1,401	1,247	1,113	994.8	890.8	798.3	717.4	644.9	580.6
42	Mo	2,045	2,110	1,870	1,662	1,479	1,318	1,177	1,053	943.1	845.8	760.5	684.0	616.2
43	Tc	2,191	1,942	2,008	1,785	1,590	1,418	1,268	1,134	1,017	912.7	821.3	739.1	666.2
44	Ru	1,726?	2,015?	1,789	1,855	1,654	1,476	1,321	1,183	1,061	952.8	857.9	772.6	696.7
45	Rh	502.6	442.7	1,436	1,690	1,757	1,569	1,405	1,259	1,131	1,016	915.2	824.7	744.2
46	Pd	538.4	474.1	418.1	1,328	1,572	1,639?	1,468	1,316	1,183	1,064	958.9	864.6	780.6
47	Ag	587.4	517.0	455.9	403.3	1,254	1,493	1,339	1,400	1,259	1,133	1,022	922.1	833.1
48	Cd	622.2	547.6	482.7	426.9	378.5	1,152	1,381	1,240	1,300	1,171	1,057	954.3	862.7
49	In	672.1	591.4	521.3	460.9	408.5	362.9	1,083	1,520?	1,174	1,233	1,114	1,006	910.3
50	Sn	715.5	629.7	555.0	490.7	434.8	386.2	344.0	1,006?	906.1	1,098	1,157?	1,046	946.5
51	Sb	766.2	674.4	594.4	525.5	465.6	413.5	368.3	328.4	293.7	850.3	1,037	937.8	989.9
52	Te	801.4	705.5	621.9	549.9	487.2	432.7	385.3	343.6	307.2	275.1	782.6	957.5	867.6
53	I	881.3	776.1	684.3	605.2	536.3	476.2	424.1	378.1	338.1	302.7	271.8	758.7	688.0
54	Xe	929.4	818.8	722.2	638.8	566.2	502.9	447.8	399.3	357.0	319.6	287.0	257.9	707.7
55	Cs	999.5	880.9	777.3	687.8	609.7	541.6	482.4	430.2	384.6	344.3	309.1	277.8	250.0
56	Ba	1,050	926.1	817.5	723.6	641.7	570.1	507.9	453.0	405.0	362.6	325.6	292.5	263.3

57	La	1,125	992.4	876.5	776.1	688.5	612.0	545.3	486.4	435.0	389.4	349.7	314.2	282.8
58	Ce	1,205	1,064	940.0	832.8	739.1	657.1	585.6	522.5	467.4	418.5	375.8	337.7	304.0
59	Pr	1,292	1,141	1,009	894.2	793.9	706.1	629.5	561.9	502.7	450.2	404.4	363.4	327.1
60	Nd	1,357	1,199	1,061	940.9	835.8	743.7	663.3	592.2	529.9	474.7	426.5	383.3	345.1
61	Pm	1,436	1,270	1,123	995.8	884.6	787.0	702.0	626.7	560.8	502.4	451.4	405.7	365.3
62	Sm	1,495	1,323	1,172	1,040	925.1	823.9	735.4	657.1	588.4	527.4	474.0	426.2	383.9
63	Eu	1,569	1,391	1,234	1,097	976.5	870.5	777.7	695.3	623.0	558.8	502.5	452.0	407.3
64	Gd	1,607	1,427	1,267	1,128	1,005	896.7	801.8	717.4	643.3	577.3	519.4	467.5	421.3
65	Tb	1,682	1,495	1,329	1,185	1,057	943.7	844.5	756.3	678.6	609.3	548.5	493.9	445.4
66	Dy	1,737	1,546	1,376	1,228	1,096	979.9	877.6	786.4	706.1	634.4	571.4	514.8	464.4
67	Ho	1,805	1,609	1,434	1,280	1,144	1,023	917.2	822.5	738.9	664.3	598.6	539.5	486.9
68	Er	1,876	1,673	1,493	1,334	1,193	1,068	958.0	859.6	772.7	695.0	626.6	565.0	510.2
69	Tm	1,956	1,746	1,559	1,394	1,248	1,118	1,003	900.8	810.1	729.0	657.6	593.2	535.8
70	Yb	2,010	1,796	1,604	1,435	1,286	1,152	1,035	929.5	836.4	753.0	679.5	613.2	554.1
71	Lu	2,090	1,869	1,670	1,496	1,340	1,202	1,080	970.3	873.5	786.7	710.2	641.2	579.5
72	Hf	2,154	1,927	1,723	1,543	1,383	1,241	1,116	1,003	903.3	813.9	735.0	663.8	600.1
73	Ta	2,232	1,997	1,787	1,601	1,436	1,289	1,159	1,042	939.0	846.3	764.5	690.6	624.6
74	W	2,306?	2,064	1,848	1,656	1,486	1,334	1,200	1,079	972.8	877.1	792.5	716.2	647.9
75	Re	2,233	2,139	1,915	1,717	1,541	1,384	1,245	1,121	1,010	911.0	823.4	744.3	673.5
76	Os	2,334	2,041	1,966	1,764	1,583	1,423	1,280	1,153	1,039	937.5	847.6	766.3	693.6
77	Ir	2,204	2,155	1,889	1,830	1,643	1,477	1,329	1,197	1,080	974.4	881.2	796.9	721.5
78	Pt	2,317	2,027	1,984	1,745	1,696	1,525	1,373	1,237	1,116	1,007	911.4	824.4	746.5
79	Au	2,447	2,141	1,877	1,841	1,759?	1,582	1,425	1,284	1,159	1,046	946.9	856.7	776.0
80	Hg	2,042	2,240	1,965	1,924?	1,698	1,626	1,465	1,321	1,192	1,077	974.7	882.2	799.3
81	Tl	2,132	2,343	2,055	1,809	1,773	1,569	1,505	1,357	1,226	1,108	1,003	908.1	823.0
82	Pb	2,235	1,957	2,158	1,900	1,677	1,645	1,461	1,401	1,266	1,144	1,037	938.9	851.2
83	Bi	2,352	2,062	1,809	2,005	1,770	1,567	1,538	1,369	1,314	1,188	1,077	975.5	884.7
84	Po	2,478	2,177	1,912	1,685	1,874	1,660	1,474	1,446	1,290	1,238	1,122	1,017	923.1
85	At	1,378	2,306	2,029	1,789	1,993?	1,765	1,568	1,395	1,369	1,224	1,175	1,066	967.7
86	Rn	755.7	1,267	2,033	1,795	1,586	1,774	1,576	1,403	1,252	1,228	1,101	1,056	959.8
87	Fr	797.2	707.6	1,227?	1,893	1,675	1,484	1,667	1,484	1,325	1,185	1,162	1,043?	1,002
88	Ra	831.2	737.5	1,306	1,970	1,750	1,552	1,380	1,555	1,389	1,242	1,114	1,090	1,037?
89	Ac	871.6	773.0	686.3	1,268	1,838	1,636	1,456	1,643	1,467	1,313	1,178	1,058?	1,033
90	Th	894.5	793.0	703.8	626.2	1,212	1,689	1,508	1,344	1,523	1,362	1,223	1,099	1,069?
91	Pa	937.6	831.0	737.2	655.7	1,160?	1,190	1,598	1,428	1,277	1,451	1,302	1,170	1,054
92	U	943.9	836.4	741.8	659.6	587.5	1,088	1,605	1,462	1,310	1,173	1,339	1,203	1,084
93	Np	977.0	865.6	767.5	682.2	607.4	541.7	1,034	1,517	1,387	1,246	1,424	1,280	1,153
94	Pu	977.7	866.1	767.8	682.3	607.3	541.5	483.9	936.9	1,399	1,285	1,158	1,326	1,195

(*continued*)

Table 14.4. (*Continued*)

Absorber	Emitter	5034 eV Pr	5230 eV Nd	5433 eV Pm	5636 eV Sm	5846 eV Eu	6057 eV Gd	6273 eV Tb	6495 eV Dy	6720 eV Ho	6949 eV Er	7180 eV Tm	7416 eV Yb
1	H	0.0	0.0	0.0	0.0	0.0	0.0	0.0	0.0	0.0	0.0	0.0	0.0
2	He	0.3	0.2	0.2	0.2	0.2	0.1	0.1	0.1	0.1	0.1	0.1	0.1
3	Li	1.2	1.0	0.9	0.8	0.7	0.6	0.6	0.5	0.5	0.4	0.4	0.3
4	Be	3.7	3.3	2.9	2.6	2.3	2.0	1.8	1.6	1.4	1.3	1.2	1.0
5	B	8.7	7.8	6.9	6.1	5.5	4.9	4.4	3.9	3.5	3.2	2.8	2.6
6	C	18.1	16.1	14.4	12.9	11.5	10.3	9.3	8.4	7.5	6.8	6.2	5.6
7	N	29.9	26.7	23.8	21.3	19.0	17.1	15.4	13.8	12.5	11.3	10.2	9.3
8	O	45.9	41.0	36.5	32.7	29.3	26.3	23.7	21.3	19.2	17.4	15.8	14.3
9	F	63.1	56.3	50.3	45.0	40.4	36.3	32.7	29.4	26.6	24.0	21.8	19.8
10	Ne	91.5	81.7	73.0	65.5	58.7	52.8	47.6	42.9	38.8	35.1	31.8	28.9
11	Na	118.0	105.4	94.3	84.6	76.0	68.4	61.7	55.6	50.3	45.5	41.3	37.5
12	Mg	157.3	140.8	126.1	113.3	101.8	91.8	82.8	74.8	67.6	61.3	55.7	50.6
13	Al	193.1	173.1	155.2	139.6	125.6	113.4	102.4	92.6	83.8	76.0	69.1	62.8
14	Si	245.6	220.5	197.9	178.3	160.6	145.1	131.2	118.7	107.6	97.7	88.8	80.9
15	P	287.4	258.4	232.3	209.5	189.0	171.0	154.8	140.2	127.2	115.6	105.2	95.9
16	S	350.5	315.6	284.1	256.7	231.9	210.0	190.3	172.6	156.8	142.6	129.9	118.5
17	Cl	392.7	354.2	319.3	288.9	261.3	236.9	215.0	195.2	177.5	161.6	147.4	134.6
18	Ar	424.6	383.5	346.2	313.6	284.0	257.9	234.3	213.0	193.9	176.7	161.4	147.5
19	K	521.0	471.2	426.0	386.4	350.4	318.5	289.8	263.7	240.3	219.2	200.4	183.3
20	Ca	602.7	545.8	494.1	448.7	407.4	370.8	337.7	307.6	280.7	256.4	234.6	214.8
21	Sc	630.2	571.4	517.9	470.9	428.0	390.0	355.6	324.3	296.2	270.8	248.0	227.3
22	Ti	687.1	623.7	566.0	515.2	468.8	427.7	390.3	356.3	325.8	298.1	273.3	250.7
23	V	90.4	81.3	73.1	559.9	510.0	465.7	425.4	388.8	355.8	325.9	299.0	274.5
24	Cr	106.1	95.5	85.9	77.6	70.1	524.2	479.4	438.4	401.6	368.2	338.1	310.7
25	Mn	119.4	107.4	96.7	87.3	78.9	71.5	64.9	58.9	435.0	399.1	366.8	337.3
26	Fe	138.6	124.6	112.2	101.3	91.6	83.0	75.3	68.4	62.2	56.7	411.0	378.2
27	Co	153.8	138.4	124.5	112.5	101.7	92.1	83.6	75.9	69.1	63.0	57.5	52.6
28	Ni	179.6	161.6	145.4	131.4	118.8	107.7	97.7	88.7	80.7	73.6	67.2	61.4
29	Cu	191.8	172.6	155.4	140.4	126.9	115.1	104.4	94.8	86.3	78.7	71.9	65.7
30	Zn	214.4	193.0	173.8	157.0	141.9	128.7	116.8	106.1	96.6	88.0	80.4	73.5
31	Ga	229.9	207.0	186.4	168.5	152.3	138.1	125.3	113.9	103.6	94.5	86.3	78.9
32	Ge	251.3	226.3	203.8	184.2	166.6	151.0	137.1	124.6	113.4	103.4	94.4	86.4
33	As	275.8	248.4	223.8	202.3	183.0	165.9	150.7	136.9	124.6	113.6	103.8	94.9

34	Se	295.0	265.8	239.5	216.6	195.9	177.7	161.4	146.7	133.5	121.8	111.3	101.8
35	Br	327.3	295.0	265.9	240.5	217.6	197.5	179.3	163.0	148.4	135.4	123.7	113.2
36	Kr	349.1	314.8	283.8	256.8	232.4	210.9	191.6	174.2	158.6	144.7	132.2	121.0
37	Rb	381.4	344.0	310.2	280.8	254.2	230.8	209.7	190.7	173.7	158.4	144.8	132.5
38	Sr	413.1	372.7	336.3	304.5	275.7	250.4	227.6	207.0	188.6	172.0	157.3	144.0
39	Y	450.6	406.7	367.1	332.5	301.2	273.6	248.7	226.2	206.2	188.2	172.1	157.5
40	Zr	484.5	437.5	395.1	358.0	324.4	294.7	268.0	243.9	222.3	202.9	185.6	169.9
41	Nb	523.2	472.7	427.0	387.1	350.8	318.9	290.1	264.1	240.8	219.8	201.1	184.2
42	Mo	555.6	502.1	453.8	411.6	373.2	339.3	308.8	281.2	256.5	234.2	214.3	196.3
43	Tc	601.0	543.5	491.4	445.8	404.4	367.9	334.9	305.0	278.3	254.2	232.7	213.2
44	Ru	628.8	569.0	514.7	467.2	424.0	385.8	351.4	320.2	292.2	267.0	244.5	224.0
45	Rh	672.1	608.4	550.7	500.1	454.0	413.3	376.6	343.3	313.4	286.5	262.4	240.5
46	Pd	705.4	638.9	578.6	525.7	477.5	434.9	396.4	361.4	330.1	301.9	276.5	253.6
47	Ag	753.2	682.6	618.5	562.2	511.0	465.6	424.5	387.2	353.8	323.6	296.6	272.0
48	Cd	780.4	707.7	641.6	583.5	530.5	483.6	441.2	402.6	368.0	336.7	308.7	283.2
49	In	824.0	747.6	678.1	617.0	561.3	511.9	467.2	426.5	390.0	357.0	327.4	300.5
50	Sn	857.3	778.2	706.3	643.0	585.2	534.0	487.6	445.3	407.3	373.0	342.2	314.2
51	Sb	897.1	814.8	739.9	674.0	613.7	560.3	511.8	467.6	428.0	392.1	359.8	330.5
52	Te	917.0	833.4	757.2	690.1	628.7	574.2	524.8	479.7	439.2	402.5	369.6	339.5
53	I	846.0	896.6	815.1	743.2	677.4	619.0	566.0	517.6	474.1	434.7	399.3	367.0
54	Xe	642.5	794.0	722.2	767.9	700.4	640.3	585.7	535.9	491.1	450.5	413.9	380.6
55	Cs	674.8	614.3	761.0	694.6	738.8	675.7	618.4	566.1	519.0	476.2	437.8	402.7
56	Ba	237.3	214.5	574.9	716?	653.7	697.2	638.4	584.7	536.3	492.3	452.7	416.6
57	La	254.9	230.3	208.2	551.2	503.5	629.7	672?	616.1	565.3	519.2	477.6	439.7
58	Ce	274.0	247.6	223.8	203.0	529.4	485.0	608.1	557.5	596.5	548.0	504.4	464.5
59	Pr	294.8	266.5	240.8	218.5	198.2	511.1	468.6	588.8	540.8	579.4	533.5	491.5
60	Nd	311.1	281.1	254.1	230.5	209.1	190.3	484.7	444.7	409?	515.4	553.3	509.9
61	Pm	329.2	297.6	269.0	244.0	221.3	201.4	183.4	461.7	424.4	390.5	494.0	455.5
62	Sm	346.1	312.9	282.9	256.6	232.8	211.9	193.0	175.9	439?	403.7	372.2	471.6
63	Eu	367.3	332.2	300.4	272.5	247.3	225.1	205.0	186.9	170.6	156.0	389.0	358.9
64	Gd	380.2	343.9	311.0	282.3	256.2	233.2	212.5	193.7	176.8	161.7	148.2	366.3
65	Tb	402.0	363.7	329.1	298.7	271.2	246.9	225.0	205.1	187.3	171.3	157.0	144.0
66	Dy	419.3	379.5	343.5	311.9	283.2	257.9	235.0	214.3	195.7	179.0	164.0	150.5
67	Ho	439.8	398.2	360.5	327.5	297.4	270.9	246.9	225.2	205.7	188.1	172.4	158.2
68	Er	461.0	417.5	378.1	343.5	312.1	284.3	259.2	236.4	216.0	197.6	181.1	166.2
69	Tm	484.3	438.8	397.5	361.3	328.3	299.1	272.7	248.8	227.4	208.0	190.7	175.0
70	Yb	501.0	454.1	411.4	374.0	340.0	309.8	282.5	257.8	235.6	215.6	197.7	181.4
71	Lu	524.2	475.2	430.7	391.6	356.1	324.5	296.1	270.2	247.0	226.0	207.3	190.3
72	Hf	543.0	492.4	446.4	406.0	369.2	336.6	307.1	280.3	256.3	234.6	215.2	197.5

(continued)

Table 14.4. *(Continued)*

Absorber \ Emitter		5034 eV Pr	5230 eV Nd	5433 eV Pm	5636 eV Sm	5846 eV Eu	6057 eV Gd	6273 eV Tb	6495 eV Dy	6720 eV Ho	6949 eV Er	7180 eV Tm	7416 eV Yb
73	Ta	565.3	512.8	465.0	423.0	384.8	350.9	320.2	292.3	267.3	244.7	224.5	206.1
74	W	586.5	532.1	482.7	439.2	399.6	364.4	332.6	303.7	277.8	254.4	233.4	214.3
75	Re	609.9	553.4	502.1	457.0	415.8	379.3	346.3	316.3	289.3	264.9	243.1	223.2
76	Os	628.2	570.3	517.5	471.1	428.7	391.2	357.2	326.3	298.5	273.4	250.9	230.4
77	Ir	653.6	593.4	538.6	490.4	446.4	407.4	372.0	339.9	311.0	284.9	261.5	240.2
78	Pt	676.5	614.3	557.7	507.9	462.4	422.1	385.5	352.3	322.4	295.4	271.1	249.1
79	Au	703.4	638.9	580.2	528.4	481.2	439.3	401.3	366.8	335.7	307.6	282.4	259.5
80	Hg	724.7	658.4	598.0	544.8	496.2	453.1	414.0	378.4	346.4	317.5	291.5	267.9
81	Tl	746.4	678.3	616.2	561.5	511.5	467.2	426.9	390.3	357.4	327.6	300.8	276.5
82	Pb	772.2	701.9	637.9	581.4	529.7	483.9	442.3	404.4	370.3	339.5	311.8	286.6
83	Bi	802.9	730.0	663.6	605.0	551.4	503.7	460.5	421.2	385.7	353.7	324.9	298.7
84	Po	838.0	762.2	693.1	632.0	576.2	526.5	481.5	440.4	403.4	370.0	339.9	312.5
85	At	878.9	799.8	727.5	663.6	605.1	553.1	505.9	462.8	424.1	389.0	357.4	328.7
86	Rn	872.2	794.1	722.6	659.4	601.5	549.9	503.1	460.4	421.9	387.0	355.7	327.1
87	Fr	911.1	830.0	755.7	689.9	629.5	575.8	526.9	482.3	442.0	405.6	372.8	342.9
88	Ra	943.6	860.2	783.7	715.8	653.5	598.0	547.4	501.2	459.5	421.7	387.7	356.6
89	Ac	986.1	899.9	820.5	750.0	685.1	627.1	574.3	526.0	482.4	442.8	407.2	374.7
90	Th	963.9	925.3	844.6	772.7	706.4	647.0	592.8	543.1	498.3	457.5	420.8	387.3
91	Pa	1,023	924.9	892.6	817.6	748.1	685.8	628.7	576.4	529.0	485.9	447.0	411.5
92	U	977.1	947.5	857.2	835.8	765.9	702.8	644.9	591.6	543.3	499.3	459.5	423.1
93	Np	1,040	940.4	908.4	824.9	811.7	746.0	685.4	629.4	578.4	531.8	489.7	451.1
94	Pu	1,078	974.9	882.2	851.5	773.5	771.8	710.4	653.3	601.0	553.1	509.6	469.6

Absorber \ Emitter		7656 eV Lu	7899 eV Hf	8146 eV Ta	8398 eV W	8652 eV Re	8912 eV Os	9175 eV Ir	9442 eV Pt	9713 eV Au	9989 eV Hg	10,269 eV Tl	10,552 eV Pb
1	H	0.0	0.0	0.0	0.0	0.0	0.0	0.0	0.0	0.0	0.0	0.0	0.0
2	He	0.1	0.1	0.1	0.0	0.0	0.0	0.0	0.0	0.0	0.0	0.0	0.0
3	Li	0.3	0.3	0.2	0.2	0.2	0.2	0.2	0.1	0.1	0.1	0.1	0.1
4	Be	0.9	0.9	0.8	0.7	0.6	0.6	0.5	0.5	0.4	0.4	0.4	0.3
5	B	2.3	2.1	1.9	1.7	1.6	1.4	1.3	1.2	1.1	1.0	0.9	0.8
6	C	5.1	4.6	4.2	3.8	3.5	3.2	2.9	2.7	2.5	2.3	2.1	1.9
7	N	8.4	7.7	7.0	6.4	5.8	5.3	4.9	4.5	4.1	3.8	3.5	3.2

8	O	13.0	11.8	10.8	9.8	9.0	8.2	7.5	6.9	6.3	5.8	5.4	4.9
9	F	18.0	16.4	14.9	13.6	12.4	11.4	10.4	9.6	8.8	8.1	7.4	6.9
10	Ne	26.3	23.9	21.8	19.9	18.2	16.7	15.3	14.0	12.9	11.9	10.9	10.1
11	Na	34.2	31.1	28.4	25.9	23.7	21.7	19.9	18.3	16.8	15.5	14.3	13.1
12	Mg	46.1	42.0	38.3	35.0	32.1	29.4	27.0	24.8	22.8	21.0	19.3	17.8
13	Al	57.2	52.2	47.7	43.6	39.9	36.6	33.6	30.9	28.4	26.2	24.1	22.3
14	Si	73.7	67.3	61.5	56.3	51.6	47.3	43.5	40.0	36.8	33.9	31.2	28.8
15	P	87.5	79.9	73.1	66.9	61.4	56.3	51.8	47.6	43.9	40.4	37.3	34.4
16	S	108.2	99.0	90.6	83.0	76.2	70.0	64.3	59.2	54.6	50.3	46.4	42.9
17	Cl	123.0	112.6	103.2	94.6	86.9	79.8	73.5	67.7	62.4	57.5	53.1	49.1
18	Ar	134.9	123.6	113.3	104.0	95.6	87.9	80.9	74.6	68.8	63.5	58.7	54.3
19	K	167.8	153.9	141.3	129.7	119.3	109.8	101.2	93.3	86.2	79.6	73.6	68.1
20	Ca	196.8	180.6	165.9	152.5	140.4	129.3	119.3	110.1	101.7	94.0	86.9	80.5
21	Sc	208.5	191.5	176.1	162.0	149.2	137.6	126.9	117.2	108.4	100.2	92.8	86.0
22	Ti	230.2	211.6	194.7	179.3	165.3	152.5	140.8	130.1	120.4	111.4	103.2	95.7
23	V	252.2	232.1	213.7	196.9	181.7	167.7	155.0	143.4	132.7	122.9	113.9	105.7
24	Cr	285.7	263.1	242.4	223.6	206.4	190.7	176.3	163.2	151.1	140.1	129.9	120.6
25	Mn	310.4	286.1	263.8	243.4	225.0	207.9	192.4	178.2	165.1	153.1	142.1	132.0
26	Fe	348.4	321.3	296.5	273.8	253.1	234.1	216.8	200.9	186.3	172.8	160.5	149.1
27	Co	48.1	345.6	319.2	294.9	272.9	252.5	233.9	216.9	201.3	186.8	173.6	161.4
28	Ni	56.3	51.6	47.4	334.8	310.0	287.0	266.0	246.8	229.1	212.8	197.8	184.0
29	Cu	60.1	55.2	50.7	46.6	42.9	39.5	277.1	257.2	238.9	222.0	206.4	192.1
30	Zn	67.3	61.7	56.7	52.1	48.0	44.2	40.8	37.7	261.0	242.7	225.8	210.2
31	Ga	72.3	66.3	60.9	55.9	51.5	47.5	43.8	40.4	37.4	34.6	32.1	221.0
32	Ge	79.1	72.5	66.6	61.2	56.4	52.0	47.9	44.3	40.9	37.9	35.1	32.6
33	As	87.0	79.8	73.3	67.3	62.0	57.1	52.7	48.7	45.0	41.7	38.6	35.8
34	Se	93.2	85.5	78.5	72.2	66.5	61.3	56.5	52.2	48.3	44.7	41.4	38.4
35	Br	103.6	95.1	87.3	80.3	74.0	68.2	62.9	58.1	53.7	49.7	46.1	42.8
36	Kr	110.8	101.7	93.4	85.9	79.1	72.9	67.3	62.2	57.5	53.2	49.3	45.7
37	Rb	121.4	111.4	102.4	94.1	86.7	79.9	73.8	68.1	63.0	58.3	54.1	50.2
38	Sr	131.9	121.1	111.2	102.3	94.2	86.9	80.2	74.1	68.5	63.4	58.8	54.6
39	Y	144.3	132.5	121.7	112.0	103.2	95.1	87.8	81.1	75.1	69.5	64.4	59.8
40	Zr	155.8	143.0	131.4	120.9	111.4	102.7	94.8	87.6	81.1	75.1	69.6	64.6
41	Nb	168.8	155.0	142.5	131.1	120.8	111.4	102.9	95.1	88.0	81.5	75.5	70.1
42	Mo	180.0	165.3	152.0	139.9	128.9	118.9	109.8	101.5	93.9	87.0	80.6	74.8
43	Tc	195.5	179.6	165.2	152.0	140.1	129.3	119.4	110.4	102.2	94.6	87.7	81.4
44	Ru	205.5	188.8	173.7	159.9	147.4	136.0	125.6	116.2	107.5	99.6	92.3	85.7
45	Rh	220.7	202.8	186.6	171.8	158.4	146.2	135.1	124.9	115.6	107.1	99.3	92.2

(continued)

Table 14.4. (*Continued*)

Absorber	Emitter	7656 eV Lu	7899 eV Hf	8146 eV Ta	8398 eV W	8652 eV Re	8912 eV Os	9175 eV Ir	9442 eV Pt	9713 eV Au	9989 eV Hg	10,269 eV Tl	10,552 eV Pb
46	Pd	232.7	213.9	196.9	181.3	167.2	154.3	142.6	131.9	122.1	113.2	105.0	97.5
47	Ag	249.8	229.6	211.4	194.7	179.7	165.8	153.3	141.8	131.3	121.7	112.9	104.8
48	Cd	260.1	239.2	220.3	202.9	187.3	172.9	159.8	147.9	137.0	127.0	117.8	109.4
49	In	276.1	254.0	233.9	215.6	199.0	183.8	169.9	157.3	145.7	135.0	125.3	116.4
50	Sn	288.8	265.7	244.8	225.7	208.4	192.5	178.0	164.8	152.7	141.6	131.4	122.1
51	Sb	303.8	279.7	257.7	237.7	219.5	202.8	187.6	173.7	161.0	149.3	138.6	128.8
52	Te	312.3	287.6	265.1	244.5	225.9	208.8	193.2	178.9	165.8	153.8	142.8	132.7
53	I	337.6	311.0	286.8	264.6	244.5	226.1	209.2	193.8	179.7	166.7	154.8	143.9
54	Xe	350.3	322.8	297.7	274.8	254.0	234.9	217.5	201.5	186.9	173.4	161.1	149.8
55	Cs	370.7	341.8	315.3	291.2	269.2	249.0	230.6	213.7	198.3	184.0	171.0	159.0
56	Ba	383.7	353.8	326.6	301.6	279.0	258.1	239.1	221.7	205.7	191.0	177.5	165.1
57	La	405.1	373.7	345.1	318.8	295.0	273.0	253.0	234.6	217.7	202.2	188.0	174.9
58	Ce	428.2	395.1	365.0	337.3	312.2	289.0	267.9	248.5	230.7	214.3	199.3	185.4
59	Pr	453.2	418.4	386.6	357.4	330.9	306.5	284.1	263.6	244.8	227.5	211.6	196.9
60	Nd	470.4	434.5	401.6	371.4	344.0	318.7	295.5	274.3	254.8	236.8	220.3	205.1
61	Pm	489.9	452.6	418.5	387.2	358.7	332.4	308.4	286.4	266.1	247.4	230.2	214.4
62	Sm	435.4	469.0	433.9	401.5	372.1	345.0	320.2	297.3	276.4	257.0	239.2	222.8
63	Eu	455.6	421.3	454.5	420.7	390.0	361.7	335.7	311.9	290.0	269.7	251.1	234.0
64	Gd	338.4	313.0	398.3	430.0	398.8	369.9	343.5	319.2	296.9	276.3	257.3	239.8
65	Tb	353.2	326.8	302.6	385.5	357.7	386.8	359.3	334.0	310.8	289.2	269.4	251.2
66	Dy	138.2	336.9	312.1	289.2	369.0	342.6	371.0	345.0	321.0	298.9	278.5	259.7
67	Ho	145.3	133.6	323.8	300.2	278.7	258.8	330.6	358.5	333.7	310.7	289.6	270.2
68	Er	152.6	140.4	129.3	311.5	289.3	268.7	249.9	319.4	297.4	322.9	301.0	280.9
69	Tm	160.7	147.9	136.2	125.6	301?	279.9	260.3	242.3	309.8	288.7	313.8	292.9
70	Yb	166.7	153.4	141.3	130.2	120.3	111.1	267.2	248.8	231.8	296?	276.4	300.7
71	Lu	174.8	160.9	148.2	136.6	126.2	116.6	107.9	258.4	240.8	224.5	209.5	268.0
72	Hf	181.5	167.1	153.9	141.9	131.1	121.1	112.1	103.8	247.8	231.1	215.7	201.5
73	Ta	189.4	174.3	160.6	148.1	136.8	126.5	117.0	108.4	100.5	239.1	223.2	208.6
74	W	197.0	181.3	167.1	154.1	142.4	131.6	121.8	112.8	104.6	97.1	230.3	215.3
75	Re	205.2	189.0	174.2	160.6	148.4	137.2	127.0	117.6	109.1	101.2	94.1	223?
76	Os	211.9	195.1	179.8	165.9	153.3	141.7	131.2	121.5	112.7	104.6	97.2	90.4
77	Ir	220.9	203.4	187.5	173.0	159.9	147.8	136.8	126.8	117.6	109.2	101.4	94.4
78	Pt	229.1	211.0	194.5	179.5	165.9	153.4	142.0	131.6	122.1	113.3	105.3	98.0
79	Au	238.7	219.9	202.7	187.1	172.9	159.9	148.1	137.2	127.3	118.2	109.9	102.2

80	Hg	246.4	227.0	209.4	193.2	178.6	165.2	153.0	141.8	131.6	122.2	113.5	105.6
81	Tl	254.4	234.4	216.2	199.5	184.5	170.6	158.0	146.5	135.9	126.2	117.3	109.2
82	Pb	263.7	243.0	224.2	206.9	191.4	177.0	163.9	152.0	141.1	131.0	121.8	113.3
83	Bi	274.9	253.3	233.7	215.8	199.5	184.6	171.0	158.5	147.1	136.6	127.0	118.2
84	Po	287.6	265.1	244.6	225.8	208.9	193.2	179.0	166.0	154.1	143.1	133.1	123.8
85	At	302.5	278.9	257.3	237.6	219.8	203.4	188.4	174.7	162.2	150.7	140.1	130.4
86	Rn	301.2	277.6	256.2	236.6	218.9	202.6	187.7	174.1	161.6	150.1	139.6	129.9
87	Fr	315.8	291.1	268.7	248.2	229.6	212.5	196.9	182.6	169.6	157.5	146.5	136.4
88	Ra	328.4	302.9	279.6	258.2	238.9	221.1	204.9	190.1	176.5	164.0	152.5	142.0
89	Ac	345.1	318.2	293.8	271.4	251.1	232.4	215.4	199.8	185.6	172.4	160.4	149.3
90	Th	356.8	329.1	303.8	280.7	259.7	240.4	222.9	206.8	192.0	178.4	165.9	154.5
91	Pa	379.1	349.8	323.0	298.4	276.2	255.7	237.0	219.9	204.2	189.8	176.5	164.4
92	U	389.9	359.7	332.2	307.0	284.1	263.1	243.9	226.3	210.1	195.3	181.7	169.2
93	Np	415.8	383.7	354.4	327.6	303.2	280.7	260.3	241.5	224.3	208.5	193.9	180.6
94	Pu	433.1	399.8	369.3	341.4	316.0	292.6	271.3	251.8	233.8	217.3	202.2	188.3

Absorber \ Emitter		10,839 eV Bi	11,131 eV Po	11,427 eV At	11,727 eV Rn	12,031 eV Fr	12,340 eV Ra	12,652 eV Ac	12,969 eV Th	13,291 eV Pa	13,615 eV U	13,944 eV Np	14,279 eV Pu
1	H	0.0	0.0	0.0	0.0	0.0	0.0	0.0	0.0	0.0	0.0	0.0	0.0
2	He	0.0	0.0	0.0	0.0	0.0	0.0	0.0	0.0	0.0	0.0	0.0	0.0
3	Li	0.1	0.1	0.1	0.1	0.1	0.1	0.1	0.1	0.0	0.0	0.0	0.0
4	Be	0.3	0.3	0.3	0.2	0.2	0.2	0.2	0.2	0.2	0.1	0.1	0.1
5	B	0.8	0.7	0.7	0.6	0.6	0.5	0.5	0.4	0.4	0.4	0.3	0.3
6	C	1.8	1.6	1.5	1.4	1.3	1.2	1.1	1.0	0.9	0.9	0.8	0.8
7	N	2.9	2.7	2.5	2.3	2.1	2.0	1.8	1.7	1.6	1.5	1.4	1.3
8	O	4.6	4.2	3.9	3.6	3.3	3.1	2.9	2.7	2.5	2.3	2.1	2.0
9	F	6.3	5.8	5.4	5.0	4.6	4.3	4.0	3.7	3.4	3.2	3.0	2.8
10	Ne	9.3	8.6	7.9	7.3	6.8	6.3	5.8	5.4	5.0	4.7	4.4	4.1
11	Na	12.1	11.2	10.4	9.6	8.9	8.2	7.7	7.1	6.6	6.2	5.7	5.3
12	Mg	16.4	15.2	14.1	13.0	12.1	11.2	10.4	9.7	9.0	8.4	7.8	7.3
13	Al	20.6	19.0	17.6	16.3	15.1	14.0	13.0	12.1	11.3	10.5	9.8	9.1
14	Si	26.7	24.7	22.8	21.1	19.6	18.2	16.9	15.7	14.6	13.6	12.7	11.9
15	P	31.8	29.5	27.3	25.3	23.5	21.8	20.3	18.8	17.5	16.3	15.2	14.2
16	S	39.7	36.7	34.0	31.6	29.3	27.2	25.3	23.6	21.9	20.4	19.1	17.8
17	Cl	45.5	42.1	39.0	36.2	33.6	31.3	29.1	27.1	25.2	23.5	21.9	20.5
18	Ar	50.3	46.6	43.2	40.1	37.3	34.7	32.2	30.0	28.0	26.1	24.4	22.7
19	K	63.1	58.5	54.3	50.4	46.9	43.6	40.6	37.8	35.2	32.9	30.7	28.7

(continued)

Table 14.4. *(Continued)*

Absorber	Emitter	10,839 eV Bi	11,131 eV Po	11,427 eV At	11,727 eV Rn	12,031 eV Fr	12,340 eV Ra	12,652 eV Ac	12,969 eV Th	13,291 eV Pa	13,615 eV U	13,944 eV Np	14,279 eV Pu
20	Ca	74.7	69.3	64.3	59.7	55.6	51.7	48.1	44.9	41.8	39.1	36.5	34.1
21	Sc	79.8	74.0	68.8	63.9	59.5	55.4	51.6	48.1	44.9	41.9	39.2	36.6
22	Ti	88.8	82.5	76.6	71.3	66.4	61.8	57.6	53.7	50.2	46.9	43.8	41.0
23	V	98.1	91.2	84.8	78.9	73.5	68.5	63.9	59.6	55.7	52.0	48.7	45.5
24	Cr	112.1	104.2	96.9	90.2	84.1	78.4	73.1	68.3	63.8	59.7	55.8	52.2
25	Mn	122.7	114.1	106.2	99.0	92.3	86.0	80.3	75.0	70.1	65.6	61.4	57.5
26	Fe	138.7	129.1	120.2	112.0	104.5	97.5	91.1	85.1	79.6	74.5	69.7	65.3
27	Co	150.2	139.8	130.3	121.5	113.4	105.8	98.9	92.4	86.4	80.9	75.8	71.0
28	Ni	171.3	159.6	148.7	138.8	129.5	121.0	113.1	105.7	98.9	92.7	86.8	81.4
29	Cu	179.0	166.8	155.5	145.1	135.5	126.6	118.4	110.8	103.7	97.2	91.1	85.4
30	Zn	195.9	182.6	170.4	159.1	148.6	138.9	129.9	121.6	113.8	106.7	100.1	93.9
31	Ga	206.0	192.1	179.3	167.4	156.5	146.3	136.9	128.2	120.0	112.5	105.6	99.1
32	Ge	30.2	206.2	192.5	179.9	168.1	157.3	147.2	137.9	129.2	121.1	113.7	106.7
33	As	33.3	30.9	28.7	26.7	181.7	169.9	159.1	149.1	139.7	131.1	123.0	115.5
34	Se	35.7	33.1	30.8	28.7	26.7	24.9	23.3	157.3	147.4	138.3	129.9	121.9
35	Br	39.7	36.9	34.3	31.9	29.8	27.7	25.9	24.2	22.6	151.6	142.3	133.7
36	Kr	42.5	39.5	36.7	34.2	31.9	29.7	27.7	25.9	24.2	22.6	21.2	19.8
37	Rb	46.6	43.3	40.3	37.5	34.9	32.6	30.4	28.4	26.5	24.8	23.2	21.8
38	Sr	50.7	47.1	43.8	40.8	38.0	35.4	33.1	30.9	28.9	27.0	25.3	23.7
39	Y	55.5	51.6	48.0	44.7	41.6	38.8	36.2	33.9	31.6	29.6	27.7	26.0
40	Zr	60.0	55.7	51.9	48.3	45.0	42.0	39.2	36.6	34.2	32.0	30.0	28.1
41	Nb	65.1	60.5	56.3	52.4	48.9	45.6	42.5	39.7	37.2	34.8	32.6	30.5
42	Mo	69.5	64.6	60.1	56.0	52.2	48.7	45.5	42.5	39.7	37.2	34.8	32.6
43	Tc	75.6	70.3	65.4	61.0	56.8	53.0	49.5	46.2	43.2	40.5	37.9	35.5
44	Ru	79.7	74.1	68.9	64.2	59.9	55.8	52.1	48.7	45.5	42.6	39.9	37.4
45	Rh	85.7	79.7	74.2	69.1	64.4	60.1	56.1	52.4	49.0	45.9	43.0	40.3
46	Pd	90.6	84.2	78.4	73.1	68.1	63.6	59.4	55.5	51.9	48.6	45.5	42.6
47	Ag	97.4	90.6	84.4	78.6	73.3	68.4	63.9	59.7	55.8	52.3	49.0	45.9
48	Cd	101.7	94.6	88.1	82.1	76.6	71.4	66.7	62.4	58.3	54.6	51.2	48.0
49	In	108.2	100.7	93.8	87.4	81.5	76.1	71.1	66.4	62.1	58.2	54.5	51.1
50	Sn	113.5	105.6	98.4	91.7	85.5	79.8	74.6	69.7	65.2	61.1	57.3	53.7
51	Sb	119.8	111.5	103.8	96.8	90.3	84.3	78.8	73.7	68.9	64.6	60.5	56.7
52	Te	123.5	114.9	107.1	99.8	93.2	87.0	81.3	76.0	71.1	66.6	62.4	58.6
53	I	133.9	124.7	116.2	108.3	101.1	94.4	88.2	82.5	77.2	72.4	67.8	63.6
54	Xe	139.4	129.8	120.9	112.8	105.3	98.3	91.9	86.0	80.5	75.4	70.5	66.3
55	Cs	148.0	137.8	128.5	119.9	111.9	104.5	97.7	91.4	85.6	80.2	74.2	70.5

56	Ba	153.7	143.2	133.5	124.5	116.3	108.6	101.6	95.1	89.0	83.4	78.2	73.4
57	La	162.8	151.7	141.5	132.0	123.3	115.2	107.8	100.8	94.4	88.5	83.0	77.9
58	Ce	172.7	161.1	150.1	140.0	130.9	122.3	114.4	107.1	100.3	94.0	88.2	82.8
59	Pr	183.5	171.0	159.5	148.9	139.1	130.1	121.7	113.9	106.7	100.0	93.9	88.1
60	Nd	191.1	178.2	166.3	155.3	145.1	135.6	126.9	118.9	111.3	104.4	98.0	91.9
61	Pm	199.8	186.3	173.9	162.4	151.8	141.9	132.9	124.4	116.6	109.3	102.6	96.3
62	Sm	207.7	193.8	180.9	169.0	158.0	147.7	138.3	129.6	121.4	113.9	106.9	100.4
63	Eu	218.2	203.6	190.1	177.6	166.1	155.4	145.5	136.3	127.7	119.9	112.5	105.7
64	Gd	223.7	208.7	194.9	182.2	170.4	159.4	149.3	139.9	131.2	123.1	115.6	108.5
65	Tb	234.3	218.8	204.4	191.0	178.7	167.3	156.7	146.8	137.7	129.2	121.4	114.0
66	Dy	242.4	226.3	211.5	197.7	185.0	173.2	162.3	152.1	142.6	133.9	125.8	118.2
67	Ho	252.2	235.5	220.1	205.9	192.7	180.4	169.1	158.5	148.7	139.6	131.2	123.2
68	Er	262.3	245.0	229.1	214.3	200.6	187.9	176.1	165.1	154.9	145.5	136.7	128.5
69	Tm	273.5	255.6	239.0	223.6	209.4	196.2	183.9	172.5	161.9	152.1	142.9	134.3
70	Yb	280.9	262.6	245.6	229.9	215.3	201.7	189.2	177.5	166.6	156.5	147.1	138.3
71	Lu	250.4	272.9	255.3	239.0	223.9	209.8	196.8	184.7	173.4	162.9	153.2	144.0
72	Hf	257.7	241.0	262.8	246.1	230.6	216.1	202.8	190.3	178.7	168.0	157.9	148.5
73	Ta	195.0	182?	233.3	254.6	238.6	223.8	210.0	197.1	185.1	174.0	163.7	154.0
74	W	201.3	188.4	176.4	225.3	211.2	230.8	216.7	203.4	191.1	179.7	169.0	159.1
75	Re	208.3	194.9	182.6	171.1	218.3	204.8	224.1	210.4	197.7	185.9	175.0	164.7
76	Os	84.2	199.9	187.2	175.5	164.6	154.4	196.9	216?	202.6	190.6	179.4	168.8
77	Ir	87.9	81.9	193.9	181.8	170.5	160.1	150.3	191.4	179.9	197.3	185.7	174.8
78	Pt	91.2	85.0	79.3	187.4	175.8	165.0	155.0	145.7	185?	174.2	191.2	180.0
79	Au	95.2	88.7	82.7	77.2	182.1	170.9	160.6	151.0	142.0	133.6	169.6	159.7
80	Hg	98.4	91.7	85.5	79.8	74.6	175.4	164.8	155.0	145.8	137.2	129.2	163.6
81	Tl	101.7	94.8	88.4	82.5	77.1	72.1	67.5	159.0	149.6	140.9	132.7	125.0
82	Pb	105.6	98.4	91.8	85.7	80.1	74.9	70.1	65.6	154.2	145.2	136.8	128.9
83	Bi	110.1	102.7	95.8	89.4	83.6	78.2	73.2	68.5	64.2	150.4	141.7	133.5
84	Po	115.4	107.6	100.4	93.7	87.6	81.9	76.7	71.8	67.3	63.2	147.3	138.8
85	At	121.5	113.3	105.7	98.7	92.3	86.3	80.8	75.7	70.9	66.6	62.5	145.0
86	Rn	121.1	112.9	105.4	98.4	92.0	86.0	80.5	75.5	70.7	66.4	62.3	58.5
87	Fr	127.1	118.5	110.6	103.3	96.6	90.3	84.6	79.2	74.3	69.7	65.5	61.5
88	Ra	132.3	123.4	115.2	107.6	100.6	94.1	88.1	82.5	77.4	72.6	68.2	64.1
89	Ac	139.2	129.8	121.1	113.2	105.8	99.0	92.7	86.8	81.4	76.4	71.8	67.4
90	Th	144.0	134.3	125.4	117.1	109.5	102.4	95.9	89.9	84.3	79.1	74.3	69.8
91	Pa	153.2	142.9	133.4	124.6	116.5	109.0	102.1	95.7	89.7	84.2	79.1	74.3
92	U	157.7	147.1	137.3	128.3	119.9	112.2	105.1	98.5	92.4	86.7	81.4	76.5
93	Np	168.3	157.0	146.6	136.9	128.1	119.8	112.2	105.2	98.6	92.6	87.0	81.7
94	Pu	175.5	163.7	152.8	142.8	133.5	124.9	117.0	109.7	102.8	96.5	90.7	85.2

[a] Heinrich (1986).

Table 14.5. Mass Absorption Coefficients for $M\alpha$ Lines[a] (? denotes significant uncertainty due to the presence of an absorption edge)

Absorber \ Emitter		833 eV La	883 eV Ce	929 eV Pr	978 eV Nd	1029 eV Pm	1081 eV Sm	1131 eV Eu	1185 eV Gd	1240 eV Tb	1293 eV Dy	1348 eV Ho	1406 eV Er
1	H	13.1	10.8	9.1	7.7	6.5	5.5	4.7	4.0	3.5	3.0	2.6	2.3
2	He	108.2	89.7	76.0	64.3	54.5	46.3	39.9	34.2	29.5	25.7	22.3	19.4
3	Li	418.8	349.6	298.4	254.0	216.4	185.1	160.4	138.2	119.5	104.4	91.3	79.7
4	Be	1,113	936.1	804.2	689.0	590.8	508.4	442.7	383.6	333.4	292.7	257.1	225.3
5	B	2,189	1,855	1,604	1,383	1,193	1,033	904.8	788.3	688.8	607.8	536.4	472.4
6	C	3,529	3,019	2,633	2,289	1,991	1,738	1,532	1,345	1,184	1,051	933.6	827.5
7	N	5,223	4,494	3,938	3,441	3,008	2,638	2,337	2,060	1,821	1,624	1,448	1,288
8	O	7,177	6,204	5,459	4,792	4,208	3,705	3,295	2,917	2,588	2,316	2,072	1,851
9	F	8,838	7,671	6,775	5,969	5,260	4,650	4,149	3,686	3,282	2,947	2,645	2,370
10	Ne	770.5	10,039?	8,893	7,860	6,950	6,162	5,515	4,914	4,390	3,952	3,558	3,197
11	Na	1,051	898.2	782.9	680.9	592.8	7,280?	6,532	5,837	5,228	4,719	4,259	3,837
12	Mg	1,477	1,264	1,103	960.5	837.0	732.1	647.3	569.8	503.3	448.6	5,292	4,778
13	Al	1,903	1,631	1,426	1,243	1,084	949.3	840.0	740.0	654.0	583.3	520.4	463.6
14	Si	2,531	2,174	1,902	1,661	1,451	1,272	1,126	993.0	878·3	783.9	699.8	623.7
15	P	3,090	2,659	2,330	2,037	1,782	1,564	1,386	1,224	1,083	967.5	864.3	770.9
16	S	3,922	3,381	2,968	2,598	2,276	2,000	1,775	1,568	1,389	1,242	1,110	991.1
17	Cl	4,562	3,940	3,463	3,036	2,663	2,343	2,082	1,842	1,634	1,462	1,308	1,168
18	Ar	5,108	4,418	3,890	3,415	3,000	2,643	2,352	2,083	1,849	1,656	1,483	1,326
19	K	6,473	5,609	4,945	4,349	3,826	3,375	3,006	2,666	2,370	2,125	1,905	1,705
20	Ca	7,715	6,697	5,913	5,207	4,587	4,053	3,614	3,209	2,856	2,563	2,300	2,060
21	Sc	8,289	7,206	6,372	5,619	4,957	4,385	3,916	3,481	3,102	2,787	2,503	2,245
22	Ti	9,261	8,064	7,140	6,306	5,571	4,935	4,411	3,926	3,503	3,150	2,833	2,543
23	V	10,252	8,940	7,926	7,010	6,201	5,500	4,923	4,387	3,918	3,528	3,176	2,853
24	Cr	11,704	10,221	9,074	8,035	7,117	6,320	5,664	5,053	4,518	4,073	3,670	3,301
25	Mn	12,791	11,186	9,943	8,815	7,818	6,952	6,236	5,571	4,987	4,499	4,059	3,654
26	Fe	12,362	12,617	11,227	9,966	8,850	7,878	7,076	6,327	5,671	5,122	4,625	4,168
27	Co	13,312	11,673	12,120?	10,771	9,576	8,535	7,673	6,869	6,163	5,572	5,036	4,544

28	Ni	1,896	13,236?	11,804	10,503	10,894	9,721	8,748	7,840	7,042	6,373	5,766	5,207
29	Cu	2,012	1,772	1,586?	10,913	9,723	8,685	9,119	8,181	7,355	6,663	6,034	5,454
30	Zn	2,285	2,009	1,795	1,602	8,655?	9,459	8,529	7,660	8,034	7,284	6,603	5,974
31	Ga	2,515	2,209	1,971	1,757	1,567	1,403	7,250?	8,015	7,220	6,552?	6,928	6,275
32	Ge	2,819	2,473	2,205	1,963	1,749	1,564	1,411	1,269	6,227	6,999	6,355	5,761
33	As	3,169	2,778	2,475	2,201	1,960	1,750	1,578	1,417	1,277	1,159	5,479	6,188
34	Se	3,466	3,037	2,705	2,405	2,139	1,909	1,720	1,543	1,389	1,260	1,143	1,036
35	Br	3,926	3,441	3,064	2,723	2,422	2,160	1,944	1,744	1,568	1,421	1,289	1,167
36	Kr	4,267	3,741	3,332	2,961	2,633	2,348	2,113	1,894	1,702	1,542	1,397	1,265
37	Rb	4,741	4,159	3,707	3,295	2,931	2,613	2,351	2,107	1,893	1,714	1,553	1,404
38	Sr	5,212	4,578	4,083	3,632	3,231	2,882	2,593	2,324	2,088	1,890	1,712	1,548
39	Y	5,758	5,064	4,521	4,025	3,584	3,198	2,879	2,581	2,319	2,099	1,901	1,718
40	Zr	6,259	5,514	4,929	4,393	3,915	3,496	3,149	2,824	2,538	2,299	2,081	1,881
41	Nb	6,819	6,018	5,387	4,808	4,289	3,835	3,456	3,102	2,789	2,527	2,289	2,069
42	Mo	7,292	6,448	5,781	5,167	4,616	4,131	3,727	3,348	3,012	2,731	2,474	2,238
43	Tc	7,929	7,026	6,310	5,649	5,054	4,529	4,090	3,678	3,312	3,005	2,724	2,466
44	Ru	8,327	7,395	6,653	5,966	5,346	4,797	4,338	3,905	3,520	3,196	2,900	2,626
45	Rh	8,917	7,937	7,154	6,426	5,768	5,184	4,694	4,230	3,818	3,470	3,151	2,856
46	Pd	9,364	8,354	7,544	6,789	6,104	5,495	4,982	4,496	4,063	3,696	3,360	3,048
47	Ag	9,991	8,934	8,084	7,288	6,564	5,919	5,374	4,856	4,394	4,002	3,642	3,307
48	Cd	10,330	9,259	8,394	7,583	6,842	6,179	5,618	5,084	4,607	4,200	3,826	3,478
49	In	10,871?	9,767	8,872	8,030	7,258	6,566	5,979	5,418	4,916	4,487	4,093	3,724
50	Sn	10,929	9,416?	9,231	8,370	7,579	6,868	6,263	5,684	5,164	4,720	4,309	3,926
51	Sb	11,874	10,274	9,017	8,768	7,953	7,219	6,592	5,992	5,451	4,988	4,559	4,158
52	Te	11,522?	10,896?	9,596	8,404	8,145	7,405	6,772	6,164	5,615	5,144	4,708	4,298
53	I	10,400	11,078?	9,803?	9,396	8,236	7,984?	7,311	6,664	6,078	5,575	5,108	4,669
54	Xe	10,805	9,576	8,529	9,163	8,821	7,759	6,882	6,907	6,308	5,792	5,313	4,862
55	Cs	11,282	10,157	9,133	8,119	8,739	8,468?	7,527	6,649	6,675	6,136	5,634	5,161
56	Ba	11,315	10,414	9,485	8,508	7,565	8,176?	7,287	7,104	6,295	5,623?	5,838	5,353
57	La	2,744?	10,772	9,973	9,047	8,106	7,220	7,883?	6,995	6,856	6,131	5,478	5,660
58	Ce	2,581	2,332?	10,386	9,562	8,650	7,755	6,963	7,575?	6,737	6,679?	5,974	5,330
59	Pr	2,748	2,481	2,268?	10,043	9,197	8,311	7,499	6,698	5,972?	6,554	6,513?	5,816

(continued)

Table 14.5. *(Continued)*

Absorber		833 eV La	883 eV Ce	929 eV Pr	978 eV Nd	1029 eV Pm	1081 eV Sm	1131 eV Eu	1185 eV Gd	1240 eV Tb	1293 eV Dy	1348 eV Ho	1406 eV Er
60	Nd	2,872	2,590	2,365	2,157?	9,528	8,696	7,895	7,083	6,334	5,691?	6,252	6,215?
61	Pm	3,013	2,714	2,477	2,256	2,362?	9,056	8,294	7,486	6,721	6,054	5,438	5,970
62	Sm	3,146	2,831	2,581	2,350	2,140	2,261?	8,607	7,833	7,071	6,391	5,754	5,158
63	Eu	3,321	2,986	2,721	2,474	2,252	2,054	1,888?	8,232	7,485	6,797	6,138	5,514
64	Gd	3,423	3,077	2,801	2,546	2,314	2,109	1,937	1,773?	7,693	7,028	6,373	5,742
65	Tb	3,610	3,243	2,951	2,680	2,435	2,218	2,035	1,861	1,706?	7,392?	6,742	6,098
66	Dy	3,762	3,379	3,073	2,789	2,533	2,305	2,113	1,931	1,769	1,631?	7,011?	6,375
67	Ho	3,948	3,545	3,223	2,925	2,654	2,414	2,211	2,020	1,848	1,703	1,569?	6,680?
68	Er	4,145	3,722	3,384	3,070	2,784	2,531	2,318	2,115	1,934	1,781	1,640	1,508?
69	Tm	4,369	3,924	3,568	3,236	2,934	2,666	2,441	2,226	2,035	1,872	1,722	1,583
70	Yb	4,541	4,079	3,709	3,364	3,051	2,771	2,536	2,312	2,112	1,942	1,786	1,640
71	Lu	4,779	4,295	3,907	3,544	3,214	2,919	2,670	2,434	2,223	2,043	1,877	1,723
72	Hf	4,986	4,485	4,081	3,703	3,358	3,050	2,790	2,543	2,321	2,133	1,959	1,797
73	Ta	5,234	4,712	4,290	3,894	3,532	3,209	2,936	2,675	2,441	2,243	2,059	1,888
74	W	5,483	4,939	4,500	4,087	3,709	3,371	3,084	2,810	2,564	2,355	2,162	1,982
75	Re	5,761	5,195	4,736	4,304	3,908	3,553	3,251	2,963	2,704	2,484	2,280	2,090
76	Os	6,003	5,419	4,944	4,496	4,085	3,715	3,401	3,100	2,830	2,599	2,386	2,187
77	Ir	6,322	5,713	5,217	4,748	4,317	3,928	3,597	3,280	2,995	2,751	2,526	2,314
78	Pt	6,627	5,996	5,480	4,991	4,541	4,135	3,788	3,456	3,156	2,900	2,663	2,440
79	Au	6,982	6,323	5,785	5,273	4,801	4,374	4,010	3,660	3,344	3,073	2,822	2,587
80	Hg	7,289	6,608	6,051	5,520	5,030	4,586	4,206	3,841	3,511	3,228	2,965	2,718
81	Tl	7,688	6,902	6,326	5,776	5,267	4,806	4,411	4,030	3,685	3,389	3,115	2,856
82	Pb	7,803	7,328	6,638	6,067	5,537	5,055	4,642	4,244	3,883	3,573	3,284	3,013
83	Bi	7,989	7,501	7,093	6,397	5,843	5,338	4,905	4,487	4,108	3,781	3,477	3,191
84	Po	8,146	7,647	7,229	6,824	6,180	5,651	5,195	4,755	4,356	4,011	3,690	3,387
85	At	8,398	7,883	7,452	7,032	6,634	5,998	5,518	5,054	4,631	4,267	3,927	3,607
86	Rn	8,146	7,647	7,228	6,820	6,432	6,072	5,532	5,069	4,648	4,284	3,945	3,624

87	Fr	8,301	7,794	7,367	6,951	6,555	6,187	5,862	5,336	4,895	4,514	4,158	3,821
88	Ra	8,374	7,865	7,435	7,015	6,615	6,242	5,914	5,589	5,093	4,698	4,330	3,981
89	Ac	8,474	7,963	7,530	7,105	6,700	6,322	5,988	5,658	5,350	4,916	4,532	4,168
90	Th	8,401	7,898	7,472	7,052	6,650	6,275	5,943	5,615	5,308	5,036	4,641?	4,270
91	Pa	8,566	8,061	7,629	7,203	6,794	6,411	6,072	5,736	5,422	5,143	4,877	4,467
92	U	8,331	7,847	7,432	7,020	6,624	6,251	5,921	5,593	5,286	5,014	4,753	4,501
93	Np	8,345	8,870	7,459	7,050	6,655	6,283	5,952	5,623	5,314	5,040	4,777	4,522
94	Pu	8,091	7,641	7,250	6,858	6,477	6,117	5,797	5,476	5,176	4,908	4,652	4,403

Absorber	Emitter	1,462 eV Tm	1,521 eV Yb	1,581 eV Lu	1,645 eV Hf	1,710 eV Ta	1,775 eV W	1,843 eV Re	1,910 eV Os	1,980 eV Ir	2,051 eV Pt	2,123 eV Au	2,195 eV Hg
1	H	2.0	1.7	1.5	1.3	1.2	1.0	0.9	0.8	0.7	0.6	0.6	0.5
2	He	17.1	15.0	13.1	11.5	10.1	8.9	7.9	7.0	6.2	5.5	4.9	4.4
3	Li	70.2	61.7	54.5	47.8	42.1	37.3	32.9	29.3	26.0	23.2	20.7	18.5
4	Be	199.3	175.9	155.7	137.3	121.4	107.7	95.6	85.2	75.9	67.8	60.7	54.5
5	B	419.7	372.1	330.7	292.8	259.9	231.7	206.3	184.7	165.1	148.0	132.9	119.8
6	C	739.5	659.6	589.5	525.1	468.8	420.1	376.0	338.3	304.1	273.8	247.1	223.6
7	N	1,155	1,034	926.9	828.4	741.8	666.7	598.4	539.8	486.4	439.0	397.0	360.0
8	O	1,665	1,495	1,345	1,206	1,084	976.7	879.2	795.3	718.5	650.1	589.3	535.7
9	F	2,139	1,927	1,739	1,564	1,409	1,274	1,150	1,043	944.5	856.9	778.6	709.4
10	Ne	2,894	2,614	2,365	2,133	1,927	1,747	1,581	1,438	1,306	1,187	1,081	987.6
11	Na	3,481	3,153	2,860	2,586	2,342	2,128	1,931	1,760	1,602	1,460	1,333	1,220
12	Mg	4,345	3,943	3,585	3,249	2,949	2,685	2,442	2,231	2,035	1,859	1,700	1,559
13	Al	416.4	373.5	4,146	3,765	3,425	3,125	2,848	2,606	2,382	2,180	1,998	1,835
14	Si	560.5	502.8	452.1	405.4	364.3	328.7	3,447?	3,160	2,893	2,652	2,435	2,241
15	P	693.1	622.1	559.6	501.9	451.2	407.2	367.1	332.7	301.2	273.3	248.4	2,523
16	S	891.7	800.8	720.7	646.7	581.6	525.1	473.6	429.3	388.7	352.7	320.7	292.5
17	Cl	1,052	945.3	851.3	764.3	687.7	621.1	560.4	508.1	460.3	417.8	379.9	346.5
18	Ar	1,195	1,075	968.4	870.0	783.3	707.8	638.9	579.5	525.2	476.8	433.7	395.7
19	K	1,537	1,383	1,248	1,122	1,010	913.5	825.0	748.7	678.8	616.5	561.0	512.0

(continued)

Table 14.5. (*Continued*)

Absorber		Emitter: 1,462 eV Tm	1,521 eV Yb	1,581 eV Lu	1,645 eV Hf	1,710 eV Ta	1,775 eV W	1,843 eV Re	1,910 eV Os	1,980 eV Ir	2,051 eV Pt	2,123 eV Au	2,195 eV Hg
20	Ca	1,859	1,675	1,512	1,360	1,226	1,109	1,002	910.1	825.5	750.2	682.9	623.4
21	Sc	2,028	1,828	1,651	1,487	1,341	1,213	1,098	997.6	905.4	823.1	749.6	684.7
22	Ti	2,299	2,075	1,876	1,690	1,526	1,383	1,251	1,137	1,033	939.5	856.1	782.3
23	V	2,582	2,332	2,110	1,903	1,720	1,560	1,412	1,285	1,167	1,063	968.7	885.6
24	Cr	2,990	2,703	2,448	2,210	1,999	1,814	1,644	1,496	1,361	1,239	1,131	1,034
25	Mn	3,313	2,998	2,718	2,456	2,223	2,019	1,831	1,668	1,518	1,383	1,263	1,156
26	Fe	3,783	3,427	3,109	2,812	2,547	2,315	2,102	1,916	1,745	1,591	1,454	1,331
27	Co	4,127	3,742	3,398	3,076	2,790	2,538	2,305	2,103	1,917	1,750	1,599	1,465
28	Ni	4,734	4,296	3,905	3,538	3,211	2,924	2,658	2,427	2,214	2,022	1,849	1,696
29	Cu	4,964	4,509	4,102	3,720	3,379	3,079	2,802	2,560	2,337	2,136	1,955	1,794
30	Zn	5,442	4,947	4,504	4,089	3,718	3,390	3,088	2,824	2,579	2,359	2,161	1,984
31	Ga	5,720	5,205	4,743	4,309	3,921	3,579	3,262	2,986	2,729	2,498	2,290	2,104
32	Ge	6,126	5,579	5,088	4,626	4,214	3,849	3,511	3,216	2,942	2,695	2,472	2,273
33	As	5,650	5,150	5,479	4,986	4,545	4,154	3,793	3,477	3,183	2,918	2,679	2,465
34	Se	4,726	5,399	4,932	4,492	4,776	4,369	3,992	3,662	3,355	3,078	2,828	2,604
35	Br	1,065	969.8	4,261	4,891	4,465	4,088	4,357	4,000	3,667	3,367	3,095	2,852
36	Kr	1,153	1,049	956.9	870.6	3,693	4,284	3,921	3,602	3,852	3,539	3,255	3,002
37	Rb	1,279	1,164	1,061	964.3	878.6	803.2	3,307	3,871	3,555	3,268	3,506	3,235
38	Sr	1,409	1,281	1,167	1,061	965.6	882.2	805.4	738.5	2,961	3,487	3,212	2,966
39	Y	1,564	1,422	1,295	1,176	1,070	977.1	891.5	817.0	748.2	686.4	2,684	3,190
40	Zr	1,713	1,556	1,417	1,287	1,170	1,068	974.4	892.6	817.0	749.1	688.1	633.7
41	Nb	1,884	1,712	1,559	1,415	1,287	1,174	1,071	980.6	897.2	822.3	754.9	695.0
42	Mo	2,038	1,853	1,687	1,531	1,393	1,271	1,158	1,061	970.1	888.9	815.8	750.8
43	Tc	2,246	2,042	1,860	1,689	1,536	1,402	1,278	1,170	1,070	980.1	899.3	827.4
44	Ru	2,394	2,178	1,984	1,802	1,640	1,497	1,364	1,249	1,142	1,046	960.0	883.1
45	Rh	2,605	2,371	2,161	1,964	1,788	1,632	1,488	1,363	1,246	1,142	1,047	963.5
46	Pd	2,782	2,534	2,312	2,102	1,914	1,748	1,594	1,460	1,336	1,224	1,123	1,033

47	Ag	3,021	2,754	2,514	2,287	2,084	1,904	1,738	1,592	1,457	1,335	1,225	1,127
48	Cd	3,180	2,902	2,651	2,414	2,201	2,012	1,837	1,684	1,542	1,413	1,297	1,194
49	In	3,409	3,113	2,846	2,594	2,367	2,165	1,978	1,814	1,662	1,524	1,400	1,288
50	Sn	3,597	3,288	3,009	2,744	2,506	2,294	2,098	1,925	1,764	1,619	1,487	1,370
51	Sb	3,814	3,490	3,196	2,918	2,667	2,443	2,235	2,053	1,882	1,728	1,589	1,463
52	Te	3,946	3,614	3,313	3,028	2,770	2,540	2,325	2,137	1,961	1,801	1,656	1,527
53	I	4,290	3,934	3,610	3,301	3,023	2,774	2,541	2,337	2,146	1,973	1,815	1,674
54	Xe	4,472	4,104	3,770	3,451	3,162	2,904	2,663	2,451	2,252	2,071	1,907	1,760
55	Cs	4,752	4,365	4,013	3,677	3,372	3,099	2,844	2,619	2,408	2,217	2,042	1,886
56	Ba	4,933	4,536	4,173	3,827	3,513	3,231	2,968	2,735	2,516	2,318	2,137	1,974
57	La	5,221	4,805	4,424	4,061	3,730	3,434	3,156	2,911	2,680	2,470	2,279	2,106
58	Ce	5,528	5,092	4,692	4,310	3,962	3,650	3,357	3,098	2,855	2,633	2,430	2,248
59	Pr	5,231	5,401?	4,981	4,579	4,212	3,883	3,574	3,300	3,043	2,808	2,594	2,400
60	Nd	5,594	5,024	5,184?	4,769	4,390	4,050	3,730	3,446	3,179	2,935	2,713	2,512
61	Pm	5,378	5,385	4,847	4,349?	4,686	4,318	3,973	3,667	3,380	3,119	2,881	2,666
62	Sm	5,712	5,137	5,164	4,636	4,170	4,395	4,053	3,749	3,463	3,201	2,962	2,745
63	Eu	4,979	5,511	4,967	4,986	4,487	4,052	4,174	3,870	3,582	3,317	3,074	2,854
64	Gd	5,194	4,682	5,198	4,672	4,707	4,252	3,837	3,898	3,615	3,354	3,113	2,895
65	Tb	5,530	4,993	4,509	4,997	4,501	4,558?	4,115	3,733	3,733?	3,469	3,225	3,003
66	Dy	5,800	5,249	4,747	4,273	4,752	4,297	4,352?	3,949	3,579	3,546?	3,301	3,077
67	Ho	6,105	5,542	5,022	4,527	4,084	4,564	4,123	3,742	3,808	3,460	3,149?	3,176
68	Er	6,399?	5,834	5,302	4,789	4,325	3,916	4,377	3,975	3,605	3,678	3,348	3,057
69	Tm	1,463	6,139?	5,601	5,073	4,590	4,160	3,762	4,232	3,839	3,489	3,569	3,259
70	Yb	1,515	1,398	5,813	5,284	4,791	4,349	3,937	3,578	4,026	3,660	3,333	3,420
71	Lu	1,591	1,467	1,355	5,552?	5,052	4,595	4,165	3,789	3,441	3,883	3,538	3,232
72	Hf	1,658	1,528	1,411	1,299	1,638	4,795	4,357	3,969	3,607	3,283	3,717?	3,397
73	Ta	1,742	1,604	1,480	1,362	1,256	1,621	4,573	4,175	3,800	3,460	3,155	3,588?
74	W	1,828	1,683	1,551	1,427	1,315	1,215	1,601	4,368	3,984	3,633	3,315	3,033
75	Re	1,926	1,773	1,634	1,502	1,383	1,277	1,179	1,594	4,180	3,820	3,490	3,194
76	Os	2,015	1,854	1,708	1,570	1,445	1,334	1,230	1,139	1,572?	3,971?	3,636	3,332
77	Ir	2,133	1,962	1,807	1,661	1,528	1,410	1,299	1,202	1,111	1,570?	3,817?	3,505

(continued)

Table 14.5. *(Continued)*

Absorber \ Emitter		1,462 eV Tm	1,521 eV Yb	1,581 eV Lu	1,645 eV Hf	1,710 eV Ta	1,775 eV W	1,843 eV Re	1,910 eV Os	1,980 eV Ir	2,051 eV Pt	2,123 eV Au	2,195 eV Hg
78	Pt	2,249	2,069	1,905	1,750	1,610	1,485	1,368	1,265	1,169	1,082	1,563?	1,442
79	Au	2,384	2,193	2,020	1,855	1,706	1,573	1,449	1,340	1,237	1,145	1,060	984.0
80	Hg	2,506	2,305	2,123	1,950	1,793	1,653	1,522	1,407	1,299	1,201	1,112	1,032
81	Tl	2,633	2,423	2,232	2,050	1,885	1,737	1,600	1,478	1,365	1,261	1,167	1,083
82	Pb	2,778	2,557	2,355	2,163	1,989	1,834	1,688	1,560	1,440	1,331	1,231	1,142
83	Bi	2,943	2,709	2,496	2,293	2,109	1,944	1,790	1,654	1,526	1,410	1,304	1,209
84	Po	3,126	2,878	2,652	2,437	2,241	2,066	1,903	1,758	1,622	1,499	1,386	1,285
85	At	3,329	3,066	2,826	2,597	2,389	2,203	2,029	1,874	1,730	1,598	1,478	1,370
86	Rn	3,346	3,083	2,843	2,613	2,404	2,217	2,042	1,887	1,741	1,608	1,487	1,378
87	Fr	3,530	3,253	3,000	2,759	2,539	2,342	2,157	1,993	1,839	1,699	1,572	1,456
88	Ra	3,678	3,391	3,129	2,877	2,649	2,444	2,251	2,080	1,920	1,774	1,641	1,520
89	Ac	3,853	3,553	3,279	3,016	2,777	2,563	2,361	2,183	2,015	1,862	1,722	1,596
90	Th	3,948	3,642	3,362	3,094	2,849	2,630	2,423	2,240	2,068	1,911	1,768	1,638
91	Pa	4,131	3,812	3,520	3,240	2,985	2,755	2,539	2,348	2,168	2,004	1,854	1,718
92	U	4,151	3,831	3,539	3,258	3,002	2,772	2,555	2,363	2,182	2,017	1,866	1,730
93	Np	4,295	3,959?	3,657	3,368	3,104	2,867	2,643	2,445	2,258	2,087	1,931	1,790
94	Pu	4,182	3,966	3,654	3,366	3,103	2,866	2,643	2,445	2,259	2,088	1,933	1,792

Absorber \ Emitter		2,271 eV Tl	2,346 eV Pb	2,423 eV Bi	2,501 eV Po	2,580 eV At	2,659 eV Rn	2,741 eV Fr	2,824 eV Ra	2,909 eV Ac	2,996 eV Th	3,028 eV Pa	3,171 eV U
1	H	0.5	0.4	0.4	0.3	0.3	0.3	0.2	0.2	0.2	0.2	0.2	0.1
2	He	3.9	3.5	3.2	2.8	2.6	2.3	2.1	1.9	1.7	1.5	1.4	1.3
3	Li	16.6	14.9	13.4	12.0	10.9	9.8	8.9	8.1	7.3	6.6	6.0	5.5
4	Be	48.8	44.0	39.6	35.8	32.3	29.3	26.6	24.1	21.9	19.9	18.2	16.6
5	B	107.7	97.2	87.9	79.6	72.1	65.6	59.6	54.3	49.4	45.1	41.2	37.7
6	C	202.0	183.2	166.3	151.2	137.7	125.7	114.7	104.8	95.8	87.6	80.4	73.8
7	N	325.8	296.1	269.1	245.0	223.4	204.3	186.6	170.7	156.3	143.1	131.4	120.7

8	O	485.8	442.4	402.9	367.5	335.7	307.4	281.3	257.7	236.2	216.5	199.2	183.1
9	F	644.9	588.5	537.2	491.0	449.3	412.2	377.8	346.7	318.3	292.3	269.2	247.8
10	Ne	899.8	823.0	752.7	689.4	632.2	581.1	533.6	490.6	451.1	414.9	382.7	352.8
11	Na	1,114	1,021	935.6	858.6	788.9	726.4	668.2	615.5	566.9	522.3	482.6	445.6
12	Mg	1,426	1,310	1,203	1,106	1,018	939.1	865.5	798.5	736.8	679.9	629.3	581.9
13	Al	1,682	1,548	1,424	1,311	1,209	1,117	1,032	953.2	881.0	814.3	754.8	699.1
14	Si	2,058	1,896	1,747	1,612	1,489	1,378	1,274	1,179	1,091	1,010	937.9	870.0
15	P	2,321	2,142	1,977	1,826	1,689	1,566	1,450	1,344	1,245	1,155	1,073	996.9
16	S	266.2	243.3	222.5	2,160	2,000	1,856	1,721	1,598	1,483	1,377	1,281	1,192
17	Cl	315.5	288.4	263.8	241.6	221.7	203.9	187.5	1,747?	1,624	1,509	1,406	1,310
18	Ar	360.3	329.4	301.3	276.1	253.4	233.1	214.3	197.3	181.8	167.5	154.9	143.1
19	K	466.3	426.5	390.2	357.6	328.2	301.9	277.6	255.7	235.5	217.1	200.7	185.5
20	Ca	568.0	519.6	475.5	435.9	400.1	368.2	338.6	311.9	287.3	264.9	244.9	226.4
21	Sc	624.0	571.1	522.8	479.3	440.1	405.1	372.6	343.2	316.3	291.6	269.7	249.3
22	Ti	713.3	653.0	598.0	548.5	503.7	463.8	426.7	393.1	362.4	334.1	309.0	285.7
23	V	808.0	740.0	677.9	622.0	571.5	526.3	484.4	446.4	411.5	379.5	351.1	324.7
24	Cr	943.9	864.9	792.7	727.6	668.8	616.2	567.3	522.9	482.2	444.8	411.6	380.7
25	Mn	1,056	967.8	887.5	815.0	749.4	690.7	636.1	586.5	541.0	499.3	462.1	427.5
26	Fe	1,217	1,116	1,024	940.7	865.3	797.9	735.1	678.1	625.7	577.6	534.8	494.8
27	Co	1,340	1,230	1,129	1,038	955.2	881.2	812.2	749.6	691.9	638.9	591.7	547.7
28	Ni	1,552	1,425	1,309	1,204	1,109	1,023	943.6	871.2	804.5	743.2	688.6	637.5
29	Cu	1,643	1,510	1,387	1,277	1,176	1,086	1,002	925.7	855.3	790.4	732.6	678.6
30	Zn	1,818	1,672	1,538	1,416	1,305	1,206	1,113	1,029	950.9	879.1	815.2	755.3
31	Ga	1,930	1,775	1,634	1,505	1,389	1,284	1,185	1,096	1,014	937.6	869.8	806.3
32	Ge	2,086	1,921	1,768	1,630	1,505	1,392	1,286	1,190	1,101	1,019	945.5	876.9
33	As	2,263	2,085	1,921	1,772	1,637	1,515	1,401	1,296	1,200	1,111	1,032	957.3
34	Se	2,392	2,206	2,034	1,877	1,735	1,606	1,486	1,376	1,275	1,181	1,097	1,018
35	Br	2,622	2,419	2,232	2,062	1,906	1,766	1,635	1,515	1,404	1,301	1,209	1,123
36	Kr	2,762	2,550	2,354	2,175	2,013	1,866	1,728	1,602	1,486	1,378	1,281	1,191
37	Rb	2,979	2,751	2,542	2,351	2,176	2,019	1,871	1,735	1,610	1,494	1,390	1,292
38	Sr	3,184	2,943	2,721	2,518	2,332	2,165	2,007	1,863	1,729	1,606	1,495	1,390

(continued)

Table 14.5. *(Continued)*

Absorber \ Emitter		2,271 eV Tl	2,346 eV Pb	2423 eV Bi	2,501 eV Po	2,580 eV At	2,659 eV Rn	2,741 eV Fr	2,824 eV Ra	2,909 eV Ac	2,996 eV Th	3,082 eV Pa	3,171 eV U
39	Y	2,941	2,720	2,932	2,715	2,516	2,337	2,168	2,014	1,870	1,737	1,618	1,506
40	Zr	2,414	2,888	2,673	2,476	2,677	2,487	2,309	2,145	1,994	1,853	1,727	1,608
41	Nb	638.7	589.2	2,195	2,643	2,453	2,280	2,468	2,295	2,134	1,984	1,850	1,723
42	Mo	689.7	635.9	586.6	541.9	1,974	2,396	2,227	2,071	2,246	2,090	1,949	1,817
43	Tc	759.8	700.4	645.8	596.4	551.5	511.2	1,820	2,219	2,066	1,923	2,092	1,951
44	Ru	810.8	747.2	688.8	635.9	587.9	544.7	504.4	467.7	1,628	1,995	1,863	1,738
45	Rh	884.5	815.0	751.2	693.4	640.9	593.7	549.6	509.4	472.4	438.2	1,495	1,844
46	Pd	948.5	873.9	805.4	743.3	686.9	636.2	588.9	545.7	505.9	469.2	436.4	406?
47	Ag	1,035	953.8	879.1	811.3	749.7	694.3	642.5	595.4	551.9	511.7	475.9	442.3
48	Cd	1,096	1,010	931.4	859.6	794.3	735.6	680.7	630.7	584.6	542.0	504.0	468.3
49	In	1,183	1,091	1,006	928.3	857.9	794.5	735.3	681.3	631.4	585.4	544.2	505.7
50	Sn	1,258	1,160	1,070	987.8	913.1	845.7	782.8	725.3	672.2	623.2	579.4	538.4
51	Sb	1,345	1,241	1,145	1,057	977.3	905.4	838.1	776.7	719.9	667.5	620.6	576.7
52	Te	1,404	1,296	1,196	1,104	1,021	946.5	876.4	812.3	753.0	698.3	649.3	603.4
53	I	1,540	1,422	1,313	1,213	1,122	1,040	963.5	893.2	828.3	768.2	714.4	663.9
54	Xe	1,620	1,496	1,382	1,278	1,182	1,096	1,016	942.0	873.7	810.5	753.9	700.8
55	Cs	1,737	1,605	1,483	1,372	1,270	1,178	1,092	1,013	939.7	872.0	811.3	754.3
56	Ba	1,819	1,682	1,555	1,439	1,333	1,237	1,147	1,064	987.6	916.7	853.1	793.3
57	La	1,942	1,797	1,662	1,539	1,426	1,324	1,228	1,140	1,058	982.4	914.5	850.7
58	Ce	2,074	1,920	1,776	1,645	1,525	1,417	1,315	1,221	1,134	1,053	980.6	912.4
59	Pr	2,216	2,052	1,900	1,761	1,633	1,517	1,408	1,309	1,216	1,130	1,052	979.3
60	Nd	2,320	2,150	1,991	1,846	1,713	1,593	1,479	1,375	1,277	1,187	1,106	1,030
61	Pm	2,461	2,279	2,111	1,956	1,815	1,687	1,566	1,455	1,352	1,257	1,171	1,090
62	Sm	2,539	2,355	2,184	2,026	1,882	1,751	1,628	1,514	1,408	1,310	1,221	1,138
63	Eu	2,644	2,455	2,280	2,119	1,970	1,835	1,707	1,590	1,480	1,378	1,286	1,199
64	Gd	2,685	2,497	2,322	2,160	2,011	1,875	1,746	1,627	1,516	1,413	1,320	1,231
65	Tb	2,789	2,597	2,417	2,251	2.098	1,958	1,825	1,703	1,588	1,481	1,384	1,292

66	Dy	2,861	2,667	2,485	2,317	2,161	2,019	1,883	1,758	1,641	1,531	1,432	1,338
67	Ho	2,957	2,759	2,573	2,401	2,241	2,095	1,956	1,827	1,707	1,594	1,491	1,394
68	Er	3,057	2,855	2,665	2,488	2,324	2,174	2,031	1,899	1,774	1,658	1,552	1,452
69	Tm	2,971	2,966	2,770	2,588	2,419	2,264	2,117	1,980	1,851	1,730	1,621	1,517
70	Yb	3,118	2,855	2,840	2,654	2,482	2,324	2,174	2,034	1,903	1,779	1,667	1,561
71	Lu	3,310?	3,031	2,777	2,757?	2,579	2,416	2,260	2,115	1,979	1,852	1,736	1,626
72	Hf	3,098	2,837	2,919	2,679	2,462	2,488	2,328	2,180	2,040	1,909	1,790	1,677
73	Ta	3,273	2,998	2,747	2,830	2,602	2,398	2,412	2,259	2,115	1,979	1,856	1,740
74	W	2,767	3,157	2,894	2,657	2,740?	2,525	2,326	2,333?	2,185	2,046	1,919	1,799
75	Re	2,916	2,673	3,055	2,805	2,579	2,378	2,455	2,265	2,091	2,120	1,989	1,864
76	Os	3,044	2,791	2,559	2,934	2,698	2,488	2,292	2,368	2,186	2,019	2,042	1,915
77	Ir	3,207	2,943	2,700	2,480	2,851	2,629	2,422	2,236	2,308?	2,132	1,976	1,830?
78	Pt	3,357	3,084	2,832	2,602	2,394	2,763?	2,547	2,350	2,170	2,005	2,075	1,922
79	Au	1,446	3,243	2,981	2,742	2,524	2,329	2,147?	2,482	2,292	2,118	1,963	2,028
80	Hg	955.9	1,441	3,105	2,860	2,635	2,432	2,243	2,071	2,399	2,216	2,055	1,904
81	Tl	1,003	931.4	1,437	2,975?	2,746	2,537	2,341	2,162	1,998	2,318	2,149	1,991
82	Pb	1,057	981.2	911.3	1,440?	1,335	2,655	2,452	2,266	2,095	1,936	2,257?	2,092
83	Bi	1,119	1,039	964.4	896.4	834?	1,351	2,577	2,385	2,206	2,040	1,892	1,754
84	Po	1,188	1,103	1,024	951.2	884.9	824.7	1,370	2,512	2,328	2,155	1,999	1,854
85	At	1,267	1,175	1,091	1,013	942.1	877.8	817.3	1,395	2,459?	2,283	2,120	1,967
86	Rn	1,275	1,183	1,097	1,019	947.3	882.4	821.3	765.2	1,340?	1,255	2,122	1,972
87	Fr	1,347	1,249	1,159	1,076	1,000	931.4	866.7	807.3	752.1	701?	1,275	2,074
88	Ra	1,406	1,304	1,210	1,123	1,044	971.7	904.0	841.9	784.1	730.4	682.2	1,279
89	Ac	1,475	1,368	1,269	1,178	1,095	1,019	948.1	882.8	822.0	765.5	714.8	667.0
90	Th	1,515	1,405	1,303	1,210	1,124	1,046	973.2	906.0	843.4	785.3	733.1	683.9
91	Pa	1,589	1,474	1,367	1,269	1,179	1,097	1,020	949.6	884.0	822.9	768.0	716.3
92	U	1,600	1,484	1,376	1,278	1,187	1,105	1,027	956.1	889.8	828.3	772.9	720.7
93	Np	1,656	1,536	1,425	1,323	1,229	1,144	1,063	989.6	920.9	857.1	799.7	745.6
94	Pu	1,657	1,537	1,426	1,324	1,230	1,145	1,064	990.4	921.6	857.7	800.1	745.9

[a] Heinrich (1986).

Table 14.6. *K* Series X-Ray Wavelengths and Energies[a]

	$K\alpha_1$		*K* edge	
Element	λ (Å)	E(keV)	λ(Å)	E (keV)
4 Be	114.00	0.109	110.0	0.111
5 B	67.6	0.183	64.57	0.192
6 C	44.7	0.277	43.68	0.284
7 N	31.6	0.392	30.9	0.400
8 O	23.62	0.525	23.32	0.532
9 F	18.32	0.677	18.05	0.687
10 Ne	14.61	0.849	14.30	0.867
11 Na	11.91	1.041	11.57	1.072
12 Mg	9.89	1.254	9.512	1.303
13 Al	8.339	1.487	7.948	1.560
14 Si	7.125	1.740	6.738	1.84
15 P	6.157	2.014	5.784	2.144
16 S	5.372	2.308	5.019	2.470
17 Cl	4.728	2.622	4.397	2.820
18 A	4.192	2.958	3.871	3.203
19 K	3.741	3.314	3.437	3.608
20 Ca	3.358	3.692	3.070	4.038
21 Sc	3.031	4.091	2.762	4.489
22 Ti	2.749	4.511	2.497	4.965
23 V	2.504	4.952	2.269	5.464
24 Cr	2.290	5.415	2.070	5.989
25 Mn	2.102	5.899	1.896	6.538
26 Fe	1.936	6.404	1.743	7.111
27 Co	1.789	6.930	1.608	7.710
28 Ni	1.658	7.478	1.483	8.332
29 Cu	1.541	8.048	1.381	8.980
30 Zn	1.435	8.639	1.283	9.661
31 Ga	1.340	9.252	1.196	10.37
32 Ge	1.254	9.886	1.17	11.10
33 As	1.176	10.544	1.045	11.87
34 Se	1.109	11.181	0.9797	12.65
35 Br	1.040	11.924	0.9204	13.47
36 Kr	0.980	12.649	0.8655	14.32
37 Rb	0.926	13.395	0.8155	15.20
38 Sr	0.875	14.165	0.7697	16.11
39 Y	0.8288	14.958	0.7277	17.04
40 Zr	0.7859	15.775	0.6888	17.999
41 Nb	0.7462	16.615	0.6530	18.99
42 Mo	0.7093	17.479	0.6198	20.00
43 Te	0.6750	18.367	0.5891	21.05
44 Ru	0.6431	19.279	0.5605	22.12
45 Rh	0.6133	20.21	0.534	23.22
46 Pd	0.5854	21.18	0.5092	24.34
47 Ag	0.5594	22.16	0.4859	25.52
48 Cd	0.5350	23.17	0.4641	26.72
49 In	0.5121	24.21	0.4437	27.94
50 Sn	0.4906	25.27	0.4247	29.19
51 Sb	0.4704	26.36	0.4067	30.49
52 Te	0.4513	27.47	0.3897	31.81
53 I	0.43333	28.61	0.3738	33.17
54 Xe	0.4163	29.78	0.3584	34.59

Table 14.6. *(Continued)*

Element	$K\alpha_1$		K edge	
	λ (Å)	E(keV)	λ(Å)	E (keV)
55 Cu	0.4003	30.97	0.3445	35.99
56 Ba	0.3851	32.19	0.3310	37.45
57 La	0.3707	33.44	0.3184	38.93
58 Ce	0.3571	34.72	0.3065	40.45
59 Pr	0.3441	36.03	0.2952	42.00
60 Nd	0.3318	37.36	0.2845	43.57
61 Pm	0.3202	38.72	0.2743	45.20
62 Sm	0.3090	40.12	0.2646	46.85
63 Eu	0.2984	41.54	0.2555	48.52
64 Gd	0.2884	42.996	0.2468	50.23
65 Tb	0.2787	44.48	0.2384	52.00
66 Dy	0.2695	45.998	0.2305	53.79
67 Ho	0.2608	47.55	0.2229	55.62
68 Er	0.2524	49.13	0.2157	57.49
69 Tm	0.2443	50.74	0.2088	59.38
70 Yb	0.2367	52.39	0.2022	61.30
71 Lu	0.2293	54.07	0.1959	63.31
72 Hf	0.2270	54.61	0.1898	65.31
73 Ta	0.2155	57.53	0.1839	67.40
74 W	0.2090	59.32	0.1784	69.51
75 Re	0.2028	61.14	0.1730	71.66
76 Os	0.1968	63.00	0.1679	73.86
77 Ir	0.1910	64.90	0.1629	76.10
78 Pt	0.1855	66.83	0.1582	78.38
79 Au	0.1851	66.99	0.1536	80.72
80 Hg	0.1751	70.82	0.1492	83.11
81 Tl	0.1701	72.87	0.1450	85.53
82 Pb	0.1654	74.97	0.1409	88.00
83 Bi	0.1608	77.11	0.1369	90.53
84 Po	0.1546	79.29	0.1331	93.12
85 At	0.1521	81.52	0.1294	95.74
86 Rn	0.1480	83.78	0.1260	98.42
87 Fr	0.1440	86.10	0.1226	101.15
88 Ra	0.1401	88.47	0.1193	103.93
89 Ac	0.1364	90.88	0.1161	106.76
90 Th	0.1328	93.35	0.1131	109.65
91 Pa	0.1293	95.87	0.1101	112.58
92 U	0.1259	98.44	0.1072	115.62

[a] Bearden (1964).

Table 14.7. *L* Series X-Ray Wavelengths and Energies[a]

	$L\alpha_1$		L_3 edge			$L\alpha_1$		L_3 edge	
Element	λ (Å)	E (keV)	λ (Å)	E (keV)	Element	λ (Å)	E (keV)	λ (Å)	E (keV)
20 Ca	36.33	0.341	35.49	0.349	57 La	2.666	4.651	2.261	5.484
21 Se	31.35	0.395	30.54	0.406	58 Ce	2.562	4.840	2.166	5.723
22 Ti	27.42	0.452	27.3	0.454	59 Pr	2.463	5.034	2.079	5.963
23 V	24.25	0.511	24.2	0.512	60 Nd	2.370	5.230	1.997	6.209
24 Cr	21.64	0.573	20.7	0.598	61 Pm	2.282	5.433	1.919	6.461
25 Mn	19.45	0.637	19.4	0.639	62 Sm	2.200	5.636	1.846	6.717
26 Fe	17.59	0.705	17.53	0.707	63 Eu	2.121	5.846	1.776	6.981
27 Co	15.97	0.776	15.92	0.779	64 Gd	2.047	6.057	1.712	7.243
28 Ni	14.56	0.852	14.52	0.854	65 Tb	1.977	6.273	1.650	7.515
29 Cu	13.34	0.930	13.29	0.933	66 Dy	1.909	6.495	1.592	7.790
30 Zn	12.25	1.012	12.31	1.022	67 Ho	1.845	6.720	1.537	8.068
31 Ga	11.29	1.098	11.10	1.117	68 Er	1.784	6.949	1.484	8.358
32 Ge	10.44	1.188	10.19	1.217	69 Tm	1.727	7.180	1.433	8.650
33 As	9.671	1.282	9.37	1.324	70 Yb	1.672	7.416	1.386	8.944
34 Se	8.99	1.379	8.65	1.434	71 Lu	1.620	7.656	1.341	9.249
35 Br	8.375	1.480	7.984	1.553	72 Hf	1.57	7.899	1.297	9.558
36 Kr	7.817	1.586	7.392	1.677	73 Ta	1.522	8.146	1.255	9.877
37 Rb	7.318	1.694	6.862	1.807	74 W	1.476	8.398	1.216	10.20
38 Sr	6.863	1.807	6.387	1.941	75 Re	1.433	8.653	1.177	10.53
39 Y	6.449	1.923	5.962	2.079	76 Os	1.391	8.912	1.141	10.87
40 Zr	6.071	2.042	5.579	2.223	77 Ir	1.351	9.175	1.106	11.21
41 Nb	5.724	2.166	5.230	2.371	78 Pt	1.313	9.442	1.072	11.56
42 Mo	5.407	2.293	4.913	2.523	79 Au	1.276	9.713	1.040	11.92
43 Tc	5.115	2.424	4.630	2.678	80 Hg	1.241	9.989	1.009	12.29
44 Ru	4.846	2.559	4.369	2.838	81 Tl	1.207	10.27	0.979	12.66
45 Rh	4.597	2.697	4.130	3.002	82 Pb	1.175	10.55	0.951	13.04
46 Pd	4.368	2.839	3.907	3.173	83 Bi	1.144	10.84	0.923	13.43
47 Ag	4.154	2.984	3.699	3.351	84 Po	1.114	11.13	0.897	13.82
48 Cd	3.956	3.134	3.505	3.538	85 At	1.085	11.43	0.872	14.22
49 In	3.772	3.287	3.324	3.730	86 Rn	1.057	11.73	0.848	14.62
50 Sn	3.600	3.414	3.156	3.929	87 Fr	1.030	12.03	0.825	15.03
51 Sb	3.439	3.605	3.000	4.132	88 Ra	1.005	12.34	0.803	15.44
52 Te	3.289	3.769	2.856	4.342	89 Ac	0.9799	12.65	0.781	15.87
53 1	3.149	3.938	2.720	4.559	90 Th	0.956	12.97	0.761	16.30
54 Xe	3.017	4.110	2.593	4.782	91 Pa	0.933	13.29	0.741	16.73
55 Cs	2.892	4.287	2.474	5.011	92 U	0.911	13.61	0.722	17.17
56 Ba	2.776	4.466	2.363	5.247					

[a] Bearden (1964).

Table 14.8. *M* Series X-Ray Wavelengths and Energies[a]

	$M\alpha_1$		M_5 edge	
Element	λ (Å)	E (keV)	λ (Å)	E (keV)
57 La	14.88	0.833	14.90	0.832
58 Ce	14.04	0.883	14.04	0.883
59 Pr	13.34	0.929	13.32	0.931
60 Nd	12.68	0.978	12.68	0.978
61 Pm	12.00	1.033	12.07	1.027
62 Sm	11.47	1.081	11.48	1.080
63 Eu	10.96	1.131	10.97	1.130
64 Gd	10.46	1.185	10.46	1.185
65 Tb	10.00	1.240	9.99	1.241
66 Dy	9.59	1.293	9.57	1.295
67 Ho	9.20	1.348	9.177	1.351
68 Er	8.82	1.406	8.799	1.409
69 Tm	8.48	1.462	8.451	1.467
70 Yb	8.149	1.521	8.11	1.528
71 Lu	7.840	1.581	7.81	1.588
72 Hf	7.539	1.645	7.46	1.661
73 Ta	7.252	1.710	7.11	1.743
74 W	6.983	1.775	6.83	1.814
75 Re	6.729	1.843	6.56	1.89
76 Os	6.490	1.910	6.30	1.967
77 Ir	6.262	1.980	6.05	2.048
78 Pt	6.047	2.051	5.81	2.133
79 Au	5.840	2.123	5.58	2.220
80 Hg	5.645	2.196	5.36	2.313
81 Tl	5.460	2.271	5.153	2.406
82 Pb	5.286	2.346	4.955	2.502
83 Bi	5.118	2.423	4.764	2.603
90 Th	4.138	2.996	3.729	3.325
91 Pa	4.022	3.082	3.602	3.442
92 U	3.910	3.171	3.497	3.545

[a] Bearden (1964).

Table 14.9. J and Fluorescent Yield (ω) by Atomic Number

Z	J (eV)[a]	ω_K[b]	ω_L[b]	ω_M[b]
1	68.2	—	—	—
2	70.8	—	—	—
3	76.7	—	—	—
4	83.9	0.00045	—	—
5	91.8	0.00101	—	—
6	100	0.00198	—	—
7	108	0.00351	—	—
8	117	0.00579	—	—
9	126	0.00901	—	—
10	135	0.0134	—	—
11	144	0.0192	—	—
12	153	0.0265	—	—
13	162	0.0357	—	—
14	172	0.0469	—	—
15	181	0.0603	—	—
16	190	0.0760	—	—
17	200	0.0941	—	—
18	209	0.115	—	—
19	218	0.138	—	—
20	228	0.163	0.00067	—
21	237	0.190	0.00092	—
22	247	0.219	0.00124	—
23	256	0.249	0.00163	—
24	266	0.281	0.0021	—
25	275	0.314	0.00267	—
26	285	0.347	0.00335	—
27	294	0.381	0.00415	—
28	304	0.414	0.00507	—
29	313	0.446	0.00614	—
30	323	0.479	0.00736	—
31	333	0.510	0.00875	—
32	342	0.540	0.0103	—
33	352	0.568	0.0121	—
34	361	0.596	0.0140	—
35	371	0.622	0.0162	—
36	380	0.646	0.0186	—
37	390	0.669	0.0213	—
38	400	0.691	0.0242	—
39	409	0.711	0.0274	—
40	419	0.730	0.0308	—
41	429	0.747	0.0345	—
42	438	0.764	0.0385	—
43	448	0.779	0.0428	—
44	457	0.793	0.0474	—
45	467	0.806	0.0523	—
46	477	0.818	0.0575	—
47	486	0.830	0.0630	—
48	496	0.840	0.0689	—
49	506	0.850	0.0750	—
50	515	0.859	0.0815	—
51	525	0.867	0.0882	—
52	535	0.875	0.0953	—
53	544	0.882	0.103	—

Table 14.9. (*Continued*)

Z	J (eV)[a]	ω_K[b]	ω_L[b]	ω_M[b]
54	554	0.888	0.110	—
55	564	0.895	0.118	—
56	573	0.900	0.126	—
57	583	0.906	0.135	0.00111
58	593	0.911	0.144	0.00115
59	602	0.915	0.153	0.00120
60	612	0.920	0.162	0.00124
61	622	0.924	0.171	0.00129
62	631	0.927	0.181	0.00133
63	641	0.931	0.190	0.00137
64	651	0.934	0.200	0.00141
65	660	0.937	0.210	0.00145
66	670	0.940	0.220	0.00149
67	680	0.943	0.231	0.00153
68	689	0.945	0.241	0.00156
69	699	0.947	0.251	0.00160
70	709	0.950	0.262	0.00163
71	718	0.952	0.272	0.00165
72	728	0.954	0.283	0.00168
73	738	0.956	0.294	0.00170
74	748	0.957	0.304	0.00172
75	757	0.959	0.315	0.00174
76	767	0.960	0.325	0.00175
77	777	0.962	0.335	0.00177
78	786	0.963	0.346	0.00177
79	796	0.964	0.356	0.00178
80	806	0.966	0.366	0.00178
81	815	0.967	0.376	0.00178
82	825	0.968	0.386	0.00177
83	835	0.969	0.396	0.01777
84	845	0.970	0.406	0.00175
85	854	0.971	0.415	0.00174
86	864	0.972	0.425	0.00172
87	874	0.972	0.434	0.00170
88	883	0.973	0.443	0.00167
89	893	0.974	0.452	0.00164
90	903	0.975	0.461	0.00161
91	912	0.975	0.469	0.00158
92	922	0.976	0.478	0.00155

[a] Berger and Seltzer (1964).
[b] Bambynek *et al.* (1972).

Table 14.10. Important Properties of Selected Coating Elements

Element	Symbol	Thermal conductivity at 300 K ($W\ cm^{-1}\ K^{-1}$)	Resistivity at 300 K ($\mu\Omega$ cm)	Melting point (K)	Boiling point (K)	Vaporization temperature at 1.3 Pa
Aluminum	Al	2.37	2.83	932	2330	1273
Antimony	Sb	0.243	41.7	903	1713	951
Barium	Ba	0.184	50.0	990	1911	902
Beryllium	Be	2.00	4.57	1557	3243	1519
Bismuth	Bi	0.079	119.0	544	1773	971
Boron	B	0.270	1.8×10^{12}	2573	2823	1628
Cadmium	Cd	0.968	7.50	594	1040	537
Calcium	Ca	2.00	4.60	1083	1513	878
Carbon	C	1.29	3500	4073	4473	2954
Chromium	Cr	0.937	13.0	2173	2753	1478
Cobalt	Co	1.00	9.7	1751	3173	1922
Copper	Cu	4.01	1.67	1356	2609	1393
Germanium	Ge	0.599	89×10^{3}	1232	3123	1524
Gold	Au	3.17	2.40	1336	2873	1738
Indium	In	0.816	8.37	430	2273	1225
Iridium	Ir	1.47	6.10	2727	4773	2829
Iron	Fe	0.802	10.0	1811	3273	1720
Lead	Pb	0.353	22.0	601	1893	991
Magnesium	Mg	1.56	4.60	924	1380	716
Manganese	Mn	0.078	5.0	1517	2360	1253
Molybdenum	Mo	1.38	5.70	2893	3973	2806
Nickel	Ni	0.907	6.10	1725	3173	1783
Palladium	Pd	0.718	11.0	1823	3833	1839
Platinum	Pt	0.716	10.0	2028	4573	2363
Rhodium	Rh	1.50	4.69	2240	2773	2422
Silicon	Si	1.48	23×10^{4}	1683	2773	1615
Silver	Ag	4.29	1.60	1233	2223	1320
Strontium	Sr	0.353	23.0	1044	1657	822
Tantalum	Ta	0.575	13.1	3269	4373	3273
Thorium	Th	0.540	18.0	2100	4473	2469
Tin	Sn	0.666	11.4	505	2610	1462
Titanium	Ti	0.219	42.0	2000	3273	1819
Tungsten	W	1.74	5.50	3669	6173	3582
Vanadium	V	0.307	18.2	1970	3673	2161
Zinc	Zn	1.16	5.92	693	1180	616
Zirconium	Zr	0.21	40.0	2125	4650	2284

Latent heat of evaporation ($kJ\ g^{-1}$)	Specific heat at 300 K ($J\ g^{-1}\ K^{-1}$)	Relative sputtering yield (atoms/600 eV Ar^+)	Relative cost/g highest purity	Evaporation technique
12.77	0.900	1.2	14	Ta, W; coil, basket; wets and alloys
1.603	0.205	—	1	Ta; basket, wets
—	0.193	—	1	Ta, Mo, W; basket, boats; wets
24.77	1.825	—	115	Ta, W, Mo; basket; wets, toxic
0.855	0.124	—	1	Ta, W, Mo; basket, boat
34.70	1.026	—	7	Carbon crucible, forms carbides
1.199	0.232	—	52	Ta, W, Mo; basket, boat, wets, toxic
—	0.653	—	1	W; basket
—	0.712	—	1	Pointed rods, resistive, evaporation
6.170	0.448	1.3	1	W; basket
6.280	0.456	1.4	6	W; basket, alloys easily
4.810	0.385	2.3	45	Ta, W, Mo; basket, boats, not wet easily
—	0.322	1.2	—	Ta, Mo, W; basket, boat, wets Ta, Mo
1.740	0.129	2.8	60	Mo, W, loop; basket, boat, wets, alloys Ta
2.030	0.234	—	21	W, Fe; basket; Mo, boat
3.310	0.133	—	51	Refractory; thick W loop
6.342	0.444	1.3	24	W; loop, coil, alloys easily
0.857	0.159	—	12	Fe; basket, boat, not wet W, Ta, Mo
5.597	1.017	—	1	Ta, W, Mo; basket, boat, wets
4.092	0.481	—	1	Ta, W, Mo, loop; basket, wets
5.115	0.251	0.9	3	Refractory
6.225	0.444	1.5	20	W; coil (heavy), alloys
0.370	0.245	2.4	37	W; loop, coil
2.620	0.133	1.6	26	Stranded with W, Ta; alloys
4.814	0.244	1.5	39	Sublimation from W, low pressure
10.59	0.703	0.5	1	BeO crucible, contaminates with SiO
2.330	0.236	3.4	22	Ta, W, Mo; coil, basket, boat; not wet easily
—	0.298	—	1	Ta, W, Mo; basket; wets
4.165	0.140	0.6	8	Refractory
2.340	0.133	—	52	W; basket, wets
2.400	0.222	—	18	Ta, Mo; basket, boat; wets
9.837	0.523	0.6	5	Ta, W; loop, cool, basket; alloys
4.345	0.133	0.6	6	Refractory
9.000	0.486	—	25	W, Mo; basket; alloys with W
1.756	0.389	—	1	Ta, W, Mo; loop, basket; wets
5.693	0.281	—	5	Ta; basket; forms oxide

References

Abraham, J. L. (1977). *SEM 1977* **2,** 119.
Akahori, H., H. Ishii, I. Nonaka and H. Yoshida (1988). *J. Electron Microsc.* **37,** 351.
Albee, A. L., and L. Ray (1970). *Anal. Chem.* **42,** 1408.
Albrecht, R. M., and B. Wetzel (1979). *SEM/1979* **3,** 203.
Allen, T. D. (1983). *SEM 1983* **4,** 1963.
Amrein, M., R. Dürr, H. Winkler, G. Travaglini, R. Wepf, and H. Gross (1989). *J. Ultrastruct. and Mol. Struct. Res.* **102,** 170–177.
Anderson, C. A., and M. F. Hasler (1966). In *Proc. 4th Intl. Conf. on X-ray Optics and Microanalysis* (R. Castaing, P. Deschamps, and J. Philibert, eds.) (Hermann, Paris), p. 310
Anderson, R. L. (1961). "Revealing Microstructures in Metals," Westinghouse Research Laboratory, Scientific Paper 425-C000-P2.
Anderson, T. F. (1951). *Trans. NY Acad. Sci. Ser. II* **13,** 130.
Aoki, M. and M. Tavassoli (1981). *J. Histochem. Cytochem.* **29,** 682.
Ardenne, M. von (1938). *Z. Tech. Phys.* **109,** 553.
Argall, K. and P. Armati (1990). *J. Electron Microsc. Tech.* **16,** 347.
Armati, P. J., D. Van Reyk and L. Van den Lubbe (1988). *J. Immunol. Methods* **110,** 267.
Armstrong, J.T. (1978). *SEM/1978* **1,** 455.
Armstrong, J. (1988). *Microbeam Analysis-1988* (San Francisco, San Francisco Press), p. 239.
Armstrong, J. T. (1988). In *Microbeam Analysis-1988* (D. E. Newbury, ed.) (San Francisco, San Francisco Press), p. 469.
Armstrong, J. T. (1991). "Quantitative Elemental Analysis of Individual Microparticles with Electron Beam Instruments." In *Electron Probe Quantitation* (K. F. J. Heinrich and D. E. Newbury, eds.) (Plenum Press, New York), p. 261.
Armstrong, J. T. and P. R. Buseck (1975). *Anal. Chem.* **47,** 2178.
Arnal, F., P. Verdier and P-D. Vincinsini (1969). *C.R. Acad. Sci Paris* **268,** 1526.
ASM Metals Handbook (1985). *Metallography and Microstructures* **5,** ASM International, Metals Park, Ohio.
ASTM, (1960). *Methods of Metallographic Specimen Preparation,* ASTM, STP 284, American Society for Testing Materials, Philadelphia, PA.
Bald, W. B. (1987). *Quantitative Cryofixation* (Bristol Adam Hilger).
Bambynek, W., B. Crasemann, R. W. Fink, H. U. Freund, H. Mark, C. D. Swift, R. E. Price, and P. V. Rao (1972). *Rev. Mod. Phys.* **44,** 716.
Barber, T. A. (1976). In *Principles and Techniques of SEM: Biological Applications, Vol. 5* (M. A. Hyatt, ed.) (Van Nostrand-Reinhold, New York).
Barbi, N. C. and D. P. Skinner (1976). *SEM/1976* **1,** 393.
Barkalow, R. H., R. W. Kraft and J. I. Goldstein (1972). *Metall. Trans.* **3,** 919.

Barnes, S. H. (1990). *Microsc. Anal. 1990*(3), 35.
Barnes, S. H., and S. Blackmore (1984). *Micron. Microsc. Acta* **15,** 187.
Barnes, S. H., and S. Blackmore (1986). *Scanning Microscopy/1986* **1,** 281.
Barnes, S. H., and S. Blackmore (1988). *Ann. Bot.* **62,** 615.
Barnett, J. R. (1988). *Microsc. Anal. 1988*(4), 11.
Bartlett, A. A., and H. P. Burstyn (1975). *SEM/1975,* 305.
Bastin, G. F., and H. J. M. Heijligers (1984). In *Microbeam Analysis-1984* (A. D. Romig, Jr. and J. I. Goldstein, eds.), p. 291.
Bastin, G. F., and H. J. M. Heijligers (1986a). *X-ray Spectrom.* **15,** 143.
Bastin, G. F., and H. J. M. Heijligers (1986b). *Quantitative Electron Probe Microanalysis of Boron in Binary Borides,* Internal Report, Eindhoven University of Technology.
Bastin, G. F., and H. J. M. Heijligers, (1990a). *Quantitative Electron Probe Microanalysis of Carbon in Binary Carbides,* Internal Report, Eindhoven University of Technology.
Bastin, G. F., and H. J. M. Heijligers (1990b). *Scanning* **12,** 225.
Bastin, G. F. and H. J. M. Heijligers (1990c). In *Proc. 12th Intl. Cong. Electron Micros.* (L. Peachey and D. B. Williams, eds.) (San Francisco Press, San Francisco), II, 216–217.
Bastin, G. F., and H. J. M. Heijligers (1991a). In *Electron Probe Quantitation* (K. F. J. Heinrich and D. E. Newbury eds.) (Plenum Press, New York), p. 163.
Bastin, G. F. and H. J. M. Heijligers (1991b). In *Electron Probe Quantitation* (K. F. J. Heinrich and D. E. Newbury, eds.) (Plenum Press, New York), p. 145.
Bastin, G. F., F. J. J. van Loo and H. J. M. Heijligers (1984a). *X-Ray Spectrom.* **13,** 91.
Bastin, G. F., H. J. M. Heijligers, and F. J. J. van Loo (1984b). *Scanning* **6,** 58.
Bastin, G. F., H. J. M. Heijligers, and F. J. J. van Loo (1986). *Scanning* **8,** 45.
Bauer, E. L. (1971). *A Statistical Manual for Chemists,* 2d ed. (Academic Press, New York), p. 189.
Bauer, V. M., and W. O. Butler (1983). *Scanning* **5,** 145.
Bear, F. E., ed., (1964). *Chemistry of Soil,* 2d ed. (Reinhold Publishing, New York).
Bearden, J. A. (1964). "X-ray Wavelengths," Report NYO 10586, U.S. Atomic Energy Commission, Oak Ridge, Tennessee.
Bearden, J. A. (1967a). *Rev. Mod. Phys.* **39,** 78.
Bearden, J. A. (1967b). "X-ray Wavelengths and X-ray Atomic Energy Levels," NSRDS-NBS 14, National Bureau of Standards, Washington, DC.
Becker, R. P., and P. P. H. De Bryun (1976). *SEM/1976* **2,** 171.
Becker, R. P., and J. S. Geoffrey (1981). *SEM/1981* **4,** 195.
Beesley, J. E. (1989). "Colloidal Gold: A New Perspective for Cytochemical Marking," *Royal Microscopical Handbook No. 17,* Oxford Univ. Press.
Beeuwkes, R. (1979). *SEM/1979* **2,** 767.
Bell, P. B., M. Lindroth, and B.-A. Fredriksson (1988). *Scanning Microsc.,* **2,** 1647.
Benada, O., and V. Pokorny (1990). *J. Electron Microsc. Tech.* **16,** 235.
Bence, A. E., and A. Albee (1968). *J. Geol.* **76,** 382.
Bence, A. E., and W. Holzwarth (1977). In *Proc. 12th MAS Conf.-1977,* p. 38.
Bence, T. (1978). Personal communication.
Bendayan, M., and M. Zollinger (1983). *J. Histochem. Cytochem.* **31,** 101.
Berger, M. (1963). *Methods in Computational Physics I* (B. Adler, S. Fernback, and M. Rotenberg, eds.) (Academic, New York).
Berger, M. J., and S. M. Seltzer (1964). Nat. Acad. Sci./Nat. Res. Council Publ. 1133, Washington, 205.
Bertin, E. P. (1975). *Principles and Practice of X-ray Spectrometric Analysis,* 2nd ed. (Plenum Press, New York).
Bethe, H. (1930). *Ann. Phys. (Leipzig),* **5,** 325.
Bethe, H. (1933). In *Handbook of Physics,* Springer, Berlin, Vol. 24, p. 273.
Birks, L. S., and E. J. Brookes (1957). *Rev. Sci. Instrum.* **28,** 709.
Bishop, H. (1966). In *Proc. 4th Intl. Conf on X-ray Optics and Microanalysis* (R. Castaing, P. Deschamps, and J. Philibert, eds.) (Hermann, Paris), p. 153.
Blaschke, R. (1980). *Proc. R. Microsc. Soc.* **15,** 280.
Blodgett, K. B., and I. Langmuir (1937). *Phys. Rev.* **51,** 964.

Bodson, M. J., W. H. Outlaw, and S. H. Silvers (1991). *J. Histochem. Cytochem.* **39,** 435.
Bolon, R. B., and E. Lifshin (1973). In *Proc. 8th Nat. Conf. Elect. Probe Anal.* EPASA, New Orleans, Paper 31.
Bolon, R. B., and M. D. McConnell (1976). *SEM/1976* **1,** 163.
Boon, M. E., and L. P. Kok (1987). *Microwave Cookbook of Pathology* (Coulomb Press, Leiden, The Netherlands).
Boone, T. (1984). *Ultramicroscopy* **14,** 359.
Boonstra, J., P. M. P. Van Bergen en Henegouwen, N. van Belzen, P. Rijken and A. J. Verkleij (1991). *J. Microsc.* **161,** 135.
Borovskii, I., and N. P. Ilin (1953). *Dokl. Akad. Nauk SSSR* **106,** 655.
Bowen, I. A., and T. A. Ryder (1978). "The Application of X-ray Microanalysis to Histochemistry. In *Electron Microprobe Analysis in Biology* (D. A. Erasmus, ed.) (Chapman Hall, London).
Boyde, A. (1973). *J. Microsc.* **98,** 452.
Boyde, A. (1974a). *SEM/1974,* 101.
Boyde, A. (1974b). *SEM/1974,* 93.
Boyde, A. (1976). *SEM/1976* **1,** 683.
Boyde, A. (1978). *SEM/1978* **2,** 203.
Boyde, A. (1979). *SEM/1979/2,* 67.
Boyde, A., and P. G. T. Howell (1977). *SEM/1977* **1,** 571.
Boyde, A., and E. Maconnachie (1979). *Scanning* **2,** 149.
Boyde, A., and E. Maconnachie (1979). *Scanning* **2,** 164.
Boyde, A., and A. D. G. Stewart (1962). *J. Ultrastruc. Res.* **7,** 159.
Boyde, A., and A. Tamarin (1984). *Scanning* **6,** 30.
Boyde, A., E. Bailey, S. J. Jones and A. Tamarin (1977). *SEM 1977* **1,** 507.
Bright, D. S., R. L. Myklebust, and D. E. Newbury, (1984). *J. Microsc.* **136,** 113.
Bright, D. S., E. B. Steel and D. E. Newbury (1991). In *Microbeam Analysis—1990* (D. G. Howitt, ed.) (San Francisco Press, San Francisco), p. 185.
Broers, A. N. (1969). *J. Phys. E* **2,** 273.
Broers, A. N. (1969a). *Rev. Sci. Instrum.* **40,** 1040.
Broers, A. N. (1973). In *Microprobe Analysis* (C. A. Anderson, ed.) (Wiley, New York), p. 83.
Broers, A. N. (1974). *SEM/1974,* 9.
Broers, A. N. (1975). *SEM/1975,* 662.
Broers, A. N., and E. Spiller (1980). *SEM/1980* **1,** 201.
Brombach, J. A. (1975). *J. Microsc. Biol. Cell* **22,** 233.
Brown, J. A., and A. Teetsov (1976). *SEM/1976* **1,** 385.
Brown, J. D., and R. H. Packwood (1982). *X-Ray Spectrom.* **11,** 187.
Brown, J. D., and L. Parobek (1972). In *Proc. 6th Int. Conf. X-ray Optics and Microanalysis* (G. Shinoda, K. Kohra and T. Ichinokawa, eds.) (Univ. of Tokyo Press, Tokyo), p. 163.
Bruining, H. (1954). *Physics and Application of the Secondary Electron Process* (Pergamon, London).
Bryce, M. R., and L. C. Murphy (1984). *Nature* **309,** 119.
Buehler, Ltd. (1973). Petrographic Sample Preparation, *AB Met. Dig.* **12/13,** 1.
Bullock, G. R. (1984). *J. Microsc.* **133,** 1.
Bullock, G. R., and P. Petrusz, eds. (1981 *et seq.*) *Techniques in Immunohistochemistry* (Academic Press, New York).
Butler, T. W. (1966). In *Proc. 6th Int. Congr. Elect. Micros., Kyoto* **1,** 193.
Caldwell, D. E., and P. W. Weiblem (1965). *Econ. Geol.* **60,** 1320.
Cailler, M., and J.-P. Ganachaud (1990a). In *Scanning Microsc. Suppl.* **4** (J. Schou, P. Kruit, and D. E. Newbury, eds.), p. 57.
Cailler, M., and J.-P. Ganachaud (1990b). In *Scanning Microsc. Suppl.* **4** (J. Schou, P. Kruit, and D. E. Newbury, eds.), p. 81.
Cameca Instruments (1974). *Users Manual for CAMEBAX.*
Cameron, I. L., and N. K. R. Smith (1980). *SEM/1980* **2,** 463.
Campbell, G. J., and M. R. Roach (1981). *Scanning* **4,** 188.

Campos, A., E. Fernandez-Segura and J. M. Garcia-Lopez (1991). *Scanning Microscopy* **5,** 175.

Carlemalm, E., R. M. Garavilo, and W. Villiger (1982). *J. Microsc.* **126,** 123.

Carlemalm, E., W. Villiger, J.-D. Acetarin, and E. Kellenberger (1985). "Low Temperature Embedding." In *The Science of Biological Specimen Preparation.* (M. Muller, R. P. Becker, A. Boyde, and J. J. Wolosewick, eds.) (SEM Inc. Chicago), p. 147.

Castaing, R. (1951). Ph.D. thesis, University of Paris.

Castaing, R. (1960). In *Advances in Electronics and Electron Physics,* Vol. **13,** (L. Marton, ed.) (Academic, New York), p. 317.

Castaing, R., and J. Deschamps (1954). *C.R. Acad. Sci.* **238,** 1506.

Castaing, R., and A. Guinier (1950). *Proc. 1st Intl. Conf. Electron Microscopy,* Delft 1949, p. 60.

Castaing, R., and J. Henoc (1966). In *Proc. 4th Int. Conf. on X-Ray Optics and Microanalysis* (R. Castaing, P. Deschamps, and J. Philibert, eds.) (Hermann, Paris), p. 120.

Champress, P. E., G. Cliff, and G. W. Lormier (1976). *J. Microsc.* **108,** 231.

Chandler, J. A. (1985). *SEM/1985,* **2,** 731.

Chang, C. C. Y., and J. V. Alexander (1981). *Biol. Cell* **40,** 99.

Chatfield, E. J. and M. J. Dillon (1978). *SEM/1978* **1,** 487.

Cheng, G., G. M. Hodges, and L. Trejdosiewicz (1980). *J. Microsc.* **137,** 9.

Claugher, D. (1986). *SEM/1986* **1,** 139.

Clay, C. S., and G. W. Peace (1981). *J. Microsc.* **123,** 25.

Cleaton-Jones, P. (1976). *J. Anat.* **122,** 23.

Cleaver, J. R. A., and K. C. A. Smith, (1973). *SEM/1973* **1,** p. 487.

Cliff, G., and G. W. Lorimar (1975). *J. Microsc.* **103,** 203.

Coates, D. G. (1967). *Philos. Mag.* **16,** 1179.

Coetzee, J., and C. F. van den Merwe (1984). *J. Microsc.* **135,** 147.

Coetzee, J., and C. F. van der Merwe (1985). *J. Microsc.* **137,** 129.

Coetzee, J., and C. F. van der Merwe (1989). *J. Electron. Microsc. Tech.* **11,** 155.

Cohen, A. L. (1979). *SEM/1979* **2,** 303.

Colby, J. W., D. R. Wonsidler, and D. K. Conley (1969). In *Proc. 4th Nat. Conf. Elect. Prob. Anal.,* EPASA, Pasadena, CA, Paper 9.

Colquhoun, W. R. (1984). *Ultrastruct. Res.* **87,** 97.

Considine, D. M. (1976). *Van Nostrand's Scientific Encyclopedia,* 5th ed. (Van Nostrand, New York), p. 710.

Cosslett, E., and R. N. Thomas (1966). In *The Electron Microprobe* (T. D. McKinley, K. F. J. Heinrich, and D. B. Wittry, eds.) (Wiley, New York), p. 248.

Cosslett, V. E. (1978). *J. Microsc.* **113,** 113.

Cosslett, V. E., and P. Duncumb (1956). *Nature* **177,** 1172.

Costello, M. J., and J. M. Corless (1987). *J. Microsc.* **112,** 17.

Crang, R. F. E., and K. L. Klomparens (1988). *Artifacts in Biological Electron Microscopy* (Plenum Press, New York).

Crawford, C. K. (1980). *SEM1980,* **4,** 11.

Crewe, A. V. (1969). *SEM/1969,* 11.

Crewe, A. V. (1979). *SEM/1979,* **2,** 31.

Crewe, A. V., D. N. Eggenberger, J. Wall and L. M. Welter (1968). *Rev. Sci. Instrum.* **39,** 576.

Crewe, A. V., M. Isaacson and D. Johnson (1971). *Rev. Sci. Instrum.* **42,** 411.

Crissman, R. S., and P. McCann (1979). *Micron* **10,** 37.

Curgenven, L., and P. Duncumb (1971). Tube Investments Res. Labs. Report 303.

Cuthill, J. R., L. L. Wyman, and H. Yakowitz (1963). *J. Met.* **15,** 763.

Dahlberg, E. P. (1976). *SEM/1976* **1,** 715.

Dalen, H., P. Scheie, R. L. Myklebust and T. Saltersdad (1983). *J. Microsc.* **131,** 35.

Danilatos, G. D. (1988). *Adv. Electronics and Electron Optics* **71,** 109.

Danilatos, G. D. (1990). *J. Microsc.* **160,** 9.

Danilatos, G. D. (1991). *J. Microsc.* **162,** 391.

Davies, G. (1983). *SEM/1983* **3,** 1163.

de Harven, E. (1987). *Ultrastruct. Pathol.* **11,** 711.
DeMay, J., and M. Moremans (1986). In *Advanced Techniques in Biological Electron Microscopy, Vol. 3* (J. K. Kochler, ed.) (Springer, Berlin), p. 229.
Deming, S. N., and S. L. Morgan (1973). *Anal. Chem.* **45,** 278.
DeNee, P. B. (1976). *SEM/1978* **1,** 479.
Dengler, L. A. (1978). *SEM/1976* **1,** 603.
Dorge, A., R. Rick, K. Gehring, and K. Thureau (1978). *Pflugers Arch.* **373,** 85–97.
Doucet, R. G., and S. A. Bradley (1989). *J. Electron. Microsc.* **12,** 58.
Drescher, H., L. Reimer, and H. Seidel (1970). *Z. Agnew. Phys.* **29,** 331.
Drier, T. M., and E. L. Thurston (1978). *SEM/1978,* **2,** 843.
Dubochet, J., J. Lepault, R. Freeman, J. A. Berriman, and J. C. Homo (1982). *J. Microsc.* **128,** 219.
Dubochet, J., M. Adrian, J. J. Chang, J. C. Homo, J. Lepault, A. W. McDowall, and P. Schultz (1988). *Quart. Rev. Biophys.* **21,** 129.
Duerr, J. S., and R. E. Ogilvie (1972). *Anal. Chem.* **44,** 2361.
Duncumb, P. (1957). In *X-ray Microscopy and Microradiography* (V. E. Cosslett, A. Engstrom, and H. H. Pattee, eds.) (Academic Press, New York), p. 617.
Duncumb, P. (1971). *Proc. EMAG,* Institute of Physics, London, Conf. Ser., Vol. 10, p. 132.
Duncumb, P., and D. A. Melford (1966). In *X-Ray Optics and Microanalysis, 4th Int. Cong. on X-Ray Optics and Microanalysis* (R. Castaing, P. Deschamps, and J. Philibert, eds.) (Hermann, Paris), p. 240.
Duncumb, P., and S. J. B. Reed (1968). In *Quantitative Electron Probe Microanalysis* (K. F. J. Heinrich, ed.) National Bureau of Standards Special Publication 298, p. 133.
Duncumb, P., and P. K. Shields (1966). In *The Electron Microprobe* (T. D. McKinley, K. F. J. Heinrich, and D. B. Wittry, eds.) (Wiley, New York), p. 284.
Durand, M., M. De Leplanque, and A. Kahane (1967). *Solid State Commun.* **5,** 750.
Dvoracheck, M., and A. Rosenfeld (1990). *J. Microsc.* **159,** 211.
Echlin, P. (1978a). *SEM/1978* **1,** 109.
Echlin, P. (1978b). *J. Microsc.* **112,** 225.
Echlin, P. (1981). *SEM/1981* **1,** 79.
Echlin, P., ed. (1984). *Analysis of Biological and Organic Surfaces* (Wiley, New York).
Echlin, P. (1989). *Beitr. Tabakforsch.* **14,** 297.
Echlin, P. (1991). *J. Microsc.* **161,** 159.
Echlin, P. (1992). *Low Temperature Microscopy and Analysis* (Plenum Press, New York).
Echlin, P., and A. Burgess (1977). *SEM/1977* **1,** 491.
Echlin, P., and P. Galle, eds. (1976). *Biological Microanalysis* (Société Française de Microscopie Electronique, Paris).
Echlin, P., and G. Kaye (1979). *SEM/1979* **2,** 21.
Echlin, P., and R. Moreton (1973). *SEM/1973* **1,** 325.
Echlin, P., and S. E. Taylor (1986). *J. Microsc.* **141,** 329.
Echlin, P., A. N. Broers, and W. Gee (1980). *SEM/1980* **1,** 163.
Echlin, P., B. Chapman, L. Stoter, W. Gee, and A. Burgess (1982). *SEM/1982* **1,** 29.
Echlin, P., B. Chapman, and W. Gee (1985). *J. Microsc.* **137,** 155.
Echlin, P., C. Lai and T. L. Mayes (1981). *SEM/1981* **2,** 489.
Edelmann, L. (1991). *J. Microsc.* **161,** 217.
Elad, E., C. N. Inskeep, R. A. Sareen, and P. Nestor (1973). *IEEE Trans. Nucl. Sci.,* **20,** 354.
Elder, H. Y., C. C. Gray, A. G. Jardine, J. N. Chapman, and W. H. Biddlecombe (1982). *J. Microsc.* **126,** 45.
Ellisman, M. H., P. L. Friedman, and W. J. Hamillon (1980). *J. Neurocytol.* **9,** 185.
Erlandsen, S. L., P. R. Gould, C. Frethem, C. L. Wells, J. Pawley, and D. W. Hamilton (1989). *Scanning* **11,** 169.
Eskelinen, S., and R. Peura (1988). *Scanning Microscopy* **2,** 1765.
Eskelinen, S., and P. Saukko (1983). *J. Microsc.* **130,** 63.
Estable-Puig, R. F., J. F. Estable-Puig, M. L. L. Slobodrian, A. Pusteria, A. B. Estable, and L. Le Blance-Laberge (1975). In *Proc. 2nd Ann. Meet. EMS Canada,* p. 40.
Evans, R. D. (1955). *The Atomic Nucleus* (McGraw-Hill, New York).

Everhart, T. E., And R. F. M. Thornley (1960). *J. Sci. Instr.* **37,** 246.
Everhart, T. E., R. F. Herzog, M. S. Chang and W. J. DeVore (1972). In *Proc. 6th Intl. Conf. on X-ray Optics and Microanalysis* (G. Shinoda, K. Kohra, and T. Ichinokawa, eds.) (Univ. Tokyo Press, Tokyo), p. 81.
Evers, P., K. Robinson, and L. Maistry (1983). *SEM/1983* **1,** 333.
Exley, R. R., B. G. Butterfield, and B. A. Meylan (1974). *J. Microsc.* **101,** 21.
Exley, R. R., B. A. Meylan and B. G. Butterfield (1977). *J. Microsc.* **110,** 75.
Falk, R. H. (1980). *SEM/1980,* **2,** 79.
Farley, A. N., and J. S. Shah (1990a). *J. Microsc.* **158,** 379.
Farley, A. N., and J. S. Shah (1990b). *J. Microsc.* **158,** 389.
Farley, A. N., And J. S. Shah (1991). *J. Microsc.* **164,** 107.
Fathers, D. J., J. P. Jakubocics, D. C. Joy, D. E. Newbury, and H. Yakowitz (1973). *Phys. Status Solidi A* **20,** 535.
Fathers, D. J., J. P. Jakubocics, D. C. Joy, D. E. Newbury, and H. Yakowitz (1974). *Phys. Status Solidi A* **22,** 609.
Fiori, C. E., and R. L. Myklebust (1979). In *Proc. Amer. Nucl. Soc. Conf.* (Mayaguez, Puerto Rico), p. 139.
Fiori, C. E., and D. E. Newbury (1978). *SEM/1978* **1,** 401.
Fiori, C. E., and C. R. Swyt (1989). In *Microbeam Analysis—1989* (P. E. Russell, ed.) (San Francisco Press, San Francisco), p. 236.
Fiori, C. E., H. Yakowitz, and D. E. Newbury (1974). *SEM/1974,* 167.
Fiori, C. E., R. L. Myklebust, K. F. J. Heinrich, and H. Yakowitz (1976). *Anal. Chem.* **48,** 172.
Fiori, C. E., C. R. Swyt, and K. E. Gorlen (1984). In *Microbeam Analysis-1984* (J. I. Goldstein and A. Romig, eds.), (San Francisco Press, San Francisco) p. 179.
Fiori, C. E., C. R. Swyt, and J. R. Ellis (1986). In *Principles of Analytical Electron Microscopy* (D. C. Joy, A. D. Romig, and J. I. Goldstein, eds.) (Plenum Press, New York), Chapter 13.
Fisher, D. W., and W. L. Baun (1967). *Norelco Reporter* **14,** 92.
Fisher, G. L., and G. D. Farningham (1972). *Quantitative Carbon Analysis of Nickel Steels with the Electron Probe Microanalyzer,* ASM Materials Engrg. Cong., Cleveland, Ohio.
Fisher, R. M., and J. C. Schwarts (1957). *J. Appl. Phys.* **28,** 1377.
Fitzgerald, R., K. Keil, and K. F. J. Heinrich (1968). *Science* **159,** 528.
Flegler, S. L. (1988). *Artifacts in Scanning Electron Microscope Operation.* In *Artifacts in Biological Electron Microscopy* (R. F. E. Crang and K. L. Klomparens, eds.) (Plenum Press, New York).
Flood, P. R. (1975). *SEM/1975* **1,** 287.
Flood, P. R. (1980). *SEM/1980* **1,** 183.
Follstaedt, D. M., J. A. Van den Avyle, A. D. Romig, Jr., and J. A. Knapp (1991). "Materials Reliability Issues in Microelectronics." In *Symposium on Electromigration, Vol. 225* (J. R. Lloyd, F. G. Yost, and P. S. Ho, eds.) (Materials Research Society), p. 225.
Fowler, R. H., and L. W. Nordheim (1928). *Proc. R. Soc. London.* **A111,** 173.
Franks, J., C. S. Clay, and G. W. Peace (1980). *SEM/1980* **1,** 155.
Freund, H. U., J. S. Hansen, S. J. Karttunen, and R. W. Fink (1972). In *Proc. 1969 Intl. Conf. on Radioactivity and Nuclear Spectroscopy* (J. H. Hamilton, and J. C. Manthuruthil, eds.) (Gordon & Breach, New York), p. 623.
Fuchs, W., J. D. Brombach, and W. Trosch (1978). *J. Microsc.* **112,** 63.
Gamliel, H. (1985). *SEM/1985,* **4,** 1649.
Gamliel, H., and A. Pollicak (1983). *SEM/1983* **2,** 929.
Garcia, R., D. Keller, J. Panitz, D. G. Bear and C. Bustamante (1989). *Ultramicroscopy* **27,** 367.
Garland, C. D., A. Lee, and M. R. Dickison (1979). *J. Microsc.* **112,** 75.
Garruto, R. M., R. Fukatsu, R. Yanagihara, D. Carleton D. Gajdusek, G. Hook, and C. E. Fiori (1984). *Proc. Nat. Acad. Sci.* (*USA*) **81,** 1875.
Gedcke, D. A., J. B. Ayers, and P. B. DeNee (1978). *SEM/1978* **1,** 581.

Geller, J. D. (1977). *SEM/1977* **1,** 281.
Gilfrich, J. V., L. S. Birks, and J. W. Criss (1978). In *X-ray Fluorescence Analysis of Environmental Samples* (T. G. Dzubay, ed.) (Ann Arbor Science Publishers, Ann Arbor, Michigan), p. 283.
Glaeser, R. M. (1979). In *Radiation Damage with Biological Specimens and Organic Materials* (J. J. Hren, J. I. Goldstein, and D. C. Joy, ed.) (Plenum Press, New York), p. 423.
Glaeser, R. M. (1980). In *Introduction to Analytical Electron Microscopy* (J. J. Hren, J. I. Goldstein, and D. C. Joy, eds.) (Plenum Press, New York), p. 423.
Glaeser, R. M., and K. A. Taylor (1978). *J. Microsc.* **112,** 127.
Glauert, A. M. (1974). "Fixation, Dehydration and Embedding of Biological Specimens." In *Practical Methods in Electron Microscopy, Vol. 3* (A. M. Glauert, ed.) (North Holland/Elsevier, Amsterdam).
Goebel, R., and W. Stocklein (1987). *Scanning Micros.* **1,** 1007.
Goldmark, P. C., and J. J. Hollywood (1951). *Proc. IRE* **39,** 1314.
Goldstein, J. I. (1967). *J. Geophys. Res.* **72,** 4689.
Goldstein, J. I., R. E. Hanneman, and R. E. Ogilvie (1965). *Trans. Met. Soc. AIME* **233,** 812.
Goldstein, J. I., J. L. Costley, G. W. Lorimer, and S. J. B. Reed (1977). *SEM/1977* **1,** 315.
Gomer, R. (1961). *Field Emission and Field Ionization* (Harvard Univ. Press, Cambridge, Massachusetts).
Good, R. H., Jr., and E. W. Muller (1956). In *Handbuch der Physik* (S. Flugge, ed.) (Springer-Verlag, Berlin), p. 176.
Goode, D., and T. K. Maugel (1986). *J. Electron. Microsc. Tech.* **5,** 263.
Goodhew, P. J. (1974). "Specimen Preparation in Materials Science." In *Practical Methods in Electron Microscopy* (A. M. Glavert, ed.) (Elsevier, Amsterdam).
Goulding, F. S. (1972). *Nucl. Instrum. Methods* **100,** 493.
Green, M. (1962). Ph.D. thesis, University of Cambridge, England.
Green, M. (1963). In *Proc. 3rd Intl. Conf. on X-ray Optics and Microanalysis* (H. A. Patee, V. E. Cosslett, and A. Engstrom, eds.) (Academic Press, New York), p. 361.
Green, M., and V. E. Cosslett (1961). *Proc. Phys. Soc.,* London, **78,** 1206.
Green, P. B., and P. Linstead (1990). *Protoplasma* **158,** 33.
Grocki, K., and R. Dermietzel (1984). *J. Microsc.* **133,** 95.
Gross, H., M. Muller, I. Wildhaber, and H. Winkler (1985). *Ultramicroscopy* **16,** 287.
Gupta, B. (1979). In *Microbeam Analysis in Biology* (C. P. Lechene and R. R. Warner, eds.) (Academic Press, New York), p. 375.
Haggis, G. H. (1982). *SEM/1982* **2,** 751.
Haggis, G. H. (1983). *J. Microsc.* **132,** 185.
Haggis, G. H. (1985). *J. Microsc.* **139,** 49.
Haggis, G. H., and E. F. Bond (1979). *J. Microsc.* **115,** 225.
Haggis, G. H., E. F. Bond, and B. Phipps (1976). *SEM/1976* **1,** 281.
Haine, M. E., and V. E. Cosslett (1961). *The Electron Microscope* (Spon, London).
Haine, M. E., and P. A. Einstein (1952). *Br. J. Appl. Phys.* **3,** 40.
Haine, M. E., and T. Mulvey (1959). *J. Phys. E.* **26,** 350.
Hall, C. E. (1966). *Introduction to Electron Microscopy* (McGraw-Hill, New York).
Hall, T. A. (1979a). *J. Microsc.* **117,** 145.
Hall, T. A. (1979b). In *Microbeam Analysis in Biology* (C. Lechene and R. R. Warner, eds.) (Academic Press, New York).
Hall, T. A. (1986). *J. Microsc.* **141,** 319.
Hall, T. A. (1988). *Ultramicroscopy* **24,** 181.
Hall, T. A., and B. F. Gupta (1974). *J. Microsc.* **100,** 177.
Hall, T. A., and B. L. Gupta (1984). *J. Microsc.* **136,** 193.
Hall, T. A. *et al.* (1966). In *The Electron Microprobe* (T. D. McKinley, K. F. J. Heinrich, and D. Wittry, eds.) (Wiley, New York), p. 805.
Halloran, B. P., and R. G. Kirk (1979). In *Microbeam Analysis in Biology* (C. Lechene and R. R. Warner, eds.) (Warner/Academic Press, New York), p. 571.
Hanaki, M., K. Tanaka, and Y. Kashima (1985). *J. Electron Microsc.* **34,** 373.

Hawes, C. R., and J. C. Horne (1985). *J. Microsc.* **137,** 35.

Hayat, M. A. (1982). *Fixation for Electron Microscopy* (Academic Press, New York).

Hayat, M. A. (1989). *Principles and Techniques of Electron Microscopy: Biological Applications* (Macmillan, London).

Hayat, M. A., ed. (1989). *Colloidal Gold: Principles, Methods and Applications, Vols. 1, 2* (Academic Press, Orlando).

Heathcock, J. F. (1988). *Microscopy and Analysis, 1988*(5), 17.

Heinrich, K. F. J. (1966a). In *Proc. 4th Intl. Conf. on X-ray Optics and Microanalysis* (R. Castaing, P. Deschamps, and J. Philibert, eds.) (Hermann, Paris), p. 159.

Heinrich, K. F. J. (1966b). In *The Electron Microprobe* (T. D. McKinley, K. F. J. Heinrich, and D. B. Wittry, eds.) (Wiley, New York), p. 296.

Heinrich, K. F. J. (1969). National Bureau of Standards Technical Note 521.

Heinrich, K. F. J. (1975). "Scanning Electron Probe Microanalysis." In *Advances in Optical and Electron Microscopy, Vol. 6* (R. Barer and V. E. Cosslett, eds.), p. 275.

Heinrich, K. F. J. (1981). *Electron Probe Microanalysis* (Van Nostrand, New York).

Heinrich, K. F. J. (1986). In *Proc. 11th Intl. Conf. on X-ray Optics and Microanalysis* (J. D. Brown and R. H. Packwood, eds.) (Univ. Western Ontario, London, Ontario, Canada), p. 67.

Heinrich, K. F. J., and D. E. Newbury, eds. (1991). *Electron Probe Quantitation* (Plenum, New York).

Heinrich, K. F. J., and H. Yakowitz (1968). *Mikrochim. Acta* **5,** 905.

Heinrich, K. F. J., and H. Yakowitz (1970). *Mikrochim. Acta* **7,** 123.

Heinrich, K. F. J., C. E. Fiori, and H. Yakowitz (1970). *Science* **167,** 1129.

Heinrich, K. F. J., R. L. Myklebust, S. D. Rasberry, and R. E. Michaelis (1971). National Bureau of Standards Special Publication 260.

Heinrich, K. F. J., D. E. Newbury, and H. Yakowitz, eds. (1976). *Use of Monte Carlo Calculations in Electron Probe Microanalysis and Scanning Electron Microscopy,* National Bureau of Standards Special Publication 460.

Heinrich, K. F. J., C. E. Fiori, and R. L. Myklebust (1979). *J. Appl. Phys.* **50,** 5589.

Hemsley, D. A. (1984). *Royal Microscopical Society Microscopy Handbook No. 7. The Light Microscopy of Synthetic Polymers* (Oxford Univ. Press, New York, Oxford).

Henderson, W. J., and K. Griffiths (1972). In *Principles and Techniques of Electron Microscopy: Biological Applications, Vol. 2* (M. A. Hayat, ed.) (Van Nostrand, Reinhold, New York).

Henglein, A. (1984). *Ultramicroscopy* **14,** 195.

Henke, B. L. (1964). In *Advances in X-Ray Analysis, Vol. 7* (Plenum Press, New York), p. 460.

Henke, B. L. (1965). In *Advances in X-Ray Analysis, Vol. 8* (Plenum Press, New York), p. 269.

Henke, B. L., and E. S. Ebisu (1974). In *Advances in X-Ray Analysis, Vol. 17* (Plenum Press, New York), p. 150.

Henke, B. L., P. Lee, T. J. Tanaka, R. L. Shimabukuro, and B. K. Fujikawa (1982). *Atomic Data and Nuclear Data Tables* **27,** 1.

Henoc, J. (1968). In *Quantitative Electron Probe Microanalysis* (K. F. J. Heinrich, ed.), National Bureau of Standards Special Publication 298, p. 197.

Henoc, J., and F. Maurice (1976). In *Use of Monte Carlo Calculations in Electron Probe Microanalysis and Scanning Electron Microscopy* (K. F. J. Heinrich, D. E. Newbury, and H. Yakowitz, eds.). National Bureau of Standards Special Publication 460, p. 61.

Henoc, J., and F. Maurice (1991). In *Electron Probe Quantitation* (K. F. J. Heinrich and D. E. Newbury, eds.) (Plenum Press, New York), p. 105.

Hermann, R., J. Pawley, T. Nagatani, and M. Müller (1988). *Scanning Microsc.* **2,** 1215.

Hewins, R. H. (1979). *Proc. 10th Lunar Sci. Conf.* (Pergamon, New York), p. 1109.

Hewins, R. H., J. I. Goldstein, and H. J. Axon (1976). *Proc. 7th Lunar Sci. Conf.* (Pergamon, New York), p. 819.

Hibi, T., K. Yada, and S. Takahashi (1962). *J. Electron. Microsc.* **11,** 244.

Hiller, J. (1947). U.S. Patent 2,418,029.

Hobbs, L. W. (1979). "Radiation Effects in Analysis of Inorganic Specimens by TEM." In

Introduction to Analytical Electron Microscopy (J. J. Hren, J. I. Goldstein, and D. C. Joy, ed.) (Plenum Press, New York).
Hobbs, P. V. (1974). *Ice Physics* (Clarendon Press, Oxford).
Hodde, K. C., D. A. Steebera, and R. M. Albrecht (1990). *Scanning Microscopy* **4,** 693.
Hodges, G. M. and K. E. Carr (1990). *Microsc. Anal. 1990*(7), 29.
Hodges, G. M., J. Southgate, and E. C. Toulson (1987). *Scanning Microscopy* **1,** 301.
Holbrook, K. A., J. R. Holbrook, and G. F. Odland (1976). *SEM/1976* **1,** 266.
Holliday, J. E. (1967). *Norelco Reporter* **14,** 84.
Holliday, J. E. (1963). In *The Electron Microprobe* (T. D. McKinley, K. F. J. Heinrich and D. B. Wittry, eds.) (Wiley, New York), p. 3.
Holy, J., C. Simerly, S. Paddock, and G. Schathen (1991). *J. Electron. Microsc. Tech.* **17,** 384.
Horisberger, M. (1985). *Tech. Immunocytochem.* **3,** 155.
Horisberger, M., and J. Rosset (1977). *SEM/1977* **2,** 75.
Horl, E. M., and E. Mugschl (1972). *Proc. 5th Cong. on Electron Microscopy, Inst. Physics,* London, Conf. Ser., **14,** 502.
Howell, P. G. T. (1975). *SEM/1975,* 697.
Hudson, B., and M. J. Markin (1970). *J. Phys. E.: Scien. Inst.* **3,** 311.
Humphreys, W. J. (1977). *SEM 1977,* **1,** 537.
Humphreys, W. J., and W. G. Henk (1979). *J. Microsc.* **116,** 255.
Humphreys, W. J., B. O. Spurlock, and J. S. Jackson (1977). In *Principles and Techniques of SEM Biological Applications, Vol. 6* (M. A. Hayat, ed.) (Van Nostrand, Reinhold, New York).
Hunger, H.-J., and L. Kuchler (1979). *Phys. Status Solidi A* **56,** K45.
Hunt, C. J. (1984). "Isothermal Freeze Fixation." In *Science of Biological Sample Preparation* (J. P. Revel, T. Barnard, and G. Haggis, eds.) (SEM Inc., Chicago), p. 123.
Ingram, M. J., and C. A. M. Hogben (1967). *Anal. Biochem.* **18,** 54.
Inoué, T. (1983). *SEM/1983* **1,** 227.
Inoué, T. (1986). In *Science of Biological Specimen Preparation for Microscopy and Analysis* (M. Muller, R. P. Becker, A. Boyde, and A. Wolosewick, eds.) (SEM Inc., Illinois), p. 245.
Inoué, T., and M. Koike (1989). *J. Microsc.* **156,** 137.
Isaacson, M. S. (1977). In *Principles and Techniques of Electron Microscopy: Biological Applications, Vol. 7.* (M. A. Hayat, ed.) (Van Nostrand, New York).
Jeng, T. W., and W. Chui (1984). *J. Microsc.* **136,** 35.
Jeszka, J. K., J. Ulanski and M. Kryszewski (1981). *Nature* **289,** 390.
Johansen, B. V., and E. Namork (1984). *J. Microsc.* **133,** 83.
Johansen, B. V., E. Namork, and G. Bukholm (1983). *J. Microsc.* **132,** 67.
Jones, D. B. (1981). *SEM/1981* **2,** 77.
Joy, D. C. (1984). *J. Micros.* **136,** 241.
Joy, D. C. (1987). *J. Micros.* **147,** 51.
Joy, D. C. (1988). Unpublished.
Joy, D. C. (1989). *Hitachi Inst. News,* #18, 3.
Joy, D. C. (1991). *J. Microsc.* **161,** 343.
Joy, D. C. (1991). Private communication concerning the Hunger–Kuchler expression.
Joy, D. C., and J. P. Jakubovics (1968). *Philos. Mag.* **17,** 61.
Joy, D. C., and S. Luo (1989). *Scanning* **11,** 176.
Joy, D. C., D. E. Newbury, and D. L. Davidson (1982). *J. Appl. Phys.* **53,** R81.
Joy, D. C., A. D. Romig, Jr., and J. I. Goldstein, eds. (1986). *Principles of Analytical Electron Microscopy* (Plenum Press, New York).
Joy, D. C., C. S. Joy, and D. A. Armstrong (1989). In *Electron Probe Microanalysis; Applications in Biology and Medicine* (K. Zierold, and M. K. Hagler, eds.) (Springer-Berlin), p. 127.
Judge, A. W. (1950). *Stereographic Photography,* 3rd ed. (Chapman and Hall, London).
Junger, E., and L. Bachman (1980). *J. Microsc.* **119,** 199.
Kahn, L. E., S. P. Frommes and P. A. Cancilla (1977). *SEM/1977* **1,** 501.

Kan, F. W. K. (1990). *J. Electron Microsc. Tech.* **14,** 21.

Kanaya, K., and S. Okayama (1972). *J. Phys. D.: Appl. Phys.* **5,** 43.

Kanaya, K., and S. Oro (1984). In *Electron Beam Interactions with Solids for Microscopy, Microanalysis, and Microlithography* (D. F. Kyser, H. Niedrig, D. E. Newbury, and R. Shimizu, eds.) (SEM, O'Hare, Illinois), p. 69.

Kanaya, K., Y. Muranaka, and H. Fujita (1982). *SEM/1982* **4,** 1379.

kanaya, K., E. Ohu, N. Osaki, and T. Oda (1988). *Micron Microsc. Acta* **19,** 163.

Kanaya, K., N. Baba, Y. Yamamoto, and K. Yonehara (1988). *Micron Microsc. Acta* **19,** 199.

Kandiah, K. (1971). *Nucl. Instrum. Methods* **95,** 289.

Kaneko, Y., H. Matsushima, M. Sekine and K. Matsumoto (1990). *J. Electron Microsc.* **39,** 426.

Kanter, H. (1961). *Phys. Rev.* **121,** 677.

Kardon, R. H., and R. G. Kessel (1979). *SEM/1979* **3,** 743.

Katoh (1978). *J. Electron Microsc.* **27,** 329.

Kazmerski, L. L., and D. M. Racine (1975). *J. Appl. Phys.* **46,** 791.

Kehl, G. L. (1949). *Principles of Metallographic Laboratory Practice,* 3rd ed. (McGraw-Hill, New York).

Kemmenoe, B. H., and G. R. Bullock (1983). *J. Microsc.* **132,** 153.

Kendall, C. E., A. Venketeshver Rao, S. A. Janezic, R. J. Temkin, M. J. Hollenburg, and P. J. Lea (1991). *J. Electron Micros. Tech.* **18,** 223.

Kennedy, J. R., R. W. Williams and J. P. Gray (1989). *J. Electron Microsc. Tech.* **11,** 117.

Kiessling, R., and N. Lange (1964). "Nonmetallic Inclusions in Steel." Special Report No. 90, The Iron and Steel Institute, London.

Kimoto, S., and H. Hashimoto (1966). In *The Electron Microprobe* (Wiley, New York), p. 480.

Kiss, J. Z., T. H. Giddings, L. A. Staehlin, and F. D. Sack (1990). *Protoplasma* **157,** 64.

Kitazawa, T., H. Shuman, and A. P. Somlyo (1983). *Ultramicroscopy* **11,** 251.

Kittel, C. (1956). *Introduction to Solid State Physics* (Wiley, New York).

Kittel, C. (1966). *Introduction to Solid State Physics,* 3rd ed. (Wiley, New York).

Knoll, M. (1935). *Z. Tech. Phys.* **11,** 467.

Koike, H., K. Ueno, and M. Suzuki (1971). *Proc. 29th Annual Meeting Electron Microscopy Soc. America* (G. W. Bailey, ed.) (Claitor's Publishing, Baton Rouge), p. 28.

Kopf, D. A., A. LeFurgey, L. A. Hawkey, B. L. Craig, and P. Ingram (1986). In *Microbeam Analysis-1986* (A. D. Romig and W. F. Chambers, eds.) (San Francisco Press, San Francisco), p. 241.

Koshikawa, T. and R. Shimizu (1974). *J. Phys. D.: Appl. Phys.* **7,** 1303.

Kotorman, L. (1980). *SEM/1980* **4,** 77.

König, H., and G. Helwig (1950). *Optik* **6,** 111.

Kramers, H. A. (1923). *Philos. Mag.* **46,** 836.

Krause, S. J., W. W. Adams, S. Kumar, T. Reilly, and T. Suziki (1987). *Proc. 4th Conf., Electron Microscopy Society of America* (G. W. Bailey, ed.) (San Francisco Press, San Francisco), p. 466.

Kubotsu, A., and M. Veda (1980). *J. Electron Microsc.* **29,** 45.

Kucera, L. J. (1986). *J. Microsc.* **142,** 71.

Kumon, H., K. Ohno, Y. Matsumura, H. Ohmori, and T. Tanaka (1983). *J. Electron Microsc.,* **32,** 20.

Kuo, H. P. (1976). Ph.D. Dissertation, Cornell University.

Kuzurian, A. M., and S. B. Leighton (1983). *SEM/1983* **4,** 1877.

Kyser, D. F. (1972). In *Proc. 6th Int. Conf. X-ray Optics and Microanalysis* (G. Shinoda, K. Kohra and T. Ichinokawa, eds.) (Univ. of Tokyo Press, Tokyo), 147.

Kyser, D. F., and K. Murata (1974). *IBM J. Res. Dev.* **18,** 352.

Kyser, D. F., and D. B. Wittry (1966). In *The Electron Microprobe* (T. D. McKinley, K. F. J. Heinrich and D. B. Wittry, eds.) (Wiley, New York), p. 691.

Lafferty, J. M. (1951). *J. Appl. Phys.* **22,** 299.

Laguitton, D., R. Rousseau, and F. Claisse (1975). *Anal. Chem.* **47,** 2174.
Lamb, J. C. and P. Ingram (1979). *SEM/1979* **2,** 459.
Lametschwandtner, A., A. Miodonski and P. Simonsberger (1980). *Microskopie* **36,** 270.
Lametschwandtner, A., U. Lametschwandtner and T. Weiger (1990). *Scanning Microscopy* **4,** 889.
Lamvik, M. K. (1991). *J. Microsc.* **161,** 171.
Lamvik, M. L. and S. D. Davilla (1988). *J. Electron Microsc. Tech.* **8,** 349.
Lander, J. J., H. Schreiber, T. M. Buck and J. R. Mathews (1963). *Appl. Phys. Lett.* **3,** 206.
Lane, W. C. (1970). In *Proc. 3rd Ann. Stereoscan Coll.* (Kent/Cambridge Scientific, Morton Grove, Illinois), p. 83.
Langmuir, D. B. (1937). *Proc. IRE* **25,** 977.
Laschi, R., G. Pasquinelli, and P. Versura (1987). *Scanning Microscopy* **1,** 1771.
Läuchli, A., R. Stelzer, R. Guggenhein and L. Henning (1978). "Precipitation Techniques as a Means of Intracellular Ion Localization by Use of Electron Probe Analysis." In *Microprobe Analysis as Applied to Cells and Tissue* (T. A. Hall, P. Echlin, and R. Kaufmann, eds.) (Academic Press, New York).
Leapman, R. D., and S. B. Andrews (1991). *J. Microsc.* **161,** 3.
Leapman, R. D., C. E. Fiori, and C. R. Swyt (1984a). *J. Microsc.* **133,** 239.
Leapman, R. D., C. E. Fiori, and C. R. Swyt (1984b). In *Analytical Electron Microscopy* (D. B. Williams and D. C. Joy, eds.) (San Francisco Press, San Francisco), p. 83.
Leaver, C., and B. Chapman (1971). *Thin Films* (Wykeham, London).
Lechene, C. (1978). *Microsc. Acta Suppl.* **2,** 228.
Liebhafsky, H. A., H. G. Pfeiffer, and P. D. Zemany (1955). *Anal. Chem.* **27,** 1257.
Liebhafsky, H. A., H. G. Pfeiffer, and P. D. Zemany (1960). In *X-ray Microscopy and X-ray Microanalysis* (A. Engström, V. Cosslett, and H. Pattee, eds.) (Elsevier/North-Holland), p. 321.
Liebmann, G. (1955). *Proc. Phys. Soc. B* **68,** 737.
Liepins, A. and E. de Harven (1978). *SEM/1978* **2,** 37.
Lifshin, E. (1974). In *Proc. 9th Ann. Conf. Microbeam Analysis Soc.,* Ottawa, Canada, p. 53.
Lifshin, E. (1975). In *Advances in X-Ray Analysis, Vol. 19* (R. W. Gould, C. S. Barrett, J. B. Newkirk, and C. O. Roud, eds.) (Kendall/Hunt, Dubuque, Iowa), p. 113.
Lifshin, E. and R. C. DeVries (1972). In *Proc. 7th Ann. Conf. Electron Probe Analysis Soc. Amer.,* San Francisco, California, p. 18.
Lifshin, E., M. F. Ciccarelli, and R. B. Bolon (1980). In *Proc. 8th Intl. Conf. on X-ray Optics and Microanalysis* (D. R. Beaman, R. E. Ogilvie, and D. B. Wittry, eds.) (Pendell, Midland, Michigan), p. 141.
Lindroth, M., and J.-E. Sundgren (1989). *Scanning* **11,** 5.
Lindroth, M., P. B. Bell, and B. A. Fredriksson (1987). *Scanning* **9,** 47.
Lindroth, M., B.-E. Fredriksson, and P. B. Bell (1991). *J. Microsc.* **161,** 229.
Linton, R. W., M. E. Framer, P. Ingram, J. R. Sommer, J. D. Shelburne (1984). *J. Microsc.* **134,** 101.
Livesey, S. S., A. A. del Campo, A. W. McDowall, and J. T. Stasy (1991). *J. Microsc.* **161,** 123.
Lohnes, R. A., and T. Demirel (1978). *SEM/1978* **1,** 643.
Long, J. V. P., and S. O. Agrell (1965). *Min. Mag.* **34,** 318.
Love, G. and V. D. Scott (1985). In *Microbeam Analysis—1985* (J. T. Armstrong, ed.), (San Francisco Press, San Francisco), p. 93.
Luftig, R. B. and P. M. McMillan (1981). *Int. Rev. Cytol. Suppl.* **12,** 309.
Lundquist, T. R., R. Alani, and P. R. Swann (1988). *Ultramicroscopy* **24,** 27.
Lyman, C. E., D. E. Newbury, J. I. Goldstein, D. B. Williams, A. D. Romig, J. T. Armstrong, P. Echlin, C. E. Fiori, D. C. Joy, E. Lifshin, and K. R. Peters (1990). *Scanning Electron Microscopy, X-Ray Microanalysis and Analytical Electron Microscopy* (Plenum Press, New York).
Lytton, D. G., E. Yuen, and K. A. Richard (1979). *J. Microsc.* **115,** 35.

Maeki, G., and M. Benoki (1977). *J. Electron Microsc.* **26,** 223.

Maisell, L. I., and R. Glang, eds. (1970). *Handbook of Thin Film Technology* (McGraw-Hill, New York).

Marinenko, R. B., K. F. J. Heinrich, and F. C. Ruegg (1979). *Microbeam Analysis—1979* (D. E. Newbury, ed.) (San Francisco Press, San Francisco), p. 221.

Marinenko, R. B., K. F. J. Heinrich, and F. C. Ruegg (1979). "Microhomogeneity Studies of NBS Standard Reference Materials, NBS Research Materials and Other Related Samples," NBS Special Publication 260–265.

Marinenko, R. B., F. Biancaniello, L. DeRoberts, P. A. Boyer, and A. W. Ruff (1981). "Preparation and Characterization of an Iron–Chromium–Nickel Alloy for Microanalysis: SRM 479a," NBS Special Publication 260–270.

Marinenko, R. B., R. L. Myklebust, D. B. Bright, and D. E. Newbury (1987). *J. Micros.* **145,** 207.

Marshall, A. T. (1975). *Micron.* **5,** 272.

Marshall, A. T. (1977). *Microsc. Acta* **79,** 254.

Marshall, A. T. (1980). In *X-Ray Microanalysis in Biology* (M. A. Hayat, ed.) (University Park Press, Baltimore), p. 167.

Marshall, A. T., and D. Carde (1984). *J. Microsc.* **134,** 113.

Marshall, A. T., and R. J. Condon (1985). *J. Microsc.* **140,** 109.

Marshall, A. T., D. Carde, and M. Kent (1985). *J. Microsc.* **139,** 335.

Marshall, D. J., and T. A. Hall (1966). In *X-Ray Optics and Microanalysis* (R. Castaing, J. Deschamps, and J. Philibert, eds.) (Hermann, Paris), p. 374.

Marton, L. L. (1941). U.S. Patent 2,233,286.

Masters, S. K., S. W. Bell, P. Ingram, D. O. Adams, and J. D. Shelburne (1979). *SEM/1979* **3,** 97.

Maugel, T. K., D. B. Bonar, W. J. Creegan, and E. G. Small (1980). *SEM/1980* **2,** 57.

McCall, J. L., and W. M. Mueller (1974). *Metallographic Specimen Preparation: Optical and Electron Microscopy* (Plenum Press, New York).

McCarthy, D. A., B. K. Pell, C. M. Molburn, S. R. Moore, J. A. Perry, D. H. Goddard, and A. P. Kirk (1985). *J. Microsc.* **137,** 57.

McCrone, W. A., and J. Delly (1973). *The Particle Atlas, Vol. 1* (Ann Arbor Science Publishers, Ann Arbor, Michigan).

McKee, A. E. (1977). *SEM/1977* **2,** 239.

McKenzie, R. L., ed. (1990). *Standard Reference Materials Catalog,* National Institute of Standards and Technology, Gaithersburg, Maryland.

McMullan, D. (1952). Ph.D. dissertation, Cambridge University.

Michaelis, R. E., H. Yakowitz, and G. A. Moore (1964). *J. Res. Nat. Bur. Std. U.S.* **68A,** 343.

Michel, M., T. Hillman, and M. Muller (1991). *J. Microsc.* **163,** 1.

Mills, A. A. (1988). *Scanning Microsc.* **2,** 1265.

Moll, S. H., F. Healy, B. Sullivan, and W. Johnson (1978). *SEM/1978* **1,** 303.

Moller, C. (1931). *Z. Phys.* **70,** 786.

Morel, F., and N. Roinel (1969). *J. Chem. Phys.* **66,** 1084.

Morgan, A. J. (1985). *X-ray Microanalysis in Electron Microscopy for Biologists: Royal Microscopical Society Handbooks No.* 5 (Oxford University Press, New York, Oxford).

Morgan, A. J., T. W. Daviey, and D. Erasmus (1978). In *Electron Probe Analysis in Biology* (D. A. Erasmus, ed.) (Chapman Hall, London), Chapter 4.

Morrison, J. C., and E. M. Buskirk (1984). *SEM/1984,* **2,** 857.

Moseley, H. G. J. (1913). *Philos. Mag.* **26,** 1024.

Moseley, H. G. J. (1914). *Philos. Mag.* **27,** 703.

Muller, M., R. P. Becker, A. Boyde, and J. L. Wolosewick, eds. (1986). *Science of Biological Specimen Preparation* (SEM Inc., Chicago).

Müller, T., R. Guggenheim, M. Duggelin, and C. Scheidegger (1991). *J. Microsc.* **161,** 73.

Munger, B. L. (1977). *SEM* 1977 **1,** 481.

Murakami, T., and A. L. Jones (1980). *SEM/1980* **1,** 221.

Murakami, T., A. Kubotsu, A. Ohtsuka, A. Akita, K. Yamamoto, and A. L. Jones (1982). *SEM/1982* **1,** 459.

Murakami, T., N. Iida, T. Taguchi, O. Ohtani, A. Kikuta, A. Ohtsuka, and T. Iroshima (1983). *SEM/1983,* 235.
Muranaka, Y., K. Hojou, and K. Kanaya (1982). In *Proc. 10th Int. Cong. EM* **3,** 459.
Murata, K. (1973). *SEM/1973,* 268.
Murata, K. (1974). *J. Appl. Phys.* **45,** 4110.
Murata, K., D. F. Kyser, and C. M. Ting (1981). *J. Appl. Phys.* **7,** 4396.
Murphey, J. A. (1978). *SEM/1978* **2,** 175.
Murphey, J. A. (1980). *SEM/1980* **1,** 209.
Murphey, J. A. (1982). *SEM/1982* **2,** 657.
Myklebust, R. L., C. E. Fiori, and K. F. J. Heinrich (1979). National Bureau of Standards Tech. Note 1106.
Myklebust, R. L., D. E. Newbury, and R. B. Marinenko (1989). *Anal. Chem.* **61,** 1612.
Myklebust, R. L., C. E. Fiori, and D. E. Newbury (1990). In *Microbeam Analysis—1990* (J. R. Michael and P. Ingram, eds.) (San Francisco Press, San Francisco), p. 147.
Nagatani, T., S. Saito, M. Sato, and M. Yamada (1987). *Scanning Microscopy* **1,** 901–909.
Nagel, D. (1968). In *Quantitative Electron Probe Microanalysis,* National Bureau of Standards Special Publication 298 (K. F. J. Heinrich, ed.), Washington, DC, 189.
Nagele, R. G., K. J. Doane, M. Lee, F. J. Wilson, and F. J. Roisen (1984). *J. Microsc.* **133,** 177.
Naguro, T., S. Inaga, and A. Iino (1990). *J. Electron Microsc.* **39,** 511.
Nakadera, T., A. Mitsushima, and K. Tanaka (1991). *J. Microsc.* **163,** 43.
Namork, E., and H. E. Meier (1989). *J. Electron Microsc. Tech.* **11,** 102.
Nation, J. L. (1983). *Stain Technol.* **58,** 347.
National Bureau of Standards Special Publication 533, 39.
Nelder, J. A., and R. Mead (1962). *Comput. J.* **7,** 308.
Neugebauer, D. C., and H. P. Zingsheim (1979). *J. Microsc.* **117,** 313.
Newbury, D. E., H. Yakowitz and R. L. Myklebust (1973). *Appl. Phys. Lett.* **23,** 488.
Newbury, D. E. (1974). *SEM/1974,* 537.
Newbury, D. E. (1976). *SEM/1976* **1,** 111.
Newbury, D. E., and R. L. Myklebust (1979). *Ultramicroscopy* **3,** 391.
Newbury, D. E., and R. L. Myklebust (1984). In *Electron Beam Interactions with Solids* (D. F. Kyser, R. Shimizu, and D. E. Newbury, eds.) (SEM, Inc., Chicago), p. 153.
Newbury, D. E., and R. L. Myklebust (1991). In *Microbeam Analysis—1991* (D. G. Howitt, ed.) (San Francisco Press, San Francisco), p. 561.
Newbury, D. E., and H. Yakowitz (1976). *SEM/1976* **1,** 151.
Newbury, D. E., R. L. Myklebust, K. F. J. Heinrich, and J. A. Small (1980).
Newbury, D. E., D. C. Joy, P. Echlin, C. E. Fiori, and J. I. Goldstein (1986). *Advanced Scanning Electron Microscopy and X-ray Microanalysis* (Plenum Press, New York), pp. 87–175.
Newbury, D. E., D. Bright, D. Williams, C. M. Sung, T. Page, and J. Ness (1986). "Application of Digital SIMS Imaging to Light Element and Trace Element Mapping" in *Secondary Ion Mass Spectrometry, Vol. 5* (A. Benninghoven, R. J. Colton, D. S. Simons, and H. W. Werner, eds.) (Springer-Verlag, Berlin), p. 261.
Newbury, D. E., R. B. Marinenko, D. S. Bright, and R. L. Myklebust (1988). *Scanning* **10,** 213.
Newbury, D. E., C. E. Fiori, R. B. Marinenko, R. L. Myklebust, C. R. Swyt, and D. S. Bright, (1990a). *Anal. Chem.* **62,** 1159A.
Newbury, D. E., C. E. Fiori, R. B. Marinenko, R. L. Myklebust, C. R. Swyt, and D. S. Bright (1990b). *Anal. Chem.* **62,** 1245A.
Niedrig, H. (1978). *Scanning* **1,** 17.
Nockolds, C. E., K. Moran, E. Dobson, and A. Phillips (1982). *SEM/1982* **3,** 907.
Nopanitaya, W., J. G. Aghajanian, and L. D. Gray (1979). *SEM/1979* **3,** 751.
Nordestgaard, B. G., and J. Rostgaard (1985). *J. Microsc.* **137,** 189.
Norenburg, J. L. and J. M. Barrett (1987). *J. Electron Microsc. Tech.* **6,** 35.
Nowell, J. A. and C. L. Lohse (1974). *SEM/1974* **1,** 267.
Nowell, J. A., and J. B. Pawley (1980). *SEM/1980* **2,** 1.

Oatley, C. W. (1972). *The Scanning Electron Microscope, Part 1, The Instrument* (Cambridge University Press, Cambridge).
Oatley, C. W. (1982). *J. Appl. Phys.* **53,** R1.
Oatley, C. W., and T. E. Everhart (1957). *J. ELectron.* **2,** 568.
Oatley, C. W., W. C. Nixon, and R. F. W. Pease (1965). In *Advances in Electronics and Electron Physics* (Academic Press, New York), p. 181.
Ohtani, O., B. Gannon, A. Ohtsuka, and T. Murakami (1982). *SEM/1982* **1,** 427.
Orloff, J. (1984). *SEM/1984* **4,** 1585.
Osatake, H., Tanaka, and T. Inoue (1985). *J. Electron Microsc. Tech.* **2,** 201.
Packwood, R. (1991). *Electron Probe Quantitation* (K. F. J. Heinrich and D. E. Newbury, eds.) (Plenum Press, New York), p. 83.
Packwood, R. H., and J. D. Brown (1981). *X-Ray Spectrom.* **10,** 138.
Pameijer, C. H. (1979). *SEM/1979* **2,** 571.
Panayi, P. N., D. C. Cheshire, and P. Echlin (1977). *SEM/1977* **1,** 463.
Panitz, J. A. (1986). In *Science of Biological Specimen Preparation for Microscopy and Analysis* (M. Muller, R. P. Becker, A. Boyde, and A. Wolosewick, eds.) (SEM Inc., O'Hare, Illinois), p. 283.
Panitz, J. A., C. L. Andrews, and D. G. Bear (1985). *J. Electron Microsc. Tech.* **2,** 285.
Parobek, L., and J. D. Brown (1978). *X-Ray Spectrom.* **7,** 26.
Pashley, D. W., M. J. Stowell, M. H. Jacobs, and T. J. Law (1964). *Philos. Mag.* **10,** 127.
Paulson, G. G., and R. W. Pierce (1972). *Proc. 30th Ann. EMSA* (Claitors Publishing, Baton Rouge, Louisiana, p. 406.
Pawley, J. B. (1974). *SEM/1974,* 27.
Pawley, J. B. (1984). *J. Microsc.* **136,** 45.
Pawley, J. B. (1988). *Proc. Eur. Reg. Cong on EM,* Inst. Phys. Conf. Ser. 93 **1,** 233.
Pawley, J. B. and S. Dole (1976). *SEM/1976* **1,** 287.
Pawley, J. B., and Norton, J. T. (1978). *J. Microsc.* **112,** 169.
Pease, D. C., and R. G. Peterson (1972). *J. Ultrastruc. Res.* **41,** 133.
Pease, R. F. W. (1963). Ph.D. Dissertation, Cambridge University.
Pease, R. F. W., and W. C. Nixon (1965). *J. Phys. E.* **42,** 281.
Pease, R. W., and J. F. Bailey (1975). *J. Microsc.* **112,** 169.
Pell, E. M. (1960). *J. Appl. Phys.* **31,** 291.
Peters, K.-R. (1980). *SEM/1980* **1,** 143.
Peters, K.-R. (1982). *SEM/1982* **4,** 1359.
Peters, K.-R. (1984). In *Electron Beam Interactions with Solids for Microscopy, Microanalysis, and Microlithography* (D. F. Kyser, H. Niedrig, D. E. Newbury, and R. Shimizu, eds.) (SEM, Inc., O'Hare, Illinois), p. 363.
Peters, K.-R. (1984a). *Proc. 2nd Pfefferkorn. Conf.,* 221.
Peters, K.-R. (1984b). *J. Microsc.* **133,** 17.
Peters, K.-R. (1985). *SEM/1985* **4,** 1519.
Peters, K.-R. (1986a). *J. Microsc.* **142,** 25.
Peters, K.-R. (1986b). In *Advanced Techniques in Biological Electron Microscopy* (J. K. Koethler, ed.) (Springer-Verlag, Berlin), p. 277.
Peters, K.-R. (1991). University of Connecticut Health Center, personal communication.
Peters, K.-R., and W. W. Carley (1988). *Scanning Microsc.* **2,** 2055.
Peters, K.-R., G. E. Palade, B. G. Schneider, and D. S. Papermaster (1983). *J. Cell Biol.* **96,** 265.
Petroff, P. M., D. V. Long, J. L. Strudel, and R. A. Logan (1978). *SEM/1978* **1,** 325.
Philibert, J. (1963). *Proc. 34rd Intl. Symp. X-Ray Optics and X-Ray Microanalysis,* Stanford University (H. H. Pattee, V. E. Cosslett, and A. Engstrom, eds.) (Academic Press, New York), p. 379.
Philibert, J., and R. Tixier (1969). *Micron* **1,** 174.
Polak, J. M., and I. M. Varndell, eds. (1984). *Immunolabelling for Electron Microscopy* (Elsevier, Amsterdam)
Pouchou, J. L., and F. Pichoir (1984). *Rech. Aerosp.* **3,** 13.
Pouchou, J. L., and F. Pichoir (1987). *Proc. 11th Int. Cong on X-ray Optics and Microanalysis* (J. D. Brown and R. H. Packwood, eds.) (Univ. of Western Ontario Press), p. 249.

Pouchou, J. L., and F. Pichoir (1991). In *Electron Probe Quantitation* (K. F. J. Heinrich and D. E. Newbury, eds.) (Plenum Press, New York), p. 31.
Pouchou, J. L., F. Pichoir, and D. Boivin (1989). *Proc. 12th Int. Conf. on X-ray Optics and Microanalysis,* Krakow, Poland, 1989 and ONERA Report, TP 157.
Pouchou, J. L., F. Pichoir, and D. Bolvin (1990). *Microbeam Analysis—1990* (J. R. Michael and P. Ingram, eds.), p. 120.
Powell, C. J. (1976a). *Rev. Mod. Phys.* **48,** 33.
Powell, C. J. (1976b). In *Use of Monte Carlo Calculations in Electron Probe Microanalysis and Scanning Electron Microscopy* (K. F. J. Heinrich, D. E. Newbury, and H. Yakowitz, eds.) National Bureau of Standards, Special Publication 460, p. 97.
Powell, C. J. (1990). In *Microbeam Analysis—1990* (J. R. Michael and P. Ingram, eds.) (San Francisco Press, San Francisco), p. 13.
Pulker, H. K. (1980). *Proc. 7th Eur. Cong. EM* **2,** 788.
Quinton, R. M. (1978). *SEM/1978* **2,** 391.
Rampley, D. N. (1976). *J. Microsc.* **107,** 99.
Rao-Sahib, T. S., and D. B. Wittry (1972). In *Proc. 6th Int'l Cong. on X-ray Optics and Microanalysis* (G. Shinoda, K. Kohra, and T. Ichinokawa, eds.) (Univ. Tokyo Press, Tokyo), p. 121.
Rasband, W. (1991). NIH Image, an image-processing program for the Macintosh computer, National Institutes of Health, Bethesda, Maryland.
Read, N. D., and C. E. Jeffree (1991). *J. Microsc.* **161,** 59.
Reed, S. J. B. (1965). *Br. J. Appl. Phys.* **16,** 913.
Reed, S. J. B. (1966). In *Proc. 4th Intl. Conf. on X-ray Optics and Microanalysis* (R. Castaing, P. Deschamps, and J. Philibert, eds.) (Hermann, Paris), p. 339.
Reed, S. J. B. (1971). *J. Phys. D.* **4,** 1910.
Reed, S. J. B. (1975). *Electron Microprobe Analysis* (Cambridge University Press, Cambridge).
Reed, S. J. B. (1990). *Microbeam Analysis—1988* (San Francisco Press, San Francisco), p. 109.
Reimer, L. (1975). In *Physical Aspects of Electron Microscopy and Microbeam Analysis* (B. M. Siegel and D. R. Beamon, eds.) (Wiley, New York), p. 231.
Reimer, L. (1976). *Elektronenmikroskopische Unter. u. Prep.-Methoden* (Springer, Berlin), p. 226.
Reimer, L., ed. (1984). *Transmission Electron Microscopy, Vol. 36* (Springer, Berlin).
Reimer, L., and C. Tollkamp (1980). *Scanning* **3,** 35.
Reimer, L., and B. Volbert (1979). *Scanning* **2,** 238.
Remond, G. (1990). *Proc. 12th Intl. Cong. Electron Microscopy, Analytical Sciences, Vol. 2* (San Francisco Press), p. 112.
Remond, G., Ph. Coutures, C. Gilles, and D. Massiot (1989). *Scanning Microscopy* **3,** 1059.
Reuter, W. (1972). In *Proc. 6th Int'l Cong. on X-ray Optics and Microanalysis* (G. Shinoda, K. Kohra, and T. Ichinokawa, eds.) (Univ. Tokyo Press, Tokyo), p. 121.
Revel, J. P., T. Barnard, and G. H. Haggis, eds. (1984). *Science of Biological Specimen Preparation* (SEM Inc., Chicago).
Rick, R., A. Dorge, and K. Thurau (1982). *J. Microsc.* **125,** 239.
Robards, A. W., and P. Crosby (1979). *SEM/1979* **2,** 325.
Robards, A. W. and U. B. Sleyter (1985). *Low Temperature Methods in Biological Electron Microscopy, Vol. 10. Practical Methods in Electron Microscopy.* (M. A. Glavert, ed.) (Elsevier, Amsterdam).
Robinson, V. N. E. (1975). *SEM/1975,* 51.
Robinson, V. N. E., and B. W. Robinson (1978). *SEM/1978* **1,** 595.
Roinel, N. (1988). *J. Electron Microsc. Tech.* **9,** 45.
Roli, J., and P. R. Flood (1978). *J. Microsc.* **112,** 359.
Romig, A. D., S. J. Plimpton, J. R. Michael, R. L. Myklebust, and D. E. Newbury (1990). In *Microbeam Analysis—1990* (J. R. Michael and P. Ingram, eds.) (San Francisco Press, San Francisco), p. 275.
Romig, A. D., Jr., D. R. Frear, P. F. Hlava, F. M. Hosking, J. J. Stephens, and P. T.

Vianco (1991). *Microbeam Analysis—1991* (D. G. Howitt, ed.) (San Francisco Press, San Francisco), p. 305.

Roos, N., and A. J. Morgan (1990). *Cryopreparation of Thin Biological Specimens for Electron Microscopy: Methods and Applications. Royal Microscopy Handbooks No. 21,* Oxford University Press, New York, Oxford).

Rose, A. (1948) In *Advances in Electronics* (A. Marton, ed.) (Academic Press, New York), p. 131.

Rosowski, J. R., S. C. Roemer, K. D. Hoagland, and W. A. Roth (1984). *SEM/1984* **1,** 29.

Rostgaard, J., and J. Tranum-Jensen (1980). *J. Microsc.* **119,** 213.

Ryan, K. P., D. H. Purse, S. G. Robinson, and J. W. Wood (1987). *J. Microsc.* **145,** 89.

Ryan, K. P., W. B. Bald, K. Neumann, P. Simonsberger, D. H. Purse, and D. N. Nicholson (1990). *J. Microsc.* **158,** 365.

Ryder, P. L. (1977). *SEM/1977* **1,** 273.

Ryder, T. A., and I. D. Bowen (1974). *J. Microsc.* **101,** 143.

Salem, S. I., and P. L. Lee (1976). *At. Data, Nucl. Data Tables* **18,** 233.

Samuels, L. E. (1982). *Metallographic Polishing by Mechanical Methods,* 3rd ed. (American Society for Metals, Metals Park, Ohio).

Sanders, S., E. Alexander, and R. Braylan (1975). *J. Cell Biol.* **67,** 476.

Sawyer, G. R., and T. F. Page (1978). *J. Mater. Sci.* **13,** 885.

Sawyer, L. C., and D. T. Grubb (1987). *Polymer Microscopy* (Chapman and Hall, London).

Scala, C., G. Cenacchi, P. Preda, R. P. Vici, R. P. Apkarian, and G. Pasquinelli (1991). *Scanning Microscopy* **5,** 135.

Schamber, F. H. (1973). *Proc. 8th Nat. Conf. on Electron Probe Analysis,* New Orleans, p. 85.

Schamber, F. H. (1977). In *X-Ray Fluorescence Analysis of Environmental Samples* (T. G. Dzubay, ed.) (Ann Arbor Science Publishers, Ann Arbor, Michigan), p. 241.

Schamber, F. H. (1978). In *Proc. 13th Nat. Conf. Microbeam Analysis Society,* Ann Arbor, Michigan, p. 50.

Schneider, G. B., S. M. Pockwinse, S. Billing, S. Gagliardi (1978). *SEM/1978* **2,** 77.

Schwabe, K. G., and L. Terracio (1980). *Cryobiology* **17,** 571.

Scopsi, L., L. I. Larson, L. Bastholm, and M. H. Nielsen (1986). *Histochem.* **86,** 35.

Scott, E. C. (1982). *J. Microsc.* **125,** 337.

Seiler, H. (1967). *Z. Angew. Phys.* **22,** 249.

Seiler, H. (1983). *J. Appl. Phys.* **54,** R1.

Seiler, H. (1984). In *Proc. 1st Pfefferkorn Conf.* (D. Kyser, H. Nedrig, R. Shimizu, and D. Newbury, eds.) (Scanning Microscopy, Inc., Chicago), p. 33.

Sela, J., and A. Boyde (1977). *J. Microsc.* **111,** 229.

Shao, Z., and A. V. Crewe (1989). *Ultramicroscopy* **31,** 199.

Shennawy, I. E. E., D. J. Gee, and S. R. Aparicio (1983). *J. Microsc.* **132,** 243.

Shimizu, R., and K. Murata (1971). *J. Appl. Phys.* **42,** 387.

Shimizu, R., Y. Kataoka, T. Ikuta, T. Koshikawa, and H. Hashimoto (1976). *J. Phys. D.: Appl. Phys.* **9,** 101.

Shiraiwa, T., N. Fujino, and J. Murayama (1972). In *Proc. 6th Int. Conf. on X-Ray Optics and Microanalysis* (G. Shinoda, K. Kohra, and T. Ichinokawa, eds.) (Univ. Tokyo Press, Tokyo), p. 213.

Shuman, H., A. V. Somlyo, and A. P. Somlyo (1976). *Ultramicroscopy* **1,** 317.

Silyn-Roberts, H. (1983). *J. Microsc.* **130,** 111.

Simonescu, N., and M. Simonescu (1976). *J. Cell. Biol.* **70,** 608.

Simpson, J. A. V., H. L. Bank, and S. S. Spicer (1979). *SEM/1979* **2,** 779.

Slayter, H. S. (1980). *SEM/1980* **1,** 171.

Small, J. A. (1981). *SEM/1981* **1,** 447.

Small, J. A., K. F. J. Heinrich, C. E. Fiori, R. L. Myklebust, D. E. Newbury, and M. F. Dilmore (1978). *SEM/1978* **1,** 445.

Small, J. A., K. F. J. Heinrich, D. E. Newbury, and R. L. Myklebust (1979). *SEM/1979* **2,** 807.

Small, J. A., R. L. Myklebust, and D. E. Newbury (1987). *J. Appl. Physics* **61,** 459.

Smith, D. G., C. M. Gold, and D. A. Tomlinson (1975). *X-Ray Spectrom.* **4,** 149.
Smith, D. G. W., ed. (1976). *Microbeam Techniques, A Short Course Handbook, Vol. 1* (Mineralogical Association of Canada, Edmonton).
Smith, K. C. A. (1956). Ph.D. Dissertation, Cambridge University.
Smith, K. C. A., and C. W. Oatley (1955). *Br J. Appl. Phys.* **6,** 391.
Soligo, D., N. Lampen, and E. de Harven (1981). *SEM/1981* **2,** 95.
Somlyo, A. V., H. Shuman, and A. P. Somlyo (1977). *J. Cell Biol.* **74,** 828.
Somlyo, A. V., H. Gonzalez, H. Shuman, G. McClellan, and A. P. Somlyo (1981). *J. Cell Biol.* **90,** 577.
Somlyo, A. P., M. Bond, and A. V. Somlyo (1985). *Nature* **314,** 622.
Spicer, S. S., B. A. Schulle, and J. D. Shelburne (1983). *SEM/1983* **4,** 1827.
Spivak, G. V., E. I. Rau, A. E. Lukianov, V. I. Petrov, and M. V. Bicov (1972). In *Proc 5th Eur. Cong. Electron Microscopy,* Manchester, England, 92.
Springer, G. (1966). *Mikrochim. Acta* **3,** 587.
Statham, P. J. (1976). *X-ray Spectrom* **5,** 16.
Statham, P. J. (1977). *Anal. Chem.* **49,** 2149.
Statham, P. J. (1979). *Mikrochem. Acta Suppl.* **8,** 229.
Stratham, P. J. and J. B. Pawley (1978). *SEM/1978* **1,** 469.
Steffens, W. L. (1978). *J. Microsc.* **113,** 95.
Steinbrecht, R. A., and K. Zierold (1987). *Cryotechniques in Biological Electron Microscopy* (Springer Verlag, Berlin).
Sternberger, L. A. (1986). *Immunocytochemistry,* 3rd ed. (Wiley, New York).
Streitwolf, H. W. (1959). *Ann. Phys. (Leipzig)* **3,** 183.
Sturgess, J. M., and M. A. Moscarello (1976). *SEM/1976* **2,** 145.
Swanson, L. W. (1975). *J. Vac. Sci. Technol.* **12,** 1228.
Swanson, L. W., and L. C. Crouser (1969). *J. Appl. Phys.* **40,** 4741.
Swanson, L. W., M. A. Gesley, and P. R. Davis (1981). *Surf. Sci.* **107,** 263.
Sylvester-Bradley, P. C. (1969). *Micropaleontology* **15,** 366.
Talmon, Y. (1982). *J. Microsc.* **125,** 227.
Talmon, Y. (1984). *Ultramicroscopy* **14,** 305.
Talmon, Y. (1987). In *Cryotechniques in Biological Electron Microscopy* (R. A. Steinbrecht and K. Zierold, eds.) (Springer Verlag, Berlin), Chapter 3.
Tanaka, K. (1974). In *Principles and Techniques of SEM: Biological Applications, Vol. 1.* (M. A. Hayat, ed.) (Van Nostrand, Rheinhold, New York).
Tanaka, K. (1989). *Biol. of the Cell* **65,** 89.
Tanaka, K., and H. Fukudome (1991). *J. Electron Microsc. Tech.* **17,** 15.
Tanaka, K., and A. Mitzushima (1984). *J. Microsc.* **133,** 213.
Tanaka, K., and T. Naguro (1981). *Biomed. Res.* **2,** 63.
Tanaka, K., A. Mitsushima, Y. Kashima, T. Nakadera, and H. Osatake (1989). *J. Electron Microsc. Tech.* **12,** 146.
Tanaka, K., A. Mitsushima, N. Yamagata, Y. Kashima, and H. Takayama (1991). *J. Microsc.* **161,** 455.
Taylor, C. M., and A. S. Radtke (1965). *Econ. Geol.* **60,** 1306.
Tegart, W. (1959). *Electrolytic and Chemical Polishing of Metals in Research and Industry* , 2nd ed. (Pergamon, New York).
Teo, B. K., and D. C. Joy (1981). *EXAFS Spectroscopy* (Plenum Press, New York).
Thomas, P. M. (1964). U.K. Atomic Energy Auth. Rept. AERE-R 4593.
Thornburg, D. D., and C. M. Wayman (1973). *Phys. Status Solidi A* **15,** 449.
Thornley, R. F. M. (1960). Ph.D. Dissertation, Cambridge University.
Thorpe, J. R., and D. M. R. Harvey (1979). *J. Ultrastruc. Res.* **68,** 186.
Thurley, K. W., and W. C. Mouel (1974). *J. Microsc.* **101,** 215.
Tovey, N. K., and K. Y. Wong (1973). In *Proc. Intl. Symp. on Soil Structure,* Swedish Geotechnical Society, Stockholm, Sweden, p. 59.
Traquair, J. A. (1987). *Can. J. Bot.* **65,** 888.
Troyon, M. (1984). *Proc. 8th European Congress on Electron Microscopy,* Budapest (A. Csandy, P. Rohlich, and D. Szabo, eds.) **1,** p. 11.
Troyon, M. (1987). *J. Microsc. Spectrosc. Electron.* **12,** 431.

Tsuji, T. (1983). *J. Microsc.* **131,** 115.
Tsuji, T., and M. A. Karasek (1985). *J. Microsc.* **137,** 75.
Tuggle, D. W., J. Z. Li, and L. W. Swanson (1985). *J. Microsc.* **140,** 293.
Tzaphlidou, M., and D. P. Matthopoulos (1988). *Micron Microsc. Acta* **19,** 137.
Ulrich, R. G., and K. J. McClung (1983). *J. Ultrastruc. Res.* **82,** 327.
Valdre, U. (1979). *J. Microsc.* **117,** 55.
Van Eekelen, C. A. G., A. Bolkestein, A. L. H. Slols, and A. M. Stadhouders (1980). *Micron* **11,** 137.
Van Essen, C. G. (1974). *J. Phys. E.* **7,** 98.
Van Steveninck, R. F. M., and M. E. Van Steveninck (1991). *Microanalysis in Electron Microscopy of Plant Cells* (J. L. Hall and C. Hawes, eds.) (in press).
Van Stevininck, R. F. M., B. Ballment, P. D. Peters, and T. A. Hall (1976). *Aust. J. Plant Physiol.* **3,** 359.
van der Voort, G. F. (1984). *Metallography: Principles and Practice* (McGraw Hill, New York).
Vaughan, J. G., ed. (1979). *Food Microscopy, Vol. 5* (Academic Press, London).
Vaz, O. W. (1986). Ph.D. Thesis, Arizona State University.
Vaz, O. W., and S. J. Krause (1986). *Proc. 4th Conf., Electron Microscopy Society of America* (G. W. Bailey, ed.) (San Francisco Press, San Francisco), p. 676.
Veneklasen, L. H., and B. M. Siegel (1972). *J. Appl. Phys.* **43,** 1600.
Volbert, B., and L. Reimer (1980). *SEM/1980* **4,** 1.
Walker, D. A. (1978). *SEM/1978* **1,** 185.
Wall, J. (1979). *Scanning Electron Microsc.* **2,** 291.
Walther, P., and J. Hentschel (1989). *Scanning Microsc.* **3,** 201.
Walther, P., J. Hentschel, P. Herler, T. Muller, and K. Zierold (1990). *Scanning* **12,** 300.
Walz, D. A., J. Penner, and M. I. Barnhart (1984). *SEM/1984* **1,** 303.
Walzthony, D., H. M. Eppenberger, and T. Wallimann (1984). *Eur. J. Cell Biol.* **35,** 216.
Ward, D. C., and T. L. Carlson (1983). *Crime Lab. Digest.* Feb. 1983, 2.
Ware, N. G., and S. J. B. Reed (1973). *J. Phys. E.* **6,** 286.
Warner, R. R. (1984). *J. Microsc.* **135,** 203.
Waterman, R. E. (1980). *SEM/1980* **2,** 21.
Watson, J. H. L., R. H. Page, and J. L. Swedo (1975). *SEM/1975* **1,** 417.
Watson, L. P., A. E. McKee, and B. R. Merell (1980). *SEM/1980* **2,** 45.
Watt, P. R., and J. Warrack (1989). *J. Microsc.* **156,** 273.
Wehner, G. K., and G. S. Anderson (1970). "The Nature of Physical Sputtering." In *Handbook of Thin Film Technology* (L. I. Maisell and R. Glang, eds.) (McGraw Hill, New York).
Weiser, B. A., M. J. Hollenberg, M. Mandlecorn, R. J. Temkin, and P. J. Lea (1991). *J. Electron Microsc. Tech.* **18,** 231.
Weiss, R. L. (1980). *SEM/1980* **4,** 123.
Wells, O. C. (1959). *J. Electron. Control.* **7,** 373–376.
Wells, O. C. (1960). *Br. J. Appl. Phys.* **11,** 199.
Wells, O. C. (1974a). *Scanning Electron Microscopy* (McGraw-Hill, New York).
Wells, O. C. (1974b). *SEM/1974,* 1.
Wepf, R., M. Amrein, U. Bürkli, and H. Gross (1991). *J. Microsc.* **163,** 51.
Wergin, W. P., and C. D. Pooley (1988). In *Artifacts in Biological Electron Microscopy* (P. R. E. Crang and K. L. Klomparens, eds.) (Plenum Press, New York), Chapter 9.
Westphal, C., and D. Frosch (1985). *J. Microsc.* **137,** 17.
Wetzel, B., G. H. Cannon, E. L. Alexander, B. W. Erickson, and E. W. Westbrook (1974). *SEM/1974* **1,** 581.
Wharton, D. A., and D. S. Murray (1990). *J. Microsc.* **158,** 81.
Whitcomb, M. J. (1981). *J. Microsc.* **121,** 289.
White, J. R., and E. L. Thomas (1984). *Rubber Chem. Technol.* **57,** 457.
Wild, P., M. Krahenbuhl, and E. M. Schraner (1989). *Histochem.* **91,** 213.
Willison, J. M. H., and A. J. Rowe (1980). In *Practical Methods in Electron Microscopy, Vol. 8* (M. A. Glavert, ed.) (North-Holland, Amsterdam).
Wilson, D. C., W. W. Ambrose, M. A. Crenshaw, and J. S. Hanker (1979). In *Proc. 37th EMSA Meet.,* p. 360.

Witcomb, M. J. (1985). *J. Microsc.* **139,** 75.

Wittry, D. B. (1957). Ph.D. Dissertation, California Institute of Technology.

Wittry, D. B. (1963). ASTM STP 349, Philadelphia, Pennsylvania, p. 128.

Wittry, D. B. (1966). In *Proc. 4th Intl. Conf. on X-ray Optics and Microanalysis* (R. Castaing, P. Deschamps, and J. Philibert, eds.) (Hermann, Paris), p. 168.

Woldseth, R. (1973). *X-ray Energy Spectrometry* (Kevex Corp., Foster City, California).

Wolf-Ingo, F., M. D. Worret, D. E. Cunningham, and R. E. Nordquist (1982). *Proc. 10th Int. E. M. Congress,* p. 253.

Wollweber, L., R. Strake, and U. Gothe (1981). *J. Microsc.* **121,** 185.

Woods, P. S., and M. E. Ledbetter (1976). *J. Cell. Sci.* **21,** 47.

Woulters, C. H., S. Hesseling, W. Th. Daems, and J. S. Ploem (1984). *J. Microsc.* **136,** 315.

Yacobi, B. G., and D. B. Holt (1990). *Cathodoluminescence Microscopy of Inorganic Solids* (Plenum Press, London).

Yakowitz, H., and K. F. J. Heinrich (1968). *Mikrochim. Acta* **5,** 183.

Yakowitz, H., and K. F. J. Heinrich (1969). *J. Res. Nat. Bur. Std. Ser. A.* **73A,** 113.

Yakowitz, H., and D. E. Newbury (1976). *SEM/1976* **1,** 151.

Yakowitz, H., D. L. Vieth, K. F. J. Heinrich, and R. E. Michaelis (1965). "Homogeneity Characterization of NBS Spectroscopic Standards II: Cartridge Brass and Low Alloy Steel," NBS Publ. 260–10.

Yakowitz, H., R. E. Michaelis, and D. L. Vieth (1969). "Preparation and Microprobe Characterization of W–20% Mo Alloy by Powder Metallurgical Methods," National Bureau of Standards Special Publication 260–16.

Yakowitz, H., C. E. Fiori, and R. E. Michaelis (1971). National Bureau of Standards Special Publication 260–20.

Yakowitz, H., A. Ruff, Jr., and R. E. Michaelis (1972). National Bureau of Standards Special Publication 260–43.

Yakowitz, H., R. L. Myklebust, and K. F. J. Heinrich (1973). National Bureau of Standards Tech. Note 796.

Yakowitz, H., C. E. Fiori, and D. E. Newbury (1974). *SEM/1974,* 167.

Yamada, N., M. Nagano, S. Murakami, M. Ikeuchi, E. Oho, N. Baba, K. Kanaya, and M. Osumi (1983). *J. Elect. Microsc.* **32,** 321.

Yarom, R., C. Maunder, M. Scripps, T. A. Hall, and V. Dubowitz (1975). *Histochemistry* **45,** 59.

Yasuda, K., A. Aiso, T. Nagatani, and M. Yamada (1989). *J. Electron Microsc. Tech.* **12,** 155.

Yonehara, K., N. Baba, and K. Kanaya (1989). *J. Electron Microsc. Tech.* **12,** 71.

Ziebold, T. O. (1967). *Anal. Chem.* **39,** 858.

Ziebold, T. O., and R. E. Ogilvie (1964). *Anal. Chem.* **36,** 322.

Zierold, K. (1980). *Microsc. Acta* **85,** 25.

Zierold, K. (1987). "Cryomicrotomy." In *Cryotechniques in Biological Electron Microscopy* (R. A. Steinbrecht and K. Zierold, eds.) (Springer Verlag, Berlin), p. 132.

Zworykin, V. K., J. Hiller, and R. L. Snyder (1942). *ASTM Bull.* **117,** 15.

Index

a-factors, 466
Aberration disks, 59
Absolute x-ray intensity, 458
Absorbed current, 187
Absorption, 349, 417, 419, 553
Absorption correction, 424, 445, 522
Absorption correction factor, 370, 371, 409, 425
Absorption edges, 316, 368
Absorption effect, 399, 480, 504
Absorption factor, 427
Absorption path, 429, 484
AC magnetic field interference, 64
Accumulation time for dot mapping, 528
Adhesives, 685
Agglomeration effects, 700
Air drying, 585
Air jet, 515
Airy disk, 55
Alumina trap, 704
Ambient temperature sublimation, 594
Analytical errors, 478
Analytical performance, 459
Analytical sensitivity, 497, 498
Analytical spatial resolution, 614, 634
Analyzing crystal, 274
Anderson–Hasler range, 131
Angular distribution, 97
Angular distribution, secondary electrons, 112
Angular distribution, tilted specimens, 99
Anisotropy, 370
Annular target, 705
Anode, 26
Anodization, 685
Antenna effects, 321
Anticontamination plates, 249
Antistatic agents, 683
Aperture alignment, 48
Aperture angle, optimum, 59
Aperture diffraction, 55
Aquatic specimens, 581
Area peak factor, 511
Area scanning, 526
Argon ion-beam etching, 640
Artifact peaks, 363
Artifacts, 330, 529, 658, 668
Astigmatism, 56
Atomic energy levels, 120
Atomic force microscope, 10
Atomic number, 84, 401, 417, 522
Atomic number contrast, 92
Atomic number correction, 419, 422, 443
Atomic number effect, 405, 406
Atomic number, average, 376, 377
Auger electrons, 1, 142
Auger process, 119
Auger yield, 119
Autofluorography, 629
Automatic qualitative analysis, 363
Autoradiography, 628, 651, 653

Background correction, 366, 381, 537
Background filtering, 368, 373
Background intensities, 516
Background measurements, 366
Background modeling, 363, 368
Background removal, 363, 398
Backscatter coefficient, 91
Backscatter detector, overhead, 24
Backscattered electron component, 200
Backscattered electron image, 694, 714
Backscattered electron signal, 674
Backscattered electrons, 1, 23, 91, 220, 256, 438, 525, 701, 714
Backscattered electrons, directionality, 97
Backscattered image, 604, 625

Backscattering, 405
Backscattering coefficient, 624
Backscattering correction factor, 420
Backstreaming, 692, 715, 719
Baffles, 693
Bandwidth, 176
Be window detector, 345
Beam current, 28, 662
Beam damage, 379, 665, 667, 673
Beam energy dependence, 110
Beam limiting aperture, 46
Beam sensitive materials, 510
Beer's Law, 372
Bence-Albee analytical technique, 466
Beryllium window, 317, 371, 372
Bethe cross section, 76, 127
Bethe range, 88
Bias condition, optimum, 31
Bias resistor, variable, 26
Bias voltage, 29, 30, 284
Bilby layer, 562
Biochamber, 727
Biological EDS analysis, 353
Biological material, 571
Biological sections, 335
Biological specimens, 460
Black level, 235
Black level suppression, 8
Blooming, 236
Bound water, 646
Bragg angle, 334
Bragg's law, 273, 357
Bremsstrahlung, 75, 116, 117
Brightness, 29
Brightness condition, optimum, 31
Brightness equation, 59
Brittle fractures, 638
BSE absorption plate, 714
BSE annular detectors, 625
BSE imaging, 253
BSE signal, 624, 643
BSE to SE conversion detector, 182
BSE_I, 221
BSE_{II}, 221
Bulk conductivity, 684
Bulk conductivity materials
  ammoniacal silver, 679
  ammoniacal silver nitrate, 682
  bifunctional mordant, 678
  chromium, 678
  chromium trioxide, 678
  dimethyl sulfoxide, 679
  galloylglucoses, 678
  gold-palladium, 740
  heavy-metal salts, 723
  hydrazine, 682
  hydrazine hydrate, 679
Bulk conductivity materials (*Cont.*)
  lead, 678
  manganese, 678
  mordant, 678
  osmium, 678, 682
  osmium tetroxide, 679
  osmium vapor, 679
  osmium–dimethyl sulphoxide, 679
  osmium–hydrazine vapor, 679
  osmium–thiocarbohydrazide, 679
  phenylenediamine, 678
  polyamines, 683
  silver, 678
  silver nitrate, 679
  sodium chloroaurate, 678
  tannic acid, 682, 683
  tannic acid–osmium, 681
  tannic acid–osmium method (TaO), 679
  tannic acid–osmium–thiocarbohydrazide, 679
  tetramethyl ethylenediamine, 678
  thiocarbohydrazides, 678
  uranyl nitrate, 678
Bulk conductivity methods, 633, 681
Bulk conductivity preparation techniques, 681
Bulk conductivity staining methods, 678
Bulk samples, 614
Bulk substrates, 565

Calculation of coating thickness, 732
Calibration curves, 515
Capping, 738
Carbon analysis, 514
Carbon fiber, 728
Carbon support films, 703
Cathode current density, 38
Cathode ray tube, 23, 152
Cathodoluminescence, 16, 144, 317, 701
Cathodoluminescence detector, 188
Celloidin–paraffin embedding, 604
Ceramics, 556
Channel plate detector, 186
Characteristic cross sections, 462
Characteristic fluorescence, 140, 419
Characteristic fluorescence correction, 429
Characteristic peak, 373, 376
Characteristic radiation, 371
Characteristic to continuum ratio, 464
Characteristic x-ray, 1, 116, 123
Charge carriers, 311
Charge-collection microscopy, 558, 560
Charge-to-voltage converter, 298
Charging, 249, 660, 672, 726
Charging effects, 677
Charging phenomena, 671
Chemical
  bonding effects, 392
  bonding shifts, 510

Chemical (*Cont.*)
 composition, 395, 417
 dehydrating agents, 643
 dehydration, 585, 586, 621
 dissolution, 640
 etching, 548
 fixation, 583, 618
 polishing, 549
 redistribution, 634
Chromatic aberration, 55
Chromatic aberration coefficient, 56
Clays, 562
Coating, 574
Coating artifacts, 700, 737
Coating frozen-hydrated material, 726
Coating materials
 for scanning electron microscopy
  alkaline earth oxides, 718
  alkaline halides, 718
  aluminum, 673, 696, 697, 702, 716, 733, 736
  calcium, 696, 697
  carbon, 674, 675, 686, 694, 733, 739
  chromium, 686, 689, 696, 705, 714, 716, 719, 720, 734, 740
  copper, 672, 673, 696, 697, 716
  germanium, 721, 734
  gold, 672, 674, 688, 692, 694, 695, 696, 697, 698, 699, 704, 708, 710, 717, 719, 720, 728, 729, 733
  gold–palladium, 694, 696, 697, 716, 725, 728, 729, 736, 739, 740
  heavy metals, 737
  iridium, 691
  molybdenum, 686, 696, 713
  nickel, 705, 716
  niobium, 689, 714, 740
  noble metals, 692, 704, 705, 707, 727
  noble-metal alloys, 716
  palladium, 694, 720
  platinum, 687, 689, 691, 704, 716, 719, 720, 721, 725, 728, 729, 735, 738, 740
  platinum films, 725
  platinum–carbon, 694, 720, 721
  platinum–iridium–carbon, 691, 695, 736, 739
  refractory metals, 707, 712
  rhenium–tungsten, 736
  rhodium, 740
  silver, 672, 673, 688, 694, 696, 697
  tantalum, 674, 686, 689, 696, 713, 719, 720, 737, 738, 740
  tantalum–tungsten, 721
  tin, 696, 697
  titanium, 696, 697
  tungsten, 674, 686, 689, 691, 692, 696, 710, 713, 720, 721, 736, 738, 740
Coating materials (*Cont.*)
 for scanning electron microscopy (*Cont.*)
  zirconium, 720
 for x-ray microanalysis
  aluminum, 673, 698, 700, 705, 729, 730, 731
  beryllium, 730
  carbon, 674, 675, 686, 694, 695, 697, 698, 701, 705, 708, 716, 720, 728, 729, 730, 736, 737, 739
  chromium, 730, 731, 734
Coating materials for x-ray microanalysis, 730
Coating techniques for x-ray microanalysis, 729
Coating thickness, 716, 732
Coaxial light microscope, 278
Coincidence losses, 289
Cold finger, 514
Cold plasma gas discharge, 581
Cold stage, 654, 715
Collection angle, 336
Collimator, 325
Colloidal gold, 630, 631, 633
Color superposition, 530, 545
Comparative radiation doses, 665
Complementary fractures, 600
Compositional contrast, 92, 191, 192
 with backscattered electrons, 191
 with secondary electrons, 195
 with specimen current, 197
Compositional mapping, 10
Compton scattering, 138
Computer x-ray analyzer, 304
Condenser lens, 46, 51
Conducting layer, 470
Conductive paint, 549, 608
Conductive polymers, 677
Conductive staining, 682
Conductive staining method, 611, 679, 680
Conductive support, 677
Conductivity, 470
Confidence level, 499
Confocal microscopes, 9
Conical lens, 46
Contaminants, 738
Contamination, 247, 255, 515, 660, 664, 697, 698, 700, 704, 705, 709, 712, 715, 720, 725, 728, 737
Contamination layer, 514
Continuous energy loss approximation, 75
Continuous energy loss expression, 76
Continuous film, 694, 698, 709, 734
Continuum, 373
Continuum cross sections, 462
Continuum fluorescence, 140, 486
Continuum fluorescence correction, 432
Continuum photon energies, 370
Continuum signal, 462

Continuum x ray, 75
Continuum x-ray production, 117
Contrast, 190
Contrast calculations, 93
Contrast components, 190
Contrast expansion, 235
Contrast formation, 215
Contrast from magnetic domains, 214
Contrast in elevated-pressure microscopy, 257
Contrast mechanisms, 92
Contrast reversal, 198
Convergence angle, 24
Convolution, 378
Cooling rate, 647, 648
Correlative microscopy, 576, 625
Corrosion casts, 604
Cosine distribution, 98, 328
Count rate effects, 533
Counting statistics, 375
Counting times, 498
Critical excitation energy, 412
Critical point drying, 590, 591, 592, 593, 622, 643
Cross section, 70, 71
  for inner shell ionization, 127
Crossover, 43, 162
Crossover point
  lower (or first), 111
  upper (or second), 111
Cryoelectron microscopy, 663
Cryofixation, 575
Cryofracturing, 597, 600, 602
Cryogens, 648
Cryomicroscopy, 666
Cryomicrotomy, 649
Cryopreparation system, 651
Cryoprotectant, 592
Cryoprotection, 647
Cryopumping, 693
Cryosectioning, 649
Crystal growth, 738
Crystal rocking, 536
Crystal spectrometer, 278
Crystallographic contrast (electron channeling), 214
Crystallographic effects, 511
Cumulative radial distribution, 101
Cumulative radiation doses, 665
Current density, 29
Current stability, 28
Curve fitting, 378
Curved crystal, 275
Cytochemical methods, 624, 625
Cytochemistry, 628

d spacing, 273, 358
Damage, 659, 668, 689
Damage (*Cont.*)
  arising during specimen observation, 661
  during examination and analysis, 659
  during preparation, 658
  to samples, 668
  to the specimen, 658
Dark level, 235
Data collection, 536
Dead layer, 290, 317, 371
Dead time, 282, 289, 302, 336
Dead time correction, 321, 536
Deconvolution, 378
Decoration artifacts, 719
Decoration effects, 737, 738
Defocusing or decollimation correction, 536
Degree of confidence, 496, 497
Dehydration, 574
Dehydration chemicals
  acetone, 586, 587, 590, 591, 622
  alcohol, 587
  carbon dioxide, 591, 592
  diethyl ether, 586
  dimethoxypropane, 587
  dimethyl sulfoxide, 592
  ethanol, 585, 587, 590, 591, 621, 633
  ethyl alcohol, 586
  Freon 113, 590, 622
  Freon 13, 591
  Freon TF, 594
  glycerol, 592
  hexamethyldisilazane, 590
  hexane, 590
  hydroxyethyl starch, 592
  methanol, 587, 591, 621, 633
  methyl, 586
  Peldri II, 594
  polyvinylpyrrolidone, 592
  t-butyl alcohol, 594
Dehydration damage, 663
Demagnification, 34, 49
Density, 44
Deposition rate, 702, 709
Depth distribution, 371, 404, 437
Depth distribution of x-ray production, 130, 133
Depth of field, 2, 163, 260
Depth of focus, 163
Derivative signal processing, 8, 238
Detectability limit, 499, 500
Detection limit, 657
Detector artifacts, 370
Detector efficiency, 362, 371
Detector resolution, 292
Detector response function, 368
Detectors for elevated pressure microscopy, 256
Diamond, 371
Differential amplification, 8, 235
Differential pumping, 38

Diffracting crystals, 392
Diffraction, 358
Diffusion pump, 692
Digital compositional mapping, 535
Digital filter, 374, 383
Digital framestores, 8
Digital image enhancement, 244
Digital image interpretation, 244
Digital imaging, 156
Diode coaters, 719, 734
Diode sputter coater, 704
Diode sputtering, 703, 709
Discontinuous films, 698
Discriminator setting, 300
Discriminator window, 358
Distribution of the continuum, 368
Divergence, 163
Divergence angle, 24
Dose, 663
Dot mapping, 525
Double apertures, 328
Double-deflection scanning, 151
Doubly ionized atoms, 363
Dry argon, 715
Dry deposition, 567
Dry fracturing, 599
Dry specimens, 581
Drying techniques, 563
Duane–Hunt limit, 117
Ductile fractures, 638
Dynamic focus correction, 170

EBIC, 561
EDS, 386
EDS decollimation, 530
Effect of tilt, 453
Effective dehydration, 590
Elastic scattering, 69, 71
Electric arc method, 686
Electric charge, 663
Electric field, 38
Electrical conductivity, 672
Electrical field buildup, 470
Electrical ground loops, 386
Electrical instabilities, 56
Electrical insulators, 572
Electromagnetic lenses, 7
Electromagnetic scan coils, 151
Electron
  backscattering, 419
  beam, 24
  beam etching, 640
  beam evaporation, 721
  beam evaporation assembly, 688
  beam method, 686
  beam-induced conductivity contrast (EBIC), 214
Electron (*Cont.*)
  channeling, 549, 560, 562
  channeling contrast, 8
  detectors, 176
  focusing, 44
  gun, 4
  gun, self-biased, 26
  lenses, 43
  microprobe, 1
  multiplier tube, 6
  probe, 24
  probe convergence angle, 66
  probe current, 21, 65
  probe diameter, 21
  probe size, 65
  range, 87
  retardation, 419
  scattering, 70
Electron-hole pairs, 296, 315
Electron-induced remote sources, 326
Electronic devices, 557
Electronic packages, 557
Electroplating, 685
Electrostatic lenses, 6
Elemental losses, 618
Elemental x-ray microanalysis, 612
Embedding chemicals
  epoxy resins, 622
  ethanol, 611
  glutaraldehyde-urea, 622
  high viscosity resins, 653
  polyvinyl alcohol, 622
Embedding procedure, 622
Emission current, 28
Emitted photons, 368
Empirical a factors, 466
Empirical method, 415, 466, 468
Energy
  deposition, 79, 83
  deposition rate, 80
  dispersive detector, 9, 378
  dispersive spectrometer, 16, 18, 64, 68
  distribution, 100, 108
  level diagram, 125
  response, 179
  space, 373
  spread, 32
  width of characteristic x rays, 123, 124
Environmental scanning microscopy, 564, 573, 678
Environmental stages, 572
Enzymatic digestion, 641
Enzymes, 620
Equipotentials, 27
Error distribution, 436
Error estimation, 386
Error histogram, 413, 490

Error matrix, 387, 389
Error propagation, 423
Escape depth of secondary electrons, 113
Escape peak, 287, 310, 349, 352
Evaporation techniques, 696
Evaporation temperature, 695
Evaporative coaters, 734
Evaporative coating, 727
Everhart–Thornley detector, 23, 177
Excitation potential, 371
Extended x-ray absorption fine structure, 136
Extraction voltage, 41
Extraneous radiation, 328

False minima, 386
False peaks, 29
Families of x-ray lines, 125
Family of x-ray lines, 343
Fano factor, 311
Faraday cup, 65, 326
Fast amplifer, 299
Fermi level, 26
Fibers, 564, 615
Field effect transistor, 297, 311
Field emission, 38
  cold, 41
  thermal, 42
Field-emission electron source, 8
Field-emission gun, 2, 38, 61
Field-emission SEM, 62
Field emitter, 6
Filament failure, 34
Filament heating current, 28
Filament power supply, 26
Filament-to-grid spacing, 31
Film adhesion, 700, 720
Film on substrate, 472, 473
Film thickness, 473, 702, 704, 718
Filter membranes, 579
Fine grain films, 703, 709, 723
First condenser lens, 46
Fixation, 574, 582, 617
Fixation chemicals
  acrolein, 583
  bicarbonate, 619
  cacodylate, 619
  calcium, 584
  carbodiimides, 633
  chymotrypsin, 581
  cobalt-molybdenum, 584
  collagenase hyaluronidase, 581
  collidine, 619
  dimethyladipimidates, 583
  DMSO, 602
  elastase, 581
  ferricyanides, 584
  ferrocyanides, 584
Fixation chemicals (*Cont.*)
  formaldehyde, 583, 584, 619
  glutaraldehyde, 583, 584, 618, 633
  lead-iron, 584
  lithium, 584
  neuraminindase, 581
  organic aldehydes, 618, 633
  osmium, 618
  osmium tetroxide, 583, 602, 611, 619
  palladium-tungsten, 584
  parabenzoquinone, 633
  permanganate, 618
  picric acids, 584
  piperazine-N,N′-bisethane sulfonic acid, 584
  piperzine-N-N′bis-ethane, 619
  platinum-carbon, 611
  sodium hypochlorite, 581
  sodium metaperiodate, 633
  tannic, 584
  thiocarbohydrazide, 583
  veronal acetate, 619
Fixative solution, 582
Flash evaporation, 695
Flashing, 42
Flat polished specimens, 417
Flotation fixation, 582
Flow proportional x-ray detector, 362
Fluorescence, 412, 417
Fluorescence effect, 399, 412, 486
Fluorescence factor, 412, 431
Fluorescence yield, 119
Focal length, 46
Focusing circle, 275
Foils, 471, 476
Foreline trap, 693
Foreshortening, 168
Formation of a thin film, 688, 689
Forward scattering, 87
Fourier transform, 373
Fracturing, 599, 623, 638
Free water, 646
Freeze dried bulk, 657
Freeze dried sections, 619, 657
Freeze dried ultra thin sections, 657
Freeze dryer, 592
Freeze drying, 593, 643, 651, 652
Freeze fracture replication, 649, 651
Freeze substitution, 652
Freeze-etch replicas, 601
Freeze-fracture-fix thaw technique, 601
Freeze-fracture-thaw technique, 610
Frequency space, 373
Fringing field, 44
Frozen dried, 649, 654, 655, 667
Frozen dried material, 726
Frozen dried sections, 596
Frozen hydrated, 649, 654, 655, 667

Frozen hydrated bulk, 657
Frozen hydrated material, 726
Frozen hydrated samples, 728, 729
Frozen hydrated sections, 657
Frozen hydrated state, 614
Full width at half maximum (FWHM), 53, 275, 310, 337, 375, 377, 391

Gamma processing, 7, 237
Gas amplification detector, 256
Gas ionization, 573
Gas-proportional counter, 280
Gaussian, 345, 375, 376, 379
Gaussian curve, 391
Gaussian distribution, 311
Gaussian image plane, 55
Gaussian intensity distribution, 53
Gaussian peak, 366
Gaussian probe diameter, 53
Gaussian profile, 385
Geiger region, 282
Generated intensity, 399, 418
Geological analysis, 466
Geological specimens, 556
Geological unknowns, 470
Geometric effects, 486, 487, 488
Geometric factors, 486
Geometrical probe size, 50
Global minimum, 386
Glow-discharge activation, 703
Glow-discharge techniques, 686
Gold absorption edges, 310
Gold electrical contact, 296
Gold labelling, 632
Goodness of fit, 379, 380
Granularity, 694
Gravimetric measurements, 735
Grid cap, 26
Grids, 623
Ground loops, 321, 338
Growth of silver film, 690
Growth of thin films, 701
Gun brightness, 6
Gun crossover, 29

Hall method, 461, 463, 465
Heat of radiation, 700
Heat sensitive samples, 691, 695
High melting point metals, 689, 696
High-quality dot maps, 527
High-resolution
coating, 683, 714, 720, 739
films, 720
images, 681, 706
imaging at high voltage, 224
imaging at low voltage, 226
High-resolution (*Cont.*)
microscopy, 219, 691
SEM, 61, 609, 610, 685, 691, 712, 724, 739
thin films, 737
High-vacuum evaporation, 688, 692
High-vacuum evaporation techniques, 694
High-vacuum thermal evaporation unit, 687
High-voltage supply, 26
Histochemical, 612, 619
Histochemical-analytical techniques, 620
Hollow magnification, 162
Homogeneity, 415, 493, 496
Homogeneity level, 497
Hydrocarbon, 247
Hydrocarbon contamination, 548
Hydrogen bonding, 646
Hyperbolic correction method, 466
Hyperbolic relationship, 466

Ice build up, 317
Ice polymorphs, 646
Image (area) scanning, 154
Image construction, 152
Image distortion, 166, 671
Image processing, 231
Image rotation, 159
Imaging strategy, 272
Immersion cooling, 648
Immersion fixation, 582
Immersion lens, 47
Immunocytochemical analysis, 653
Immunocytochemical methods, 612, 629
Immunocytochemical studies, 630
Immunocytochemistry, 651
In situ chemical analysis, 575
In-hole spectrum, 327
Incident beam current, 65
Incident electron energy, 369
Incipient charging, 253
Inclined spectrometer, 278
Inclusions, 554
Incomplete charge collection, 313
Indirect backscattered electrons, 203
Inelastic scattering, 69, 73
Inelastic scattering cross-section, 664
Inert dehydration, 622
Inhomogeneous particles, 492
Inner-shell ionization, 119
Inner-shell ionization process, 116
Instrument drift, 496
Instrumental factors, 495
Instrumental response function, 392
Instrumentation, 616
Integral intensity, 511
Integrated circuit (IC), 557
Intensity modulation, 152
Intensity of characteristic x-rays, 127

Intensity ratios (*k* values), 365
Intensity, x-ray continuum, 118
Interaction volume, 79, 83, 84, 86
Interfamily calculations, 457
Interfering peaks, 508
Internal fluorescence peak, 319
Internal space charging, 726
Interpretation of images, 654, 669
Intrinsic Ge, 292
Intrinsic semiconductor, 292
Ion beam sputter coater, 708
Ion beam sputtered tungsten, 711
Ion beam sputtering, 707, 709, 711, 721, 739
Ion milling, 564
Ion pumps, 693
Ion-beam etching, 606
Ionization of inner shells, 75
Ionization potential, mean, 76
Iron meteorites, 500
Irregular surfaces, 492
Isolated cells, 615
Iteration, 383, 434

J values, 422
Johann optics, 277
Johansson, 275

k ratio, 15, 442, 517
k value, 399
k_{AB}, 478
Kanaya–Okayama electron range, 89, 131
Kernel, 245
KLM markers, 342
Kramers' constant, 369
Kramers' Law, 376

Lack of background correction
  continuum image artifact, 534
  detection limits, 534
Lanthanum hexaboride, 7, 35
Laplacian, 245
Lateral range effects, 229
Lateral spatial distribution, 101
Laurate, 505
Layered synthetic microstructure, 279, 505
LDE crystal, 506
Lead stearate (STE), 505
Lens aberrations, 6, 30
Lens coil, 46
Lens excitation, 55
Lens gap, 44, 46
Lens system, 4
Level of homogeneity, 496, 497
Lifetime, gun, 32
Light element analysis, 503, 513, 519
Light element detection, 289
Light element mass absorption coefficient, 519
Light microscopy, 576
Light optical analogy, 203
Limit of detection, 17, 341, 396
Line markers, 336
Linear fitting procedure, 381
Linear interpolation, 376
Liquid cryogens, 647
Liquid dehydration, 591
Liquid nitrogen reservoir, 294, 323
Liquid nitrogen trap, 693
Liquid samples, 615
Liquid water, 645
Lithium fluoride, 332
Live time, 302
Live-time corrections, 365
Living organisms, 571
Local minimum, 386
Localization artifacts, 621
Lorentzian, 289, 391
Lorentzian tails, 376
Low energy x-rays, 503
Low loss electrons, 101
Low magnification operation, 163
Low resolution SEM, 704
Low temperature
  coating, 725
  embedding, 652
  for high spatial resolution, 723
  fracturing, 649
  methods, 645
  microanalysis, 656
  miroscopy, 645
  procedures, 653
  SEM, 654
  specimen preparation, 644
  x-ray microanalysis, 654, 657
Low vacuum evaporation, 701
Low voltage microscopy, 254
Low voltage operation, 60
Low voltage SEM, 43, 61, 676

Macromolecules, 624
Magnetic contrast, 549
Magnetic field, 43
Magnetic field strength, 44
Magnetic flux, 44
Magnetic quantum number, 121
Magnetic shielding, 64
Magnification, 22, 157
Main amplifier, 285
Marshall–Hall method, 119, 461
Mass absorption coefficient, 136, 371, 409, 518
Mass depth, 403
Mass distance units, 78
Mass effect, 479, 488
Mass loss, 460, 463, 664, 666, 667
Mass thickness contrast, 229

Mass thickness, average, 734
Material contrast, 192
Matrix correction, 413, 417, 538
Matrix correction procedures, 366
Matrix effects, 400, 417
Matrix inversion, 382
Maximum brightness, 30
Maximum probe current, 53
Mean free path, 71
Measured intensity, 400, 418
Measurement
  background, 516
  of phi rho z curves, 439
Mechanical polishing, 549
Mechanical stability, 675
Mechanical vibration, 64
Membrane filters, 685
Mercuric iodide, 292
Metal stains, 643
Metallographic methods, 636
Microanalysis, 63
Microchemical analysis, 547, 552
Microdroplets, 616
Microorganisms, 615
Microphony, 320, 338
Microstructural analysis, 547
Microwave-enhanced processing, 584
Migration, 249
Minimizing the radiation dose, 157
Minimum concentration, 499
Minimum dose microscopy, 246
Minimum probe size, 53
Moiré effects, 174
Moiré fringes, 174
Monte Carlo
  calculations, 454
  electron-trajectory simulation, 81
  method, 389, 474
  simulations, 335, 401
Morphological spatial resolution, 634
Moseley's relation, 123
Multicellular organisms, 578
Multichannel analyzer, 294
Multicomponent film, 475
Multiple beam interferometry, 735
Multiple linear least-squares curve fitting, 378
Multiple marking technique, 631
Multiple scattering, 712
Multiple scattering, Monte Carlo, 81
Multiple specimen holder, 609

Natural decoration effect, 738
Natural energy distribution, x-rays, 391
Natural line width, 310
Natural peak width, 313
Negative bias, 178
Newer fixatives, 584

Noble gas, 715
Noble metal, 707, 727
Noise, 216, 373
Nonconducting samples, 671, 676
Nonlinear amplification, 8
Nonlinear methods, 383
Nonnormal electron beam incidence, 453
Nonobservable damage, 662
Nonoptimal operation, 574
Normal beam incidence, 97
Normalization, 457, 487
Normalized chi squared, 381, 389
Nucleation, 701
Nucleation density, 703, 723
Nuclepore polycarbonate filter, 568
Number component, 190, 199

Objective aperture, real, 48
Objective lenses, 46
Organelles, 615
Organic buffers, 619
Organic samples, 571, 573
Osmium tetroxide, 583
Oversaturating, 34
Overscanning, 489
Overvoltage, 349, 402
OVH crystal, 506
Oxidation states, 289
Oxide standards, 451, 470
Oxygen absorption, 452
Oxygen plasma etching, 606, 640

P-10 gas, 282, 362
Packwood–Brown phi rho z equation, 441, 520
PAP approach, 442
Paraffin wax, 597
Parallax, 261
Particle
  dispersal, 567
  shapes, 487
  size, 483
  substrates, 565
  transfer, 566
Particles, 471, 564, 615
Path length, 409
Pauli exclusion principle, 121
PDH equation, 426
Peak
  centroids, 386
  deconvolution, 352, 366
  modelling, 537
  overlap, 379, 502
  positions, 356
  recognition, 363
  searching, 363
  shape alterations, 512
  shifts, 513

Peak (*Cont.*)
stripping, 363, 378
unraveling, 378
Peak to background, 366, 367
Peak-to-background method, 119, 487, 490. 492
Peak-to-background ratio, 328, 333, 359, 362
Peak-to-background values, 17
Peak-shift measurements, 289
Peelback procedure, 637
Penning sputter coater, 712, 721
Penning sputtering, 712, 714, 739
Phase structure, 551
Phi rho *z*, 133, 417
Phi rho *z* curves, 417, 436, 437, 440, 473
Phi rho *z* emitted, 445
Phi rho *z* function, 134
Phi rho *z* method, 413, 417, 425, 436, 519
Phi rho *z* thin film method, 474
Phi rho *z* versus rho *z* curve, 403
Philibert–Duncumb–Heinrich (PDH) equation, 426
Philibert–Duncumb–Heinrich absorption correction, 519
Phonon excitation, 74
Photoelectric absorption, 135, 296, 319
Photomultiplier, 7
Picture element, 159
Picture point, 159
Pileup continuum, 302
Pinhole lens, 46
Pixel, 159
Plasma magnetron coater, 705, 740
Plasma magnetron sputter, 722, 725
Plasma magnetron sputter coater, 706, 711, 715
Plasma–ion etching, 557
Plasma-magnetron, 715
Plasma-magnetron coater, 719
Plasmon excitation, 74
Plastics, 635
Pointed tungsten filaments, 34
Poisson counting statistics, 382
Poisson statistics, 494
Pole piece, 326
Polepieces, 44
Polished cut surfaces, 636
Polishing, 560
Polishing abrasive, 554
Polyethylene glycol, 597, 603
Polymer composite, 643
Polymer functional groups, 641, 642
Polymeric specimens, 460
Polymers, 572, 635, 636, 637, 639
Poor contrast sensitivity, 534
Positive bias, 179
Preamplifier, 282, 297
Precipitation methods, 621
Precipitation techniques, 620
Precision, 493
Preparative procedures, 578
Preparative techniques, 579
Primary drying, 651
Primary radiation, 139
Principal quantum number n, 121
Probe current, 28
Probe forming lens, 46
Processing time, 299
Projection distortion
gnomonic projection, 166
image foreshortening, 167
Proportional counter, 274
PROZA method, 442
PROZA program, 404
Pulse height analysis, 282, 285
Pulse height analyzer, 507
Pulse height shift, 505
Pulse pileup, 299, 304
Pulse pileup rejector, 339
Pulse shaping linear amplifier, 297
Pulse shifts, 508
Pulsed optical feedback (POF), 298
Pumping speeds, 692
Pyroantimonate, 620
Pyrolitic graphite, 566
Pyroxene, 452
Pyroxene microanalysis, 450

Quadrature addition, 311
Qualitative analysis, 341, 359, 362, 365, 612
Qualitative stereo microscopy, 260
Quantitative analysis, 10, 365, 613
Quantitative stereo microscopy, 263
Quantitative x-ray microanalysis, 387, 554
Quantum numbers, 120
Quartz, 332
Quartz crystal thin film monitor, 698, 708, 733, 735

R factor, 406, 420
Radiant heat load, 727
Radiation damage, 663, 666, 671
Radiation doses, 666
Radiation effects, 664
Radiation resistance, 667
Radiation sensitive, 9
Radio frequency etching, 606
Radio frequency sputtering, 703
Radiographic techniques, 629
Range of fluorescence radiation, 141
Range of homogeneity, 496
Range, low beam energy, 90
Range, tilted specimen, 89
Rapid chemical dehydration, 587
Rapid scanning, 253

Rate meter, 274
Ratio method, 478
Real time, 302
Real time digital imaging, 244
Reference spectrum, 379
Refractory metals, 707
Relative accuracy, 365
Relative contributions of SE_I and SE_{II}, 115
Relative errors, 488
Relative sputtering yield, 702
Relative transition probabilities, 127
Remote excitation, 485
Replica techniques, 604
Replicas, 639
Residual carbon level, 516
Resistive heating technique, 686
Resolution, 258, 332, 377
Resolution improvements, the secondary electron signal, 227
Resolution, EDS, WDS, 357
Response function, 384
Reverse bias, 294
Richardson equation, 26
Right-hand rule, 44
Rose criterion, 216, 218
Rotary coating, 601
Rotary pump, 692, 704
Rotary shadowing, 706
Rough surfaces, 471, 485, 492
Rowland circle, 277
Rubbers, 636
Rutherford scattering, 72

S factor, 406, 421
Safety, 693
Safety precautions, 591
Sample contamination, 649
Sample dehydration, 585
Sampling depth, backscattered electrons, 104, 106
Sands, 562
Satellite lines, 289, 363
Saturation, gun, 28
Scan coils, 22, 46
Scan generator, 151
Scanning action, 150
Scanning system, 22
Scanning transmission electron microscope, 5, 267
Scanning tunneling microscope, 10
Scattering angles, 72
Schottky barrier, 560
Schottky emitter, 42
Scintillator backscatter detectors, 181
Scintillator detector, 256
Second condenser lens, 46
Second crossover point E2, 254
Secondary drying, 651
Secondary-electron coefficient, δ, 108
Secondary-electron emission, 108
Secondary-electron excitation, 75
Secondary-electron flux, 729
Secondary-electron image, 714
Secondary-electron imaging, 714
Secondary-electron signal, 674
Secondary-electrons, 1, 23, 107, 176, 201, 223, 256
Secondary plus backscattered electron component, 201
Secondary radiation, 139
Sectioning, 596, 623, 636
SE_I, 114, 223
SE_{II}, 114, 223
Self-cleaning apertures, 326
SEM at elevated pressures (environmental SEM), 255
SEM autoradiography, 629
SEM operating conditions, 429
SEM–STEM, 739
Sensitivity, 493
Sensitivity factor, 477
Separation of contrast components, 210
Sequential recording, 530
Sequential simplex, 384
Severe charging, 254
Shadowing, 738
Shape factors, 488
Shape function, 379
Short wavelength limit, 470
Shottky effect, 35
Shrinkage, 585, 591
Si escape peak, 363
Si(Li) detector, 16, 292
Siegbahn notation, 124
Signal differentiation, 240
Signal mixing, 242
Signal quality, 215
Signal variations, 495
Signal-to-noise ratio, 216
Silicate minerals, 468
Silicon, 371, 372
Silicon absorption edge, 310, 338
Silicon dead zone, 317, 371
Silicon fluorescence peak, 338
Silicon internal fluorescence peak, 310
Single channel analyzer, 274, 285
Single scattering, Monte Carlo, 81
Single tissue culture, 588
Small grain size films, 691
Small particles, 483
Small specimens, 578
Soils, 562
Solid angle, 30, 332
  of collection, 325

Solid cryogens, 647
Solid-state diode detector, 183
Soluble elements, 618, 619
Solvent evaporation drying, 622
Source size, 32
Spatial resolution, 415, 739
Specialized coating methods, 720
Specific dyes, 619
Specimen
  borne contamination, 515
  cleaning, 579
  coating, 254
  composition dependence, 108
  contamination, 623
  current, 65, 186, 187, 207
  current amplifier, 65
  heating, 146
  preparation, 682
  selection, 578
  separation, 547
  stabilization, 633
  stub, 608
  support, 606, 623, 684
  thickness, 478
  tilt, 111, 453
  topography, 549
Spectral artifacts, 296, 336
Spectral distribution, 369
Spectral filtering, 363
Spectral interferences, 17
Spectral overlap, 379, 389
Spectral resolution, 376
Spectral unraveling, 387
Spectrometer defocusing, 336
Spectrometer efficiency, 457
Spectrum processing, 366
Spherical aberration, 54
Spherical aberration coefficient, 55
Spherical particles, 483
Spin quantum numbers, 121
Spot size, 6, 21
Sputter coating, 716, 728, 729
Sputter shadowing, 706
Sputtering rate, 702
Sputtering techniques, 715
Sputtering yield, 716
Stage scanning, 536
Staining of polymers, 641
Staining techniques, 612, 624
Stainless steel, 413
Stains
  Alcian blue, 620
  celloidin-paraffin embedding, 604
  chlorosulfonic acid, 642
  colloidal gold, 630, 631
  dental-impression plastic, 604
  diamino-benzididine, 627
Stains (*Cont.*)
  diaminobenzidine-osmium staining, 626
  epon, 618
  fibrin, 604
  fibrinogen, 604
  hydrazine, 642
  iron carbonyl, 628
  lead, 619
  mercury trifluoroacetate, 642
  modified lead-Gomori method, 626
  osmium, 627, 628
  osmium tetroxide, 641, 642
  periodic acid-thiocarbohydrazide-osmium staining, 626
  phosphotungstic acid, 641, 642
  polyethylene glycol, 603
  pyroantimonate, 620
  ruthenium tetroxide, 641, 642
  Sevriton, 604
  silicone rubber, 605
  silver, 628
  silver methenamine staining, 626
  silver sulfide, 641, 642
  thorotrast, 628
  thrombin, 604
  tin chloride, 642
  uranium, 628
  uranyl acetate, 641
  water soluble resins, 633
  xylene, 604
Standard deviation, 493, 494
Standard for x-ray microanalysis, 398
Standard mapping, 537
Standard Reference Materials, 159, 171, 555
Standardless analysis, 456
Standards, 419, 513, 613
Statistical equations, 493
Statistical quality, 543
Statistical scatter, 375
Statistics in compositional mapping, 539
STEM in SEM, 478
Stereo microscopy, 260
Stereographic pairs, 7, 261
Stigmator, 7, 46
Stopping power, 78, 405, 664
Stray magnetic fields, 64
Stray radiation, 325, 336, 338
Stretched polypropylene, 283
Structural studies, 575
Student *t* factor, 497, 498
Sublimation pumps, 693
Substrates, 607
Sum peaks, 299, 310, 348
Surface
  barrier contact, 371
  charging, 726
  coating, 515

Surface (*Cont.*)
conductivity, 682
contamination, 505, 514
etched fracture, 655
etching, 639, 719
roughness, 415
Synthetic multilayer crystals, 506
Synthetic multilayer diffraction structures, 392
System peaks, 349

Take off angle, 174, 409, 411, 553
Target current, 187
Target material, 716
TEM, 576
Theoretical resolution of the microscope, 60
Thermal damage, 673, 719, 739
Thermal drift, 38
Thermal effects, 673
Thermal evaporation, 686, 692, 709, 711, 739
Thermal noise, 311
Thermal wave contrast, 214
Thermionic emission, 25
Thermionic source, 43
Thermocouples, 648
Thick sections, 614
Thickness measurement during coating, 733
Thin conductive films, 685
Thin film, 473, 712
Thin foil, 335, 478
Thin-foil substrate, 566
Thin materials, 471
Thin sections, 596, 615
Thin window, 346
Thin-lens equation, 48
Thiocarbohydrazide, 678, 679
Threshold current, 217
Threshold equation, 30, 217
Tilt, 86
Tilt correction, 170
Tilt dependence, 95
Tilted specimen, 453
Tilted surfaces, 480
Time constant, 299
Tissue distortion, 588
Tissue extraction, 585
Top hat digital filter, 374
Topographic analysis, 547
Topographic contrast, 6, 97, 198
Topography, 263
Tortuous path filters, 568
Trace element analysis, 391, 499
Trace level, 341
Tracer technique, 439
Trajectory component, 190, 199
Transmission electron microscope, 5
Transmitted electron signal, 267
Triode electron gun, 30
Triode sputtering, 703
Tungsten evaporation, 34
Tungsten filament, 26, 35, 61
Tungsten sources, 334

Ultrahigh vacuum electron beam evaporation, 721
Ultramicrotome, 596
Ultraplaning, 602
Ultrathin films, 713, 722
Ultrathin-window EDS, 294
Uncoated biological specimens, 676
Unsupported films, 476
Untreated specimens, 675

Vapor fixation, 582
Vertical spectrometer, 277
Vibration, 64
Video printer, 24
Viewing screen, 23
Virtual ground, 297
Virtual objective aperture, 48
Virtual source, 38
Visible light photons, 674
Vitrified state, 647
Voltage contrast, 7, 214, 249, 558, 560

Wavelength dispersive spectrometer, 14, 18, 64, 68, 126, 273, 362, 656
Wavelength shift, 510
Waxes, 636
WDS, 376, 392
WDS defocusing, 530
WDS proportional counters, 365
WDS qualitative analysis, 359
WDS spectrometer, 359
Wehnelt cylinder, 26
Weighting factor, 382
Weights of lines, 344
Wet specimens, 572
Wilkinson type, 305
Windowless detector, 317
Windowless EDS, 294, 346, 359, 504
Wood samples, 599
Work function, 25
Working curves, 473
Working distance, 47

X-ray
absorption, 135, 401, 407, 408, 729
absorption edge, 136
absorption energy, 122
analytical studies, 666
analyzer, 294
area scanning, 525
continuum background, 483
cross sections, 463

X-ray (*Cont.*)
detectors, 656
dot mapping, 292
fluorescence, 139, 326, 401, 412
focusing, 496
generation, 402
generation curves, 444
intensity ratios, 398
lines, 124
microanalysis, 395, 644, 662, 663, 664, 667, 676, 683, 694
peak to background ratio, 129
peaks, 124
photon energy, 369
production, 404
production in thick targets, 128
production in thin foils, 128
range, 131
range, Andersen and Hasler, 131
spectra, 12
spectrometer defocusing, 292
take off angle, 371

Y-modulation, 243

Z contrast, 92
ZAF, 417, 554
ZAF factors, 405
ZAF method, 413, 419, 434
Ziebold relation, 501
Zoom capability, 158